Dan LaDuke

7   USW

Bemidji, MN 56601

751-3625 H

751-3456 W

electricity
one-seven

Hayden Electricity One—Seven Series

Harry Mileaf, Editor-in-Chief

electricity
one-seven

HARRY MILEAF EDITOR-IN-CHIEF

1-7

revised second edition

Hayden Book Company

A DIVISION OF HAYDEN PUBLISHING COMPANY, INC.
HASBROUCK HEIGHTS, NEW JERSEY / BERKELEY, CALIFORNIA

ISBN   0-8104-5952-3
Library of Congress Catalog Card Number   75-45504

21 22 23 24 25 26 27 28   PRINTING

85 86 87 88   YEAR

preface

This book combines a series of volumes designed specifically to teach electricity. The series is logically organized to fit the learning process. Each volume covers a given area of knowledge, which in itself is complete, but also prepares the student for the ensuing volumes. Within each volume, the topics are taught in incremental steps and each topic treatment prepares the student for the next topic. Only *one* discrete topic or concept is examined on a page, and *each* page carries an illustration that graphically depicts the topic being covered. As a result of this treatment, neither the text nor the illustrations are relied on solely as a teaching medium for any given topic. Both are for *every* topic, so that the illustrations not only complement but reinforce the text. In addition, to further aid the student in retaining what he has learned, the important points are summarized in text form on the illustration. This unique treatment allows the book to be used as a convenient review text. Color is used not for decorative purposes, but to accent important points and make the illustrations meaningful.

In keeping with good teaching practice, all technical terms are defined at their point of introduction so that the student can proceed with confidence. And, to facilitate matters for both the student and the teacher, key words for each topic are made conspicuous by the use of italics. Major points covered in prior topics are often reiterated in later topics for purposes of retention. This allows not only the smooth transition from topic to topic, but the reinforcement of prior knowledge just before the declining point of one's memory curve. At the end of each group of topics comprising a lesson, a summary of the facts is given, together with an appropriate set of review questions, so that the student himself can determine how well he is learning as he proceeds through the book.

Much of the credit for the development of this series belongs to various members of the excellent team of authors, editors, and technical consultants assembled by the publisher. Special acknowledgment of the contributions of the following individuals is most appropriate: Frank T. Egan, Jack Greenfield, and Warren W. Yates, principal contributors; Peter J. Zurita, Steven Barbash, Solomon Flam, and A. Victor Schwarz, of the publisher's staff; Paul J. Barrotta, Director of the Union Technical Institute; Albert J. Marcarelli, Technical Director of the Connecticut School of Electronics; Howard Bierman, Editor of *Electronic Design;* E. E. Grazda, Editorial Director of *Electronic Design;* and Irving Lopatin, Editorial Director of the Hayden Book Companies.

HARRY MILEAF
Editor-in-Chief

contents

ELECTRICITY ONE

CONTENTS

CONTENTS

CONTENTS

ELECTRICITY FOUR

CONTENTS

ELECTRICITY FIVE

CONTENTS

CONTENTS

CONTENTS

CONTENTS

electricity
one-seven

electricity
one

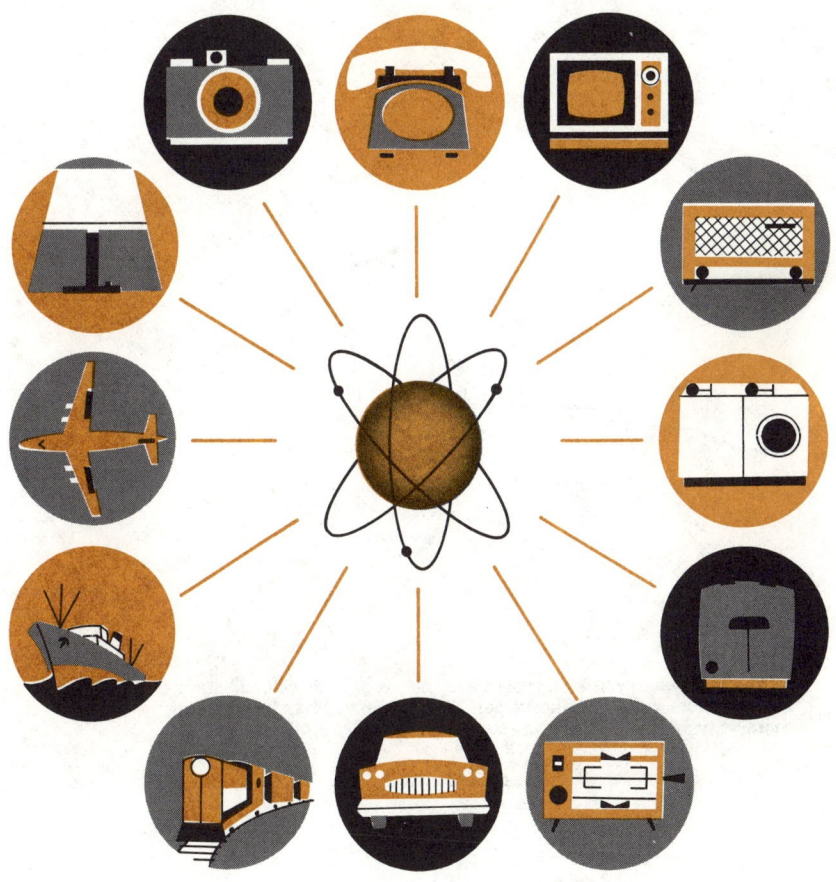

the importance of electricity

Electricity is one of the most important forms of energy used in the world today. Without it, there would be no convenient lights, no radio or television communications, no telephone service; and people would have to go without the many household appliances that are taken for granted. In addition, the field of transportation would not be as it is today without electricity. Electricity is used in all types of vehicles. When you stop and think about it, electricity is used everywhere.

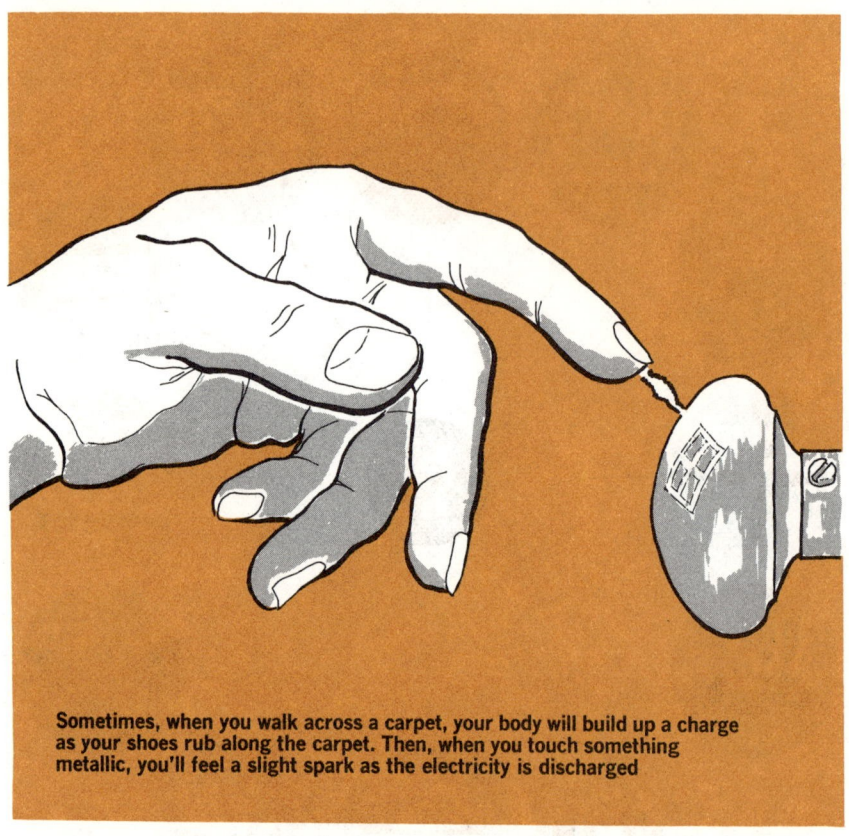

Sometimes, when you walk across a carpet, your body will build up a charge as your shoes rub along the carpet. Then, when you touch something metallic, you'll feel a slight spark as the electricity is discharged

early history

Although electricity first became useful during modern times, it was discovered 2000 years ago by the Greeks. They noticed that when a material that we now call amber was rubbed with some other materials, it became *charged* with a mysterious force. The charged amber attracted such materials as dried leaves and wood shavings. The Greeks called the amber *elektron,* which is how the word electricity came about.

Around 1600, William Gilbert classified materials that acted like amber as *electriks,* and those materials that did not, as *nonelectriks.*

In 1733, a Frenchman, Charles DuFay, noticed that a charged piece of glass *attracted* some charged objects, but *repelled* other charged objects. He concluded that there were two types of electricity.

Around the middle 1700's, Benjamin Franklin called these two kinds of electricity *positive* and *negative.*

what is electricity?

In Benjamin Franklin's time, scientists thought that electricity was a fluid that could have positive and negative charges. But today, scientists think electricity is produced by very tiny particles called *electrons* and *protons*. These particles are too small to be seen, but they exist in all materials. To understand how they exist, you must first understand the structure of all matter.

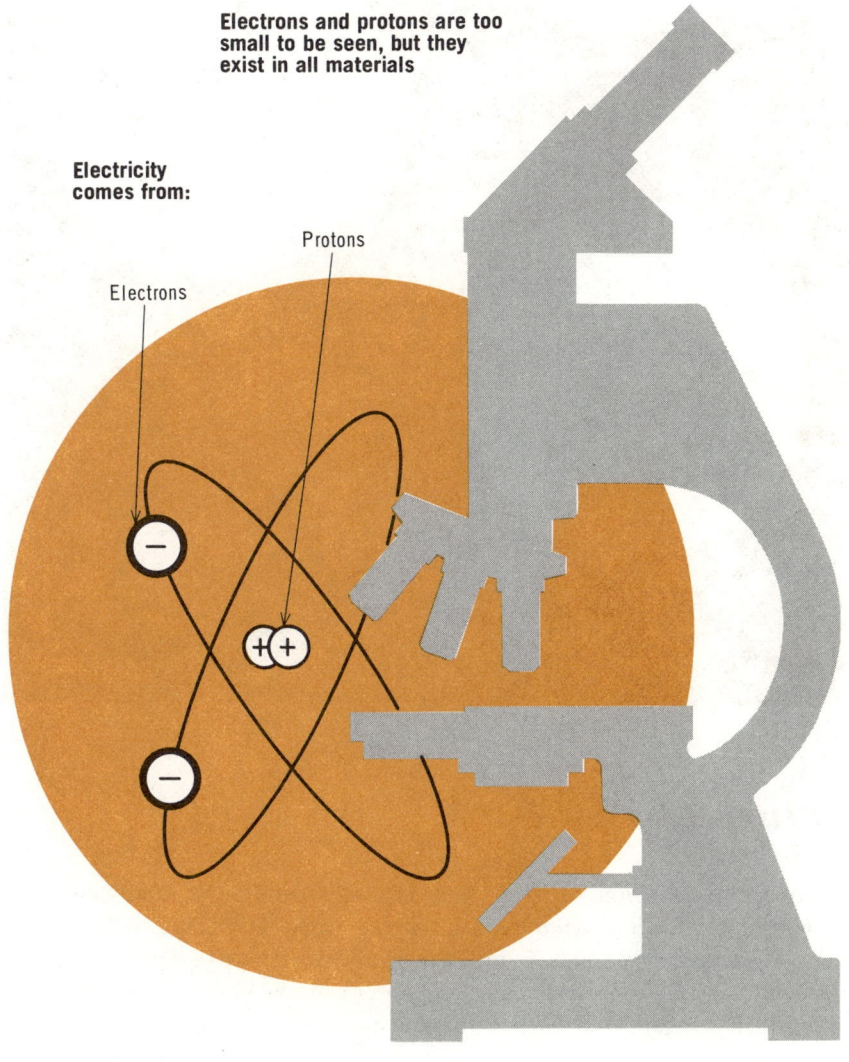

Electrons and protons are too small to be seen, but they exist in all materials

Electricity comes from:

Protons

Electrons

MATTER CAN BE:

A SOLID

A LIQUID

A GAS

what is matter?

Anything that you can see, feel, or use is matter. Actually, matter is anything that has *weight* and occupies *space*. It can be in the form of a solid, liquid, or gas. Stone, wood, and metal are forms of matter (solids), and so are water, alcohol, and gasoline (liquids), as well as oxygen, hydrogen, and carbon dioxide (gases).

the elements

The *elements* are the basic materials that make up all matter. Oxygen and hydrogen are elements, and so are aluminum, copper, silver, gold, and mercury. Actually, there are just a little over 100 known elements. Ninety-two of these elements are natural, and the others are man-made. In the last few years, a number of new ones were made, and it is expected that there are many more still to be produced.

Everything we see about us is made from the elements. But the elements themselves cannot be produced by the simple chemical combination or separation of other elements.

THE NATURAL ELEMENTS

Atomic Number	Name	Symbol	Atomic Number	Name	Symbol	Atomic Number	Name	Symbol
1	Hydrogen	H	32	Germanium	Ge	62	Samarium	Sm
2	Helium	He	33	Arsenic	As	63	Europium	Eu
3	Lithium	Li	34	Selenium	Se	64	Gadolinium	Gd
4	Beryllium	Be	35	Bromine	Br	65	Terbium	Tb
5	Boron	B	36	Krypton	Kr	66	Dysprosium	Dy
6	Carbon	C	37	Rubidium	Rb	67	Holmium	Ho
7	Nitrogen	N	38	Strontium	Sr	68	Erbium	Er
8	Oxygen	O	39	Yttrium	Y	69	Thulium	Tm
9	Fluorine	F	40	Zirconium	Zr	70	Ytterbium	Yb
10	Neon	Ne	41	Niobium	Nb	71	Lutetium	Lu
11	Sodium	Na		(Columbium)		72	Hafnium	Hf
12	Magnesium	Mg	42	Molybdenum	Mo	73	Tantalum	Ta
13	Aluminum	Al	43	Technetium	Tc	74	Tungsten	W
14	Silicon	Si	44	Ruthenium	Ru	75	Rhenium	Re
15	Phosphorus	P	45	Rhodium	Rh	76	Osmium	Os
16	Sulfur	S	46	Palladium	Pd	77	Iridium	Ir
17	Chlorine	Cl	47	Silver	Ag	78	Platinum	Pt
18	Argon	A	48	Cadmium	Cd	79	Gold	Au
19	Potassium	K	49	Indium	In	80	Mercury	Hg
20	Calcium	Ca	50	Tin	Sn	81	Thallium	Tl
21	Scandium	Sc	51	Antimony	Sb	82	Lead	Pb
22	Titanium	Ti	52	Tellurium	Te	83	Bismuth	Bi
23	Vanadium	V	53	Iodine	I	84	Polonium	Po
24	Chromium	Cr	54	Xenon	Xe	85	Astatine	At
25	Manganese	Mn	55	Cesium	Cs	86	Radon	Rn
26	Iron	Fe	56	Barium	Ba	87	Francium	Fr
27	Cobalt	Co	57	Lanthanum	La	88	Radium	Ra
28	Nickel	Ni	58	Cerium	Ce	89	Actinium	Ac
29	Copper	Cu	59	Praseodymium	Pr	90	Thorium	Th
30	Zinc	Zn	60	Neodymium	Nd	91	Protactinium	Pa
31	Gallium	Ga	61	Promethium	Pm	92	Uranium	U

THE ARTIFICIAL ELEMENTS

Atomic Number	Name	Symbol	Atomic Number	Name	Symbol	Atomic Number	Name	Symbol
93	Neptunium	Np	97	Berkelium	Bk	101	Mendelevium	Mv
94	Plutonium	Pu	98	Californium	Cf	102	Nobelium	No
95	Americium	Am	99	Einsteinium	E	103	Lawrencium	Lw
96	Curium	Cm	100	Fermium	Fm			

the compound

Actually, there are many more materials than there are elements. The reason for this is that the elements can be combined to produce materials that have characteristics that are completely different from the elements. Water, for example, is a *compound* that is made from the element hydrogen and the element oxygen. And, ordinary table salt is made from the element sodium and the element chlorine.

Notice how, although hydrogen and oxygen are gases, they can produce water as a liquid.

COMPOUNDS MAY BE PRODUCED BY COMBINING:

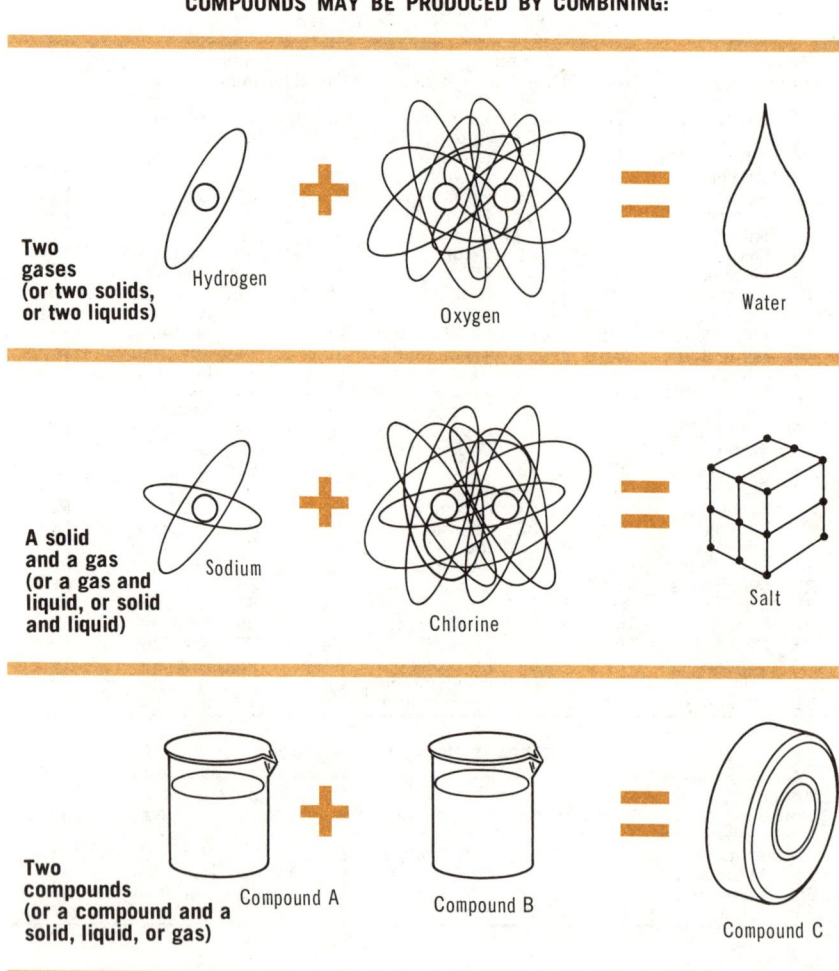

Two gases (or two solids, or two liquids)   Hydrogen   +   Oxygen   =   Water

A solid and a gas (or a gas and liquid, or solid and liquid)   Sodium   +   Chlorine   =   Salt

Two compounds (or a compound and a solid, liquid, or gas)   Compound A   +   Compound B   =   Compound C

the molecule

The molecule is the smallest particle that a *compound* can be reduced to before it breaks down into its elements. For example, if we took a grain of table salt and kept breaking it in half till it got as small as it possibly could and yet still be salt, we would have a *molecule* of salt. If we then broke it in half again, the salt would change into its elements.

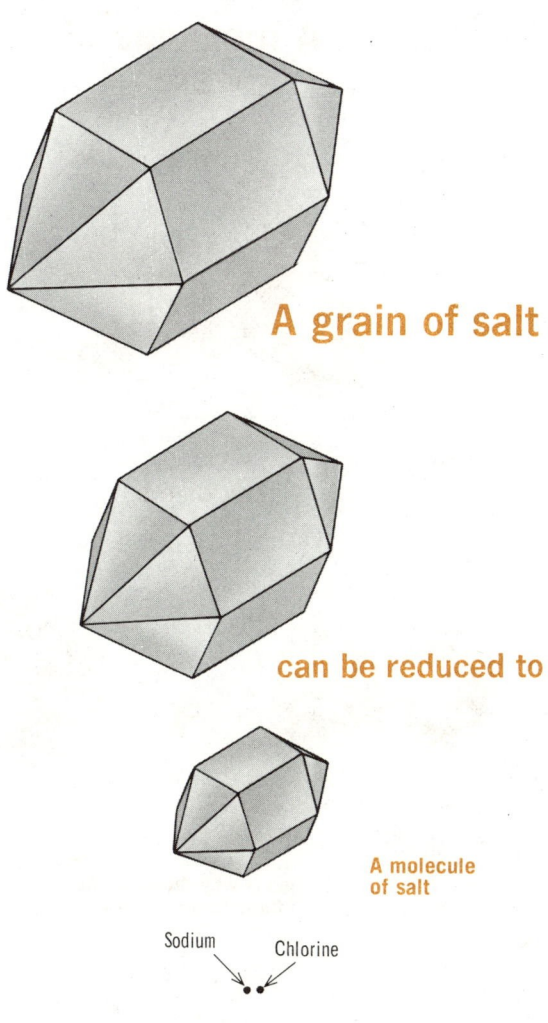

A grain of salt

can be reduced to

A molecule of salt

Sodium   Chlorine

the atom

The atom is the smallest particle that an *element* can be reduced to and still keep the properties of that element. If a drop of water was reduced to its smallest size, a molecule of water would be produced. But if that molecule of water was reduced still further, *atoms* of hydrogen and oxygen would appear.

A molecule of water

becomes

Two atoms
of hydrogen

One atom
of oxygen

This is why the chemical
formula for water is
H_2O

THE CARBON ATOM

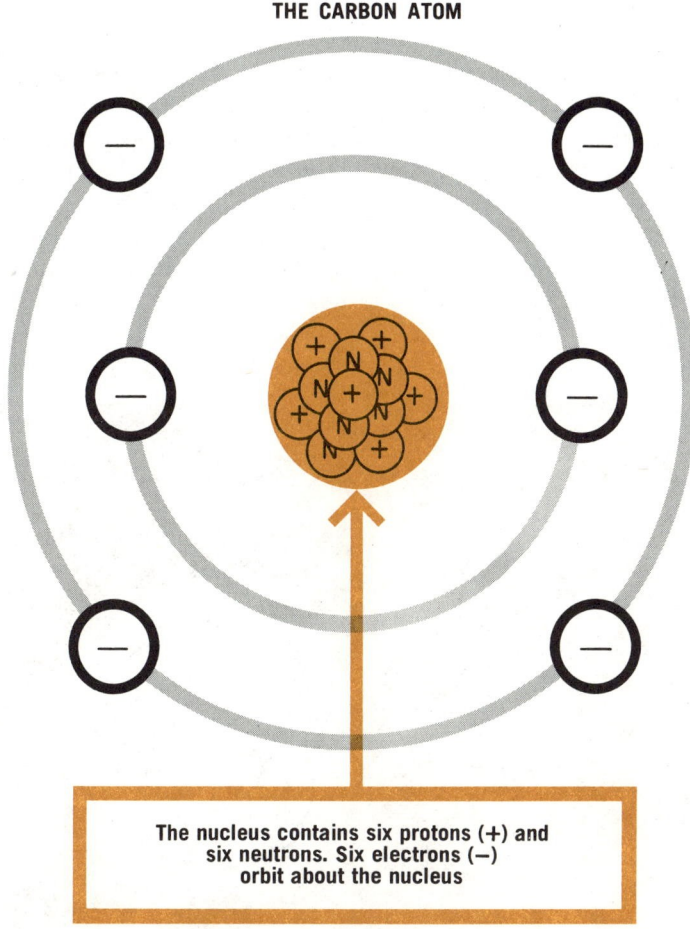

The nucleus contains six protons (+) and
six neutrons. Six electrons (−)
orbit about the nucleus

the structure of the atom

If the atom of an element is broken down any further, that element
would no longer exist in the particles that remain. The reason for this
is that these smaller particles are present in all the atoms of the different elements. The atom of one element differs from the atom of another
element only because it contains different numbers of these subatomic
particles.

Basically, an atom contains three types of subatomic particles that
are of interest in electricity: *electrons, protons,* and *neutrons.* The protons and neutrons are located in the center, or *nucleus,* of the atom,
and the electrons travel about the nucleus in *orbits.*

the nucleus

The *nucleus* is the central part of the atom. It contains the *protons* and *neutrons* of an atom. The number of protons in the nucleus determines how one atom of an element differs from another. For example, the nucleus of a hydrogen atom contains one proton, oxygen has eight, copper has 29, silver has 47, and gold has 79. As a matter of fact, this is how the different elements are identified by *atomic numbers,* as shown in the table on page 1-5. The atomic number is the number of protons that each atom has in its nucleus.

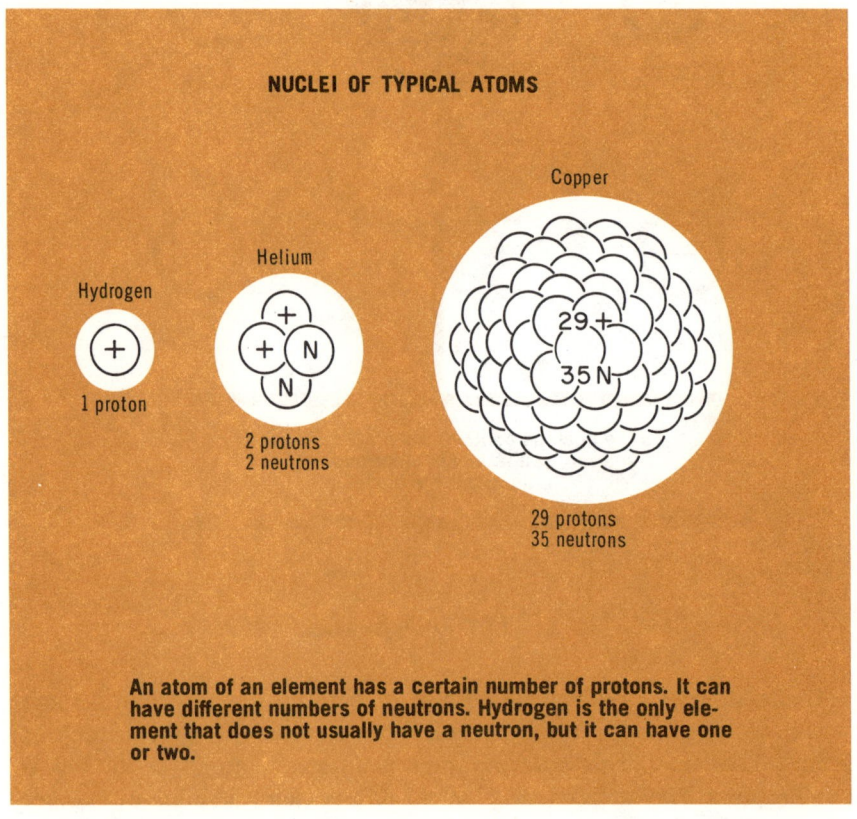

NUCLEI OF TYPICAL ATOMS

Hydrogen
1 proton

Helium
2 protons
2 neutrons

Copper
29 protons
35 neutrons

An atom of an element has a certain number of protons. It can have different numbers of neutrons. Hydrogen is the only element that does not usually have a neutron, but it can have one or two.

Although a neutron is actually a particle by itself, it is generally thought of as an electron and proton combined, and is *electrically neutral.* Since neutrons are electrically neutral, they are not too important to the electrical nature of atoms.

the proton

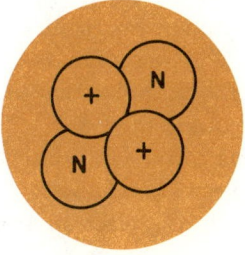

Since the nucleus of an atom contains neutrons, which are neutral,
and protons, which are positive,
the nucleus of any atom is always
positive

The positive lines of force of a proton go
straight out in all directions

The proton is very small. It is estimated to be 0.07 trillionth of an inch in diameter. The proton is one-third the diameter of an electron, but it has almost 1840 times the *mass* of an electron; the proton is almost 1840 times *heavier* than the electron. It is extremely difficult to dislodge a proton from the nucleus of an atom. Therefore, in electrical theory, protons are considered permanent parts of the nucleus. Protons do not take an active part in the flow or transfer of electrical energy.

The *proton* has a *positive* electrical *charge*. The lines of force of this charge go straight *out* in all directions from the proton.

the electron

As explained earlier, the electron is three times larger in diameter than the proton, or about 0.22 trillionth of an inch; but it is about 1840 times lighter than the proton. The electrons are *easy to move*. They are the particles that actively participate in the flow or transfer of electrical energy.

Electrons revolve in orbits around the nucleus of an atom, and have *negative* electrical *charges*. The lines of force of these charges come straight *in* to the electron from all sides.

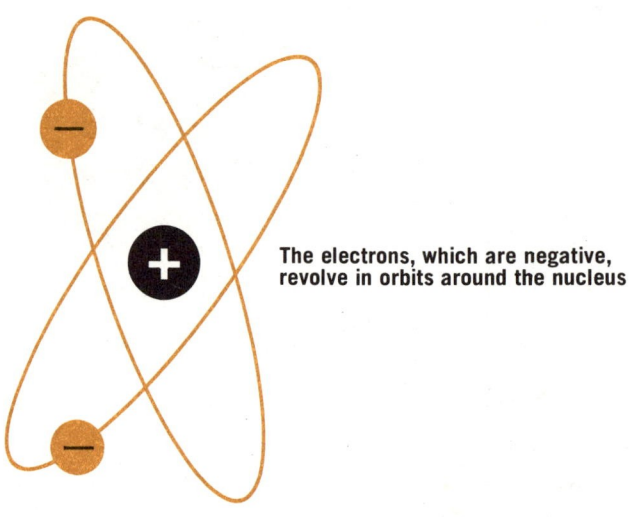

The electrons, which are negative, revolve in orbits around the nucleus

The negative lines of force of an electron come straight in from all directions

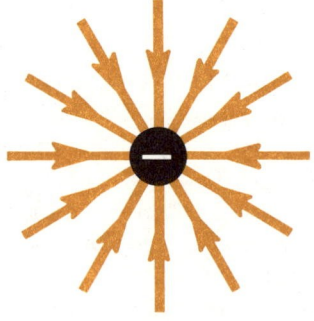

summary

☐ Electricity is produced by tiny particles called electrons and protons. ☐ Matter is anything that has weight and occupies space. It can be in the form of a solid, liquid, or gas. ☐ The basic materials that make up matter are the elements. ☐ There are ninety-two natural elements. All others are man-made. ☐ The elements can be combined to form compounds, with characteristics that are completely different from the elements from which they are produced.

☐ The molecule is the smallest particle that a compound can be reduced to before it breaks down into its elements. ☐ The atom is the smallest particle that an element can be reduced to and still keep the properties of that element. ☐ The atom of one element differs from the atom of another element only because it contains different numbers of subatomic particles. ☐ The three basic types of subatomic particles that are of interest in electricity are electrons, protons, and neutrons.

☐ The nucleus is the central part of the atom. ☐ The number of protons in the nucleus determines how the elements differ from each other. The different elements are identified by the atomic number. The atomic number is the number of protons in the nucleus. ☐ The proton has a positive charge, is smaller but 1840 times heavier than the electron, and is in the nucleus of the atom. It is difficult to dislodge from the nucleus. ☐ The electron has a negative charge, and is larger but 1840 times lighter than the proton. It revolves around the nucleus in orbits, and is easy to move. ☐ The neutron is electrically neutral, and is in the nucleus.

review questions

1. What particles produce electricity?
2. There are how many natural occurring elements?
3. Define *atomic number* of an element.
4. The proton has a _____ charge, and the electron has a _____ charge.
5. What particles are found in the nucleus of an atom? In the orbits?
6. What is the smallest particle that retains the characteristics of the compound? Of the element?
7. Is salt an element or a compound? Oxygen? Water?
8. Which is heavier, and by how much: a proton or an electron?
9. Which has a greater diameter: a proton or an electron? By how much?
10. What is the electrical charge of a neutron?

the law of electrical charges

The *negative* charge of an electron is *equal* but *opposite* to the *positive* charge of a proton.

The charges on an electron and a proton are called *electrostatic charges*. The lines of force associated with each particle produce *electrostatic fields*. Because of the way these fields act together, charged particles can attract or repel one another. The law of electrical charges is that particles with *like charges repel* each other, and those with *unlike charges attract* each other.

A proton (+) repels another proton (+).
An electron (−) repels another electron (−).
A proton (+) attracts an electron (−).

Because protons are relatively heavy, the repulsive force they exert on one another in the nucleus of an atom has little effect.

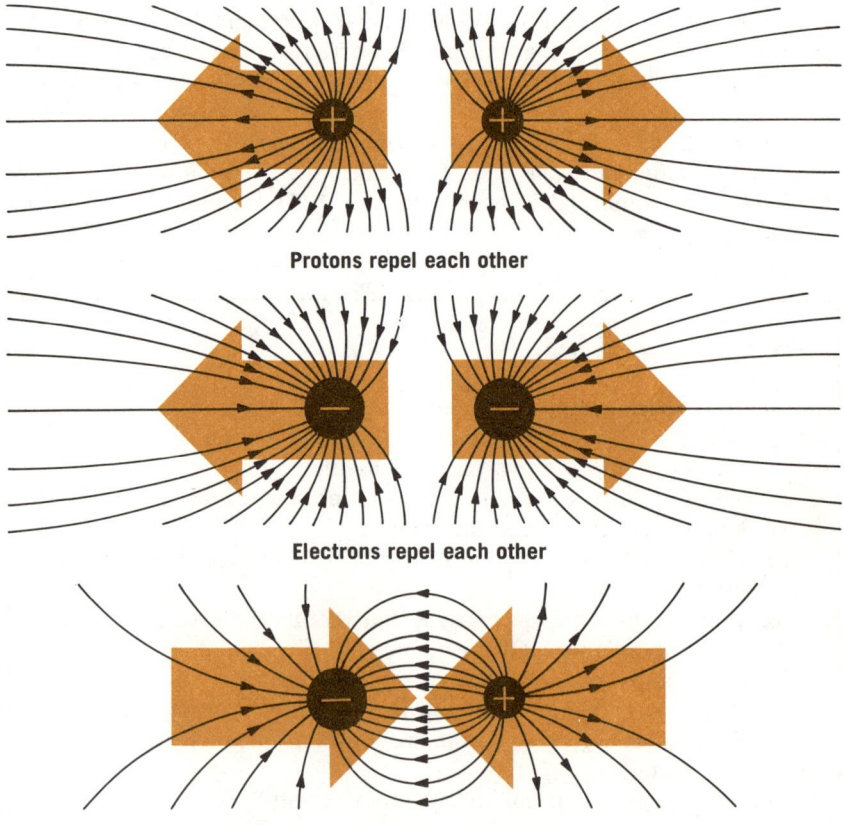

Protons repel each other

Electrons repel each other

Electrons and protons attract each other

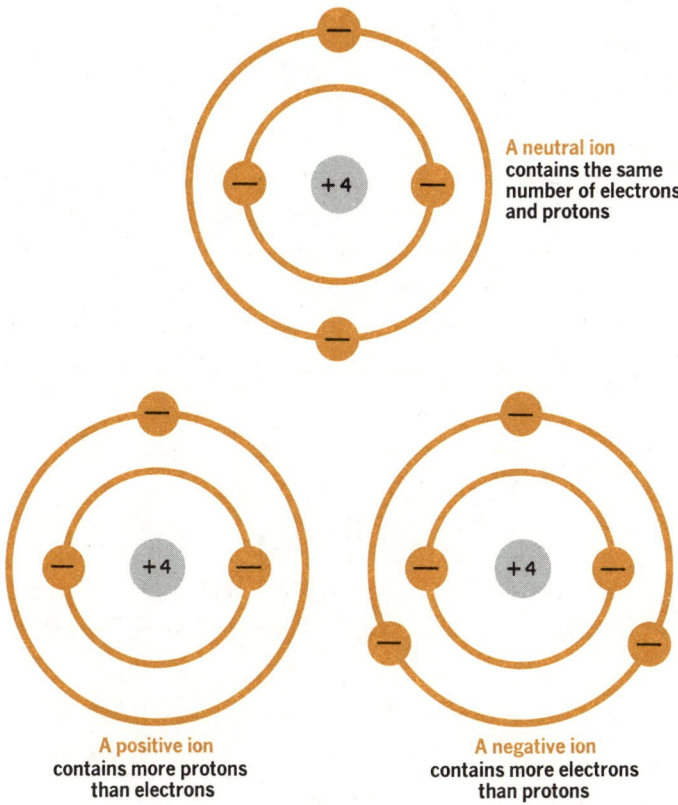

A neutral ion
contains the same
number of electrons
and protons

A positive ion
contains more protons
than electrons

A negative ion
contains more electrons
than protons

atomic charges

Normally, an atom contains the same number of electrons and protons, so that the *equal* and *opposite* negative and positive charges cancel each other to make the atom electrically *neutral*. But, as was explained earlier, since it is the number of protons in the nucleus that gives the atom of an element its properties, the number of electrons can be changed.

The drawings on this page show beryllium atoms, which have four protons in the nucleus. When the atom also has four electrons, the equal number of positive charges (protons) and negative charges (electrons) cancel each other, and the atom has no charge. If the beryllium atom has only three electrons, it will have more protons (+) than electrons (−), so the atom will have a *positive charge*. When the atom has five electrons, it will have more electrons (−) than protons (+). The atom will then have a *negative charge*.

Charged atoms are called *ions*. A positively charged atom is a *positive ion*, and a negatively charged atom is a *negative ion*.

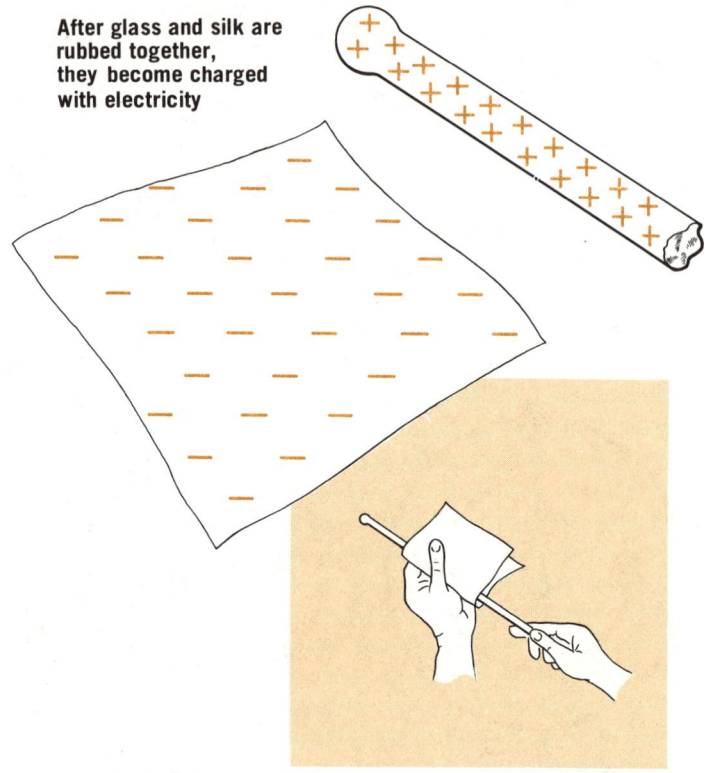

After glass and silk are
rubbed together,
they become charged
with electricity

charged materials

If a large number of atoms in a piece of neutral material loses or gains electrons, that material will become charged. Atoms can be made to do this in a number of ways, as will be explained later. The method that the ancient Greeks discovered was by *friction*. For example, if we rub a glass rod with a piece of silk, the glass rod will give up electrons to the silk. The glass rod will become positively charged, and the silk will become negatively charged.

The reason these charges result is because the glass rod has surface electrons that are easily dislodged by friction. This same thing will happen when any two materials are rubbed together, as long as one material can give up electrons easily, and the other material will accept those electrons readily. If you experiment with a few different combinations, you will find that some work well while others do not. If you comb your hair with a rubber or plastic comb, you will find that the comb will become charged, and will attract pieces of paper. Along the same lines, if you rub a rubber rod with fur, the rod will be charged negatively because it picks up electrons from the fur.

charging by contact

Suppose you had a rubber rod that was charged negatively by a piece of fur. Using this charged rubber rod, you could now *charge* other materials such as copper just by *touching* them. This method is called charging by contact, and it works because the negative charge of the rod tries to *repel electrons* from the rod's surface.

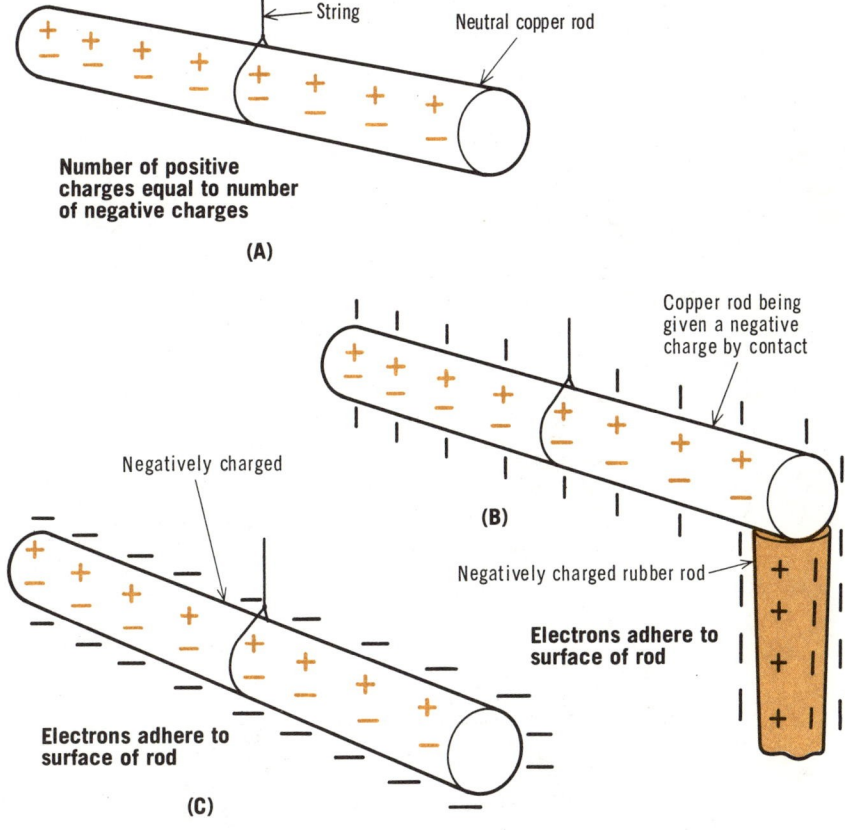

Number of positive
charges equal to number
of negative charges

(A)

Copper rod being
given a negative
charge by contact

Negatively charged

(B)

Negatively charged rubber rod

Electrons adhere to
surface of rod

Electrons adhere to
surface of rod

(C)

The electrons on the surface of the rubber rod will go onto the surface of the *suspended* copper rod to give it a negative charge. If a positive glass rod is used instead of a negative rubber rod, electrons would be attracted from the surface of the copper rod to give it a positive charge.

charging by induction

Because electrons and protons have *attracting* and *repelling* forces, an object can be charged without being touched by the charged body. For example, if the negatively charged rubber rod is brought *close* to a piece of aluminum, the *negative* force from the rubber rod will *repel* the *electrons* in the aluminum rod to the other end. One end of the rod will then be negative, and the other positive. If we move the rubber rod away, the electrons in the aluminum rod will redistribute themselves to neutralize the rod. If you want the aluminum to remain charged,

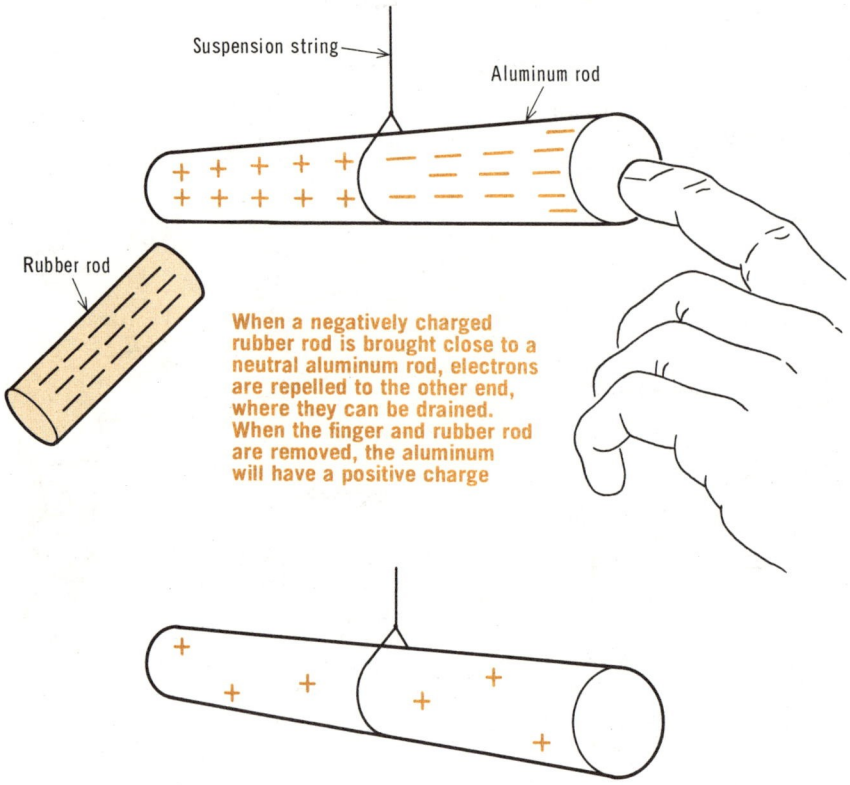

Suspension string

Aluminum rod

Rubber rod

When a negatively charged rubber rod is brought close to a neutral aluminum rod, electrons are repelled to the other end, where they can be drained. When the finger and rubber rod are removed, the aluminum will have a positive charge

bring the rubber rod close again and then touch the negative end with your finger. Electrons will leave the rod through your body. (Very small charges are involved here, and so you will not feel it.) Then, when your finger is taken away before the rubber rod is removed, the aluminum rod will remain charged. This method is called *charging by induction*.

neutralizing a charge

After glass and silk are rubbed together, they become charged with electricity. But, if the glass rod and silk are brought together again, the attraction of the positive ions in the rod pulls the electrons back out of the silk until both materials become electrically neutral.

Unlike charges

Electrons leap across gap

The arc is due to the force of attraction

Lightning is an arc discharge

A wire can also be connected between the charged bodies to *discharge* them. But, if the charges on both materials are strong enough, they could discharge through an arc, like lightning.

attraction and repulsion

If you give glass rods a positive charge by rubbing them with silk, and give rubber rods a negative charge by rubbing them with fur, and then experiment with the glass, rubber, silk, and fur, without letting them touch, you would find that:

Like charges repel.

Unlike charges attract.

If you pivot the negative rubber rod to swing freely, it would be repelled by another negative rubber rod or a negative piece of silk if it were brought close to it. On the other hand, the negative rubber rod would be attracted toward the positive glass rod or the positive piece of fur. The two positive glass rods would repel each other just the way the two negative rubber rods did.

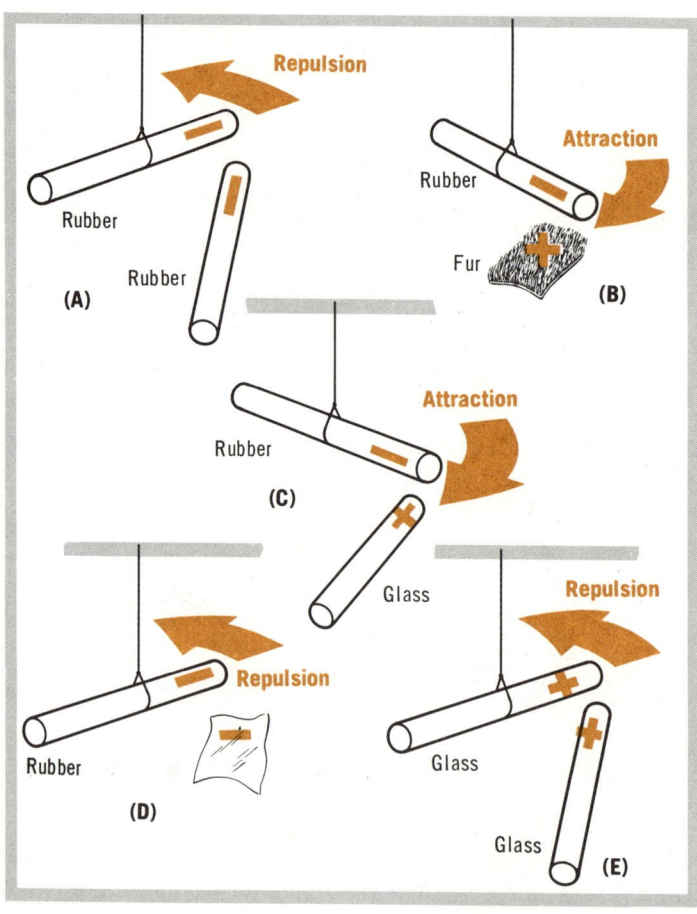

electrostatic fields

The attracting and repelling forces on charged materials occur because of the *electrostatic lines of force* that exist around the charged materials.

In a *negatively charged* object, the lines of force of the excess electrons add to produce an electrostatic field that has lines of force coming *into* the object from all directions.

In a *positively charged* object, the lack of electrons causes the lines of force on the excess protons to add to produce an electrostatic field that has lines of force going *out* of the object in all directions.

These electrostatic fields either *aid* or *oppose* each other to *attract* or *repel*.

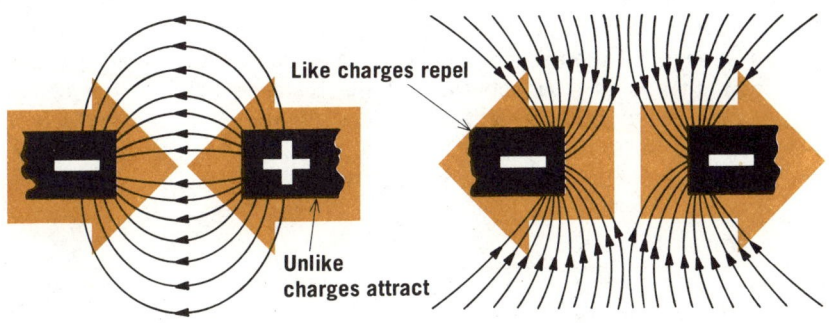

Like charges repel

Unlike charges attract

The strength of the attraction or repulsion force depends on two factors: (1) the amount of *charge* that is on each object, and (2) the distance between the objects. The greater the electric charges on the objects, the greater will be the electrostatic force. And the closer the charged objects are to each other, the greater the electrostatic force. The attraction or repulsion force gets weaker if either charge is reduced, or if the objects are moved farther apart.

During the 18th century, a scientist named Coulomb experimented with electrostatic charges and came up with a law of electrostatic attraction, which is commonly referred to as *Coulomb's law of electrostatic charges*. The law is: the force of electrostatic attraction or repulsion is directly proportional to the product of the two charges, and inversely proportional to the square of the distance between them. Of course, the more surplus electrons that a charged object has, the greater its negative charge will be. And the greater its lack of electrons, the greater its positive charge. This is explained further on page 1-56.

summary

☐ The negative electrostatic charge of an electron is equal and opposite to the positive charge of a proton. ☐ Electrostatic fields are produced by the lines of force associated with the charges. ☐ Like charges repel. A proton (+) repels another proton (+). An electron (−) repels another electron (−). ☐ Unlike charges attract. A proton (+) attracts an electron (−).

☐ An atom is neutral if it contains the same number of protons and electrons. ☐ If an atom contains less electrons than protons, it has a positive charge. ☐ If an atom contains more electrons than protons, it has a negative charge. ☐ Atoms that have a positive or negative charge are called ions. ☐ Objects can be charged by friction, contact, or induction. ☐ Neutralizing a charged object can be done by bringing it in contact with an object of opposite charge.

☐ Lines of force enter a negatively charged object, by convention. ☐ Lines of force leave a positively charged object, by convention. ☐ The electrostatic fields created by the lines of force aid or oppose each other to attract or repel. ☐ The strength of the attraction or repulsion force depends on the amount of charge that is on each object, and the distance between the objects. ☐ Coulomb's Law of Electrostatic Charges relates the forces of attraction and repulsion: Force is directly proportional to the product of the two charges, and inversely proportional to the square of the distance between them.

review questions

1. If an electron were brought in the vicinity of a proton, would the proton repel or attract the electron?
2. Why don't protons in a nucleus repel each other with sufficient force to split the nucleus?
3. Do the protons in a nucleus have any repulsive force on each other?
4. What is the polarity of the charge of an object that has less electrons than protons?
5. Name three ways in which a material can be charged.
6. If a rubber rod is rubbed with a piece of fur, what is the polarity of the rubber? What is the polarity of the fur?
7. How can a charged object be neutralized?
8. Do the lines of force enter or leave an electron?
9. Does the force of repulsion between two electrons increase or decrease with distance? If the distance is doubled, what is the magnitude of the new force as compared to the old?
10. State Coulomb's Law.

electron orbits

As you have seen, electricity is produced when electrons leave their atoms. To understand the different ways of accomplishing this, it would be helpful to know more about the nature of the different *electron orbits* about the nucleus of an atom.

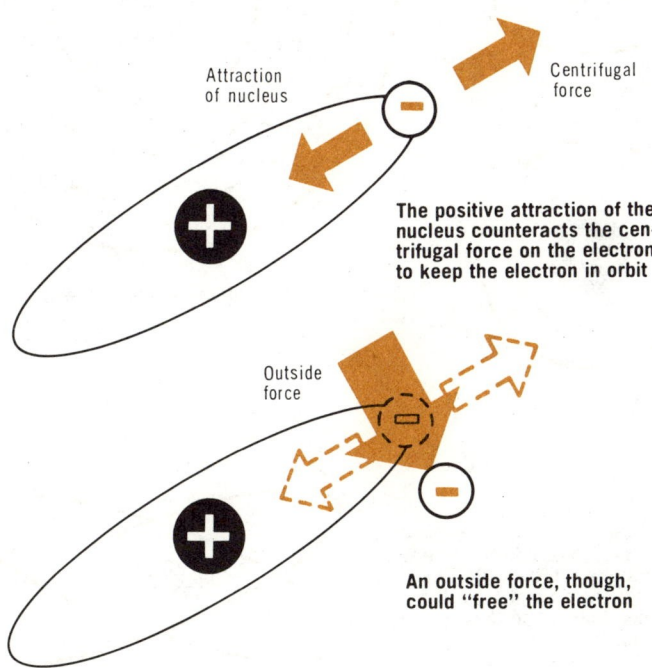

Attraction
of nucleus

Centrifugal
force

The positive attraction of the nucleus counteracts the centrifugal force on the electron to keep the electron in orbit

Outside
force

An outside force, though, could "free" the electron

The electrons *revolve* at high speed in their orbits about the atom's nucleus. Because of the electron's great speed, centrifugal force tends to pull the electron out of orbit. But the *positive attraction* of the *nucleus* keeps the electron from breaking away. However, if a sufficient outside force were applied to aid the centrifugal force, the electron could be "freed."

When a force is applied to an atom, it gives energy to the orbiting electron, and it is the amount of energy absorbed by the electron that determines whether or not it will be freed. There are a number of ways that energy can be applied, which you will learn about later: friction, chemicals, heat, pressure, magnetism, and light.

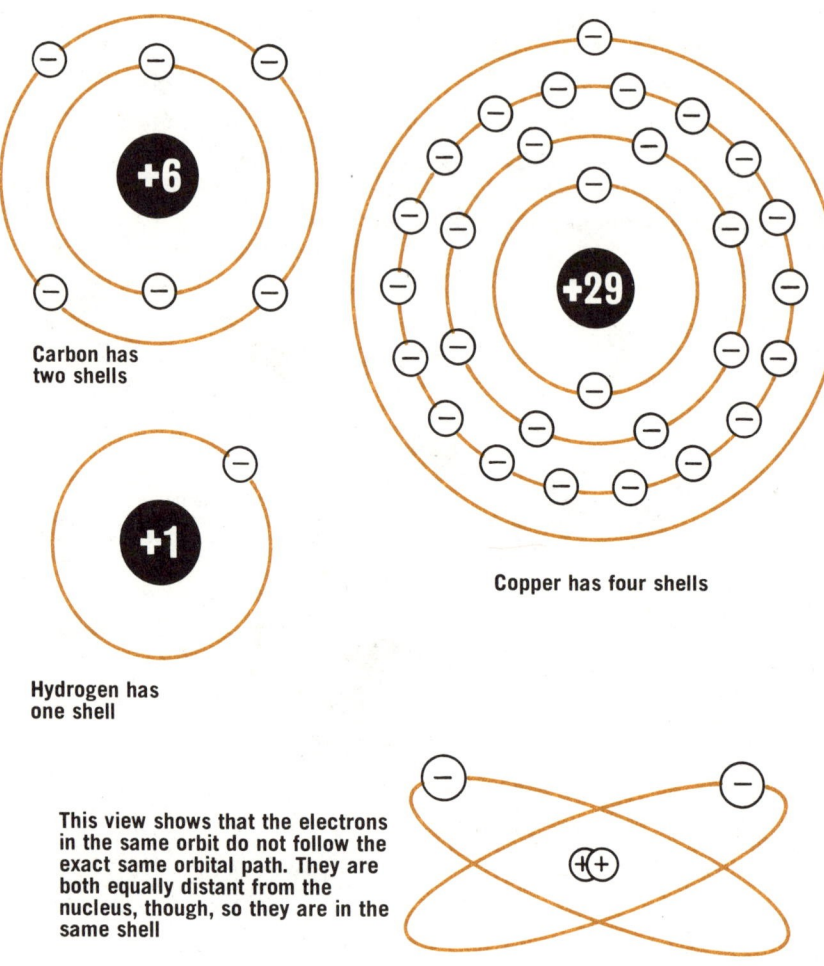

Carbon has
two shells

Hydrogen has
one shell

Copper has four shells

This view shows that the electrons
in the same orbit do not follow the
exact same orbital path. They are
both equally distant from the
nucleus, though, so they are in the
same shell

orbital shells

Electrons that orbit close to the nucleus are difficult to free because they are close to the positive force that holds them. The farther the electrons are from the nucleus, the weaker the positive force becomes. As you might have noticed in some of the earlier diagrams, the more electrons that an atom has, the more orbits there are. The orbital paths are commonly called *shells*.

The atoms of all the known elements can have up to seven shells. The table on page 1-25 lists the 103 elements, showing the number of electrons in each shell for each atom.

the elements
and their atomic shells

ELECTRON SHELLS

Atomic No.	Element	1	2	3	4	5	Atomic No.	Element	1	2	3	4	5	6	7
1	Hydrogen, H	1					53	Iodine, I	2	8	18	18	7		
2	Helium, He	2					54	Xenon, Xe	2	8	18	18	8		
3	Lithium, Li	2	1				55	Cesium, Cs	2	8	18	18	8	1	
4	Beryllium, Be	2	2				56	Barium, Ba	2	8	18	18	8	2	
5	Boron, B	2	3				57	Lanthanum, La	2	8	18	18	9	2	
6	Carbon, C	2	4				58	Cerium, Ce	2	8	18	19	9	2	
7	Nitrogen, N	2	5				59	Praseodymium, Pr	2	8	19	20	9	2	
8	Oxygen, O	2	6				60	Neodymium, Nd	2	8	19	21	9	2	
9	Fluorine, F	2	7				61	Promethium, Pm	2	8	18	22	9	2	
10	Neon, Ne	2	8				62	Samarium, Sm	2	8	18	23	9	2	
11	Sodium, Na	2	8	1			63	Europium, Eu	2	8	18	24	9	2	
12	Magnesium, Mg	2	8	2			64	Gadolinium, Gd	2	8	18	25	9	2	
13	Aluminum, Al	2	8	3			65	Terbium, Tb	2	8	18	26	9	2	
14	Silicon, Si	2	8	4			66	Dysprosium, Dy	2	8	18	27	9	2	
15	Phosphorus, P	2	8	5			67	Holmium, Ho	2	8	18	28	9	2	
16	Sulfur, S	2	8	6			68	Erbium, Er	2	8	18	29	9	2	
17	Chlorine, Cl	2	8	7			69	Thulium, Tm	2	8	18	30	9	2	
18	Argon, A	2	8	8			70	Ytterbium, Yb	2	8	18	31	9	2	
19	Potassium, K	2	8	8	1		71	Lutetium, Lu	2	8	18	32	9	2	
20	Calcium, Ca	2	8	8	2		72	Hafnium, Hf	2	8	18	32	10	2	
21	Scandium, Sc	2	8	9	2		73	Tantalum, Ta	2	8	18	32	11	2	
22	Titanium, Ti	2	8	10	2		74	Tungsten, W	2	8	18	32	12	2	
23	Vanadium, V	2	8	11	2		75	Rhenium, Re	2	8	18	32	13	2	
24	Chromium, Cr	2	8	13	1		76	Osmium, Os	2	8	18	32	14	2	
25	Manganese, Mn	2	8	13	2		77	Iridium, Ir	2	8	18	32	15	2	
26	Iron, Fe	2	8	14	2		78	Platinum, Pt	2	8	18	32	16	2	
27	Cobalt, Co	2	8	15	2		79	Gold, Au	2	8	18	32	18	1	
28	Nickel, Ni	2	8	16	2		80	Mercury, Hg	2	8	18	32	18	2	
29	Copper, Cu	2	8	18	1		81	Thallium, Tl	2	8	18	32	18	3	
30	Zinc, Zn	2	8	18	2		82	Lead, Pb	2	8	18	32	18	4	
31	Gallium, Ga	2	8	18	3		83	Bismuth, Bi	2	8	18	32	18	5	
32	Germanium, Ge	2	8	18	4		84	Polonium, Po	2	8	18	32	18	6	
33	Arsenic, As	2	8	18	5		85	Astatine, At	2	8	18	32	18	7	
34	Selenium, Se	2	8	18	6		86	Radon, Rn	2	8	18	32	18	8	
35	Bromine, Br	2	8	18	7		87	Francium, Fr	2	8	18	32	18	8	1
36	Krypton, Kr	2	8	18	8		88	Radium, Ra	2	8	18	32	18	8	2
37	Rubidium, Rb	2	8	18	8	1	89	Actinium, Ac	2	8	18	32	18	9	2
38	Strontium, Sr	2	8	18	8	2	90	Thorium, Th	2	8	18	32	19	9	2
39	Yttrium, Y	2	8	18	9	2	91	Protactinium, Pa	2	8	18	32	20	9	2
40	Zirconium, Zr	2	8	18	10	2	92	Uranium, U	2	8	18	32	21	9	2
41	Niobium, Nb	2	8	18	12	1	93	Neptunium, Np	2	8	18	32	22	9	2
42	Molybdenum, Mo	2	8	18	13	1	94	Plutonium, Pu	2	8	18	32	23	9	2
43	Technetium, Tc	2	8	18	14	1	95	Americium, Am	2	8	18	32	24	9	2
44	Ruthenium, Ru	2	8	18	15	1	96	Curium, Cm	2	8	18	32	25	9	2
45	Rhodium, Rh	2	8	18	16	1	97	Berkelium, Bk	2	8	18	32	26	9	2
46	Palladium, Pd	2	8	18	18	0	98	Californium, Cf	2	8	18	32	27	9	2
47	Silver, Ag	2	8	18	18	1	99	Einsteinium, E	2	8	18	32	28	9	2
48	Cadmium, Cd	2	8	18	18	2	100	Fermium, Fm	2	8	18	32	29	9	2
49	Indium, In	2	8	18	18	3	101	Mendelevium, Mv	2	8	18	32	30	9	2
50	Tin, Sn	2	8	18	18	4	102	Nobelium, No	2	8	18	32	31	9	2
51	Antimony, Sb	2	8	18	18	5	103	Lawrencium, Lw	2	8	18	32	32	9	2
52	Tellurium, Te	2	8	18	18	6									

shell capacity

If you study the table on page 1-25 briefly, you will notice that each shell can only hold a certain number of electrons. The shell closest to the nucleus (the first shell) cannot hold more than 2 electrons; the second shell cannot hold more than 8 electrons; the third, no more than 18; the fourth, no more than 32; and so on.

If you look again at the table on page 1-25, you will see that up to atomic number 10, the second shell built up to 8 electrons. Since this is the limit for the second shell, a third shell had to be started. From atomic numbers 11 through 18, the third shell built up to 8, and then a fourth shell started. Then, from numbers 19 through 29, the third shell built up to its maximum of 18.

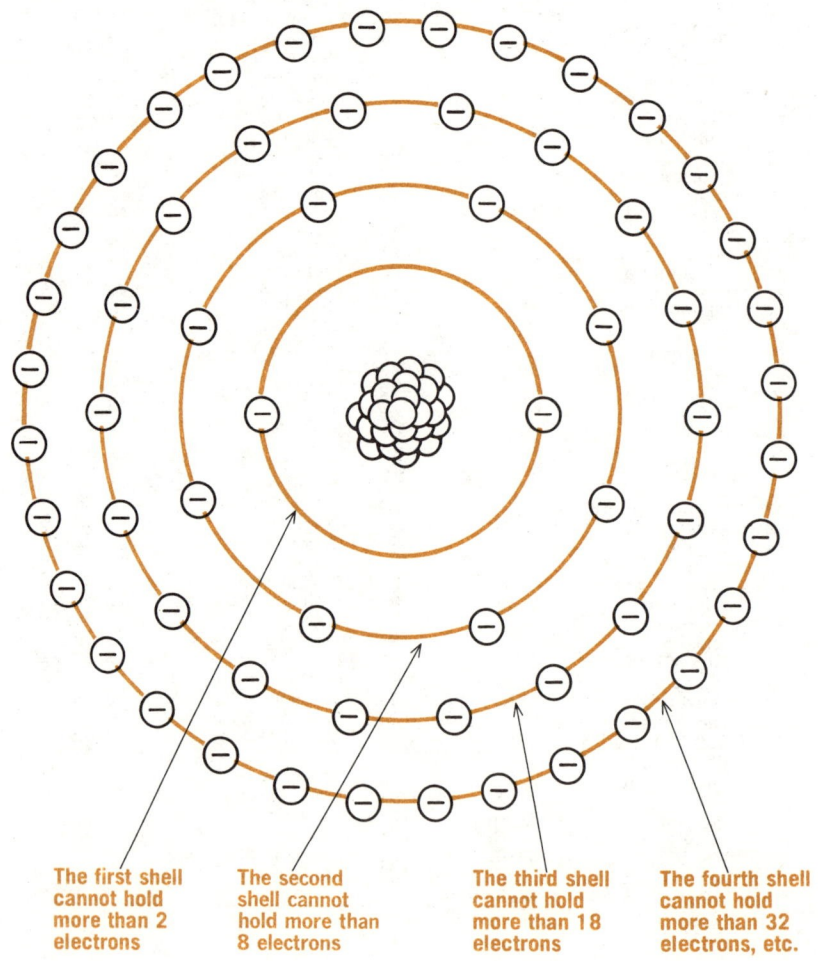

The first shell cannot hold more than 2 electrons

The second shell cannot hold more than 8 electrons

The third shell cannot hold more than 18 electrons

The fourth shell cannot hold more than 32 electrons, etc.

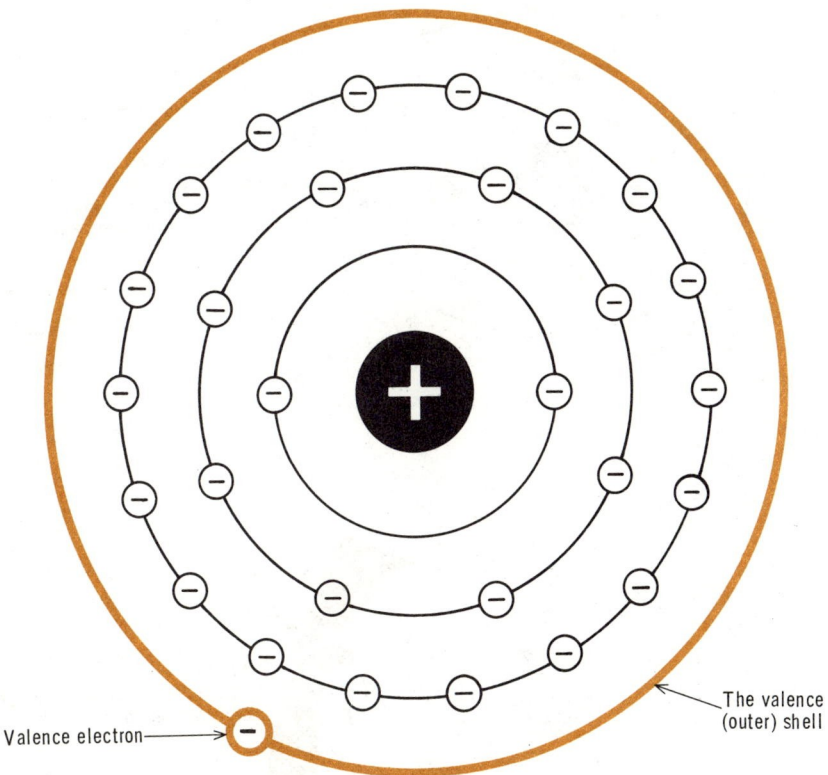

Valence electron⟶

The valence
(outer) shell

**The outer shell is called the valence shell, and the
electrons in that shell are called valence electrons**

the outer (valence) shell

As you can notice in the table on page 1-25, although the third shell can hold up to 18 electrons, it did not take on any more than 8 electrons until the fourth shell started. This is also true of the fourth shell. It will not take on any more electrons than 8 until a fifth shell starts, even though the fourth shell can hold up to 32 electrons. This shows that there is another rule. *The outer shell of an atom will have no more than 8 electrons.* The outer shell of an atom is called the *valence shell,* and its electrons are called *valence electrons.* The number of electrons in the valence shell of an atom is important in electricity, as you will see later.

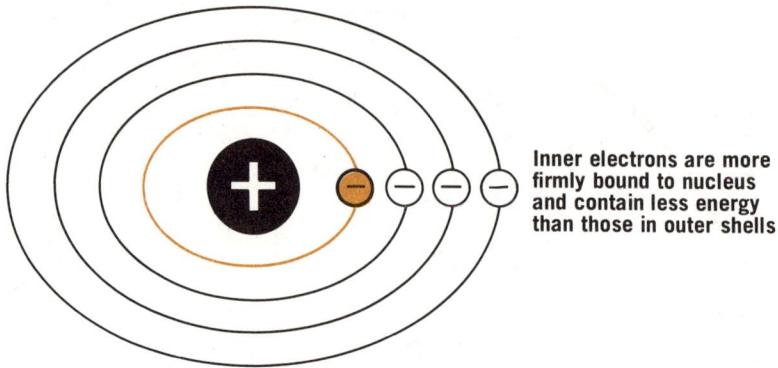

Inner electrons are more firmly bound to nucleus and contain less energy than those in outer shells

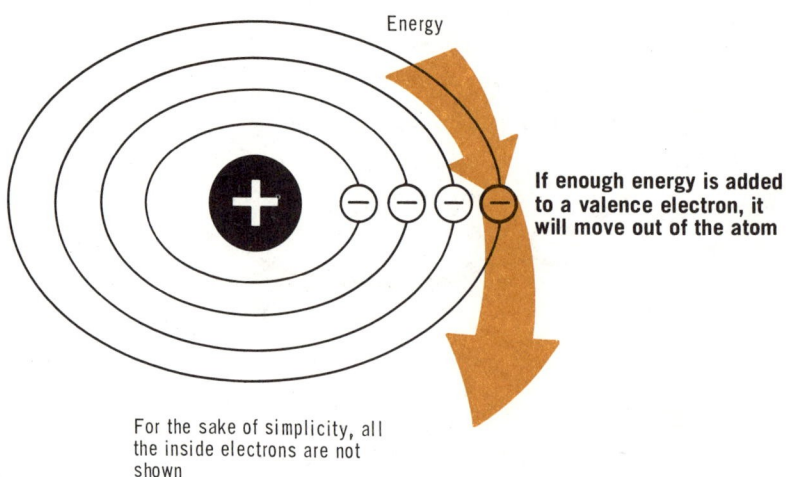

Energy

If enough energy is added to a valence electron, it will move out of the atom

For the sake of simplicity, all the inside electrons are not shown

electron energy

Although every electron has the same negative charge, not all electrons have the same *energy level*. The electrons that orbit close to the nucleus contain less energy than those that orbit further away. The further an electron orbits from the nucleus, the greater its energy.

If enough energy is added to an electron, that electron will move *out of its orbit* to the next higher orbit. And, if enough energy is added to a *valence* electron, that electron will move *out of its atom,* since there is no next higher orbit.

producing electricity

Electricity is produced when *electrons are freed* from their atoms. Since the valence electrons are farthest from the attractive force of the nucleus and also have the highest energy level, they are the electrons that are most easily set free. When enough force or energy is applied to an atom, the valence electrons will become free. However, the energy supplied to a valence shell is distributed amongst the electrons in that shell. Therefore, for a given amount of energy, the more valence electrons there are, the *less* energy each electron will get.

Energy is applied to the valence shell, and is distributed amongst the valence electrons

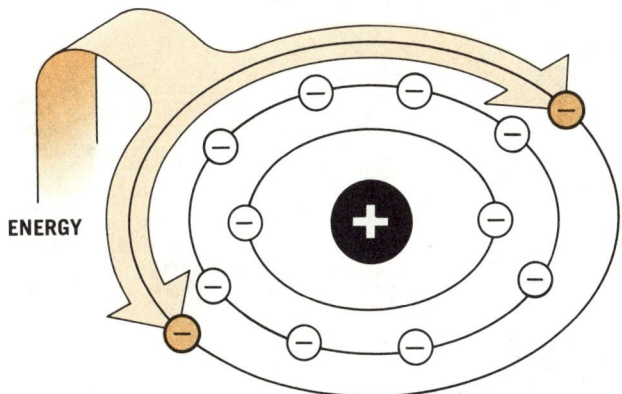

ENERGY

Two electrons share the energy equally

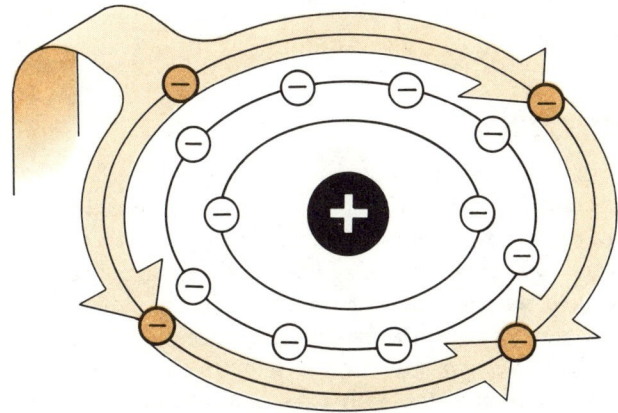

Four electrons share the energy equally, but each has less energy gained than any electron above

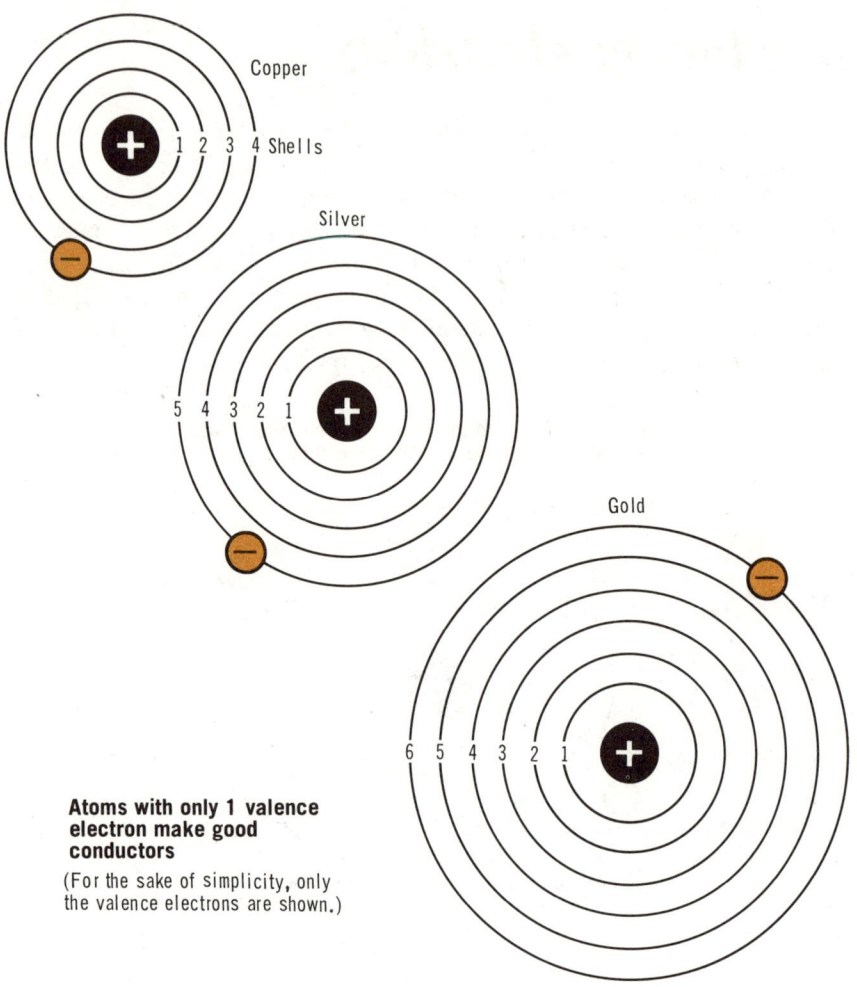

Copper

1 2 3 4 Shells

Silver

5 4 3 2 1

Gold

6 5 4 3 2 1

Atoms with only 1 valence electron make good conductors

(For the sake of simplicity, only the valence electrons are shown.)

conductors

The valence shell can contain up to 8 valence electrons. Since valence electrons share any energy applied to them, the atoms that have less valence electrons will more easily allow those electrons to be freed. Materials that have electrons that are easily freed are called *conductors*. The atoms of conductors have only 1 or 2 valence electrons. The ones with only *1 valence electron* are the best electrical conductors.

If you look at the atomic table on page 1-25, you can pick the good conductors. They all have one electron in their outer shell. Most *metals* are good conductors. The ones you are probably most familiar with are: copper (No. 29), silver (No. 47), and gold (No. 79).

An atom that is more than half filled, but has less than 8 electrons, tries to become stable by filling its valence shell

Valence shell

These atoms make good insulators because it is very difficult to free an electron from their valence shell

insulators

Insulators are materials from which electrons are very difficult to free. The atoms of insulators have their valence shells filled with 8 electrons or are more than half filled. Any energy applied to such an atom will be distributed amongst a relatively large number of electrons. But, in addition to this, these atoms resist giving up their electrons because of a phenomenon known as *chemical stability*.

An atom is completely *stable* when its *outer shell* is completely *filled* or when it has 8 valence electrons. A stable atom resists any sort of activity. In fact, it will not combine with any other atoms to form compounds. There are six naturally stable elements: helium, neon, argon, krypton, xenon, and radon. These are known as the *inert gases*.

All atoms that have less than 8 valence electrons tend to attain the stable state. Those that are less than half filled (the conductors), tend to release their electrons to empty the unstable shell. But those that are more than half filled (the insulators), strive to collect electrons to fill up the valence shell. So, not only is it difficult to free their electrons, but these atoms of the insulators also oppose the production of electricity with their tendency to catch any electrons that may be freed. Those atoms that have 7 valence electrons most actively try to be filled, and are excellent electrical insulators.

semiconductors

Semiconductors are those materials that are neither good conductors nor good insulators. In other words, they can conduct electricity better than insulators can, but not as well as conductors can.

Semiconductors do not conduct electricity as well as good conductors because they have more than 1 or 2 valence electrons, and so do not give up the electrons so easily. By the same token, semiconductors have less than 7 or 8 valence electrons, which is what insulators have. Thus, they do not resist giving up electrons as much as the insulators do, and therefore will allow some conduction.

An atom that has 4 valence electrons is a semiconductor

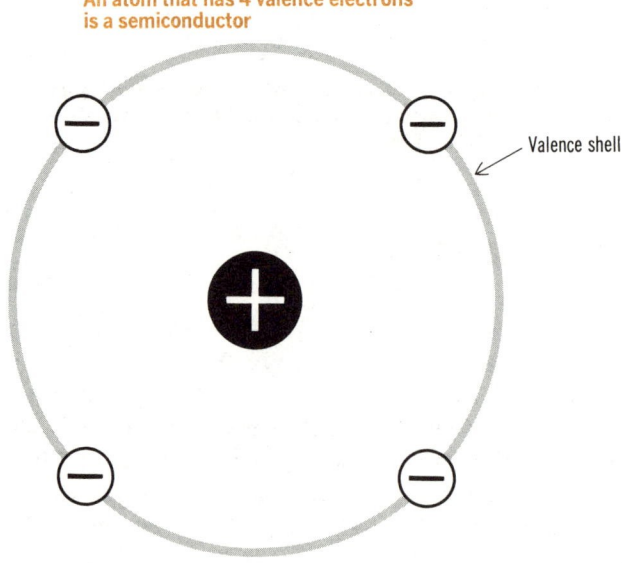

Valence shell

These atoms give up electrons less easily than good conductor atoms, but more easily than good insulator atoms

Generally, good conductors have their valence shells less than half filled, and good insulators have their valence shells more than half filled. So, a semiconductor is a material whose atoms are *half filled*. Those atoms have 4 valence electrons.

A review of the table on page 1-25 shows that carbon, silicon, germanium, tin, and lead are all semiconductors, since they have 4 electrons in their outer (valence) shell.

atomic bonds

Until now, you have studied the characteristics of individual atoms in relation to conductors, semiconductors, and insulators. Actually, when you work with materials, the characteristics of the *molecules* are what become important. As you recall, atoms join together to form molecules. Because of the nature of the way atoms bond together, the characteristics of the molecules they form can be quite different from those of the atoms that form the molecule.

Two hydrogen atoms join together when each of them shares its electrons with the other to form a stable hydrogen molecule

You remember that atoms tend to strive for *chemical stability*, the condition in which the outer or valence shell is completely filled. Atoms, then, tend to bond together in such a way that this stability is brought about.

Hydrogen is a good example of the simple way this happens. As the table on page 1-25 shows, the hydrogen atom has only one shell and one electron. The table also shows that the maximum number of electrons this first shell can hold is 2. Thus, the hydrogen atoms will combine to become stable with each atom having 2 valence electrons. This is done by the two atoms joining together and *sharing* their electrons. So, even though there is a total of only 2 electrons for both atoms, when the *electron pair is shared*, each atom sees 2 electrons, and a stable molecule is produced. This is known as *electron pair* or *covalent bonding*.

Other atoms join in similar ways, with each determined by the number needed for stability. Take chlorine, for example, which has 7 valence electrons. As the table on page 1-25 shows, 8 electrons are needed in its valence shell for stability. If two atoms each permit one electron to be shared as a pair, each atom in the completed molecule will have 8 valence electrons. For oxygen, which has 6 valence electrons, three atoms will join, with one forming an *electron pair bond* with each

atomic bonds (cont.)

of the other two atoms. You can see, then, that the number of atoms that must join together to form a *stable molecule* depends on the particular atoms involved. With atoms that have fewer valence electrons, the molecular structure becomes more and more complex.

On the other hand, those atoms of elements whose valence shells are already filled do not have to join with other atoms. In these elements, the atoms *are* the molecules.

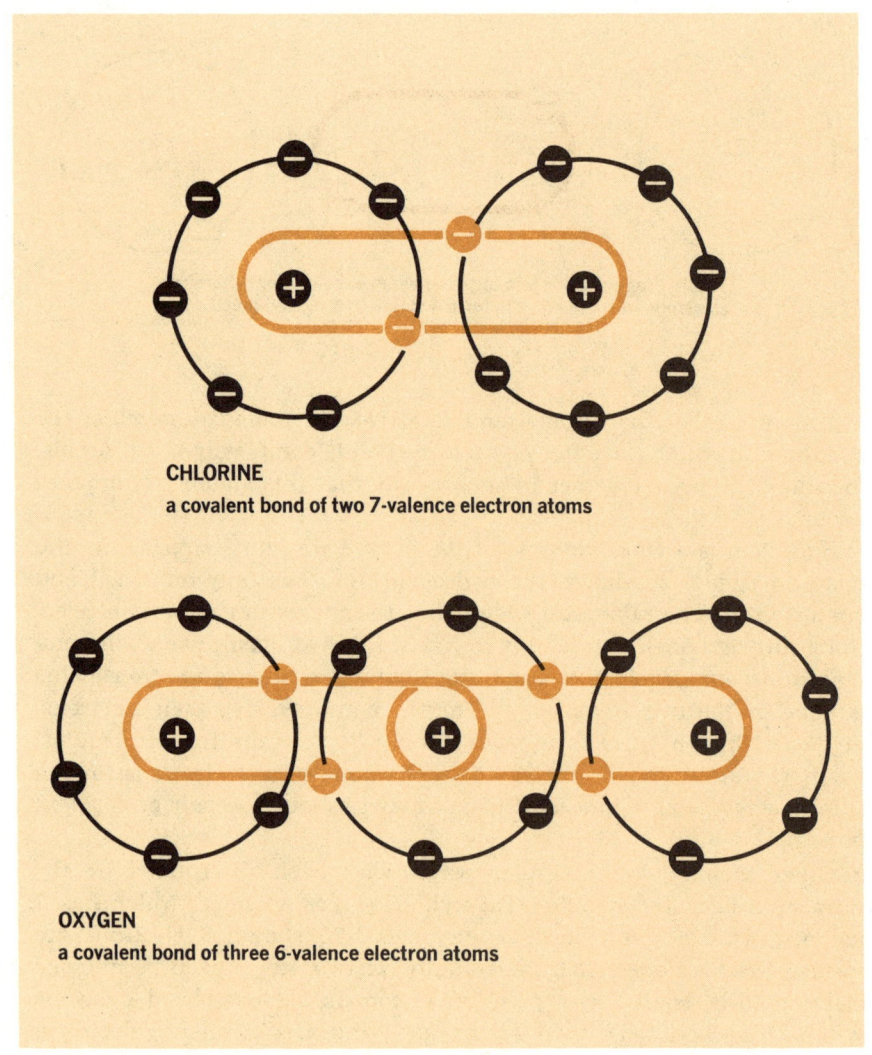

CHLORINE
a covalent bond of two 7-valence electron atoms

OXYGEN
a covalent bond of three 6-valence electron atoms

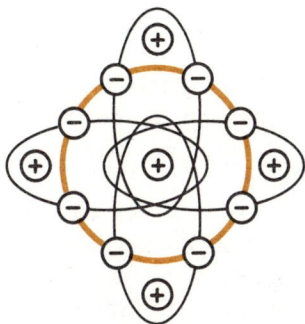

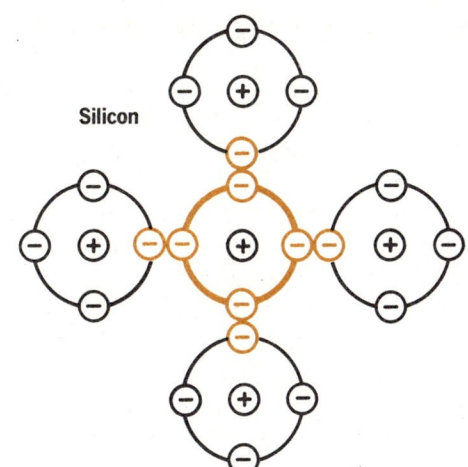

Silicon

Each of the four atoms shares an electron with the center atom, so that only the center atom is stable with 8. This is how the five atoms look with the electrons orbiting

covalent bonds

Semiconductor materials are a good example of how atoms must join together in a complex way to produce a stable molecular arrangement. Silicon, one of the better-known semiconductors, has atoms with 4 valence electrons: 8 are needed for stability. This means that each valence electron of any atom must pair up and be shared with another valence electron from another atom. Therefore, a total of five atoms is needed to make only one of them stable. This arrangement could be considered a molecule, but actually it is not, because the other four atoms are not stable. Each of those atoms will have to join with others in the same way to become stable, but in each case there will still be some unstable atoms that must join with others. A continuous atomic structure must be formed, which is known as a *crystal lattice*. There are no real discrete molecules, but there are groups of atoms that can be considered as molecular groupings.

Covalent bonds cause the atomic structure of semiconductors to repeat the same geometric pattern throughout the material— the crystal lattice

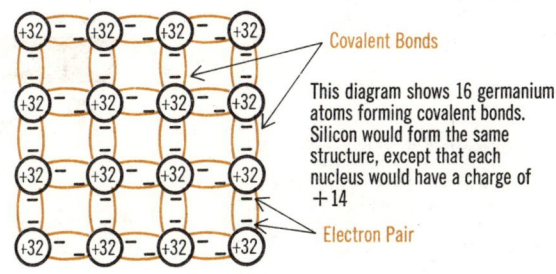

Covalent Bonds

This diagram shows 16 germanium atoms forming covalent bonds. Silicon would form the same structure, except that each nucleus would have a charge of $+14$.

Electron Pair

electrovalent bonds

The previous pages discussed bonding primarily as it relates to *elements*. When the atoms of different elements combine to form *compounds*, the bonds can be similar and also quite different, depending on the number of valence electrons in the different atoms.

Water is an example of a compound that uses covalent bonding similar to the ordinary bonds of the atoms of the same elements. However, since two different atoms are involved with a compound, with different amounts of valence electrons, it appears different. With water, two hydrogen atoms combine with one oxygen atom to form a water molecule. Oxygen has 6 valence electrons, and hydrogen has 1. Each hydrogen electron pairs up with an oxygen electron, so that the hydrogen atoms each stabilize with an *electron pair*, and the oxygen atom sees 8.

Another form of bonding that is common with compounds is *electrovalent bonding*.

The combination of the elements sodium and chlorine is an example of this. Sodium has 1 valence electron, which it tends to give up to become stable; and chlorine has 7 valence electrons, and needs 8 to become stable. When the atoms of sodium and chlorine are brought together, the sodium atom gives up its electron to the chlorine atom. The sodium atom then becomes a *positive ion* and the chlorine atom becomes a *negative ion*. They attract each other and bond together to form sodium chloride. Because ionic attraction forms the bond, this is also known as *ionic bonding*.

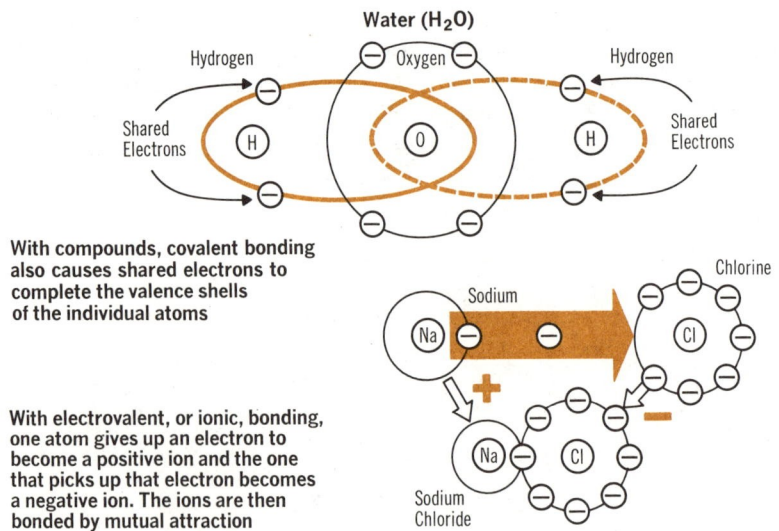

With compounds, covalent bonding also causes shared electrons to complete the valence shells of the individual atoms

With electrovalent, or ionic, bonding, one atom gives up an electron to become a positive ion and the one that picks up that electron becomes a negative ion. The ions are then bonded by mutual attraction

metallic bonds

In the strictest sense, the *metallic bond* of atoms is a form of covalent bonding, wherein the valence electrons are *shared* by the various atoms that join together. However, the covalent bonds you just studied dealt with atoms whose valence shells were more than half filled, and shared *electron pairs* to complete a stable bond.

With metals, which are good conductors, the atoms have only 1 valence electron. And these atoms seem to *lose* their valence electrons to attain stability. With elements such as copper, silver, gold, etc., the atoms join together, with each atom continually giving up its electron to a neighboring atom and taking on another electron from another neighboring atom. This process of giving up and taking on electrons is continuous, and the individual valence electrons are *free* to wander aimlessly and randomly from atom to atom in and out of orbits. This does not mean that the electrons wander free of the atoms. It means only that groups of atoms continuously *share* all of their valence electrons.

With metallic bonding, the valence electrons move from atom to atom, so that all the atoms share their valence electrons

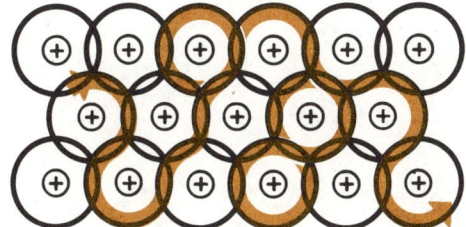

The electrons are free to wander randomly from atom to atom

Another way to look at this is that any group of atoms can form a molecular structure with all their valence electrons orbiting all of the atoms in the group. This is described as a *metallic lattice*, as opposed to the crystal lattices formed by *electron pair* bonds. Some theories describe the metallic sharing of valence electrons as producing an *electron cloud* which surrounds their bonded atoms. As you can see, there are no discrete molecules formed by metallic bonding.

the effect of atomic bonds

Earlier you learned how the number of valence electrons in the atoms of different elements determined whether those atoms made good conductors, insulators, or semiconductors, depending on the stability of the element. But now you have learned that when the atoms bond together, the valence structure of the material can change. The table on page 1-25 shows that there are more elements available with these valence shells less than half filled, which could mean that most elements are good conductors. But this is not true because of the way the atoms of most elements are bonded. Most covalent bonds produce stable atomic arrangements or molecules. Except for those elements that use metallic bonds, most elements are not good conductors. As a matter of fact, if the elements were pure, most of them would be perfect insulators.

It is important to keep in mind, however, that under practical conditions most materials contain impurities. Also, stable bonds can be broken by heat energy, so that conduction can usually take place.

Compounds are good examples of how bonds can change electrical characteristics. Copper, which is a good conductor, becomes a good insulator when combined with oxygen to produce a stable copper oxide molecule, with 8 valence electrons.

Those good conductors that use metallic bonding also end up with the different materials having different conduction characteristics. Although they all have 1 valence electron in their atoms, some metals have a denser grouping of atoms, and so have more valence electrons available in a given amount of material. This is why copper is a better conductor than gold, and silver is better than copper.

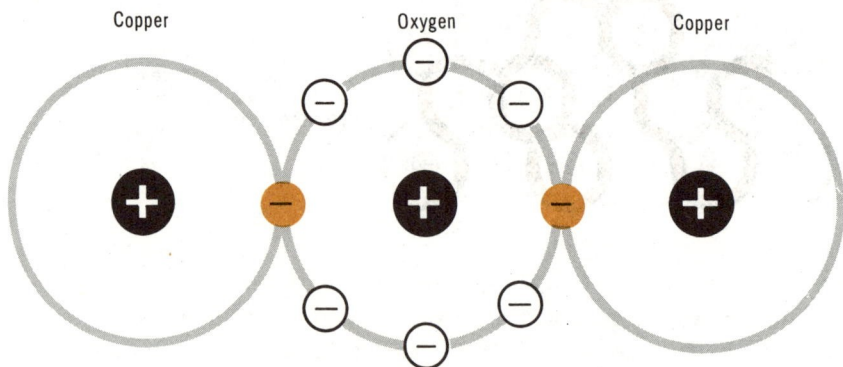

Copper   Oxygen   Copper

Copper is a good conductor because it has only one valence electron. But when two copper atoms combine with one oxygen atom, they produce a molecule of copper oxide (Cu_2O); eight valence electrons now exist to make the molecule stable. Copper oxide, then, is a good insulator in its pure form

summary

☐ Electrons revolve at high speed in their orbits about the atom's nucleus. This high speed gives rise to a centrifugal force that tends to pull the electron out of orbit. The positive attraction of the nucleus prevents this. A large outside force can free the electron from the atom. ☐ The positive attractive force is greater on the electrons in orbit that are closer to the nucleus; hence, they are more difficult to free.

☐ Electrons orbit in one of seven shells. ☐ The innermost shell can hold a maximum of 2 electrons; the second, 8; the third, 18; the fourth, 32; etc. ☐ The outermost shell of an atom is the valence shell. ☐ All electrons in this outer shell are called valence electrons. ☐ The valence shell never contains more than 8 electrons. ☐ An atom with a completely filled valence shell is very stable, and chemically inactive. ☐ Electrons in orbit farther away from the nucleus contain more energy. If sufficient energy is added to an electron, it will move to the next outer orbit; if sufficient energy is added to a valence electron, it will be freed. ☐ The flow of free electrons constitutes electric current.

☐ Conductors are materials that have one or two valence electrons, which are easily freed. ☐ Insulators are materials that have five or more valence electrons, which are difficult to free. ☐ Semiconductors are materials that have more free electrons than insulators, but less than conductors. Impurities can make them better conductors. ☐ Bonding is the force that keeps elements together to form compounds. The electrons in the compounds form stable octets.

review questions

1. What prevents an electron from being freed due to its centrifugal force?
2. What is a shell, and how many shells are there?
3. What is a "valence" electron?
4. What is the maximum number of valence electrons in an atom?
5. What is a "free" electron?
6. Why are compounds good insulators? Name two good insulators.
7. Name two semiconductors, and two conductors.
8. What are the characteristics of semiconductors?
9. Why are impurities added to compounds?
10. Is an element that contains six valence electrons a good conductor? Two valence electrons?

how electricity is produced

Until now, the discussion dealt with the general idea of applying a force or energy to the electrons to move them out of their orbits, but no mention was made as to how this can be done. It can be done in a variety of ways, all of which fall into six accepted categories.

friction

chemicals

ELECTRICITY
CAN BE
PRODUCED
BY

heat

pressure

magnetism

light

electricity from friction

This is the method that was first discovered by the ancient Greeks, which was described earlier in the book. An electric charge is produced when two pieces of material are rubbed together, such as silk and a glass rod, or when you comb your hair. Did you ever walk across a carpet and get a shock when you touched a metal doorknob? Your shoe soles built up a charge by rubbing on the carpet, and this charge was transferred to you and was discharged to the knob. These charges are called *static electricity;* and results when one material *transfers* its *electrons* to another.

TRIBOELECTRICITY

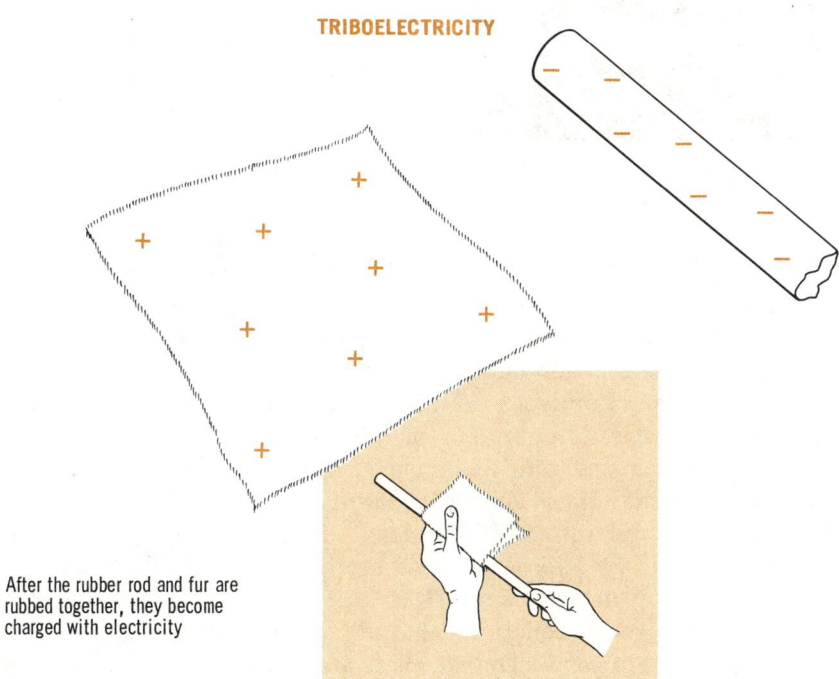

After the rubber rod and fur are rubbed together, they become charged with electricity

It is something that is still not completely understood. But, one theory is: On the *surface* of a material, there are many atoms that cannot combine with other atoms as they do inside the material; the surface atoms, then, have some free electrons. This is the reason why insulators such as glass and rubber can produce charges of static electricity. The heat energy produced by the rubbing friction is supplied to the surface atoms to release the electrons. This is known as the *triboelectric* effect.

ELECTROCHEMISTRY

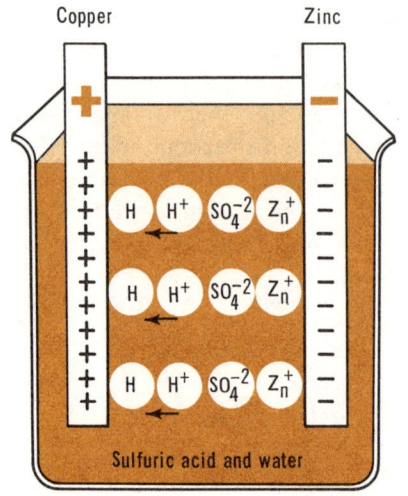

Copper   Zinc

Sulfuric acid and water

The "Wet" Cell

The solution, known as the electrolyte, pulls positive ions from the zinc bar and free electrons from the copper bar

The "dry" flashlight battery uses an electrolytic paste instead of a fluid solution

electricity from chemicals

Chemicals can be combined with certain metals to cause a chemical action that will transfer electrons to produce electric charges. This is how the ordinary *battery* works. This process works on the principle of *electrochemistry*. One example of this is the basic *wet cell*. When sulfuric acid is mixed with water (to form the *electrolyte*) in a glass container, the sulfuric acid breaks down into separate chemicals of hydrogen (H) and sulfate (SO_4), but because of the nature of the chemical action, the hydrogen atoms are positive ions (H^+) and the sulfate atoms are negative ions (SO_4^{-2}). The number of positive and negative charges are equal, so that the entire solution has no net charge. Then when copper and zinc bars are added, they *react* with the solution.

The zinc combines with the sulfate atoms; and since those atoms are negative, positive zinc ions (Zn^+) are given off by the zinc bar. The electrons from the zinc ions are left behind in the zinc, so that the zinc bar has a surplus of electrons, or a *negative charge*. The zinc ions combine with and neutralize the sulfate ions so that the solution now has more positive charges. The positive hydrogen ions attract free electrons from the copper bar to again neutralize the solution. But, now the *copper* bar has a lack of electrons, so it gets a *positive charge*.

Batteries and cells are covered in greater detail in Volume 6.

electricity from pressure

When pressure is applied to some materials, the *force* of the *pressure* is passed through the material to its atoms, where it drives the electrons out of orbit in the direction of the force. The electrons leave one side of the material and accumulate on the other side. Thus, positive and negative charges are built up on opposite sides. When the pressure is released, the electrons return to their orbits. The materials are cut in certain ways to control the surfaces that will be charged. Some materials will react to a *bending* pressure, while others will respond to a *twisting* pressure.

Piezoelectricity is the name given to the effect of *pressure* in causing electrical charges. *Piezo* is a name derived from the Greek word for pressure. This effect is most noticeable in crystal materials, such as Rochelle salts and certain ceramics like barium titanate. These piezo crystals are used in some microphones and phonograph pickups.

PIEZOELECTRICITY

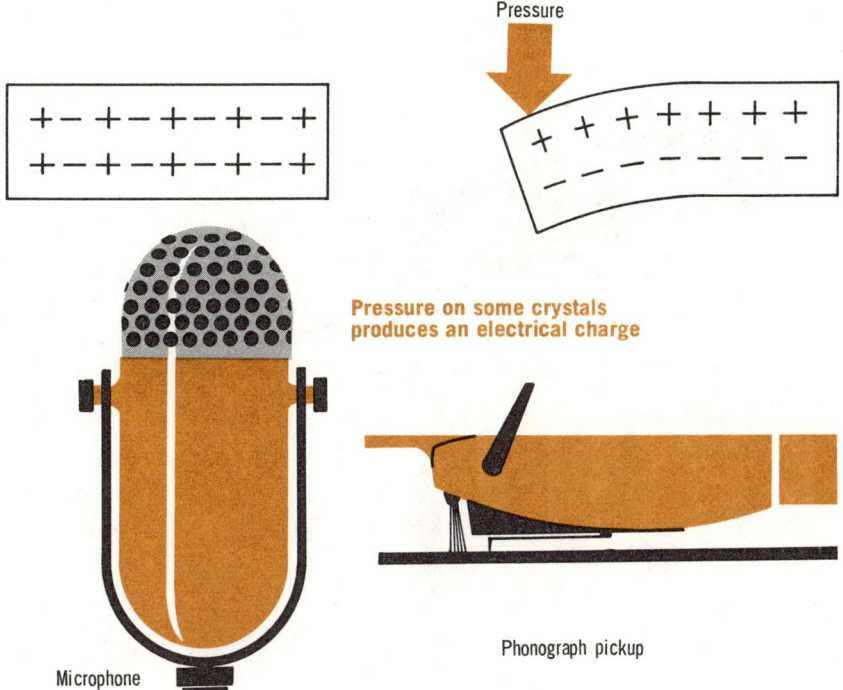

Pressure

Pressure on some crystals produces an electrical charge

Phonograph pickup

Microphone

electricity from heat

Because some materials readily give up their electrons, and other materials accept electrons, a transfer of electrons can take place when two *dissimilar metals*, for example, are joined. With particularly active metals, the heat energy of normal room temperature is enough to make those metals release electrons. Copper and zinc, for example, will act this way. Electrons will leave the copper atoms and enter the zinc atom. The zinc, then, gets a surplus of electrons and becomes negatively charged. The copper, having lost electrons, takes on a positive charge.

The charges developed at room temperature are small, though, because there is not enough heat energy to free more than a relatively few electrons. But, if heat is applied to the junction of the two metals to provide more energy, more electrons will be freed. This method is called *thermoelectricity*. The more heat that is applied, the greater the charge that will be built up. When the heat is removed, the metals will cool, and the charges will dissipate. The device described is called a *thermocouple*. When a number of thermocouples are joined together, a *thermopile* is made.

THERMOELECTRICITY

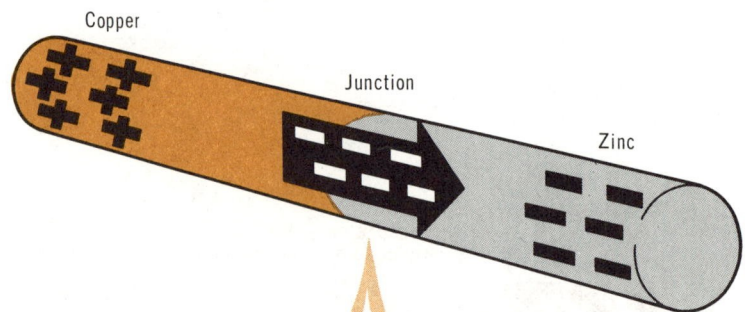

Copper

Junction

Zinc

Heat energy causes the copper to release electrons to the zinc

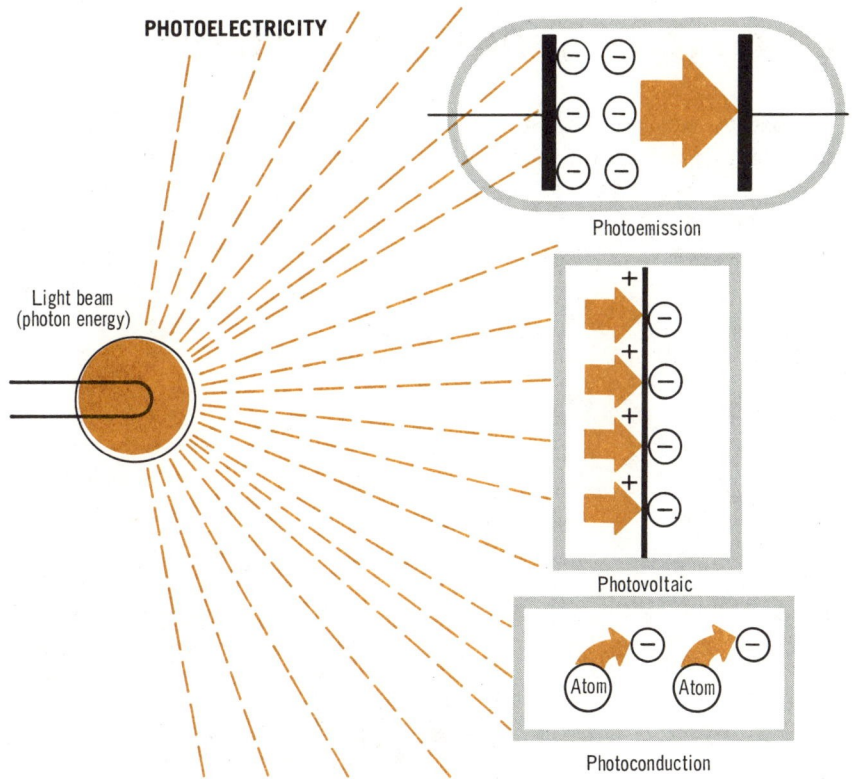

PHOTOELECTRICITY

Light beam
(photon energy)

Photoemission

Photovoltaic

Photoconduction

Atom Atom

electricity from light

Light in itself is a form of energy, and is generally considered by many scientists to be made of small particles of energy called *photons.* When the photons in a light beam strike a material, they release their energy. With some materials, the energy from the photons can cause atoms to release electrons. Materials such as potassium, sodium, cesium, lithium, selenium, germanium, cadmium, and lead sulfide react this way to light. This *photoelectric effect* can be used in three ways:

1. *Photoemission:* The photon energy from a beam of light could cause a surface to release electrons in a vacuum tube. A plate would then collect the electrons.
2. *Photovoltaic:* The light energy on one of two plates that are joined together causes one plate to release electrons to the other. The plates build up opposite charges, like a battery.
3. *Photoconduction:* The light energy applied to some materials that are normally poor conductors causes free electrons to be produced in the materials, so that they become better conductors.

electricity from magnetism

Most of you are probably familiar with magnets, and have toyed with them at one time or another. You may have seen that, in some cases, magnets attract each other, and in other cases, they repel each other. The reason for this is that magnets have force fields that interact with one another. (This is explained more fully later in this book.)

ELECTROMAGNETISM

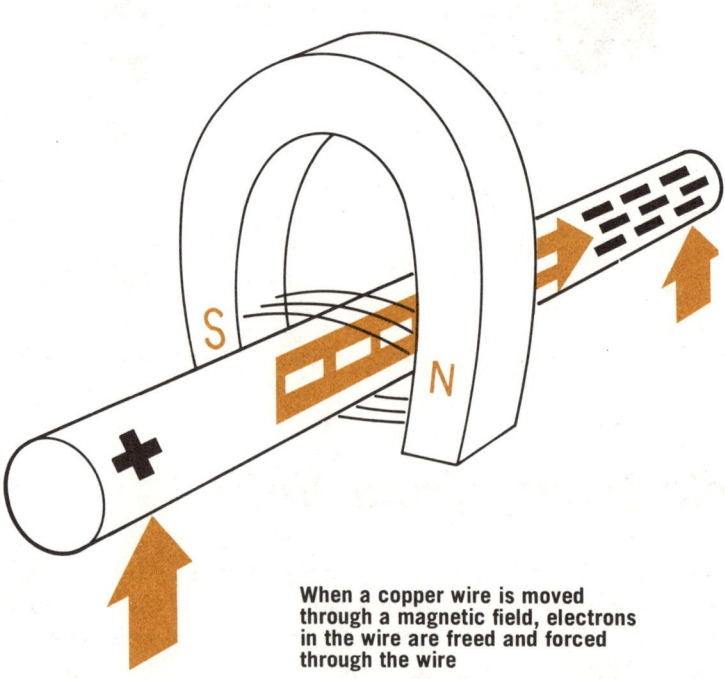

When a copper wire is moved through a magnetic field, electrons in the wire are freed and forced through the wire

The force of a magnetic field can also be used to move electrons. This is known as *magnetoelectricity,* and is the basis for how an electric *generator* produces electricity. When a good conductor, such as copper, is passed through a *magnetic* field, the *force* of the field will provide enough energy to cause the copper atoms to free their valence electrons. The electrons will all be moved in a certain direction, depending on how the wire crosses the magnetic field. Actually it is not necessary to only move the conductor through the magnetic field; the same effect will be obtained if the field is moved across the conductor. All that is really required is relative motion between any conductor and a magnetic field. (Magnetoelectricity is further examined at the end of the book.)

summary

☐ Electrons can be made to move out of their orbits by applying a force or energy to them. ☐ Six accepted ways of doing this are triboelectric effect, electrochemistry, piezoelectricity, thermoelectricity, photoelectric effect, and magnetoelectricity.

☐ Triboelectric effect causes electrons that are on the surface of a material to be released by rubbing. The energy comes from heat friction. ☐ Electrochemistry enables chemicals to be combined with certain metals to cause a chemical action that will transfer electrons to produce charges. ☐ Piezoelectricity is the effect of pressure causing electrical charges. Its effects are most noticeable in crystals. ☐ Thermoelectricity is the effect of heat being applied to two dissimilar metals, producing opposite charges on the two metals. ☐ Photoelectric effect causes the atoms of certain materials to release electrons when light energy, in the form of photons, strikes them.

☐ Photoemission: Photon energy releases electrons of one surface in a vacuum tube. Another surface in the tube collects the electrons. ☐ Photovoltaic: Light energy on one of two joined plates causes electrons to be released to the other. The plates then act like a battery. ☐ Photoconduction: Light energy applied to some materials causes them to become better conductors. ☐ Magnetoelectricity is the effect of the force of a magnetic field, which can be used to move electrons.

review questions

1. What causes electrons to move out of their orbits?
2. What is the effect of applying a pressure to a Rochelle salt crystal?
3. What is the triboelectric effect?
4. In thermoelectricity, heat is applied at the junction of two _____ metals.
5. What is the difference between a thermocouple and a thermopile?
6. What are photons?
7. Name and describe three ways in which the photoelectric effect can be used.
8. Is it always necessary for the conductor to move in order for its electrons to be freed by a magnetic force?
9. Describe the basic wet cell. It operates on the principle of _____.
10. The electric generator operates on the principle of _____.

what is electric current?

The book, till now, dealt with what electricity is, and how electric charges are produced. The subjects mostly covered what is called *static electricity*, which is an electric *charge at rest*. But, a static electric charge cannot usually perform a useful function. In order to use electrical energy to do some kind of work, the electricity must be set in "motion." This is done when an electric current is produced. This electric current is developed when many *free electrons* in a wire are moved in the same direction.

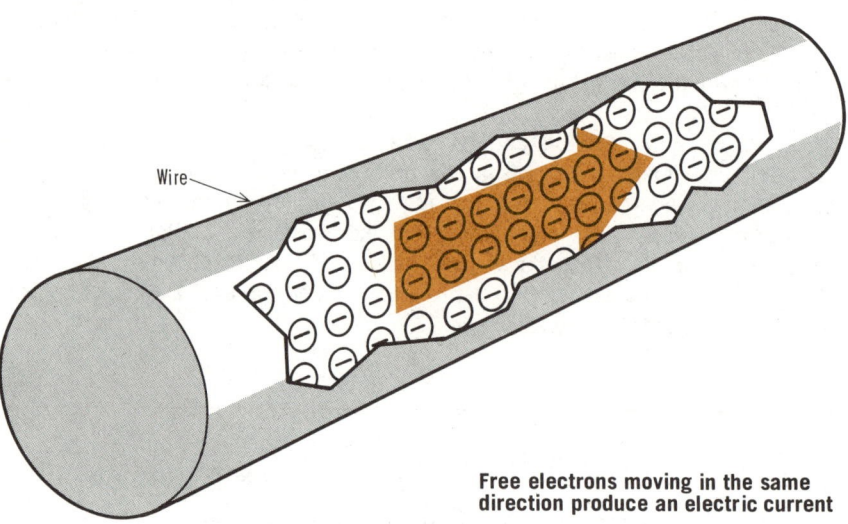

Wire

Free electrons moving in the same direction produce an electric current

As you will learn later, every electron contains energy that can cause certain effects. Ordinarily, electrons are moving in various directions, so that their effects cancel. But when we get the electrons to move in the same direction, that is, form a current flow, their effects can add, and the energy they release can be put to work. Also, the more electrons there are moving in the same direction, the greater is the current flow, and the more electron energy is made available to do work. Therefore, the larger or smaller electric currents are caused by the greater or fewer number of electrons set in motion in the same direction.

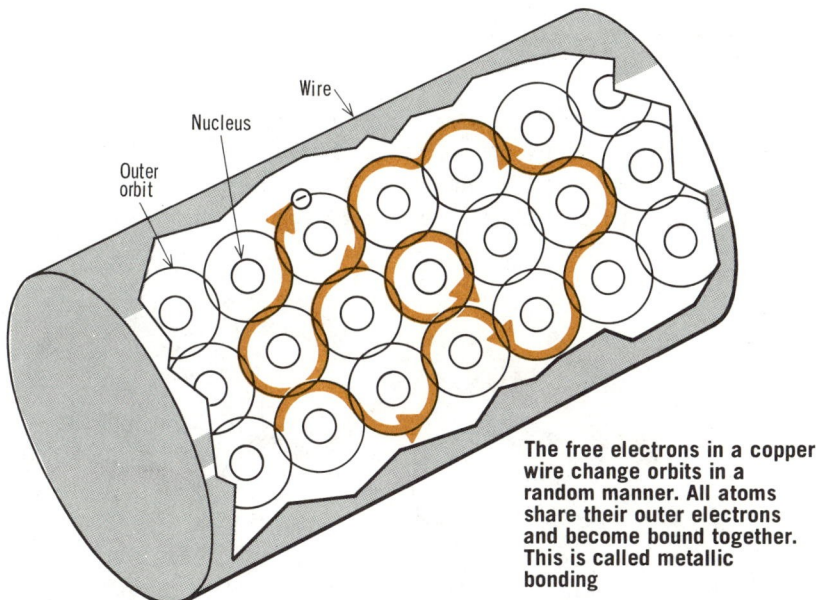

The free electrons in a copper
wire change orbits in a
random manner. All atoms
share their outer electrons
and become bound together.
This is called metallic
bonding

free electrons

In order to understand how electrons produce electric current, it
would help to review how the atoms of a good conductor, such as
copper, are bound together to form the solid metal.

In a copper wire, the atoms each have 1 valence electron, which is
loosely held in orbit. And the atoms are close together so that the
outer orbits overlap. While in motion, the electron of one atom can
come under the influence of another atom and enter its orbit. At about
the same time, the electron in the second atom is displaced and moves
into the orbit of another atom. Most of the outer electrons continually
change orbits this way in a *random* manner, so that the valence electrons
are not really associated with any one atom. Instead, all of the atoms
share all of the valence electrons, and so are bonded together. The electrons are "free" to wander randomly. The action is continuous, so that
every atom always has an electron, and vice versa. Therefore, no electric charge results, but the conductor has a lot of free electrons.

In order to produce an electric current, the free electrons in the copper wire must be made to move in the same direction instead of just
randomly. This can be done by putting electrical charges on each end
of the copper wire; a negative charge at one end, and a positive charge
at the other end.

electron movement

Since the electrons are negative, they are repelled by the negative charge and are attracted by the positive charge. Because of this, they cannot change to orbits that would make them move against the forces of the electrical charges. Instead, they *drift* from orbit to orbit toward

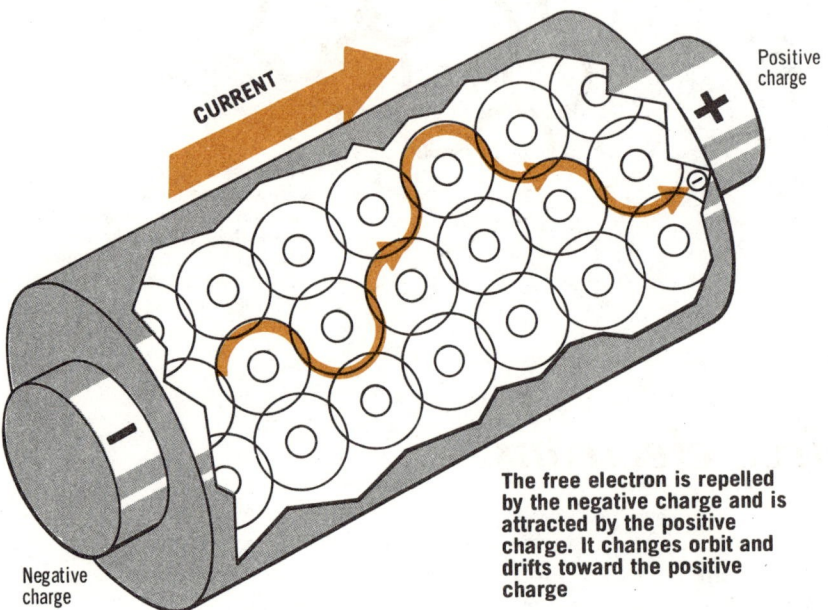

CURRENT

Positive charge

Negative charge

The free electron is repelled by the negative charge and is attracted by the positive charge. It changes orbit and drifts toward the positive charge

the positive charge, causing an *electric current* to go in that direction.

You can see on the diagram that the density of the atoms in the copper wire is such that the valence orbits of the individual atoms overlap, so that the electrons find it easy to move from one atom to the next. The path that the electron moves in depends on the direction of the orbits that the electron encounters while it drifts to the positive charge. You can see that it does not follow a straight line. But as the charges on each end are made stronger, they control each electron more, causing it to follow a straighter path, and thus move faster through the wire.

The strength of the charge at each end of the wire also determines how many electrons change from a random drifting motion to a more directional drift through the wire. Small charges cause only a small number of electrons to drift to the positive charge. But the stronger the negative and positive charges are made, the more and more electrons will be repelled on a straighter path through the wire.

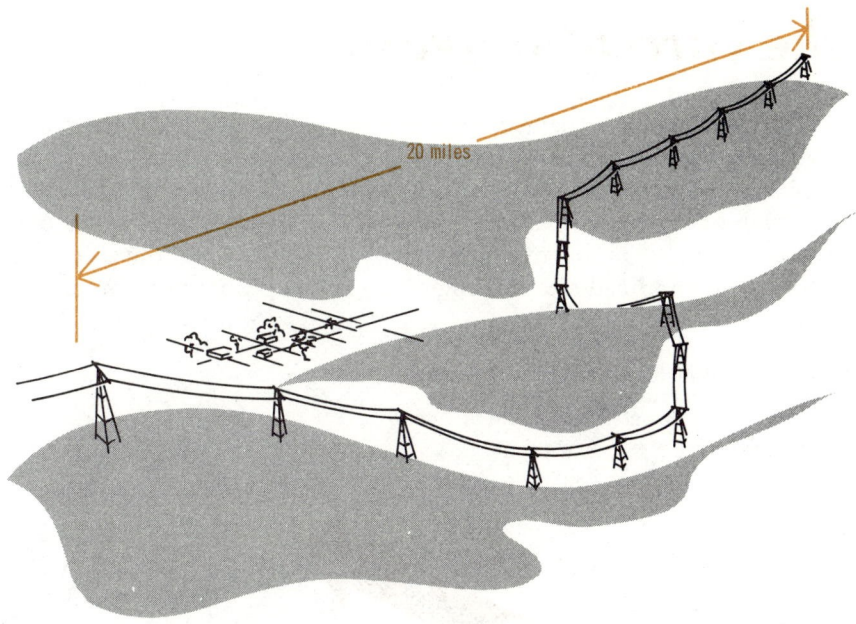

20 miles

If a free electron had to drift down a wire 20 miles long, it could take over 30 days. Yet an electric current travels that distance in a fraction of second

current flow

Although it is sometimes easier to think of the drifting free electrons as being the electric current, it is important to remember this is not completely so. The free *electron* movement *produces* the *current*. This is made clearer when you compare the speed of an electron to the speed of current. The speed of the electron can vary, depending on the conducting material and the number of electrical charges being used. But the speed of the current is always the same.

The free electron that wanders randomly, travels relatively fast because it is only under the influence of the atomic orbital forces; its speed can be a few *hundred miles per second.*

The free electron that drifts under the influence of the electrostatic charges has to oppose some of the atomic orbital forces, and so it is slowed down considerably; it can travel at speeds measured in *inches per second.* This is very slow when you realize that electric current travels at the speed of light: *186,000 miles per second.*

the current impulse

The electric *current* is actually the *impulse* of electrical energy that one electron transmits to another as it changes orbit. When energy is applied to an electron so that it leaves its orbit, it will have to encounter an orbit of another atom as it leaves. The reason for this is that all of the outer orbits overlap and obstruct the free travel of the electron. As the freed electron enters the new orbit, its negative charge

As each electron leaves its orbit and enters another orbit, it repels an electron out of orbit to repeat the action from atom to atom throughout the wire

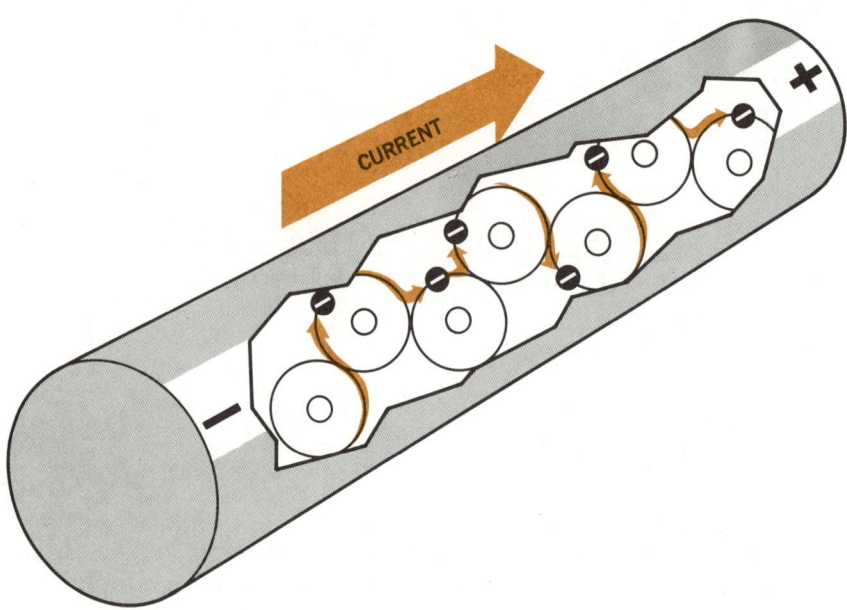

The impulse that is transferred from one electron to the next down the line is the electric current

reacts with the negative charge of the electron already in that orbit. The first electron *repels* the other out of orbit, transmitting its energy to it. The second electron repeats the performance of the first as it encounters the next orbit; and this process continues through the wire. The impulse of energy that is transferred from electron to electron is the electric current.

the speed of electric current

Since the atoms are close, and the orbits overlap, the electron that is freed does not have to travel far to encounter a new orbit. And the moment it enters the new orbit, it transfers its energy to the next electron to free it. The action is almost instantaneous. The same is true for all the following electrons, so that even though each electron is moving relatively slowly, the impulse of electrical energy is transferred down the line of atoms at a very great speed: *186,000 miles per second.* The free electrons are considered *current carriers.*

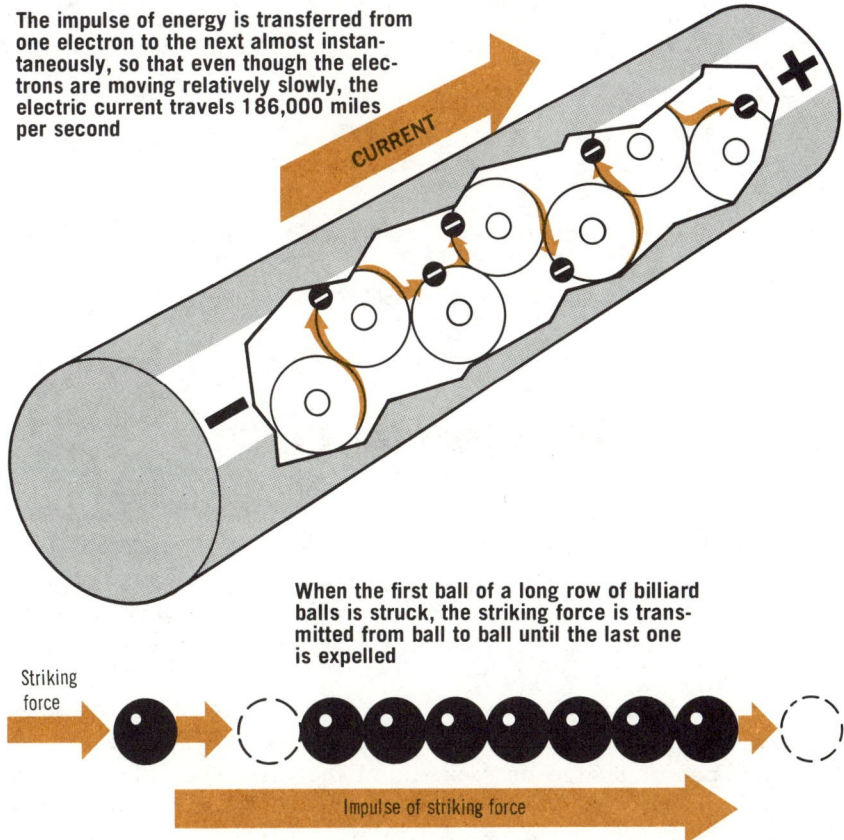

The impulse of energy is transferred from one electron to the next almost instantaneously, so that even though the electrons are moving relatively slowly, the electric current travels 186,000 miles per second

When the first ball of a long row of billiard balls is struck, the striking force is transmitted from ball to ball until the last one is expelled

Striking force

Impulse of striking force

A good analogy of this *impulse transfer* is a long row of billiard balls. When a ball is hit into one end of the row of balls, its striking force is transmitted from ball to ball until the ball at the other end is knocked free. The last ball is released at almost the same time that the first ball is struck.

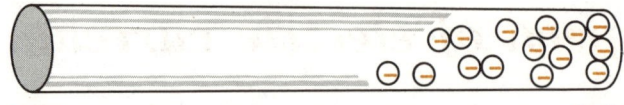

A negative charge put at one end of a wire will repel electrons to the other end until an equal charge is built up to stop electrons from flowing. But, if a power source applies opposite charges to each end of the wire, electrons will continue to flow

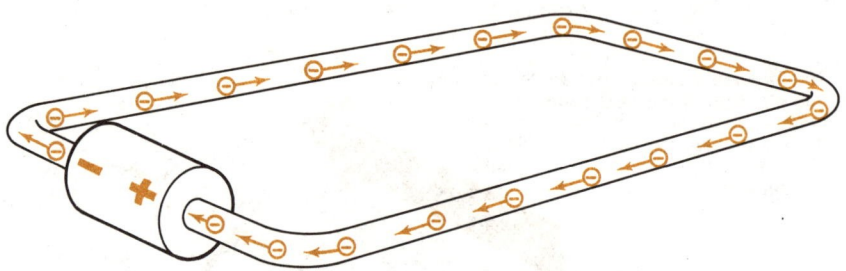

An electron leaves the negative side of the battery for each electron that enters the positive side

A COMPLETE OR CLOSED CIRCUIT

a complete (closed) circuit

If a negative charge is put at one end of a wire, that charge would repel free electrons to the other end of the wire. Current would flow only for a moment, until enough electrons accumulate at the other end to produce an equal negative charge that would prevent more electrons from coming. This would be *static electricity* because everything came to rest.

In order to have an electric current, the free electrons must continue to flow; for this to happen, an electrical energy source must be used to apply opposite charges to *each* end of the wire. Then, the negative charge would repel the electrons through the wire. At the positive side, electrons would be attracted into the *source;* but for each electron attracted into the source, an electron would be supplied by the negative side into the wire. Current would, therefore, continue to flow through the wire as long as the energy source continued to apply its electrical charges. This is called a *complete* or *closed circuit.* A battery is a typical electrical source.

A complete or closed circuit is needed for current to flow.

an open circuit

If the wire were broken at any point, electrons would build up at the end of the wire that is connected to the negative side of the battery; and electrons would be attracted away from the end of the wire that is connected to the positive side of the battery. A charge would be built up across the opening to stop the electrons from moving. Current would stop flowing.

An open circuit will not conduct current.

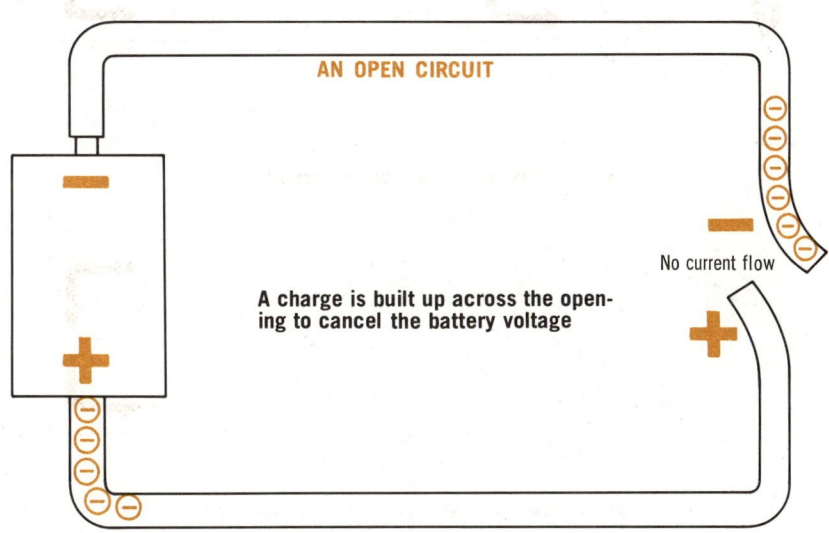

AN OPEN CIRCUIT

No current flow

A charge is built up across the opening to cancel the battery voltage

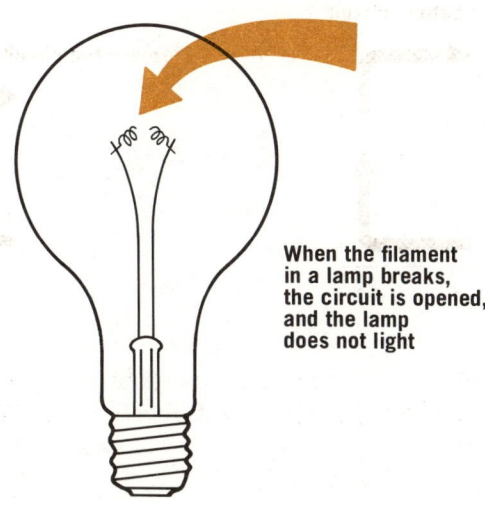

When the filament in a lamp breaks, the circuit is opened, and the lamp does not light

the electrical energy source

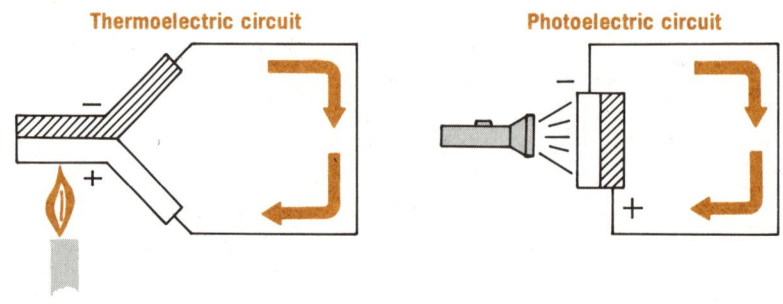

Thermoelectric circuit   Photoelectric circuit

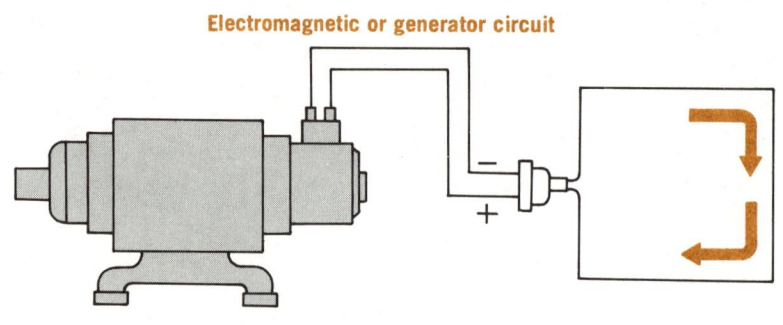

Electromagnetic or generator circuit

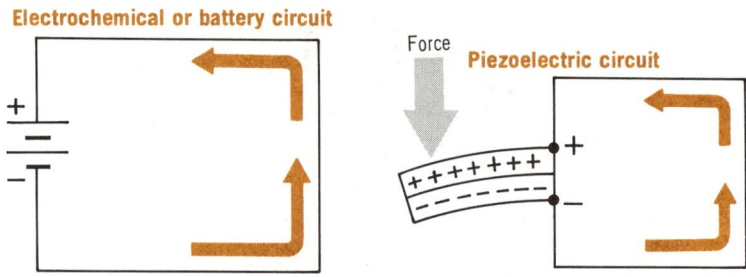

Electrochemical or battery circuit

Force   Piezoelectric circuit

Any one of the five types of sources described on pages 1-42 through 1-46 can be used to cause current to flow through a wire. The battery and the generator are the most common. The electrical outlet in your home is supplied by a distant generator.

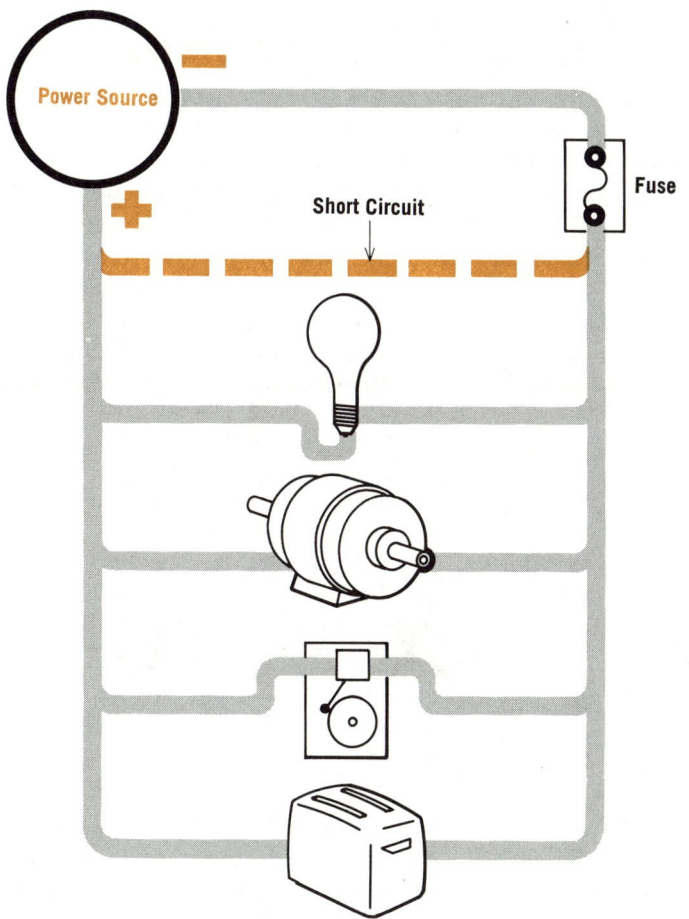

putting electricity to work

Actually, when a good conducting wire is put directly across the *terminals* of a battery or generator, a *short circuit* is produced. Much more current flows than the battery or generator can supply. The battery or generator can burn out and the wire will become very hot. This is why protective fuses are used. They melt when too much current flows, and "open" the circuit.

The wire is used to carry current to other things to make them work. For example, it carries current to heat the filament of a lamp to make it light; it provides electrical energy to turn a motor, ring a bell, heat a toaster, and so on. Some of these uses are explained at the end of this book.

electrical units of measurement

You can see now that two conditions are needed to get current flow: (1) electrical charges to move the free electrons, and (2) a complete circuit to allow electric current to flow. Different amounts of electrical charges can be used, and different amounts of current can flow. There are certain units of measurements that indicate the different values.

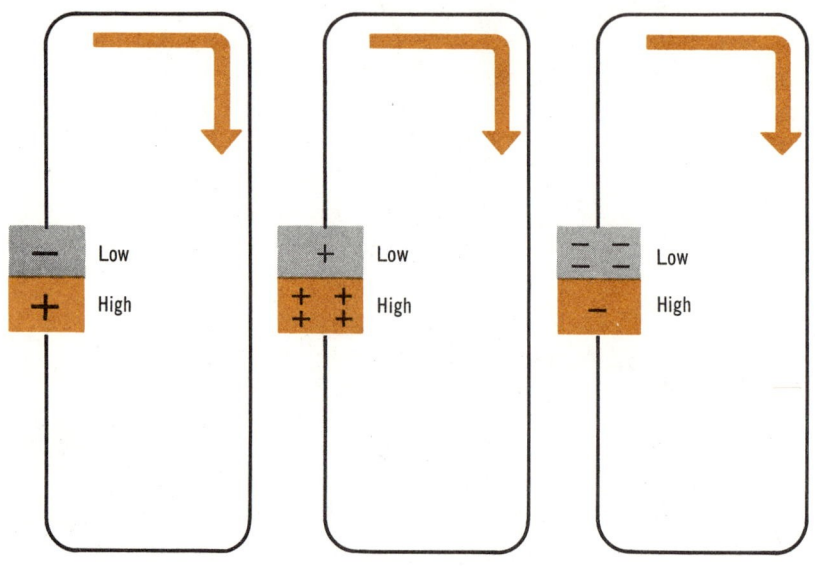

Electron current flows from a low potential to a high potential, or between any difference of potential

The electrical charge that an object gets is called an electric potential, because the electrons that are displaced accumulate potential energy that can be used to move other electrons. Since two charges are needed to complete a circuit, it is the *difference of potential* between these two charges that provides the electric force. *Negative* is considered to be a *low potential,* and *positive,* a *high potential. Electron current* in a wire always goes from the *low to high potential.* This also means that electron current will flow from a low positive potential to a high positive potential, and between two different negative potentials as well.

electromotive force (voltage)

The electrical *charge* that an object gets is determined by the *number of electrons* that the object lost or gained. Because such a vast number of electrons moves, a unit called the *coulomb* is used to indicate the charge. If an object has a negative charge of 1 coulomb, it has gained 6.28×10^{18} (billion billion) extra electrons. This is 6,280,000,000,-000,000,000 electrons.

When two charges have a difference of potential, the electric force that results is called an *electromotive force* (emf). The unit used to indicate the strength of the emf is the *volt*. When a difference of potential causes 1 coulomb of current to do 1 joule of work, the emf is 1 volt. Some typical *voltages* you will probably work with are: 1.5 volts for a flashlight battery, 6 volts for the older auto batteries, 12 volts for the newer auto batteries, 115 volts for the home, 220 volts for industrial power, and so on. Voltages actually vary from *microvolts* (millionths of a volt) to *megavolts* (millions of volts). The terms potential, electromotive force (emf), and voltage are often used interchangeably.

UNITS OF VOLTAGE

A charge of 1 coulomb = 6.28×10^{18} electrons
An emf of 1 volt (v) = 1 coulomb doing 1 joule of work
1 microvolt (μv) = 1/1,000,000 volt
1 millivolt (mv) = 1/1000 volt
1 kilovolt (kv) = 1000 volts
1 megavolt (megav) = 1,000,000 volts

CONVERSION OF UNITS

volts (v) $\times$ 1000 = millivolts (mv)
volts (v) $\times$ 1,000,000 = microvolts (μv)
millivolts (mv) $\times$ 1000 = microvolts (μv)
volts (v) $\div$ 1000 = kilovolts (kv)
volts (v) $\div$ 1,000,000 = megavolts (megav)
megavolts (megav) $\times$ 1000 = kilovolts (kv)

---

millivolts (mv) $\div$ 1000 = volts (v)
microvolts (μv) $\div$ 1,000,000 = volts (v)
microvolts (μv) $\div$ 1000 = millivolts (mv)
kilovolts (kv) $\times$ 1000 = volts (v)
megavolts $\times$ 1,000,000 = volts (v)
kilovolts $\div$ 1000 = megavolts

---

1 v = 1000 mv = 1,000,000 μv = 0.001 kv = 0.000001 megav

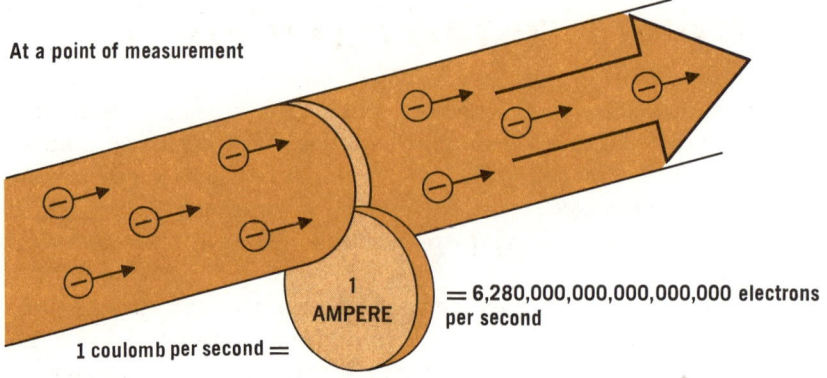

At a point of measurement

1
AMPERE

1 coulomb per second =

= 6,280,000,000,000,000,000 electrons
per second

amount of current (ampere)

The quantity of current flowing in a wire is determined by the number of electrons that pass a given point in one second. As mentioned previously, a coulomb is 6.28×10^{18} electrons. If 1 *coulomb* passes a point in 1 *second,* then 1 *ampere* of current is flowing. The unit term ampere came from the name of another 18th century scientist, A. M. Ampere. Current is also measured in *microamperes* (millionths of an ampere) and *milliamperes* (thousandths of an ampere).

UNITS OF CURRENT

1 ampere (a) = 1 coulomb/sec
1 milliampere (ma) = 1/1000 ampere
1 microampere (µa) = 1/1,000,000 ampere

CONVERSION OF UNITS

amperes (a) × 1000 = milliamperes (ma)
amperes (a) × 1,000,000 = microamperes (µa)
milliamperes (ma) × 1000 = microamperes (µa)

milliamperes (ma) ÷ 1000 = amperes (a)
microamperes (µa) ÷ 1,000,000 = amperes (a)
microamperes (µa) ÷ 1000 = milliamperes (ma)

0.5 a = 500 ma = 500,000 µa

summary

☐ Metallic bonds are important in the study of electricity. ☐ Electrons in a metal are loosely held in their orbits, and with a little force can be made to enter an overlapping orbit of another atom. ☐ The valence electrons are free to wander from atom to atom. ☐ When a force causes the electrons to move in a specific direction, an electric current is created. ☐ The force which moves the electrons is an electromotive force, emf. It is also called voltage and potential.

☐ The speed of a free electron in random motion can be a few hundred miles per second. Under an emf, its speed will be slowed considerably. ☐ Although the actual speed of an electron under the influence of an emf is slow, the impulse of energy that is transferred from electron to electron is at 186,000 miles per second. This is the rate of current flow. ☐ A current will not flow in a circuit unless there is complete path or loop. Otherwise, the circuit is said to be an open circuit. When a circuit is closed, there is a complete path. ☐ The battery and the generator are the most commonly used sources of force (emf) to transfer electrons. ☐ To prevent an excessive current flow, fuses are used to open the circuit to protect it.

☐ Electron current in a wire always flows from a low to a high potential. ☐ The basic unit of emf is the volt; other units are the microvolt (μv), the millivolt (mv), the kilovolt (kv), and the megavolt (megav). ☐ The amount of 6.28×10^{18} electrons is the coulomb. ☐ The basic unit of electron current is the ampere (a), which is one coulomb per second; other units are the micro-ampere (μa), and the milliampere (ma).

review questions

1. What is electric current, and in what units is it measured?
2. What is a metallic bond?
3. Do orbits of different atoms in a conductor ever overlap?
4. Do electrons under the influence of an emf travel from atom to atom at the rate of 186,000 miles per second?
5. How fast does electron *current* travel? Why is it different from the electron's speed?
6. What is meant by coulomb of charge? Ampere?
7. What is meant by potential difference, voltage, and emf?
8. How many volts are there in 2500 megavolts? 2500 milli-volts?
9. What is a microampere? Milliampere?
10. How does a fuse prevent excessive current from flowing in a circuit?

effects of electricity

Except for friction, electricity can be used to produce the same effects described on page 1-40 that were used to produce electricity.

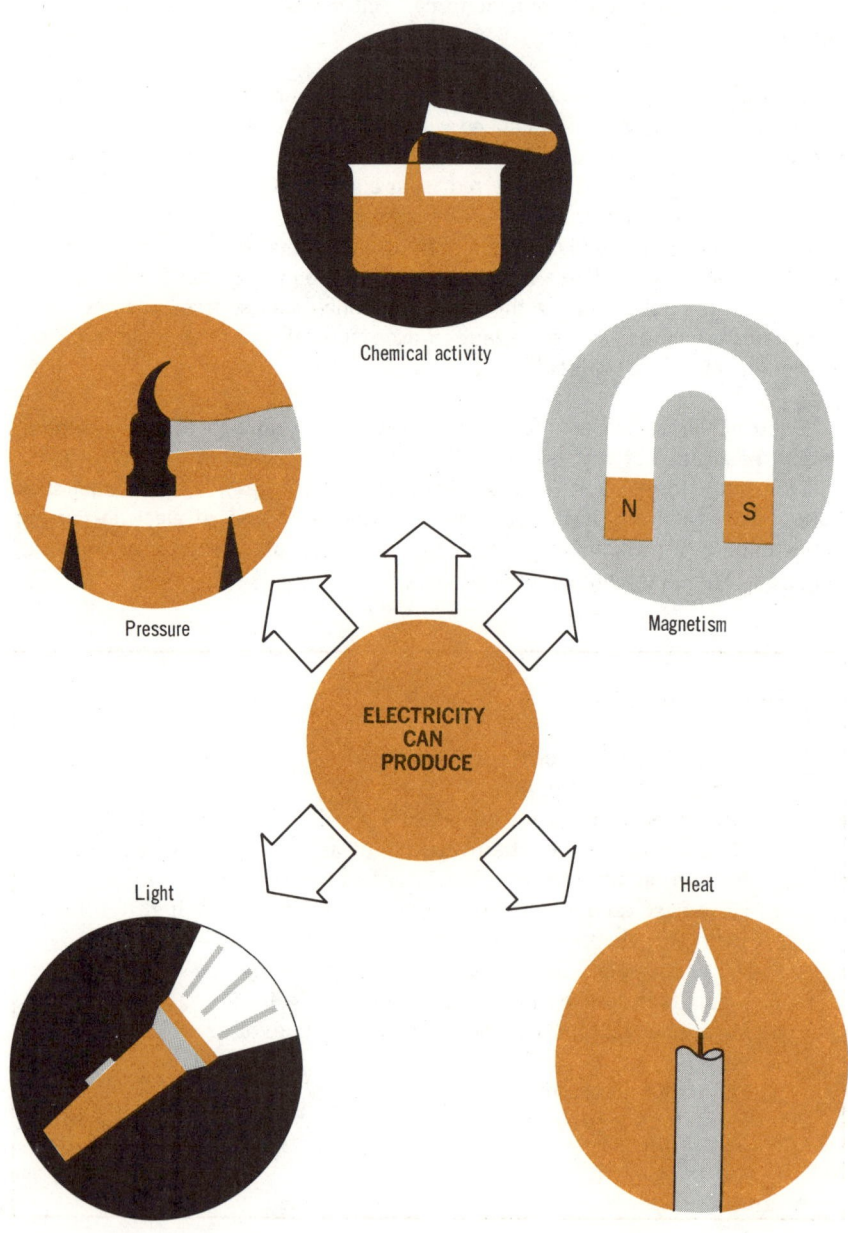

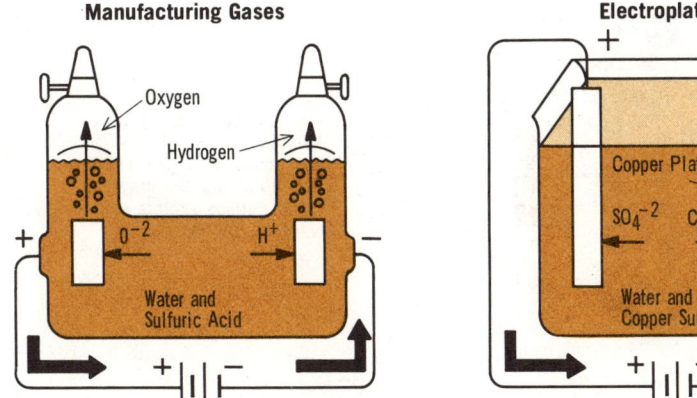

Manufacturing Gases

Oxygen

Hydrogen

O^{-2}

H^+

Water and
Sulfuric Acid

+ | | | –

Electroplating

+ –

Copper Plating

SO_4^{-2} Cu^{+2}

Water and
Copper Sulfate

+ | | | –

The electric potential and current decompose the electrolytes into ions

electricity causes chemical activity

Since the electrical charge is the basic binding force that causes the chemical bonding of compounds, an electric potential or current could be used to alter normal chemical effects. In electrochemistry, this process is called *electrolysis.* For example, if an electric current is passed through water (H_2O) that contains a small amount of sulfuric acid, the water molecules will be broken down into hydrogen and oxygen atoms. The oxygen atoms, though, do not give up the hydrogen atoms' electrons that it previously shared. As a result, the hydrogen atoms become positive ions (H^+), and the oxygen atoms, negative ions (O^{-2}). The ions are attracted to oppositely charged *electrodes.*

At the negative electrode, the positive hydrogen ions pick up electrons, become neutral, and escape the water as a gas. At the positive electrode, the negative oxygen ions give up electrons, become neutral, and escape the water as a gas. The gases can then be bottled. Since an electron comes in at the negative electrode to replace each electron that a hydrogen ion picks up, and one goes out at the positive electrode for each electron that an oxygen atom gives up, current continues to flow until all the water becomes gaseous hydrogen and oxygen.

Another application of electrolysis is *electroplating.* If the water were mixed with copper sulfate ($CuSO_4$), the copper sulfate would break down into positive copper ions (Cu^{+2}) and negative sulfate ions (SO_4^{-2}). The copper ions would go to the negative electrode and pick up electrons. But since copper is a metal, it adheres to the electrode. After a while, the electrode becomes completely plated with copper. Silver and gold plating can also be done this way.

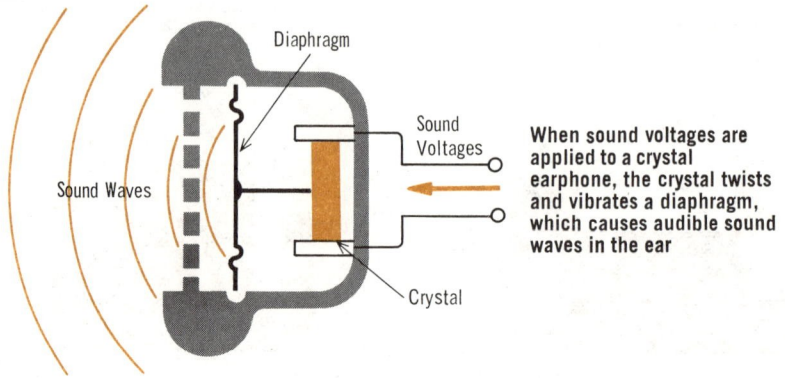

Diaphragm

Sound Voltages

Sound Waves

Crystal

When sound voltages are applied to a crystal earphone, the crystal twists and vibrates a diaphragm, which causes audible sound waves in the ear

electricity causes pressure

Just as a force or pressure that bends or twists some crystals to produce a piezoelectric charge, the application of a voltage will cause the crystal to bend or twist. If an electric potential is placed across a slab of Rochelle salt crystal, the force of the electric field will exert a piezoelectric pressure on the atomic structure and distort the shape of the crystal. This is the way a crystal earphone works, and it is also one method used to "cut" phonograph records.

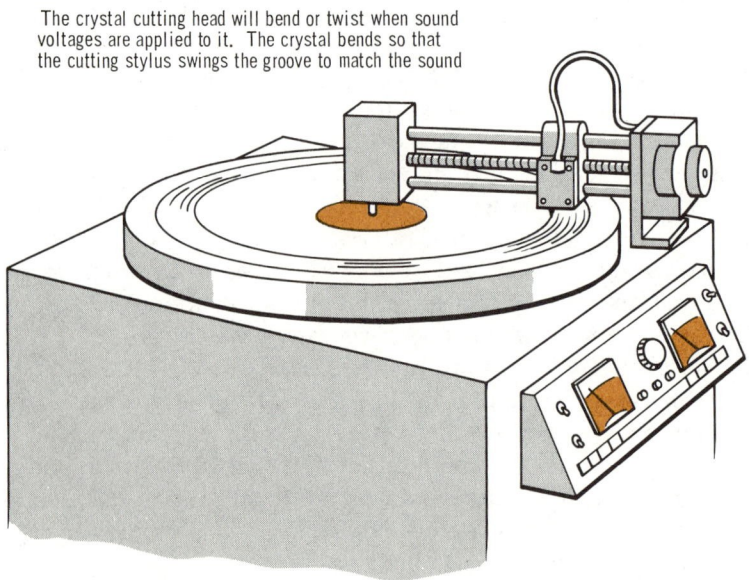

The crystal cutting head will bend or twist when sound voltages are applied to it. The crystal bends so that the cutting stylus swings the groove to match the sound

electricity causes heat

Whenever electric current flows through a wire, it produces some *heat*. The reason for this is that some energy is used up in causing the current to flow. This energy is given off in the form of heat. Since it is easiest to cause current flow in good conductors, less heat is produced in good conductors. A poor conductor, such as nichrome, produces a great deal of heat when it conducts current. Copper is about 60 times as good a conductor as nichrome.

The heating effects of electricity are used in many appliances: toasters, irons, electric dryers, electric blankets, heaters, etc.

You should remember, though, that even good conductors produce some heat.

electricity causes light

When many of the poor conductors become hot from conducting current, they glow red and even white hot. Because of this glow, they give off light as well as heat. This is the way the ordinary *incandescent lamp* works.

Light can also be produced by electricity without much heat, by such methods as *fluorescence, phosphorescence,* and *electroluminescence.*

Electroluminescence is produced by some solid materials when they conduct current. The amount of light they give off, though, is relatively slight, and so they are used for display purposes. Many gases, when they conduct current, become ionized and produce light radiations. Neon, argon, and mercury vapor are some examples. They are used in the "neon" signs we see atop our local stores.

Phosphorescence occurs when an electron beam strikes some *phosphors* and other types of materials. The television picture tube works this way.

Fluorescence combines electroluminescence and phosphorescence. A gas, such as mercury vapor, carrying electric current becomes ionized. It emits ultraviolet radiation. The radiation strikes a phosphorescent coating that gives off "white" light.

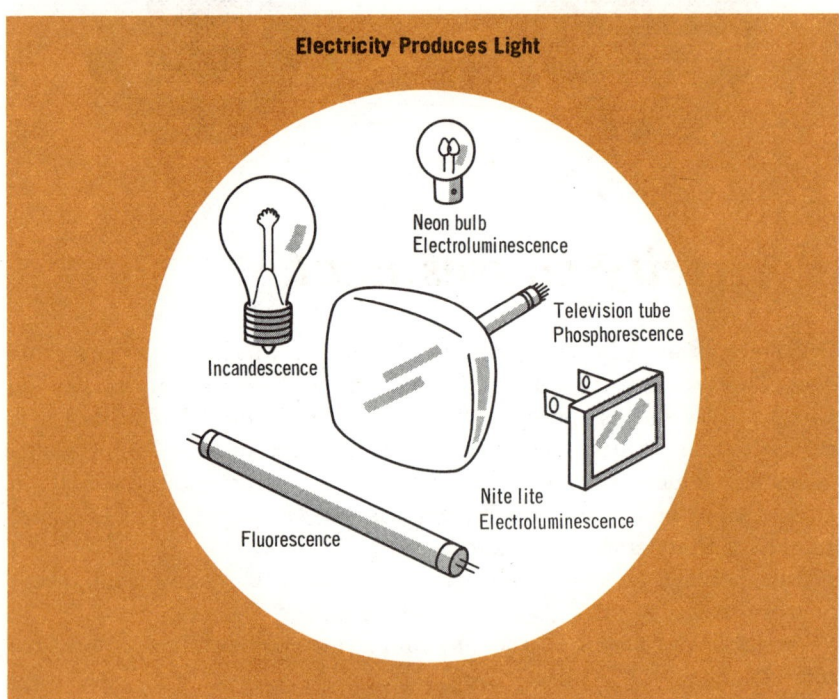

Electricity Produces Light

Neon bulb
Electroluminescence

Television tube
Phosphorescence

Incandescence

Nite lite
Electroluminescence

Fluorescence

Electricity Produces Magnetism

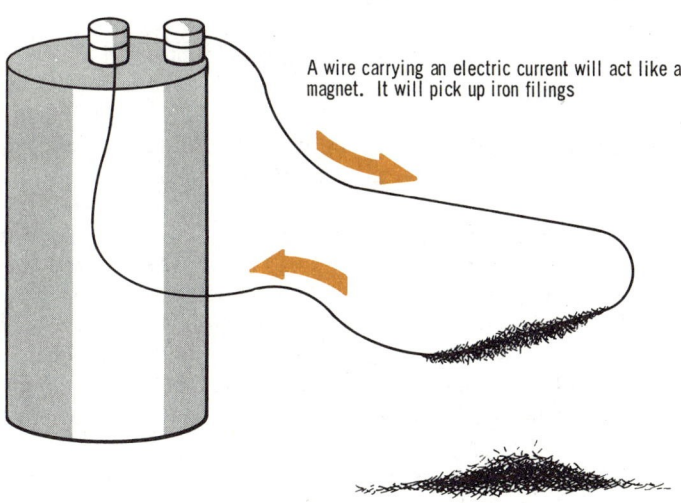

A wire carrying an electric current will act like a magnet. It will pick up iron filings

electricity causes magnetism

Magnetism, too, can be caused by electricity, just as electricity is produced with magnetism. Any conductor that carries an electric current will act like a magnet. This is called *electromagnetism*. Magnetism and electromagnetism are explained more fully in the following pages.

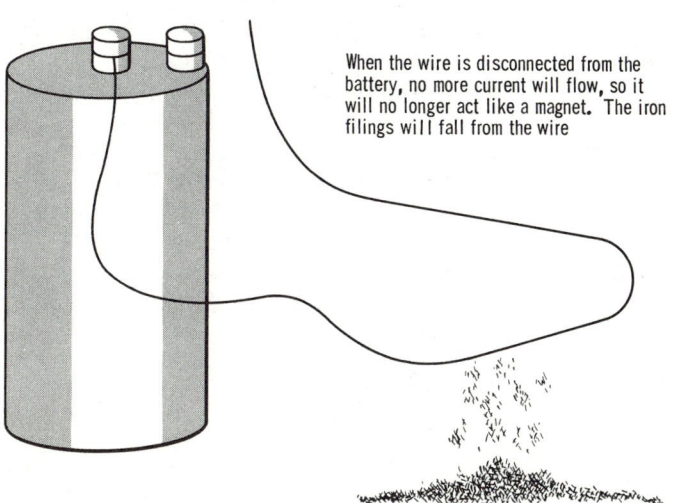

When the wire is disconnected from the battery, no more current will flow, so it will no longer act like a magnet. The iron filings will fall from the wire

summary

☐ Electricity can cause chemical, piezo, thermo, photo, and magnetic effects. ☐ In electrochemistry, the decomposition of chemicals caused by electric current gives rise to electrolysis and electroplating. ☐ Electrolysis is the decomposition of a chemical compound caused by passing a current through its solution. ☐ Electroplating is one application of electrolysis.

☐ If a voltage is applied to certain crystals, a piezoelectric force will exert a pressure that will deform the crystal. ☐ When electricity flows through a poor conductor, it produces heat. ☐ This thermal effect is used in toasters, irons, electric dryers, etc. ☐ When electricity flows, it can be made to produce light. This can be accompanied by considerable heat, as in the common incandescent lamp; or by little heat, as in such methods as fluorescence, phosphorescence, and electroluminescence.

☐ Electroluminescence is produced by gases and some solid materials when they conduct current. "Neon" signs are produced when gases are used for conductors. ☐ Phosphorescence occurs when radiation or an electron beam strikes some phosphors and other types of materials. The television picture tube works on this principle. ☐ Fluorescence combines electroluminescence and phosphorescence. A gas is ionized and emits ultraviolet radiation. The radiation strikes a phosphorescent coating that gives off "white" light. ☐ Any conductor that carries an electric current will act like a magnet. This is electromagnetism.

review questions

1. Name the five effects of electric current.
2. What is electroplating? When a liquid is decomposed into gases by an electric current, what is this called?
3. Which produces more heat: a good or a poor conductor? Name a metal that gives a high thermal effect.
4. What is incandescence?
5. What are electroluminescence and phosphorescence?
6. What process combines electroluminescence and phosphorescence?
7. "Neon" signs are made with ＿＿＿＿＿＿ as a conductor.
8. How does incandescence differ from the other photoelectric effects?
9. A crystal earphone works on what principle? What other kind of device can use this principle?
10. What is the difference betwen magnetoelectricity and electromagnetism?

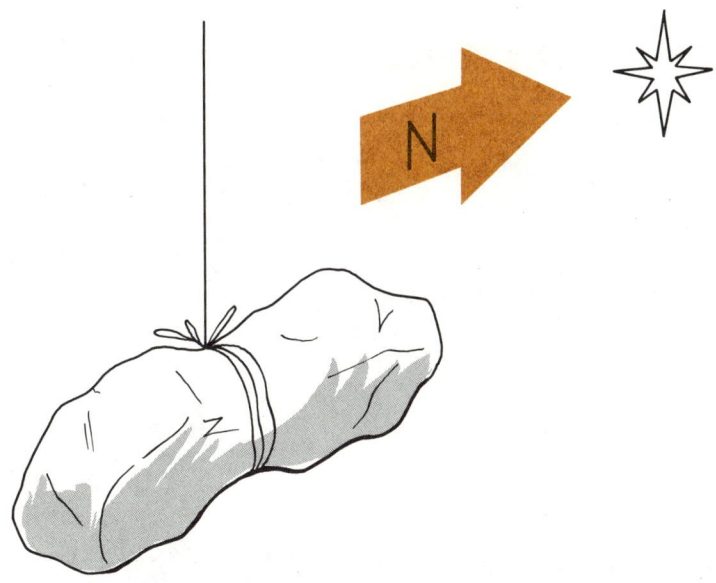

Lodestone was a natural magnet discovered by the Greeks in Asia Minor over 2000 years ago

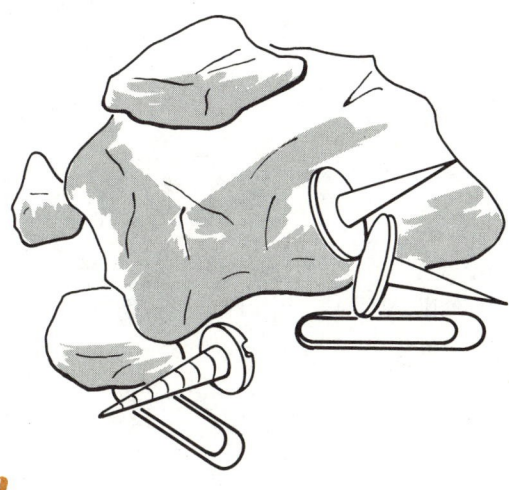

magnetism

Magnetism was first discovered over 2000 years ago by the ancient Greeks when they noticed that a certain kind of stone was attracted to iron. Since this stone was first found in Magnesia in Asia Minor, the stone was called *magnetite*. Later, when it was discovered that this stone would align itself north and south when suspended on a string, it was referred to as the *leading stone* or *lodestone*. Lodestone, therefore, is a *natural magnet* that will attract magnetic materials.

magnetism and the electron

Although the forces of electricity and magnetism are related, they are completely different. Magnetic forces and electrostatic forces have no effect on one another as long as there is *no motion*. But, if the force field of either is set in motion, then something happens to cause the two forces to interact. Since the electron is the smallest particle of matter, a theory has been developed to explain the relationship between electricity and magnetism. This is the *electron theory of magnetism*.

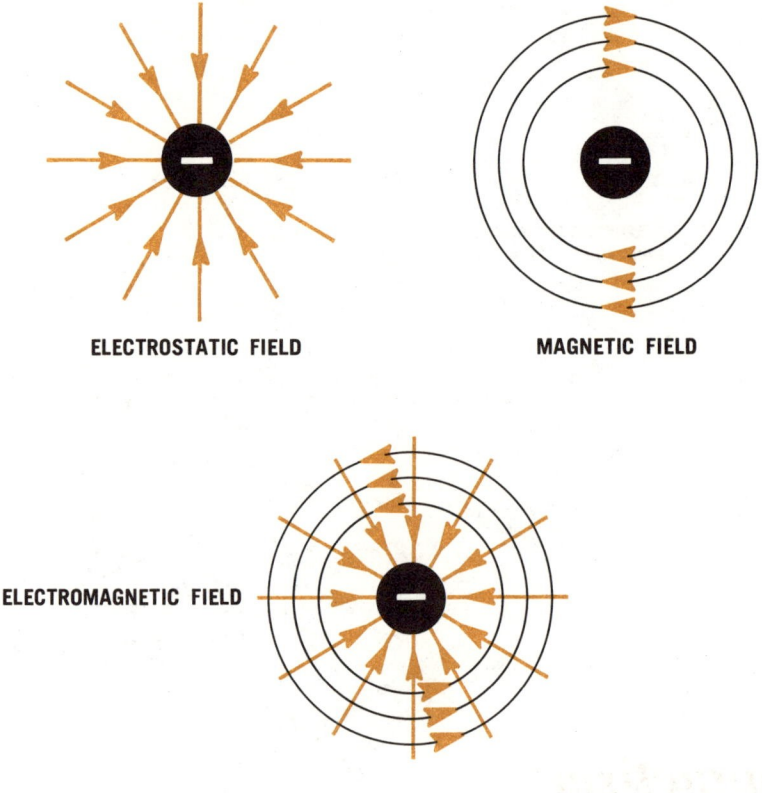

ELECTROSTATIC FIELD   MAGNETIC FIELD

ELECTROMAGNETIC FIELD

We know that the electron contains a negative charge. This charge produces a force field that comes straight in to the electron from all directions. But scientists claim that a *spinning charge* also produces a *magnetic field*. Because of its orbital spin, the electron also contains a magnetic field. But this field exists in concentric circles around the electron. The electrostatic lines of force, then, and the magnetic lines of force are at *right angles* to one another at any one point. The two fields combined are called the *electromagnetic field*.

the magnetic molecule

Actually, iron, nickel, and cobalt are the only naturally magnetic metals; but since all materials contain electrons, you might ask why everything does not have magnetic properties? The answer to this is that the electrons in atoms tend to pair off in orbits with opposite spins, so that their magnetic fields are opposite and cancel each other. But then this might lead one to believe that those elements that have an odd number of electrons are magnetic. If those atoms could be isolated, this could be so; but when atoms combine to form molecules, they ordinarily arrange themselves to produce a total of 8 valence electrons, and in the process, the orbital spins of the electrons cancel the magnetic fields in most materials.

For some reason, though, this orderly process does not occur in iron, nickel, and cobalt. When the atoms of these metals combine, they become ions and share their valence electrons in such a way that many of the electron spins *do not cancel, but add.* This produces regions in the metal called *magnetic domains,* which are referred to as *magnetic molecules.* These magnetic molecules act just like little magnets.

Although iron, nickel, and cobalt are the only naturally magnetic materials, compounds can be manufactured with a controlled process to give them good magnetic properties.

NONMAGNETIC ATOM

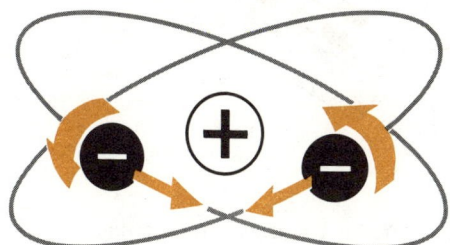

The opposite spins of paired electrons cancel their magnetic effects

In a magnetic molecule, the electrons do not pair off with opposite spins, and the molecule has magnetic properties

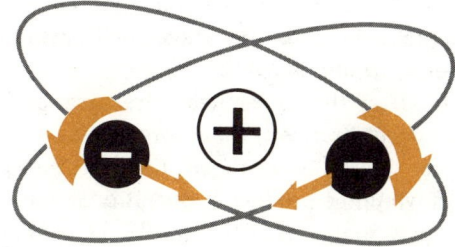

magnetic materials

The naturally magnetic materials are called *ferromagnetic* materials. Ferrous stands for iron, and each of the materials acts like iron with respect to magnetism.

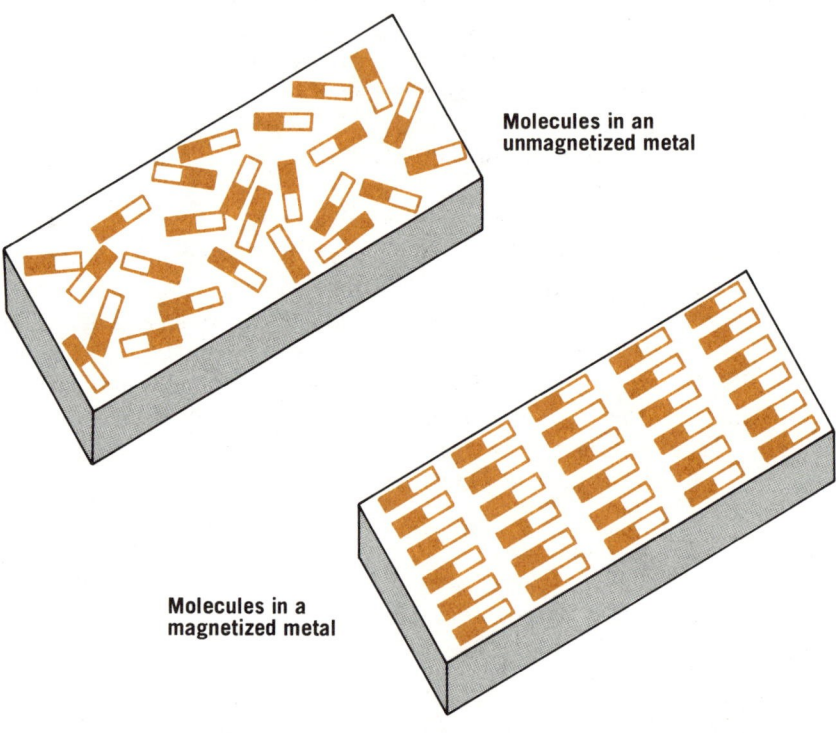

Molecules in an
unmagnetized metal

Molecules in a
magnetized metal

Since the magnetic materials contain magnetic molecules, it might seem as though they would always act as magnets. But they do not. The reason for this is that under ordinary circumstances, the magnetic molecules are scattered and oriented in a random manner, so that the magnetic fields of the molecules cancel each other. The metal is considered *unmagnetized.*

If all of the molecules were arranged so that they were pointing in the *same direction,* their force fields would *add.* The metal would then be considered *magnetized.* If *all* of the molecules were aligned, a *strong* field would be produced. But, if only *some* of the molecules were aligned, a *weak* magnetic field would be produced. So a magnetic material could also be *partially magnetized.*

how to magnetize iron

Since a magnetic material can be magnetized by *aligning* its molecules, the best way to do it is by applying a magnetic force. The force would act against the magnetic field of each molecule and force it into alignment. This can be done in two ways: (1) by magnetic stroking, and (2) by an electric current.

When a magnet is stroked across the surface of an unmagnetized piece of iron, the field of the magnet aligns the molecules to magnetize the iron.

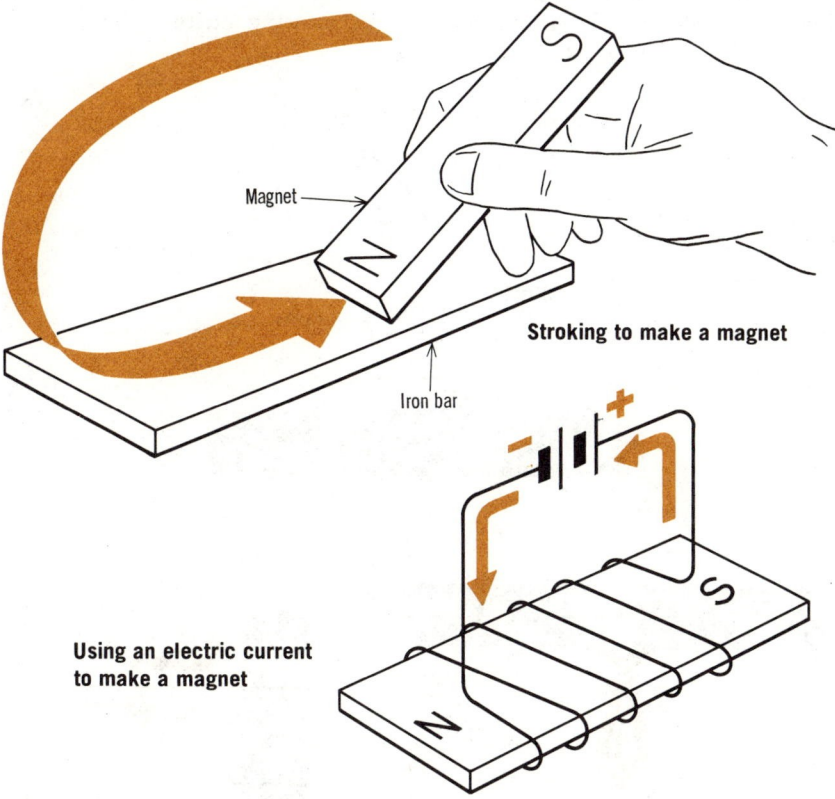

Magnet

Stroking to make a magnet

Iron bar

Using an electric current
to make a magnet

When an unmagnetized piece of iron is put in a coil of wire, and the wire is connected across a battery, the electric current produces a magnetic field that magnetizes the iron. This is explained later.

When a magnetized material keeps its magnetic field for a long time, it is called a *permanent magnet*. If it loses its magnetism fast, it is called a *temporary magnet*. Hard iron or steel makes a good permanent magnet. Soft iron is used for temporary magnets.

how to demagnetize a magnet

In order to *demagnetize* a magnet, the molecules must again be *disarranged* so that their fields *oppose* each other.

If the magnet is struck hard, the force would jar the molecules and they would rearrange themselves. Sometimes, repeated strikes are needed.

If the magnet were heated, the *heat energy* would cause the molecules to vibrate sufficiently and rearrange themselves.

If the magnet were placed in a rapidly *reversing magnetic field*, the molecules would become disarranged trying to follow the field. A rapidly reversing field can be produced with alternating current. This topic is covered in Volume 3.

DEMAGNETIZING A MAGNET

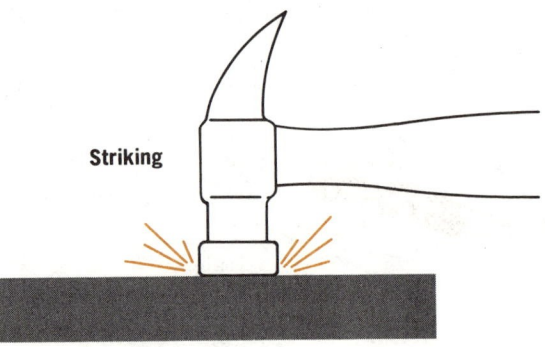

Striking

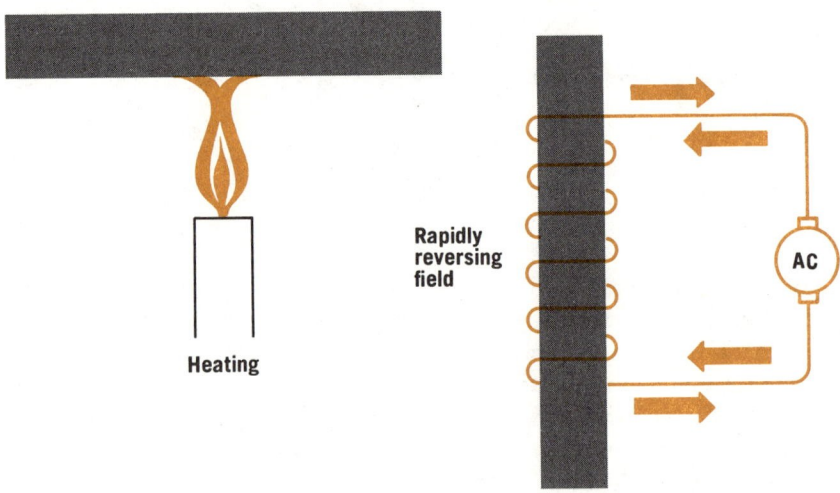

Heating

Rapidly reversing field

AC

the earth's magnetic field

Since the earth itself is a large spinning mass, it too produces a magnetic field. The earth acts as though it has a bar magnet extending through its center, with one end near the north geographic pole, and the other end near the south pole.

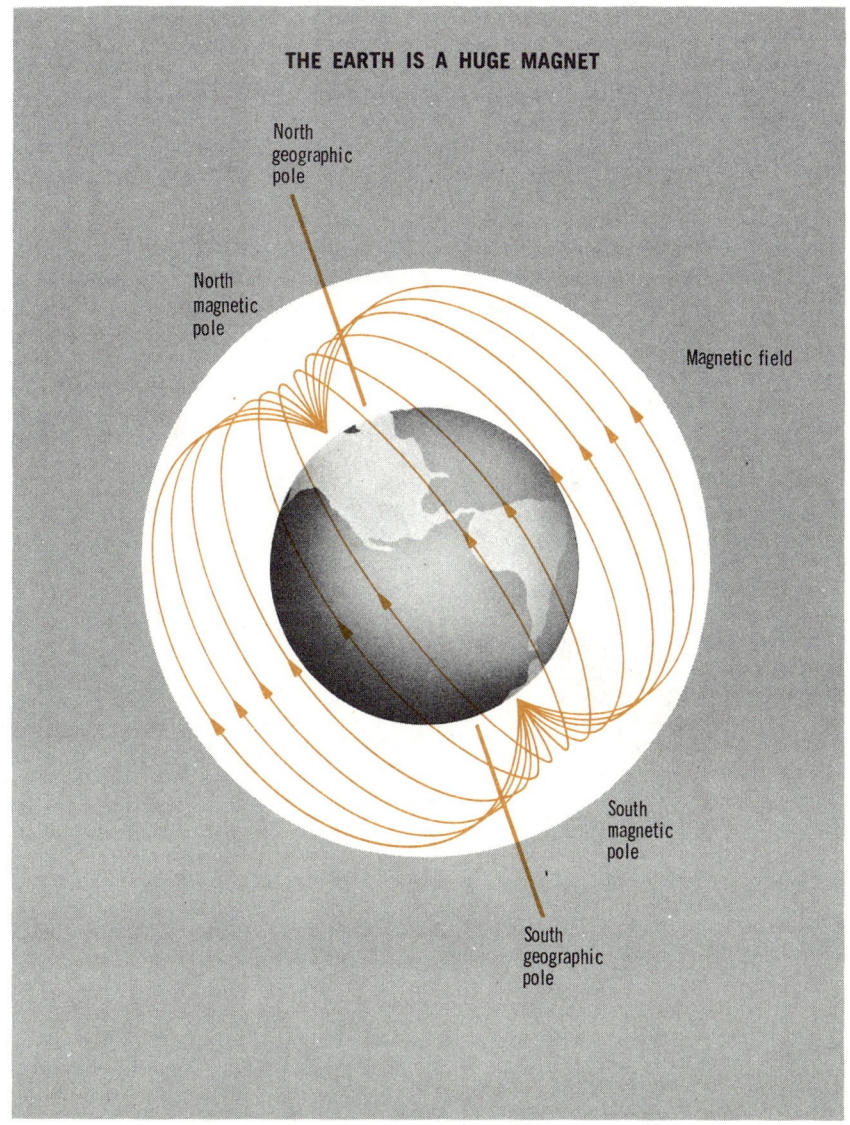

THE EARTH IS A HUGE MAGNET

North geographic pole

North magnetic pole

Magnetic field

South magnetic pole

South geographic pole

magnetic polarities

In order to set rules on how magnets affect one another, *polarities* are assigned to the ends of magnets. The polarities are called *north* (N) and *south* (S). The north end of a magnet is determined by hanging that magnet by a string to allow it to swing freely. The magnet will then

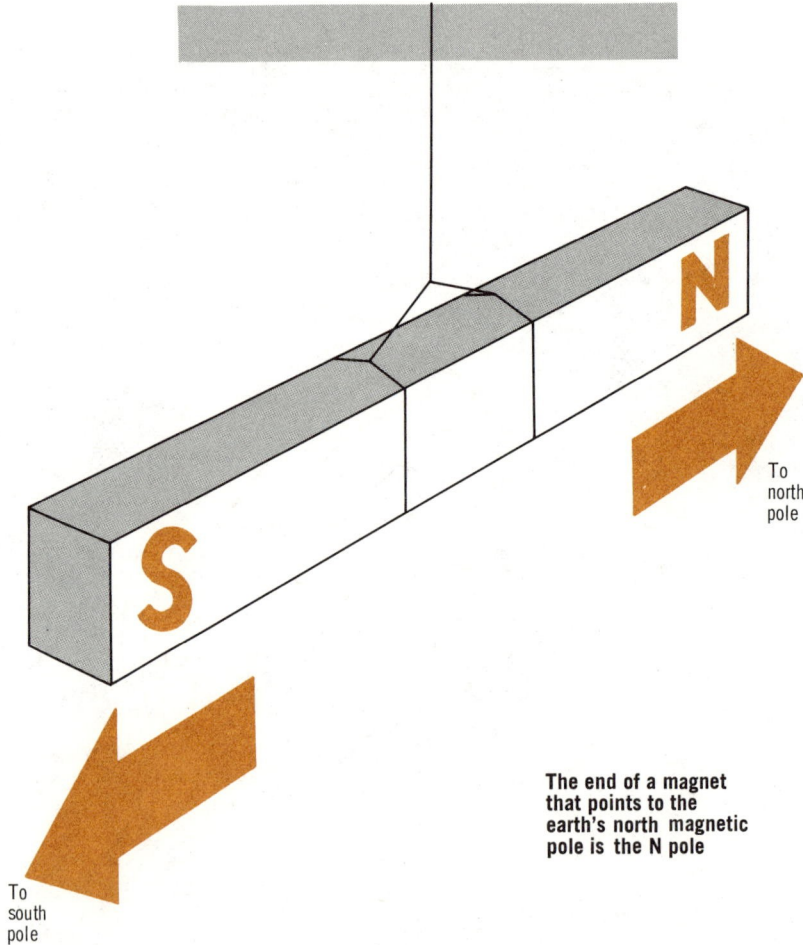

To north pole

To south pole

The end of a magnet that points to the earth's north magnetic pole is the N pole

align itself with the earth's magnetic field. The end of the magnet that points to the earth's north magnetic pole is called the N pole of the magnet. The other end of the magnet is called the S pole. The magnet will always align itself this way. The reason will be explained later.

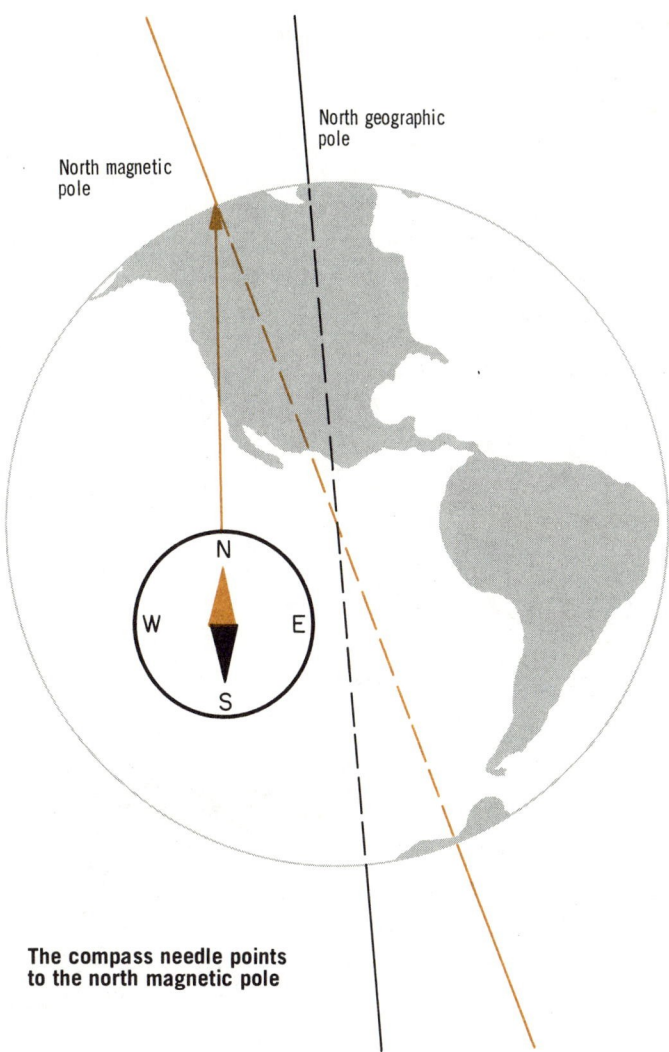

North geographic
pole

North magnetic
pole

N

W   E

S

**The compass needle points
to the north magnetic pole**

the magnetic compass

Since a magnet will align itself with its N pole pointing north, this
can be used to determine directions. A compass is made with a tiny
light magnet that is freely pivoted so that it will easily keep itself
aligned to the earth's magnetic north pole. Regardless of how the compass is turned, the magnetic needle will point north.

attraction and repulsion

Since a magnet always lines itself up with the earth's magnetic north pole, there seems to be some definite laws governing magnetic effects. These are the laws of attraction and repulsion. The attraction and repulsion laws of magnetism are the same as those of electric charges, except N and S polarities are used instead of negative and positive. The laws are: *Like poles repel, unlike poles attract.*

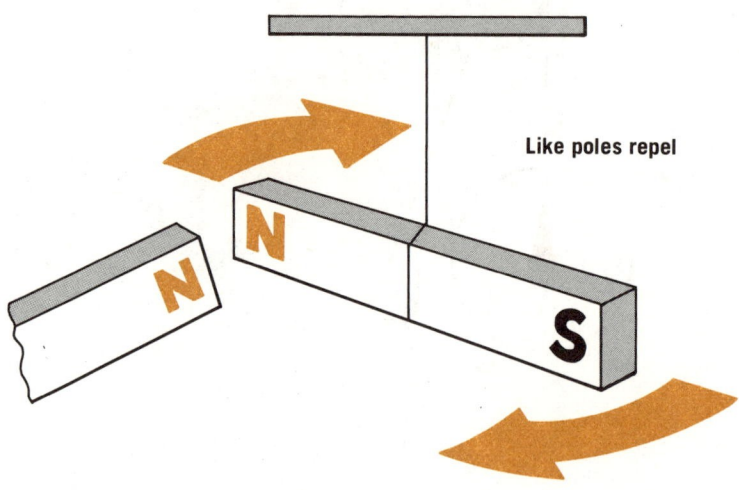

Like poles repel

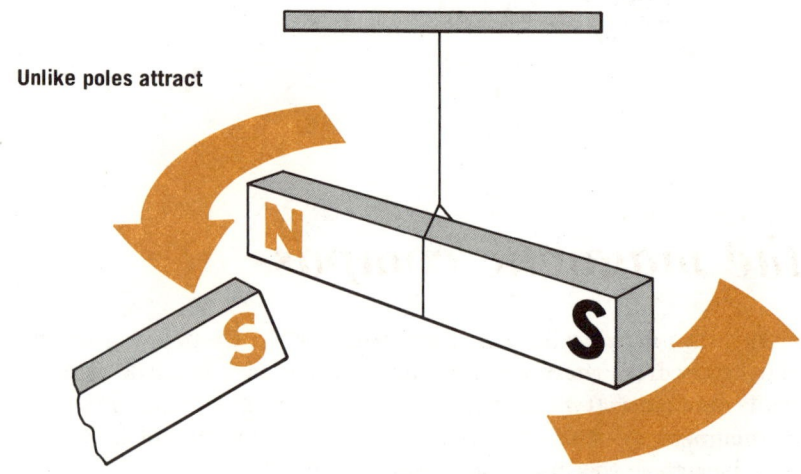

Unlike poles attract

the magnetic field

As you can see by the attraction and repulsion of the magnetic poles, there are forces coming out of the magnetic poles to cause those actions. But the actions do not only take place at the poles. The magnetic *force* actually surrounds the magnet in a *field*. This can be seen when a compass is moved around the bar magnet. In each position around the bar magnet, one end of the compass needle will point to the opposite pole on the bar.

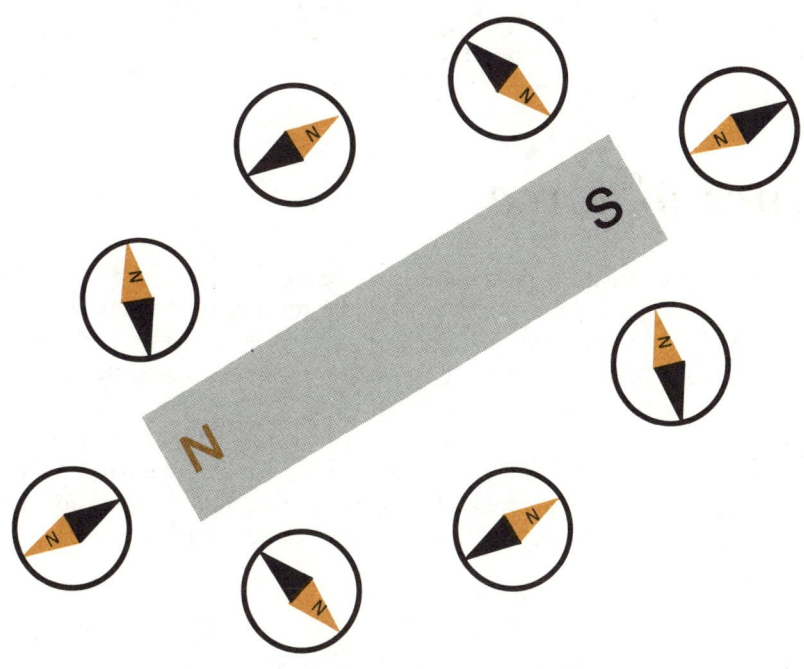

**The compass
shows how the
magnetic force
surrounds a magnet**

The compass can also be used to see how far the magnetic field extends away from the magnet. By withdrawing the compass slowly, you will reach a point where the compass needle is no longer affected by the magnetic field of the magnet, but will again be attracted by the earth's north magnetic pole.

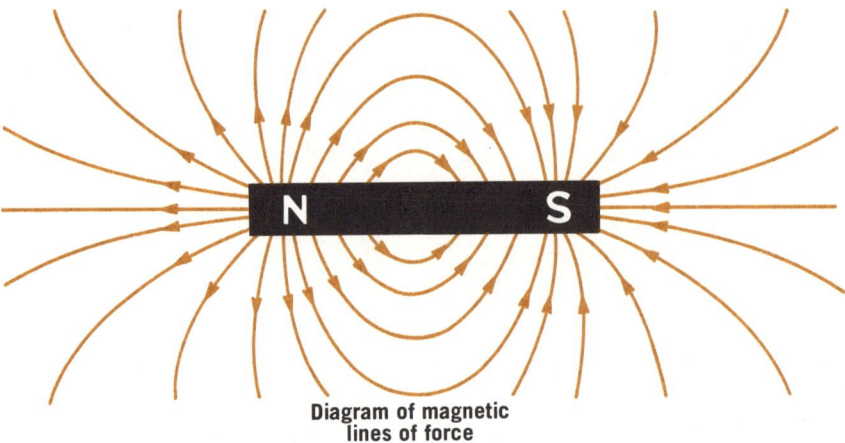

**Diagram of magnetic
lines of force**

lines of force

The magnetic field of a magnet is made up of *lines of force* that extend out into space from the N pole of the magnet to the S pole. These lines of force do not cross, and they become wider apart away from the magnet. The closer the lines of force are and the greater the number of force lines, the stronger the magnetic field.

The existence of the lines of force can be demonstrated by sprinkling iron filings on a flat surface, and then placing a bar magnet on them. The iron filings will arrange themselves along the lines of force to show the magnetic field. The lines of force are also called *flux lines*.

**Fine iron filings show up the
lines of force of a bar magnet**

interaction of magnetic fields

When two magnets are brought together, their fields interact. Magnetic lines of force *will not cross* one another. This fact determines how fields act together.

If the lines of force are going in the *same direction,* they will *attract* each other and join together as they approach each other. This is why unlike poles attract.

If the lines of force are going in *opposite directions,* they *cannot combine.* And, since they cannot cross, they apply a force against each other. This is why like poles repel.

The interaction of the flux lines can also be shown with iron filings.

Unlike poles attract

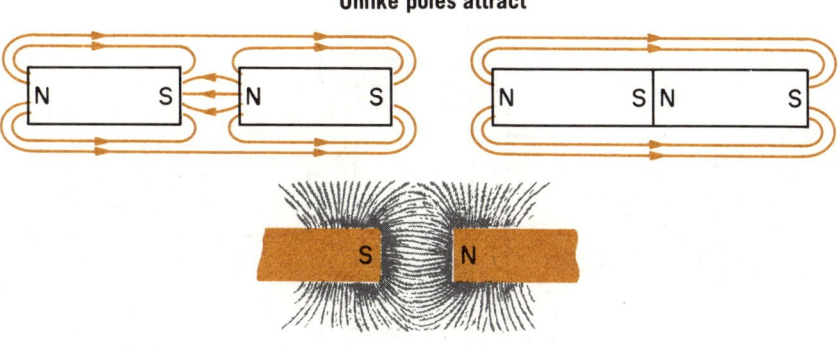

Like poles repel

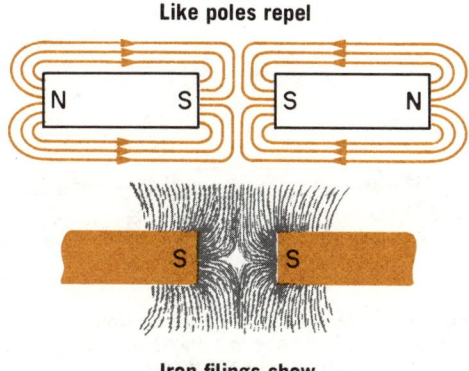

Iron filings show
the flux lines

magnetic shielding

Magnetic flux lines can pass through all materials, even those that have no magnetic properties. Some materials, though, resist the passage of flux lines somewhat. This property is known as *reluctance*. Magnetic materials have a very low reluctance to flux lines. The lines of flux will be attracted through a magnetic material even if it has to take a longer path. This characteristic allows us to shield things from magnetic lines of force by enclosing them with a magnetic material. This is the way antimagnetic watches are made.

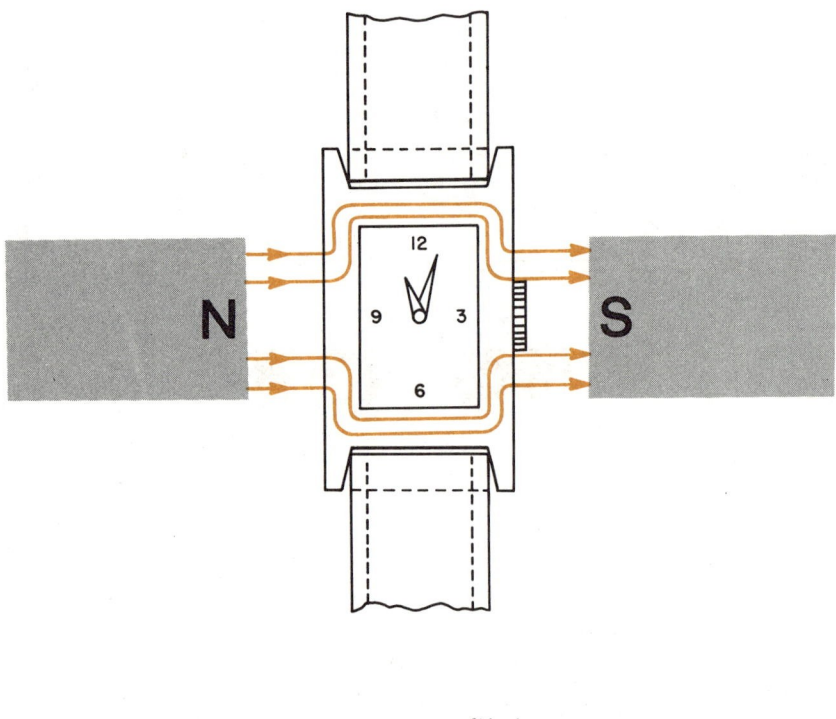

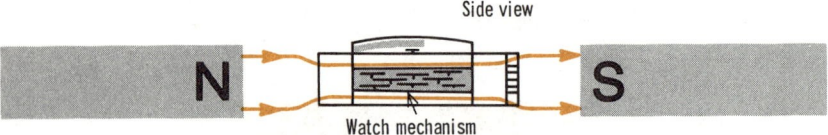

Side view

Watch mechanism

Magnetic materials shield a watch from magnetic fields. The shield becomes only temporarily magnetized. The flux lines cannot reach this watch mechanism

summary

☐ The interaction of electricity and magnetism to form an electromagnetic field is explained by the electron theory of magnetism. ☐ Atoms of certain metals combine so that their valence electrons are shared in such a way to form magnetic domains or molecules. These are examples of magnetic materials. ☐ A material with its magnetic domains or molecules aligned is said to be magnetized. ☐ A magnetic material can be magnetized by applying a magnetic force either by stroking or by an electric current. ☐ A magnetic material can be demagnetized by being heated, struck, or by being placed in a rapidly reversing magnetic field.

☐ The earth produces a magnetic field. ☐ North (N) and south (S) polarities are assigned to magnets. ☐ The N pole of a freely pivoted magnet will point to the earth's north pole. The other pole is the S pole. A compass utilizes this principle. ☐ The laws of repulsion and attraction for magnets are: like poles repel, and unlike poles attract.

☐ Magnetic force surrounds the magnet in a magnetic field. ☐ The magnetic field is made up of lines of force that extend into space from the N pole of the magnet to the S pole. ☐ Lines of force are called flux lines. ☐ Lines of force do not cross. ☐ The closer the lines of force and the greater the number of force lines, the stronger the magnetic field. ☐ Lines of force in the same direction attract and join; this is why unlike poles attract. ☐ Lines of force in opposite directions repel and cannot combine; this is why like poles repel.

review questions

1. What are magnetic domains or molecules?
2. How can a magnet be demagnetized?
3. What is a ferromagnetic material?
4. What is meant by the reluctance of a material? Why should the reluctance of a shield be smaller than the shielded material?
5. What is an electromagnetic field?
6. How can a ferromagnetic metal be magnetized?
7. What are the laws of repulsion and attraction for magnetic poles?
8. What are flux lines?
9. Which pole of a compass needle points to the earth's north pole?
10. What is the difference between permanent and temporary magnets?

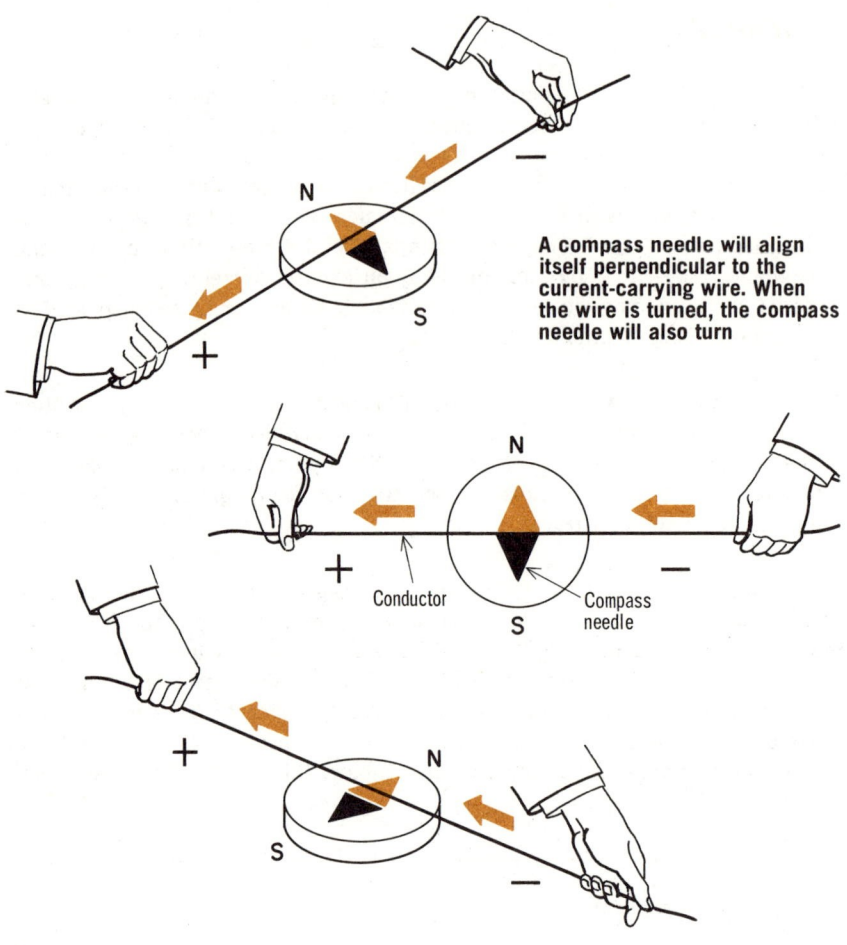

A compass needle will align itself perpendicular to the current-carrying wire. When the wire is turned, the compass needle will also turn

what is electromagnetism?

Since the electron produces its own magnetic field because of its orbital spin, the accumulation of surplus electrons in an object might seem to be able to produce a magnetic field. But, again, with static charges, electrons with opposite spins pair off to cancel their magnetic effects. Static electricity, then, has no magnetic field.

Electrons moving through a wire under a force causing a *current flow,* though, cannot pair off with opposite spins. On the contrary, since they are drifting in the same direction, their magnetic fields tend to add.

In 1819, Hans Christian Oersted discovered that an electric current produced a magnetic field when he noticed how a wire carrying a current affected a compass.

electromagnetism in a wire

Since the magnetic field around an electron forms a loop, the fields of the electrons combine to form a series of loops around the wire. The direction of the magnetic field depends on the direction that the current flows. A compass moved around the wire will align itself with the flux lines.

A *left-hand rule* can be used to determine the direction of the magnetic field. If you wrap your fingers around the wire with your thumb pointing in the direction of electron current flow, your fingers will point in the direction of the magnetic field.

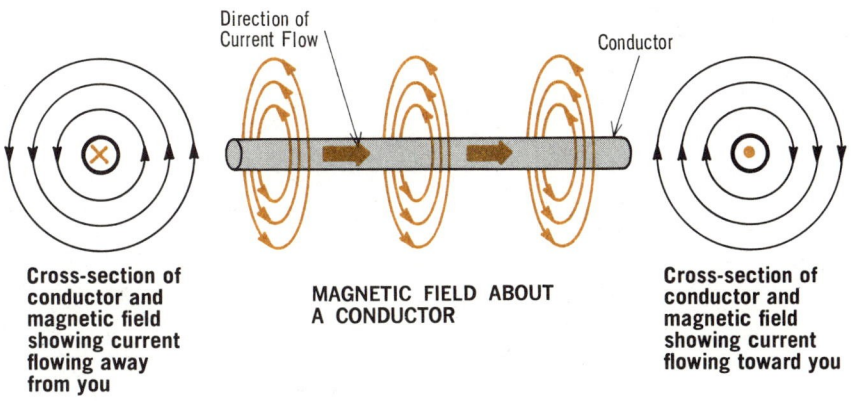

Cross-section of conductor and magnetic field showing current flowing away from you

MAGNETIC FIELD ABOUT A CONDUCTOR

Cross-section of conductor and magnetic field showing current flowing toward you

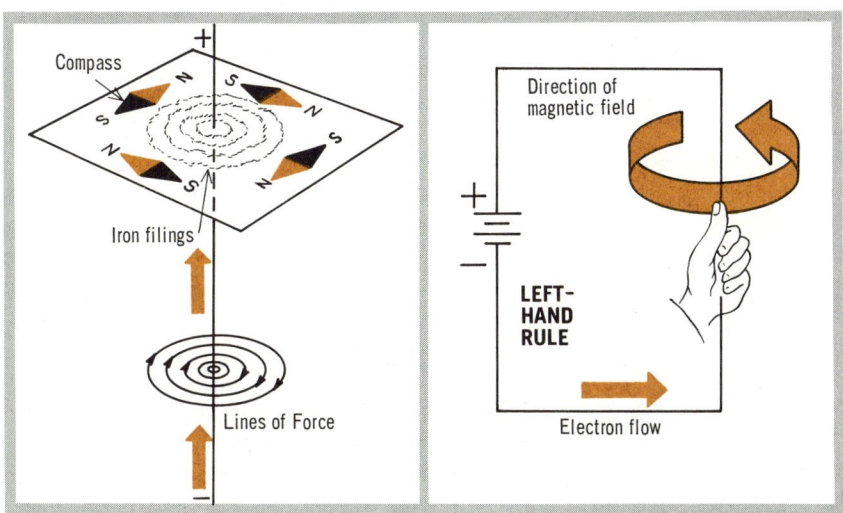

field intensity

The more current that flows through a wire, the stronger the magnetic field will be. As with the magnet's magnetic field, the flux lines are closer together near the wire, and they move further apart as they move away from the wire. The field, then, is stronger near the wire, and becomes weaker with distance.

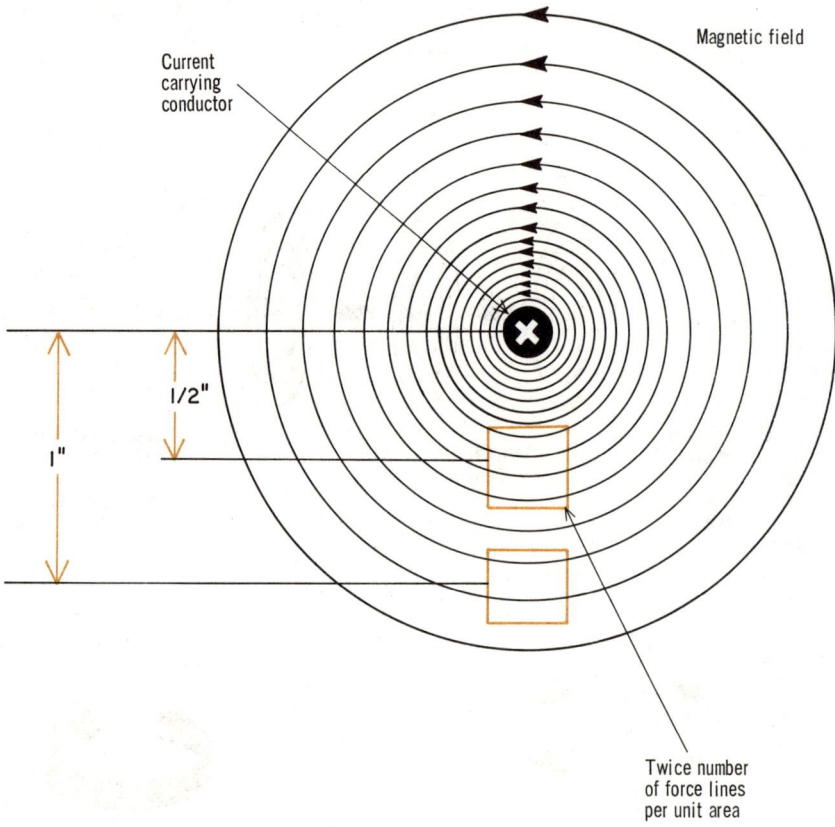

The decrease in the number of lines of force per unit area is in inverse proportion to the distance from the conductor. That is, at a distance of one inch from the conductor, for instance, there is one-half the density of force as at a distance of one-half inch.

field interaction

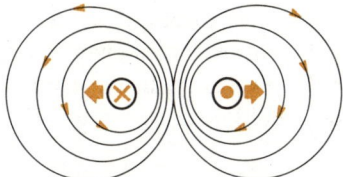

**Opposite currents cause
opposite fields that repel**

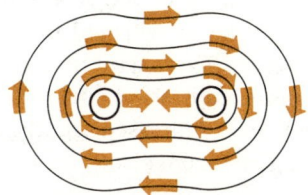

**Currents in the same direction
cause fields to add and attract**

If two wires that are carrying current in opposite directions are brought close together, their magnetic fields will oppose one another, since the flux lines are going in opposite directions. The flux lines cannot cross, and the fields move the wires apart.

When wires that are carrying current in the same direction are brought together, their magnetic fields will aid one another, since the flux lines are going in the same direction. The flux lines join and form loops around both wires, and the fields bring the wires together. The flux lines of both wires add to make a stronger magnetic field. Three or four wires put together in this way would make a still stronger field.

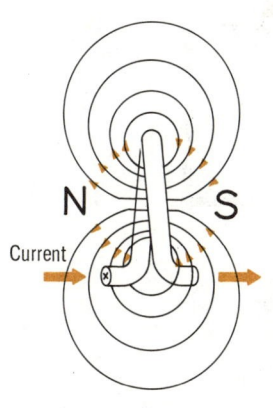

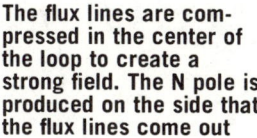

When a wire is being formed into a loop, all of the circular fields enter one side of the loop and leave the other side

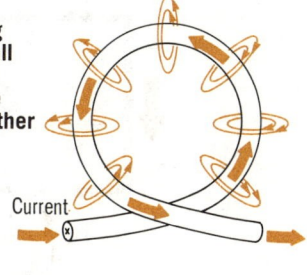

The flux lines are compressed in the center of the loop to create a strong field. The N pole is produced on the side that the flux lines come out

If the wire is twisted to form a *loop*, the magnetic fields around the wire will all be arranged so that they each flow into the loop on one side, and all come out on the other side. In the center of the loop, the flux lines are compressed to create a dense and stronger field. This produces magnetic poles, with north on the side that the flux lines come out, and south on the side that they go in.

electromagnetism in a coil

If a number of loops are wound in the same direction to form a *coil*, more fields will add to make the flux lines through the coil even more dense. The magnetic field through the coil becomes even stronger. The more loops there are, the stronger the magnetic field becomes. If the coil is compressed tightly, the fields would join still more to produce a still stronger *electromagnet*.

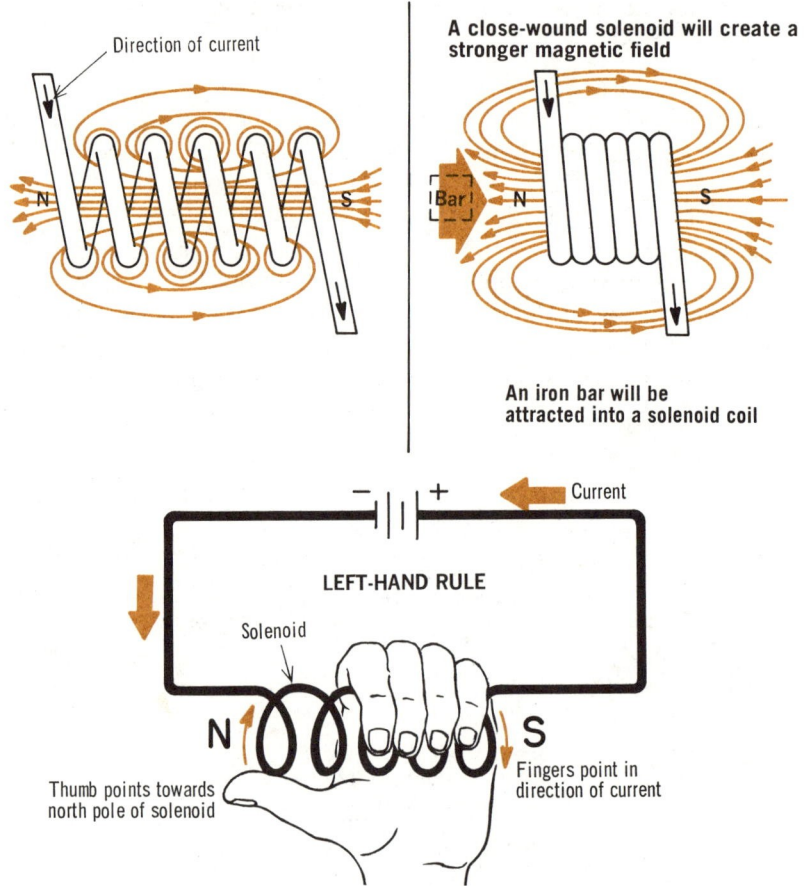

Direction of current

A close-wound solenoid will create a stronger magnetic field

An iron bar will be attracted into a solenoid coil

LEFT-HAND RULE

Current

Solenoid

N   S

Thumb points towards north pole of solenoid

Fingers point in direction of current

A helically wound coil that is made to produce a strong magnetic field is call a *solenoid*. The flux lines in a solenoid act the same as in a magnet. They leave the N side and go around to the S pole. When a solenoid attracts an iron bar, it will draw the bar *inside* the coil.

There is a *left-hand rule* for solenoids too. If you wrap your fingers around the coils in the direction of electron current flow, your thumb will point to the N pole.

the magnetic core

The magnetic field of a coil can be made stronger still by keeping an *iron core* inside the coil of wire. Since the soft iron is magnetic and has a low *reluctance,* it allows more flux lines to be concentrated in it than would air. The more flux lines there are, the stronger the magnetic field. Soft iron is used as a core in an electromagnet because hard iron would become permanently magnetized.

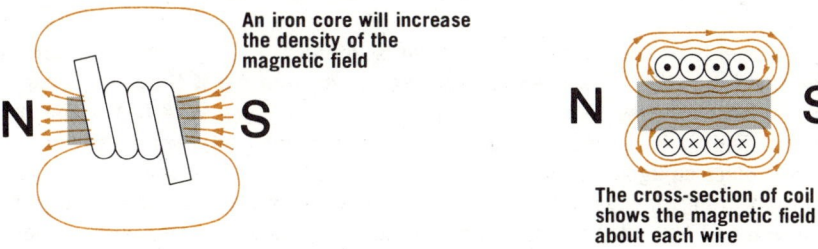

An iron core will increase the density of the magnetic field

The cross-section of coil shows the magnetic field about each wire

magnetomotive force

The magnetizing force that is caused by current flowing in a wire is called the *magnetomotive force* (mmf). The mmf depends on the current flowing in the coil and the number of turns in the coil. If the current is doubled, the mmf will be doubled. Also, if the number of turns (loops) in the coil is increased, the mmf will be increased.

The mmf, then, is determined by a term called ampere-turns, which is the electric current multiplied by the number of turns of the coil.

The magnitude of the mmf determines the number of lines of flux there will be in the field, or how strong the field will be. As the mmf is increased, the number of flux lines also increases. But there will be a point at which the mmf being increased will no longer produce more flux lines. This is known as the *saturation point.*

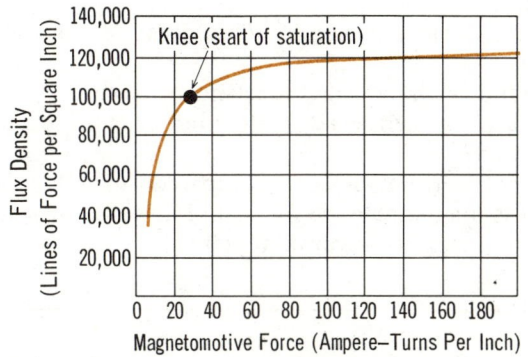

summary

☐ Electrons produce magnetic as well as electrostatic fields. But normally, electrons cancel each other's magnetic effects. ☐ Electrons moving through a wire because of current flow produce a magnetic field because the electrons' fields aid each other. ☐ The direction of the magnetic field depends on the direction of the current flow, according to the left-hand rule: If your fingers are wrapped around the wire, with your thumb in the direction of the electron current flow, they will indicate the direction of the magnetic field. ☐ The stronger the electron current, the stronger is the magnetic field. ☐ The field is stronger near the wire, and becomes weaker with distance.

☐ The fields of two current-carrying wires interact. ☐ If the currents are in the same direction, the wires are attracted to each other; if they are in opposite directions, they repel. ☐ A loop of current-carrying wire has a concentrated field at its center. ☐ Many loops can be helically wound to form a strong electromagnet. This electromagnet is called a coil or solenoid. ☐ The left-hand rule for a solenoid gives the direction of the poles: If your fingers are wrapped around the wire in the direction of the electron current flow, your thumb will point to the N pole.

☐ Soft iron cores are used to concentrate and strengthen the lines of flux. ☐ The magnetizing force that is caused by current flowing in a wire is called the magnetomotive force, mmf. ☐ Ampere-turns is the unit of mmf. It is equal to the number of loops or turns of wire multiplied by the current in amperes. ☐ When an increasing mmf no longer increases the number of flux lines, the coil is saturated.

review questions

1. State the rule to determine the direction of the magnetic field around a current-carrying conductor.
2. How does the current magnitude affect the field magnitude?
3. How does the number of lines of force vary with distance?
4. What happens when two current-carrying wires are brought together?
5. What is a coil or solenoid? What does it do for the magnetic field?
6. State the left-hand rule for solenoids.
7. Why are cores inserted in electromagnets? Are steel cores ever used?
8. How do electromagnets act compared to regular magnets?
9. What is magnetomotive force, and how is it measured?
10. What is meant by saturation point?

putting electricity and magnetism to work

Because of the various effects that electricity has, and the way it is related to magnetism, these two types of energy can be applied in many ways to do work.

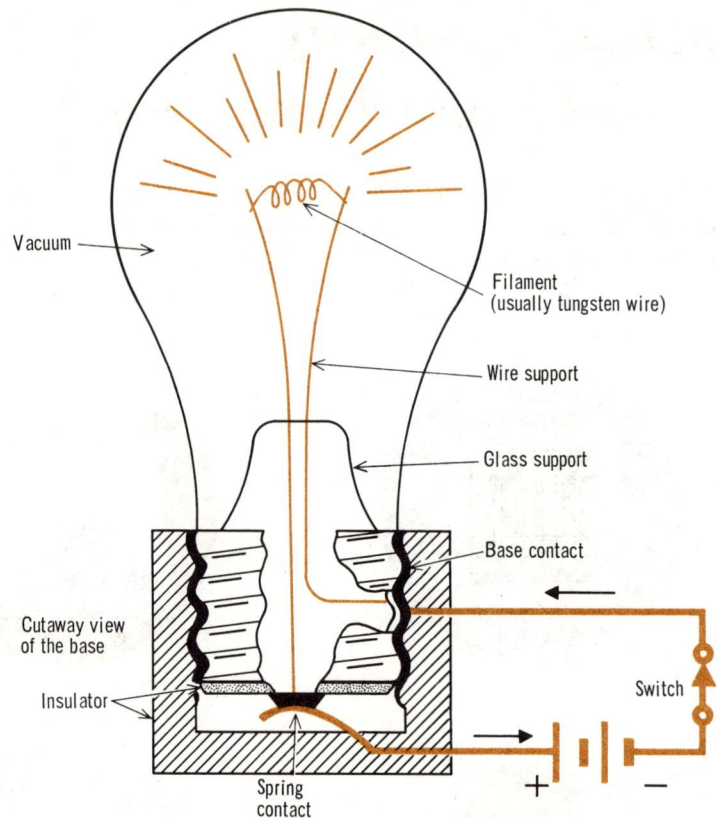

Vacuum

Filament
(usually tungsten wire)

Wire support

Glass support

Base contact

Cutaway view
of the base

Insulator

Switch

Spring
contact

+   −

the incandescent light

As explained earlier, the incandescent lamp works because electricity that flows through a poor conductor causes it to glow red or white hot; and light is given off. That poor conductor in the light bulb is called the *filament,* and is usually made of tungsten wire.

The filament is held up by two wire supports that go down through a piece of glass support to the base. One wire goes to the brass threads, and the other wire goes to a metal button at the bottom. The threaded brass and the button are good conductors and are separated by an insulator.

When the switch is closed, the circuit is complete. Electron current flows from the negative side of the battery, through the switch and base contact, and up one support wire. The current then goes through the filament, down the other wire support to the button; and then through the spring contact to the positive side of the battery.

The air is taken out of the bulb to keep the filament from burning up.

the electric heater

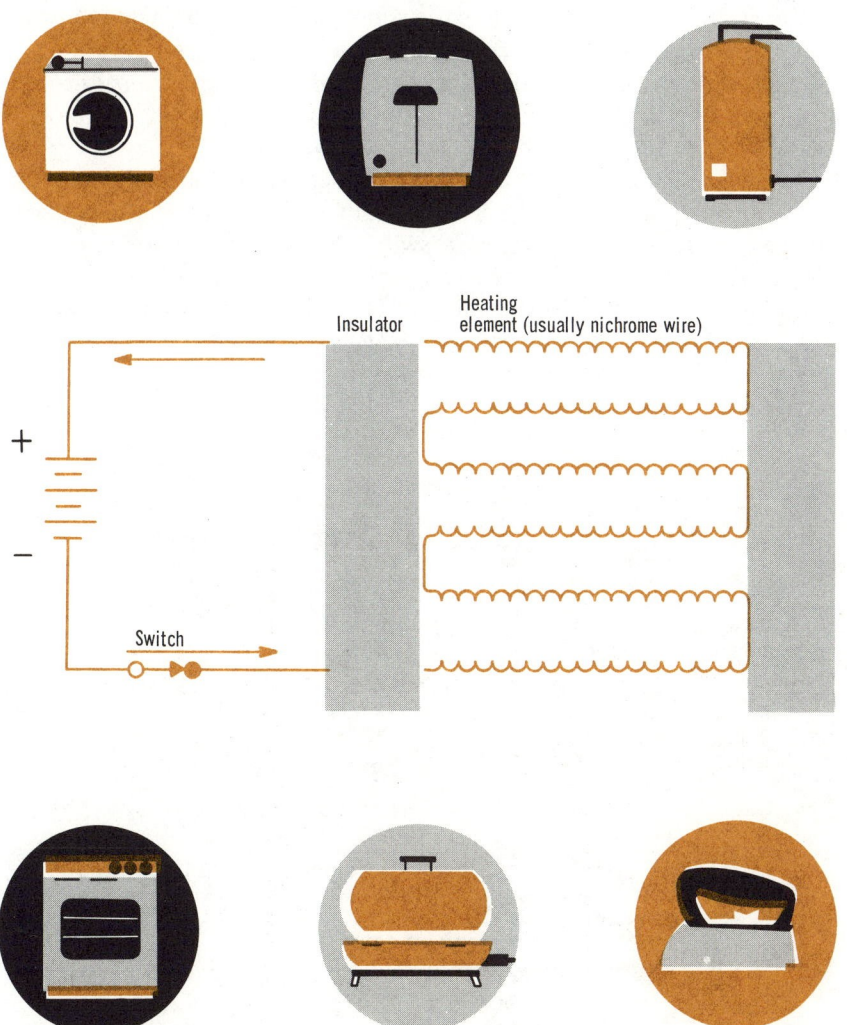

The electric heater is made similar to the incandescent light bulb, except that the material used for the heating element may not glow too brightly. When the switch is closed, the current from the battery goes through the heating element to heat it. It usually gives off a soft red glow, and heats the surrounding air. The heating element, which is usually nichrome wire, is supported by insulation.

Electric clothes dryers, hot water heaters, stoves, irons, toasters, grills, etc. work in similar ways.

the electromagnetic relay

The *electromagnetic relay* uses the action of a magnetic field to attract a movable contact against a fixed contact to close another circuit.

When the switch is closed, the battery sends electron current through a coil that is wrapped around a soft iron core. The current in the coil produces a magnetic field that magnetizes the core. The magnetic field from the core attracts the movable contact, which is also made of magnetic material, and pulls it down firmly against the fixed contact. The two contacts close just like a switch to energize the other circuit. When the main switch is opened, the current stops going through the coil, and the magnetic field collapses. The movable contact is released, and springs away from the fixed contact to open the other circuit.

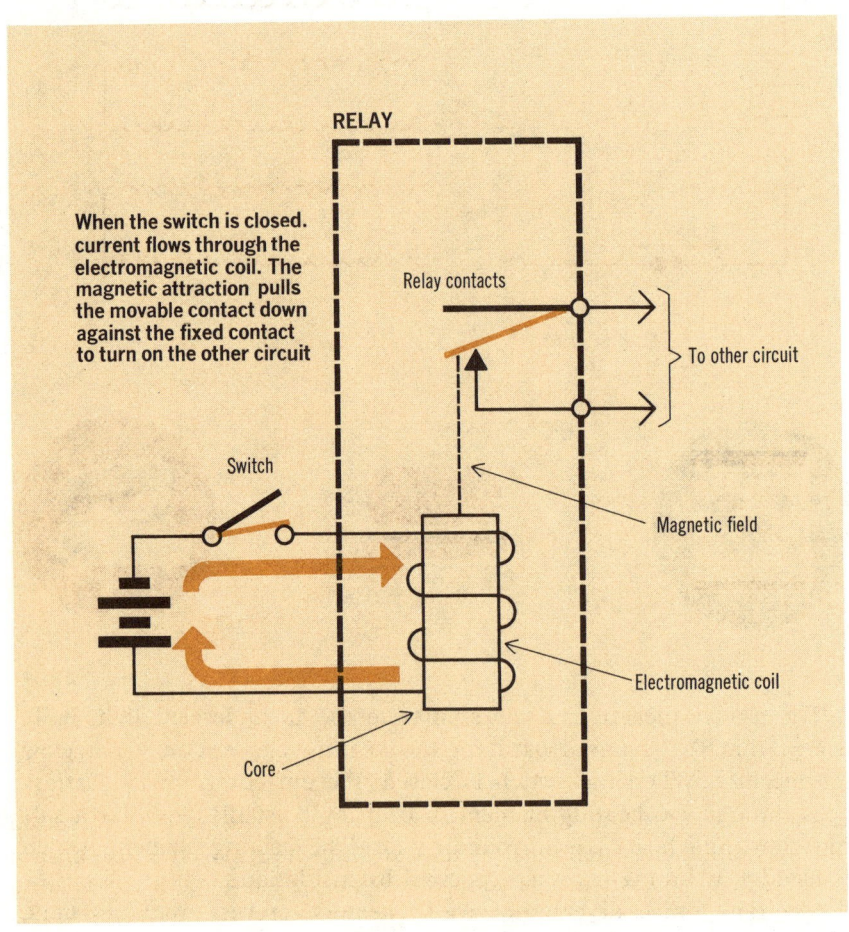

RELAY

When the switch is closed, current flows through the electromagnetic coil. The magnetic attraction pulls the movable contact down against the fixed contact to turn on the other circuit

Relay contacts

To other circuit

Switch

Magnetic field

Electromagnetic coil

Core

the electromagnetic relay (cont.)

Sometimes it is desirable to want the relay to stay energized when the operating switch is released. A special relay, called a *holding relay*, is used, which has an extra set of contacts to keep current flowing

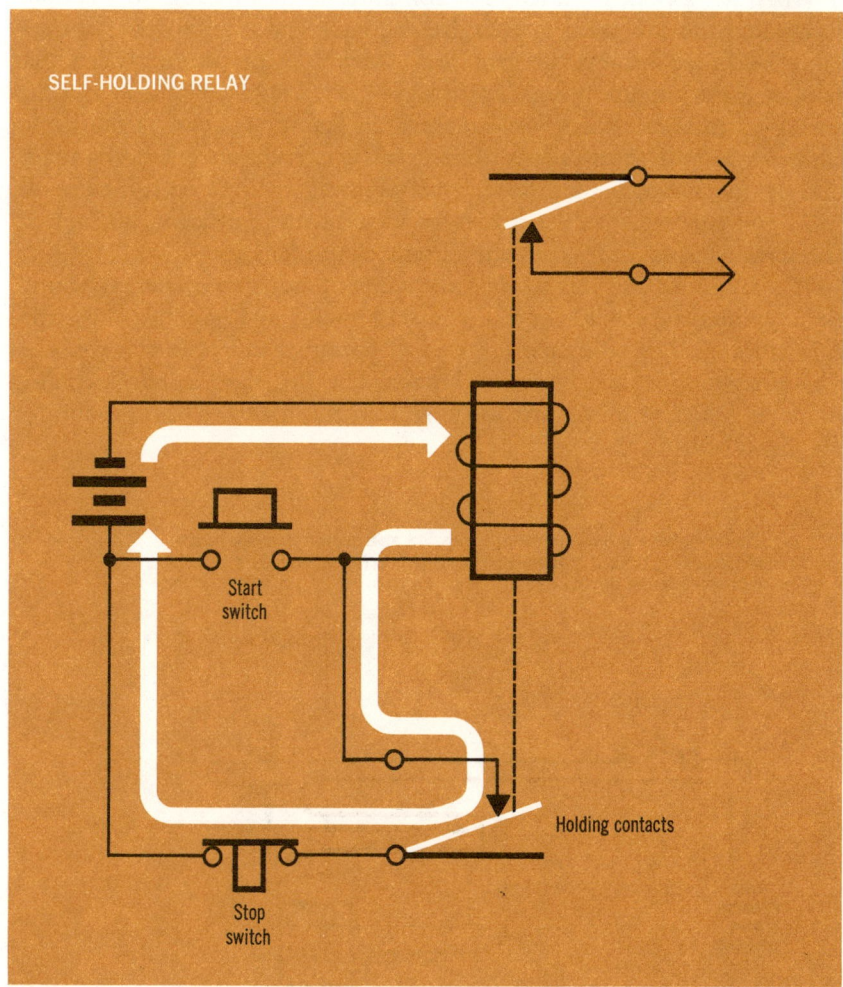

SELF-HOLDING RELAY

Start switch

Stop switch

Holding contacts

through the relay even after the main switch is released. With this kind of relay, another switch must be used to shut the relay off. It is called a *stop switch*, and when it is pushed it opens the circuit to the *holding contacts*.

the buzzer

The *buzzer* operates similarly to the relay, except that the contacts are specially arranged so that the *electromagnetic coil* cannot stay energized continuously. A movable armature which normally stays pressed against an adjustment screw forms the relay contacts.

When the switch is closed, electron current goes through the contact screw and the armature to energize the coil. The coil produces a magnetic field that attracts the armature away from the screw. When that happens, the circuit is opened, the magnetic field collapses, and the springy armature swings back against the screw. As soon as the armature touches the screw again, the circuit is closed, current flows again, and the magnetic field pulls the armature away again to open the circuit.

So you can see that the armature moving up and down, on and off the screw, opens and closes the circuit continuously.

The armature chatters against the screw, producing a buzzing sound. By adjusting the screw, you can control how far and how fast the armature vibrates, thus adjusting the kind of buzzing sound you will get.

A *bell* works similarly to the buzzer. The bell uses a long swinging hammer attached to the armature, so that the hammer can be vibrated against a bell.

The armature and adjustment screw act as a switch that opens and closes every time the armature is attracted to the coil. The armature vibrates up and down against the screw to produce a buzzing sound

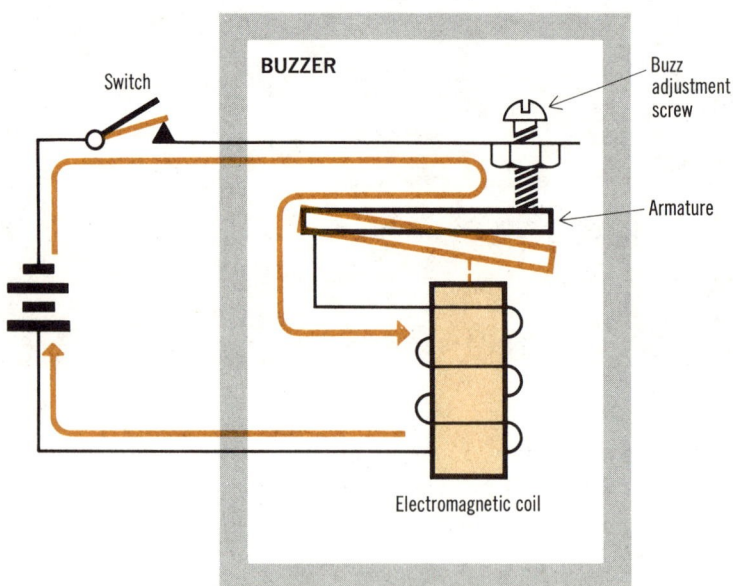

Switch

BUZZER

Buzz adjustment screw

Armature

Electromagnetic coil

the basic telegraph set

The diagram below shows how a basic telegraph set works. The telegraph *keys* are used as *switches* that open and close the circuit to form dots and dashes.

If the "east" station wants to receive from the "west" station, he must hold his key down. Then, when the west station puts his key down, current will flow through the relays and energize them. The relays will close their armature contacts to connect the sounder batteries to the sounder relays. The sounder relay armatures vibrate when they are energized just like the buzzer described previously. If the west station's key is held down for a *short* time, he sends a *dot*. But, if the key is held down *longer*, he sends a *dash*. Both sounder relays work together, so the west station can also hear what he *transmits*.

When the east station wants to send a message, he opens his key to open the entire circuit. Then, when the west station hears his sounder stop, he knows that the east station wants to send; the west station then closes his key and listens.

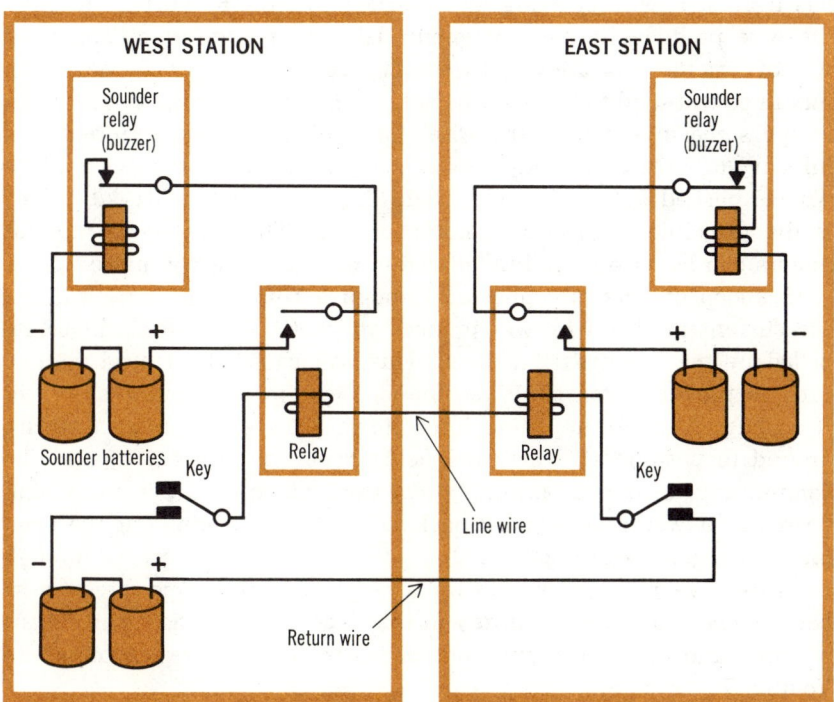

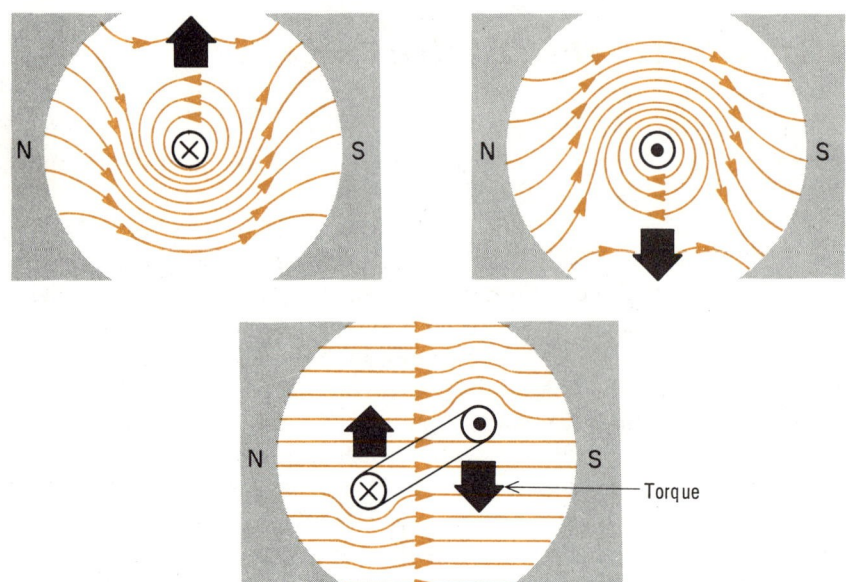

the electric motor

The electric motor works because of the effect that a magnetic field has against a wire carrying an electric current. The current through the wire produces its own magnetic field around the wire. This field will distort the flux lines that exist between two magnetic poles. The flux lines will tend to move to the side of the wire where they are going in the same direction as the wire's lines of force. The distorted flux lines try to straighten, and so exert a repelling force on the wire. The wire is pushed out of the field where the flux lines are weakest. This is the principle of electric motor operation. The *right-hand rule* for motors can be used to find the direction that the wire will move.

If a loop of wire is connected through a *commutator* to a battery, the current in that wire will produce magnetic fields that will be repelled by the magnet's flux lines. This will cause the looped wire to turn, or produce a *torque*. When the loop gets to the position shown in B on page 1-99, the repelling force stops, but inertia carries the wire around to position C, where the field's repulsion turns it again. The commutator is needed because, when the loop reaches position B and passes it, it would be repelled back to position A again. But the commutator segments are split at that point, so that the current through the wire is reversed, and the wire is repelled in the same direction as before. The rotor (or armature) in this type of motor has many turns of wire and many commutator segments. Motors are examined in Volume 7.

the electric motor (cont.)

RIGHT-HAND RULE

To determine the direction in which a current-carrying wire will move in a magnetic field, use the right hand motor rule. Point the index-finger in the direction of the magnetic field and the second finger in the direction of current flow. The thumb will indicate the direction of motion

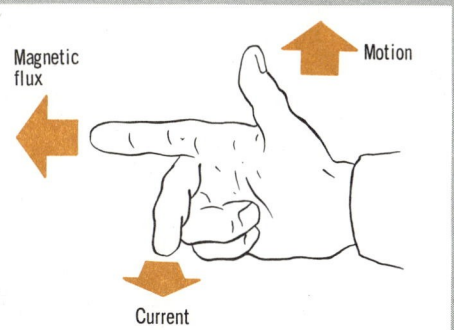

Magnetic flux

Motion

Current

(A)

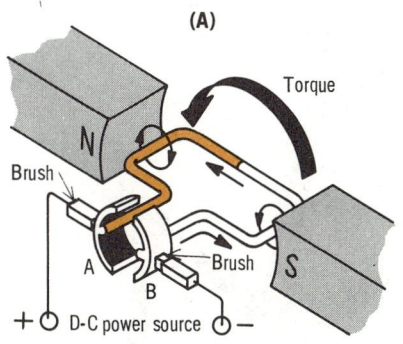

Torque

N

Brush

A

Brush

B

+ ○ D-C power source ○ −

Current through a wire causes a field that repels the loop with a torque

(B)

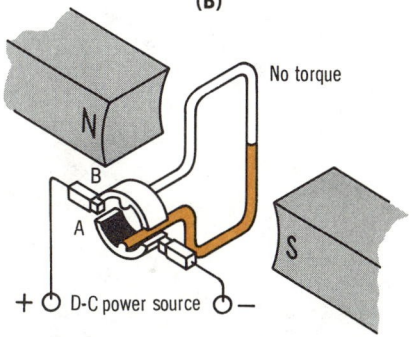

No torque

N

B

A

S

+ ○ D-C power source ○ −

At this point, the field's repulsion stops, but inertia carries the loop toward diagram (C)

(C)

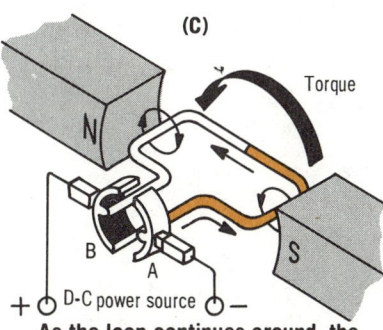

Torque

N

B

A

S

+ ○ D-C power source ○ −

As the loop continues around, the commutator reverses the current direction so that the magnetic field repels the wire in the same direction

(D)

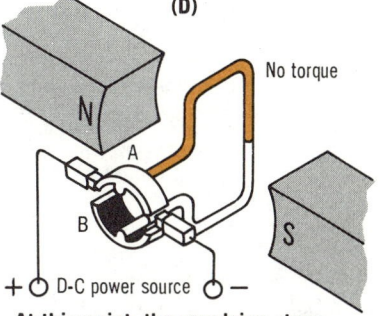

No torque

N

A

B

S

+ ○ D-C power source ○ −

At this point, the repulsion stops again, but inertia carries the loop toward position A to repeat the cycle

the meter

A basic type of meter can use a solenoid coil and a movable core to measure the current that is flowing. Whenever current flows through the coil, a magnetic field is built up, which attracts the core into it. However, the other end of the core is pivoted with a spring that tends to hold it back. The distance that the core moves, depends on the

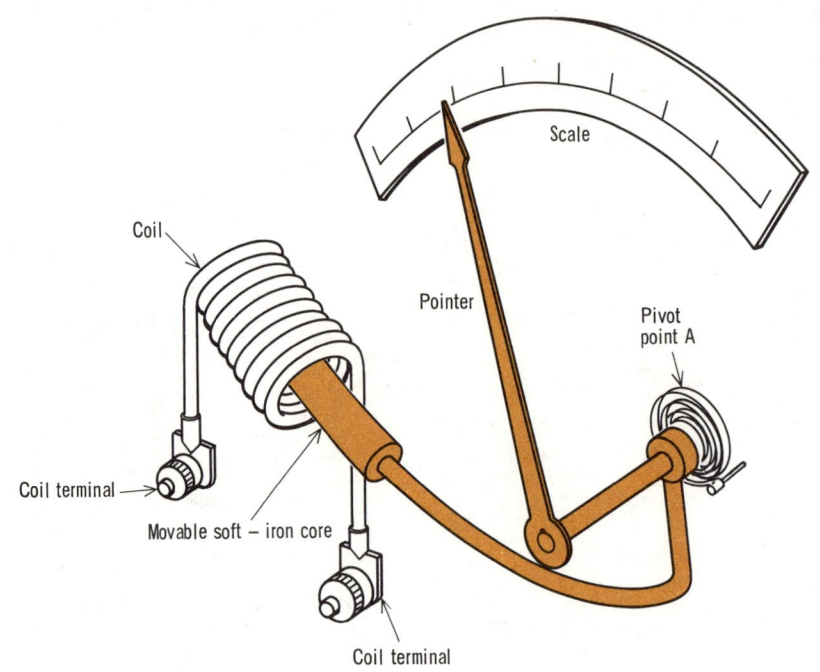

Current through the coil produces a magnetic field that attracts the core in and swings the pointer. The greater the current through the coil, the stronger the magnetic field, and the more the core will be attracted into the coil to swing the pointer further

strength of the magnetic field; and the strength of the field is determined by the amount of current flowing through the coil. Therefore, the core moves farther into the coil when more current flows.

The pivot of the core also contains a pointer that swings along a scale to indicate the current that is being measured. Meters are examined in Volume 5.

the basic generator

The generator's operation is opposite to that of the motor. Instead of putting a current in the rotor windings to produce a magnetic field, the rotor is *turned mechanically,* usually with a motor.

Then when the rotor windings pass through the lines of flux, the magnetic energy forces current to flow in the wire. When the wire goes down the field, current flows in one direction; but when the wire goes up the field, the current flows in the other direction. The commutator, though, switches the wires outside the generator while the rotor turns, to keep the current flow in the *same direction* through the meter at all times. Therefore, it is called a *direct-current* (d-c) generator. If a commutator is not used, the current coming out of the generator will change direction as the loop turns. This is called *alternating current* (ac). It is examined in Volume 3.

Current is induced in a wire that is moved through a magnetic field. The direction of the current flow depends on the direction the wire is moved

The commutator keeps the current flowing out of the generator in the same direction at all times

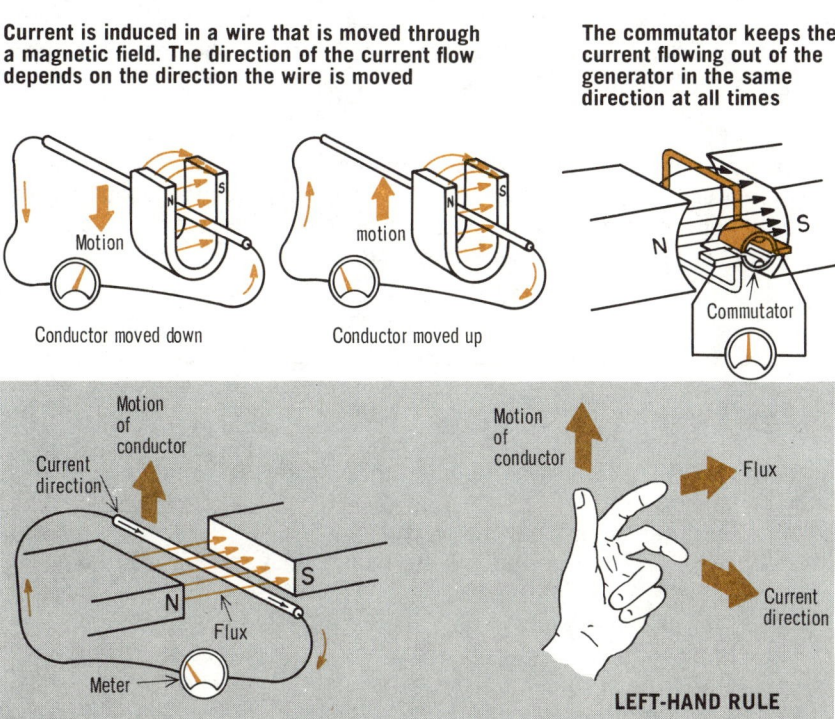

Conductor moved down

Conductor moved up

Commutator

Motion of conductor

Current direction

Flux

Meter

Motion of conductor

Flux

Current direction

LEFT-HAND RULE

The *left-hand rule* for generators can be used to determine the direction of current flow that will result when a wire is moved through a magnetic field.

Generators are examined in detail in Volume 6.

summary

☐ Some electrical energy applications are: the incandescent lamp and the electric heater. ☐ The incandescent lamp gives off light when current flows through a poor conductor. The conductor or filament, is usually made of tungsten. ☐ The electric heater uses the same principle as the incandescent lamp, but the heating element does not glow as brightly. The heating element is usually nichrome. ☐ Other devices using heating elements are irons, toasters, grills, etc.

☐ Some magnetic energy applications are: the electromagnetic bell, the relay, the basic telegraph set, and the electric motor. ☐ The electromagnetic bell uses the action of a magnetic field to vibrate an armature so that a hammer strikes a bell. ☐ A relay is an electromagnet that opens or closes contacts. ☐ A basic telegraph set consists of keys that act as switches to open and close the circuit by relay action to form dots and dashes. ☐ The electric motor operates because of the effect of a magnetic field on a current-carrying wire. The interaction causes the rotation of an armature. The right-hand motor rule gives the direction that the rotor loop will rotate.

☐ The basic meter is an instrument that uses a solenoid coil to attract a movable core to measure current. ☐ The basic generator produces electricity. Its operation is opposite to that of the motor. The left-hand generator rule gives the direction of the current generated.

review questions

1. How does an incandescent lamp differ from an electric heater?
2. State the right-hand motor rule.
3. What do the electromagnetic bell and buzzer have in common?
4. Why is a commutator used in certain motors?
5. Is a commutator always necessary for a generator to work?
6. State the left-hand generator rule.
7. How can a meter measure current?
8. What is the difference between direct current and alternating current?
9. Describe a basic telegraph set.
10. What is the difference between a magnetic circuit breaker and a fuse?

electricity
two

using electricity

In itself, electricity is nothing more than an interesting phenomenon. To be of practical use, it must be made to perform some work or function. Generally, this requires that the electricity be controlled, and often converted to other forms of energy. The physical means for accomplishing this transition from phenomenon to practical use is the *electric circuit.*

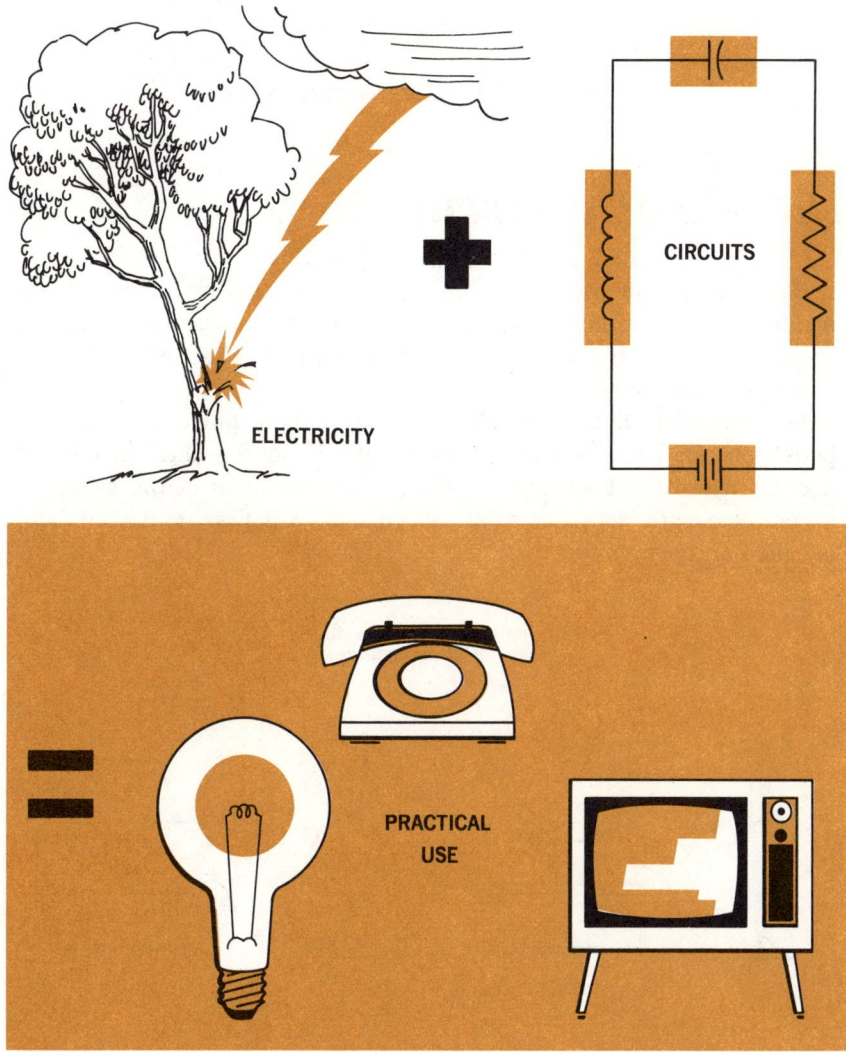

ELECTRICITY

CIRCUITS

PRACTICAL
USE

**Three elements are necessary
to have an electric circuit**

Connecting
Wires

Connecting
Wires

| Power Source | Closed Circuit | Load |

| Power Source | Open Circuit | Load |

A complete path is necessary for current to flow

the electric circuit

Basically, an electric circuit consists of (1) a power source; (2) connecting wires, or conductors; and (3) a device that uses the electrical energy of the source to accomplish some purpose. The device that uses the energy is called the *load*.

For current to flow in an electric circuit, there must be a complete path from the negative terminal of the power source, through the connecting wires and load, back to the positive terminal of the source. If a complete path does not exist, no current will flow, and the circuit is called an *open circuit*.

**The load is a device that
uses electrical energy**

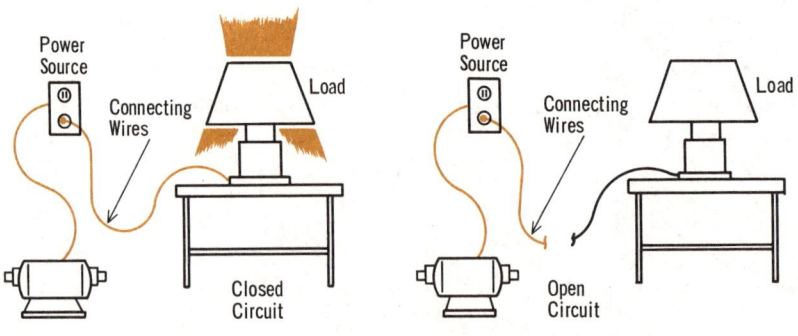

Power
Source

Load

Connecting
Wires

Closed
Circuit

Power
Source

Load

Connecting
Wires

Open
Circuit

A complete path is necessary for current to flow

the switch

An electric circuit has to provide a complete path for current flow only when electrical energy is needed by the load. At all other times, the circuit is kept "open," and no current flows.

All switches perform the same basic function of opening and closing electric circuits

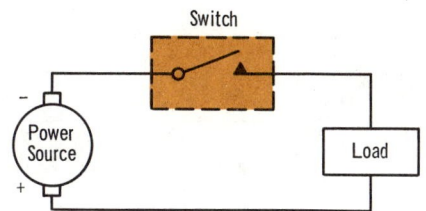

The opening and closing of an electric circuit is normally accomplished by a *switch*. In its simplest form, a switch consists of two pieces of conducting metal that are connected to the circuit wires. These two pieces of metal are arranged so that they may easily be made to either touch each other or be separated. When they touch, a complete path for current flow exists and the circuit is closed. When they are separated, no current can flow, and the circuit is open.

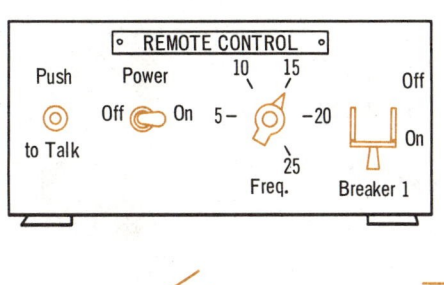

Many types of electrical switches are used today. Each type has its own schematic representation, so that you can look at a circuit diagram and know what type of switch is being used

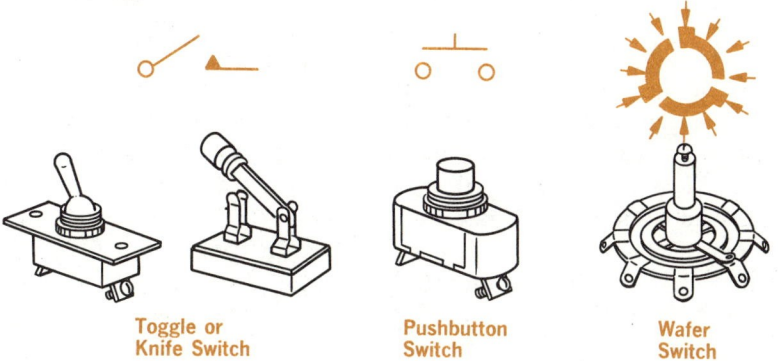

Toggle or Knife Switch

Pushbutton Switch

Wafer Switch

the load

In a simple electric circuit, the load is the device that takes the electrical energy from the power source and uses it to perform some useful function. To do this, the load may convert the electrical energy to another form of energy, such as light, heat, or sound, or it may merely change or control the amount of energy delivered by the source.

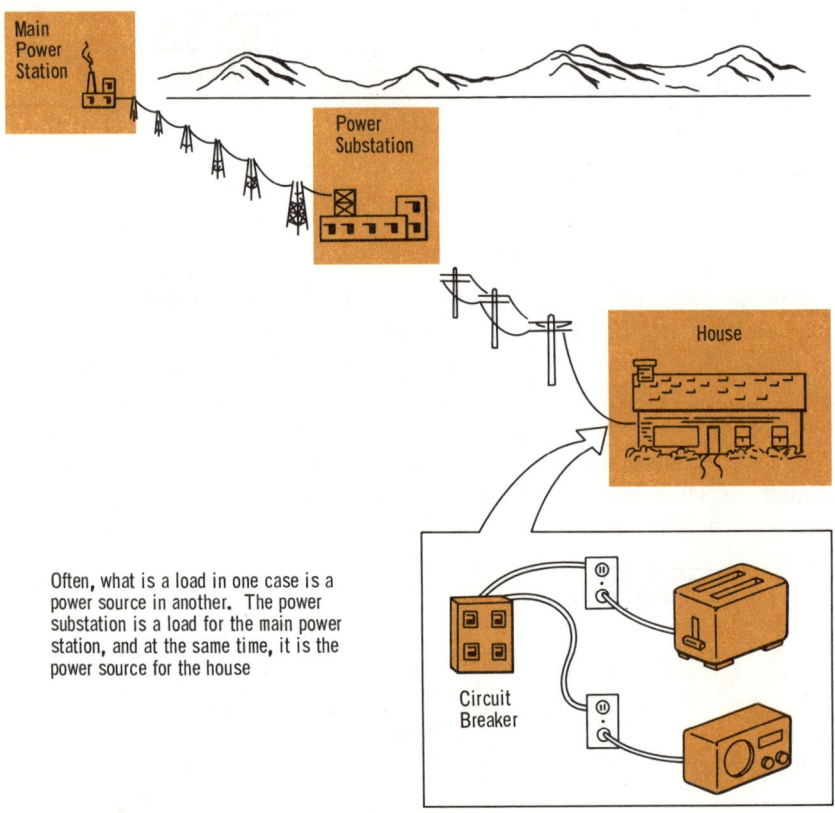

Often, what is a load in one case is a power source in another. The power substation is a load for the main power station, and at the same time, it is the power source for the house

A light bulb is a load, and so is a motor, toaster, heater, and so forth. The type of load used determines the amount of energy taken from the power source. Because of this, the term "load" is also often used to mean the *power* delivered by the source. In this case, when someone says the load is increased or decreased, it means that the source is supplying more or less power. (This is described in detail later.) Keep in mind that the term load stands for two things: (1) the *device* that takes power from the source, and (2) the *power* that is taken from the source.

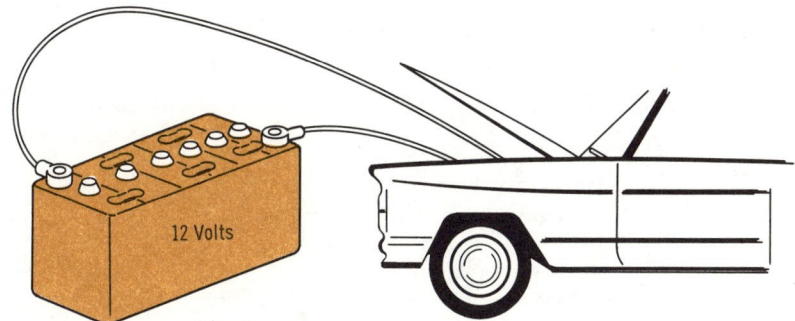

One of the most widely used voltage sources today is the battery

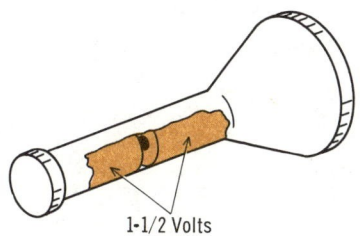

There is a wide variety of sizes, shapes, and types of batteries in use today

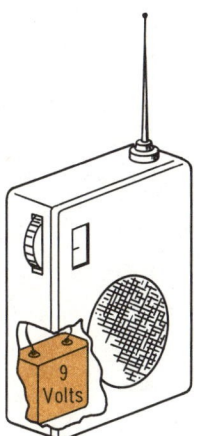

the power source

The power source produces electrical energy by chemical, magnetic, or other means. (This is explained in Volume 1.) This energy is usually in the form of a *difference of electrical potential* between the output terminals of the source, which is called an *electromotive force*. Usually, the abbreviation emf is used in place of the term electromotive force. Emf is measured in volts, and so the source producing it is called a *voltage source*. The polarity of the voltage source determines in which direction the current flows in the circuit; and the amount of voltage supplied by the source determines how much current will flow. Power sources are covered in detail in Volume 6.

the direct-current circuit

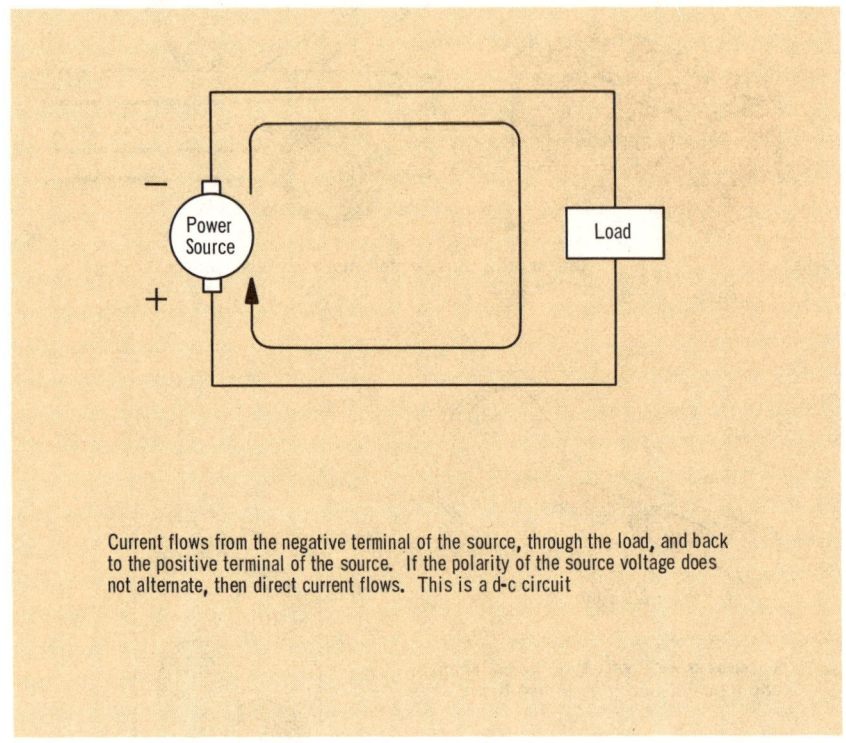

Current flows from the negative terminal of the source, through the load, and back to the positive terminal of the source. If the polarity of the source voltage does not alternate, then direct current flows. This is a d-c circuit

Since electron current always flows out of the negative terminal of the power source, the current flow in a circuit will always be in the same direction, if the polarity of the source voltage always remains the same. This type of current flow is called direct current, and the source is called a direct-current source. Any circuit that uses a direct-current source is then a direct-current circuit. For simplicity, direct current is generally abbreviated as dc, and we then speak of d-c sources, d-c voltage, d-c current, and d-c circuits. The three types of sources most often used for d-c circuits are the battery, the d-c generator, and the electronic power supply. Regardless of the type of d-c source used, the theory of operation of all d-c circuits is the same. D-c theory is covered in this volume.

When the voltage polarity of the power source changes, or alternates, the direction of current flow will also alternate. This type of current is called *alternating current* (ac), and is covered in Volume 3.

what controls current flow?

Generally, electric circuits are designed for some specific amount of current flow. If too little current flows in the circuit, the load will not operate properly, or maybe not at all. If too much current flows, the voltage source or load could be damaged. There are only two factors that determine how much current will flow in a d-c circuit. The first is the amount of voltage that is supplied by the power source, and the second is how well the load and the wires conduct current.

Circuit Current is Too Low   **Circuit Current is Too High**

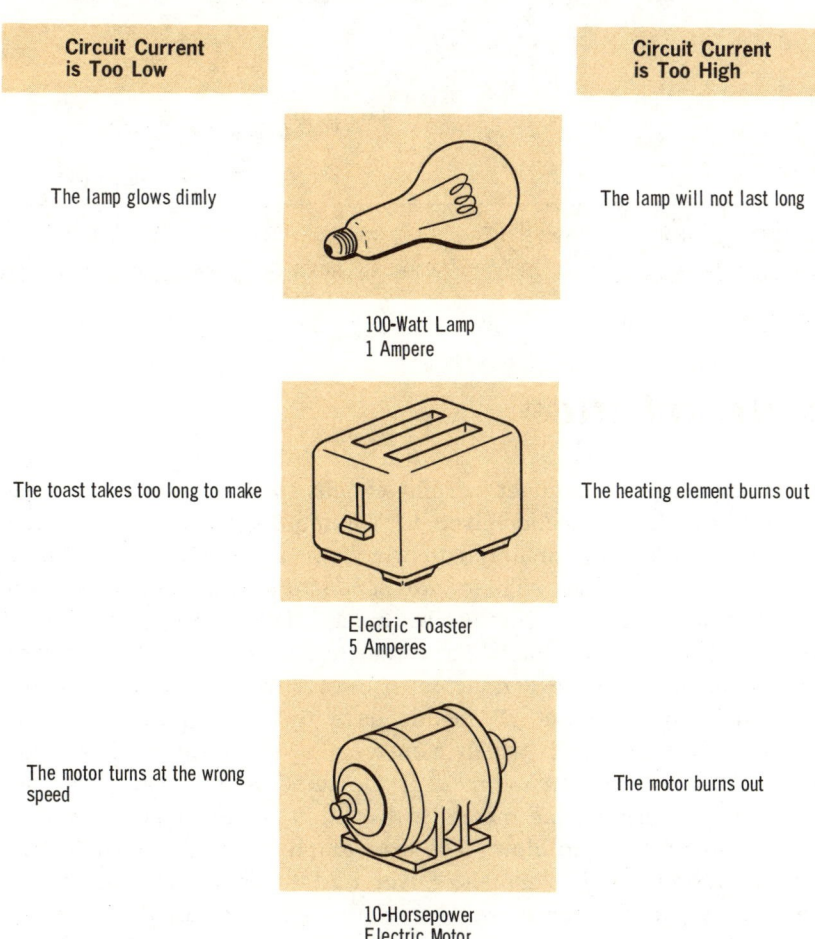

The lamp glows dimly   The lamp will not last long

100-Watt Lamp
1 Ampere

The toast takes too long to make   The heating element burns out

Electric Toaster
5 Amperes

The motor turns at the wrong speed   The motor burns out

10-Horsepower
Electric Motor
75 Amperes

The current flowing in any circuit must be controlled if the circuit is to work properly

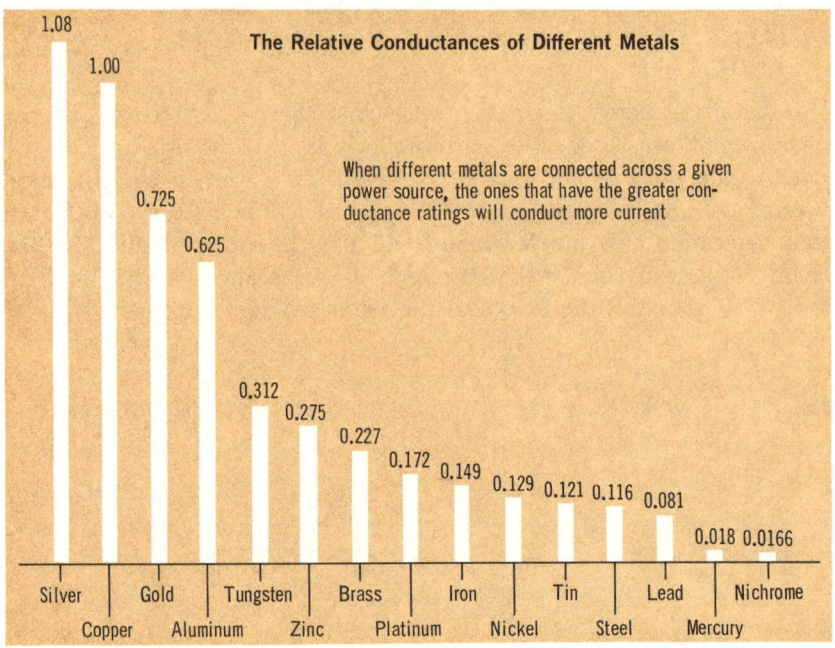

The Relative Conductances of Different Metals

When different metals are connected across a given power source, the ones that have the greater conductance ratings will conduct more current

1.08 1.00 0.725 0.625 0.312 0.275 0.227 0.172 0.149 0.129 0.121 0.116 0.081 0.018 0.0166

Silver Copper Gold Aluminum Tungsten Zinc Brass Platinum Iron Nickel Tin Steel Lead Mercury Nichrome

conductance

Not all materials conduct current equally well. If you recall some of the basic electric theory given in Volume 1, you know that there are two basic kinds of material in which we are interested, as far as electricity is concerned. These are *conductors* and *insulators*. Conductors allow current to flow easily, and insulators oppose the flow of current. The reason for this is that conductors have many *free electrons*.

Most metals are good conductors. However, some metals are better than others because not all metals have the same number of free electrons. The ease with which a metal allows current to flow is described by the term *conductance*. If the same voltage source is used with different metals, the metals with a high conductance rating will allow more current to flow. The bar graph gives the conductance ratings of some similar metals. Silver has the greatest conductance; but since copper is used more often than silver because it is cheaper, copper is given a conductance rating of 1, and the other metals are rated in comparison to copper. For example, tungsten, which is used in light bulbs, has only 0.312 the conductance of copper. Therefore, copper will allow more than 3 times as much current to flow than will tungsten if they were both connected across the same power source.

summary

☐ The electric circuit is the physical means for accomplishing the transition of electricity from phenomenon to practical use. ☐ The electric circuit consists of (1) a power source, (2) connecting wires or conductors, and (3) a device that uses the electrical energy of the source. ☐ The device that uses the electrical energy is the load. ☐ For current flow, there must be a complete path from the negative terminal of the power source, through the connecting wires and load, back to the positive terminal of the source. ☐ A switch is used to open and close the path between the source and the load. An open switch means that there is no continuous path through the circuit; the circuit is therefore open. A closed switch means that there is a continuous path; the circuit is then closed.

☐ The power source produces electrical energy by chemical, magnetic, or other means. ☐ The power source discussed in this volume is a direct-current source: a battery, a d-c generator, or an electronic power supply. The theory of circuit operation is the same regardless of the source type used. ☐ The circuit is usually designed to operate with a specific current flow. Too low a value may cause the load to be inoperative; too high a value may damage the voltage source or load. ☐ Two factors determine how much current will flow in a d-c circuit: (1) the amount of voltage supplied, and (2) how well the load and the wires conduct current.

☐ Two materials of interest in electricity are conductors and insulators. Conductors allow current to flow; insulators oppose it. ☐ The ease with which a metal allows current to flow is described as conductance. ☐ All metals are given a conductance rating which shows how well they conduct electricity, as compared to copper. ☐ The best conductor is silver, but copper is used more often because it is cheaper.

review questions

1. What are the three basic elements of an electric circuit?
2. What is meant by an *open* circuit?
3. What controls a circuit so that it is open or closed?
4. What is meant by a *load*?
5. Can a battery ever be considered a load? Explain.
6. What do a battery, d-c generator, and electronic power supply have in common?
7. What controls the current flow in a circuit?
8. What is meant by *conductance rating*?
9. Conductance ratings are based on what metal?
10. What can happen if there is too much current in a circuit?

resistance

The term *conductance* is used to describe how well a material *allows* current to flow. Another way to look at this is that the *low-conductance* materials oppose or *resist* the flow of electric current. Some materials, then, offer more *resistance* to the flow of electrons than other materials. This is actually the way that materials are rated in the field of electricity.

If you were to cut a piece of each of the more common metals to a standard size, and then connect the pieces to a battery, one at a time, you would find that different amounts of current would flow. Each metal offers a different resistance to the movement of electrons.

The Relative Resistances of Different Metals As Compared to Copper

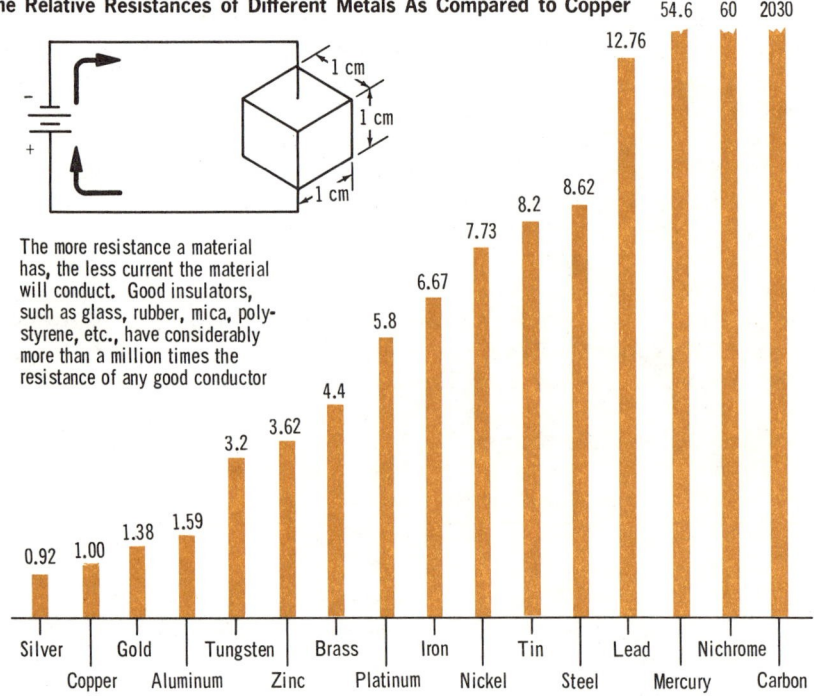

The more resistance a material has, the less current the material will conduct. Good insulators, such as glass, rubber, mica, polystyrene, etc., have considerably more than a million times the resistance of any good conductor

The standard size that is usually used to test the resistance of metals is a 1-centimeter cube. The bar graph shows the resistance of some common metals as compared to copper. Silver is a better conductor than copper because it has less resistance. Nichrome has 60 times more resistance than copper, and so copper will conduct 60 times as much current as nichrome if they were connected to the same battery, one at a time.

Point of
Current
Measurement

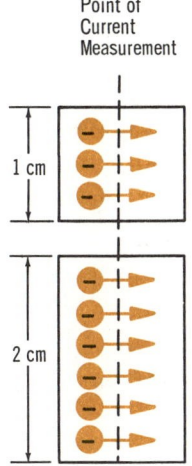

When a conductor is made thicker, it
will conduct more current and will
have less resistance

**Conductors with greater cross-sectional
area have more free electrons available,
and, therefore, have less resistance**

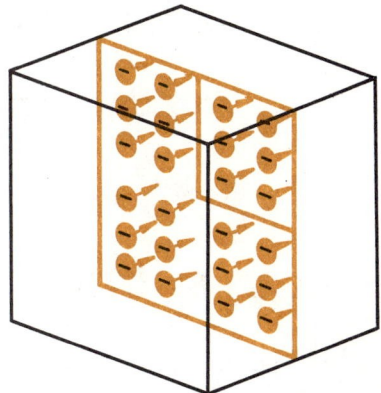

how resistance can be decreased

Actually, the resistance of any material depends on the number of free electrons the material has. If you think back to what was explained in Volume 1, you should recall that electric current is rated in amperes; and 1 ampere is 6,280,000,000,000,000,000 free electrons passing any given point in a wire in 1 second. Therefore, a good conductor must have a number of free electrons available to allow several amperes to flow. Since current is a measurement of electrons passing a point in a wire, we can make more free electrons available by using a *thicker* piece of metal so that *more current* will flow.

A piece of copper 2 centimeters high and 1 centimeter wide will have twice as many free electrons available along the point at which current is being measured than a piece of copper only 1 centimeter high and 1 centimeter wide. The copper that is twice as high will conduct twice as much current. If you use copper that is 2 centimeters wide, you will double the current and halve the resistance again. When you increase the width or height of a piece of metal, you are increasing its *cross-sectional area*. The *greater* the cross-sectional *area* of a wire, the *lower* its *resistance*.

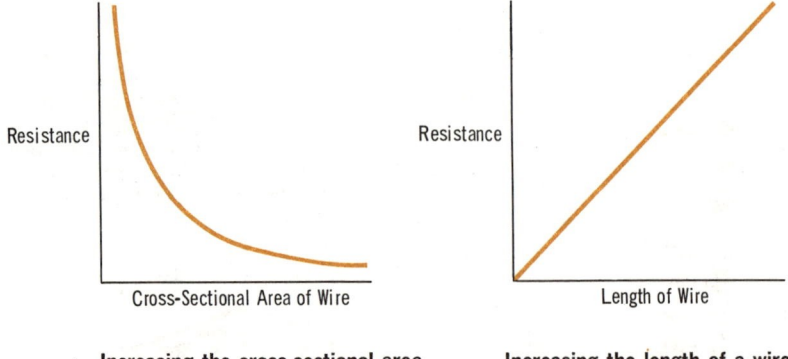

Increasing the cross-sectional area
of a wire decreases its resistance

Increasing the length of a wire
increases its resistance

how resistance can be increased

By increasing the cross-sectional area of a conductor, you make more free electrons available for current, and lower the resistance of the conductor. This might lead you to think that by changing the length of a piece of copper, you could accomplish the same thing. But this isn't so. Although a longer piece of copper has more free electrons in the whole piece, the extra free electrons are not made available along the line of current measurement. Actually, each length of conductor has a certain amount of resistance. When you add an *extra length* of copper, you are adding *more resistance*. The longer a wire is, the more resistance it has.

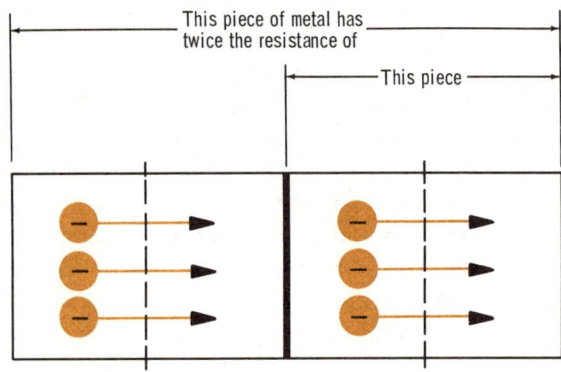

This piece of metal has twice the resistance of

This piece

how resistance is controlled

You can see now that the resistance of a piece of wire can be increased by making it longer, or decreased by making it shorter. You can also lower the resistance by increasing its cross-section, or raise the resistance by decreasing its cross-section.

If you double the length, you will double the resistance. Because of this relationship, we say that the resistance of a wire is *directly proportional* to its *length*.

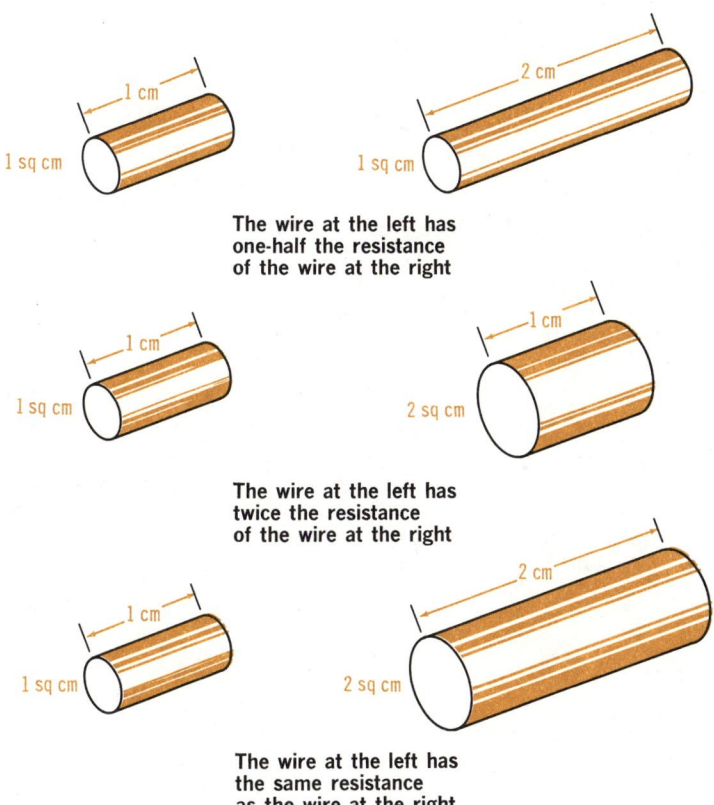

The wire at the left has
one-half the resistance
of the wire at the right

The wire at the left has
twice the resistance
of the wire at the right

The wire at the left has
the same resistance
as the wire at the right

If you double the cross-sectional area of wire, you will halve its resistance. As a result, we say that the resistance of a wire is *inversely proportional* to its *cross-section*.

Therefore, by choosing the proper metal for a conductor, and making it with a certain cross-section and length, you can produce any kind of resistance effect you want.

effect of temperature

Actually, the relative values of resistance that were given before apply to the metals when they are at about room temperature. At higher or lower temperatures, the resistances of all materials change. In most cases, when the temperature of a material goes up, its resistance goes up too. But with some other materials, increased heat causes the resistance to go down. The amount that the resistance is

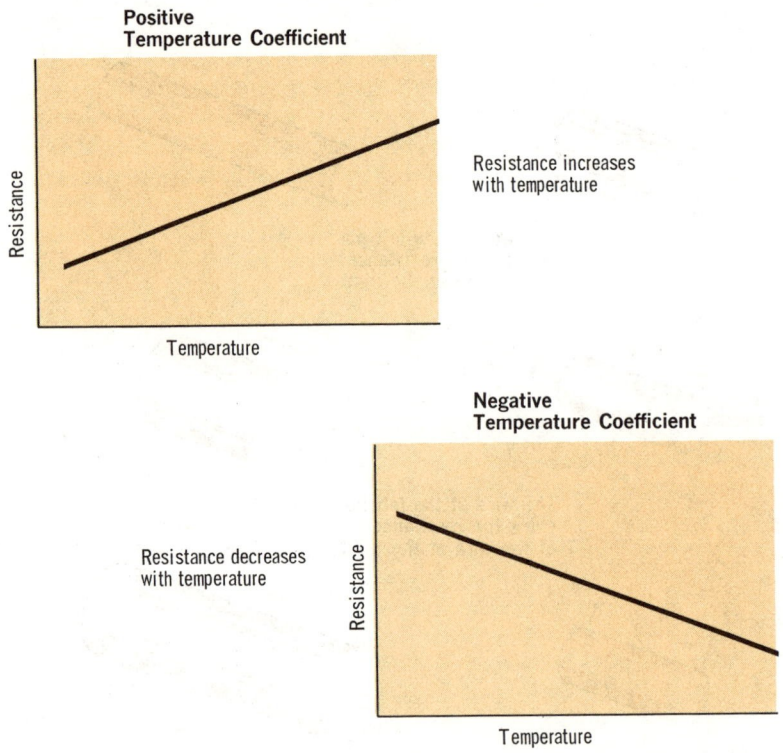

affected by each degree of temperature change is called the *temperature coefficient*. And the words positive and negative are used to show whether the resistance goes up or down with temperature.

When a material's resistance goes *up* as temperature is increased, it has a *positive* temperature coefficient.

When a material's resistance goes *down* as temperature is increased, it has a *negative* temperature coefficient.

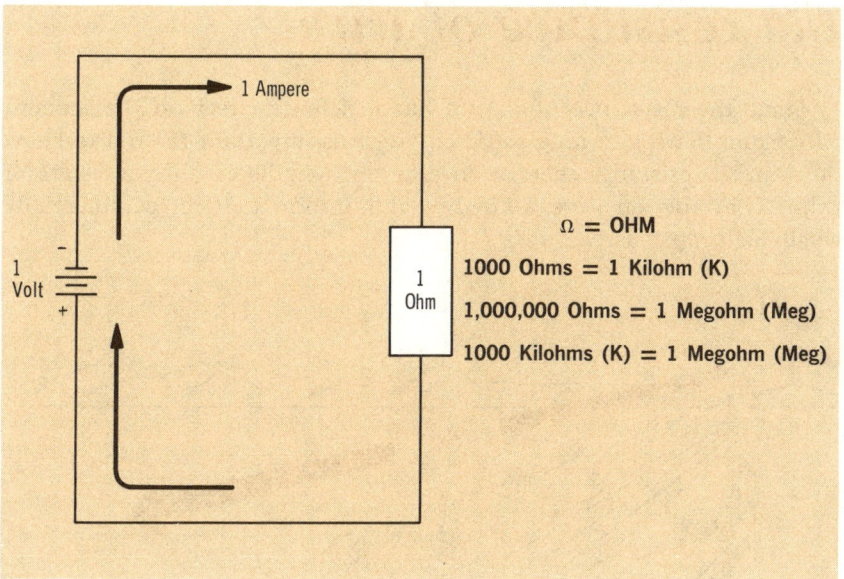

the unit of resistance

During the early 1800's, a German scientist named Georg Simon Ohm performed a number of experiments with electricity, and made some of the first important discoveries about the nature of electrical resistance. In his honor, the unit of resistance is called the *ohm*.

A conductor is said to have a resistance of 1 ohm when an emf of 1 volt causes a current of 1 ampere to flow through that conductor. If, of course, 1 volt causes only ½ ampere to flow, then the conductor has twice the resistance, or 2 ohms. By using this relationship, the exact resistance of all types, sizes, and shapes of conductors can be found. Resistance can vary from fractions of an ohm to kilohms (1000 ohms) and megohms (1,000,000 ohms). The Greek letter omega (Ω) is often used as the symbol for the ohm.

CONVERSION OF UNITS
ohms ÷ 1000 = kilohms (K)
ohms ÷ 1,000,000 = megohms (Meg)
kilohms (K) ÷ 1000 = megohms (Meg)
kilohms × 1000 = ohms
megohms × 1,000,000 = ohms
megohms × 1000 = kilohms (K)
500,000 ohms = 500 kilohms = 0.5 megohm
or
500,000 Ω = 500 K = 0.5 Meg

the resistance of wire

Since the resistance of a wire has a definite effect on the amount of current flowing in an electric circuit, it is important for you to know how much resistance there is in different lengths of different sizes of wire. The table on page 2-17 gives this information for commercially available copper wire.

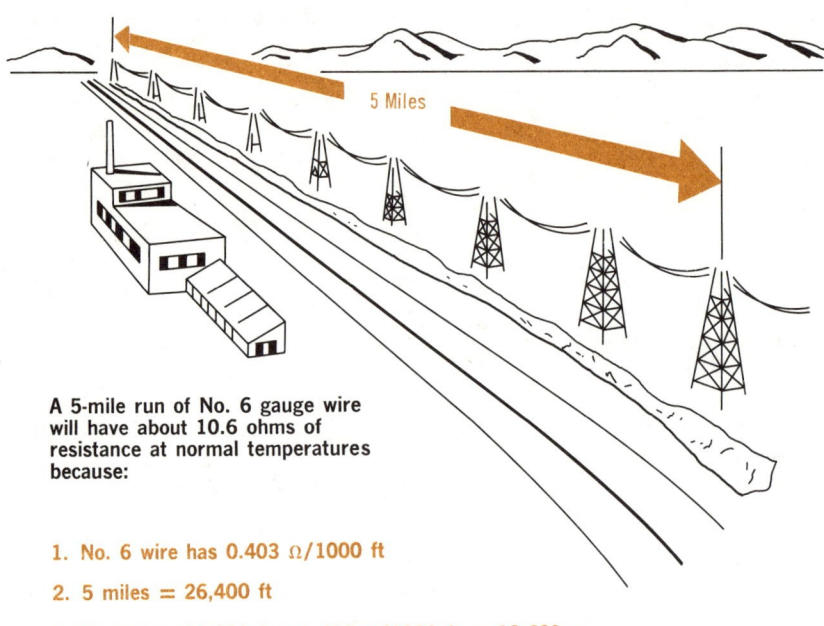

5 Miles

A 5-mile run of No. 6 gauge wire will have about 10.6 ohms of resistance at normal temperatures because:

1. No. 6 wire has 0.403 Ω/1000 ft
2. 5 miles = 26,400 ft
3. Therefore, 26,400 ft x 0.403 Ω/1000 ft = 10.639 Ω

The table was originally called the Brown and Sharpe Wire Gauge, but is now known as the American Standard Wire Gauge. The different diameters of copper wire are identified by gauge numbers. There are 40 gauges shown, ranging from 0000 through 36. The higher the gauge number, the smaller the diameter and cross-sectional area of the wire. Therefore, the *higher* numbered *gauges* of wire have *more resistance*. The table gives the diameter for each gauge and the typical resistance for 1000 feet of wire. Resistances are given for both the normal temperature of 70°F, and a high temperature of 167°F.

the resistance of wire (cont.)

AMERICAN STANDARD WIRE GAUGES

Dimensions and Typical Resistances of Commercial Copper Wire

B & S Gauge No.	Diameter of Bare Wire (Inches)	Ohms per 1000 ft		Current Capacity (Amperes)	
		70°F	167°F	Rubber Insulation	Other Insulation
0000 (4/0)	0.460	0.050	0.060	160-248	193-510
000 (3/0)	0.410	0.062	0.075	138-215	166-429
00 (2/0)	0.365	0.080	0.095	120-185	145-372
0	0.325	0.100	0.119	105-160	127-325
1	0.289	0.127	0.150	91-136	110-280
2	0.258	0.159	0.190	80-118	96-241
3	0.229	0.202	0.240	69-101	83-211
4	0.204	0.254	0.302	60-87	72-180
5	0.182	0.319	0.381	52-76	63-158
6	0.162	0.403	0.480	45-65	54-134
7	0.144	0.510	0.606	45-65	54-134
8	0.128	0.645	0.764	35-48	41-100
9	0.114	0.813	0.963	35-48	41-100
10	0.102	1.02	1.216	25-35	31-75
11	0.091	1.29	1.532	25-35	31-75
12	0.081	1.62	1.931	20-26	23-57
13	0.072	2.04	2.436	20-26	23-57
14	0.064	2.57	3.071	15-20	18-43
15	0.057	3.24	3.873	15-20	18-43
16	0.051	4.10	4.884	6	10
17	0.045	5.15	6.158	6	10
18	0.040	6.51	7.765	3	6
19	0.036	8.21	9.792	3	6
20	0.032	10.3	12.35	–	–
21	0.028	13.0	15.57	–	–
22	0.025	16.5	19.63	–	–
23	0.024	20.7	24.76	–	–
24	0.020	26.2	31.22	–	–
25	0.018	33.0	39.36	–	–
26	0.016	41.8	49.64	–	–
27	0.014	52.4	62.59	–	–
28	0.013	66.6	78.93	–	–
29	0.011	82.8	99.52	–	–
30	0.010	106	125.50	–	–
31	0.009	134	158.20	–	–
32	0.008	165	199.50	–	–
33	0.007	210	251.60	–	–
34	0.006	266	317.30	–	–
35	0.005	337	400.00	–	–
36	0.005	423	504.50	–	–

how much current can wire carry?

The resistance of a wire determines how much current will flow through the wire when it is connected across a voltage source. But there are two other factors that should also be considered when it comes to current flow. One is that although the wire resistance will *allow* a certain amount of current to flow, no more current can flow than the power source is able to supply. This is why a wire is rarely put directly across a source. Sources have a *safe limit* of maximum current they can supply before they *burn out*. This is explained in detail in Volume 6.

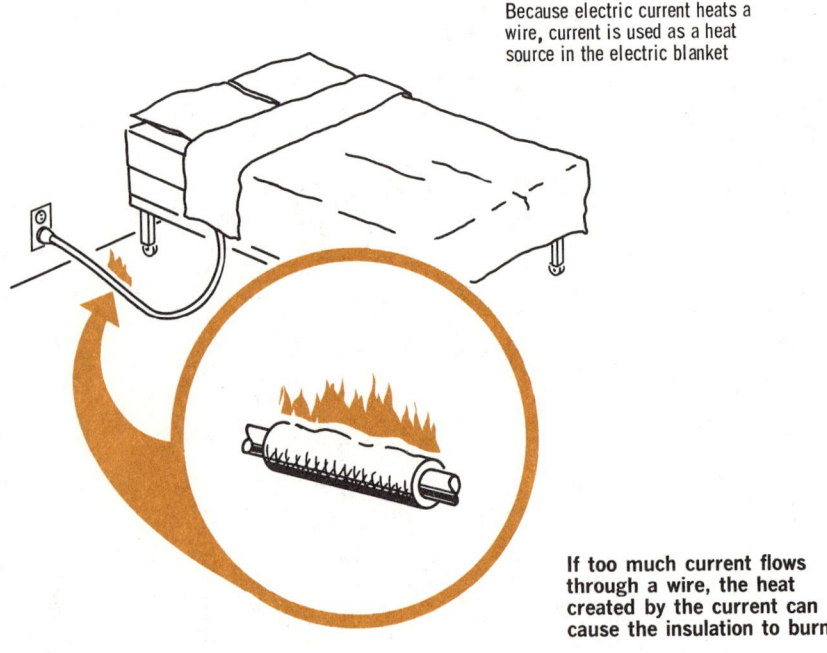

Because electric current heats a wire, current is used as a heat source in the electric blanket

If too much current flows through a wire, the heat created by the current can cause the insulation to burn

The second factor to consider is how much current the wire can *safely* carry. You should remember from what was explained in Volume 1 that electric *current heats* a wire. If the wire becomes too hot, its insulation covering will burn, and the wire can deteriorate. This is the reason why the AWS table on page 2-17 gives the maximum range of current that each gauge of wire can safely carry. As the table shows, if another insulation material which is better than rubber is used, the wire can carry more current.

the resistance of the load and of the power source

Up to this point, you have studied the resistance of ordinary wires, and how they affect current flow. But circuits must have a load if they are to perform a function with electricity. In most cases, the resistance of the load is much greater than the resistance of the circuit wiring. For example, a motor uses a large number of turns of wire, so the wiring of a motor is actually very long. And a heater, toaster, and electric iron use high-resistance conductors to produce heat. Therefore, the load resistance has more effect on how much current flows than the resistance of the circuit wires.

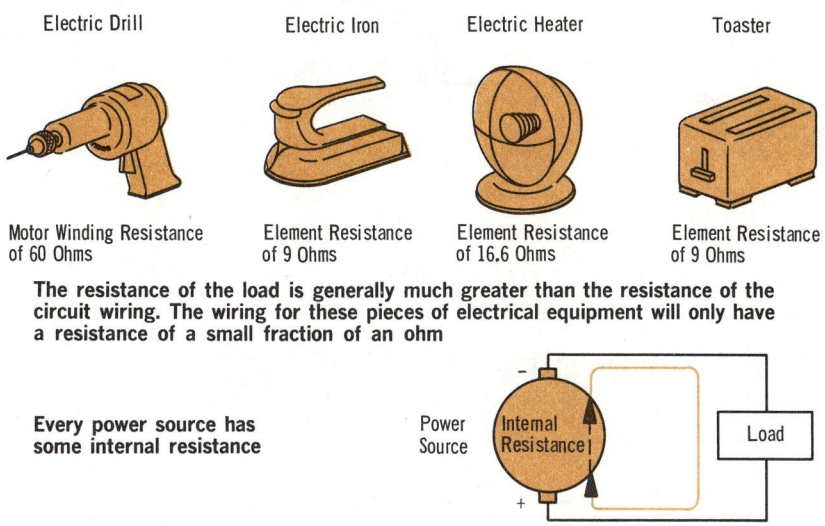

Electric Drill   Electric Iron   Electric Heater   Toaster

Motor Winding Resistance of 60 Ohms   Element Resistance of 9 Ohms   Element Resistance of 16.6 Ohms   Element Resistance of 9 Ohms

The resistance of the load is generally much greater than the resistance of the circuit wiring. The wiring for these pieces of electrical equipment will only have a resistance of a small fraction of an ohm

Every power source has some internal resistance

Power Source   Internal Resistance   Load

Some quantity of resistance is present in everything that conducts current. When current flows in a circuit, electrons leave the negative terminal of the power source, go around the wires and through the load, and reenter the power source through the positive terminal. The power source does not continually provide electron current from the negative terminal without replenishing its supply of electrons. The current that enters the positive terminal of the source must be conducted through the source to complete the circuit. Therefore, current flows from positive to negative *inside the source*. And every source has some *internal resistance* that also opposes current flow. This internal resistance is usually very low compared to the load.

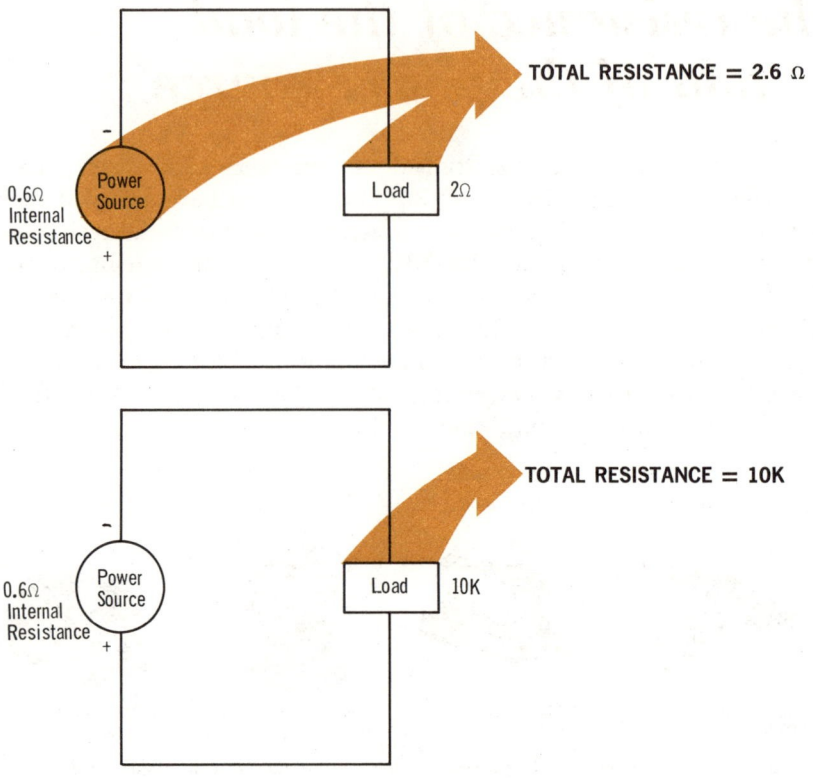

If the internal resistance of the power source is very much smaller than the load resistance, the total resistance of the circuit is that of the load alone. The resistance of long transmission lines, however, must be considered

total circuit resistance

The total resistance of a circuit is the sum of the individual resistances of the power source, the circuit wiring, and the load. As you know, the load resistance is generally much higher than the resistance of either the source or the wiring. In such cases, you can ignore the smaller resistances and consider the load resistance as the total resistance. But if the load resistance is small too, then those resistances should be added to get the true total resistance. Very rarely does the circuit wiring have enough resistance to be considered. But with long electric power, telephone, or telegraph lines, the resistance is large enough so that it must be considered.

summary

☐ Materials are rated in the field of electricity by their resistance to the flow of electrons. ☐ Good conductors have a low resistance, and good insulators have a high resistance. ☐ The resistance of a wire is determined by the type of material and the dimensions of the wire. The greater the cross-sectional area of the wire, the lower is its resistance. The longer the wire is, the greater is its resistance. In other words, the resistance of a wire is inversely proportional to its cross-sectional area and directly proportional to its length.

☐ The temperature coefficient of a material describes the effect that temperature has on the resistance of the material. A positive temperature coefficient means the resistance increases with an increase in temperature; a negative coefficient means the resistance decreases with an increase in temperature. ☐ The unit of resistance is the ohm. 1 ohm exists when an emf of 1 volt causes a current of 1 ampere to flow.

☐ The American Standard Wire Gauge lists resistance and current capacity characteristics for different sizes of copper wire. The higher the AWG number, the smaller is the wire diameter and therefore the greater is the resistance. ☐ Although a particular wire gauge can allow a high current flow, the actual current in a circuit is determined by: (1) the safe limit the source can supply before burning out, and (2) the safe limit the wire can carry. ☐ The total resistance of a circuit is the sum of the individual resistances of the power source, the circuit wiring, and the load. ☐ Usually, the load resistance is so much greater than that of the source or the wiring that it can be considered the total resistance.

review questions

1. Would a metal with a conductance rating of 0.99 be a good or poor insulator?
2. A certain material has a resistance of 15 ohms. If its cross-sectional area were tripled, what would be its resistance?
3. Define *temperature coefficient.*
4. Does the length of a wire affect its temperature coefficient?
5. Define *internal resistance* of a power source.
6. Does copper have a positive or a negative temperature coefficient?
7. What must be considered when calculating total circuit resistance?
8. What is the unit and symbol for resistance?
9. What is the American Standard Wire Gauge?
10. If a copper wire is heated, what happens to its resistance?

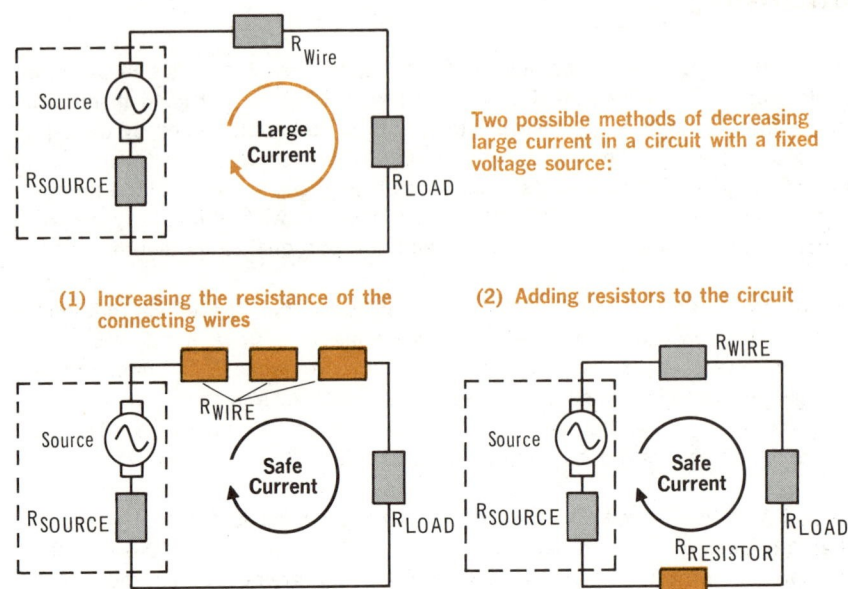

Two possible methods of decreasing large current in a circuit with a fixed voltage source:

(1) Increasing the resistance of the connecting wires

(2) Adding resistors to the circuit

resistors

Very often, if a load is connected to an existing source of voltage, too much current might flow in the circuit. This could happen if the resistance of the load was very low or the output voltage of the source was very high. The current could be decreased by reducing the source voltage, but usually this is impossible, or at least impractical. As you already know, the only other way to decrease the current is to add resistance to the circuit. This might be done by increasing the resistance of the voltage source, the load, or the connecting wires. However, the resistances of the source and the load are built right into these devices and cannot be changed. This leaves the connecting wires; but their resistance is so low that miles and miles of wire might be needed just to add a few hundred ohms to a circuit. Of course, connecting wires with a higher resistance could be used, and as a matter of fact, this has been done in the past for certain applications. However, if this was always done, it would greatly increase the number of different types of wires used to interconnect electric circuits. What is needed, then, is a method of easily adding various amounts of resistance to a circuit without drastically changing either its physical size or the materials used to make it. *Resistors* are the electric circuit components used to accomplish this.

how resistors are used

Resistors are used for *adding* resistance to an electric circuit. Basically, they are materials that offer a high resistance to current flow. The materials most often used for resistors are *carbon*, and special *metal alloys*, such as nichrome, constantan, and manganan. A resistor is connected into a circuit in such a way that the circuit current flows through it as well as through the load and the source. The *total circuit resistance* is then the sum of the individual resistances of the *load*, the *source*, the connecting *wires*, and the *resistor*. You can see from this that by just adding the proper resistor to the circuit, the resistance of the circuit can be changed to almost any value.

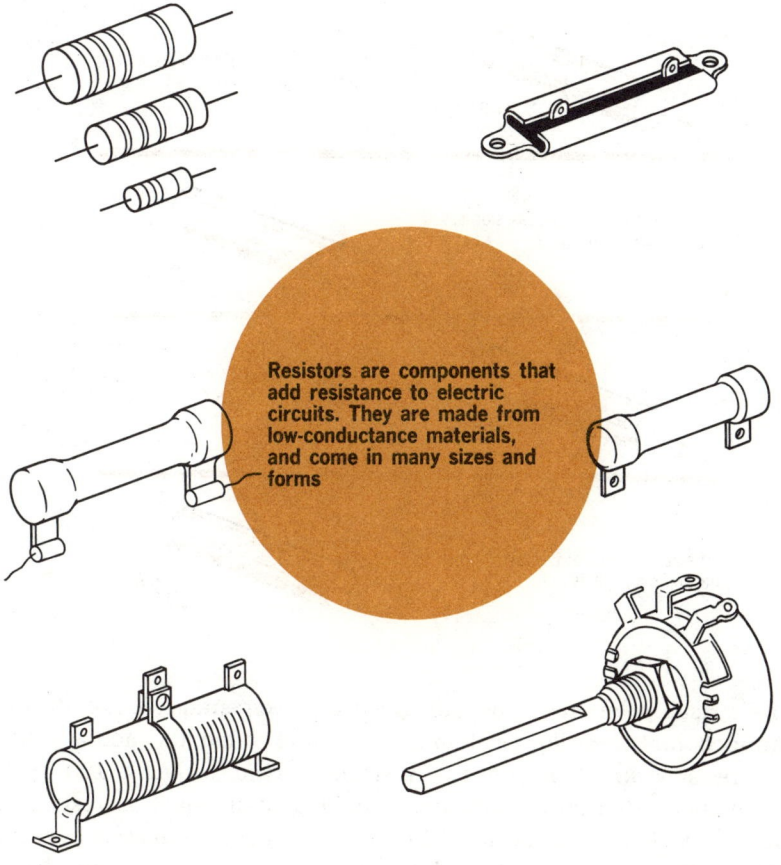

Resistors are components that add resistance to electric circuits. They are made from low-conductance materials, and come in many sizes and forms

tolerance

The basic characteristic of any resistor is the number of ohms of resistance it has. This is called the *resistor value,* and is normally marked on the resistor in some manner. However, the value marked on a resistor is really only a "nominal" value. The actual value may be somewhat higher or lower. The reason for this is that resistors are normally mass produced. And, as with all mass produced items, variations occur during manufacture. To account for these variations, resistors are marked with a *tolerance.*

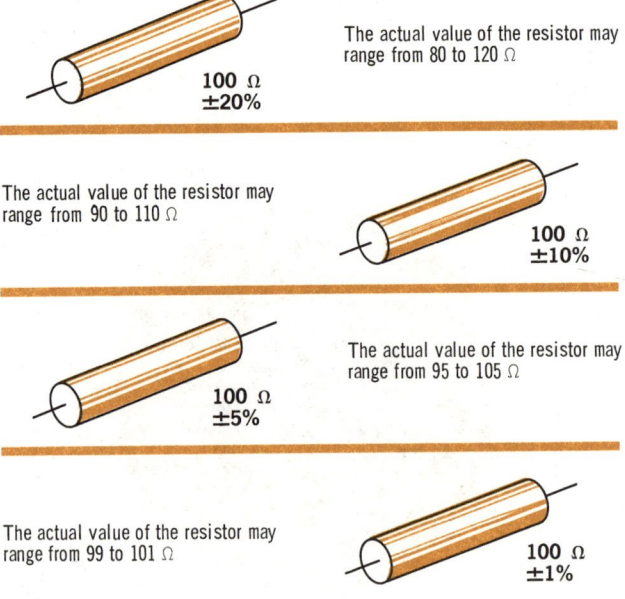

The resistor tolerance indicates how much above or below the nominal value the actual value of the resistor may be. Typical tolerance values are 20, 10, 5, and 1 percent

The actual value of the resistor may range from 80 to 120 Ω

100 Ω
±20%

The actual value of the resistor may range from 90 to 110 Ω

100 Ω
±10%

100 Ω
±5%

The actual value of the resistor may range from 95 to 105 Ω

The actual value of the resistor may range from 99 to 101 Ω

100 Ω
±1%

The resistor tolerance is usually given as a percentage, and indicates how much higher or lower than the nominal value the actual value of the resistor *might* be. Thus, a resistor marked as being 100 ohms and having a 10 percent tolerance would actually have a resistance value of anywhere from 90 to 110 ohms. The most common tolerances are 20, 10, 5, and 1 percent. And as you can probably guess, the lower the tolerance, the more expensive is the resistor.

current rating

You already know that when current flows in a wire it generates heat in the wire. The cause of this is the wire's resistance. The higher this resistance is, the more heat is produced. In a resistor, the resistance is concentrated in a small area, and so the heat generated by current flow is also concentrated in a small area. As a result, a resistor can become very hot when connected in a circuit. This means that it must be able to either withstand the heat generated, or give up the heat to the surrounding air. If it can do neither, it will eventually be damaged or destroyed. Even if the heat is not severe enough to damage the resistor, it may cause a large *change* in the *resistance*, since, as you remember, the resistance of all materials changes as their temperatures change.

Every resistor has a *maximum current rating*, and should not be used in a circuit where more than this maximum current will flow. Otherwise, the resistor may burn out. The current rating of a resistor is normally given as the *wattage rating* of the resistor. This will be explained later.

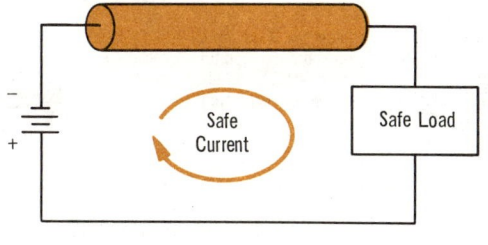

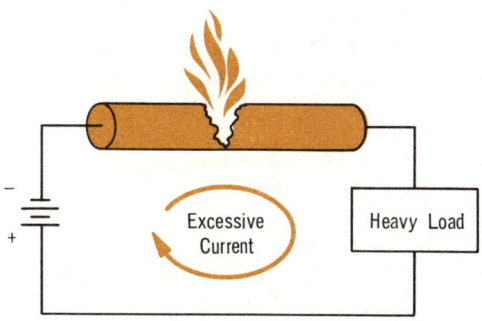

All resistors are designed for a certain maximum current. If this current is exceeded, the heat generated in the resistor will cause a change in the resistance and possibly even destroy the resistor. The wattage rating determines the maximum current a resistor can carry

resistor types

Based on what you now know about resistors, you may think that choosing a resistor for a circuit is a simple matter. It is just a matter of picking one that has the right resistance value and proper tolerance, and which also can carry the circuit current without burning out.

Although these are important considerations, they are not the only ones. There are many more, such as the cost of the resistor, how sturdy or rugged it is, how it is mounted in a circuit, and whether or not age or long use would cause changes in its resistance value. You can see, then, that you must consider many things when you pick a resistor. However, not all of these things are important in every case.

ALL-PURPOSE RESISTOR

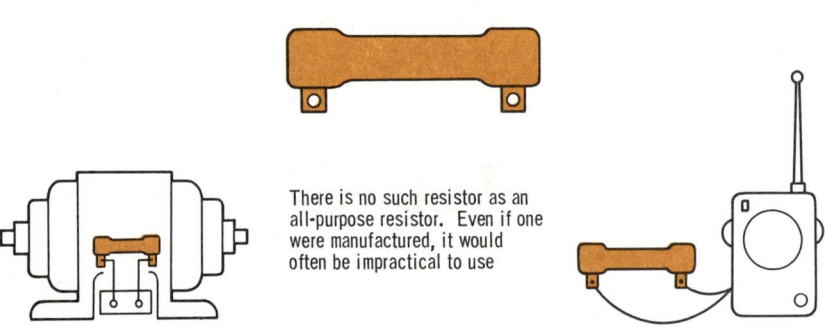

There is no such resistor as an all-purpose resistor. Even if one were manufactured, it would often be impractical to use

For example, the cost of the resistors in a small table radio was very important to the manufacturer, but the tolerances were not. If a single type of resistor was made that could be used in all circuits and under all conditions, it would be very expensive and would have many features that would often be unnecessary. Instead, different types of resistors are made, with each type best suited for certain uses.

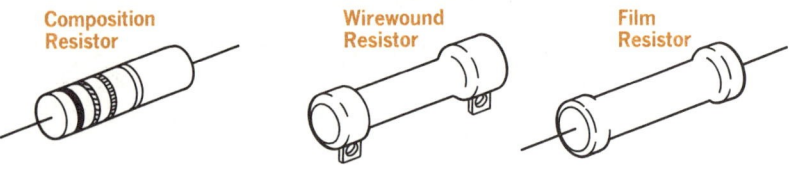

Composition Resistor   Wirewound Resistor   Film Resistor

Most of the resistors used today are one of two basic types: *composition resistors,* or *wirewound resistors.* However, a third type, called *film resistors,* is being used more and more.

composition resistors

In most cases where a resistor is used, the requirements are not severe, and what is needed is a resistor that will do the job as cheaply as possible. The type of resistor most often used in these cases is the *composition resistor.* The most common type of composition resistor consists essentially of a *powdered-carbon* resistance element, a tubular *plastic case* for sealing and protecting the resistance element, and *wire leads* for connecting the resistor into a circuit. As you can see from the bar graph on page 2-10, carbon has a resistance 2030 times that of copper. It only takes small amounts of carbon, therefore, to obtain high resistance. The powdered carbon is mixed with an insulating material, called a *binder,* and the value of the resistor depends on the relative amounts of carbon and binder material used.

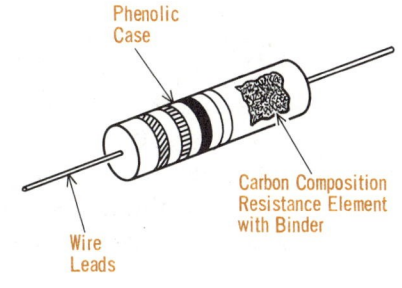

Phenolic Case

Carbon Composition Resistance Element with Binder

Wire Leads

Always be careful when soldering composition resistors. Heat from the soldering iron may be conducted along the leads to the resistance element and cause a significant change in the resistor's value

Advantages	Disadvantages
Small	Overheats with high currents
Rugged	Large temperature coefficient
Low Cost	Wide tolerance range

Composition resistors are made with resistance values of from less than 10 ohms to more than 20 million ohms (20 Megs), and with tolerances of 20, 10, and 5 percent. They cannot carry high currents without overheating, and have large temperature coefficients. They have the advantages, however, of small size, ruggedness, and low cost. In general, composition resistors are used in applications where large currents are not involved and narrow tolerances not required.

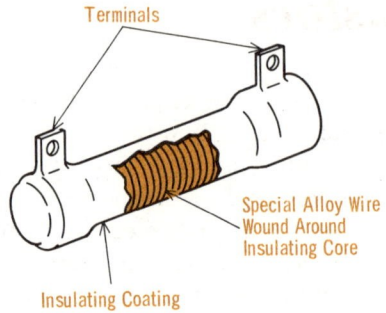

Terminals

Special Alloy Wire
Wound Around
Insulating Core

Insulating Coating

Power-type wirewound resistors can carry high currents and dissipate large amounts of heat

Precision-type wirewound resistors have very small tolerances and keep within these tolerances during use. This property of maintaining a small tolerance under all operating conditions is called the STABILITY of a resistor

wirewound resistors

The two main shortcomings of composition resistors are their limited current-carrying ability, and the difficulty in making them with small tolerances. Both of these limitations can be overcome, but at an increase in cost, by using resistance elements of special resistance wire instead of powdered carbon. Long lengths of wire are usually needed to obtain the necessary resistance values; to keep the resistor as small as possible, the wire is wound around a core. Resistors made in this way are called *wirewound resistors*.

There are two basic types of wirewound resistors: the *power* type, and the *precision* type. The power type is used for circuits having large currents, whereas the precision type is used when resistances with very small tolerances are required. Essentially, both types are made by winding special alloy *wire* around an insulating *core*, and then applying a ceramic, plastic, or other insulating *coating*. The ends of the winding are attached to metal caps at each end of the core. The caps have some form of *terminal* for connecting the resistor to a circuit. The high currents used with power type resistors generate large amounts of heat, which must be transferred to the surrounding air or dissipated. These resistors are therefore often large, since the more surface area a body has, the more heat it can transfer. The power-type wirewound resistors are made with resistance values of a few ohms to thousands (K) of ohms, with tolerances of 10 or 20 percent. Precision-type wirewound resistors are made with low resistance values, as low as 0.1 ohm and with tolerances as small as 0.1 percent. To obtain such small tolerances, costly materials and construction methods are used, and as a result, precision-type resistors are very expensive.

film resistors

Film resistors can be considered a compromise between composition resistors and precision-type wirewound resistors. They have some of the *accuracy* and *stability* of the wirewound type, but are smaller, more rugged, and cheaper.

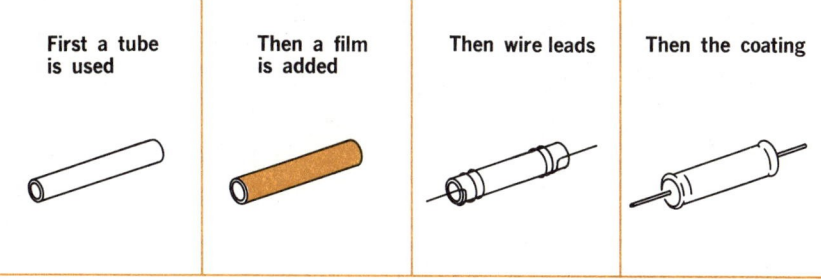

| First a tube is used | Then a film is added | Then wire leads | Then the coating |

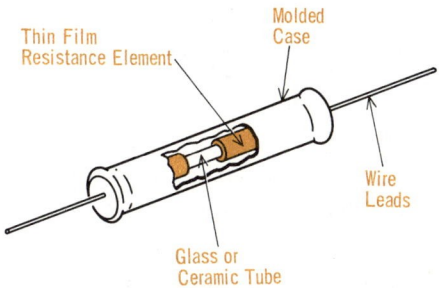

Thin Film
Resistance Element

Molded
Case

Wire
Leads

Glass or
Ceramic Tube

Film resistors are named according to the film material used. Thus, there are carbon film resistors, boron-carbon film resistors, metal film resistors, and metallic oxide film resistors

Film resistors are usually made by depositing, by a special process, a *thin film* of resistance material on a glass or ceramic tube. Leads for connecting the resistor into a circuit are connected to caps at the end of the tube. An insulating coating is then molded around the unit for protection.

The resistance of a film resistor is determined by the material used for the film, and its *thickness*. In general, the thicknesses used range between 0.00001 and 0.00000001 inch. You can see why these resistors are often called *thin film resistors*.

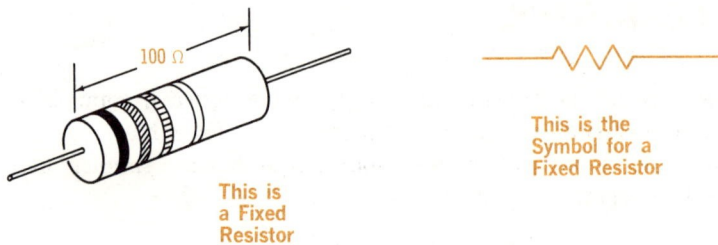

This is
a Fixed
Resistor

This is the
Symbol for a
Fixed Resistor

The value of a fixed resistor is set and cannot be varied

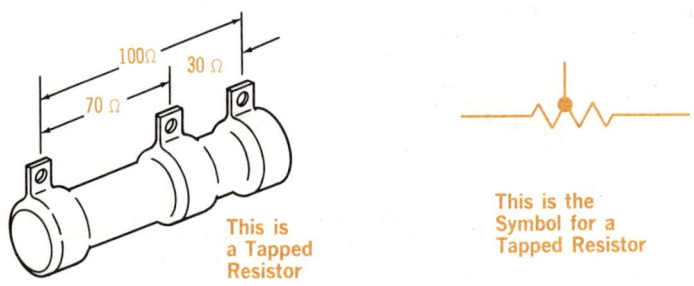

This is
a Tapped
Resistor

This is the
Symbol for a
Tapped Resistor

You can obtain more than one resistance value from a tapped resistor. However, each of these values is fixed

fixed resistors

You have now learned how resistors are classified according to the materials used for their resistance elements. But there is another way of classifying resistors. This is by whether their resistance value is *fixed* and unchangeable, or whether it can be *varied*. The types of resistors described earlier have two leads, each connected to one end of the resistance element; when these resistors are connected into a circuit, their entire resistance value is added to the circuit. You can see, therefore, that a *fixed* resistor has only one resistance value. However, there is a special type of fixed resistor that has more than one value. This type has, in addition to the terminals at the ends of the resistance element, one or more other terminals somewhere between the ends of the resistance element. By connecting different terminals into a circuit, different values of resistance can be obtained. Each of these different resistances, however, is still in itself a fixed resistance. This type of resistor is called a *tapped* resistor.

Fixed resistors can be either of the composition, wirewound, or film type.

adjustable resistors

From the preceding pages, you can see that a fixed resistor has no flexibility as far as its value of resistance is concerned. It has one value, which cannot be changed or varied. The tapped resistor offers some flexibility, since more than one value of resistance can be obtained from it. However, the number of resistance values you can obtain from a tapped resistor is usually limited to three or four. What is desirable in many applications is a resistor from which you can obtain a *range of resistance values,* from zero up to some maximum; for example, a resistor that you could adjust for any value from 0 to 100 ohms, or maybe 0 to 25K. One type of resistor that gives this flexibility is the *adjustable resistor.* An adjustable resistor is similar to a tapped, fixed wirewound resistor, except that all or part of the winding is exposed. A *movable clamp,* with a terminal attached, makes contact with the winding and can be moved to any position along the length of the winding. The resistance between the movable terminal and either of the end terminals then depends on the position of the movable clamp.

These resistors are not built to be frequently adjusted. They are normally set to the required resistance value when they are installed in a circuit, and then left at this value.

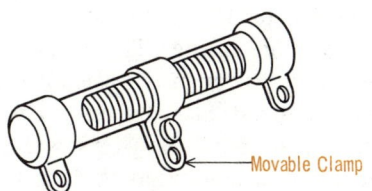

Movable Clamp

**This is an
Adjustable Resistor**

**This is the
Symbol for an
Adjustable Resistor**

With an adjustable resistor, you can obtain any resistance value covered by the range of the resistor

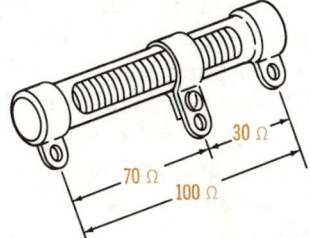

30 Ω

70 Ω

100 Ω

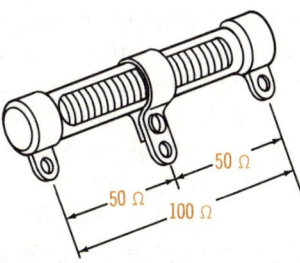

50 Ω

50 Ω

100 Ω

Though a wide range of resistance values is possible, these resistors are made in such a way that frequent adjustment is impractical

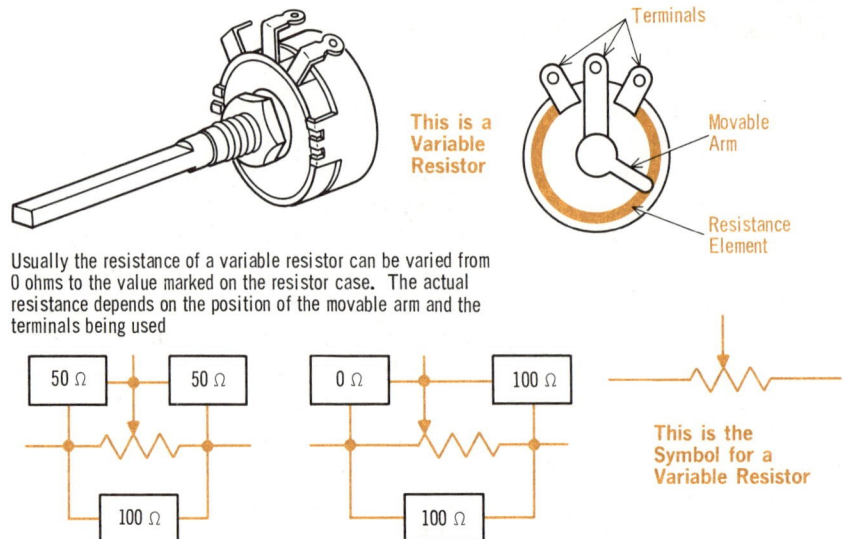

Usually the resistance of a variable resistor can be varied from 0 ohms to the value marked on the resistor case. The actual resistance depends on the position of the movable arm and the terminals being used

variable resistors

In many electrical devices a resistance value must be changed frequently. Examples of this are the volume control on your radio, the brightness control on your television set, or an electric light dimmer or motor speed control. This cannot be done using an adjustable resistor, since it would be difficult and time consuming. The resistors used must be continuously variable over a certain range of resistances, the same as adjustable resistors, but they must also be very *easy* to vary and built to withstand *frequent* adjustments. Resistors which do this are called *variable* resistors. Usually, a variable resistor consists of a circular resistance element contained in a housing or case. The element can be wirewound, composition, or film. A movable contact can slide across the element while maintaining electrical contact with it.

The movable contact is turned by a shaft. The resistance between movable contact and the ends of the element depends, therefore, on the position of the shaft. Both ends of the resistance element and the movable contact are connected to external terminals. When all three of these terminals are connected into a circuit, the resistor is called a *potentiometer*. When only the center terminal and one of the other terminals are used in a circuit, the resistor is called a *rheostat*. Sometimes, rheostats are made without the end terminal that will not be used. So remember: potentiometers and rheostats are both variable resistors. The only difference is the way they are used in a circuit.

resistor color code

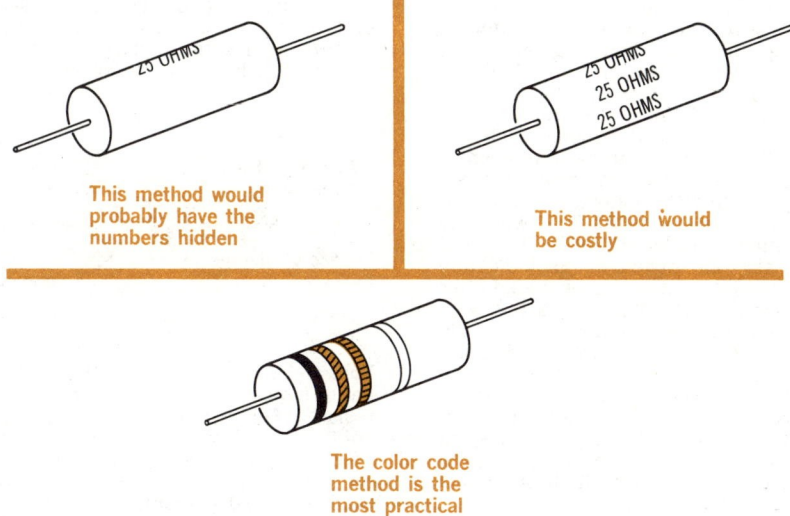

This method would
probably have the
numbers hidden

This method would
be costly

The color code
method is the
most practical

It is not always practical to mark the resistance value using numbers on axial-lead, fixed composition resistors. Instead, colored bands are used. The relationship of these bands to the resistance value is called the resistor color code

All resistors have their resistance value marked on them in some way. At first, you might suppose that this would always be done using numbers; for example, 50 ohms or 1000 ohms. The larger power resistors, precision resistors, and variable resistors are marked this way, but this is impractical for small fixed, composition resistors. These resistors are often too small to be marked that way. Also, they are tubular in shape and have axial leads, and can therefore be physically mounted in a circuit in any position. If their resistance values were marked in numbers, there would be a good chance the numbers would be hidden once the resistors were connected in a circuit. Of course, the numbers could be marked all around the resistor, but this would be difficult and costly from a manufacturing standpoint. This problem has been solved by using a series of *colored bands* around the resistors to indicate their resistance values. The positions of the bands and their color, making up what is called a *color code*, indicate the resistance values. A single standard color code has been adopted by the United States Armed Forces and the Electronic Industries Association (EIA) for fixed composition, axial-lead resistors.

standard resistor color code

First Significant Figure: The color of the first band indicates the first digit of the resistor value. For example, using the Color Code Table, if this band is yellow, the first digit is 4.

Multiplying Value: The color of the third band indicates how much the first two digits are to be multiplied to obtain the resistance value. For example, using the Color Code Table, if this band is green, the first two digits are multiplied by 100,000. This band can also be looked on as indicating the number of zeros to be added after the second digit. When used this way, the number of zeros shown in the Significant Figures column of the Color Code Table is the number of zeros to add. For example, if this band is orange, add three zeros after the second digit. But if the band is black, no zeros are added. If the third band is gold or silver, the multiplication factor must be used.

Second Significant Figure: The color of the second band indicates the second digit of the resistor value. For example, using the Color Code Table, if this band is black, the second digit is 0.

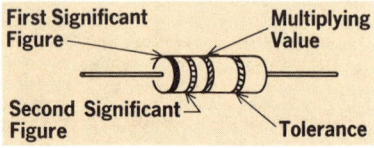

Tolerance: The color of the fourth band indicates the tolerance of the resistor. For example, if this band is gold, the resistor tolerance is ±5 percent. If there is no tolerance band on a resistor, the tolerance is automatically ±20 percent.

COLOR CODE TABLE

Color	Significant Figures	Multiplying Value	Tolerance
Black	0	1	—
Brown	1	10	—
Red	2	100	—
Orange	3	1000	—
Yellow	4	10,000	—
Green	5	100,000	—
Blue	6	1,000,000	—
Violet	7	10,000,000	—
Gray	8	100,000,000	—
White	9	1,000,000,000	—
Gold	—	0.1	±5%
Silver	—	0.01	±10%
No Color	—	—	±20%

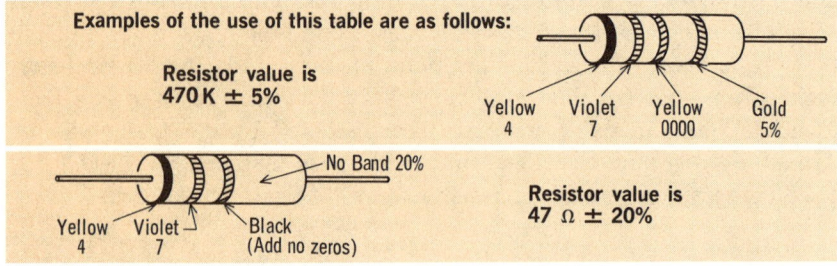

Examples of the use of this table are as follows:

Resistor value is
470 K ± 5%

Yellow   Violet   Yellow   Gold
4   7   0000   5%

No Band 20%

Yellow   Violet   Black
4   7   (Add no zeros)

Resistor value is
47 Ω ± 20%

summary

☐ Resistors are inserted into a circuit to lower the current flow by adding resistance. ☐ There are two basic types of resistors: composition and wirewound. Film resistors, a third type, are now becoming more and more popular. ☐ Composition resistors are usually made with a powdered carbon resistance element. They have high temperature coefficients, low current capacity, and large tolerances. Their advantages include ruggedness, small size, and low cost.

☐ Wirewound resistors are usually made from special resistance wire, wound about a core. They have large current capacity, and can be of the power or precision types. The power types have large current capacity; the precision types have very small tolerances. ☐ Film resistors are usually made by depositing a thin film of resistance material on a glass or ceramic tube. They have some of the advantages of both composition and wirewound resistors.

☐ Adjustable resistors have a movable clamp that is set to a required resistance value when installed in a circuit. They are not made for frequent adjustment. ☐ Variable resistors are similar to adjustable resistors, except that they are continuously variable over a range of resistances. ☐ If all three terminals of a variable resistor are connected in a circuit, the resistor is called a potentiometer. ☐ If only the center and one of the other terminals are connected, the resistor is called a rheostat. ☐ The resistor color code for fixed composition, axial-lead resistors indicates both the nominal value and tolerance of the resistor. The first two color bands indicate the significant figures; the third band is the multiplier; and the fourth band is the tolerance.

review questions

1. Describe the construction of the three types of fixed resistors.
2. State the advantages and disadvantages of composition resistors.
3. State the advantages and disadvantages of wirewound resistors.
4. What are the advantages of film resistors?
5. Define *nominal value* of a resistor.
6. Define *tolerance* of a resistor.
7. What is the difference between a rheostat and a potentiometer?
8. How do the *tapped, variable,* and *adjustable* resistors differ?
9. Describe the resistor color code.
10. What is meant by *significant figures,* as used in the color code?

ohm's law

As you learned earlier, since voltage causes current to flow in a closed circuit and resistance opposes the flow of current, a relationship exists between voltage, current, and resistance. This relationship was first determined in a series of experiments made by Georg Simon Ohm, who, as you remember from page 2-15, is the person for whom the unit of resistance was named.

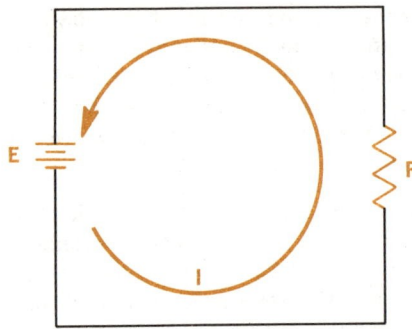

It is common practice to abbreviate current as I, voltage as E, and resistance as R

George Simon Ohm proved that current, I, in a d-c circuit is directly proportional to the voltage, E, and inversely proportional to the resistance, R

This means:
1. If you raise E, I will go up
2. If you lower E, I will go down
3. If you raise R, I will go down
4. If you lower R, I will go up

Ohm found that if the resistance in a circuit was kept constant, and the source voltage was increased, there would be a corresponding increase in current. Likewise, a decrease in voltage would cause a decrease in current. Stated another way, Ohm found that in a d-c circuit, current is directly proportional to voltage. Ohm also discovered that if the source voltage was held constant, while the circuit resistance was increased, the current would decrease. Similarly, a decrease in resistance resulted in an increase in current. In other words, current is inversely proportional to resistance. This relationship between the current, voltage, and resistance in a d-c circuit is known as Ohm's Law, and can be summarized as follows: *In a d-c circuit, the current is directly proportional to the voltage and inversely proportional to the resistance.*

equations

Strictly speaking, Ohm's Law is a statement of proportion and not a mathematical equation. However, if the current is in amperes, the voltage in volts, and the resistance in ohms, then Ohm's Law can be expressed by the equation:

$$I = E/R$$

which states that current (I) equals voltage (E) divided by resistance (R). Two variations of this equation that are very useful in analyzing d-c circuits are

$$R = E/I$$

which states that resistance (R) equals voltage (E) divided by current (I) and

$$E = IR$$

which states that voltage (E) equals current (I) times resistance (R).

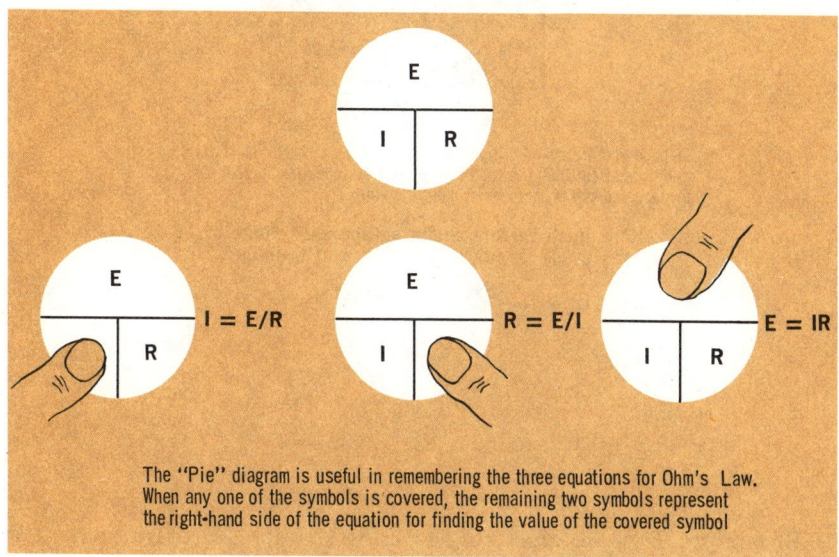

The "Pie" diagram is useful in remembering the three equations for Ohm's Law. When any one of the symbols is covered, the remaining two symbols represent the right-hand side of the equation for finding the value of the covered symbol

You can see that with these three equations, when *two* of the three circuit elements (current, voltage, and resistance) are known, the third can easily be found.

It is very important for you to *memorize* these three equations. You will be using them over and over while you study and work with circuits.

calculating current

At one time or another, you will probably have to calculate the current flowing in a circuit. You know that this can be done using Ohm's Law, so the first step is to decide which of the equations for Ohm's Law applies. A good practice to always follow at this point is to think in terms of *knowns* and *unknowns*. In any equation, the unknown is the term whose value you are trying to find. It is always the term to the left of the equal sign. The knowns are all the other terms of the equation. They are to the right of the equal sign.

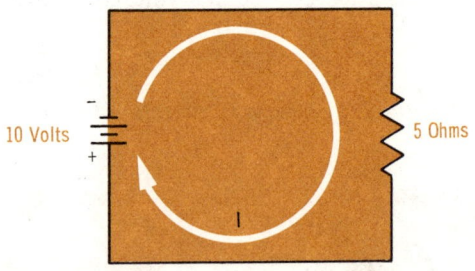

Suppose you are asked to find the current in the circuit. The first step is to study the circuit diagram and then phrase the question in your own mind as simply as possible, perhaps as the following:

How much current would an applied voltage of 10 volts cause through a resistance of 5 ohms?

Since current (I) is the unknown, you would use the equation

$$I = E/R$$
$$= 10 \text{ Volts}/5 \text{ Ohms}$$
$$= 2 \text{ Amperes}$$

In our problem, we are trying to find the value of the current, and therefore I is the unknown. As shown on page 2-37, the equation for Ohm's Law in which I is the *unknown* is

$$I = E/R$$

Therefore, this is the equation to use when calculating current in a circuit by using Ohm's Law.

calculating current (cont.)

The circuit diagram at the left shows a 20-ohm resistor used as the load in a circuit having a 100-volt battery as a voltage source. If the resistor has a maximum current rating of 8 amperes, will its rating be exceeded when the switch is closed?

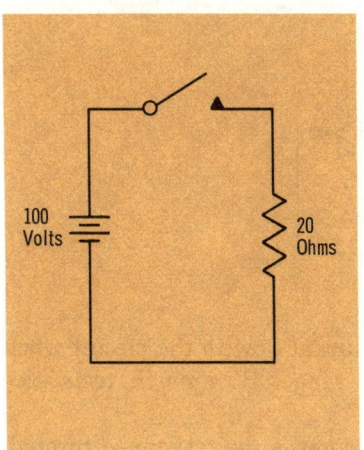
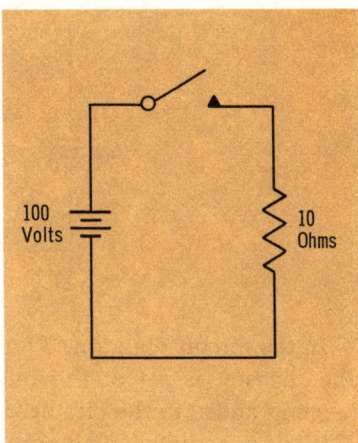

After reading the question and studying the diagram, you should see that you are really being asked two questions: (1) How much current would an applied voltage of 100 volts cause through 20 ohms of resistance, and (2) is this unknown current greater than 8 amperes? To answer the first question, the unknown is the current, and so the equation $I = E/R$ is used.

$$I = E/R = 100 \text{ volts}/20 \text{ ohms} = 5 \text{ amperes}$$

The second question can then be answered by a simple comparison. Since only 5 amperes of current flows, the 8-ampere current rating of the resistor is not exceeded.

What if a 10-ohm resistor, also with a maximum rating of 8 amperes, is used instead?

The equation $I = E/R$ is used again.

$$I = E/R = 100 \text{ volts}/10 \text{ ohms} = 10 \text{ amperes}$$

The resulting current of 10 amperes does exceed the 8-ampere current rating. The resistor will probably burn out.

calculating resistance

You calculate resistance by Ohm's Law by the equation:

$$R = E/I$$

You would use this equation to choose the proper size resistor to be connected into a circuit, or to determine the resistance of a resistor or other load already in a circuit.

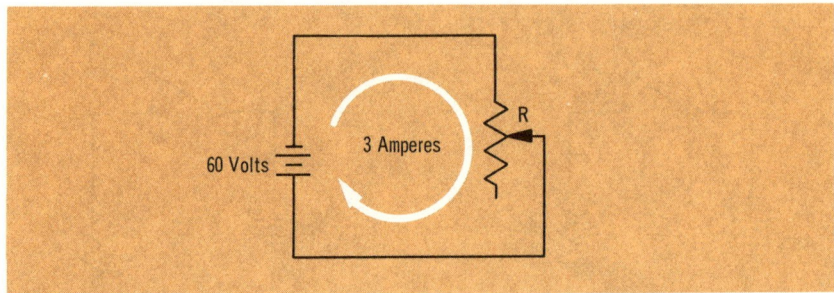

In the circuit diagram, 3 amperes of current flow in the circuit when the rheostat is set at the middle of its range. How much resistance is being added to the circuit?

What is actually being asked here is: What is the resistance through which an applied voltage of 60 volts will cause a current of 3 amperes to flow? Since resistance is the unknown, the equation $R = E/I$ is used.

$$R = E/I = 60 \text{ volts}/3 \text{ amperes} = 20 \text{ ohms}$$

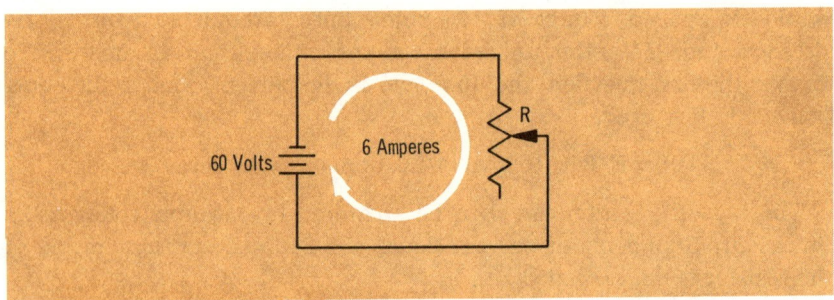

In the circuit, how much resistance would the rheostat have to add to the circuit to increase the current to 6 amperes? Again, resistance is the unknown, and the equation $R = E/I$ is used.

$$R = E/I = 60 \text{ volts}/6 \text{ amperes} = 10 \text{ ohms}$$

Thus, to double the current, the resistance must be halved.

calculating voltage

You calculate voltage by Ohm's Law using the equation:

$$E = IR$$

If the light bulb in the circuit diagram has a resistance of 100 ohms and 1 ampere of current flows in the circuit when the switch is closed, what is the voltage output of the battery?

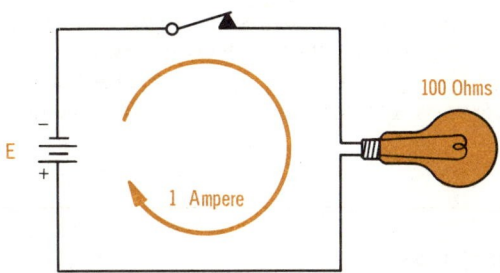

After studying the circuit diagram, you can see that what is being asked is: How much voltage will cause a current of 1 ampere to flow through a resistance of 100 ohms? Voltage is the unknown, so the equation $E = IR$ is used.

$$E = IR = 1 \text{ ampere} \times 100 \text{ ohms} = 100 \text{ volts}$$

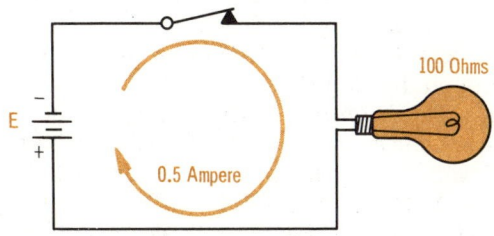

If the battery shown in the circuit wears out so that only 0.5 ampere flows in the circuit, what is the battery output voltage?

Again the voltage equation, $E = IR$, is used:

$$E = IR = 0.5 \text{ ampere} \times 100 \text{ ohms} = 50 \text{ volts}$$

You can see from this that the current was cut in half when the source voltage was reduced by one half.

summary

☐ Ohm's Law describes the relationship between voltage, current, and resistance in a d-c circuit. The Law states: In a d-c circuit, the current is directly proportional to the voltage and inversely proportional to the resistance.

☐ Ohm's Law can be expressed mathematically by three equations: $I = E/R$; this equation is used when current I is the unknown quantity in the circuit. ☐ $R = E/I$; this equation is used when resistance R is the unknown quantity in the circuit. ☐ $E = IR$; this equation is used when voltage E is the unknown quantity in the circuit.

☐ A Pie diagram is useful in remembering any of the three equations for Ohm's Law. When any one of the symbols is covered, the remaining two symbols represent the right-hand side of the equation (the known quantities). The symbol covered is the left-hand side of the equation (the unknown quantity).

review questions

1. State Ohm's Law and give its three equations.
2. Draw a circuit with a 15-volt battery power source and a 5-ohm resistive load. What is the current flowing in the circuit?
3. In what two ways can the current in Question 2 be doubled?
4. If the resistance of a circuit is increased to four times its original value, what would have to be done to the source voltage to maintain the original current flow?
5. A 100-ohm *film* resistor has a maximum current rating of 2 amperes. What is the maximum value of source voltage that can be applied to the circuit?
6. In Question 5, what effect would an applied voltage of 500 volts have on the circuit?
7. What is a *Pie diagram*? Illustrate its use.
8. If the resistance of a circuit is decreased to 1/4 of its original value, what happens to the current if the source voltage is unchanged?
9. If the resistance of a circuit is decreased to 1/4 of its original value, what circuit change can be made to maintain the original circuit current?
10. Doubling the resistance of a circuit has what effect on the current, if the source voltage is held constant? Halving the source voltage has what effect on the current, if the resistance is held constant? Doubling both the source voltage and resistance has what effect on the current?

power

As you now know, the purpose of the power source in an electric circuit is to supply electrical energy to the load. The load uses this energy to perform some useful function. But another way you can put it is that the load uses the energy from the source to do *work*. In doing the work, the load uses up the energy; this is why batteries "wear out" and have to be recharged or replaced. The amount of work done by a load depends on the amount of energy provided to the load and how fast the load can use the energy. In other words, with equal amounts of energy available, some loads can do more work in the same time than can others. Thus, some loads do work faster than others.

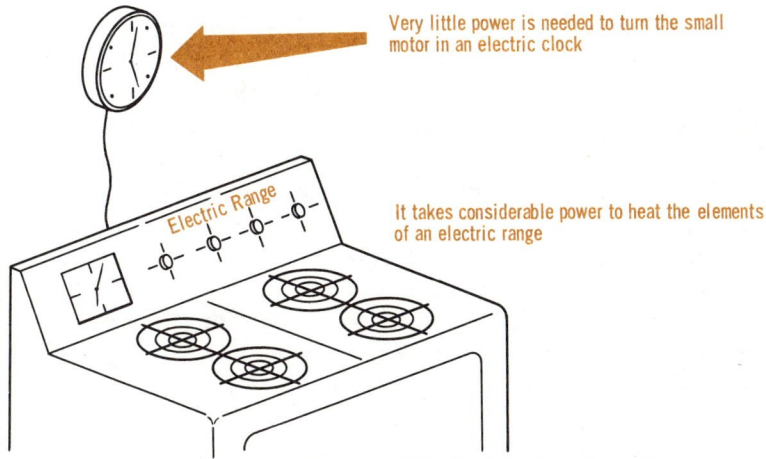

Very little power is needed to turn the small motor in an electric clock

It takes considerable power to heat the elements of an electric range

Power is the rate at which work is done. The faster work is done, the more power is used

The term *power* is used to describe how fast a load can do work. It can be defined as follows: *Power is the amount of work that can be done by a load in some standard amount of time, usually one second.* An important point you should remember is that the work done in an electric circuit can be *useful* work or it can be *wasted* work. In both cases, the rate at which the work is done is still measured in power. The turning of an electric motor is useful work, as is the heating of the element in an electric oven. The heating of the connecting wires or resistors in a circuit on the other hand, represents wasted work, since no useful function is performed by the heat. When power is used for wasted work, it is said to be *dissipated*.

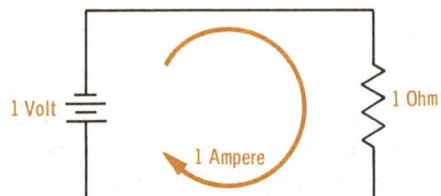

1 watt of power is used when 1 ampere flows through a potential difference of 1 volt. Therefore, in the circuit, 1 watt of power is used

the unit of power

Since power is the rate at which work is done, it has to be expressed in the units of *work* and *time*. You undoubtedly know that the basic unit of time is the second. However, you might not know the unit of work. For our purposes here, the unit of work will only be defined. You can find a description of how it is derived in many elementary physics books.

The unit of electrical work is the *joule*. This is the amount of work done by one *coulomb* flowing through a potential difference of 1 volt. Thus, if 5 coulombs flow through a potential difference of 1 volt, 5 joules of work are done. You can see that the time it takes these coulombs to flow through the potential difference has no bearing on the amount of work done. Whether it takes one second or one year, 5 joules of work are still done.

It is more convenient when working with circuits to think of amperes of current rather than coulombs; and as you remember from Volume 1, 1 ampere equals 1 coulomb passing a point in 1 second. Using amperes, then, one joule of work is done in one second when 1 ampere moves through a potential difference of 1 volt. This rate of 1 joule of work in 1 second is the basic unit of *power,* and is called a watt. Therefore, *a watt is the power used when one ampere of current flows through a potential difference of one volt.*

Mechanical power is usually measured in units of *horsepower,* abbreviated *hp.* You will sometimes find it necessary to convert from watts to horsepower and vice versa. To convert from horsepower to watts, multiply the number of horsepower by 746. And to convert from watts to horsepower divide the number of watts by 746.

CONVERSION OF UNITS
1000 watts (w) = 1 kilowatt (kw)
1,000,000 watts (w) = 1 megawatt (Megaw)
1000 kilowatts (kw) = 1 megawatt (Megaw)
1 watt (w) = 0.00134 horsepower (hp)
1 horsepower (hp) = 746 watts (w)

equations

From the definition of the watt, you know that 1 watt of power is used when 1 ampere flows through a difference of 1 volt. Then, if 2 amperes flow through a difference of 1 volt, 2 watts of power must be used. Or, if 1 ampere flows through a difference of 2 volts, 2 watts must be used. In other words, the number of watts used is equal to the number of amperes of current times the potential difference. This is expressed in equation form as:

$$P = EI$$

where P is the power used, in watts; E is the potential difference, in volts; and I is the current, in amperes.

The equation is sometimes called Ohm's Law for power, because it is similar to Ohm's Law. This equation allows you to find the power used in a circuit or load when you know the values of current and voltage. Two other useful forms of the equation are:

$$E = P/I$$

which is used when you know the power and current and want to find the voltage; and:

$$I = P/E$$

which is used to find the current when you know the power and voltage. You can see, then, that with these three equations you can calculate the power, voltage, or current in a circuit as long as you know the value of the other two.

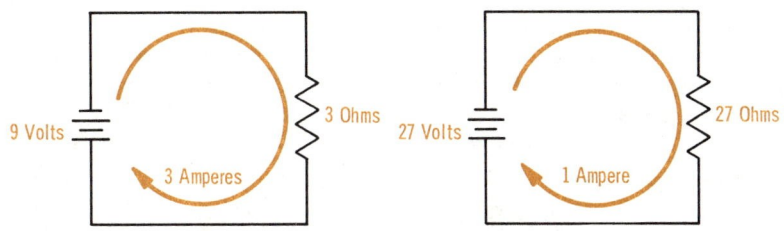

Since the equation for power is $P = EI$, both of these circuits use the same power:

$$P = EI = 9 \text{ volts} \times 3 \text{ amperes} = 27 \text{ watts}$$

$$P = EI = 27 \text{ volts} \times 1 \text{ ampere} = 27 \text{ watts}$$

resistance equations

There will be times when you have to find the power in a circuit and all you know are the voltage and resistance. You could first apply Ohm's Law to find the circuit current, but this takes time. It is easier to use an equation that gives power in terms of *voltage* and *resistance*. Since the power equations and Ohm's Law are similar, an equation can be easily derived.

You know that P = EI and that I = E/R. So if you replace the I in the power equation with its Ohm's Law equivalent, E/R, you get:

$$P = E \times (E/R) = E^2/R$$

With this equation, all you have to know are the resistance and voltage and you can calculate the power. The term E^2 is pronounced "E squared," and means E multiplied by itself.

In the same way that the equation $P = E^2/R$ was derived, an equation can be obtained giving power in terms of *current* and *resistance*. Such an equation would be used when you know the current and resistance and are asked to find the power. To derive this equation, you use E = IR. When these are combined, you get:

$$P = IR \times I = I^2R$$

Two steps are needed to find the power used in the first circuit below using the equations on page 2-45. How would you do it now in only one step? Since power is the unknown and voltage and resistance are the knowns, you have to use the equation that relates power to voltage and resistance. The equation is

$$P = E^2/R = (100)^2/10 = (100 \times 100)/10 = 1000 \text{ watts}$$

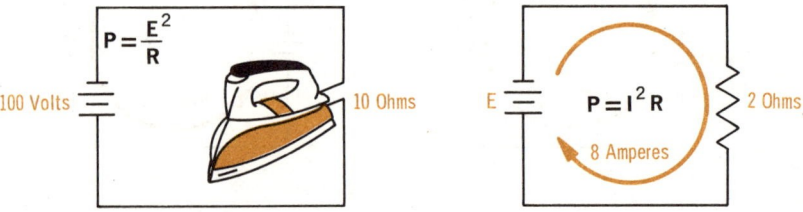

What is the power consumed in the second circuit? Now you would use the equation $P = I^2R$, since the current and resistance are known:

$$P = I^2R = 8 \times 8 \times 2 = 128 \text{ watts}$$

power losses

The power consumed in a circuit indicates how much work is done in that circuit. But you will remember that this work is not always useful. Much of it may be wasted, or lost. Power used to perform wasted work is, therefore, lost or *dissipated* power. In terms of the power source, lost power represents electrical energy that is not being used productively. And as you know, the production of electrical energy, whether by a battery or an electrical generator, costs money. It is important, therefore, that power losses in any electric circuit be kept as small as possible.

The most common loss of power in an electric circuit is the heat produced when current flows through a resistance. The exact relationship between the three quantities of heat, current, and resistance is given by the power equation:

$$P = I^2R$$

where P is the rate at which the heat is produced. You can see from the equation that you can decrease the amount of heat produced by lowering either the current or the resistance.

This I^2R heating, as it is often called, takes place in the circuit wires as well as in resistors. It is normally very small in the wires, since material and wire sizes are used that have low resistance values. In a resistor, little can be done about the I^2R heating inasmuch as the circuit current and the resistance value of the resistor cannot usually be changed without affecting the operation of the circuit.

In some electrical appliances, such as toasters and irons, the I^2R heating is needed and so does not represent a power loss.

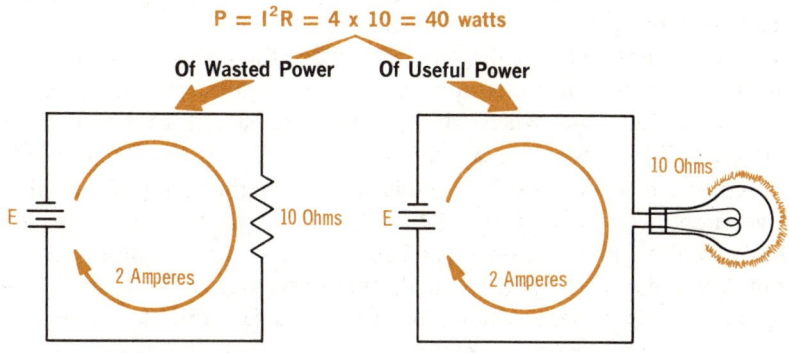

$P = I^2R = 4 \times 10 = 40$ watts

Of Wasted Power　Of Useful Power

I^2R heating usually represents wasted power. Sometimes, though it does useful work

1 Watt

1/4 Watt

10 Watts

1 Watt

Often, power ratings are not marked on resistors, but indicated by their physical size. However, the sizes used for particular wattages vary not only between different types of resistors but also between manufacturers, so it might be difficult to judge the power rating. Check the manufacturer's list of characteristics

power rating of resistors

If too much current flows through a resistor, the heat caused by the current will damage or destroy the resistor. This heat is caused by I^2R heating, which you know is power loss expressed in watts. Therefore, every resistor is given a *wattage*, or *power*, rating to show how much I^2R heating it can take before it burns out. This means that a resistor with a power rating of 1 watt will burn out if it is used in a circuit where the current causes it to dissipate heat at a rate greater than 1 watt.

If you know the power rating of a resistor and want to find the maximum current it can carry, you can use an equation derived from $P = I^2R$:

$$P = I^2R \text{ becomes } I^2 = P/R, \text{ which becomes } I = \sqrt{P/R}$$

Using this equation, you can find the maximum current that can be carried by a 1-ohm resistor with a power rating of 4 watts:

$$I = \sqrt{P/R} = \sqrt{4/1} = \sqrt{4} = 2 \text{ amperes}$$

If such a resistor conducts more than 2 amperes, it will dissipate more than its rated power and will burn out.

Power ratings assigned by resistor manufacturers are usually based on the resistors being mounted in an open location where there is free air circulation, and where the temperature is not higher than 40°C (104°F). Therefore, if a resistor is mounted in a small, crowded, enclosed space, or where the temperature is higher than 40°C, there is a good chance it will burn out even before its power rating is exceeded. Also, some resistors are designed to be attached to a chassis or frame, which will carry away the heat. If these types of resistors are mounted in the air, their heat will not be carried away and they will become too hot. In some cases, special mounting clamps, called heat sinks and cooling fins, are used to carry the heat away. Sometimes fan blowers are used to cool the resistors to increase their power rating.

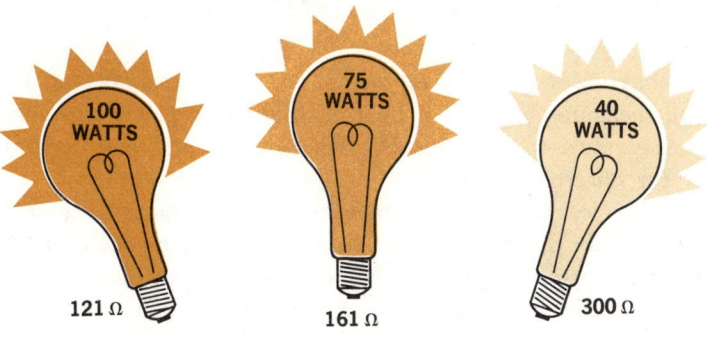

The wattage, or power, rating of a lamp tells how bright the lamp will be when used in a circuit. Actually, it is a measure of the I^2R heating of the lamp filament, which depends on the resistance of the filament

incandescent lamp power rating

From Volume 1, you know that an incandescent lamp is made by enclosing a resistance element, called a filament, in a glass bulb. When the lamp is connected into a circuit, current flows through the filament, and I^2R heating takes place. The heat is so severe that the filament becomes white-hot, and gives off light. The *more* the filament is *heated,* the *more* light the lamp gives off. You can see then that a convenient way of rating electric lamps is according to the I^2R heating they produce. This is exactly what is done by commercial lamp manufacturers. They stamp on each lamp the value of I^2R heating, in watts, that will be produced when the lamp is connected to a standard source of power. And, then, when you buy a lamp according to its *wattage rating,* you are really selecting it according to its light output.

You may wonder what the physical difference is between a 40-watt lamp and a 100-watt lamp. You should see from the equation $P = EI$ that the higher wattage bulb either has more current flowing through it or is connected to a higher source of voltage, or maybe both. However, you know that in most cases, such as in your own home, the source voltage is set by the local power company and cannot be changed. This means that the 100-watt lamp must conduct more current. To do this, it must have a lower resistance than the 40-watt lamp. Thus, the wattage and, therefore, the light output of an electric lamp depends on the resistance of the lamp filament. The *higher* the *resistance,* the *lower* the rated *wattage;* and the *lower* the *resistance,* the *higher* the rated *wattage.*

typical power ratings

As you have seen, the power ratings used for resistors and electric lamps are measures of the I^2R heating that takes place. Although the power rating always measures I^2R heating, its *practical* meaning is different for different devices.

Many other electrical devices are selected on the basis of their power ratings, especially those appliances that use heat to do their job. These include irons, toasters, heaters, ovens, etc. For most of these appliances, the larger the power rating, the more heat is produced. This means, for example, that a 1500-watt electric heater puts out more heat than a 1000-watt heater, and can therefore heat a larger area. Appliances with the highest power ratings, however, are not always the best. Toasters with power ratings of 10,000 watts or more could be made, but they would not toast your bread, they would burn it to a crisp almost instantly.

The power rating of working equipment such as motors, though, are not based on I^2R losses. They are based mostly on the power they can use to do mechanical work. A *1-horsepower* motor uses *746 watts* of electrical power, plus whatever power is dissipated because of I^2R losses. A ¼-hp motor needs at least 186.5 watts of electrical power.

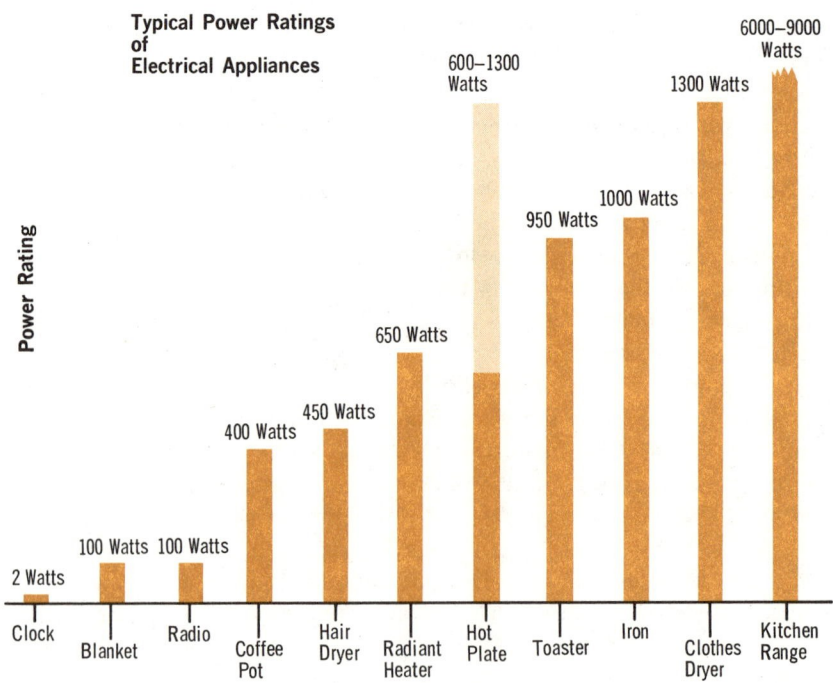

Typical Power Ratings of Electrical Appliances

Power Rating

Clock — 2 Watts
Blanket — 100 Watts
Radio — 100 Watts
Coffee Pot — 400 Watts
Hair Dryer — 450 Watts
Radiant Heater — 650 Watts
Hot Plate — 600–1300 Watts
Toaster — 950 Watts
Iron — 1000 Watts
Clothes Dryer — 1300 Watts
Kitchen Range — 6000–9000 Watts

the kilowatt-hour

Practically all of the electricity used in this country is produced, or generated, by large electric companies. From the power stations where it is generated, the electricity is distributed to the users by a complicated arrangement of wires, cables, and other devices. This distribution system ends at the individual factories, stores, or homes where the electricity is to be used. Since the electric companies are in the business of selling electricity, they must have some way of knowing *how much* electricity is used by each of their customers. Otherwise, they would have no way of knowing how much money to charge. They do this by supplying a meter to every user. The meter is usually located where the electricity enters a home, apartment, or factory, and measures the electricity used.

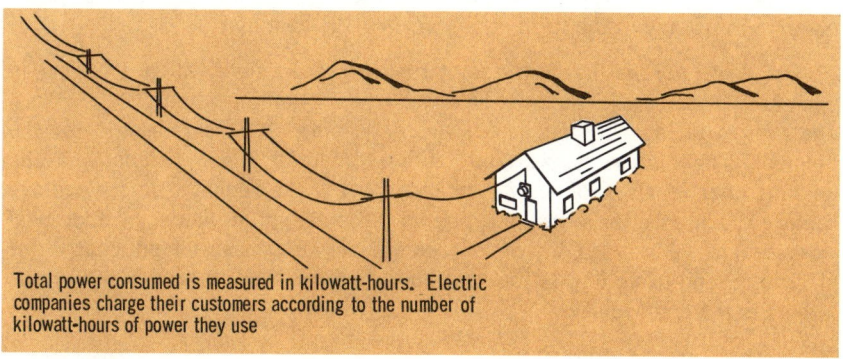

Total power consumed is measured in kilowatt-hours. Electric companies charge their customers according to the number of kilowatt-hours of power they use

Electricity itself cannot be measured, since as you know, it is really only a phenomenon. Current and voltage can be measured, but as you will learn later, to charge a user on the basis of current or voltage alone is impractical. Instead, every customer is charged on the basis of how much *work* is done by the electricity he uses.

You remember that the rate at which work is done is measured in watts. So to find the *total work* done, which is the same as the total power consumed, you multiply the rate of doing work (watts) by the total time during which the work is done. Thus, if a 100-watt lamp burns for one hour, the total work is 100 watts times 1 hour, or 100 watt-hours. This, then, is the way the electric companies measure and charge for electricity.

The watt-hour is a small unit. If it was used to indicate total power consumed, large numbers would result. So instead, units of *kilowatt-hours* are used. Each kilowatt-hour is equal to 1000 watt-hours.

summary

☐ Power describes how fast a load can do work. It is the amount of work that can be done in some standard amount of time. ☐ The unit of electrical work is the joule, which is the amount of work done by 1 coulomb flowing through a potential difference of 1 volt. The unit of time used is the second. ☐ One basic electrical power unit is the rate of 1 joule of energy consumed per second. It is called the watt. The watt can also be expressed as the power used when one ampere of current flows through a potential difference of 1 volt. ☐ The unit for mechanical power is the horsepower, hp. ☐ One horsepower equals 746 watts.

☐ Power can be calculated in terms of voltage, current, and resistance by three equations: $P = EI$, $P = I^2R$, and $P = E^2/R$. ☐ The three equations can be rearranged to solve for voltage, current, and resistance: $E = P/I$ and $E = \sqrt{PR}$; $I = P/E$ and $I = \sqrt{P/R}$; and $R = E^2/P$ and $R = P/I^2$. ☐ Work done in an electric circuit can be useful or wasted work. When power is used for wasted work, it is said to be dissipated. ☐ Power losses (dissipated power) are usually in the form of heat, called I^2R heating.

☐ The power, or wattage, ratings of resistors are usually at 104°F, and in open air. These conditions must be noted when mounting a resistor. ☐ Wattage ratings of incandescent lamps are determined by the resistance of the filament. The higher the wattage rating, the lower the resistance. ☐ The amount of light given off by an incandescent lamp depends on its wattage rating. The higher the wattage rating, the more light it yields. ☐ Electrical work or energy is used by electric companies to measure and charge for electricity. The basic unit is the watt-hour, but the kilowatt-hour (1000 times larger) is more convenient.

review questions

1. Define the following: *joule, watt, watt-hour,* and *kilowatt-hour.*
2. How many joules are there in 1 kilowatt-hour?
3. Draw a "Pie" diagram for P, E, and I.
4. A certain motor consumes 1492 watts. What is the power rating in hp? In kw?
5. If it was desired to obtain more light in a room, should a bulb with a smaller or greater wattage rating be used? Why?
6. What is meant by I^2R *losses?*
7. What is the equation for finding P, if E and I are given?
8. What is the equation for finding E, if P and R are given?
9. What is the equation for finding I, if R and P are given?
10. What is the equation for finding R, if P and E are given?

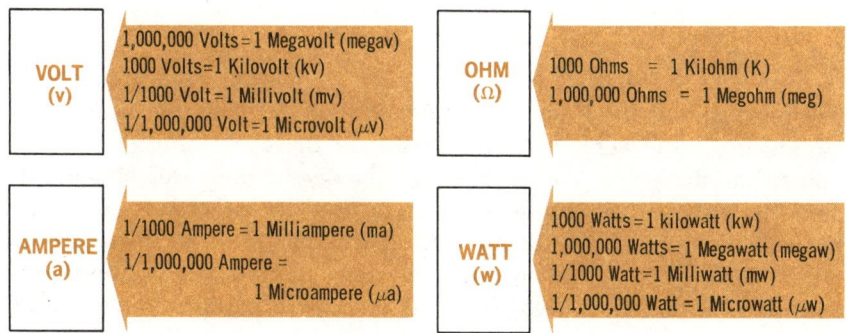

VOLT (v)	1,000,000 Volts = 1 Megavolt (megav) 1000 Volts = 1 Kilovolt (kv) 1/1000 Volt = 1 Millivolt (mv) 1/1,000,000 Volt = 1 Microvolt (μv)
OHM (Ω)	1000 Ohms = 1 Kilohm (K) 1,000,000 Ohms = 1 Megohm (meg)
AMPERE (a)	1/1000 Ampere = 1 Milliampere (ma) 1/1,000,000 Ampere = 1 Microampere (μa)
WATT (w)	1000 Watts = 1 kilowatt (kw) 1,000,000 Watts = 1 Megawatt (megaw) 1/1000 Watt = 1 Milliwatt (mw) 1/1,000,000 Watt = 1 Microwatt (μw)

Often, the basic units of the volt, ohm, ampere, and watt are either too large or too small for practical use. Fractional or multiple values of the units are used instead:

mega = 1,000,000; kilo = 1000; milli = 1/1000; and micro = 1/1,000,000

basic electrical units

Based on what you have learned up to now, you can see that in every d-c circuit there are four electrical quantities that you will work with most often. These are (1) the source *voltage* applied to the circuit, (2) the *resistance* present in the circuit, (3) the *current* that flows in the circuit, and (4) the *power* consumed by the circuit. The units in which each of these quantities are normally expressed make up what is called the practical system of units. These units are:

Quantity	*Unit*
Voltage	Volt
Resistance	Ohm
Current	Ampere
Power	Watt

The *volt* is defined in terms of electrical work as follows: When an emf moves 1 coulomb of electrons (6.28 billion billion electrons) to do 1 joule of work, the emf has a potential difference of 1 volt.

The *ampere* is defined in terms of coulombs of charge. A current of 1 ampere flows when 1 coulomb passes a given point in 1 second. You can see that the ampere is a measure of the rate of flow.

The *ohm* is defined in terms of the volt and the ampere. A material has a resistance of 1 ohm when an emf of 1 volt causes a current of 1 ampere to flow through it.

The *watt* is also defined in terms of the volt and the ampere. It is the power used when 1 ampere of current flows through a potential difference of 1 volt.

series circuits

Thus far, you have learned that the basic elements of the electric circuit are the power source, the load, and the connecting wires. You know that the power source provides energy in the form of voltage, and that this voltage causes current to flow in the circuit. You also know that the resistance in the circuit, whether it is the resistance of the load or of the wires, opposes the current flow. In addition, you learned how the load uses electrical energy to perform some useful function. This is the way the basic circuit, its parts, and their purpose were presented. Once you understood the basic circuit, the relationships that exist between voltage, current, resistance, and power were explained.

Throughout the presentation of this material, all the circuits described had one basic thing in common. That is, there was *one* and only one *path* for the current to follow as it flowed from the negative terminal of the power source, through the circuit, and back to the positive terminal of the source. There was *never* any point in a circuit where the current could divide and take more than one path. As a result, exactly the same current flowed through every part or device in the circuit.

This type of one-path circuit is called a *series circuit*.

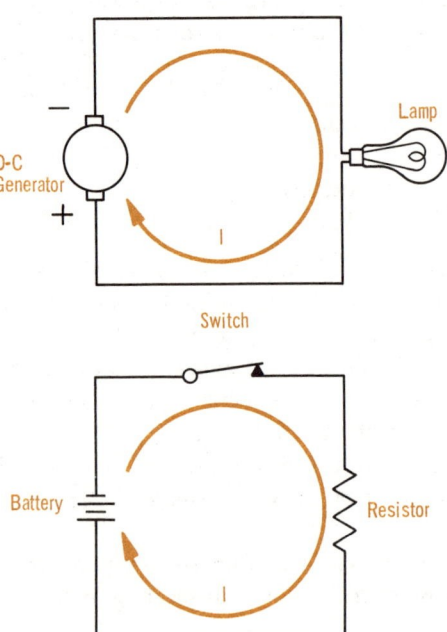

In a series circuit, the same current flows through every part. It makes no difference how many parts or devices there are. As long as the identical current passes through each, it is a series circuit

series loads

Up until now, you have studied circuits that have a single load, such as one resistor or one lamp. The resistance of this single load has been the total circuit resistance, and the power consumed in the circuit was the power used by this one load. In actual practice, however, you will very often find that a circuit has more than one load. It may have two resistors, or a resistor and a lamp, or maybe even five resistors and two lamps. In fact, there is almost no limit to the number of individual loads that a circuit may have.

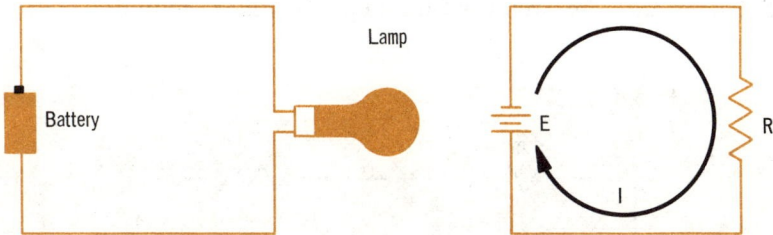

Previously, we have examined circuits that have a single load. The total circuit resistance was the resistance of the load, and the total circuit power was the power used by the load

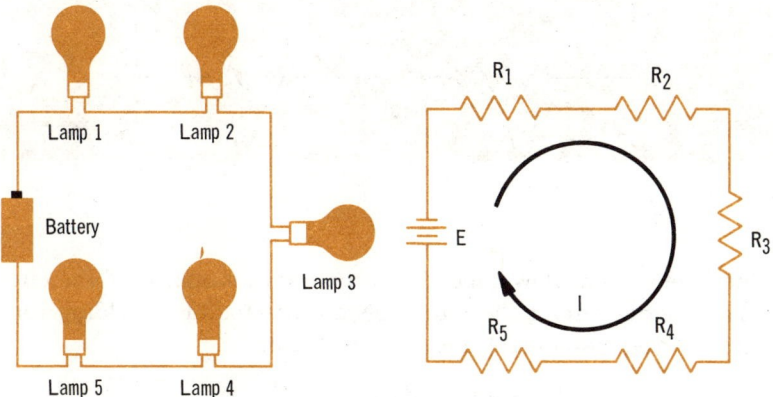

More than one load can be used in a circuit. If the total current (I) flows through each of them, they are series loads

If the loads are connected in the circuit in such a way that the total circuit current passes through each, then they are connected in *series*. They are *series loads*.

how series loads affect current

The current that flows in a circuit depends on the source voltage and the *total circuit resistance*. When there is only one load in a circuit, it usually provides the total circuit resistance. However, when series loads are used, the total circuit resistance is the *sum* of the resistances of each individual load. Thus, if a circuit has five loads connected in series, and each load is a 10-ohm resistor, the total circuit resistance is 5 × 10 or 50 ohms.

To find the current in a circuit that contains series loads, you first determine the total circuit resistance by adding the resistances of the individual loads. Then you use Ohm's Law ($I = E/R$) to find the current.

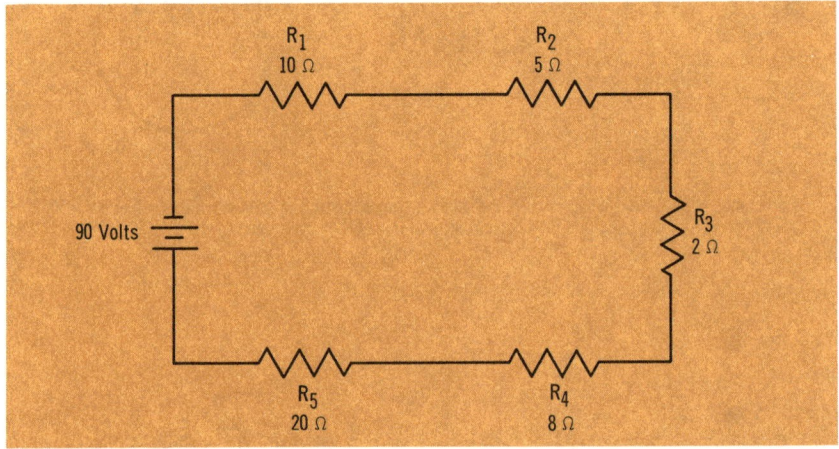

Since the current flow depends on the total resistance of the circuit, and for a series circuit the total resistance is found by adding the resistances of the individual loads, for this circuit:

$$R_{TOT} = R_1 + R_2 + R_3 + R_4 + R_5$$
$$= 10 + 5 + 2 + 8 + 20$$
$$= 45 \text{ ohms}$$

Once the total resistance is known, Ohm's Law can be used to find the circuit current:

$$I = E/R_{TOT}$$
$$= 90 \text{ volts}/45 \text{ ohms}$$
$$= 2 \text{ amperes}$$

series power sources

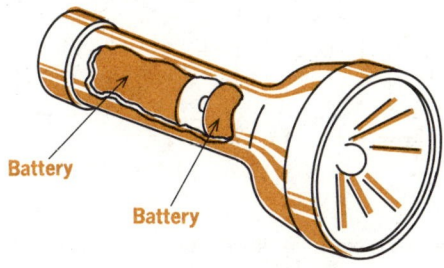

Battery

Battery

Frequently, the voltage required for operation of a device or circuit is larger than that of any available power source. In these cases, power sources can be used in series to give the needed voltage. An example of this is the common flashlight

You have probably at one time or another bought batteries for your car, your flashlight, your portable radio, or the flash unit for your camera. And as a result, you know that there are batteries with 1½-volt outputs, 6-volt outputs, 9-volt outputs, and 12-volt outputs, just to name a few. But you have probably never seen a 3-volt battery, a 15-volt battery, or a 29-volt battery. The reason is that it is unprofitable for battery manufacturers to mass produce batteries with every possible voltage output. So instead, they make certain standard sizes that can be *combined* to get most of the required voltage.

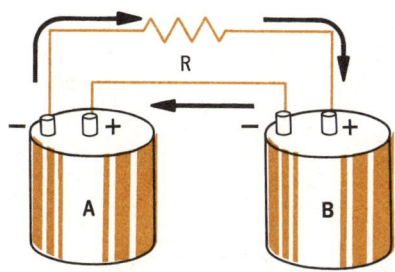

When two or more batteries are used in a circuit to produce a greater voltage than either battery can alone, then the batteries are connected in series. This produces *series power sources.*

When two batteries are connected in series, the negative terminal of one is connected to the positive terminal of the other. The other two terminals are connected to the circuit. As shown, current leaves the negative terminal of battery A, flows through the circuit, and enters the positive terminal of battery B. It then leaves the negative terminal of B and returns to the positive terminal of A.

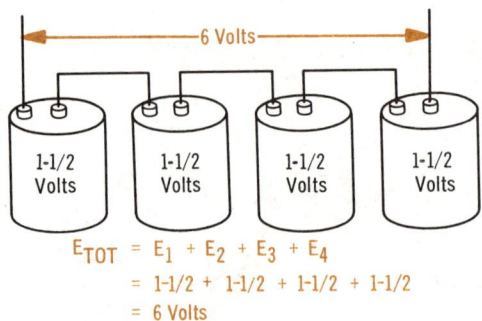

$$E_{TOT} = E_1 + E_2 + E_3 + E_4$$
$$= 1\text{-}1/2 + 1\text{-}1/2 + 1\text{-}1/2 + 1\text{-}1/2$$
$$= 6 \text{ Volts}$$

For power sources connected in series, the total voltage output is the sum of the individual source voltages

how series power sources affect current

When power sources are connected in series, the *total* voltage is equal to the *sum* of the *individual* source voltages. To find the current in a circuit containing power sources connected in series, you must therefore first find the total source voltage. Then you can use Ohm's Law ($I = E/R$) to calculate the current.

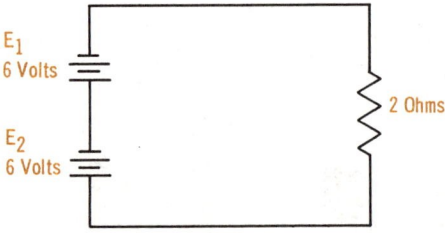

For the above circuit with series power sources, the total source voltage is

$$E_{TOT} = E_1 + E_2$$
$$= 6 + 6$$
$$= 12 \text{ volts}$$

To calculate the current,

$$I = E_{TOT}/R$$
$$= 12 \text{ volts}/2 \text{ ohms}$$
$$= 6 \text{ amperes}$$

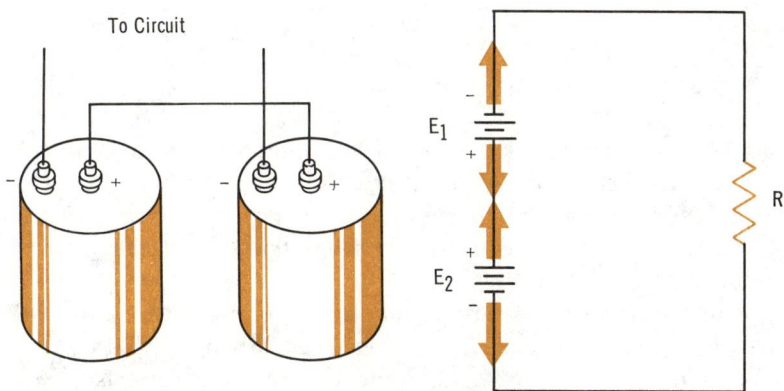

If the battery polarities are not connected in the same direction, they will be in series opposing, and will subtract from each other. Therefore,

$$E_{TOT} = E_1 - E_2$$

series-opposing power sources

If power sources are connected so that their polarities are not in the same direction, they will oppose each other. Power sources connected in this way are said to be *series-opposing*. The *total* voltage of series-opposing sources is the *difference* between the *individual* voltages:

$$E_{TOT} = E_{LARGER} - E_{SMALLER}$$

The polarity of the total source voltage will be the same as that of the larger battery, but if both batteries have the same voltage, then E_{TOT} will be zero, and no current will flow.

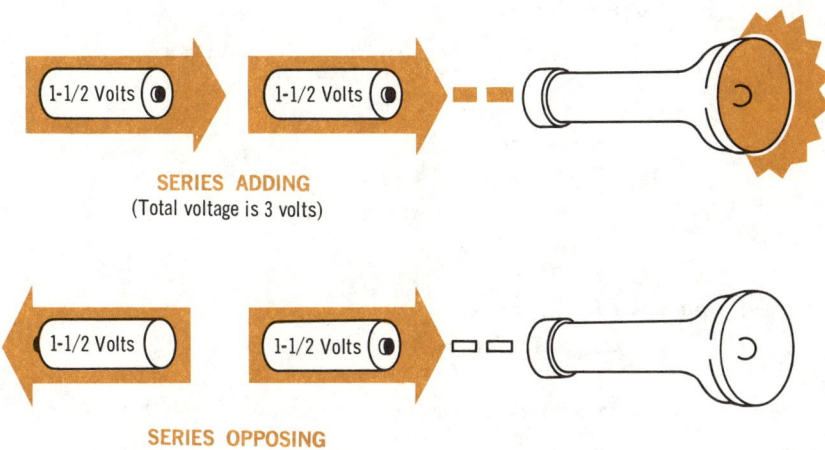

SERIES ADDING
(Total voltage is 3 volts)

SERIES OPPOSING
(Total voltage is 0 volts)

power consumption

Power is the rate at which a load does work. And when there is only one load in a circuit, the power used by that load is the total power consumed in the circuit. When a circuit has a number of loads connected in series, each of the individual loads uses power. Therefore, the total power consumed in the circuit is the sum of the power consumptions of each load.

The total power in a series circuit can be found in two ways. One is to calculate the power used by each load, and then add these together. The second, and easier way, is to find the total circuit resistance, and then calculate the power that is used by the total resistance.

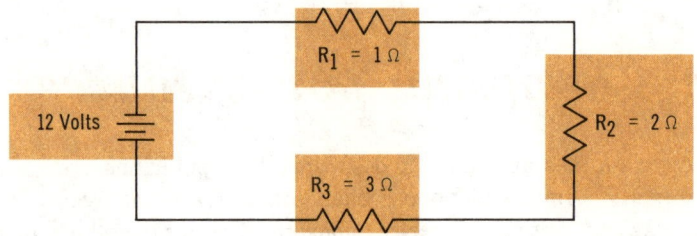

To find the power used by each of the loads in this circuit, you must first know the circuit current. But before you can find the current, you have to find the total circuit resistance.

Calculating Total Circuit Resistance:

$$R_{TOT} = R_1 + R_2 + R_3 = 1 + 2 + 3 = 6 \text{ ohms}$$

Calculating Circuit Current:

$$I = E/R_{TOT} = 12 \text{ volts}/6 \text{ ohms} = 2 \text{ amperes}$$

Calculating Power Used by Each Load:

$$P = I^2R$$
$$P_1 = 2 \times 2 \times 1 = 4 \text{ watts}$$
$$P_2 = 2 \times 2 \times 2 = 8 \text{ watts}$$
$$P_3 = 2 \times 2 \times 3 = 12 \text{ watts}$$

Calculating Total Circuit Power:

$$P_{TOT} = P_1 + P_2 + P_3 = 4 + 8 + 12 = 24 \text{ watts}$$

The total circuit power could also be found more simply by the equation:

$$P_{TOT} = EI = 12 \text{ volts} \times 2 \text{ amperes} = 24 \text{ watts}$$

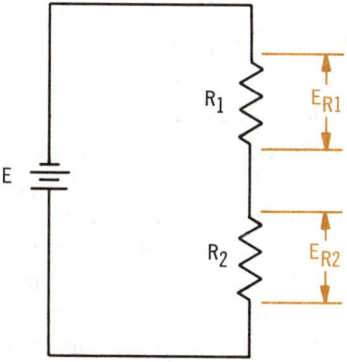

The total voltage supplied by a source is dropped across the circuit resistance. You can consider the voltage drop either as a loss of energy of the electron flow, or as the work done by the current when it flows through resistance. The energy lost is usually given off as heat. The total voltage drop equals the source voltage:

$$E_{SOURCE} = E_{R1} + E_{R2}$$

the voltage drop

You remember from Volume 1 that in a battery, a difference of potential is set up between the positive and negative terminals. This is done by chemically causing an *excess of electrons* at the *negative terminal* and a corresponding *lack of electrons,* or positive charges, at the *positive terminal.* When a wire or any conducting material is connected between the terminals, the difference of potential causes a force field, which we will call a force, to be sent down the wire at the speed of light. Electrons then flow from the negative terminal through the wire and back to the positive terminal under the pressure of the force. For every electron that leaves the negative terminal, another is produced chemically by the battery to take its place. Likewise, for every electron that arrives at the positive terminal and neutralizes a positive ion, another ion is produced by the battery to replace it. The battery thus keeps its difference of potential constant, although electrons are flowing.

Every electron at the negative battery terminal has been given energy by the battery. When the electron moves around the circuit it gives up the energy, so that when it arrives at the positive terminal it has lost all the energy the battery had given it. The electron loses its energy by giving it to the circuit resistance, usually in the form of heat.

Since the difference of potential across the battery terminals is normally given in volts, the energy lost by the electrons in the circuit resistance is also expressed in volts. Thus, if a resistor was connected across a 10-volt battery, 10 volts would be lost, or *dropped,* by the current flowing through the resistor. If two or more resistors were connected across the battery, some voltage would be dropped across each resistor, but the total voltage dropped would still be only 10 volts. Therefore, the total voltage drop in a circuit always equals the source voltage.

calculating voltage drops

In a series circuit, the *total* voltage dropped across all of the loads is equal to the source voltage. This is true whether there is one load or fifty loads. You can see, then, that for a fixed source voltage, the more loads there are, the less voltage will be dropped across each load.

Since the *voltage drop* across any load is the energy given to the load, the voltage that is dropped depends on the *current* flowing through the load and the *resistance* of the load. The greater the current, or the higher the resistance, the more voltage is dropped. And, the smaller the current, or the lower the resistance, the less voltage is dropped. This is shown by the equation $E = IR$. You will recognize this as one of the equations of Ohm's Law. It shows the relationships between the current, voltage, and the resistance of individual circuit components, as well as entire circuits.

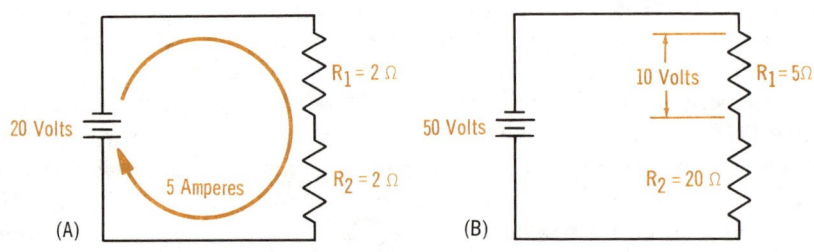

(A)   (B)

For circuit A the voltage drop across R_1 is

$$E_{R1} = IR_1 = 5 \text{ amperes} \times 2 \text{ ohms} = 10 \text{ volts}$$

Since you know that the drop across R_1 is 10 volts, and that the total drop must equal the source voltage, then the remaining voltage must be dropped across R_2. This can be found by:

$$E_{R2} = E_{TOT} - E_{R1} = 20 - 10 = 10 \text{ volts}$$

The voltage drops across R_1 and R_2, then, are the same: 10 volts each. But this makes sense since they both have the same resistance and carry the same current.

As shown in circuit B, sometimes you may know the resistance of a load and the voltage drop across it, and you are asked to find the current. You would do this by using the equation $I = E/R$. Therefore, the current through R_1 is

$$I = E_{R1}/R_1 = 10 \text{ volts}/5 \text{ ohms} = 2 \text{ amperes}$$

polarities

You remember that all voltages and currents have polarity as well as magnitude. In a series circuit there is only one current, and its polarity is from the negative battery terminal, through the circuit, to the positive battery terminal. The voltage drops across the loads also have polarities. The easiest way to find out what these polarities are, is to use the direction of the electron current as a basis. Then, where the electron current enters the load, the voltage is negative; and where it leaves the load, the voltage is positive. This holds true no matter how many loads there are in the circuit or what type they may be.

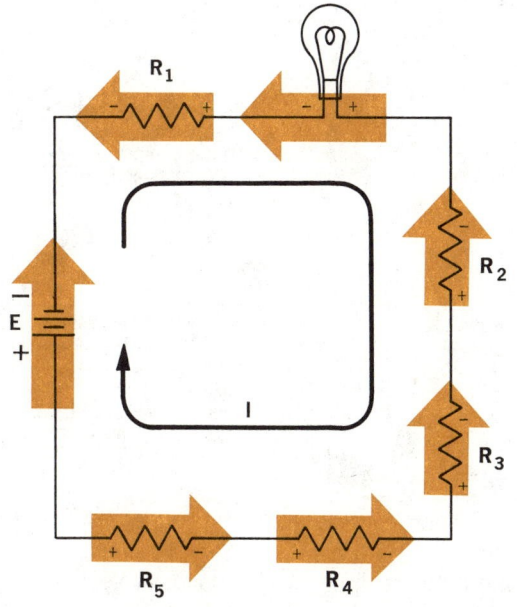

The voltage drops oppose the voltage from the source

Current enters the negative side of a load and leaves by the positive side. Current flow is from negative to positive within loads, and from positive to negative within power sources. This can be explained by the energy of the flowing charges. Inside of the source, their energy is increased, while inside the loads, their energy is decreased

The drop across the load, then, is opposite to that of the source. The voltage drops oppose the source voltage, and reduce it for the other loads. The reason for this is that each load uses energy, leaving less energy for the other loads.

polarity and voltage of a point

Outside of a power source, current always flows from negative to positive. You may have wondered while studying the illustration on page 2-63 whether that was always true, inasmuch as the polarities seem to show that between the loads the current was flowing from positive to negative. Actually, each pair of polarities applies only to the load it is near. The plus and minus for one load has no relationship to the plus and minus of any other load. If you look at the illustration, this will probably be clear to you.

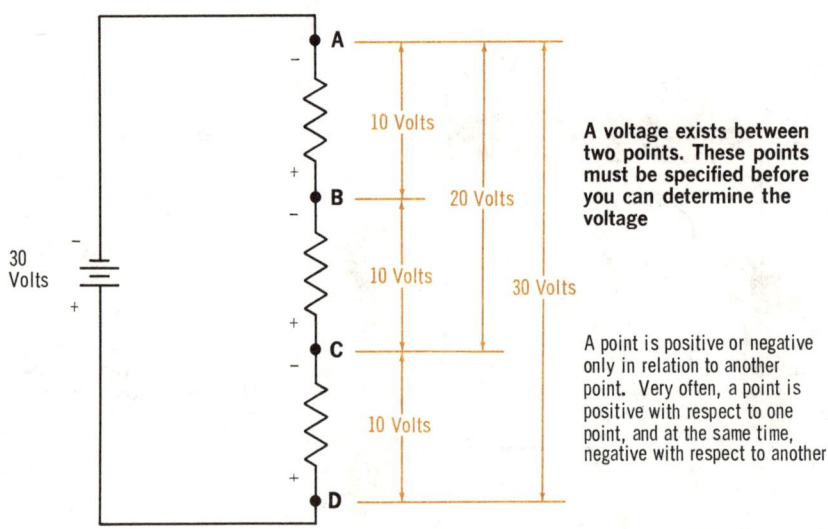

A voltage exists between two points. These points must be specified before you can determine the voltage

A point is positive or negative only in relation to another point. Very often, a point is positive with respect to one point, and at the same time, negative with respect to another

Point B has both a plus and a minus because it is positive with respect to point A, but negative with respect to point C. You can see, then, that to say a point is positive or a point is negative has no meaning. A point has to be positive or negative with respect to some other point.

In the same way, a voltage does not exist at a point. Voltage is a *difference of potential* between two points. This is shown in the illustration. If 10 volts is dropped across each resistor, the voltage between points A and B, B and C, and C and D are all 10 volts. But the voltage between A and C is the sum of the voltage drops between A and B, and B and C; or 20 volts. Similarly, the voltage between points A and D is 30 volts. You can see that to specify a voltage, you have to identify the *two points*.

the potentiometer circuit

You can see from the illustration on page 2-64 that in a circuit with series loads, different voltages exist between different points in the circuit. Later you will find that this is very useful, since it allows different values of voltage to be obtained from one source. Obviously, the more series loads there are, the more values of voltage there are in the circuit. But to get many voltages in this way requires a very large number of series loads.

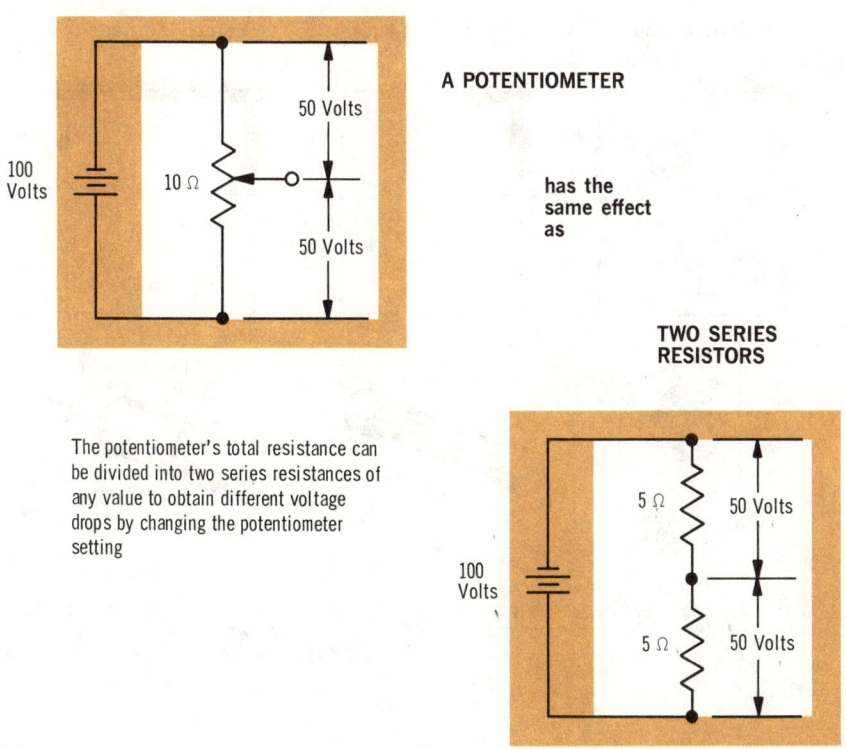

A POTENTIOMETER

has the same effect as

TWO SERIES RESISTORS

The potentiometer's total resistance can be divided into two series resistances of any value to obtain different voltage drops by changing the potentiometer setting

One way to get many voltages and still keep the number of loads to a minimum, is to use a potentiometer as the load. You remember from page 2-32 that a potentiometer has three terminals, and all three are connected into a circuit. The terminal that is connected to the movable arm of the potentiometer, therefore, actually divides the total resistance into two series resistances, and each resistance has its own voltage drop across it. By varying the resistance of the potentiometer, any value of resistance, and thus any voltage drop, can be obtained.

open circuits

A circuit has to provide a *complete* path from the negative terminal of the power source to the positive terminal in order for current to flow. A series circuit has only one path, and if it is broken, no current flows, and the circuit becomes an *open* circuit.

In a series circuit, all of the current flows through every point in the circuit. Therefore, if the circuit is open at ANY point, all current flow stops

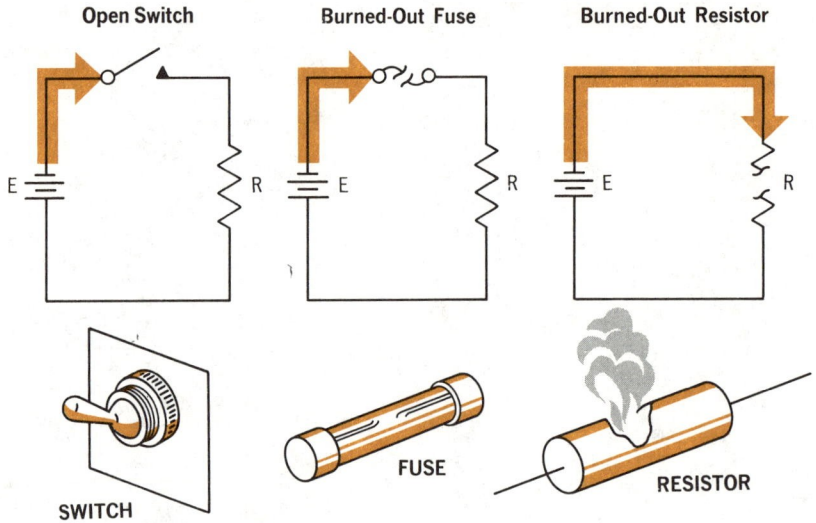

Open Switch   **Burned-Out Fuse**   **Burned-Out Resistor**

SWITCH   FUSE   RESISTOR

Circuits can be open by switches and fuses, as well as by damaged or broken parts or open connections

Since all current stops flowing when a series circuit opens, everything in a circuit that depends on the current is also affected

Circuits may be opened deliberately, such as by switches, or they may be opened as a result of some defect or trouble, such as a broken wire or a burned-out resistor. Since too much current in a circuit can damage the power source and the load, fuses are usually used in circuits to protect against damage. A fuse does its job by opening the circuit before high currents can do any harm.

Inasmuch as *no current flows* in an *open* series circuit, there are no voltage drops across the loads. There is also no power used by the loads, and so the total power consumed in the circuit is zero.

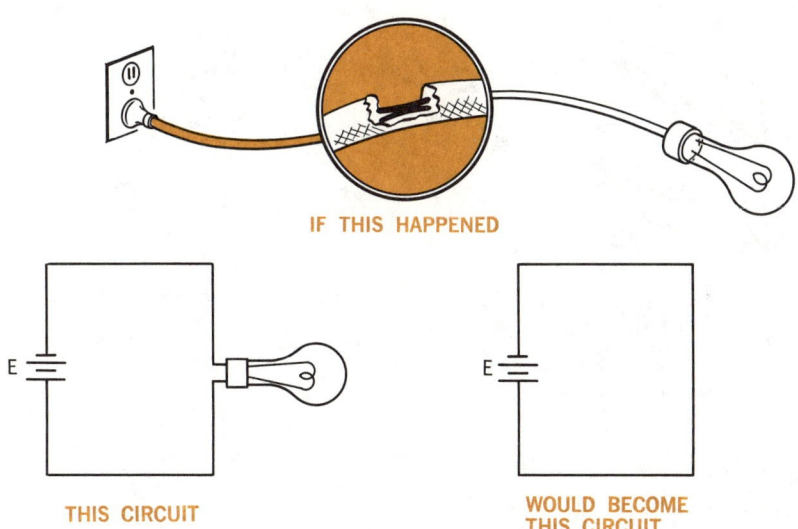

IF THIS HAPPENED

THIS CIRCUIT

WOULD BECOME
THIS CIRCUIT

A short circuit exists when current can flow from the negative terminal of the power source through connecting wires and back to the positive terminal of the power source without going through any load

short circuits

In a d-c circuit, the resistance is the only thing that opposes current flow. Therefore, if there was no resistance in a circuit, or if the resistance suddenly became zero, a very large current will flow. This condition of *no resistance* and very *high current* is called a *short circuit*.

From a practical standpoint, the resistance of a circuit cannot be reduced completely to zero. Even if a piece of silver wire with a large cross-sectional area was connected directly between the output terminals of a power source, there would still be some resistance in the circuit. This would consist of the resistance of the wire and the internal resistance of the power source. These resistances would be so low, though, that they would not limit the current flow very much. For example, if the combined resistance of the wire and the power source were 0.5 ohm and the source voltage was 100 volts, from Ohm's Law, the current would be: $I = E/R = 100$ volts$/0.5$ ohm $= 200$ amperes.

A short circuit is therefore said to exist whenever the resistance of a circuit becomes so low that the circuit current increases to the point where it can damage the circuit components. Current from short circuits can damage power sources, burn the insulation on wires, and start fires from the intense heat it produces in conductors. Fuses and other circuit breakers are the major means of protecting against the dangers of short circuits.

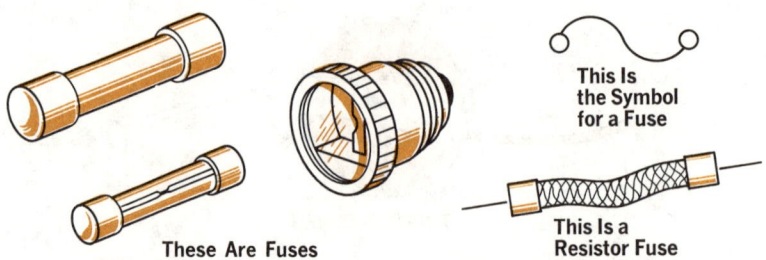

These Are Fuses

This Is
the Symbol
for a Fuse

This Is a
Resistor Fuse

Fuses open circuits before the high currents caused by short circuits can do any damage. The maximum current a fuse can carry before it melts and opens the circuit is called the rating. Fuses are normally rated in amperes, such as a 1-ampere fuse, or a 5-ampere fuse. The rating is usually marked on the fuse. Resistor fuses also have a resistance rating.

the fuse

You know that fuses open circuits to prevent the high currents caused by short circuits from doing damage. A fuse, therefore, has to do three things: (1) it has to know, or sense, when a short circuit exists, (2) it has to open the circuit before any damage is done, and (3) it has to have no effect on the circuit during normal operation, that is, when no short circuit exists.

Basically, most fuses are pieces of soft metal conductors or elements, contained in a housing. The fuse is connected into a circuit so that the fuse element is in series with the loads and power source. This means that the total circuit current flows through the fuse element. The element has very little resistance, and so has practically no effect on the circuit under normal conditions.

When a short circuit occurs, the current flowing through the fuse increases greatly; this causes the I^2R heating of the fuse element to increase. The fuse element has a low melting point, which means that it melts at a lower temperature than ordinary wire conductors. And when the heat caused by the short-circuit current reaches the melting point of the fuse element, the element melts and opens the circuit. The current that a fuse can carry before it melts depends on the material used for the element and its cross-sectional area. When a fuse melts, it is no longer any good, and must be replaced.

Sometimes *resistors* are made in such a way that they also act as *fuses*. A wire is used in the resistor which melts when it gets too hot. This happens when enough current flows to exceed the power rating of the resistor. In other cases, a special *fuse wire* is used to make the connection in the circuit that has to be protected. It is a plain length of insulated, thin wire that opens when it gets too hot.

circuit breakers

The trouble with fuses of any kind is that they must burn out to accomplish their purpose. This means they must be replaced, and must be stocked nearby so that you will have them for replacement when they are needed. Circuit breakers are another kind of protective device. They have the same function as fuses, but work in a different way so that they do not have to go bad to work. The *electromagnetic circuit breaker* uses the principle of magnetism to open a circuit when too much current flows. A set of switch contacts inside the circuit breakers is normally kept closed by an armature. A spring tries to pull one contact away, but the armature holds it in place. When too much current flows through the coil, a magnetic field builds up and attracts the armature to the coil. This releases the contact, which is pulled away to open the circuit. The circuit breakers are made with a toggle lever to reset the contacts and armature so that the unit can still be used.

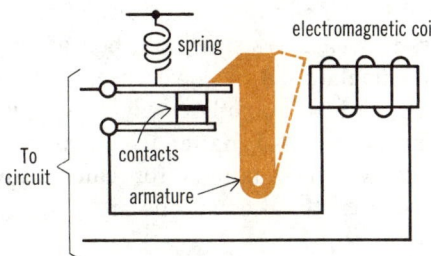

Electromagnetic Circuit Breaker

When too much current flows in the circuit breaker the magnetic field of the coil pulls in the armature. This allows the spring to pull the contacts apart to open the circuit

A *thermal circuit breaker* uses heat from the high current to work. One of the contact arms in this type of breaker is made of a temperature-sensitive metal that bends when it gets too hot. When it bends far enough, it will release the contact pulled by the spring. This kind of circuit breaker can also be reset after the bent metal cools and straightens.

Thermal Circuit Breaker

The thermal contact bends from the heat when too much current goes through it. This allows the other contact to be pulled **away** by the spring. Some thermal breakers do not use a spring to pull away the other contact. With these, when the thermal contact cools, it moves back in position to close the circuit again automatically

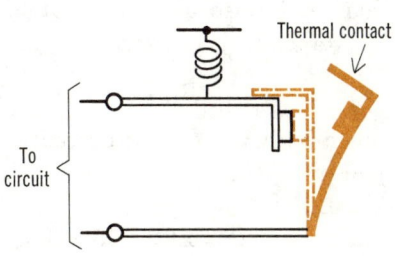

solved problems

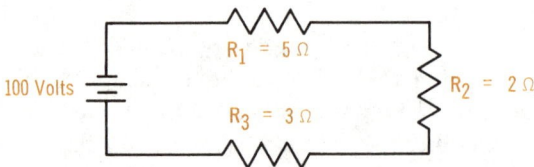

Problem 1. In the circuit, what is the current and the total power consumed?

Since in a series circuit, $I = E/R_{TOT}$, you have to find the total resistance before you can calculate the circuit current.

$$R_{TOT} = R_1 + R_2 + R_3 = 5 + 2 + 3 = 10 \text{ ohms}$$

With the total resistance, you find the current by Ohm's Law:

$$I = E/R_{TOT} = 100 \text{ volts}/10 \text{ ohms} = 10 \text{ amperes}$$

You can find the power consumed in the circuit in a variety of ways. One is by calculating the power used by each resistor, and then adding them. Another way is by directly calculating the power used by the total circuit resistance of 10 ohms. And still another way is by just using the source voltage and circuit current. No matter which method you use, the first step is to decide which equations for calculating power you should use. You remember that these equations are

$$P = EI \qquad P = I^2R \qquad P = E^2/R$$

If you decide to find the total power by first finding the power used by each individual resistor, the equation $P = I^2R$ should be used. The reason for this is that you already know the current through each resistor and the resistor value. Thus,

$$P_{TOT} = I^2R_1 + I^2R_2 + I^2R_3$$
$$= (100 \times 5) + (100 \times 2) + (100 \times 3) = 1000 \text{ watts}$$

To find the total power using the total circuit resistance of 10 ohms, you would also use the equation $P = I^2R$, since the current I and total resistance R_{TOT} are known.

$$P_{TOT} = I^2R_{TOT} = 100 \times 10 = 1000 \text{ watts}$$

Since both the circuit current and the source voltage are known, the power consumed in the circuit can also be found using the equation $P = EI$.

$$P = EI = 100 \text{ volts} \times 10 \text{ amperes} = 1000 \text{ watts}$$

solved problems (cont.)

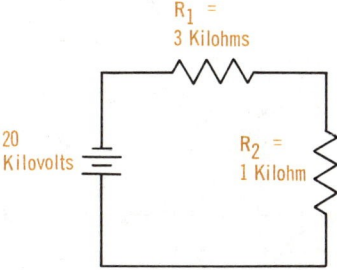

Problem 2. *In this circuit, what is the current?*

As you can see, the values of the source voltage and the resistances are given in multiples of the basic units. Whenever this happens, always convert to the basic units before trying to solve the problem.

You will remember from page 2-53 that 1 kilovolt equals 1000 volts. The source voltage in the circuit is therefore 20 × 1000, or 20,000 volts. Also you should recall that 1 kilohm equals 1000 ohms. Thus, the value of resistor R_1 is 3 × 1000, or 3000 ohms; and the value of resistor R_2 is 1 × 1000, or 1000 ohms.

Once you have converted the given values into the basic units, Ohm's Law is used to find the current.

$$I = E/R_{TOT} = E/(R_1 + R_2)$$
$$= 20,000 \text{ volts}/ (3000 \text{ ohms} + 1000 \text{ ohms}) = 5 \text{ amperes}$$

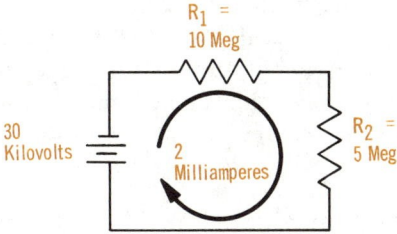

Problem 3. *Convert the quantities given on this circuit to the basic units.*

You know that 1 kilovolt equals 1000 volts. Therefore, 30 kilovolts is 30 × 1000, or 30,000 volts. One Meg equals 1,000,000 ohms. Therefore, 10 Meg is 10 × 1,000,000, or 10,000,000 ohms. And 5 Meg is 5 × 1,000,000, or 5,000,000 ohms. One milliampere equals 1/1000 ampere. Therefore, 2 milliamperes equals 2 × 1/1000, or 2/1000 (0.002) ampere.

solved problems (cont.)

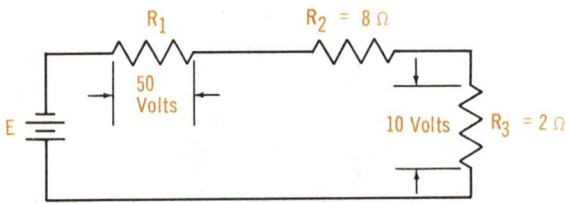

Problem 4. *What is the source voltage in this circuit?*

You know that from Ohm's Law the equation for voltage is $E = IR$. And so, after examining the circuit diagram, you can see that one way to find the source voltage is to first find the circuit current and then the resistance value of R_1. You could then use $E = IR_{TOT}$ to calculate the source voltage. This would require the following steps:

1. Using the equation $I = E_3/R_3$ to find the current through R_3, which is also the total circuit current, since this is a series circuit.

$$I = E_3/R_3 = 10 \text{ volts}/2 \text{ ohms} = 5 \text{ amperes}$$

2. Using the equation $R_1 = E_1/I$ to find the resistance of R_1.

$$R_1 = E_1/I = 50 \text{ volts}/5 \text{ amperes} = 10 \text{ ohms}$$

3. Using the equation $E = IR_{TOT}$ to calculate the source voltage.

$$E_{TOT} = IR_{TOT} = I(R_1 + R_2 + R_3) = 5(10 + 8 + 2) = 100 \text{ volts}$$

Another slightly easier way to solve this problem is to first find the circuit current, and then calculate the voltage drop across R_2. You would then know the voltage drop across each individual load in the circuit. And you remember that one of the basic laws of series circuits is that the sum of the individual voltages drops is equal to the applied source voltage.

Thus, to find the circuit current, you apply the equation $I = E/R$ to resistor R_3.

$$I = E_3/R_3 = 10 \text{ volts}/2 \text{ ohms} = 5 \text{ amperes}$$

Since it is a series circuit, the same current flows through resistor R_2. To find the voltage drop across R_2, then, you use the equation $E_2 = IR_2$:

$$E_2 = IR_2 = 5 \text{ amperes} \times 8 \text{ ohms} = 40 \text{ volts}$$

Now that you know the voltage drops across R_1, R_2, and R_3, the source voltage is simply the sum of the three voltage drops:

$$E_{SOURCE} = E_{R1} + E_{R2} + E_{R3} = 50 + 40 + 10 = 100 \text{ volts}$$

solved problems (cont.)

Problem 5.   In the circuit, how much power is consumed by resistor R when the switch is opened?

Obviously, no power is used, since for power to be used, current must flow; and when the switch is opened, there is no current flow anywhere in the circuit.

Problem 6.   If 5 amperes is the most current that the fuse can carry before it melts and opens the circuit, what is the lowest value at which R can be set?

The lowest value of R is the value which allows 5 amperes to flow, or

$$R = E/I = 200 \text{ volts}/5 \text{ amperes} = 40 \text{ ohms}$$

For any lower value of resistance, the current will be greater than 5 amperes and the fuse will "blow."

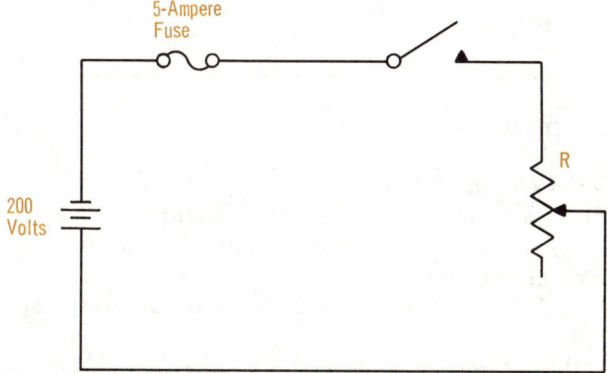

Problem 7.   If the fuse does "blow" and opens the circuit, what will be the voltage drop across R?

No voltage will be dropped across R after the fuse blows, since there will be no current flow in the circuit.

Problem 8.   If R is a 1000-watt resistor, and it is set for 20 ohms, there will be 10 amperes of current flow. This will cause R to dissipate 2000 watts of power. Will R burn out?

No. Current in the circuit will actually not reach 10 amperes since the fuse will blow at 5 amperes and open the circuit.

summary

☐ A series circuit has only one path for the current. ☐ If a series circuit is broken, the circuit is open, and no current flows. ☐ Series loads are connected so that the total current passes through each. ☐ The total circuit resistance for series loads is the sum of the individual resistances.

☐ Series power sources consist of two or more batteries connected to produce a greater voltage than either battery can alone. ☐ The total voltage for series power sources is the sum of the individual power sources. ☐ For power sources in series-aiding, the battery polarities are in the same direction, and the voltages add. ☐ In series-opposing, the battery polarities are in opposite directions, and the voltages subtract.

☐ The total power consumed in a series circuit is the sum of the powers used by the individual loads. ☐ The total voltage dropped across all of the loads in a series circuit is equal to the source voltage. ☐ The polarity of a voltage drop across a load is opposite to that of the source. ☐ Voltage-drop polarity is determined by the current direction. The end of a load where current enters is negative; where it leaves positive. ☐ Polarity is always with respect to a reference point.

review questions

1. Define *series circuit. Series load.*
2. The total power consumed in a series circuit is 100 watts. If there are two loads, and one consumes 35 watts, what does the other consume?
3. In the circuit of Question 2, which load has the greater current?
4. With two batteries of 9 volts, and 1.5 volts, show how you can obtain supply voltages of 10.5, 7.5, −10.5, and −1.5 volts.

For Questions 5 to 10, consider a circuit consisting of five resistors, whose values are 5, 4, 3, 2, and 1 ohms, in series with a 30-volt battery.

5. What is the total resistance of the circuit?
6. What is the circuit current?
7. What is the voltage drop across the 3-ohm resistor?
8. What is the total power consumed in the circuit?
9. What is the power consumed by the 4-ohm resistor?
10. Answer Questions 6 to 9 with the circuit open.

parallel circuits

All of the circuits you have studied so far have been series circuits. The current flowing at every point in these circuits was the same. Thus, once you found the current at any point, you knew what the current was everywhere else in the circuit. It would greatly simplify the analysis of d-c circuits if all circuits had this feature, but this is not the case.

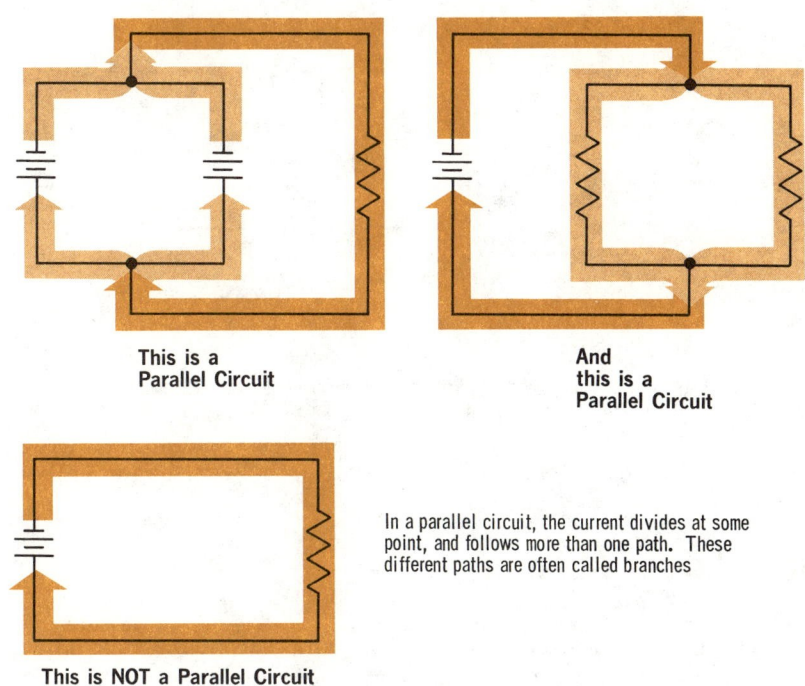

This is a
Parallel Circuit

And
this is a
Parallel Circuit

In a parallel circuit, the current divides at some point, and follows more than one path. These different paths are often called branches

This is NOT a Parallel Circuit

There is a large number of circuits in which the current is not the same at every point. In these circuits, there can be any number of *different currents*. All of the currents have the same polarity, since they are d-c circuits, but their magnitudes can vary greatly. These circuits are called *parallel circuits*, and can be defined as follows: *A parallel circuit is one in which there are one or more points where the current divides and follows different paths.*

When circuit components are connected in such a way that they provide different current paths, the components are said to be *connected in parallel*.

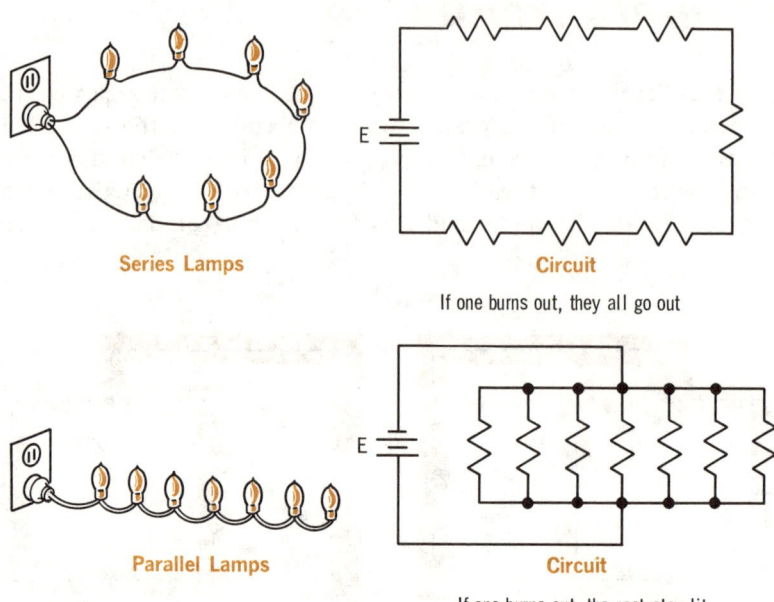

Series Lamps　　　　　　Circuit

If one burns out, they all go out

Parallel Lamps　　　　　　Circuit

If one burns out, the rest stay lit

If you have Christmas tree lamps, you are familiar with one of the basic features of parallel loads. That is, unlike series loads, if one load or branch becomes open, current still flows through other loads

parallel loads

In a series circuit, every load has the same current flowing through it; and this is the same current that both leaves and enters the power source. Very often, however, loads are connected in a circuit so that the *current* from the power source is *divided* between the loads, with only a *portion* of the current flowing through each load. The loads are then said to be connected in parallel, and the circuit is a parallel circuit.

In the parallel circuit, each load provides a separate path for current flow. The separate paths are called branches, and the current flowing in each branch is called *branch current*. Since the current that leaves the power source is divided between the branches, it is obvious that the current in any branch is less than the power source current.

If one branch of a parallel circuit is broken, or opened, current still flows in the circuit since there are still one or more complete paths for current flow through the other branches.

voltage drop across parallel loads

You will remember from series circuits that a portion of the source voltage is dropped across each series load. And the sum of these individual voltage drops is equal to the source voltage. When loads are connected in parallel, voltage is also dropped across each load. Instead of a portion of the source voltage being dropped across each load as in a series circuit, however, the *entire source voltage* is dropped across each. The reason for this is that all parallel loads are connected together directly across the power source.

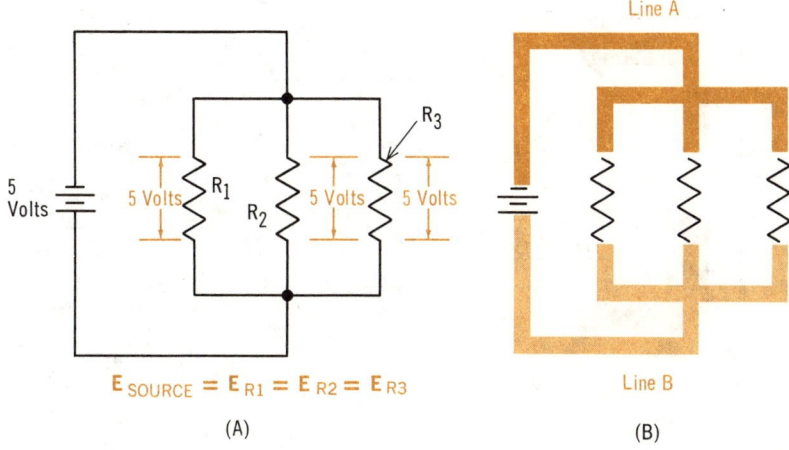

$$E_{SOURCE} = E_{R1} = E_{R2} = E_{R3}$$

(A)   (B)

Examining the circuit diagram (A) above, you can see that the entire source voltage is dropped across each branch, if you consider the lines that represent the connecting wires as lines of equal potential. In other words, the lines that connect the load and the power source have the same potential all along their length. This is shown on diagram (B).

Line A is at the potential of the negative battery terminal all along its length. Line B is at the potential of the positive battery terminal all along its length. The difference of potential between the battery terminals is 5 volts, so potential difference across each load must also be 5 volts.

effect of parallel loads on current

In a parallel circuit, the current leaves the power source, divides at some point to flow through the branches, and then recombines and flows back into the power source. There are thus *two* types of current in a parallel circuit: the current that leaves and reenters the power source, called the *total current;* and the *branch currents.* Since the total current divides to become the branch currents, the *sum* of the branch currents must be equal to the total current.

$$I_{TOT} = I_1 + I_2 + I_3$$

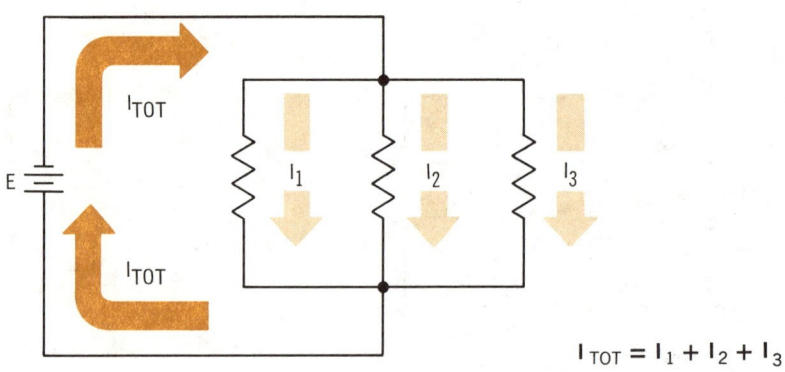

$$I_{TOT} = I_1 + I_2 + I_3$$

The total current in a parallel circuit is equal to the sum of the branch currents

Branch currents are determined by the resistance of the branch and the voltage across it. Since all branches have the same voltage across them, the more resistance there is in a branch, the less current there will be. And likewise, the less resistance there is, the greater the current will be. The total current in a parallel circuit depends on the source voltage and the total resistance of the circuit. As you will soon see, adding parallel loads to a circuit decreases the total circuit resistance, and thus increases the total current.

calculating branch currents

Each branch in a parallel circuit carries a separate current. Inside of a branch, though, the current is the same at every point. Every branch, therefore, has a voltage across it that is equal to the source voltage, a resistance, and a current that is the same at every point. You will recognize this as being similar to a series circuit. And as a matter of fact, this is exactly how you calculate the currents in the branches of parallel circuits. You take *one* branch at a time, consider it a series circuit, and use the equations for Ohm's Law.

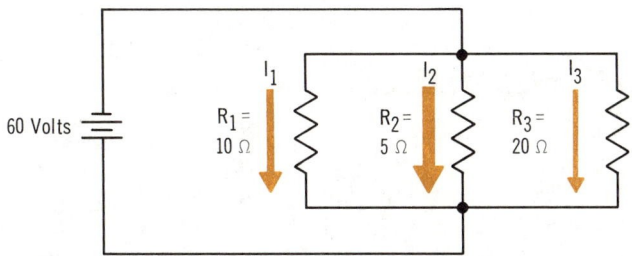

To calculate the branch currents of a parallel circuit, consider each branch as a separate series circuit, and then apply Ohm's Law. Thus, you would handle the above circuit as three separate circuits

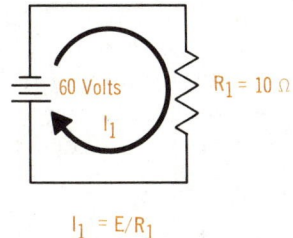

$$I_1 = E/R_1$$

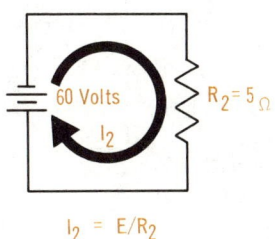

$$I_2 = E/R_2$$

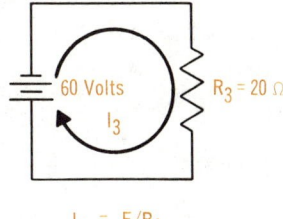

$$I_3 = E/R_3$$

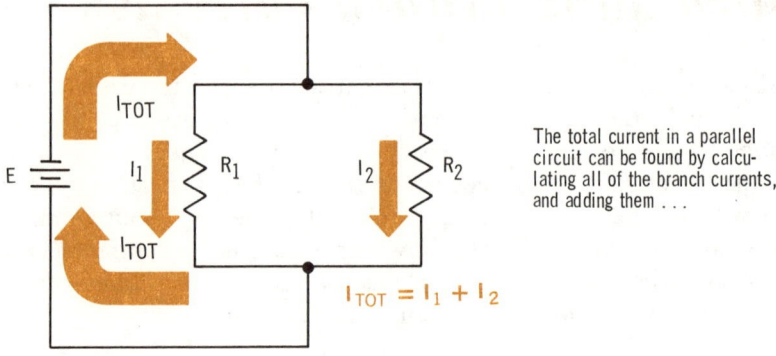

The total current in a parallel circuit can be found by calculating all of the branch currents, and adding them . . .

$$I_{TOT} = I_1 + I_2$$

calculating total current

Since the total current in a parallel circuit is equal to the sum of the branch currents, one way of finding the total current is to calculate all of the branch currents and then add their values. As an example of this, in the illustration on page 2-79, the total current in the circuit is the sum of the currents in the three branches, or 6 amperes plus 12 amperes plus 3 amperes, for a total of 21 amperes. Often it is easier to find the total current in a parallel circuit by calculating the total resistance in the circuit and then using the equation for Ohm's Law:

$$I_{TOT} = E/R_{TOT}$$

Based on what you know about series circuits, you may think that this would always be the easiest method to use. This is not true, though, since finding the total resistance of a parallel circuit is sometimes just as much, or more, work than calculating all of the branch currents.

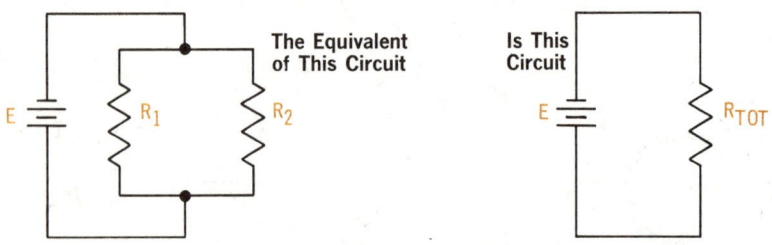

. . . Or the total current can be found by calculating the total resistance of the circuit, and then finding the total current as you would for series circuits

calculating total resistance

In a series circuit, the total resistance is simply the sum of all the individual resistances. The more resistances there are, the greater is the total resistance. It is obvious, therefore, that the total resistance is larger than any of the individual resistances.

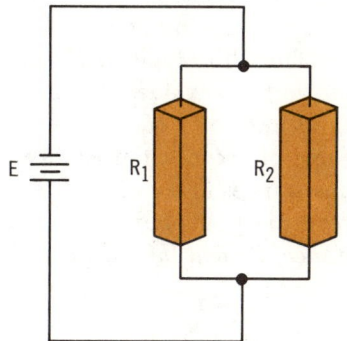

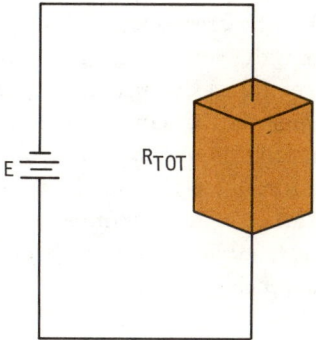

How is the total value of parallel resistances smaller than the individual resistances? This will be clear if you recall that the resistance of a material can be lowered by increasing its cross-sectional area

And this is effectively what is done when resistances are connected in parallel. Each resistance path allows more total current to flow

In a parallel circuit, the relationship between the total resistance and the individual resistances is completely different, and, as a matter of fact, almost opposite. For parallel circuits, the total resistance is *not* the sum of the individual resistances; the *more* resistances there are, the *lower* is the total resistance; and the *total* resistance is *smaller* than any of the individual resistances. The reason for this is that each new branch resistor draws more current from the source to increase the total current. And an increase in total current can only occur from a decrease in total resistance. You can probably see from this that calculating the total resistance for a parallel circuit is quite different than for a series circuit.

There are various ways to find the total resistance of parallel resistances. The best method to use in any particular case usually depends on how many individual resistances there are, and whether or not their values are equal.

equal resistances in parallel

The simplest arrangement of parallel resistances is when two are in parallel, and their values are *equal*. If you think of the two resistances as two equal-sized pieces of the same material, you know that the two together have twice the cross-sectional area of either one. They could, therefore, be replaced in the circuit with one piece of the material having the same length but double the cross-sectional area.

If you glance again at page 2-13, you will remember that if you double the cross-sectional area of a piece of material and do not change its length, you cut its resistance in half. Thus, the total resistance of *two* equal resistances in parallel is *one-half* the value of either resistance. In the same way, the total resistance of *three* equal resistances in parallel is *one-third* the value of any one of the resistances. We can therefore state a general rule for equal resistances in parallel: *The total resistance of any number of parallel resistances all having the same value is equal to the value of one resistance divided by the number of resistances.*

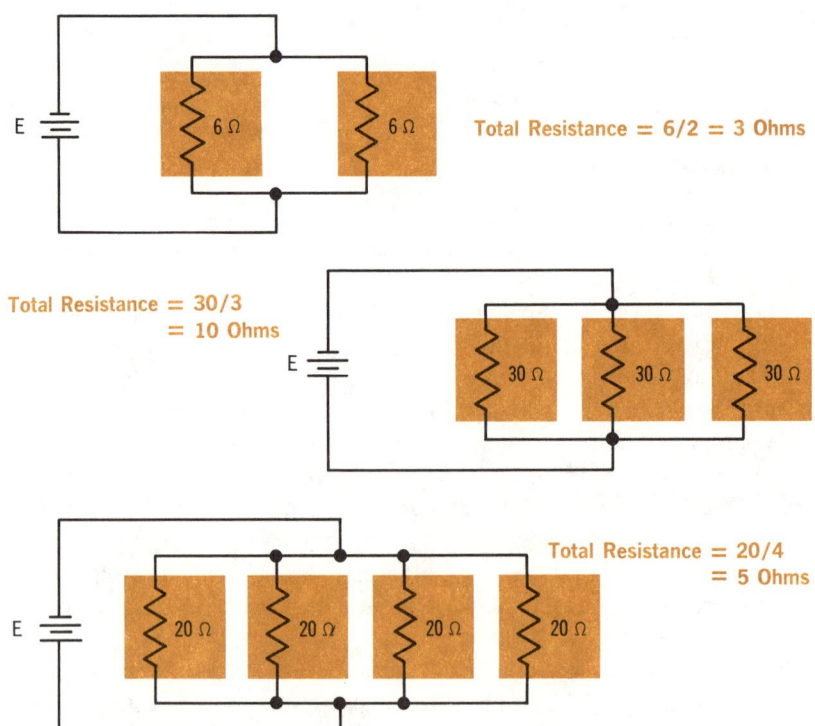

equal resistances in parallel (cont.)

Usually, if the parallel resistances are not equal, this equal-resistance rule cannot be used. However, if the different resistance values are *multiples* of one another, a variation of the equal-resistance rule can be used because any one resistance can be considered as two or more other resistors in parallel. For example, a 4-ohm resistor can be thought of as two 8-ohm resistors in parallel, or three 12-ohm resistors in parallel, or ten 40-ohm resistors in parallel. Therefore, when the parallel resistors have values that are multiples of each other, the lower valued resistors can be thought of as combinations to produce the same effective circuit with equal resistances. Then the equal-resistance rule can be used. This method also applies when the resistor values are not multiples of each other, but can be divided by a common multiple. As an example, 10- and 15-ohm resistors can both be changed to 30-ohm equivalents.

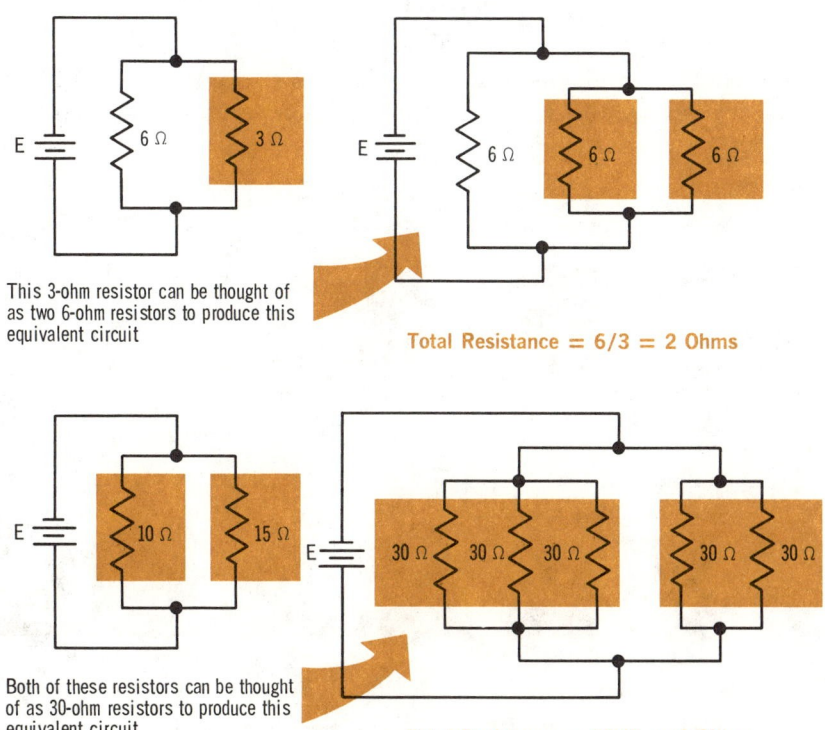

This 3-ohm resistor can be thought of as two 6-ohm resistors to produce this equivalent circuit

Total Resistance = 6/3 = 2 Ohms

Both of these resistors can be thought of as 30-ohm resistors to produce this equivalent circuit

Total Resistance = 30/5 = 6 Ohms

two unequal resistances in parallel

When there are two resistances in parallel, but their values are not the same and cannot be converted to the same multiple, the total resistance cannot be found with the equal-resistance rule. Instead, you must use a method called the *product/sum*, or product over the sum, method. To use this method, you first multiply the values of the two resistances to get their *product*. Then you add the values of the two resistances to get their *sum*. Finally, you divide the product by the sum, and the result is the total resistance. For example, let's use a set of resistor values that were used on page 2-83 to show how this method will produce the same results. If the values of two parallel resistances were 6 ohms and 3 ohms, you first multiply their values ($6 \times 3 = 18$) to get a product of 18. You then add their values ($6 + 3 = 9$) to get a sum of 9. Finally, you divide the product by the sum ($18/9 = 2$) to obtain the total resistance of 2 ohms. As an equation, this method can be expressed as:

$$R_{TOT} = \frac{\text{product}}{\text{sum}} = \frac{R_1 \times R_2}{R_1 + R_2}$$

The product/sum method can also be used for two parallel resistances whose values are the same or have the same multiple. However, this might often cause unnecessary work.

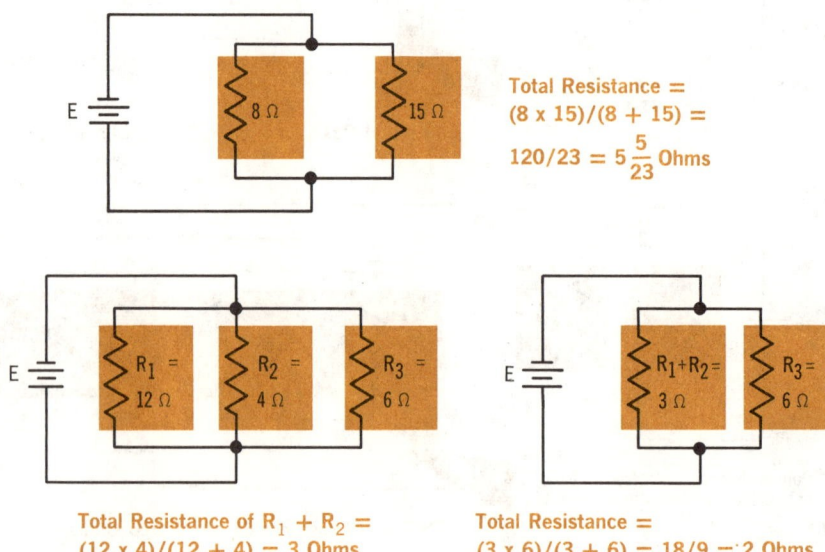

Total Resistance =
(8 x 15)/(8 + 15) =
$120/23 = 5\frac{5}{23}$ Ohms

Total Resistance of $R_1 + R_2$ =
(12 x 4)/(12 + 4) = 3 Ohms

Total Resistance =
(3 x 6)/(3 + 6) = 18/9 = 2 Ohms

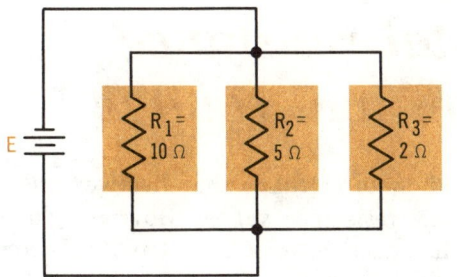

$$R_{TOT} = \cfrac{1}{\cfrac{1}{R_1} + \cfrac{1}{R_2} + \cfrac{1}{R_3}}$$

$$R_{TOT} = \cfrac{1}{\cfrac{1}{10} + \cfrac{1}{5} + \cfrac{1}{2}}$$

$$= \cfrac{1}{\cfrac{2}{20} + \cfrac{4}{20} + \cfrac{10}{20}}$$

$$= \cfrac{1}{\cfrac{16}{20}} = \cfrac{20}{16}$$

$$= 1\text{-}1/4 \text{ Ohms}$$

The reciprocal method can be used to calculate the total resistance of any combination of parallel resistances. In the circuit,

three or more unequal resistances in parallel

The product/sum method you have just learned is a special case of a much more general method for calculating the total resistance of a parallel circuit. The general method can be used for *any number* of resistances, and works whether the resistances are *equal* or *unequal*. The method is actually derived from the equations for Ohm's Law.

This general method is called the *reciprocal method*. The reciprocal of a number is 1 divided by that number. Thus, the reciprocal of 2 is ½, or 0.5; the reciprocal of 4 is ¼, or 0.25; and the reciprocal of 10 is 1/10, or 0.1.

To use the reciprocal method, you first calculate the reciprocal of the value of each resistance. Then add these reciprocals to get the total reciprocal. And finally, find the reciprocal of the total reciprocal. The reciprocal method is described by the equation:

$$R_{TOT} = \cfrac{1}{\cfrac{1}{R_1} + \cfrac{1}{R_2} + \cfrac{1}{R_3} + \ldots + \text{etc.}}$$

calculating total resistance from current and voltage

The various methods you know to calculate the effective resistance of parallel circuits involve only resistance values. However, you can also use Ohm's Law to find the effective resistance of parallel circuits. If the total circuit current and the applied voltage are known, the total resistance can easily be found using Ohm's Law in the form:

$$R_{TOT} = E/I_{TOT}$$

where E is the source voltage and I_{TOT} is the total current.

If only the source voltage is known, the total current can be found by first calculating each of the branch currents, and then adding them. Ohm's Law is then used to find the total resistance.

When the source voltage and total circuit current are both *not* known, Ohm's Law still can be used to find the effective resistance of a parallel circuit. Assume any source voltage you want, and calculate hypothetical branch currents with this voltage. By adding these branch currents, you find the total hypothetical current. Using the assumed voltage and the total current that you found, use the equation $R_{TOT} = E/I_{ASSUMED}$ to find the total resistance.

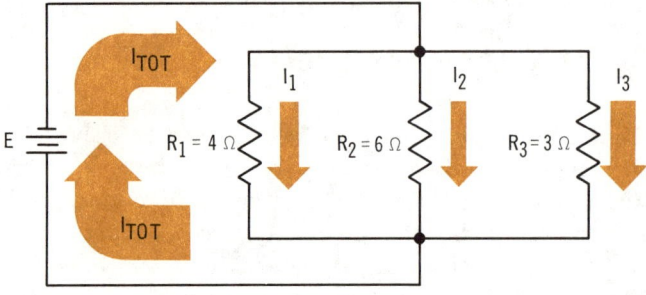

For the circuit, assume a source voltage of 12 volts. The voltage across each branch is, therefore, also assumed to be 12 volts. Therefore:

$$I_{TOT} = E/R_1 + E/R_2 + E/R_3 = 3 + 2 + 4 = 9 \quad \text{amperes}$$

Then the total resistance can be found by:

$$R_{TOT} = E_{ASSUMED}/I_{TOT} = 12 \text{ volts}/9 \text{ amperes} = 1.33 \text{ ohms}$$

You should obtain the same result by using the reciprocal method.

summary

☐ A parallel circuit is one in which there are one or more points where the current divides and follows different paths. ☐ Parallel loads are connected in such a way that the total current is divided between the loads. ☐ The entire source voltage is dropped across each parallel load. ☐ The separate paths in a parallel circuit are called branches. ☐ The current flowing in each branch is called the branch current. ☐ The total current of a parallel circuit is equal to the sum of the branch currents. ☐ Adding parallel loads to a circuit decreases the total circuit resistance, and thus increases total current.

☐ The total resistance of any number of parallel resistances all having the same value is equal to the value of one resistance divided by the number of resistances. ☐ The total resistance of two unequal resistances in parallel can be found by the product/sum method: $R_{TOT} = (R_1 \times R_2)/(R_1 + R_2)$.

☐ The reciprocal method for finding total resistance can be used for any number of resistances, equal or unequal:

$$R_{TOT} = \frac{1}{1/R_1 + 1/R_2 + 1/R_3 + \ldots +}$$

☐ When the source voltage and total current are not known, the total resistance can be found by assuming a convenient source voltage, solving for the total assumed current, and then solving for the total resistance by $R_{TOT} = E_{ASSUMED} \div I_{TOT}$.

review questions

1. What is a parallel circuit?
2. Find R_{TOT} for six 10-ohm resistors in parallel.
3. Find R_{TOT} for six 10-ohm resistors in series.
4. Doubling the source voltage of a parallel circuit has what effect on the branch currents? On the total circuit resistance?
5. What is the *reciprocal method* equation for finding R_{TOT}?
6. What is the *product/sum* equation for total resistance?
7. What equation should be used for finding the equivalent resistance of two equal resistors in parallel? Of two unequal resistors?
8. When can the reciprocal method be used for finding R_{TOT}?
9. A four-branch parallel circuit has branch currents of 2, 3, 5, and 10 amperes. What is the total current?
10. If the resistance of the 2-ampere branch of Question 9 is 10 ohms, what are the values of the other resistances, and the total resistance?

parallel power sources

Remember that a parallel circuit is a circuit that has more than one path for current flow. So far, the only types of parallel circuits you have studied are those having parallel loads. In these circuits, the current leaving the power source is the total circuit current, and this current is divided between the branches of the circuit.

Another type of parallel circuit is one that has parallel power sources. In this type circuit, each power source supplies *part* of the current that flows through the load. Thus, the current through the load is the *total circuit current,* and the current through each power source is *branch current.*

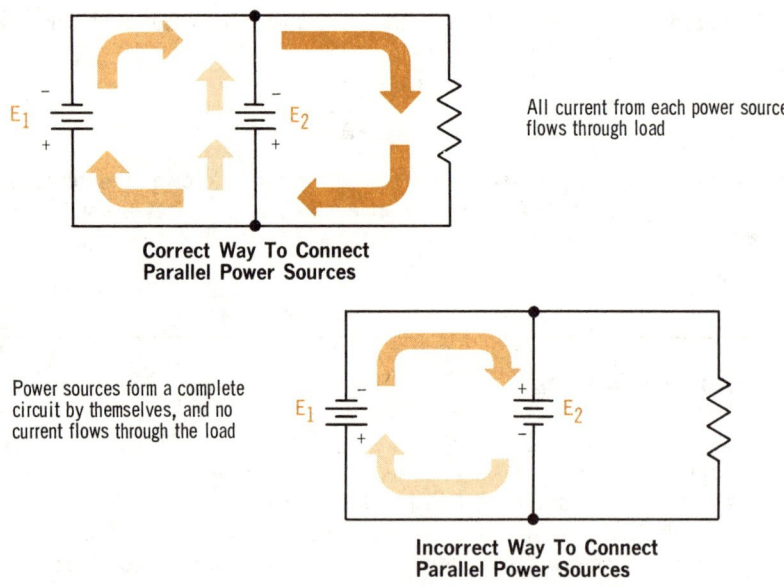

All current from each power source flows through load

**Correct Way To Connect
Parallel Power Sources**

Power sources form a complete circuit by themselves, and no current flows through the load

**Incorrect Way To Connect
Parallel Power Sources**

Each power source connected in parallel supplies part of the total circuit current. This current may be supplied to only one load, or it may be divided among parallel loads

Circuits that have parallel power sources can also have parallel loads. In these circuits, the total current is the *sum* of the currents flowing in the branches.

Only power sources with the *same output voltage* should ever be connected in parallel. If sources with different voltages were put in parallel, some current would flow from the source with the higher voltage into the one with the lower voltage. This would be, *wasted* current, since it would not flow through the load.

effect of parallel power sources on current

When two power sources with the same output voltage are connected in parallel, the voltage output of the parallel combination is the *same* as the *individual* sources. There is no increase or decrease in voltage. Therefore, the current through the loads, whether there are one or more of them, is the same as if only one power source was used. Why bother putting power sources in parallel? One answer is that very often power sources cannot supply the *total* current needed by a circuit. But by putting sources in parallel, each only has to supply a *portion* of the circuit current. For example, consider a load of 2 ohms that needs a current of 10 amperes to operate properly. From Ohm's Law, the voltage required:

$$E = IR = 10 \text{ amperes} \times 5 \text{ ohms} = 50 \text{ volts}$$

But suppose that the only 50-volt power sources available can supply 8 amperes at most. In this case, two 50-volt power sources would have to be connected in parallel.

Batteries are often connected in parallel to *extend* their *life*. When current is drawn from a battery, the battery discharges. And the more current that is drawn, the quicker the battery discharges. When batteries are put in parallel, each supplies only a portion of the circuit current. They, therefore, discharge more slowly, and last longer. Power sources and their current limitations are covered in Volume 6.

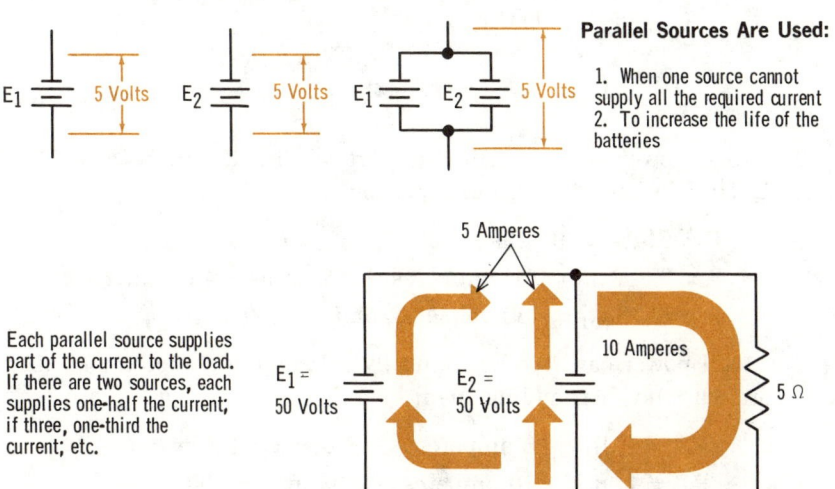

E_1 5 Volts E_2 5 Volts E_1 E_2 5 Volts

Parallel Sources Are Used:

1. When one source cannot supply all the required current
2. To increase the life of the batteries

Each parallel source supplies part of the current to the load. If there are two sources, each supplies one-half the current; if three, one-third the current; etc.

5 Amperes

$E_1 =$ 50 Volts $E_2 =$ 50 Volts   10 Amperes   5 Ω

power consumption

You have learned that in a series circuit, the total power consumed is equal to the sum of the power used by each individual load. You also know that the total power can be calculated directly, using the values of the total circuit current, total circuit resistance, and source voltage, if any two of these values are known. These *same* relationships are true for power in a parallel circuit. It can be calculated *directly* from total current, total resistance, and source voltage; or it can be found *indirectly* by taking the sum of the individual power consumptions of the loads.

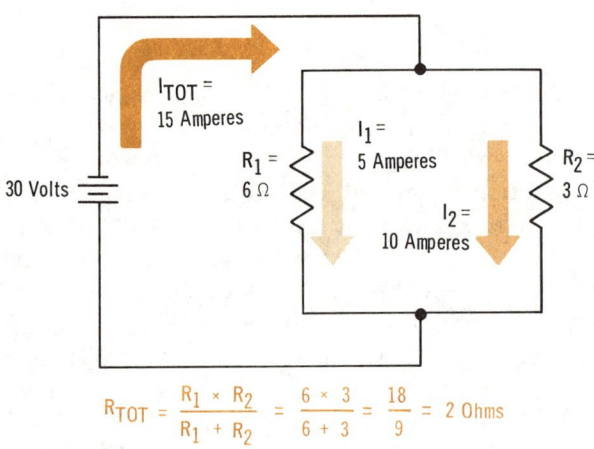

$$R_{TOT} = \frac{R_1 \times R_2}{R_1 + R_2} = \frac{6 \times 3}{6 + 3} = \frac{18}{9} = 2 \text{ Ohms}$$

The total power consumed in the parallel circuit can be calculated directly from either of the three equations:

$$P = EI_{TOT} = 30 \text{ volts} \times 15 \text{ amperes} = 450 \text{ watts}$$
$$P = I^2_{TOT} R_{TOT} = (15 \text{ amperes})^2 \times 2 \text{ ohms} = 450 \text{ watts}$$
$$P = E^2/R_{TOT} = (30 \text{ volts})^2/2 \text{ ohms} = 450 \text{ watts}$$

The total power can also be found by calculating the power used in each branch, and then adding them:

$$P_{R1} = I_1^2 R_1 = (5 \text{ amperes})^2 \times 6 \text{ ohms} = 150 \text{ watts}$$
$$P_{R2} = I_2^2 R_2 = (10 \text{ amperes})^2 \times 3 \text{ ohms} = 300 \text{ watts}$$
$$P_{TOT} = P_{R1} + P_{R2} = 150 + 300 = 450 \text{ watts}$$

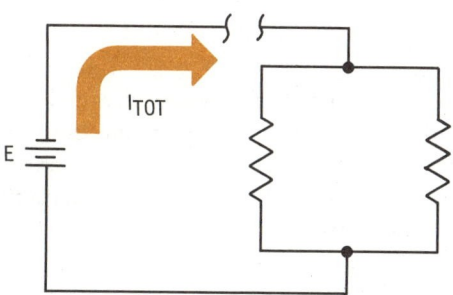

Breaking a parallel circuit where the total current flows opens the entire circuit, stopping all current flow. With no current, there is no power consumed in the circuit and no voltage drops across the loads

Breaking a parallel circuit where branch current flows opens only that branch. Total current and current in the other branches still flow. The values of the total currents, however, will change. The reason is that when a branch is opened, its resistance is no longer part of the circuit, and the total resistance is increased

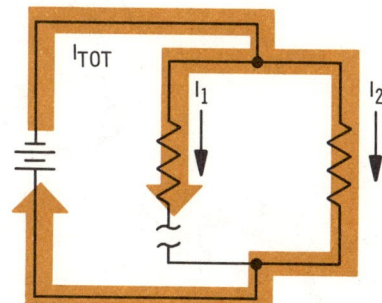

open circuits

If a series circuit is broken at any point, no current flows. The reason for this is that there is only one path for current flow in a series circuit, and that path must be complete or the circuit is open. A parallel circuit, however, has more than one current path. Thus, even if one of the paths is opened, current will still flow in the circuit as long as one or more of the other paths provide a *complete* circuit from the negative terminal of the power source to the positive terminal. This does not mean, however, that you cannot stop current flow in a parallel circuit by opening it at one point. What it does mean is that the behavior of a parallel circuit that is open at some point depends on where the opening, or break, is.

If the circuit is opened at a point through which *total* circuit current flows, the entire circuit is open, and all current flow stops. If, however, it is opened at a point where only a *branch* current flows, then only that branch is open, and current continues to flow in the rest of the circuit. You can see from this that for a fuse to do its job in a parallel circuit, it must be connected at a place where total circuit current flows, or else each branch must have a fuse.

short circuits

When a parallel circuit becomes shorted, the same effects happen that occur when a series circuit is shorted. You remember that these effects include a sudden and very large increase in circuit current, heating of the connecting wires and possible burning of the insulation, and the possible burning out of the power source.

If any load in a parallel circuit becomes shorted, the resistance of the circuit drops to practically zero. The reason is that each load is connected across the power source terminals. This can be shown by the equation:

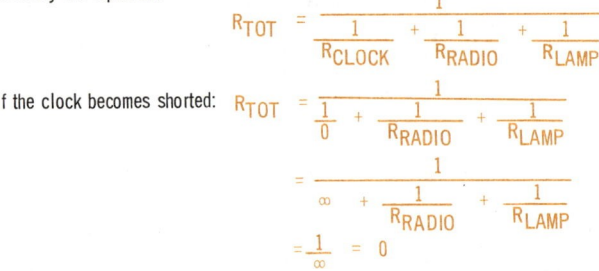

$$R_{TOT} = \cfrac{1}{\cfrac{1}{R_{CLOCK}} + \cfrac{1}{R_{RADIO}} + \cfrac{1}{R_{LAMP}}}$$

If the clock becomes shorted:

$$R_{TOT} = \cfrac{1}{\cfrac{1}{0} + \cfrac{1}{R_{RADIO}} + \cfrac{1}{R_{LAMP}}}$$

$$= \cfrac{1}{\infty + \cfrac{1}{R_{RADIO}} + \cfrac{1}{R_{LAMP}}}$$

$$= \frac{1}{\infty} = 0$$

Parallel circuits are usually more likely to develop damaging short circuits than are series circuits. The reason for this is that each parallel load is connected *directly* between the power source terminals. And so if any one of the loads becomes shorted, it drops the resistance between the power source terminals to practically zero. But if a series load becomes shorted, the resistance of the other loads in series with it keep the circuit resistance from dropping to zero.

solved problems

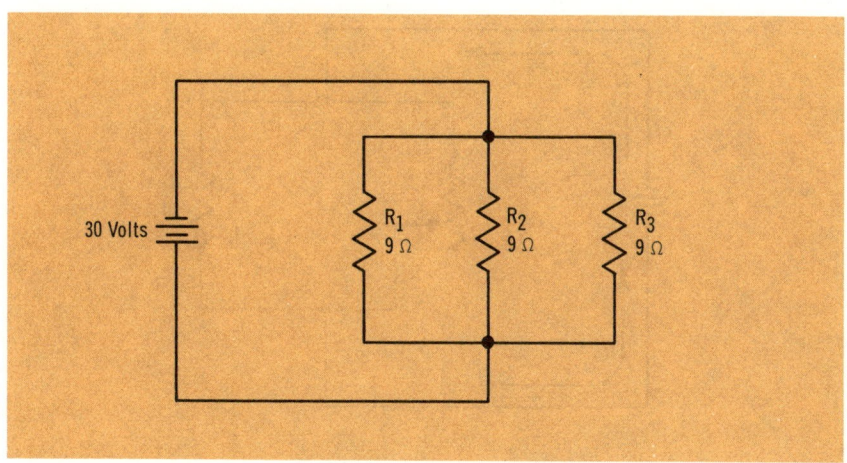

Problem 9. *What is the total current in this circuit?*

One way to find the total current would be to calculate the three branch currents and then add their values. However, since the three resistances are equal, it is easier in this case to find the total resistance and then calculate the total current from that. The total resistance of equal parallel resistances is given by:

$$R_{TOT} = \frac{\text{value of one resistance}}{\text{number of resistances}} = \frac{9}{3} = 3 \text{ ohms}$$

The total current can be found using Ohm's Law:

$$I_{TOT} = E/R_{TOT} = 30 \text{ volts}/3 \text{ ohms} = 10 \text{ amperes}$$

Problem 10. *What is the current through R_2?*

R_2 is one of the branches in the circuit since it provides a separate path for current. And you know that the current in a branch depends on the resistance of the branch and the voltage across it. So, using Ohm's Law:

$$I_2 = E/R_2 = 30 \text{ volts}/9 \text{ ohms} = 3.33 \text{ amperes}$$

This problem could also have been solved by just looking at the circuit. Since the three branch resistances are equal the total current must divide equally between them. That means one-third of the total current flows through each branch. With the total current being 10 amperes, the branch current through R_2 is one-third of 10, or 3.33 amperes.

solved problems (cont.)

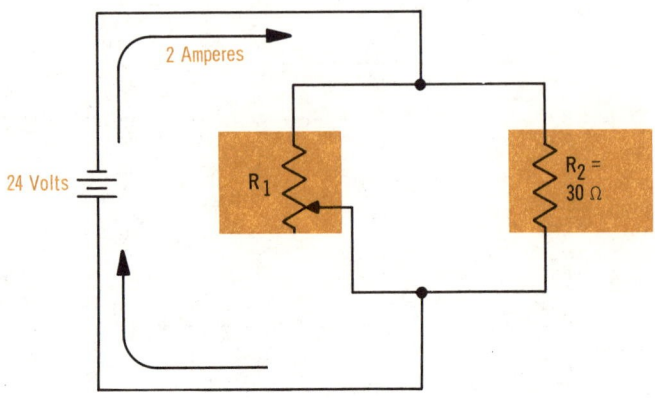

Problem 11.   What is the total resistance in the circuit?

Since both branch resistances are not known, you cannot use the reciprocal method or the product/sum method to find the total resistance. But the applied voltage and the total current are known, and so you can use Ohm's Law to calculate the total resistance:

$$R_{TOT} = E/I_{TOT} = 24 \text{ volts}/2 \text{ amperes} = 12 \text{ ohms}$$

Problem 12.   If R_1 was set to have a value of 60 ohms, what would the total current be then?

The total resistance of the parallel combination has to be found before the total current can be calculated. There are two unequal resistances, so the product/sum method can be used:

$$R_{TOT} = (R_1 \times R_2)/(R_1 + R_2) = (60 \times 30)/(60 + 30)$$
$$= 1800/90 = 20 \text{ ohms}$$

With the total resistance and the applied voltage known, the total current can then be calculated by Ohm's Law:

$$I_{TOT} = E/R_{TOT} = 24 \text{ volts}/20 \text{ ohms} = 1.2 \text{ amperes}$$

Problem 13.   How much of this total current of 1.2 amperes flows through R_2?

The voltage across R_2 is 24 volts and its resistance is 30 ohms. The current through R_2 is, therefore, found by the equation:

$$I = E/R_2 = 24 \text{ volts}/30 \text{ ohms} = 0.8 \text{ ampere}$$

solved problems (cont.)

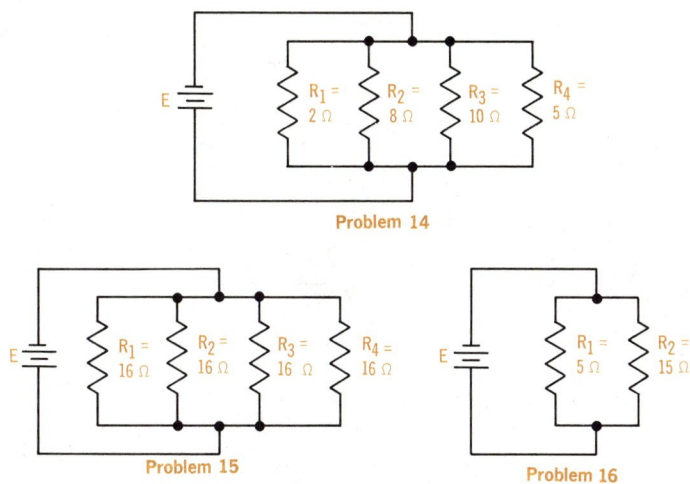

Problem 14

Problem 15

Problem 16

Problem 14. **What is the total resistance of the circuit?**

There are more than two resistances in this parallel combination, so the reciprocal method is used:

$$R_{TOT} = \cfrac{1}{\cfrac{1}{R_1} + \cfrac{1}{R_2} + \cfrac{1}{R_3} + \cfrac{1}{R_4}} = \cfrac{1}{\cfrac{1}{2} + \cfrac{1}{8} + \cfrac{1}{10} + \cfrac{1}{5}}$$

$$= \cfrac{1}{\cfrac{20}{40} + \cfrac{5}{40} + \cfrac{4}{40} + \cfrac{8}{40}} = \cfrac{1}{\cfrac{37}{40}} = 40/37 = 1\text{-}3/37 \text{ ohms}$$

Problem 15. **What is the total resistance of the circuit?**

The reciprocal method could be used for this circuit, but since all the resistances have the same value, it is much easier to use the equal resistance method. Thus,

$$R_{TOT} = \frac{\text{value of one resistance}}{\text{number of resistances}} = \frac{16}{4} = 4 \text{ ohms}$$

Problem 16. **What is the total resistance of the circuit?**

The product/sum method can be used here since there are two resistances with different values:

$$R_{TOT} = (R_1 \times R_2)/(R_1 + R_2) = (5 \times 15)/(5 + 15) = 3.75 \text{ ohms}$$

This problem can also be solved by the equal resistance rule if R_1 is converted to three 15-ohm resistors. Then, there would be a total of four 15-ohm resistors in the circuit, so 15/4 equals 3.75 ohms.

solved problems (cont.)

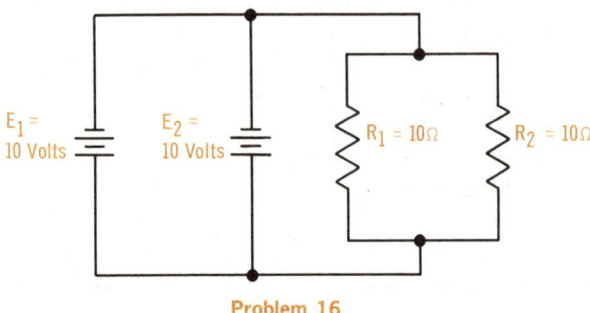

Problem 16

Problem 17. **What is the total current in this circuit?**

By now, you should be able to just look at these two parallel resistances and mentally calculate their total value. They are equal, and so their total value is one-half the value of either, or 5 ohms.

Once you know the total resistance, all you need is the source voltage and you can calculate the total current from Ohm's Law. You remember that if equal power sources are connected in parallel, their combined output voltage is the same as their individual voltages. Thus, the source voltage for this circuit is 10 volts. The total circuit current is, then:

$$I_{TOT} = E/R_{TOT} = 10 \text{ volts}/5 \text{ ohms} = 2 \text{ amperes}$$

And since parallel power sources each supply only a part of the total circuit current, each battery in this circuit is supplying 1 ampere.

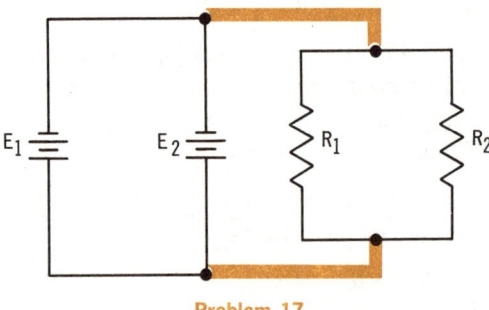

Problem 17

Problem 18. **Where would a fuse have to be connected in the above circuit so that it could open the entire circuit if it blew?**

The fuse would have to be connected at some point where total circuit current flowed. The path of the total current is shown by the colored lines in the circuit.

comparison of series and parallel circuits

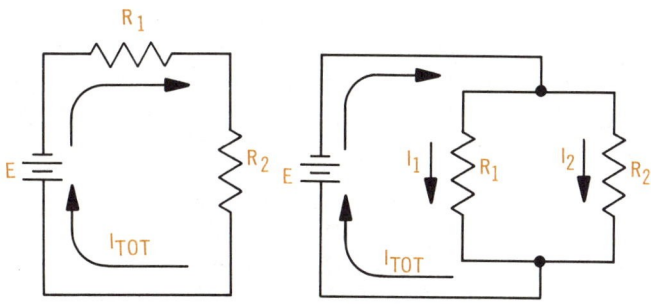

	Series Circuit	Parallel Circuit
Current	There is only one path for current to flow.	There is more than one path for current to flow.
	The current at every point in the circuit is the same.	The total current is equal to the sum of the branch currents.
Voltage	The sum of the voltage drops across the individual loads is equal to the source voltage.	The voltage across each branch is the same as the source voltage.
Resistance	The total resistance is equal to the sum of the individual resistances.	The total resistance is equal to the reciprocal of the sum of the reciprocals of the individual resistances.
Power	The total power consumed is equal to the sum of the power consumptions of the individual loads.	

summary

☐ When power sources are connected in parallel, each source supplies part of the circuit current. ☐ Only power sources with the same output voltage should ever be connected in parallel. ☐ When two power sources with the same output voltage are connected in parallel, the voltage output of the parallel combination is the same as the individual sources.

☐ Total power consumption in a parallel circuit is found the same way as for a series circuit: $P_{TOT} = EI_{TOT}$, $P_{TOT} = I^2_{TOT}/R_{TOT}$, and $P_{TOT} = E^2/R_{TOT}$. ☐ The total power is equal to the sum of the powers consumed in the branch circuits: $P_{TOT} = P_1 + P_2 + P_3 + \ldots + $ etc.

☐ In a parallel circuit, even if one of the current paths is opened, current will still flow in the circuit as long as one or more of the other paths provides a complete circuit. ☐ If a parallel circuit is opened at a point through which the total circuit current flows, the entire circuit is open, and all current flow stops. ☐ For a fuse to protect a parallel circuit, it should be placed in the circuit where total current flows, or else each branch must have a fuse.

review questions

1. Why shouldn't power sources having different output voltages be connected in parallel?
2. If three 10-volt batteries are connected in parallel, what is their total output voltage?

For Questions 3 to 8, consider a circuit consisting of three parallel resistors, whose values are 20, 50, and 80 ohms, connected across a 150-volt battery.

3. What is the total resistance of the circuit?
4. What is the total current and the branch currents?
5. What is the total power supplied by the battery, and what is the power consumed by each resistor?
6. Where should a fuse be placed to protect the circuit?
7. Would a 2-ampere fuse protect the battery against a short?
8. If the 20-ohm resistor became open-circuited, what would be the value of the total circuit current?
9. A circuit consists of ten equal-value resistors in parallel. If the total power dissipated by the circuit is 40 watts, and if the circuit is powered by a 10-volt battery, what is the value of each resistor?
10. For the circuit of Question 9, if the battery output falls to 5 volts, what is the total power dissipated by the circuit?

series-parallel circuits

From the material given so far, it should be easy for you to recognize both series and parallel circuits. But there is another type of circuit that has branches, like parallel circuits, and series loads or components,

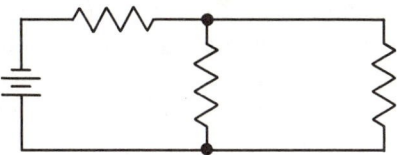

This is a Series-Parallel Circuit

Series-parallel circuits are a combination of both series circuits and parallel circuits. They can be fairly simple and have only a few components, but they can also have many components and be quite complicated

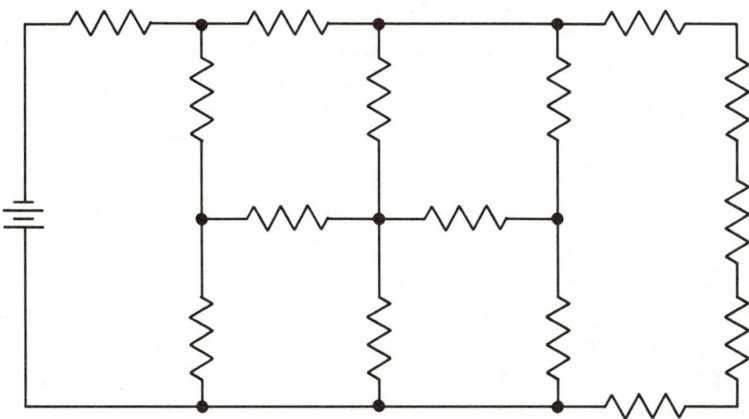

And This is a Series-Parallel Circuit

like series circuits. This is called a *series-parallel* circuit since it is a combination of the others. The methods you will use to analyze series-parallel circuits are mostly combinations of those you already know for series circuits and parallel circuits.

analyzing series-parallel circuits

In any d-c circuit, there are certain basic factors in which you will be interested. From what you have learned about series and parallel circuits, you know that these factors are (1) the total current from the power source and the current in each part of the circuit, (2) the source voltage and the voltage drops across each part of the circuit, and (3) the total resistance and the resistance of each part of the circuit. Once you know these circuit factors, you can easily calculate others, such as total power or the power consumed in some part of the circuit.

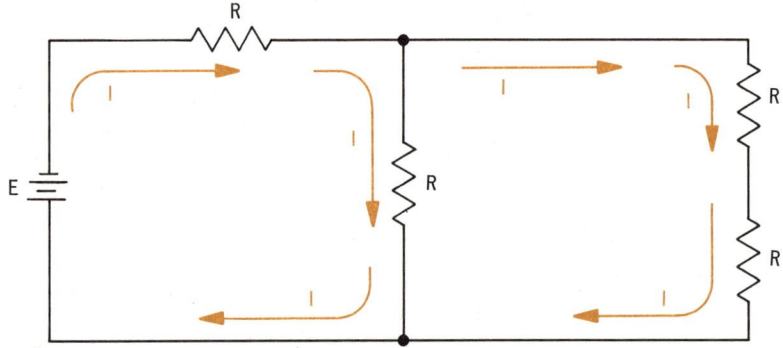

Every branch and every load in a series-parallel circuit has current, voltage and resistance. The circuit has also total current, total voltage, and total resistance. To find any of these, you have to use the rules for series circuits, as well as those for parallel circuits

To find the various currents, voltages, and resistances in series and parallel circuits is fairly easy. You know the rules of series and parallel circuits; and when working with either type, you use only the rules that apply to that type. In a series-parallel circuit, on the other hand, some parts of the circuit are *series-connected* and some parts are *parallel-connected*. Thus, in some parts of a series-parallel circuit, you have to use the rules for series circuits, and in other parts the rules for parallel circuits apply.

You can see then, that before you can analyze or solve a problem involving a series-parallel circuit you have to be able to *recognize* which parts of the circuit are series-connected and which parts are parallel-connected. Sometimes this is obvious, if the circuit is simple. Many times, however, you will have to *redraw* the circuit, putting it into a form that is easier for you to recognize.

redrawing series-parallel circuits

In a series circuit, the current is the same at all points. In a parallel circuit, there are one or more points where the current divides and flows in separate branches. And in a series-parallel circuit, there are both separate branches and series loads. You can see, then, that the easiest way to find out whether a circuit is a series, parallel, or series-parallel circuit is to start at the negative terminal of the power source and *trace* the path of current through the circuit and back to the positive terminal of the power source. If the current does not divide anywhere, it is a series circuit. If the current does divide into separate branches, but there are no series loads, it is a parallel circuit. And if the current divides into separate branches and there are also series loads, it is a series-parallel circuit. When tracing the circuit in this way, remember that there are *two types* of series loads. One type consists of two or more resistances in one branch of the circuit. The other type is any resistance through which the total circuit current flows. You can see these two types in the illustration.

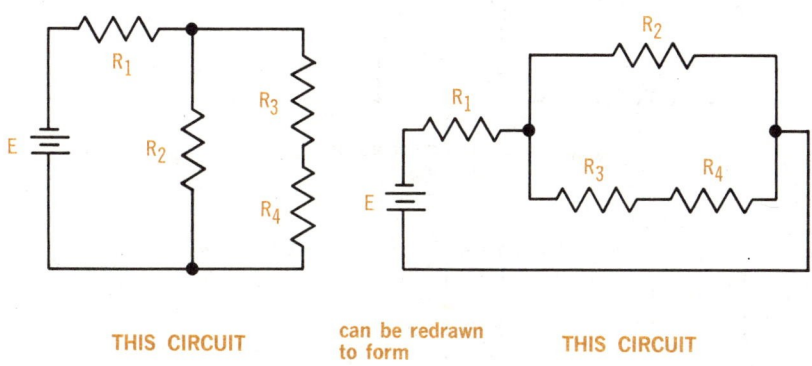

THIS CIRCUIT   can be redrawn to form   THIS CIRCUIT

The analysis of a series-parallel circuit can be often simplified if the circuit is redrawn to that the branches and series loads can be recognized quickly

Often after determining that a circuit is series-parallel, you will find it very helpful to redraw the circuit so that the branches and the series loads are more easily recognized. This will be especially helpful when you have to find the total resistance of the circuit. Examples of how circuits can be redrawn are shown on this and the next page.

redrawing
series-parallel circuits (cont.)

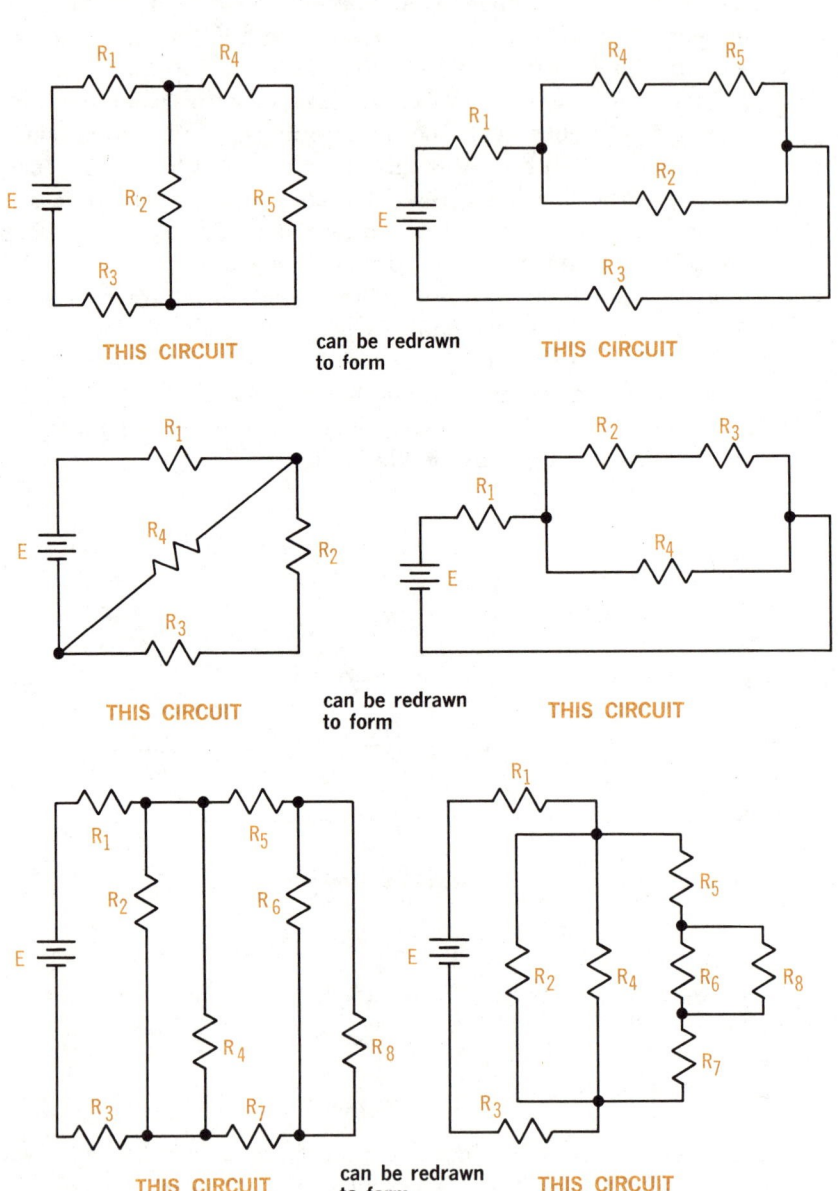

THIS CIRCUIT   can be redrawn to form   THIS CIRCUIT

THIS CIRCUIT   can be redrawn to form   THIS CIRCUIT

THIS CIRCUIT   can be redrawn to form   THIS CIRCUIT

reducing series-parallel circuits

Very often, all that is known about a series-parallel circuit is the applied voltage and the values of the individual resistances. To find the voltage drop across any of the loads or the current in any of the branches, you usually have to know the total circuit current. But to find the total current you have to first know the total resistance of the circuit. To find the total resistance, you reduce the circuit to its *simplest*

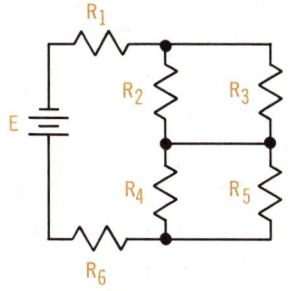

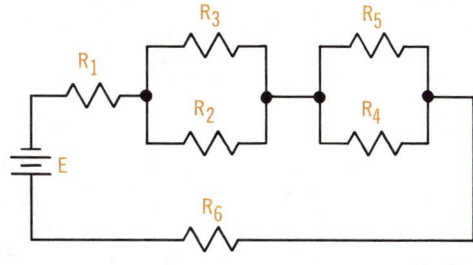

1. If necessary, redraw the circuit so that all parallel combinations of resistances and series resistances are easily recognized

2. For each parallel combination of resistances, calculate its effective resistance

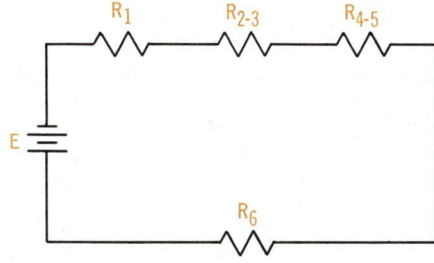

3. Replace each of the parallel combinations with one resistance, whose value is equal to the effective resistance of that combination. This gives you a circuit with all series loads

4. Find the total resistance of this circuit by adding the resistances of all the series loads

form, which is usually *one resistance* that forms a series circuit with the *voltage source*. This simple series circuit has the equivalent resistance of the series-parallel circuit it was derived from, and so also has the same total current. There are four basic steps in reducing a series-parallel circuit.

reducing
series-parallel circuits (cont.)

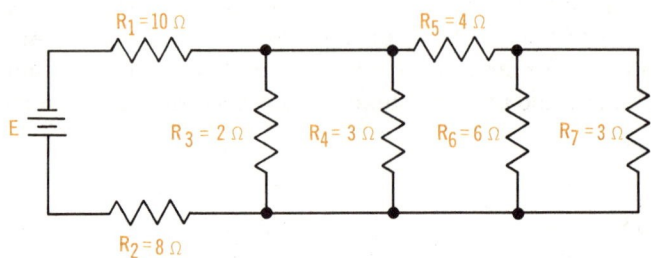

What is the simplest equivalent series circuit for this series-parallel circuit? The first step is to redraw the circuit so that you can easily see the parallel portions and the series portions. (The following figure references are to diagrams on page 2-105.) When this is done, you can see that there are three main branches: one is R_3, the second is R_4, and the third is R_5 and the parallel combination of R_6 and R_7. You have to reduce the three branches so they have one resistance each. (See Fig. A.)

Start by finding the effective resistance of R_6 and R_7, and putting this in the circuit in place of R_6 and R_7 (See Fig. B):

$$R_{6\text{-}7} = (R_6 \times R_7)/(R_6 + R_7) = (6 \times 3)/(6 + 3) = 18/9 = 2 \ \text{ohms}$$

You now have two series resistances in the third branch. So you reduce this branch to one resistance by adding the two series resistances and replacing them with a single resistance whose value is equal to their sum. (See Fig. C.)

The three branches now contain one resistance each. These three parallel resistances can be replaced with one resistance, whose value can be found by using the reciprocal method. (See Fig. D.)

$$R_{3\text{-}4\text{-}5\text{-}6\text{-}7} = \cfrac{1}{\cfrac{1}{R_3} + \cfrac{1}{R_4} + \cfrac{1}{R_{5\text{-}6\text{-}7}}} = \cfrac{1}{\cfrac{1}{2} + \cfrac{1}{3} + \cfrac{1}{6}}$$

$$= \cfrac{1}{\cfrac{3}{6} + \cfrac{2}{6} + \cfrac{1}{6}} = \cfrac{1}{\cfrac{6}{6}} = 1 \ \text{ohm}$$

The circuit has now been reduced to a series circuit having three series resistances. It can be reduced further to only one resistance by adding the values of the three series resistances. (See Fig. E.)

reducing
series-parallel circuits (cont.)

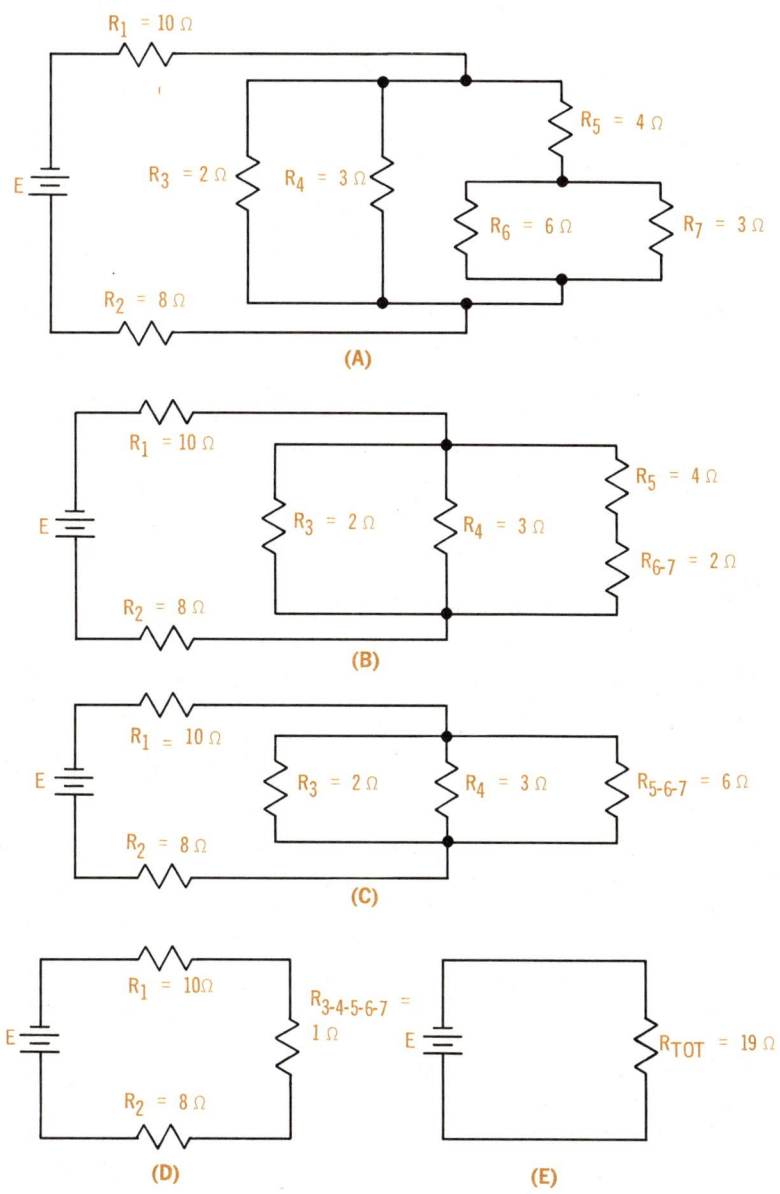

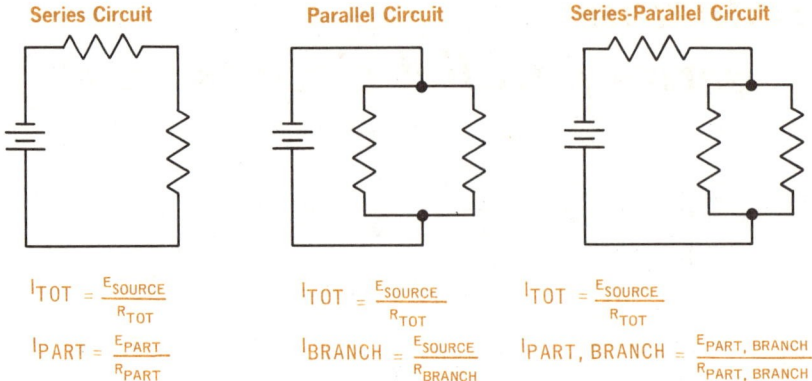

Series Circuit

$$I_{TOT} = \frac{E_{SOURCE}}{R_{TOT}}$$

$$I_{PART} = \frac{E_{PART}}{R_{PART}}$$

Parallel Circuit

$$I_{TOT} = \frac{E_{SOURCE}}{R_{TOT}}$$

$$I_{BRANCH} = \frac{E_{SOURCE}}{R_{BRANCH}}$$

Series-Parallel Circuit

$$I_{TOT} = \frac{E_{SOURCE}}{R_{TOT}}$$

$$I_{PART, BRANCH} = \frac{E_{PART, BRANCH}}{R_{PART, BRANCH}}$$

current

In any d-c circuit, the total current is equal to the power source voltage divided by the total resistance. For series circuits, this is the only current. Therefore, if you know the total current, you also *know* the current through every part of the circuit. In parallel circuits, the current divides and follows more than one path. Therefore, if you just know the total circuit current, you *do not know* the current in every part of the circuit.

Branch currents are usually calculated by applying Ohm's Law to the voltage across the branch and the branch resistance. In series-parallel circuits, the current also divides, following more than one path. So, like parallel circuits, the branch currents have to be found using Ohm's Law. There is an important difference, however. In both cases you use Ohm's Law in the form $I = E/R$. But for parallel circuits, the voltage across every branch in the circuit is the same, and is equal to the source voltage. Therefore, if the source voltage and branch resistance are known, all branch currents can be found.

In series-parallel circuits, the voltage across every branch is usually *not* the same. So, very often, the voltage must be calculated before the branch current can be found. You may think from this that if you calculated all of the circuit voltages first, you could then use them to find all the currents in the circuit. This is *not* the case, though. The approach you will usually use is to first find the total current in the circuit. Then, using the current, you will calculate the voltage across some part or branch of the circuit. With this voltage you will calculate the current through that part or branch. You will then use the current you find to calculate the voltage across some other part or branch. Using this method, you will eventually find all of the currents and voltages in the circuit.

voltage

In a series circuit, the sum of the voltage drops around the circuit equals the applied voltage. And in a parallel circuit, the voltage across each branch is the same as the applied voltage. No such simple relationship exists for a series-parallel circuit between the applied voltage and the voltages throughout the circuit. However, you know that the voltage dropped across any resistance or group of resistances is equal to the current through the resistance times the value of the resistance. This relationship is true in any d-c circuit, whether series, parallel, or series-parallel. And generally, this is the method used to find voltages in a series-parallel circuit.

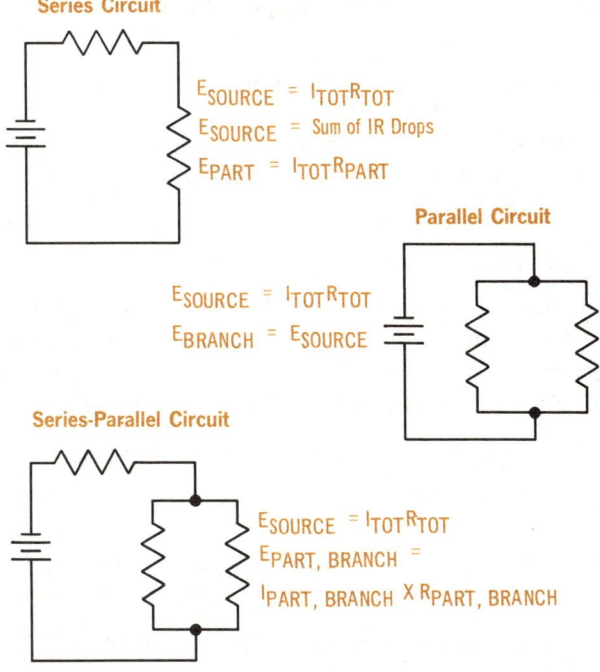

Series Circuit

$$E_{SOURCE} = I_{TOT}R_{TOT}$$
$$E_{SOURCE} = \text{Sum of IR Drops}$$
$$E_{PART} = I_{TOT}R_{PART}$$

Parallel Circuit

$$E_{SOURCE} = I_{TOT}R_{TOT}$$
$$E_{BRANCH} = E_{SOURCE}$$

Series-Parallel Circuit

$$E_{SOURCE} = I_{TOT}R_{TOT}$$
$$E_{PART, BRANCH} = I_{PART, BRANCH} \times R_{PART, BRANCH}$$

Remember that you usually cannot calculate all of the currents or all of the voltages in a series-parallel circuit by using only the total current and applied voltage. You have to work around the circuit load by load and branch by branch, finding the current through and voltage across each load or branch before moving on to the next. Of course, as you acquire more experience and practice you will develop shortcuts that will enable you to eliminate some of the work.

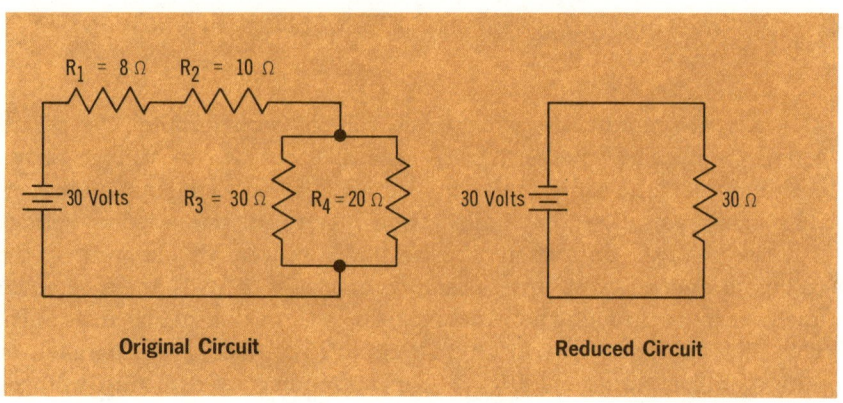

Original Circuit   Reduced Circuit

calculating voltage and current

In the series-parallel circuit, calculate the current through the resistances and voltage drops across them. First reduce the circuit to its simplest form. Start by reducing the parallel combination of R_3 and R_4 to one equivalent resistance by the product/sum method:

$$R_{3\text{-}4} = (R_3 \times R_4)/(R_3 + R_4) = (30 \times 20)/(30 + 20)$$
$$= 600/50 = 12 \text{ ohms}$$

Now the original circuit has been reduced to a series circuit with three resistances: 8, 10, and 12 ohms. The completely reduced circuit, then, has one resistance of 30 ohms.

The total current in this circuit can be found by using Ohm's Law:

$$I = E/R = 30 \text{ volts}/30 \text{ ohms} = 1 \text{ ampere}$$

Referring back to the original circuit, you can see that this 1-ampere current flows through resistances R_1 and R_2, and then separates through R_3 and R_4. Since you know the current through R_1 and R_2, calculate the voltage drops across them using Ohm's Law.

$$E_{R1} = IR_1 = 1 \text{ ampere} \times 8 \text{ ohms} = 8 \text{ volts}$$
$$E_{R2} = IR_2 = 1 \text{ ampere} \times 10 \text{ ohms} = 10 \text{ volts}$$

If 8 volts are dropped across R_1 and 10 volts across R_2, 12 volts remain across the combination of R_3 and R_4. The current through each of these can now be found.

$$I_{R3} = E/R_3 = 12 \text{ volts}/30 \text{ ohms} = 0.4 \text{ ampere}$$
$$I_{R4} = E/R_4 = 12 \text{ volts}/20 \text{ ohms} = 0.6 \text{ ampere}$$

You know that both branch currents must add up to the total circuit current of 1 ampere, so you can check your results by adding them.

solved problems

Problem 19. Redraw the circuit so that the series and parallel portions are shown more clearly.

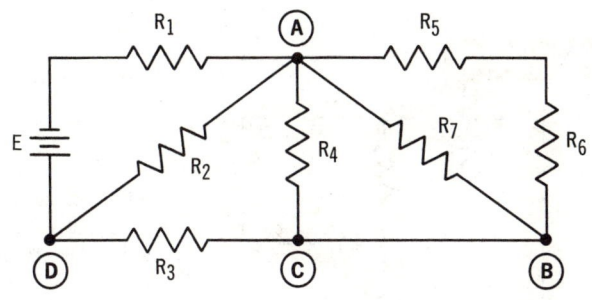

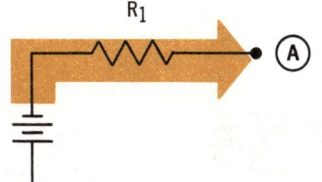

Current leaves the negative terminal of the battery, and this total current flows through R_1 to point A

Current divides at point A, and follows four separate paths. One of these paths is through R_2 to point D, and from there to the positive terminal of the battery

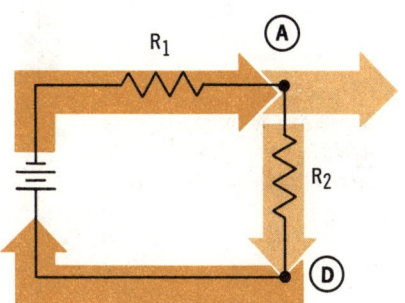

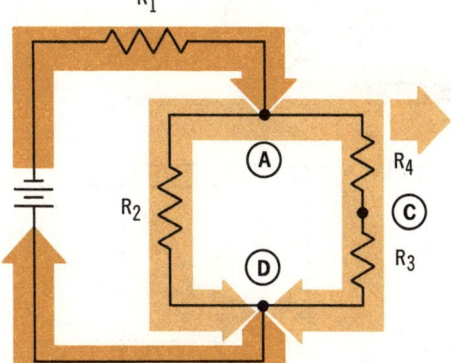

A second path from point A is through R_4 to point C, and through R_3 to point D, and from there to the positive terminal of the battery

The circuit is further simplified on page 2-110.

solved problems (cont.)

The circuit from page 2-109 is further simplified:

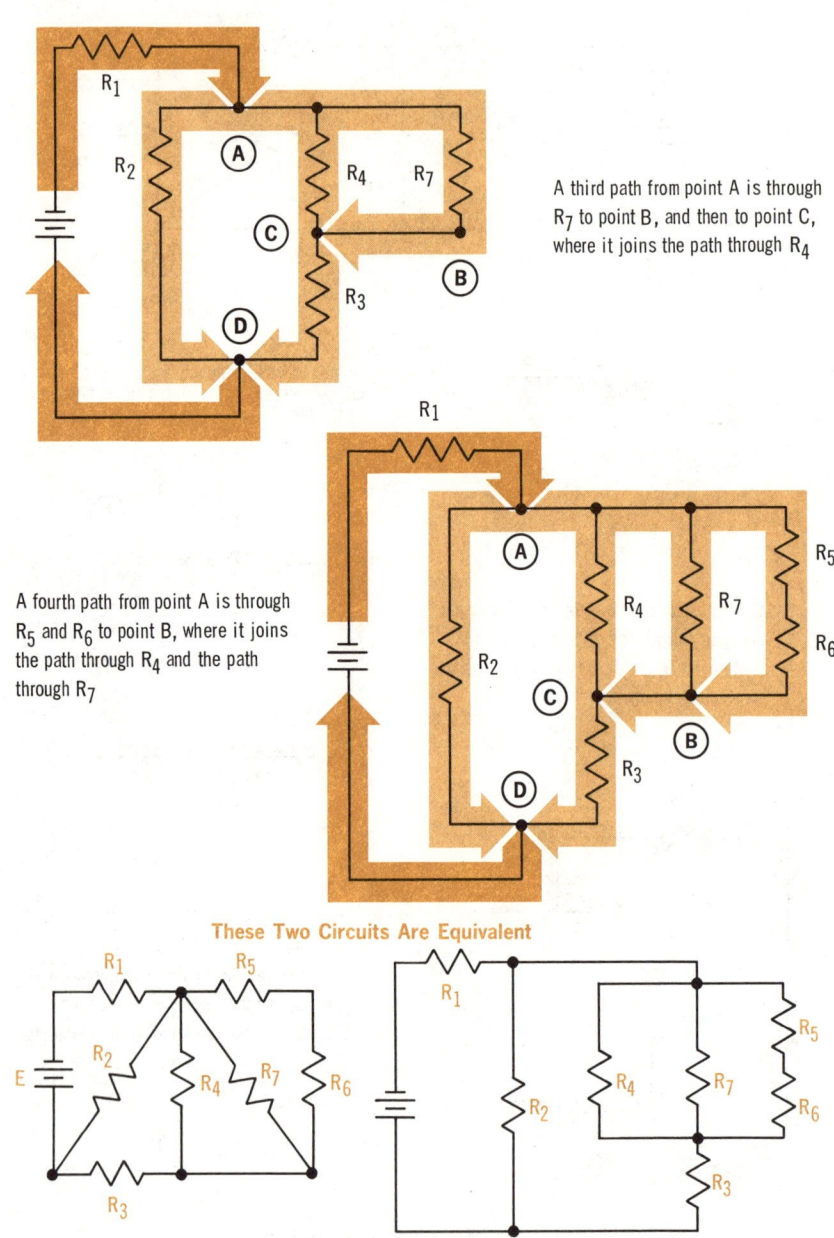

A third path from point A is through R_7 to point B, and then to point C, where it joins the path through R_4

A fourth path from point A is through R_5 and R_6 to point B, where it joins the path through R_4 and the path through R_7

These Two Circuits Are Equivalent

solved problems (cont.)

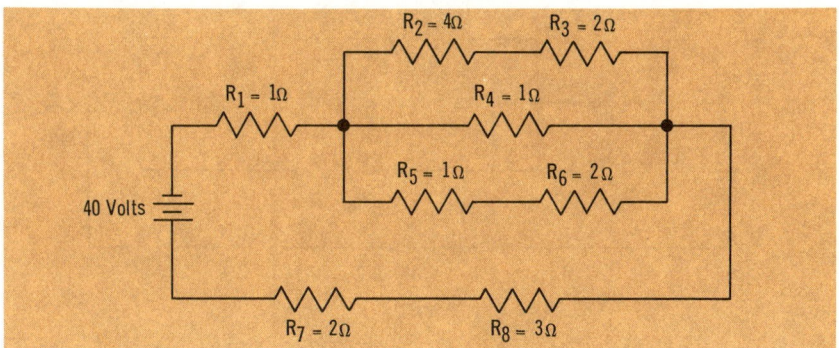

Problem 20. *What is the total current flowing in the circuit?*
To do this, reduce the circuit to only one resistance that is equivalent to the total value of all the series and parallel resistances in the circuit. You can see that there are three branches in the circuit. One of them is through R_2 and R_3, the second is through R_4, and the third is through R_5 and R_6. Each of these branches must first be reduced. R_2 and R_3 are in series; so their total resistance is $4 + 2 = 6$ ohms. R_5 and R_6 are also in series, and their total resistance is, therefore, $1 + 2 = 3$ ohms.

The resistances of the three branches are then 6 ohms, 1 ohm, and 3 ohms, and the effective resistance of the three branches can be found by the reciprocal method for calculating parallel resistances:

$$R_{EFF} = \cfrac{1}{\cfrac{1}{6} + \cfrac{1}{1} + \cfrac{1}{3}} = \cfrac{1}{\cfrac{1}{6} + \cfrac{6}{6} + \cfrac{2}{6}}$$

$$= \cfrac{1}{\cfrac{9}{6}} = 0.667 \text{ ohms}$$

Now replace the parallel resistances with the 0.667-ohm resistance, and the result is a series circuit with four resistances. Since it is a series circuit, the total resistance is 6.667 ohms. This reduces the original circuit to a simple one having 40 volts across 6.667 ohms. The total current, then, is found by Ohm's Law:

$$I = E/R = 40 \text{ volts}/6.667 \text{ ohms} = 6 \text{ amperes}$$

solved problems (*cont.*)

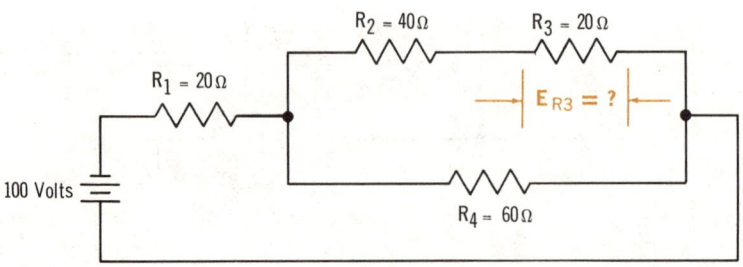

Problem 21.   *What is the voltage drop across R_3 in the circuit?*

To find the voltage drop across R_3, you have to know the current through it. The current is not given, and so you must first calculate the total circuit current, and then find how much of the total current flows through R_3.

Start by reducing the circuit to a single equivalent resistance. There are two parallel branches: one is R_2 and R_3, and the other is R_4. You know the total resistance of the R_4 branch, and you find the total of the R_2–R_3 branch by adding the resistances: $40 + 20 = 60$ ohms. Both branches have total resistances of 60 ohms, so the effective resistance of the parallel combination is one-half the value of either branch, or 30 ohms. The resistance of the entire circuit is the 30 ohms plus the 20 ohms of R_1 or 50 ohms. The total circuit current can be found by Ohm's Law:

$$I_{TOT} = E/R_{TOT} = 100 \text{ volts}/50 \text{ ohms} = 2 \text{ amperes}$$

This total current flows through R_1, and drops some voltage across it.

$$E_{R1} = IR_1 = 2 \text{ amperes} \times 20 \text{ ohms} = 40 \text{ volts}$$

This means that the difference between 40 volts and the power source voltage, or 60 volts, must be across the two parallel branches.

With 60 volts across each branch, the current through the branch containing R_2 and R_3 can be found.

$$I = E/R_{2\text{-}3} = 60 \text{ volts}/60 \text{ ohms} = 1 \text{ ampere}$$

This current flows through both R_2 and R_3, since they are in series. You know the current through R_3, and can therefore calculate the voltage drop across it using Ohm's Law:

$$E_{R3} = IR_3 = 1 \text{ ampere} \times 20 \text{ ohms} = 20 \text{ volts}$$

summary

☐ A series-parallel circuit can be identified by tracing the path of current. If the current divides into separate branches, and there are also series loads, the circuit is a series-parallel circuit. Any resistance through which the total current flows is in the series part of the circuit. ☐ Any resistance through which only part of the total current flows is in the parallel part of the circuit. ☐ In a series-parallel circuit, all the laws for series and parallel circuits are obeyed.

☐ Series-parallel circuits are often reduced to their simplest form to solve for electrical quantities. The simplest form is usually one resistance that forms a series circuit with the voltage source. ☐ Four basic steps in reducing series-parallel circuits are: Redraw the circuit, if necessary, so that all parallel combinations of resistances and series resistances are easily recognized. ☐ Calculate the effective resistance of each parallel combination of resistances. ☐ Replace each parallel combination with one resistance equal to the effective resistance of the combination, resulting in a series circuit. ☐ Find the total resistance of the circuit by adding the resistances of all series loads.

☐ The total voltage dropped around a series-parallel circuit equals the source voltage. ☐ The branch currents through a parallel part of a circuit can be found by calculating the voltage drop across the parallel part, and then applying Ohm's Law. This voltage drop is equal to the total current times the series-equivalent resistance of the parallel part. ☐ The power dissipated by each load of a series-parallel circuit is found by the standard power equations. ☐ The total power is equal to the power developed by the equivalent total resistance of the circuit, or to the sum of the powers developed by each load.

review questions

1. Define *series, parallel,* and *series-parallel circuits.*

For Questions 2 to 10, consider the circuit on page 2-104, with all 4-ohm resistors and a 100-volt power source.

2. What is the total circuit resistance?
3. What is the total circuit current?
4. What is the current through R_2?
5. What is the voltage across R_1?
6. What is the voltage across R_3, and across R_7?
7. What is the current through R_4?
8. What is the current through R_6?
9. What is the power consumed by R_5?
10. What is the total power consumed in the circuit?

law of proportionality

You have learned how to apply Ohm's Law in simple and complex circuits to calculate current, voltage, or resistance. For many circuits it might be convenient to reduce resistances and trace current the way the earlier lessons taught. But as you become experienced and adept at the calculations, you will probably notice that the *relative values* of resistances in a circuit will allow some short-cut methods of calculations. The law of proportionality is an example of this.

Take the series circuit on this page. The source voltage is given, and so are both resistor values. Ordinarily, if you want to find the voltage drop across R_1, you would use Ohm's Law to find first the current flow in the circuit, and then again use Ohm's Law to calculate the voltage drop. But the law of proportionality allows you to find the voltage drop across R_1 with only one calculation.

Since the same current is flowing through both resistors, the voltage drop across each is directly proportional to the value of that resistor. If one resistor is twice the value of the other, it will have twice the voltage drop; if it is three times the value, it will have three times the voltage drop; and so on. So a short-cut method is to arrange an equation that will show what percentage the resistance of the resistor in question is of the overall total circuit resistance, and then multiply that percentage by the total voltage to get the voltage drop.

Suppose you wanted to find the voltage drop across R_1. The equation

$$\frac{R_1}{R_1 + R_2}$$

will show the proportion of resistance that R_1 has of the total circuit resistance. If you multiply that by the total voltage, you will find E_1:

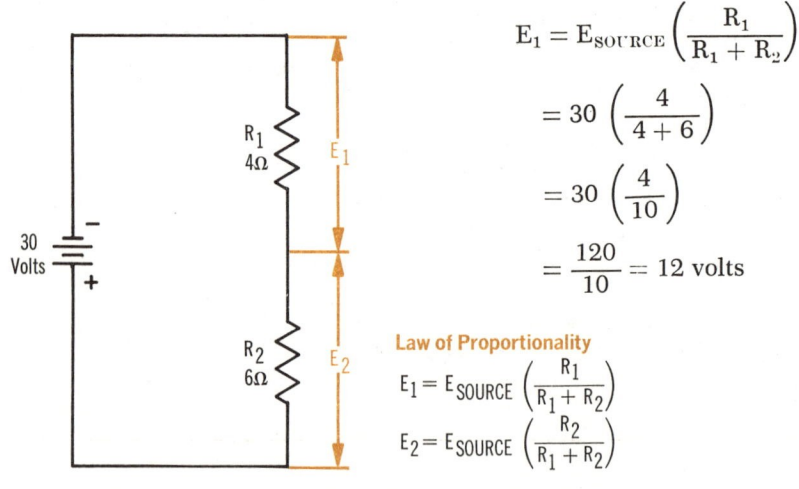

$$E_1 = E_{SOURCE}\left(\frac{R_1}{R_1 + R_2}\right)$$

$$= 30\left(\frac{4}{4+6}\right)$$

$$= 30\left(\frac{4}{10}\right)$$

$$= \frac{120}{10} = 12 \text{ volts}$$

Law of Proportionality

$$E_1 = E_{SOURCE}\left(\frac{R_1}{R_1 + R_2}\right)$$

$$E_2 = E_{SOURCE}\left(\frac{R_2}{R_1 + R_2}\right)$$

law of proportionality (cont.)

The voltage across R_2 could be found the same way, with the equation:

$$E_2 = E_{SOURCE} \left(\frac{R_2}{R_1 + R_2} \right)$$

As a double check on the answer, here is how E_1 would be found with ordinary Ohm's Law:

$$R_{TOT} = R_1 + R_2 = 4 + 6 = 10 \text{ ohms}$$

$$I = \frac{E_{SOURCE}}{R_{TOT}} = \frac{30 \text{ volts}}{10 \text{ ohms}} = 3 \text{ amperes}$$

$$E_1 = IR_1 = 3 \times 4 = 12 \text{ volts}$$

As shown in the drawing on this page, the law of proportionality can also be applied to finding currents in parallel circuits. Since the voltage that will be dropped across two parallel resistors is the same, then the current that flows through any one resistor will be *inversely* proportional to the value of that resistor as compared to the value of the other parallel resistor. This means that if one resistor has twice the value of another, it will have half the current, and so on. Notice that the voltage proportionality in the series circuit was *directly* proportional to the value, whereas here, the current value is *inversely* proportional. Because of this the equation here is slightly different. The value of I_1 depends on the proportionality of R_2 and vice versa. Let's find I_1 in this circuit:

$$I_1 = I_{TOT} \left(\frac{R_2}{R_1 + R_2} \right)$$

$$= 4 \left(\frac{6}{4 + 6} \right)$$

$$= 4 \left(\frac{6}{10} \right)$$

$$= \frac{24}{10} = 2.4 \text{ amperes}$$

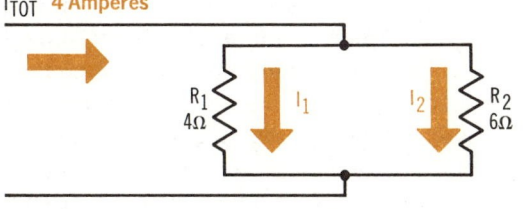

I_{TOT} **4 Amperes**

R_1 4Ω I_1 I_2 R_2 6Ω

Law of Proportionality

$$I_1 = I_{TOT} \left(\frac{R_2}{R_1 + R_2} \right)$$

$$I_2 = I_{TOT} \left(\frac{R_1}{R_1 + R_2} \right)$$

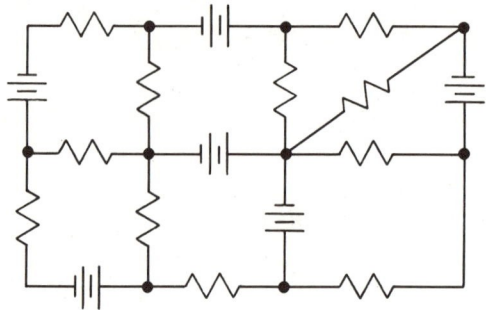

The voltages and currents in this circuit cannot be found by Ohm's Law. There are too many branches and too many power sources. With Kirchhoff's Laws, though, all of the voltages and currents can be calculated

kirchhoff's laws

In all of the circuits examined so far, Ohm's Law has described the relationships between the current, voltage, and resistance. However, all of the circuits covered have been *relatively simple*. There are many circuits that are so complex that they cannot be solved by Ohm's Law. These circuits have *many branches* or *many power sources,* and Ohm's Law would be either impractical or impossible to use on them. Methods are needed, therefore, for solving complex circuits. Any methods used, though, must not violate Ohm's Law, since Ohm's Law is the very basis of d-c circuit theory.

Methods for solving complex circuits have been developed, and are based on the experiments of a German physicist, Gustav Kirchhoff. About 1857, Kirchhoff developed two conclusions as a result of his experiments. These conclusions, known as *Kirchhoff's Laws,* can be stated as follows:

Law No. 1: The sum of the voltage drops around any closed loop is equal to the sum of the emf's in that loop.

Law No. 2: The current arriving at any junction point in a circuit is equal to the current leaving that point.

In using Kirchhoff's Laws, it makes no difference which is called the first law and which is called the second. Since, in use, law No. 1 above is usually applied first, we will call it Kirchhoff's first law. Law No. 2 will, therefore, be his second law.

These two laws may seem obvious to you based on what you already know about circuit theory. In spite of their apparent simplicity, though, they are powerful tools when it comes to solving difficult and complex circuits. Although the laws themselves are simple, the mathematics for applying them becomes more difficult as the circuits become more complex. Because of this, the discussion here will be limited to the use of the laws for solving only the relatively minor complex circuits.

kirchhoff's voltage law

Kirchhoff's first law is also known as his *voltage law*. You will often see it written in many different forms, but no matter what form it is in, it is expressing the same fact. It gives the relationship between the *voltage drops* around any closed loop in a circuit and the voltage sources in that loop. The *totals* of these two quantities are *always equal*. This can be given in equation form as: $\Sigma E_{SOURCE} = \Sigma IR$, where the symbol Σ, which is the Greek letter *sigma,* means "the sum of."

These Circuits ARE Loops

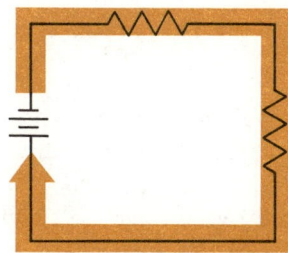 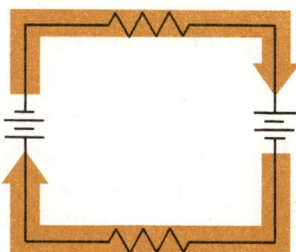

Kirchhoff's Voltage Law can only be applied to closed loops. A closed loop must satisfy two conditions:

1. It must have one or more voltage sources
2. It must have a complete path for current flow from any point, around the loop, and back to that point

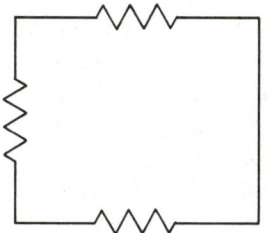 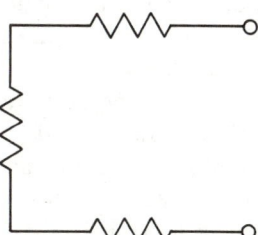

These Circuits ARE NOT Loops

You will recall that in a simple series circuit, the sum of the voltage drops is equal to the applied voltage. This is actually Kirchhoff's voltage law applied to the simplest possible case, that is, where there is only one loop and one voltage source in that loop.

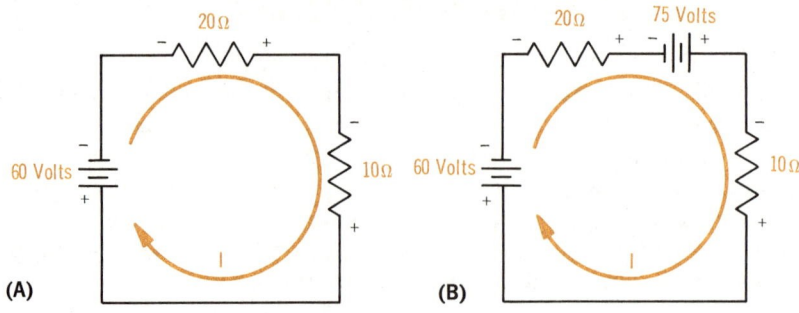

(A)   (B)

applying kirchhoff's voltage law

For a simple series circuit, Kirchhoff's voltage law corresponds to Ohm's Law. To find the current in circuit (A) by using Kirchhoff's voltage law, simply use the equation: $\Sigma\,E_{SOURCE} = \Sigma\,IR$. There is only one voltage source, or emf, in the loop, and two voltage, or IR, drops. So the equation becomes:

$$60 = 20I + 10I$$
$$60 = 30I$$
$$I = 60/30 = 2 \text{ amperes}$$

In the above problem, the direction of the current flow was known before the problem was solved. When there is more than one voltage source, the direction of current might not be known. In this case, you assume a direction at the beginning of the problem. All the sources that would aid the current in this assumed direction are then positive, and all that would oppose current in this direction are negative. The answer to the problem will be positive if you assumed the correct direction of current flow, and negative if you assumed the wrong direction. In either case, you will get the right magnitude of current.

For example, what is the current in circuit (B)? If you assume that the current is flowing in the direction shown, the equation for Kirchhoff's voltage law is

$$\Sigma\,E_{SOURCE} = \Sigma\,IR$$
$$60 - 75 = 20I + 10I$$
$$-15 = 30I$$
$$I = -15/30 = -0.5 \text{ ampere}$$

The result is negative, and so the current is actually 0.5 ampere in the direction opposite to what we assumed.

kirchhoff's current law

Kirchhoff's second law is called his *current law*. Like the voltage law, it too is often stated in different ways. No matter how it is stated, though, its meaning does not change. And the law is: at any junction point in a circuit, the current *arriving* is *equal* to the current leaving. This should be obvious to you from what you learned in Volume 1. Current cannot *collect* or *build up* at a point. For every electron that arrives at a point, one must leave. If this was not so, potential would build up, and current would eventually stop when the potential equaled that of the power source. Thus, if 10 amperes of current arrives at a junction that has two paths leading away from it, the 10 amperes will divide among the two paths, but the total 10 amperes must leave the junction. You are already familiar with the most obvious application of Kirchhoff's second law from parallel circuits. That is, that the sum of the branch currents is equal to the total current entering the branches as well as to the total current leaving the branches.

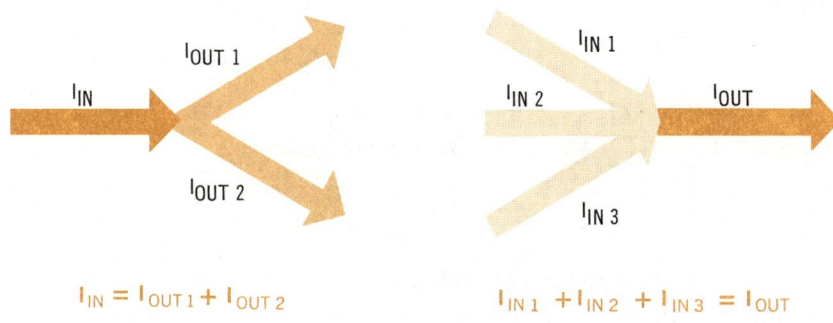

$$I_{IN} = I_{OUT\,1} + I_{OUT\,2}$$

$$I_{IN\,1} + I_{IN\,2} + I_{IN\,3} = I_{OUT}$$

Kirchhoff's current law states that current cannot collect at a point. The current that leaves a point must be equal to the current that enters a point. Thus, if you assign a positive polarity to current entering a point and a negative polarity to the current leaving a point, the algebraic sum of the currents at any point is zero:

$$\Sigma I_{IN} - \Sigma I_{OUT} = 0$$

$$\text{or} \quad \Sigma I_{IN} = \Sigma I_{OUT}$$

Normally, Kirchhoff's current law is not used by itself, but together with the voltage law in solving a problem. This is shown on the following pages.

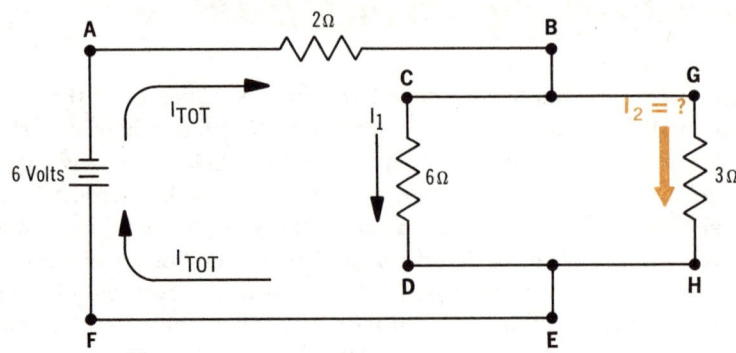

applying kirchhoff's laws

Find the current through the 3-ohm resistance in the circuit using Kirchhoff's Laws. There are two loops in this circuit: one is ABCDEFA, and the other is ABGHEFA. First apply Kirchhoff's voltage law to both loops:

$$2I_{TOT} + 6I_1 = 6 \qquad (1)$$
$$2I_{TOT} + 3I_2 = 6 \qquad (2)$$

Now, since $I_{TOT} = I_1 + I_2$, if you substitute $(I_1 + I_2)$ in place of I_{TOT} in Eqs. (1) and (2), and simplify, you will get:

$$8I_1 + 2I_2 = 6 \qquad (3)$$
$$2I_1 + 5I_2 = 6 \qquad (4)$$

You now have two equations and two unknowns, and must eliminate I_1 to find I_2. One way to do this is to multiply Eq. (4) by four, and subtract Eq. (3) from the result:

$$8I_1 + 20I_2 = 24$$
$$\underline{- (8I_1 + 2I_2 = 6)}$$
$$18I_2 = 18$$

You now have an equation with only I_2, the current you are looking for. So the current I_2 through the 3-ohm resistor is

$$18I_2 = 18$$
$$I_2 = 18/18 = 1 \text{ ampere}$$

You could have solved this problem by just using Ohm's Law, but it was solved by Kirchhoff's Laws to show you the techniques used in solving complex circuits when Ohm's Law cannot be used.

the principle of superposition

When there is more than one power source in a circuit or loop, the current is affected by *each* of the sources. You have examined two ways of finding the current in such cases. One is to determine the combined voltage of the sources, and then use Ohm's Law to find the current. The other way was to use Kirchhoff's voltage law around the loop. A third method that you can use is based on the fact that the current at any point is the *sum* of the currents caused by *each* of the power sources. Therefore, if you can calculate the current that would exist if there was only one power source, and do this for each individual source, the sum of these currents will be the total current with all the sources acting in the circuit. This is called the principle of *superposition*. There are four steps involved in applying the principle:

1. Replace all power sources with a short circuit, except one, and assume a direction of current flow.
2. Calculate the current you want with the one source in the circuit.
3. Do this for each power source in the circuit.
4. Add the individual currents you found. Currents in the direction of the assumed current flow are positive. Those in the opposite direction are negative. If the total current turns out negative, the assumed direction was wrong.

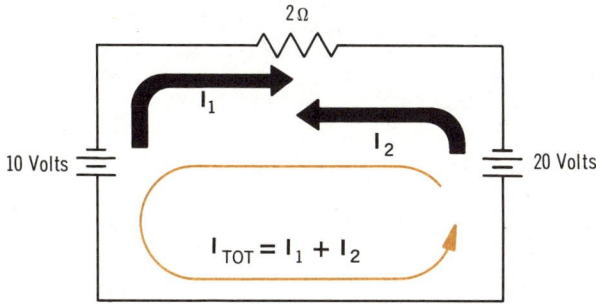

For the circuit shown, we assume a direction for the total current. Thus,

$$I_1 = E_1/R = -10 \text{ volts}/2 \text{ ohms} = -5 \text{ amperes}$$

$$I_2 = E_2/R = 20 \text{ volts}/2 \text{ ohms} = 10 \text{ amperes}$$

$$I_{TOT} = I_1 + I_2 = -5 + 10 = 5 \text{ amperes}$$

an example of superposition

Problem: Find the currents through each resistor, and calculate their voltage drops.

$$R_{TOT} = R_2 + \frac{R_1 R_3}{R_1 + R_3}$$

$$= 20 + \frac{20 \times 20}{20 + 20} = 20 + \frac{400}{40}$$

$$= 20 + 10 = 30 \text{ ohms}$$

$$I_2 = E_2/R_{TOT} = 30/30 = 1 \text{ ampere}$$

Since $R_1 = R_3$, the current $I_1 = 0.5$ ampere and $I_3 = 0.5$ ampere.

$$R_{TOT} = R_1 + \frac{R_2 R_3}{R_2 + R_3}$$

$$= 20 + \frac{20 \times 20}{20 + 20} = 20 + \frac{400}{40}$$

$$= 20 + 10 = 30 \text{ ohms}$$

$$I_1 = E_1/R_{TOT} = 30/30 = 1 \text{ ampere}$$

Since $R_3 = R_2$, the current $I_3 = -0.5$ ampere and $I_2 = 0.5$ ampere.

Answers: $I_{1TOT} = I_1 + I_1 = (0.5) + (1.0) = 1.5$ amperes

$I_{2TOT} = I_2 + I_2 = (1.0) + (0.5) = 1.5$ amperes

$I_{3TOT} = I_3 + I_3 = (0.5) + (-0.5) = \text{zero current}$

Since the currents I_1 and I_2 are in the same direction in both calculations, they are both positive currents. I_3, however, is in opposite directions in both calculations; so one of them is negative.

$$E_1 = I_1 R_1 = 1.5 \times 20 = 30 \text{ volts}$$

$$E_2 = I_2 R_2 = 1.5 \times 20 = 30 \text{ volts}$$

$$E_3 = I_3 R_3 = (0) \times 20 = \text{zero volts}$$

thevenin's theorem

Earlier in the book you learned how to reduce a complex circuit to its simplest form, so that you could make the basic Ohm's Law calculations that could be applied to the actual overall circuit. Often, in many circuits, the circuit data is needed only for a load resistor at the output, and it would be convenient if all of the calculations did not have to be made. Also, if the load resistor value still were not decided, it would be impossible to make any calculations because each different value of load resistor would change the currents and voltage drops throughout the circuit.

A late 19th century scientist, Leon Thevenin, pondered this problem and developed a theory that any complex circuit with a two terminal output could be reduced to a simple *equivalent circuit* across whose output any load resistor would function in the same way as the original circuit. *Thevenin's theorem* states that the equivalent circuit need contain only an equivalent voltage source, $E_{THEVENIN}$, and an equivalent series resistance, $R_{THEVENIN}$.

As a proof that this is possible, take the circuit shown on this page. We want to find the current in R_L. First disconnect R_L. The rest of the circuit is to be thevenized. If you connect a voltmeter across the output terminals, you will measure Thevenin's equivalent voltage, E_{THEV}. You can also calculate it using Ohm's Law or the law of proportionality. You will find that $E_{THEV} = 4.5$ volts. Next connect an ammeter to read Thevenin's current. Since the ammeter will short out R_2, only R_1 will be left in the circuit to give $I_{THEV} = E/R = 6/2 = 3$ amperes.

Now to find Thevenin's equivalent resistance, divide E_{THEV} by I_{THEV}, and you will get 1.5 ohms, R_{THEV}. So Thevenin's equivalent circuit will be as shown, with E_{THEV} in series with R_{THEV} and R_L. Simple Ohm's Law can now be used to find the current through R_L, no matter what value you decide to use. This same problem was solved with Kirchhoff's Laws a few pages earlier.

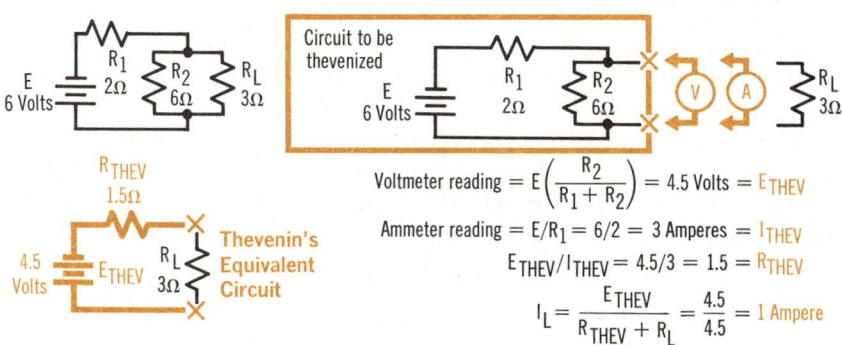

Voltmeter reading $= E\left(\dfrac{R_2}{R_1 + R_2}\right) = 4.5$ Volts $= E_{THEV}$

Ammeter reading $= E/R_1 = 6/2 = 3$ Amperes $= I_{THEV}$

$E_{THEV}/I_{THEV} = 4.5/3 = 1.5 = R_{THEV}$

$I_L = \dfrac{E_{THEV}}{R_{THEV} + R_L} = \dfrac{4.5}{4.5} = 1$ Ampere

thevenizing

Actually, the discussion on the previous page only showed the proof that Thevenin's theorem works. To actually thevenize a circuit, you would not use a voltmeter or ammeter, nor would you make *all* calculations that were made. Thevenin's theorem requires that you:

1. First determine E_{THEV} across the output terminals with R_L disconnected.

2. *Look back into* the output terminals to see what equivalent resistance R_{THEV} exists with the source voltage *shorted out.*

E_{THEV} is found just the way it was on the previous page, either by Ohm's Law or by the law of proportionality.

To find R_{THEV}, when the source voltage is shorted out, resistors R_1 and R_2 become connected in parallel when viewed from the output terminals.

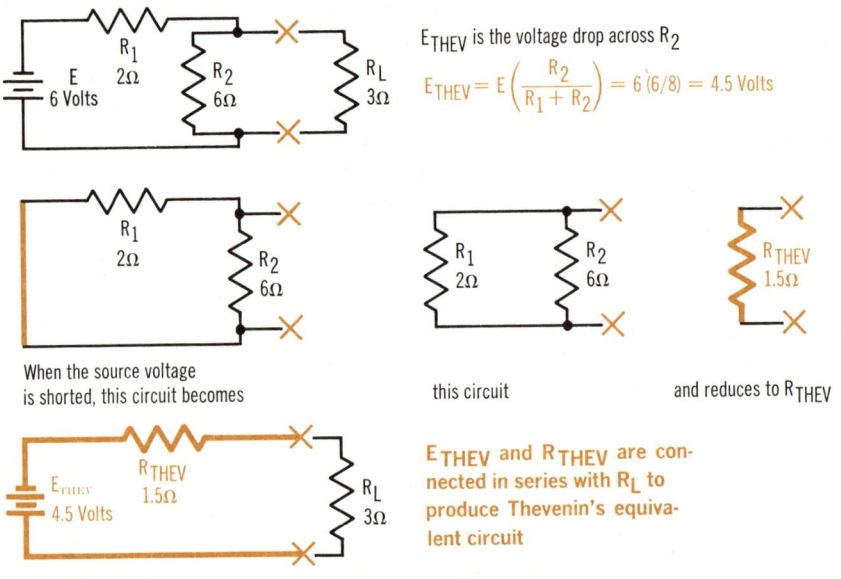

E_{THEV} is the voltage drop across R_2

$$E_{THEV} = E\left(\frac{R_2}{R_1 + R_2}\right) = 6\,(6/8) = 4.5 \text{ Volts}$$

When the source voltage is shorted, this circuit becomes

this circuit

and reduces to R_{THEV}

E_{THEV} and R_{THEV} are connected in series with R_L to produce Thevenin's equivalent circuit

So, using the parallel resistance equation, you will find that R_{THEV} = 1.5 ohms, and Thevenin's equivalent circuit is the same as previously shown.

To show you the real value of Thevenin's equivalent circuit, you can now test a variety of different R_L values, and determine the current, voltage, and power with simple Ohm's Law in each case. If you used the original circuit to do this, complex computations would have to be repeated for each value of R_L tested.

a sample problem

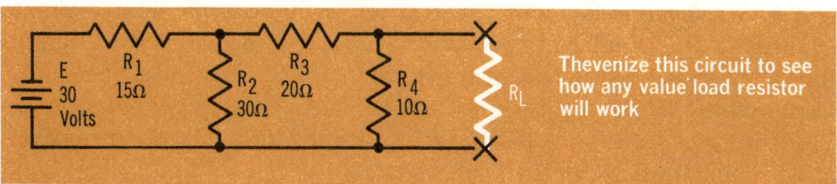

Thevenize this circuit to see how any value load resistor will work

First find the voltage drop across R_4, which will be E_{THEV}

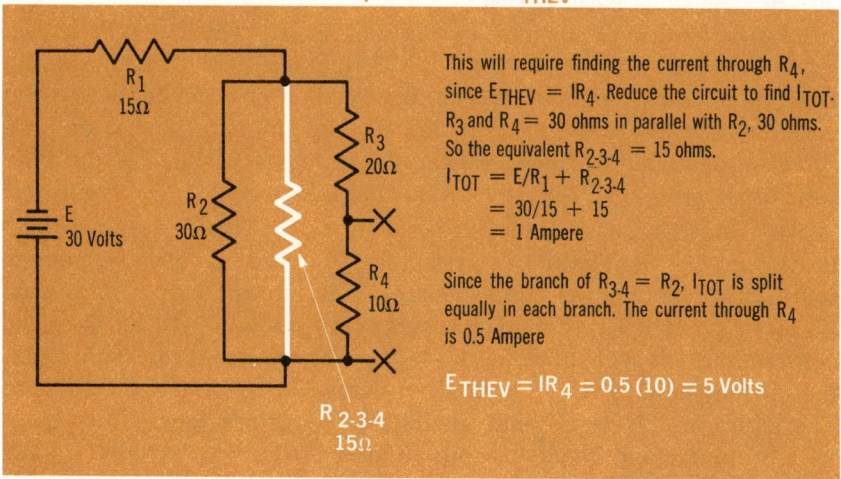

This will require finding the current through R_4, since $E_{THEV} = IR_4$. Reduce the circuit to find I_{TOT}. R_3 and $R_4 = 30$ ohms in parallel with R_2, 30 ohms. So the equivalent $R_{2\text{-}3\text{-}4} = 15$ ohms.

$$I_{TOT} = E/R_1 + R_{2\text{-}3\text{-}4}$$
$$= 30/15 + 15$$
$$= 1 \text{ Ampere}$$

Since the branch of $R_{3\text{-}4} = R_2$, I_{TOT} is split equally in each branch. The current through R_4 is 0.5 Ampere

$$E_{THEV} = IR_4 = 0.5 \, (10) = 5 \text{ Volts}$$

Now short out the source voltage and find Thevenin's equivalent resistance, R_{THEV}, looking back from the output terminals

This circuit becomes   this circuit   and reduces to this

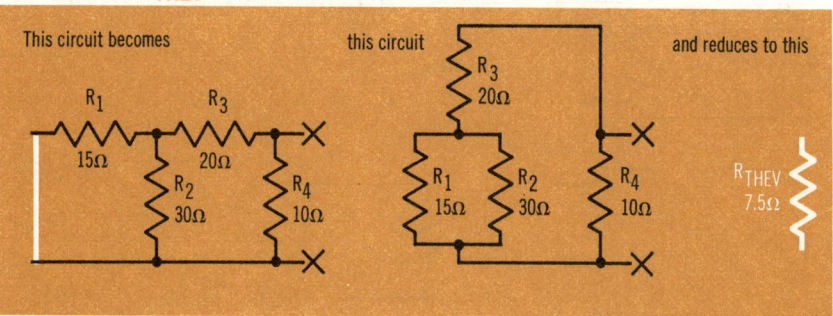

Draw Thevenin's equivalent circuit

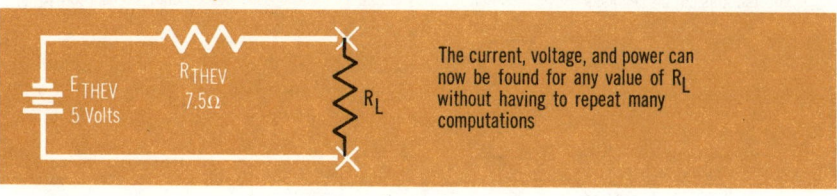

The current, voltage, and power can now be found for any value of R_L without having to repeat many computations

norton's theorem

Norton's theorem is another method of reducing a circuit to a simple equivalent. It is similar in concept to Thevenin's theorem, but differs in that it uses an equivalent constant current source instead of an equivalent voltage source. It also uses an equivalent resistance, but whereas Thevenin's resistance was put in series with the load, Norton's resistance is placed in parallel with the load.

To show how Norton's theorem accomplishes the same as Thevenin's, let's use the same basic circuit.

First find Norton's constant current source, I_N, which will be shown as an arrow in a circle showing the current direction. Disconnect R_L from the output terminals, and connect a short-circuit wire across the terminals. Calculate the current flowing into the wire. Since the wire shorts out resistor R_2, then resistor R_1, which is 2 ohms, is the only resistance across the 6-volt source. So Norton's current, I_N, is 3 amperes.

Next, find Norton's equivalent resistance, R_N. This is found exactly the same way as Thevenin's resistance, by shorting out the voltage source and reducing the circuit looking back into the output terminals.

Once I_N and R_N are found, the equivalent circuit is drawn with the constant current source feeding R_N in *parallel* with the loads. Notice that the current flow in R_L and the drop across R_L are the same in both Thevenin's and Norton's equivalent circuits; Ohm's Law shows them to be 1 ampere and 3 volts in both circuits.

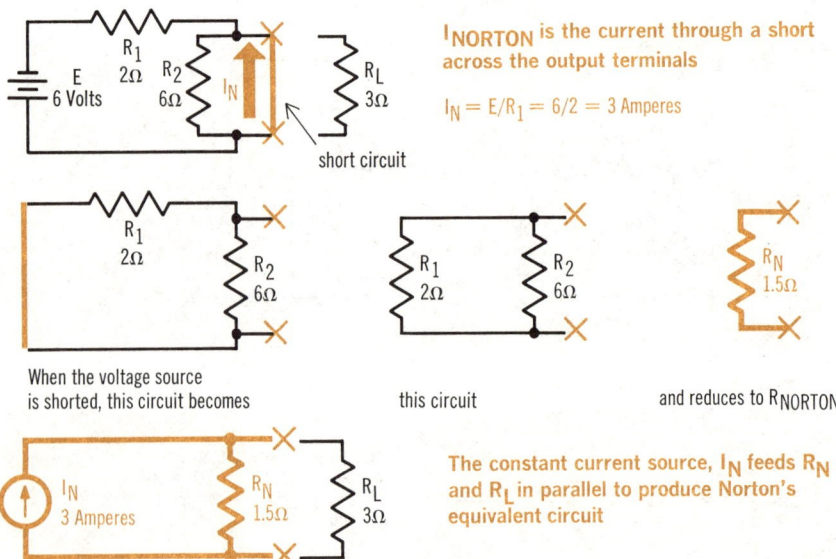

INORTON is the current through a short across the output terminals

$I_N = E/R_1 = 6/2 = 3$ Amperes

short circuit

When the voltage source is shorted, this circuit becomes

this circuit

and reduces to R NORTON

The constant current source, I_N feeds R_N and R_L in parallel to produce Norton's equivalent circuit

internal resistance
of power sources

You will remember from page 2-19 that every power source has some *internal resistance* that opposes current flow. Normally this resistance is very small and has little effect on circuit operation. For this reason, the internal resistance of the power source has been neglected in all circuits throughout this volume. When the internal resistance of a source must be considered in a circuit, it is usually represented as a resistance in series with the source. And for the most part, this is the effect it has on a circuit: that of an *additional resistance* in the circuit in series with the power source.

Every resistance in a circuit has voltage dropped across it when current flows. Since the internal resistance of the source is inside the source, its voltage drop is also *internal*. This internal drop *subtracts* from the source output voltage. And since the amount of any voltage drop follows the equation $E = IR$, the higher the internal resistance of a source or the more current it conducts, the greater will be its internal voltage drop, and the lower will be its output voltage.

For some power sources, the current output is limited by the internal voltage drop. If you try to draw more than a certain amount of current from these sources, the increased current causes an increase in the internal voltage drop, which lowers the output voltage and, therefore, decreases the current.

When power sources are connected in parallel, their internal resistances are also in parallel. So their effective internal resistance is lower than that of any of the individual sources.

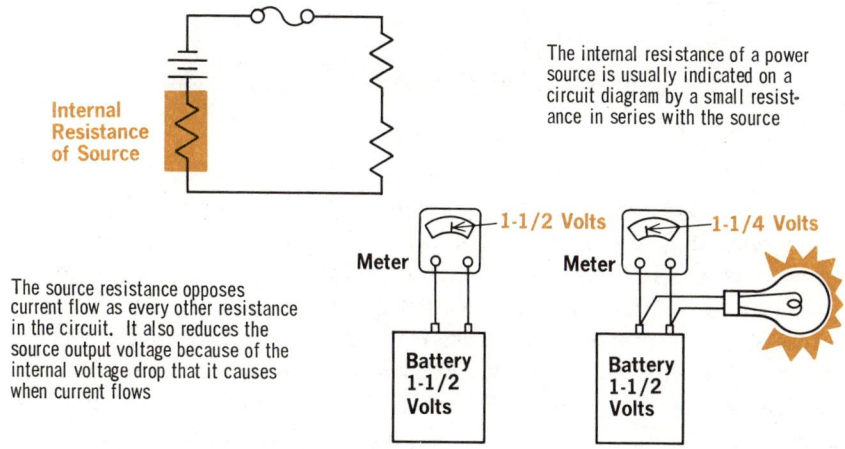

Internal
Resistance
of Source

The internal resistance of a power source is usually indicated on a circuit diagram by a small resistance in series with the source

The source resistance opposes current flow as every other resistance in the circuit. It also reduces the source output voltage because of the internal voltage drop that it causes when current flows

1-1/2 Volts

Meter

Battery
1-1/2
Volts

1-1/4 Volts

Meter

Battery
1-1/2
Volts

d-c circuit failures

For a circuit to operate correctly, *every part* has to do its job. The *power source* has to supply the required voltage, the *connecting wires* have to provide low-resistance connections between the circuit parts without overheating or short-circuiting, and the *loads* have to do their job without drawing too much current from the power source. Auxiliary devices such as switches also have to operate properly. You can see, then, that if any part in a circuit goes bad, or *fails,* the entire circuit fails, since it cannot operate in the way it was designed.

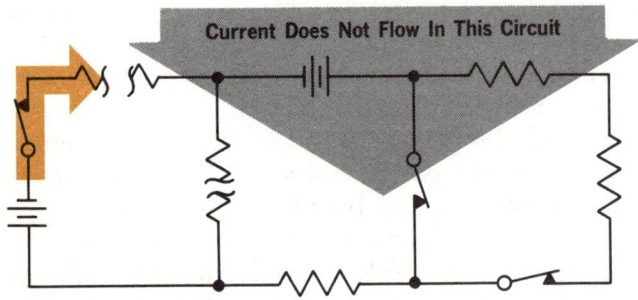

Current Does Not Flow In This Circuit

When an electrical part opens, this does not necessarily mean that the circuit is open. Likewise, when a part becomes shorted, it does not always short the entire circuit

The exact effect that a faulty part has on a circuit depends on what the part is and how it fails, as well as the type of circuit and the location of the part in the circuit

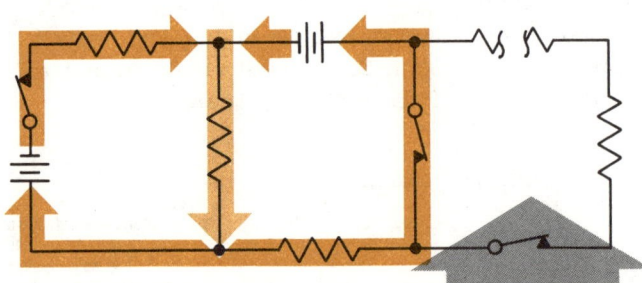

Current Does Not Flow In This Branch of the Circuit

When a part fails, the way in which it affects the circuit depends on what type of a part it is, how it fails, what type of circuit it is used in, and where it is connected in the circuit. Thus, the same type of part, failing in the same way, can cause different effects in different circuits. For example, if a resistor became open in a series circuit, it would open the entire circuit. But if the same resistor opened in one branch of a parallel circuit, it would only open that one branch. Total circuit current would still flow, although it would decrease.

power source failures

A common failure in d-c circuits is failure of the power source. Sources can fail by losing their output entirely, or by having their output voltage decrease to a value less than what is considered normal.

In the field of electricity, the two most common d-c power sources are the d-c generator and the battery. D-c generators can fail because of some mechanical defect within the generator, or because of electrical troubles, such as open or shorted wires.

The most common type of battery failure is a decrease in output voltage caused by the discharging of the battery. Any battery gradually discharges as current is drawn from it over a period of time. When it has discharged to the point where its output voltage is less than that needed to operate the circuit, the battery either has to be recharged or replaced. Batteries can also develop internal short circuits, and then must be replaced. However, this does not happen frequently. D-c generators and batteries are described in detail in Volume 6.

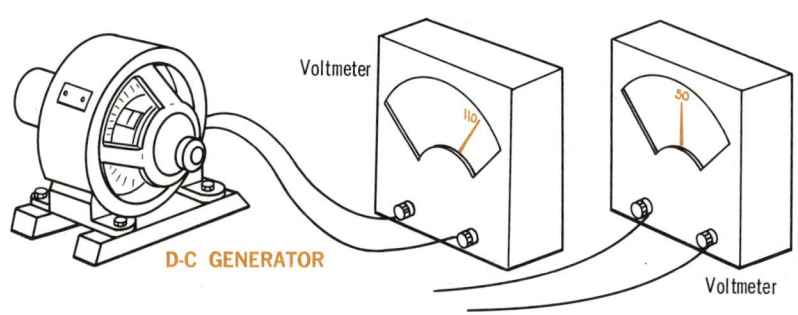

Voltmeter

D-C GENERATOR

Voltmeter

D-c generators and batteries fail whether by losing their outputs completely, or by having their output voltages decrease considerably

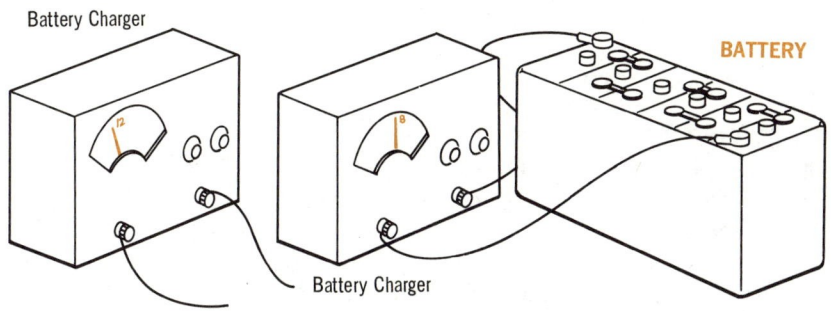

Battery Charger

BATTERY

Battery Charger

resistor failures

Resistors are one of the most common causes of failure in a circuit. This is not because they are exceptionally delicate or prone to damage, but rather because so many are used. In electric circuits, resistors fail in two main ways: (1) they burn out and *open,* and (2) their resistance value *changes.* Resistors can change in value so that their resistances go up or down. If the value increases greatly, it will act like an open resistor. If the value decreases greatly, it will act like a shorted resistor. These failures are usually the result of the *heat* generated in the resistor by the current. If the heat is not severe, but is generated for a long period of time, the resistor can change in value, while if the heat is very severe, the resistor will more likely burn out in a short time. Often, a resistor failure is caused by some *other failure* in the circuit which results in an increase in circuit current. The increased current then overheats the resistor and causes it to fail.

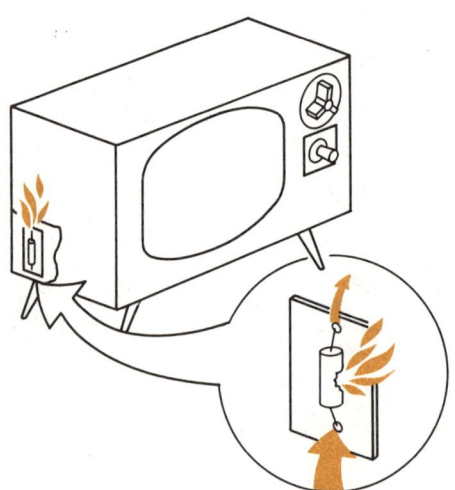

Most resistor failures result from the heat generated in the resistor by the current flow. This heat can cause resistors to change in value, as well as to burn out and become open

Since most resistor failures are caused by too much heat, a defective resistor can frequently be detected by its charred and discolored look. This is especially true for composition resistors.

Adjustable and variable resistors develop the same troubles as do fixed resistors, but in addition, they can fail by losing their ability to be adjusted or varied. This can happen as a result of some mechanical defect, such as a broken sliding contact, or because of an electrical trouble, like a dirty contact.

failure of other loads

Resistors are not the only type of load that can fail in a circuit. Actually, any device used as a load can develop trouble and cause a circuit failure. The loads that use some sort of *heating element* to accomplish their function are a frequent cause of failure. Examples of this type of load are toasters, irons, electric heaters, and lamps. During their normal operation, these loads experience severe heating, which causes the material in the element to *expand*. When the current is turned off, the element cools, and *contracts* to its original size. The amount of this expansion and contraction is not great, but it occurs each time the device is turned on and off. This alternate expansion and contraction *fatigues* the material, and eventually the element *breaks* and opens the circuit.

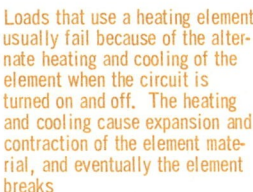

Loads that use a heating element usually fail because of the alternate heating and cooling of the element when the circuit is turned on and off. The heating and cooling cause expansion and contraction of the element material, and eventually the element breaks

failure of auxiliary devices

The two circuit auxiliary devices you are familiar with are the *switch* and the *fuse*. Both of these can fail and affect circuit operation.

Since the purpose of a switch is to open and close a circuit, any switch must have two positions: one to allow current to flow, and the other to stop current flow. The switch is operated mechanically to one position or the other. If a switch *sticks* in one position or *breaks* so that it stays in one position, its basic switching action is lost. It then keeps the circuit either closed or open at all times. Another type of switch failure occurs when the switch *contacts* become dirty. You remember that a switch should have no effect on a circuit when it is closed. It should, therefore, have almost zero resistance. But if the contacts are dirty, the dirt can act as an *additional resistance* in the circuit and cause a decrease in circuit current.

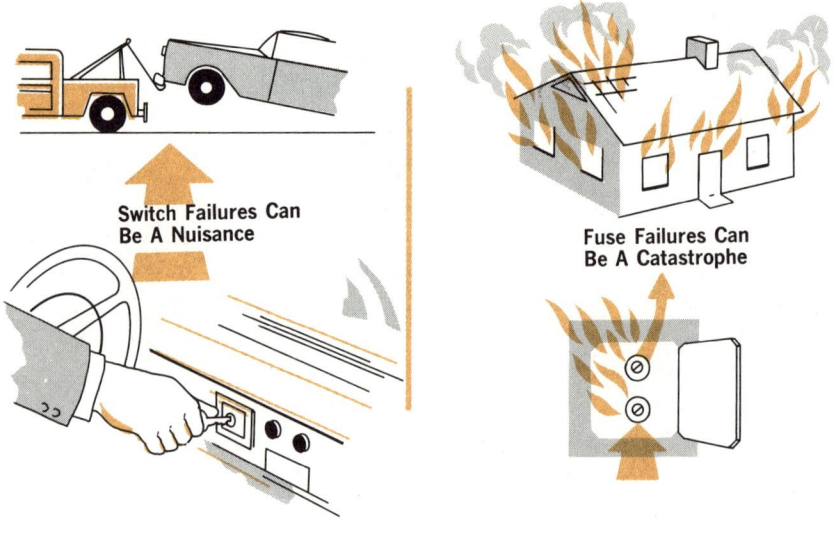

Switch Failures Can
Be A Nuisance

Fuse Failures Can
Be A Catastrophe

Fuses do not fail very often, but if they do, the results can be catastrophic. If, due to imperfections in the fuse material, a fuse should blow at a *lower* current level than it is supposed to, the circuit will be opened *unnecessarily*. This is inconvenient, but not serious. But, if a fuse does not blow when it should, the *high damaging currents* that it is supposed to prevent will flow in the circuit. These currents can burn out the power source or load, and possibly even start a fire.

summary

☐ Proportionality is an easier way of finding series voltage drops and parallel currents. ☐ Kirchhoff's voltage law is: The sum of the voltage drops around any closed loop is equal to the algebraic sum of the emf's in that loop. Mathematically, $\Sigma E_{SOURCE} = \Sigma IR$. ☐ Kirchhoff's current law is: The current arriving at any point in a circuit is equal to the current leaving that point. Mathematically, $\Sigma I_{IN} = \Sigma I_{OUT}$.

☐ In using Kirchhoff's voltage law, assume a direction for the loop current and traverse the loop in the direction of the assumed current, returning to the starting point. ☐ The emf's in the circuit are positive if they tend to aid the current in the assumed direction, and are negative if they tend to oppose the current. ☐ If the value of current found using Kirchhoff's laws is negative, the assumed direction of current flow was wrong.

☐ The principle of superposition is used to solve for current when there is more than one power source. The current through a particular load is found by adding the currents due to each source. The current caused by each source is calculated by replacing the other sources by their internal resistance, or by a short circuit if the internal resistance is negligible. ☐ Thevenin's and Norton's theorems are used to produce simple equivalent circuits. ☐ Power sources fail by a loss or decrease of output voltage. ☐ Resistors can fail by burning out and opening, or by changing value. ☐ Wires can fail by breaking. ☐Switches can fail by locking in one position, or by having dirty contacts. ☐ Fuses can fail by opening prematurely or failing to open.

review questions

For Questions 1 to 5, consider circuit B on page 2-118.

1. Solve for the current using the principle of superposition.
2. Which end of the 20-ohm resistor is negative?
3. What is the voltage drop across each resistor?
4. What is the total power supplied by the two batteries?
5. Answer Questions 1 to 3 with the 75-volt battery terminals reversed.

For Questions 6 to 10, consider the circuit on page 2-121.

6. Solve for the current using Kirchhoff's voltage law.
7. What is the polarity of the voltage drop across the resistor?
8. What is the value of the voltage drop across the resistor?
9. What is the power consumed by the resistor?
10. Answer Questions 6 to 9 with the 20-volt battery terminals reversed.

electricity
three

what is alternating current?

In Volume 1, electric current was described as the movement of free electrons in a conductor that is connected between a difference of potential. The current flows as long as the difference of potential is present. If the polarity of the potential difference never changes, the current always flows in *one direction* and is called *direct,* or d-c *current.* D-c current and d-c circuits were described in detail in Volume 2.

There is a type of electric current that does *not* always flow in the same direction. Instead, it *alternates,* flowing first in one direction, and then reversing and flowing in the other. This type of current is called *alternating,* or a-c *current.*

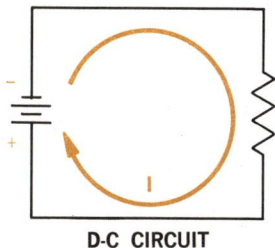

Unlike d-c current, which always flows in the same direction, a-c current periodically changes its direction

D-C CIRCUIT

A-c current flows in one direction, then . . .

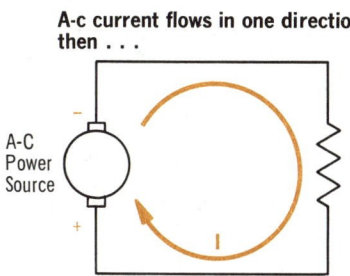

It reverses and flows in the other direction

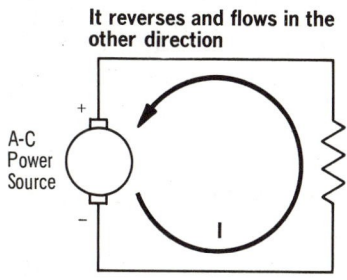

A-C CIRCUIT

Since, in any circuit, current flows from the negative terminal of the power source to the positive terminal, it is obvious that for a-c current to flow, the polarity of the power source must alternate, or change direction. Power sources that do this are called *a-c power sources.* Circuits that use a-c power sources, and that therefore have a-c current flowing in them, are called *a-c circuits.* Similarly, the power consumed in an a-c circuit is *a-c power.*

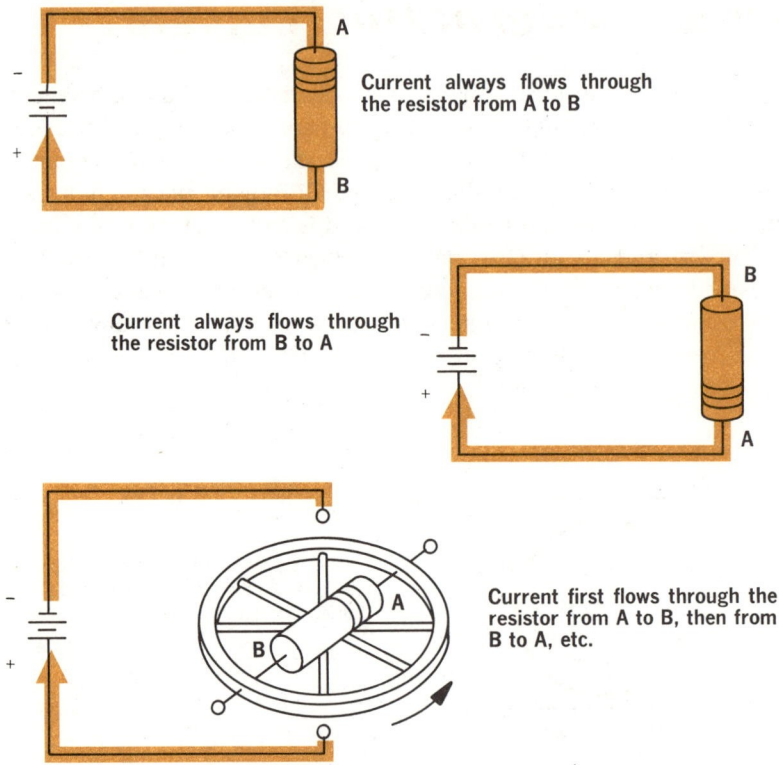

Current always flows through
the resistor from A to B

Current always flows through
the resistor from B to A

Current first flows through the
resistor from A to B, then from
B to A, etc.

Heat is developed in this resistor regardless of the direction in which the current flows. Alternating current, therefore, has the same effect on the resistor as does direct current

is alternating current useful?

When you are first introduced to alternating current you may wonder whether it can be put to any *practical* use. Since it reverses in direction, it may seem that anything it would do while flowing in one direction, it would undo when it reversed and flowed in the opposite direction. This is not the case, however.

In a circuit, the flowing electrons themselves do not do any useful work. It is the *effects* they have on the loads they flow through that are important. And these effects are the same regardless of the direction of the current. For example, when current flows through a resistance, heat is *always* produced. It makes no difference whether the current is always in one direction, always in the opposite direction, or sometimes in one direction and sometimes in the other.

why is alternating current used?

Direct current was the first source of electric power to be widely used. But as more became known about the characteristics of alternating current, it gradually replaced dc as the primary source of electric power in use. Today about 90 percent of all the electric power consumed through the world is ac. In the United States, this percentage is even higher. There are some sections in our older cities where the electric power is still dc. These sections, though, are gradually being changed to ac.

What are the reasons for this changeover? Why should nine times as much a-c power be used as dc? Basically, there are two reasons for it. One is that, for the most part, ac can do everything that dc can, and in addition, can be *sent* from the point where it is produced to the point where it will be used *more easily* and *cheaply* than dc can. The second reason for the widespread use of ac is the fact that it can do certain things and be used for certain applications for which dc is unsuited.

You should not get the impression that dc will soon be obsolete and all electricity used will be ac. There are many applications, particularly inside of electrical equipment, where only dc can perform the desired function.

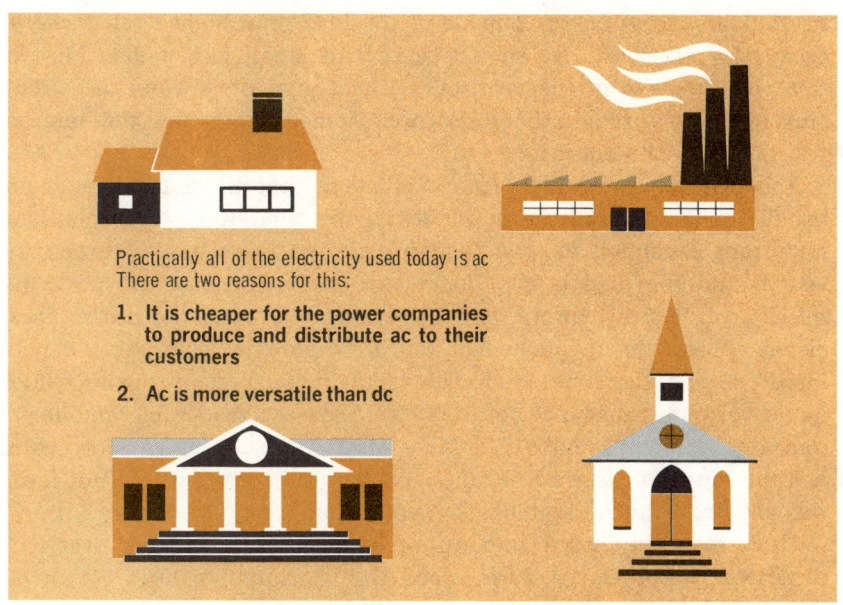

Practically all of the electricity used today is ac
There are two reasons for this:

1. It is cheaper for the power companies to produce and distribute ac to their customers

2. Ac is more versatile than dc

Electric power stations cannot be built near every home, factory, and farm. They must usually be located near the available sources of natural energy, such as large rivers

The electric power the stations produce, then, has to be sent long distances to the users. Large power losses result if dc is sent over long distances. With ac, these losses are greatly reduced

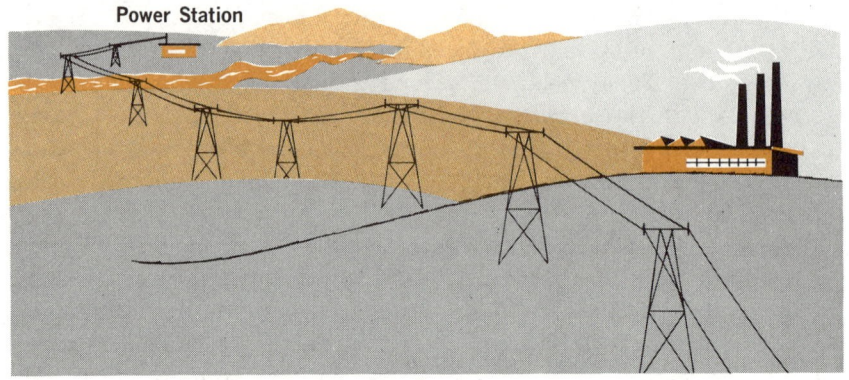

Power Station

electric power transmission

In an ideal electric circuit, *all* of the energy produced by the power source would be converted by the load into some *useful* form, such as light or heat. In practice, though, it is impossible to build an ideal circuit. Some of the energy from the power source is used in the circuit interconnecting wires, and some is also used within the power source itself. This use of energy outside of the load represents *wasted energy* or *wasted power,* and so should be kept as small as possible. Most of this power *loss* is in the form of the *heat* generated when the circuit current flows through the resistance of the wiring and the internal resistance of the source.

You probably remember from Volume 2 that these resistances are usually very low, and the power losses are, therefore, very small. An important exception to this, though, is when the wiring between the source and the load is *very long,* such as is the case in the transmission of electric power from the power stations to the users. These electric power lines, with which you are undoubtedly familiar, can be hundreds of miles long. And even large-diameter, copper wire, which has a very low resistance for 1000 feet, has a considerable amount of total resistance if hundreds of miles of it are used. Silver wire, which has the lowest resistance of all, could be used, but it would not lower the total resistance significantly, and would cost too much.

How, then, can large amounts of electric power be sent over long distances without large losses along the transmission lines? With dc, this cannot be done. But with ac, it is relatively easy.

transmission power loss

In the transmission of electric power, a portion of the power is converted into heat along the length of the transmission line. From Volume 2, you should recall that this heat loss is *directly proportional* to the resistance and to the *square* of the current. This is shown by the formula for *power loss:*

$$P = I^2R$$

The heat, or power loss (P) can thus be reduced by lowering either the current (I) carried by the transmission line, the resistance (R) of the wire, or both. However, the resistance has much less effect on the power loss than the current because the current is "squared."

In the transmission of a-c power, the current carried by the transmission line is relatively small

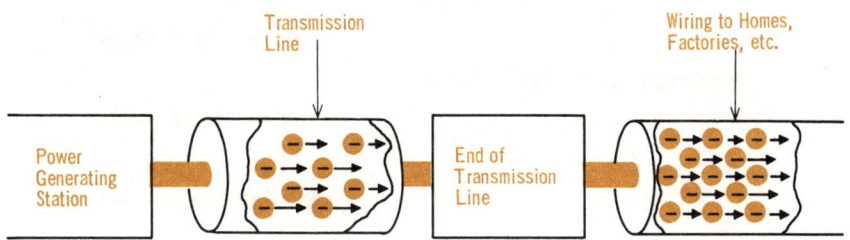

At the end of the transmission line, this small current is converted to the large currents required by the users of power. This keeps the I^2R (power) losses low in the transmission lines

If the resistance was doubled, the power loss would double; but if the current is doubled, the power loss would be quadrupled. So the best way to reduce the power losses is to *lower* the *current*. But large currents are required by the users of the electric power at the end of the transmission line. What is needed, then, is a method of having low currents in the transmission line, but large currents available at the end of the line. This is possible with a-c power. Relatively low currents are sent along the transmission lines, and when they reach the point where they are to be used, they are converted to high currents.

transmission of a-c power

It may seem unreasonable that electric power can be transmitted with *low current* in the transmission line, and yet at the end of the line, the power is available with a *high current*. To understand this, you should recall the voltage–current equation for electric power:

$$P = EI$$

You can see from this equation that the *identical* power (P) can be produced by *many combinations* of currents (I) and voltages (E). For example, you can get 1000 watts of power from a voltage of 100 volts and a current of 10 amperes, from a voltage of 200 volts and a current of 5 amperes, or from a voltage of 1000 volts and a current of 1 ampere.

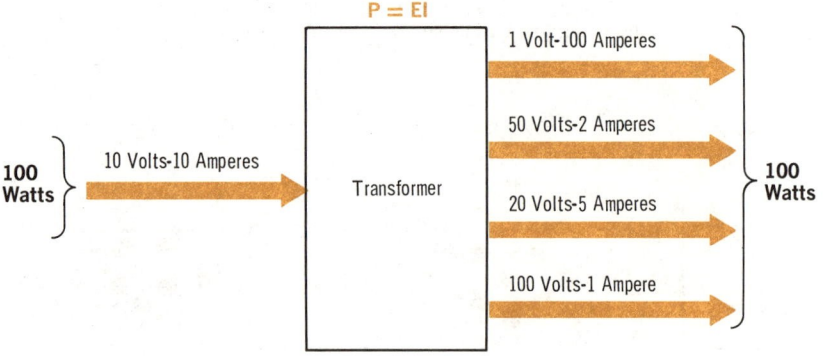

A-c power can be converted to various combinations of voltage and current. The value of the power after the conversion is the same as its value before conversion, since P = EI

A million watts of power can, therefore, be sent over a transmission line in many ways. It could be sent at a voltage of 1000 volts, in which case the current would then be 1000 amperes, and the power loss along the line would be very high. Or it could be sent at a voltage of 100,000 volts with a current of only 10 amperes, and the power loss would be much less. At the end of the transmission line, the transmitted combination of voltage and current can then be converted to any other combination of voltage and current that produces a total of a million watts.

The devices used to convert a-c power from one combination of voltage and current and another are called *transformers*. They will be examined later in this volume.

Electrical communications as we know it today is based on the characteristics of ac

other characteristics of ac

The fact that it can be sent over long transmission lines cheaply and easily is not the only advantage that ac has over dc. Ac has other characteristics that dc does not have which make it ideally suited for certain uses. One of these characteristics is the fact that it *varies* in value, or magnitude. It does not rise immediately to a maximum value and stay at that value until the circuit is opened, as dc does. This changing level is similar to the way the level of sound waves in the air change. Thus, in the field of electrical communications, ac is used to reproduce and transmit sound electrically.

Another important characteristic of ac is the fact that electrical energy *radiates* from any circuit in which alternating current is flowing. This characteristic is the basis of radio communications, and is widely used in the field of electronics.

the a-c power source

The purpose of any electric power source is to provide a *voltage,* or difference of potential, between its output terminals, and to maintain this voltage when the circuit is closed and *current* flows. In d-c power sources, the polarity of the output voltage *never* changes. One terminal is *always negative,* and the other is *always positive.* Therefore, the circuit current is always in the same direction; from the negative terminal of the source, through the load, and back to the positive terminal of the source. A-c sources, on the other hand, *constantly change polarity.* At any instant, one terminal is negative and the other one is positive. At some later instant, the terminal that was negative becomes positive, and the one that was positive becomes negative. This reversal in polarity is continuous, and each time it happens, the circuit current changes direction, since it must always flow from the negative to the positive terminal.

A-c power sources are called *a-c generators,* or *alternators.*

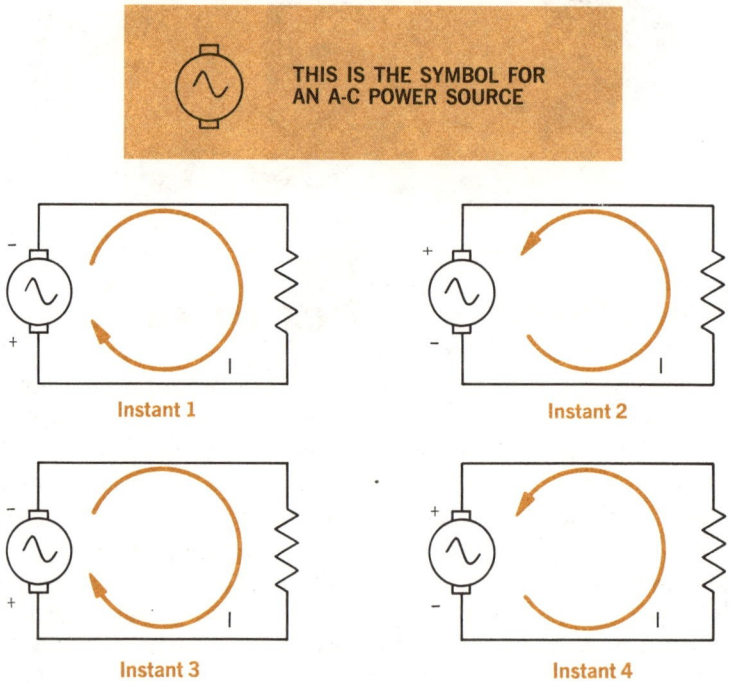

The polarity of a-c power sources changes continuously. Each time the polarity reverses, the circuit current also changes direction

a simple a-c generator

A-c generators combine physical motion and magnetism to produce an a-c voltage. You learned from Volume 1 that if a conductor is moved through a magnetic field in such a way that it passes through the flux lines, a force is applied to the free electrons in the conductor, causing them to move. Since the force thus causes current flow, it can be considered an emf, or a voltage. This is the basic principle on which a-c generators work.

The simplest type of a-c generator is shown. It consists of a single *loop* of wire, which is placed between the poles of a permanent magnet and made to *rotate*. As the loop rotates, it passes through, or *cuts*, the magnetic lines of force, and a voltage is developed. In a practical generator, this loop is actually a series of loops contained on a *rotor* or an *armature*.

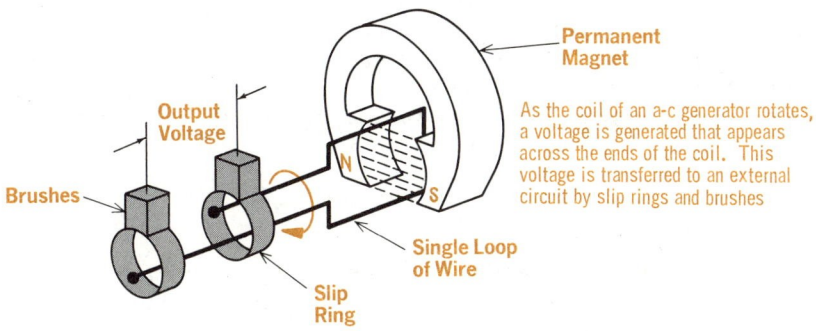

Permanent Magnet

Output Voltage

Brushes

As the coil of an a-c generator rotates, a voltage is generated that appears across the ends of the coil. This voltage is transferred to an external circuit by slip rings and brushes

Single Loop of Wire

Slip Ring

The voltage produced exists between the two ends of the loop. *Slip rings* and *brushes* are used to apply the voltage to an external circuit. The slip rings are smooth rings made of conducting material. One ring is connected to each end of the loop, and both rings rotate as the loop rotates. The output voltage of the generator thus exists between the two slip rings. The brushes are in contact with the slip rings, one brush for each ring. The brushes do not move, but stay in contact with the slip rings by sliding along their surface as they rotate. In this way, the output voltage of the generator is between the brushes and can easily be applied to a circuit.

You can see from the description that something must turn the loop for the generator to work. This "something" could be flowing water, a gasoline engine, steam created by burning coal, or even steam produced by a nuclear reactor.

A-c generators are covered in detail in Volume 6.

summary

☐ Alternating current flows first in one direction, and then reverses and flows in the other. ☐ A-c power sources cause a-c current to flow. ☐ Circuits that use a-c power sources are called a-c circuits, and the power consumed by them is a-c power. ☐ Ninety percent of all electric power consumed throughout the world is ac.

☐ There are two basic reasons why ac is used: it can be transmitted more easily and cheaply than can dc; and it can do certain things and be used for certain applications for which dc is unsuited. ☐ Heat generated in transmission lines is wasted power. ☐ The transmission of a-c power at a relatively low current keeps power losses to a minimum. ☐ The same power level can be produced by many combinations of voltage and current.

☐ Electrical energy radiates from any circuit in which alternating current flows. This is the basis of radio communications. ☐ A-c power sources are called a-c generators or alternators. ☐ The simplest type of a-c generator consists of a single loop of wire placed between the poles of a magnet and made to rotate. The cutting of magnetic lines of force develops a voltage. ☐ A rotating series of loops is called a rotor, or an armature. ☐ Slip rings and brushes are used to apply the output voltage from a generator to an external circuit.

review questions

1. What is *a-c current?* *D-c current?*
2. Is d-c electric power still used in the United States?
3. What is meant by *power losses?*
4. How can power losses be minimized?
5. What is the effect on the power loss if the resistance of a transmission line is doubled?
6. What is the effect on the power loss if the current through a transmission line is doubled
7. What device is used to convert a-c power from one combination of voltage and current to another? Why is it used?
8. What are a-c power sources called?
9. What are *slip rings* and *brushes?*
10. Which would produce the greater power loss: the transmission of power at 100 volts and a current of 10 amperes, or at 200 volts and 5 amperes?

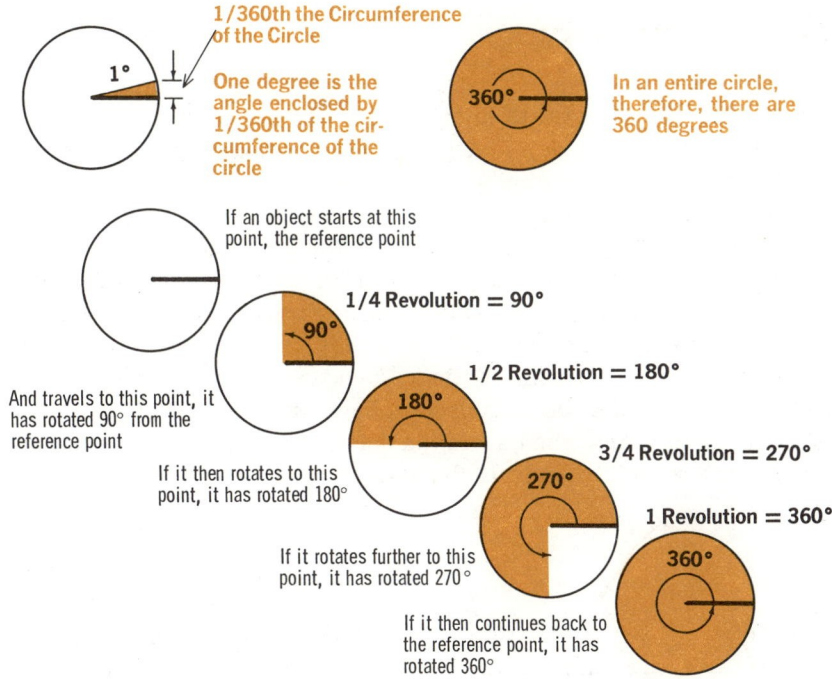

1/360th the Circumference of the Circle

1°

One degree is the angle enclosed by 1/360th of the circumference of the circle

360°

In an entire circle, therefore, there are 360 degrees

If an object starts at this point, the reference point

1/4 Revolution = 90°

90°

And travels to this point, it has rotated 90° from the reference point

1/2 Revolution = 180°

180°

If it then rotates to this point, it has rotated 180°

3/4 Revolution = 270°

270°

If it rotates further to this point, it has rotated 270°

1 Revolution = 360°

360°

If it then continues back to the reference point, it has rotated 360°

angular motion

The loop of wire in the simple generator that was described on page 3-9 rotated within the magnetic field. And, as you know, rotation is movement in a *circle*, such as the turning of an airplane propeller. You will find later that it is often necessary to talk about the voltage that results from each position of the loop of wire as it rotates. To do this, you have to understand *angular motion* and the way it is expressed.

Angular motion is motion in a circle. It is usually described by dividing the circumference of a circle into 360 equal lengths. If lines are drawn from the ends of any one of these lengths to the center of the circle, the distance between the lines is called *1 degree*. Since this can be done for each of the 360 equal lengths, there are *360 degrees* in the *circle*. A line from the circumference of the circle to the center is called a *radius*. And so the distance between any two radii of a circle is measured in *degrees*. This distance is always measured in a counter-clockwise direction from one radius to the other. In practical use, one radius corresponds to the body or object that is rotating. The other radius is a *reference point* from which the position of the first radius is measured.

a-c waveforms

Very often, it is useful to know how a current or voltage *changes* with time. The easiest way to do this, and a way that gives a picture representation of the current or voltage, is to construct a *waveform* on graph paper. A waveform shows the *magnitude* and *direction* of the current or voltage at every instant of *time*. To make a waveform, you label the two axes as shown. One axis, usually the vertical one, is the *current* or *voltage axis*, and is divided into suitable divisions of current or voltage. The other axis is usually the *time axis*, and is divided into suitable divisions of time, such as seconds. With the axes labeled, you then plot the current or voltage at each unit of time as a dot on the graph. And when all of the dots are connected by a continuous line, the resulting figure is the waveform.

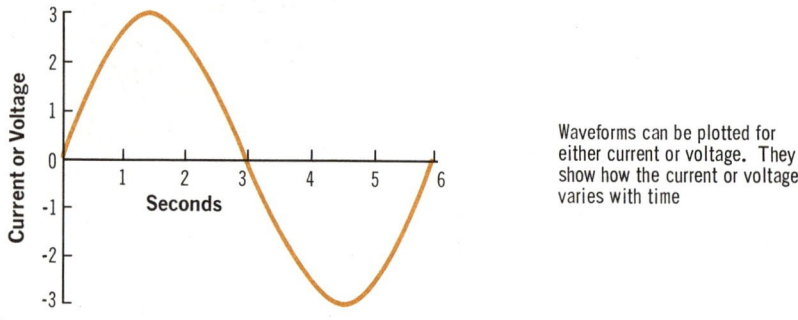

Waveforms can be plotted for either current or voltage. They show how the current or voltage varies with time

When you are dealing with a-c power sources, you may be interested in knowing how the output voltage of a generator varies as the position of the armature changes during its rotation. In this case, instead of labeling the horizontal axis in units of time, you would divide it into *degrees of rotation*. The waveshape would then show the magnitude and polarity of the voltage for each position of the armature.

For a-c power sources, waveforms can also show how the output voltage varies with the position of the generator armature

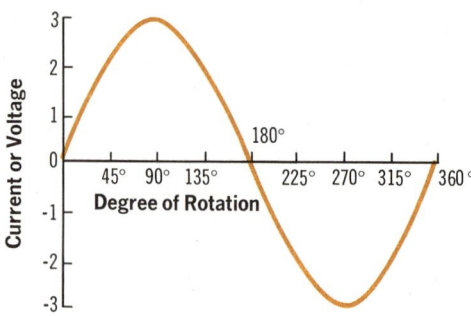

a-c vs. d-c waveforms

The polarity and magnitude of a d-c current or voltage never change. Therefore, the waveform of a 2-volt d-c voltage would be a straight line.

An a-c current or voltage changes in both magnitude and polarity. This can be seen from the waveform of an a-c current. Where the waveform is above the zero current line, the current is flowing in one direction, which is called the positive direction in this case. Where the

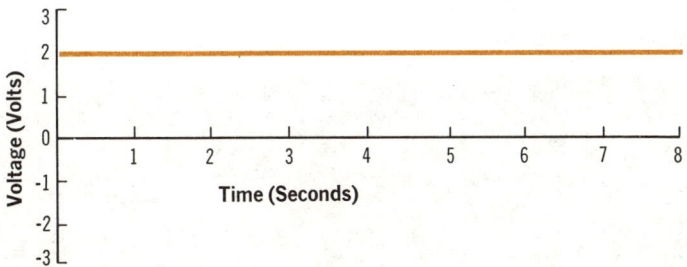

waveform is below the zero current line, as shown in color, the current has reversed and is flowing in the opposite direction. You can see that the current represented by the waveform first flowed in one direction, then reversed and flowed in the other direction, and started to reverse again, all within 8 seconds time.

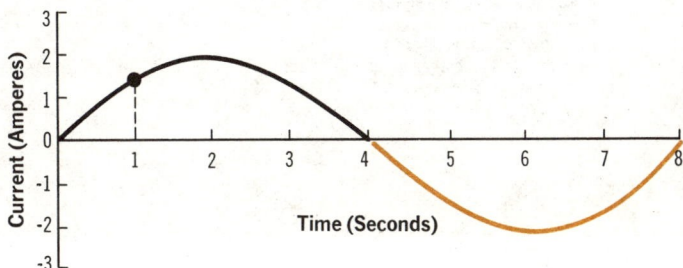

The distance from any point on the waveform to the time axis is the magnitude of the current at that point in time. Thus at 1 second, the current is 1½ amperes, as shown by the dot on the waveform. Similarly, at 8 seconds there is no distance between the waveform and the time axis, and so the current is zero.

the sine wave

The voltage produced by the simple a-c generator described previously has a characteristic waveform that is important throughout a-c circuit theory. This waveform describes the output voltage of the generator during *one full revolution* of the armature. The voltage starts at *zero,* when the armature is not cutting any magnetic lines of force. As the armature turns, the voltage increases from zero to a *maximum* value in one direction. It then decreases until it reaches *zero* again. At zero, the voltage reverses polarity, and increases until it reaches a *maximum* at this opposite polarity. It then decreases until it reaches *zero* again. At this point, the generator armature has completed one *full revolution.*

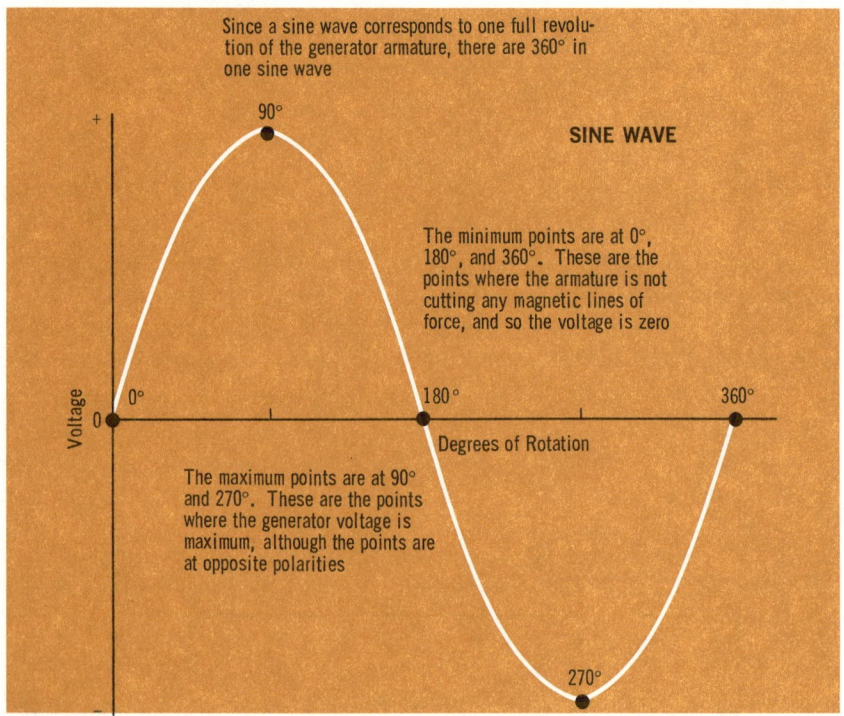

Since a sine wave corresponds to one full revolution of the generator armature, there are 360° in one sine wave

90°

SINE WAVE

The minimum points are at 0°, 180°, and 360°. These are the points where the armature is not cutting any magnetic lines of force, and so the voltage is zero

0°   180°   360°

Degrees of Rotation

The maximum points are at 90° and 270°. These are the points where the generator voltage is maximum, although the points are at opposite polarities

270°

Voltage

For every revolution that the armature makes, the voltage varies in this same way. The waveform that describes this variation in voltage for one full revolution of the armature is called a *sine wave.* It gets its name from the fact that the voltage generated at any point in the armature's travel is proportional to the sine of the angle between the magnetic field and the direction of motion of the armature.

generation of a sine wave

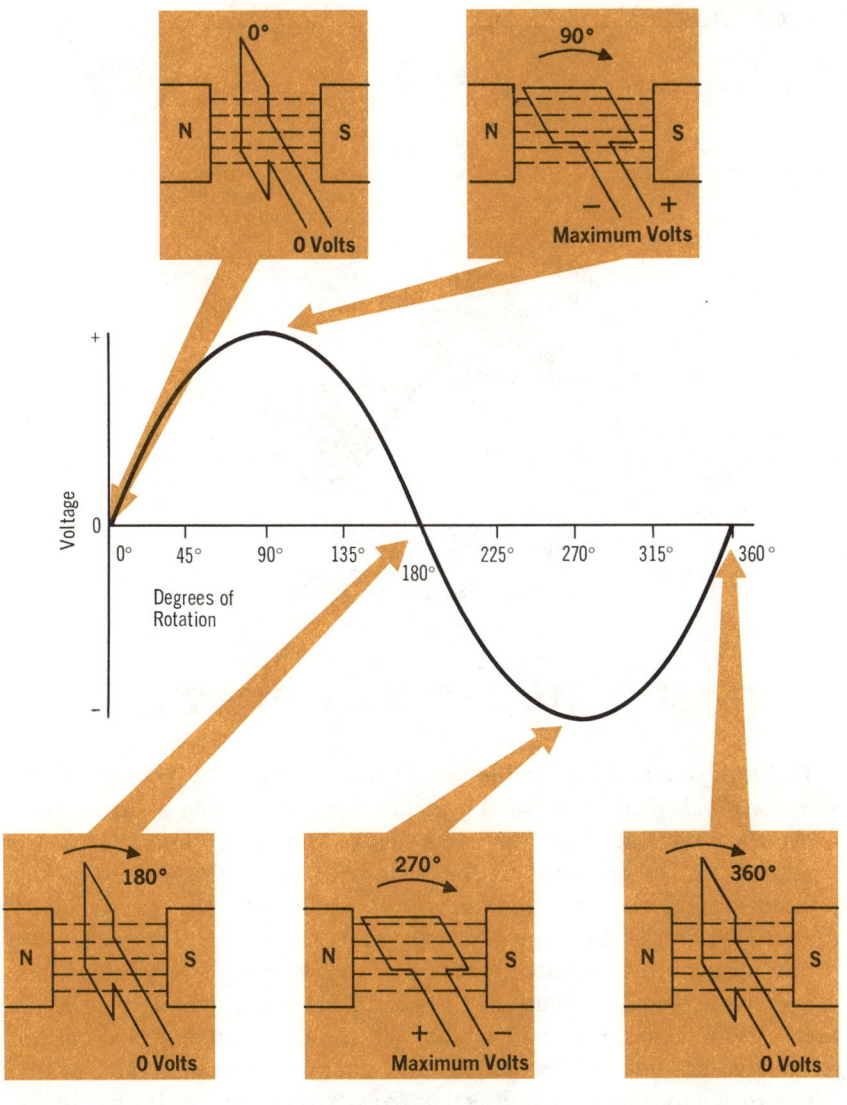

As the generator armature rotates, the magnitude and polarity of the voltage produced follows the pattern of the sine wave

**A-C Waveforms Are Symmetrical
About the Horizontal Axis**

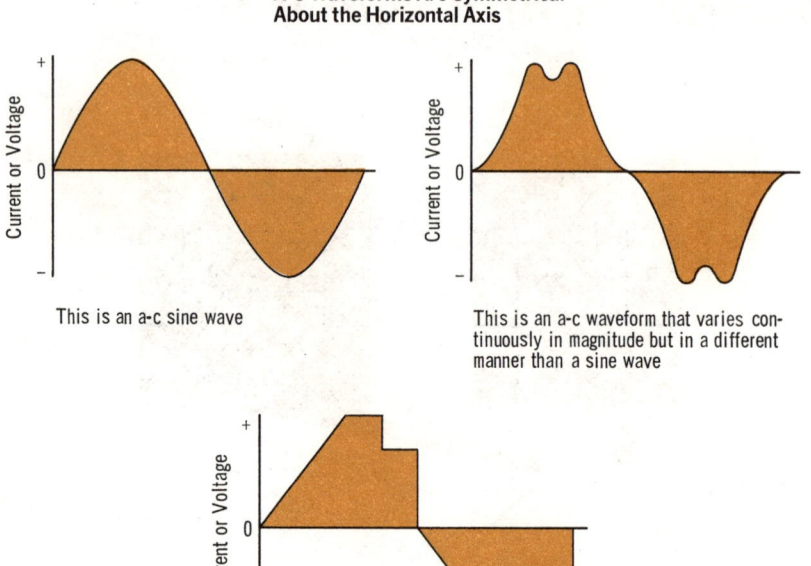

This is an a-c sine wave

This is an a-c waveform that varies continuously in magnitude but in a different
manner than a sine wave

This is an a-c waveform that does not
vary continuously in magnitude. The
magnitude does not change where the
waveform is a horizontal straight line

symmetry of a-c waveforms

Perhaps you noticed on the previous page that the portion of the
sine wave *below* the horizontal axis has the *same* shape as the *portion*
above the axis. Both have the same height and width, and vary in the
same way. In other words, if the negative part of the waveform was
pivoted around the axis and brought up alongside the positive part,
the two halves of the waveform would be identical. This *symmetry*
of the positive and negative portions of the waveform is characteristic
of a-c voltages and currents. If the waveform is not symmetrical about
the horizontal axis, it is not *pure* ac. We can thus define an a-c current
or voltage as one which periodically changes direction, and which
varies in magnitude in one direction exactly the same as it does in the
other direction.

In working with ac, you will become familiar with other waveforms
besides the sine wave. Two of the most common, and two with which
you should become thoroughly familiar, are the *square wave* and the
sawtooth wave.

the square wave

A very common type of waveform in which the current or voltage does not vary in magnitude continuously is called the *square wave*. In a square wave, the current or voltage increases *instantly* from *zero* to some *maximum* value. It then does not vary, but *stays* at this maximum value for a period of time. After this period of time, the current or voltage does three things *instantly:* (1) it *decreases* to zero, (2) it *reverses* direction, and (3) it *increases* to its maximum value in this opposite direction. It stays at this negative maximum value for a time, and then *decreases instantly* to zero. The waveform is thus made up of a series of *straight lines,* as shown.

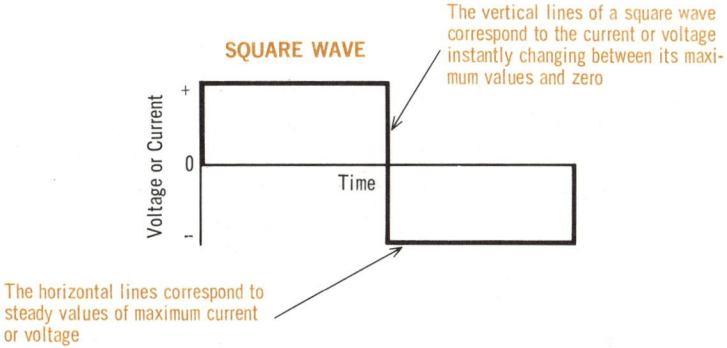

SQUARE WAVE

The vertical lines of a square wave correspond to the current or voltage instantly changing between its maximum values and zero

The horizontal lines correspond to steady values of maximum current or voltage

If a square wave was greatly expanded with respect to the time axis, the vertical lines would actually be sloping lines, because the current or voltage cannot change in value instantly. There would be a definite time required for the changes

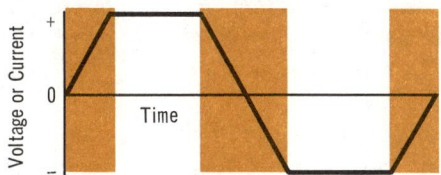

Actually, the current or voltage does *not* change between its maximum values and zero instantly. However, these changes are so fast, that for all practical purposes they can be considered as happening instantly. You will find that this is very often true in the field of electricity. Many things happen so quickly that they can be considered, and will be called *instantaneous,* even though they are actually not.

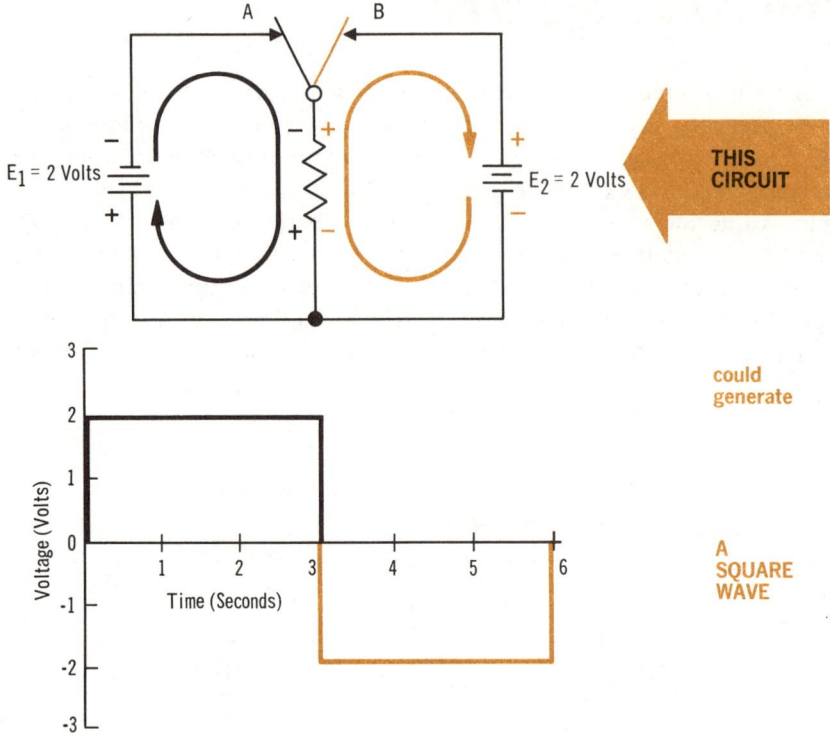

THIS
CIRCUIT

could
generate

A
SQUARE
WAVE

generation of a square wave

If a switch in the circuit is alternately flipped between A and B, the waveform of the voltage across the resistor will be a square wave. For example, when the switch is first set in position A, the current flows from battery E_1 through the resistor, and instantly 2 volts are dropped across the resistor. This voltage stays at 2 volts as long as the switch is at A.

If after 3 seconds the switch is flipped to B, the voltage across the resistor caused by the current from E_1 drops to zero. Current from E_2 then flows through the resistor in the opposite direction, and causes 2 volts to be dropped across the resistor. This 2 volts is opposite in polarity to that previously caused by E_1.

If after 3 seconds, the switch is moved away from position B, the current from E_2 stops and the voltage across the resistor drops to zero. The voltage across the resistor has then completed one full square wave. If the switch was put back to position A and the process repeated, another square wave would be generated.

the sawtooth wave

You have probably noticed that waveforms get their names from their shapes. Thus, a square wave is square, or possibly rectangular, and a sine wave is sinusoidal. There is another very common waveform, which once you hear its name, you will probably have a good idea of its shape. This waveform is called the *sawtooth* wave, and looks very much like a tooth on a common wood saw.

To understand how a sawtooth wave is produced, you first have to know what a *linear* increase in current or voltage is. You already know that an instantaneous change in current or voltage is represented on a waveform by a *straight vertical line*. Curved lines on a waveform, such as a sine wave, show that the current or voltage is changing in a *nonlinear* way. This means that in each *equal increment*, or piece, of time, the current or voltage changes by a *different amount*. For example, in the first second current may rise from 0 to 5 amperes, or increase 5 amperes; in the next second it may go from 5 to 8 amperes, or increase 3 amperes; and in the next second it may rise to 10 amperes, or increase 2 amperes. In equal increments of 1 second, the current thus rose 5, 3, and 2 amperes. This is a *nonlinear* change in current.

To change linearly, the current or voltage must change by *equal amounts* in *equal time intervals*. This means that in the above example, the current would have to go from 0 to 5 amperes in the first second, from 5 to 10 amperes in the second second, and from 10 to 15 amperes in the third second. Its linear increase would then be 5 amperes per second. On a waveform, a linear change in current or voltage is represented by a *straight sloping* line.

The sawtooth wave starts at zero, and linearly increases to its maximum value in one direction. It then instantly drops to zero, reversing its direction, and increases to its maximum value in this direction. The instant it reaches its maximum value, it begins to linearly decrease to zero again.

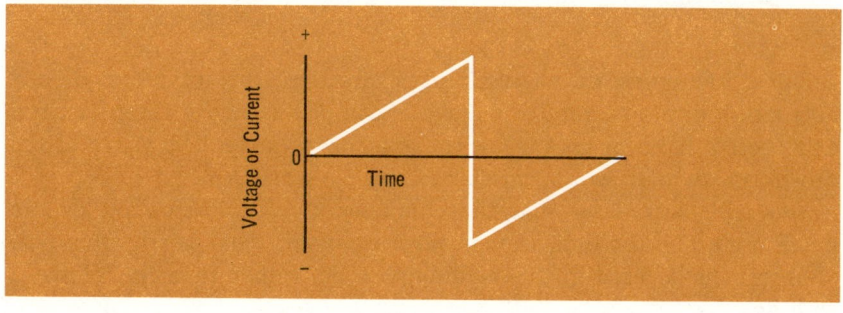

fluctuating dc

You know that every current has a *magnitude* and a *direction*. For direct current, both the magnitude and direction are constant and never change. For alternating current, on the other hand, both of them do change, with the direction reversing periodically, and the magnitude varying between zero and some maximum value in both directions. There is another type of current in which the *magnitude varies* but the *direction never changes*. This current is called *fluctuating dc*, since it can be considered a direct current that fluctuates, or changes, in value.

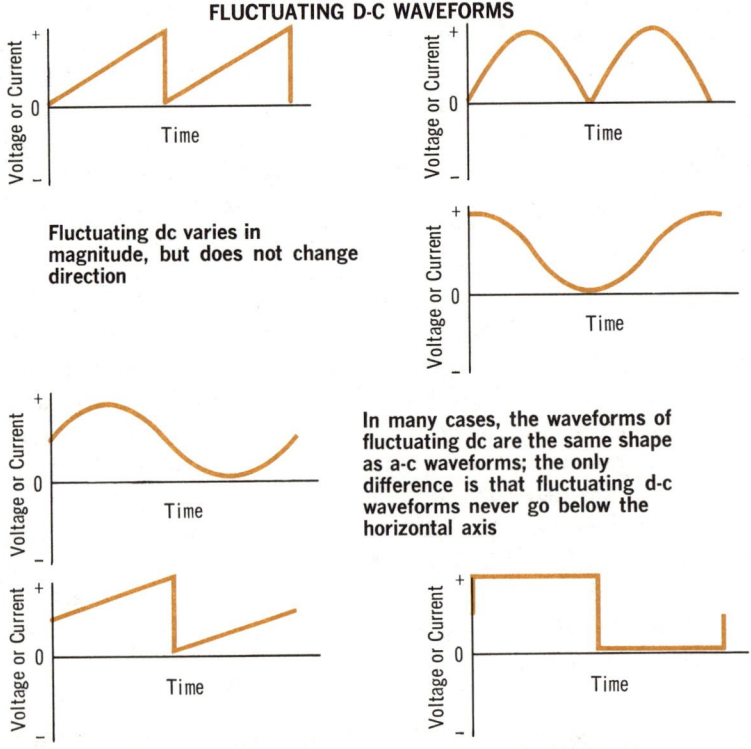

FLUCTUATING D-C WAVEFORMS

Fluctuating dc varies in magnitude, but does not change direction

In many cases, the waveforms of fluctuating dc are the same shape as a-c waveforms; the only difference is that fluctuating d-c waveforms never go below the horizontal axis

Since fluctuating dc never changes direction, the waveform of any fluctuating direct current or voltage is completely above the horizontal axis (zero). It never goes below the axis, or negative. The shape of the waveform, though, can be similar to any a-c waveform.

You will learn later that the characteristics of fluctuating dc are much more like those of ac than they are of dc. That is why this type of current is covered in this volume with ac, rather than in Volume 2 with dc.

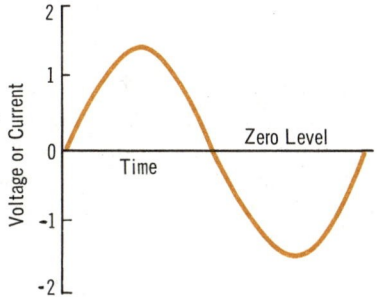

An a-c voltage or current changes in magnitude in one direction the same way that it changes in magnitude in the other direction. It, therefore, varies around a zero voltage or current level.

the a-c component

Fluctuating dc is similar to regular dc in that it does not change direction. It is also similar to ac, since it varies in magnitude. Some types of fluctuating dc can be considered as *combinations* of ac and dc. In actual electric circuits, this is often the case. A d-c voltage or current is combined with an a-c voltage or current, and fluctuating dc is produced. When this is done, the dc varies in magnitude in an a-c way. The a-c variation is called the a-c component, and the dc is called the *d-c reference level.*

The waveform for such a voltage or current is identical to an a-c waveform, except that it is entirely *above* the horizontal axis. The d-c reference for this type of waveform is the horizontal line that half of the waveform is above and the other half is below. Thus, the a-c component varies *around* the d-c reference.

The a-c component can be removed from its d-c reference and be converted to a pure a-c voltage or current that varies around zero by devices called *transformers* and *capacitors.* These devices will be covered later.

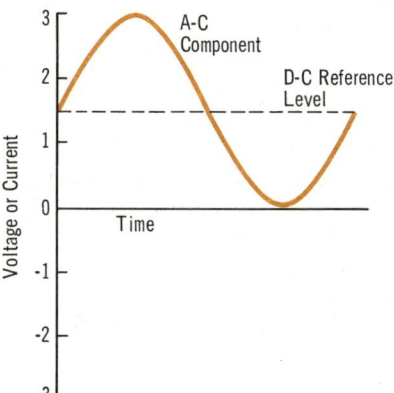

When a d-c voltage or current varies in an a-c way, it does not change direction. However, it fluctuates around the d-c reference level in exactly the same way that ac fluctuates around the zero level

summary

☐ Angular motion is motion in a circle. ☐ A circle is divided into 360 degrees. ☐ A waveform shows the magnitude and direction of current and voltage at every instant of time. ☐ The direction and magnitude of a d-c current or voltage never change. ☐ The voltage produced by the simple a-c generator has the sine wave as its characteristic waveform. ☐ The sine wave has symmetrical positive and negative portions above and below the zero reference level.

☐ If a waveform is not symmetrical about the horizontal axis, it is not pure ac. ☐ An a-c current or voltage is one which periodically changes direction, and which varies in magnitude in one direction exactly the same as it does in the other direction. ☐ The square wave increases instantly from zero to a maximum value. It stays at this maximum value for a period of time, decreases instantly to zero, and increases to a maximum value in this opposite direction. It stays at this negative maximum value for a period of time and then decreases instantly to zero.

☐ The sawtooth wave increases linearly from zero to a maximum value. It then instantly drops to zero, reversing its direction, and increases to a maximum value in this opposite direction. When it reaches this negative maximum value, it linearly decreases to zero again. ☐ Fluctuating dc changes in value, but never goes below the horizontal axis. ☐ A fluctuating d-c voltage can be considered to be made up of an a-c component varying above and below a d-c reference level. ☐ The a-c component can be removed from the d-c reference level by devices such as transformers and capacitors.

review questions

1. What is *angular motion*? How is it expressed?
2. With respect to the reference radius, in what direction does the rotating radius rotate?
3. What does a waveform show? Which axis is usually the time axis?
4. What is a *sine wave*? Why is it so called?
5. Is a d-c wave symmetrical above and below the horizontal axis?
6. What is a *square wave*? A *sawtooth wave*?
7. Can a square wave be generated by a battery? What additional components are necessary besides the battery?
8. What is *fluctuating dc*? Does it ever go below the horizontal axis?
9. Draw the waveform of a fluctuating d-c voltage.
10. What is meant by the *a-c component*? *D-c reference level*?

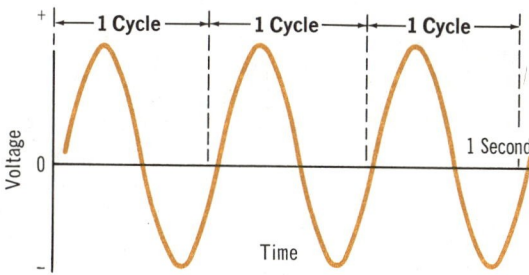

The frequency of a voltage or current is the number of cycles generated each second. The frequency of this voltage is, therefore, 3 Hz

frequency

When an a-c voltage or current goes through 360 degrees, it is said to have completed one *cycle*. To do this, the waveform starts at zero, goes to a maximum positive, then drops to zero; then it continues on to a maximum negative, and returns again to zero. When it starts to repeat itself, it has completed one cycle and is starting the next cycle. The *frequency* of the a-c wave is the number of cycles it completes in 1 second. The more cycles it completes in 1 second, the higher is the frequency.

Frequency is expressed in *cycles per second*. Until the 1960s, this used to be abbreviated *cps*, which seemed natural. Then, though, the electrical engineering profession decided to honor a great scientist named Heinrich Hertz. They abbreviated his name *Hz*, and said it should be used in place of cps. So, a frequency of 10 cycles per second is abbreviated as 10 Hz, and is pronounced 10 hertz.

In the United States the frequency of ordinary house power is 60 hertz, and you will usually see a statement on electrical appliances that they can only use 60-Hz power. Frequency of electrical power in other countries may vary from 25 Hz to 125 Hz. The other major standard frequency in Europe and South America is 50 Hz. In special applications, such as aircraft or military equipment, the power frequency can vary from 400 to 1000 Hz.

CONVERSION OF UNITS

1 kilohertz (kHz) = 1000 Hz	1 megahertz (MHz) = 1,000,000 Hz
Hz X 1000 = kHz	kHz/1000 = Hz
Hz X 1,000,000 = MHz	MHz/1,000,000 = Hz
kHz X 1000 = MHz	MHz/1000 = kHz

Since the frequency of the voltage is 3 Hz, the wavelength is

Wavelength (meters)

= 300,000,000 ÷ frequency

= 300,000,000 ÷ 3

= 100,000,000 meters

As the frequency decreases, the wavelength becomes longer

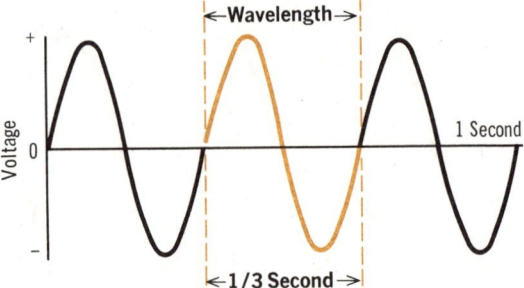

wavelength

You remember from Volume 1 that, although the individual electrons that make up electric current move through a wire quite slowly, the electric field, or impulse, that causes the current flow, moves down a wire at about 186,000 miles per second. Since the current travels at a definite speed, it can only travel a certain *distance* during any period of *time*. And since frequency is actually a measure of the number of *cycles* for a given period of *time*, you can calculate how far current can travel during 1 cycle of a-c voltage. This distance is called the *wavelength*. It is the *distance* that current can move in the time it takes to complete 1 full cycle of a-c voltage.

With a 60-Hz voltage, for example, 1 cycle takes 1/60 of a second. And, since current travels 186,000 miles in one second, it can only travel 3100 miles during 1/60 of a second, or 1 cycle of a 60-Hz a-c voltage. The wavelength of a 60-Hz voltage, therefore, is 3100 miles. Since the wavelength of a-c voltage depends on its frequency and the speed at which the electric impulse moves along a wire, it can be calculated by the equation:

Wavelength = velocity of current/frequency

The velocity of current, as far as basic electricity is concerned, is the speed of light: 186,000 miles per second. However, for many years, it has been standard practice to use the metric system for the speed of light. In the metric system, the basic unit for distance is the *meter;* 1 meter equals 39.37 inches, so that the speed of light is 300,000,000 meters per second. The equation for wavelength then becomes:

Wavelength (meters) = 300,000,000/frequency

The wavelength for 1 cycle of a 60-Hz voltage is now said to be 5,000,000 meters.

You can see that *wavelength* is just another way of expressing *frequency*. The use of wavelength is not too important in electrical power but it is often used in the field of communications.

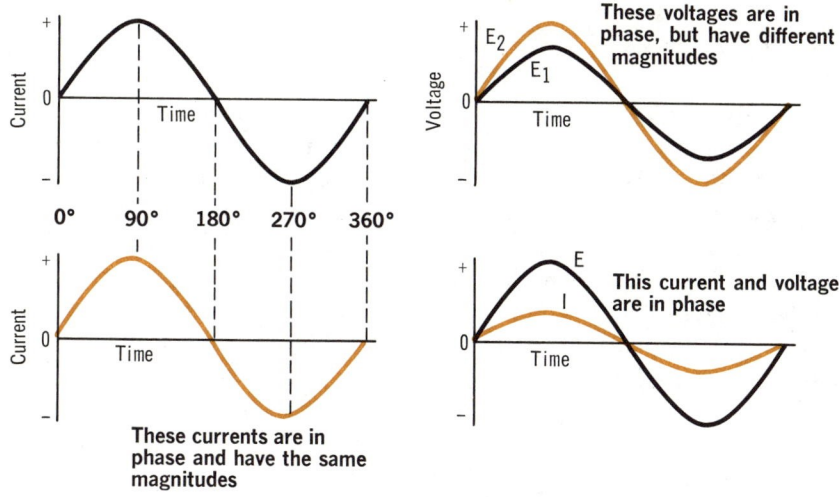

Current and voltages are in phase when they both reach their maximum values and their minimum values at the same time

phase

The output of a simple a-c generator varies as a sine wave. Therefore, if two such generators are started, they each will generate one full sine wave output after one full revolution. If the generators are started at the *same time* and turn at exactly the *same speed*, the two waveforms will *begin* simultaneously and *end* simultaneously. They will also reach their maximum values and pass through zero at the same time. The two waveforms are thus "in step" with each other, and the voltages they represent are said to be *in phase*. From this you can see that the term *phase* is used to indicate the *time* relationship between alternating voltages and currents.

When two currents or voltages are in phase, it does not mean that their *magnitudes* are the same. The peak magnitudes are reached at the same time, but they can have different values.

Although phase is usually used to compare the time relationship of two waveforms, it can also be used to indicate a point in time for one waveform. As shown on page 3-15, a full cycle can be represented by degrees. These degrees are often called *phase angles*. The phase of the positive peak is 90 degrees, and the negative peak is 270 degrees. The sine wave is zero at the phase angles of 0, 180, and 360 degrees. In this way, any point on a sine wave can be referred to as some phase angle.

phase difference

If two identical generators are started at the same time and turn at the same speed, their maximum and minimum output values will occur simultaneously, and so the two outputs will be *in phase*. But if one generator is started after the other, its maximum and minimum output values will occur *after* the corresponding values for the other generator. The two outputs in this case are *out of phase*. Or, to put it another way, a *phase difference* exists between the two outputs. The amount of the phase difference depends on how far behind one output is from the other.

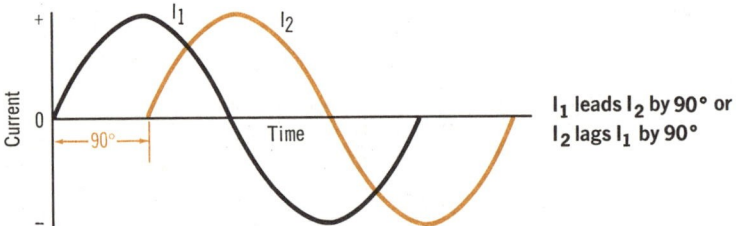

I₁ leads I₂ by 90° or
I₂ lags I₁ by 90°

When maximum and minimum points of one voltage or current occur before the corresponding points of another voltage or current, the two are out of phase. When such a phase difference exists, one of the voltages or currents leads, and the other lags

E₁ leads E₂ by 180° or
E₂ lags E₁ by 180°

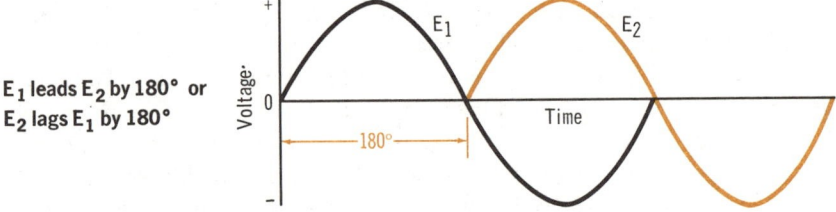

Phase difference can be expressed in fractions of a cycle. Then, if one output begins when the other has just completed one-half of a cycle, the phase difference is one-half cycle. Usually though, for more accuracy, phase difference is given in *degrees*. And since one full sine wave corresponds to 360 degrees, a phase difference of one-half cycle is a 180-degree phase difference; a one-quarter cycle difference is a 90-degree phase difference, etc.

The terms *lead* and *lag* are used to describe the relative positions in time of two voltages or currents that are out of phase. The one that is *ahead* in time is said to *lead*, while the one *behind lags*.

other a-c terms

You are already familiar with most of the terms normally used to describe a-c voltages and currents and their waveforms. However, in addition to *cycle, frequency, wavelength,* and *phase,* there are other a-c terms that are used very often, which you should know. For example, ½ of a full cycle is often referred to as an *alternation.*

Another term is *amplitude.* The amplitude of an a-c current or voltage is the *maximum value* that the current or voltage reaches. It is the same in both the positive and negative directions. On a waveform, the amplitude is the distance from 'the horizontal axis to either the highest point of the wave above the axis, or to the lowest point below the axis. The amplitude is often also called the peak value.

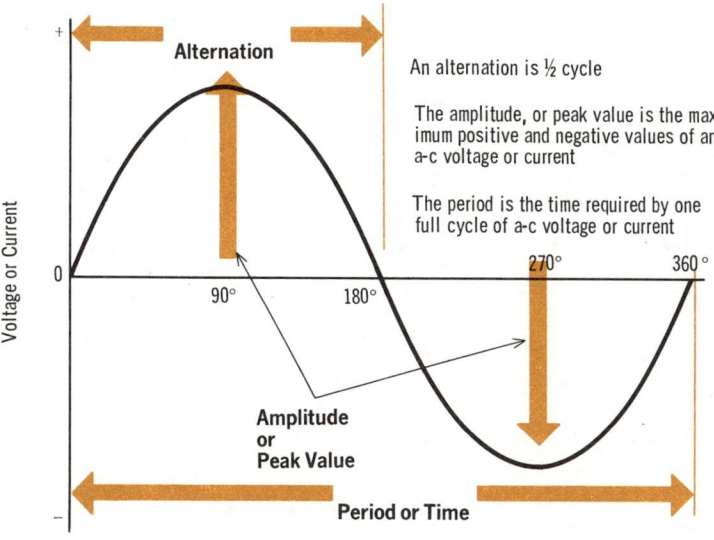

Alternation

An alternation is ½ cycle

The amplitude, or peak value is the maximum positive and negative values of an a-c voltage or current

The period is the time required by one full cycle of a-c voltage or current

Voltage or Current

90° 180° 270° 360°

Amplitude
or
Peak Value

Period or Time

Another term you should know is *period.* The period of an alternating quantity, such as a-c voltage or current, is the *time* it takes to complete *one full cycle* of the quantity. If you know the frequency, the period can easily be calculated. For example, for a 60-Hz voltage, 60 cycles are generated in one second. It therefore takes 1/60 of a second to generate one cycle. Thus, you obtain the period by dividing the frequency into 1.

$$\text{Period} = \frac{1}{\text{frequency}}$$

The period is in seconds, and the frequency is in cycles per second, Hz.

values of a-c voltage and current

To specify the value of a d-c voltage or current is no problem, since d-c values do not change. The values of a-c voltages and currents, though, are constantly changing, and so there is a problem when it comes to specifying them. Before you can give the value of an a-c voltage or current, you usually have to determine what type of value is needed. And this in turn depends on how you want to use the value. The most obvious value to you is probably the *peak value,* which you know is the *amplitude* or maximum value of the voltage or current. Another value sometimes used is the peak-to-peak value, which is twice the peak value. On a waveform, the peak-to-peak value is the distance from the maximum positive value to the maximum negative value.

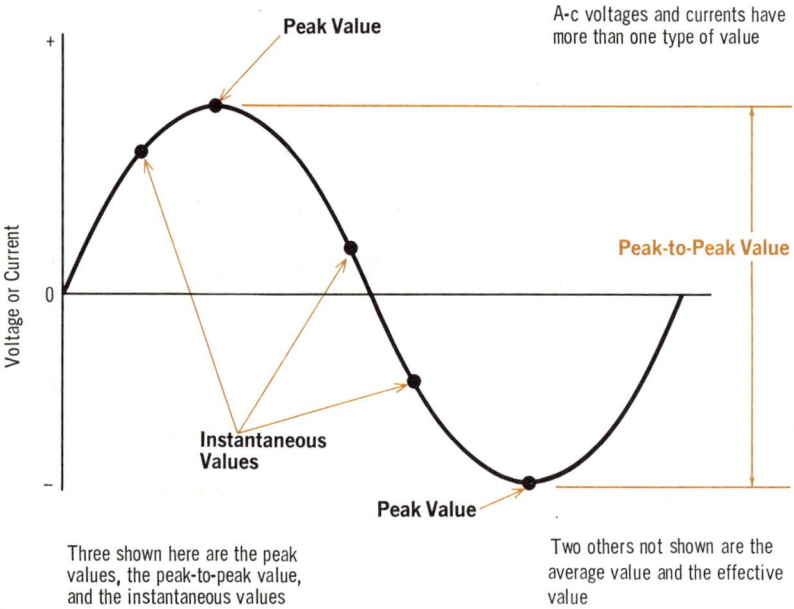

A-c voltages and currents have more than one type of value

Three shown here are the peak values, the peak-to-peak value, and the instantaneous values

Two others not shown are the average value and the effective value

Occasionally, you might be interested in the *instantaneous value* of voltage or current. This is the value at one particular instant of time; and depending on the instant you select, it can range anywhere from the zero to the peak value.

In most cases, none of these values (peak, peak-to-peak, or instantaneous) are satisfactory for giving the actual values of a-c voltages and currents. Instead, two other values, called the *average* and the *effective* values, are generally used.

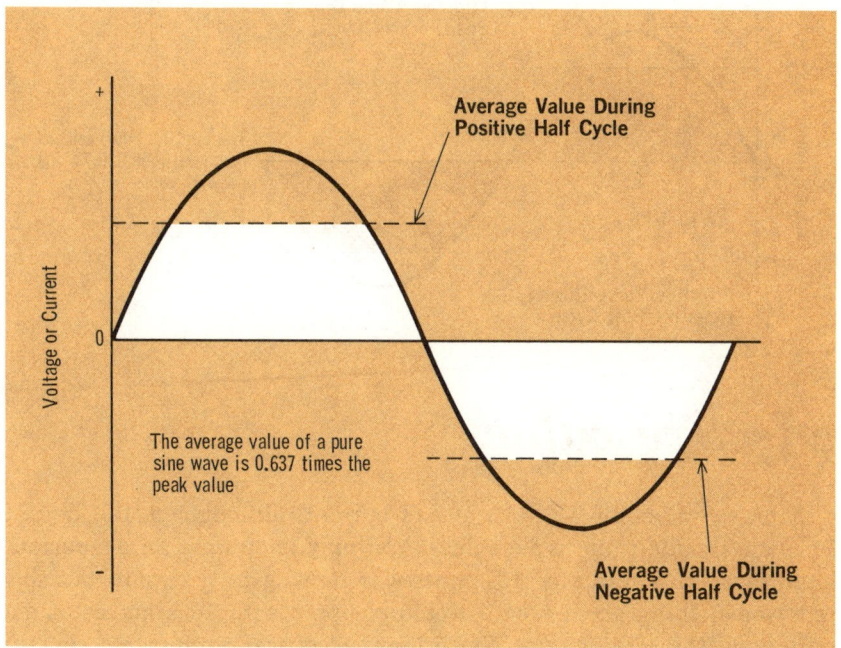

average values

The average value of an a-c voltage or current is the *average* of all the instantaneous values during *one half cycle,* or alternation. Since during a half cycle, the voltage or current increases from zero to the peak value and then decreases back to zero, the average value must be somewhere between zero and the peak value. For a pure sine wave, which is the most common waveform in a-c circuits, the average value is 0.637 times the peak value. For voltage, this is expressed by the equation:

$$E_{AV} = 0.637\ E_{PK}$$

As an example, if the peak voltage in a circuit is 100 volts, the average voltage will be:

$$E_{AV} = 0.637\ E_{PK} = 0.637 \times 100 = 63.7\ \text{volts}$$

The equation for average current in terms of peak current is identical to that for voltage.

You should be careful never to confuse average value, which is the average of *one-half cycle,* with the average over a *complete cycle.* Since both half cycles are identical, except that one is positive and the other negative, the average over a *complete cycle* would be *zero.*

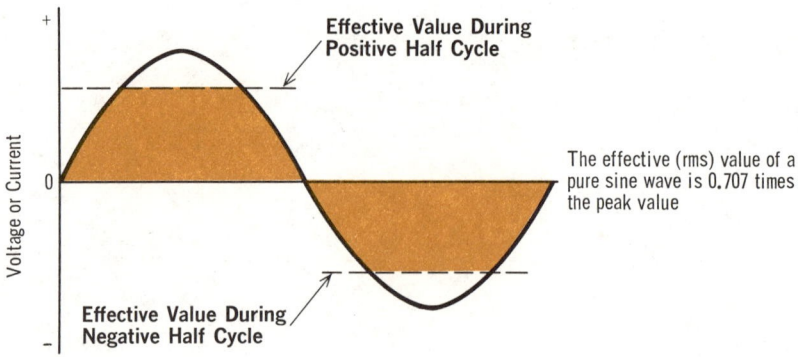

The effective (rms) value of a pure sine wave is 0.707 times the peak value

effective values

Average values of a-c voltage and current, although useful, do not have any relationship to d-c values. You may know that an a-c current with an average value of 10 amperes is flowing in a circuit, but this tells you nothing about how it would compare with 10 amperes of d-c current in the same circuit. Since many electrical equipments contain both a-c and d-c circuits, it is very useful to be able to express a-c currents and voltages in values related to dc. This is possible with the use of *effective values.*

The effective value of an a-c voltage or current is the value that will cause the same amount of *heat* to be produced in a circuit containing only resistance that would be caused by a d-c voltage or current of the same value. Thus, an alternating current with an effective value of 1 ampere generates the same heat in a 10-ohm resistor as does a direct current of 1 ampere. The effective value is also called the *root-mean-square,* or *rms,* value because of the way it is derived. It is equal to the square root of the average value of the squares of all the instantaneous values of current or voltage during one-half cycle.

In a pure sine wave, the effective value is 0.707 times the peak value. The equations for the effective values of current and voltage are, therefore,

$$I_{EFF} = 0.707 \; I_{PK} \qquad E_{EFF} = 0.707 \; E_{PK}$$

Therefore, for a peak voltage of 100 volts, an a-c voltage would have an rms value of 70.7 volts. This means that a resistor that is put across a 100-volt a-c source will produce the same heat as if it were placed across a 70.7-volt d-c source.

The effective value is the one that is usually used to rate a-c voltage and currents. The home line voltage of 110 volts is the *rms* value. And so is the 220-volt industrial power voltage.

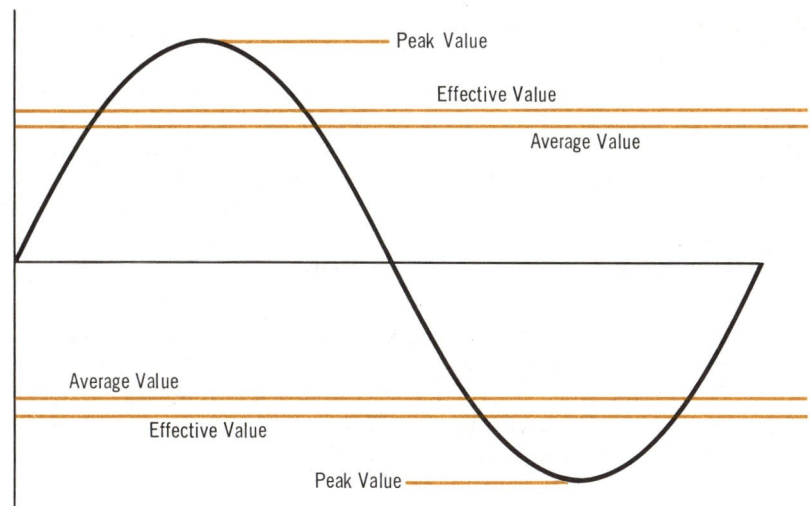

converting values

In working with a-c circuits you will often have to *convert* given or measured values of a-c voltage or current to other values. For example, you may have to convert an average value to a peak value, or maybe an effective value to an average value. For all conversions between peak, average, and effective values, there are six basic equations that apply. By using the proper equation, you can easily convert from any one of these values to any other. The six equations are given below for converting voltage and current values.

CONVERTING A-C VOLTAGES AND CURRENTS		
To Convert From	**To**	**Use the Equations**
Peak	Average	$E_{AV} = 0.637\ E_{PK}$ $\qquad I_{AV} = 0.637\ I_{PK}$
Peak	Effective	$E_{EFF} = 0.707\ E_{PK}$ $\qquad I_{EFF} = 0.707\ I_{PK}$
Average	Peak	$E_{PK} = 1.57\ E_{AV}$ $\qquad I_{PK} = 1.57\ I_{AV}$
Average	Effective	$E_{EFF} = 1.11\ E_{AV}$ $\qquad I_{EFF} = 1.11\ I_{AV}$
Effective	Peak	$E_{PK} = 1.414\ E_{EFF}$ $\qquad I_{PK} = 1.414\ I_{EFF}$
Effective	Average	$E_{AV} = 0.9\ E_{EFF}$ $\qquad I_{AV} = 0.9\ I_{EFF}$

In some cases, you may have to convert from or to peak-to-peak values. To do this it is best to use the equations for the peak value, and remember that the peak-to-peak value is twice the peak value, and conversely that the peak value is one half the peak-to-peak value.

summary

☐ The number of cycles generated in one second is called the frequency of a voltage or current, and is expressed in cycles per second, Hz. ☐ Most of the electric power generated in the United States has a frequency of 60 Hz. Other countries use different power frequencies, ranging from 25 to 125 Hz. ☐ Frequency can also be expressed in kilohertz, kHz; and megahertz, MHz. ☐ The distance a wave travels during one cycle is called its wavelength. ☐ The wavelength is equal to the wave's velocity divided by its frequency.

☐ Phase is used to indicate the time relationship for a-c voltages and current. ☐ Phase difference expresses the time relationship between two voltages, currents, or combinations of both, with respect to each other. This difference is expressed in parts of a cycle, or degrees. ☐ When two waveforms are in step with each other, they are said to be in phase.

☐ An alternation is one-half of a complete cycle. ☐ The time it takes to complete one full cycle is the period of the waveform. Period equals 1/frequency. ☐ The average value for one alternation (1/2 cycle) of a sine wave is equal to 0.637 times the peak value. The average for a complete cycle is zero. ☐ For a pure sine wave, the effective, or rms, value is equal to 0.707 times the peak value.

review questions

1. What is the frequency most commonly used for electric power in the United States?
2. Frequency is measured in what units?
3. What is meant by *phase? Phase difference?*
4. Define *amplitude* and *alternation.*
5. What is the effective value of a voltage whose maximum amplitude is 200 volts?
6. What is the rms value of a voltage whose peak-to-peak voltage is 200 volts?
7. What is the average value of the output voltage of a 6-volt battery? The rms value?
8. The average value of a sine-wave current is 5 amperes. What is its peak, effective, and peak-to-peak values?
9. One alternation of a sine wave takes 4/1000 of a second. What is the frequency of the wave?
10. What is the maximum value of a sine wave whose effective value is 50 volts? Whose peak-to-peak value is 50 volts? Whose average value for one alternation is 50 volts?

resistive a-c circuits

In a d-c circuit, the only property that opposes or reduces current flow is resistance. Resistance also affects the current flow in a-c circuits, although it is not always the only property that does. You will learn later that a-c circuits have other properties, which, like resistance, affect the current and voltage throughout a circuit. The *simplest* type of a-c circuit, though, contains *only resistance*. And like a d-c circuit, this resistance includes the individual resistances of the loads, the power source, and the connecting wires.

An a-c resistive circuit is one that contains only resistance

Electric appliances used to produce heat, such as toasters, irons, frying pans, and ranges, are examples of a-c resistive circuits when they are connected to the standard home power source

In actual practice, no a-c circuit can contain only resistance. The other properties that affect the voltage and current are always present in some amount. However, when their effects are very small compared with the effects of the resistance, they can be considered as being not present. The circuit is then a completely *resistive* circuit.

current and voltage

When an a-c voltage is applied across a resistance, an a-c current flows through the resistance. The magnitude of the *current* at any instant is *directly proportional* to the magnitude of the *voltage* at that instant, and is *inversely proportional* to the value of the *resistance*. This is the same relationship that exists between the current, voltage, and resistance in a d-c circuit, and so *Ohm's Law* also applies to the *instantaneous* values of current and voltage in an a-c circuit.

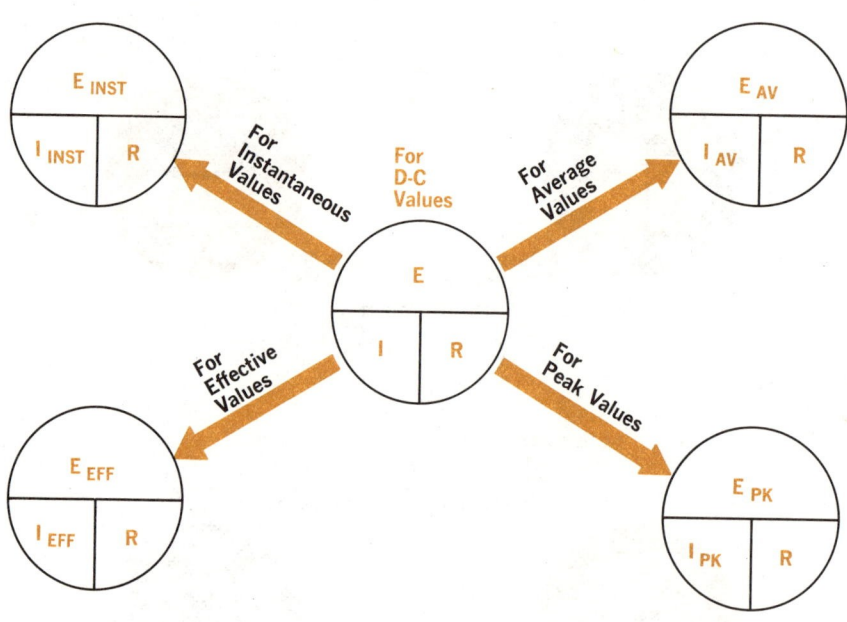

In a-c circuits that contain only resistance, the relationship between the current, voltage, and resistance is that of Ohm's Law

Since the average, effective, and peak values of a-c current and voltage are derived from the instantaneous values, Ohm's Law also applies to them. This means that to find resistances, currents, and voltages in a-c resistive circuits, you use the same rules and methods that you learned for d-c circuits in Volume 2.

phase relationships

Since the instantaneous values of current and voltage in an a-c circuit that contains only resistance follow Ohm's Law, this means . that at any instant when the voltage is zero, the current is also zero. And when the voltage is maximum, the current must also be maximum,

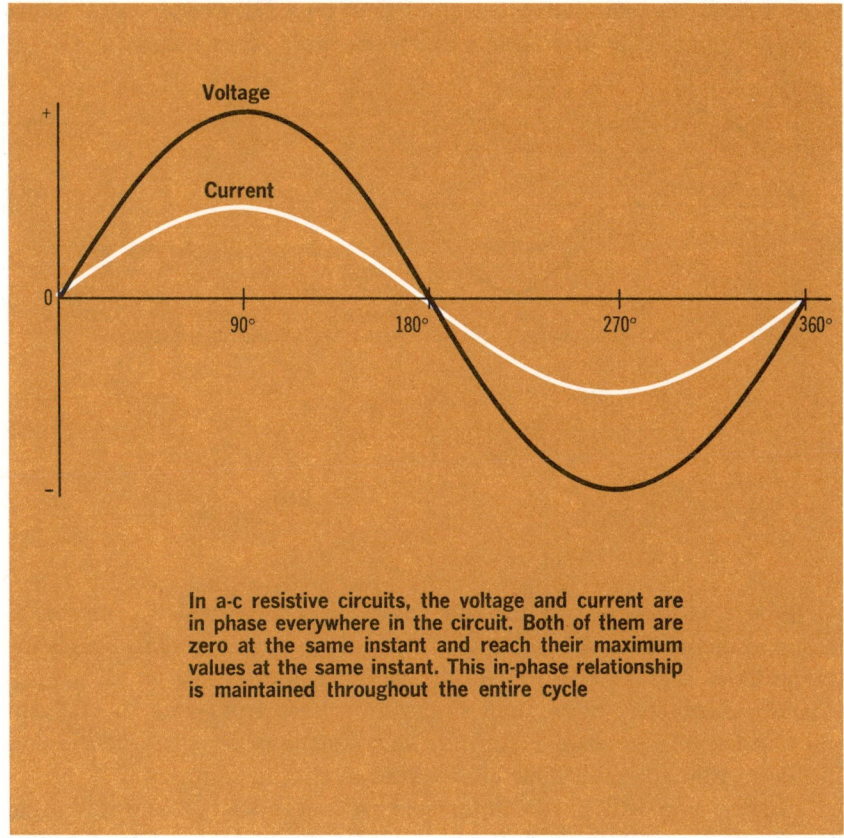

In a-c resistive circuits, the voltage and current are in phase everywhere in the circuit. Both of them are zero at the same instant and reach their maximum values at the same instant. This in-phase relationship is maintained throughout the entire cycle

since the resistance is constant. When the voltage reverses and becomes negative, the current also reverses, inasmuch as it always flows from negative to positive. Thus, at every *instant* of time the current is exactly *in step* with the applied voltage.

In an a-c resistive circuit, therefore, the current and voltage are *in phase*. This is true not only of the total circuit current and the source voltage, but is so for the voltage and current in *every part* of the circuit.

power

The power consumed in any circuit depends on the voltage and current in the circuit. And since in a purely resistive circuit, a-c voltages and currents follow Ohm's Law, it would seem that the power in such a circuit would be calculated in the same way as the power in a d-c circuit. Essentially, this is true. The power in an a-c resistive circuit does follow the standard d-c equation $P = EI$. However, a-c voltages have different types of values, and so, therefore, does a-c power.

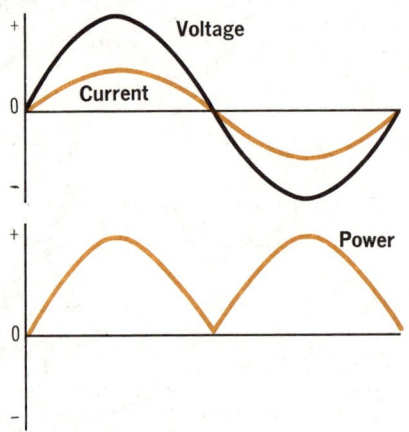

The value of each point on the power waveform is equal to the product of the voltage and current at that instant. The second half cycle of the power waveform is positive since it is produced by the product of two negative quantities

At any particular instant, the power in an a-c resistive circuit is equal to the product of the voltage and current at that instant. This is the *instantaneous power* and is found by the equation $P_{INST} = E_{INST}I_{INST}$. The instantaneous power can range anywhere from zero, if the current and voltage at that instant are zero, to the peak power, if the current and voltage are at their peak values at that instant.

Usually, you will not be interested in instantaneous power, but in the power used during a full cycle. This is often referred to either as *power* or *average power*. To find this, you would use the *effective* or *rms* values of voltage and current, since these values are the ones that give the same power loss effect as the d-c equivalent, as you learned earlier. The formula, then, for finding this power dissipated in a circuit is

$$P = E_{EFF}I_{EFF}$$

Another formula that can be used if you know the peak value is

$$P = \frac{E_{PK}I_{PK}}{2}$$

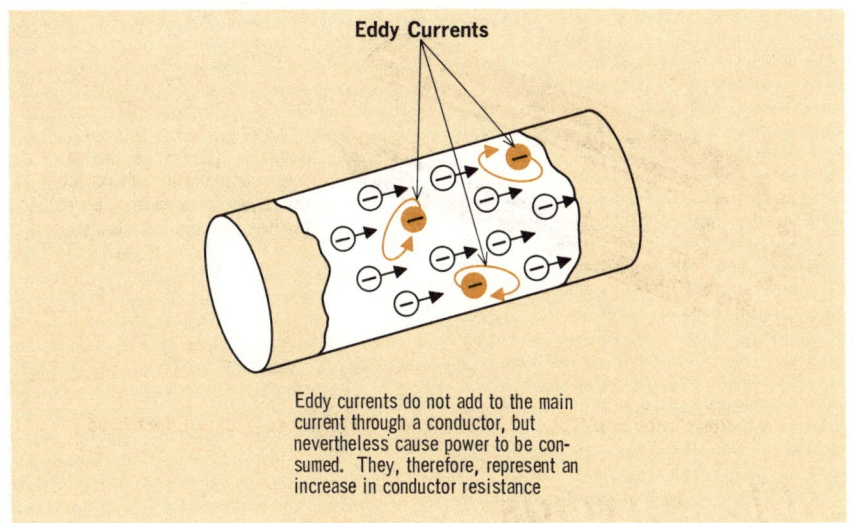

Eddy Currents

Eddy currents do not add to the main
current through a conductor, but
nevertheless cause power to be consumed. They, therefore, represent an
increase in conductor resistance

eddy currents and skin effect

In d-c circuits, resistance is a physical property of the conductors
that opposes current flow. The resistance is directly proportional to
conductor length and inversely proportional to conductor cross-
sectional area. This d-c, or *ohmic*, resistance opposes a-c current the
same as it does d-c current. However, when a-c current flows in a
conductor, the resistance offered to the current by the conductor is
somewhat *greater* than the resistance that would be offered to d-c current by the same conductor. There are two reasons for the increase
in resistance, and both are due to the fact that when an a-c current
flows in a conductor, it causes *voltages* to be set up inside of the
conductor. How this is done will be explained later in this volume.
The voltages set up in the conductor cause small independent currents,
called *eddy currents,* to exist. These eddy currents flowing through
the resistance of the conductor consume power, and therefore represent a *power loss,* or an increase in resistance, in the circuit.

Besides producing eddy currents, the voltages set up in a conductor
by a-c current repel the flowing electrons towards the *surface* of the
conductor. See illustration on page 3-38. More current, therefore, flows
near the conductor surface than at the center of the conductor. This
has the effect of *decreasing* the *cross-sectional area* of the conductor,
and as you have learned, a decrease in cross-sectional area causes an
increase in resistance. The concentration of current flow near the surface of a conductor is called *skin effect.*

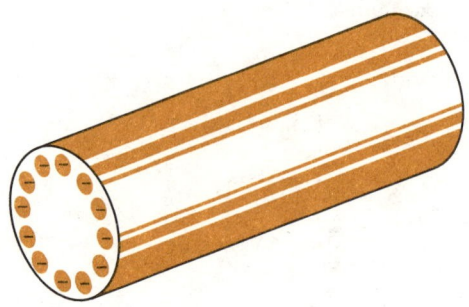

Skin effect results in a higher concentration of current near the conductor surface than at the conductor. Effectively, this reduces the cross-sectional area of the conductor

A-C Resistance = D-C Resistance + Eddy Current loss + Skin Effect Loss

eddy currents
and skin effect (cont.)

The losses in a-c conductors due to eddy currents and skin effect are *directly proportional* to the *frequency* of the current flowing in the conductor. The higher the frequency, the greater are the eddy current and skin effect losses. These losses only become significant when the frequency is very high. For this reason, they will be considered as zero, and therefore neglected in the remainder of the volume. Power cables are usually made of *stranded wires* to reduce the skin effect loss. This works because the surface areas of all the individual wire strands added together is greater than the surface area of a solid conductor.

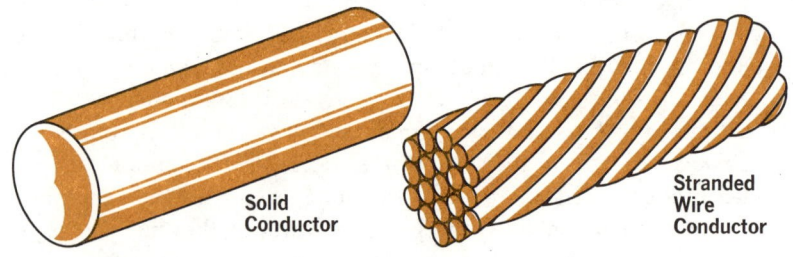

Solid
Conductor

Stranded
Wire
Conductor

Skin effect loss is reduced in power cables by the use of stranded wire. The total surface area of the wires is greater than that of a single conductor

solved problems

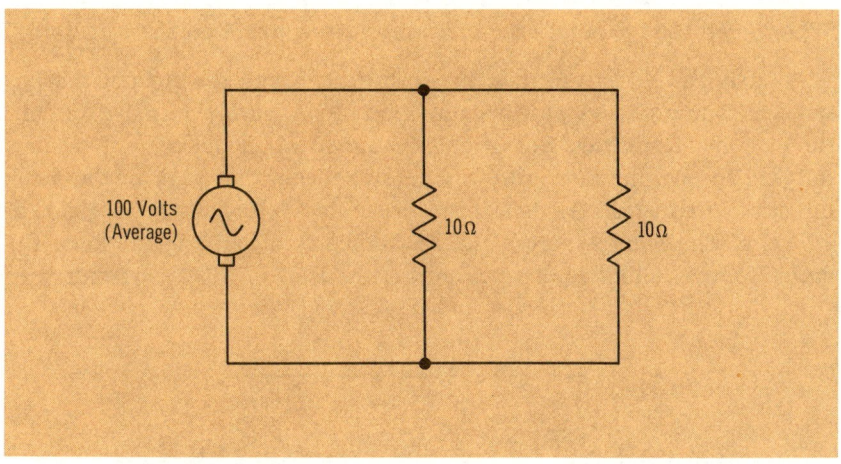

Problem 1. What is the effective value of the total current in the circuit?

The effective value of the current is asked for, but the source voltage given is the average value. The average value could be converted to its equivalent effective value, and the effective current then calculated directly. Or, the current could be solved for the average current and the conversion to the effective value could then be made. Since only one conversion is required, it will be made first, although either method would be equally satisfactory in this problem.

The equation for converting average voltage to effective voltage can be taken from the table on page 3-31.

$$E_{EFF} = 1.11 \, E_{AV} = 1.11 \times 100 \text{ volts} = \text{approx. } 110 \text{ volts}$$

The total resistance of the two 10-ohm resistors in parallel is 5 ohms, and so you know the effective value of the applied voltage and the circuit resistance. Since it is a purely resistive circuit, Ohm's Law can be used to find the effective current.

$$I_{EFF} = E_{EFF}/R = 110 \text{ volts}/5 \text{ ohms} = 22 \text{ amperes}$$

Problem 2. What is the peak value of the circuit current?

The effective current is 22 amperes. To convert this to its equivalent peak current, you use the appropriate equation from the table on page 3-31.

$$I_{PK} = 1.414 \, I_{EFF} = 1.414 \times 22 \text{ amperes} = 31.11 \text{ amperes}$$

solved problems (cont.)

Problem 3. *The current in this circuit has a waveform as shown. What is the voltage across the resistance at instant A on the waveform?*

Point A on the waveform is the instant at which the current reaches its peak value in the positive direction. In a purely resistive circuit, the current and voltage are in phase; and so when the current reaches its peak in one direction, the voltage also reaches its peak in the same direction. At instant A, therefore, the source voltage is at its peak of 10 volts. The voltage across the resistance is also 10 volts, since the entire source voltage is dropped across the load in a series circuit.

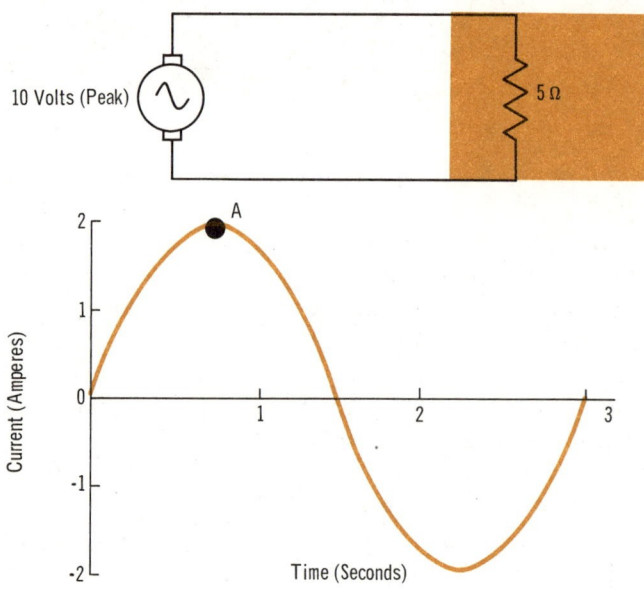

Problem 4. *What is the power consumed in the circuit?*

The values of current and voltage could be converted to their equivalent effective values, and the power then calculated from the equation $P = E_{EFF}I_{EFF}$. However, since both values are peak, it is easier to use the equation that expresses power in terms of peak voltage and current. The equation is

$$P = \frac{E_{PK}I_{PK}}{2} = \frac{10 \text{ volts} \times 2 \text{ amperes}}{2} = 10 \text{ watts}$$

summary

☐ Ohm's Law applies to instantaneous values of current and voltage in an a-c circuit, just as in a d-c circuit. ☐ In an a-c resistive circuit, the current and voltage are in phase. ☐ The average power dissipated in a resistor in an a-c circuit is $P = E_{EFF}I_{EFF}$ and $P = E_{PK}I_{PK}/2$.

☐ Instantaneous power is found by the equation: $P_{INST} = E_{INST}I_{INST}$. It can range from zero to a maximum, which is reached when the current and voltage are both at their peak values at the same time, as is the case in a pure resistive circuit.

☐ The losses in a-c conductors due to eddy current and skin effect are directly proportional to the frequency of the current. Because these losses only become significant at high frequencies, they are neglected and considered to be zero in this volume. ☐ Power cables are usually constructed of stranded wires to reduce skin effect loss.

review questions

1. What are *eddy currents*? How are they related to frequency?
2. What is *skin effect*? Is it present in d-c circuits?
3. How is skin effect overcome in power cables? Why does this work?
4. What is the average power dissipated in an a-c circuit where the maximum voltage is 20 volts, and the maximum current is 40 amperes?
5. What is the average power dissipated in an a-c circuit where the effective voltage is 70 volts and the effective current is 30 amperes?
6. The instantaneous current through a resistor is 14 amperes, and the instantaneous voltage across the resistor is 28 volts. What is the instantaneous power developed?
7. The peak value of the voltage in an a-c circuit is 100 volts, and the effective current is 7 amperes. What is the average power dissipated?
8. The peak power developed across a resistor is 1500 watts, when the effective voltage across the resistor is 500 volts. What is the average current over one alternation?
9. The average value of one alternation of a sine-wave voltage is 75 volts, and the peak value of the current is 3 amperes. What is the average power developed?
10. The average power developed across a resistor is 2000 kilowatts, and the peak-to-peak voltage is 200 kilovolts. What is the value of the effective current through the resistor?

nonresistive a-c circuits

In a d-c circuit, resistance is the *only* thing that opposes current flow. Therefore, a d-c circuit *without* resistance, or with a very low resistance, is a *short circuit*. Damagingly high currents will flow in such a circuit, and no useful function can be accomplished. In a-c circuits, on the other hand, resistance is *not* the only thing that opposes current. Two other circuit properties, called *inductance* and *capacitance*, oppose the flow of a-c current. Therefore, if either of these properties are present, a-c current is *limited*, even though the circuit resistance may be zero.

IDENTICAL CURRENTS WILL FLOW

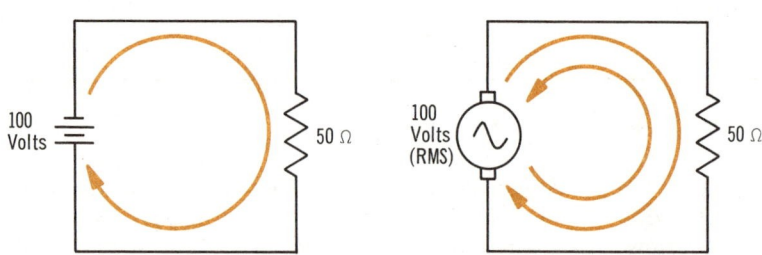

IDENTICAL CURRENTS WILL NOT NECESSARILY FLOW

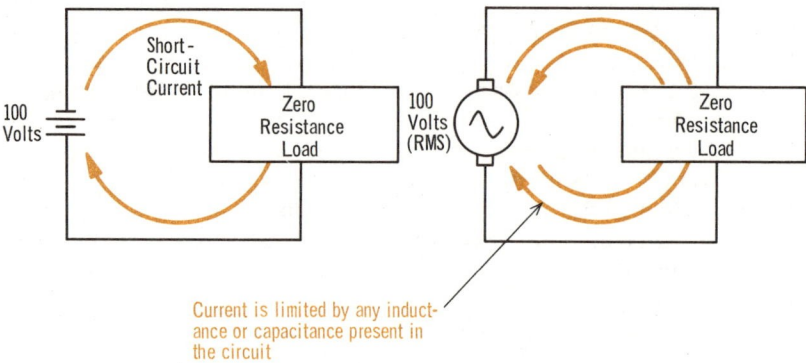

Current is limited by any inductance or capacitance present in the circuit

The nature and characteristics of inductance and capacitance are described in the remainder of this volume. Inductance is covered first, followed by capacitance. Before examining inductance, though, some of the basic principles of electromagnetism and its effects will be reviewed, since inductance is basically an electromagnetic phenomenon.

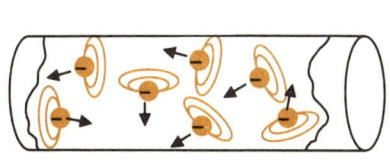 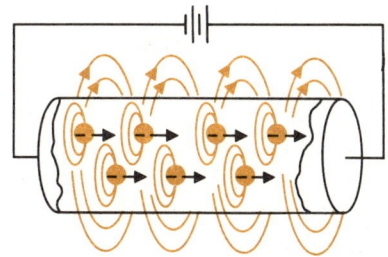

With no voltage applied and therefore no current, the magnetic fields of the free electrons cancel each other. There is no field outside the conductor

With a voltage applied, the free electrons move in the same direction, and their magnetic fields combine. The overall field extends outside the conductor

the magnetic field around a conductor

An electric current is made up of many free *electrons* moving in the *same direction* in a wire. Each moving electron sets up its own magnetic field, and since the electrons are moving in the same direction, the fields of the individual electrons *combine* to produce one overall magnetic field.

In a conductor that has no voltage applied to it, the current is zero. The free electrons in the conductor are moving and creating their individual magnetic fields, but their motion is *random*. At each instant, for every electron that is moving in one direction, there is another moving in the opposite direction. This causes the individual magnetic fields to oppose, or *cancel,* each other. As a result, there is no magnetic field outside of the conductor.

If a voltage is applied to the conductor, many of the free electrons begin moving in the same direction. Their individual magnetic fields then combine, and an overall field is produced. This magnetic field extends outside of the conductor, with each line of force making a circle, or loop, around the conductor. If the voltage applied to the conductor is increased, the current also increases. More electrons will then contribute to the overall magnetic field, and so its strength will be greater. The strength of a magnetic field is usually indicated by the number of individual lines of force, and their distance apart. Strong fields have many lines, and they are closely spaced.

The direction of the magnetic field around a current-carrying conductor is given by the *left-hand rule,* which you learned in Vol. 1: If you wrap your left hand around a conductor with your thumb pointing in the direction of the current flow, the magnetic field surrounding the conductor is in the direction of your curled fingers.

induced emf

When an electron *moves* through a *magnetic field,* a force is exerted on the electron as a result of the interaction between the magnetic field of the electron and the field through which it is moving. When a piece of conductor is moved through a magnetic field, therefore, a force is exerted on *each* of the free electrons in the conductor. Effectively, these forces add together, and the effect is an emf generated, or *induced,* in the conductor. The direction of the induced emf depends on the direction of motion of the conductor relative to the direction of the magnetic field, and can be found by using the *right-hand rule,* which you studied in Volume 1. The right-hand rule states that if the thumb, forefinger, and middle finger of the right hand are held at right angles to one another, with the thumb pointing in the direction in which the conductor is moving and the forefinger pointing in the direction of the magnetic field, then the middle finger points in the direction of the induced emf.

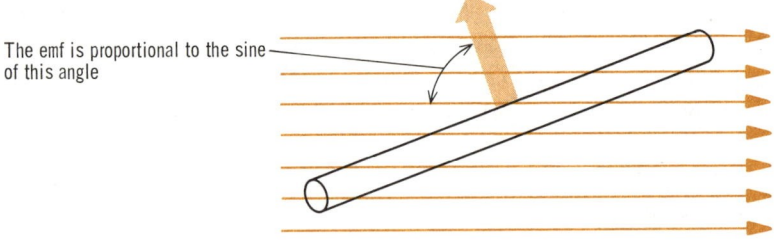

The emf is proportional to the sine of this angle

The magnitude of the induced emf is *directly proportional* to the strength of the magnetic field, the length of the conductor, and the speed at which the conductor moves through the magnetic field. Increasing any of these three factors will cause an increase in the induced emf. The magnitude of the induced emf also depends on the direction in which the conductor moves through the field. If the conductor moves at right angles to the direction of the field, the emf is maximum. If the direction of the conductor is parallel to the direction of the field, no emf is induced. See illustration on page 3-45. And if the conductor moves neither at a right angle nor parallel to the direction of the field, the emf is proportional to the sine of the angle between the direction of the field and the direction of motion of the conductor.

The above description was for the case of a *moving conductor* in a *stationary magnetic field.* As you will learn later, exactly the same results can be obtained by having a *moving magnetic field* and a *stationary conductor.*

factors determining induced emf

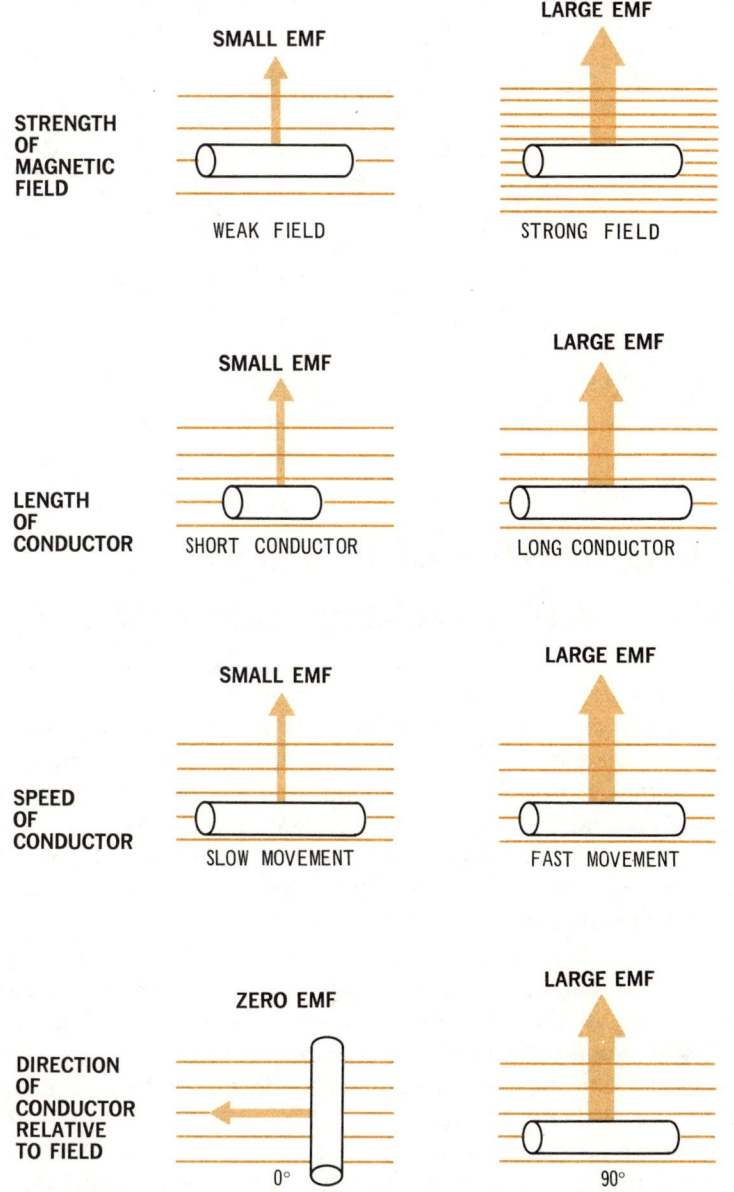

STRENGTH OF MAGNETIC FIELD

SMALL EMF — WEAK FIELD

LARGE EMF — STRONG FIELD

LENGTH OF CONDUCTOR

SMALL EMF — SHORT CONDUCTOR

LARGE EMF — LONG CONDUCTOR

SPEED OF CONDUCTOR

SMALL EMF — SLOW MOVEMENT

LARGE EMF — FAST MOVEMENT

DIRECTION OF CONDUCTOR RELATIVE TO FIELD

ZERO EMF — 0°

LARGE EMF — 90°

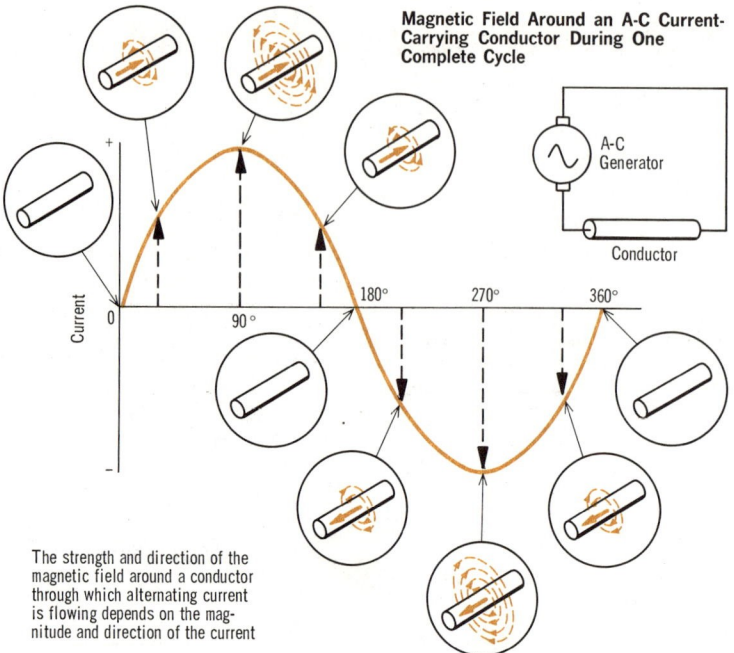

Magnetic Field Around an A-C Current-
Carrying Conductor During One
Complete Cycle

The strength and direction of the
magnetic field around a conductor
through which alternating current
is flowing depends on the magnitude and direction of the current

the magnetic field created
by an alternating current

When a d-c voltage is applied to a conductor, the current goes from
zero to its maximum value almost instantly. The magnetic field around
the conductor also goes from zero to its maximum strength almost instantly, and *stays* at this strength as long as the *current flows*. When
the circuit is opened, the current drops to zero, and the magnetic field
around the conductor also decreases to zero. Usually, you picture the
decrease of a magnetic *field* as the *collapse* of the lines of force back
into the electrons that produced them.

When an alternating current flows in a conductor, the current constantly varies in magnitude. This means that the number of free
electrons moving in the same direction also varies. As a result, the
magnetic field around the conductor *constantly changes* in *strength*.
The greater the current, the stronger is the field. Similarly, the less
the current, the weaker is the field.

Since alternating current periodically changes direction, the magnetic field it produces also reverses direction. At any instant, the direction of the magnetic field is determined by the direction of current flow.

self-induction

As the alternating current in a conductor goes through one complete cycle, the magnetic field around the conductor builds up and then collapses. It then builds up in the opposite direction, and collapses again. When the magnetic field begins building up from zero, the lines of force, or *flux lines,* expand from the center of the conductor outward. As they expand outward, they can be thought of as cutting through the conductor. Remember that an emf is induced in any wire that moves in a magnetic field. In this case, it is the *field* that moves, but

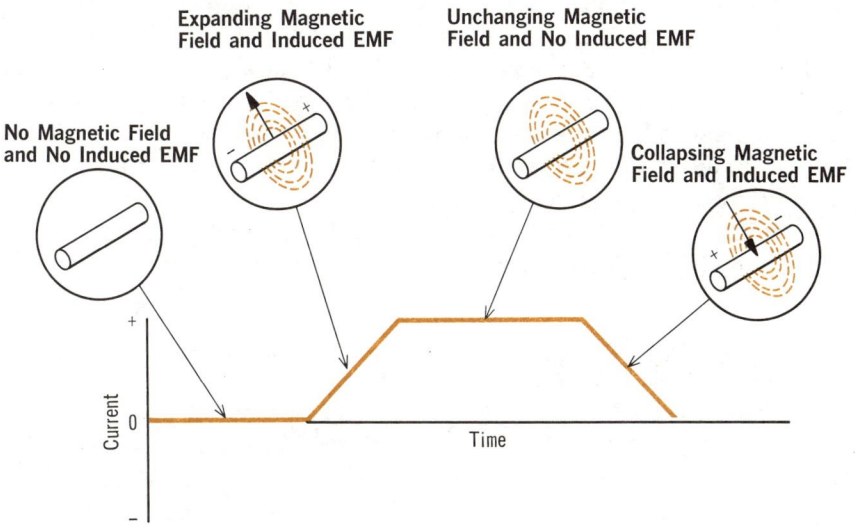

Any change in the current through a conductor causes self-duced emf in the conductor. If the current is zero or constant, no emf is induced

the effect is the same as if the wire was moving and the field was stationary. All that is required is relative motion between the magnetic field and the electron. Therefore, as the magnetic field expands outward through the conductor, it tends to produce a current flow of its own. Similarly, when the magnetic field collapses, the flux lines cut through the conductor again, and again an emf is induced.

You can see then that any change in current causes an expansion or collapsing of the magnetic field around a conductor, which in turn induces an emf in the conductor. This is called *self-induction.*

magnitude of self-induced emf

The emf induced in a conductor by a change in the current through the conductor has *magnitude* and *polarity,* the same as all emf's. One factor that determines the magnitude of the induced emf is the rate at which the magnetic field expands or collapses. This, in turn, depends on how rapidly the current changes. For pure alternating current, which varies as a sine wave, the *frequency* is a measure of how fast the current changes. Therefore, the magnitude of the induced emf depends on the frequency of the current. For a given value of current, the higher the frequency, the more rapidly the current changes, and thus the larger the emf induced. Similarly, the lower the frequency, the smaller will be the induced emf.

The magnitude of the induced emf also depends on the value of the *current.* Larger currents produce stronger magnetic fields. And when a strong field collapses, more flux lines cut the conductor, and a greater emf is induced. Therefore, for a given frequency, higher amplitude currents produce greater induced emf's.

Summarizing, the magnitude of the self-induced emf is proportional to the amplitude and frequency of the current.

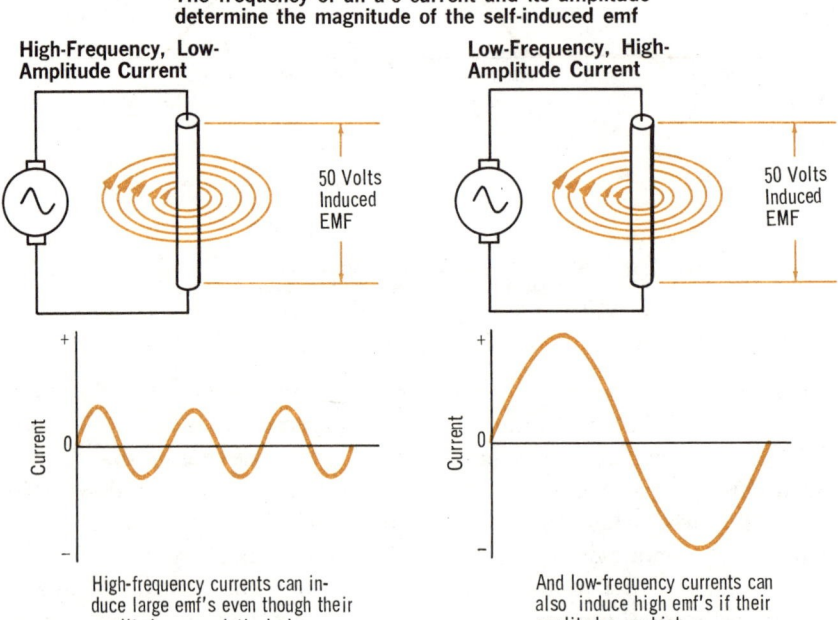

The frequency of an a-c current and its amplitude determine the magnitude of the self-induced emf

High-Frequency, Low-Amplitude Current

50 Volts Induced EMF

Low-Frequency, High-Amplitude Current

50 Volts Induced EMF

Current

Current

High-frequency currents can induce large emf's even though their amplitudes are relatively low

And low-frequency currents can also induce high emf's if their amplitudes are high

polarity of self-induced emf

Every emf has a polarity, and a self-induced emf is no exception. You might think that the polarity of an induced emf should always be in the same direction as the current that causes it. However, if you think in terms of a direct current flowing in a conductor, you will see that this cannot be so. When a direct current increases from zero to its maximum value, the magnetic field it creates around the conductor induces an emf in the conductor. If this emf was in the same direction as the current, it would cause an increase in the current. This increase in current would cause more induced emf, which would increase the current further. This sequence would continue until so much current flowed that something in the circuit would burn out. You know that this does not happen, and so the polarity of the induced emf is not always in the same direction as the current that causes it. The polarity of an induced emf is given by the right-hand rule, but this is hard to visualize when you are dealing with *self-induction.*

The direction of the self-induced emf was first explained by the German physicist H. F. E. Lenz, in what is now called *Lenz's Law: A change in current produces an emf whose direction is such that it opposes the change in current.* In other words, when a current is decreasing, the induced emf is in the same direction as the current and tries to keep the current from decreasing. And when a current is increasing, the polarity of the induced emf is opposite to the direction of the current and tries to prevent the current from increasing. The relationship between the induced emf and the applied voltage that causes the current flow is such that the two voltages are always *180 degrees out of phase.*

The magnitude of the cemf induced at any point depends on the rate of change of the flux lines. So, at 90 and 270 degrees, where I momentarily becomes steady at its peak, the rate of change is zero, and so is the cemf

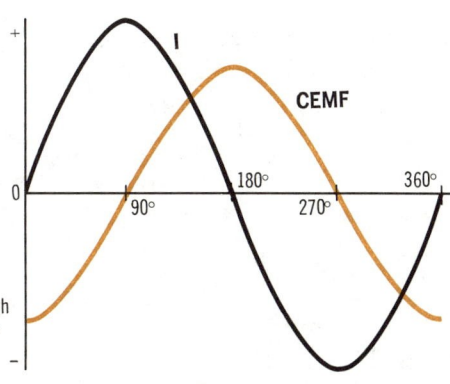

At 0, 180, and 360 degrees, when I passes through zero and changes direction, the rate of change is greatest, and so is the cemf. So I and cemf are 90 degrees out of phase

polarity
of self-induced emf (cont.)

When the applied voltage is at its maximum in one direction, the induced emf is at its maximum in the opposite direction. And when the applied voltage is increasing or decreasing in one direction, the induced emf is increasing or decreasing in the opposite direction. Since the action of the induced emf is to oppose the applied voltage, it is often called the *back emf* or *counter emf*, and is generally abbreviated as *cemf*.

The fact that the back emf always opposes the applied voltage, but sometimes opposes and sometimes aids the current flow may seem confusing. This seeming contradiction is caused by the phase relationship between the applied voltage and the current, and will be explained later in this volume.

The Back EMF Opposes Any Change in the Current

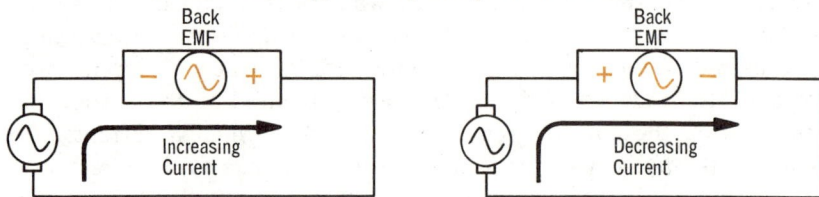

The polarity of the back emf is opposite to the circuit current when current is increasing, and the same as the circuit current when current is decreasing

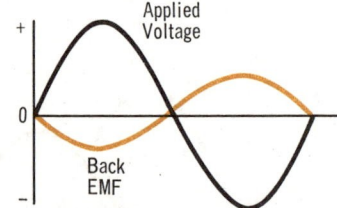

The back emf always opposes the applied voltage. Back emf is also known as counter emf or cemf

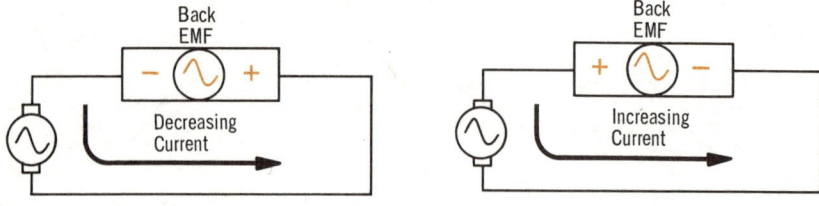

self-induction
from an energy standpoint

Self-induction can also be explained from the standpoint of *energy*. When this is done, the magnetic field that surrounds a current-carrying conductor is considered as *exchanging* energy with the circuit. When circuit current *increases*, energy is *removed* from the circuit and stored in the magnetic field. This is why the magnetic field becomes stronger. The removal of circuit energy shows up as a decrease in the potential along the conductor, and corresponds to the back emf opposing the source voltage.

When current stops increasing, the magnetic field becomes constant and stops removing energy from the circuit; and all of the energy supplied by the power source is used by the circuit current. The magnetic field holds all of the energy it has taken from the circuit until the current begins *decreasing*.

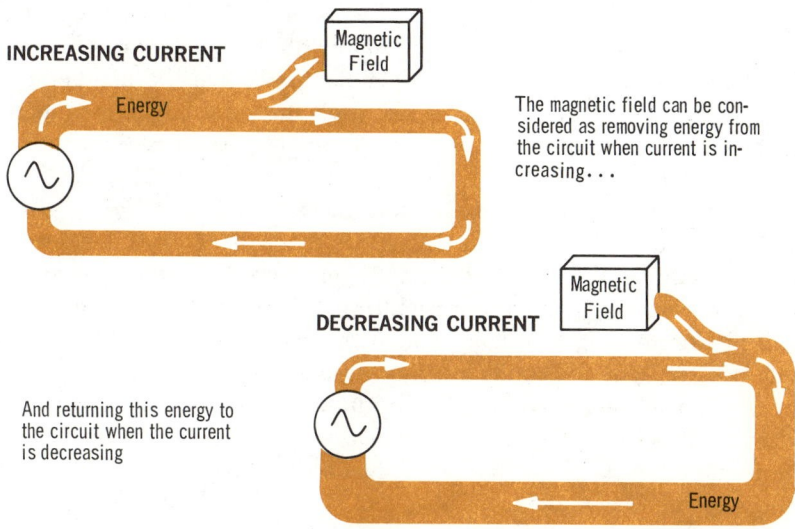

INCREASING CURRENT

Magnetic Field

Energy

The magnetic field can be considered as removing energy from the circuit when current is increasing...

Magnetic Field

DECREASING CURRENT

And returning this energy to the circuit when the current is decreasing

Energy

When the current starts decreasing, the magnetic field starts collapsing and returns the stored energy to the circuit. It returns the energy in the form of increased potential along the conductor. This corresponds to the self-induced emf being in the same direction as the source voltage, and thus adding to it. From the energy standpoint, therefore, self-induction is a removal of energy from a circuit when current increases, and a returning of this energy to the circuit when current decreases.

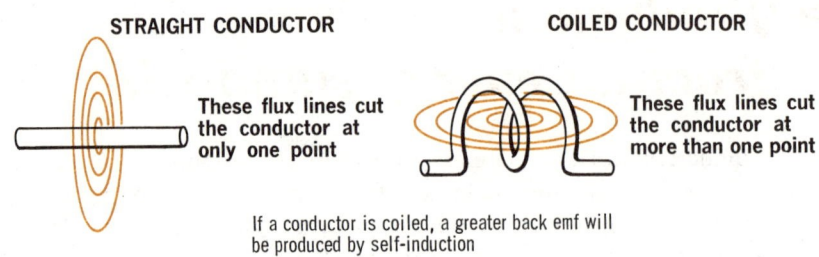

STRAIGHT CONDUCTOR   **COILED CONDUCTOR**

These flux lines cut
the conductor at
only one point

These flux lines cut
the conductor at
more than one point

If a conductor is coiled, a greater back emf will
be produced by self-induction

effect of conductor shape on self-induction

You now know that self-induction in a conductor opposes any change in the current through the conductor, and that the amount, or magnitude, of the self-induction is determined by the amplitude and frequency of the current. However, there is one other factor that affects self-induction. This is the *physical shape* of the conductor. Until now, you have considered only *straight* conductors. When the magnetic field in this type of conductor builds up, each flux line cuts the conductor in one place, and the back emf produced is determined by the total number of flux lines.

If a conductor is *coiled*, into adjacent loops, the situation is entirely different. For one thing, the length of the conductor is longer, so a greater cemf is induced. But, in addition, there is another reason why the cemf will go up even more with a coiled wire. Changes in current still produce a magnetic field around the conductor, but now each flux line does not cut the conductor at only one point. First of all, the flux lines add to create a stronger field, and as each flux line expands outward, it cuts the conductor in the loop in which it is produced, and it also cuts *adjacent loops* of the conductor. The more the flux lines add and expand, the more loops they cut. Each flux line, therefore, generates a back emf at more than one point along the conductor. The polarity of all these back emf's are such that they *add* to produce a total counter emf much greater than would be generated in a straight conductor by the same change in current. When the flux lines collapse, the situation is the same. Each line cuts more than one loop of the conductor, and generates a cemf in every loop it cuts. The cemf's again add to produce a large total back emf.

You can see that for a given current, the amount of back emf produced in a conductor depends on the number of individual flux lines generated, the shape of the conductor, and the frequency of the current.

summary

☐ If either inductance or capacitance is present in an a-c circuit, the current flow is limited. ☐ An induced emf is developed when a conductor cuts magnetic lines of force. Either a moving conductor in a stationary magnetic field, or a moving magnetic field and a stationary conductor will produce an induced emf.

☐ When a varying current flows in a conductor, it creates a magnetic field which induces an emf in the conductor. This is self-induction. ☐ The magnitude of the self-induced emf is proportional to the amplitude and frequency of the current. ☐ The direction of a self-induced emf is explained by Lenz's Law. Lenz's Law states: A change in current produces an emf whose direction is such that it opposes the change in current. ☐ Self-induced emf is sometimes called back emf, or counter emf, and is abbreviated cemf.

☐ Self-induction can be considered an exchange of energy when the magnitude of the current through a conductor changes. ☐ The physical shape of a conductor also determines the magnitude of self-induction. ☐ Conductors are sometimes coiled to increase the number of flux lines cut by the conductor, and this results in larger self-induced emf.

review questions

1. What properties, in addition to resistance, limit the current in an a-c circuit?
2. What rule gives the direction of the magnetic field around a current-carrying conductor? State the rule.
3. For a conductor moving through a magnetic field, what rule gives the direction of the induced emf when the direction of the magnetic field is known? State the rule.
4. What factors determine the magnitude of the induced emf?
5. What is *self-induction*?
6. On what two factors does the magnitude of a self-induced emf depend?
7. What is mean by *counter emf*? By *back emf*?
8. State Lenz's Law.
9. What determines the direction of the self-induced emf? How does Lenz's Law explain this?
10. How can self-induction be explained from an energy standpoint?

inductance

Since, for a given amplitude and frequency of current, the back emf developed in a conductor depends on the shape of the conductor, the exact relationship between the current, back emf, and the shape of the conductor can be expressed mathematically: When the number of flux lines produced by the current is multiplied by a constant that is determined by the shape of the coil, the product equals the back emf produced. The equation for this is

$$E_{cemf} = L \times \text{number of flux lines}$$

The constant, L, which depends on the conductor shape, is called the *inductance* of the conductor.

The inductance of *straight* conductors is usually *very low,* and for our purposes can be considered zero. The inductance of *coiled conductors,* though, can be *high,* and plays an important role in the analysis of a-c circuits.

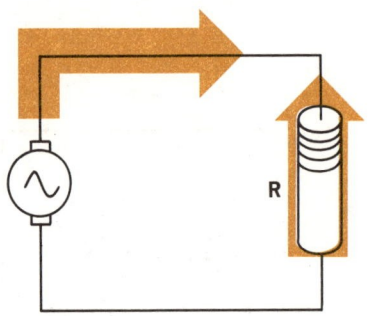

Resistors oppose all current in a circuit. They have resistance, which is abbreviated R

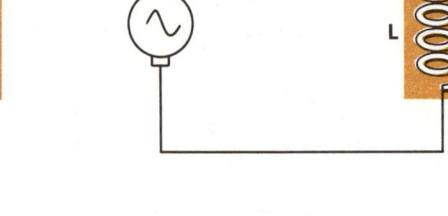

Inductors oppose changes in current in a circuit. They have inductance, which is abbreviated L

Although inductance is actually a physical characteristic of a conductor, it is often defined in terms of its effect on current flow. This definition of inductance is as follows: *Inductance is that property of an electrical circuit that tends to oppose any change of current through the circuit.* You can see from this definition that inductance has no effect on a steady d-c current. It only opposes changes in current. Coiled conductors are usually used in a-c circuits to deliberately introduce inductance into the circuit. Such a coiled conductor is called an *inductor.*

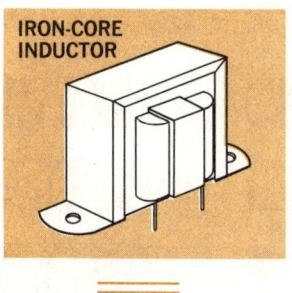 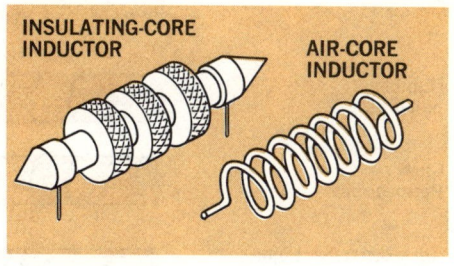

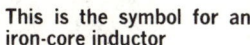

This is the symbol for an iron-core inductor

This is the symbol for an insulating-core inductor or an air-core inductor

inductors

Basically, all inductors are made by winding a length of conductor around a core. The conductor is usually solid copper wire coated with enamel insulation; and the core is made either of magnetic material, such as powdered iron, or of insulating material. When an inductor is wound around an insulating core, the core is used only for a support, since it has no magnetic properties. If heavy wire is used in making the inductor, a core is actually not needed; the rigid loops of wire support themselves. When a magnetic core is not used, the inductor is usually referred to as an *air-core inductor*.

Inductors with set values of inductance that cannot be changed are called fixed inductors. Inductors whose inductance can be varied over some range are called variable inductors. Usually, variable inductors are made so that the core can be moved into and out of the winding. The position of the core then determines the inductance value.

Inductors are also frequently called *chokes* or *coils*. All three of these terms mean the same thing, and you should be familiar with each of them.

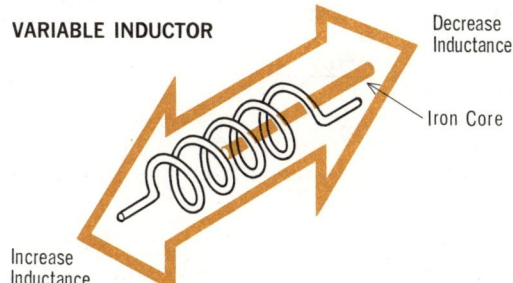

VARIABLE INDUCTOR

Decrease Inductance

Iron Core

Increase Inductance

This is the symbol for a variable inductor

FACTORS AFFECTING INDUCTANCE

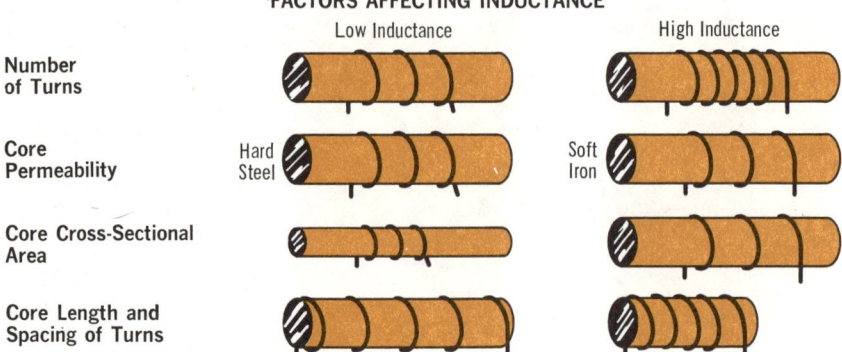

High inductance is obtained with a reasonably small inductor by winding the wire around the core in many layers

factors determining inductance

The physical characteristics, or geometry, of both the *core* and the *windings* around the core affect the inductance produced. Inductors with *magnetic cores* have much greater inductances than those with *insulating* or *air* cores. The reason is that all of the flux lines produced by an inductor pass through the core, and in doing so, magnetize the core, if it is made of magnetic material. The flux lines of the magnetic field of the core then *add* to and *strengthen* the flux lines created by the winding, and therefore, a greater cemf is produced. For a given number of turns in the inductor winding, a core with a *larger cross-sectional area* will produce more flux lines. Also, the *longer* the core is for a given number of turns, the *less* flux lines it will produce. The inductance, therefore, is directly proportional to the cross-sectional area of the core, and inversely proportional to its length.

The number and spacing of the individual turns of wire in an inductor also greatly affect the inductance. The *more turns* there are, the *greater* is the inductance. And the *closer together* they are, also the greater is the inductance. The relationships between inductance and all the physical factors that affect it are expressed by the equation:

$$L = \frac{0.4\pi N^2 \mu A}{l}$$

where N is the number of turns; μ is the core permeability, which is high for magnetic materials and low for other materials; A is the core cross-sectional area; and l is the core length.

For every magnetic core material there is a point where the core becomes *saturated;* then even large changes in current cannot increase flux, and very little cemf is produced.

inductance values and cemf

Inductance is actually a measure of how much counter emf is generated in a circuit or component for a certain change in current through that circuit or component. In other words, it is the amount of cemf produced for a unit change of current. The unit of inductance is the *henry*, named in honor of the American physicist Joseph Henry, who shares the honor of the discovery of induction with Michael Faraday. The henry is defined as follows: *A conductor or coil has an inductance of one henry if a current that changes at the rate of one ampere per second produces a cemf of 1 volt.* Therefore, the greater the inductance, the greater the henrys. The abbreviation for the henry is *h*.

The henry is a fairly large unit. For this reason, inductance is often given in the smaller units of the millihenry and the microhenry. One millihenry is 1/1000 of a henry, and one microhenry is 1/1,000,000 of a henry. The millihenry is abbreviated *mh*; the microhenry *μh*.

Since the amount of cemf produced is part of the definition of the henry, you can actually calculate the cemf that an inductor generates in a circuit when you know its *inductance* value, and the *amplitude* and *frequency* of the current. One form of the equation for cemf is

$$\text{cemf} = -L(\Delta I / \Delta t)$$

The minus sign indicates that the cemf is opposite in polarity to the applied voltage. The ΔI term, pronounced "delta I," is the change in current that takes place in a time of Δt, which is the change in time. As an example of the use of the equation, consider the cemf developed by an inductor having an inductance of 10 henrys when the current changes from 5 to 3 amperes in 1 second.

$$\text{cemf} = -L \frac{\Delta I}{\Delta t} = -10 \left(\frac{5-3}{1} \right) = -20 \text{ volts}$$

You can see that by changing either the inductance (L) or the rate of change of the current ($\Delta I / \Delta t$), which is the frequency, various values of cemf can be obtained. The table below shows how the cemf would rise as the rate of change of the current was increased.

L (HENRYS)	I (AMPERES)	t (SECONDS)	cemf (VOLTS)
1	1	1	1
1	1	1/2	2
1	1	1/4	4
1	1	1/10	10
1	1	1/20	20
1	1	1/50	50
1	1	1/100	100
1	1	1/500	500
1	1	1/1000	1000

inductive d-c circuits

In a d-c circuit, the only *changes* in current occur when the circuit is closed and when it is opened. If the circuit contains only resistance, these changes can be considered as *instantaneous*. Thus, when the circuit is closed, the current instantly increases from zero to its maximum value. And when the circuit is opened, the current instantly drops to zero. If inductance is added to a d-c circuit, say in the form of an inductor, the current can no longer change instantly. When the circuit is closed, the current tries to increase instantly, but is opposed by the back emf generated by the inductor. So instead of increasing instantly, it takes a definite amount of time for the current to reach its maximum value. The greater the inductance, the larger is the back emf produced, and so the longer it takes the current to reach its maximum.

The situation is identical when the circuit is opened and the current tries to decrease to zero instantly. The back emf opposes the change, and so the current decreases to zero gradually. The waveform of any d-c current change through an inductor has the same *basic shape*, regardless of the values of current and inductance. This waveform shows that the current changes rapidly at first, and then gradually changes less and less, until it reaches its maximum value if it is an increasing current, or zero if it is a decreasing current. Because of its shape, this waveform is called an *exponential* waveform, or curve.

D-C Current Through an Inductance

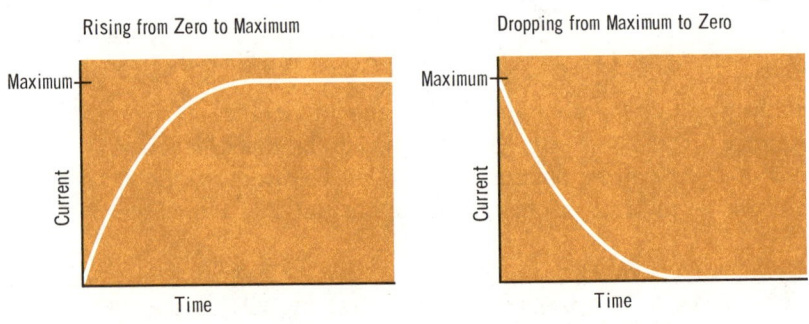

Rising from Zero to Maximum

Dropping from Maximum to Zero

In a d-c circuit containing inductance, the current follows an exponential waveform when it changes from zero to maximum (circuit closed)

The current follows an exponential waveform when it changes from maximum to zero (circuit opened)

At all times when the current is not changing, the inductance has no effect on the circuit

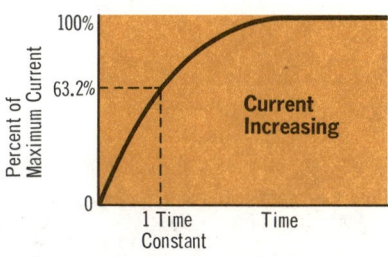

In 1 time constant the current increases
from zero to 63.2% of its maximum value

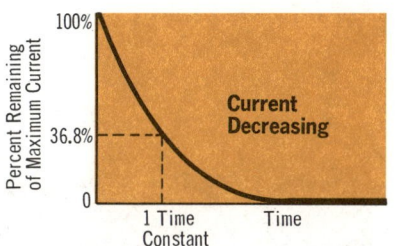

In 1 time constant the current falls 63.2%
from its maximum value, leaving 36.8%

inductive time constant

In a d-c circuit containing inductance as well as resistance, the current changes gradually between zero and maximum, and between maximum and zero. Regardless of the values of inductance and resistance in the circuit, these changes always follow a *similar* pattern. Initially, the change is greatest, and then it gradually tapers off until the current reaches its constant value, which is either its maximum or zero. During these changes, a relationship exists between the *values* of current reached, and the *time* it takes to reach them. This relationship is expressed by a quantity called the *time constant*.

The time constant is defined as the time required for the current to either increase to 63.2 *percent* of its maximum value, or to decrease 63.2 percent from its maximum value. In any d-c circuit, the time constant depends on the value of the *inductance* and the *resistance*. The value of the time constant is directly proportional to the inductance and inversely proportional to the resistance. If these two quantities are known, the time constant can be calculated from the equation

$$\text{Time constant} = \text{inductance}/\text{resistance} \quad \text{or} \quad t = L/R$$

With this equation, if the inductance is in henrys and the resistance in ohms, the time constant will be in seconds. In actual practice, time constants are usually very short. For this reason, they are often expressed in *milli*seconds and *micro*seconds, where one millisecond is one one-thousandth (1/1000) of a second, and one microsecond is one-millionth (1/1,000,000) of a second. Millisecond is often abbreviated *millisec,* and microsecond is abbreviated μsec.

Once the time constant for a circuit is known, you can easily figure how long it takes for the current to go from zero to maximum, or from maximum to zero. This is because, as you can see from the illustration on page 3-60, it takes a time equal to five time constants for the current to rise to its maximum or drop to zero.

rising and falling currents

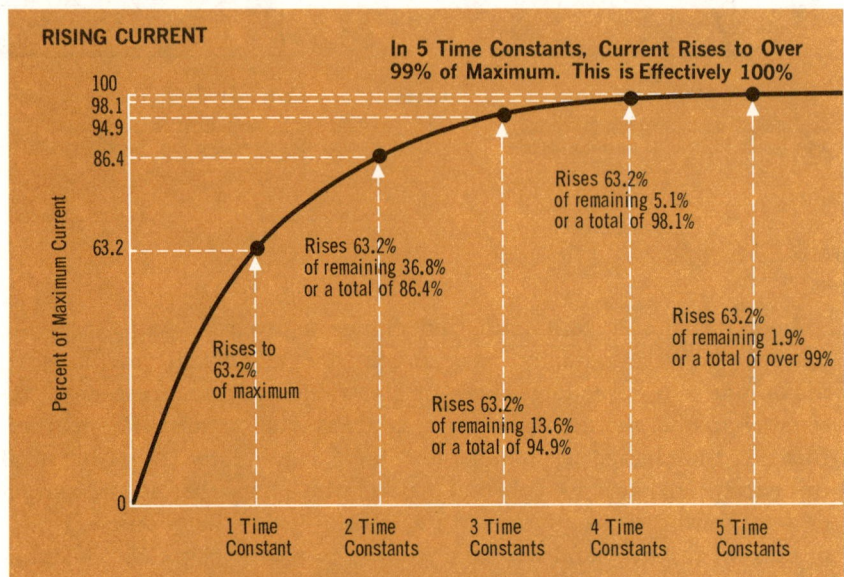

In each time constant, the current increases to a value
63.2% closer to its maximum value

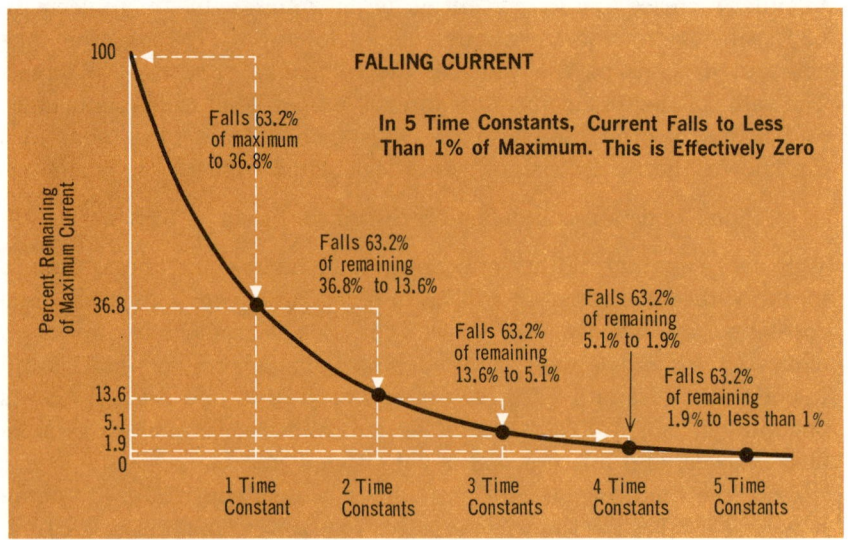

In each time constant, the current decreases to a value
63.2% closer to zero

effect of increasing
or decreasing inductance

The amount of inductance in a d-c circuit determines *how long* it takes the current to build up to its maximum value when the circuit is closed, as well as how long it takes the current to drop to zero when the circuit is opened. If there is no inductance in the circuit, the current changes are for all practical purposes *instantaneous*. The effect

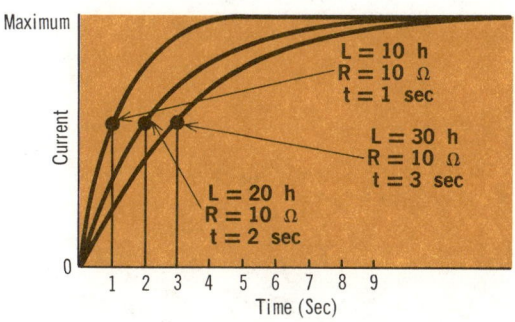
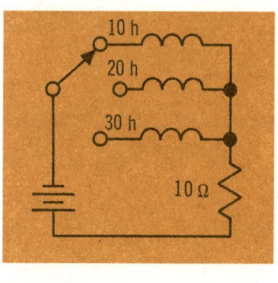

Increasing the inductance in a d-c circuit increases the circuit time constant, and, therefore, increases the time it takes the current to change between its zero and maximum values

of adding inductance is to create a *delay* in the time it takes the current to change. The more inductance added, the longer it takes the current to change. The exact relationship between the inductance and the time required to change is given by the equation for the inductive time constant, $t = L/R$. For example, if the resistance of a circuit is 10 ohms and the inductance is 2 henrys, the circuit time constant is

$$t = L/R = 2 \text{ henrys}/10 \text{ ohms} = 0.2 \text{ second}$$

And since the current changes from zero to maximum or vice versa in 5 time constants, it takes 5×0.2 second, or 1 second, for such a change. If the inductance is increased to 4 henrys, the circuit time constant becomes

$$t = L/R = 4 \text{ henrys}/10 \text{ ohms} = 0.4 \text{ second}$$

And 5 time constants equal 5×0.4 second, or 2 seconds. Thus, doubling the inductance, doubles the time it takes the current to change between its two values. Similarly, if the inductance is tripled, the time will be tripled. And if the inductance is cut in half, the time will also be halved.

inductive a-c circuits

Unlike a d-c circuit, in which the current only changes when the circuit is opened and when it is closed, in an a-c circuit the current changes *frequently.* For *sinusoidal-type ac,* the current changes almost constantly, varying in magnitude continuously, and reversing direction periodically. Since this is the most common type of ac, and the type which forms the basis of a-c circuit theory, the rest of this volume will apply to ac that varies as a sine wave, unless otherwise stated.

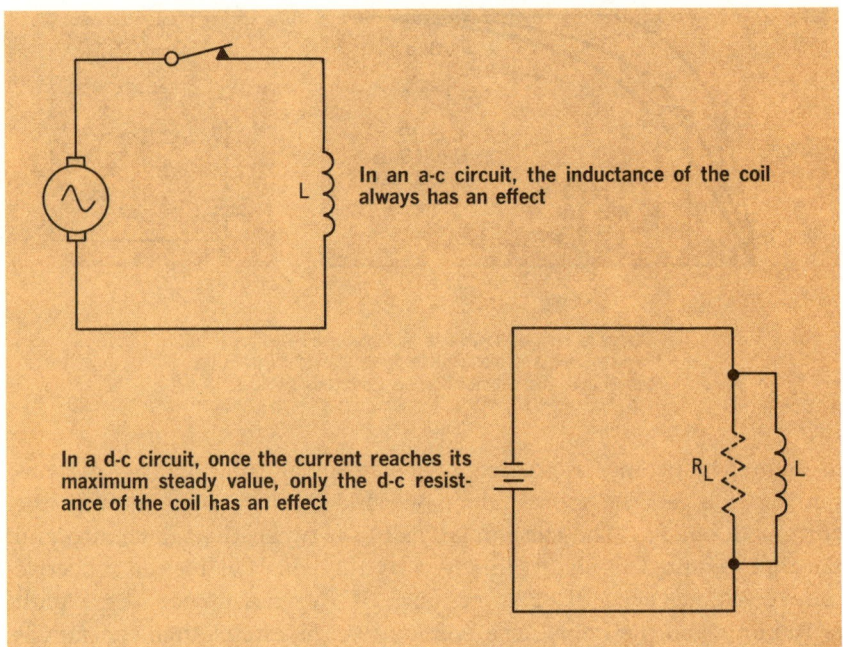

In an a-c circuit, the inductance of the coil always has an effect

In a d-c circuit, once the current reaches its maximum steady value, only the d-c resistance of the coil has an effect

Inductance affects the operation of pure d-c circuits only at the instant they are opened and the instant they are closed. In an a-c circuit, on the other hand, the current is always changing, and the inductance is always opposing the change. The inductance, therefore, has a *constant* effect on circuit operation. Thus, from the time the switch in the circuit is closed until it is opened again, inductance affects the circuit operation.

The circuit shown contains only inductance. Actually, this can never be, since the source, the connecting wires, and the inductor all must have some resistance. However, if these resistances are very small and have a much smaller effect on the circuit current than does the inductance, the circuit can be considered as containing only inductance.

a-c voltage and current relationships

In any a-c circuit that contains only inductance, there are three varying quantities with which you will normally be interested. These are (1) the applied voltage, (2) the induced back emf, and (3) the circuit current. You will remember that if an a-c circuit has only *resistance*, the voltage and current in the circuit are in phase. If the circuit has inductance instead of resistance, this is no longer true. The back emf and time delay generated by the inductance destroys the in-phase relationship between the applied voltage and the circuit current. The applied voltage and the current are, therefore, out of phase. Furthermore, the back emf is out of phase with both the applied voltage and the current.

The phase relationships in an inductance can most easily be understood by considering first the current and the back emf. You know two things about the current and the back emf. One is that the cemf is maximum when the *rate of change* of current is the greatest, and is zero when the current is not changing. The reason for this is that when the current changes rapidly, the flux lines expand or collapse rapidly, and so there is a greater relative motion between the conductor and the magnetic field. The second relationship is that the direction of the cemf is such that it always opposes the current change.

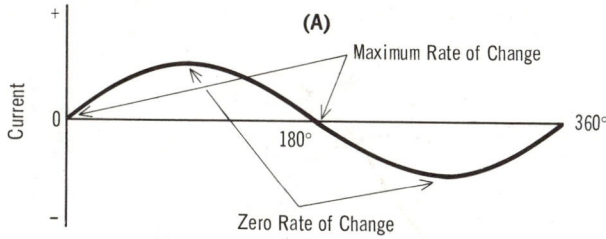

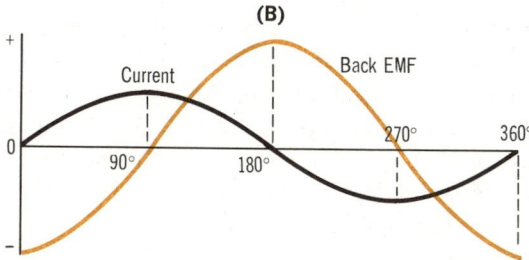

a-c voltage and current relationships (cont.)

Waveform A, on page 3-63, shows one cycle of a-c current. The rate of change is greatest where the slope of the waveform is greatest. You can see that this occurs at those points where the waveform passes through zero; or at 0, 180, and 360 degrees. This means that the most cemf is generated at 0, 180, and 360 degrees, as shown on the waveforms in B on page 3-63. Around 90 and 270 degrees, the current changes very little; as a matter of fact, at exactly 90 and 270 degrees, where the current changes from rising to falling, the current is momentarily steady. Therefore, the flux lines do not change at those points,

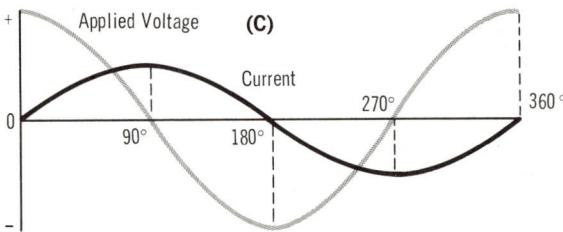

and no cemf is induced. Since at 0 degrees the current is passing through zero in a positive direction, the cemf must be maximum in the negative direction, inasmuch as it always opposes the increase in current. Similarly, when the current begins to decrease, at 90 degrees,

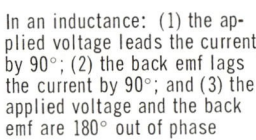

In an inductance: (1) the applied voltage leads the current by 90°; (2) the back emf lags the current by 90°; and (3) the applied voltage and the back emf are 180° out of phase

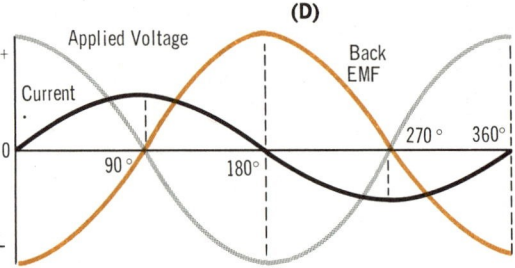

the cemf must be increasing in the positive direction to aid the current flow. As shown, therefore, the cemf follows *Lenz's Law* by lagging the current by 90 degrees. You know that the applied voltage is 180 degrees out of phase with the cemf, and so the applied voltage must lead the current by 90 degrees. This is shown on the waveforms in C on this page. The relationships between all three quantities (current, cemf, and applied voltage) are shown on the waveforms in D on this page.

inductive reactance

In a d-c circuit with a fixed applied voltage, the amount of current that flows depends on the resistance of the circuit, which opposes the flow of current. In an a-c circuit that has only resistance, the same is true. But, in an a-c circuit that only has inductance, the amount of current that flows is determined by the cemf, which counteracts the applied voltage to oppose current flow. The cemf acts just like a resistance to limit current flow. But cemf is discussed in terms of volts, so it cannot be used in Ohm's Law to compute current. However, the effect of cemf can be given in terms of ohms. This effect is called *inductive reactance*, and is abbreviated X_L. Since the cemf generated by an inductor is determined by the inductance (L) of the inductor, and the frequency (f) of the current, the inductive reactance must also depend on these things. The inductive reactance can be calculated by the equation:

$$X_L = 2\pi fL$$

where X_L is the inductive reactance, in ohms; the value of 2π is approximately 6.28; f is the frequency of the current, in cycles per second, Hz; and L is the inductance, in henrys. The quantity $2\pi f$ together actually represents the *rate of change* of the current.

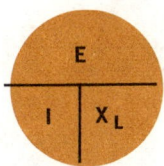

In a purely inductive circuit, inductive reactance has the same effect as resistance in either a d-c circuit or an a-c resistive circuit

You can see from the equation that the higher the frequency or the greater the inductance, the more will be the inductive reactance. Likewise, the lower the frequency or the inductance, the smaller will be the inductive reactance.

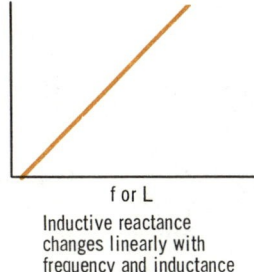

f or L

Inductive reactance changes linearly with frequency and inductance

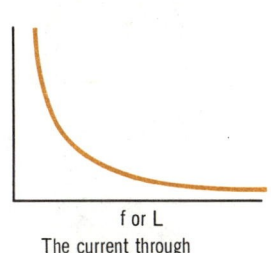

f or L

The current through an inductance changes nonlinearly with frequency

inductive reactance (cont.)

In a-c circuits that contain only inductance, the inductive reactance is the only thing that limits current flow. Such circuits can be solved by Ohm's Law merely by using the inductive reactance in place of the resistance. Therefore, $I = E/X_L$. An important difference, though, is that a particular value of inductive reactance applies only to *one specific frequency*. Thus, if you calculated the inductive reactance in a circuit for a frequency of 30 Hz and then used this value to find the circuit current, the result would hold only if the frequency was constant. If the frequency changed, the inductive reactance would change, and so also would the circuit current. With only resistance, as you know, the frequency has no effect on the basic resistance.

Inductive Reactance Is Directly Proportional to Inductance and Frequency

$$X_L = 2\pi fL$$

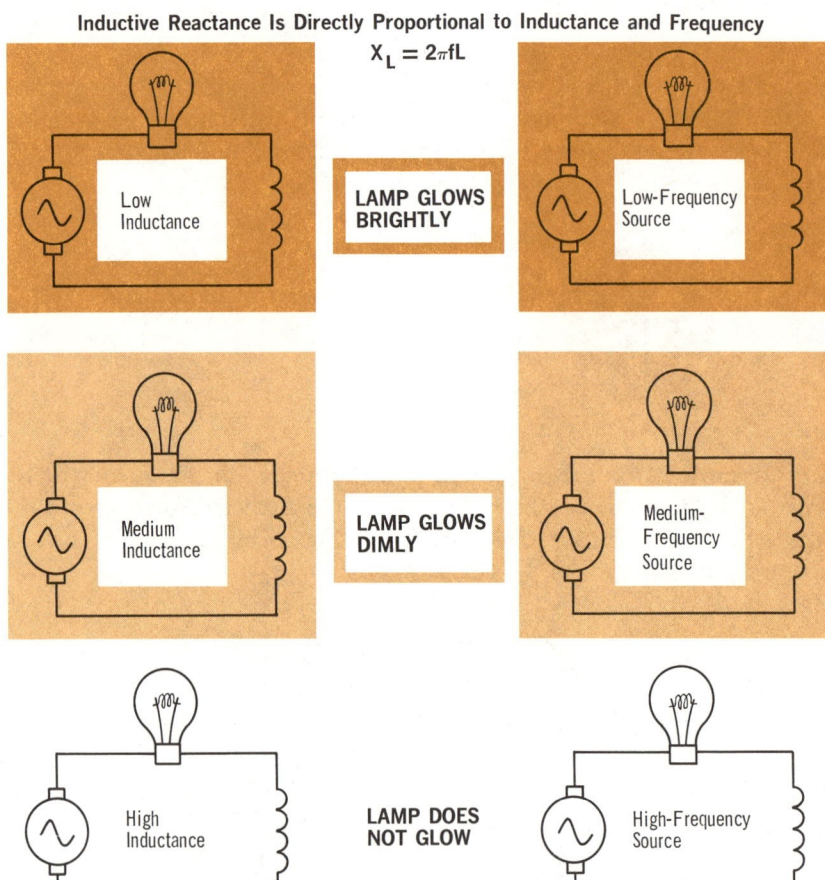

| Low Inductance | **LAMP GLOWS BRIGHTLY** | Low-Frequency Source |

| Medium Inductance | **LAMP GLOWS DIMLY** | Medium-Frequency Source |

| High Inductance | **LAMP DOES NOT GLOW** | High-Frequency Source |

series and parallel inductors

To meet circuit requirements, it is frequently necessary to connect inductors in series or in parallel. When this is done the total inductance is calculated in exactly the *same way* as you calculate the *total resistance* of series and parallel resistors. For series inductors, this means that the total inductance is the *sum* of the *inductances* of the individual inductors. In the form of an equation, this is expressed as:

$$L_{TOT} = L_1 + L_2 + L_3 + \ldots + \text{etc.}$$

For parallel inductors, all of the methods described in Volume 2 for finding the effective resistance of parallel resistances can be used. The general method, which you remember applies in all cases, is the *reciprocal method*. For parallel inductors, this method is expressed as:

$$L_{TOT} = \cfrac{1}{\cfrac{1}{L_1} + \cfrac{1}{L_2} + \cfrac{1}{L_3} + \ldots + \text{etc.}}$$

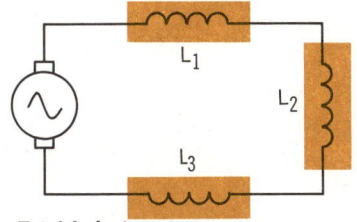

Total Inductance:
$$L_{TOT} = L_1 + L_2 + L_3$$
Total Inductive Reactance:
$$X_{L\,TOT} = X_{L1} + X_{L2} + X_{L3}$$

The total inductance or inductive reactance of series and parallel inductors is calculated the same as the total resistance of series and parallel resistance combinations

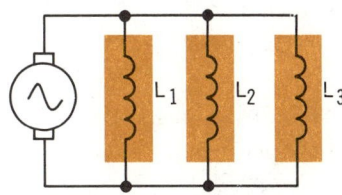

Total Inductance:
$$L_{TOT} = \cfrac{1}{\cfrac{1}{L_1} + \cfrac{1}{L_2} + \cfrac{1}{L_3}}$$

Total Inductive Reactance:
$$X_{L\,TOT} = \cfrac{1}{\cfrac{1}{X_{L1}} + \cfrac{1}{X_{L2}} + \cfrac{1}{X_{L3}}}$$

The total inductive reactance of series and parallel combinations of inductors can be found using the same methods used to find the total inductance. Or, the total inductive reactance can be found by first determining the total inductance and then calculating the total inductive reactance of this single inductance.

The methods just described for finding the total inductance of series and parallel inductors assume that the inductors are physically placed so that the flux lines from each inductor do not cut the windings of the other inductors. This cutting of the windings of one inductor by the flux lines of another is called *mutual induction*.

the Q of a coil

Up till now, you have treated the coil as though it were a pure inductor having only inductive reactance. However, you know that any coil of wire also has resistance, even though it might be a very small value.

For any given inductance value, the inductor can be made in different ways, to produce different amounts of resistance. If a coil is wound with heavy wire on a core to produce, say, 1 henry, it might have a very small resistance. But, if a fine wire is used without a core, many more turns might be needed to produce 1 henry, and this coil would have a much higher resistance than the other coil.

The resistance of a coil affects how the coil actually works in an a-c circuit. If it has a very low resistance compared to its inductive reactance, it will work like a pure inductor. But the higher the resistance compared to the inductive reactance, the less the coil works like a pure inductor.

There is a way of rating the merit of a coil as an inductor. It is called the Q *of a coil*, and is derived by dividing the inductive reactance by the resistance:

$$Q = \frac{X_L}{R_L}$$

The resistance of the coil is made up of both d-c resistance and a-c resistance. As the coil is used at higher frequencies, the a-c resistance increases. But X_L also increases with frequency. So, over most of its operating range, the Q of a coil will remain relatively constant. It might vary, however, depending on how high the d-c resistance is compared to the a-c resistance.

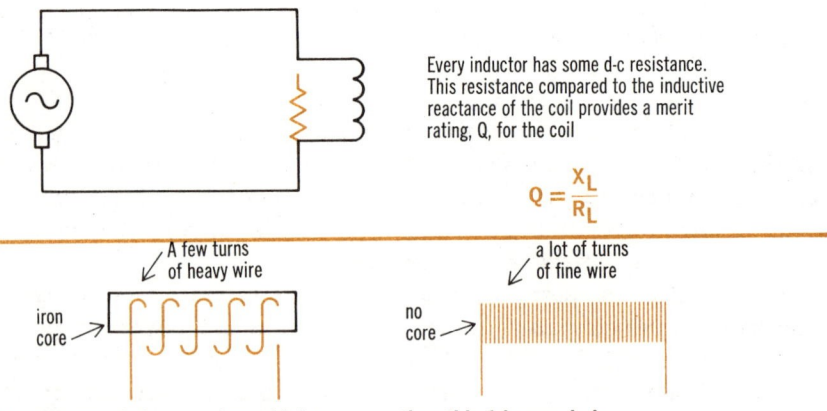

Every inductor has some d-c resistance. This resistance compared to the inductive reactance of the coil provides a merit rating, Q, for the coil

$$Q = \frac{X_L}{R_L}$$

A few turns of heavy wire   a lot of turns of fine wire

iron core   no core

This 1 henry choke . . . has a higher Q . . . than this 1 henry choke

solved problems

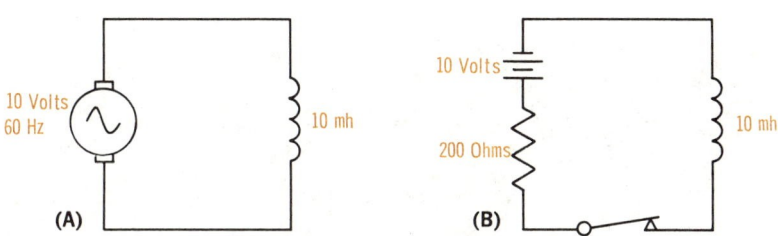

(A)   **(B)**

Problem 5. In which of these circuits is the current the greatest?

In circuit B, the applied voltage is 10 volts and the total circuit d-c resistance is 200 ohms. Theoretically, the inductor has no effect on the d-c current once the current has reached its maximum, or steady-state value. You can, therefore, consider the 10-millihenry inductor as a short circuit to the d-c current since it has negligible resistance. The current in circuit B is calculated simply by Ohm's Law:

$$I = E/R = 10 \text{ volts}/200 \text{ ohms} = 0.05 \text{ ampere}$$

(In actual practice, inductors do have some d-c resistance, since they are made of turns of wire. A 10-mh inductor, though, can be considered as having an insignificant d-c resistance in the problem.)

Circuit A is a purely inductive circuit, and so Ohm's Law can be used to find the current by using the inductive reactance of the inductor in place of resistance. First, then, you must calculate X_L:

$$X_L = 2\pi fL = 6.28 \times 60 \text{ Hz} \times 0.01 \text{ h} = 3.768 \text{ ohms}$$

Notice that the inductance value of 10 mh was converted to the basic unit of the henry before the equation was solved. This must always be done. If the frequency had been given in kilohertz or megahertz, it also would have had to be converted to its basic unit (Hz).

Once X_L is known, use Ohm's Law to find the circuit current. Actually, since you know the voltage applied to both circuits, and the opposition to current flow in each, you should realize that more current will flow in circuit A than in circuit B. This is obvious, inasmuch as Ohm's Law says that for the same voltage, less resistance will allow more current to flow. If you wanted to calculate the current in circuit A, though, it would be:

$$I = E/X_L = 10 \text{ volts}/3.77 \text{ ohms} = 2.65 \text{ amperes}$$

In the above problem, an important assumption was made that the 10 volts given for the source voltage of circuit A was an rms value. If it were not, the current in circuit A would bear no relationship to the current in B, and therefore the two currents could not be compared.

solved problems (cont.)

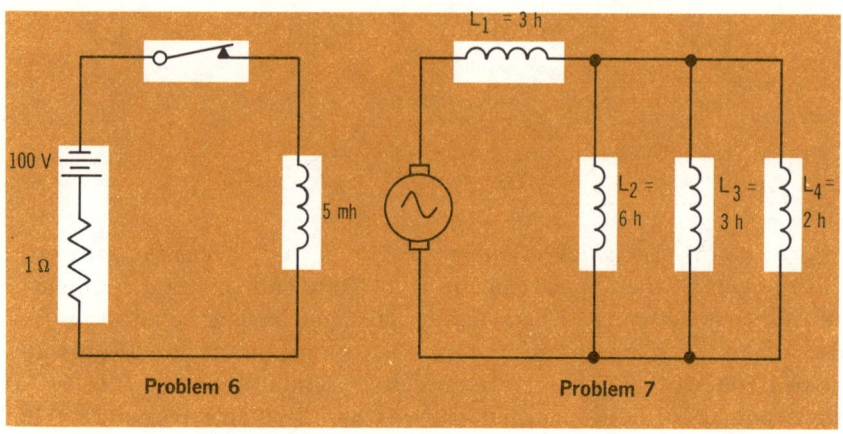

Problem 6   Problem 7

Problem 6. How long will it take the current in the circuit to reach its full value?

Once the switch is closed, it will take a time equal to 5 time constants for the current to reach its full value. One time constant, t, equals the inductance, in henrys, divided by the resistance, in ohms. Therefore,

$$t = L/R = 0.005 \text{ henry}/1 \text{ ohm} = 0.005 \text{ second}$$

Thus, $5t = 0.025 \text{ second}$

Problem 7. What is the total inductance in the circuit?

A series-parallel combination of inductances is reduced to a single equivalent inductance, the same as a series-parallel combination of resistances would be reduced, by applying the reciprocal method.

$$L_{2\text{-}3\text{-}4} = \cfrac{1}{\cfrac{1}{L_2} + \cfrac{1}{L_3} + \cfrac{1}{L_4}} = \cfrac{1}{\cfrac{1}{6} + \cfrac{1}{3} + \cfrac{1}{2}}$$

$$= \cfrac{1}{\cfrac{1}{6} + \cfrac{2}{6} + \cfrac{3}{6}} = \cfrac{1}{\cfrac{6}{6}} = 1 \text{ henry}$$

This reduces the original circuit to two series inductances. The total circuit inductance is now found by simply adding L_1 and $L_{2\text{-}3\text{-}4}$.

$$L_{TOT} = L_1 + L_{2\text{-}3\text{-}4} = 3 + 1 = 4 \text{ henrys}$$

The total inductance of the four series-parallel inductances is, therefore, 4 henrys.

solved problems (cont.)

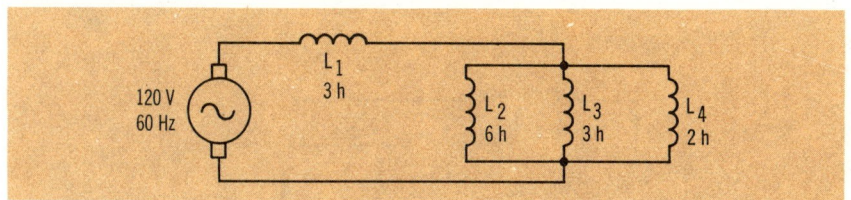

Problem 8. *What is the total current flow in this circuit?*

To find the total current, you must first find the total inductive reactance of the circuit, then use Ohm's Law. The total reactance can be found in two ways. First, you can use the equation $X_2 = 2\pi fL$ for each inductor, and then add up the individual reactances to get the total. Another way is to use the total inductance of the circuit and find its reactance. This circuit is the same one used in Problem 7, where the total inductance was found to be 4 henrys. So, we will find the total reactance by:

$$X_2 = 2\pi fL$$
$$= 2 \times 3.1414 \times 60 \times 4$$
$$= 1{,}507.87 \text{ ohms}$$

Now use Ohm's Law to find the total current:

$$I_{TOT} = \frac{E}{X_L}$$
$$= \frac{120}{1{,}507.87}$$
$$= 79.6 \text{ ma}$$

Problem 9. *What is the voltage drop across L_1?*

There are two ways of doing this problem also. First, since we know the current through L_1, we could find the value of X_{L_1} and use Ohm's Law, $E = I_{TOT}X_L$. Or, since we found in Problem 6 that the equivalent inductance of L_2, L_3, and L_4 was 1 henry, we could use the law of proportionality (Vol. 2) to find the voltage drop across L_1. Normally, we would use the reactance values for the proportionality law, but since the reactance of an inductor is *directly* proportional to its inductance, either value can be used.

$$E_{L1} = E \left(\frac{L_1}{L_1 + L_{2\text{-}3\text{-}4}} \right)$$
$$= 120 \,(3/4)$$
$$= 90 \text{ volts}$$

solved problems (cont.)

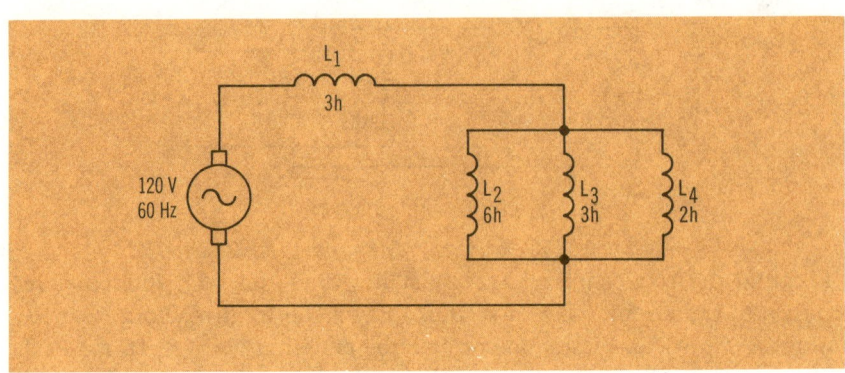

Problem 10.   *How much current is flowing through L_2, L_3, and L_4?*

Since you know that the voltage drop across L_1 is 90 volts, and the source voltage is 120 volts, then the remaining 30 volts is dropped across the $L_{2\text{-}3\text{-}4}$ parallel circuit. Thus, if you find the reactance of each inductor, you can calculate the current through each with Ohm's Law.

$$X_{L2} = 2\pi fL_2 = 2 \times 3.1414 \times 60 \times 6 = 2,261.8$$
$$X_{L3} = 2\pi fL_3 = 2 \times 3.1414 \times 60 \times 3 = 1,130.9$$
$$X_{L4} = 2\pi fL_4 = 2 \times 3.1414 \times 60 \times 2 = 753.9$$

Now:

$$I_{L2} = \frac{E_2}{X_{L2}} = \frac{30}{2,261.8} = 13.27 \text{ ma}$$

$$I_{L3} = \frac{E_3}{X_{L3}} = \frac{30}{1,130.9} = 26.54 \text{ ma}$$

$$I_{L4} = \frac{E_4}{X_{L4}} = \frac{30}{753.9} = 39.81 \text{ ma}$$

Problem 11.   *Is there another way of finding I_{L2}, I_{L3}, and I_{L4}?*

Yes, there is: using a version of the law of proportionality. Since you know the total current in the circuit, and you know that this current will be split up *inversely* proportional to the value of inductance, you can proportionalize the branch circuit to determine the current. Notice that L_2, L_3, and L_4 are 6, 3, and 2 henrys, respectively. They are all divisible by 6. L_2 is 1/6; L_3 is 2/6; and L_4 is 3/6. So if we divided I_{TOT} by 6, we can then proportion it by branch.

$$I_{L2} = (1/6)I_{TOT} = (1/6)(79.6 \text{ ma}) = 13.27 \text{ ma}$$
$$I_{L3} = (2/6)I_{TOT} = (2/6)(79.6 \text{ ma}) = 26.54 \text{ ma}$$
$$I_{L4} = (3/6)I_{TOT} = (3/6)(79.6 \text{ ma}) = 39.81 \text{ ma}$$

summary

☐ Counter emf voltage can be calculated by: $E_{cemf} = L \times$ number of flux lines, where L is the inductance of the conductor. ☐ Inductance is that property of an electrical circuit that tends to oppose any change in current through the circuit. ☐ Inductance is measured in henrys. Other units are the millihenry, mh; and the microhenry, μh. ☐ Coiled conductors that are inserted into a circuit to deliberately add inductance are called inductors. ☐ Inductors are sometimes called chokes or coils. ☐ A coil is said to be saturated when even large changes in current cannot increase the flux, and very little cemf is produced.

☐ Counter emf can be found by: cemf $= -L(\Delta i / \Delta t)$. ☐ When an inductor is placed in a d-c circuit, the current will increase in the shape of an exponential waveform. The reverse happens when the circuit is opened. ☐ The time required for the current to increase to 63.2 percent of its maximum value or to decrease to 63.2 percent from its maximum value, is defined as the time constant.

☐ The inductive time constant is equal to: $t = L/R$. ☐ Steady state is reached after five time constants. ☐ Inductance in an a-c circuit has a constant effect on circuit operation: the applied voltage leads the current by 90 degrees, and the cemf lags the current by 90 degrees.

review questions

1. Define *inductive time constant*.
2. How many time constants does it take for the current to reach 63.2 percent of its maximum value in a d-c circuit?
3. In what units are time constants measured?
4. What is the total inductive reactance of three coils in series whose reactances are 5, 5, and 10 ohms?
5. What would be the total inductive reactance for the coils in Question 4 if they were in parallel?
6. What is the inductive reactance of a 10-millihenry coil at frequencies of 100 Hz, 1000 Hz, 10 kHz, 100 kHz, and 1 MHz?
7. What is the phase relationship between current through, and the voltage across, an inductor? Why is this so?
8. What is the value of the time constant for a circuit consisting of a 5000-ohm resistor and a coil having an inductive reactance of 1.884 ohms at 60 Hz?
9. It takes 10 seconds for the current to drop to zero after the switch is opened in a d-c circuit consisting of a battery, switch, resistor, and coil. What is the circuit time constant?
10. If the value for the resistor of Question 9 is 100 ohms, how many 40-henry inductors are needed to provide the inductance? Are they connected in series or in parallel?

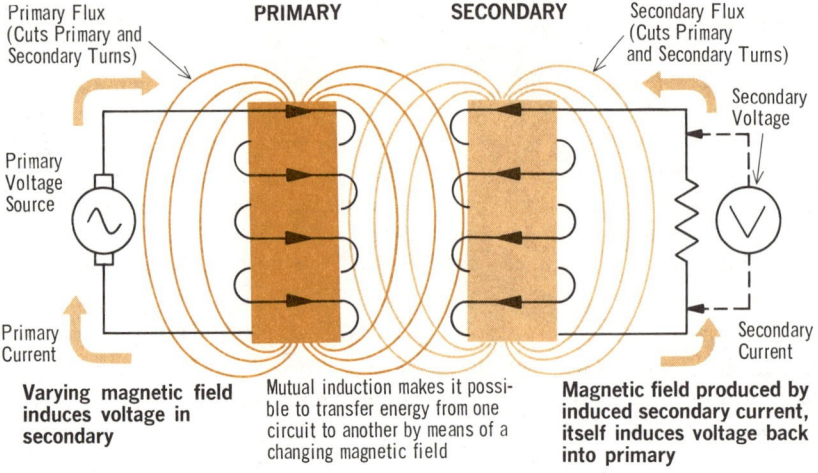

Primary Flux
(Cuts Primary and
Secondary Turns)

PRIMARY   **SECONDARY**

Secondary Flux
(Cuts Primary
and Secondary Turns)

Secondary
Voltage

Primary
Voltage
Source

Primary
Current

Secondary
Current

Varying magnetic field induces voltage in secondary

Mutual induction makes it possible to transfer energy from one circuit to another by means of a changing magnetic field

Magnetic field produced by induced secondary current, itself induces voltage back into primary

mutual induction

Self-induction in a coil or conductor is actually an induced back emf, which is generated when the magnetic field caused by the current flow cuts the coil or conductor. If the flux lines from the expanding and contracting magnetic field of one coil were to cut the windings of another nearby coil, a voltage would also be induced in that coil. The amount of emf induced in this way depends on the relative *positions* of the two coils. Also, the more turns of the second coil that are cut by the flux lines from the first, the greater will be the emf that is induced. This inducing of an emf in a coil or conductor by magnetic flux lines generated in another coil or conductor is called *mutual induction*. The coil in which the flux originates is called the *primary*, or primary winding, and the one in which the emf is induced is called the *secondary*. Similarly, the current that flows through the primary is the *primary current*, and if the secondary is connected to a load so that current flows, this current is called *secondary current*.

When current flows in the secondary, it sets up its own magnetic field, which induces a voltage back into the primary winding. Thus, when mutual induction occurs between two coils, there are four voltages present. These are (1) the applied voltage in the primary, (2) the self-induced emf in the primary, (3) the induced emf in the secondary, and (4) the emf induced back into the primary by the secondary current. Therefore, the actual or effective overall, inductance of two coils that are mutually coupled is complex because of the complex interactions between the magnetic fields. This effective inductance of two coils that are mutually coupled is called *mutual inductance*.

mutual inductance

Mutual inductance can be considered as the amount or degree of mutual induction that exists between two coils or windings. The mutual inductance of any two coils depends on the *flux linkage* between the coils, which in turn depends on their positions relative to each other. The degree of flux linkage is expressed by a factor called the *coefficient of coupling*. When all of the flux lines from each coil cut, or link, the other coil, the coefficient of coupling is 1, which is the *maximum* value. If only some of the flux lines from each coil cut the other, the coefficient of coupling has some value less than 1. You can see then that when no mutual inductance exists between two coils, the coefficient of coupling between the two is 0. When the value of the coefficient of coupling is close to 1, the two coils are said to have tight coupling; and when the value is much less than 1, the coils have loose coupling. The term *critical coupling* is used to describe the dividing line between loose and tight coupling.

Mutual Inductance Between Coils Depends Upon Coefficient of Coupling

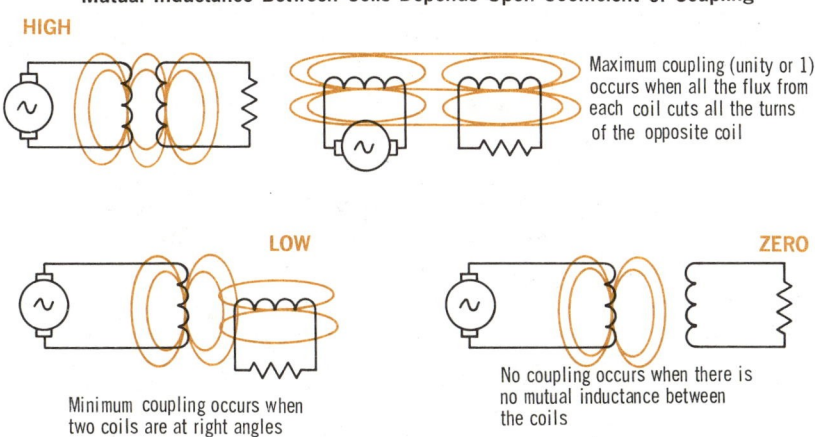

HIGH

Maximum coupling (unity or 1) occurs when all the flux from each coil cuts all the turns of the opposite coil

LOW

Minimum coupling occurs when two coils are at right angles

ZERO

No coupling occurs when there is no mutual inductance between the coils

When the coefficient of coupling between two coils is known, the total inductance of the coils is found by multiplying the values of inductance of the coils, taking the square root of the result, and multiplying it by the coefficient of coupling. As an equation, this is given by:

$$M = k \sqrt{L_1 \times L_2}$$

where M is the total inductance of the mutually coupled coils, in henrys; k is the coefficient of coupling; and L_1 and L_2 are the individual inductances of the coils, in henrys.

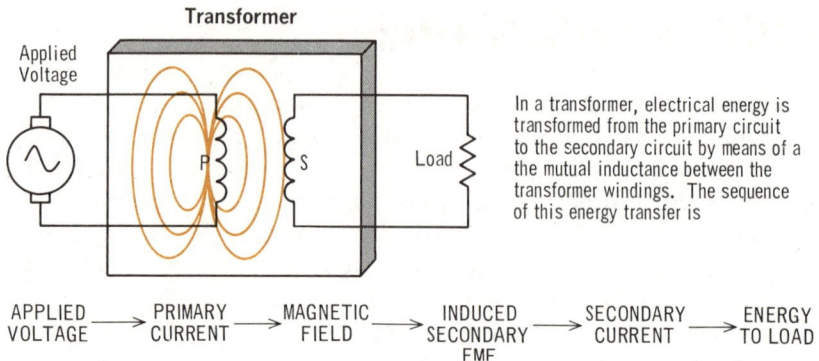

In a transformer, electrical energy is transformed from the primary circuit to the secondary circuit by means of a the mutual inductance between the transformer windings. The sequence of this energy transfer is

APPLIED → PRIMARY → MAGNETIC → INDUCED → SECONDARY → ENERGY
VOLTAGE   CURRENT   FIELD   SECONDARY   CURRENT   TO LOAD
                                   EMF

the transformer

When mutual induction exists between two coils or windings, a change in current through one induces a voltage in the other. Devices which make use of this principle are called *transformers*. Every transformer has a primary winding and one or more secondary windings. The primary winding receives electrical energy from a power source and couples this energy to the secondary winding by means of a changing magnetic field. The energy appears as an emf across the secondary winding, and if a load is connected to the secondary, the energy is transferred to the load.

By means of transformers, electrical energy can be transferred from one circuit to another, with no physical connection between the two. The energy transfer is accomplished completely by the magnetic field. The transformer thus acts as a coupling device. Transformers are also indispensable in a-c power distribution, since they can convert electrical power at a given current and voltage into the equivalent power at some other current and voltage.

Transformers Can Have More Than One Secondary Winding

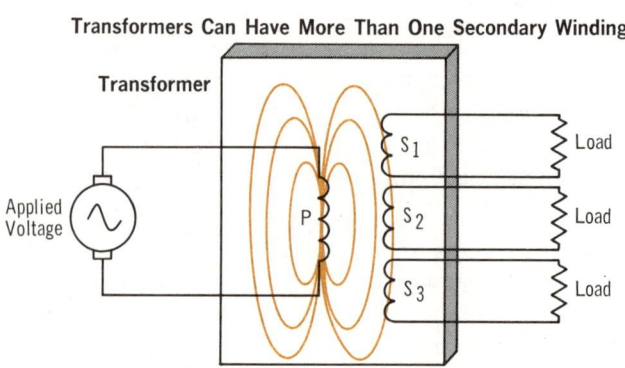

transformer operation with open secondary

You will remember that when mutual induction exists between two coils, not only does the current in the primary coil induce a voltage in the secondary coil, but the resulting current in the secondary coil, in turn, induces a voltage *back* into the primary coil. This occurs in a transformer, and it is in large part responsible for how a transformer works. However, for ease of understanding, transformer operation will first be described for the case of a transformer having an *open secondary*. No current flows in the secondary in this case, and so no voltage is induced back into the primary. After the description of how a transformer with an open secondary works, a complete transformer, with current flowing in both the primary and the secondary, will be described.

In a transformer having an open secondary, the primary operates essentially the same as an inductor. This means that the primary current *lags* the applied voltage by 90 degrees and at the same time *leads* the back emf by 90 degrees. The applied voltage and the back emf are thus of *opposite* polarity. Most transformers are designed to have large back emf's in the primary when the secondary is open, so the primary current is therefore *very low*. As the magnetic field around the primary winding, caused by the changing current through the primary, alternately expands and collapses, it cuts the turns of the secondary winding, thereby inducing a voltage in the secondary. The voltage induced in the secondary is *maximum* when the rate of *change* of the primary current is *greatest* (0, 180, and 360 degrees), and *zero* when the primary current is *not changing* (90 and 270 degrees). When this relationship is plotted, as on page 3-78, you can see that the secondary voltage lags the primary current by 90 degrees. Since the primary applied voltage leads the primary current, the secondary voltage is therefore 180 degrees out of phase with the primary voltage. It is also in phase with the back emf in the primary.

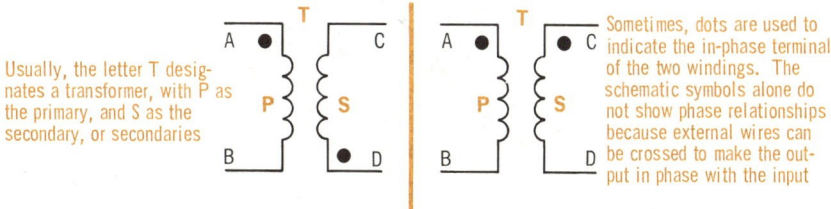

Usually, the letter T designates a transformer, with P as the primary, and S as the secondary, or secondaries

Sometimes, dots are used to indicate the in-phase terminal of the two windings. The schematic symbols alone do not show phase relationships because external wires can be crossed to make the output in phase with the input

voltage and current

PHASE RELATIONSHIPS IN TRANSFORMER WITH UNLOADED SECONDARY

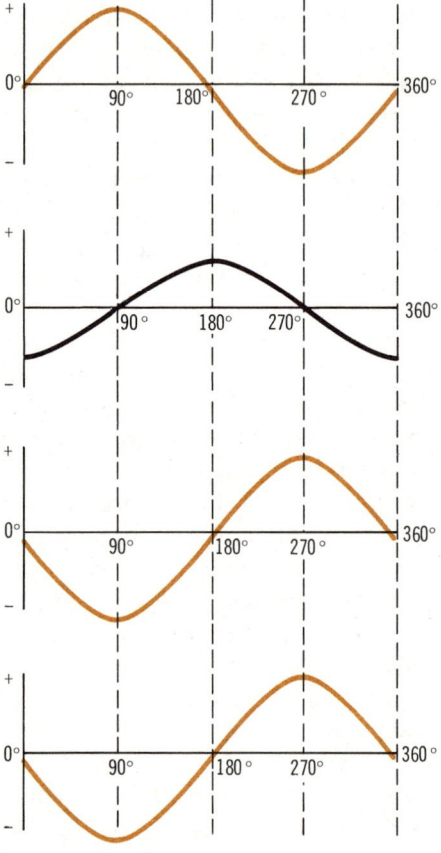

Voltage Applied to Primary

- Lead primary current by 90°
- 180° out of phase with back emf
- 180° out of phase with secondary voltage

Primary Current

- Lags applied voltage by 90°
- Leads back emf by 90°
- Leads secondary voltage by 90°

Back EMF Induced in Primary

- Lags primary current by 90°
- 180° out of phase with applied voltage
- In phase with secondary voltage

Secondary Voltage

- Lags primary current by 90°
- 180° out of phase with applied voltage
- In phase with back emf

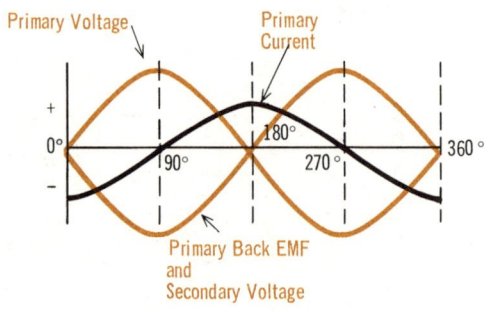

Primary Voltage

Primary Current

Primary Back EMF and Secondary Voltage

transformer operation with loaded secondary

When a *load* is connected to the secondary of a transformer, current flows in the secondary. As in any inductance, the *current* through the secondary winding *lags* the secondary *voltage* that produces it by 90 degrees. Therefore, since the secondary voltage, as you remember, lags the primary current by 90 degrees, and the secondary current lags the secondary voltage by 90 degrees, then the secondary current is 180 degrees out of phase with the primary current. As the secondary current changes, it generates its own magnetic field, whose flux lines *oppose* those of the magnetic field created by the primary current. This *reduces* the strength of the primary magnetic field, and as a result, *less back emf* is generated in the primary. With less back emf to oppose

When there is no current flowing in the secondary of a transformer, very little current flows in the primary. But, when current flows in the secondary, the current in the primary goes up in direct proportion to the secondary current

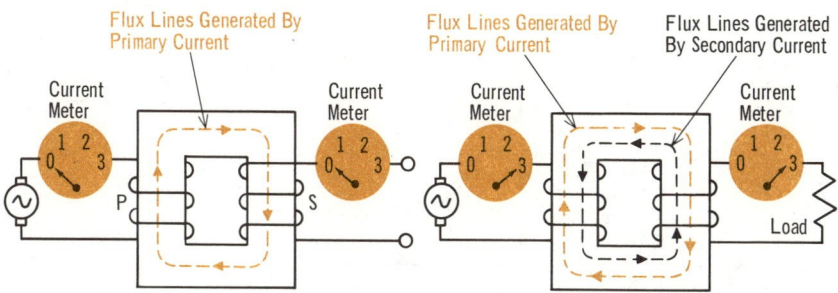

With an open secondary, only flux lines caused by primary current exist. The primary current is limited by the back emf, which is normally quite large

Secondary current produces flux lines that oppose those caused by primary current. This causes a decrease in back emf, and a resulting increase in primary current

the applied voltage, the primary current *increases*. The amount of increase is in *direct proportion* to the current flowing in the secondary. Thus, when secondary current in a transformer increases, the primary current automatically also increases. And when secondary current decreases, the primary current does likewise. You can see, then, that if the secondary of a transformer should become *shorted*, too much current would flow in the primary as well as in the secondary. Not only would the transformer burn out, but there is a possibility that the source supplying power to the primary would also be damaged.

voltage and current

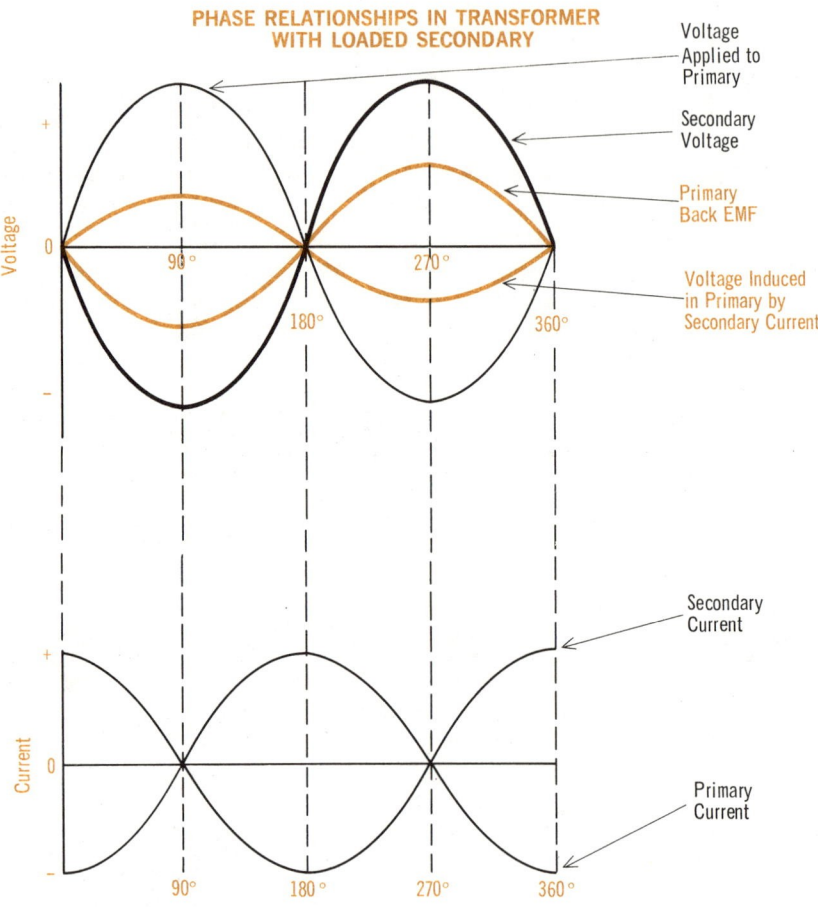

PHASE RELATIONSHIPS IN TRANSFORMER
WITH LOADED SECONDARY

Voltage Applied to Primary

Secondary Voltage

Primary Back EMF

Voltage Induced in Primary by Secondary Current

Secondary Current

Primary Current

The secondary current is 180° out of phase with the primary current. The secondary current induces a voltage in the primary that opposes the back emf of the primary. This causes the back emf of the primary to decrease, and allows more primary current to flow

effect of load on phases

The 90-degree phase difference (between the voltage and current in both the primary and the secondary) actually only exists when the secondary current is very small. When the resistance of the secondary load is decreased, secondary current increases. This is often called *increasing the load*. This makes the secondary circuit become more resistive. With a more resistive load, the phase difference between the voltage and current becomes smaller, since they are in phase in a purely resistive circuit. The greater the secondary current, the smaller is the phase angle.

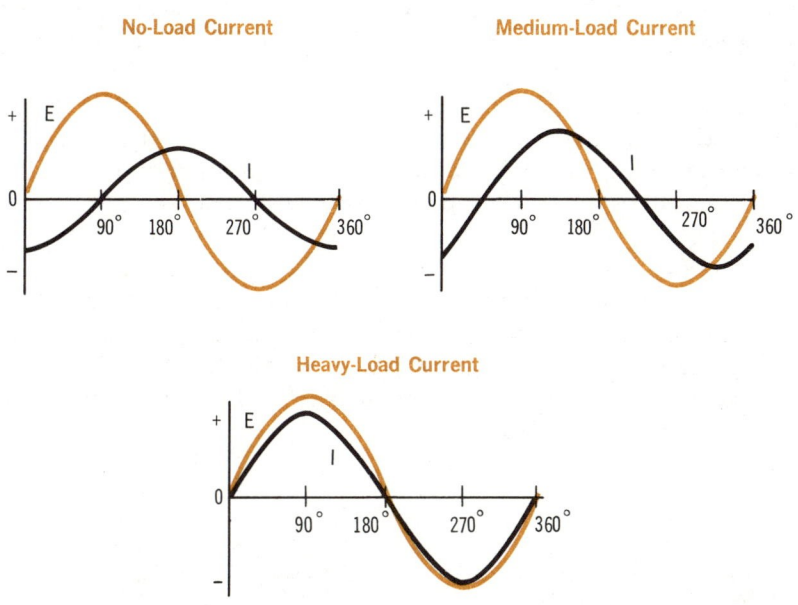

As the secondary current increases, it also reduces the back emf in the primary, and the more primary current flows. Since the cemf decreases, the inductive reactance of the primary has been reduced. The primary circuit, therefore, also becomes more resistive as current increases. As a result, the phase angle between the primary voltage and current is also reduced.

In summary, then, when very little current flows in the secondary of a transformer that contains a resistive load, the voltage and current in both the primary and secondary are 90 degrees out of phase. But as the secondary load increases, the phase angle between primary voltage and current, and secondary voltage and current becomes smaller (closer to 0 degrees). This is more fully explained in Volume 4.

primary and secondary power

A transformer actually transfers electric *power* from the primary to the secondary circuit. The primary circuit *draws* the power from the *source*, and the secondary *delivers* the power to the *load*. The power transferred from the primary to the secondary is determined by the current flowing in the secondary, which, in turn, depends on the power required by the load. If the load requires a great deal of power, such as would be the case for a low-resistance load, a high current will flow in the secondary. As you learned previously, this high current will cause a decrease in the back emf of the primary, and the primary current will increase; this will produce the stronger magnetic field necessary for the high secondary current. The transformer thus regulates the power transferred from the source to the load in response to the load requirements.

In an *ideal* transformer, the power in the primary circuit equals the power in the secondary circuit. Since power equals current times voltage, the equation for the relationship between the primary power (P_p) and the secondary power (P_s) in an ideal transformer is

$$E_p \times I_p = E_s \times I_s$$

Thus, assuming that primary and secondary voltages are equal, which is the case when the transformer primary and secondary windings have the *same* number of *turns*, the primary current will automatically adjust to the same value as the secondary current so that the primary and secondary powers are equal.

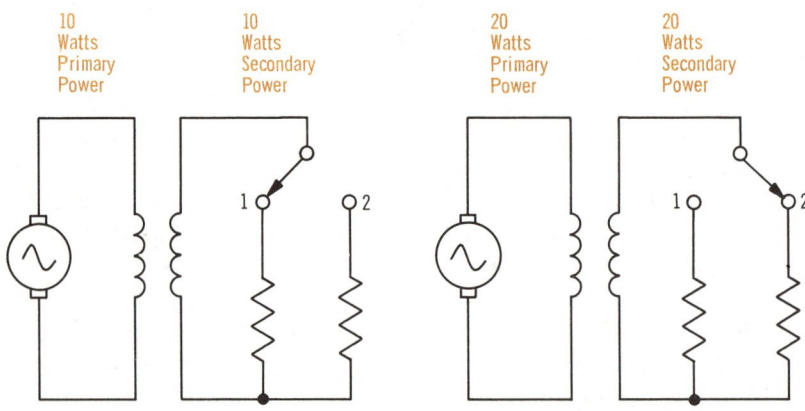

10 Watts Primary Power   10 Watts Secondary Power   20 Watts Primary Power   20 Watts Secondary Power

In an ideal transformer, the power in the primary circuit equals the power in the secondary circuit

turns ratio vs. voltage and current

As was mentioned, an important use of the transformer in power transmission is to convert power at some values of current and voltage into the *same power* at some *other values* of current and voltage. Basically, this can be done because with a given applied voltage in the primary, the secondary voltage depends on the number of *turns* in the *secondary* winding as compared to the number of *turns* in the *primary* winding. When the secondary winding has *more* turns than the primary, the secondary voltage is *greater* than the primary voltage. A step up in voltage, therefore, takes place, and the transformer is called a *step-up transformer*. Likewise, if the secondary winding has *fewer* turns than the primary, the secondary voltage is lower than the primary voltage, and the transformer is called a *step-down transformer*.

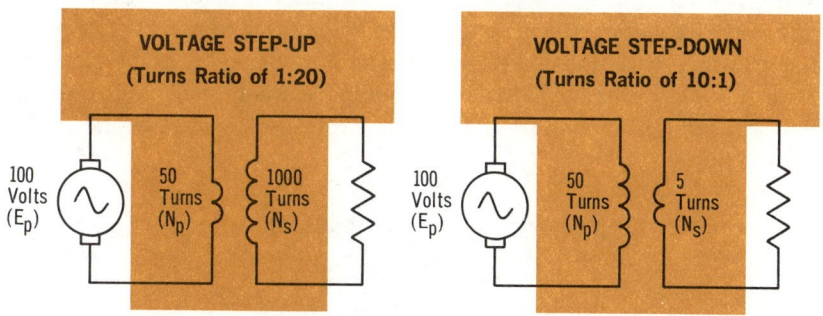

VOLTAGE STEP-UP
(Turns Ratio of 1:20)

100 Volts (E_p) 50 Turns (N_p) 1000 Turns (N_s)

$E_S = E_p (N_S/N_p) = 100 \times (1000/50) = 2000$ volts

VOLTAGE STEP-DOWN
(Turns Ratio of 10:1)

100 Volts (E_p) 50 Turns (N_p) 5 Turns (N_s)

$E_S = E_p (N_S/N_p) = 100 \times (5/50) = 10$ volts

The reason for this increase or decrease in voltage is simple if you remember that the voltage induced in any coil is actually the sum of the many voltages induced in each loop that the flux lines cut. The more loops, therefore, the more individual voltages are induced, and the greater is their sum. The exact relationship in an ideal transformer between the primary and secondary voltages (E) and their number of turns (N) is given by the equation:

$$E_p/N_p = E_s/N_s \qquad \text{or} \qquad E_p/E_s = N_p/N_s$$

Therefore, the secondary voltage is equal to:

$$E_s = E_p \left(\frac{N_s}{N_p} \right)$$

turns ratio vs. voltage and current (cont.)

The relative number of turns in the windings (N_p/N_s) is called the *turns ratio* of the transformer, and is usually expressed as a proportion; for example, $10:1$, $50:1$, $1:20$, etc. You can see from the equation that if the secondary has *twice the turns* as the primary (turns ratio of $1:2$), the secondary voltage is *twice* the *primary voltage.* Similarly, if the secondary only has *half the turns* of the primary (turns ratio of $2:1$), the secondary voltage is *half* the *primary voltage.*

Since the turns ratio determines the relationship between primary and secondary voltages, and since ideally, the power in the primary is equal to the power in the secondary, then there must also be a relationship between the turns ratio and the primary and secondary *currents.* From the power equation ($P = EI$), you can see that for the primary and secondary powers to be equal, the winding with the *higher voltage,* and therefore more turns, must have a *lower current* in it. Similarly, the winding with the *lower voltage,* and therefore fewer turns, must have a *higher current.* This relationship between the turns ratio and the primary and secondary currents is expressed by the equation:

$$I_p \times N_p = I_s \times N_s \qquad \text{or} \qquad I_s/I_p = N_p/N_s$$

Therefore, the primary current is equal to:

$$I_p = I_s \left(\frac{N_s}{N_p} \right)$$

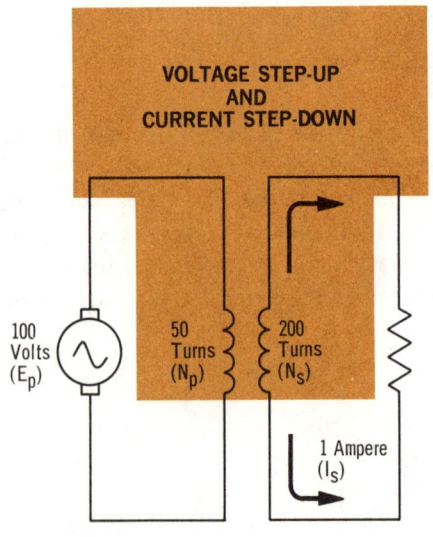

**VOLTAGE STEP-UP
AND
CURRENT STEP-DOWN**

100 Volts (E_p)

50 Turns (N_p)

200 Turns (N_s)

1 Ampere (I_s)

$I_p = I_s \, (N_s/N_p) = 1 \times (200/50) = 4$ amperes

$E_s = E_p \, (N_s/N_p) = 100 \times (200/50) = 400$ volts

$P_p = E_p \times I_p = 100 \times 4 = 400$ watts

$P_s = E_s \times I_s = 400 \times 1 = 400$ watts

turns ratio vs. voltage and current (cont.)

The equation expressing the relationship between the turns ratio and the primary and secondary currents tells us that the current in the primary times the number of turns in the primary winding is equal to the current in the secondary times the number of turns in the secondary winding. The current times the number of turns is often called the *ampere-turns*. And so, the ampere-turns of the primary equals the ampere-turns of the secondary.

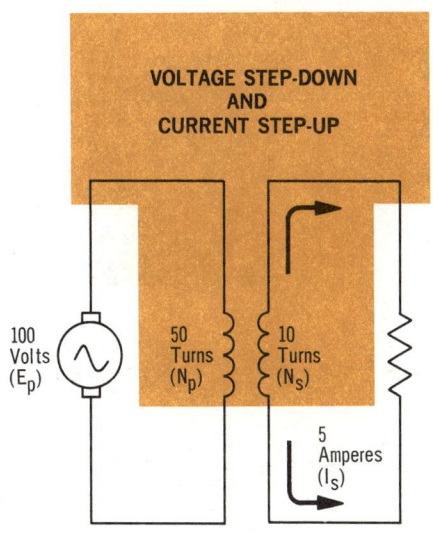

VOLTAGE STEP-DOWN AND CURRENT STEP-UP

100 Volts (E_p)

50 Turns (N_p)

10 Turns (N_s)

5 Amperes (I_s)

$I_p = I_s (N_s/N_p) = 5 \times (10/50) = 1$ ampere

$E_s = E_p (N_s/N_p) = 100 \times (10/50) = 20$ volts

$P_p = E_p \times I_p = 100 \times 1 = 100$ watts

$P_s = E_s \times I_s = 20 \times 5 = 100$ watts

If you compare the relationships of the turns ratio with voltages and currents, that is

$$E_p/E_s = N_p/N_s$$

and

$$I_s/I_p = N_p/N_s$$

you can see that the primary-secondary current ratio is the *opposite* of the primary-secondary voltage ratio. A transformer with a 1 : 50 step-up in voltage, therefore, has a 50 : 1 step-down in current. You can see, then, that power at one current and voltage can be converted to the same power at any other current and voltage by a transformer having the appropriate turns ratio.

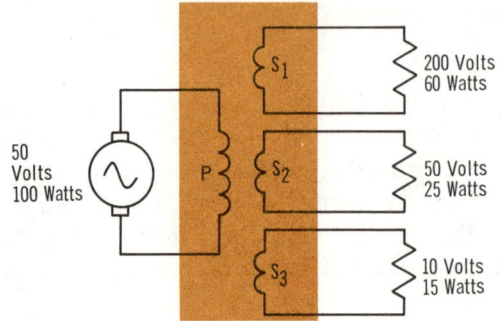

In transformers with multiple secondary windings, the power in the primary is equal to the sum of the powers in the individual secondaries:

$$P_P = P_{S_1} + P_{S_2} + P_{S_3} + \ldots + \text{etc.}$$

Multiple Secondary Windings
Connected to Common Load

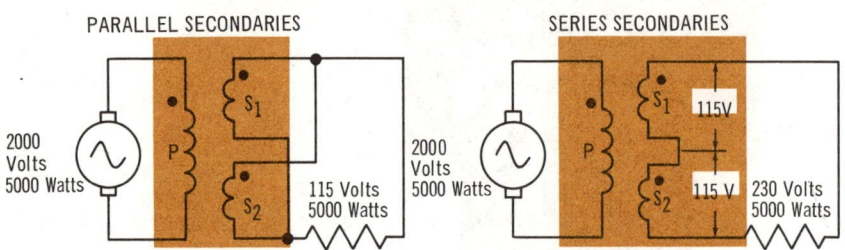

multiple-secondary transformers

There is a type of transformer that has a *single* primary winding but *more than one* secondary winding. All of the secondary windings may be step-up or step-down windings, or some may be step-up and some step-down. The voltage induced in each secondary winding is *independent of* the other windings, and is determined, as in any simple transformer, by the primary voltage and the ratio of the number of turns of the secondary to those of the primary winding.

In some applications, the secondary windings are connected to individual, independent circuits. In the transmission of electric power, though, it is common to use transformers having two secondaries, and to connect the secondary windings either in series or parallel. Both secondaries then deliver power to the same load. When the windings are connected in series, if their polarities are in the same direction, their voltages add; they conduct the same current. If their polarities are opposing, their voltages will subtract. Secondary windings are usually connected in parallel so that greater current can be supplied to a load with less loss.

summary

☐ Mutual induction exists between two coils when the flux lines generated in one coil cut the turns of the other. ☐ Mutual induction between two coils is given by: $M = k\sqrt{L_1L_2}$. ☐ Transformers make use of the mutual inductance between two coils, or windings. ☐ With an open secondary, the secondary transformer voltage is 180 degrees out of phase with the primary voltage. It is in phase with the back emf of the primary. ☐ When a load is connected to the secondary of a transformer, current flows in the secondary. The secondary current, in turn, creates a magnetic field, which induces an emf back in the primary. ☐ Shorting the secondary can cause the primary to draw excessive current, which may damage the transformer.

☐ The value of the secondary load has an effect on the phase angles between primary voltage and current, and secondary voltage and current. ☐ In an ideal transformer, the power in the primary equals the power in the secondary. ☐ The ratio of the number of turns in the primary to the number of turns in the secondary is called the turns ratio. ☐ The ratio of the primary to the secondary voltage is equal to the turns ratio.

☐ The ratio of the secondary to the primary current is equal to the turns ratio. ☐ The ampere-turns of the primary is equal to the ampere-turns of the secondary. ☐ In multiple-secondary transformers, the total power in the primary is equal to the sum of the powers of the secondaries.

review questions

1. What is *mutual induction?* What are its units?
2. What is meant by *coupling coefficient?*
3. What is the mutual inductance of two coils, each having an inductance of 14 millihenrys, when the coefficient of coupling is 0? 1/4? 1/2? 3/4? 1?
4. What is a *transformer?*
5. What is the *turns ratio* of a transformer?
6. What is the phase relationship between the primary and secondary voltages of a transformer with an open-circuited secondary?
7. In an ideal transformer, what is the relationship between the power in the primary and the power developed by the secondary?
8. The primary voltage of a transformer with a turns ratio of 1:9 is 120 volts. What is the voltage across the secondary?
9. A transformer whose primary draws 2000 watts has five equal secondaries. How much power is developed by each secondary?
10. What is meant by *ampere-turns?*

types of transformers

There are many types of transformers in use today. Each type is designed, and therefore best suited, for a particular use. They vary not only in physical characteristics, such as size and shape, but in electrical characteristics and efficiency as well. A complete description of all the various types of transformers used is beyond the scope of this volume. However, there are certain basic categories into which all transformers fall. These categories will be described on the following pages.

TRANSFORMERS VARY IN SIZE AND SHAPE

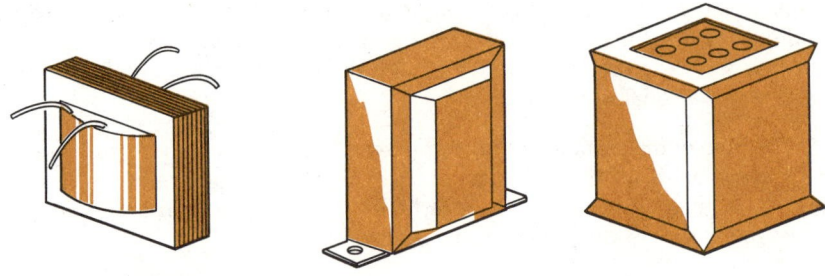

TRANSFORMERS VARY IN CORE MATERIAL

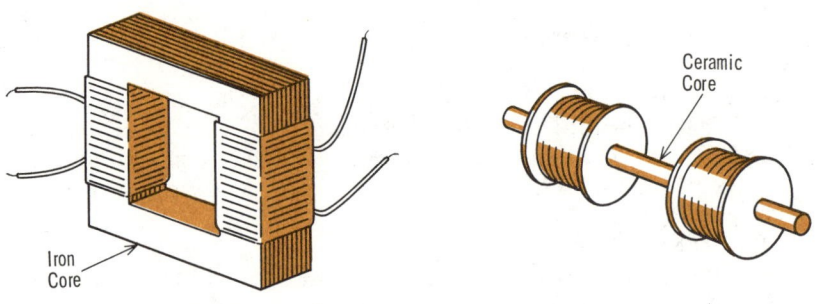

Ceramic Core

Iron Core

TRANSFORMERS VARY IN TURNS RATIO

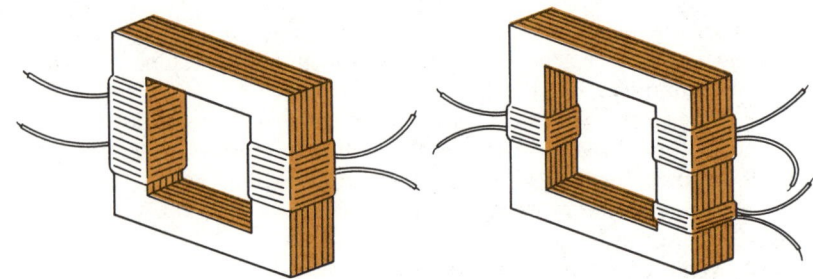

air-core transformers

The two broadest categories of transformers are *iron-* and *air-core transformers*. They obtain their names from the materials between the primary and secondary windings through which the flux lines travel. The windings of air-core transformers are around *insulating* forms, and the flux lines follow a path in *air* between the windings. Air does not offer as good a path for the flux lines as does iron, and so the coupling between the primary and secondary windings is less than that obtained if an iron core was used. The maximum coefficient of coupling possible

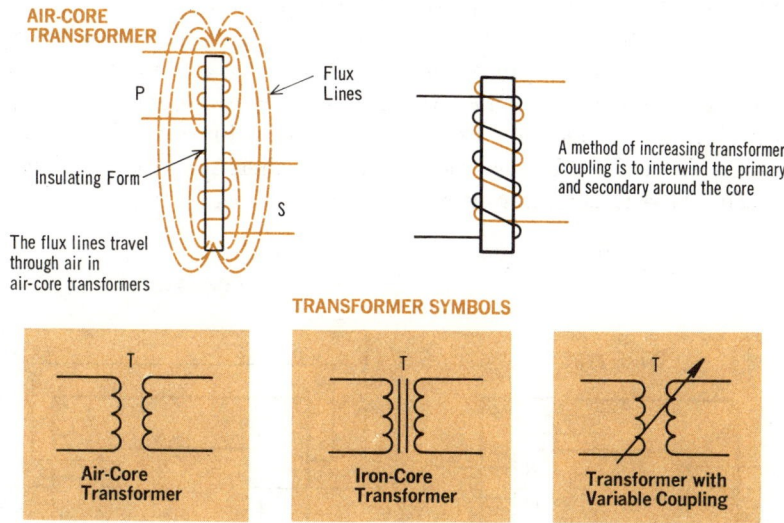

AIR-CORE TRANSFORMER

P

Flux Lines

Insulating Form

S

The flux lines travel through air in air-core transformers

A method of increasing transformer coupling is to interwind the primary and secondary around the core

TRANSFORMER SYMBOLS

T

Air-Core Transformer

T

Iron-Core Transformer

T

Transformer with Variable Coupling

with air-core transformers is about 0.65, which is much less than that provided by iron-core transformers. As a result, the simple primary–secondary power ratios do not apply to air-core transformers, since these formulas are based on almost perfect coupling. This is explained more fully for transformer efficiency, covered later. Generally, air-core transformers are used only at high frequencies, where iron-core transformers are impractical because of the losses that occur in them.

There is a type of radio transformer in which the coupling can be varied by changing the positions of the primary and secondary windings relative to each other. When the windings are at right angles to one another, there is zero coupling. When they are parallel, there is maximum coupling (1.00). Another method of changing the coupling is to pass a magnetic shield between the windings of an iron-core transformer.

iron-core transformers

The purpose of the iron core in a transformer is to provide an easy path for the flux lines that couple the windings, and to permit more flux lines to increase the coupling. The core also reduces *flux leakage*. The cores are made in different ways and of different materials to control the coupling and efficiency of the transformer.

Silicon steel is often used because of its high *permeability*, and is *laminated* to reduce *eddy current* losses. Generally, two types of construction are employed. The *core-type* is a rectangular stack of laminated strips in which the windings can be wound on opposite sides, on the same side, on adjacent sides, or in any combination of the above. The *shell-type* has an added laminated leg down the center of the core on which both windings are wound together for *tight coupling*.

Solid cores made of *ferrite*, which is a fused, powdered-iron composition, can also be used for high frequency transformers. Ferrite is not a good electrical conductor, and so has an inherently low eddy-current loss. This type of core, however, is brittle.

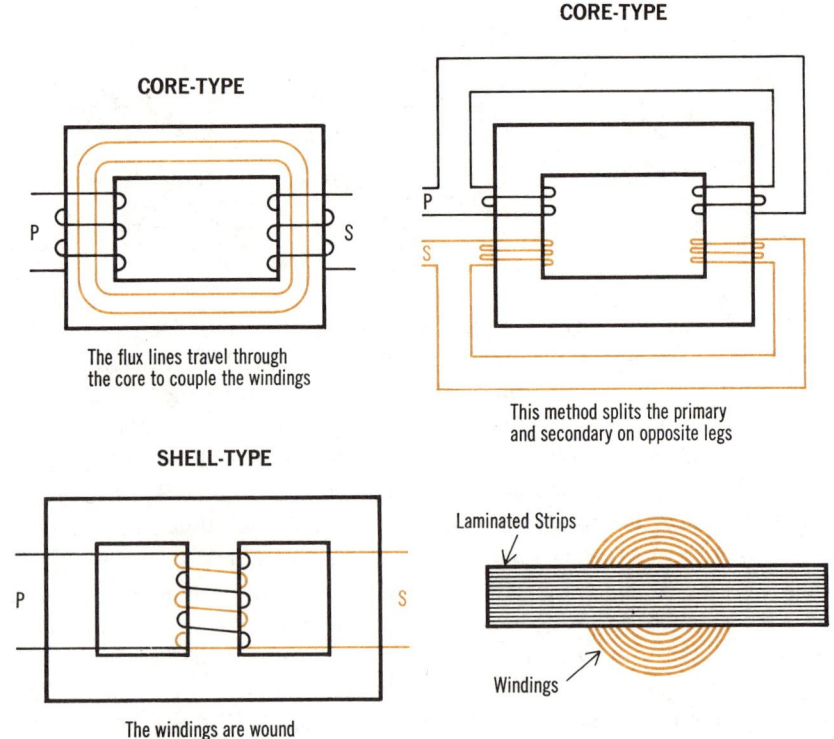

CORE-TYPE

The flux lines travel through
the core to couple the windings

CORE-TYPE

This method splits the primary
and secondary on opposite legs

SHELL-TYPE

The windings are wound
on the center leg

Laminated Strips

Windings

the autotransformer

There is a special type of iron-core transformer which physically has only *one* winding. Functionally, though, the one winding serves as *both* the primary and secondary. This type of transformer is called an *autotransformer*. When an autotransformer is used to step up the voltage, part of the single winding acts as the primary, and the entire winding acts as the secondary. When an autotransformer is used to step down the voltage, the entire winding acts as the primary, and part of the winding acts as the secondary.

The single winding of an autotransformer serves as both the primary and the secondary

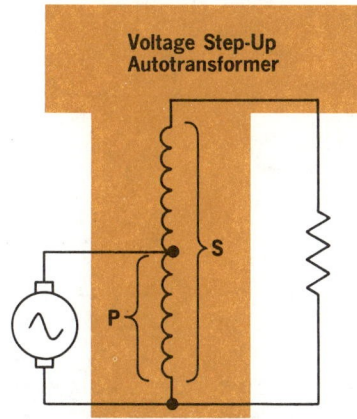

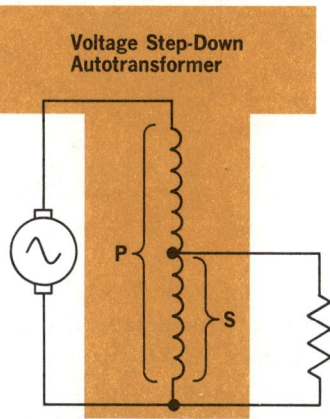

When the voltage is stepped up, part of the winding is the primary, and the entire winding is the secondary

When the voltage is stepped down, the entire winding acts as the primary, and part of the winding acts as the secondary

The action of an autotransformer is basically the same as the standard two-winding transformer. Power is transferred from the primary to the secondary by the changing magnetic field, and the secondary in turn regulates the current in the primary to set up the required condition of equal primary and secondary power. The amount of step-up or step-down in voltage depends on the turns ratio between the primary and secondary, with each winding considered as separate, even though some turns are *common* to both the primary and secondary.

A disadvantage of the autotransformer is the lack of *isolation* between the primary and secondary circuits. This results from the primary and secondary both using some of the same turns. Despite this disadvantage, the autotransformer is used in many circuits because it is inexpensive.

the adjustable transformer

Transformers can be made similar to variable resistors, so that their output voltages can be set to a specific value. Such transformers are used in power line regulators, where it is important to get an exact line voltage. Such a transformer generally uses a sliding contact that can be positioned to obtain just the right turns ratio for the proper output voltage when the value of the input voltage may not be reliable. This can be done with a regular transformer, or with an autotransformer.

The output voltage of either of these transformers can be adjusted from 0 to some maximum value

Another method to obtain the proper output voltage uses a special transformer with a tapped winding. Either the primary or secondary can be tapped, and a switch picks out the turns ratio. But this method does not allow as precise control as does the transformer with the sliding contact. However, transformers can be made using both these methods. For example, a switch with a tapped primary for a *coarse* adjustment, and a sliding contact on the secondary for a *fine*, or *vernier*, adjustment.

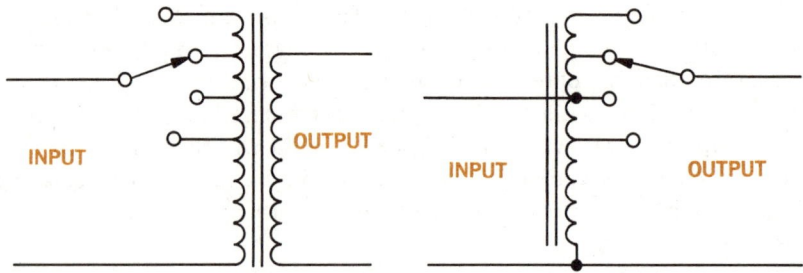

The output voltage of either of these transformers can be adjusted within a range to get close to the desired value

transformer losses

You remember that in an ideal transformer the power in the secondary is exactly *equal* to the primary power. This is true for a transformer with a coefficient of coupling of 1.0 (complete coupling) and no internal losses. In practice, such a transformer cannot be made. The degree to which any transformer approaches these ideal conditions is called the *efficiency* of the transformer. Mathematically, the efficiency is equal to the output (secondary) power divided by the input (primary) power. Or,

$$\text{Efficiency } (\%) = \frac{\text{output power}}{\text{input power}} \times 100$$

You can see from this equation that when the output and input power are equal, the efficiency is 100 percent. The smaller the output power is in relation to the input power, the lower is the efficiency.

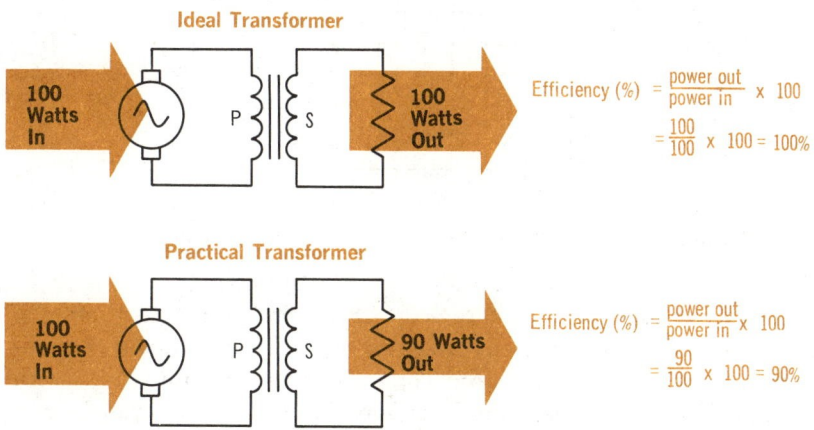

In any practical transformer, the output power is less than the input power and so the efficiency is less than 100%. The lost, or wasted, power is the result of transformer losses, and less than complete coupling

Since transformer losses reduce the efficiency of a transformer, and therefore represent wasted power, these losses are usually kept to a minimum. This is especially true in the design of iron-core transformers that must deliver large amounts of power. The most common type of transformer losses are described on the following pages.

copper loss and leakage

Transformer windings are usually made of many turns of copper wire. As with any wire, these windings have *resistance*. The *more turns* there are to the windings, the longer is the effective length of the wire, and so the *greater* is the *resistance*. When the primary and secondary currents flow through the windings, power is *dissipated* in the form of heat. These I²R losses are called *copper losses*, and are proportional to the square of the current and to the resistance. Copper losses can be minimized by winding the transformer primary and secondary with wire that has a large cross-sectional area, but this increases the size and weight of the transformer.

COPPER LOSSES

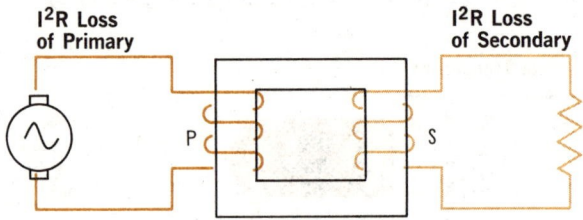

Copper losses result from the resistance of the wire used for the transformer windings

A source of inefficiency in iron-core transformers results from the fact that not all of the flux lines produced by the primary and secondary windings travel through the iron core. Some of the lines *leak* from the windings out into space, and therefore do not link the primary and secondary. This leakage of the flux lines represents *wasted* energy.

FLUX LEAKAGE

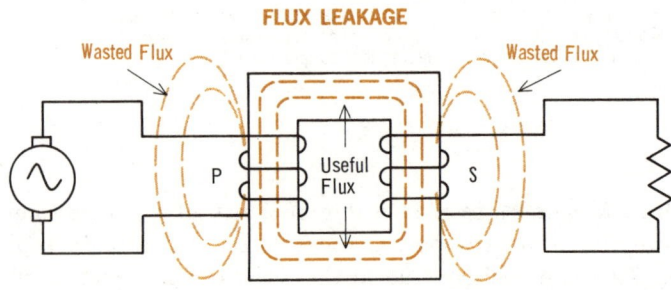

Leakage is caused by the fact that not all of the flux lines travel through the iron core

hysteresis loss

In an iron-core transformer, the core is magnetized by the magnetic field created by the current through the windings. The direction in which the core is magnetized is the same as the direction of the magnetic field that causes it to become magnetized. Thus, each time the magnetic field around the windings expands and collapses, the direction in which the core is magnetized also changes. You will remember from Volume 1 that each molecule of iron behaves as a small individual magnet. To magnetize a piece of iron, all or most of these individual magnets must be aligned in the *same* direction. Therefore, every time the direction of magnetization of the core reverses, the molecules of the core turn around to become aligned in the new direction of the flux lines. However, the molecules do not follow the reversals of the magnetic field exactly.

When the core is initially magnetized, the molecules are aligned in the direction of the field. But when the magnetic field drops to zero, the molecules do not return to their original random orientations. As a result, although the magnetizing force has dropped to zero, the core still retains some of its magnetization. The magnetic field has to reverse direction and apply a magnetizing force in the opposite direction before the core returns to its unmagnetized state. The molecules then reverse and orient themselves in the new direction of the field. The lagging behind the magnetizing force by the molecules is called *hysteresis*. The energy that has to be supplied to the molecules to cause them to turn around and, in effect, try to catch up with the magnetic field, is called the *hysteresis loss* of the core. The more energy that is required, the greater is the hysteresis loss.

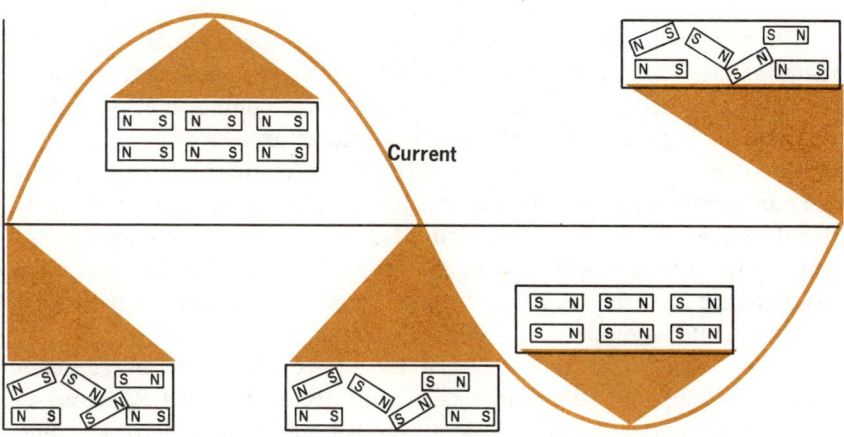

Current

hysteresis loop

Hysteresis losses depend mainly on the type of core material used. Materials that retain a large part of their magnetization after the magnetizing force is removed have high hysteresis losses. These materials are said to have a high *permanence*.

For a given core material, hysteresis losses are directly proportional to the frequency of the current in the transformer. The higher the frequency, the more times per second the molecules of the core material must reverse their alignment; and so the larger is the total energy required for this purpose. This relationship between hysteresis loss and frequency is one of the main reasons why iron-core transformers cannot be used in applications involving high frequencies.

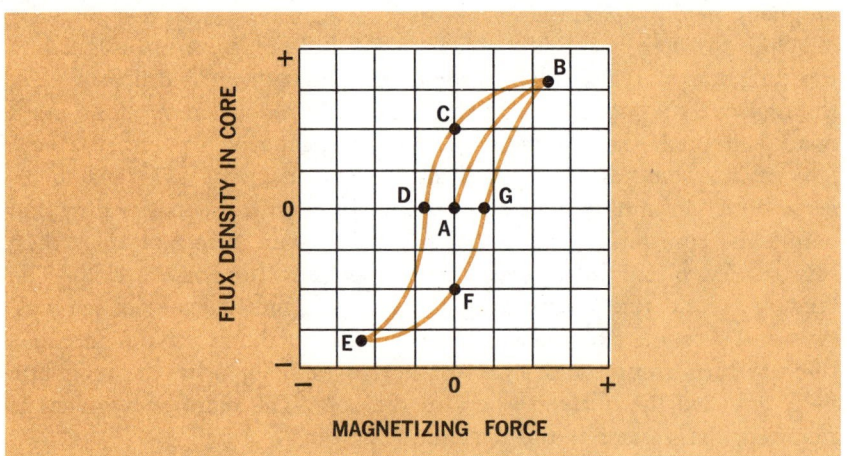

A hysteresis loop is a curve that shows how the magnetization of a material lags behind the magnetizing force. It can, therefore, be used to indicate hysteresis loss. On the curve shown, point A corresponds to no magnetizing force, and therefore no core flux.

When the magnetizing force is first applied in the positive direction, the curve travels to point B, which corresponds to core flux in the positive direction. When the magnetizing force drops to zero (point C), you can see that there is still flux in the core in the positive direction. The magnetizing force has to reverse direction and reach point D before the material is demagnetized (zero flux density).

You can follow the remainder of the first cycle of the magnetizing force (D to E), and the complete second cycle (EFGBCDE), and see how the magnetization of the core lags behind the magnetizing force.

eddy current loss

Since the iron core of a transformer is a *conducting* material, the magnetic field of the transformer *induces* a voltage in the *core*. This voltage then causes small *currents* to flow within the core. These currents are called *eddy currents*. Eddy currents can be considered as short-circuit currents, inasmuch as the only resistance they encounter is the small resistance of the core material. Like hysteresis losses, eddy currents remove energy from the transformer windings, and so represent wasted power.

Eddy currents in a transformer core are reduced by dividing the core into many flat sections, or *laminations*, with the laminations insulated from each other by means of an insulating coating applied to both sides of the lamination. The eddy currents can then only flow in the individual laminations. And since the laminations have very small cross-sectional areas, the resistance offered to the eddy currents is greatly increased.

Unlaminated Core

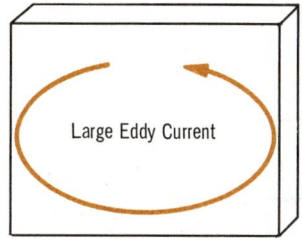

In an unlaminated core, the eddy currents are limited only by the small resistance of the core material. The currents are, therefore, large

In a laminated core, the eddy currents are confined to the small cross-sectional area of the individual laminations. This increases the resistance to the eddy currents, thereby keeping them small

Laminated Core

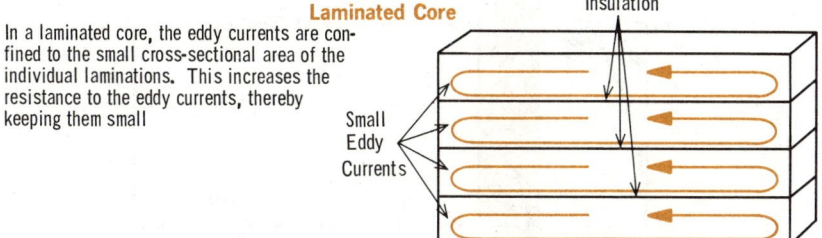

The power loss due to eddy currents is proportional to both the *frequency* and *magnitude* of the current in the transformer. Eddy current losses, like hysteresis losses, therefore limit the use of iron-core transformers in applications involving high frequencies.

saturation loss

When the current in the primary of an iron-core transformer increases, the flux lines generated follow a path through the core to the secondary winding, and back through the core to the primary winding, As the current first begins to increase, the number of flux lines in the core increases *rapidly*. The more the current rises, the greater is the number of flux lines existing in the core. When the current has risen to the point where there are a *great number* of flux lines in the core (*high flux density*), additional increases in current produce only a *few* additional flux lines. The core is then said to be *saturated*. Any further increases in primary current after core saturation has been reached results in wasted power, since the magnetic field cannot couple the additional power to the secondary.

CORE SATURATION

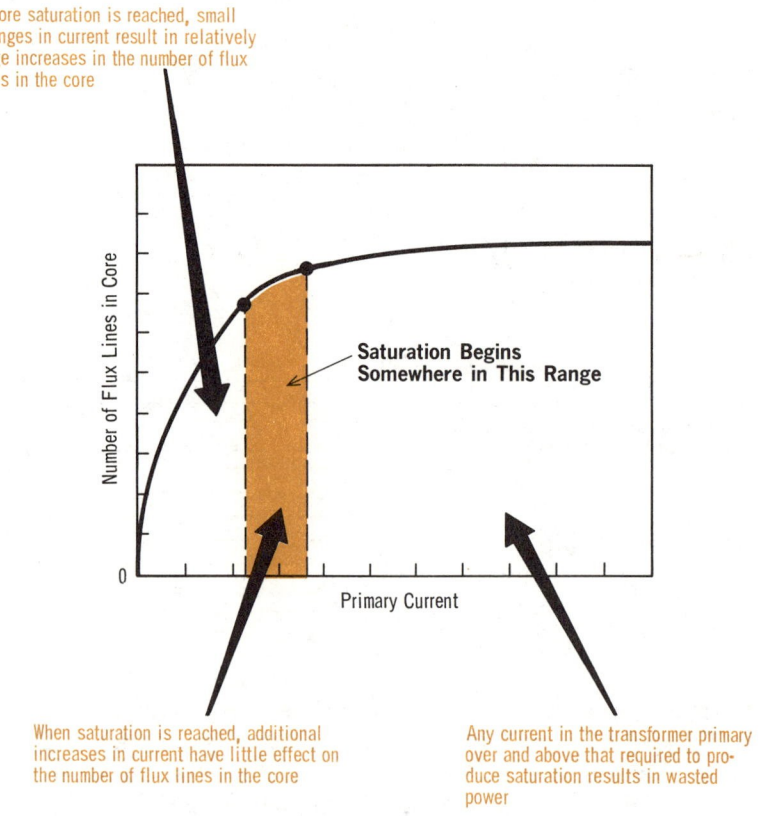

Before saturation is reached, small changes in current result in relatively large increases in the number of flux lines in the core

Saturation Begins Somewhere in This Range

Number of Flux Lines in Core

0

Primary Current

When saturation is reached, additional increases in current have little effect on the number of flux lines in the core

Any current in the transformer primary over and above that required to produce saturation results in wasted power

summary

□ Air-core transformers are wound around insulating forms, and the flux lines pass through the air between the windings. □ The maximum coefficient of coupling possible for air-core transformers is 0.65. □ Iron-core transformers have as their core iron, which provides a better path for the flux lines. □ The autotransformer has only one winding, which is tapped. □ A disadvantage of the autotransformer is the lack of isolation between the primary and secondary circuits.

□ Variable transformers have sliding contacts or different taps. This permits the transformer to have various voltages, depending on where the slide is set or which tap is used. □ The efficiency of a transformer indicates the ratio of the output power to the input power. □ The difference between the transformer's input power and output power is due to the transformer losses. □ Transformer losses include: copper loss, leakage, hysteresis loss, eddy current loss, and saturation loss.

□ Copper loss is due to the heat dissipated in the windings by current flow. These are sometimes called I^2R losses. □ Leakage is due to some lines of flux not linking both primary and secondary. This leakage represents wasted energy. □ Hysteresis loss depends on the type of core material used. It is directly proportional to frequency. □ Eddy current loss is due to circulating currents or short-circuit currents that flow in the core material. These losses are reduced by using insulated laminations for the core. □ Saturation loss is due to the inability of the core material to contain the increased number of flux lines when high current flows.

review questions

1. How many separate windings does an autotransformer have?
2. Does the autotransformer have isolation between the primary and secondary? Is this an advantage or disadvantage? Why?
3. When are iron-core transformers used?
4. What is an *air-core transformer*?
5. What is meant by *transformer efficiency*?
6. What is the efficiency of an ideal transformer?
7. What is a *hysteresis loop*? What does it indicate?
8. What is meant by I^2R *losses* in a transformer?
9. What is the difference between *hysteresis loss* and *eddy current loss* in a transformer?
10. What is meant by a *saturated core*?

RL circuits

Inductance and inductive circuit components, such as inductors and transformers, have been described so far from the standpoint of their effects on circuits that contain *no* resistance. In any circuit, though, some resistance is present, even if it is only the resistance of the connecting wires and the transformer or inductor windings. When the circuit resistance is *very low* compared to any inductive reactance present, the resistance can often be ignored. Frequently, though, resistance is deliberately added to a circuit that contains inductance. Both the resistance and the reactance then affect the circuit current and must be considered in any analysis of the circuit.

RL Circuits Contain Resistors

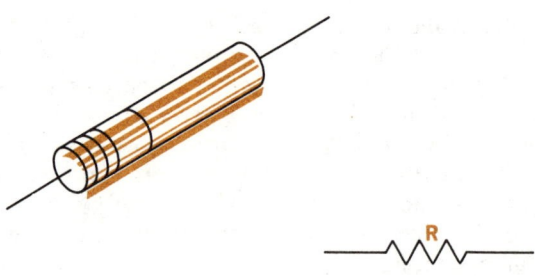

Circuits that contain both resistance (R) and inductance (L) are called *RL circuits*. The relationships between the resistance and the inductance in RL circuits and their joint effect on circuit operation are covered in Volume 4.

RL Circuits Contain Inductors

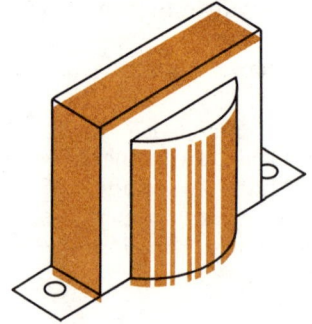

capacitance and the capacitor

Capacitance may loosely be defined as the property of an electric circuit that enables it to store electric energy by means of an *electrostatic field* and to *release* this energy sometime later. Devices that introduce capacitance into circuits are called *capacitors*. Physically, a capacitor exists whenever an *insulating* material separates *two conductors* that have a difference of potential between them. Capacitors are items that are manufactured for deliberately adding capacitance

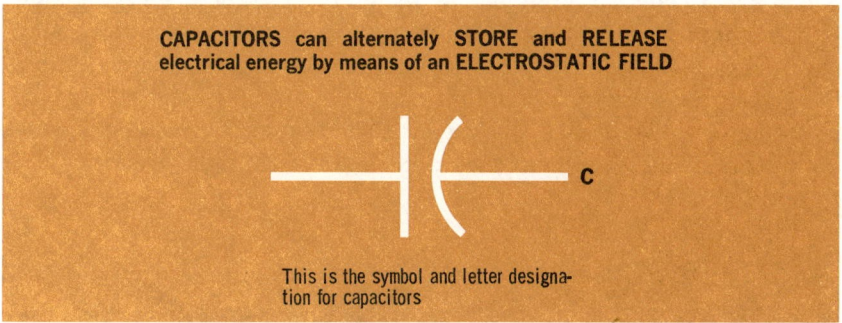

CAPACITORS can alternately STORE and RELEASE electrical energy by means of an ELECTROSTATIC FIELD

C

This is the symbol and letter designation for capacitors

to a circuit. Capacitance, however, can also be a byproduct of the arrangement and location of parts in an electric circuit or system. In this case, the capacitance they introduce is usually unwanted. In a capacitor, the electric energy is stored in the form of an electrostatic field between the two conductors, or *plates,* as they are usually called.

Capacitors used to be and are still sometimes referred to as *condensers*. The term capacitor, however, is more correct.

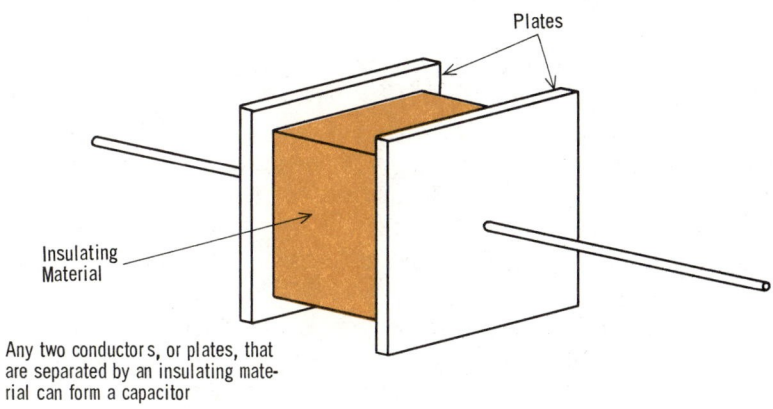

Plates

Insulating
Material

Any two conductors, or plates, that are separated by an insulating material can form a capacitor

charging a capacitor

When a capacitor has a potential difference between its plates, it is said to be *charged*. To produce a potential difference, or in other words, to charge a capacitor, *free electrons* are made to *accumulate* on one plate, and at the same time free electrons are *removed* from the other plate. One plate thus has an *excess* of free electrons, and the other plate has a *lack* of them. Since electrons are negative, the plate with the excess electrons has an overall *negative* charge, while the plate from which electrons were removed has an overall *positive* charge. A difference of potential, or a voltage, therefore exists between the plates. This is shown for the case of a simple capacitor that has two metal plates separated by air, which acts as an electrical insulator. Before the capacitor is charged, both plates are electrically *neutral*. As you remember from Volume 1, this means that they contain an *equal number* of positive charges (protons) and negative charges (electrons).

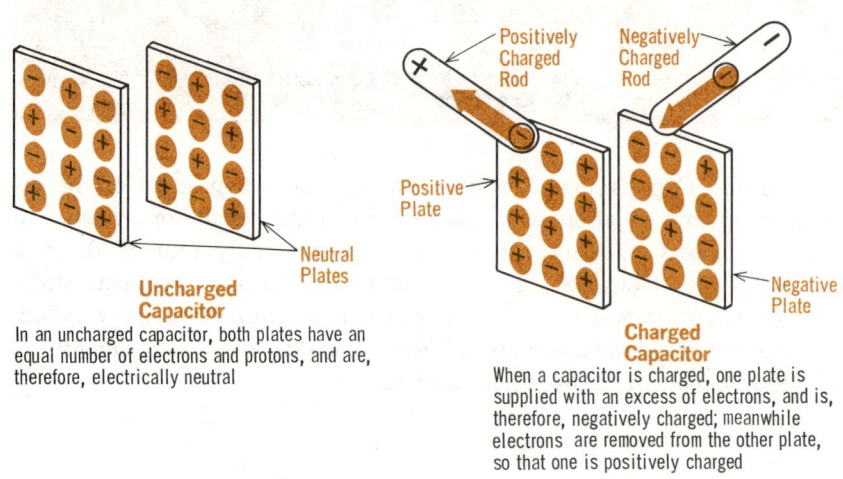

Uncharged Capacitor
In an uncharged capacitor, both plates have an equal number of electrons and protons, and are, therefore, electrically neutral

Charged Capacitor
When a capacitor is charged, one plate is supplied with an excess of electrons, and is, therefore, negatively charged; meanwhile electrons are removed from the other plate, so that one is positively charged

If a rubber rod that has been charged negatively by being rubbed with cat's fur is touched to one plate, electrons from the rod will be drawn into the plate. The plate then has a negative charge, since it has more electrons than protons. If at the same time, a glass rod that has been charged positively by being rubbed with silk is touched to the other plate, electrons will be attracted from the plate by the rod. This plate then has a positive charge, because it has more protons than electrons.

In actual circuits, of course, capacitors are not charged by rubber and glass rods. Practical charging sources, such as batteries and generators, are used.

the electric field

When a capacitor is charged, one plate is negative and the other is positive. Since electrons are *attracted* by a *positive* potential, the excess electrons on the negative plate would tend to move to the positive plate were it not for the fact that the two plates are separated by an insulating material. Even though the electrons cannot flow to the positive plate, the *electrostatic force* that attracts them still exists. This force is called the *electric field,* and can be pictured as lines of electric force that exist between the two capacitor plates. The more a capacitor is charged, the stronger is the electric field. As the charges are increased, so also does the electric field and the force of attraction between the plates. In actual practice, a capacitor can be *overcharged* to the point where the attraction between the positive plate and the negative plate is so great that the electrons will be *pulled* through the insulation to the positive plate. When this happens, the insulation is said to have *broken down.*

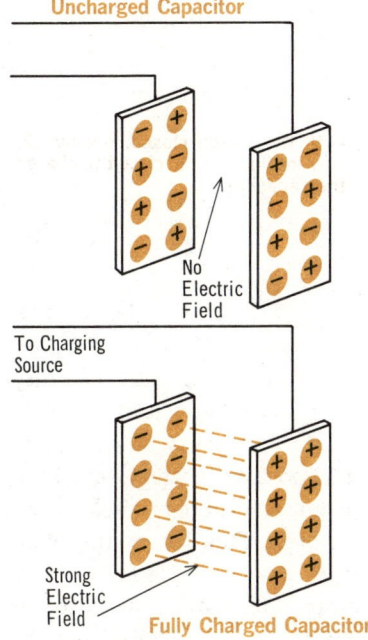

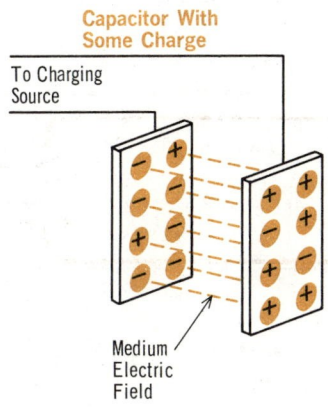

As a capacitor charges, an electric field is set up between the plates. The more the capacitor charges, the stronger is the electric field. The electrical energy stored by a capacitor is contained in the electric field, and shows up as a voltage across the plates

The electric field between the plates of a capacitor can be considered as *stored energy,* and it shows up as a *voltage* across the plates. The voltage will drop if the stored energy is released when the capacitor is discharged. This is covered later.

charging a capacitor in a d-c circuit

For a capacitor to become charged and thus store electrical energy, it must have a difference of potential, or voltage, applied to the plates. If this voltage is supplied by a battery, one plate of the capacitor is connected to the positive battery terminal, and the other plate, to the negative battery terminal. If a switch is placed in the circuit, as shown, no voltage is applied to the capacitor as long as the switch is open. Consequently, the capacitor plates are neutral, and no energy is stored.

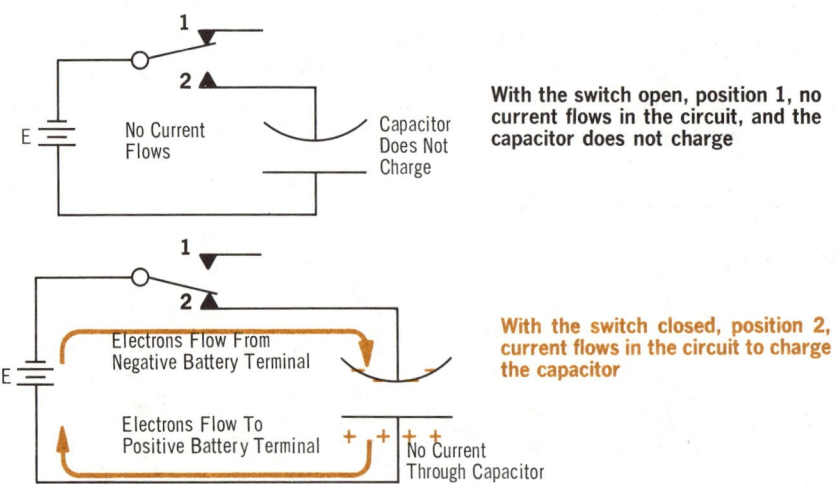

With the switch open, position 1, no current flows in the circuit, and the capacitor does not charge

With the switch closed, position 2, current flows in the circuit to charge the capacitor

When the switch is closed, electrons flow from the negative battery terminal, which has a negative potential, to the capacitor plate to which it is connected. The plate thus acquires an excess of electrons, or a negative charge. At the same time, the positive battery terminal, which has a positive potential, attracts an equal number of electrons from the capacitor plate to which it is connected. This leaves the plate with a lack of electrons, or a positive charge.

While the capacitor is charging, therefore, electrons flow through the circuit wires and enter and leave the battery. In other words, current flows in the circuit. You should note carefully, though, that although current flows in the circuit, it does not flow through the capacitor. The current enters and leaves the capacitor, but the insulation between the capacitor plates prevents current from flowing through the capacitor.

charging a capacitor
in a d-c circuit (cont.)

As electrons flow into the negative plate of a capacitor and out of the positive plate, the increasing electrostatic field causes a voltage to be built up across the capacitor. This voltage starts at zero when the circuit is first closed, and increases as more and more electrons enter the negative plate and a like number leave the positive plate. The voltage building up across the capacitor has a polarity *opposite* to that of the battery supplying the current. As a result, the voltage across the capacitor *opposes* the battery voltage. The total circuit voltage, therefore, consists of two *series-opposing* voltages.

As the voltage across the capacitor increases, the *effective* circuit voltage, which is the *difference* between the battery voltage and the capacitor voltage, decreases. This in turn causes a decrease in circuit current. When the voltage across the capacitor *equals* the battery voltage, the effective voltage in the circuit is zero, and so current flow stops. At this point, the capacitor is *fully* charged, and no further current can flow in the circuit.

The current that flows while the capacitor is charging is called the *charging current*. From this you can see that the charging current and the voltage across the capacitor behave in *opposite* ways. When charging begins, the current is maximum and the capacitor voltage is zero. And as the capacitor voltage increases, the current decreases. Finally, when the capacitor voltage reaches its maximum value, the current is zero.

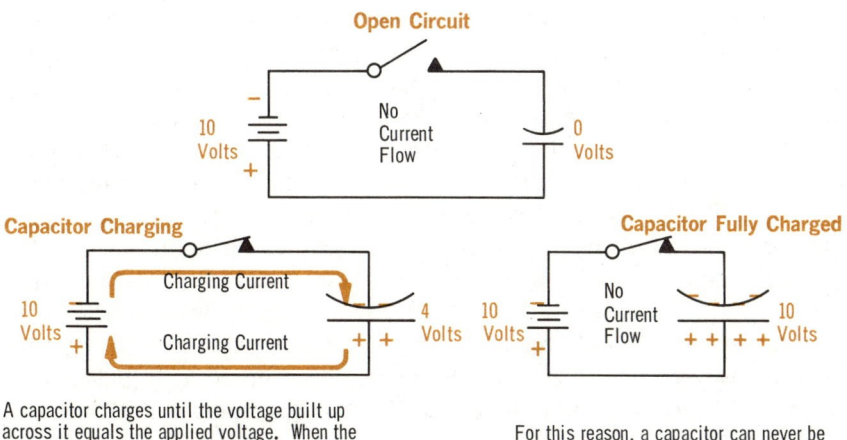

Open Circuit

10 Volts — No Current Flow — 0 Volts

Capacitor Charging

10 Volts — Charging Current — Charging Current — 4 Volts

Capacitor Fully Charged

10 Volts — No Current Flow — 10 Volts

A capacitor charges until the voltage built up across it equals the applied voltage. When the two voltages are equal, the capacitor is fully charged, and current stops

For this reason, a capacitor can never be charged to a higher voltage than the source supplying the charging current

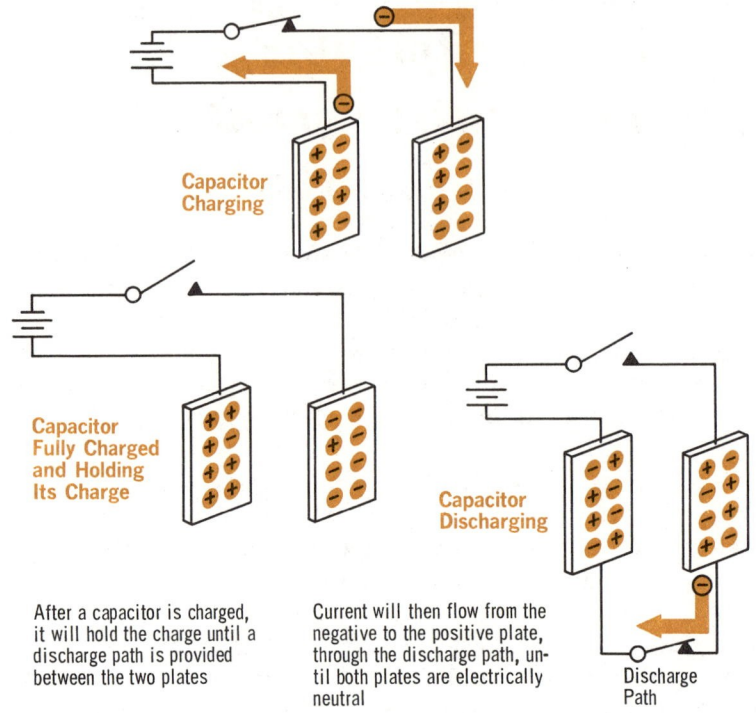

Capacitor
Charging

Capacitor
Fully Charged
and Holding
Its Charge

Capacitor
Discharging

Discharge
Path

After a capacitor is charged, it will hold the charge until a discharge path is provided between the two plates

Current will then flow from the negative to the positive plate, through the discharge path, until both plates are electrically neutral

discharging a capacitor

Once a capacitor is charged, it will theoretically hold the charge indefinitely. In practice, though, when a charged capacitor is removed from its charging source it will *eventually* lose its charge. This loss of charge takes time, however, and so for practical circuit applications, a capacitor can be considered as holding its charge until it is *deliberately* made to return the electrical energy stored in it. The recovering of this energy is referred to as *discharging the capacitor.*

To discharge a capacitor, all that is required is a *conducting path* between the capacitor plates. The free electrons on the negative plate will then flow to the positive potential of the positive plate. In this way, the positive plate acquires enough electrons to make it electrically neutral, and the negative plate loses enough electrons to make it neutral also. With both plates neutral, the capacitor has no voltage across it, and is said to be discharged. The flow of electrons from the negative to the positive plate during discharging makes up what is called *discharge current.* The path followed by this current is known as the *discharge path.*

discharging a capacitor in a d-c circuit

After a capacitor in a d-c circuit has charged to the value of the source voltage, current stops flowing in the circuit. Thus, the circuit is, in effect, an *open* circuit. It remains this way until the capacitor discharges. The capacitor *cannot* discharge through the power source,

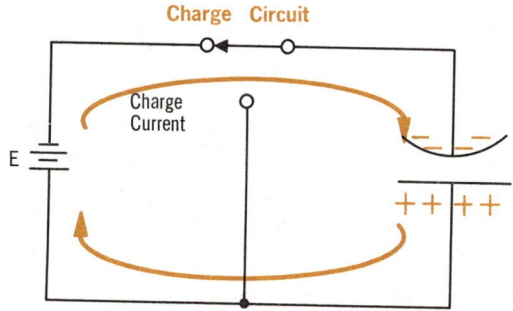

Charge Circuit

since the polarity of the source voltage is such that it opposes the capacitor voltage. For the capacitor to discharge, therefore, some other discharge path must be provided.

When such a path is available, the capacitor will *completely* discharge through it. During discharge, the capacitor and the discharge path can be considered as an independent circuit, with the capacitor

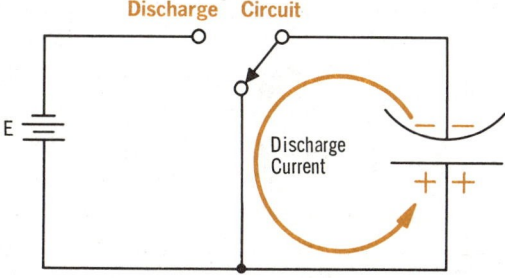

Discharge Circuit

supplying the voltage for the circuit. At the instant the capacitor begins to discharge, the voltage applied to this circuit, which is the voltage across the capacitor, is maximum. So also, therefore, is the current. As the capacitor discharges, both the voltage and current in the discharge circuit decrease, until finally they both reach zero when the capacitor is fully discharged.

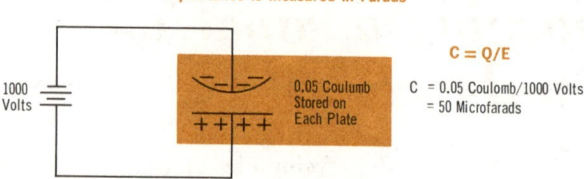

Capacitance is Measured in Farads

1000 Volts

0.05 Coulomb Stored on Each Plate

+ + + +

C = Q/E

C = 0.05 Coulomb/1000 Volts
   = 50 Microfarads

unit of capacitance

Capacitance is a measure of *how well* a capacitor can *store* electrical charge. More exactly, the capacitance is proportional to the *amount of charge* (in coulombs) that is stored in the capacitor for each volt applied across it. Actually, the net charge stored in a capacitor is zero, since the two plates have equal charges of *opposite* polarities. Therefore, "the charge stored in a capacitor" means the charge stored on either one of the capacitor plates. If two capacitors each have 100 volts applied, and one stores 5 coulombs of charge on each plate, while the other stores 2 coulombs, the one storing 5 coulombs has the greater capacitance.

Also if both capacitors stored 5 coulombs, but one capacitor needed 100 volts applied to it, and the other only needed 50 volts, the one that required less voltage to get the same charge has a greater capacitance.

The unit of capacitance is the *farad,* named in honor of the scientist Michael Faraday. A capacitor has a capacitance of one farad, when with one volt applied to its plates, it stores one coulomb of charge on each of its plates. The equation for capacitance is

$$C = Q/E$$

where C is the capacitance given in farads, Q is the charge on one plate measured in coulombs, and E is the voltage applied to the capacitor.

In practice, the farad is an enormous storage capability. For this reason, fractional units of the farad are practically always used. These fractional units are the microfarad (μf) and the micromicrofarad ($\mu\mu$f), known as the picofarad (pf).

CONVERSION OF UNITS
1 farad (f) = 1,000,000 microfarads (μf) = 1,000,000,000,000 picofarads (pf)
1 microfarad (μf) = $\dfrac{1}{1,000,000}$ farad (f) = 1,000,000 picofarads (pf)
1 picofarad (pf) = $\dfrac{1}{1,000,000,000,000}$ farad (f)
1 picofarad (pf) = $\dfrac{1}{1,000,000}$ microfarad (μf)
1 picofarad (pf) = 1 micromicrofarad ($\mu\mu$f)

capacitive time constant

When a capacitor is connected to a source of d-c voltage, it charges very rapidly. If there was *no resistance* in the charging circuit, the capacitor would become fully charged almost *instantly*. Resistance has the effect of causing a delay in the time required for charging. And since, as you know, every circuit has some resistance, it takes some definite amount of time for a capacitor to become charged. The exact time required depends on both the resistance (R) in the charging circuit, and the capacitance (C) of the capacitor. The relationship between these two quantities and the charging time is expressed by the equation:

$$t = RC$$

where t is the *capacitive time constant,* representing the time required for the capacitor to charge to 63.2 percent of its fully-charged voltage. In each succeeding time constant, the voltage across the capacitor increases an additional 63.2 percent of the remaining voltage. Thus, after the second time constant (2t) the capacitor has charged to 86.4 percent of its maximum voltage, after 3t to 94.9 percent, after 4t to 98.1 percent, and after 5t to more than 99 percent. The capacitor is considered to be *fully charged* after a period of *five* time constants.

In a similar manner, the capacitive time constant also shows the time required for the voltage across a discharging capacitor to drop to various percentages of its maximum value.

You have probably noticed a similarity between the capacitive time constant and the inductive time constant, which was described earlier. The similarity is that the voltage across a capacitor builds up and drops off in exactly the same way and at the same rate that current through an inductor rises and falls. This can easily be seen by comparing the curves shown on the following page with those for the inductive time constant given on page 3-60.

UNITS OF CAPACITIVE TIME CONSTANT		
If R Is In	**And C Is In**	**Then t Is In**
Ohms	Microfarads	Microseconds
Kilohms	Microfarads	Milliseconds
Megohms	Microfarads	Seconds
Ohms	*Picofarads	Microseconds/1,000,000
Kilohms	Picofarads	Microseconds/1,000
Megohms	Picofarads	Microseconds

*Also known as micromicrofarad.

capacitive time constant (cont.)

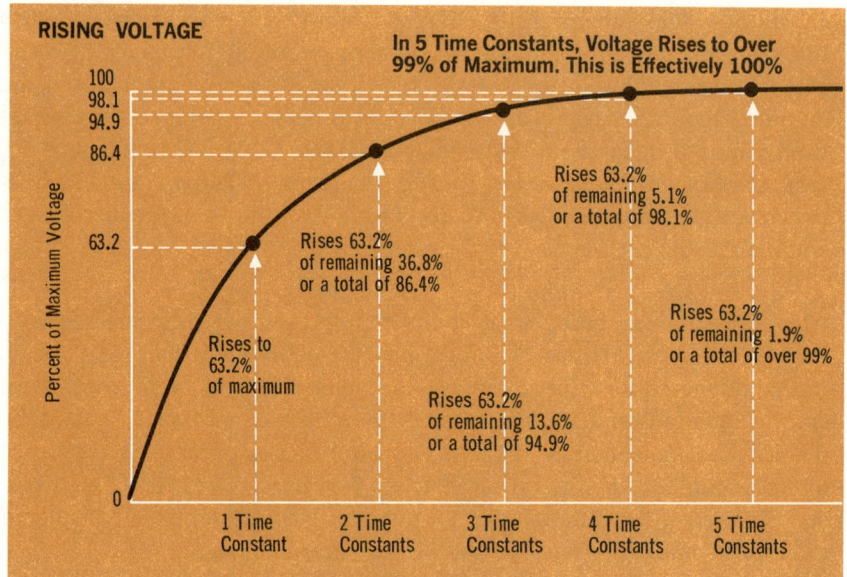

In each time constant, the voltage increases to a value
63.2% closer to its maximum value

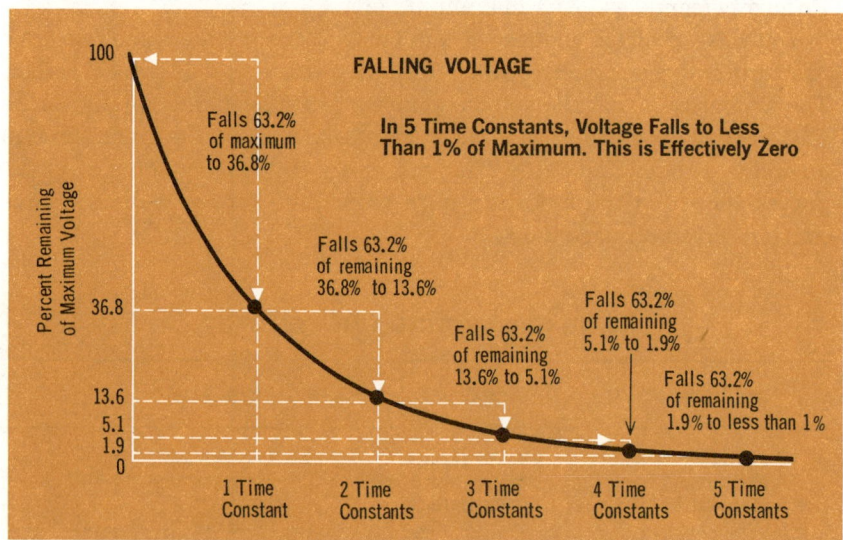

In each time constant, the voltage decreases to a value
63.2% closer to zero

summary

☐ Capacitance exists whenever two conductors are separated by an insulating material, and have a difference of potential between them. ☐ The basic unit of capacitance is the farad; however, practical units are the microfarad and the picofarad (also known as the micromicrofarad). ☐ Capacitors are devices that add capacitance into circuits. ☐ Capacitors can be charged by connecting batteries or generators across their plates.

☐ When a difference of potential, or a voltage, exists between the plates of a capacitor, it is said to be charged. ☐ A charged capacitor has an electrostatic force, or electric field, existing between the plates. The electric field stores energy. ☐ In a d-c circuit, although electrons flow through the circuit wires when the capacitor is being charged, they do not flow across the insulation between the capacitor plates. ☐ When a charged capacitor is allowed to lose its charge, the path followed by the electrons on discharging is known as the discharge path.

☐ The equation for capacitance is $C = Q/E$, where Q is the charge stored by the capacitor, and E is the voltage across the capacitor. ☐ The capacitive time constant in an RC circuit is expressed by the equation: $t = RC$. ☐ It takes five time constants to charge a capacitor from zero to more than 99 percent of its maximum charge. ☐ The waveform of the voltage build-up across a capacitor is similar to that of the current increase through an inductor.

review questions

1. How is energy stored in a capacitor?
2. Express 2 farads in picofarads; in microfarads.
3. Express 2 picofarads in microfarads. Express 2 farads in microfarads.
4. What is the capacitance of a capacitor that stores 5000 microcoulombs when the voltage across it is 1000 volts?
5. What is the *capacitive time constant*?
6. It takes 10 seconds for a capacitor to charge to the maximum voltage when it is applied in series with a resistor across a battery. What is the time constant of the circuit? How long does it take for the capacitor to discharge completely?
7. For the circuit of Question 6, what is the value of the capacitor if the resistor has a value of 5 megohms?
8. If R is in ohms and C is in microfarads, in what units is the time constant, t?
9. If t is in seconds and C is in microfarads, in what units is R?
10. If R is in megohms and t is in microseconds, in what units is C?

what controls
capacitance values?

There are three factors that determine the capacitance of a capacitor. They are: (1) the *surface area* of the plates; (2) the *distance* between the plates; and (3) the insulating material, or *dielectric* as it is called, used between the plates. The amount of voltage applied to a capacitor, or the quantity of charge stored in it, have *no* effect on capacitance. This, of course, assumes that the capacitor is not connected to excessively high voltages, which could damage or destroy it.

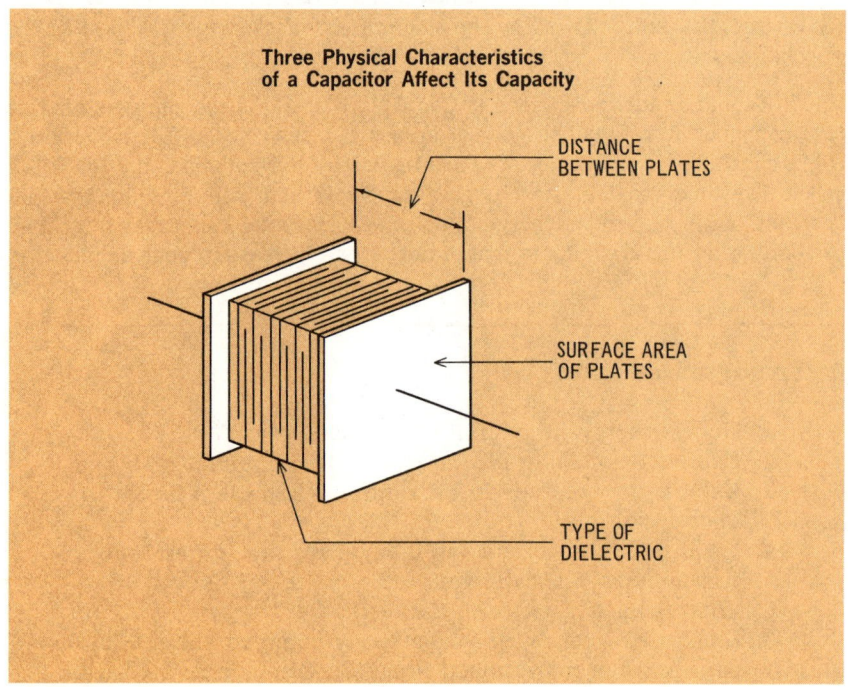

Three Physical Characteristics of a Capacitor Affect Its Capacity

DISTANCE BETWEEN PLATES

SURFACE AREA OF PLATES

TYPE OF DIELECTRIC

There are other factors which can affect capacitance, and which would have to be considered when selecting capacitors for certain applications. These factors include the frequency of the applied voltage, the temperature of the capacitor, and the age of the capacitor. Generally, these additional factors have only slight effects on capacitance, and can be ignored. At other times, though, they are very important and must be taken into consideration. More detailed information in this area is given later.

effect of plate surface area

If two capacitors are made with the same dielectric and the same separation between their plates, but with different sized plates, the one with the *most* plate *surface area* will have the *greatest* capacitance. The reason for this is that the larger the plates are, the more charge can be stored on them. A large negative plate has more room to accumulate free electrons than does a smaller one. And a large positive plate has more free electrons to give up than does a smaller one. If the separation between plates and the dielectric is held constant, the capacitance of a capacitor is *directly proportional* to the surface area of the plates. Doubling the surface area results in a doubling of the capacitance. Likewise, if the surface area is cut in half the capacitance is halved. These characteristics only apply, of course, if the plates are completely aligned and parallel to each other.

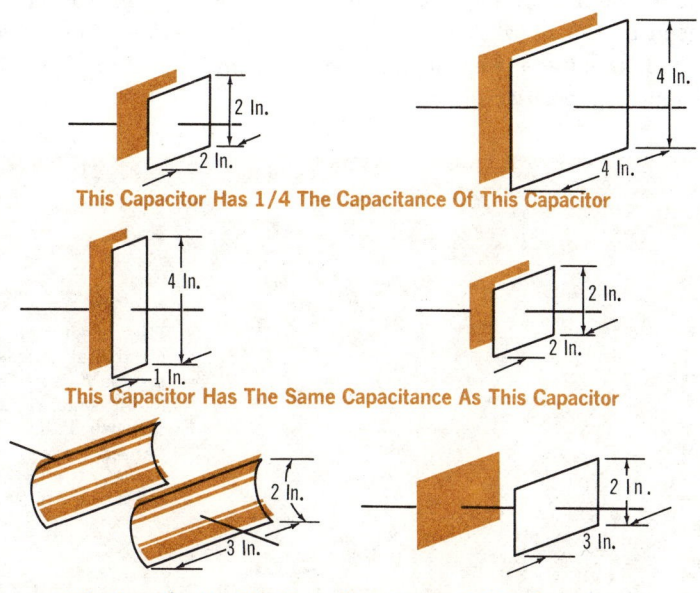

This Capacitor Has 1/4 The Capacitance Of This Capacitor

This Capacitor Has The Same Capacitance As This Capacitor

This Capacitor Has The Same Capacitance As This Capacitor

It was pointed out previously that the farad represents an enormous amount of stored charge, and that is why the units of microfarad and picofarad are used. As an example of the impractical nature of the farad, consider the plate area required for a 1-farad capacitor. If the separation between the plates is 1 millimeter and air is used as the dielectric, each plate would have to be approximately 6½ miles high by 6½ miles wide.

effect of plate separation

As you learned earlier, the capacitor charges because electrons accumulate on the negative plate, and leave the positive plate. Actually, the charges on each plate affect one another greatly. The greater the negative charge built up on the negative plate, the more electrons will be *repelled* off the positive plate; and the greater the charge on the positive plate, the more electrons will be *attracted* to the negative plate. In this way, the electrostatic field that is built up between the plates aids the source voltage. The stronger the field is for a given charge, the more it will aid the applied voltage.

If you think back to Volume 1, you should recall *Coulomb's Law of Electric Charges:* The strength of the electrostatic lines of force is inversely proportional to the *square* of the distance between the charges. This means that if two charged plates are moved apart so that the space between them is doubled, the strength of the electrostatic field will be reduced to 1/4 of what it was. And a weaker field will give *less aid* to the applied voltage; less electrons will accumulate on the negative plate and leave the positive plate.

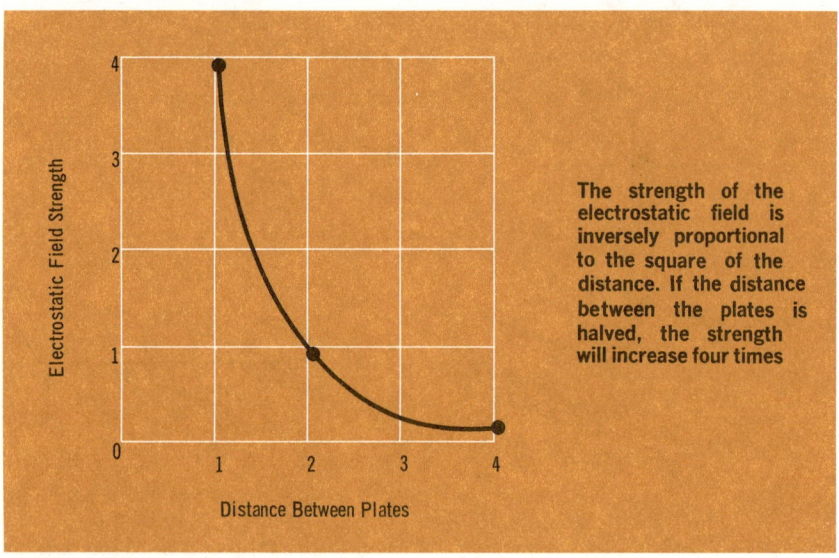

The strength of the electrostatic field is inversely proportional to the square of the distance. If the distance between the plates is halved, the strength will increase four times

Capacitance, therefore, is *inversely proportional* to the distance between the capacitor plates. This means that the smaller the distance between the plates, the greater is the capacitance. Conversely, the larger the distance between plates, the smaller is the capacitance.

effect of plate separation (cont.)

There is a *limit*, though, on how small the separation between plates can be made. If the plates are too close together, electrons will be *torn loose* from the negative plate by the potential on the positive plate. This *breakdown* can damage, and even destroy, the capacitor. Actually, the smallest spacing possible before breakdown occurs depends on the voltage across the capacitor, and the dielectric between the plates. Where high voltages are involved, breakdown will occur even in capacitors with large separations; low voltages will allow smaller separations. Some commercially available capacitors, however, have spacings as small as 0.0005 inch, and can withstand voltages of hundreds of volts.

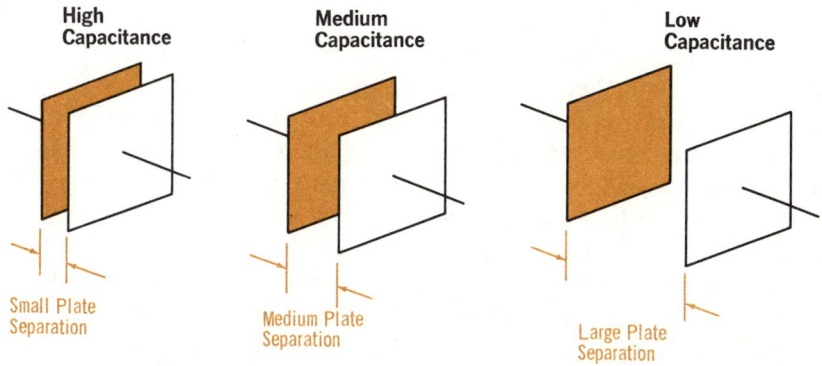

The capacitance of a capacitor is inversely proportional to the distance between the plates

Without considering the effect of the dielectric, the following equation relates the capacitance to the area of the plates and the distance between them:

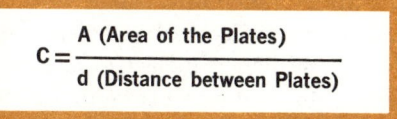

$$C = \frac{A \text{ (Area of the Plates)}}{d \text{ (Distance between Plates)}}$$

Using this equation, you can see that these two capacitors have the same capacitance value

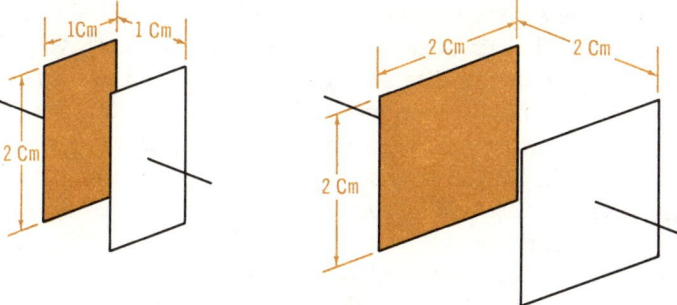

effect of the dielectric

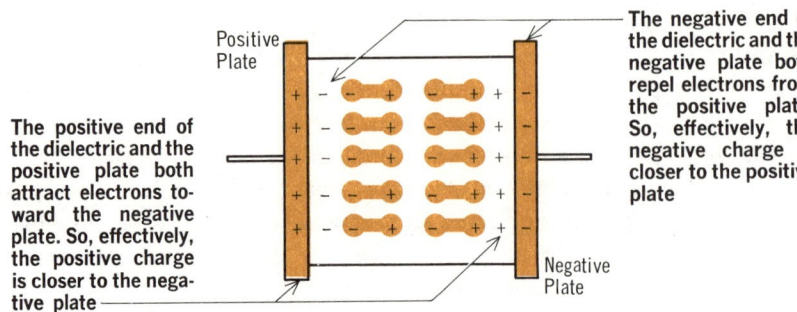

The positive end of the dielectric and the positive plate both attract electrons toward the negative plate. So, effectively, the positive charge is closer to the negative plate

Positive Plate

The negative end of the dielectric and the negative plate both repel electrons from the positive plate. So, effectively, the negative charge is closer to the positive plate

Negative Plate

Polar molecules align with the electrostatic field, effectively moving the charges on the plates closer together to increase capacitance

The insulating material used between the plates of a capacitor is called a *dielectric*. The dielectric can be air, glass, paper, or even a vacuum. In fact, any *insulating* material will work, but some produce a high capacitance, and, therefore, are good dielectrics, while others yield a low capacitance, and so are poor dielectrics.

The difference between good and poor dielectrics can be explained on the basis of how easily the molecules of the dielectric are affected by electrostatic forces. Many insulating materials have molecules whose orbiting electrons are unsymmetrically located. One side of the molecule has more electrons than the other, and so in effect, the molecule has a negative side and a positive side. These are called *polar molecules*. They have a net charge of zero because there is an equal number of electrons and protons; but because of a lopsided accumulation of electrons, the molecule has a positive pole and a negative pole.

When such a dielectric is inserted between the plates of a charged capacitor, one side of the dielectric is exposed to the positively charged plate, and the other side to the negatively charged plate. The polar molecules in the dielectric are attracted to the charged plates. As a result, the molecules *align* themselves with their positive poles towards the negative plate and their negative poles towards the positive plate. An electrostatic force then exists between each end of the dielectric and the nearest plate. The direction of these forces is such that the negative end of the dielectric repels electrons from the positive plate, and in doing so aids the negative plate, which is doing the same thing. Similarly, the positive end of the dielectric attracts electrons to the negative plate, and so aids the positive plate. Effectively, then, the dielectric moves the charges on the capacitor plates *closer together*, and, as you have learned, this results in increased capacitance.

dielectric constant and strength

Since the dielectric material has a significant effect on the capacitance value of a capacitor, it is often convenient to compare various materials from the standpoint of how well they perform as dielectrics. For this purpose, a rating system is used in which the dielectric properties of materials are related to those of *air*. Actually, the system is based on the dielectric properties of a pure *vacuum*. However, there is so little difference between an air dielectric and a vacuum dielectric, that for all practical purposes they can be thought of as identical.

In this system, air is assigned the value of 1. The other materials are assigned values which indicate how much more effective they are as dielectric materials than is air. For example, a capacitor with a dielectric material whose value is 5 will have five times the capacitance of a similar capacitor having an air dielectric. The value of each material is called the *dielectric constant* of that material. The dielectric constant is usually designated by the letter K. Therefore, to see the relationship of all the physical properties of a capacitor compared to its capacity, the following equation can be used:

$$C = K \frac{A \text{ (plate area)}}{d \text{ (plate spacing)}}$$

In addition to its dielectric constant, each material also has a property called its *dielectric strength*. This indicates the *maximum* voltage that can be applied across the dielectric *safely*. If the voltage is exceeded, the molecules of the dielectric material break down, and arcing occurs between the dielectric and the capacitor plate. Dielectric strength is expressed as a maximum allowable voltage for a specific thickness of material.

DIELECTRIC CONSTANTS		DIELECTRIC STRENGTHS	
Material	Dielectric Constant (K)	Material	Dielectric Strength (Volts/0.001 in.)
Air	1	Air	80
Resin	2.5	Fiber	50
Hard Rubber	2.8	Glass	200
Dry Paper	3.5	Castor Oil	370
Glass	4.2	Bakelite	500
Bakelite	4.5 to 7.5	Porcelain	750
Mica	5 to 9	Paper (paraffined)	1200
Porcelain	5.5	Paper (beeswaxed)	1800
Mycalex	8	Mica	2000
Titanium Dioxide Compounds	90 to 170		

capacitive d-c circuits

A d-c circuit that contains only capacitance behaves quite differently than a d-c circuit containing only resistance or inductance. Both resistance and inductance *oppose* circuit current, although in different ways. Capacitance, on the other hand, *prevents* d-c current flow, once the initial charging current has charged the capacitor. Capacitors are, therefore, said to *block* dc. You can see, then, that there are not a great many applications for capacitors in d-c circuits. This does not mean that capacitors are never used in d-c circuits. There are many d-c applications for which capacitors are well suited. One of these is to prevent arcing across the contacts of electric switches, as shown. When a switch opens a circuit in which a large current is flowing, the current attempts to continue flowing. This causes arcing across the switch contacts, which can eventually ruin the switch. A capacitor connected across the switch contacts provides a path for current flow until the switch is fully open and the danger of arcing is passed.

Capacitors are also used with batteries to supply large current for a very short time, which the battery alone could not supply. With the battery connected across the capacitor, the battery could supply a safe low level of current over a long period to charge the capacitor. Then, if the capacitor is connected to, say a photoflash lamp, it could discharge very quickly and supply a higher value of current than the battery could in such a short time.

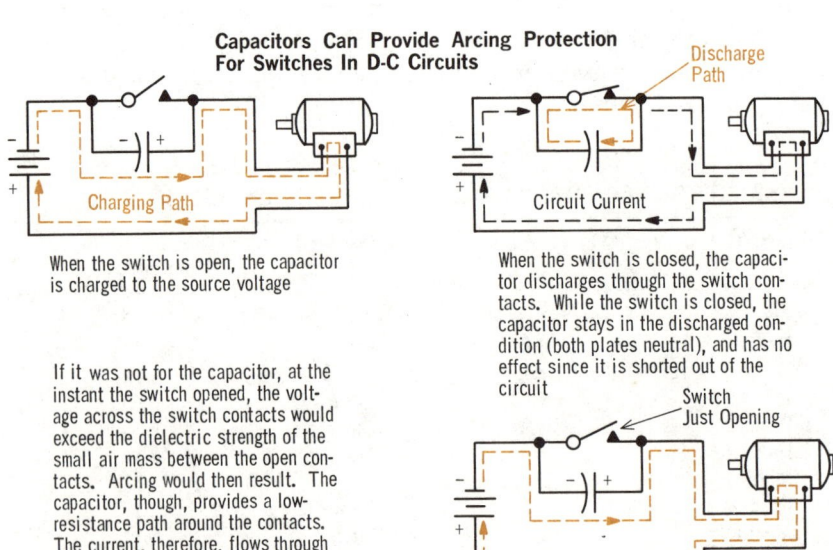

Capacitors Can Provide Arcing Protection For Switches In D-C Circuits

Charging Path

When the switch is open, the capacitor is charged to the source voltage

If it was not for the capacitor, at the instant the switch opened, the voltage across the switch contacts would exceed the dielectric strength of the small air mass between the open contacts. Arcing would then result. The capacitor, though, provides a low-resistance path around the contacts. The current, therefore, flows through the capacitor instead of across the air gap until the capacitor is charged

Discharge Path

Circuit Current

When the switch is closed, the capacitor discharges through the switch contacts. While the switch is closed, the capacitor stays in the discharged condition (both plates neutral), and has no effect since it is shorted out of the circuit

Switch Just Opening

summary

☐ The capacitance of a capacitor is determined by three factors: the surface area of the plates, the distance between the plates, and the insulating material used between the plates. ☐ The effects of the three factors determining capacitance are expressed in the equation: $C = KA/d$. The equation shows that capacitance is directly proportional to the surface area of the plates, and inversely proportional to the distance between the plates. ☐ Other factors, such as frequency of applied voltage and temperature and age of the capacitor can affect capacitance. Usually, though, their effects are slight.

☐ The insulating material used between the plates of a capacitor is called the dielectric. ☐ A good dielectric increases capacitance by effectively moving the plates of the capacitor closer together. ☐ The dielectric constant, K, indicates how effective a material is as a dielectric. ☐ A system of comparing the dielectric constants of various materials is used in which air serves as a reference with a dielectric constant of 1. ☐ The maximum voltage that can be applied across a dielectric is called the dielectric strength.

☐ In d-c circuits, charged capacitors prevent d-c current flow. Capacitors are therefore said to block dc. ☐ Capacitors are also used to supply large currents for short intervals.

review questions

1. What are the three factors that determine the capacitance of a capacitor?
2. Why are dielectrics used in capacitors?
3. What is meant by the *dielectric strength* of a material?
4. What is meant by *dielectric constant*?
5. Which has the larger dielectric constant: air or hard rubber? Air or glass?
6. What happens to the capacitance of a capacitor if its plate area is doubled, and the distance between plates is halved?
7. If, in a capacitor, the dielectric constant is doubled without changing the plate area, how can the plates be adjusted so that the capacitance remains the same?
8. If the voltage applied across a capacitor is doubled, what happens to the capacitance?
9. How are capacitors used in a photoflash lamp?
10. Why are capacitors said to block dc in a d-c circuit?

capacitive a-c circuits

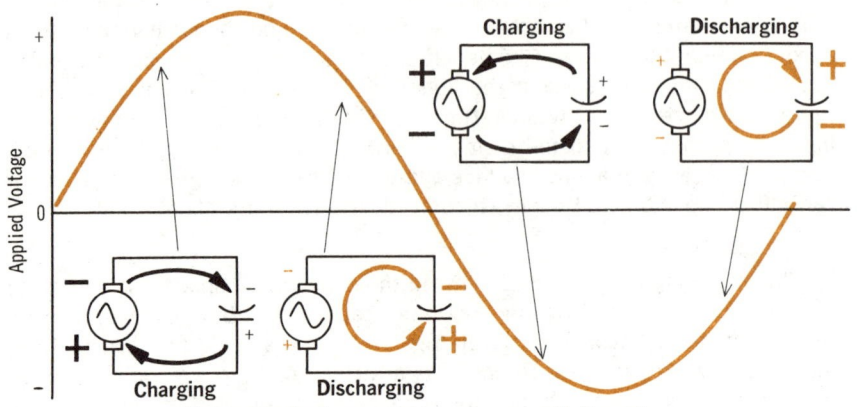

When a capacitor is connected to an a-c source, it alternately charges and discharges, first in one direction and then the other

A capacitor blocks dc. Therefore, once it is fully charged by an applied d-c voltage, no further current will flow in the circuit unless some means is provided for discharging the capacitor. In an a-c circuit, the applied voltage, as well as the current it produces, periodically changes direction. As a result, a capacitor in an a-c circuit is first charged by the voltage being applied in one direction. Then, when the applied voltage starts to decrease, less current flows, but the capacitor is still being charged in the same direction. As a result, as the applied voltage continues to drop, the voltage developed across the capacitor becomes greater. The capacitor, then acts as the source, and starts *discharging*. The capacitor becomes fully discharged when the applied voltage drops to zero and reverses its direction. Then the capacitor starts charging again, but in the same direction in which it was previously discharging. This continues until the applied voltage again starts to drop, and the events repeat themselves. This alternate charging and discharging, first in one direction, and then in the other, occurs during every cycle of the applied ac. An a-c current, therefore, flows in the circuit *continuously*. It can be said, then, that although a capacitor blocks dc, it *passes* ac.

The charging a path for a capacitor in an a-c circuit is from the negative terminal of the source to one capacitor plate, and from the other plate to the positive terminal of the source. The discharge path for the capacitor is from the negative plate, through the power source, and back to the positive plate. Remember, though, that in an a-c circuit, the terms positive plate and negative plate refer to a specific instant of time.

voltage relationships

When an a-c voltage source is connected across a capacitor, the amount of *charge* on the capacitor plates follows the sine-wave pattern of the applied voltage. At the instant the applied voltage is zero, no current flows, and so no charging potential exists for the capacitor; both plates are neutral. As the voltage source begins its sinusoidal rise from zero, the current flows, and the charge on the capacitor plates starts building. The charge continues building as the applied voltage increases. When the applied voltage reaches its peak value, the charge on the plates is maximum. This relationship between the applied voltage and the charge on the capacitor plates continues for the complete cycle of applied voltage. They both decrease and reach zero simultaneously, and then increase to a maximum in the opposite direction. Finally, both decrease again, reaching zero simultaneously.

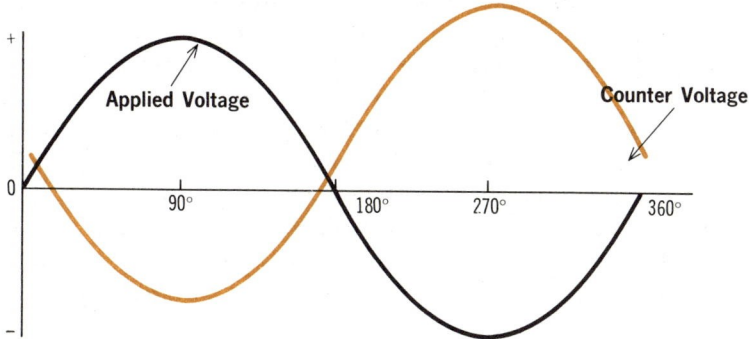

The counter voltage developed across a capacitor by the accumulation of charges on the capacitor plates is 180° out of phase with the voltage applied across the capacitor

The charges on the plates produce an electrostatic field that shows up as a voltage across the capacitor. This charge voltage follows the applied voltage exactly. If you measure the waveform pattern across the source and compare it to the pattern across the capacitor, you will see that they are the same. Actually, this has to be so, because when you measure one, you automatically must measure the other, since they are considered across one another. Since the charge on the capacitor plates follows exactly the same sine wave pattern as the applied voltage, you might think they are in phase. However, the charge on the capacitor actually *opposes* the applied voltage, so they are out of phase. Since the capacitor develops its own voltage, it acts as a voltage source that tries to cause current to flow into the battery. The capacitor voltage, therefore, is often called a *counter voltage* because it acts 180 degrees out of phase with the applied voltage.

voltage
and current relationships

When an a-c voltage source is connected across a capacitor, *maximum* current flows in the circuit the instant the source voltage begins its sinusoidal rise from zero. At first it may seem strange to you that maximum current flows when the source voltage is at its lowest value. But if you remember what you learned earlier for d-c circuits, this current is actually the movement of free electrons from the negative terminal of the source to one capacitor plate and from the other plate to the positive terminal of the source; as a result, maximum current would occur at this time, since the plates are neutral and present *no opposing* electrostatic forces to the source terminals. Therefore, as you can see by Ohm's Law, if the opposition to current flow is very, very low, a small applied voltage can cause considerable current to flow.

As the source voltage rises, however, the charges on the capacitor plates, which result from the current flow, build up. The charge voltage, then, presents an *increasing opposition* to the lower voltage and so the current *decreases*. When the source voltage reaches its peak value, the charge voltage across the capacitor plates is maximum. This charge is sufficient to completely *cancel* the source voltage, and so current flow in the circuit stops. As the source voltage begins to decrease, the electrostatic charge on the capacitor plates becomes greater than the potential of the source terminals, and so the capacitor starts to discharge.

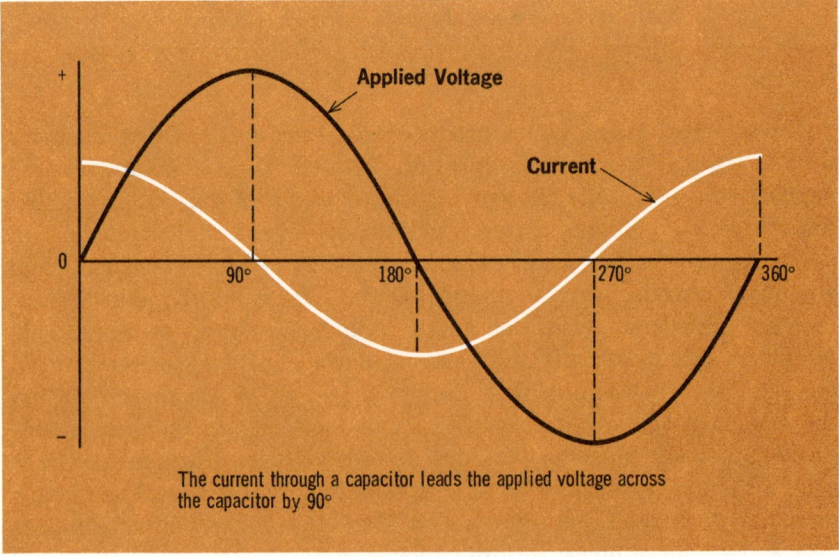

The current through a capacitor leads the applied voltage across the capacitor by 90°

voltage
and current relationships (cont.)

Electrons flow from the negative plate to the negative source terminal (which is becoming less negative), while an equal number of electrons flow from the positive source terminal (which is becoming less positive) to the positive plate. This direction of electron flow is *opposite* to the direction taken by the electrons during capacitor charging. Thus, at the point that the applied voltage passes through its maximum value and begins decreasing, the current in the circuit passes through zero and changes direction. As can be seen from the graph this constitutes

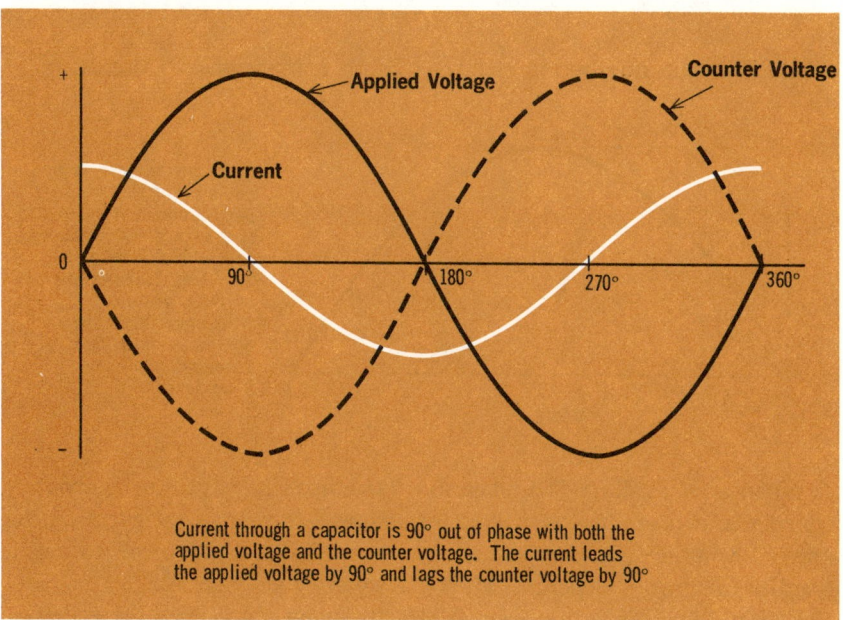

Current through a capacitor is 90° out of phase with both the applied voltage and the counter voltage. The current leads the applied voltage by 90° and lags the counter voltage by 90°

a 90-degree phase difference, with the current *leading* the applied voltage. This 90-degree difference is maintained throughout the complete cycle of applied voltage. When the applied voltage has dropped to zero, the circuit current has increased to its maximum in the opposite direction; and when the voltage reverses direction, the current begins decreasing. Therefore, the *voltage* applied to a capacitor is said to *lag* the *current* through the capacitor by 90 degrees. Or, the *current* through a capacitor *leads* the applied *voltage* by 90 degrees. And, since the counter voltage is 180 degrees out of phase with the applied voltage, the current *lags* the counter voltage by 90 degrees.

effect of frequency on capacitor current

Current flow in an a-c capacitive circuit is made up of free electrons flowing between the terminals of the voltage source and the plates of the capacitor. The electrons flow in one direction while the capacitor is charging, and in reverse direction during the time the capacitor discharges and charges in the opposite direction. The *amount* of *current* flow, you will remember from Volume 1, is determined by the *number* of *electrons* that flow past any point in the circuit in a unit *time*. Put another way, current is equal to the rate of electron flow. As an equation, this can be shown as

$$I = Q/t$$

where Q is the number of electrons being transported in time *t*.

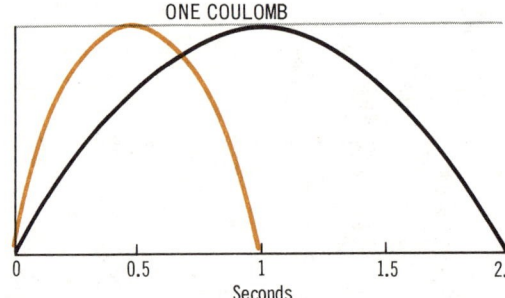

ONE COULOMB

Seconds

Since an ampere is one coulomb of electrons per second, a frequency that allows one coulomb to flow in 0.5 second produces 2 amperes of current, while a frequency that allows one coulomb to flow in 1 second produces 1 ampere

When an a-c voltage is applied to a capacitor, the capacitor alternately charges and discharges (in both directions) during every cycle of the applied voltage. For any given capacitor, the time required to reach maximum charge depends on the frequency of the applied voltage. The reason for this is that, as you already know, the charge on a capacitor follows the applied voltage exactly. Each time the applied voltage is maximum, the capacitor is at peak charge. And the higher the frequency, the faster the applied voltage rises from zero to maximum. So it follows that the higher the frequency, the faster the capacitor will reach its peak charge.

The amount of charge stored is independent of frequency; it depends only on the capacitance of the capacitor and the amplitude of the applied voltage. You can see, then, that two *identical capacitors* connected to *identical voltage* sources, which have *different frequencies*, will have *equal* values of Q but *different* values of *t* in the equation $I = Q/t$.

effect of frequency on capacitor current (cont.)

The currents through the two capacitors will therefore be different; although the same *number* of electrons flow, they flow *faster* in one capacitor than in the other, producing a greater current. The *higher frequency* (lower t) causes the *greater current,* and the *lower frequency* (higher t) causes the *lesser current.*

In summary, current through a capacitor is *directly proportional* to the frequency of the applied voltage. High frequencies result in large currents, whereas low frequencies produce small currents in the same capacitor.

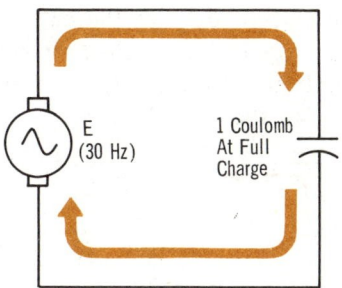 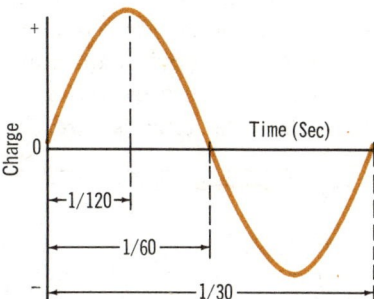

Assume that the capacitors in the circuits above and below are identical, and the source voltages have the same amplitude, but have the different frequencies as shown. Both capacitors would reach a maximum charge of 1 coulomb, and would discharge this amount during the cycle. For the charge to build up from zero to 1 coulomb with the 30-Hz voltage, it would take 1/120 of a second. The formula for current (I = Q/t) gives a current flow of 120 amperes. However, for the capacitor to build up a 1 coulomb charge with the 60-Hz voltage, it would take 1/240 of a second. This is 240 amperes of current flow.

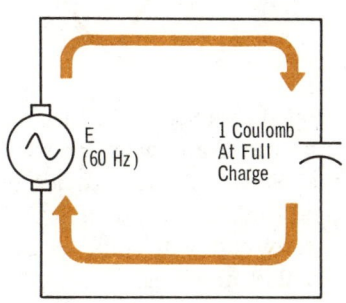 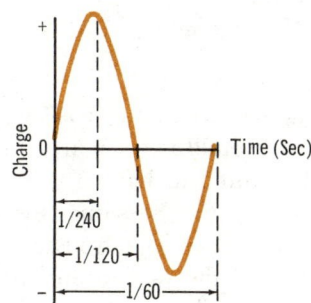

effect of capacitance
on capacitor current

As you learned, capacitance is the amount of charge stored in a capacitor for each volt applied across the capacitor; or $C = Q/E$. Since this is the case, the *total charge* stored is equal to the capacitance times the applied voltage; or $Q = CE$. Thus, if equal voltages are applied across two capacitors that have different values of capacitance, the charges stored in the capacitors will be different. The capacitor with the larger capacitance will store the larger charge. This means that with the larger capacitance, more free electrons will flow during the charge and discharge cycles. If the times required for fully charging the two capacitors were the same, which would be the case if the applied voltages had the same frequency, the current through the capacitor with the larger capacitance would be greater than the current through the other capacitor. You can see from this that the current through a capacitor is *directly proportional* to the capacitance.

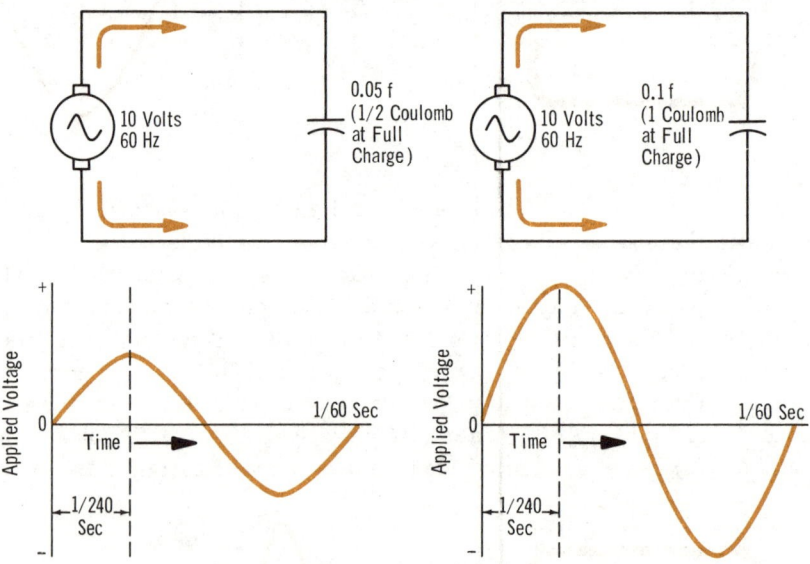

Using the equation $Q = CE$, you can see that the capacitor in the first circuit will build up a charge of ½ coulomb, and the capacitor in the second circuit will have a 1-coulomb charge. Since the frequencies of both sources are the same, the charges must both build up in 1/240 second. Using $I = Q/t$, you find that the first circuit will conduct 120 amperes and the second circuit will conduct 240 amperes.

capacitive reactance

A capacitor offers *opposition* to the flow of a-c electric current similar to a resistor or an inductor. You know that the amount of a-c current that a capacitor conducts depends on the *frequency* of the applied voltage and the *capacitance*. The amplitude of the applied voltage, of course, also controls the value of current, but if the voltage amplitude were kept constant, the current would change whenever the frequency

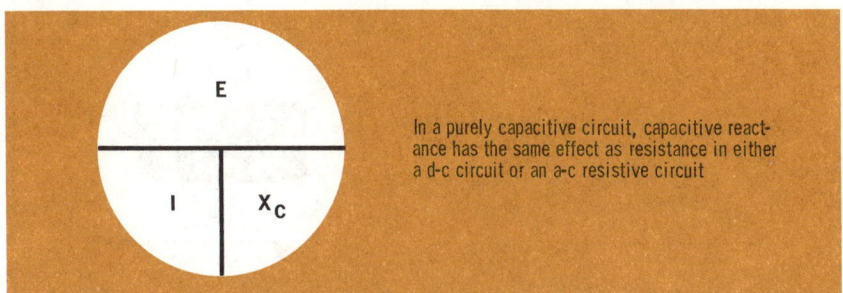

In a purely capacitive circuit, capacitive reactance has the same effect as resistance in either a d-c circuit or an a-c resistive circuit

or the capacitance was changed. As you saw earlier, current flow could be calculated using certain equations, but it is much more convenient to use Ohm's Law. But capacitance and frequency, in themselves, cannot be used directly in Ohm's Law. What is needed is some characteristic that can be expressed in ohms, just as resistance and inductive reactance. The *opposition* to current flow in a capacitor is used, since the opposition also depends on frequency and capacitance. However, since current flow is *directly proportional* to frequency and capacitance, the opposition to current flow must be *inversely proportional* to these quantities.

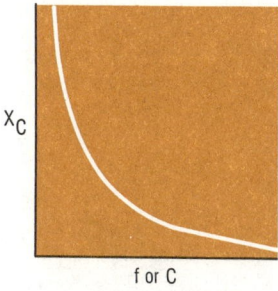

The decrease in capacitive reactance with increases in either frequency or capacitance is nonlinear. Thus, each additional increase in frequency or capacitance results in a smaller decrease in X_C

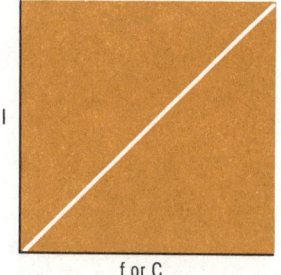

Current through a capacitor, on the other hand, increases linearly with increasing frequency or capacitance

capacitive reactance (cont.)

The opposition offered to the flow of current by a capacitor is called *capacitive reactance*, and is abbreviated X_C. Capacitive reactance can be calculated by:

$$X_C = \frac{1}{2\pi fC}$$

where 2π is approximately 6.28; f is the frequency of the applied voltage in cycles per second, Hz; C is the capacitance in farads. You can see that the higher the frequency, or the greater the capacitance,

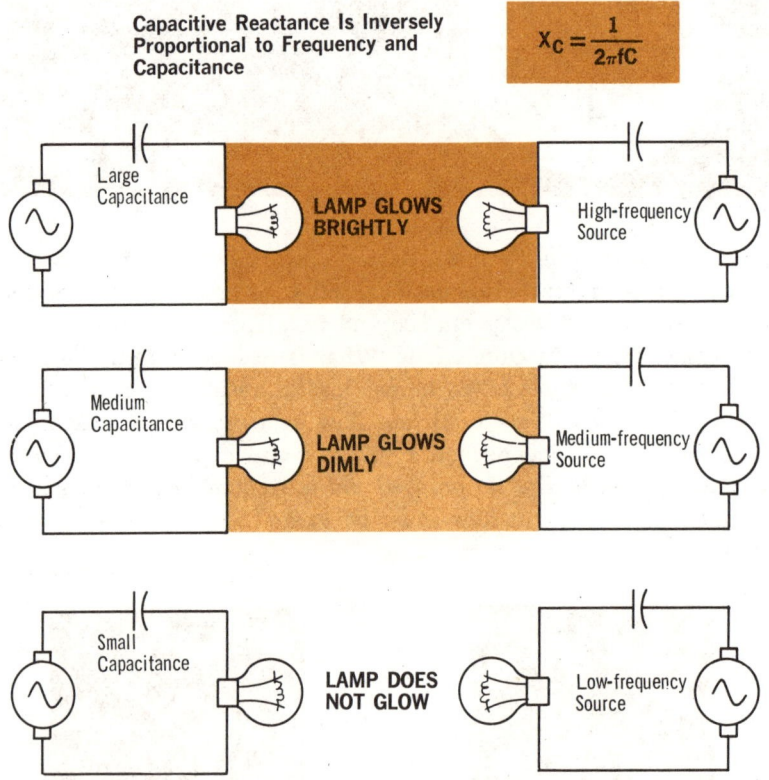

Capacitive Reactance Is Inversely Proportional to Frequency and Capacitance

$$X_C = \frac{1}{2\pi fC}$$

Large Capacitance

LAMP GLOWS BRIGHTLY

High-frequency Source

Medium Capacitance

LAMP GLOWS DIMLY

Medium-frequency Source

Small Capacitance

LAMP DOES NOT GLOW

Low-frequency Source

the less will be the capacitive reactance. Like its inductive counterpart, inductive reactance, capacitive reactance is expressed in *ohms,* and acts just like a resistance in limiting a-c current flow. When you know the capacitive reactance, you can find current by:

$$I = E/X_C$$

leakage resistance and Q

The capacitor also has a Q *rating* similar to a coil. But since the capacitor is constructed to act as an insulator to dc, the Q rating of a capacitor is not always as meaningful or as easy to compute as with a coil.

Theoretically, the a-c current flow in a capacitor depends on the capacitive reactance of the capacitors, and is primarily in the form of charge and discharge currents. Current flow between the capacitor plates should be prevented by the dielectric. But you know that there is no perfect insulator. So, a certain amount of current will *leak* between the plates. The capacitor, then, has a *leakage resistance* between the plates, which is usually very high. Often, when a capacitor is used at high voltages, a corona discharge can occur, which would lower the leakage resistance.

In a good capacitor, the leakage resistance is generally measured in megohms, and capacitive reactance is generally low in comparison for it to act like a pure capacitor. One way to look at it is the leakage resistance acting as a shunt on the capacitor, and the higher the value of leakage resistance, the less it will affect the action of the capacitor.

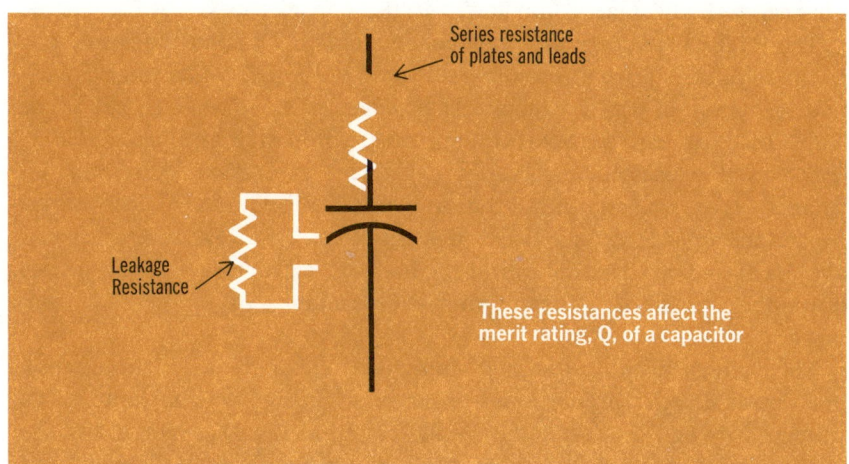

Series resistance of plates and leads

Leakage Resistance

These resistances affect the merit rating, Q, of a capacitor

However, there is also the resistance of the plates and wire in the capacitor, which acts like a small *series* resistance. This must be low compared to X_C, and its merit rating would be similar to that of a coil.

You can see, then, that the Q of a capacitor is not as straightforward as that of a coil. Generally, the Q of a capacitor is found as the *reciprocal* of the power factor, which is covered later.

series capacitors

Putting Capacitors In Series Decreases Capacitance

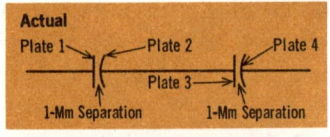

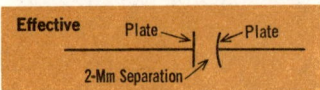

Series capacitors act as a single capacitor with a separation between their plates equal to the sum of the separations of the individual capacitors. The total capacitance is, therefore, less than that of any of the individual capacitors

The total capacitance of series capacitors is calculated the same as the total resistance of parallel resistors. The reciprocal method:

$$C_{TOT} = \frac{1}{\frac{1}{C_1} + \frac{1}{C_2} + \frac{1}{C_3} + \cdots + etc.,}$$

or any of the shorter, special methods can be used. The total capacitive reactance, though, is simply the sum of the individual reactances:

$$X_{C\,TOT} = X_{C1} + X_{C2} + X_{C3} + \ldots + etc.$$

To obtain a desired value of capacitance, capacitors are often connected in series. The total capacitance of the series combination is *less* than the capacitance of any individual capacitor. The reason is that the series combination acts as a *single* capacitor, with a separation between its plates equal to the *sum* of the *separations* of the individual capacitors. And, as you have previously learned, the larger the separation between the plates, the lower is the capacitance.

The total capacitance of *series capacitors* is calculated the same way as the total resistance of *parallel resistors* (Volume 2). The total capacitive reactance of series capacitors, however, is not found in the same way. Since capacitive reactance is inversely proportional to capacitance, if the total capacitance of series capacitors goes down, then the reactance goes up. Actually, the individual reactances of the series capacitors are added just like series resistors. The total reactance can also be found by first determining the total capacitance of the series combination, and then calculating the reactance of the total capacitance.

The voltage drop across each individual capacitor of a series combination is *directly proportional* to the *reactance* of that capacitor. The voltage drop can be calculated by the equation $E = IX_C$, where I is the series circuit current, which is the *same* through each capacitor of the combination, and X_C is the reactance of the particular capacitor. Since voltage drop is directly proportional to capacitive reactance, which, in turn, is inversely proportional to capacitance, the voltage drop is also *inversely proportional* to the *capacitance*. Thus, in a series combination, the capacitor with the lowest capacitance has the largest voltage drop across it.

parallel capacitors

When capacitors are connected in *parallel*, the total capacitance is equal to the *sum* of the individual capacitances. The reason for this is that parallel capacitors act as a single capacitor having a *plate area* equal to the sum of the plate areas of the individual capacitors. With a larger plate area, therefore, the capacitance is increased. As a result, the total capacitance of parallel capacitors is found by taking the sum of the individual capacitances, just as with series resistors. The total capacitive reactance of parallel capacitors, on the other hand, goes in the other direction. The total reactance, then, is found by treating the individual reactances like parallel resistors; or it can be found by first determining the total capacitance and then finding the reactance of this total capacitance.

As is the case with parallel resistors and parallel inductors, the same voltage is dropped across each capacitor of a parallel combination, but the current through each differs for different capacitance values.

Putting Capacitors in Parallel Increases Capacitance

Actual   **Effective**

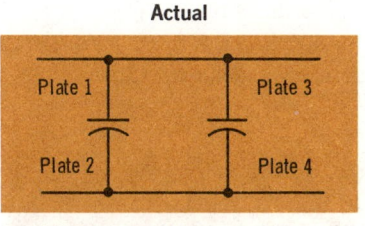

Parallel capacitors act as a single capacitor with a plate area equal to the sum of the plate areas of the individual capacitors

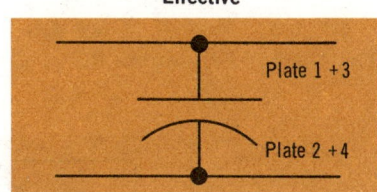

The total capacitance is, therefore, more than that of any of the individual capacitors

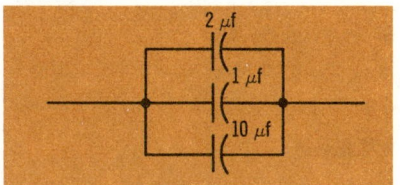

The total capacitance of parallel capacitors is equal to the sum of the capacitances of the individual capacitors:

$$C_{TOT} = C_1 + C_2 + C_3 + \ldots + \text{etc.}$$

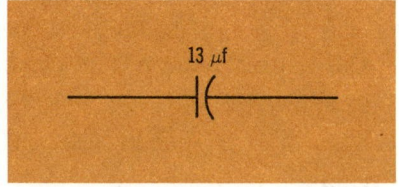

The total capacitive reactance, though, is found by treating the individual reactances as parallel resistances:

$$X_{C\,TOT} = \frac{1}{\frac{1}{X_{C1}} + \frac{1}{X_{C2}} + \frac{1}{X_{C3}} + \ldots + \text{etc.}}$$

solved problems

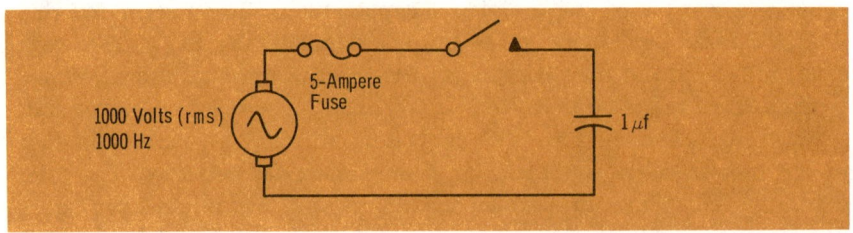

Problem 12. *Will the fuse in the circuit blow when the switch is closed?*

The fuse will blow if the current is greater than 5 amperes. Find the current with Ohm's Law after X_C is calculated.

$$X_C = \frac{1}{2\pi fc} = \frac{1}{(6.28 \times 1000 \times 0.000001)} = \frac{1}{0.00628} = 159 \text{ ohms}$$

Using Ohm's Law, then:

$$I = \frac{E}{X_C} = \frac{1000 \text{ volts}}{159 \text{ ohms}} = 6.3 \text{ amperes}$$

The fuse will therefore blow, since the current exceeds its rating.

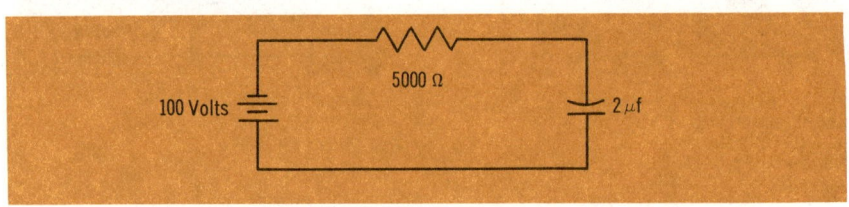

Problem 13. *How long will the current flow in the circuit?*
Current will flow until the capacitor is fully charged, which will take 5 time constants. One time constant equals resistance times capacitance:

$$t = RC = 5000 \text{ ohms} \times 0.000002 \text{ farads} = 0.01 \text{ second}$$

Therefore,

$$5t = 5 \times 0.01 = 0.05 \text{ second}$$

Problem 14. *If the voltage in the above circuit is increased to 200 volts, how long will the current flow?*

With a higher voltage applied, the capacitor will charge to a higher voltage, but it will still do it in 0.05 second.

solved problems (cont.)

Problem 15.   *What is the total capacitance of the circuit below?*

The capacitors are in series, so you use the same methods for finding the total resistance of parallel resistors. Since there are two unequal capacitors, the product/sum method is the best to use:

$$C_{TOT} = \frac{(C_1 \times C_2)}{(C_1 + C_2)} = \frac{(10\ \mu f \times 5\ \mu f)}{(10\ \mu f + 5\ \mu f)} = \frac{50\ \mu f}{15\ \mu f} = 3.33\ \mu f$$

Notice that it was not necessary to convert the values of the capacitors from microfarads to the farad. The reason for this is that only capacitance was involved in the problem, and all of the capacitances were in the same units.

Problem 16.   *In the circuit, which capacitor has the most charge on its plates?*

Since current is the flow of charge, and at any instant is the same in all parts of a series circuit, both capacitors must have the identical charge at any instant.

Problem 17.   *Which capacitor has the most voltage across it?*

This problem could be solved by finding the capacitive reactance of the two capacitors, using it to calculate the total circuit current, and then using the current and the reactance of the individual capacitors to find the voltage drop across each. The easiest way to solve the problem is to rearrange the basic equation for capacitance, $C = Q/E$.

$$E = Q/C$$

And since you know that the charge (Q) on both capacitors is the same, this equation shows that the one with the smallest capacitance (C) must have the largest voltage (E) across it. This fact holds true whenever capacitors are connected in series. The one with the smallest capacitance always has the largest voltage across it. C_2, then will drop twice as much voltage as C_1.

solved problems (cont.)

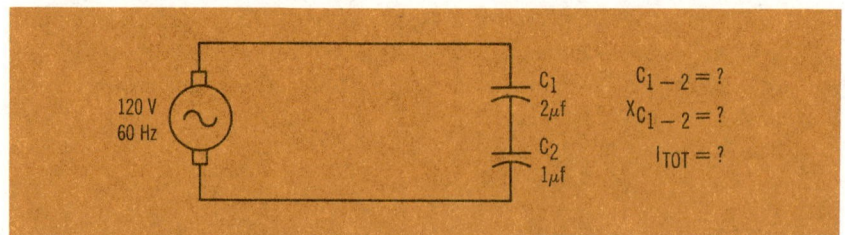

Problem 18.   What is the current flow in the circuit?

To find the total current flow, I_{TOT}, in this circuit, you must first determine the total capacitive reactance. This can be done either by finding the individual capacitive reactances and adding them, or by first finding the total capacitance and then determining its reactance. Let's do it the latter way.

Remember that capacitors placed in series reduce the total capacitance. They act like parallel resistors, and the parallel equation is used to make the calculation:

$$C_{1-2} = \frac{C_1 C_2}{C_1 + C_2}$$

$$= \frac{2 \times 1}{2 + 1} = 2/3$$

$$= 0.66 \ \mu f$$

Now, to find the total capacitive reactance,

$$X_{C_{1-2}} = \frac{1}{2\pi fC}$$

$$= \frac{1}{2 \times 3.1414 \times 60 \times 0.6 \times 10^{-6}}$$

$$= \frac{1}{.00022618} = 3978.78 \text{ ohms}$$

Now you can use Ohm's Law to find the current.

$$I_{TOT} = E/X_{C_{1-2}} = 120/3978.78$$

$$= 30.16 \text{ ma}$$

Problem 19.   What is the voltage drop across C_1?
Using the proportionality law:

$$E_{C1} = E(C_2/C_{1+2})$$

$$= 120 \ (2/3) = 80 \text{ volts}$$

solved problems (cont.)

Since series capacitors act like parallel resistors, the C_2 proportionality was used to find the C_1 voltage drop.

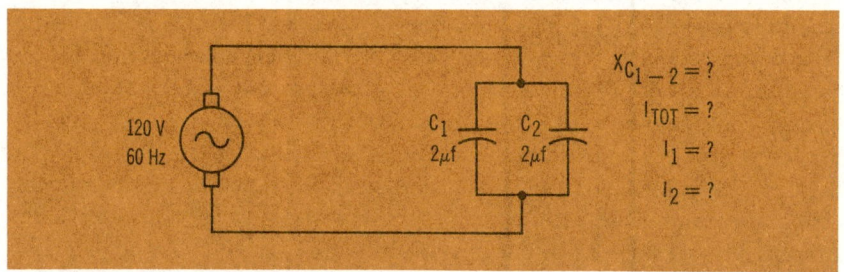

Problem 20. What is the total current flow in the circuit?

To determine the total current, you must first determine the total capacitive reactance. This can be done either by first finding the individual reactances or by first finding the total capacitance. To solve the following problems, let's use the first method:

$$X_C = \frac{1}{2\pi f C_1} = \frac{1}{2 \times 3.1414 \times 60 \times 2 \times 10^{-6}}$$
$$= 1326.26 \text{ ohms}$$

$$X_{C2} = \frac{1}{2\pi f C_2} = \frac{1}{2 \times 3.1414 \times 60 \times 1 \times 10^{-6}}$$
$$= 2652.52 \text{ ohms}$$

Now:

$$X_{C_{1-2}} = \frac{X_{C1} X_{C2}}{X_{C1} + X_{C2}} = \frac{(1326.26)(2652.52)}{1326.26 + 2652.52}$$
$$= 884.17 \text{ ohms}$$

To find the total current:

$$I_{TOT} = E/X_{C_{1-2}} = 120/884.17$$
$$= 135.72 \text{ ma}$$

Problem 21. What are the branch currents, I_{C_1}, and I_{C_2}?
These can now be found simply with Ohm's Law:

$$I_{C_1} = E/X_{C_1} \qquad\qquad I_{C_2} = E/X_{C_2}$$
$$= 120/1326.26 \qquad\qquad = 120/2652.52$$
$$= 90.48 \text{ ma} \qquad\qquad = 45.25 \text{ ma}$$

Problem 22. What is the circuit's total capacitance?
Since parallel capacitances act like series resistances:

$$C_{1-2} = C_1 + C_2 = 1 + 2 = 3 \ \mu f$$

capacitance in fluctuating d-c circuits

You learned that fluctuating dc has characteristics of both ac and dc. It consists of a component, the *a-c component,* which varies in amplitude but not direction, as well as another component, the *d-c reference.* The d-c reference is a value of dc around which the a-c component varies. When a fluctuating d-c voltage is applied to a capacitor, the capacitor charges almost instantly to the value of the d-c reference. It then charges further and discharges according to the a-c component. The capacitor never fully discharges, since it always maintains some charge as a result of the d-c reference. The maximum voltage across the capacitor is equal to the value of the d-c reference plus the peak value of the a-c component. The minimum voltage is equal to the d-c reference minus the peak value of the a-c component.

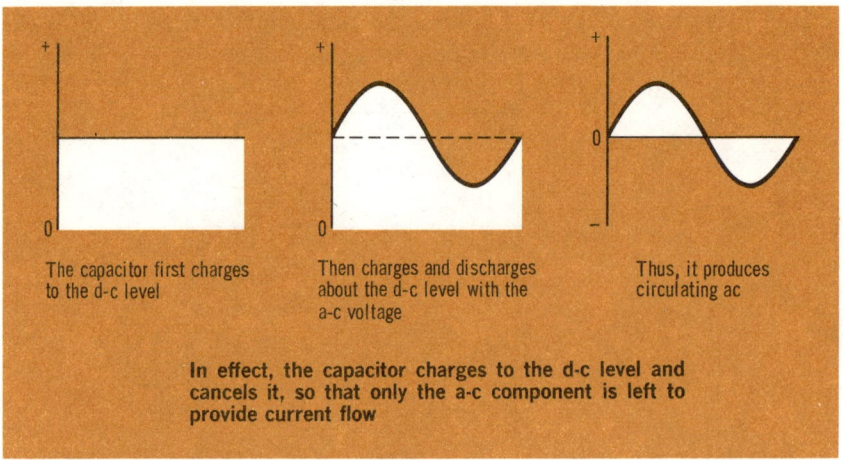

The capacitor first charges to the d-c level

Then charges and discharges about the d-c level with the a-c voltage

Thus, it produces circulating ac

In effect, the capacitor charges to the d-c level and cancels it, so that only the a-c component is left to provide current flow

When the a-c component is *increasing,* the capacitor is *charging,* and the current flows in the corresponding direction. When the a-c component *decreases,* the capacitor *discharges partially,* and so the current *reverses* direction. Thus, although the applied voltage never changes direction, the circuit current does each time the capacitor starts to discharge. The circuit current, then, acts as if the applied voltage consisted only of the a-c component. In effect, the a-c component is *separated* from the d-c reference; an a-c current that varies around *zero* is produced. If a resistor were added in series with the capacitor, the a-c component would be dropped across it as voltage that varied around zero. Thus, the capacitor *blocks* the dc, and *passes* the ac.

capacitance in fluctuating d-c circuits (cont.)

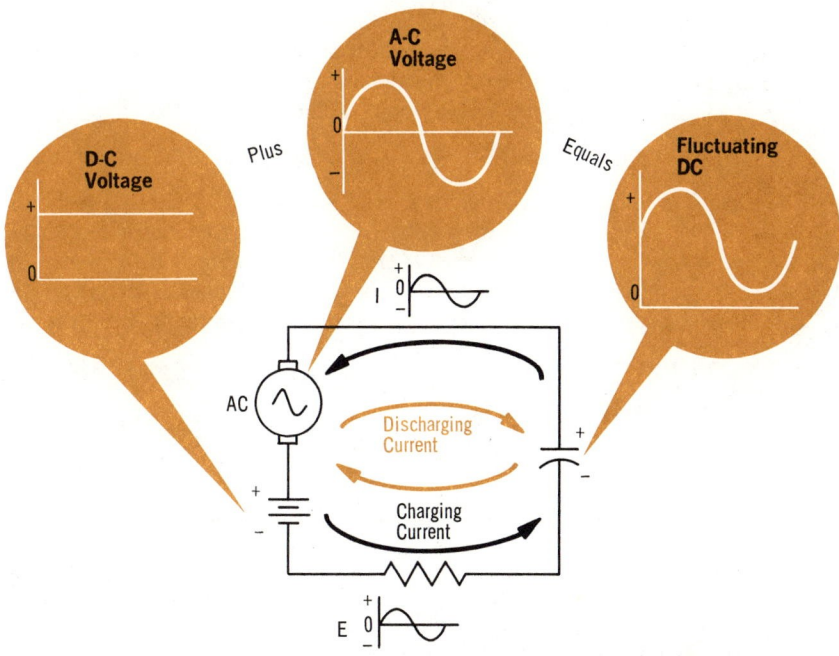

A fluctuating d-c voltage source varies in amplitude, but never reverses polarity. The current through a capacitor connected to such a source, though, does change direction

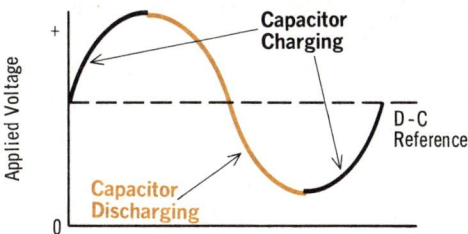

Capacitor charges when the applied voltage is increasing, and discharges when voltage is decreasing. Current, therefore, follows the a-c component

Current variations follow the a-c component, but are around zero instead of around the d-c reference. In effect, then, the capacitor separates the a-c component

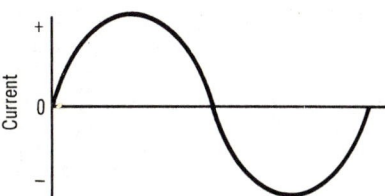

summary

☐ In an a-c circuit, the voltage across a capacitor constantly changes **polarity**. ☐ A-c current is passed by a capacitor because the capacitor alternately charges and discharges during every cycle of the applied ac.

☐ The capacitor voltage is often called the counter voltage, because it acts 180 degrees out of phase with the applied voltage. ☐ The voltage across a capacitor lags the current through the capacitor by 90 degrees. Or, the current through a capacitor leads the applied voltage by 90 degrees. ☐ Current through a capacitor is directly proportional to the frequency of the applied voltage.

☐ The opposition offered to the flow of current by a capacitor in an a-c circuit is called capacitive reactance, and is abbreviated X_C. ☐ Capacitive reactance can be found by the equation: $X_C = 1/(2\pi fC)$. X_C is expressed in ohms. ☐ Leakage current in a capacitor is due to the leakage resistance, which is caused by the dielectric not being a perfect insulator. ☐ The total capacitance of capacitors in series is $C_{TOT} = \dfrac{1}{1/C_1 + 1/C_2 + 1/C_3 + \ldots + \text{etc.}}$. ☐ The total capacitance of capacitors in parallel is $C_{TOT} = C_1 + C_2 + C_3 + \ldots + \text{etc.}$ ☐ Capacitors in fluctuating d-c circuits block the dc while passing the ac.

review questions

1. How does a capacitor block dc?
2. What are the voltage and current relationships between the capacitor and applied voltages? Between the capacitor current and applied voltage?
3. What happens to the capacitor current if the frequency of the applied voltage is decreased?
4. What is the opposition to the flow of a-c current offered by a capacitor called? What are its units?
5. What is the equation for finding X_C?
6. What is meant by *leakage current* in a capacitor, and what causes it?
7. What is the total capacitance of three capacitors in series, whose values are 10, 10, and 5 microfarads?
8. What is X_C for each capacitor in Question 7, and the total capacitive reactance at 100 Hz? 10 kHz? 1 MHz?
9. What is the total capacitance of three capacitors in parallel, whose values are 10, 10, and 5 microfarads?
10. What is X_C for each capacitor in Question 9, and the total capacitive reactance at 100 Hz? 10 kHz? 1 MHz?

power factor

In d-c circuits, or purely resistive a-c circuits, the power consumed is equal to the voltage times the current, or P = EI. In circuits containing reactance, though, the relationship between voltage, current, and power consumed is not so simple. The reason for this is that much of the power drawn from the source by inductors and capacitors instead of being consumed, is *stored* temporarily, and then returned to the source. The power is stored in the magnetic field of inductors and the electrostatic field of capacitors. If you were to measure the voltage and current in an inductive or capacitive circuit and then multiply them together, you would obtain the *apparent power*. This is the power supplied to the circuit by the source, but is not the power consumed by the circuit, which is the *true power*. To convert apparent power to true power, you must multiply the apparent power by the *cosine* of the phase angle (θ) between the voltage and current in the circuit. Mathematically, this is expressed as:

$$\text{True power (P)} = \text{EI} \cos \theta$$

The value of cosine θ is called the *power factor* of the circuit. In a purely resistive circuit, the voltage and current are *in phase*, and so the phase angle between them is *zero*. Since the cosine of θ is 1, the apparent power *equals* the true power in resistive circuits. In a purely inductive or capacitive circuit, the phase angle between the voltage and current is 90 degrees. The cosine of 90 degrees is zero, and so the true power equals the apparent power times zero; the true power, then is zero. This means that no power is consumed in the circuit. All of the power drawn from the source is returned to it.

Actually, all circuits contain some resistance. In reactive circuits, the effect of resistance is to *reduce* the phase angle between the voltage and current. The cosine of the angle is no longer zero, and some of the power drawn from the load is consumed in the circuit resistance. The true power then depends on the relative amounts of resistance and reactance in the circuit. This is covered in detail with RL and RC circuits in Volume 4.

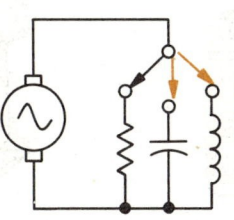

In a purely resistive circuit, the power factor equals 1.0, and so the apparent power equals the true power

In a purely inductive or capacitive circuit, the power factor equals 0, and so no power is consumed by the circuit

FIXED CAPACITOR

**This is the Symbol
For A Fixed Capacitor**

types of capacitors

A wide variety of capacitor types is used today. Essentially, they fall into either of two broad categories: *fixed* capacitors, or *variable* capacitors. Fixed capacitors have their capacitance value *permanently* set by their construction, and this value cannot be changed. The most common get their names from the dielectric materials used. Examples of these are *paper* capacitors, *mica* capacitors, and *ceramic* capacitors. The plates of these capacitors are usually made of metal foil. To keep their physical size to a minimum, techniques such as the use of a series of plates separated by dielectric material, or the rolling of the plates and the dielectric material into a tubular shape, are used in the construction of fixed capacitors.

Typical Fixed Capacitors

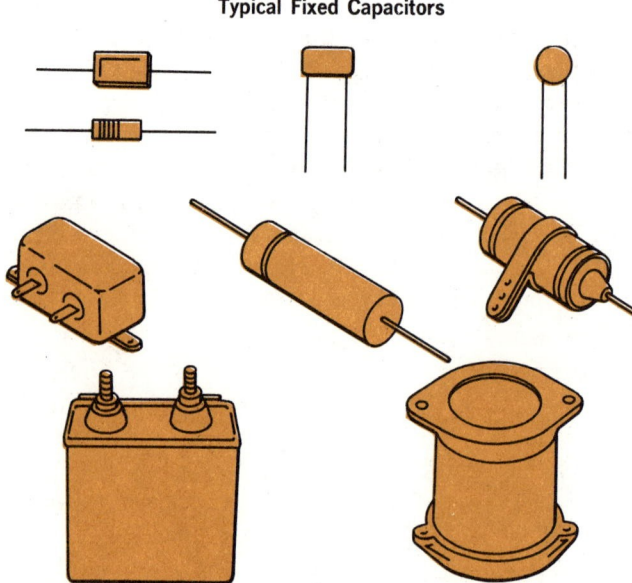

capacitor construction

Generally, the way a capacitor is made depends to some degree on the value of capacitance it must have, and the voltage it must withstand. Small capacitance values and low working voltages require simple construction, but as the need for more plate area and better dielectric strength increases, the construction becomes more complex.

The simple small ceramic capacitor merely has metallic film deposited on a ceramic sleeve to form the plates and the dielectric. When more capacitance is needed, interleaved sets of plates can be used, with the alternate plates connected together to act as a single large plate. Dielectric layers between the plates keep them apart. Mica capacitors use mica layers, and other kinds of capacitors use glass or vitreous enamel. When even more capacitance is needed, long sheets of metal foil, kept apart by sheets of paper dielectric, are rolled up to form the basic paper capacitor. Many capacitors, after they are made, are kept encased in phenolic or plastic jackets to protect them from humidity.

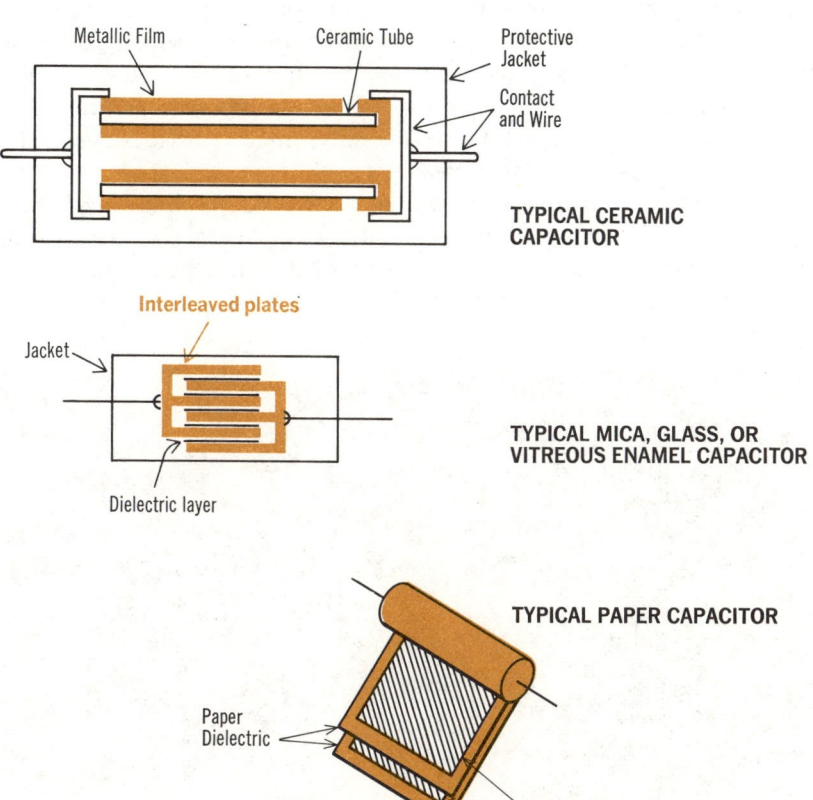

Metallic Film   Ceramic Tube   Protective Jacket

Contact and Wire

TYPICAL CERAMIC CAPACITOR

Interleaved plates

Jacket

TYPICAL MICA, GLASS, OR VITREOUS ENAMEL CAPACITOR

Dielectric layer

TYPICAL PAPER CAPACITOR

Paper Dielectric

Metal Foil

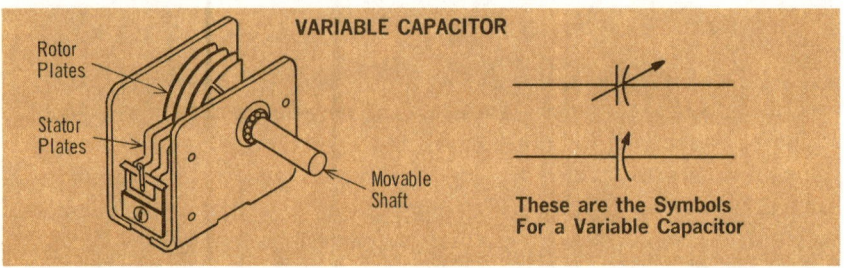

VARIABLE CAPACITOR

Rotor Plates

Stator Plates

Movable Shaft

These are the Symbols For a Variable Capacitor

types of capacitors (cont.)

Variable capacitors are made so that their capacitance values can be *continuously adjusted* over a certain range. The most common type consists of two sets of interleaving metal plates. One set of these plates, the *rotor* plates, is mounted on a shaft, which when rotated, moves the plates in between the other set of plates, the *stator* plates. Effectively, one set of plates acts as one of the capacitor plates, and the other set as the other plate. The more the two sets of plates are interleaved, the greater is the effective *area* of the capacitor plates, and so the greater is the capacitance. Air serves as the dielectric between the plates. Another variable capacitor consists of two metal plates whose *separation* can be varied by a screw adjustment. Generally, mica is used between the plates of this type of variable capacitor.

Most capacitors are marked with a d-c working voltage (WVDC), the *maximum* d-c voltage that can safely be applied across the capacitor. If this voltage is exceeded, breakdown of the dielectric might result, and the capacitor would become useless.

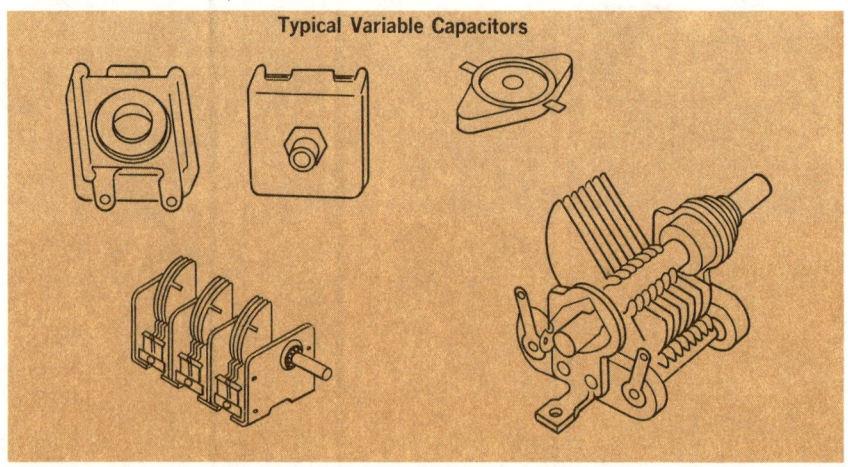

Typical Variable Capacitors

the electrolytic capacitor

The capacitors previously described are generally limited to values less than 1 μf because of size and cost considerations. For larger values, special capacitors, called *electrolytic capacitors*, are used. Capacitances as large as several thousand microfarads can be obtained with electrolytic capacitors at a reasonable size and cost. There are two basic types of construction used for electrolytic capacitors: the *wet type* and the *dry type*. The wet type is now practically obsolete, and so only the dry type will be described.

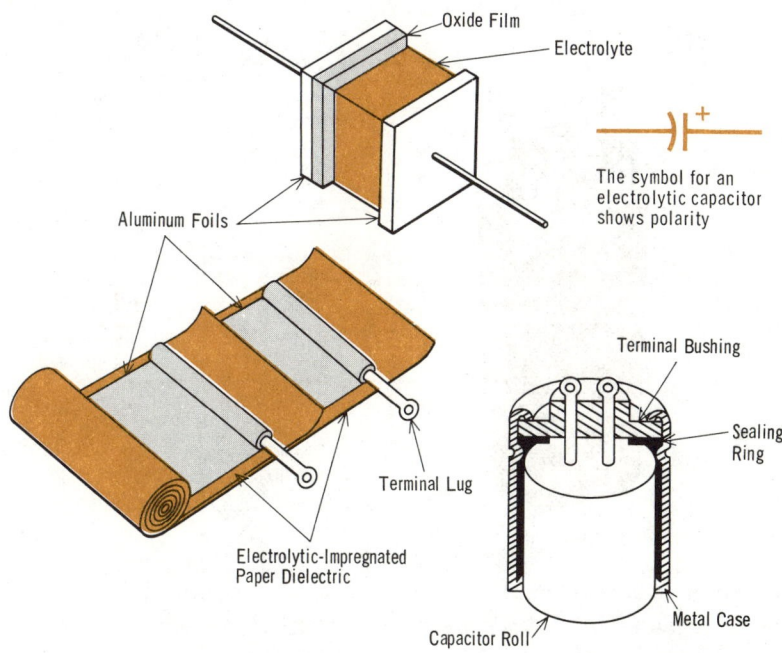

The symbol for an electrolytic capacitor shows polarity

Essentially, a dry-type electrolytic capacitor consists of two aluminum sheets separated by a layer of paper saturated with a liquid chemical called the *electrolyte*. All three sheets are rolled together and sealed in a container. When the capacitor is first made, a d-c voltage is applied between the two aluminum sheets. The resulting current causes a thin *oxide layer* to be formed on one aluminum sheet, and as a result, the capacitor is *polarized*.

The aluminum sheet with the oxide layer is positive, and serves as the positive capacitor plate. The oxide layer has insulating properties, and serves as the dielectric for the capacitor. The electrolyte is the negative capacitor plate, and the second aluminum sheet merely provides a connection between the electrolyte and external circuits.

the electrolytic capacitor (cont.)

Since the electrolytic capacitor is polarized, it can only be used in circuits containing *fluctuating d-c voltages*. Furthermore, the positive and negative terminals of electrolytic capacitors must be connected to circuit points of the same polarity (negative to negative, and positive to positive). The reason for this is that the electrolytic capacitor acts as a capacitor in only one direction of applied voltage. In the other direction, the electrolytic capacitor conducts like a low-value resistor.

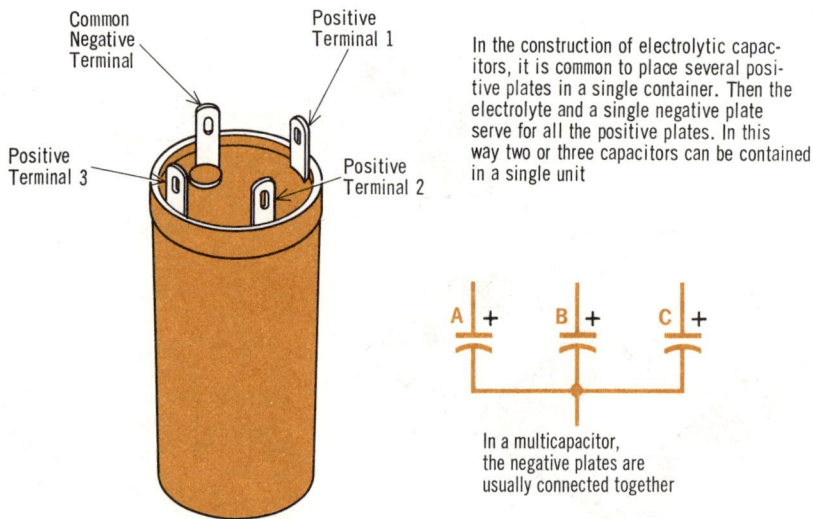

In the construction of electrolytic capacitors, it is common to place several positive plates in a single container. Then the electrolyte and a single negative plate serve for all the positive plates. In this way two or three capacitors can be contained in a single unit

In a multicapacitor, the negative plates are usually connected together

As a matter of fact, if an electrolytic capacitor is connected backwards, it could "explode" because of the high current that would flow. Special electrolytic capacitors are available for use in a-c circuits. In these capacitors, two positive plates are used, thus permitting a reversal in polarity of the applied voltage. Essentially, this type is made of two electrolytics connected back to back.

Wet-type electrolytic capacitors are similar in construction to the dry type. The principal difference is that they use liquid electrolytes. The electrolytic capacitor has a much lower leakage resistance than the regular capacitors.

In order to save space electrolytic capacitors are commonly made in multiple units. Two, three, or more capacitors can be made in one case. Usually, multiple capacitors have a common negative plate, but the positive sides can be connected independently. Each capacitor section has its own capacitance value and voltage rating.

capacitor color codes

All capacitors are marked in some way with their capacitance value. In addition, other important characteristics, such as working voltage and tolerance, are sometimes indicated. Where practical, this information is printed on the capacitor. In many cases, though, this is impractical, and a system of *colored markings* is used instead. The interpretation of these colored markings make up capacitor *color coding systems*. These are somewhat similar to the resistor color code system described in Volume 2. However, capacitor color codes are not too well standardized, so it would be wise to get the code from the specific manufacturers.

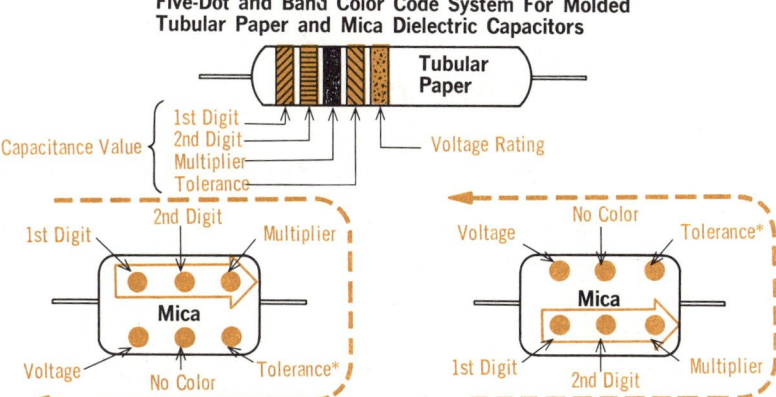

Five-Dot and Band Color Code System For Molded Tubular Paper and Mica Dielectric Capacitors

Sometimes a six-color system is used. When it is, the third color represents the third digit, and the rest of the code is the same

COLOR CODE SYSTEM					
Color	First Digit	Second Digit	Multiplier	Tolerance	Voltage Rating (Volts)
Black	0	0	1	±20	100
Brown	1	1	10	±1	200
Red	2	2	100	±2	300
Orange	3	3	1000	±3**	400
Yellow	4	4	10,000	±4**	500
Green	5	5	100,000	±5	600
Blue	6	6	1,000,000	±6	700
Violet	7	7	10,000,000	±7	800
Gray	8	8	100,000,000	±8	900
White	9	9	1,000,000,000	±9	1000
Gold	—	—	0.1	±5	2000
Silver	—	—	0.01	±10	—

* No color for tolerance dot indicates tolerance of ±20%.
** Multiply by 10 for tubular paper capacitors.

Color Code System For Some Ceramic Capacitors

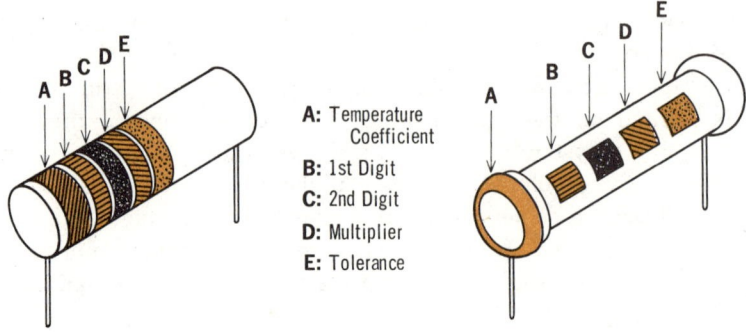

A: Temperature Coefficient
B: 1st Digit
C: 2nd Digit
D: Multiplier
E: Tolerance

capacitor color codes (cont.)

The temperature coefficient band is wider than the other bands for identification. Sometimes this band is last. When it is, though, the rest of the code remains the same. Also, the tolerance band can be left off. If six bands are used, there is a third digit. No tolerance band means 20%.

The temperature coefficient indicates the change in temperature in parts per million that will take place for every degree rise in the capacitor's operating temperature above 20°C. A minus means that capacitance will decrease, while a plus means that it will increase. These capacitors are all rated at 500 WVDC, unless otherwise indicated.

				Tolerance		Tempera-
Color	First Digit	Second Digit	Multiplier	More than 10 pf (%)	Less than 10 pf (pf)	ture Coefficient
Black	0	0	1.0	±20	±2	0
Brown	1	1	10	±1	—	−30
Red	2	2	100	±2	—	−80
Orange	3	3	1000	±3	—	−150
Yellow	4	4	10,000	±4	—	−220
Green	5	5	—	±5	±0.5	−330
Blue	6	6	—	±6	—	−470
Violet	7	7	—	±7	—	−750
Gray	8	8	0.01	±8	±0.25	+30
White	9	9	0.1	±10	±1	+120 to −750
Gold	—	—	0.1	±5	—	—
Silver	—	—	0.01	±10	—	—

COLOR CODE SYSTEM

RC Circuits Contain

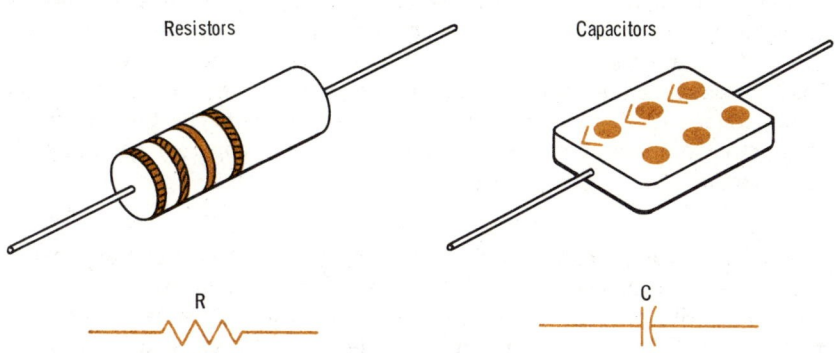

Resistors   Capacitors

R   C

RC circuits

Capacitors and capacitance have been described so far from the standpoint of their effects on circuits that contain *no resistance*. As you know, though, every circuit contains some resistance, even if it is only the resistance of the circuit wiring and the internal resistance of the source. When the circuit resistance is very low compared to the capacitive reactance, the resistance can usually be ignored, as has been done in this volume. However, when the resistance becomes appreciable, it must be considered.

Circuits that contain both resistance (R) and capacitance (C) are called *RC circuits*. The relationships between resistance and capacitance in RC circuits and their joint effect on circuit operation are covered in Volume 4.

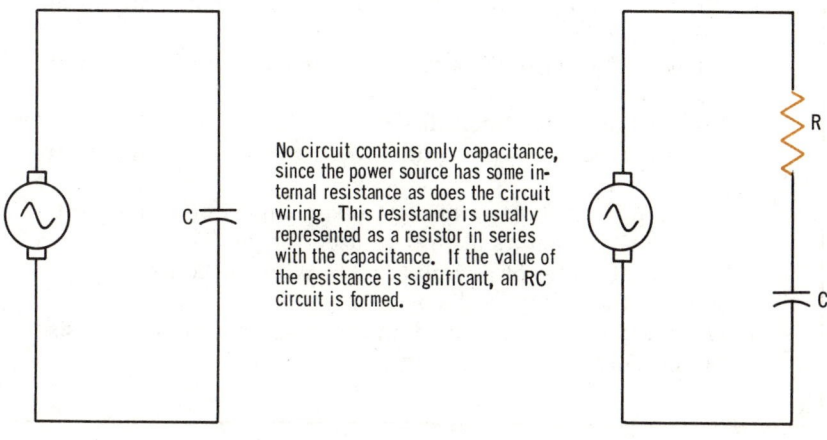

No circuit contains only capacitance, since the power source has some internal resistance as does the circuit wiring. This resistance is usually represented as a resistor in series with the capacitance. If the value of the resistance is significant, an RC circuit is formed.

summary

☐ The power factor of a circuit is the cosine of the phase angle (θ) between the voltage and the current. It can vary between 0 and 1, depending on the circuit. ☐ The actual power consumed in a circuit is called true power. ☐ Apparent power is the power supplied to the circuit by the source. It is equal to the voltage of the source times the current delivered by the source. ☐ True power in a circuit can be found by multiplying apparent power by the power factor.

☐ Capacitors can be either fixed or variable. ☐ Some types of fixed capacitors are paper, mica, and ceramic. ☐ Variable capacitors include the rotating-plate type, with air as the dielectric; and the screw-adjustment type, with mica as the dielectric. ☐ The maximum d-c voltage a capacitor can safely withstand is abbreviated WVDC.

☐ The electrolytic capacitor is polarized, and must be inserted in a circuit with the proper polarity. Special electrolytic capacitors, with high values of capacitance, are available for use in a-c circuits. ☐ Capacitor color codes permit identification of important characteristics. These characteristics include capacitance value, tolerance, voltage rating, and temperature coefficient.

review questions

1. What is meant by the *power factor* of a circuit?
2. The apparent power in a circuit is 50 volt-amperes, and the true power is 40 watts. What is the power factor?
3. Why are paper, mica, and ceramic capacitors so called? Are these fixed or variable capacitors?
4. What is meant by the abbreviation *WVDC* that is found on a capacitor?
5. What is an *electrolytic capacitor*? Can it ever be used in an a-c circuit?
6. How are capacitors identified?
7. What is meant by *rotor* and *stator plates*? Do fixed capacitors have these?
8. Why are the polarities of electrolytic capacitors important? Can these capacitors always be used for dc? For ac?
9. The power factor of a circuit is 0.5. What is the phase angle between the voltage and current?
10. What is the power factor in a pure capacitive circuit? Resistive circuit?

electricity
four

This volume describes circuits that contain:

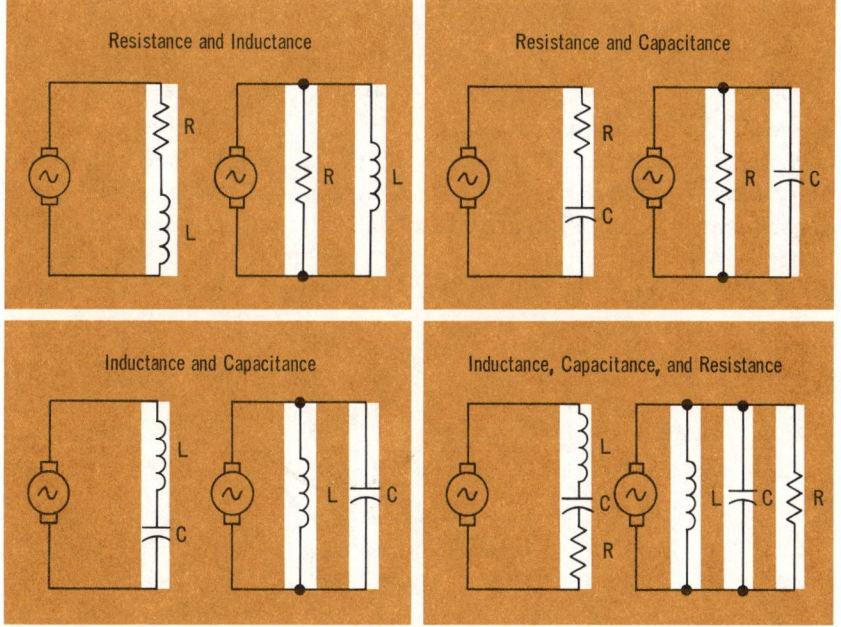

introduction

In Volume 3, you learned the properties of alternating current and how it is used. You also learned about inductance and capacitance, and were shown how to analyze a-c circuits that contain resistance, inductance, or capacitance. Volume 3 was limited, however, to circuits having only *one* of these circuit elements. The operation of circuits having two of the elements, such as resistance and capacitance or resistance and inductance, or even all three of the elements, was not covered. These more complex circuits are the major subject of this volume. You will learn how to analyze and solve circuits that contain both series and parallel combinations of resistance and inductance; resistance and capacitance; inductance and capacitance; and inductance, capacitance, and resistance. You will also learn some of the unique applications of these circuits.

Much of the material in this volume is based on the *phase relationships* between the voltages and currents in the various circuits. These relationships can be both explained and understood more easily if *vectors* are used to describe them. Therefore, before the actual circuits are described, a basic description of vectors and their use in electricity will be given.

review of LCR phase angles

You will remember from Volume 3 that the term *phase angle* is used to describe the *time* relationship between a-c voltages and currents, as well as to specify a *position* or *point* in time of one a-c voltage or current. For example, if two a-c voltages are of opposite polarity at every instant of time, they are 180 degrees out of phase, or the phase angle between them is 180 degrees. Similarly, if a current reaches its maximum amplitude after one-quarter of its cycle, or 90 degrees, maximum amplitude is said to occur at a phase angle of 90 degrees.

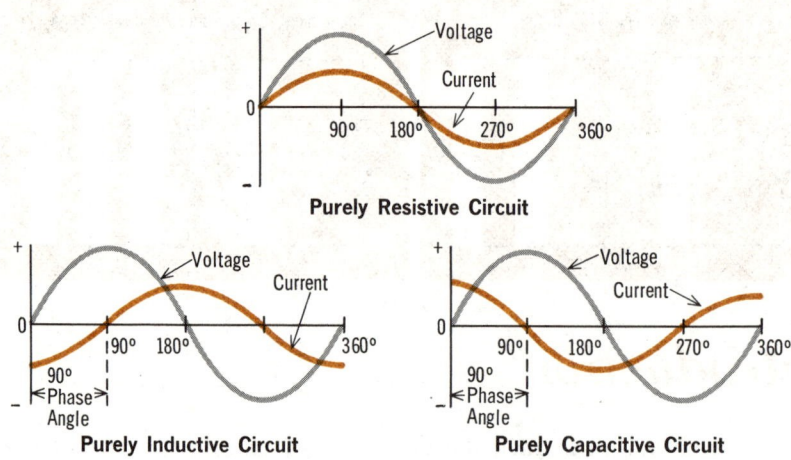

Purely Resistive Circuit

Purely Inductive Circuit   **Purely Capacitive Circuit**

In a purely resistive circuit, the phase angle between the voltage and current is zero. In a purely inductive circuit, the phase angle is 90°, with the voltage leading. In a purely capacitive circuit, the phase angle is 90°, with the current leading

You also learned in Volume 3 that there are very definite phase relationships between the *applied voltage* and the *circuit current* in purely resistive, purely inductive, and purely capacitive circuits. These relationships can be summarized as follows:

1. In a purely *resistive* circuit, the voltage and current are *in phase*.
2. In a purely *inductive* circuit, the applied *voltage leads* the *current* by 90 degrees.
3. In a purely *capacitive* circuit, the *current leads* the applied *voltage* by 90 degrees.

You already know how these relationships can be described using voltage and current *waveforms*. However, there is another, and often easier, way to express these same relationships. This is by using *vectors*.

what is a vector?

Every physical quantity has *magnitude*. The terms "5 apples," "5 days," and "5 ohms" all express physical quantities, and each is *completely* described by the number 5. There are some quantities, however, that are *not* completely described if only their magnitudes are given. Such quantities have a *direction* as well as magnitude, and without the direction these quantities are *meaningless*. For example, if

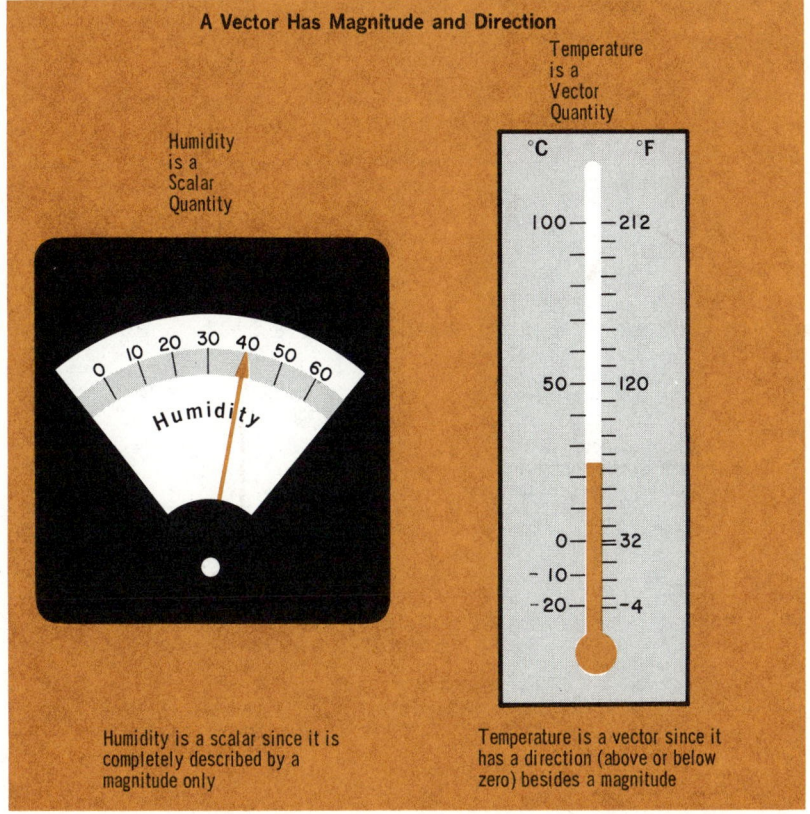

A Vector Has Magnitude and Direction

Humidity is a Scalar Quantity

Temperature is a Vector Quantity

Humidity is a scalar since it is completely described by a magnitude only

Temperature is a vector since it has a direction (above or below zero) besides a magnitude

someone were to ask you how to get from Chicago to Milwaukee, and you told them to drive 80 miles, this would be meaningless. But if you said to drive 80 miles *north,* your directions would be complete. The quantity "80 miles north" thus has a magnitude of 80 and a direction of north.

Quantities that have only magnitude are called *scalars*. Those that have magnitude and direction are called *vectors*.

graphical representation

Graphically, a vector is represented by a *straight line* with an *arrowhead* at one end. The *length* of the line is proportional to the *magnitude* of the vector quantity, and the *arrowhead* indicates the *direction*. When a vector is drawn on a graph, the direction of the vector is usually represented by the *angle* it makes with the *horizontal* axis. The horizontal axis thus serves as the *reference line* from which the direction of the vector is measured. You will learn later that in actual practice this reference point can be a direction, such as north, or a position in time, such as the zero-degree point of a sine wave.

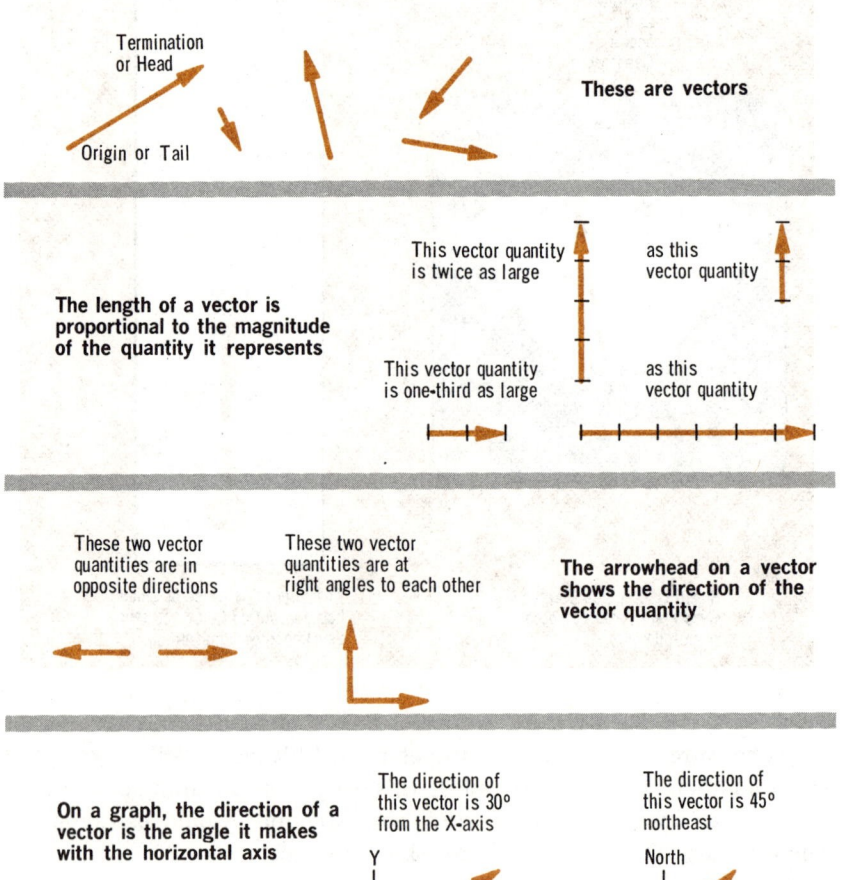

working with vectors

Since vectors represent physical quantities, they can be added, subtracted, multiplied, and divided the same as other physical quantities. In your work with electricity, you will normally only be interested in the *addition* and *subtraction* of vectors. Because of this, vector multiplication and division will not be covered in this volume.

Vector Quantities Can Be Added and Subtracted

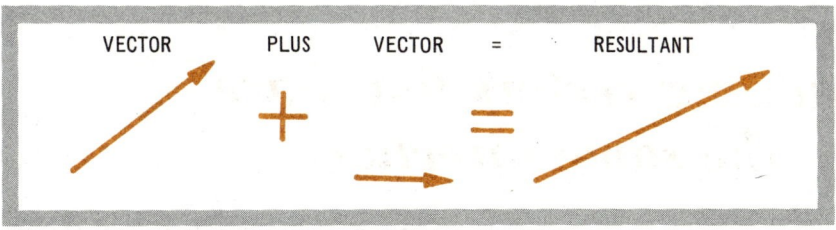

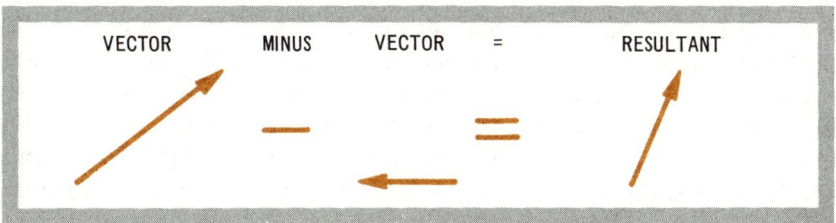

When vectors are added or subtracted, the result is also a vector, and is called the resultant vector, or just a resultant

The methods used to add and subtract scalar quantities *cannot* be used for vector quantities. The fact that vectors have direction, as well as magnitude, makes it necessary to add and subtract them *geometrically*. As a result, the adding and subtracting of vectors is a combination of geometry and *algebraic* addition and subtraction.

There are various methods you can use when adding or subtracting vectors. The particular method you select will depend on the relative directions of the vectors involved.

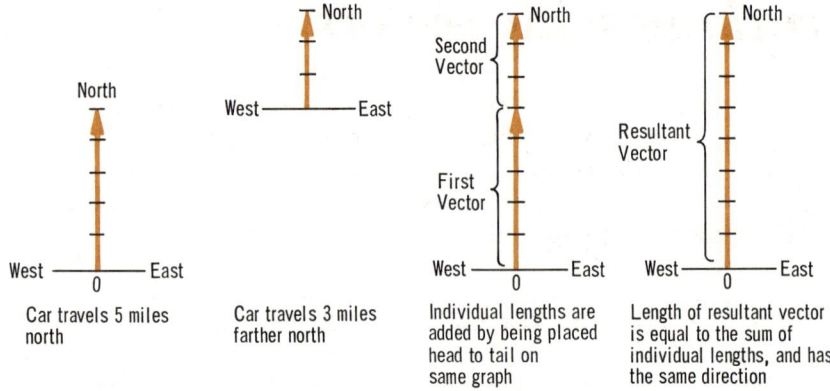

Car travels 5 miles north

Car travels 3 miles farther north

Individual lengths are added by being placed head to tail on same graph

Length of resultant vector is equal to the sum of individual lengths, and has the same direction

adding vectors that have the same direction

To add vector quantities that have the *same* direction, you simply *add* the individual *magnitudes*. This gives you the magnitude of the resultant. Its direction is the *same* as the direction of the individual vectors. An example of this would be a car that traveled 5 miles north from a starting point. This motion could be represented by a vector pointing north, and having a length corresponding to 5 miles. If the car then traveled 3 miles farther, still going north, the additional motion could be represented by another vector. This vector would also point north, but would have a length corresponding to 3 miles. The total motion of the car, or its final position, can be found by adding the lengths of the two vectors and assigning the resultant a direction of north. The resultant then shows that the car has traveled a total of 8 miles north of the starting point.

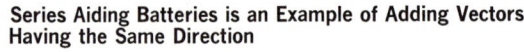

Series Aiding Batteries is an Example of Adding Vectors Having the Same Direction

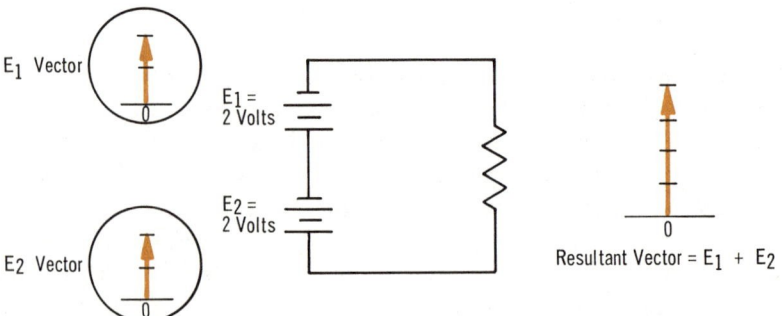

E_1 Vector

$E_1 = 2$ Volts

$E_2 = 2$ Volts

E_2 Vector

Resultant Vector = $E_1 + E_2$

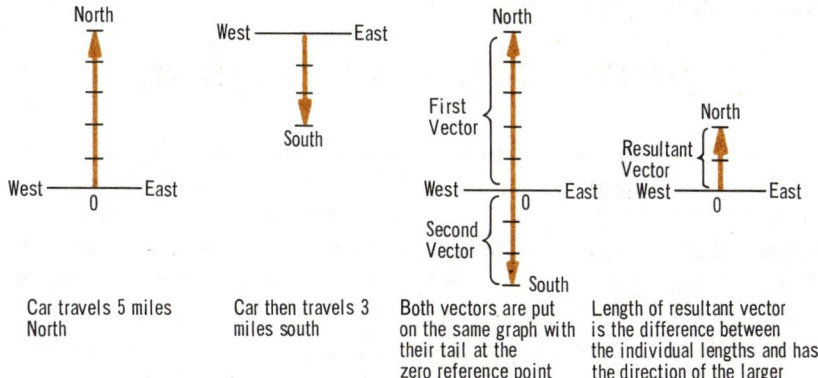

Car travels 5 miles North | Car then travels 3 miles south | Both vectors are put on the same graph with their tail at the zero reference point | Length of resultant vector is the difference between the individual lengths and has the direction of the larger

adding vectors that have opposite directions

To add vector quantities that have *opposite* directions, *subtract* the one with the *smaller* magnitude from the one with the *larger* magnitude. This gives the magnitude of the resultant. The *direction* of the resultant is the same as that of the *larger* vector. You can see from this that if one vector is added to another that has the same magnitude but opposite direction, the resultant is zero.

An example of the addition of vectors that have opposite directions is shown. If a car travels north for 5 miles, the vector representing the motion points north, and has a length corresponding to 5 miles. If the car then travels 3 miles south, the vector representing this additional motion points south, and has a length corresponding to 3 miles. The resultant of these two motions is found by adding the two vectors. Since their directions are opposite, this is done by subtracting the smaller vector (3 miles) from the larger vector (5 miles), and assigning the resultant the direction of the larger vector (north). The resultant then shows that after the two motions, the car is at a position 2 miles north of the starting point.

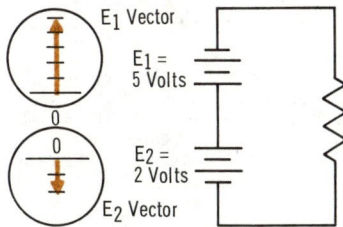

Series Opposing Batteries is an Example of Adding Vectors That Have Opposite Directions

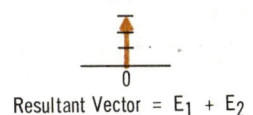

Resultant Vector = $E_1 + E_2$

adding vectors
by parallelogram method

When vectors are neither in the same nor in opposite directions they can be added *graphically* and their resultant found by means of the *vector parallelogram method.* You will remember from mathematics that a parallelogram is a *four-sided* figure whose opposite sides are *equal* in length and *parallel* to each other. To add two vectors using the parallelogram method, the vectors must first be placed "tail to tail." A parallelogram, then, is constructed by using the vectors. The *diagonal* of a parallelogram so constructed from the tail of the vectors is the *resultant* of the two vectors. The length of the diagonal is the length of the resultant, and the angle between the diagonal and the horizontal axis (θ) represents the direction of the resultant.

The parallelogram method can be used to add more than two vectors. To do this, you first find the resultant of any two of the vectors. You then use this resultant with one of the remaining vectors, and find the resultant of that combination. This resultant is then used with another vector, and so on, until the overall resultant has been found for all the vectors.

You can see that the parallelogram method is primarily a graphical method for adding vectors. It requires the use of a ruler and a protractor, and at best is only fairly accurate, unless great care is taken. For this reason, you will find that you very rarely use the parallelogram method to find numerical solutions to vector addition problems when working with electricity. It is nevertheless important to understand, since it is often used to analyze and describe relationships between vector quantities.

Constructing a Vector Parallelogram

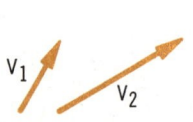 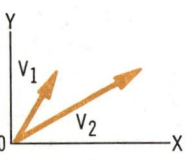 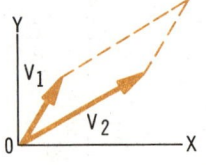 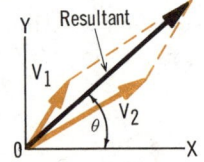

(A) To add two vectors

(B) First place them tail to tail

(C) Then construct a parallelogram using the vectors as two of the sides

(D) The diagonal of the parallelogram is the resultant. The length of the diagonal is the magnitude and the angle θ is the direction of the resultant

adding vectors
by triangle method

Another method for *graphically* adding vectors is called the *vector triangle method*. As its name implies, the triangle method involves the construction of a triangle to determine the resultant vector. To use the method for adding two vectors, the vectors are first placed "head to tail." A line is then drawn connecting the *tail* of the *first* vector with the *head* of the *second*. This is the resultant of the two vectors.

Constructing a Vector Triangle

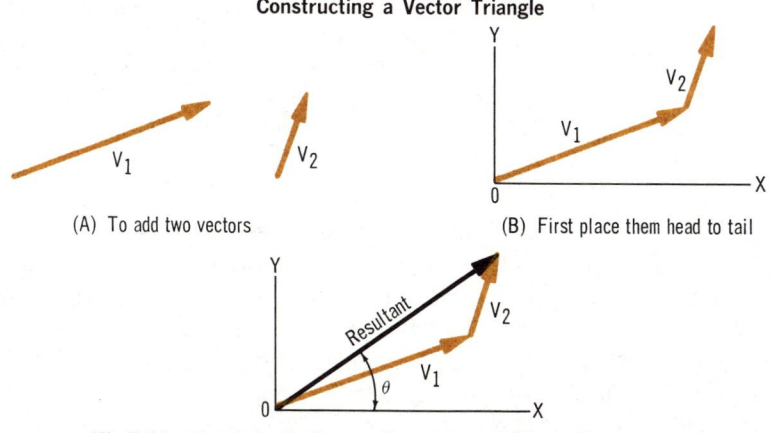

(A) To add two vectors

(B) First place them head to tail

(C) The line that closes the figure to form a triangle is the resultant. The length of the line is the magnitude, and the angle θ is the direction of the resultant

The triangle method can be used when there are more than two vectors involved. To do this, all of the vectors are connected "head to tail." A line is then drawn from the tail of the first vector to the head of the last vector. This line is the total resultant of all the individual vectors.

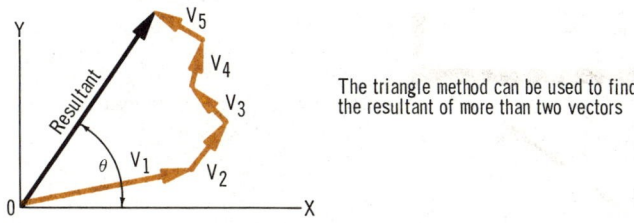

The triangle method can be used to find the resultant of more than two vectors

Like the parallelogram method, the triangle method is used mainly to analyze and describe relationships between vector quantities, rather than to obtain numerical solutions to vector additions.

adding vectors
that are 90 degrees apart

A very special case of vectors that are neither in the same nor in opposite directions are those that are *90 degrees apart*. As you will learn later, these are especially important in electricity because of their relationship to the 90-degree phase difference that exists between the voltage and current in inductive and capacitive circuits.

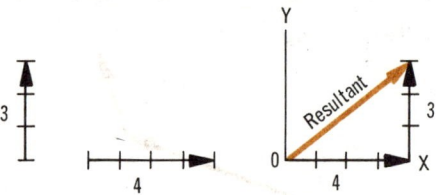

Two vectors that are 90° apart can be added graphically using the vector triangle method. The resultant and the two vectors form a right triangle, so the magnitude of the resultant can be calculated by the Pythagorean Theorem

Graphically, two vectors that are 90 degrees apart can be added using either the parallelogram method or the triangle method. When vector addition is done, the resultant is the *hypotenuse* of a *right triangle* whose other two sides are the vectors being added. Because of the properties of the right triangle, summarized on pages 4-11 and 4-12, the magnitude and direction of the resultant can easily be calculated. As shown, if the magnitudes of two vectors are known, and the vectors are 90 degrees apart, the magnitude of the resultant can be calculated by the *Pythagorean Theorem* $(c^2 = a^2 + b^2)$. The direction of the resultant, which is the angle (θ) between it and the horizontal axis, can then be found using the *trigonometric relationship* between the sine of the angle and the sides of the triangle.

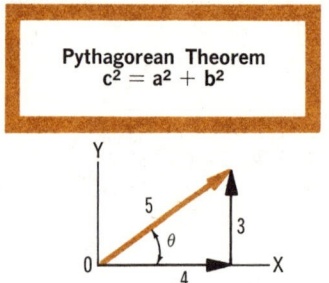

Pythagorean Theorem
$$c^2 = a^2 + b^2$$

$(\text{Resultant})^2 = a^2 + b^2$

$\text{Resultant} = \sqrt{a^2 + b^2}$

$= \sqrt{9 + 16} = \sqrt{25} = 5$

The direction of the resultant can be found using the trigonometric relationship:

$$\sin \theta = \frac{\text{opposite}}{\text{hypotenuse}} = \frac{3}{5} = 0.6$$

From a table of trigonometric functions, therefore: $\theta = 36.9°$

It might be advisable at this time for you to review right triangles and their solution. Such material is available in any standard trigonometry text, or in books covering mathematics for electricity.

properties of the right triangle

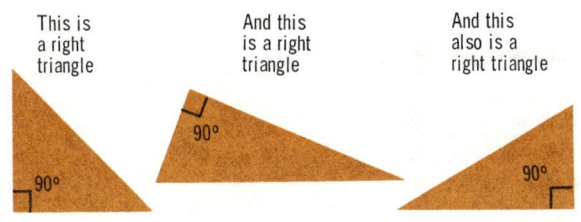

This is a right triangle

And this is a right triangle

And this also is a right triangle

A right triangle is one that contains a 90° angle

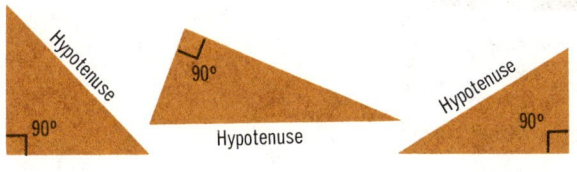

Hypotenuse

Hypotenuse

Hypotenuse

The side of the triangle opposite the 90° angle is called the hypotenuse

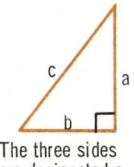

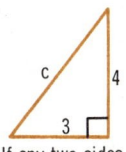

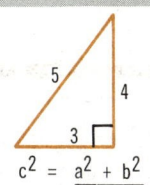

The three sides are designated a, b, and c, with c being the hypotenuse

If any two sides are known, the third side can be found by the Pythagorean Theorem

$$c^2 = a^2 + b^2$$
$$c = \sqrt{a^2 + b^2}$$
$$= \sqrt{16 + 9}$$
$$= \sqrt{25}$$
$$= 5$$

The lengths of the three sides of a right triangle are related by the Pythagorean Theorem:
$$c^2 = a^2 + b^2$$
where c is the length of the hypotenuse, and a and b are the lengths of the other two sides

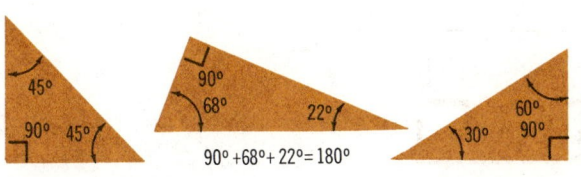

$$90° + 45° + 45° = 180°$$

$$90° + 68° + 22° = 180°$$

$$90° + 30° + 60° = 180°$$

The sum of the three angles in any right triangle is 180°

The properties of the right triangle are continued on the next page.

properties of the right triangle (cont.)

Sine of either angle	$=$	length of side opposite angle
		length of hypotenuse

EXAMPLE:

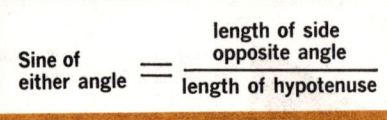

$\sin \theta = 4/5$

$= 0.8$

The two angles of a right triangle smaller than 90° are related to the lengths of the sides by the trigonometric relationships of sine, cosine, and tangent

Cosine of either angle	$=$	length of side adjacent to angle
		length of hypotenuse

EXAMPLE:

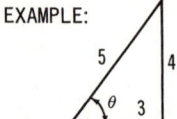

$\cos \theta = 3/5$

$= 0.6$

Once the value of the sine, cosine, or tangent of an angle is known, the size of the angle, in degrees, can be found in standard tables of trigonometric functions

Tangent of either angle	$=$	length of side opposite angle
		length of side adjacent angle

EXAMPLE:

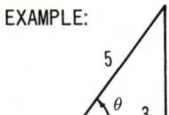

$\tan \theta = 4/3$

$= 1.33$

Since $\cos \theta = a/c$

$$a = c \times \cos \theta$$

and since $\sin \theta = b/c$

$$b = c \times \sin \theta$$

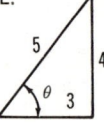
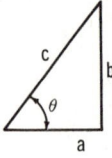

If the length of the hypotenuse and the angle between it and one of the other sides is known, the length of the other two sides can be found using the trigonometric relationships of sine or cosine

subtracting vectors

To subtract one vector from another, *add* the *negative equivalent* of the vector to be subtracted to the other vector. This means that once the vector to be subtracted has been converted to its negative equivalent, the two vectors are merely added using whichever of the methods of *vector addition* that is appropriate.

The negative equivalent of a vector is obtained by *rotating* the vector *180 degrees*. The magnitude of the vector is thus the same, but its direction is reversed.

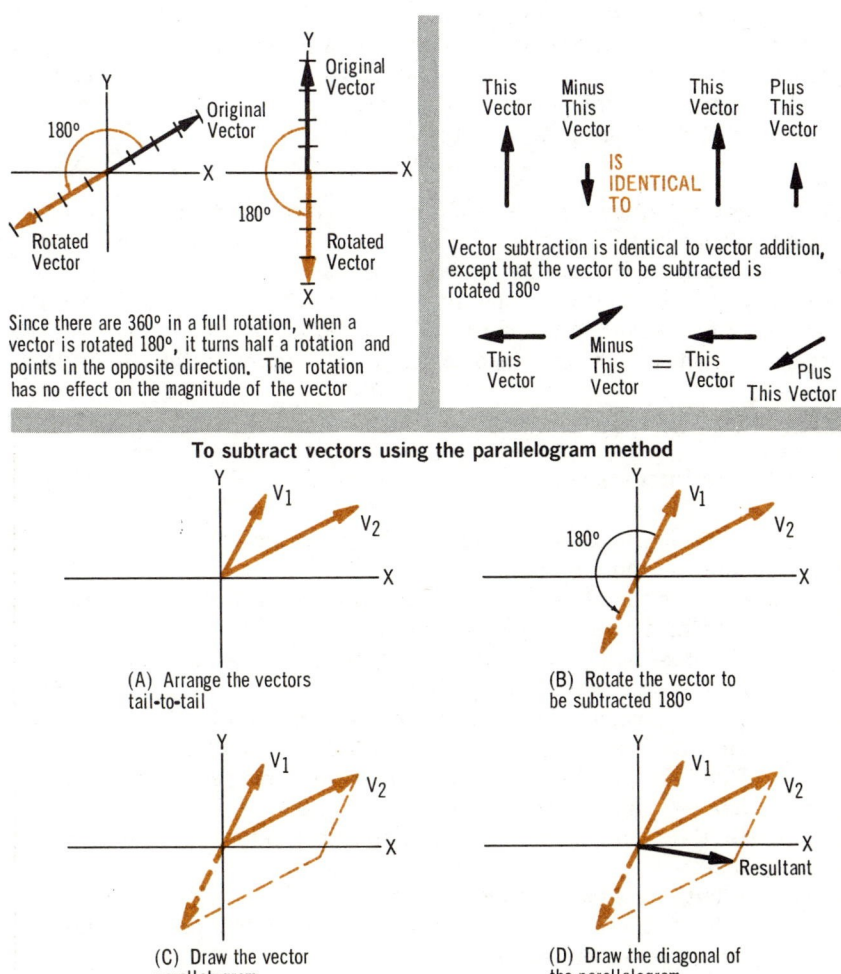

Since there are 360° in a full rotation, when a vector is rotated 180°, it turns half a rotation and points in the opposite direction. The rotation has no effect on the magnitude of the vector

Vector subtraction is identical to vector addition, except that the vector to be subtracted is rotated 180°

To subtract vectors using the parallelogram method

(A) Arrange the vectors tail-to-tail

(B) Rotate the vector to be subtracted 180°

(C) Draw the vector parallelogram

(D) Draw the diagonal of the parallelogram

summary

☐ Phase angles are used to describe the time relationships between a-c voltages and currents. They also specify the position or point in time of one voltage or current. ☐ In a purely resistive circuit, the voltage and current are in phase. ☐ In a purely inductive circuit, the applied voltage leads the current by 90 degrees. ☐ In a purely capacitive circuit, the current leads the applied voltage by 90 degrees. ☐ Quantities that have magnitude only are called scalars; those having both magnitude and direction are called vectors. ☐ Vectors are often used to express voltage and current relationships.

☐ Vectors that have the same direction can be added by simply adding the magnitude of the individual vectors. ☐ The sum of the vectors that have opposite directions is a vector whose magnitude is the difference of the two vectors, and whose direction is the same as that of the larger vector. ☐ The parallelogram method is used, in general, to graphically add two vectors. ☐ Another method for graphically adding vectors is the triangle method. This is also called the "head-to-tail" method.

☐ The resultant of two vectors that are 90 degrees apart can be calculated using the Pythagorean Theorem: $c^2 = a^2 + b^2$. ☐ The trigonometric functions of the angles of a right triangle are sine = opposite/hypotenuse; cosine = adjacent/hypotenuse; tangent = opposite/adjacent. ☐ To subtract vectors, the negative of the vectors to be subtracted is added to the other vector. ☐ The negative equivalent of a vector is obtained by rotating the vector 180 degrees.

review questions

1. What is the phase relationship between the voltage and current in a purely inductive circuit?
2. What is the phase relationship between the voltage and current in a purely capacitive circuit?
3. What is the phase relationship between the voltage and current in a purely resistive circuit?
4. How are vectors having the same direction added?
5. How are vectors having opposite directions added?
6. Does the parallelogram method for adding vectors use the head-to-tail method?
7. Is the triangle method for adding vectors a graphical method?
8. What is the *Pythagorean Theorem*?
9. What is meant by the *sine, cosine,* and *tangent* of an angle?
10. How are vectors subtracted?

separating vectors into components

Every vector can be separated, or *resolved,* into two other vectors that are 90 degrees apart, and which when added will produce the original vector. These two vectors are called the *components* of the original vectors. One of them is in the horizontal direction, and is called the *horizontal component;* the other is in the vertical direction, and is called *vertical component.* In effect, the two components divide the magnitude of the original vector, and show how much of the magnitude is in the horizontal direction and how much is in the vertical direction.

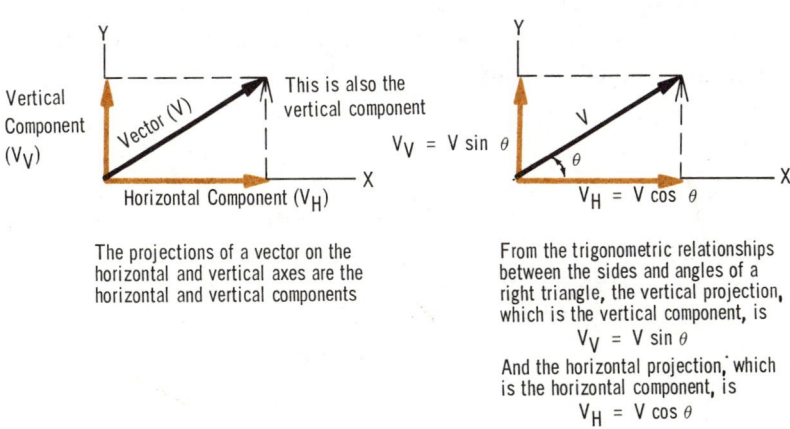

The projections of a vector on the horizontal and vertical axes are the horizontal and vertical components

From the trigonometric relationships between the sides and angles of a right triangle, the vertical projection, which is the vertical component, is
$$V_V = V \sin \theta$$
And the horizontal projection, which is the horizontal component, is
$$V_H = V \cos \theta$$

As illustrated, a vector and its two components form a right triangle. You will remember that, in a right triangle, if you know the length of the hypotenuse and one of the angles besides the right angle, you can calculate the lengths of the other two sides using the sine and cosine functions. Therefore, if you know the magnitude and direction of a vector, you can use the sine and cosine functions to find its components. The horizontal component is found by:

Horizontal component = original vector $\times \cos \theta$

And the vertical component by:

Vertical component = original vector $\times \sin \theta$

where θ is the angle between the original vector and the horizontal axis.

adding vectors
by components

You may wonder why it would ever be desirable to resolve vectors into their components. The primary reason is that it makes the addition of vectors much easier. When vectors are to be added, they can first be resolved into their components. The individual horizontal components can then be added to find the *total horizontal component*. Similarly, the vertical components can be added to find the *total vertical component*. These total components are the components of the resultant vector, and are 90 degrees apart. So the resultant can be found by adding the total components using the standard methods for solving right triangles.

The previous method is not used for adding vectors that are in the same or opposite directions or are 90 degrees apart, inasmuch as relatively simple methods already exist for adding these vectors.

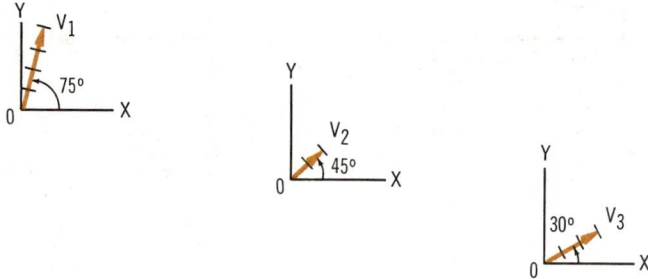

As an example of adding vectors by components, the three vectors shown will be resolved into their components and then added. The horizontal component of each vector is found using the equation $V_H = V \cos \theta$. Therefore,

$$V_{H1} = 4 \times \cos 75° = 4 \times 0.2588 = 1.14$$

$$V_{H2} = 2 \times \cos 45° = 2 \times 0.7071 = 1.41$$

$$V_{H3} = 3 \times \cos 30° = 3 \times 0.866\ = 2.6$$

The vertical component of each vector is found next, using the equation $V_V = V \sin \theta$.

$$V_{V1} = 4 \times \sin 75° = 4 \times 0.9659 = 3.86$$

$$V_{V2} = 2 \times \sin 45° = 2 \times 0.7071 = 1.41$$

$$V_{V3} = 3 \times \sin 30° = 3 \times 0.5000 = 1.5$$

adding vectors by components (cont.)

The total components are now found by adding the individual components. The horizontal component, V_H, is

$$V_H = V_{H1} + V_{H2} + V_{H3}$$
$$= 1.14 + 1.41 + 2.6 = 5.15$$

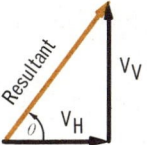

The total vertical component, V_V, is

$$V_V = V_{V1} + V_{V2} + V_{V3}$$
$$= 3.86 + 1.41 + 1.5 = 6.77$$

These total components are now added vectorially to find the magnitude of the resultant vector. Since the components and the resultant form a right triangle, the Pythagorean Theorem is used.

$$c^2 = a^2 + b^2$$
$$c = \sqrt{a^2 + b^2} = \sqrt{(5.15)^2 + (6.77)^2}$$
$$= \sqrt{26.5 + 45.8} = \sqrt{72.3} = 8.5$$

The magnitude of the resultant is thus 8.5 units. Before the vector can be drawn, though, its direction must be determined. This is done by using the relationship between the tangent of the angle enclosed by the resultant and the horizontal axis, and the two components.

Once the tangent of the angle is known, the angle itself can be found from a table of trigonometric functions. In this case, the angle is approximately 53 degrees. The sum of the three original vectors, therefore, is a vector that makes an angle of 53 degrees with the horizontal axis, and that has a magnitude of 8.5 units.

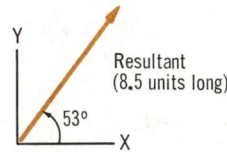

vectors associated with sine-wave phase angles

As was pointed out at the beginning of this volume, vectors are very useful for expressing the phase difference, or phase angle, between a-c voltages and currents. When used in this way, the a-c voltage or current is the physical quantity represented by the vector. As with all vectors, the length of, say, a voltage vector corresponds to the amplitude of the voltage. The direction of the vector, however, does not represent the direction of the voltage. Actually, the direction of the vector

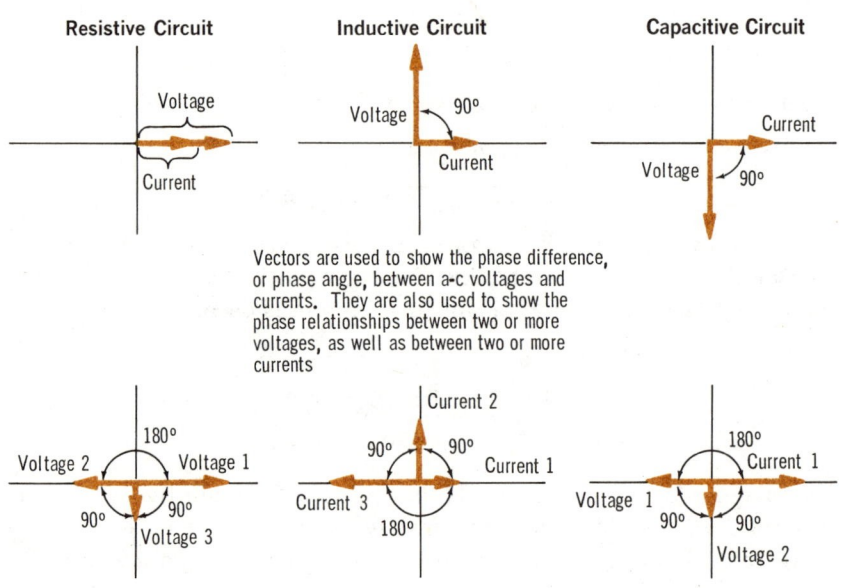

Vectors are used to show the phase difference, or phase angle, between a-c voltages and currents. They are also used to show the phase relationships between two or more voltages, as well as between two or more currents

is *meaningless* in itself, and only becomes significant when it is compared to the direction of another voltage or current vector. Thus, the difference in direction between a-c voltage and current vectors is what is important. This difference is expressed in degrees, and represents the *phase difference* between the two vector quantities.

A-c vectors are called *rotating vectors*, since they represent sine-wave quantities, which as you have learned are based on the rotation of the armature of an a-c generator.

graphical representation of rotating vectors

A-c, or rotating, vectors show both the amplitudes and the phase relationships between sine-wave voltages and currents. The *length* of the vector shows amplitude, while the *angle* between the vectors shows the phase. Also, the position of the vectors shows which is *leading* and which is *lagging*.

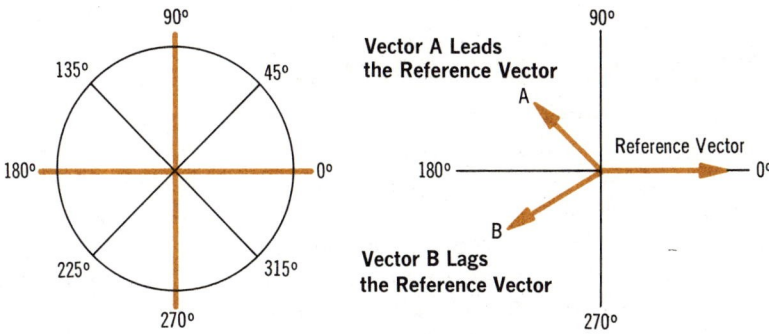

A-c vectors can have a direction corresponding to any angle from 0 to 360° measured from the reference direction of 0°

In any graph of a-c vectors, one vector serves as the reference vector, and points in the 0° direction. Vectors above the horizontal axis lead the reference vector, and those below lag it

As you can see, a vector graph is divided into 360 degrees, corresponding to one full sine-wave cycle. The starting, or *reference*, point of zero degrees is horizontally to the right, and the phase angles of other vectors are compared to this reference. Vectors that *lead* the reference vector are *above* the horizontal axis, and those that *lag* are *below*.

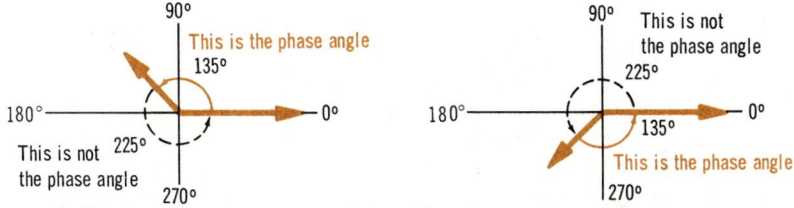

The angle between any vector and the reference vector is measured in the direction that will make it less than 180°. Thus, the phase angle between the reference vector and a vector that leads it is measured in a counterclockwise direction starting from the reference vector. The angle between the reference vector and a vector that lags it is measured in a counterclockwise direction starting from the lagging vector

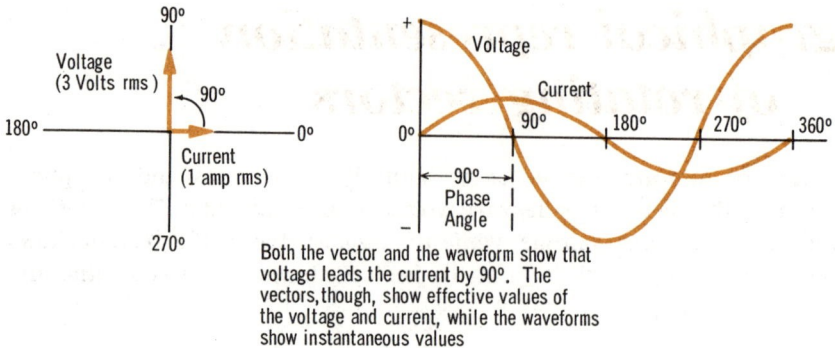

Both the vector and the waveform show that voltage leads the current by 90°. The vectors, though, show effective values of the voltage and current, while the waveforms show instantaneous values

vectors vs. waveforms

Both a-c vectors and waveforms show the amplitudes and phase relationships of a-c voltages and currents. One important difference, though, is in the *type* of amplitudes shown. A waveform shows all of the *instantaneous* values of current or voltage throughout the complete sine-wave cycle. A vector, on the other hand, shows only *one* value, since its length is fixed. This value can be the peak, average, or effective value, depending on the particular situation in which the vector is used. If the length of the vector is proportional to the *peak* amplitude, its *vertical component* is proportional to the instantaneous amplitude for any given phase angle. Therefore, if a vector corresponding to the peak amplitude of a voltage or current is rotated through 360 degrees, as shown below, the magnitude of its vertical component will trace out the waveform that corresponds to the vector.

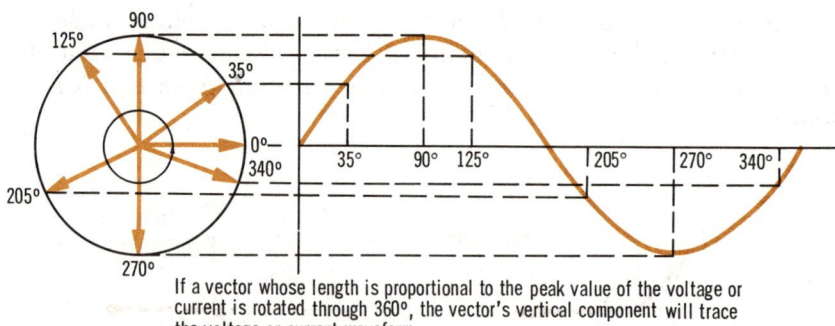

If a vector whose length is proportional to the peak value of the voltage or current is rotated through 360°, the vector's vertical component will trace the voltage or current waveform

You should not think that just because vectors show only a fixed value of voltage or current that their use is thereby limited. They can represent average, effective, or peak amplitudes, and in solving practical problems, these are the values you will usually be working with.

vectors of purely resistive, inductive and capacitive circuits

The following are vector representations, and the corresponding waveforms, of the currents and voltages in purely resistive, inductive, and capacitive circuits.

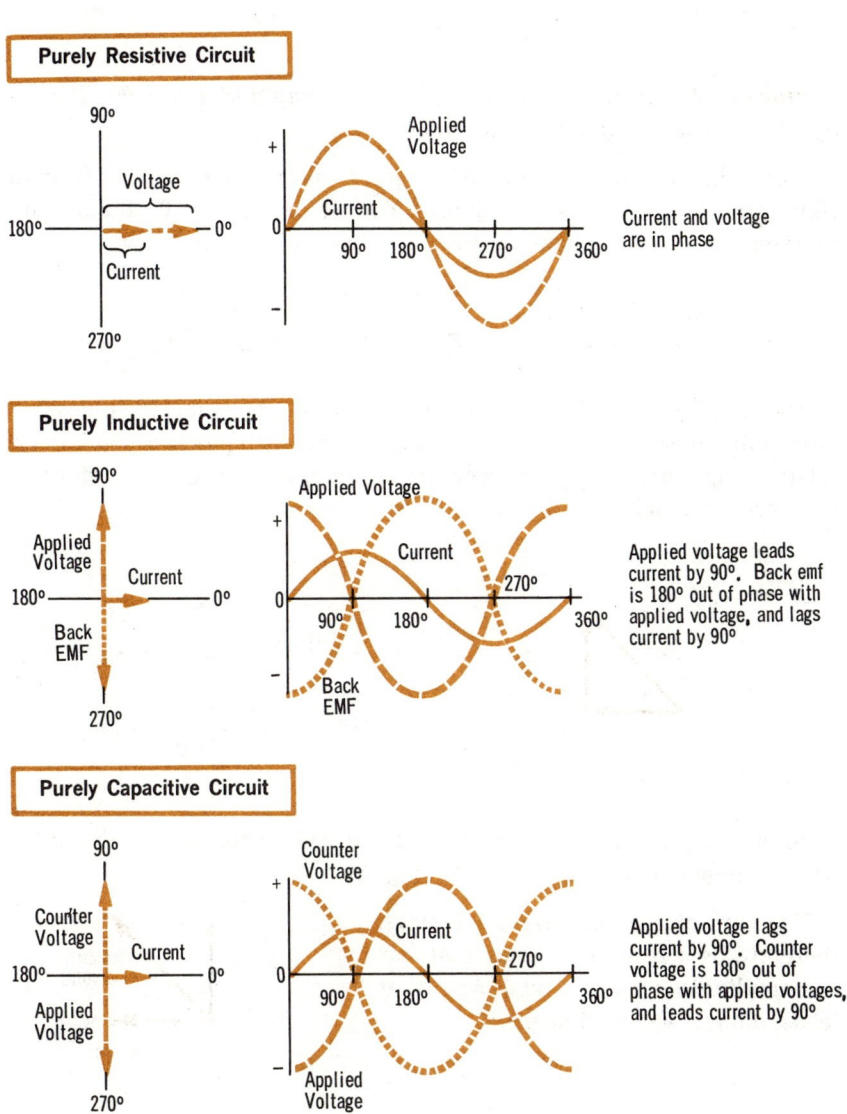

Purely Resistive Circuit

Current and voltage are in phase

Purely Inductive Circuit

Applied voltage leads current by 90°. Back emf is 180° out of phase with applied voltage, and lags current by 90°

Purely Capacitive Circuit

Applied voltage lags current by 90°. Counter voltage is 180° out of phase with applied voltages, and leads current by 90°

solved problems

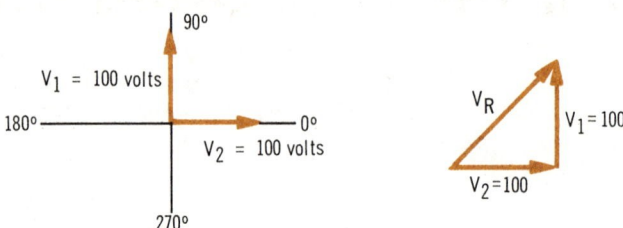

Problem 1. *If two voltages with equal amplitudes are 90 degrees apart, as shown, what is their sum?*

Since the two voltages are 90 degrees apart, their vectors form a right triangle with the resultant. The Pythagorean Theorem can, therefore, be used to calculate the magnitude of the resultant.

$$V_R^2 = V_1^2 + V_2^2$$
$$V_R = \sqrt{V_1^2 + V_2^2} = \sqrt{(100)^2 + (100)^2} = \sqrt{20{,}000} = 141 \text{ volts}$$

The amplitude of the resultant is therefore 141 volts. Its phase relationship must now be found. This is done using the trigonometric relationship between the tangent of the angle θ, and the two sides of the vector triangle.

$$\tan \theta = \frac{\text{opposite side}}{\text{adjacent side}}$$
$$= \frac{V_1}{V_2} = \frac{100}{100} = 1$$

From a table of trigonometric functions, it can be found that the angle whose tangent is equal to 1 is 45 degrees.

The sum of the two original vectors, therefore, is a resultant vector that has an amplitude of 141 volts, and that leads voltage V_2 by 45 degrees.

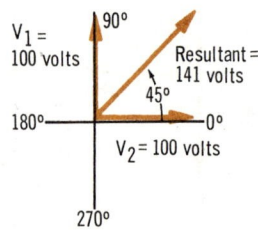

solved problems (cont.)

Problem 2. Draw the current and voltages in the following circuit on a vector diagram.

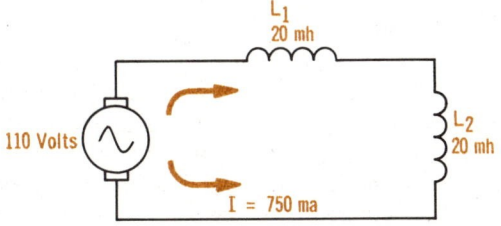

It is a series circuit, so the current throughout the circuit is the same. There will, therefore, be only one current vector, and it should be made the reference vector. The applied voltage will be one voltage vector, and since there are two coils and each develops a back emf, there will be two voltage vectors for the back emf, or a total of three voltage vectors.

Back emf is always 180 degrees out of phase with the applied voltage, so the back emf vectors and the applied voltage vector must have opposite directions. Furthermore, applied voltage in a purely inductive circuit leads the current by 90 degrees. Therefore, if the current is the reference vector and so has a direction of 0 degrees, the applied voltage vector must have a direction of 90 degrees. And since the back emf vectors are opposite in direction to the applied voltage vector, they must have a direction of 270 degrees.

The amplitudes of the applied voltage and the current are stated in the problem; but the amplitudes of the back emf's are not. However, since the inductances of the coils are equal, you do know that the back emf's must also be equal; so their vectors must be equal in length.

Based on the above reasoning, the vector diagram would be drawn as follows:

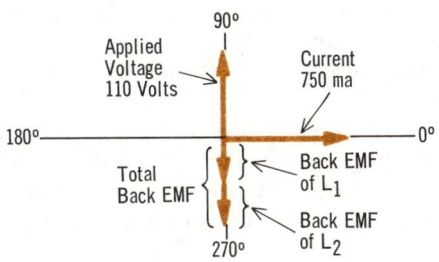

summary

☐ Every vector can be separated, or resolved, into two components that are 90 degrees apart. These are called the horizontal component and the vertical component. ☐ Component vectors in the same direction can be added arithmetically. ☐ The magnitude of the resultant vector can be found from the components by using the Pythagorean Theorem.

☐ A-c vectors are called rotating vectors; they represent sine-wave quantities. ☐ A-c vectors show both the amplitude and the phase relationships between sine-wave voltages and currents. ☐ The length of a rotating vector indicates the amplitude of the a-c voltage or current. ☐ The angle between two rotating vectors indicates the phase difference between the two vector quantities. ☐ A vector graph is divided into 360 degrees, corresponding to one full sine-wave cycle. ☐ Phase angles of vectors are compared to the reference, or zero-degree, point on the right part of the horizontal axis.

☐ A waveform shows all of the instantaneous current and voltage values throughout a complete sine-wave cycle. ☐ A vector shows only one value; it can be the peak, average, or effective value of the amplitude.

review questions

1. Illustrate and explain what is meant by the *horizontal* and *vertical components* of a vector.
2. What is the magnitude of the vertical component of a horizontal vector whose magnitude is 100? What is the magnitude of the horizontal component?
3. For Question 2, what is the phase angle between the vector and the vertical axis?
4. The horizontal component of a vector is 5 and the vertical component is 12. What is the magnitude of the vector? What is its direction with respect to the horizontal axis?
5. The magnitude of a vector is 20 and its horizontal component is 16. What is the value of the vertical component?
6. What is meant by a *rotating vector*? Why is it so called?
7. What does a rotating vector indicate?
8. What amplitude of a voltage or current does a waveform indicate?
9. Explain the difference between a *vector* and a *waveform*.
10. On a vector diagram, show the phase difference between the applied voltage and the current in a purely inductive circuit. In a purely capacitive circuit. In a purely resistive circuit.

RL circuits

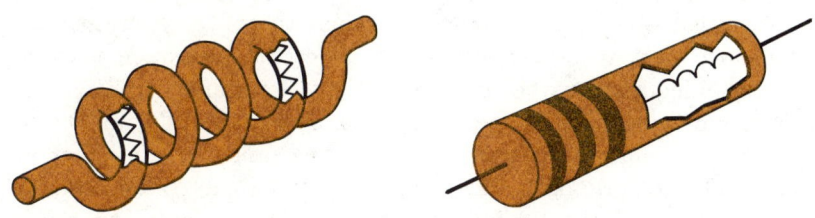

An RL circuit is one that contains both *resistance* (R) and *inductance* (L). In Volume 3, you learned the characteristics of a-c circuits having only resistance or only inductance. Many of these characteristics are *modified* when resistance and inductance are *both* present. As a result, different methods and equations must be used to solve RL circuit problems. You will find that the basic reason for the differences between RL circuits and purely resistive or inductive circuits is that the *phase relationships* in the *resistive* portions of RL circuits are *different* than the phase relationships in the *inductive* portions. Both of these relationships, though, affect the overall operation of the circuit, and must be considered when solving RL circuit problems. The nature of the phase relationships that exist in series and parallel RL circuits, as well as methods of solving circuit problems taking into account these phase relationships, are described on the following pages.

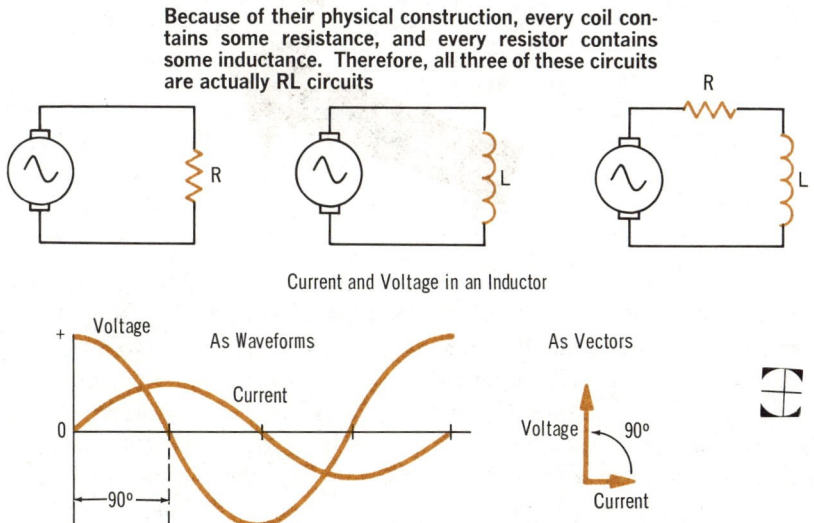

Because of their physical construction, every coil contains some resistance, and every resistor contains some inductance. Therefore, all three of these circuits are actually RL circuits

Current and Voltage in an Inductor

impedance

In resistive circuits, resistance provides the only opposition to current flow. And, in inductive circuits, all of the opposition is provided by the inductance in the form of inductive reactance. Resistance, you will remember, is "built into" a load, and is essentially unaffected by the circuit voltage or current. Inductive reactance, on the other hand, is directly proportional to the frequency, and so its value depends on the frequency of the applied voltage. Furthermore, although a voltage drop occurs when current flows through either a resistance or an inductive reactance, the *phase relationship* between the current and the voltage drop is different for a resistance than it is for a reactance. Since the voltage drop is a measure of the opposition to current, resistance and inductive reactance can be considered as differing in phase. Actually, only quantities that vary in *time* can differ in *phase;* and inasmuch as neither resistance nor inductive reactance do, strictly speaking, they cannot really differ in phase. But, since their *effects* on current flow have a *phase relationship*, we think of resistance and inductive reactance themselves as having a phase relationship.

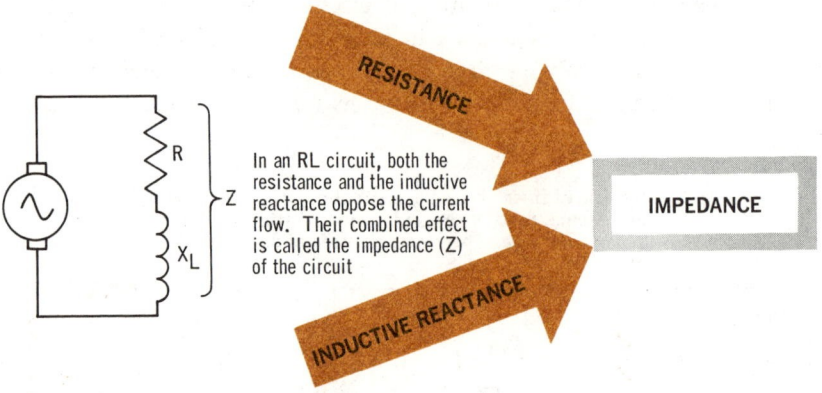

In an RL circuit, both the resistance and the inductive reactance oppose the current flow. Their combined effect is called the impedance (Z) of the circuit

In effect, then, although both resistance and inductive reactance oppose current flow, some of their characteristics and effects are different. For this reason, the *total* opposition to current flow in RL circuits is not expressed in terms of either resistance or inductive reactance. Instead, a quantity called the *impedance* is used. The impedance of an RL circuit is calculated from the values of resistance and inductive reactance, and takes into account the differences between them. Impedance is measured in ohms, and is usually designated by the letter Z. The methods used to calculate impedance depend on whether the resistance and inductive reactance are in *series* or in *parallel*.

series *RL circuits*

When the resistive and inductive components of a circuit are connected in such a way that the *total* circuit current flows through each, the circuit is a *series* RL circuit. It is important to realize that the current flowing at *every* point in the circuit is the *same*. As you will learn, any analysis of series RL circuits is based on this fact.

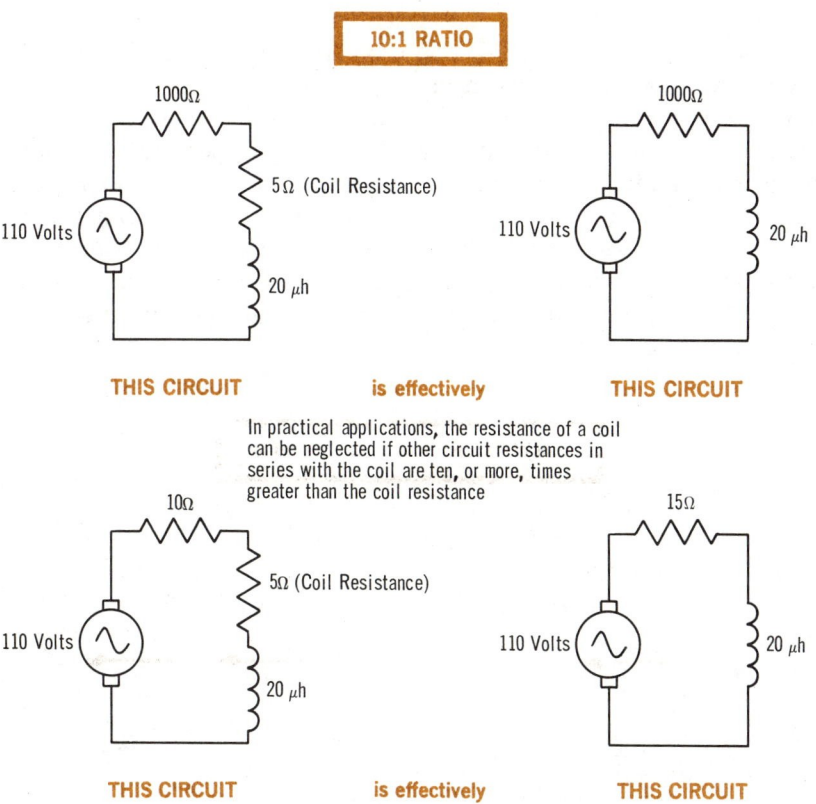

10:1 RATIO

1000Ω

5Ω (Coil Resistance)

110 Volts

20 μh

1000Ω

110 Volts

20 μh

THIS CIRCUIT   **is effectively**   **THIS CIRCUIT**

In practical applications, the resistance of a coil can be neglected if other circuit resistances in series with the coil are ten, or more, times greater than the coil resistance

10Ω

5Ω (Coil Resistance)

110 Volts

20 μh

15Ω

110 Volts

20 μh

THIS CIRCUIT   **is effectively**   **THIS CIRCUIT**

As you continue your studies of electricity and electric circuits, you will find that this 10:1 ratio applies to many situations where the relative effects of two quantities are involved

A series RL circuit may consist of one or more resistors, or resistive loads, connected in series with one or more coils. Or, since the wire used in any coil has some resistance, a series RL circuit may consist of only one or more coils, with the resistance of the coils, which is *effectively* in series with the inductance, supplying the circuit resistance.

series *RL* circuits (cont.)

If a resistor or other resistive load is connected in series with one or more coils, its resistance is usually much greater than the resistance of the coils. When it is *ten*, or more, *times greater*, the effect of the coil resistance can be ignored. In this volume, unless otherwise indicated, we will assume that this is the case, and will consider coils as having zero resistance.

You will notice that in the discussion on series RL circuits, the phase of the circuit *current* is used as the phase *reference* for all other circuit quantities. This is done for convenience, inasmuch as the current is the same throughout the circuit. Since current is used as the reference, the current vector on a vector diagram has a phase angle of *0 degrees*, which means that it is horizontal and points to the right. Any circuit quantity that is in phase with the current, therefore, also has a phase angle of 0 degrees. You should be aware, though, that other circuit quantities could be used as the phase reference. Current is chosen in the series RL circuit only because it is common to all the parts in the circuit.

Use of Current as Phase Reference

I

In series RL circuits, the current is used as the phase reference for all other circuit quantities. It, therefore, has a phase angle of 0°

The phase of all other circuit quantities are then determined relative to the current

R and E$_R$ are in phase with I

Quantities that are in phase with the current also have phase angles of 0°

voltage

When voltage is applied to a series RL circuit, the current causes a voltage drop across both the resistance and inductance. As you learned in Volume 3, the voltage drop across a *resistance* is *in phase* with the current that causes it, while the voltage drop across an *inductance leads* the current by 90 degrees. Since in a series RL circuit the current through the resistance and inductance is the same, the voltage drop across the resistance (E_R) is in phase with the circuit current, while the voltage drop across the inductance (E_L) leads the current by 90 degrees. With the current as a reference, therefore, E_L *leads* E_R by 90 degrees. The amplitude of the voltage drop across the resistance is proportional to the current and the value of resistance ($E = IR$). The amplitude of the voltage across the inductance is proportional to the current and the value of inductive reactance ($E = IX_L$).

A situation that is new to you, though, arises here. Until now, the *sum* of the individual voltage drops around a series circuit was equal to the applied voltage in all the circuits you have learned. This was according to Kirchhoff's Voltage Law. But if you were to measure the *applied* voltage and then the voltage *drops* in a series a-c circuit, you would find that the *sum* of the voltage drops appears *greater* than the applied voltage. This is because when the voltage drops are not in phase, the *vector sum* of the voltage drops, rather than their arithmetic sum must be used for Kirchhoff's Law to work. Thus, the applied voltage can be represented vectorially as the sum of the two vectors: one, the E_R vector, is at 0 degrees, since it is in phase with the circuit current; the other, the E_L vector, leads both E_R and I by 90 degrees.

The relationship between the applied voltage and the voltage drops in a series RL circuit is such that the applied voltage equals the VECTOR SUM of the voltage drops

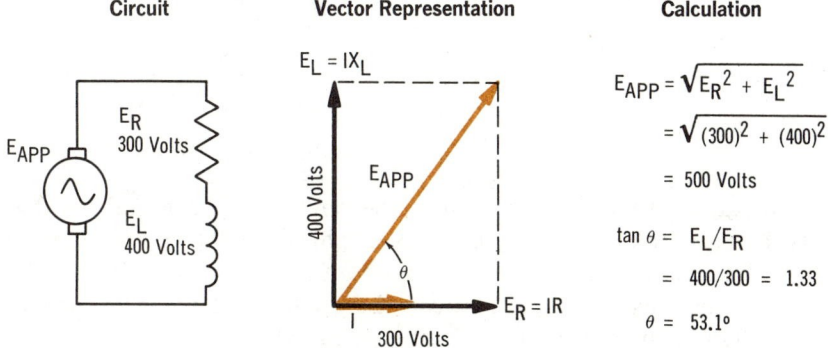

Circuit	Vector Representation	Calculation

$$E_{APP} = \sqrt{E_R{}^2 + E_L{}^2}$$
$$= \sqrt{(300)^2 + (400)^2}$$
$$= 500 \text{ Volts}$$

$$\tan \theta = E_L/E_R$$
$$= 400/300 = 1.33$$
$$\theta = 53.1°$$

voltage (cont.)

Since the two voltage vectors are 90 degrees apart, they can be added vectorially by the *Pythagorean Theorem* to find the applied voltage:

$$E_{APP} = \sqrt{E_R^2 + E_L^2}$$

Graphically the applied voltage is the hypotenuse of a right triangle whose other two sides are the circuit voltage drops. The angle between the applied voltage and E_R is the same as the phase angle between the applied voltage and the current (I). The reason for this is that E_R and I are in phase. The value of θ can be found from:

$$\tan \theta = E_L / E_R$$

or

$$\cos \theta = E_R / E_{APP}$$

With I used as the reference vector, as shown at the left below, it may seem that the applied voltage changes in phase as the voltage drops change with different values of resistance and inductive reactance. Actually, it only appears this way because the current is used as the 0-degree reference.

I as Reference Vector E_{APP} as Reference Vector

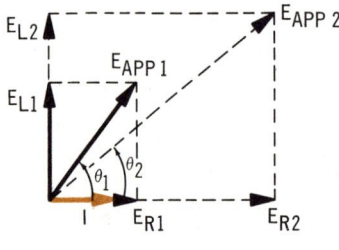

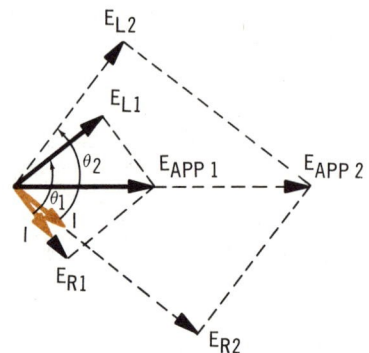

If the applied voltage was used as the reference, you would see that it is the current that actually changes in phase, as shown on the right illustration above. Therefore, to avoid confusion, always consider the phase angle θ as the angle *between* the applied voltage and current, rather than as the phase angle of either one of them.

voltage waveforms

You saw from the calculations on the previous pages that although the individual voltage drops in the circuit were 300 and 400 volts, the applied voltage, or total voltage drop, was 500 volts instead of 700 volts. The reason for this was that the individual voltage drops were out of phase. If they were in phase, they would have reached their maximum amplitudes at the same time, and could be added directly. But since they were out of phase, all of their *instantaneous values* had to be added, and then the *average* or *effective* value of the resulting voltage found. This is what vector addition accomplishes. Whether the average or effective value of the applied voltage is found, depends on what values you are using. If the voltage drops are given in effective values, the applied voltage you find will also be an effective value. Similarly, if you are working with average values, the applied voltage will be an average value.

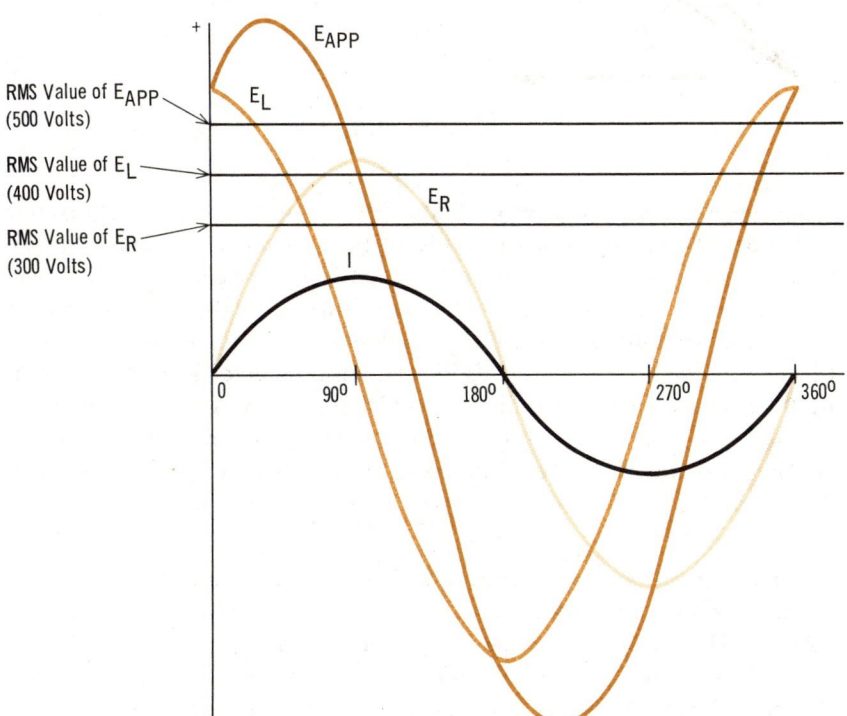

Every point on the applied voltage waveform (E_{APP}) is the algebraic sum of the instantaneous values of the E_R and E_L waveforms. The rms value of the applied voltage waveform is equal to the vector sum of the rms values of the E_R and E_L waveforms

impedance and current

In a series RL circuit, the impedance is the VECTOR SUM of the resistance and the inductive reactance.

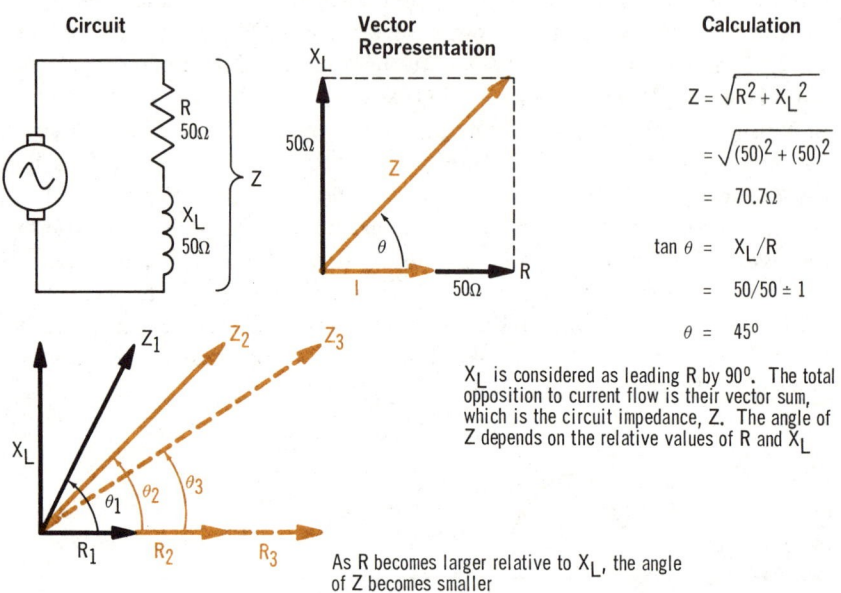

Circuit   **Vector Representation**   **Calculation**

$$Z = \sqrt{R^2 + X_L^2}$$

$$= \sqrt{(50)^2 + (50)^2}$$

$$= 70.7\Omega$$

$$\tan \theta = X_L/R$$

$$= 50/50 \doteq 1$$

$$\theta = 45^0$$

X_L is considered as leading R by 90°. The total opposition to current flow is their vector sum, which is the circuit impedance, **Z.** The angle of Z depends on the relative values of R and X_L

As R becomes larger relative to X_L, the angle of Z becomes smaller

As was explained, it is convenient in RL circuits to consider resistance and inductive reactance as differing in phase, and to use the term impedance (Z) to represent the *total* opposition to current flow. Since, in a series RL circuit, the same current flows through both the resistor and inductor, and the voltage drop across the resistor is in phase with the current, while the voltage drop across the inductor leads the current by 90 degrees, then the inductive reactance is considered to lead the resistance by 90 degrees. The *vector sum* of the reactance and resistance, the impedance, can therefore be calculated by the Pythagorean Theorem.

$$Z = \sqrt{R^2 + X_L^2}$$

Since R and X_L are 90 degrees apart, their vector sum, Z, will be at an angle somewhere *between* 0 and 90 degrees relative to the circuit current. The exact angle depends on the *comparative* values of R and X_L. If R is greater, Z will be closer to 0 degrees; and if X_L is greater, Z will be closer to 90 degrees. The angle can be found from:

$$\tan \theta = X_L/R \qquad \text{or} \qquad \cos \theta = R/Z$$

The phase angle of Z is the same as the phase angle of the applied voltage, described previously.

impedance and current (cont.)

The 10-to-1 rule you learned before also applies to impedance. If either X_L or R is 10 times greater than the other, the phase angle of Z can be considered as 0 or 90 degrees, depending on which is larger. Essentially, what the 10-to-1 rule signifies is that if R is 10 or more times greater than X_L, the circuit will operate almost as if X_L were not present; the opposite is true if X_L is 10 or more times larger than R.

The relationships between I, E, and Z in RL circuits are similar to the relationships between I, E, and R in d-c circuits. Because of this, the equations for Ohm's Law can be used in solving RL circuits by using the impedance (Z) in place of the resistance. These equations are often called *Ohm's Law for a-c circuits*. They are:

$$I = E/Z \qquad E = IZ \qquad Z = E/I$$

You will find later that these equations also apply to circuits that contain capacitance as well as inductance and resistance.

In a series RL circuit, the current is the same at every point, and lags the applied voltage by an angle between 0 and 90°

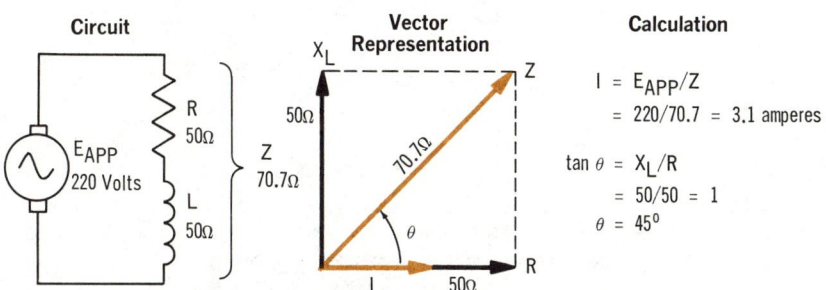

Circuit	Vector Representation	Calculation

Calculation:

$I = E_{APP}/Z$

$= 220/70.7 = 3.1$ amperes

$\tan \theta = X_L/R$

$= 50/50 = 1$

$\theta = 45^0$

The calculated angle is the phase angle of the current. It is the same as the phase angle of impedance found for the same circuit on the previous page. The reason for this is that the values of X_L and R determine the angle of the impedance, which in turn determines how inductive or resistive the current is

As in any series circuit, the current in an RL series circuit is the *same* at *every point*. As a result, the current through the resistance is *in phase* with the current through the inductance, since they are actually the same current. If the applied voltage and impedance in an RL series circuit are known, the current can be calculated by Ohm's Law: $I = E_{APP}/Z$, where E_{APP} equals the applied voltage and Z equals the vector sum of the resistance and inductive reactance ($\sqrt{R^2 + X_L^2}$).

impedance and current (cont.)

Since the current in a series circuit is common to all parts of the circuit, it is used as the 0-degree reference. And the angle between it and the impedance determines whether the current is more inductive or resistive. As you learned before, the angle of Z is somewhere between 0 and 90 degrees, depending on the relative values of the inductive reactance and resistance.

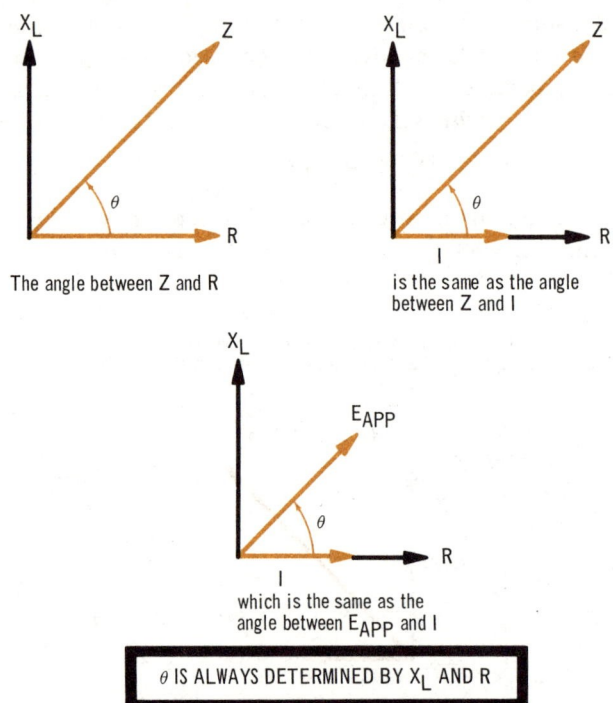

The angle between Z and R

is the same as the angle between Z and I

which is the same as the angle between E_{APP} and I

θ IS ALWAYS DETERMINED BY X_L AND R

The *greater* the inductive reactance is compared to the resistance, the *larger* is the phase angle, and the more I tends to act as an inductive current. Similarly, the *smaller* the inductive reactance is compared to the resistance, the *smaller* is the angle, and the more I tends to act as a resistive current. When Z and I are exactly in phase ($Z = R$), the current is purely resistive; and when Z leads I by 90 degrees ($Z = X_L$), the current is purely inductive. The terms resistive and inductive, when applied to the current, refer to the phase relationship between the current and the *applied* voltage. The closer the current is to being in phase with the applied voltage, the more resistive it is; and the closer it is to lagging the applied voltage by 90 degrees, the more inductive it **is.**

power

In resistive circuits, *all* of the power delivered by the source is *dissipated* by the load; but if you recall what you learned from Volume 3, in an RL circuit, only a portion of the input power is dissipated. The part delivered to the inductance is returned to the source each time the magnetic field around the inductance collapses. There are, therefore, two types of power in an RL circuit. One is the *apparent power*. The other type is the *true power* actually consumed in the circuit. The true power depends on how much was returned to the source by the inductance; this in turn depends on the phase angle between the circuit current and the applied voltage. True power is calculated by multiplying the apparent power by the cosine of the phase angle between the voltage and current:

$$P_{TRUE} = E_{APP}I \cos \theta = I^2 Z \cos \theta$$

Since from the vector diagram of impedance, $R = Z \cos \theta$, the equation can also be written as $P_{TRUE} = I^2 R$. This shows that true power is that used by the circuit resistance.

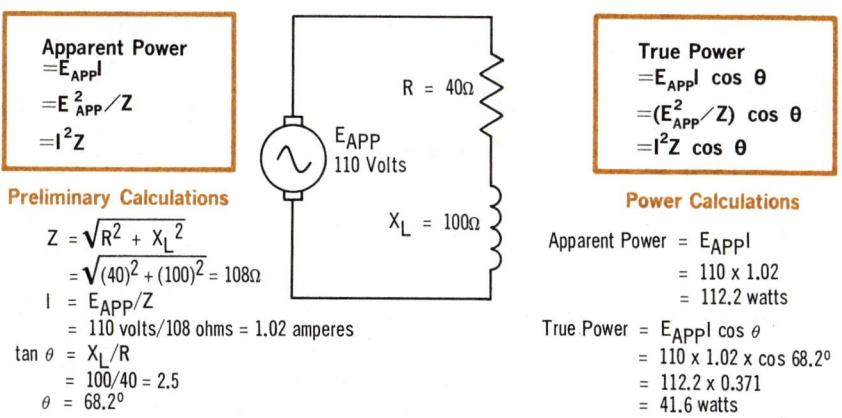

Apparent Power
$$= E_{APP}I$$
$$= E_{APP}^{2} / Z$$
$$= I^2 Z$$

Preliminary Calculations

$Z = \sqrt{R^2 + X_L^2}$
    $= \sqrt{(40)^2 + (100)^2} = 108\Omega$

$I = E_{APP}/Z$
    $= 110 \text{ volts}/108 \text{ ohms} = 1.02 \text{ amperes}$

$\tan \theta = X_L/R$
    $= 100/40 = 2.5$

$\theta = 68.2°$

$R = 40\Omega$

E_{APP}
110 Volts

$X_L = 100\Omega$

True Power
$$= E_{APP}I \cos \theta$$
$$= (E_{APP}^{2} / Z) \cos \theta$$
$$= I^2 Z \cos \theta$$

Power Calculations

Apparent Power $= E_{APP}I$
    $= 110 \times 1.02$
    $= 112.2 \text{ watts}$

True Power $= E_{APP}I \cos \theta$
    $= 110 \times 1.02 \times \cos 68.2°$
    $= 112.2 \times 0.371$
    $= 41.6 \text{ watts}$

The value of cosine θ can vary between 0 and 1, and, as you remember from Volume 3, it is called the circuit *power factor*. Small power factors (close to 0) are undesirable, since they mean that the power source has to deliver more power than is used. The power factor is found by:

$$\text{Power factor} = \frac{\text{true power}}{\text{apparent power}}$$

the Q of a coil

As you already know, every coil has some resistance, and so acts as a series RL circuit when it is connected to a source of voltage. Physically, it is impossible to measure *separately* the voltage drop across the coil resistance and the drop across the coil inductance. However, mathematically you can assume that the resistance and the inductance are both separate quantities in series with each other, and the two voltage drops and their phase angle can then be calculated. It is obvious that the smaller the coil resistance, the more the coil acts as a *perfect inductor,* that is, one with inductance, but zero resistance. You will find that it is sometimes desirable to compare coils on the basis of how close they approach the theoretically perfect coil. This is done by calculating the *ratio* of inductive reactance to resistance. Such a ratio is called the merit rating, or Q, of the coil. As an equation,

$$Q = X_L/R$$

You can see that the higher the inductive reactance, or the lower the resistance, the greater the Q.

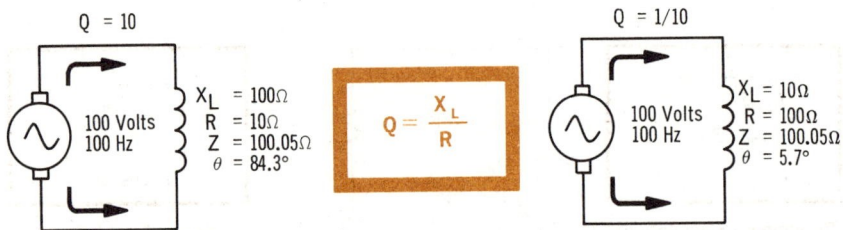

The high-Q coil produces a greater phase angle, and so is a better inductor

A coil with a high Q has a *large phase angle* (close to 90°) between the voltage across it and the current through it. Such a coil will develop a strong magnetic field, and therefore a large cemf, for a given applied voltage. A coil with a low Q has a *small phase angle* between its voltage and current, and because of the I^2R losses caused by its relatively large resistance, will develop a weaker magnetic field and a smaller cemf for a given applied voltage. The Q of a coil becomes an important factor to consider when you work with LC circuits, which are covered later.

With some coils, eddy currents might cause a-c resistance to rise with frequency, as does X_L, to keep the Q somewhat constant. But, since not all coils respond with the same eddy current effects, the Q of different coils should be compared at different frequencies.

effect of frequency

As you know, the *relative values* of X_L and R determine the phase angle of the impedance and the current, as well as the circuit power factor. When X_L is much larger than R, the circuit is very inductive and the power factor is close to zero. When R or X_L is 10 or more times larger than the other, the circuit can be considered purely resistive or purely inductive, and the power factor considered as one or

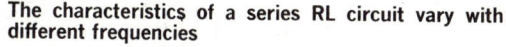

The characteristics of a series RL circuit vary with different frequencies

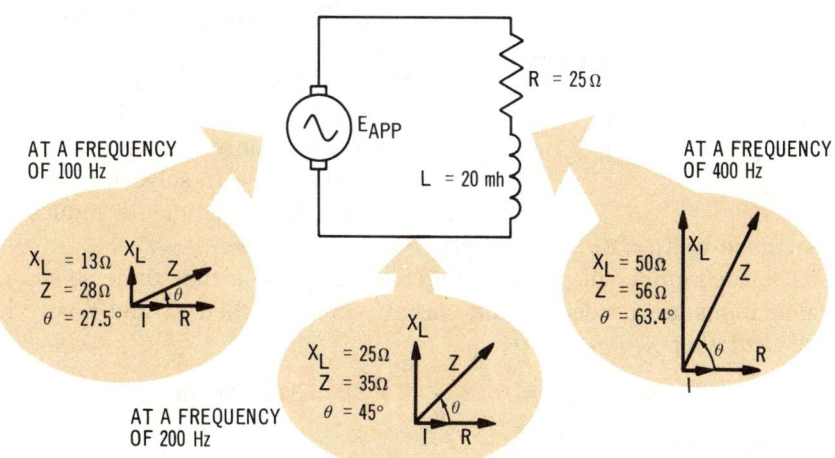

zero. However, since the value of X_L increases with frequency, so might the relative value of X_L and R. As a result, the same circuit might have different properties if the frequency is varied. A very low frequency can make the circuit almost purely resistive; while a very high frequency can cause it to be almost purely inductive. Of course, Z also changes with the relative value of X_L and R.

Very often a graph is made to show circuit impedance changes with frequency. Such a graph is called the frequency response curve of the circuit. The frequency response curve of the circuit above is shown

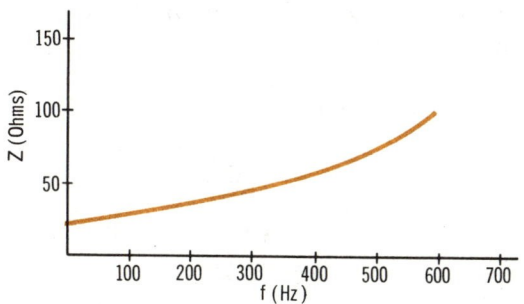

solved problems

Problem 3.   What is the current in this circuit?

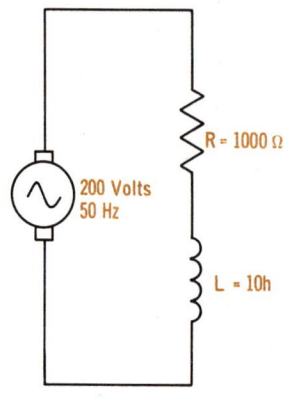

The first step in any problem is to inspect the information to help you determine whether the quantity asked for can be found directly, or if other calculations are required.

The quantity to be solved for is current. The basic equation for current is $I = E/Z$. The voltage is given, so you know that you have to calculate Z. The equation for impedance is $Z = \sqrt{R^2 + X_L^2}$, and from the circuit, you can see that only R is given. Therefore, you also have to solve for the reactance, X_L. The equation for inductive reactance is $X_L = 2\pi fL$, and both f and L are given.

The preliminary inspection, therefore, shows you that three separate calculations are needed to find current: first X_L, then Z, and finally I.

Calculating X_L:

$$X_L = 2\pi fL = 6.28 \times 50 \times 10 = 3140 \text{ ohms}$$

Calculating Z:

$$Z = \sqrt{R^2 + X_L^2} = \sqrt{(1000)^2 + (3140)^2} = 3295 \text{ ohms}$$

Calculating I:

$$I = E/Z = 200 \text{ volts}/3295 \text{ ohms} = 0.061 \text{ ampere}$$

Problem 4.   In the above circuit, what is the phase angle between the applied voltage and the current?

The phase angle can be calculated from either of the equations: $\tan \theta = E_L/E_R$ or $\tan \theta = X_L/R$. Since the voltages across the resistance and inductance are not known, but the resistance and inductance are, $\tan \theta = X_L/R$ is used.

$$\tan \theta = \frac{X_L}{R}$$

$$= \frac{3140}{1000}$$

$$= 3.14$$

$$\theta = 72.3°$$

solved problems (*cont.*)

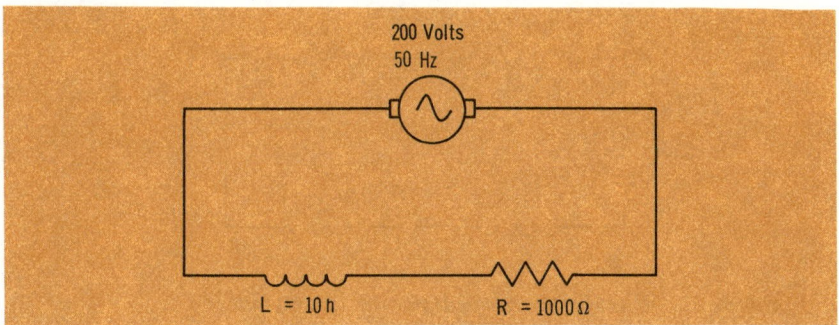

Problem 5. *In the circuit, what is the phase angle of the impedance?*
The phase angle of the impedance is always identical to the phase angle of the applied voltage and the current. This phase angle was calculated in Problem 4. The angle of the impedance is, therefore, also 72.3°.

Problem 6. *In the circuit, what are the voltage drops across R and L?*
Both of the voltage drops can be found by Ohm's Law, the drop across the resistance being $E_R = IR$, and the drop across the inductance, $E_L = IX_L$. The values of I, R and X_L are known, so the problem can be solved directly, with no preliminary calculations necessary.

Calculating E_R:

$$E_R = IR$$
$$= 0.061 \text{ ampere} \times 1000 \text{ ohms} = 61 \text{ volts}$$

Calculating E_L:

$$E_L = IX_L$$
$$= 0.061 \text{ ampere} \times 3140 \text{ ohms} = 192 \text{ volts}$$

Problem 7. *In the circuit, what would be the voltage drops across R and L if the source frequency were raised to 1000 Hz?*
The first step is to find the value that X_L would attain at 1000 Hz by:

$$X_L = 2\pi fL = 6.28 \times 1000 \times 10 = 62{,}800 \text{ ohms}$$

Normally, the next step would be to calculate the impedance and then use E/Z to find the circuit current. Then you would use Ohm's Law to find the voltage drops. But a much simpler method would be to use the 10-to-1 rule for X_L and R, and you can see immediately that the entire source voltage will be dropped across L.

solved problems (cont.)

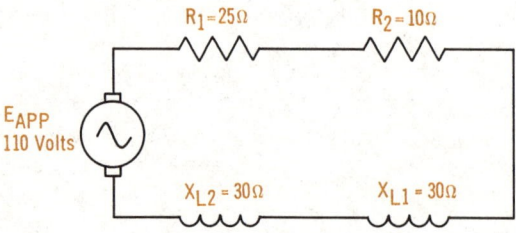

Problem 8. *What is the impedance of the circuit?*

This circuit is solved in the same manner as one containing only one resistance and one inductance, except that you first must calculate the total resistance and total inductive reactance. Since the resistances and inductances are in series, this is done by simple addition.

$$R_{TOT} = R_1 + R_2$$
$$= 25 + 10 = 35 \text{ ohms}$$
$$X_{L\ TOT} = X_{L1} + X_{L2}$$
$$= 30 + 30 = 60 \text{ ohms}$$

The impedance can then be found by the standard equation:

$$Z = \sqrt{R_{TOT}^2 + X_{L\ TOT}^2}$$
$$= \sqrt{(35)^2 + (60)^2} = \sqrt{4825}$$
$$= 69.5 \text{ ohms}$$

Problem 9. *How much power is consumed in the above circuit?*

Power consumed, or dissipated, is true power, so the true power must be found. The applied voltage is given and the impedance has just been found. The most appropriate equation to use, therefore, is $P_{TRUE} = (E^2_{APP}/Z) \cos \theta$. Before this equation can be applied, though, the phase angle, θ, must be found.

$$\tan \theta = \frac{X_{L\ TOT}}{R_{TOT}} = 60/35 = 1.71$$
$$\theta = 59.7°$$

The true power can now be calculated.

$$P_{TRUE} = \frac{E^2_{APP}}{Z} \cos \theta$$
$$= \frac{(110)^2}{69.5} \cos 59.7°$$
$$= 174.1 \times 0.505 = 87.9 \text{ watts}$$

summary

☐ The total opposition to current flow in an RL circuit is called the impedance. Impedance is measured in ohms, and is designated by Z. ☐ In a series RL circuit, the current through each component is the same. Because of this, the current is used as the phase reference for all other circuit quantities. ☐ Kirchhoff's voltage equation is valid for a series RL circuit, but the voltages across the resistor and inductor must be added vectorially.

☐ The applied voltage in a series RL circuit is found by $E_{APP} = \sqrt{E_R^2 + E_L^2}$. ☐ The value of the phase angle (θ) is found by: $\tan \theta = E_L/E_R$; or $\cos \theta = E_R/E_{APP}$. ☐ The magnitude of the impedance is the vector sum of the reactance and the resistance: $Z = \sqrt{R^2 + X_L^2}$. ☐ The phase angle between R and X_L is found by: $\tan \theta = X_L/R$; or $\cos \theta = R/Z$. ☐ Either X_L or R can be considered negligible if one is 10 times as large as the other.

☐ Ohm's Law for a-c circuits can be expressed as: $E = IZ$. Other forms of the equation are $I = E/Z$ and $Z = E/I$. ☐ The actual power dissipated in a circuit is called true power. It is equal to the apparent power times the power factor: $P_{TRUE} = E_{APP}I \cos \theta = I^2 Z \cos \theta$. ☐ The Q of a coil is the ratio of the reactance at a particular frequency to the resistance of the coil.

---

review questions

1. In a vector diagram of the voltages in a series RL circuit, what circuit quantity is used as the reference vector? Why?
2. What is the 10-to-1 rule for R and X_L in a series RL circuit?
3. What is the resistance of a coil having a Q of 65, when the inductive reactance is 325 ohms?
4. What is the resistance in a series RL circuit when the impedance is 130 ohms and the inductive reactance is 50 ohms?
5. What is the phase angle for the circuit in Question 4?
6. If an applied voltage of 100 volts causes 5 amperes of current in a series RL circuit, what is the circuit impedance?
7. The power dissipated in a circuit is 500 watts, the impedance is 10 ohms, and the phase angle is 60 degrees. What is the value of the current in the circuit?
8. What is the apparent power of a circuit which dissipates 500 watts, and has a power factor of 0.25?
9. Between what values can the power factor be found? Why?
10. What is the voltage across the coil in a series RL circuit when the applied voltage is 100 volts, and the voltage across the resistor is 80 volts?

parallel RL circuits

In a parallel RL circuit, the resistance and inductance are connected in parallel across a voltage source. Such a circuit thus has a *resistive branch* and an *inductive branch*. The circuit current *divides* before entering the branches, and a portion flows through the resistive branch, while the rest flows through the inductive branch. The currents in the branches are therefore *different*. The analysis of parallel RL circuits

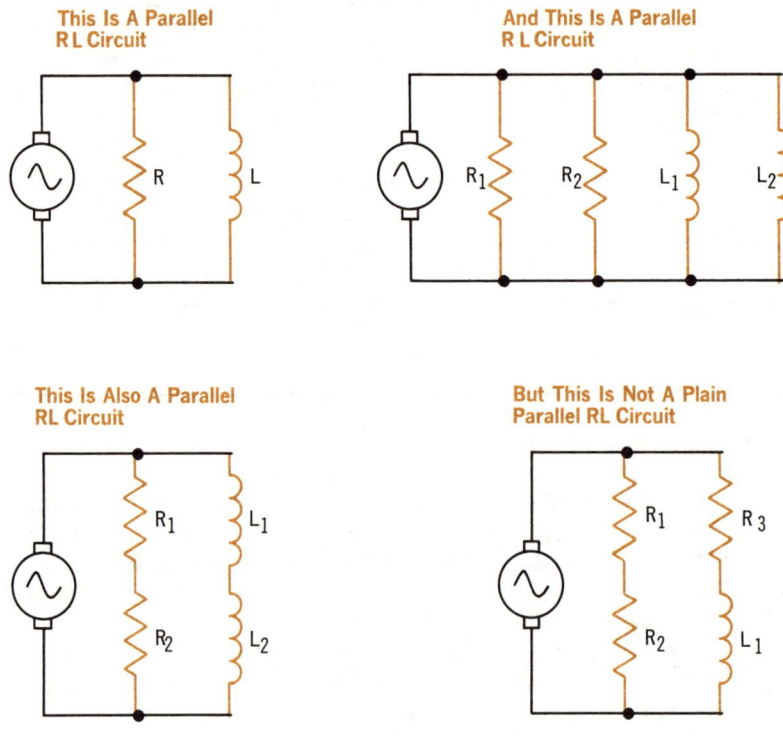

A parallel RL circuit has one or more resistive branches and one or more inductive branches. Each resistive branch is purely resistive, and each inductive branch is purely inductive. If any branch contains both resistance and inductance, the circuit is a series-parallel RL circuit. These will be covered later

and the methods used to solve them are different than the analysis and solution of series RL circuits. It is important, therefore, that you be able to recognize series and parallel RL circuits so that you can use the proper techniques and methods for solving them.

voltage

In a simple parallel RL circuit, there is one *resistive branch* and one *inductive branch*. Both branches are connected directly across the voltage source, and so have the full source voltage applied to them. Since the source voltage across both branches is the *same,* the voltages must be *in phase.* You can see, therefore, that if you know the applied voltage, you automatically know the voltage across each branch. Likewise, if you know the voltage across either branch, you also know the voltage across the other branch, as well as the applied voltage.

The voltage across every branch of a parallel RL circuit is the same as the applied voltage

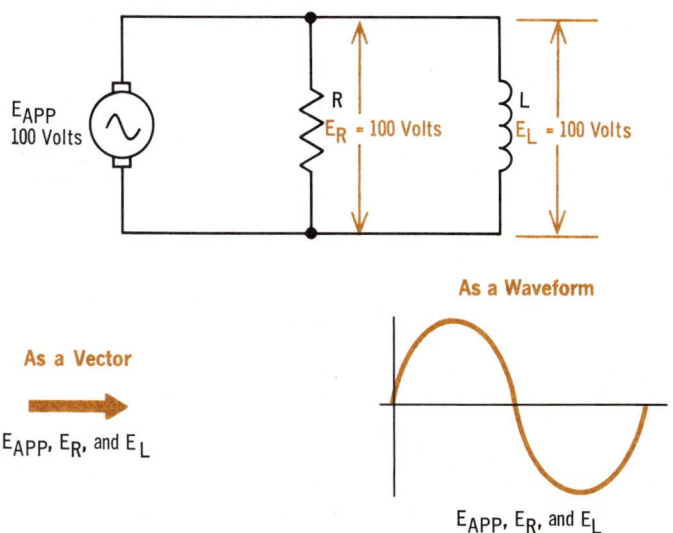

As a Vector

E_{APP}, E_R, and E_L

As a Waveform

E_{APP}, E_R, and E_L

If you know one voltage in a parallel RL circuit, you automatically know the other voltages, since they are all identical

You will remember that in *series* RL circuits, *current* was the common quantity, since it was the same in both the resistive and inductive parts of the circuit. In *parallel* RL circuits, *voltage* is the common quantity, inasmuch as the same voltage is applied across the resistive and inductive branches. The currents in the branches are not the same. The voltage, therefore, is used as the zero reference to compare the other angles.

branch current

The current in each branch of a parallel RL circuit is independent of the other branches, and can be calculated by Ohm's Law

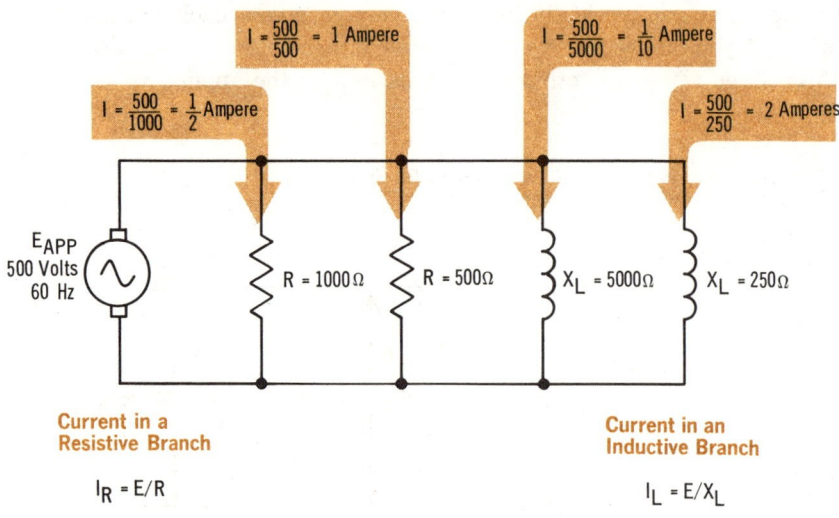

$I = \frac{500}{1000} = \frac{1}{2}$ Ampere

$I = \frac{500}{500} = 1$ Ampere

$I = \frac{500}{5000} = \frac{1}{10}$ Ampere

$I = \frac{500}{250} = 2$ Amperes

E_{APP}
500 Volts
60 Hz

R = 1000 Ω   R = 500 Ω X_L = 5000 Ω X_L = 250 Ω

**Current in a
Resistive Branch**

$I_R = E/R$

**Current in an
Inductive Branch**

$I_L = E/X_L$

As in all parallel circuits, the current in each branch of a parallel RL circuit is *independent* of the currents in the other branches. If one of the branches opened, it would have no effect on the other branch currents. The current in each branch depends only on the voltage across that branch and the opposition to current flow, in the form of either resistance or inductive reactance, contained in the branch. The branch voltages are the same, so it is the value of resistance or inductive reactance that determines the relative amount of current that flows in the branch. Each branch of a parallel RL circuit can be considered as a small separate series circuit. Ohm's Law can then be used to find the individual branch currents. In resistive branches, therefore, the current is equal to the branch voltage, which is the same as the applied voltage, divided by the resistance. In inductive branches, the current equals the branch voltage divided by the inductive reactance. Thus,

$$I_R = \frac{E}{R} \qquad I_L = \frac{E}{X_L}$$

Line current in a parallel RL circuit is equal to the VECTOR SUM
of the currents in the resistive and inductive branches

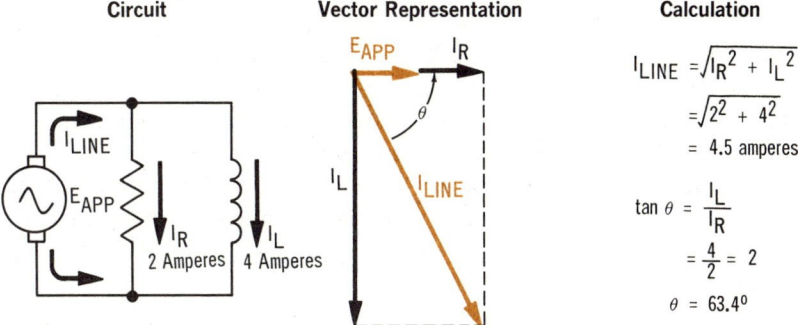

Circuit   Vector Representation   Calculation

$$I_{LINE} = \sqrt{I_R^2 + I_L^2}$$
$$= \sqrt{2^2 + 4^2}$$
$$= 4.5 \text{ amperes}$$

$$\tan \theta = \frac{I_L}{I_R}$$
$$= \frac{4}{2} = 2$$
$$\theta = 63.4^0$$

line current

In purely resistive parallel circuits, the total circuit current, or line
current as it is called, is simply the *arithmetic sum* of the individual
branch currents. In parallel RL circuits, though, there is a *phase difference* between the current in the resistive branch and the current in the
inductive branch. Because of the phase difference, the individual
branch currents must be added *vectorially* to find the line current. The
nature of the phase difference between the two currents is such that
the current in the *resistive* branch *leads* the current in the *inductive*
branch by 90 degrees. The reason for this is that the voltages across the
branches are in phase, and the current in the resistive branch is in
phase with the voltage, while the current in the inductive branch lags
the voltage by 90 degrees.

Because the two currents are 90 degrees out of phase, their vector
sum, which is the line current, can be calculated by the Pythagorean
Theorem, using the equation:

$$I_{LINE} = \sqrt{I_R^2 + I_L^2}$$

Line current is the vector sum of
the resistive and inductive
currents

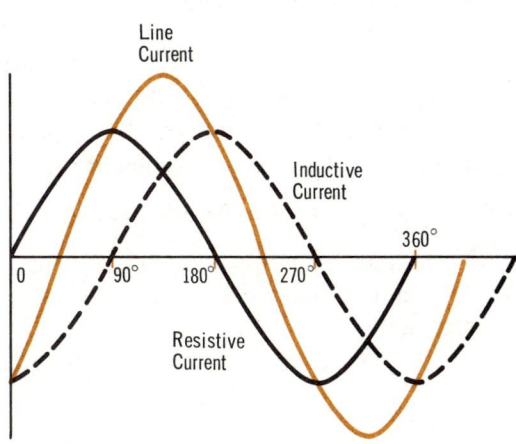

Resistive current leads the inductive current by 90°

line current (cont.)

The phase angle between the line current and the applied voltage is somewhere between 0 and 90 degrees, with the current lagging the voltage, as in all RL circuits. The actual angle depends on whether there is more inductive current or resistive current. If there is more inductive current, the phase of the line current will be closer to 90 degrees. It will be closer to 0 degrees if there is more resistive current.

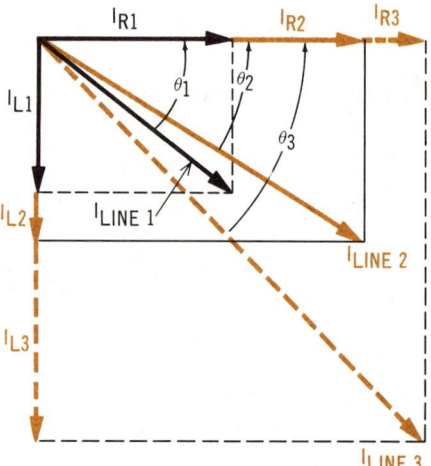

The line current in a parallel RL circuit will have a phase angle between 0 and 90°, lagging. The value depends on the relative values of the inductive and resistive currents in the branches

If either the resistive or inductive current is 10 times or more greater, the line current can be considered to have a phase angle of 0 or 90 degrees, as the case may be. From the vector diagram, you can see that the value of the phase angle can be calculated from the equation:

$$\tan \theta = I_L/I_R$$

Other very useful equations for calculating the phase angle can be derived by substituting the relationships $I_L = E/X_L$ and $I_R = E/R$ into the above equation. The equations derived in this way are

$$\tan \theta = R/X_L \quad \text{and} \quad \cos \theta = Z/R$$

If the impedance of a parallel RL circuit and the applied voltage are known, the line current can also be calculated using Ohm's Law for a-c circuits:

$$I_{LINE} = E/Z$$

current waveforms

Since the branch currents in a parallel RL circuit are out of phase, their vector sum rather than their straight arithmetic sum equals the line current. This is the same type of situation that exists for the voltage drops in a series RL circuit. By adding the currents vectorially, you are adding all of their instantaneous values, and then finding the average or effective value of the resulting current. This can be seen from the current waveforms shown. They are the waveforms for the circuit solved on the previous pages.

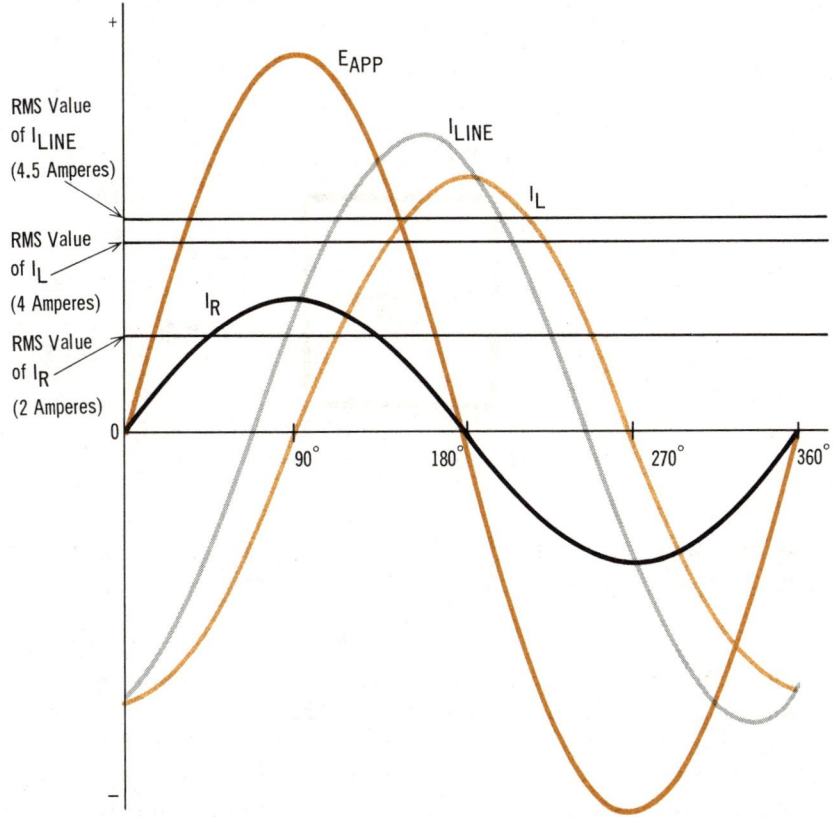

Every point on the line current waveform (I_{LINE}) is the algebraic sum of the instantaneous values of the I_R and I_L waveforms. The rms value of the line current waveform is equal to the vector sum of the rms values of the I_R and I_L waveforms

impedance

The impedance, Z, of a parallel RL circuit is the total opposition to current flow offered by the resistance of the resistive branch and the inductive reactance of the inductive branch. The impedance of a parallel RL circuit is calculated similarly to a parallel resistive circuit. However, since X_L and R are vector quantities, they must be added *vectorially*. As a result, the equation for the impedance of a parallel RL circuit is

$$Z = \frac{RX_L}{\sqrt{R^2 + X_L^2}}$$

where the quantity in the denominator is the vector addition of the resistance and the inductive reactance. If there are more than one resistive or inductive branches, R and X_L must equal the *total* resistance or reactance of these parallel branches.

In a Parallel RL Circuit:

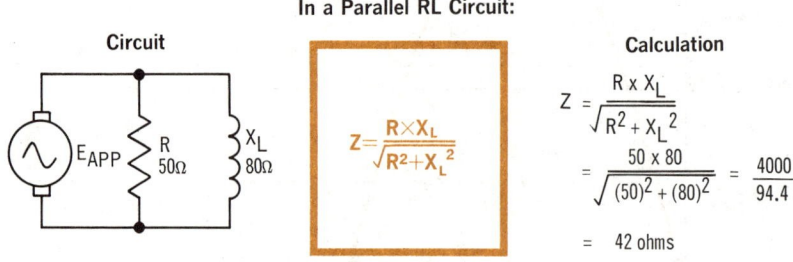

Circuit

$$Z = \frac{R \times X_L}{\sqrt{R^2 + X_L^2}}$$

Calculation

$$Z = \frac{R \times X_L}{\sqrt{R^2 + X_L^2}}$$

$$= \frac{50 \times 80}{\sqrt{(50)^2 + (80)^2}} = \frac{4000}{94.4}$$

$$= 42 \text{ ohms}$$

The impedance of a parallel RL circuit is always less than the resistance or inductive reactance of any of the branches

If the circuit line current and the applied voltage are known, the impedance can also be calculated by the equation:

$$Z = E_{APP}/I_{LINE}$$

The impedance of a parallel RL circuit is always *less* than the resistance or reactance of any one branch. The branch of a parallel RL circuit that offers the *most opposition* to current flow has the *lesser effect* on the phase angle of the current. For example, if X_L is larger than R, the resistive branch current is greater than the inductive branch current, so the line current is also more resistive (closer to 0°). This is the opposite of a series RL circuit. When either X_L or R is 10 or more times greater than the other, for practical purposes, a parallel RL circuit can be considered as a simple series circuit consisting of only X_L or R, whichever is *smaller*.

power

In parallel RL circuits, the relationships between the applied voltage, the circuit current, and the circuit power are similar to those of series RL circuits, previously described. Because of the phase difference between the branch currents, the line current and the applied voltage are out of phase. As a result, the value of power obtained by multiplying the applied voltage by the line current is only the *apparent power*. Part of this apparent power is *returned* to the source by the inductive branch. So, to find the power actually dissipated in the circuit, the *true power*, the apparent power has to be multiplied by the cosine of the phase angle (θ) between the applied voltage and line current. The value of cosine θ is the circuit *power factor*.

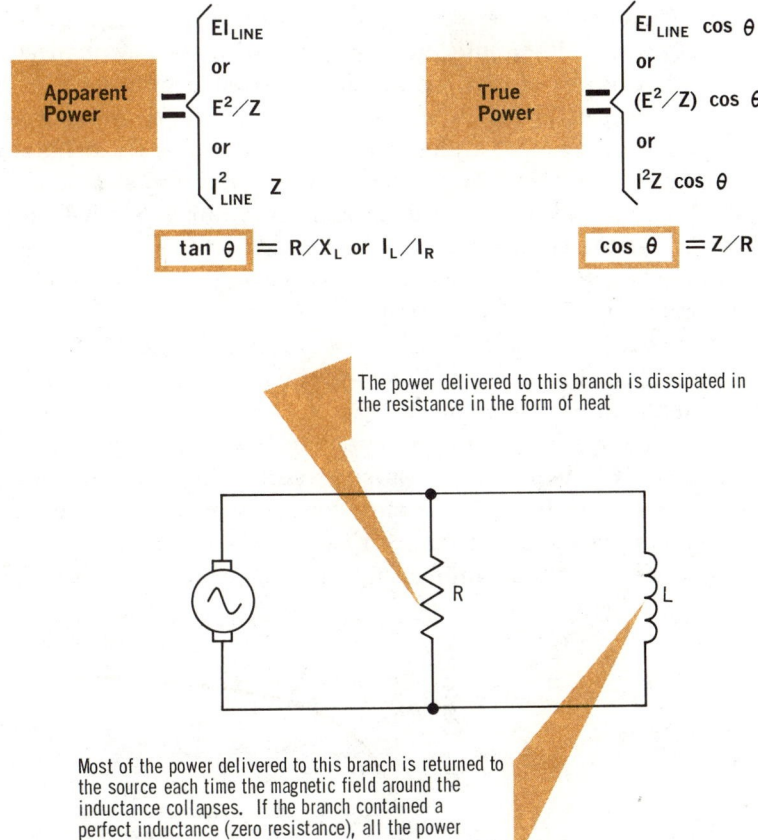

$$\text{Apparent Power} = \begin{cases} EI_{LINE} \\ \text{or} \\ E^2/Z \\ \text{or} \\ I^2_{LINE}\,Z \end{cases}$$

$$\boxed{\tan\theta} = R/X_L \text{ or } I_L/I_R$$

$$\text{True Power} = \begin{cases} EI_{LINE}\ \cos\theta \\ \text{or} \\ (E^2/Z)\ \cos\theta \\ \text{or} \\ I^2 Z\ \cos\theta \end{cases}$$

$$\boxed{\cos\theta} = Z/R$$

The power delivered to this branch is dissipated in the resistance in the form of heat

Most of the power delivered to this branch is returned to the source each time the magnetic field around the inductance collapses. If the branch contained a perfect inductance (zero resistance), all the power would be returned

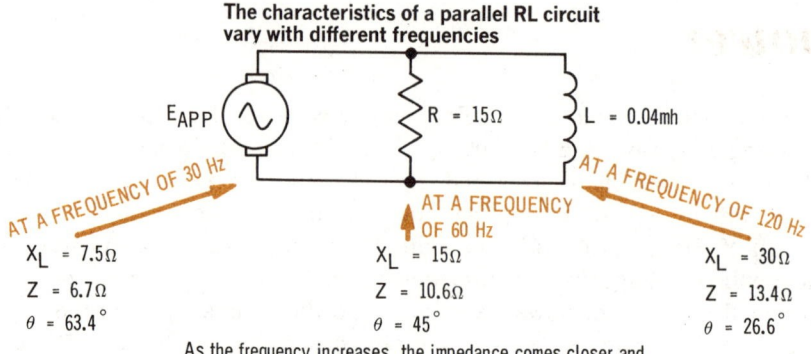

The characteristics of a parallel RL circuit vary with different frequencies

E_{APP} R = 15Ω   L = 0.04mh

AT A FREQUENCY OF 30 Hz
X_L = 7.5Ω
Z = 6.7Ω
θ = 63.4°

AT A FREQUENCY OF 60 Hz
X_L = 15Ω
Z = 10.6Ω
θ = 45°

AT A FREQUENCY OF 120 Hz
X_L = 30Ω
Z = 13.4Ω
θ = 26.6°

As the frequency increases, the impedance comes closer and closer to the value of the resistance

effect of frequency

You will recall that the frequency of the applied voltage has a significant effect on the characteristics of a series RL circuit. The same thing is true of parallel RL circuits, but the *effects* of frequency changes are *different*. In a series circuit, an increase in frequency caused an increase in the values of X_L and Z, and made the circuit more inductive. Increasing the frequency of a parallel RL circuit also causes an increase in the values of X_L and Z. However, whereas in a *series* circuit a larger X_L makes the circuit *more inductive*, increasing X_L in a *parallel* circuit makes the circuit *more resistive*. The reason for this is that the larger X_L is, the smaller is the inductive branch current, and so the larger is the relative value of the resistive branch current.

If the frequency is decreased, the opposite is true. X_L becomes smaller, causing an increase in inductive branch current, thereby making the circuit more inductive. At very *low frequencies*, therefore, a parallel RL circuit will be almost purely *inductive;* while at very *high frequencies*, it will be almost purely *resistive*. If the frequency is such that either X_L or R are 10 or more times larger than the other, the branch containing the larger one can be neglected, and the circuit treated as a series circuit containing only the smaller of the two.

The frequency response curve of the above circuit

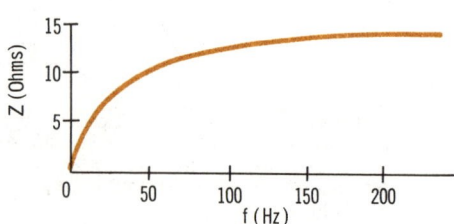

solved problems

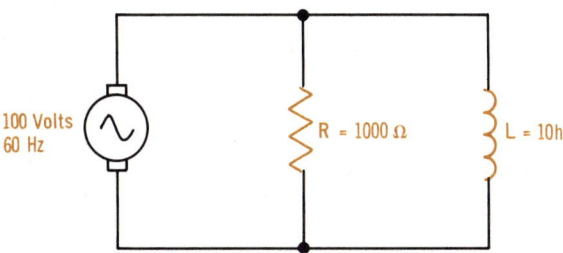

Problem 10. *Calculate the line current in this circuit in two ways: first using Ohm's Law, and then using the branch currents.*

To calculate the line current using Ohm's Law, the equation $I_{LINE} = E/Z$ is used. Before this can be done, though, the impedance must be found. The equation for the impedance is $Z = RX_L/\sqrt{R^2 + X_L^2}$. R is given in the problem, but X_L is not. So X_L must be found first. This means that to solve the problem, X_L must first be calculated, then Z, and finally the line current.

Calculating X_L:
$$X_L = 2\pi fL = 6.28 \times 60 \times 10 = 3768 \text{ ohms}$$
Calculating Z:
$$Z = \frac{RX_L}{\sqrt{R^2 + X_L^2}} = \frac{1000 \times 3768}{\sqrt{(1000)^2 + (3768)^2}} = 966 \text{ ohms}$$
Calculating I_{LINE}:
$$I_{LINE} = E/Z = 100 \text{ volts}/966 \text{ ohms} = 0.104 \text{ ampere}$$

To calculate the line current using the branch currents, you must first find the two branch currents. The vector sum of these two currents will then be the line current.

Calculating Branch Currents:
$$I_R = E/R = 100 \text{ volts}/1000 \text{ ohms} = 0.1 \text{ ampere}$$
$$I_L = E/X_L = 100 \text{ volts}/3768 \text{ ohms} = 0.0265 \text{ ampere}$$
Calculating I_{LINE}:
$$I_{LINE} = \sqrt{I_R^2 + I_L^2} = \sqrt{(0.1)^2 + (0.0265)^2} = 0.104 \text{ ampere}$$

Both methods, therefore, result in the same value for the line current, which they naturally should. You will find that many, if not most, electrical problems can be solved more than one way. When solving such problems, it is good practice, if time permits, to solve the problems in two ways. The two answers should be the same, and will serve as a check on the accuracy of your calculations.

solved problems (cont.)

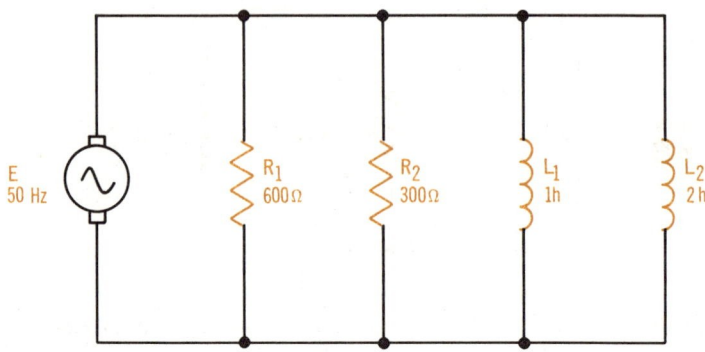

Problem 11.   *In this circuit, what is the phase angle between the applied voltage and the line current?*

The phase angle can be calculated from the equation $\tan \theta = I_L/I_R$, $\tan \theta = R/X_L$, or $\cos \theta = Z/R$. Since the applied voltage is not known, the equation $\tan \theta = I_L/I_R$ cannot be used in this problem. Of the other two equations, $\tan \theta = R/X_L$ is the easiest to use in this particular case. However, before it can be used to solve for the phase angle, the total values of R and X_L must be determined.

Calculating R_{TOT}: The two resistances are in parallel, so the product/sum method is used to find their total.

$$R_{TOT} = \frac{R_1 R_2}{R_1 + R_2} = \frac{600 \times 300}{600 + 300} = 200 \text{ ohms}$$

Calculating $X_{L\ TOT}$: The individual values of X_{L1} and X_{L2} must first be found.

$$X_{L1} = 2\pi fL = 6.28 \times 50 \times 1 = 314 \text{ ohms}$$
$$X_{L2} = 2\pi fL = 6.28 \times 50 \times 2 = 628 \text{ ohms}$$

The two reactances are in parallel, so the product/sum method can also be used to find their total.

$$X_{L\ TOT} = \frac{X_{L1} X_{L2}}{X_{L1} + X_{L2}} = \frac{314 \times 628}{314 + 628} = 209 \text{ ohms}$$

Calculating θ:

$$\tan \theta = \frac{R_{TOT}}{X_{L\ TOT}} = \frac{200}{209} = 0.957$$

θ can now be found from a table of trigonometric functions to be:

$$\theta = 43.7°$$

comparison of series and parallel RL circuits

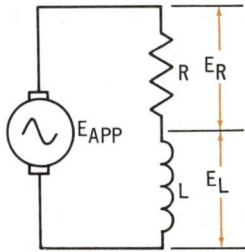

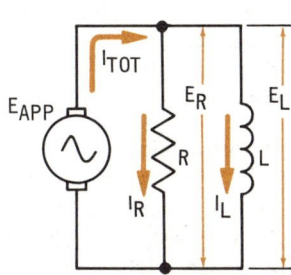

	Series RL Circuit	Parallel RL Circuit
Current	Current is same everywhere in circuit. Current through R and L are, therefore, in phase.	Current divides between resistive and inductive branches. $$I_{TOT} = \sqrt{I_R^2 + I_L^2}$$ $$I_R = E_{APP}/R \qquad I_L = E_{APP}/X_L$$ Current through R leads current through L by 90°.
Voltage	Vector sum of voltage drops across R and L equals applied voltage. $$E_{APP} = \sqrt{E_R^2 + E_L^2}$$ Voltage across L leads voltage across R· by 90°.	Voltage across each branch is same as applied voltage. Voltages across R and L are, therefore, in phase. $$E_R = E_L = E_{APP}$$
Impedance	It is the vector sum of resistance and inductive reactance. $$Z = \sqrt{R^2 + X_L^2}$$	It is calculated in same way as parallel resistances, except that vector addition is used. $$Z = RX_L / \sqrt{R^2 + X_L^2}$$
Phase Angle (θ)	It is the angle between circuit current and applied voltage. $$\tan \theta = E_L/E_R = X_L/R$$ $$\cos \theta = R/Z$$	It is the angle between applied voltage and line current. $$\tan \theta = I_L/I_R = R/X_L$$ $$\cos \theta = Z/R$$
Power	Power delivered by source is apparent power. Power actually consumed in circuit is true power. Power factor determines what portion of apparent power is true power. $$P_{APPARENT} = E_{APP}I \qquad P_{TRUE} = E_{APP}I \cos \theta \qquad P.\,F. = \cos \theta$$	
Effect of Increasing Frequency	X_L increases, which in turn causes circuit current to decrease. Phase angle increases, which means that circuit is more inductive.	X_L increases, inductive branch current decreases, and so line current also decreases. Phase angle decreases, which means that circuit is more resistive.
Effect of Increasing Resistance	Phase angle decreases, which means that circuit is more resistive.	Phase angle increases, which means that circuit is more inductive.
Effect of Increasing Inductance	Phase angle increases, which means that circuit is more inductive.	Phase angle decreases, which means that circuit is more resistive.

summary

☐ In a simple parallel RL circuit, there is one resistive branch and one inductive branch. ☐ The circuit current in a parallel RL circuit divides before entering the branches, and a portion flows through the resistive branch, while the rest flows through the inductive branch. ☐ The current through each branch is independent of the current in the other. ☐ Both branches are connected directly across the voltage source. ☐ Voltage is used as the zero reference vector since it is the common circuit quantity in a parallel RL circuit.

☐ The current in the resistive branch of a parallel RL circuit is found by $I_R = E/R$. ☐ The current in the inductive branch is found by $I_L = E/X_L$. ☐ The line current is found by adding the branch currents vectorially: $I_{LINE} = \sqrt{I_R{}^2 + I_L{}^2}$. ☐ The line current can also be found by: $I_{LINE} = E/Z$. ☐ The phase angle of a parallel RL circuit is found by: $\tan \theta = I_L/I_R$; or $\tan \theta = R/X_L$; or $\cos \theta = Z/R$. ☐ The impedance of a parallel RL circuit is found by adding the parallel oppositions vectorially. The equation for impedance is $Z = RX_L/(\sqrt{R^2 + X_L{}^2})$. ☐ The impedance can also be found by: $Z = E_{APP}/I_{LINE}$.

☐ The power equations for parallel RL circuits are identical to those for series RL circuits. ☐ Since inductive reactance increases with an increase in frequency, the parallel RL circuit becomes more resistive as the frequency is increased.

---

review questions

1. What is meant by *line current* in a parallel RL circuit?

For Questions 2 to 10, consider a parallel RL circuit with an applied voltage of 100 volts, a resistor with a resistance of 10 ohms, and an inductor with an inductive reactance of 20 ohms.

2. What is the current through the resistor; inductor?
3. What is the voltage across the resistor; inductor?
4. What is the impedance of the circuit?
5. What is the phase angle of the circuit?
6. What is the apparent power? The true power? The power factor?
7. What is the value of the line current?
8. By how much does the frequency have to be multiplied for the parallel circuit to become, effectively, a resistive circuit?
9. Answer Questions 3 to 5 for the condition where the frequency is doubled.
10. Answer Questions 5 to 7 for the condition where the frequency is tripled.

RC circuits

A circuit that contains *resistance* (R) and *capacitance* (C) is called
an *RC circuit*. The methods you use to solve RC circuits depend on
whether the resistance and capacitance are in *series* or in *parallel*. This
is similar to what you have just learned for RL circuits. Actually, the
conditions that exist in RC circuits and the methods used for solving
them are quite similar to those for RL circuits. The principal difference
is one of *phase relationship*, since as you will remember from Volume
3, the phase relationship between the current and voltage in a capacitive circuit is different from that in an inductive circuit.

RC Circuits

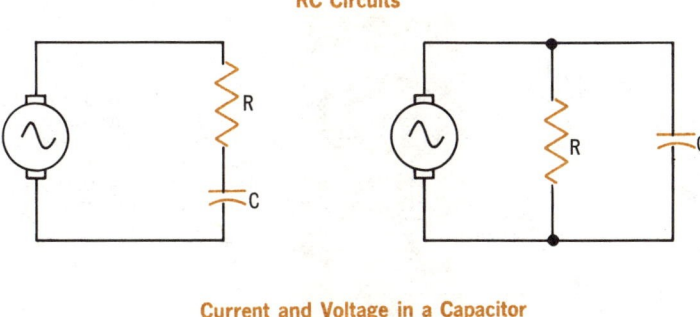

Current and Voltage in a Capacitor

AS WAVEFORMS   AS VECTORS

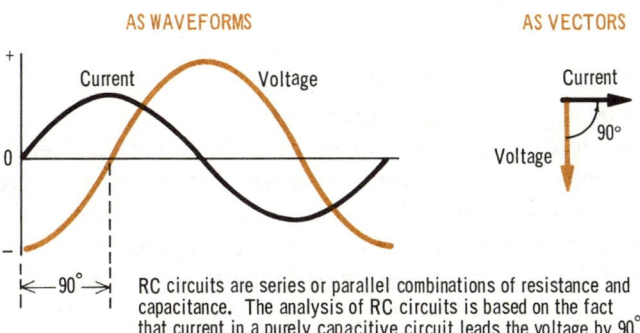

RC circuits are series or parallel combinations of resistance and
capacitance. The analysis of RC circuits is based on the fact
that current in a purely capacitive circuit leads the voltage by $90°$

An RC circuit is usually considered as one that contains resistors
and capacitors. However, any practical circuit has some resistance in
the circuit wiring, as well as some capacitance between wires or between the wiring and surrounding metal parts. RC circuits, therefore,
exist even when no resistors or capacitors are used. However, in these
cases, the values of resistance and capacitance are usually very small.
In this volume, therefore, the resistance and capacitance of the circuit
wiring will be considered negligible.

series RC circuits

In a series RC circuit, one or more resistances are connected in series with one or more capacitances, so that *total circuit current* flows through *each* individual component. For the discussion on the following pages of the voltage, impedance, and current in series RC circuits, the case of a *single resistance* in series with a *single capacitance* will be considered, unless otherwise stated. When there is more than one resistance or capacitance, the analysis of the circuit is the same, except that the single resistance or capacitance then becomes the total resistance or capacitance.

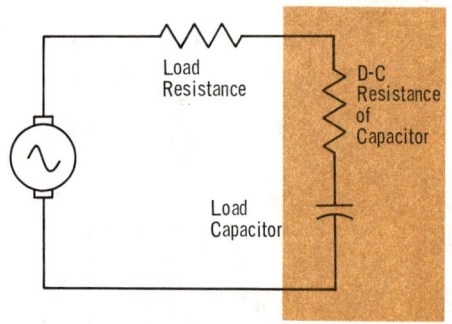

Every capacitor has leakage, which is caused by the d-c resistance of the capacitor. Normally, a capacitor's d-c resistance is very high, so, essentially, the capacitor acts as an ideal capacitor, passing ac and blocking dc

You will recall from Volume 3 that every capacitor has some *leakage,* made up of a small amount of current that flows through the dielectric. Effectively, the leakage current destroys the 90-degree relationship between the voltage across the capacitor and the current through it, so that the current actually leads the voltage by some phase angle less than 90 degrees. For most capacitors, though, the leakage current is so small that for all practical purposes the phase angle can be considered as being 90 degrees. In this volume, therefore, we will consider capacitors as having no leakage, and the capacitor current as leading the voltage by 90 degrees.

In the description of series RC circuits on the following pages, the circuit current will be used as the phase reference for all other circuit quantities, just as it was for series RL circuits. Again, the choice of current is for convenience, since it is the same in all parts of the circuit. With current used as the reference, the vectors of all quantities that are in phase with the current will have the same direction as the current vector: 0 degrees.

voltage

When current flows in a series RC circuit, the voltage drop across the resistance (E_R) is in phase with the current, while the voltage drop across the capacitance (E_C) lags the current by 90 degrees. Since the current through both is the same, E_R leads E_C by 90 degrees. The amplitudes of the two voltage drops can be calculated by:

$$E_R = IR$$
$$E_C = IX_C$$

Like series RL circuits, the *vector sum* of the voltage drops *equals* the applied voltage. As an equation:

$$E_{APP} = \sqrt{E_R^2 + E_C^2}$$

The relationship between the applied voltage and the voltage drops in a series RC circuit is such that the applied voltage equals the **VECTOR SUM** of the voltage drops

Circuit   **Vector Representation**   **Calculation**

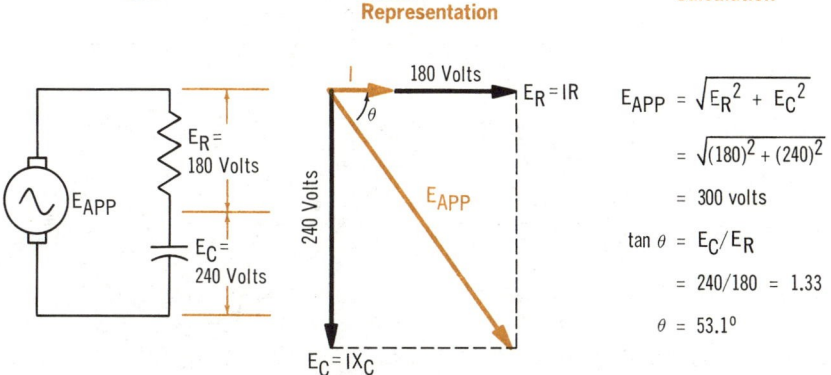

If one of the voltage drops changed as a result of a change in either R or X_C, the angle of the applied voltage vector would also appear to change. Actually, it is the current that changes phase; this is the same as was pointed out for series RL circuits. To avoid confusion, always consider the phase angle, θ, as the angle between I and E_{APP}

Graphically, the applied voltage is the hypotenuse of a right triangle whose two sides are the voltage drops E_R and E_C. The angle (θ) of this vector triangle between the applied voltage and E_R is the same as the phase angle between the applied voltage and the current. The reason for this is that E_R and I are in phase. The value of θ can be calculated from:

$$\tan \theta = E_C/E_R \quad \text{or} \quad \cos \theta = E_R/E_{APP}$$

voltage waveforms

The waveforms of the voltages in a series RC circuit are similar to those you have seen for a series RL circuit. They show how the applied voltage waveform is the sum of all the instantaneous points of the two voltage drop waveforms. They also show that the average and effective values of the applied voltage waveform equal the vector sum of the average and effective values of the voltage drop waveforms. This is illustrated for the circuit solved on the previous page.

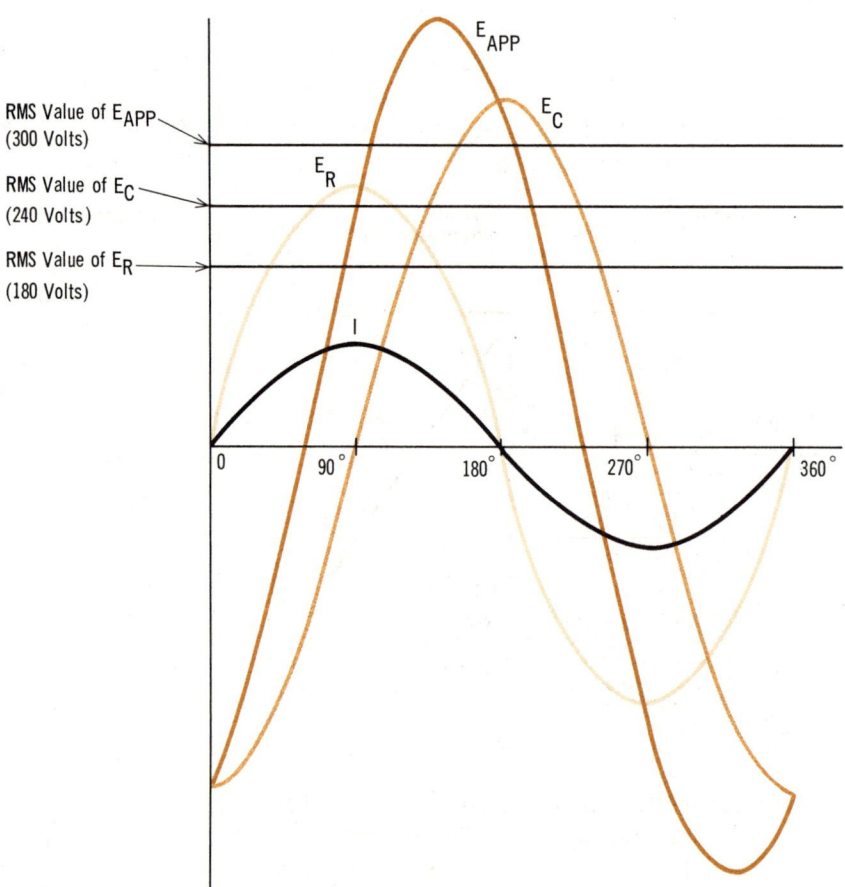

Every point on the applied voltage waveform (E_{APP}) is the algebraic sum of the instantaneous values of the E_R and E_C waveforms

impedance

The impedance of a series RC circuit is the *total* opposition to current flow offered by the circuit resistance and capacitive reactance. It is calculated in the same way as the impedance of a series RL circuit, except that capacitive reactance is used in place of inductive reactance. The equation for the impedance of a series RC circuit is, therefore,

$$Z = \sqrt{R^2 + X_C^2}$$

The vector addition takes into account the 90-degree phase difference between the voltage drop across the resistance and that across the capacitance.

In a series RC circuit, the impedance is the **VECTOR SUM** of the resistance and capacitive reactance

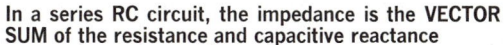

Circuit	Vector Representation	Calculation

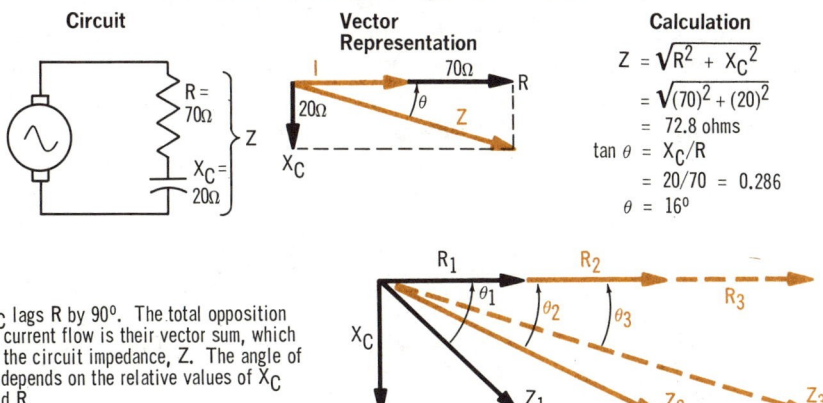

$$Z = \sqrt{R^2 + X_C^2}$$
$$= \sqrt{(70)^2 + (20)^2}$$
$$= 72.8 \text{ ohms}$$
$$\tan \theta = X_C/R$$
$$= 20/70 = 0.286$$
$$\theta = 16^0$$

X_C lags R by 90°. The total opposition to current flow is their vector sum, which is the circuit impedance, Z. The angle of Z depends on the relative values of X_C and R

As R becomes larger relative to X_C, the angle of Z becomes smaller

Since R and X_C are 90 degrees apart, with R leading, the phase angle of Z is somewhere between 0 and 90 degrees. The exact angle depends on the *relative* values of R and X_C. If R is greater, Z will be closer to 0 degrees, and if X_C is greater, Z will be closer to 90 degrees. The value of the angle can be calculated from either of the equations:

$$\tan \theta = X_C/R \qquad \text{or} \qquad \cos \theta = R/Z$$

The phase angle of Z is the same as the phase angle between the applied voltage and the current. So if you know one, you automatically know the other.

Just as it does in series RL circuits, the 10-to-1 rule applies to the impedance of series RC circuits. This means that if either X_C or R is 10 or more times greater than the other, the circuit will operate essentially as if only the larger of the two quantities was present.

current

The amplitude of the current in a series RC circuit can be calculated from Ohm's Law if you know the applied voltage and the impedance. Thus,

$$I = E_{APP}/Z \quad \text{where} \quad Z = \sqrt{R^2 + X_C^2}.$$

Since the current is the *same throughout* the circuit, it is used as the phase reference. So the angle between it and the impedance determines whether the current is more resistive or more capacitive. You will recall that this angle is somewhere between 0 and 90 degrees, with its exact value depending on the *relative* values of the resistance and the capacitive reactance. The *larger* X_C is compared to R, the closer the angle is to *90 degrees,* and the more *capacitive* is the current; similarly, the *smaller* X_C is compared to R, the closer the angle is to *0 degrees,* and the more *resistive* is the current. If X_C is 10 or more times larger than R, the current can be considered as purely capacitive, and thus to lead the applied voltage by 90 degrees; while if R is 10 or more times greater than X_C, you can consider the current as purely resistive, and thus being in phase with the applied voltage.

You can calculate the circuit power using the same equations you learned for RL circuits.

In a series RC circuit, the current is the same at every point, and leads the applied voltage by an angle between 0 and 90°

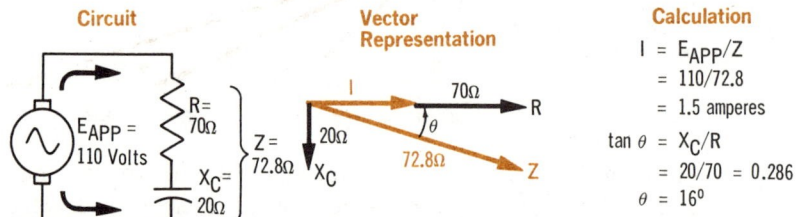

Circuit   Vector Representation   Calculation

$$I = E_{APP}/Z$$
$$= 110/72.8$$
$$= 1.5 \text{ amperes}$$
$$\tan \theta = X_C/R$$
$$= 20/70 = 0.286$$
$$\theta = 16°$$

The angle calculated here is the phase angle of the current. You will notice that it is the same as the angle of the impedance found for the same circuit on the previous page. The reason for this is that the value of X_C and R determine the angle of the impedance, which in turn determines if the current is capacitive or resistive

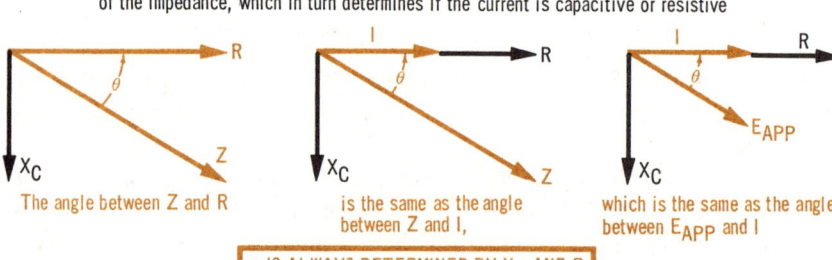

The angle between Z and R   is the same as the angle between Z and I,   which is the same as the angle between E_{APP} and I

θ IS ALWAYS DETERMINED BY X_C AND R

The characteristics of a series RC circuit vary with different frequencies

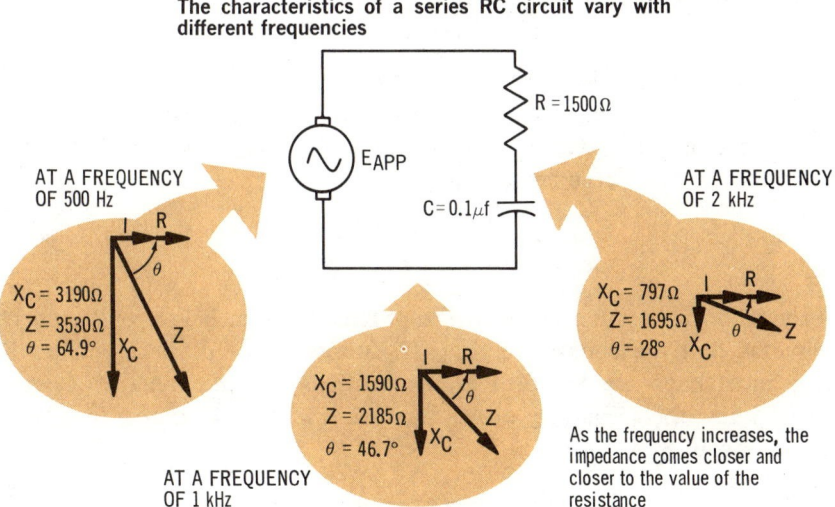

effect of frequency
============

Since the value of X_C in a series RC circuit *changes* with *frequency,* all the properties of the circuit that are affected by X_C also change with frequency. These frequency-dependent properties include the *impedance,* the amplitude and phase angle of the *current,* and the circuit *power factor.* Since the value of X_C is inversely proportional to frequency, an increase in frequency causes a decrease in X_C, while a decrease in frequency causes X_C to increase. As a result, when the frequency goes up, the impedance decreases, the circuit current increases and becomes more resistive, and the power factor goes closer to 1. Conversely, when frequency goes down, the impedance increases, the circuit current decreases and becomes more capacitive, and the power factor goes closer to 0.

This is the frequency response curve of the circuit. At very low frequencies, Z is practically infinite; and the higher the frequency becomes, the lower is the impedance, approaching but never actually reaching the value of R

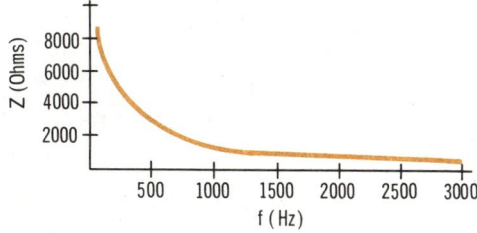

solved problems

Problem 12.   *What is the applied voltage in the circuit below?*
The applied voltage is the unknown quantity that has to be found.
Therefore, the first step is to consider the equations for calculating the
applied voltage. These equations are

$$E_{APP} = IZ$$

and $$E_{APP} = \sqrt{E_R{}^2 + E_C{}^2}$$

From the information given in the circuit diagram below, it should be
obvious that the equation $E_{APP} = IZ$ cannot be used, since to find Z,
you must know the value of X_C. In turn, to find the value of X_C, you
must know the frequency of the applied voltage, and that information
is not given.

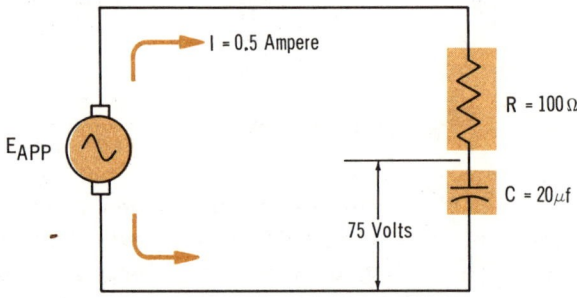

The applied voltage can only be found, therefore, by the equation
$E_{APP} = \sqrt{E_R{}^2 + E_C{}^2}$. Before this can be done, however, the voltage drop
across the resistance has to be calculated.
Calculating E_R:

$$E_R = IR = 0.5 \text{ ampere} \times 100 \text{ ohms} = 50 \text{ volts}$$

Calculating E_{APP}:

$$E_{APP} = \sqrt{E_R{}^2 + E_C{}^2} = \sqrt{(50)^2 + (75)^2} = \sqrt{8125} = 90 \text{ volts}$$

Problem 13.   *What is the impedance of the circuit?*
Before the applied voltage was calculated, the impedance could not
be determined. The reasons for this were that in the equation for
impedance, $Z = E_{APP}/I$, the value of E_{APP} was not known, while in the
equation $Z = \sqrt{R^2 + X_C{}^2}$, the value of X_C was not known. Now that the
applied voltage, E_{APP}, has been calculated, however, the impedance can
easily be found:

$$Z = E_{APP}/I = 90 \text{ volts}/0.5 \text{ ampere} = 180 \text{ ohms}$$

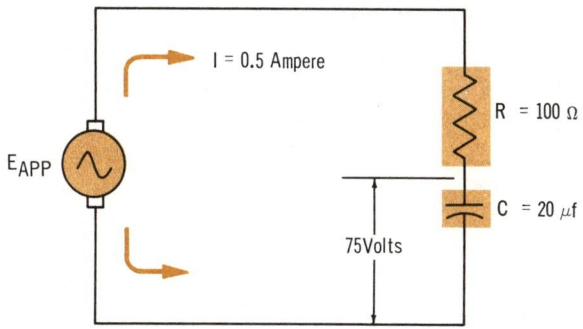

solved problems (cont.)

Problem 14. *What is the phase angle in the circuit?*

As you know, the three commonly used equations for calculating the phase angle, θ, are

$$\cos \theta = R/Z \qquad \tan \theta = X_C/R \qquad \tan \theta = E_C/E_R$$

The second equation cannot be used in this problem since the value of X_C is not known. Enough information is known, however, to use either of the other two equations.

$$\cos \theta = R/Z = 100/180 = 0.555$$

$$\theta = 56.3°$$

$$\tan \theta = E_C/E_R = 75/50 = 1.500$$

$$\theta = 56.3°$$

Problem 15. *What is the frequency of the power source in the circuit?*

The only way the frequency can be found is if the value of X_C is known. This can be calculated from the equation for the voltage drop across C, since both I and E_C are known. Thus,

$$E_C = IX_C$$

so, $X_C = E_C/I = 75/0.500 = 150$ ohms

With the value of X_C now known, the frequency can be found by the equation for calculating capacitive reactance:

$$X_C = \frac{1}{2\pi fC}$$

so, $f = \dfrac{1}{2\pi CX_C} = \dfrac{1}{6.28 \times 0.00002 \times 150} = 53$ Hz

solved problems (cont.)

Problem 16. *What capacitance should the capacitor have if the lamp is to dissipate its rated wattage?*

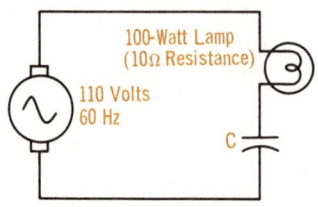

100-Watt Lamp
(10Ω Resistance)

110 Volts
60 Hz

C

To solve this problem, you will find it convenient to change the basic equations to new forms.

Since you know the lamp's wattage rating and resistance, you first have to find the current that will cause the lamp to dissipate its rated wattage. You start with the equation that relates power (P), current (I), and resistance (R), or, $P = I^2R$. This equation can be changed to a form that will allow you to solve for the current as follows:

$$P = I^2R \qquad \text{or} \qquad I^2 = P/R, \text{ which becomes } I = \sqrt{P/R}$$

Now you can calculate the current required by the lamp:

$$I = \sqrt{P/R} = \sqrt{100/10} = \sqrt{10} = 3.16 \text{ amperes}$$

It is a series circuit, so 3.16 amperes must flow through the entire circuit. With the applied voltage of 110 volts, the circuit impedance that will allow 3.16 amperes to flow can be found by Ohm's Law in the form:

$$Z = E_{APP}/I = 110 \text{ volts}/3.16 \text{ amperes} = 34.8 \text{ ohms}$$

The circuit impedance, which is the vector sum of the resistance of the lamp and the reactance of the capacitor, must therefore be 34.8 ohms. You know the lamp resistance (R) and the circuit impedance (Z). To find the capacitive reactance (X_C), the equation for impedance, $Z = \sqrt{R^2 + X_C^2}$, can be changed as follows:

$$Z = \sqrt{R^2 + X_C^2} \text{ which becomes } Z^2 = R^2 + X_C^2$$

Transposing, $X_C^2 = Z^2 - R^2$ which becomes $X_C = \sqrt{Z^2 - R^2}$

Therefore, $X_C = \sqrt{(34.8)^2 - (10)^2} = \sqrt{1111} = 33.3$ ohms

The capacitor must thus have a capacitive reactance of 33.3 ohms. To find the value of capacitance that has a reactance of 33.3 ohms with a 60-Hz applied voltage, you start with the basic equation for capacitive reactance and change it as follows:

$$X_C = \frac{1}{2\pi fC} \qquad \text{or} \qquad 2\pi fCX_C = 1$$

Solving for C:
$$C = \frac{1}{2\pi fX_C}$$

Therefore,

$$C = \frac{1}{6.28 \times 60 \times 33.3} = \frac{1}{12,547} = 79.7 \text{ microfarads}$$

summary

☐ In a series RC circuit, since the circuit current flows through both the resistance and the capacitance, it is used as the phase reference. ☐ The voltage drop across the resistance is in phase with the current, while the voltage drop across the capacitance lags the current by 90 degrees. ☐ The voltage drop across the resistance can be found by $E_R = IR$. ☐ The voltage drop across the capacitance can be found by $E_C = IX_C$.

☐ The vector sum of the voltage drops around a series RC circuit equals the applied voltage: $E_{APP} = \sqrt{E_R^2 + E_C^2}$. ☐ The phase angle can be found by: $\tan \theta = E_C/E_R$; or $\tan \theta = X_C/R$; or $\cos \theta = E_R/E_{APP}$; or $\cos \theta = R/Z$. ☐ The equation for impedance in a series RC circuit is $Z = \sqrt{R^2 + X_C^2}$. ☐ The current is the same through the circuit, and is given by: $I = E_{APP}/Z$.

☐ The power equations for a series RC circuit are similar to those for series RL circuits. ☐ Since the capacitive reactance, X_C, decreases with an increase in frequency, as the frequency increases, a series RC circuit becomes more resistive.

---

review questions

1. On a vector diagram of the voltages in a series RC circuit, what circuit quantity is taken as the reference vector? Why?
2. What is the 10-to-1 rule for resistance and capacitive reactance for a series RC circuit?
3. What is meant by *leakage current*? What effect does leakage current have on the phase relationship for the voltage and current of a capacitor?
4. What is the resistance in a series RC circuit when the impedance is 130 ohms, and the capacitive reactance is 50 ohms?
5. What is the phase angle for the circuit of Question 4?
6. The applied voltage across a series RC circuit is 100 volts, and causes a current of 5 amperes to flow. What is the magnitude of the impedance of the current?
7. The power dissipated in a circuit is 500 watts, the impedance is 10 ohms, and the phase angle is 60 degrees. What is the value of the current in the circuit?
8. What is the value of the apparent power of a circuit that dissipates 500 watts, and has a power factor of 0.25?
9. Between what values can the power factor of a series RC circuit be found? Why?
10. What is the voltage across the capacitor in a series RC circuit when the applied voltage is 100 volts, and the voltage across the resistor is 80 volts?

parallel RC circuits

In a parallel RC circuit, one or more resistive loads and one or more capacitive loads are connected in parallel across a voltage source. There are, therefore, *resistive branches,* containing only resistance, and *capacitive branches,* containing only capacitance. The current that leaves the voltage source divides among the branches, so there are different currents in different branches. The current is, therefore, not a common quantity, as it is in series RC circuits.

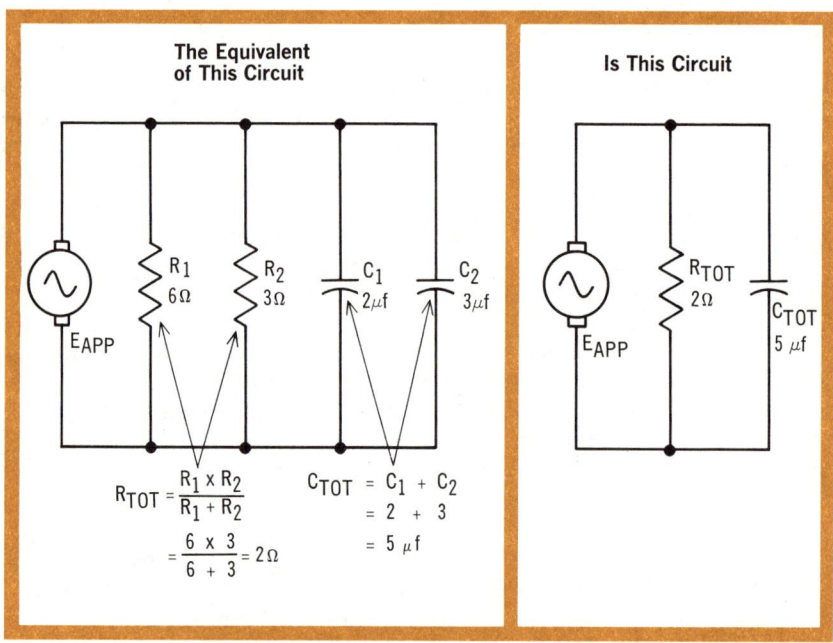

When calculating the total circuit quantities of applied voltage, line current, impedance, and power, resistive and capacitive branches should first be reduced to their simple single equivalents

The description of parallel RC circuits contained on the following pages will cover parallel circuits that contain only a single resistive branch and a single capacitive branch. Circuits that contain more than one resistive or capacitive branch are identical, except that when solving them for overall circuit quantities such as impedance or line current, the resistive or capacitive branches must first be reduced to their equivalent single resistive or capacitive branch.

voltage

In a parallel RC circuit, as in any parallel circuit, the applied voltage is *directly across* each branch. The branch voltages are, therefore, equal to each other, as well as to the applied voltage, and all three are *in phase*. So if you know any one of the circuit voltages, you know all of them. Since the voltage is *common* throughout the circuit, it serves as the common quantity in any vector representation of parallel RC circuits. This means that on any vector diagram, the reference vector will have the same direction, or phase relationship, as the circuit voltage. The two quantities that have this relationship with the circuit voltage, and whose vectors therefore have a direction of zero degrees, are the circuit resistance and the current through the resistance.

In a parallel RC circuit, each branch voltage is the same as the applied voltage

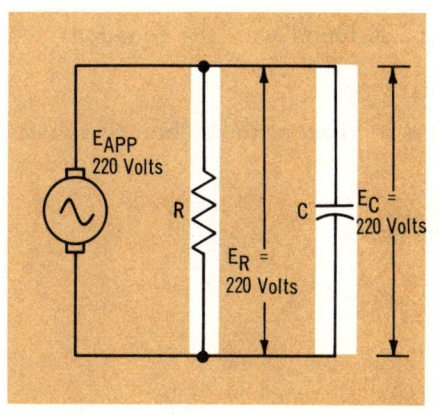

Phase relationships between the various quantities in a parallel RC circuit are expressed in relation to how they differ phasewise from the circuit voltage. The reason for this is that the voltage is the same throughout the circuit, and so provides a basis for expressing phase differences

As a Vector

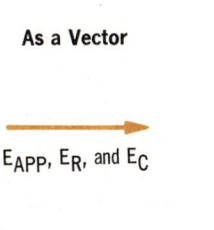

As a Waveform

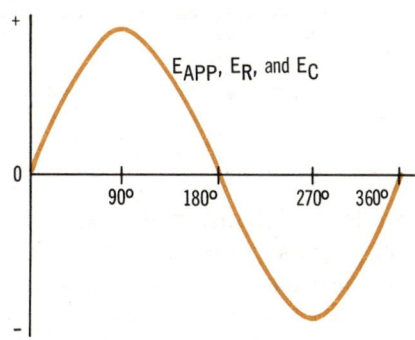

branch current

The current in each branch of a parallel RC circuit is *independent* of the current in the other branches. Current within a branch depends only on the voltage across the branch, and the resistance or capacitive

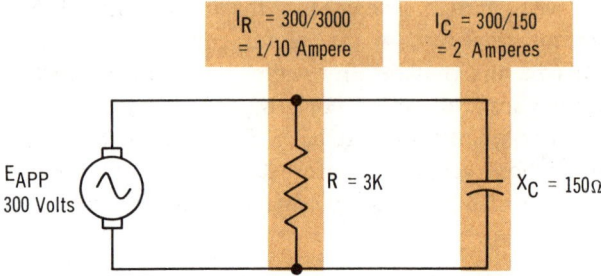

reactance contained in it. The current in the resistive branch is calculated from the equation:

$$I_R = E_{APP}/R$$

The current in the capacitive branch is found with the equation:

$$I_C = E_{APP}/X_C$$

The current in the resistive branch is in phase with the branch voltage,

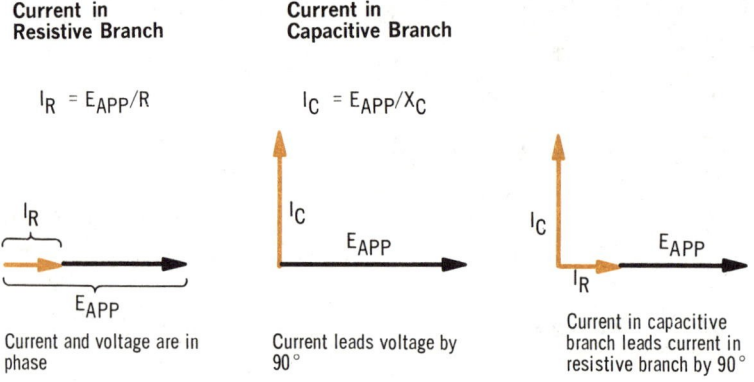

Current in Resistive Branch	Current in Capacitive Branch	
$I_R = E_{APP}/R$	$I_C = E_{APP}/X_C$	
Current and voltage are in phase	Current leads voltage by 90°	Current in capacitive branch leads current in resistive branch by 90°

while the current in the capacitive branch leads the branch voltage by 90 degrees. Since the two branch voltages are the same, the current in the *capacitive* branch (I_C) must *lead* the current in the *resistive* branch (I_R) by 90 degrees.

Line current in a parallel RC circuit is equal to the VECTOR SUM of the currents in the resistive and capacitive branches

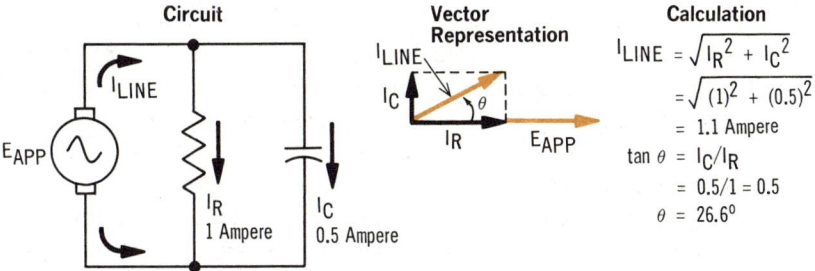

Circuit	Vector Representation	Calculation

$$I_{LINE} = \sqrt{I_R^2 + I_C^2}$$
$$= \sqrt{(1)^2 + (0.5)^2}$$
$$= 1.1 \text{ Ampere}$$
$$\tan \theta = I_C/I_R$$
$$= 0.5/1 = 0.5$$
$$\theta = 26.6^\circ$$

line current

Since the branch currents in a parallel RC circuit are out of phase with each other, they have to be added *vectorially* to find the line current. The two branches are 90 degrees out of phase, so their vectors form a right triangle, whose hypotenuse is the line current. The equation for calculating the line current, I_{LINE}, is

$$I_{LINE} = \sqrt{I_R^2 + I_C^2}$$

If the impedance of the circuit and the applied voltage are known, the line current can also be calculated from Ohm's Law.

$$I_{LINE} = E/Z$$

The line current is the vector sum of the resistive and capacitive currents

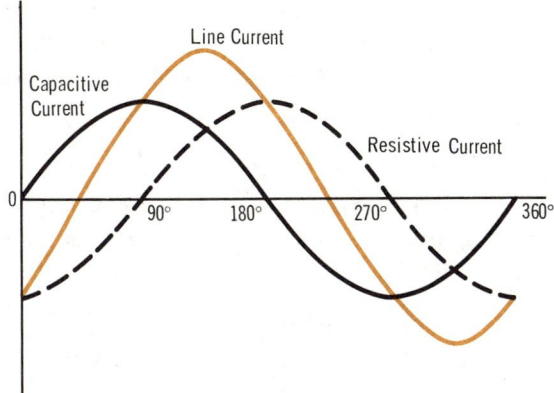

Capacitive current leads the resistive current by 90°

line current (cont.)

Inasmuch as the current in the resistive branch of a parallel RC circuit is in phase with the applied voltage, while the current in the capacitive branch leads the applied voltage by 90 degrees, the sum of the two branch currents, or line current, leads the applied voltage by some phase angle less than 90 degrees but greater than 0 degrees. The exact angle depends on whether the capacitive current or resistive current is greater. If there is more capacitive current, the angle will be closer to 90 degrees; while if the resistive current is greater, the angle is closer to 0 degrees. In cases where one of the currents is 10

The line current in a parallel RC circuit will have a phase angle somewhere between 0 and 90°, leading. The value of θ depends on the relative values of the capacitive and resistive currents in the branches

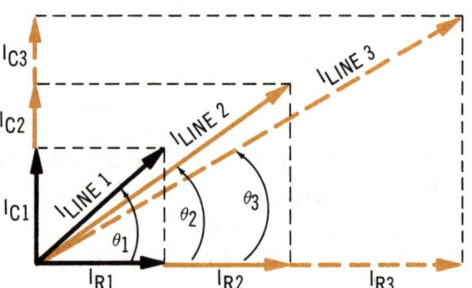

or more times greater than the other, the line current can be considered to have a phase angle of 0 degrees if the resistive current is the larger, or 90 degrees if the capacitive current is the larger. The value of the phase angle can be calculated from the values of the two branch currents with the equation:

$$\tan \theta = I_C/I_R$$

By substituting the quantities $I_C = E/X_C$ and $I_R = E/R$ in the above equation, two other useful equations for calculating the phase angle, θ, can be derived. They are:

$$\tan \theta = R/X_C \qquad \cos \theta = Z/R$$

Once you know the line current and the applied voltage in a parallel RC circuit, you can find the circuit power using the same equations you learned for parallel RL circuits. These are:

$$P_{APPARENT} = E_{APP}I_{LINE}$$
$$P_{TRUE} = E_{APP}I_{LINE} \cos \theta$$

where $\cos \theta$ is the power factor.

current waveforms

Since the branch currents in a parallel RC circuit are out of phase, their *vector sum* rather than their arithmetic sum equals the line current. This is the same condition that exists for the voltage drops in a series RC circuit. By adding the currents vectorially, you are adding their *instantaneous* values at every point, and then finding the average or effective value of the resulting current. This can be seen from the current waveforms shown. They are the waveforms for the circuit solved on page 4-69.

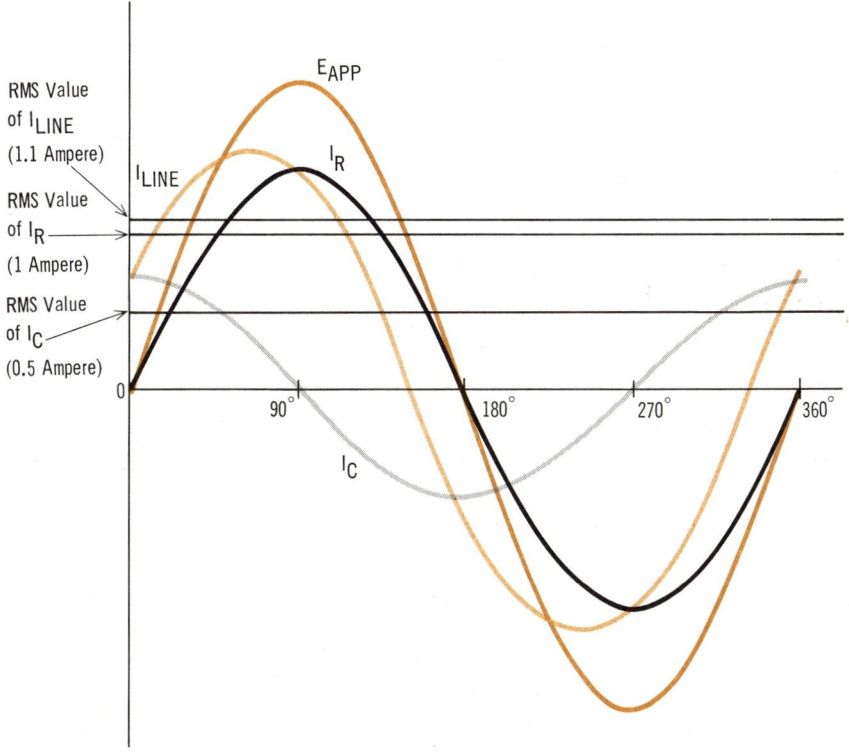

Every point on the line current waveform (I_{LINE}) is the algebraic sum of the instantaneous values of the I_R and I_C waveforms. The rms value of the line current waveform is thus shown to be equal to the vector sum of the rms values of the I_R and I_C waveforms.

In a Parallel RC Circuit:

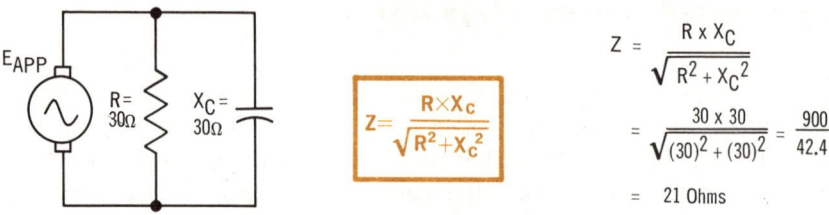

$$Z = \frac{R \times X_C}{\sqrt{R^2 + X_C^2}}$$

$$= \frac{30 \times 30}{\sqrt{(30)^2 + (30)^2}} = \frac{900}{42.4}$$

$$= 21 \text{ Ohms}$$

When resistance and capacitive reactance are equal, the impedance is not one-half the value of either one, or 15 ohms, as you might suppose based on your knowledge of parallel resistances

impedance

The impedance of a parallel RC circuit represents the total opposition to current flow offered by the resistance of the resistive branch and the capacitive reactance of the capacitive branch. Like the impedance of a parallel RL circuit, it can be calculated with an equation that is similar to the one used for finding the total resistance of two parallel resistances. However, just as you learned for parallel RL circuits, two vector quantities cannot be added *directly; vector addition* must be used. Therefore, the equation for calculating the impedance of a parallel RC circuit is

$$Z = \frac{RX_C}{\sqrt{R^2 + X_C^2}}$$

where $\sqrt{R^2 + X_C^2}$ is the vector addition of the resistance and capacitive reactance.

In cases where you know the applied voltage and the circuit line current, the impedance can be found simply by using Ohm's Law in the form:

$$Z = E_{APP}/I_{LINE}$$

The impedance of a parallel RC circuit is always *less* than the resistance or capacitive reactance of the individual branches.

The relative values of X_C and R determine how capacitive or resistive the circuit line current is. The one that is the *smallest,* and therefore allows *more* branch current to flow, is the determining factor. Thus, if X_C is smaller than R, the current in the capacitive branch is larger than the current in the resistive branch, and the line current tends to be more capacitive. The opposite is true if R is smaller than X_C. When X_C or R is 10 or more times greater than the other, the circuit will operate for all practical purposes as if the branch with the larger of the two did not exist.

effect of frequency

As in all RL and RC circuits, the frequency of the applied voltage determines many of the characteristics of a parallel RC circuit. Frequency affects the value of the capacitive reactance, and so also affects the circuit impedance, line current, and phase angle, since they are determined to some extent by the value of X_C. The higher the frequency of a parallel RC circuit, the lower is the value of X_C. This means that for a given value of R, the impedance is also lower, making the line current larger and more capacitive. Conversely, the lower the frequency, the greater is the value of X_C, the larger is the impedance, and the

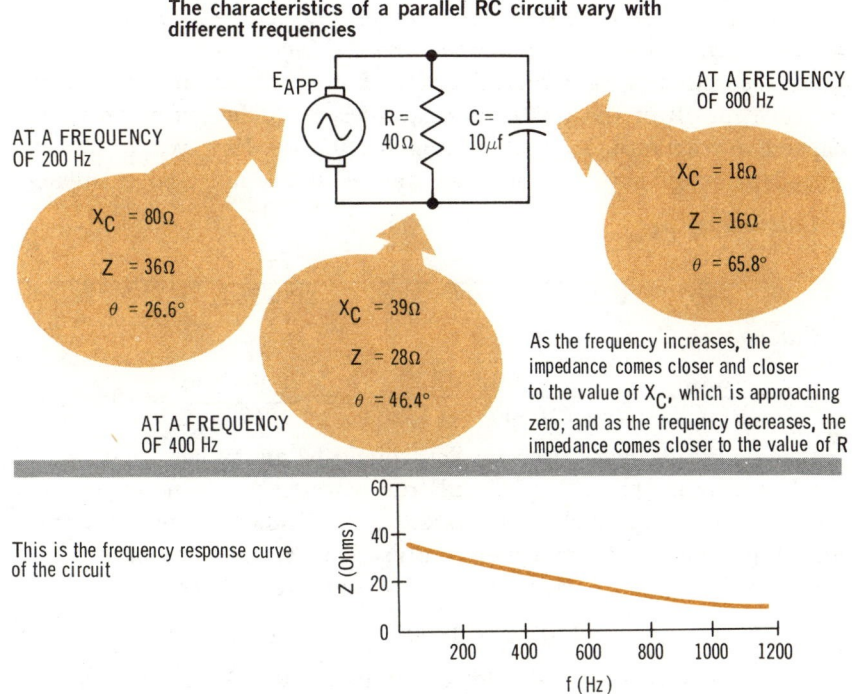

The characteristics of a parallel RC circuit vary with different frequencies

E_{APP} R = 40Ω   C = 10μf

AT A FREQUENCY OF 200 Hz

X_C = 80Ω

Z = 36Ω

θ = 26.6°

AT A FREQUENCY OF 400 Hz

X_C = 39Ω

Z = 28Ω

θ = 46.4°

AT A FREQUENCY OF 800 Hz

X_C = 18Ω

Z = 16Ω

θ = 65.8°

As the frequency increases, the impedance comes closer and closer to the value of X_C, which is approaching zero; and as the frequency decreases, the impedance comes closer to the value of R

This is the frequency response curve of the circuit

Z (Ohms): 60, 40, 20, 0

f (Hz): 200, 400, 600, 800, 1000, 1200

smaller and more resistive is the line current. The same circuit, therefore, can have a *small, resistive* line current at *low frequencies*, and a *large, capacitive* line current at *high frequencies*. Furthermore, since according to the 10-to-1 rule, you can neglect the branch with the smaller current when one of them is 10 or more times larger than the other, depending on the frequency, a parallel RC circuit can act as a simple *series* resistive or capacitive circuit.

solved problems

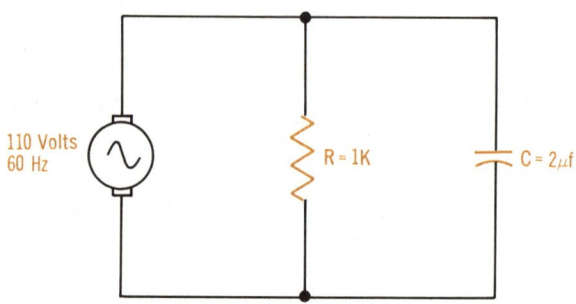

Problem 17. In the circuit, find the impedance, the branch currents, the line current, the phase angle (θ), and the true power.

An inspection of the circuit diagram shows that you are being asked to solve for all of the unknown quantities, with the exception of the capacitive reactance, X_C, and the apparent power. However, to find the impedance, you have to know X_C, and so it too must be calculated.

Calculating X_C:

$$X_C = \frac{1}{2\pi fC} = \frac{1}{6.28 \times 60 \times 0.000002} = \frac{1}{0.000754} = 1326 \text{ ohms}$$

Calculating Z:

$$Z = \frac{RX_C}{\sqrt{R^2 + X_C^2}} = \frac{1000 \times 1326}{\sqrt{(1000)^2 + (1326)^2}} = \frac{1,326,000}{1661} = 798 \text{ ohms}$$

Calculating Branch Currents: You should be aware here that to find the line current, phase angle, and true power it is not necessary that you calculate the branch currents, since you know the applied voltage and impedance, and could therefore use Ohm's Law. The branch currents are being calculated only because they were asked for.

$$I_R = E/R = 110 \text{ volts}/1000 \text{ ohms} = 0.11 \text{ ampere}$$

$$I_C = E/X_C = 110 \text{ volts}/1326 \text{ ohms} = 0.083 \text{ ampere}$$

Calculating I_{LINE}:

$$I_{LINE} = \sqrt{I_R^2 + I_C^2} = \sqrt{(0.11)^2 + (0.083)^2} = \sqrt{0.019} = 0.14 \text{ ampere}$$

Calculating θ:

$$\tan \theta = R/X_C = 1000/1326 = 0.754 \qquad \theta = 37°$$

Calculating True Power:

$$P_{TRUE} = EI_{LINE} \cos \theta$$

$$= 110 \times 0.14 \times 0.798 = 12.3 \text{ watts}$$

comparison of series and parallel RC circuits

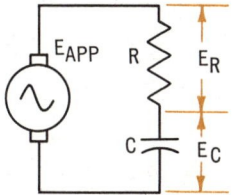 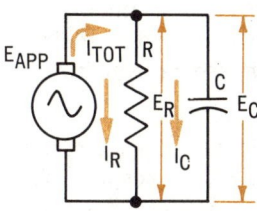

	Series RC Circuit	Parallel RC Circuit
Current	It is the same everywhere in circuit. Currents through R and C are, therefore, in phase.	It divides between resistive and capacitive branches. $$I_{TOT} = \sqrt{I_R{}^2 + I_C{}^2}$$ $$I_R = E_{APP}/R \qquad I_C = E_{APP}/X_C$$ Current through C leads current through R by 90°.
Voltage	Vector sum of voltage drops across R and C equals applied voltage. $$E_{APP} = \sqrt{E_R{}^2 + E_C{}^2}$$ Voltage across C lags voltage across R by 90°.	Voltage across each branch is same as applied voltage. Voltages across R and C are, therefore, in phase. $$E_R = E_C = E_{APP}$$
Impedance	It is the vector sum of resistance and capacitive reactance. $$Z = \sqrt{R^2 + X_C{}^2}$$	It is calculated the same as parallel resistances, except that vector addition is used. $$Z = RX_C / \sqrt{R^2 + X_C{}^2}$$
Phase Angle (θ)	It is the angle between circuit current and applied voltage. $$\tan \theta = E_C/E_R = X_C/R$$ $$\cos \theta = R/Z$$	It is the angle between line current and applied voltage. $$\tan \theta = I_C/I_R = R/X_C$$ $$\cos \theta = Z/R$$
Power	Power delivered by source is apparent power. Power actually consumed in circuit is true power. Power factor determines what portion of apparent power is true power. $$P_{APPARENT} = E_{APP}I \qquad P_{TRUE} = E_{APP}I \cos \theta \qquad P.\,F. = \cos \theta$$	
Effect of Increasing Frequency	X_C decreases, which in turn causes circuit current to increase. Phase angle decreases, which means that circuit is more resistive.	X_C decreases, capacitive branch current increases, and so line current also increases. Phase angle increases, which means that circuit is more capacitive.
Effect of Increasing Resistance	Phase angle decreases, which means that circuit is more resistive.	Phase angle increases, which means that circuit is more capacitive.
Effect of Increasing Capacitance	Phase angle increases, which means that circuit is more capacitive.	Phase angle decreases, which means that circuit is more resistive.

summary

☐ In a parallel RC circuit, the applied voltage is the same across each branch. It is therefore used as the phase reference. ☐ The current through each branch of a parallel RC circuit is independent of the current through the other branches. ☐ The current in the resistive branch is found by $I_R = E_{APP}/R$. ☐ The current in the capacitive branch is found by $I_C = E_{APP}/X_C$. ☐ The line current is found by adding the branch currents vectorially. $I_{LINE} = \sqrt{I_R^2 + I_C^2}$. It can also be found by Ohm's Law for a-c circuits: $I = E/Z$.

☐ The phase angle of a parallel RC circuit is given by: $\tan \theta = I_C/I_R$; or $\tan \theta = R/X_C$; or $\cos \theta = Z/R$. ☐ The impedance is found by adding the parallel oppositions vectorially. The equation for impedance is $Z = RX_C/(\sqrt{R^2 + X_C^2})$.

☐ The power equations for a parallel RC circuit are identical to those for series RC circuits. ☐ Since capacitive reactance, X_C, decreases with increasing frequency, the parallel RC circuit approaches a pure capacitive circuit as the frequency is increased.

review questions

1. What is meant by *line current* in a parallel RC circuit?

For Questions 2 to 10, consider a parallel RC circuit with an applied voltage of 100 volts, a resistor with a resistance of 20 ohms, and a capacitor with a capacitive reactance of 10 ohms.

2. What is the current through the resistor? Through the capacitor?
3. What is the voltage across the resistor? Across the capacitor?
4. What is the impedance of the circuit?
5. What is the phase angle of the circuit?
6. What is the apparent power? The true power? The power factor?
7. What is the value of the line current?
8. By how much does the frequency have to be multiplied for the parallel circuit to become, effectively, a capacitive circuit?
9. Answer Questions 3 to 5 for the condition where the frequency is doubled.
10. Answer Questions 5 to 7 for the condition where the frequency is tripled.

LCR circuits

You have learned the fundamental properties of resistive, inductive, and capacitive circuits, as well as circuits that contain resistance and inductance, and resistance and capacitance. You will now learn about circuits that contain all three of the basic properties of inductance (L), capacitance (C), and resistance (R). These circuits are called *LCR circuits,* and may consist of either series or parallel combinations of inductance, capacitance, and resistance. You will find that everything you have learned previously about resistive, inductive, and capacitive circuits is involved in the analysis of LCR circuits. In addition, though, some entirely new properties and characteristics are involved.

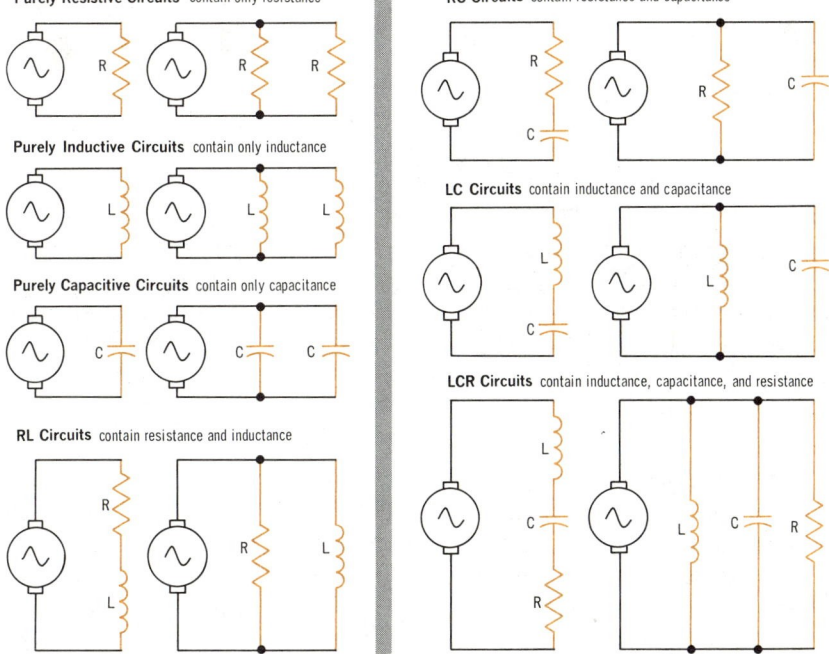

Purely Resistive Circuits contain only resistance

Purely Inductive Circuits contain only inductance

Purely Capacitive Circuits contain only capacitance

RL Circuits contain resistance and inductance

RC Circuits contain resistance and capacitance

LC Circuits contain inductance and capacitance

LCR Circuits contain inductance, capacitance, and resistance

The description of LCR circuits given on the following pages is divided into two parts: one covering series circuits, and the other parallel circuits. For both the series and parallel types, pure *LC circuits* are covered first. These are circuits that have inductance and capacitance, but no resistance. After LC circuits are thoroughly described, resistance will be included, and practical LCR circuits analyzed.

series LC circuits

A series LC circuit consists of an *inductance* and a *capacitance* connected in *series* with a voltage source. There is *no resistance* in the circuit. Of course, this is impossible in actual practice, since every circuit contains some resistance. However, since the circuit resistance of the wiring, the coil winding, and the voltage source is usually so small, it has little or no effect on circuit operation.

A Series LC Circuit

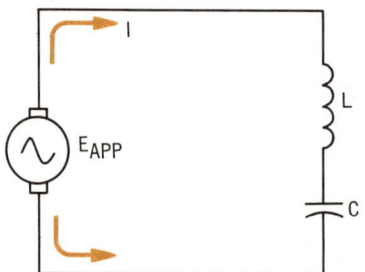

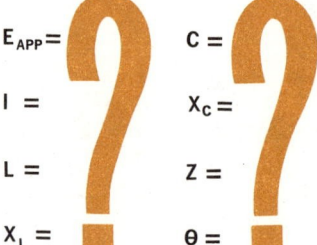

In a series LC circuit, the inductance and capacitance are connected in series, so the total circuit current flows through both. Since the inductance and capacitance are in series, their reactances, X_L and X_C, are also in series

The quantities in series LC circuits that you will normally be interested in are the applied voltage, E_{APP}; the current, I; the inductance, L; the inductive reactance, X_L; the capacitance, C; the capacitive reactance, X_C; the impedence, Z; and the phase angle, θ

As in all series circuits, the *current* in a series LC circuit is the *same* at all points. So the current in the inductance is the same as, and therefore in phase with, the current in the capacitance. Because of this, on any vector diagram of a series LC circuit the direction of the *current vector* is the *reference*, or 0-degree, *direction*. All other quantities, such as the applied voltage and the voltage drops in the circuit, are expressed on the basis of their phase relationship to the circuit current. Here again, as in RL and RC circuits, current is chosen as the phase reference for convenience; not because the phase angle of the current is fixed and does not change. The phase angle of the current depends on the circuit properties, and so it does change. What is important is the *phase difference* between the current and the applied voltage; and vectorially this is the same, whether current or voltage is used as the phase reference.

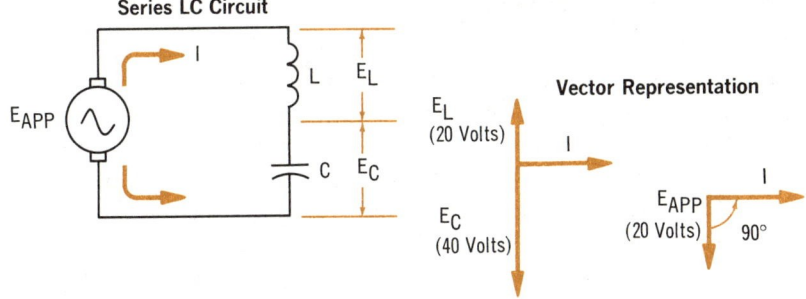

The applied voltage is equal to the vector sum of the voltage drop across the inductance (E_L) and the drop across the capacitance (E_C). If E_L is larger than E_C, the current is purely inductive, so the applied voltage leads the current by 90°; and if E_C is larger than E_L, the current is purely capacitive, so the applied voltage lags the current by 90°

voltage

When a-c current flows in a series LC circuit, the voltages dropped across the inductance and the capacitance depend on the circuit current and the values of X_L and X_C. The voltages can be found by:

$$E_L = IX_L \quad \text{and} \quad E_C = IX_C$$

The voltage across the *inductance leads* the current through it by 90 degrees, while the voltage across the *capacitance lags* the current through it by 90 degrees. Since the *current* through both is the *same*, the voltage across the *inductance leads* that across the *capacitance* by 180 degrees. You will remember that in a series RL or RC circuit, because of the phase differences, the *vector sum* of the voltage drops equals the applied voltage. This is also so in a series LC circuit. There is a difference, though, in how vector addition is used for LC circuits.

In RL and RC circuits, the voltage vectors are 90 degrees out of phase, so the Pythagorean Theorem is used to add them. However, the method used to add vectors that differ by 180 degrees, you remember, is to subtract the smaller vector from the larger and assign the resultant the direction of the larger. When applied to LC series circuits, this means that the applied voltage is equal to the *difference* between the voltage drops across the inductance and capacitance, with the *phase angle* between the applied voltage and the circuit current determined by the *larger* of the voltage drops. As an example, suppose that the voltage drop across the inductance was 50 volts and that across the capacitance 40 volts. The applied voltage would be 10 volts (50 − 40), and would lead the current by 90 degrees, since the larger voltage drop was across the inductance.

voltage (cont.)

A unique property of series LC circuits is that one of the voltage drops, either E_L or E_C, is always *greater* than the *applied voltage*. Moreover, in some cases, both of the voltage drops are greater than the applied voltage. The reason for this, as you will learn more about later, is that the reactances of the inductance and capacitance play a *dual role* in the circuit. They act *together* in opposing the circuit current, whereas they act *independently* in causing their voltage drops.

When the reactances act together, their phase relationship is such that they tend to cancel a portion of each other. The total opposition they offer to the current is therefore less than either would offer individually, so a larger current flows than either would allow by itself. When this current flows, though, it causes a voltage drop across the full value of each individual reactance. In effect, the voltage source "sees" a circuit with little opposition to current flow, and so puts out a relatively large current. But in flowing around the circuit, the current meets opposition which the source does not "see."

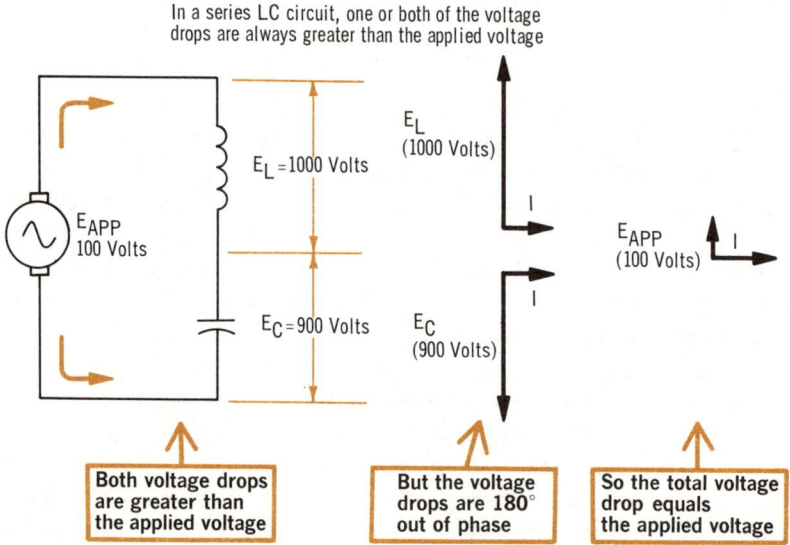

In a series LC circuit, one or both of the voltage drops are always greater than the applied voltage

$E_L = 1000$ Volts

E_{APP} 100 Volts

$E_C = 900$ Volts

E_L (1000 Volts)

E_C (900 Volts)

E_{APP} (100 Volts)

Both voltage drops are greater than the applied voltage

But the voltage drops are 180° out of phase

So the total voltage drop equals the applied voltage

An important point that you should remember is that although one or both of the voltage drops is greater than the applied voltage, they are *180 degrees out of phase*. One of them effectively cancels a portion of the other, so that the total voltage drop is always equal to the applied voltage.

voltage waveforms

You saw on the previous page a circuit that had individual voltage drops of 1000 volts and 900 volts, and yet the voltage source was only 100 volts. The reason for this was that the voltage drops were 180 degrees out of phase. Because of their phase difference, the two voltage drops had to be added vectorially. And since their phase difference was such that one voltage, in effect, *canceled* a portion of the other, the vector addition was done by *algebraic addition*. Thus, one voltage was considered positive, and the other negative, and the two quantities were added algebraically. As shown, the waveforms of the two voltage drops can also be added algebraically, and the average or effective value of the resulting waveform is the average or effective value of the applied voltage.

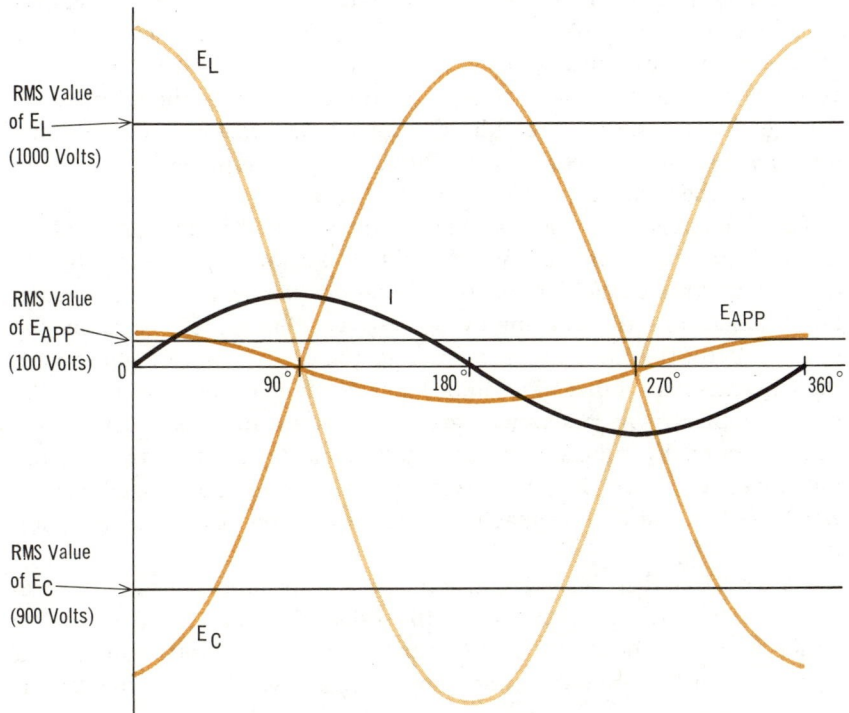

Every point on the applied voltage waveform (E_{APP}) is the algebraic sum of the instantaneous values of the E_L and E_C waveforms

Circuit	Vector Representation	Calculation

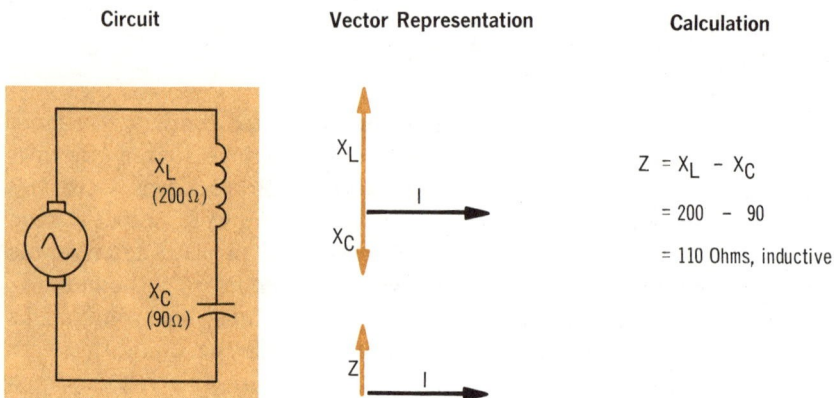

$Z = X_L - X_C$

$= 200 - 90$

$= 110$ Ohms, inductive

impedance

You will recall from series RL and RC circuits that it was convenient to consider resistance, inductive reactance, and capacitive reactance as vector quantities because of the phase relationships between the voltage drops they cause and the circuit current. The direction of these vector quantities was then the same as the direction of their respective voltage drops. And as a result, the impedance vector had the same direction as the applied voltage. These same techniques are used to determine the impedance of series LC circuits.

The inductive reactance, like the voltage across the inductance (E_L), leads the circuit current (which is the 0-degree reference) by 90 degrees, and the capacitive reactance, like the voltage across the capacitance (E_C), lags the current by 90 degrees. Since the current is the same in both, the inductive reactance is 180 degrees out of phase with the capacitive reactance. The impedance is then the vector sum of the two reactances. The reactances are 180 degrees apart, so their vector sum is found by subtracting the smaller one from the larger. If the inductive reactance is the larger, the equation for impedance is $Z = X_L - X_C$. And if the capacitive reactance is the larger, the equation is $Z = X_C - X_L$.

Unlike RL and RC circuits, in which the impedance was a combination of resistance and reactance, the impedance in an LC circuit is either purely inductive or purely capacitive. It is purely inductive if X_L is the larger reactance, and purely capacitive if X_C is the larger. Usually, the type of the impedance is specified directly after the impedance value. For example, an impedance of 50 ohms, capacitive; or 10 ohms, inductive. If the impedance is capacitive, the current is purely capacitive; and if the impedance is inductive, the current is purely inductive.

current

The same current flows through both the inductance and capacitance in a series LC circuit. If the *inductive* reactance (X_L) is the *greater* of the two circuit reactances, the current is purely inductive, and *lags* the applied voltage by 90 degrees. And if the *capacitive* reactance (X_C) is the *larger* reactance, the current is purely capacitive, and *leads* the applied voltage by 90 degrees. As far as the current and applied voltage are concerned, therefore, a series LC circuit is either a purely inductive or purely capacitive circuit: inductive if the impedance, which is the net reactance, is inductive, and capacitive if the impedance is capacitive. The magnitude of the current is related to the applied voltage and impedance by Ohm's Law for a-c circuits. So if the applied voltage and impedance are known, the current can be calculated from the equation:

$$I = E_{APP}/Z$$

The thought may have occurred to you: "What happens if the two reactances are *equal?*" This is entirely possible, and as a matter of fact is very often the case. If the two reactances were equal in a purely

The magnitude of the current in series
LC circuits is determined by Ohm's Law

The phase angle between the current
and applied voltage depends on the
relative values of X_L and X_C

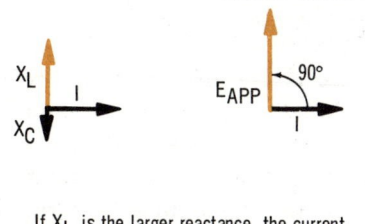

If X_L is the larger reactance, the current
lags the voltage by 90°

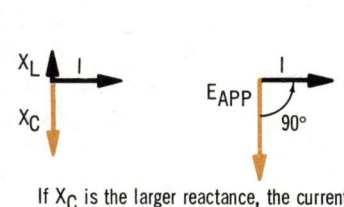

If X_C is the larger reactance, the current
leads the voltage by 90°

LC circuit, which we are considering now, the impedance would be *zero*. So with no opposition, an infinitely large current would flow. Of course this is never the case, since every circuit has some resistance. In actual circuits, when the two reactances are equal, large currents do flow, with the magnitude of the current limited only by the circuit resistance. This condition is called *resonance,* and will be covered later.

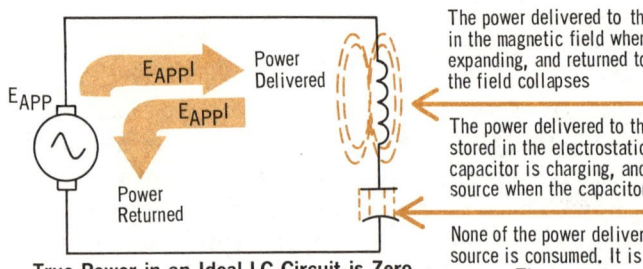

The power delivered to the inductance is stored in the magnetic field when the field is expanding, and returned to the source when the field collapses

The power delivered to the capacitance is stored in the electrostatic field when the capacitor is charging, and returned to the source when the capacitor discharges

True Power in an Ideal LC Circuit is Zero

None of the power delivered to the circuit by the source is consumed. It is all returned to the source. The true power, which is the power consumed, is thus zero

power

In a series LC circuit, power is delivered by the source to both the inductance and capacitance. The energy represented by the power delivered to the inductance is *stored* in the *magnetic field* around the inductance, and that delivered to the capacitance is *stored* in the *electrostatic field* between the capacitor plates. Since theoretically there is no resistance in the circuit, all of the stored energy is *returned* to the source each time the magnetic field around the inductance collapses and the capacitor discharges. There is thus a constant interchange of power, or energy, between the source and the circuit, but no power *consumption*. Inasmuch as true power is the power actually consumed, or dissipated, in a circuit, the *true power* in a pure LC circuit is *zero*. The apparent power, which is the total power delivered by the source, is the sum of the powers delivered to the inductance and capacitance. It is equal to the applied voltage times the circuit current, or:

$$P_{\text{APPARENT}} = E_{\text{APP}}I$$

The circuit power factor is equal to the cosine of the phase angle between the applied voltage and the current, as it is in any a-c circuit. Since the phase angle in a series LC circuit is always 90 degrees, and the cosine of 90 degrees is zero, the power factor in an LC circuit is zero. True power in an a-c circuit equals the apparent power times the power factor; this is another way of showing that the true power in a series LC circuit is always zero.

$$P_{\text{TRUE}} = E_{\text{APP}}I \cos \theta$$
$$= E_{\text{APP}}I \times 0 = 0$$

Actually, all circuits contain some resistance, of course, so the power factor is never zero because of the power dissipated by the resistance. When the resistance is very low, though, for all practical purposes the power factor can be considered zero.

effect of frequency

In series RL and RC circuits, a definite relationship exists between the *impedance,* and the *frequency* of the applied voltage. *Increasing* the frequency of an *RL* circuit, or *decreasing* the frequency of an *RC* circuit, results in an *increase* in *impedance.* Similarly, *decreasing* the frequency in an *RL* circuit, or *increasing* the frequency in an *RC* circuit, causes a *decrease* in *impedance.*

In series LC circuits, however, no similar clear-cut relationship exists between the impedance, and the frequency. The impedance is controlled by the frequency, but an increase or decrease in the frequency depends on the *relative values* of the inductive and capacitive reactance. For example, in one circuit, an increase in frequency may

The effect of frequency changes on series LC circuits depends on the relative valves of X_L and X_C

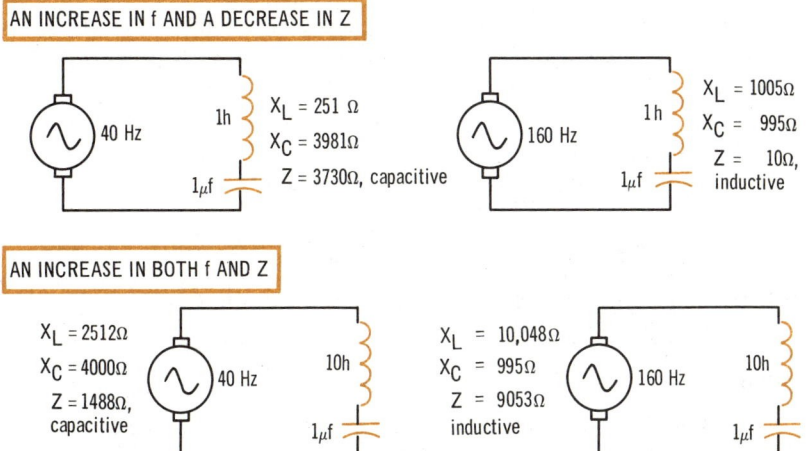

AN INCREASE IN f AND A DECREASE IN Z

40 Hz 1h X_L = 251 Ω X_C = 3981Ω 1µf Z = 3730Ω, capacitive

160 Hz 1h X_L = 1005Ω X_C = 995Ω 1µf Z = 10Ω, inductive

AN INCREASE IN BOTH f AND Z

X_L = 2512Ω X_C = 4000Ω Z = 1488Ω, capacitive 40 Hz 10h 1µf

X_L = 10,048Ω X_C = 995Ω Z = 9053Ω inductive 160 Hz 10h 1µf

Although the impedance is controlled by the frequency, no relationship can be specified. As shown, the identical increase in frequency can cause an increase in impedance in one circuit, and a decrease in another

cause an increase in impedance, while in another circuit, the same increase in frequency could cause a decrease in impedance. In the same way, a decrease in capacitance can cause an increase in impedance in one circuit and a decrease in another. The situation can even occur where a certain increase, say in inductance, will cause a decrease in impedance, but a further increase in the inductance will cause the impedance to increase. It is, therefore, usually best to actually calculate the new impedance whenever a change occurs in the frequency of a series LC circuit.

solved problems

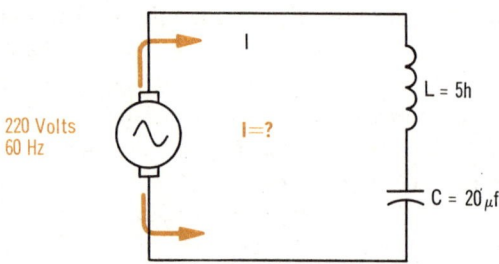

Problem 18.　　*What is the current in the circuit?*

According to Ohm's Law for a-c circuits, the equation for current is $I = E_{APP}/Z$. The voltage is given, but the impedance is not, and must therefore be calculated. Before you can calculate the impedance, though, the two reactances, X_L and X_C, must be determined. Thus, the sequence for solving this problem is to first calculate X_L and X_C, then Z, and finally I.

Calculating X_L and X_C:

$$X_L = 2\pi fL = 6.28 \times 60 \times 5 = 1884 \text{ ohms}$$

$$X_C = \frac{1}{2\pi fC} = \frac{1}{6.28 \times 60 \times 0.00002} = 133 \text{ ohms}$$

Calculating Z: X_L is larger than X_C, so the equation for Z is

$$Z = X_L - X_C = 1884 - 133 = 1751 \text{ ohms, inductive}$$

Calculating I:

$$I = E_{APP}/Z = 220 \text{ volts}/1751 \text{ ohms} = 0.125 \text{ ampere}$$

Problem 19.　　*What is the phase angle between the applied voltage and current in the above circuit?*

The phase angle in all purely LC circuits is 90 degrees; thus, in this circuit, the angle must be 90 degrees. And since the impedance was inductive, the applied voltage leads the current.

Problem 20.　　*What are the voltage drops across the reactances?*

Each voltage drop is independent, and depends only on the circuit current and the value of reactance. So,

$$E_L = IX_L = 0.125 \times 1884 = 236 \text{ volts}$$

$$E_C = IX_C = 0.125 \times 133 = 17 \text{ volts}$$

Notice that the voltage drop across the inductance (E_L) is greater than the applied voltage of 220 volts.

solved problems (cont.)

Problem 21. Which of the following circuits are inductive and which are capacitive?

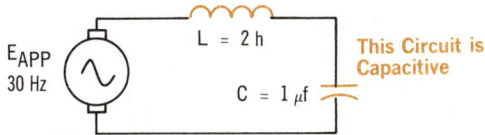

$X_L = 2\pi fL = 6.28 \times 30 \times 2 = 377$ ohms

$X_C = \dfrac{1}{2\pi fC} = \dfrac{1}{6.28 \times 30 \times 0.000001} = 5300$ ohms

$Z = X_C - X_L = 5300 - 377 = 4923$ ohms, capacitive

$X_L = 2\pi fL = 6.28 \times 60 \times 5 = 1884$ ohms

$X_C = \dfrac{1}{2\pi fC} = \dfrac{1}{6.28 \times 60 \times 0.00000001} = 265{,}000$ ohms

$Z = X_C - X_L = 265{,}000 - 1884 \cong 263K$, capacitive

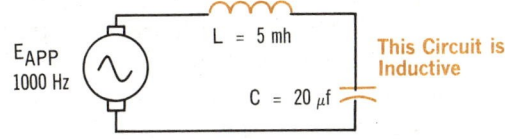

$X_L = 2\pi fL = 6.28 \times 1000 \times 0.005 = 31$ ohms

$X_C = \dfrac{1}{2\pi fC} = \dfrac{1}{6.28 \times 1000 \times 0.000020} = 8$ ohms

$Z = X_L - X_C = 31 - 8 = 23$ ohms, inductive

$X_L = 2\pi fL = 6.28 \times 200 \times 0.2 = 251$ ohms

$X_C = \dfrac{1}{2\pi fC} = \dfrac{1}{6.28 \times 200 \times 0.0000001} = 7900$ ohms

$Z = X_C - X_L = 7900 - 251 = 7649$ ohms, capacitive

summary

☐ A series LC circuit consists of an inductance and capacitance in series, with no resistance in the circuit. ☐ Since the current is the same throughout the circuit, it is used as the phase reference. ☐ The voltage across the inductor in a series LC circuit leads the current through it by 90 degrees. It is equal to: $E_L = IX_L$. ☐ The voltage across the capacitor lags the current through it by 90 degrees. It is equal to: $E_C = IX_C$. ☐ Because the voltages are 180 degrees out of phase with each other, the total, or applied, voltage is equal to the arithmetic difference of the two. ☐ The phase angle is determined by the larger of the two voltage drops.

☐ In a series LC circuit, either E_L or E_C, or in some cases both, are greater than the applied voltage. ☐ The impedance of a series LC circuit is either purely inductive or purely capacitive, depending upon the magnitudes of the reactances. Usually, the type of impedance is specified after the impedance value. ☐ Ohm's Law for a-c circuits applies in series LC circuits: $I = E_{APP}/Z$.

☐ In a pure series LC circuit, the true power is zero. ☐ The power factor is also zero. ☐ Because both X_L and X_C are dependent on frequency, there is no definite relationship between impedance and the frequency. The impedance depends on the relative values of the inductive and capacitive reactances in the circuit.

review questions

1. In a series LC circuit, can either the voltage across L and/or C ever be greater than the applied voltage?

For Questions 2 to 10, consider a series LC circuit with an applied voltage of 100 volts; a capacitor with a voltage drop of 140 volts; and an inductor with an inductive reactance of 20 ohms.

2. What is the current in the circuit?
3. What is the impedance of the circuit?
4. What is the apparent power of the circuit?
5. What is the true power of the circuit?
6. What is the phase angle of the circuit?
7. What is the power factor of the circuit?
8. What is the capacitive reactance?
9. Answer Questions 2 to 8, where the frequency of the applied voltage is doubled. (Hint: The voltage across the capacitor is no longer 140 volts.)
10. Answer Questions 2 to 8, where the frequency of the applied voltage is halved.

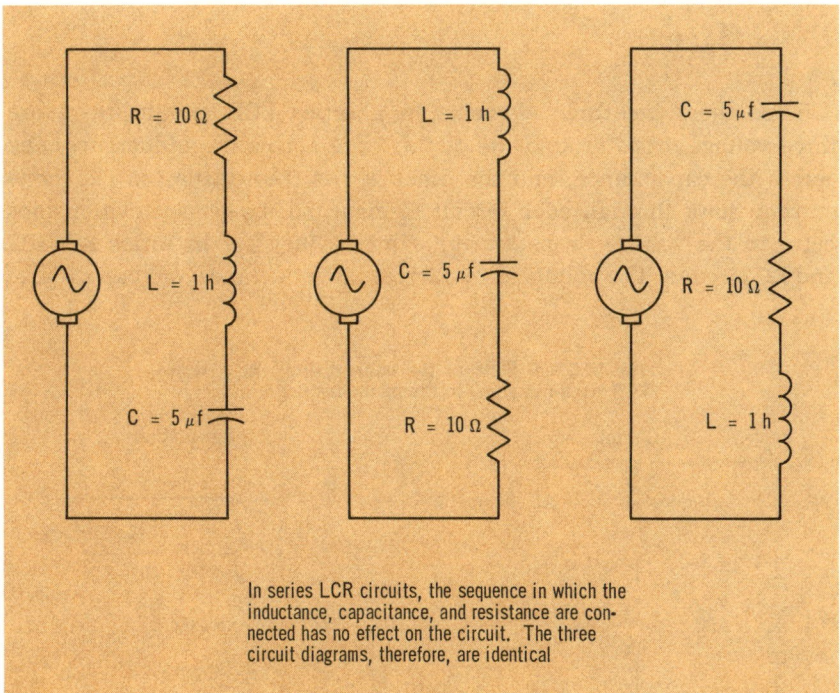

In series LCR circuits, the sequence in which the inductance, capacitance, and resistance are connected has no effect on the circuit. The three circuit diagrams, therefore, are identical

series LCR circuits

Any practical series LC circuit contains some resistance. When the resistance is very small compared to the circuit reactances, it has almost no effect on the circuit and can be considered as being zero, which is what was done on the preceding pages. When the resistance is *appreciable*, though, it has a significant effect on the circuit operation, and therefore must be considered in any circuit analysis. It makes no difference whether the resistance is the result of the circuit wiring or coil windings, or whether it is in the form of a resistor connected into the circuit. As long as it is appreciable, it affects the circuit operation and so must be considered. As a general rule, if the total reactance of the circuit is *not* 10 times or more greater than the resistance, the resistance *will* have an effect.

Circuits in which the inductance, capacitance, and resistance are all connected in series are called series LCR ciruits. You will see that the fundamental properties of series LCR circuits, and the methods used to solve them, are similar to those that you have learned for series LC circuits. The differences are caused by the effects of the resistance.

voltage

Since there are three elements in a series LCR circuit, there are three voltage drops around the circuit: one across the inductance, one across the capacitance, and the other across the resistance. The same current flows through each circuit element, so the phase relationships between the voltage drops are the same as they are in series LC, RL, and RC circuits. The voltage drops across the inductance and capacitance

The VECTOR SUM of the voltage drops in a series LCR circuit is equal to the applied voltage

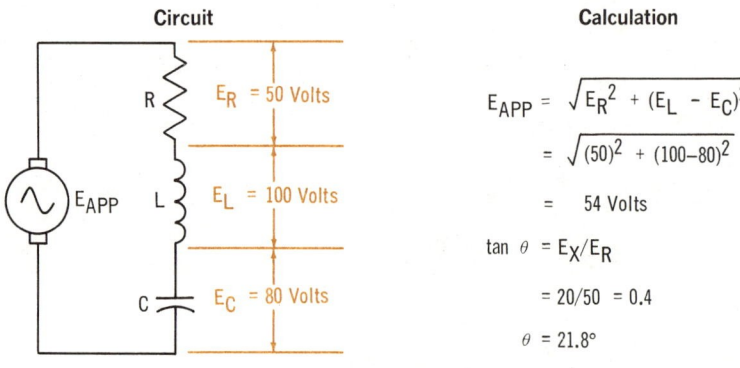

Circuit

Calculation

$$E_{APP} = \sqrt{E_R^2 + (E_L - E_C)^2}$$

$$= \sqrt{(50)^2 + (100{-}80)^2}$$

$$= 54 \text{ Volts}$$

$$\tan \theta = E_X/E_R$$

$$= 20/50 = 0.4$$

$$\theta = 21.8°$$

are 180 degrees out of phase, with the inductive voltage drop (E_L) leading the resistive voltage drop (E_R) by 90 degrees, and the capacitive voltage drop (E_C) lagging the resistive voltage drop (E_R) by 90 degrees.

The vector sum of the three voltage drops is equal to the applied voltage. However, to calculate this vector sum, a *combination* of the methods you have learned for LC, RL, and RC circuits must be used. You first have to calculate the combined voltage drop of the two reactances. This value is designated E_X, and is found, as in pure LC circuits, by subtracting the smaller reactive voltage drop from the larger. The result of this calculation is the net reactive voltage drop, and is either inductive or capacitive, depending on which of the individual voltage drops was larger. As an equation, the net reactive voltage drop can be written:

$$E_X = E_L - E_C$$

if E_L is larger than E_C; or

$$E_X = E_C - E_L$$

if E_C is larger than E_L.

voltage (cont.)

Once the net reactive voltage drop is known, it is added vectorially to the voltage drop across the resistance, using the Pythagorean Theorem. The equation for this vector addition is

$$E_{APP} = \sqrt{E_R{}^2 + E_X{}^2}$$

The vector addition of all three voltage drops can be put into one equation by substituting in the above equation the values of E_X given on the previous page. Thus,

$$E_{APP} = \sqrt{E_R{}^2 + (E_L - E_C)^2}$$

if E_L is larger than E_C; or

$$E_{APP} = \sqrt{E_R{}^2 + (E_C - E_L)^2}$$

if E_C is larger than E_L.

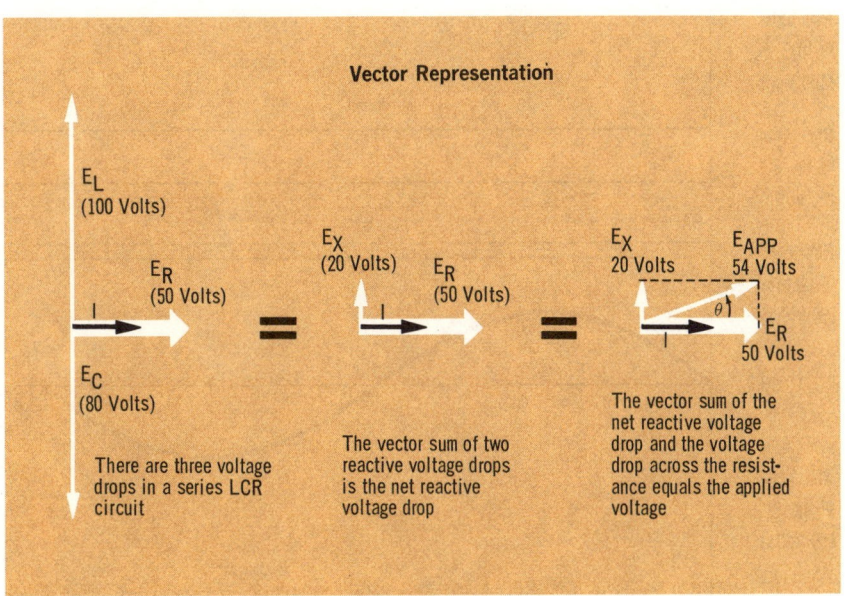

Vector Representation

E_L
(100 Volts)

E_R
(50 Volts)

E_C
(80 Volts)

There are three voltage drops in a series LCR circuit

E_X
(20 Volts)

E_R
(50 Volts)

The vector sum of two reactive voltage drops is the net reactive voltage drop

E_X
20 Volts

E_{APP}
54 Volts

θ

E_R
50 Volts

The vector sum of the net reactive voltage drop and the voltage drop across the resistance equals the applied voltage

As you can see from the vector diagrams shown, the angle between the applied voltage (E_{APP}) and the voltage across the resistance (E_R) is the same as the phase angle between the applied voltage and the circuit current. The reason for this is that E_R and I are in phase. The value of the phase angle can be found from:

$$\tan \theta = E_X/E_R$$

voltage waveforms

The voltage waveforms in a series LCR circuit are a combination of those in series RL, RC, and LC circuits. The applied voltage waveform is the sum of the instantaneous values of three voltage waveforms, all 90 degrees out of phase, rather than of two voltages less than 90 degrees out of phase, as in RL and RC circuits, or of two voltages 180 degrees out of phase, as in LC circuits. Because of their different phase relationship, the vector addition of the three voltage drops in an LCR circuit has to be done in two steps; first the two reactive voltage drops, and then their resultant and the resistive voltage drop. When they are represented as waveforms, though, the three voltage drops can be shown to add simultaneously to produce the applied voltage waveform.

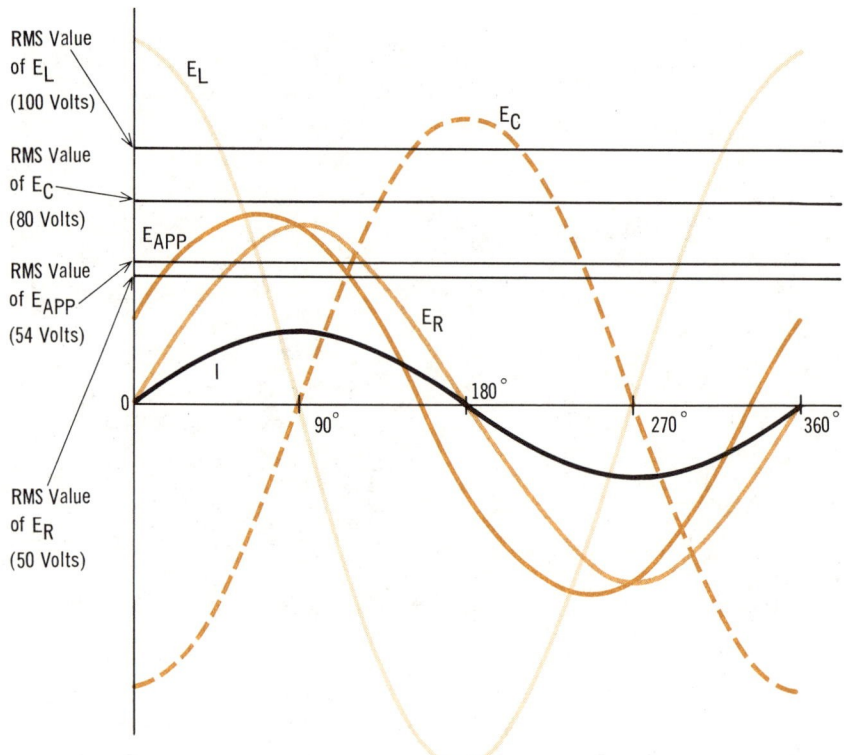

Every point on the applied voltage waveform (E_{APP}) is the algebraic sum of the instantaneous values of the E_L, E_C, and E_R waveforms

The impedance of a series LCR circuit is the **VECTOR SUM** of the inductive reactance, the capacitive reactance, and the resistance

Circuit

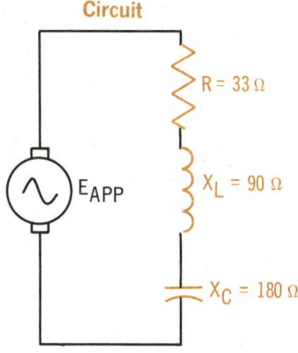

E_{APP}

$R = 33 \, \Omega$

$X_L = 90 \, \Omega$

$X_C = 180 \, \Omega$

Calculation

$$Z = \sqrt{R^2 + (X_C - X_L)^2}$$
$$= \sqrt{(33)^2 + (180 - 90)^2}$$
$$= 96 \text{ Ohms, capacitive}$$

$$\tan \theta = X/R$$
$$= (180 - 90)/33 = 2.72$$
$$\theta = 69.8°$$

impedance

The impedance of a series LCR circuit is the vector sum of the inductive reactance, the capacitive reactance, and the resistance. This vector addition is the same type as you have just learned for adding the voltage drops around a series LCR circuit. The two reactances are 180 degrees out of phase, so the net reactance, designated X, is found by subtracting the smaller reactance from the larger. Therefore,

$$X = X_L - X_C$$

if X_L is larger than X_C; or

$$X = X_C - X_L$$

if X_C is larger than X_L.

The impedance is then the vector sum of the *net reactance* and the *resistance*, and is calculated by the Pythagorean Theorem:

$$Z = \sqrt{R^2 + X^2}$$

If the equations for the net reactance and impedance are combined, the impedance can be calculated from a single equation, which is

$$Z = \sqrt{R^2 + (X_L - X_C)^2}$$

if X_L is larger than X_C; or

$$Z = \sqrt{R^2 + (X_C - X_L)^2}$$

if X_C is larger than X_L.

impedance (cont.)

When X_L is greater than X_C, the net reactance is inductive, and the circuit acts essentially as an *RL circuit*. This means that the impedance, which is the vector sum of the net reactance and resistance, will have an angle between 0 and 90 degrees. Similarly, when X_C is greater than X_L, the net reactance is capacitive, and the circuit acts as an *RC circuit*. The impedance, therefore, has an angle somewhere between 0 and 90 degrees. In both cases, the value of the impedance angle depends on the relative values of the net reactance (X) and the resistance (R). The angle can be found by:

$$\tan \theta = X/R$$

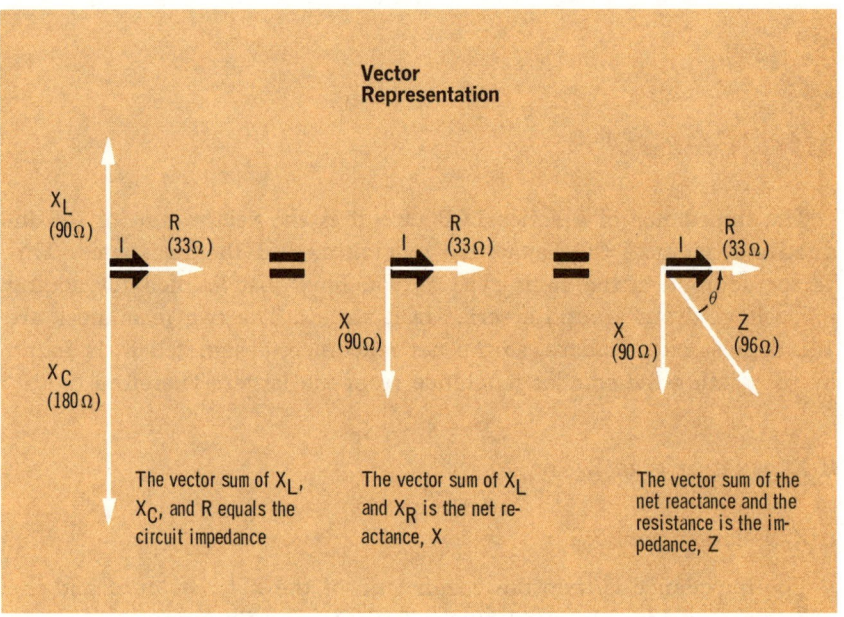

Vector Representation

The vector sum of X_L, X_C, and R equals the circuit impedance

The vector sum of X_L and X_R is the net reactance, X

The vector sum of the net reactance and the resistance is the impedance, Z

A point you should notice about the impedance of a series LCR circuit is that its value depends on the resistance and the *relative* values of X_L and X_C. High reactances do not necessarily mean a high impedance. A circuit can have very large reactances, but if their difference, or X, is small, the impedance will be low for a given value of resistance. And if R is greater than X, the impedance will be more resistive. The 10-to-1 rule applies to X and R, as it does to X_L or X_C and R in an RL or RC circuit.

The current in a series LCR circuit is calculated from Ohm's Law for a-c circuits:

$$I = E_{APP}/Z$$

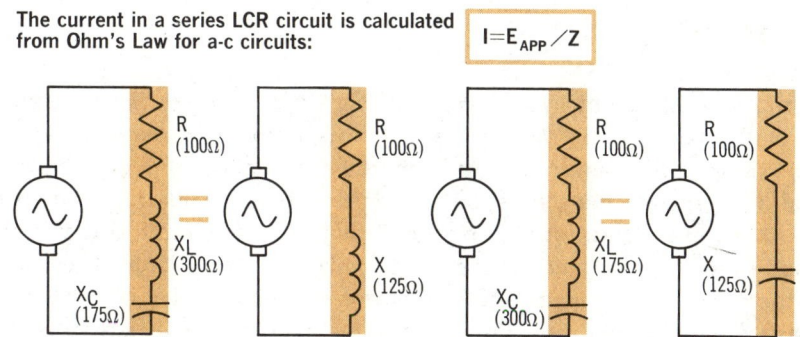

When X_L is greater than X_C, the current acts as an RL circuit, so the current lags the voltage

When X_C is greater than X_L, the circuit acts as an RC circuit, so the current leads the voltage

current

The same current flows in every part of a series LCR circuit. If the impedance and the applied voltage are known, the magnitude of the current can be calculated by Ohm's Law for a-c circuits:

$$I = E_{APP}/Z$$

The current always leads the voltage across the capacitance by 90 degrees, lags the voltage across the inductance by 90 degrees, and is in phase with the voltage across the resistance. The phase relationship between the current and the applied voltage, however, depends on the circuit *impedance*. If the impedance is *inductive* (X_L greater than X_C), the current is inductive, and *lags* the applied voltage by some phase angle less than 90 degrees. And if the impedance is *capacitive* (X_C greater than X_L), the current is capacitive, and *leads* the applied voltage by some phase angle also less than 90 degrees. The angle of the lead or lag is determined by the relative values of the net reactance and the resistance according to the equation:

$$\tan \theta = X/R$$

The greater the value of X, or the smaller the value of R, the larger is the phase angle, and the more reactive (or less resistive) is the current. Similarly, the smaller the value of X, or the larger the value of R, the more resistive (or less reactive) is the current. If either R or X is 10 or more times greater than the other, the circuit will act essentially as though it was purely resistive or reactive, as the case may be.

Other useful equations for calculating the phase angle can be derived from the vector diagrams for impedance and applied voltage. Two of these equations are

$$\cos \theta = R/Z \qquad \tan \theta = E_X/E_R$$

power

In a *pure* LC circuit, you will recall that the true power is zero, since all of the power delivered by the source is returned to it. In a series LCR circuit, the power delivered to the inductance and capacitance is also returned to the source, but in addition, power is *dissipated* by the *resistance* in the form of I^2R heating. This power represents true power, since by definition, true power is the power dissipated, or "used-up," in the circuit. The amount of true power depends on the value of the resistance and the current flow. As was pointed out previously, the impedance of a series LCR circuit, and therefore the circuit current, is determined in large part by how *close* the values of X_L and X_C are. The closer they are, the lower is the impedance, the greater is the circuit current, and the larger is the true power for a given resistance. You can see, then, that anything that affects the relative values of X_L and X_C will also affect the power dissipated in the circuit.

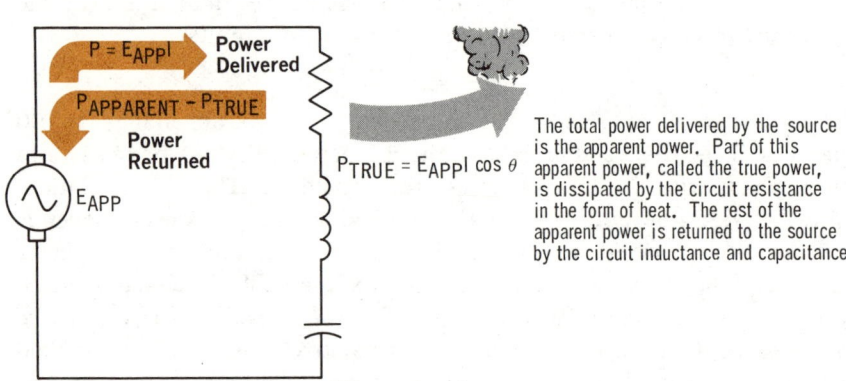

The total power delivered by the source is the apparent power. Part of this apparent power, called the true power, is dissipated by the circuit resistance in the form of heat. The rest of the apparent power is returned to the source by the circuit inductance and capacitance

The value of the true power in a series LCR circuit can be calculated from the standard equation for power in a-c circuits:

$$P_{TRUE} = E_{APP}I \cos \theta$$

The equation can also be written as $P_{TRUE} = I^2Z \cos \theta$; and since from a vector diagram for impedance, $R = Z \cos \theta$, true power can also be expressed as:

$$P_{TRUE} = I^2R$$

The apparent power, which is the total power *delivered* by the source, is simply equal to the applied voltage times the circuit current, or

$$P_{APPARENT} = E_{APP}I$$

effect of frequency

You will recall that in series LC circuits, although frequency affects the characteristics of the circuit, there is no clear-cut relationship between an increase or decrease in frequency and a corresponding increase or decrease in circuit impedance or current. The reason for this is that when frequency is increased, X_L also increases, but X_C decreases; and when frequency is decreased, X_C increases while X_L decreases. The impedance, on the other hand, varies as the *difference* between X_L and X_C. So, whether the impedance increases or decreases depends on what the *relative* values of X_L and X_C were before the frequency change took place.

As you will learn later, for every series LCR circuit there is one frequency, called the *resonant frequency*, at which X_L and X_C are equal. The frequency is determined by the values of the inductance and capacitance, and is unaffected by the circuit resistance. Any change in frequency *away* from the resonant frequency will result in an increase in the net reactance and the impedance, and a resulting decrease in current. A change which brings the frequency *closer* to the resonant frequency will have the opposite effect. Net reactance and impedance will decrease, so circuit current will increase. This is covered later.

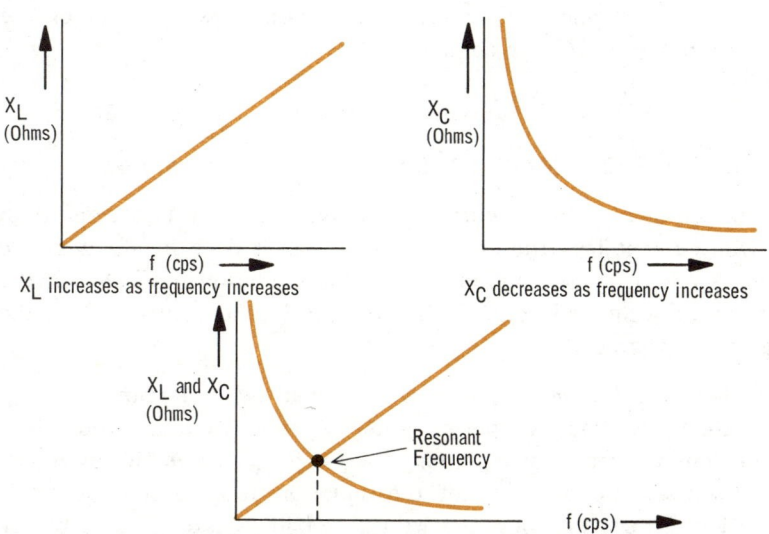

X_L (Ohms)

f (cps) ➡

X_L increases as frequency increases

X_C (Ohms)

f (cps) ➡

X_C decreases as frequency increases

X_L and X_C (Ohms)

Resonant Frequency

f (cps) ➡

For every combination of inductance and capacitance, the X_L and X_C curves intersect at one point, where the values of X_L and X_C are equal, and which corresponds to the resonant frequency. Any change in frequency from this point results in an increase in impedance. Conversely, if the frequency comes closer to this point, impedance decreases

solved problems

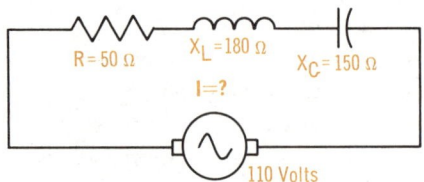

Problem 22. *What is the current in the circuit?*

The current is calculated from the equation $I = E_{APP}/Z$. The applied voltage is given, but the impedance has to be determined before the current can be calculated.

Calculating Z: X_L is larger than X_C, so the equation for Z is

$$Z = \sqrt{R^2 + (X_L - X_C)^2} = \sqrt{(50)^2 + (180 - 150)^2} = 58 \text{ ohms, inductive}$$

Calculating I:

$$I = E_{APP}/Z = 110 \text{ volts}/58 \text{ ohms} = 1.9 \text{ amperes}$$

Problem 23. *What is the relationship between the current and the applied voltage?*

With the equation that is known, the phase angle, θ, between the current and the applied voltage can be calculated with either of the equations: $\tan \theta = X/R$, or $\cos \theta = R/Z$.

$$\tan \theta = \frac{X}{R} = \frac{X_L - X_C}{R} = \frac{30}{50} = 0.6 \qquad \theta = 31°$$

$$\cos \theta = R/Z = 50/58 = 0.862 \qquad \theta = 31°$$

The impedance of the circuit is inductive, since X_L is greater than X_C, so the current lags the applied voltage, as it does in all inductive circuits. A complete description of the phase relationship between the current and the applied voltage, therefore, is that the current lags the voltage by 31 degrees.

Problem 24. *How much power is consumed in the circuit?*

Consumed power is true power, and you know that the true power can be calculated from the equation $P_{TRUE} = E_{APP}I \cos \theta$. However, you also know that another form of the equation for the true power is $P_{TRUE} = I^2R$. Since you know the values of both I and R, you can use this equation. Very often, by using this equation you will avoid the necessity of finding the value of θ in trigonometric tables.

$$P_{TRUE} = I^2R = (1.9)^2 \times 50 = 181 \text{ watts}$$

solved problems (cont.)

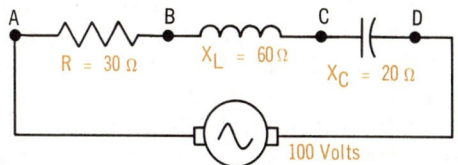

Problem 25. *What are the voltage drops between AB, BC, CD, BD, and AC?*

Before the voltage drops can be found, the circuit current has to be calculated; but before this can be done, the impedance must be determined. X_L is larger than X_C, so the impedance is calculated from the equation:

$$Z = \sqrt{R^2 + (X_L - X_C)^2} = \sqrt{(30)^2 + (60 - 20)^2}$$
$$= \sqrt{2500} = 50 \text{ ohms, inductive}$$

The current can now be found by using Ohm's Law:

$$I = E_{APP}/Z = 100 \text{ volts}/50 \text{ ohms} = 2 \text{ amperes}$$

Knowing the current, you can now calculate the voltage drops:

Calculating Voltage Drop AB: The voltage drop across the resistance is equal to the current times the resistance.

$$E_R = IR = 2 \text{ amperes} \times 30 \text{ ohms} = 60 \text{ volts}$$

Calculating Voltage Drop BC: The voltage drop across the inductance is equal to the current times the inductive reactance.

$$E_L = IX_L = 2 \text{ amperes} \times 60 \text{ ohms} = 120 \text{ volts}$$

Calculating Voltage Drop CD: The voltage drop across the capacitance is equal to the current times the capacitive reactance.

$$E_C = IX_C = 2 \text{ amperes} \times 20 \text{ ohms} = 40 \text{ volts}$$

Calculating Voltage Drop BD: Voltage drop BD is the net voltage, E_X, across the two reactances. It is equal to the difference between the two individual voltages E_L and E_C.

$$E_X = E_L - E_C = 120 - 40 = 80 \text{ volts}$$

Calculating Voltage Drop AC: Voltage drop AC is the vector sum of the voltage drops across the resistance and the inductance. These two voltages are 90 degrees out of phase, so they can be added vectorially by the Pythagorean Theorem:

$$E_{AC} = \sqrt{E_R^2 + E_L^2} = \sqrt{(60)^2 + (120)^2} = 134 \text{ volts}$$

summary

☐ A series LCR circuit consists of an inductor, a resistor, and a capacitor in series. ☐ The net reactance of the circuit determines whether the circuit behaves like an RL or RC circuit. ☐ The total reactive voltage of a series LCR circuit is given by: $E_X = E_L - E_C$, for E_L greater than E_C; and $E_X = E_C - E_L$, for E_C greater than E_L. ☐ The applied voltage is equal to the vector sum of the resistor voltage and the total reactive voltage: $E_{APP} = \sqrt{E_R + E_X{}^2}$. ☐ The phase angle is given by $\tan \theta = E_X/E_R$; or $\tan \theta = X/R$; or $\cos \theta = R/Z$.

☐ The impedance of a series LCR circuit is the vector sum of the resistance and net reactance: $Z = \sqrt{R^2 + X^2}$, where X is the net reactance, and is equal to $X_L - X_C$, or $X_C - X_L$. ☐ The current is equal to: $I = E_{APP}/Z$. It leads or lags the applied voltage depending on the net reactance of the circuit.

☐ Unlike a pure LC circuit, the LCR circuit does dissipate power in the resistor: $P_{TRUE} = E_{APP}I \cos \theta = I^2Z \cos \theta = I^2R$. ☐ The apparent power is equal to: $P_{APPARENT} = E_{APP}I$. ☐ The true power can also be found by $P_{TRUE} = P_{APPARENT} \times$ power factor. ☐ The frequency at which the net reactance is zero, or where X_L equals X_C, is known as the resonant frequency.

review questions

1. Can the voltage across the inductor or the capacitor in a series LCR circuit ever be greater than the applied voltage?
2. Can the voltage across the resistor in a series LCR circuit ever be greater than the applied voltage?

For Questions 3 to 8, consider a series LCR circuit with an applied voltage of 200 volts, an impedance of 100 ohms, an inductive reactance of 50 ohms, and a capacitive reactance of 130 ohms.

3. What is the value of the resistance?
4. What is the phase angle of the circuit?
5. What is the current in the circuit?
6. What is the apparent power? The true power?
7. What would be the impedance if the frequency were doubled?
8. What would be the impedance if the frequency were halved?
9. The applied voltage across a series LCR circuit is 200 volts, the voltage across the resistor is 160 volts, and the voltage across the inductor is 300 volts. What values can exist across the capacitor?
10. Answer Question 9, where the voltage across the inductor is 100 volts. (*Hint:* There is only one solution.)

series resonance

Resonance was briefly described as a condition that occurs when the inductive reactance and capacitive reactance of a series LCR circuit are *equal*. When this happens, the two reactances, in effect, *cancel* each other, and the impedance of the circuit is equal to the resistance. Current, therefore, is opposed only by the resistance, and if the resistance is relatively low, *very large currents* can flow. Remember, though, that the two reactances cancel each other only as far as their opposition to current is concerned. They are still present in the circuit, and because of the large current that flows when they are equal, extremely high voltages drops exist across them.

Series Resonance Occurs When:
X_L EQUALS X_C

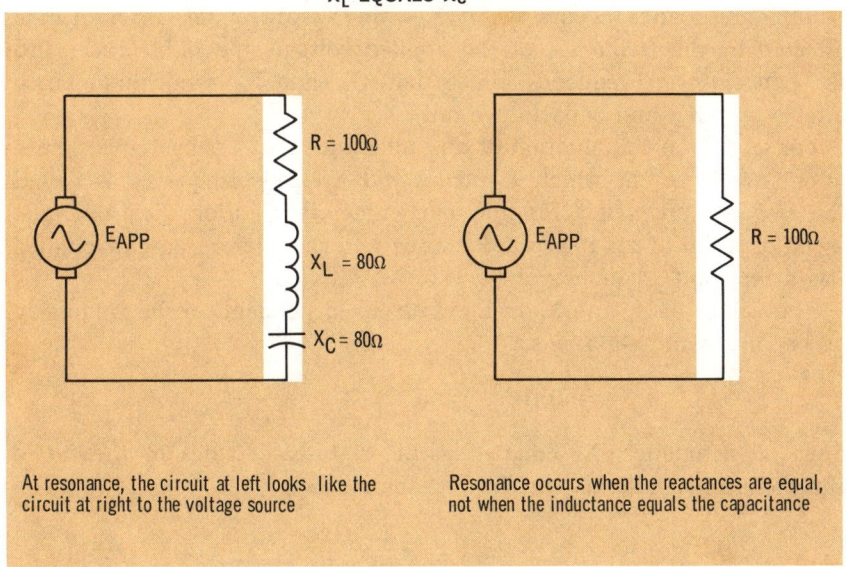

R = 100Ω
X_L = 80Ω
X_C = 80Ω
E_{APP}
R = 100Ω
E_{APP}

At resonance, the circuit at left looks like the circuit at right to the voltage source

Resonance occurs when the reactances are equal, not when the inductance equals the capacitance

The two identifying characteristics of resonance in a series LCR circuit are *low impedance* and *large current*. Actually, for any given circuit, the impedance is at its minimum and the current is at its maximum at resonance.

The type of resonance that is being described is really called *series resonance*. This is to distinguish it from another type of resonance called *parallel resonance* that occurs in parallel LCR circuits. Parallel resonance is covered later.

factors that determine resonance

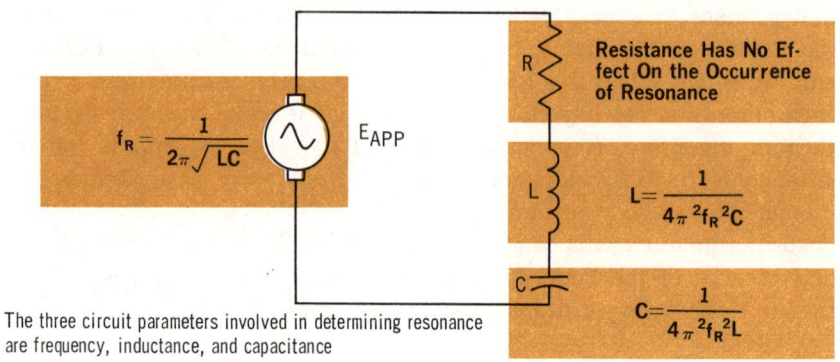

The three circuit parameters involved in determining resonance are frequency, inductance, and capacitance

Since resonance occurs when X_L and X_C are equal, resonance is affected by the *frequency* of the applied voltage, the *inductance,* and the *capacitance.* Frequency affects both X_L and X_C, while inductance affects only X_L, and capacitance only X_C.

For any given combination of an inductance and a capacitance, there is one frequency at which X_L *will equal* X_C. This frequency is called the *resonant frequency* for that particular combination. For example, a combination of a 1-henry inductance and a 4-microfarad capacitance has a resonant frequency of 80 Hz.

The values of X_L and X_C in a circuit are dependent on the frequency according to the equations:

$$X_L = 2\pi fL \qquad \text{and} \qquad X_C = \frac{1}{2\pi fC}$$

Since at resonance, X_L equals X_C, the right-hand sides of these two equations must also be equal at resonance. So,

$$2\pi fL = \frac{1}{2\pi fC}$$

When this equations is solved for f, the result is

$$f = \frac{1}{2\pi \sqrt{LC}}$$

With this equation you can find the resonant frequency, usually designated f_R, of any combination of L and C. You can also determine the value of inductance that will resonate with a particular value of capacitance at a certain frequency, and vice versa, with these equations:

$$L = \frac{1}{4\pi^2 f_R^2 C} \qquad C = \frac{1}{4\pi^2 f_R^2 L}$$

impedance at and off resonance

In most practical applications of series resonant circuits, the values of the inductance and capacitance are set, and the frequency is the variable quantity that determines whether or not a circuit is at resonance. At the resonant frequency, X_L and X_C effectively cancel each other, and the circuit impedance equals the value of the resistance.

$$Z = \sqrt{R^2 + (X_L - X_C)^2} = \sqrt{R^2 + 0} = R$$

The circuit is, therefore, completely *resistive*.

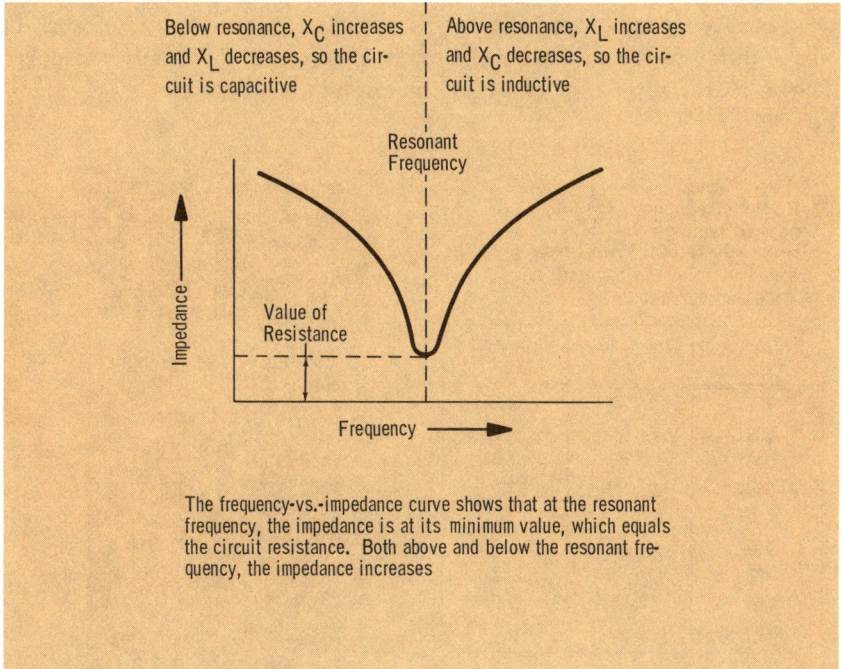

Below resonance, X_C increases and X_L decreases, so the circuit is capacitive

Above resonance, X_L increases and X_C decreases, so the circuit is inductive

Resonant Frequency

Impedance

Value of Resistance

Frequency

The frequency-vs.-impedance curve shows that at the resonant frequency, the impedance is at its minimum value, which equals the circuit resistance. Both above and below the resonant frequency, the impedance increases

If the frequency is varied above or below the resonant frequency, the net reactance, which is the difference between X_L and X_C, is no longer zero. The net reactance, therefore, has to be added to the resistance, and so the impedance increases. The further the frequency is varied from the resonant frequency, the greater the net reactance becomes, and the higher the impedance becomes. A characteristic of series resonant circuits is that when the frequency is *above* the resonant frequency, X_L is greater than X_C, so the circuit is *inductive;* and when the frequency is *below* the resonant frequency, X_C is greater than X_L, and the circuit is *capacitive*.

current at and off resonance

The *impedance* of a series LCR circuit is *minimum* at resonance, so the *current* must, therefore, be *maximum*. Both *above* and *below* the resonant frequency, circuit impedance increases, which means that *current decreases*. The further the frequency is from the resonant frequency, the greater is the impedance, and so the smaller the current becomes. At any frequency, the current can be calculated from Ohm's Law for a-c circuits, using the equation $I = E/Z$. Since at the resonant frequency the impedance equals the resistance, the equation for current at resonance becomes $I = E/R$.

If the frequency of an LCR circuit is varied and the values of current at the different frequencies are plotted on a graph, the result is a curve known as the *resonance curve* of the circuit.

If the frequency of the applied voltage is varied from 100 to 600 kHz, these are the values of impedance and current . . .

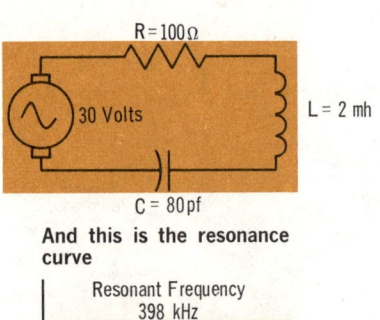

Frequency (kHz)	Impedance (Ω)	Current (ma)
100	18,634	1.61
200	7,433	4.0
300	2,862	10.48
350	1,270	23.6
398	100	300
400	122	246
450	1,240	24.2
500	2,302	13.0
600	4,221	7.1

And this is the resonance curve

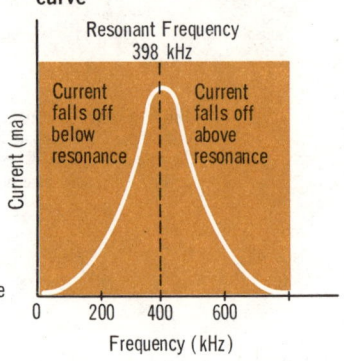

This is the characteristic shape of the resonance curve. Actual resonance curves are not always symmetrical

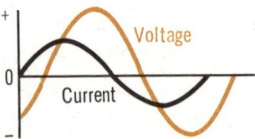

Below resonance, impedance is capacitive, so current leads voltage

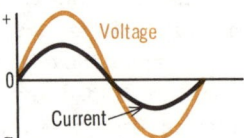

At resonance, impedance equals resistance, so current and voltage are in phase

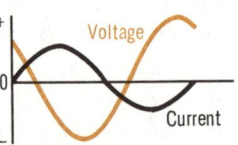

Above resonance, impedance is inductive, so voltage leads current

the resonance band

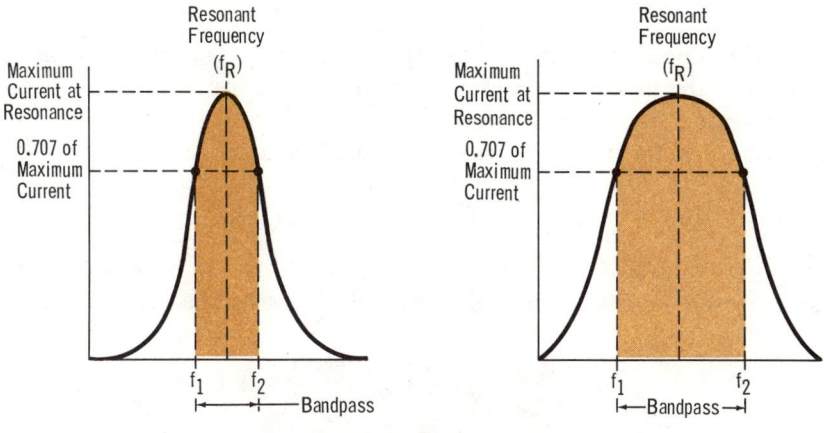

RESONANCE BAND, OR BANDPASS EQUALS $f_2 - f_1$

The range of frequencies at which the current has a value of 0.707
or greater times its value at the resonant frequency is called the
resonance band, or bandpass

From the resonance curve shown on the previous page, you can
see that although maximum current occurs only at the resonant frequency, there is a small *range* of frequencies on either side of resonance
where the current is almost as large as it is at resonance. For practical
purposes, within this range, or band, of frequencies, the circuit is at
resonance. This band of frequencies is called the *resonance band*, or
bandpass, of the circuit, and serves as a basis for rating or comparing
circuits that are used for their resonant properties. Normally, the bandpass is considered as that range of frequencies between which the
circuit current is 0.707, or greater, times its value at resonance. Thus,
if the current in a particular circuit is 2 amperes at resonance, the
bandpass of the circuit is that portion of the resonance curve between
the two points corresponding to 1.414 amperes (0.707 × 2). You can
see from this that the width of the bandpass depends on the *shape* of
the resonance curve. Curves, which because of their shape have a *wide
separation* between the points corresponding to 0.707 of resonance
current, have a *wide bandpass*. Those with a *small separation* between
the 0.707 points have a *narrow bandpass*.

The resonance band or bandpass of a circuit is also often referred
to as the *bandwidth*. All three terms mean the same thing.

effect of resistance on resonance band

The resonant frequency of a series LCR circuit depends only on the values of the inductance and capacitance. The resistance of the circuit has nothing to do with the resonant frequency. This is obvious from the equation for calculating the resonant frequency:

$$f_R = \frac{1}{2\pi\sqrt{LC}}$$

Any two values of L and C whose product is the same will have the same resonant frequency, regardless of the resistance of the circuit.

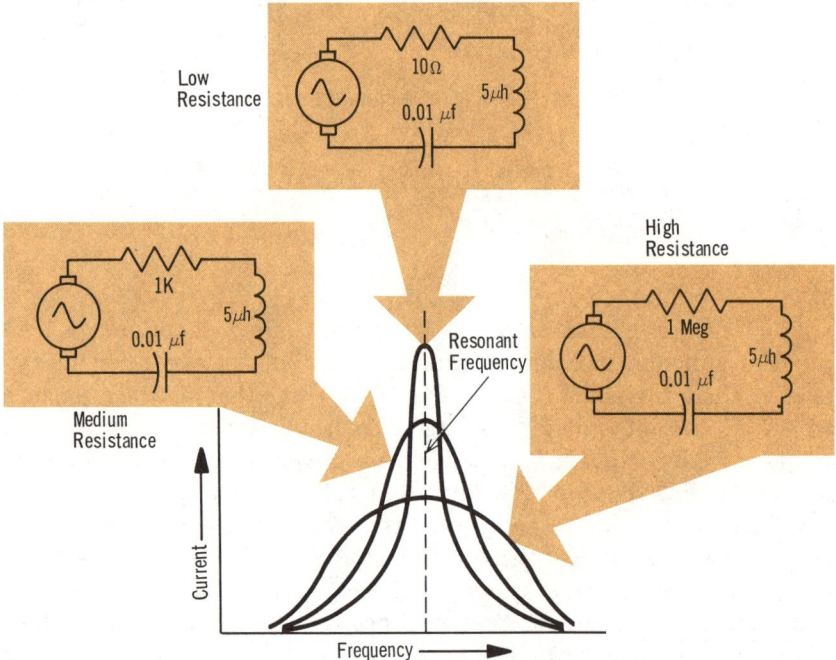

The circuit resistance determines the height and flatness of the resonance curve

Although the resistance plays no part in determining the resonant frequency, it does affect the current that flows at resonance. Since the impedance at resonance is equal to the circuit resistance, resonant current is limited only by the values of the resistance. If the resistance is small, very heavy current flows at resonance. And if the resistance is large, the current will be small, even though the resonant frequency is the same.

effect of resistance on resonance band (cont.)

Since the resonance curve of a series LCR circuit shows how the current varies with frequency, the shape of the curve depends on the value of the circuit resistance. As shown, the greater the resistance, the *lower* is the maximum height of the curve. The reason for this is that the high point on the curve corresponds to the current at resonance. In addition, the greater the resistance, the *flatter* is the curve. The reason for this is that off resonance, the shape of the curve, or the circuit current, is controlled by both the net reactance and the resistance. The net reactance depends on the frequency, but the resistance remains the same for all frequencies. So the higher the resistance, the greater is the relative control it has on the current, and the more it tends to limit the current to a constant value.

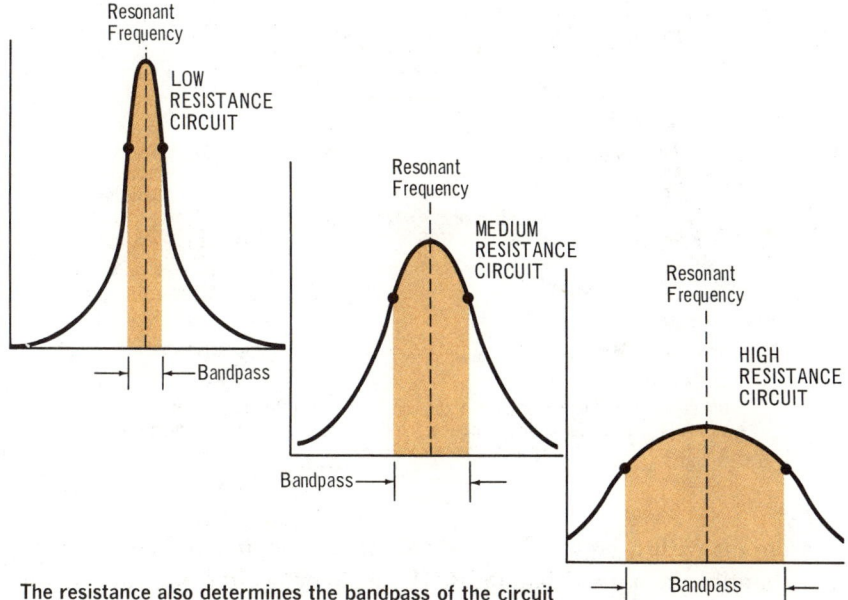

The resistance also determines the bandpass of the circuit

The flatter a resonance curve is, the further apart are the points corresponding to 0.707 of peak, or resonant, current. This means that the flatter the curve, the wider the bandpass; and since resistance determines the flatness of the resonance curve, the *higher* the *resistance, the wider* the *bandpass.* Similarly, the *smaller* the *resistance,* the *narrower* the *bandpass.*

quality

At resonance, the voltage drop across the resistance of a series resonant circuit equals the applied voltage, while the voltage drops across the inductance and capacitance are usually many times higher. The ratio of the voltage drop at resonance across either the inductance or capacitance to that across the resistance is used to express the *quality* of a series resonant circuit. This ratio is called the Q of the circuit, and is equal to either E_L/E_R or E_C/E_R. Since E_L and E_C are equal at resonance, either voltage can be used to find the Q of a circuit. If E_L is used, the equation $Q = E_L/E_R$ can also be written in the form $Q = IX_L/IR$, which can then be reduced to what is the standard equation for Q.

$$Q = X_L/R$$

You can see from this equation, that the lower the resistance, the higher the Q, or quality, of the resonant circuit.

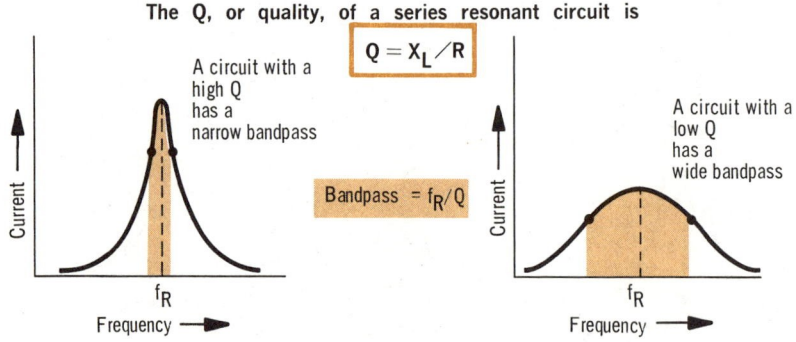

The Q, or quality, of a series resonant circuit is

A circuit with a high Q has a narrow bandpass

$$Q = X_L/R$$

Bandpass $= f_R/Q$

A circuit with a low Q has a wide bandpass

f_R
Frequency ➤

f_R
Frequency ➤

Notice that the equation for the Q of the series resonant LCR circuit is the same as you learned earlier in this volume for the Q of a coil. This shows why the d-c resistance of a coil is important in resonant circuits

A circuit with a high Q has a high, steeply sloping resonance curve, and therefore a narrow bandpass. If the Q of a circuit is known, the bandpass can be calculated by the equation:

$$\text{Bandpass (Hz)} = f_R/Q$$

The result of the equation is the total number of cycles above and below the resonant frequency that make up the resonance band. For example, if for a resonant frequency of 100 Hz this equation gives a result of 10 Hz, the resonance band extends from 95 Hz (100 − 5) to 105 Hz (100 + 5).

practical uses

You may wonder at this point whether series resonant circuits serve any practical use. The answer is yes; not only are they used in practical applications, but they are used extensively. Practically all applications of series resonant circuits make use of their property of allowing a large current to flow at frequencies in the resonance band, and providing a high opposition to current flow at frequencies outside of the resonance band. They are thus used because of their ability to *discriminate* against frequencies outside of the resonance band. As shown, a series combination of inductance, capacitance, and resistance can be connected in a circuit in such a way that it causes the circuit to respond only to those frequencies within the resonance band. Or, the same LCR combination can be connected so that the circuit responds only to those frequencies outside of the resonance band. The operation of the LCR combination is *identical* in both cases, but its effect on the overall circuit is different.

Another use of series resonant circuits is to achieve a *gain in voltage*. You know that at resonance, the voltage across X_L and X_C can be many times greater than the applied voltage. In certain applications, the voltage across either X_L or X_C is used as an output voltage to perform some function. Since this voltage is larger than the applied voltage, which is the input voltage, a voltage gain has been accomplished in the circuit.

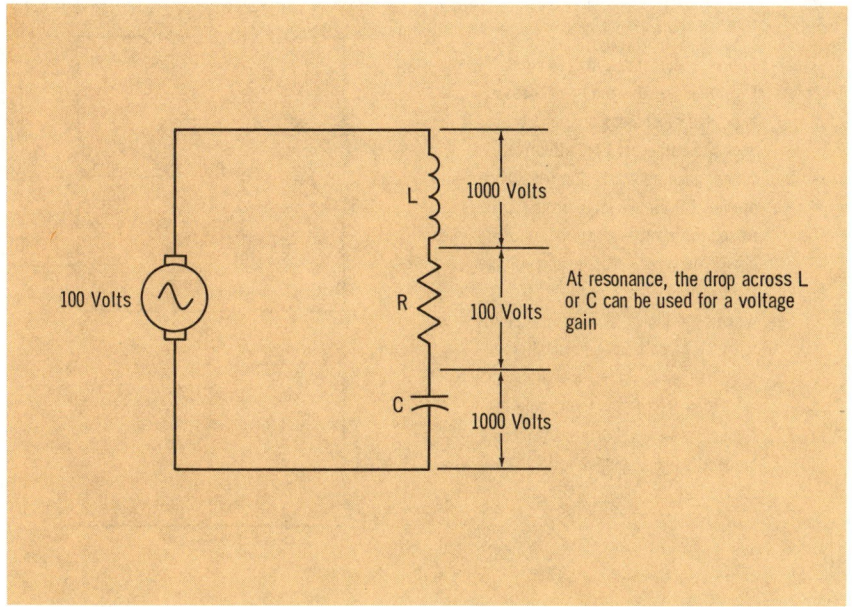

100 Volts

L — 1000 Volts

R — 100 Volts

C — 1000 Volts

At resonance, the drop across L or C can be used for a voltage gain

practical uses (cont.)

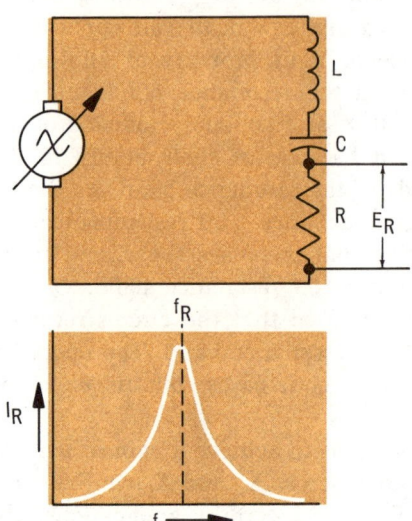

In this circuit, the current through the resistance, I_R, produces the output voltage. At resonance, X_L cancels X_C, so the circuit current, which is also I_R, is maximum. Above and below resonance, the circuit impedance increases, and I_R drops sharply. Thus, only the frequencies at or around resonance produce a useful output voltage

In this circuit, the voltage across R_3, E_{R3}, produces the output. At resonance, the reactances of L and C are opposite and equal, and effectively cancel each other. So the impedance of the LCR circuit is 10 ohms. This makes the parallel combination of R_2 and R_3 10 ohms. Most of the voltage is then dropped across R_1. Off resonance, though, the impedance of the LCR circuit can be much higher than 1000 ohms, so R_3 can drop much more than R_1. As a result, only the frequencies away from resonance will produce a useful output voltage

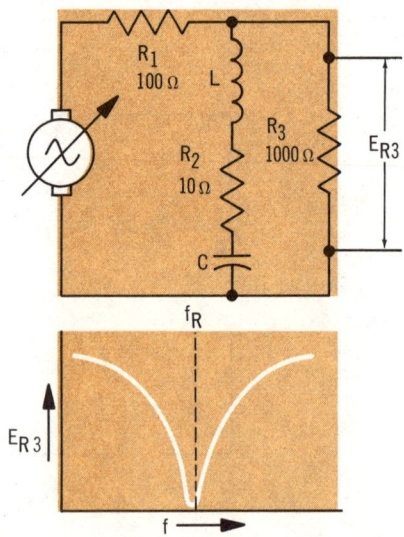

solved problems

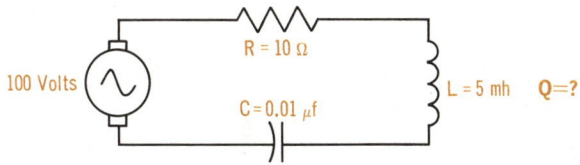

Problem 26. What is the Q of the circuit?

The Q can be calculated from the inductive reactance and resistance, using the equation $Q = X_L/R$. Or it can be found from the voltage drop across L or C and the drop across R, using either the equation $Q = E_L/E_R$ or $Q = E_C/E_R$. No matter which method is used, though, one of the reactances has to be found first; and to do this, you have to know the resonant frequency (f_R) of the circuit. Depending on the method you choose for finding Q, the following are the calculations you have to perform:

If you use $Q = X_L/R$
  (1) Find f_R
  (2) Find X_L
  (3) Find Q

If you use $Q = E_L/E_R$
  (1) Find I
  (2) Find f_R
  (3) Find X_L
  (4) Find E_L
  (5) Find Q

You can see from this comparison that the problem can be solved more easily using the equation $Q = X_L/R$.

Calculating the Resonant Frequency, f_R:

$$f_R = \frac{1}{2\pi\sqrt{LC}} = \frac{1}{6.28 \times \sqrt{0.005 \times 0.00000001}} = 22{,}500 \text{ Hz} = 22.5 \text{ kHz}$$

Calculating X_L: The inductive reactance at the resonant frequency (f_R) is the value to be calculated.

$$X_L = 2\pi f_R L = 6.28 \times 22{,}500 \times 0.005 = 707 \text{ ohms}$$

Calculating the Q:

$$Q = X_L/R = 707/10 = 70.7$$

Problem 27. What is the bandwidth of the circuit?

Since the Q of the circuit is known, the bandwidth can be easily calculated from the equation:

$$\text{Bandpass (Hz)} = f_R/Q = 22{,}500/70.7 = 318 \text{ Hz}$$

This means that for practical purposes the circuit can be considered as resonant at frequencies between 22,341 and 22,659 Hz.

solved problems (cont.)

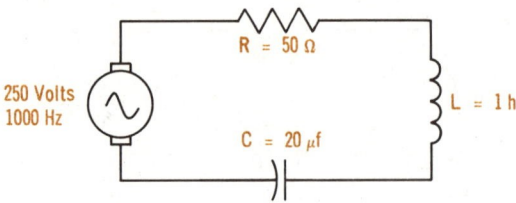

Problem 28.   *Is the circuit inductive or capacitive?*

A circuit is inductive if X_L is larger than X_C, and capacitive if X_C is larger than X_L. So you could solve this problem by calculating X_L and X_C, and seeing which is larger. This would require two separate calculations. The problem can be solved with one calculation if you remember that when the frequency of a series LCR circuit is below the resonant frequency, the circuit is capacitive; and if the frequency is above the resonant frequency, the circuit is inductive. So all you have to do is calculate the resonant frequency and compare it to the actual circuit frequency. The resonant frequency, then, is

$$f_R = \frac{1}{2\pi\sqrt{LC}} = \frac{1}{6.28 \times \sqrt{1 \times 0.00002}} = \frac{1}{0.028} = 38 \text{ Hz}$$

The circuit frequency of 1000 Hz is higher than the resonant frequency of 38 Hz, so the circuit is *inductive.*

Problem 29.   *What is the current in the circuit?*

The current is found using the same methods that apply to any series LCR circuit. This means that X_L and X_C must first be found, then the impedance, Z, and finally the current with the equation $I = E/Z$.

Calculating X_L and X_C:

$$X_L = 2\pi fL = 6.28 \times 1000 \times 1 = 6280 \text{ ohms}$$

$$X_C = \frac{1}{2\pi fC} = \frac{1}{6.28 \times 1000 \times 0.00002} = 8 \text{ ohms}$$

Calculating Z:

$$Z = \sqrt{R^2 + (X_L - X_C)^2} = \sqrt{(50)^2 + (6280 - 8)^2} = 6272 \text{ ohms}$$

Notice here that the impedance has almost exactly the same value as X_L. The reason for this is that X_L is so much larger than X_C that the effect of X_C is negligible. In practical work, if X_L is more than 10 times larger than X_C, the value of X_C could actually be neglected, and the impedance considered equal to X_L.

Calculating I:

$$I = E/Z = 250 \text{ volts}/6272 \text{ ohms} = 0.040 \text{ ampere}$$

summary

☐ Series resonance in a series LCR circuit occurs at the frequency at which the net reactance is zero. ☐ At resonance, the impedance is minimum, and current is maximum. ☐ The resonant frequency is $f_R = 1/(2\pi \sqrt{LC})$. ☐ At resonance, the series LCR circuit is completely resistive. ☐ Above resonance, X_L is greater than X_C, and the circuit is inductive. ☐ Below resonance, X_C is greater than X_L, and the circuit is capacitive. ☐ The resonance curve indicates values of current at different frequencies.

☐ The bandpass of a series LCR circuit is that range of frequencies between which the circuit current is 0.707, or greater, times its values at resonance. ☐ The resonance band, or bandpass, of a circuit is also called the bandwidth. ☐ Resistance in a series LCR circuit affects the flatness of the bandpass. The higher the resistance, the wider the bandpass; the smaller the resistance, the narrower the bandpass.

☐ The Q of a circuit is the ratio of the voltage drop across the inductor or capacitor at resonance to the voltage across the resistor. $Q = E_L/E_R = E_C/E_R$. Other equations for Q are $Q = X_L/R$ and $Q = X_C/R$. ☐ The bandpass of a series LCR circuit is given by: Bandpass $= f_R/Q$. ☐ The deviation above and below the resonant frequency is $1/2$ the bandpass.

review questions

1. What is the resonant frequency of a series LCR circuit having an L of 15 μh, a C of 15 μf, and an R of 10 ohms?
2. For the circuit of Question 1, what is the value of Q?
3. Derive the equation for the bandpass of a series LCR circuit in terms of the resistance and inductance. (*Hint:* Substitute for Q in the bandpass equation.)
4. What is the bandwidth of a series LCR circuit with R equal to 6.28 ohms and L equal to 50 millihenrys?
5. What is the power factor for a series resonant circuit?
6. If 50 volts is applied to a series LCR circuit with R equal to 5 ohms, X_L equal to 23.2 ohms, and X_C equal to 49.6 ohms, what current will flow at resonance?
7. If 50 volts is applied to a series LCR circuit whose Q is 10, what is the voltage across the capacitor at resonance?
8. In a series LCR circuit, what determines the flatness of the bandpass curve?
9. At resonance, is a series LCR circuit inductive?
10. A series LCR circuit has a current of 20 amperes at its resonant frequency of 100 kHz. What is the bandwidth if at 104 kHz the current is 14.14 amperes?

parallel LC circuits

A parallel LC circuit consists of an inductance and a capacitance connected in parallel across a voltage source. The circuit thus has two branches: an *inductive branch,* and a *capacitive branch.* In an *ideal* parallel LC circuit, which we will consider here, there is *no resistance* in either branch. This is of course impossible; but in actual practice, the resistance can be made so small as to be negligible.

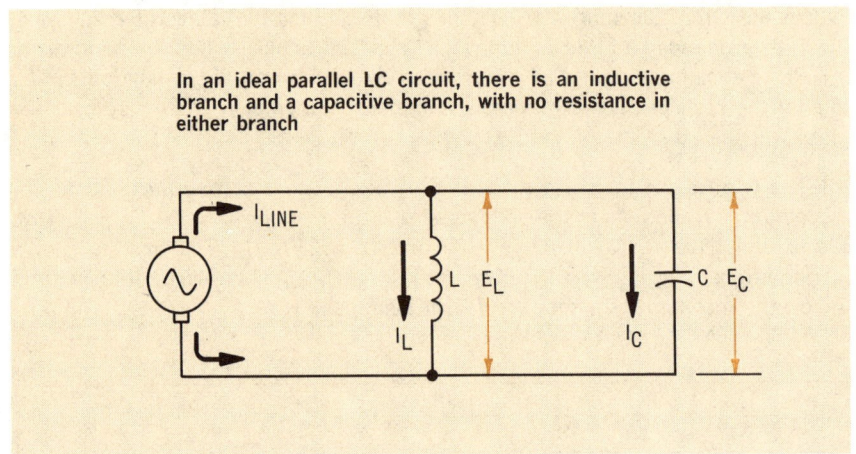

In an ideal parallel LC circuit, there is an inductive branch and a capacitive branch, with no resistance in either branch

Parallel LC circuits can have more than one inductive or capacitive branch, as well as more than one of both. However, once these circuits are reduced to their equivalent two-branch circuit, their analysis is the same as that of a simple parallel LC circuit.

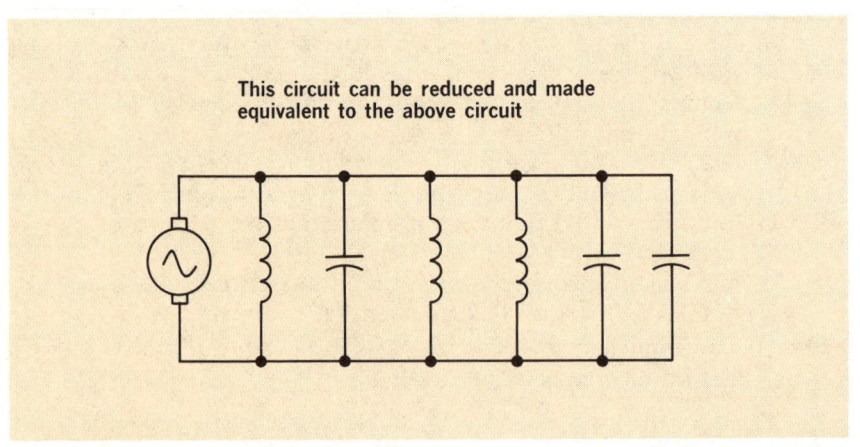

This circuit can be reduced and made equivalent to the above circuit

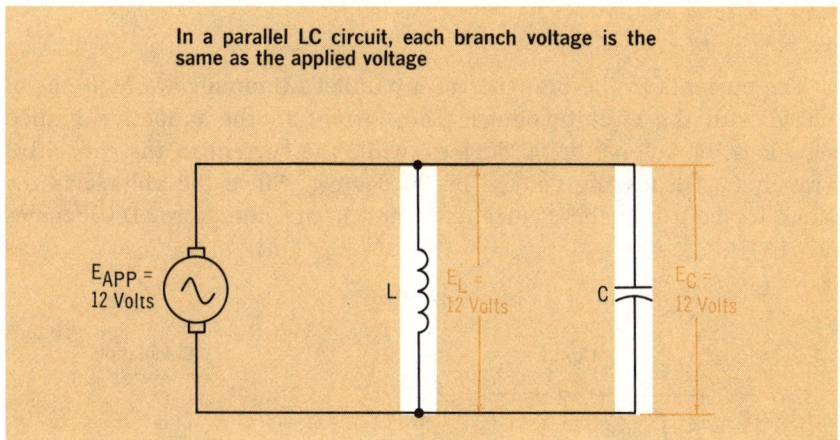

In a parallel LC circuit, each branch voltage is the same as the applied voltage

$E_{APP} = $ 12 Volts   L $E_L = $ 12 Volts   C $E_C = $ 12 Volts

voltage

The voltages across the branches of a parallel LC circuit are the same as the applied voltage, as they are in all parallel circuits. Since they are actually the *same voltage*, the branch voltages and source voltage are equal to each other and in phase. Because of this, the voltage is used as the zero-degree *phase reference*, and the phases of the other circuit quantities are expressed in relation to the voltage. The amplitude of the voltage in a parallel LC circuit is related to the circuit impedance and the line current by Ohm's Law. Thus,

$$E = I_{LINE}Z$$

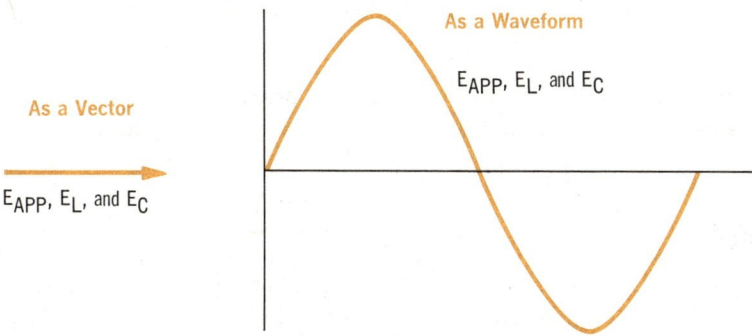

Since all the circuit voltages are the same, the voltage is used as the zero-degree reference

As a Waveform

E_{APP}, E_L, and E_C

As a Vector

E_{APP}, E_L, and E_C

current

The currents in the branches of a parallel LC circuit are *both* out of phase with the circuit voltage. The current in the inductive branch (I_L) lags the voltage by 90 degrees, while the current in the capacitive branch (I_C) leads the voltage by 90 degrees. Since the voltage is the same for both branches, currents I_L and I_C are, therefore, 180 degrees

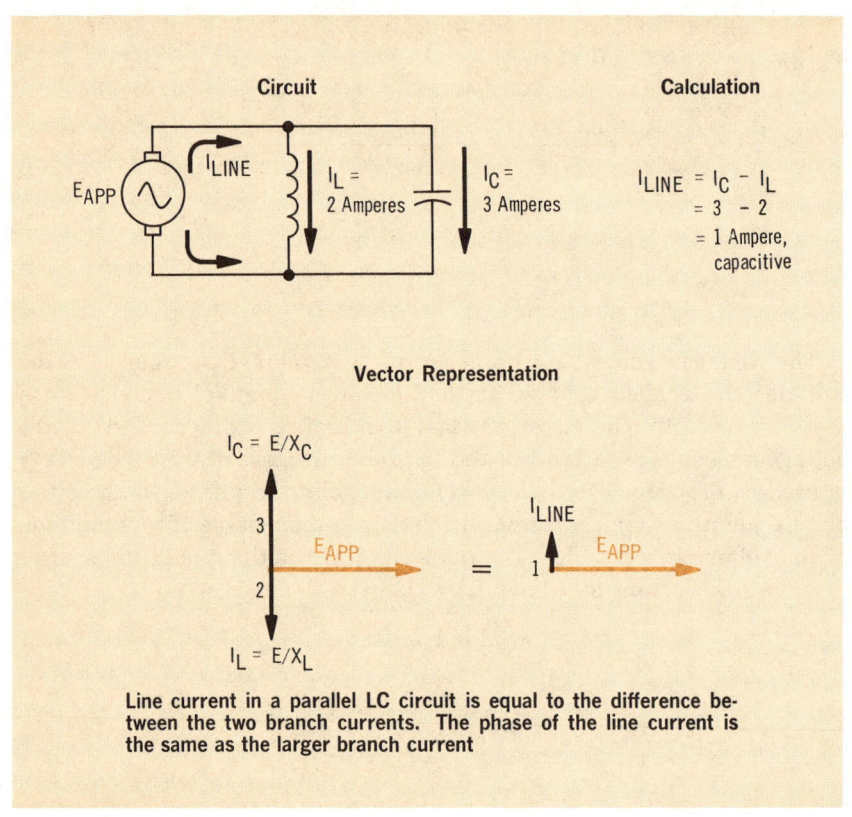

Line current in a parallel LC circuit is equal to the difference between the two branch currents. The phase of the line current is the same as the larger branch current

out of phase. The amplitudes of the branch currents depend on the value of the reactance in the respective branches, and can be found from:

$$I_L = E/X_L$$

and $$I_C = E/X_C$$

With the branch currents being 180 degrees out of phase, the line current is equal to their vector sum. This vector addition is done by subtracting the smaller branch current from the larger.

current (cont.)

The line current for a parallel LC circuit, therefore, has the *phase characteristics* of the *larger* branch current. Thus, if the inductive branch current is the larger, the line current is inductive, and lags the applied voltage by 90 degrees. And if the capactive branch current is the larger, the line current is capacitive, and leads the applied voltage by 90 degrees. In equation form, therefore, the line current is

$$I_{LINE} = I_L - I_C \text{ (if } I_L \text{ is larger than } I_C)$$

$$I_{LINE} = I_C - I_L \text{ (if } I_C \text{ is larger than } I_L)$$

If the impedance of the circuit is known, the line current can also be found by Ohm's Law:

$$I_{LINE} = E/Z$$

A unique property of the line current in a parallel LC circuit is that it is always *less* than one of the branch currents, and sometimes less than both. This is in contrast to all other parallel circuits you have studied, in which the line current was always greater than any one of the branch currents. The reason that the line current is less than the branch currents is because the two branch currents are 180 degrees out of phase. As a result of the phase difference, some *cancellation* takes place between the two currents when they combine to produce the line current. You will find later that this property is the basis of *parallel resonance*.

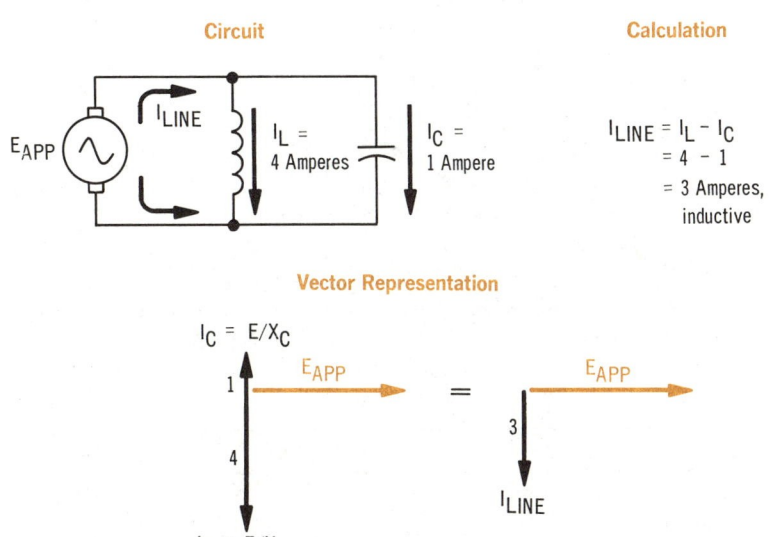

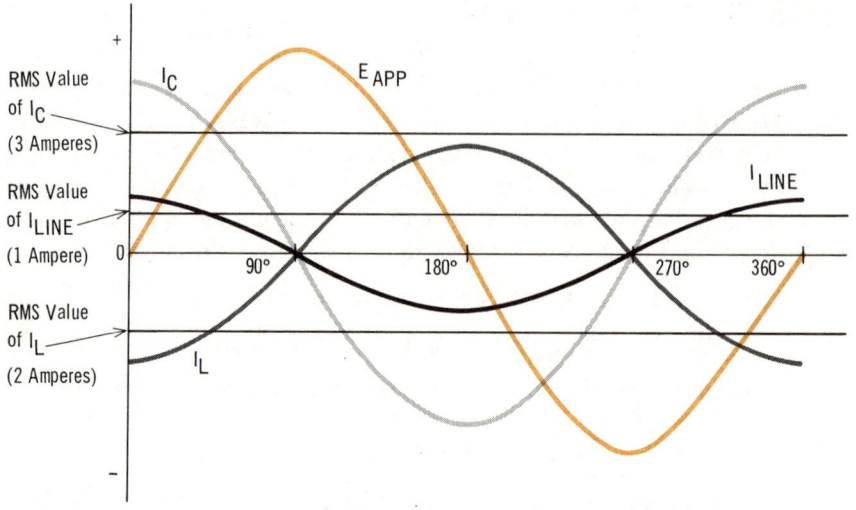

current waveforms

The waveforms of the currents in a parallel LC circuit are similar to the voltage waveforms you have seen for a series LC circuit. All of the instantaneous values of two waveforms 180 degrees out of phase are added to produce the resulting waveform, which in this case is the line current waveform. The current waveforms for the two circuits solved on the previous pages are shown.

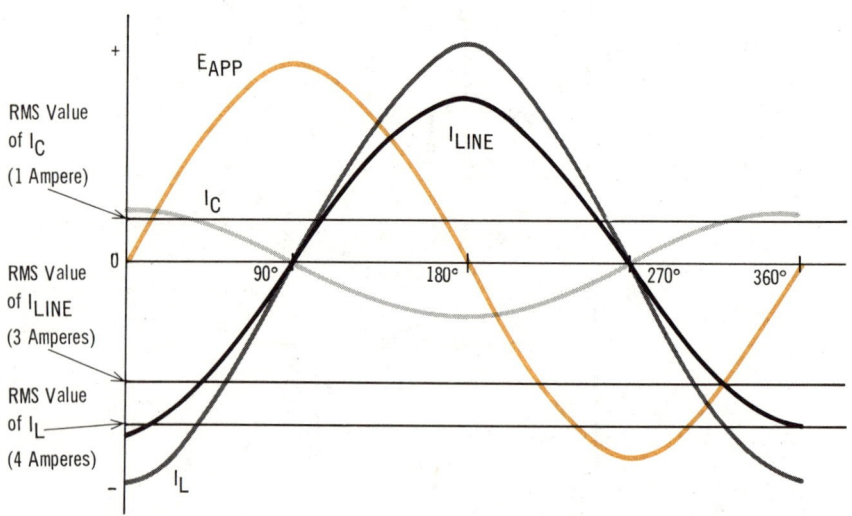

impedance

The impedance of a parallel LC circuit can be found with:

$$Z = \frac{X_L \times X_C}{X_L - X_C} \quad \text{(for } X_L \text{ larger than } X_C \text{)}$$

and
$$Z = \frac{X_L \times X_C}{X_C - X_L} \quad \text{(for } X_C \text{ larger than } X_L \text{)}$$

When using these equations, the impedance will have the phase characteristic of the *smaller* reactance.

For mathematical simplicity, a single equation can be used to find Z regardless of whether X_L or X_C is larger:

$$Z = \frac{X_L \times X_C}{X_L + X_C}$$

This is the same equation used for parallel resistances, but since X_L and X_C are 180 degrees out of phase, to use this equation, X_L is always a positive (+) quantity, and X_C is always a negative (−) quantity. When the *relative values* of X_L and X_C make Z *negative*, the impedance is *capacitive*. Similarly, when Z is *positive*, the impedance is *inductive*. Remember that X_C is not *actually* a negative quantity. It is assumed so only for this impedance equation.

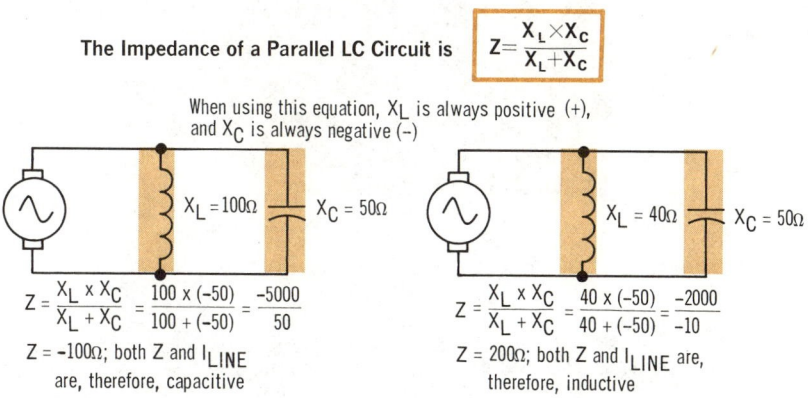

The Impedance of a Parallel LC Circuit is $Z = \frac{X_L \times X_C}{X_L + X_C}$

When using this equation, X_L is always positive (+), and X_C is always negative (−)

$Z = \frac{X_L \times X_C}{X_L + X_C} = \frac{100 \times (-50)}{100 + (-50)} = \frac{-5000}{50}$

$Z = -100\Omega$; both Z and I_{LINE} are, therefore, capacitive

$Z = \frac{X_L \times X_C}{X_L + X_C} = \frac{40 \times (-50)}{40 + (-50)} = \frac{-2000}{-10}$

$Z = 200\Omega$; both Z and I_{LINE} are, therefore, inductive

You can see that because of the difference in sign, the *closer* X_L and X_C are in value, the *larger* is the impedance. When X_L and X_C are equal, the impedance is infinitely large. As you will learn later, the circuit is then at resonance.

If the line current and applied voltage are known, the impedance can also be found by Ohm's Law:

$$Z = E_{APP}/I_{LINE}$$

effect of frequency

The frequency of the applied voltage affects the values of X_L and X_C in a parallel LC circuit. Since the value of the impedance is based on X_L and X_C, the frequency also affects the impedance. But because X_L and X_C change in *opposite* directions for a given change in frequency, no general relationships can be given for the effects of frequency on impedance, as was done for RL and RC circuits. However, as was mentioned previously in series LC circuits, for every combination of inductance and capacitance, there is one frequency called the *resonant frequency*, where the value of X_L *equals* that of X_C. And, as you learned on the previous page, when X_L and X_C are equal in a parallel LC circuit, the impedance approaches an infinitely high value. At frequencies above and below this resonant frequency, X_L and X_C have different values, and the impedance is lower.

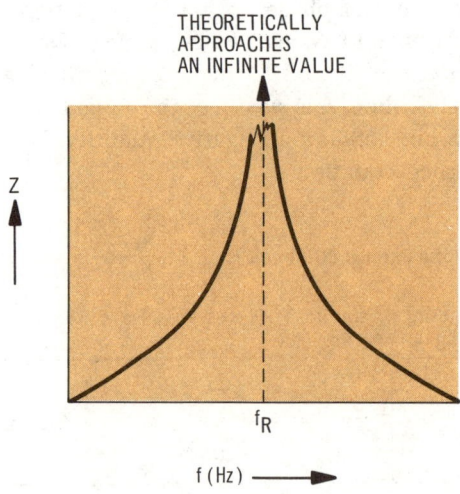

THEORETICALLY
APPROACHES
AN INFINITE VALUE

Z

f_R

f (Hz) ⟶

The frequency response curve of a parallel LC circuit would look like this. The high, or infinite impedance, point on the curve occurs at the resonant frequency

If the frequency changes away from this point, the impedance decreases. And if the frequency changes toward this point, the impedance increases

It can be said, therefore, that in parallel LC circuits, a change in frequency *towards* the resonant frequency will cause the *impedance* to *increase* and the *line current* to *decrease*. Similarly, a change in frequency away from the resonant frequency causes the *impedance* to *decrease* and the *line current* to *increase*.

solved problems

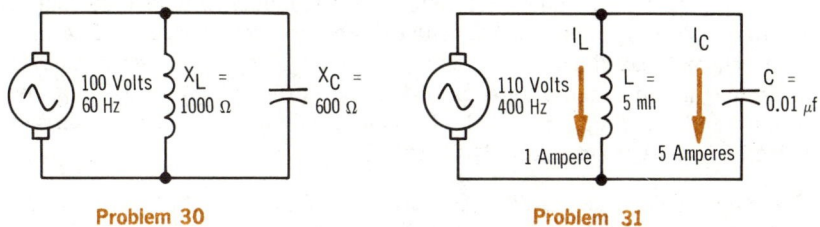

Problem 30   Problem 31

Problem 30. What is the line current in the circuit?

The problem can be solved in either of two ways. One is by finding the two branch currents, and then using them to calculate the line current. The other is by finding the impedance, and then using Ohm's Law to determine the line current. Both methods are equally suitable, so we will solve the problem both ways.

Finding I_{LINE} By the Branch Currents:

$$I_L = E_{APP}/X_L = 100/1000 = 0.1 \text{ ampere}$$

$$I_C = E_{APP}/X_C = 100/600 = 0.166 \text{ ampere}$$

Since I_C is larger than I_L, the equation for I_{LINE} is

$$I_{LINE} = I_C - I_L = 0.166 - 0.1 = 0.066 \text{ ampere}$$

I_C was the larger branch current, so the line current of 0.066 ampere is capacitive, which means it leads the applied voltage by 90 degrees.

Finding I_{LINE} By the Impedance:

$$Z = \frac{X_L \times X_C}{X_L + X_C} = \frac{1000 \times (-600)}{1000 + (-600)} = -1500 \text{ ohms, capacitive}$$

The line current, therefore, is

$$I_{LINE} = 100/1500 = 0.066 \text{ ampere}$$

Since the impedance was negative, the line current is capacitive, and so leads the applied voltage by 90 degrees.

Problem 31. In the circuit, which is larger? X_L or X_C?

This problem could be solved by calculating the values of X_L and X_C, and comparing them. But since you only have to determine which is larger, the problem can be solved by inspection. The voltages across both branches are the same, so the larger reactance will allow the least branch current to flow. Since I_L (1 ampere) is smaller than I_C (5 amperes), X_L must be larger than X_C.

summary

☐ Because the voltage across each branch of a parallel LC circuit is the same, the voltage is used as the phase reference. ☐ The inductive branch current is found by: $I_L = E/X_L$; and the capacitive branch current by: $I_C = E/X_C$. ☐ The line current has the characteristic of the larger branch current, and is equal in magnitude to the difference of the two currents: $I_{LINE} = I_L - I_C$, for I_L larger than I_C; and $I_{LINE} = I_C - I_L$, for I_C larger than I_L. ☐ The line current can also be found by Ohm's Law: $I_{LINE} = E/Z$.

☐ The impedance of a parallel LC circuit is $Z = X_L X_C/(X_L - X_C)$, for X_L greater than X_C; or $Z = X_L X_C/(X_C - X_L)$, for X_C greater than X_L. ☐ The parallel LC circuit is either capacitive or inductive, depending on the smaller of the two parallel reactances. ☐ Impedance in a parallel LC circuit can also be found by: $Z = X_L X_C/(X_L + X_C)$. When using this equation, X_C is always assigned a negative (−) sign. The resulting sign of Z then determines whether the impedance is inductive (+) or capacitive (−).

☐ At the resonant frequency, where X_L equals X_C, the impedance approaches an infinitely high value. ☐ At frequencies above and below the resonant frequency, X_L and X_C have different values, and the impedance is lower than at resonance. ☐ At resonance, the line current in a parallel LC circuit will be minimum, and will increase as the frequency is changed above and below resonance.

---

review questions

1. The line current in a parallel LC circuit is 20 amperes, and the current through the inductor is 30 amperes. What is the current through the capacitor?
2. If the current through the inductor in Question 1 is 15 amperes, what is the current through the capacitor?
3. For a parallel LC circuit, X_L is 20 ohms, and X_C is 10 ohms. What is Z?
4. Answer Question 3, where the values for X_L and X_C are interchanged.
5. Is the circuit of Question 3 inductive or capacitive?
6. Is the circuit of Question 4 inductive or capacitive?
7. Can the line current of a parallel LC circuit ever be greater than either or both branch currents?
8. At what frequency will an ideal parallel LC circuit approach an open circuit?
9. What is the line current for the circuit of Question 8?
10. Will an increase in frequency above the resonant frequency in a parallel LC circuit cause the circuit to become more capacitive or more inductive?

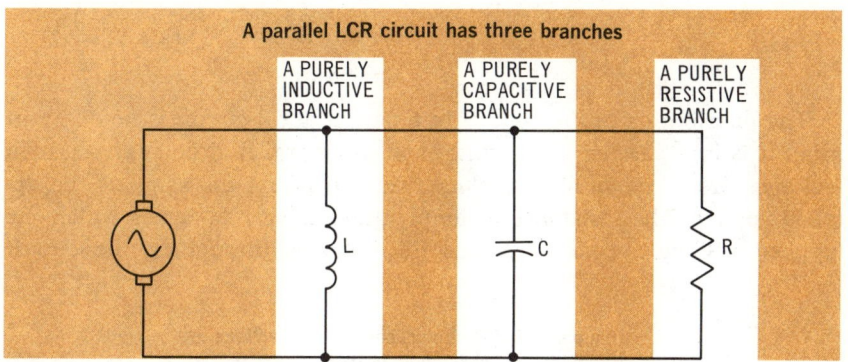

A parallel LCR circuit has three branches

A PURELY INDUCTIVE BRANCH

A PURELY CAPACITIVE BRANCH

A PURELY RESISTIVE BRANCH

parallel LCR circuits

A parallel LCR circuit is essentially a parallel LC circuit having a resistance in parallel with the inductance and capacitance. There are thus *three branches* in the circuit: a purely inductive branch, a purely capacitive branch, and a purely resistive branch. You have already learned how to analyze and solve parallel circuits that contain any two of these branches. You will now learn how to analyze circuits that have all three.

When you solve parallel LCR circuit problems, you are essentially:

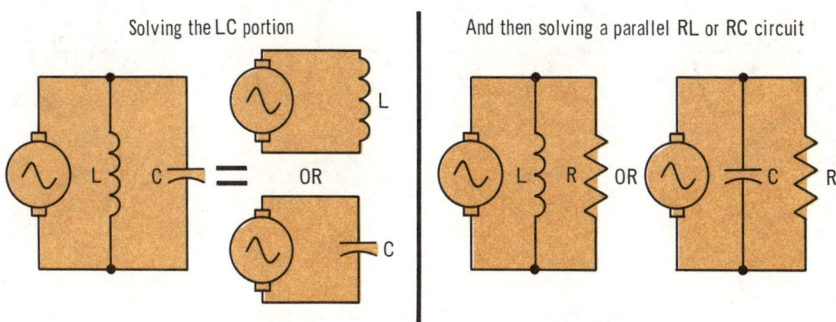

Solving the LC portion

And then solving a parallel RL or RC circuit

Actually, you will find that the solution of a parallel circuit is basically the solution of a parallel LC circuit, and then the solution of either a parallel RL circuit or a parallel RC circuit. The reason for this is that, as you will recall from the previous pages, a parallel combination of L and C appears to the source as a pure L or a pure C. So by solving the LC portion of a parallel LCR circuit first, you, in effect, reduce the circuit to an equivalent RL or RC circuit.

voltage

The distribution of the voltage in a parallel LCR circuit is no different than it is in a parallel LC circuit, or in any parallel circuit. The branch voltages are all *equal* and *in phase,* since they are the same as the applied voltage. The resistance is simply another branch across which the applied voltage appears. Because the voltages throughout the circuit

In a parallel LCR circuit, each branch voltage is the same as the applied voltage

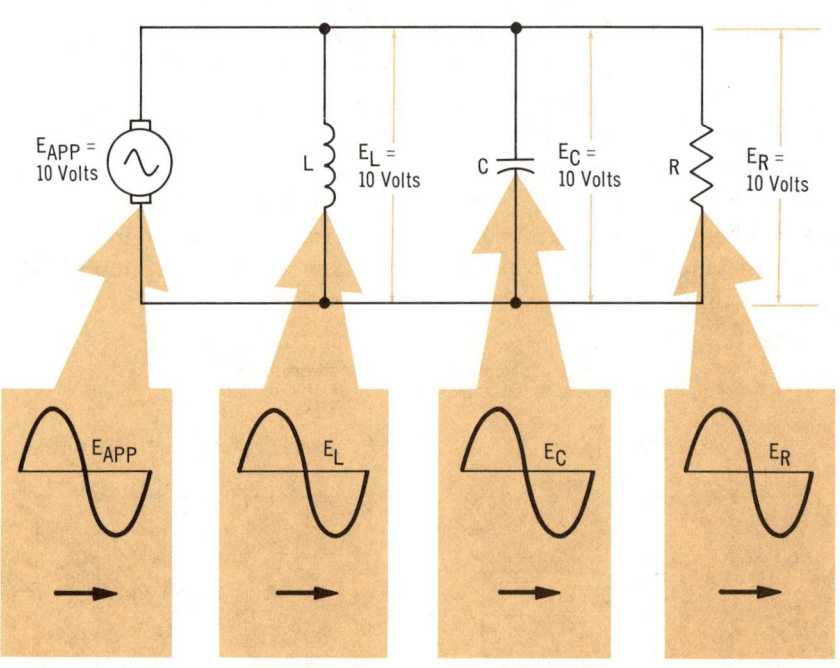

Since the circuit voltages are all the same, the voltage is used as the zero-degree phase reference

are the same, the applied voltage is again used as the zero-degree *phase reference,* as it was in parallel LC circuits. The phase angles of all other circuit quantities, then, are expressed in relation to the applied voltage. The amplitude of the applied voltage is related to the circuit impedance by Ohm's Law:

$$E_{APP} = I_{LINE}Z$$

current

The three branch currents in a parallel LCR circuit are an *inductive current* (I_L), a *capacitive current* (I_C), and a *resistive current* (I_R). Each is independent of the other, and depends only on the applied voltage and the branch resistance or reactance. The amplitudes of the branch currents are equal to:

$$I_L = E_{APP}/X_L \qquad I_C = E_{APP}/X_C \qquad I_R = E_{APP}/R$$

The three branch currents all have *different* phases with respect to the branch voltages. I_L lags the voltage by 90 degrees, I_C leads the voltage by 90 degrees, and I_R is in phase with the voltage. Since the voltages are the same, I_L and I_C are 180 degrees out of phase with each other, and both are 90 degrees out of phase with I_R. Because I_R is in phase with the voltage, it has the same zero-reference direction as the voltage. So I_C leads I_R by 90 degrees, and I_L lags I_R by 90 degrees.

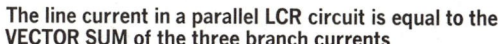

The line current in a parallel LCR circuit is equal to the VECTOR SUM of the three branch currents

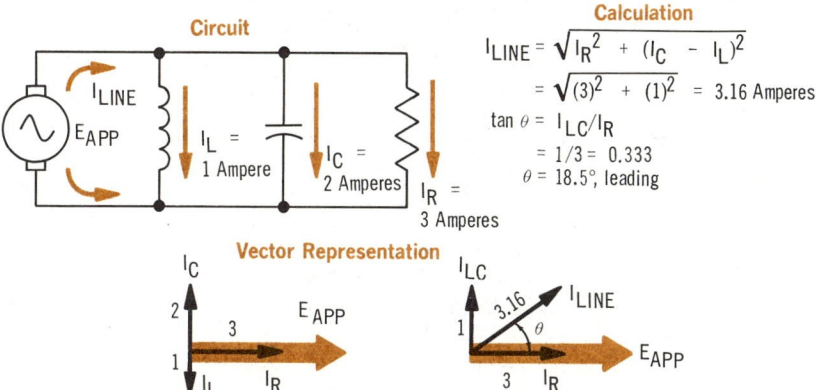

The reactive branch that has the largest current determines whether the line current leads or lags the applied voltage. The relative amplitudes of the total reactive and resistive currents then determine the angle of lead or lag.

The line current (I_{LINE}) is the vector sum of the three branch currents, and so can be calculated by adding I_L, I_C, and I_R vectorially. The different phase relationships between the three branch currents make it necessary to perform this addition in two steps. First, the two reactive currents are added, using the same methods learned for parallel LC circuits. The total of the currents, called I_{LC}, then is

$$I_{LC} = I_L - I_C \qquad \text{(if } I_L \text{ is larger than } I_C)$$

$$I_{LC} = I_C - I_L \qquad \text{(if } I_C \text{ is larger than } I_L)$$

current (cont.)

To find the line current, the quantity I_{LC} is then added to I_R, using the Pythagorean Theorem. Therefore,

$$I_{LINE} = \sqrt{I_R{}^2 + I_{LC}{}^2}$$

When the equations for these two additions are combined, they give the standard equation for line current in terms of the branch currents. This is

$$I_{LINE} = \sqrt{I_R{}^2 + (I_L - I_C)^2} \qquad \text{(if } I_L \text{ is larger than } I_C)$$

$$I_{LINE} = \sqrt{I_R{}^2 + (I_C - I_L)^2} \qquad \text{(if } I_C \text{ is larger than } I_L)$$

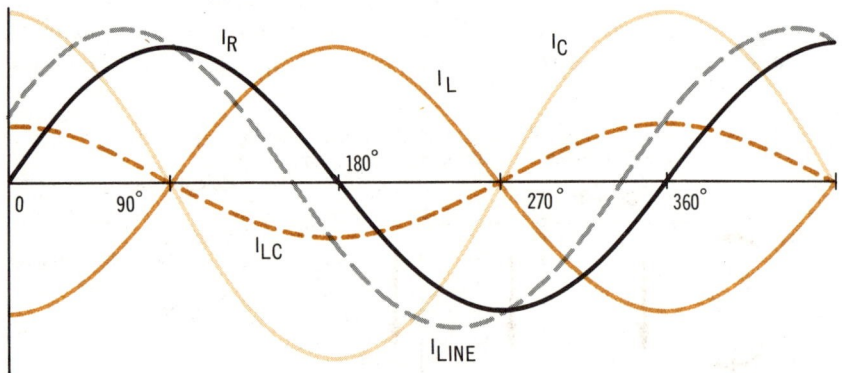

To find the line current (I_{LINE}), first add the inductive current (I_L) to the capacitive current (I_C) algebraically, to find the total reactive current (I_{LC}). Then use the Pythagorean Theorem with I_{LC} and the resistive current, I_R

Whether the line current leads or lags the applied voltage depends on which of the reactive branch currents, I_L or I_C, is the *larger*. If I_L *is larger*, I_{LINE} *lags* the applied voltage; and if I_C *is larger*, I_{LINE} *leads* the applied voltage. The exact angle of lead or lag is found by the equation:

$$\tan \theta = I_{LC}/I_R$$

Whether the angle is leading or lagging depends on which branch current, I_L or I_C, is the larger.

current waveforms

The waveforms of the currents in a parallel LCR circuit are similar to the waveforms for the voltages in a series LCR circuit. The instantaneous values of three out-of-phase waveforms combine to form one resulting waveform, which, in this case, is the circuit line current. Two of the waveforms are 180 degrees out of phase, and so their instantaneous values are always of opposite polarity. The third waveform is 90 degrees out of phase with the other two, but in phase with the applied voltage waveform. Representative waveforms of a parallel LCR circuit are shown.

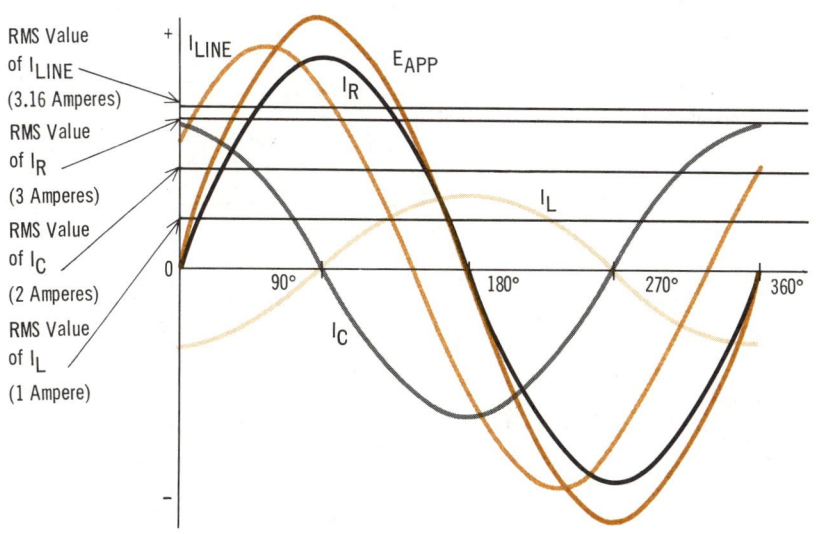

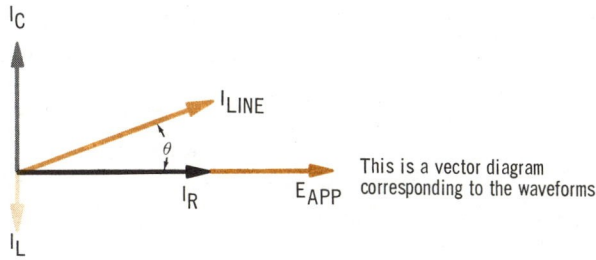

This is a vector diagram corresponding to the waveforms

impedance

To determine the impedance of a parallel LCR circuit, you must first find the net reactance (X) of the inductive and capacitive branches. Then, using X, you can find the impedance (Z) the same as you would in a parallel RL or RC circuit:

$$X = \frac{X_L \times X_C}{X_L + X_C}$$

$$Z = \frac{XR}{\sqrt{X^2 + R^2}}$$

Remember that X_L is a positive quantity, and X_C is negative. Therefore, both X and Z will also be either negative (capacitive) or positive (inductive).

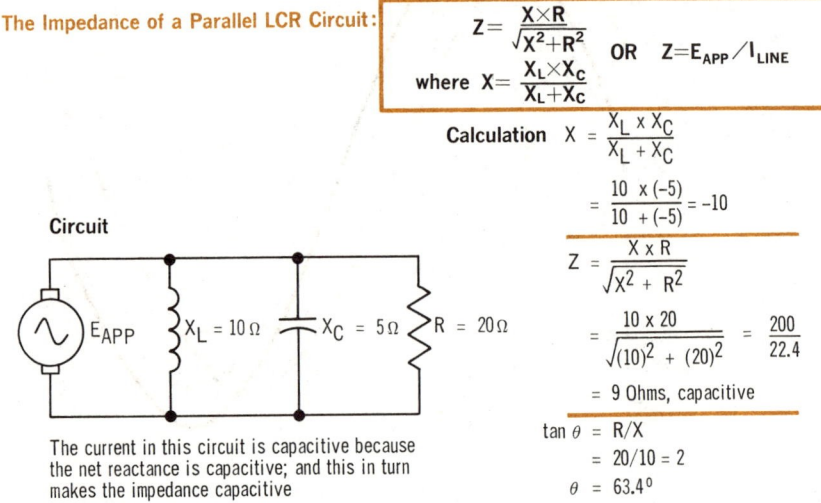

The Impedance of a Parallel LCR Circuit:

$$Z = \frac{X \times R}{\sqrt{X^2 + R^2}} \quad \text{OR} \quad Z = E_{APP}/I_{LINE}$$

$$\text{where } X = \frac{X_L \times X_C}{X_L + X_C}$$

Calculation $X = \dfrac{X_L \times X_C}{X_L + X_C}$

$$= \frac{10 \times (-5)}{10 + (-5)} = -10$$

$$Z = \frac{X \times R}{\sqrt{X^2 + R^2}}$$

$$= \frac{10 \times 20}{\sqrt{(10)^2 + (20)^2}} = \frac{200}{22.4}$$

$$= 9 \text{ Ohms, capacitive}$$

Circuit

E_{APP} $X_L = 10\,\Omega$ $X_C = 5\,\Omega$ $R = 20\,\Omega$

The current in this circuit is capacitive because the net reactance is capacitive; and this in turn makes the impedance capacitive

$$\tan\theta = R/X$$
$$= 20/10 = 2$$
$$\theta = 63.4^\circ$$

Whenever Z is inductive, the line current will lag the applied voltage. Similarly, when Z is capacitive, the line current will lead the applied voltage. The exact angle of lead or lag depends on the relative values of X and R. It can be found by the equations:

$$\tan\theta = R/X \qquad \text{or} \qquad \cos\theta = Z/R$$

Again, the 10-to-1 rule given for the other circuits applies here.

If the line current and the applied voltage are known, the impedance can also be found by Ohm's Law:

$$Z = E_{APP}/I_{LINE}$$

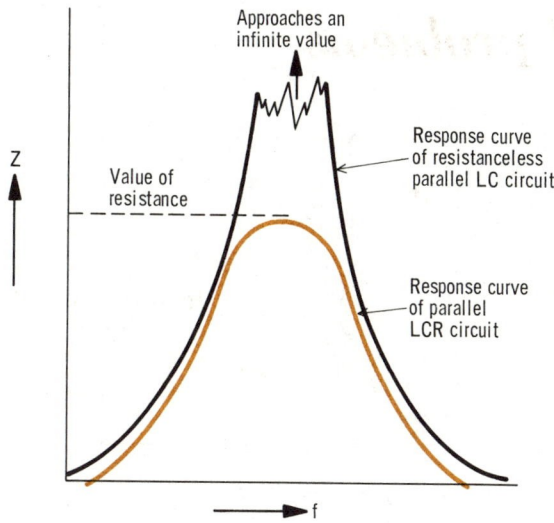

The impedance of a parallel LCR circuit can never be larger than the resistance, no matter how close to the resonant frequency the circuit frequency approaches

effect of frequency

The effects of frequency on parallel LCR circuits are similar to those for parallel LC circuits. A change in frequency causes changes in the values of both X_L and X_C, but in *different* directions. And this, in turn, results in a change in the circuit impedance. However, the exact manner in which the impedance changes depends on the *relative values* of X_L and X_C.

A definite relationship between the frequency and the impedance can only be stated in relation to the *resonant frequency*, which, you recall, is the frequency that results in X_L and X_C having equal values. This relationship is that any change in frequency *towards* the *resonant* frequency causes an *increase* in *impedance*; whereas a change in frequency *away* from the *resonant* frequency results in a *decrease* in *impedance*. This is the same relationship that exists in a parallel LC circuit. However, there is one important difference.

In an LC circuit, there is theoretically no limit to how large the impedance can go as the frequency approaches the resonant frequency. But in an LCR circuit, the impedance can *never* be larger than the value of the resistance. In effect, the resistance destroys an important characteristic of the parallel inductance and capacitance: the ability to present a very high impedance to the voltage source. This will be covered later.

solved problems

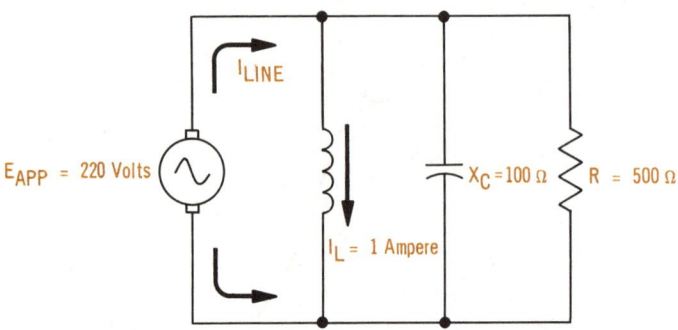

Problem 32.   *What is the line current in the circuit?*
This problem could be solved in two ways. One is by finding all of the branch currents, and using them to calculate the line current. The other is by using the equation $X_L = E_{APP}/I_L$ to find the reactance of the inductive branch, and then calculating the circuit impedance and using it to find the line current. Here, we will use the branch currents to find I_{LINE}.
Calculating I_C and I_R: Since I_L is given, only I_C and I_R have to be found.

$$I_C = E_{APP}/X_C = 220/100 = 2.2 \text{ amperes}$$

$$I_R = E_{APP}/R = 220/500 = 0.44 \text{ ampere}$$

Calculating I_{LINE}:

$$I_{LINE} = \sqrt{I_R^2 + (I_C - I_L)^2} = \sqrt{(0.44)^2 + (1.2)^2} = 1.3 \text{ amperes}$$

The current leads the applied voltage, since I_C is larger than I_L.

Problem 33.   *What is the phase angle between the current and the applied voltage in the circuit?*

$$\tan \theta = I_{LC}/I_R = 1.2/0.44 = 2.73 \qquad \theta = 69.9°$$

Since I_C is larger than I_L, the current leads the applied voltage by 69.9°.

Problem 34.   *How much power is consumed in the circuit?*
Power consumed is true power, since the values of the circuit resistance and current I_R are known, the true power can be calculated from the equation:

$$P_{TRUE} = I_R^2 R = (0.44)^2 \times 500 = 97 \text{ watts}$$

summary

☐ To solve for the impedance of a parallel LCR circuit, the parallel LC circuit is solved first, and then the net reactance is combined with the resistor as a parallel RL or RC circuit. The parallel LC circuit is characterized by the smaller of the two reactances. ☐ The voltage across a parallel LCR circuit is used as the phase reference.

☐ The currents in the three branches of a parallel LCR circuit are independent, and are found by: $I_L = E_{APP}/X_L$; $I_C = E_{APP}/X_C$; and $I_R = E_{APP}/R$. ☐ The two reactive currents are added as in a parallel LC circuit: $I_{LC} = I_L - I_C$, for I_L larger than I_C; and $I_{LC} = I_C - I_L$, for I_C larger than I_L. ☐ The line current is found by: $I_{LINE} = \sqrt{I_R^2 + I_{LC}^2}$. ☐ The phase angle is found by: $\tan \theta = I_{LC}/I_R$; or $\tan \theta = R/X$; or $\cos \theta = Z/R$. It is leading or lagging, depending on the relative magnitudes of the inductive and capacitive currents.

☐ The impedance of a parallel LCR circuit is found by finding the net reactance, X, of the L and C branches and combining it vectorially with R. The equation for the impedance is $Z = XR/(\sqrt{X^2 + R^2})$. ☐ The circuit will behave as an RL or RC circuit, depending on which reactance is the smaller. ☐ The impedance can also be found by $Z = E_{APP}/I_{LINE}$. ☐ The effect of frequency on a parallel LCR circuit is similar to that on parallel LC circuits, except that the maximum impedance can never be larger than the value of the resistance.

review questions

For Questions 1 to 5, consider a parallel LCR circuit with a resistance of 25 ohms, an inductive reactance of 50 ohms, and a capacitive reactance of 75 ohms.

1. What is the circuit impedance?
2. What is the net reactance of the circuit?
3. If the frequency is doubled, what is the impedance?
4. If the frequency is halved, what is the impedance?
5. If the capacitance is 2 microfarads, what is the inductance?

For Questions 6 to 8, consider a parallel LCR circuit with an applied voltage of 100 volts, and a line current of 20 amperes.

6. If the phase angle is 30 degrees, what is the resistance?
7. What is the power factor of the circuit?
8. What is the apparent power of the circuit?
9. For a parallel LCR circuit, the resistive, inductive, and capacitive currents are 10, 22.5, and 15 amperes, respectively. What is the line current?
10. What is the phase angle of the circuit of Question 9?

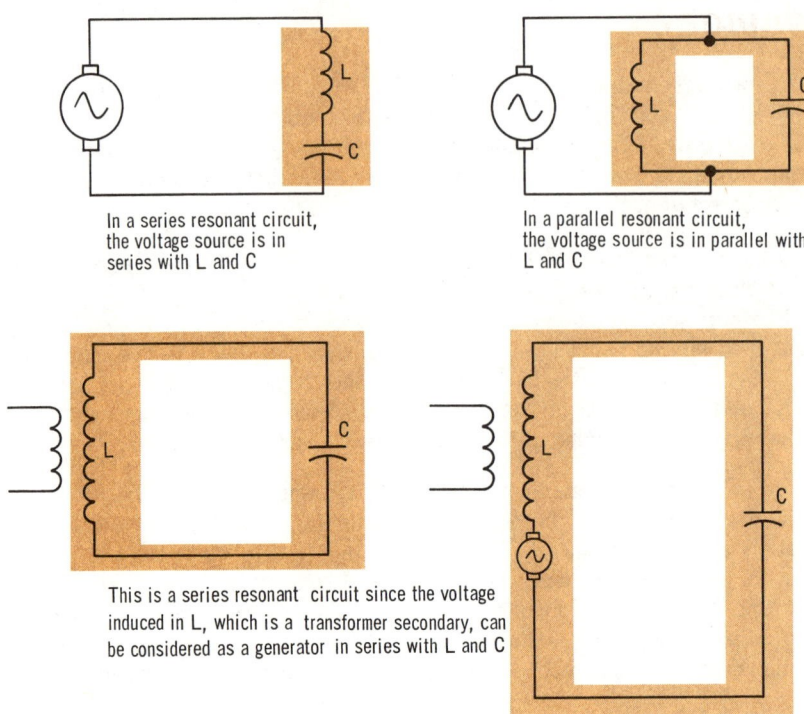

In a series resonant circuit, the voltage source is in series with L and C

In a parallel resonant circuit, the voltage source is in parallel with L and C

This is a series resonant circuit since the voltage induced in L, which is a transformer secondary, can be considered as a generator in series with L and C

parallel resonance

In parallel LC circuits, parallel resonance is the equivalent of series resonance in series LCR circuits. However, the *characteristics* of parallel resonance are quite *different* from those of series resonance.

For any given values of inductance and capacitance, the *frequency* at which parallel resonance takes place is *identical* to the frequency at which series resonance would take place for the same values of L and C. Therefore, parallel resonance can also be found by:

$$f_R = \frac{1}{2\pi\sqrt{LC}}$$

This being the case, the other equations you learned for finding L or C can also be used here:

$$L = \frac{1}{4\pi^2 f_R{}^2 C} \quad \text{and} \quad C = \frac{1}{4\pi^2 f_R{}^2 L}$$

Since it is sometimes difficult to distinguish series from parallel resonant circuits, think of a series resonant circuit as one with the voltage source in series with L and C; and a parallel resonant circuit as one with the voltage source in parallel with L and C.

the tank circuit

The properties of a parallel resonant circuit are based on the action that takes place between the parallel inductance and capacitance which is often called a *tank circuit*, because it has the ability to *store* electrical energy.

The action of a tank circuit is basically one of *interchange* of energy between the inductance and capacitance. If a voltage is *momentarily* applied across the tank circuit, C charges to this voltage. When the applied voltage is removed, C discharges through L, and a magnetic field is built up around L by the discharge current. When C has discharged, the field around L collapses, and in doing so induces a current that is in the same direction as the current that created the field. This current, therefore, charges C in the opposite direction. When the field around L has collapsed, C again discharges, but this time in the direction opposite to before. The discharge current again causes a magnetic field around L, which when it collapses, charges C in the same direction in which it was initially charged.

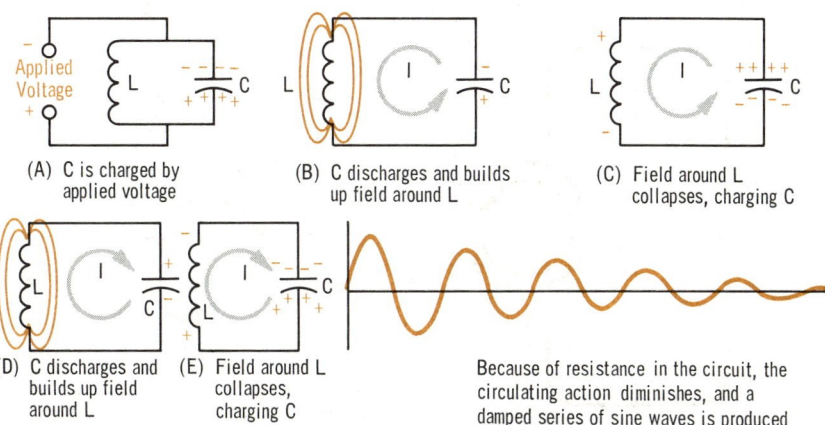

(A) C is charged by applied voltage

(B) C discharges and builds up field around L

(C) Field around L collapses, charging C

(D) C discharges and builds up field around L

(E) Field around L collapses, charging C

Because of resistance in the circuit, the circulating action diminishes, and a damped series of sine waves is produced

This interchange of energy, and the circulating current it produces, would continue *indefinitely* producing a series of *sine* waves, if we had an *ideal* tank circuit with no resistance. However, since some resistance is always present, the circulating current gradually diminishes as the resistance dissipates the energy in the circuit in the form of heat. This causes the sine-wave current to be *damped out*. If a voltage was again momentarily applied across the circuit, the interchange of energy and accompanying circulating current would begin again.

current and impedance at resonance

When an a-c voltage is applied to any parallel LC circuit having zero resistance, the currents in the inductive and capacitive branches are equal to:

$$I_L = E/X_L \qquad I_C = E/X_C$$

At resonance, X_L equals X_C, so the two currents are also equal. And since in a parallel LC circuit the two currents are 180 degrees out of phase, the line current, which is their vector sum, must be *zero*. Thus, the only current is the circulating current in the tank circuit. No line current flows. And if no line current flows, this means that the circuit has *infinite impedance* as far as the voltage source is concerned.

At resonance, an ideal parallel resonant circuit has zero line current and infinite impedance

$$Z = E_{APP}/I_{LINE}$$
$$= E_{APP}/0$$
$$= \text{Infinity}$$

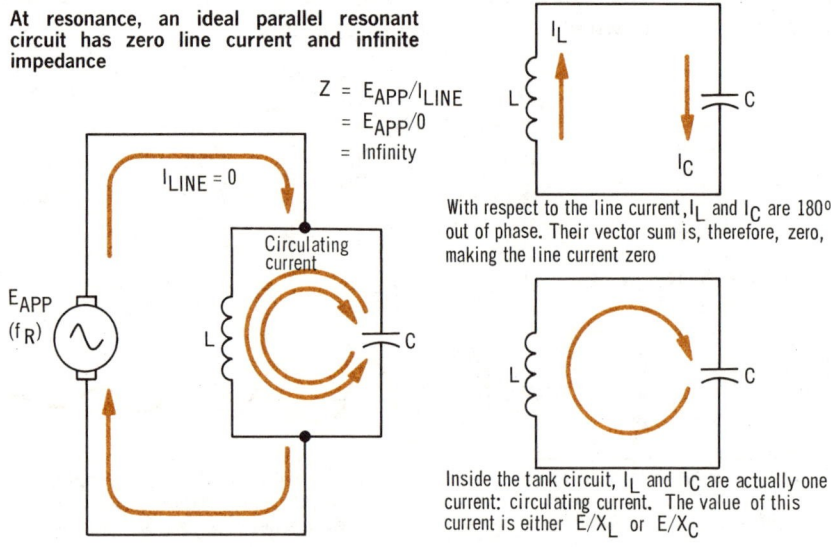

With respect to the line current, I_L and I_C are 180° out of phase. Their vector sum is, therefore, zero, making the line current zero

Inside the tank circuit, I_L and I_C are actually one current: circulating current. The value of this current is either E/X_L or E/X_C

These two conditions of zero line current and infinite impedance are characteristic of *ideal* parallel resonant circuits at resonance. In practical circuits, which contain some resistance, the theoretical conditions of zero line current and infinite impedance are not realized. Instead, practical parallel resonant circuits have *minimum line current* and *maximum impedance* at resonance. You will recognize that this is exactly *opposite* to series resonant circuits, which have maximum current and minimum impedance at resonance.

current and impedance
off resonance

In the ideal parallel resonant circuit at resonance, the branch currents, I_L and I_C, are equal, so the line current is zero and the circuit impedance is infinite. Above and below the resonant frequency, one of the reactances (X_L or X_C) is larger than the other. The two *branch currents* are therefore *unequal*, and the *line current*, which equals their vector sum (or arithmetic difference), has some value *greater than zero*. And since line current flows, the circuit *impedance is no longer infinite*. The further frequency is from the resonant frequency, the greater is the difference between the values of the reactances. As a result, the larger is the line current, and the smaller is the circuit impedance.

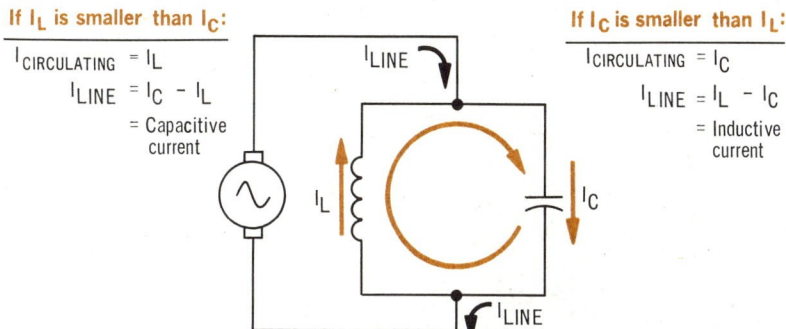

If I_L is smaller than I_C:

$I_{CIRCULATING} = I_L$
$I_{LINE} = I_C - I_L$
$\quad = $ Capacitive current

If I_C is smaller than I_L:

$I_{CIRCULATING} = I_C$
$I_{LINE} = I_L - I_C$
$\quad = $ Inductive current

The circulating current in the tank circuit equals the smaller of the two branch currents. The line current is equal to the difference between the two branch currents, and has the same phase relationship as the larger of the two

The impedance at any frequency can be calculated from the equation you learned previously for the impedance of a parallel LC circuit. That is,

$$Z = \frac{X_L \times X_C}{X_C + X_L}$$

Where X_L and X_C are the reactances at the particular frequency involved. Also, as you learned for parallel LC circuits, the impedance is always greater than at least one of the reactances, instead of always being less than both, as it is for other types of parallel a-c circuits.

current and impedance
off resonance (cont.)

At frequencies off resonance, the line current is equal to the difference (vector sum) between the values of the branch currents. Circulating current still flows in the tank circuit, and is equal to the
smaller of the two branch currents. Thus, if I_L is 5 amperes and I_C
is 3 amperes, the circulating current is 3 amperes and the line current
is 2 amperes. In effect, the line current is that portion of the larger
branch current that does not take part in the circulating current of
the tank circuit. Since branch current I_L is inductive and branch current I_C is capacitive, the line current is *inductive* if I_L is *larger* (which
means that X_L is smaller than X_C), and *capacitive* if I_C is *larger* (which
means that X_C is smaller than X_L). This is the opposite of series
resonant circuits, which you remember are inductive when X_L is
larger, and capacitive when X_C is larger.

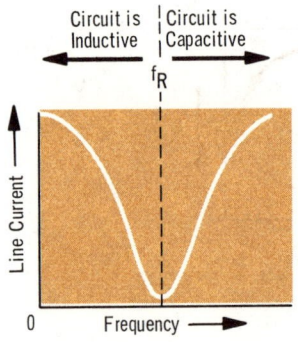

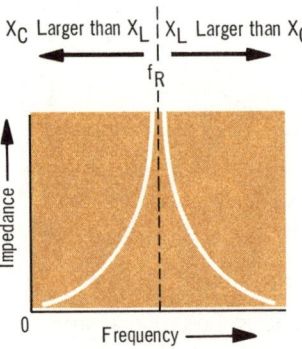

Curves showing how the line current and impedance of an ideal
parallel resonant circuit change as the frequency is varied are shown.
Notice that the characteristic shape of the *current*-vs.-frequency curve
for the parallel resonant circuit is the same as the *impedance*-vs.-
frequency curve for series resonant circuits. Similarly, the impedance-
vs.-frequency curve for a parallel circuit is the same as the current-vs.-
frequency curve for series circuits.

practical
parallel resonant circuits

Practical parallel resonant circuits differ from the ideal parallel circuit just described in one major respect: practical circuits contain *resistance*. This resistance is contained in the inductance, the capacitance, and the interconnecting wires. Normally, however, only the resistance of the inductance is large enough to be considered. In analyzing a circuit, this resistance is considered to be in *series* with the *inductance*. A practical parallel resonant circuit, therefore, consists of a purely capacitive branch, and an inductive branch that is actually a series RL circuit. You will remember from what you have learned about the 10:1 ratio, that if the inductive reactance of the inductance is 10 times or more greater than its resistance, the resistance can generally be neglected. The circuit could then be analyzed in the same manner as the ideal parallel resonant circuit. For purposes of explanation, though, we will include the effects of the resistance regardless of its value relative to X_L.

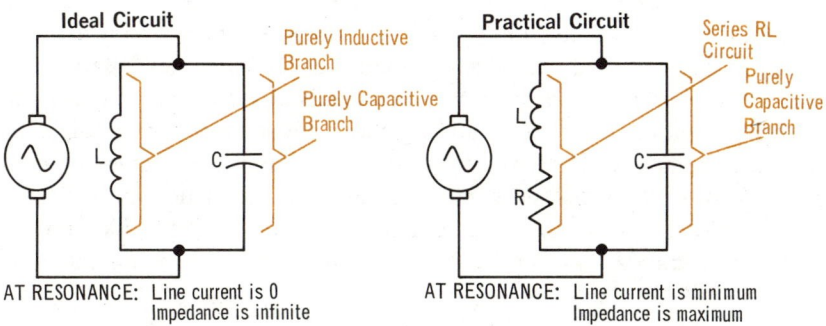

Ideal Circuit — Purely Inductive Branch — Purely Capacitive Branch

AT RESONANCE: Line current is 0
Impedance is infinite

Practical Circuit — Series RL Circuit — Purely Capacitive Branch

AT RESONANCE: Line current is minimum
Impedance is maximum

The principal effect of the resistance in a parallel resonant circuit is that it causes the current in the inductive branch to lag the applied voltage by a phase angle of *less* than 90 degrees, instead of exactly 90 degrees as is the case in the ideal circuit. As a result, the two branch currents are not 180 degrees out of phase. For simplicity, resonance can still be considered as occurring when X_L equals X_C, but now when the two branch currents are added vectorially, their sum is *not* zero. This means that at resonance some line current flows. And since there is line current, the impedance cannot be infinite, as it is in the ideal circuit. Thus, at resonance, practical parallel resonant circuits have *minimum* line current and *maximum* resistance, instead of zero line current and infinite impedance, as do ideal circuits.

practical
parallel resonant circuits (cont.)

At resonance, the *line current* in a practical parallel resonant circuit is *in phase* with the *applied voltage*. The reason for this is that since X_L and X_C are equal, the reactive (inductive) component of the current in the RL branch is equal to, and cancels, the current (capacitive) in the capacitive branch. Only the *resistive* component of the current in the RL branch, therefore, flows in the line. Since the capacitive branch contains no resistance, the current in it is equal to:

$$I_C = E/X_C$$

The current in the inductive branch is calculated the same as in any series RL circuit, and so is equal to:

$$I_L = E/\sqrt{R^2 + X_L^2}$$

The line current can then be found by adding the two branch currents *vectorially*. However, the branch currents differ in phase by less than 180 degrees, but more than 90 degrees. Therefore, to add them vectorially you cannot use the Pythagorean Theorem, since it applies only to quantities 90 degrees apart, and you cannot subtract them arithmetically, since this only applies to quantities 180 degrees apart. They can, however, be added by first being resolved into their vertical and horizontal components, then adding the components, and finally finding the resultant of the total components. Addition of vectors by components was described earlier. At resonance, the calculation of the line current, as well as the circulating current in the tank and the circuit impedance, can be done much more easily by using the Q of the circuit.

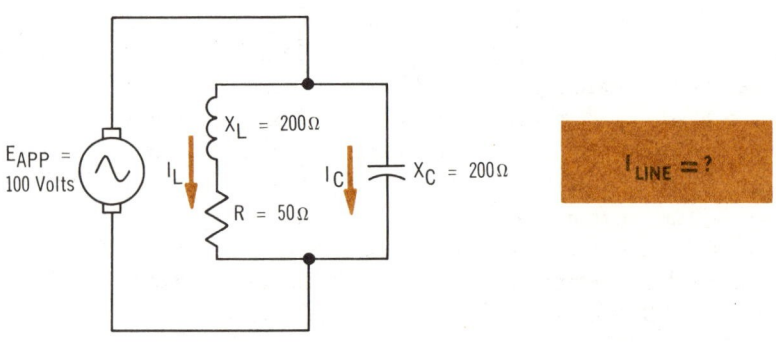

Branch currents are added vectorially by components to find the line current in practical parallel resonant circuits

practical
parallel resonant circuits (cont.)

The branch currents, I_L and I_C, of the circuit on page 4-138 are

$$I_L = \frac{E}{\sqrt{R^2 + X_L^2}} = \frac{100}{\sqrt{(50)^2 + (200)^2}} = 0.486 \text{ ampere}$$

$$\tan \theta = X_L/R = 200/50 = 4 \qquad \theta = 76°; \text{ current lagging voltage}$$

$$I_C = E/X_C = 100/200 = 0.5 \text{ ampere}$$

The horizontal (H) and vertical (V) components of the branch currents are

$$I_{L(H)} = I_L \cos \theta = 0.486 \times 0.242 = 0.118 \text{ ampere}$$

$$I_{L(V)} = I_L \sin \theta = 0.486 \times 0.97 = 0.47 \text{ ampere}$$

$$I_{C(H)} = 0; \text{ no resistive component} \qquad I_{C(V)} = 0.5 \text{ ampere}$$

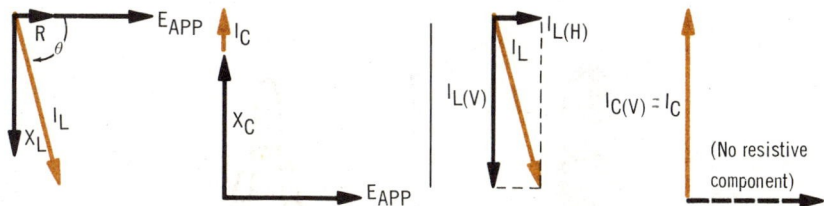

The total horizontal and vertical components are

$$I_{TOT(H)} = I_{L(H)} + I_{C(H)} = 0.118 + 0 = 0.118 \text{ ampere}$$

$$I_{TOT(V)} = I_{C(V)} - I_{L(V)} = 0.5 - 0.47 = 0.03 \text{ ampere}$$

The minus sign of equation $I_{TOT(V)}$ indicates that $I_{L(V)}$ vector points down, or lags the voltage.

The resultant of the total components, which is the line current, is

$$I_{LINE} = \sqrt{I_{TOT(H)}^2 + I_{TOT(V)}^2} = \sqrt{(0.118)^2 + (0.03)^2} = 0.123 \text{ ampere}$$

The phase angle between the line current and the applied voltage is

$$\tan \theta = \frac{I_{TOT(V)}}{I_{TOT(H)}} = \frac{0.03}{0.118} = 0.254$$

$$\theta = 14.3°; \text{ current leading voltage}$$

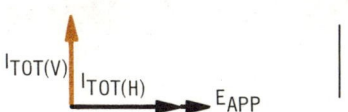

the resonance band

You will remember that for every series resonant circuit there is a *range* of frequencies above and below the resonant frequency at which, for practical purposes, the circuit can be considered as being at resonance. This range of frequencies was called the *resonance band* or *bandpass,* and consisted of all the frequencies at which the circuit current was 0.707 or more times its value at resonance. Parallel resonant circuits also have a resonance band; but it is defined in terms of the *impedance-vs.-frequency curve,* and consists of all the frequencies that produce a circuit impedance 0.707 or more times the impedance at resonance.

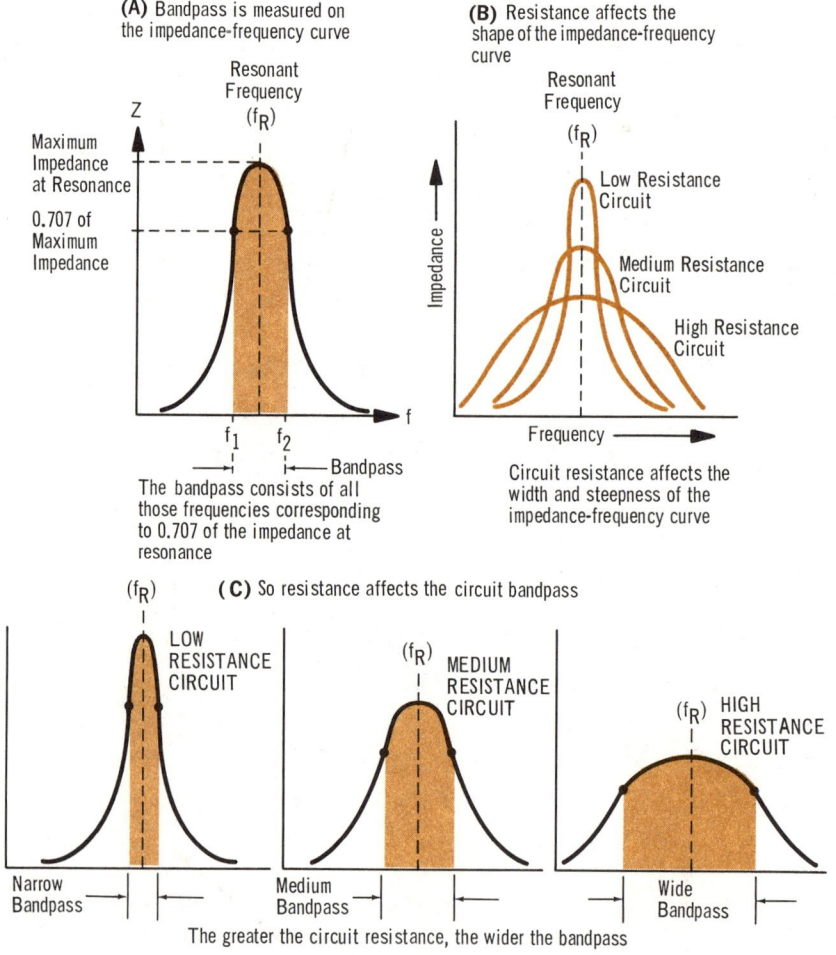

(A) Bandpass is measured on the impedance-frequency curve

Resonant Frequency (f_R)

Maximum Impedance at Resonance

0.707 of Maximum Impedance

f_1 f_2

Bandpass

The bandpass consists of all those frequencies corresponding to 0.707 of the impedance at resonance

(B) Resistance affects the shape of the impedance-frequency curve

Resonant Frequency (f_R)

Low Resistance Circuit

Medium Resistance Circuit

High Resistance Circuit

Circuit resistance affects the width and steepness of the impedance-frequency curve

(C) So resistance affects the circuit bandpass

(f_R) LOW RESISTANCE CIRCUIT

(f_R) MEDIUM RESISTANCE CIRCUIT

(f_R) HIGH RESISTANCE CIRCUIT

Narrow Bandpass

Medium Bandpass

Wide Bandpass

The greater the circuit resistance, the wider the bandpass

the Q of a parallel resonant circuit

The Q, or quality, of a series resonant circuit, you will recall, is determined by the ratio of the voltage across either X_L or X_C to the applied voltage. For parallel resonance, the Q also measures the quality of a circuit. However, in parallel resonant circuits, Q is not determined on the basis of voltage, but rather on the basis of current. The Q of a parallel resonant circuit is defined as the *ratio* of the current in the *tank* to the *line* current. Thus,

$$Q = I_{TANK}/I_{LINE}$$

where I_{TANK} is the circulating current in the tank. Mathematically, this equation can be converted to the form:

$$Q = X_L/R$$

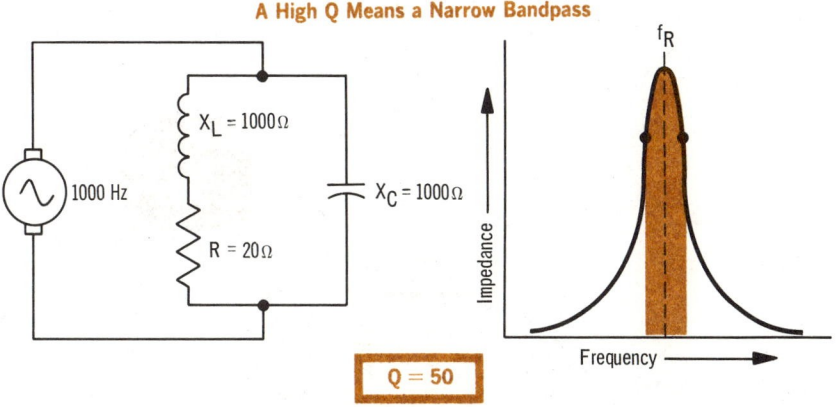

A High Q Means a Narrow Bandpass

The Q of a parallel resonant circuit is its ability to discriminate between resonant and nonresonant frequencies. The Q is the ratio of I_{TANK} to I_{LINE}, but can be calculated more easily from the equation $Q = X_L/R$. In terms of its Q, the bandpass of a parallel resonant circuit is Bandpass $= f_R/Q$.

You will recognize this as the same equation used for the Q of a series resonant circuit. As a result, resistance has the same effect on the Q of a parallel resonant circuit as it does on a series resonant circuit. The *lower* the *resistance*, the *higher* is the Q of the circuit, and the narrower is its bandpass. Conversely, the *greater* the *resistance*, the *lower* is the Q, and the *wider* is the *bandpass*.

the Q of a
parallel resonant circuit (cont.)

If the Q of a circuit is known, the bandpass can be calculated by the equation:

$$\text{Bandpass (Hz)} = f_R/Q$$

Circuits with high Q's, therefore, can discriminate between resonant and nonresonant frequencies better than circuits with low Q's. In practical applications, when the Q of a parallel resonant circuit is 10 or greater, this resistance can be neglected, and the circuit considered as an almost ideal parallel resonant circuit.

A Low Q Means a Wide Bandpass

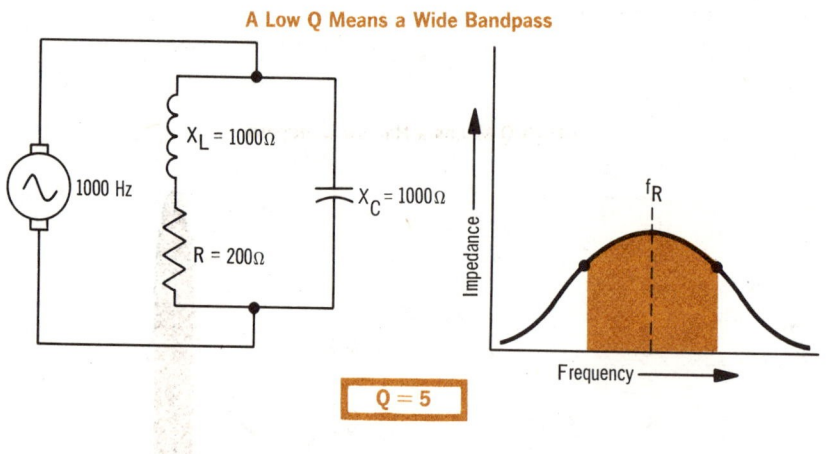

Once the Q is known, the values at resonance of the other important circuit quantities can easily be found. These quantities are the impedance (Z), the tank current (I_{TANK}), and the line current (I_{LINE}). The equations for impedance in terms of Q are

$$Z = QX_L$$
or $$Z = QX_C$$

The equation for I_{TANK} in terms of Q is

$$I_{TANK} = QI_{LINE}$$

The equations for I_{LINE} in terms of Q are

$$I_{LINE} = E/QX_L$$
or $$I_{LINE} = E/QX_C$$

controlling Q

For the most part, resonant circuits are desiged to have high Q's, so the circuits can be most efficient. Also the Q that results usually depends on the inherent design of the coil, i.e., the internal resistance of the coil that develops from the way the coil is wound.

However, in some instances there is a need for the Q to be a specific value, which may be lower than the Q that would ordinarily result from a well-designed tank circuit. A good example of this is when a specific bandpass might be wanted at the resonant frequency, and this specific bandpass would require a specific Q.

Since the basic equation for Q is $Q = X_L/R$, one method of controlling Q would be to make R the specific value needed. For example, if X_L is 1000 ohms, and you wanted a Q of 5, then you would have to make sure that R would be 200 ohms. This would require you to add a resistor in series with the inductor that, when added to the a-c resistance of the windings, would produce a total resistance of 200 ohms. Q, then, would be:

$$Q = \frac{X_L}{R_L + R_{\text{series}}}$$

This would also be true for series resonant circuits.

Another way to control the Q, which is easier in parallel resonant circuits, is to use a shunt resistor across the tank. Since Q in a parallel resonant circuit depends on $I_{\text{TANK}}/I_{\text{LINE}}$, a shunt resistor across the tank would increase I_{LINE} to lower the Q. The value of I_{LINE}, of course, depends on tank impedance, Z_{TANK}. By manipulating the equation, $Z = QX_L$, which you studied earlier, you can see that $Q = Z/X_L$. By adding R_{SHUNT} to the circuit, the impedance Z will change, and the equation becomes:

$$Q = \frac{\left|\dfrac{Z_{\text{TANK}} \times R_{\text{SHUNT}}}{Z_{\text{TANK}} + R_{\text{SHUNT}}}\right|}{X_L}$$

which reduces to

$$Q = \frac{Z_{\text{TANK}} \times R_{\text{SHUNT}}}{X_L(Z_{\text{TANK}} + R_{\text{SHUNT}})}$$

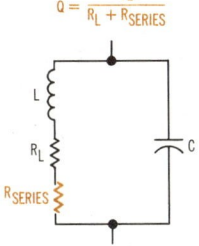

$$Q = \frac{X_L}{R_L + R_{\text{SERIES}}}$$

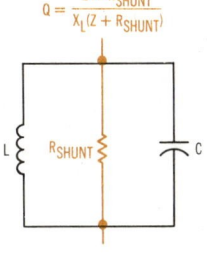

$$Q = \frac{I_{\text{TANK}}}{I_{\text{LINE}}}$$

$$Q = \frac{Z \times R_{\text{SHUNT}}}{X_L(Z + R_{\text{SHUNT}})}$$

The Q of a tank circuit, and therefore its bandpass, can be controlled by placing the proper value resistor in the choke leg or another value resistor across the tank. R_{SERIES} is usually a low value, and R_{SHUNT} is usually a high value in comparison. The shunt method is the most popular

solved problems

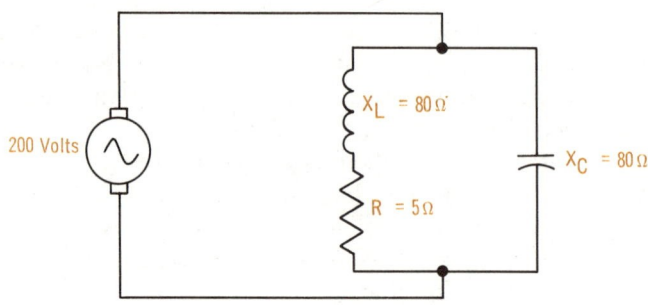

Problem 35.   *Is this circuit at resonance?*

Remember that here X_L and X_C are equal only at resonant frequency. Since X_L and X_C both equal 80 ohms, the circuit must be at resonance.

Problem 36.   *What is the value of the current flowing in the tank?*

One way to solve this problem is to use the equation $I_{TANK} = QI_{LINE}$. Of course Q and I_{LINE} must be determined first:

Calculating Q:

$$Q = X_L/R = 80/5 = 16$$

Calculating I_{LINE}:

$$I_{LINE} = E/Z = E/QX_L = 200/(16 \times 80) = 0.156 \text{ ampere}$$

Calculating I_{TANK}:

$$I_{TANK} = QI_{LINE} = 16 \times 0.156 = 2.5 \text{ amperes}$$

Thus, with only 0.156 ampere in the line, 2.5 amperes flow in the tank.

Problem 37.   *If the resonant frequency is 2000 Hz, what is the circuit bandwidth?*

Using the Q of the circuit, calculate bandwidth from:

$$\text{Bandwidth (Hz)} = f_R/Q = 2000/16 = 125 \text{ hertz}$$

The bandwidth, therefore, extends from about 1938 Hz (2000 − 62.5) to about 2063 Hz (2000 + 62.5).

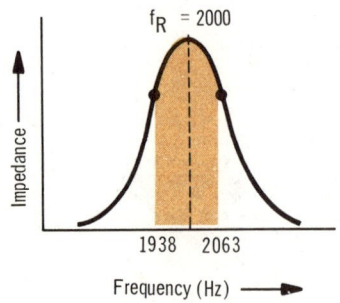

comparison of series and parallel resonant circuits

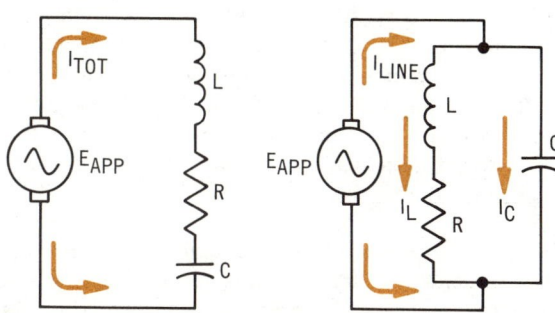

PROPERTIES AT RESONANCE

	Series Resonant Circuit	Parallel Resonant Circuit
Resonant Frequency (f_R)	$\dfrac{1}{2\pi \sqrt{LC}}$	$\dfrac{1}{2\pi \sqrt{LC}}$
Reactances	$X_L = X_C$	$X_L = X_C$
Impedance	Minimum; $Z = R$	Maximum; $Z = QX_L$
Current (I_{TOT} or I_{LINE})	Maximum; I_{TOT}	Minimum; I_{LINE}
Q, Quality	$E_L/E_{APP} = X_L/R$	$I_{TANK}/I_{LINE} = X_L/R$
Bandwidth	f_R/Q	f_R/Q

PROPERTIES OFF RESONANCE

	Series Resonant Circuit		Parallel Resonant Circuit	
	Above f_R	Below f_R	Above f_R	Below f_R
Reactances	$X_L > X_C$	$X_C > X_L$	$X_L > X_C$	$X_C > X_L$
Impedance	Increases	Increases	Decreases	Decreases
Phase Angle Between E_{APP} and I_{TOT} or I_{LINE}	I lags E	I leads E	I leads E	I lags E
Inductive or Capacitive Circuit	Inductive	Capacitive	Capacitive	Inductive

summary

☐ The properties of a parallel resonant circuit are based on the interchange of energy between an inductor and capacitor connected in parallel across a voltage source. Such a circuit is often called a tank circuit. ☐ For a given inductor and capacitor, the frequency at which parallel resonance occurs can be found by: $f_R = 1/(2\pi\sqrt{LC})$. ☐ Theoretically, a parallel resonant circuit has infinite impedance and zero line current. Practical parallel resonant circuits, however, exhibit maximum impedance and minimum line current at resonance. ☐ For frequencies above or below the resonant frequency, the line current of a parallel resonant circuit increases progressively, while its impedance decreases.

☐ Practical parallel resonant circuits contain resistance, as well as inductance and capacitance. Normally this resistance is the resistance of the inductor wire, and can be considered as being in series with the inductance. ☐ The principal effect of the resistance is that at resonance the two branch currents are not 180 degrees out of phase, so line current flows in the circuit. ☐ The magnitudes of the branch currents can be found by: $I_C = E/X_C$ and $I_L = E/(\sqrt{R^2 + X_L^2})$. ☐ The line current can be determined by adding the line currents vectorially.

☐ The bandpass of a parallel resonant circuit consists of all frequencies that produce a circuit impedance 0.707, or more, times the impedance at resonance. ☐ The bandpass can be found from the Q of a circuit by: Bandpass (cps) $= f_R/Q$. ☐ The Q, or quality, of the circuit is the ratio of the circulating current in the tank (I_{TANK}) to the line current (I_{LINE}), or: $Q = I_{TANK}/I_{LINE}$.

---

review questions

1. What is an *ideal parallel resonant circuit?*
2. How does a practical parallel resonant circuit differ from an ideal one?
3. What is the resonant frequency of a 50-microfarad capacitor and 50-millihenry coil connected in parallel?
4. How is the circulating current produced in a tank circuit?
5. What is meant by *damping* in a tank circuit?
6. Draw an impedance-vs.-frequency curve for a parallel resonant circuit.
7. How is *bandwidth* of a parallel resonant circuit defined?
8. What is meant by the Q of a parallel resonant circuit?
9. How is Q related to resistance? To bandwidth?
10. If a parallel resonant circuit has a frequency of 1000 Hz and a Q of 10, what is its bandwidth?

tuning

Until now, you studied resonant circuits in which fixed values of L and C were used to produce the desired resonant frequency. In actual practice this is impractical, because coils and capacitors cannot be made precisely enough to get the exact tuned frequency. These parts have tolerance values, and rarely do any two parts have the exact same value. Also, quite often, it is important to have the resonant circuit function at different frequencies, or even over different ranges of frequencies. To do this, either the inductor or capacitor or both must be variable, so that they can be tuned for the precise frequency; and when tuning ranges are required, different inductors and/or capacitors can be switched in or out of the tank circuit.

The simplest form of tuning can be used with a tank circuit that is designed to work at only one resonant frequency. The tuning adjustment, then, is needed only to compensate for the tolerance variations of the parts. In these cases, the basic L and C components can be fixed, and a small *trimmer capacitor* can be added to make minor *fine tuning* adjustments.

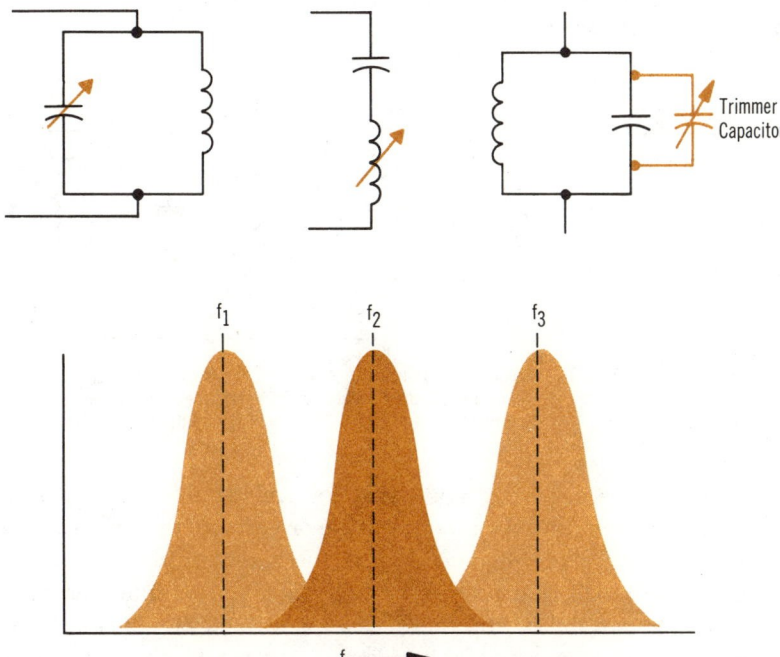

If variable capacitors or inductors are used in resonant circuits, the resonance point and bandpass frequencies can be changed to a variety of frequencies by a simple adjustment.

tuning (cont.)

For *broadband tuning*, either the inductor or capacitor should be variable. Generally speaking, there are more standard variable capacitors available than variable inductors, so it is usually easier to design a tunable circuit with a variable capacitor. Also, it is easier to design a variable capacitor to give a wider range of control than a variable inductor; so *broadband resonant circuits* are more often capacitor tuned. Many variable capacitors used in radios have built-in trimmer capacitors to adjust the minimum-to-maximum tracking of the capacitor and, thus, tuning variability.

This does not mean that variable inductors are not used for broadband tuning. They are, particularly since they allow delicate, small adjustments to be made more easily. Since variable inductors use plungers that move in and out of the coil, complicated mechanical arrangement is needed when tuning is made frequently, usually from the front panel of a piece of equipment. A variable capacitor does that job much more simply. But for tuning adjustments that are made infrequently, the variable inductor will work as well as or better than the variable capacitor. With high frequencies, the tank circuit may not have an actual capacitor across the coil. Instead, the innerwiring capacities of the variable inductor resonate with the coil.

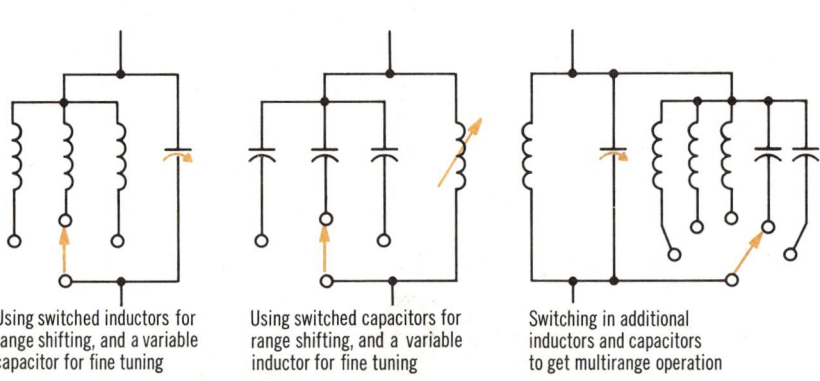

Using switched inductors for range shifting, and a variable capacitor for fine tuning

Using switched capacitors for range shifting, and a variable inductor for fine tuning

Switching in additional inductors and capacitors to get multirange operation

Sometimes, both the coil and capacitor are variable, one acting as a coarse adjustment, and the other as a fine or vernier adjustment. In some circuits, the range of frequencies that the resonant circuit must work through is so broad that no one variable capacitor or inductor can provide enough variability. In these cases, either different inductors or capacitors can be switched in and out of the tuned circuits to provide a series of frequency ranges, or additional components can be added to those already in the circuit to shift the range of operation constantly.

filter circuits

As used in electrical circuits, the term filter means to offer large opposition to, or *reject*, voltages and currents of certain frequencies, and at the same time offer little opposition to, or *pass*, voltages and currents of other frequencies. Circuits which have this capability are called *filter circuits*. Filter circuits are divided into various groups, according to the frequencies they pass and reject. One such group passes all frequencies up to a certain frequency, and rejects all above that frequency. This type of filter is called a *low-pass filter*. Another group of filters rejects all frequencies up to a certain frequency, and passes all those higher than that frequency. These are called *high-pass filters*. Other filters pass certain ranges, or bands, of frequencies and reject all frequencies outside of the band. Such filters are called *band-pass* filters. Still others reject a band of frequencies, and pass all frequencies outside of the band. They are *band-reject* filters. These, then, are the principal types of filter circuits.

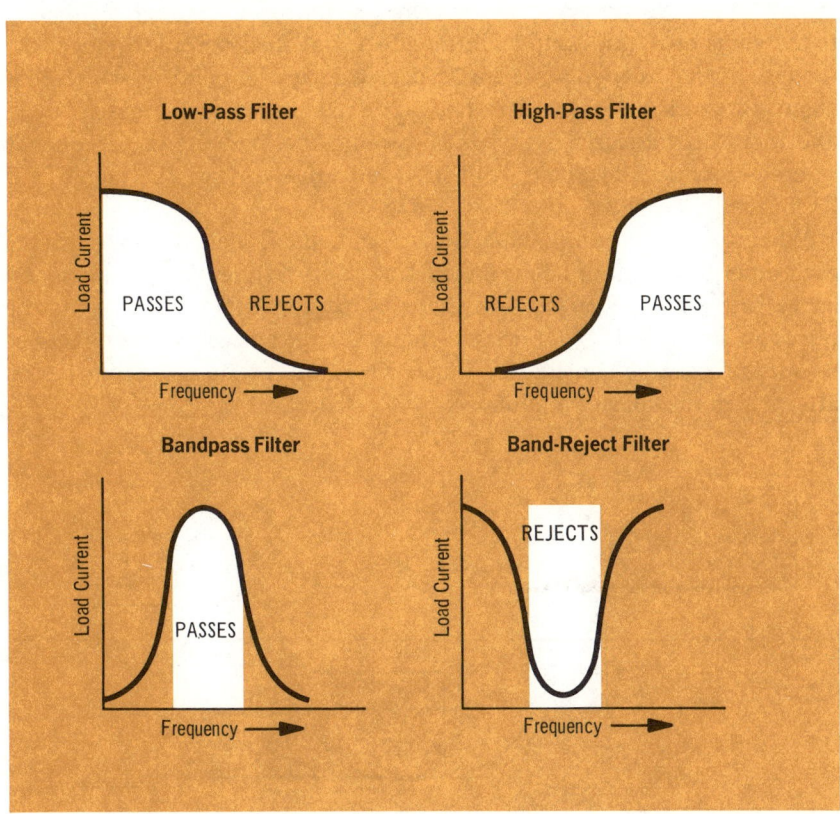

low-pass filters

The *low-pass filter* relies on the varying impedance of an inductor and/or capacitor to interfere with a signal as its frequency increases. In its simplest form a low-pass filter can be merely an inductor in series with the output load. At low frequencies, the reactance, X_L, of the inductor is small, so that very little signal is dropped across the inductor, leaving most of the signal to reach the output load. As the frequency increases, X_L increases; and when X_L reaches a value that is 1/10 of R, the drop across L will reduce the signal drop across R. The more the frequency rises, the greater the drop across L and the less the drop across R. At the frequency where X_L is equal to R, half the signal will be lost at the output. At and beyond the frequency where X_L is ten or more times greater than R, for all practical purposes, all of the signal will drop across L, leaving no signal output across R.

When a capacitor is used as a low-pass filter, it must be used differently since its reactance is inversely proportional to frequency. The capacitor is connected in parallel with the output load to bypass the signal. At low frequencies, though, its reactance, X_C, is high compared to the value of R, and so it has little effect. Its effect becomes noticeable, though, at the frequency where X_C decreases to the point at which it is about 10 times R. From there on, as the frequency is increased, more and more of the signal is bypassed around the load. At the frequency where X_L reduces to about 1/10 of R, effectively all of the signal will be shunted around the output.

When the inductor and capacitor are combined, their effects reinforce each other so that at or about the critical cutoff frequencies the passband curve drops more sharply and there is more effective filtering at the rejection frequencies. The design, though, differs because the effects of resonance must be considered. When $X_L = X_C$, the signal output peaks, after which the sharp drop follows.

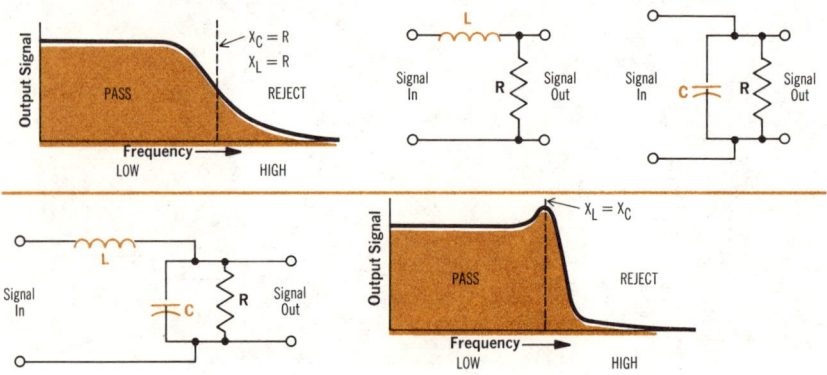

high-pass filters

The *high-pass filter* works in a manner similar to that of the low-pass filter in that it also relies on the varying reactance characteristics of the inductor and capacitor. However, since the high-pass filter is designed to block the lower frequencies, the parts are connected into the circuit differently. The inductor, for example, is *shunted* across the output load. At the lower frequencies, X_L will be very low compared to the load resistance, and will effectively short out the load. As the frequency increases and X_L rises, the shunting effect of L is proportionately reduced. For all practical purposes, when X_L reaches about 1/10 the value of R, some output signal starts to develop; and as X_L rises further with frequency, the output continues to climb until X_L is 10 or more times greater than R. Then the circuit works as though the inductor did not exist.

A capacitor would have to be connected in *series* with the output. Its reactance, X_C, would be very high at low frequencies, and so most of the signal would be dropped across C, leaving little left for the output. But with the higher frequencies, as X_C decreases to where it is only ten times the value of R, some signal would start developing at the output, and would rise with frequency as X_C continued to decrease until X_C reached 1/10 the value of R.

When L and C are combined in a high-pass filter, they are placed with C in series with the output, and L shunting the output. As with the low-pass filter, they sharpen the rejection slope and peak at resonance.

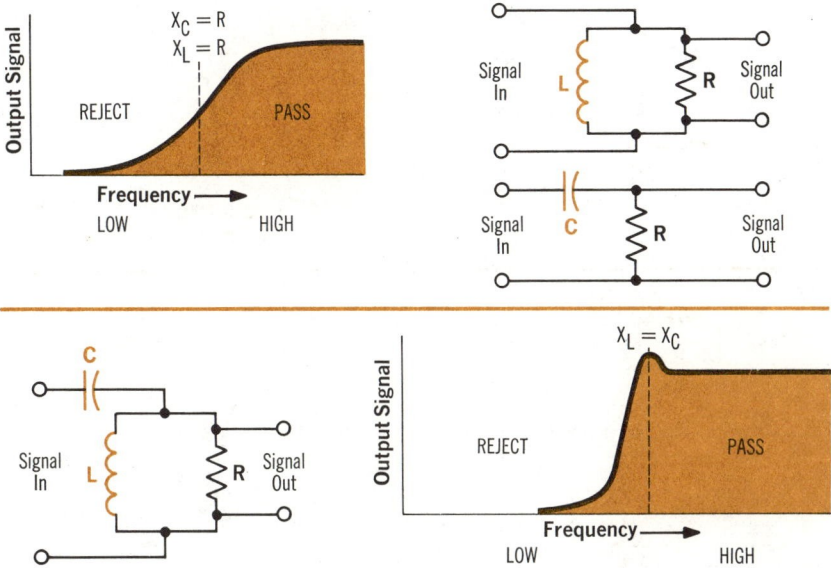

bandpass and band-reject filters

Whereas low- and high-pass filters work either above or below a certain point, the *bandpass filter* functions differently within a specific *range* of frequencies from the way it does at frequencies either above or below that range.

The ordinary resonant tank circuit in the output is the best example of a typical bandpass filter. The resonant frequency, f_R, of the tank will establish the center point of the band to be passed. At resonance, the tank will present a high load impedance at the output. But off resonance, the impedance will drop, offering less and less output load for a signal to develop. The width of the bandpass will depend on the Q of the tank; by controlling the Q, either a narrow or wide bandpass can be obtained.

The series-resonant circuit can function as a bandpass filter when it is connected in series with the output. The series resonant circuit has a very low impedance at resonance, and so allows most of the signal to reach the output. When above and below resonance, though, the high impedance of the series circuit (X_C at the lower frequencies and X_L at the higher frequencies) drops the signals before they reach the output.

Band-reject filters function in a manner exactly opposite that of bandpass filters. The resonant tank circuit placed in series with the output will use its high impedance at resonance to prevent the signal from reaching the output. And the series-resonant circuit shunted across the output will use its low impedance at resonance to bypass the output. In either case, frequencies above or below the resonant band will be only slightly affected.

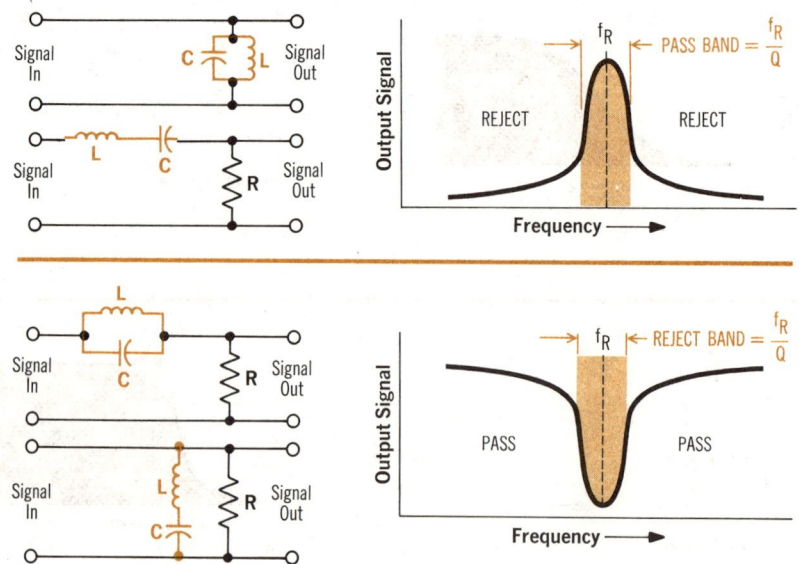

filter networks

Aside from the simple LC combinations shown on the previous pages, inductors and capacitors can be combined in a variety of ways to form many simple or intricate *filter networks*. The basic circuit patterns can be repeated to attain as high a degree of filtering as necessary.

Simple LC low- and high-pass filters are actually called *half-section filters* because they represent only a part of the more sophisticated filter circuits that are in use today. Typical filter sections generally form T or pi networks, so named because of the shape they take schematically. A low-pass T *filter* has two inductors in series with the signal, and a capacitor shunting the line from between the two inductors. A high-pass T filter has two capacitors in series with the signal line, and an inductor shunting the line from between the two capacitors. These T networks function in the same manner as *half-section filters*, except that the extra component will give added attenuation of the unwanted frequencies.

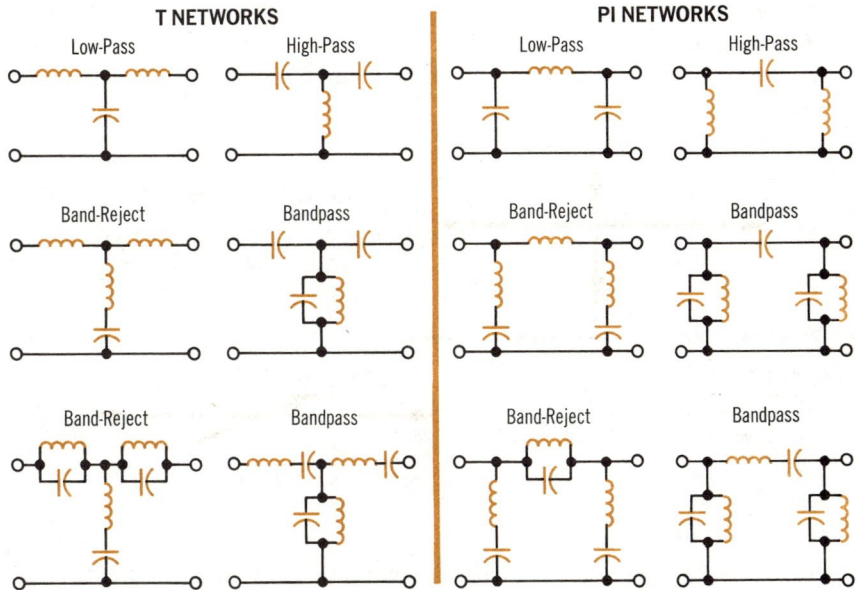

The *pi networks* are so called because they resemble the Greek letter pi (π). They use two shunt components, one on either side of the series component, and like the T filter, give better attenuation of unwanted frequencies.

For bandpass and band-reject networks, the individual inductors and/or capacitors can be replaced with resonant circuits to form a variety of T and pi networks.

transformer vectors

You recall that the theory of operation of ordinary transformers was given in Volume 3. Transformers were described then strictly on the basis of the current and voltage waveforms that exist in the primary and secondary. Transformer voltages and currents can also be represented by *vectors*, as shown below. When this is done, the *primary voltage* is used as the zero-degree *phase reference*, and the other voltage and current vectors are shown in relation to it.

In a transformer with a very small secondary current flowing, the primary current lags the primary voltage by 90 degrees, and the secondary current lags the secondary voltage by 90 degrees. Furthermore, the primary and secondary voltages are 180 degrees out of phase. On a vector diagram, therefore, the four vectors representing the currents and voltages are *mutually perpendicular*, as shown. As the secondary current increases, assuming that the secondary load is resistive, the current in the secondary begins to go resistive. As a result, it no longer lags the secondary voltage by 90 degrees. The more current that flows in the secondary, the more resistive it becomes, and the smaller the angle between the secondary voltage and current becomes. You will re-

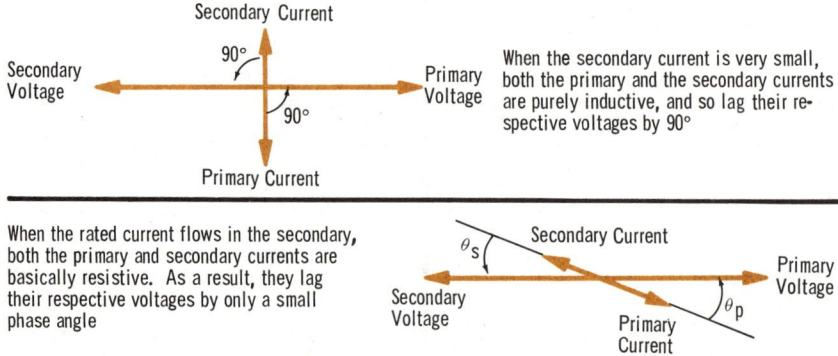

Secondary Current

Secondary Voltage   90°   Primary Voltage   When the secondary current is very small, both the primary and the secondary currents are purely inductive, and so lag their respective voltages by 90°

90°

Primary Current

When the rated current flows in the secondary, both the primary and secondary currents are basically resistive. As a result, they lag their respective voltages by only a small phase angle

θ_s Secondary Current   Primary Voltage

Secondary Voltage   Primary Current θ_p

member that when the secondary current increases, the magnetic field created in the secondary winding causes the primary current to also increase by effectively reducing the inductive reactance of the primary winding. With the primary inductive reactance reduced, the primary also becomes more resistive, and so the angle between its current and voltage also decreases. Thus, the vector diagram of a transformer carrying its rated current shows that the primary and secondary voltages are 180 degrees out of phase, with the two currents lagging their respective voltages by the same small angle.

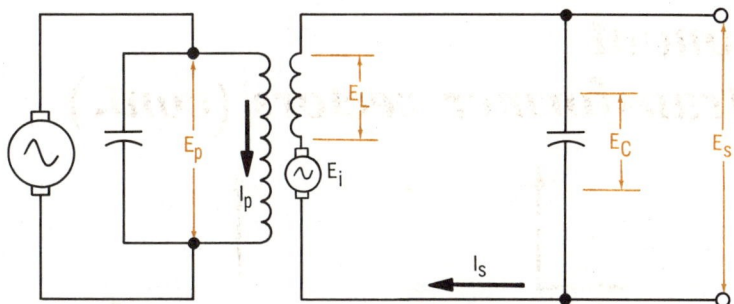

In tuned transformers, the output, or secondary voltage (E_s) is not actually the voltage induced in the secondary winding (E_i). Because the secondary circuit is a series resonant LC circuit, the voltages developed across the inductor and capacitor are greater and differ in phase with the induced voltage. The voltage across the capacitor is actually the output voltage and it is 90° ahead of the primary voltage at resonance

resonant transformer vectors

It was repeated on the previous page, as you learned in Volume 3, that the transformer primary and secondary voltages are 180 degrees out of phase. But at resonance, this is not completely true because the voltage induced in the secondary is not the voltage across the coil. The voltage *induced* in the coil can be considered as an *applied* voltage to a series LC resonant circuit, so that the voltages *developed* across the coil and capacitor are much greater than that induced in the coil. The output of the transformer is actually the voltage developed across the capacitor, which has a different phase angle than the induced voltage. Let's work it out.

In the primary tank circuit, the current in the coil (I_p) *lags* the voltage across the coil (E_p) because the current is *inductive*. The primary current then induces a voltage in the secondary circuit (E_i) that is 180 degrees out of phase with the primary voltage, just as in a regular transformer. But since this is a tuned circuit, the induced voltage is applied as a source voltage to the inductor and capacitor in series. This causes a current to flow in the secondary circuit (I_s). But at resonance, the circuit is *resistive,* so the secondary current (I_s) is in phase with the induced voltage (E_i).

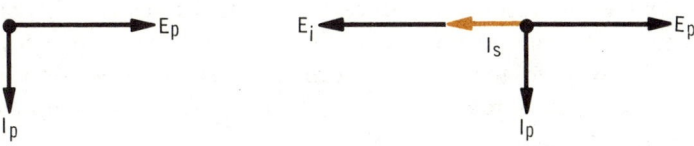

resonant transformer vectors (cont.)

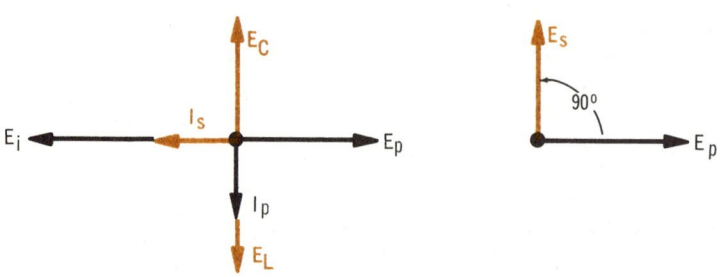

Voltages are then dropped across the inductor and capacitor, but because of the phase relationships of current and voltage in these components, the drop across the inductor (E_L) *leads* the current (I_s) by 90 degrees, and the drop across the capacitor (E_C) *lags* the current (I_s) by 90 degrees; this is the same as in an ordinary series LC circuit. The capacitor voltage (E_C), then, also lags the induced voltage (E_i) by 90 degrees. And since the induced voltage (E_i) and the primary voltage (E_p) are 180 degrees apart, the capacitor voltage (E_C) *leads* the primary voltage (E_p) by 90 degrees. Since the output, or secondary voltage, is actually the voltage developed across the capacitor, then the secondary voltage (E_s) leads the primary voltage (E_p) by 90 degrees in a tuned transformer at resonance.

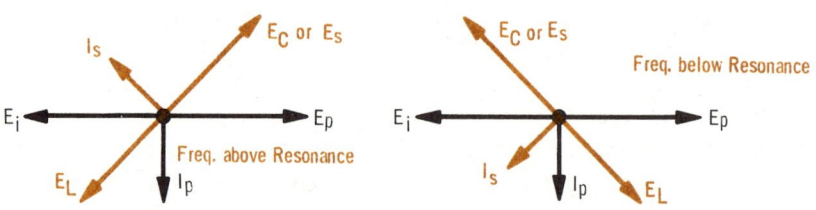

Off resonance, though, the secondary circuit is no longer resistive, so the current (I_s) will not be in phase with the induced voltage (E_i). As a result, the voltages developed across the inductor and capacitor will be more or less than 90 degrees away from E_i, depending on how far the tuning is off resonance. Since it is a series resonant circuit, the lower the frequency is off resonance, the more capacitive it becomes, and E_s will go toward 180 degrees away from E_p. For the higher frequencies, the circuit becomes inductive, and E_s will approach the same phase as E_p.

impedance matching

In an electric circuit, power is delivered by the source to the load, or loads. For the case of a simple circuit with a single resistive load, the power delivered is $P = EI = I^2R$, where R is the resistance of the load. The power, therefore, depends on the current and the resistance of the load. Since the current increases when the load resistance decreases, you may think that the smaller the load resistance was made, the larger the power would be. This would be the case, were it not for the internal resistance of the source. The source resistance is in series with the load resistance, so it also affects the current, and thus the power delivered to the load. The combined effect of the load and source resistances is such that maximum power is delivered to the load when the load impedance equals the source impedance. When the two impedances are equal, they are said to be matched. As you will find later in your electrical work, very often a low-impedance source must be matched to a high-impedance load, or vice versa. This could be done by using a source having the desired impedance, but generally this is impractical. A very common way to match the impedance of a load to the impedance of the source is by using a transformer. You will recall from Volume 3 that the impedance ratio between the primary and secondary of a transformer depends on the turns ratio of the transformers, according to the equation:

$$\sqrt{Z_p/Z_s} = N_p/N_s$$

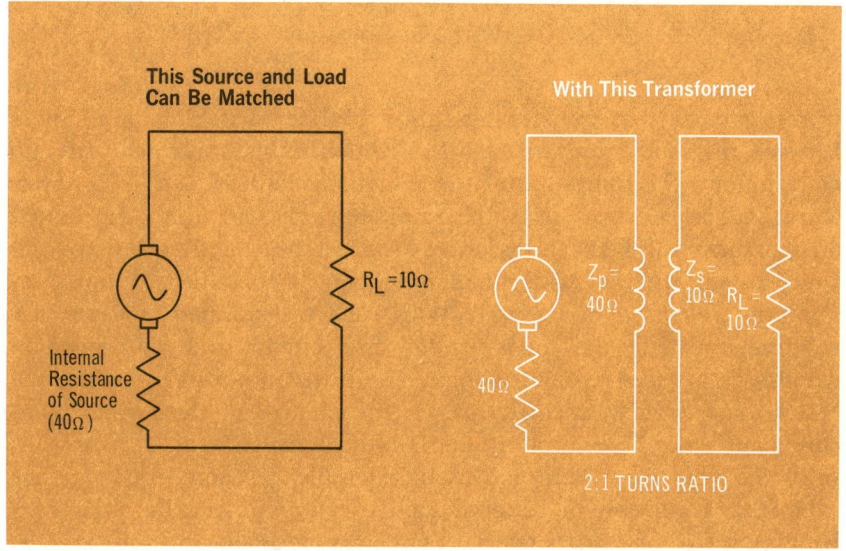

This Source and Load Can Be Matched

With This Transformer

$R_L = 10\,\Omega$

Internal Resistance of Source $(40\,\Omega)$

$Z_p = 40\,\Omega$ $Z_s = 10\,\Omega$ $R_L = 10\,\Omega$

$40\,\Omega$

2:1 TURNS RATIO

impedance matching (cont.)

Therefore, by using a transformer with the appropriate turns ratio, you can obtain any desired values of primary and secondary impedances. If the transformer primary is connected to the source, it serves as the load for the source, and if it has the same impedance as the source, maximum power is transferred to the primary. Similarly, if the secondary is connected to the load, it serves as the source for the load. And if the secondary impedance equals the load impedance, maximum power is transferred to the load.

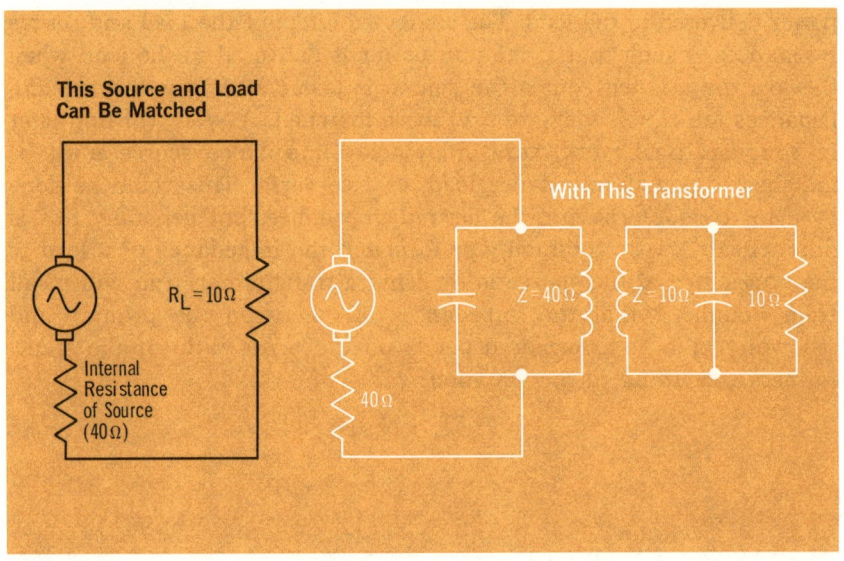

You can see that transformers are useful for matching impedances between the source and the load. In addition, transformers can step up or step down the source voltage to the value the load needs. However, since both impedance and voltage matching depend to a great extent on the turns ratio, the transformer is often designed to *compromise*. Quite often, though, better matching can be obtained if *tuned* or *resonant transformers* are used. With such transformers, the *primary* winding is usually part of a *parallel* tank circuit, and the secondary winding is a *series* resonant circuit. The turns ratio can be set for the best voltage match, since the proper impedances of the resonant circuits can be set by controlling the Q's of the coils. In addition to this, since the *primary* winding is part of a parallel resonant circuit, it has a *high* tank *current* for better inductive coupling. And the *secondary* winding is part of a series resonant circuit, so it produces a *voltage gain*.

tuned transformers

In many high-frequency applications, the tuning of resonant transformers is handled in particular ways to control the way the circuits respond to the signal frequencies involved. Because of the amplification factors involved in the primary tank circuit, the transformer coupling, and the secondary tank circuit, the bandpass action of the entire tuned transformer can provide a variety of bandwidth curves, depending on how the transformers are tuned and coupled.

When both tank sides of the transformer are tuned to the same frequency, the bandwidth of the circuit can have a sharply sloping curve that is moderately flat at the top as long as both windings are properly coupled. The exact width of the curve will depend on the Q's of both resonant circuits. If the transformer is *overcoupled*, the passband will widen and the top of the curve will dip in proportion to the amount of overcoupling. *Undercoupling*, on the other hand, will reduce the height of the curve as well as its rate of slope.

If the primary and secondary resonant circuits are tuned above and below the desired center frequency, bandpass widening can be controlled to give more bandwidth with sharp sloping curves. This is called *stagger-tuning*. As the primary and secondary tuning go further apart, the height of the curve will drop and the top of the curve will dip, with peaks showing at the primary frequency, f_p, and the secondary frequency, f_s. If only the primary *or* the secondary is tuned off the center frequency, f_c, the bandwidth curve will skew in that direction.

Combinations of off-tuning and degrees of coupling can produce varieties of bandwidth shapes.

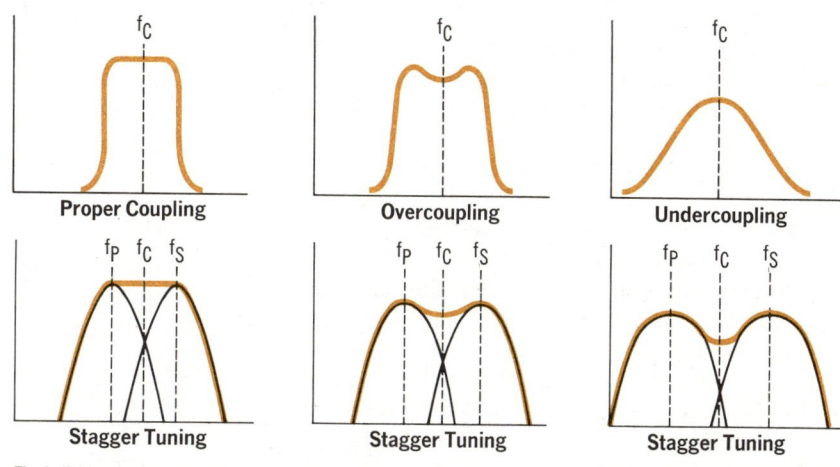

Proper Coupling	Overcoupling	Undercoupling
Stagger Tuning	Stagger Tuning	Stagger Tuning
The individual tank curves combine to form one flat curve	Tuning the primary and secondary further apart dips the curve	Setting f_P and f_S further apart will widen the curve and increase the dip

summary

☐ Series-parallel RL, RC, and LCR circuits contain both series and parallel combinations of resistance, capacitance, and inductance. ☐ When solving series-parallel circuits, vector additions of voltages, currents, and impedances usually have to be performed by first resolving the various quantities into their components. ☐ Tuned circuits are resonant circuits whose resonant frequency can be changed. This is accomplished by making either the capacitor or inductor in the circuit variable.

☐ Filter circuits reject voltages or currents of certain frequencies, while passing voltages or currents of other frequencies. ☐ Lowpass filters pass all frequencies up to a certain point, and reject all higher frequencies. ☐ Highpass filters reject all frequencies up to a certain point, and pass all higher frequencies. ☐ Other types of filters are bandpass, and band-reject filters. ☐ Transformer voltages and currents can be represented by vectors. When this is done, the primary voltage is used as the phase reference. ☐ In resonant transformers, the primary and secondary voltages are not 180 degrees out of phase. Instead, the secondary voltage leads the primary voltage by 90 degrees at resonance.

☐ Transformers are often used to match the impedance of a source to that of a load. Such impedance matching is accomplished by using a transformer having the appropriate turns ratio. ☐ The impedance ratio between a transformer primary and secondary depends on the transformer turns ratio according to the equation: $\sqrt{Z_p/Z_s} = N_p/N_s$.

---

review questions

1. What is a *series-parallel RL, RC,* or *LCR circuit?* Draw a a schematic diagram of one such circuit.
2. Draw another type of series-parallel RL, RC, or LCR circuit.
3. What is a *tuned circuit?* Draw a schematic diagram of such a circuit.
4. Why are tuned circuits used?
5. What is a *filter circuit?*
6. Draw the frequency response curve of a band-reject filter. Of a low-pass filter.
7. Describe a simple low-pass filter and its operation.
8. A capacitor connected in series with a load is what kind of filter?
9. Draw a vector diagram for a typical transformer.
10. What is meant by *impedance matching?* Why is it important?

**electricity
five**

the need for test equipment

Whether you design, install, operate, or repair electrical equipment, you must know how to measure many different electrical quantities, such as frequency, power, power factor, impedance, distortion, sensitivity, current, voltage, and resistance. No doubt, you are not familiar with some of these quantities, but by now in your study of electricity you are familiar with *current, voltage, resistance,* and *power.* And, these four quantities are, in most cases, the most important quantities you will have to measure.

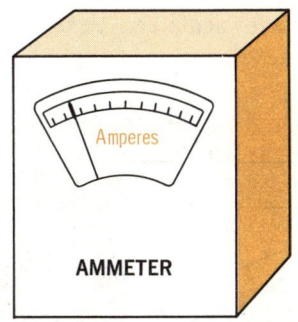

AMMETER

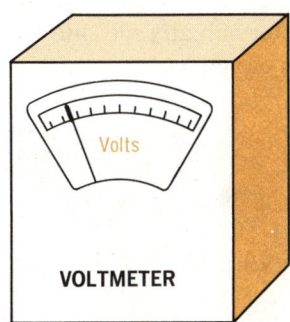

VOLTMETER

It is important that you know how to use the four basic meters, and to have a thorough understanding of how they work

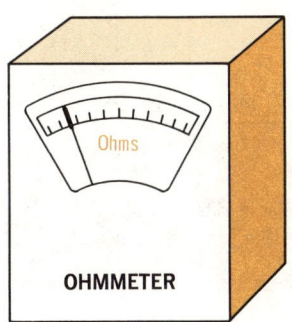

OHMMETER

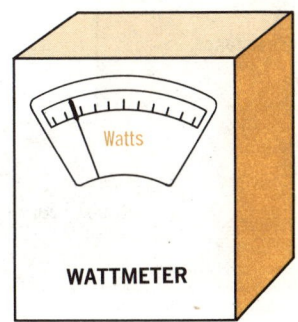

WATTMETER

Meters are the most common instruments used to measure current, voltage, resistance, and power. It is important that you know how to connect a meter into a circuit to make a particular measurement, but it is just as important, however, that *you understand how the meter works* so that you can properly interpret the measurement.

the basic meter

Meters, except for the few that operate on electrostatic principles, can only measure the amount of current flowing through them. However, they can be calibrated to indicate almost any electrical quantity. For example, you know that according to Ohm's law, the current that flows through a meter is determined by the voltage applied to the meter and the resistance of the meter:

$$I = E/R$$

Therefore, for a given meter resistance, different values of applied voltage will cause specific values of current to flow. As a result, although a meter actually measures current, the meter scale can be calibrated in units of voltage.

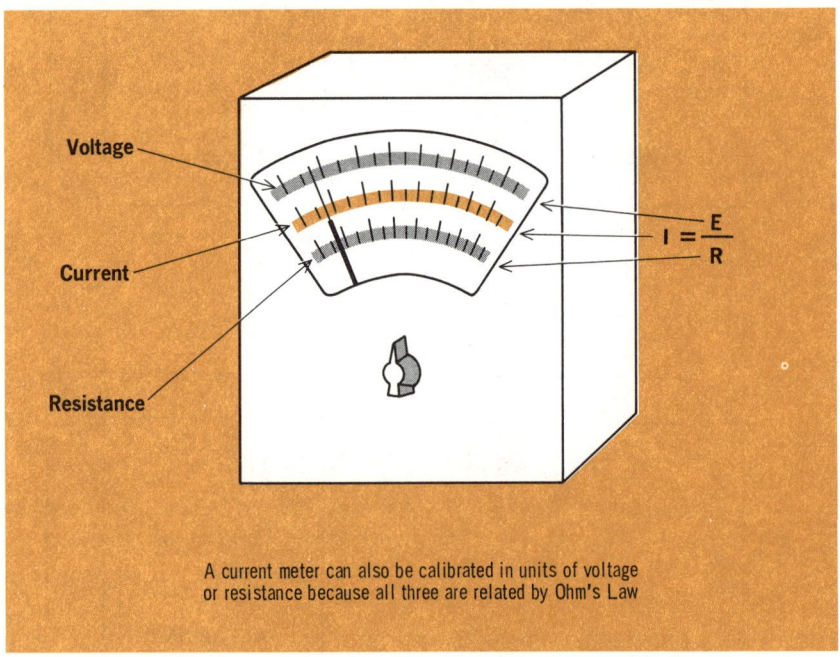

A current meter can also be calibrated in units of voltage or resistance because all three are related by Ohm's Law

Similarly, for a given applied voltage, different values of resistance will cause specific values of current to flow. Therefore, the meter scale can also be calibrated in units of resistance, rather than in units of current. The same holds true for power, since power is proportional to current:

$$P = EI \qquad or \qquad P = I^2R$$

the current meter

As you learned in Volumes 1 and 2, when current flows through a wire, it produces two effects:

1. It creates a *magnetic field* that surrounds the wire.
2. It generates *heat* in the wire.

The amount of current flowing through the wire determines both the strength of the magnetic field and the amount of heat produced. These effects are used in the two basic types of current meters: the electromagnetic current meter, and the thermal current meter. From their names, you can see that the electromagnetic meter makes use of the magnetic field to measure the amount of current flow and the thermal meter makes use of the heat produced by the current flow to measure the amount of current flow.

The electromagnetic current meter makes use of the magnetic field around a current-carrying wire to measure current

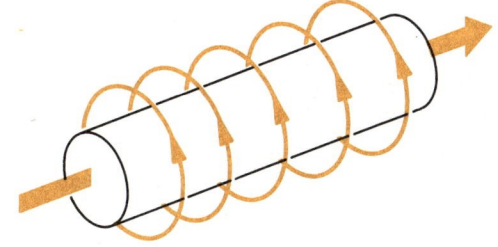

The thermal current meter makes use of the heat produced by the current flowing through a wire to measure current

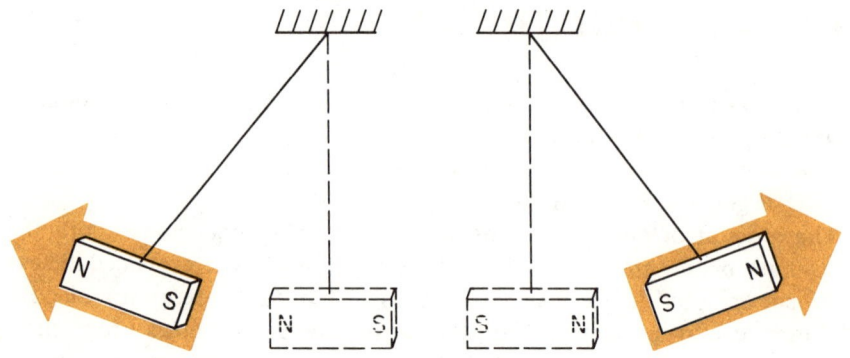

Like magnetic poles repel each other

review of electromagnetism

The electromagnetic current meter is, by far, the one used most often to measure current, voltage, resistance, and power. This type of meter is easy to understand if you know the basic magnetic principles upon which the meter operates. These principles were taught to you in Volume 1, but let's review them briefly again.

Magnetic fields interact in certain ways. For example, the like poles of two iron magnets will repel each other and the unlike poles will attract each other. The same is true for the like and unlike poles of electromagnets. Furthermore, an iron magnet and an electromagnet will repel each other if they are positioned so that their like poles are facing each other, and they will attract each other if their unlike poles are facing each other.

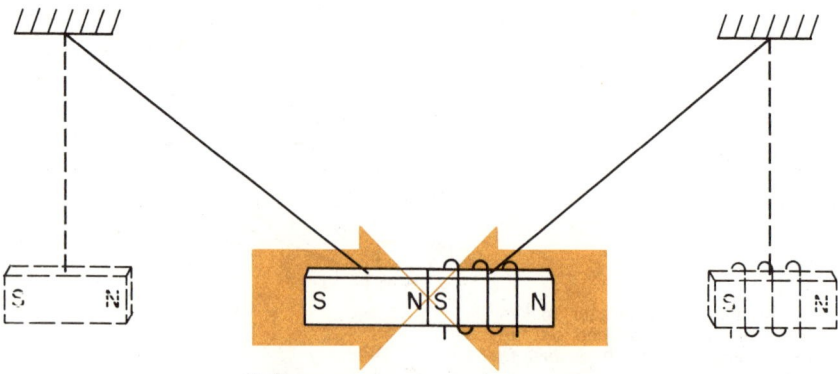

Unlike magnetic poles attract each other

attraction

If you place a soft iron bar close to a magnetized solenoid, the iron bar will become magnetized. The magnetic lines of force set up in the iron will line up in the same direction as those of the solenoid. As a result, the poles set up in the iron bar will also be in the same direction. Therefore, the poles of the solenoid and iron bar that face each other are opposite. Since opposite poles attract each other, the iron bar will be drawn into the coil. The plunger-type moving iron meter, which will be discussed later, operates on this exact principle.

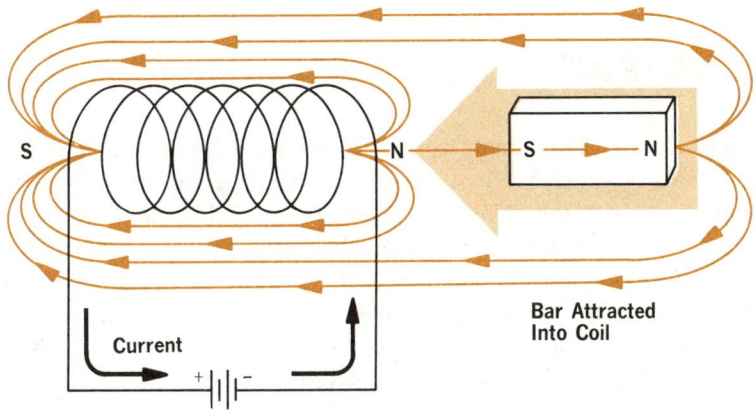

Bar Attracted
Into Coil

Because unlike poles attract each other, the iron bar will be magnetized with the polarities shown, and will be attracted into the coil

Once the bar moves into the coil, the magnetic field is concentrated within the bar

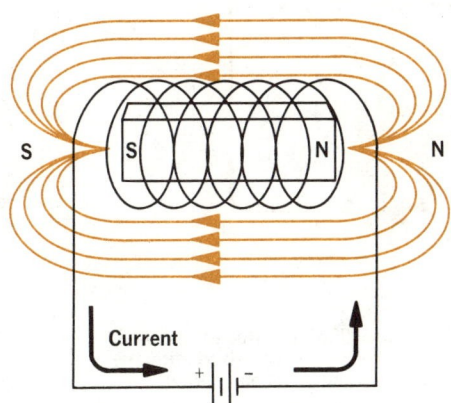

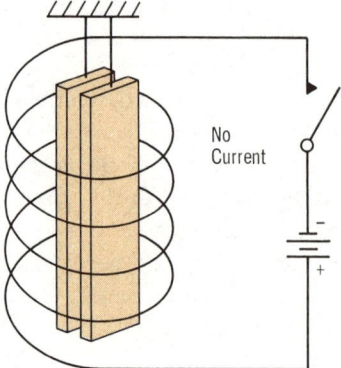

With no current flowing through the coil, no magnetic field is set up and the iron bars do not move

No Current

repulsion

Now, let's place two soft iron bars inside of a solenoid coil side by side to see what happens. When the coil is energized, both bars will become magnetized with the same polarity. Because of this, they have like poles facing each other; and since like poles repel, the bars will move apart.

When the current through the coil reverses, the polarity of the bars also reverses, but there are still like poles facing each other. This principle of like poles repelling each other is used in the repulsion-type moving-iron meter, which will be examined later.

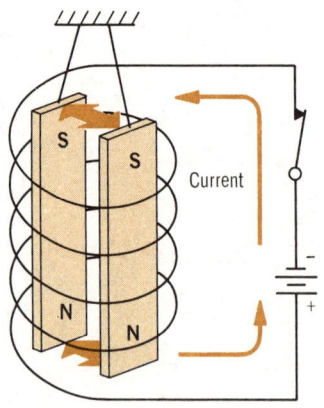

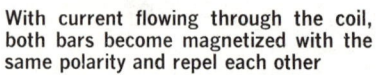

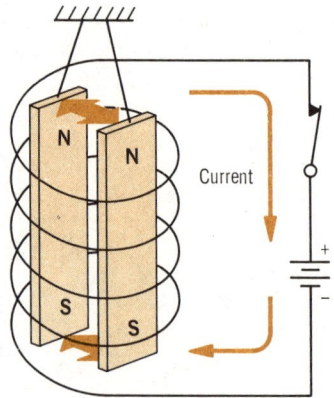

With current flowing through the coil, both bars become magnetized with the same polarity and repel each other

With current flowing through the coil in the opposite direction, the north and south poles of each bar reverse, but the bars still repel each other

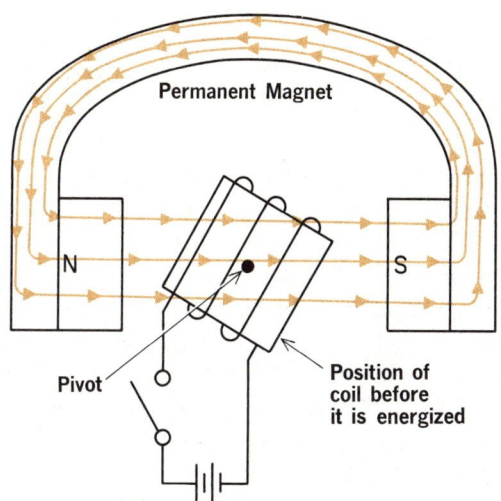

Before the coil is energized, it remains in its normal position (inclined slightly). The permanent magnetic field has no effect on the position of the coil

repulsion (cont.)

In the two examples of magnetic field interaction you have studied thus far, the coil remained stationary and the iron magnet moved. But, if the coil was pivoted between the poles of a stationary permanent magnet and was then energized, the magnetic fields of the permanent magnet and the coil would interact, causing the coil to pivot. Moving-coil meters, which will be discussed later, operate on this principle.

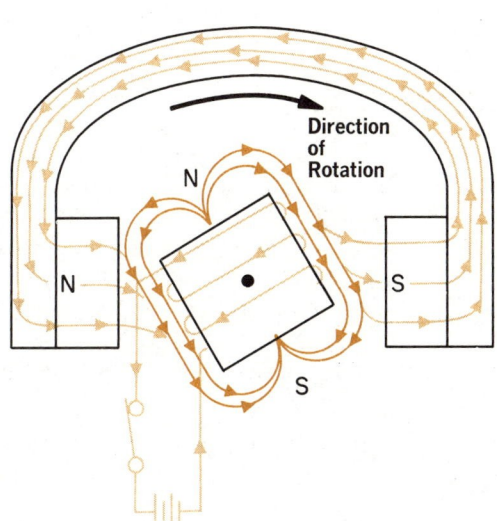

When the coil is energized, a magnetic field is set up within the coil. The field is such that the like poles of the magnet and the coil oppose each other, causing the coil to rotate on its pivot

how current affects
a magnetic field

You have seen how one magnetic field interacts with another. So far, however, there has not been any discussion on what controls the amount of interaction. You know that current flowing through a coil produces the magnetic field that surrounds the coil. The *strength* of the *magnetic field* is proportional to the *amount of current* flowing through the coil. As the current increases, the strength of the magnetic field increases, and as the current decreases, the strength of the magnetic field decreases. For example, if a current through a certain coil is increased from 1 to 1.6 amperes, the magnetic field around the coil will be stronger for 1.6 amperes than it was for 1 ampere.

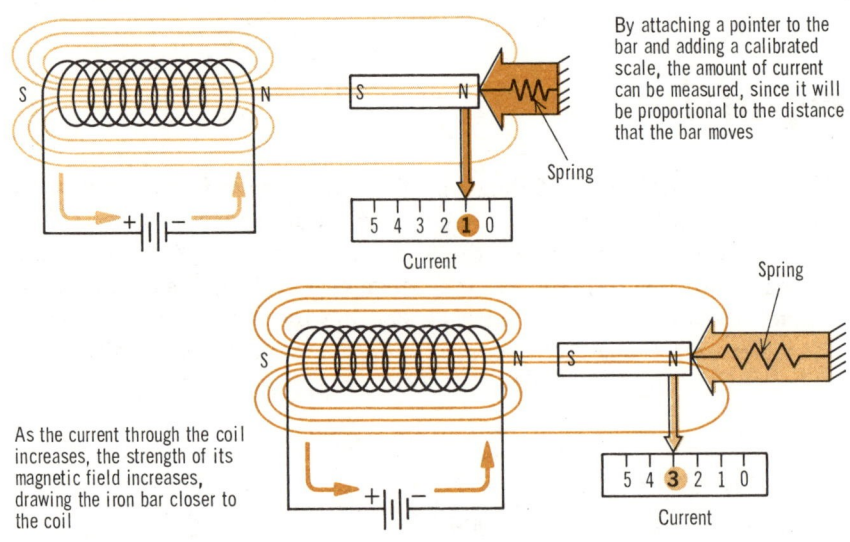

By attaching a pointer to the bar and adding a calibrated scale, the amount of current can be measured, since it will be proportional to the distance that the bar moves

Spring

Current

Spring

As the current through the coil increases, the strength of its magnetic field increases, drawing the iron bar closer to the coil

Current

Now suppose a spring is attached to the iron bar so that it holds back the bar. The magnetic field, therefore, will have to overcome the spring tension. The stronger the field, the more the spring tension will be overcome. Therefore, the greater the current flowing through the coil, the greater will be the magnetic field and the further the iron bar will be drawn into the coil. The greater the current flowing through a coil enclosing two iron bars, page 5-6, the further each iron bar will be repelled from the other. Similarly, the moving coil, page 5-7, will rotate further as the current through the coil increases. All electromagnetic current meters operate on the principle that the strength of the magnetic field about a coil is *proportional* to the amount of current flowing through it.

types of
electromagnetic current meters

You have learned how magnetic fields can be used to cause motion between magnetized objects; and that the amount of motion is proportional to the strength of the magnetic field, which, in turn, is proportional to the current that produces the field. Now you will learn how these effects are used to measure electric current.

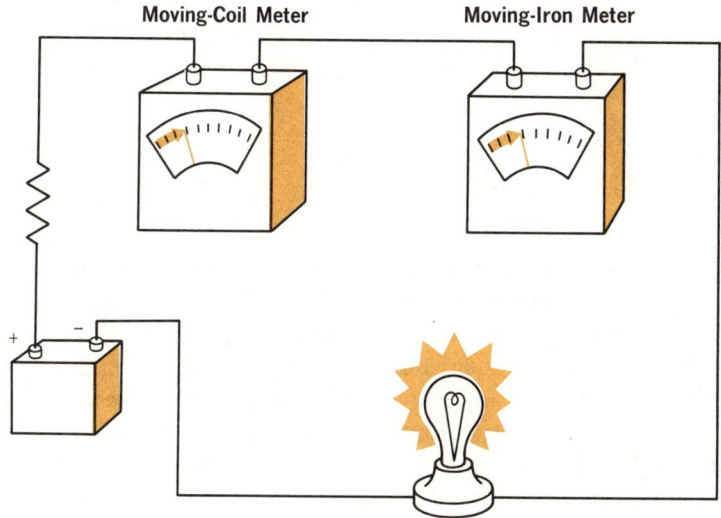

Both meter movements make use of slightly different magnetic principles to measure the current flowing through the lamp

There are two basic types of electromagnetic current meter movements in use today: the *moving-coil* type and the *moving-iron* type. Both types operate on electromagnetism, but each type uses magnetic fields in a slightly different way to indicate the amount of current flowing in a circuit. Also, each type has certain advantages and disadvantages, which will be discussed later.

You can see that it is not easy to tell the difference between the different types of meters just by looking at them or by using them. From the outside, they appear the same, and they are generally used in the same manner to take current measurements. But, when you know how each type works, you can easily identify them when you examine their movements.

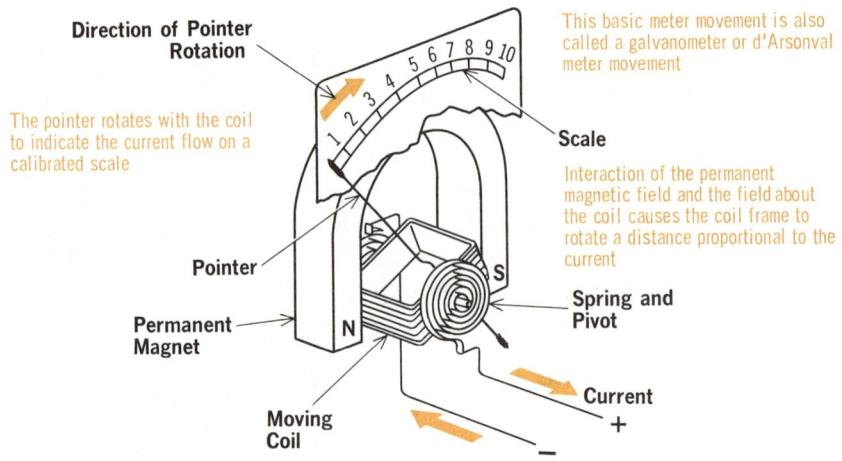

the moving-coil meter movement

In 1882, Arsene d'Arsonval, a Frenchman, invented the *galvanometer*, named in honor of Italian scientist Galvani. The meter was basically a device that used a stationary permanent magnet and a *moving coil*. Although the early galvanometer was very accurate, it could only measure very small currents and was very delicate. Over the years, many improvements were made that extended the range of the meter and made it very rugged. (To this day, the basic moving-coil meter galvanometer movement is often referred to as the d'Arsonval meter movement.)

Because it is very accurate and rugged, the moving-coil movement is by far the most common meter used today. It is the basic meter movement used to measure current, voltage, resistance, and a wide variety of other electrical quantities. Therefore, a thorough understanding of the moving-coil meter is a "must" for anyone studying electricity.

The moving-coil meter operates on the electromagnetic effect described on page 5-7. In its simplest form, the moving-coil meter uses a coil of very fine wire wound on a light aluminum frame. A permanent magnet surrounds the coil. The aluminum frame is mounted on pivots to allow it and the coil to rotate freely between the poles of the permanent magnet. When current flows through the coil, it becomes magnetized, and the polarity of the coil is such that it is repelled by the field of the permanent magnet. This causes the coil frame to rotate on its pivots, and the distance it rotates depends on how much current flows through the coil. Therefore, by attaching a pointer to the coil frame and by adding a scale calibrated in units of current, the amount of current flowing through the meter can be measured.

moving-iron meter movements

In the review of electromagnetism you saw how two soft iron bars repelled each other when placed inside an energized electromagnetic coil. This effect is used in moving-iron meters to measure electric current. There are three types of moving-iron meters: (1) the *radial-vane* type, (2) the *concentric-vane* type, and (3) the *plunger* type.

Basically, the radial-vane meter movement consists of two *rectangular* pieces of soft iron, called *vanes*, surrounded by a coil. One vane is fixed and the other is free to rotate along one edge on pivots. They are called *radial vanes* because the pivoted vane rotates like a radial line in a circle. A pointer is attached to the vane that rotates. When current flows through the coil, it sets up a magnetic field around the coil, and this magnetic field, in turn, induces a magnetic field of the same polarity in both vanes. This causes the vanes to repel each other, and the moving vane, with its pointer, rotates a distance proportional to the current flowing in the coil. Just as in the moving-coil meter, the pointer moves across a scale calibrated in units of current to indicate the current flowing through the meter.

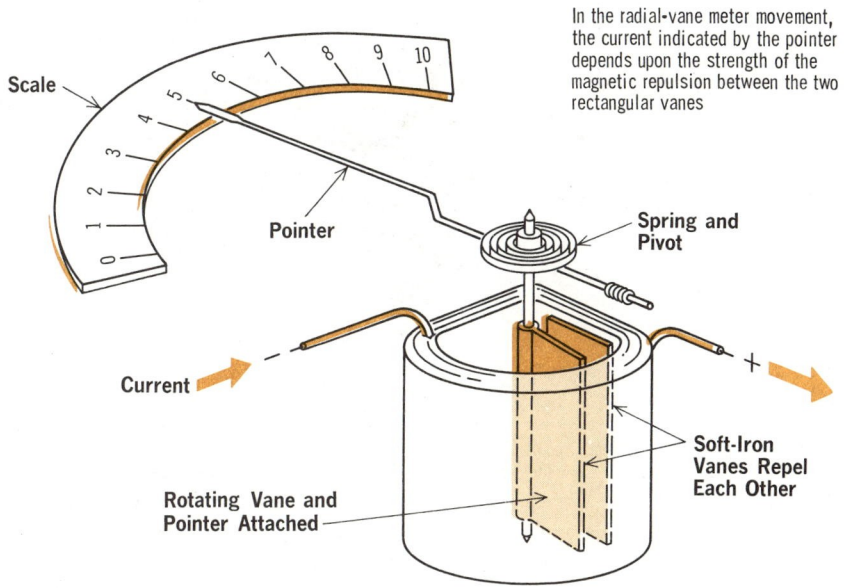

In the radial-vane meter movement, the current indicated by the pointer depends upon the strength of the magnetic repulsion between the two rectangular vanes

Scale

Pointer

Spring and Pivot

Current

Rotating Vane and Pointer Attached

Soft-Iron Vanes Repel Each Other

The strength of the magnetic repulsion, in turn, depends on the current flowing through the coil

the concentric-vane
meter movement

The *concentric-vane* meter movement operates similarly to the radial-vane meter movement. The only differences between the two meters are the shapes of the vanes and their position with respect to each other.

The vanes in the concentric-vane meter are *semicircular* in shape, and one semicircular vane is placed parallel to the other vane. The vanes are like segments of two different-size circles that have a common center; that is, they are *concentric*. The *inner vane* is free *to rotate* about this center point.

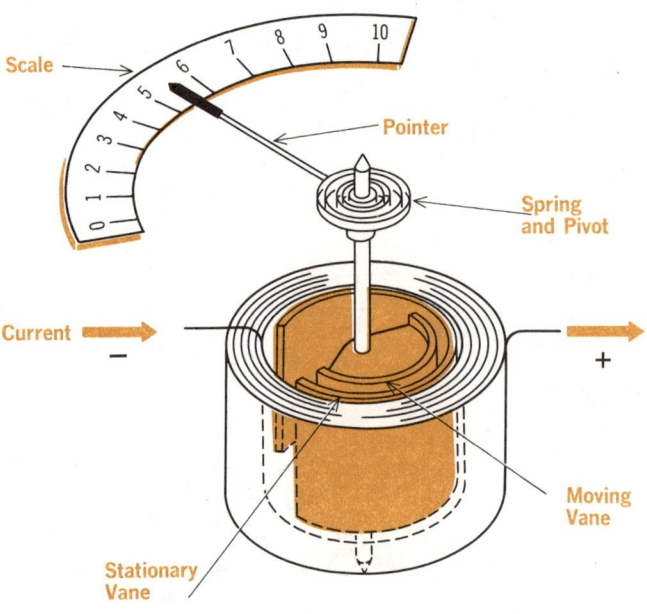

The magnetic repulsion between two semicircular vanes causes the inner vane to rotate and move the pointer, which indicates the current flowing through the coil

As with the radial-vane meter movement, current flow through the coil sets up magnetic fields of the same polarity in both vanes. The distance that the free vane rotates depends on the current flowing through the coil, and the pointer indicates the amount of current on a calibrated scale.

the moving-plunger meter movement

The *plunger* type meter movement consists basically of a movable soft iron core placed partially within a fixed coil. The core is connected to a pivoted arm that allows it to swing in and out of the coil. A pointer is connected to the pivot point so that it swings with the core plunger. When current flows through the coil, a magnetic field is set up about the coil. As explained in the review of electromagnetism, this causes the core to become magnetized, drawing it further into the coil. The distance that the core is pulled into the coil depends on the amount of current flowing through the coil. Since the pointer is attached to the plunger's pivot, its movement across a calibrated scale is used to indicate the amount of current flowing through the coil.

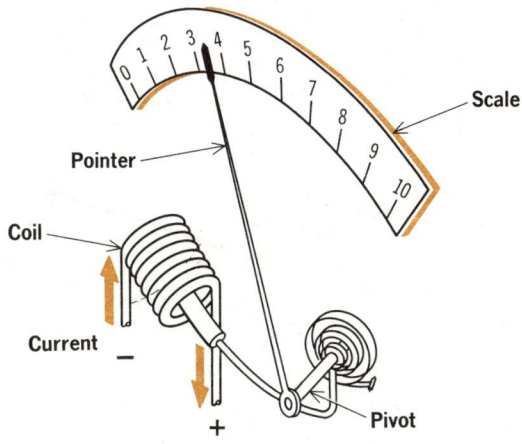

Current through the coil sets up a magnetic field that draws the iron core into the coil. The amount of current is indicated by the pointer, which is attached to the pivot point

The plunger type meter was the first moving-iron meter invented. However, it is used seldomly now because it is not as accurate or as sensitive as the other types of moving-iron meters that were developed later. Therefore, it is not covered in detail in this book.

the hot-wire ammeter

In Volume 2, you learned that a wire becomes heated when current flows through it. You also learned that the amount of heat increases as the current increases. The hot-wire ammeter makes use of this effect to measure current.

A wire expands when it becomes heated. The more the wire is heated, the more it expands. Therefore, the greater the current through the wire, the more it will expand. If a second wire and spring are attached to

When current flows through the wire, the wire is heated in proportion to the current flow. As a result, the wire expands in proportion to how hot it gets

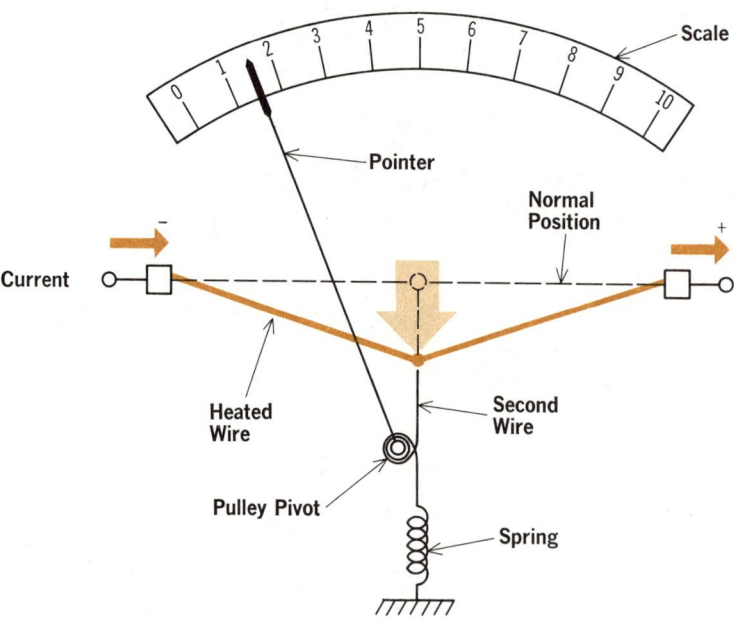

When the wire expands, it becomes less taut and the spring and second wire pull it down. The spring and wire are attached to the pulley pivot so that they will also swing the pointer to indicate the current flow

the original current-carrying wire, they will pull the current-carrying wire from its normal position whenever it expands due to heat. If a pointer is attached to the second wire, the pointer will also move as the current-carrying wire expands. The distance that the pointer moves is used to indicate the amount of current flowing through the wire.

the thermocouple meter

Previously, you studied meters that operated either on electromagnetic effects or on thermal effects. The thermocouple meter makes use of *both* effects to measure current. Basically, it is a combination hot-wire ammeter and moving-coil meter, with a device known as a *thermocouple* added.

If you will recall what you learned in Volume 1 about thermoelectricity, you know that a thermocouple consists of two dissimilar metals, which when joined produce an emf when their junction is heated.

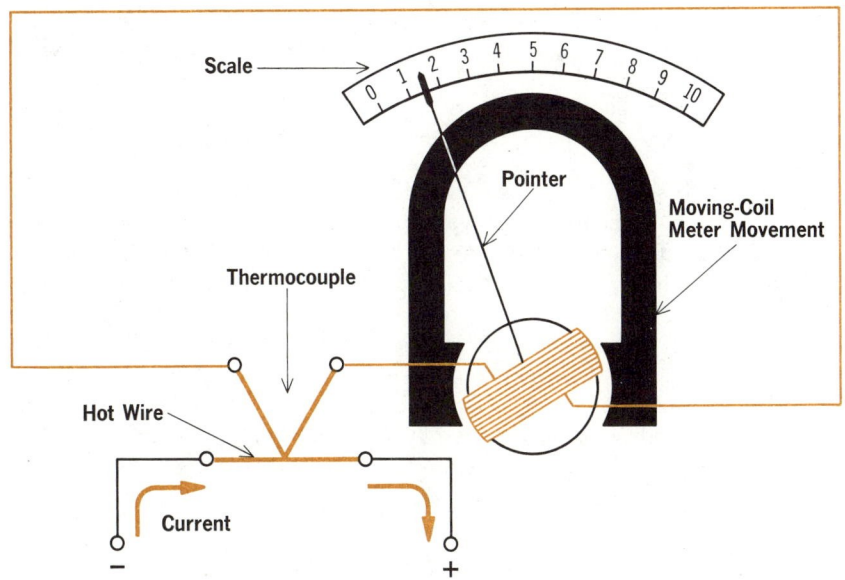

Current through the hot wire heats both the hot wire and the thermocouple junction. A small d-c voltage is produced at the free ends of the thermocouple, which causes the moving-coil meter to indicate the current flowing through the hot wire

In this meter movement, a heater, usually a hot wire, is also attached to the junction of the thermocouple. As you can see from the figure, the current to be measured is passed through the thermocouple heater, or hot wire. As in the case of the hot-wire ammeter, the wire heats up to a temperature that depends on the amount of current flow. The wire then heats the thermocouple junction, and the thermocouple in turn, generates a small d-c voltage. This voltage causes a current to flow through the moving-coil movement to indicate the amount of current flowing through the hot wire.

summary

☐ Practically all current meters measure only the current flowing through them. Nevertheless, they can be used to measure voltage, resistance, and power, by calibrating the meter scale in these quantities. ☐ There are two basic types of current meters: the electromagnetic type and the thermal type. ☐ The electromagnetic type, the one used most often, operates on the basis of interacting magnetic fields. ☐ The thermal current meter uses the heat produced in a wire by current flow.

☐ The moving-coil meter is the most common type of meter used today. It is also called the galvanometer, in honor of Italian scientist Galvani. ☐ The basic moving-coil meter consists of a coil of fine wire mounted on an aluminum frame, and positioned between the poles of a permanent magnet. The coil frame is free to rotate. ☐ When a current produces a magnetic field around the meter coil, the coil rotates, since it is repelled by the field of the permanent magnet. ☐ A pointer is attached to the coil frame of the moving-coil meter. When the coil rotates, it moves the pointer past a calibrated scale.

☐ In moving-iron meters, the current-carrying coil is stationary. ☐ The radial-vane and concentric-vane types of moving-iron meters use soft iron vanes positioned within the coil. When the coil produces a magnetic field, it causes one of the vanes to rotate. The amount the vane rotates depends on the strength of the coil's magnetic field, and therefore on the amount of current flowing through it. ☐ In the moving-plunger type, a movable core swings into and out of the coil under the influence of the coil's magnetic field. ☐ The hot-wire ammeter is a thermal-type meter. It operates on the principle that a wire expands when it is heated by a current flowing through it. ☐ The thermocouple meter makes use of both the electromagnetic and thermal effects of current.

review questions

1. Why can a current meter measure voltage and resistance?
2. What are the two basic types of current meters?
3. If the current through a coil is increased, what happens to the magnetic field around the coil?
4. Draw a basic moving-coil meter movement.
5. What is another name for the moving-coil meter?
6. What are the three types of moving-iron meter movements?
7. Draw a basic moving-plunger meter movement.
8. Draw a basic hot-wire ammeter.
9. What is a *thermocouple*?
10. Briefly describe a thermocouple meter.

meter movement parts

While all of the basic meter movements you have just examined operate on different electrical principles, their basic construction is similar. All contain the following basic parts: (1) a coil, (2) a pointer, (3) a scale, (4) pivots, (5) bearings, (6) springs, (7) retaining pins, (8) a zero-adjust screw, and (9) a damping mechanism.

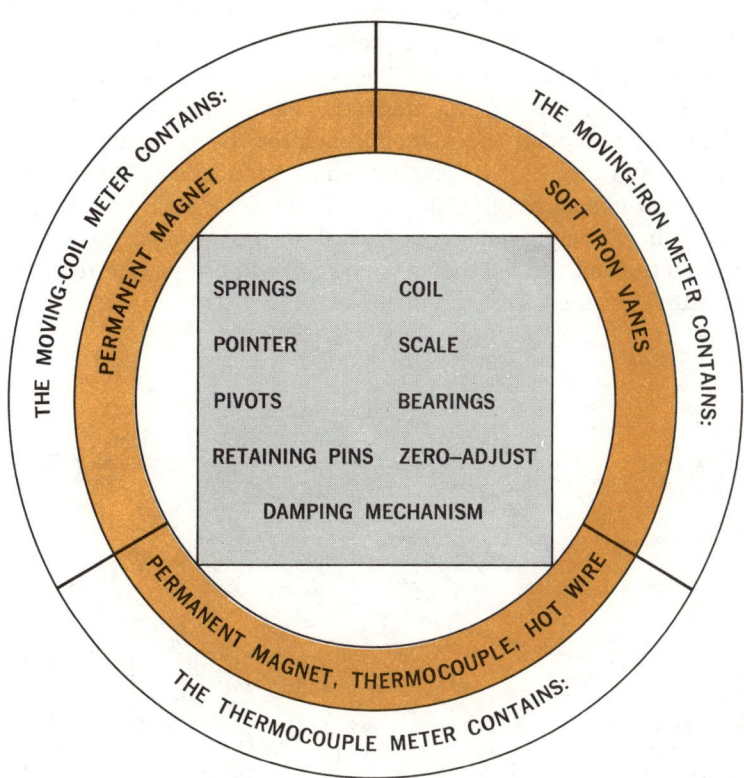

Except for the coil, scale, and damping mechanism, these parts are essentially the same in all meters. The moving-coil meter movement, in addition to containing the above parts, has a permanent magnet, while the moving-iron meter movement has soft iron vanes. A thermocouple meter is identical to the moving-coil meter except that it has a thermocouple and hot wire added to it.

Now you will learn how moving-coil, moving-iron, and thermocouple meter movements are constructed. You will also learn the function of each major part of these movements.

moving-coil (d'arsonval) meter movement

The moving-coil meter movement shown earlier was simplified to help you become familiar with the basic principles of its operation. Actually, as you will now see, the moving-coil meter movement contains many more parts than were shown earlier.

All the major parts of a moving-coil meter are now shown. You will recall that the permanent magnet provides the uniform magnetic field in which the moving coil rotates. The current to be measured is fed to the moving coil, and produces a magnetic field about the coil. This magnetic field interacts with the magnetic field of the permanent magnet, causing the coil to rotate. The pointer, which is attached to the coil, also rotates and swings across the calibrated scale to indicate the amount of current flowing. The greater the current flow, the stronger the magnetic field around the coil will be, the more the coil will rotate, and the further the pointer will swing across the meter scale.

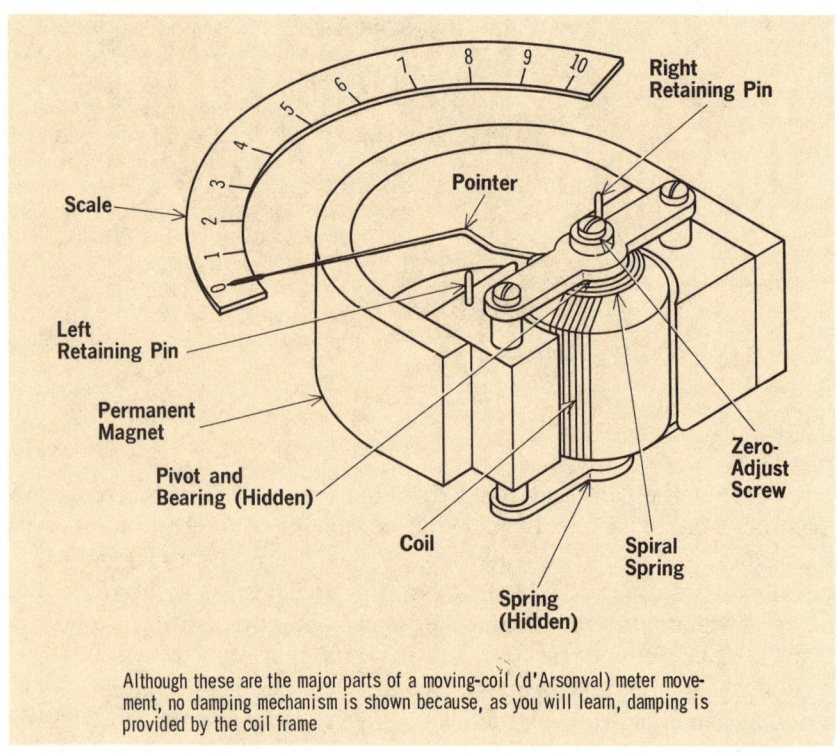

Although these are the major parts of a moving-coil (d'Arsonval) meter movement, no damping mechanism is shown because, as you will learn, damping is provided by the coil frame

permanent magnets

The moving-coil meter movement uses a horseshoe shaped permanent magnet. The moving coil is placed within the magnetic field between the magnet's two poles. However, if a *simple* horseshoe magnet were used, many of the magnetic lines of force would not cut through the moving coil. But, you will recall from your study of magnetism in Volume 1, that magnetic lines of force travel the path of least resistance. You will also recall that soft iron offers less resistance to lines of force than does air. Therefore, soft iron *pole pieces* are attached to the poles of the magnet to concentrate the lines of force between the magnetic poles.

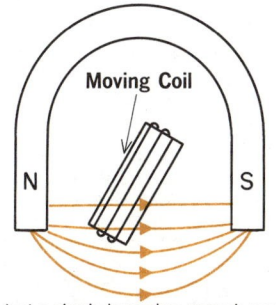

With just a simple horseshoe magnet, many of the lines of force would not cut the coil

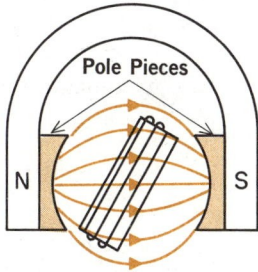

Pole pieces concentrate the lines of force between the two poles. Therefore, more lines of force cut the coil

To further concentrate the lines of force between the magnet poles, a circular, soft-iron core is placed between the pole pieces. The core not only causes a very strong, uniform magnetic field between the poles, but it also acts as a *keeper* to help the permanent magnet retain its magnetism. The moving coil rotates around the soft iron core, which is fixed in place.

A horseshoe magnet with pole pieces and a soft iron core has a highly concentrated, uniform magnetic field across the gap. Therefore, most of the lines of force cut the coil

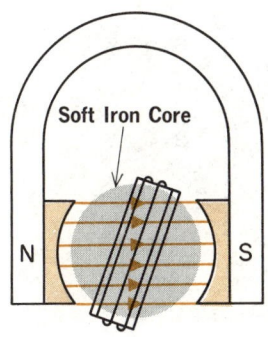

moving-iron meters

On page 5-17, it was pointed out that the moving-coil and the moving-iron meter movements contain essentially the same parts except for the permanent magnet of the moving-coil movement and the iron bars of the moving-iron movement. We will now study the moving-iron movement and its iron vanes, and proceed to study all the parts that are common to both meters.

You will recall that a moving-iron meter contains two vanes mounted within a coil. One vane is fixed and the other, with a pointer attached, is free to rotate. Current through the coil induces a magnetic field of the same polarity in both vanes. The free vane, therefore, is repelled by the fixed vane. It rotates a distance that depends on the strength of the magnetic field, and, therefore, on the strength of the current. The pointer, which is attached to the free vane, also rotates and swings across a calibrated scale to indicate the amount of current flowing.

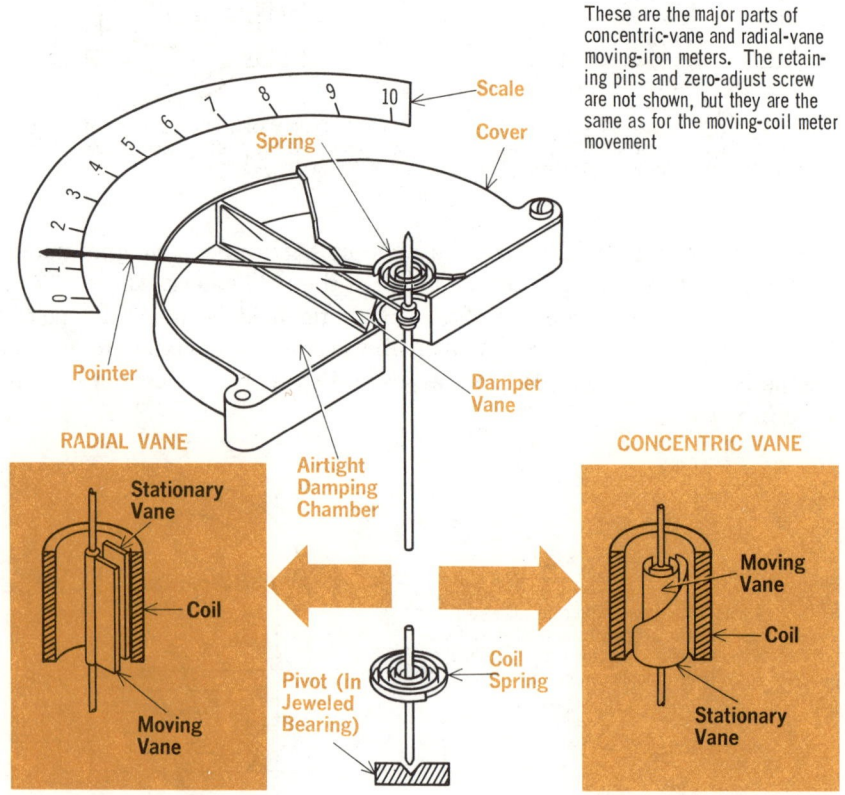

These are the major parts of concentric-vane and radial-vane moving-iron meters. The retaining pins and zero-adjust screw are not shown, but they are the same as for the moving-coil meter movement

iron vanes

The differences between the concentric-vane and radial-vane iron meters are the shape of the vanes and the physical placement of the vanes with respect with each other.

The concentric-vane meter contains two *semicircular* soft iron vanes. One vane is essentially inside the other, which is why the meter is called a *concentric*-vane meter. The outside vane, which is tapered at one edge, is fixed; the inner vane, which has square edges, is pivoted. When current flows through the coil, the lines of force cut through both vanes,

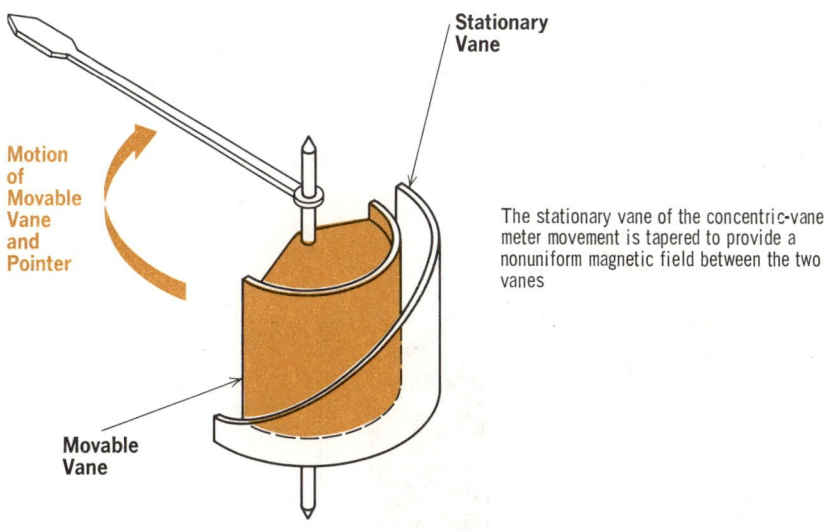

Stationary Vane

Motion of Movable Vane and Pointer

Movable Vane

The stationary vane of the concentric-vane meter movement is tapered to provide a nonuniform magnetic field between the two vanes

but the distribution of the lines of force is not the same in one vane as in the other. The lines of force are distributed uniformly through the movable (inner) vane because its dimensions are uniform, but they are not distributed evenly in the stationary (outer) vane because of its tapered edge. Less lines of force pass through the tapered end than through the rest of the vane because the tapered end is smaller than the rest of the vane, and, therefore, has a higher reluctance.

When both vanes become magnetized in the *same* polarity, they will repel each other, causing the movable vane to rotate on its pivot. The strongest repulsion will occur in the area where the stationary vane is *not* tapered since there will be more magnetic lines than in the tapered area. This means that the movable vane will swing towards the tapered end of the stationary vane since there are fewer lines of force there.

iron vanes (cont.)

The radial-vane meter contains two rectangular iron vanes: one fixed, and the other free to move. The radial-vane meter operates on exactly the same principle as the concentric-vane meter, except that, because its vanes are of the same shape and size, a uniform magnetic field is present between the movable and the fixed vanes. As you can see, then, the two types of meters are identical except for the shape and orientation of the vanes. Both types of movements are enclosed in iron cases to prevent outside magnetic fields from affecting meter readings.

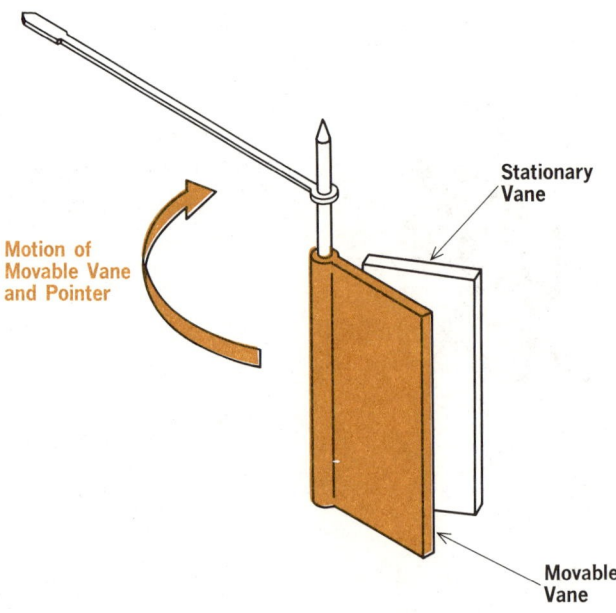

Both the stationary vane and the movable vane of the radial-vane meter movement have the same dimensions to provide a uniform magnetic field between the vanes

Just like the moving-coil meter movement, the concentric-vane and radial-vane moving-iron movements contain springs, pivots, bearings, etc. to control the action of the pointer. These parts serve the same function in both types of movements, as you will see.

coils

In both the moving-coil and moving-iron meter movements, the current being measured flows through the coil. Except for this similarity, the coils in each type of movement are different.

On page 5-10, you learned that when current flows through the coil of the moving-coil meter movement, a magnetic field is produced that causes the coil to rotate. For the coil to rotate easily, it must be as light as possible. To make the coil light, it is wound on an aluminum frame. Furthermore, the coil is made from very fine wire, and when compared with the coil in the other meter movements, contains very few turns to keep it as light as possible.

When you studied the moving-iron meter movements, you learned that the coil in this type of movement remains stationary, and the magnetic field about the coil moves an iron vane. Because this iron vane is relatively heavy, a strong magnetic field is required to move it. Therefore, the coil of a moving-iron meter movement contains many turns of wire to produce this strong magnetic field.

The iron vanes of the moving-iron meter movement are placed inside of the coil. However, the shape of the coil is different for the concentric-vane meter movement and the radial-vane movement. The coil for the concentric-vane movement is constructed so that it can accept semi-circular vanes and the coil for the radial-vane movement is constructed so that it can accept rectangular plates.

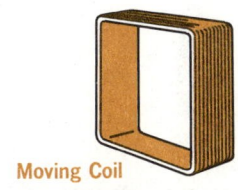

The coil of the moving-coil meter movement consists of several turns of fine wire wound on an aluminum frame

Moving Coil

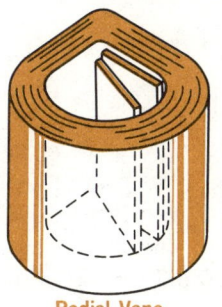

The coils on both the radial-vane and the concentric-vane moving-iron meter movements contain many turns of wire. The shape of the coil differs to accommodate the particular shape of the vane; that is, radial or concentric

Radial Vane

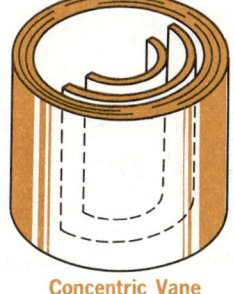

Concentric Vane

pointers

When a *pointer* is attached to the moving element of the meter movement, and a calibrated *scale* is put behind the pointer tip, the pointer will swing with the meter movement according to the amount of current flowing, and will stop at the proper place on the scale to show the current flow.

Because of the delicate nature of the meter movement, and the need to have it respond to current flow without having to overcome external forces, the pointer is generally made of thin aluminum to keep it light.

The tips of the pointer usually come in three shapes: spade, knife, and lance. The *spade* pointer is used when high visibility is needed, particularly at a distance. However, the broad width of the tip makes it difficult to differentiate scale index marks that are close together on high-accuracy scales. The thin, sliver-like *knife pointer* is best for high accuracy readings, but it is difficult to see at a distance. The *lance pointer* is a compromise shape that gives good accuracy with acceptable visibility.

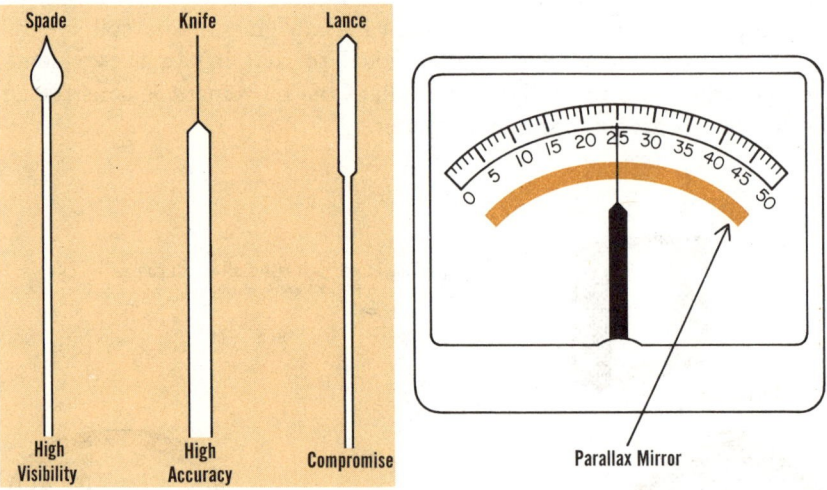

When knife-edge pointers are used for high accuracy, *parallax error* must be considered. This is the error in reading when the pointer *appears* to be pointing off the true reading because it is not being read exactly head on. Good instruments contain a parallax mirror to show the error. The reading should be made when the mirrored reflection of the pointer is directly in line with the pointer. In this way, the meter will not be read at an angle.

counterweights and retaining pins

Although the pointer in a meter is very light, the extremely delicate sensitivity of the meter movement is such that the pointer must be properly *balanced* so that it does not interfere with the accuracy of the movement.

When the meter is manufactured, counterweights are attached to the pointer assembly, and meticulously adjusted to balance the pointer on the pivot so that the weight of the pointer will neither aid nor oppose the motion of the meter movement.

To limit the range over which the meter movement can travel, and to protect it from being overdriven, retaining pins are placed on both sides of the pointer to keep the pointer from going too far off the bottom or top ends of the scale. The pins thus limit any undue pressure being applied to the delicate meter movement spring.

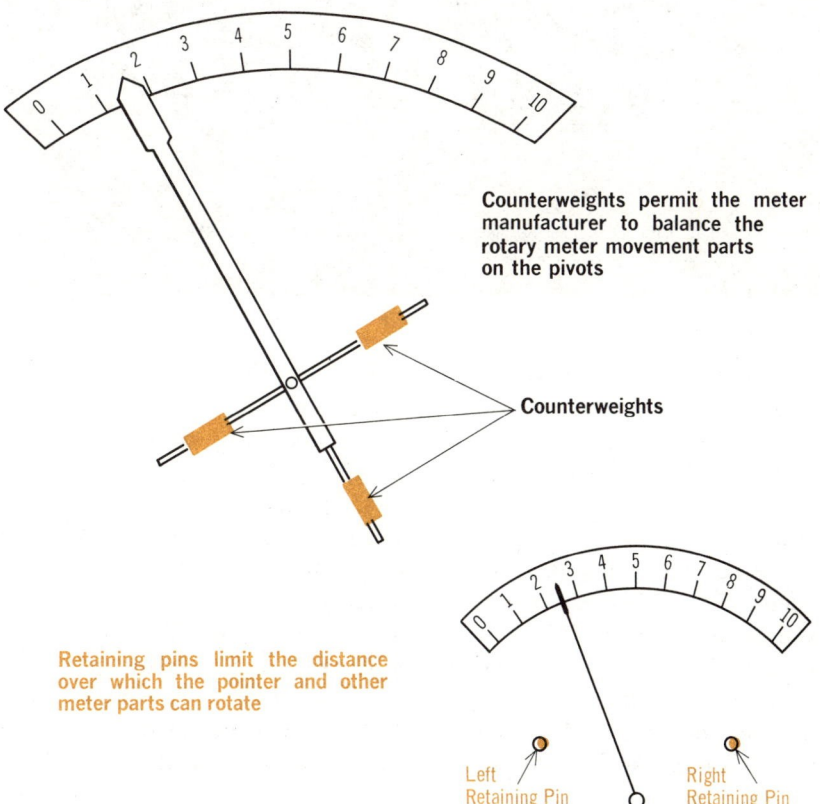

Counterweights permit the meter manufacturer to balance the rotary meter movement parts on the pivots

Counterweights

Retaining pins limit the distance over which the pointer and other meter parts can rotate

Left Retaining Pin   Right Retaining Pin

springs

When a meter is connected into a circuit, the pointer must indicate the amount of current flowing in the circuit. When the meter is disconnected from the circuit or when the current stops flowing, the pointer must return to zero. This action is controlled by the meter movement springs. The springs, then, control the rotating action of the moving coil in the moving-coil meter and the movable vane in the moving-iron meter. Therefore, they must be manufactured very precisely to insure meter accuracy.

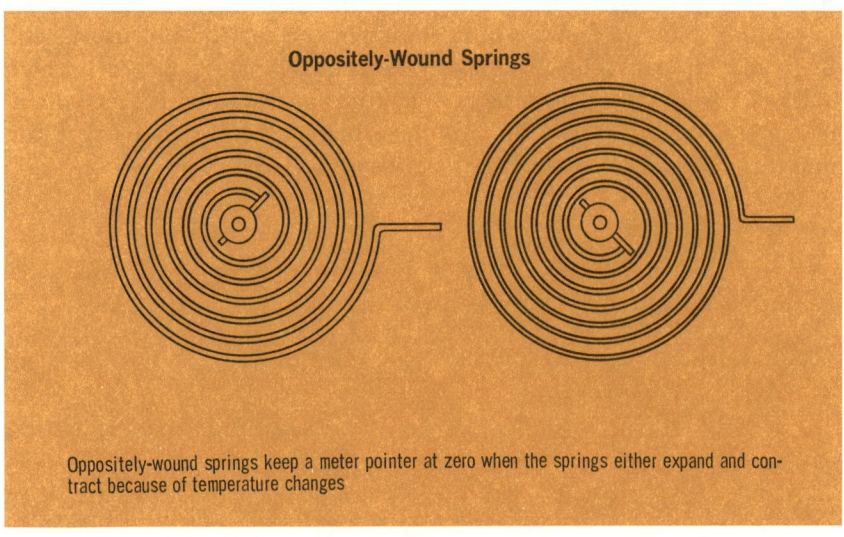

Oppositely-Wound Springs

Oppositely-wound springs keep a meter pointer at zero when the springs either expand and contract because of temperature changes

Two springs are used with each of the meter movements, and these springs are oppositely wound. This is done for a very important reason. You know that metal expands as temperature increases and contracts as temperature decreases. If both springs were wound in the same direction, both would expand or contract in the same direction as the temperature changed. This would cause the pointer to rotate off zero with no current flowing, and would cause an error in the meter reading when a current is measured. However, because the springs are oppositely wound, as one expands because of an increase in temperature, it tries to move the pointer in one direction, and as the other expands it tries to move the coil in the opposite direction. Therefore, the two actions cancel, and the pointer remains at zero. When temperature decreases the oppositely-wound springs contract an equal amount, and the pointer remains at zero.

zero-adjust

You learned that the springs that control the rotation of the meter pointer are oppositely wound to keep the pointer at zero when no current is flowing through the meter. If the two springs were perfectly matched in every respect, then the pointer would rest exactly on zero with no current flowing. In actual practice, however, it is impossible to construct two perfectly matched springs that will expand or contract equally with temperature changes. Also, springs tend to lose tension as they age. Therefore, the springs will not always keep the pointer exactly at the zero scale reading with no current flowing. To correct this, most meters have a *zero-adjust screw* that permits setting the pointer to zero.

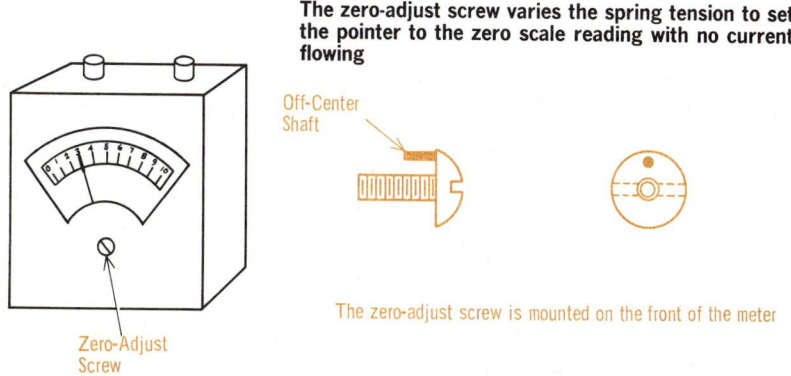

The zero-adjust screw varies the spring tension to set the pointer to the zero scale reading with no current flowing

Off-Center Shaft

The zero-adjust screw is mounted on the front of the meter

Zero-Adjust Screw

The zero-adjust screw is located on the front of the meter. As the screw is turned, it increases or decreases the tension on one of the springs, depending upon the direction in which it is turned. By turning the screw, the tension on its associated spring can be adjusted until it is equal to the tension on the other spring. And, when the tension on both springs are equal, the pointer will rest at the zero scale reading.

When the zero-adjust screw is turned to the left, spring tension is reduced, and the pointer moves up-scale to the right

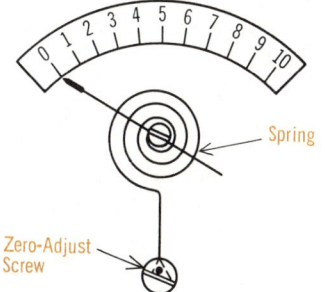

Spring

When the zero-adjust screw is turned to the right, spring tension is increased, and the pointer moves off-scale to the left

Zero-Adjust Screw

summary

☐ Although their electrical operation is different, all of the basic meter movements have similar construction. ☐ The basic meter parts are a coil, a pointer, a scale, pivots, bearings, springs, retaining pins, a zero-adjust screw, and a damping mechanism. ☐ The moving-coil meter movement also contains a permanent magnet, while the moving-iron meter movement has soft iron vanes. ☐ A thermocouple meter is identical to a moving-coil meter, except that it also contains a thermocouple and hot wire.

☐ The permanent magnet of a moving-coil meter has soft iron pole pieces, as well as a soft iron core to insure that the magnetic lines of force are concentrated between the magnetic poles. ☐ The vanes of moving-iron meters are constructed of soft iron. These movements are enclosed in iron cases to prevent outside magnetic fields from affecting meter readings. ☐ The coil of a moving-coil meter consists of a few turns of very fine wire wound on an aluminum frame to make the coil as light as possible. ☐ The coil of a moving-iron meter contains many turns of wire to produce a strong magnetic field to move the relatively heavy iron vane.

☐ Since the pointer of a meter must move with the moving coil or the iron vane, it is usually made of very thin, lightweight aluminum. ☐ Counterweights are usually attached to the pointer, and provide the means for balancing the moving portion of the meter. ☐ Springs are used to control the moving action of a meter. ☐ Two springs are used, and they are wound in opposite directions. This is done to compensate for the expansion and contraction that occurs in springs as a result of temperature changes. ☐ The tension of the two meter springs is controlled by the zero-adjust screw. The screw is adjusted for equal tension on both springs, so that with no current through the meter, the pointer indicates zero on the meter scale.

review questions

1. What basic parts are common to most meter movements?
2. What are *pole pieces*? Why are they used?
3. What are the differences between the vanes of the concentric-vane and radial-vane meters?
4. Must the coil of a moving-coil meter produce a strong magnetic field? Why?
5. Why must the coil of a moving-coil meter be lightweight?
6. What are *counterweights*? Why are they used?
7. How do the springs affect the accuracy of a meter?
8. Why are two springs used?
9. If both springs lost most of their tension, how would this affect meter operation?
10. What is the function of the *zero-adjust screw*?

pivots

Remember that the meter movement is a delicately balanced and sensitive instrument, and should not be unduly interfered with by outside forces. The movement is so sensitive that even a lightweight, unbalanced pointer will cause inaccuracies. Also, the movement must overcome friction as it rotates. This friction must be kept extremely low, especially in highly sensitive movements that must respond to very small currents.

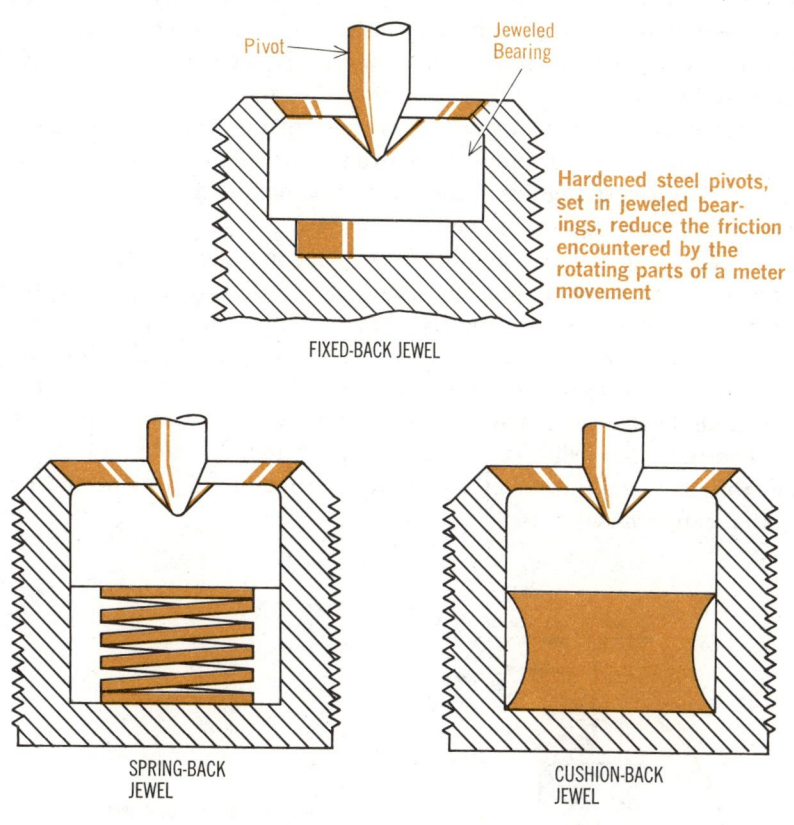

Hardened steel pivots, set in jeweled bearings, reduce the friction encountered by the rotating parts of a meter movement

FIXED-BACK JEWEL

SPRING-BACK JEWEL

CUSHION-BACK JEWEL

Most meter movements use hardened, finely tipped steel *pivots*, set in *jewel* bearings to keep rotating friction at a minimum. Three types of jewel bearing assemblies are used. The *fixed jewel* type is a rigid unit. Any shock can damage either the pivot or the jewel. The *cushion-back* jewel costs a little more, but the resiliency of the cushion allows it to absorb some shocks without damage. The *spring-back jewel* costs still more, but is the most shock-resistant assembly manufactured today.

taut band movements

The rotating contact of the pivot in the jewel still produces some friction to retard the rotation of the movement, regardless of how well and how precisely the pivots are made, and how fine a jewel bearing you use. The pivot must be held loosely in the bearing in order to rotate as freely as possible in the jewel assembly. But if it is made too loose for the least possible friction, there will be too much end play and shifting of the movement. Sensitive meters using pivots and jewels are often set for minimum friction and its accompanying end play, but such meters can function properly in only one position. At any other position, the shifting meter movement will cause parts to touch and interfere with the overall accuracy.

The *taut band meter movement* was designed to produce a *suspended* movement that did not employ the friction-producing pivots and bearings, so that a very high sensitivity could be obtained with a meter that could be kept in any reasonable position.

The taut band meter uses thin metal ribbons, kept taut, to suspend the meter movement. A tension spring exerts the pressure to keep the ribbon bands taut so that the movement is kept in proper alignment, regardless of its orientation. With such an arrangement, the taut bands flex and twist to allow the movement to rotate. The opposition of the taut bands to the movement rotation is relatively constant and is taken into consideration when the scale is calibrated. Its accuracy is more definable than the unpredictable friction variations of a pivot and bearing that can wear with use.

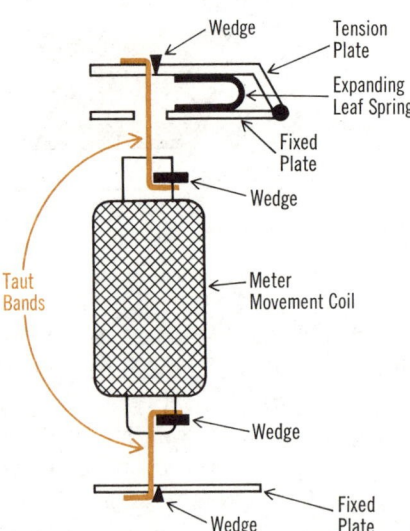

This is a partial representation of the taut band meter. Only the main parts are shown. The metal ribbon bands are locked in place with wedges. The expanding leaf spring pushes up on the tension plate, which in turn pulls the bands taut so that they hold the meter movement coil suspended and free to rotate

scales for moving-coil meters

Moving-coil meter movements have a *linear* scale; that is, a scale in which the space between numbers is equal. The distance that the pointer deflects across the scale is *directly proportional* to the amount of current flowing through the meter coil.

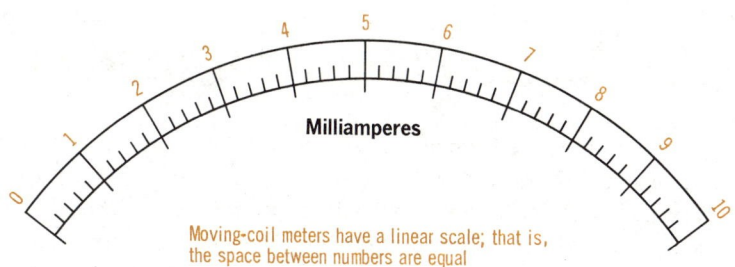

Moving-coil meters have a linear scale; that is, the space between numbers are equal

When the full rated current of a moving-coil meter movement flows through the coil, the pointer deflects full scale; when one-half the rated meter current flows through the coil the pointer will move one-half the distance across the scale, and so on. The reason for this is that the magnetic flux produced by the coil increases in direct proportion to the current; and so the interaction of the fields also increases proportionally to give a linear reading. This is not true for moving-iron type meters, as you will soon learn.

Moving-Coil Meters Have A Linear Scale

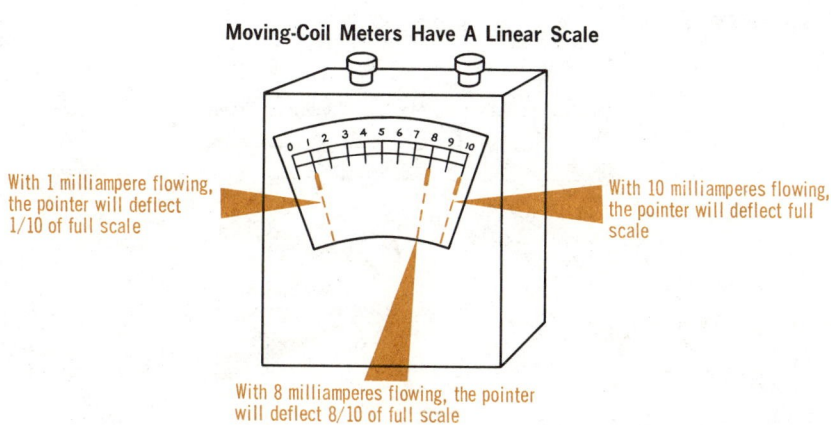

With 1 milliampere flowing, the pointer will deflect 1/10 of full scale

With 10 milliamperes flowing, the pointer will deflect full scale

With 8 milliamperes flowing, the pointer will deflect 8/10 of full scale

The pointer deflection is directly proportional to the amount of current flowing in the meter coil

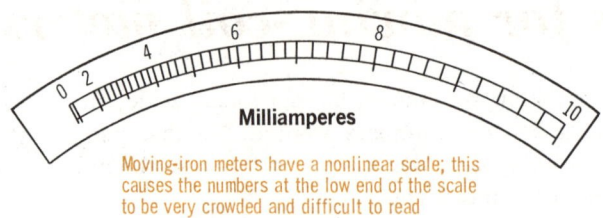

Milliamperes

Moving-iron meters have a nonlinear scale; this
causes the numbers at the low end of the scale
to be very crowded and difficult to read

scales for moving-iron meters

The scale for a moving-coil meter, as you learned, is linear. If the amount of current through the meter doubles, the distance the pointer deflects doubles; if the current through the meter triples, the distance the pointer deflects triples. This relationship does not hold for moving-iron meter movements, however. Instead, the deflection increases with the *square* of the current. If the current through the meter doubles, the strength of the magnetic field about each vane doubles, and, therefore, the repulsion of each vane becomes twice as great. Since the repulsion of each vane is now twice as great, the combined repulsion of the two vanes becomes *four times* as great. If the current is *tripled,* the repulsion of each vane becomes three times as great, and the combined repulsion of the two vanes becomes *nine times* as great. Therefore, the deflection varies as the *square* of the current, rather than in a linear manner.

Since deflection is nonlinear, the scale of a moving-iron meter has to be nonlinear. The numbers at the low end of the scale are crowded, and they are farther and farther apart toward the high end of the scale where deflection is greater.

Moving-Iron Meters Have a Nonlinear Scale

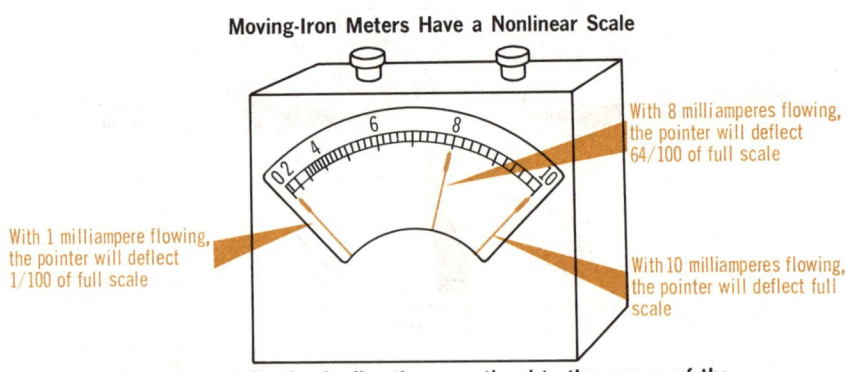

With 8 milliamperes flowing,
the pointer will deflect
64/100 of full scale

With 1 milliampere flowing,
the pointer will deflect
1/100 of full scale

With 10 milliamperes flowing,
the pointer will deflect full
scale

The pointer deflection is directly proportional to the square of the
current flowing in the meter coil

edgewise scales

The standard stock meter design uses the broad round or rectangular face that is higher and wider than it is deep. The large face area of the stock meter permits room for the scales to be highly legible, and also provides room for multiscales on its face.

For equipment where panel space is at a premium, the *edgewise reading meter* can be used. These meters run deep, but take up only a small rectangular area for the scale. The edgewise meters can be designed for either horizontal or vertical use, and are often made available so that they can be stacked. With such an arrangement, a few edgewise meters will take up the same panel space as one stock meter.

Stock meters are more accurate than edgewise meters because their pointers arc in a plane that is parallel to the scale face. With an edgewise meter, the arc of the pointer causes the scale face to curve to the same sweeping arc as the pointer. The curved scale increases *parallax error* possibilities from one end of the scale to the other.

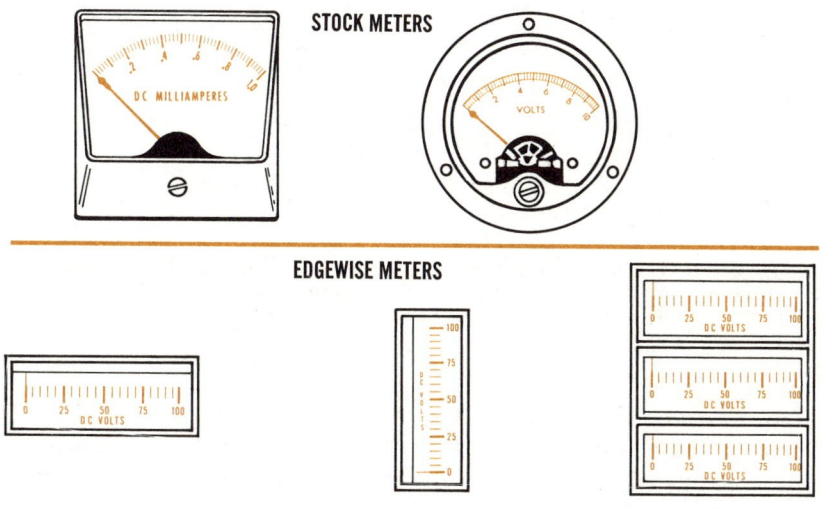

STOCK METERS

EDGEWISE METERS

There are two types of edgewise meters: the projected pointer, and the projected scale. The *projected pointer* movement is the more common type, in which the pointer sweeps across the fixed scale face as in stock meters. The *projected scale* meter uses a *fixed pointer*, and a circular lightweight scale is attached to the meter movement. The scale moves by the fixed pointer as the movement rotates, and the reading is always taken in the same position at the fixed pointer. Because of the extra weight carried by the projected scale meter, it is not a sensitive, accurate instrument.

damping

You learned that all of the rotating meter parts are as light as possible and rotate on pivots set in bearings to keep friction to a minimum. Keeping friction to a minimum permits measuring small currents, but creates a major problem when reading the meter.

When a meter is in a circuit, the pointer should move across the scale and stop immediately at the correct reading. However, because of the very little friction of the rotating parts the pointer does not come to rest immediately at the correct reading; it overshoots because of inertia, and then the spring pulls it back, it overshoots slightly again, and so on. As a result, the pointer tends to swing back and forth, or vibrate, about the correct reading point many times before coming to rest.

UNLESS A METER IS DAMPED

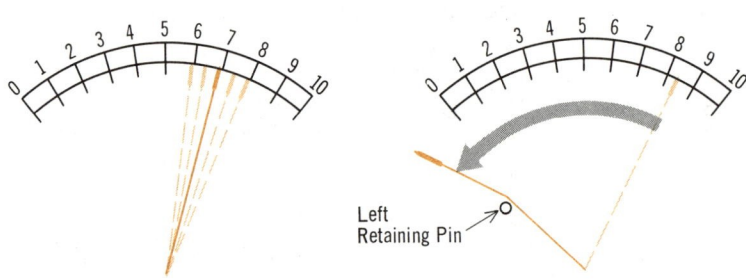

the pointer will vibrate several times about the correct reading before it comes to rest

the pointer might return so quickly to zero so that it bends about the left retaining pin

To overcome this problem, the meter movement must be *damped*. Damping can be thought of as a *braking* action on the rotating parts. It almost completely eliminates the vibrating action of the pointer, resulting in quick, correct pointer indication.

Damping also eliminates another problem. When a meter is removed from an external circuit or when the circuit is deenergized, the pointer returns to zero. Because of the very low friction of the rotating parts, the springs tend to pull the parts back to zero very quickly; so quickly in fact, that the pointer could bend as it overshoots and strikes the left retaining pin. This is particularly true when the meter returns to zero from near full-scale deflection. However, you will learn how damping overcomes this problem by applying a braking action to the pointer as it returns to zero.

damping a moving-coil meter

Moving-coil meter movements make use of the aluminum frame on which the coil is wound to provide damping. Since aluminum is a conductor, the frame acts as a one-turn coil. When the coil assembly and pointer rotate to register current, the aluminum frame cuts through the field flux lines of the permanent magnet. Small currents, called eddy currents, are induced in the frame, which set up a magnetic field about the frame.

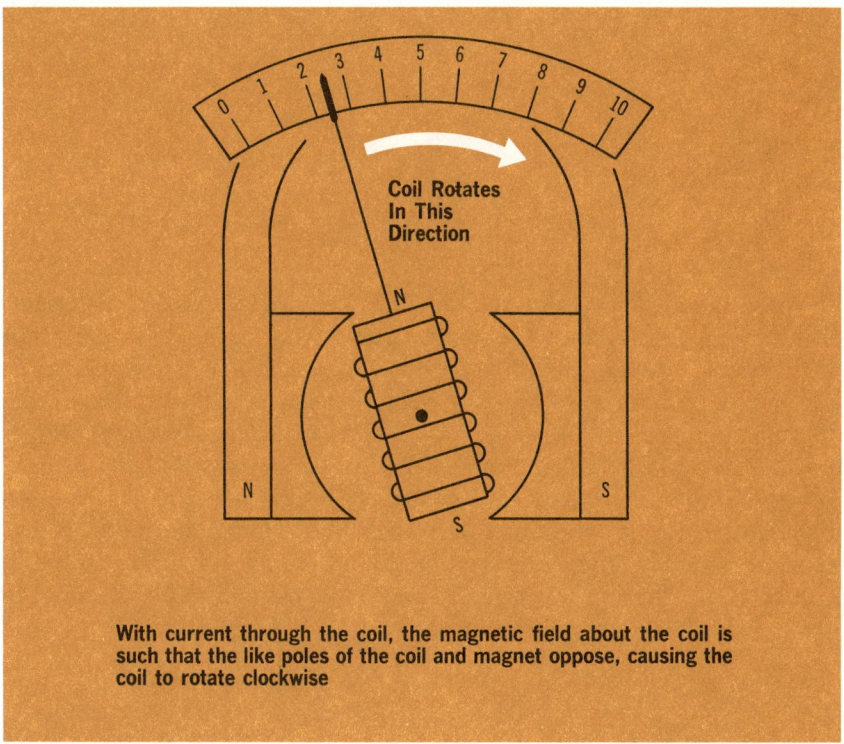

Coil Rotates In This Direction

With current through the coil, the magnetic field about the coil is such that the like poles of the coil and magnet oppose, causing the coil to rotate clockwise

The polarity of this magnetic field is opposite to that of the magnetic field about the coil. Therefore, the magnetic field about the frame *opposes* the magnetic field about the coil. This action reduces the overall field of the moving coil so that it swings more slowly. In effect, the faster the coil swings, the more the aluminum frame's field slows it down. This causes the coil and pointer to rotate relatively slowly and smoothly to the correct reading without vibrating. As soon as the coil assembly and pointer comes to rest, no further eddy currents are induced in the frame and its magnetic field disappears.

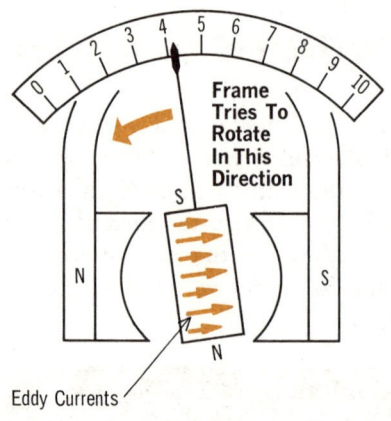

The eddy currents produced in the aluminum frame, which acts like a 1-turn coil, set up a magnetic field opposite to that of the coil. Therefore, the like poles of the frame and the magnet attract, thereby slowing the rotation of the coil assembly

Eddy Currents

damping
a moving-coil meter (cont.)

When the meter is disconnected from the circuit, or when the circuit is deenergized, essentially the same action occurs. The coil assembly and pointer start to rotate very rapidly toward zero, but now, since it is rotating in the opposite direction, the eddy currents produced in the frame flow in the opposite direction. This results in a magnetic field with a polarity opposite to the one previously. As the springs pull the coil assembly and pointer closer and closer to zero, the like poles of the permanent magnet and the frame repel each other more and more so that the coil assembly is again braked or slowed down, and slowly comes to rest at zero. Thus, the pointer is prevented from striking the left retaining pin and, perhaps, bending around it.

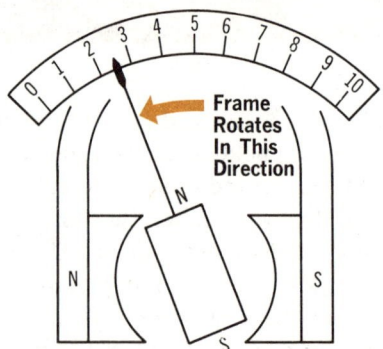

As the springs bring the coil assembly back toward zero, a magnetic field is again set up around the frame, but in the opposite direction. As the coil assembly nears zero, the like poles of the magnet and the frame oppose, thereby braking the coil so that it slowly comes to rest at zero

damping a moving-iron meter

You learned how electrical damping is accomplished in moving-coil meters. Eddy currents were produced in the aluminum frame of the coil as the coil rotated in the field of the permanent magnet, and these eddy currents produced a magnetic field that damped the movement of the pointer. This type of damping, however cannot be used with a moving-iron meter because the coil is stationary. *Air damping* is used instead.

There are two different methods of obtaining air damping. In the method used most often, a light aluminum vane, a *damping vane,* is attached to the same shaft as the pointer and moving vane and enclosed in an airtight *damping chamber.* As the moving vane is repelled, the damping vane rotates with it. Since the damping vane is contained in an airtight chamber, it compresses the air ahead of it as it moves. This decreases the speed at which the vane moves, and, therefore, decreases the speed of the moving vane and pointer. Because of the slower speed, there is less tendency of the pointer to overshoot. When the pointer comes to a stop, the air in front of the damping vane is no longer compressed, and the damping action ceases.

Radial-vane and concentric-vane meters are usually damped in this manner. Occasionally, however, small radial-vane meters are damped by enclosing the coil and vanes in an airtight metal case rather than using a separate damping vane and chamber. As the moving vane is repelled, it compresses the air ahead of it to slow the vane and the pointer. The same principle is used in both methods, however, using a separate damping vane and chamber usually provides the most effective damping.

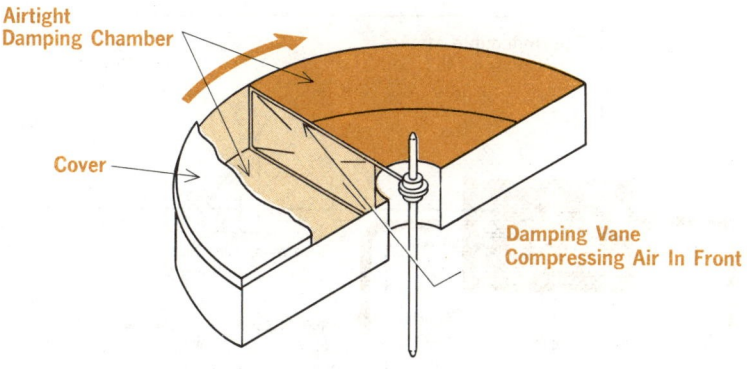

The shaft movement, which rotates the pointer and moving vane, is retarded as air is compressed in front of the damping vane. This slows the pointer so that there is less overshoot

thermocouple meter scale

You learned that a thermocouple meter consists of a device called a *thermocouple* and of a *moving-coil meter*. When current flows through the heating element of the thermocouple, a voltage is produced across the two free ends of the thermocouple. This voltage causes a d-c current to flow through the moving-coil meter and the pointer deflects an amount that is proportional to the amount of current flowing through the thermocouple heating element.

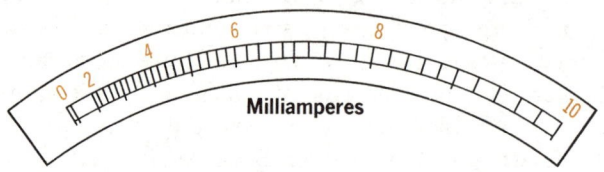

Milliamperes

The pointer deflection is directly proportional to the square of the current flowing in the meter coil

The amount of deflection is proportional to the amount of heat produced in the heater wire. The amount of heat produced in the heater element is, in turn, proportional to the square of the current flowing in the heating wire ($P = I^2R$). Therefore, thermocouple meters, like moving-vane meters, have a *square law scale*, and the lower readings on the scale are very crowded.

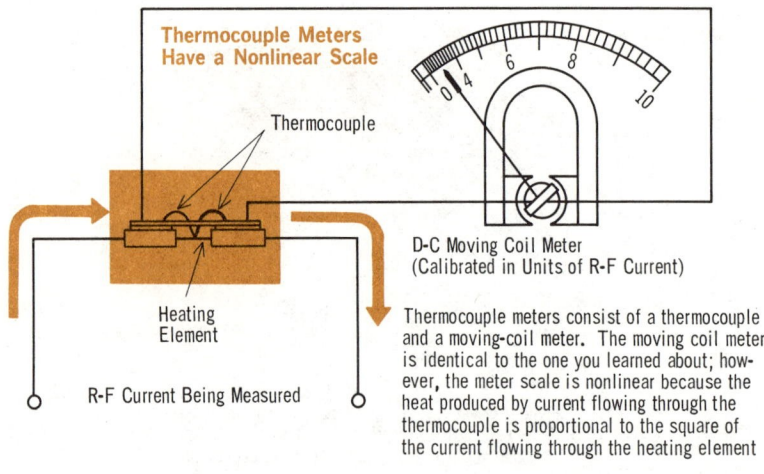

Thermocouple Meters Have a Nonlinear Scale

Thermocouple

Heating Element

R-F Current Being Measured

D-C Moving Coil Meter
(Calibrated in Units of R-F Current)

Thermocouple meters consist of a thermocouple and a moving-coil meter. The moving coil meter is identical to the one you learned about; however, the meter scale is nonlinear because the heat produced by current flowing through the thermocouple is proportional to the square of the current flowing through the heating element

thermocouple construction

The voltage that is generated at the free ends of the thermocouple is dependent upon the *temperature difference* between the free (cool) ends and the junction (hot) ends. Incorrect meter readings will result if this temperature difference is affected by the ambient temperature of the area where the meter is being used or by heat given off by currents flowing through the meter. To nullify these unwanted effects, each free end of the thermocouple is attached to the center of a copper strip. The ends of the copper strips are placed very close to the *heater wire* ends so that the free ends of the thermocouple have the same temperature as the copper strips. The ends of the copper strips are insulated from the ends of the heater element by very thin sheets of mica. Since the copper strips attain the same temperature as the *ends* of the heater wire, then the free ends of the thermocouple will attain the same temperature as the free ends of the heater wire. The temperature of the free ends of the thermocouple is, therefore, dependent on the current flowing through the heater wire, and not on other factors such as ambient temperature.

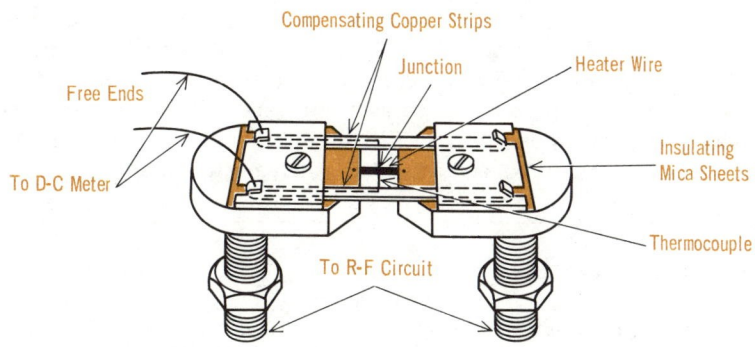

Regardless of outside temperature effects, the temperature difference between the free ends and the junction ends of this compensated thermocouple will remain constant for a given r-f current

When current flows through the heater wire, the temperature becomes much greater at its center, that is, at the point where the wire and the thermocouple junction are joined, than it does at the ends of the heater wire. For any given current, this *temperature difference* between the thermocouple junction and the free ends of the thermocouple will remain constant regardless of outside temperature effects.

summary

☐ The rotating parts of a meter movement are connected to a shaft that has hardened steel pivots attached to each end. The pivots are set into jewelled bearings to keep friction to a minimum. ☐ Retaining pins are used at both ends of a meter scale to prevent the pointer from moving off scale. ☐ Moving-coil meters have linear scales. This is because the distance that the pointer deflects across the scale is directly proportional to the amount of current flowing through the meter coil. ☐ Moving-iron meters have square-law scales. This is because the deflection of the pointer varies with the square of the current through the coil. ☐ The numbers at the low end of a square-law scale are very close together. They become farther and farther apart toward the high end. ☐ Thermocouple meters, like moving-iron meters, have square-law scales.

☐ Damping provides a braking action on a meter pointer to prevent it from vibrating about a reading before coming to rest. It also prevents the pointer from returning too rapidly to zero when the meter is disconnected from a circuit. ☐ In moving-coil meters, damping is provided by the aluminum frame of the movement. ☐ The flux lines of the moving coil cut the aluminum frame and set up eddy currents in the frame. These produce a magnetic field of their own that opposes the field of the coil, slowing down the coil.

☐ In moving-iron meters, damping is usually provided by a damping vane moving in an airtight chamber. ☐ The damping vane moves with the moving vane and pointer. As it moves, it compresses the air ahead of it, and in so doing, is slowed down. ☐ The voltage generated by a thermocouple depends on the temperature difference between the free ends and the hot junction. A thermocouple meter is designed so that this temperature is affected only by the current being measured, and not by outside temperature effects.

review questions

1. What is a *linear scale*? A *square-law scale*?
2. Why are linear scales used on moving-coil meters?
3. Why are square-law scales used on moving-iron meters?
4. If the current through a moving-coil meter doubles, by how much does the pointer deflection increase?
5. In a meter, what is meant by *damping*?
6. Why is damping necessary?
7. How is damping provided in a moving-coil meter?
8. How is damping provided in a moving-iron meter?
9. What effect would too much damping have on a meter?
10. How are the effects of ambient temperature compensated for in a thermocouple meter?

internal resistance
and sensitivity

Every meter coil has a certain amount of d-c resistance. The amount of resistance depends upon the number of turns on the coil and the size of the wire used to wind the coil. In Volume 1, you learned that the strength of the magnetic field about a coil increases as the number of turns on the coil increases. Therefore, if more windings are placed on a meter coil, a small current can create a magnetic field strong enough to cause the coil to deflect full scale.

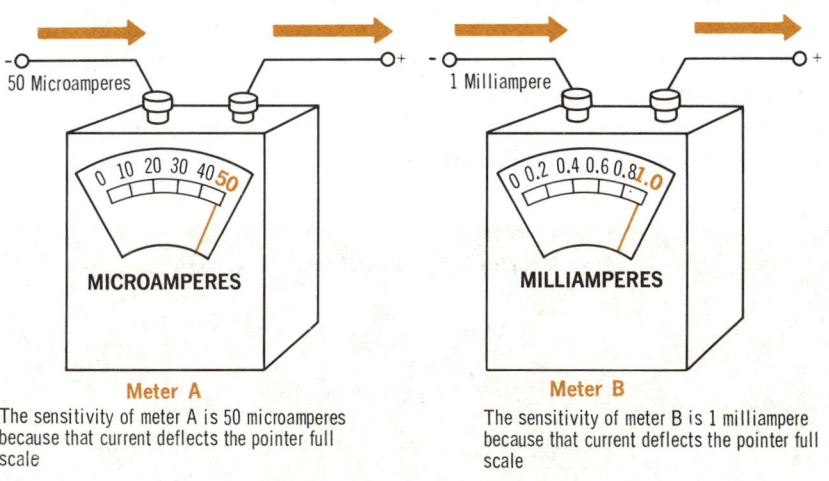

Meter A
The sensitivity of meter A is 50 microamperes because that current deflects the pointer full scale

Meter B
The sensitivity of meter B is 1 milliampere because that current deflects the pointer full scale

Meter A has a higher sensitivity than meter B because it requires less current for full-scale deflection

The amount of current necessary to cause the meter pointer to deflect full scale is the *meter sensitivity;* it is an important characteristic of any meter. Typical current meter sensitivities vary from about 5 microamperes (0.000005 amperes) to about 10 milliamperes (0.010 amperes). Some common values are 5, 50, and 100 microamperes; and 1, and 10 milliamperes.

You have probably recognized that the sensitivity of a meter movement is the *maximum* current that the movement can measure. Any current greater than this value will very likely damage the meter. Too much current might cause the pointer to rotate past full-scale deflection and bend about the right retaining pin. Or, too much current might cause the coil to burn out. A very heavy current overload sometimes causes both types of damage.

meter accuracy

The accuracy of a meter is specified as the *percentage of error at full-scale deflection*. For example, if the specified accuracy of a 100-milliampere meter is specified as ±2 percent, not only might the meter be off by ±2 milliamperes at a 100-milliampere reading, but it might be off by as much as ±2 milliamperes for any reading below full-scale deflection. Therefore, the accuracy of a meter becomes progressively poorer as the pointer moves farther and farther from full-scale deflection towards zero.

For example, at a meter reading of 50 milliamperes, the meter, since it could still be off by ±2 milliamperes, is only really accurate to ±4 percent. At a meter reading of 10 milliamperes, the meter could still be off by ±2 milliamperes, which is a true reading accuracy of ±20 percent. This, of course, is all based on a full-scale reading of 100 milliamperes. By using certain meter range circuits though, this situation can be improved, as you will learn later.

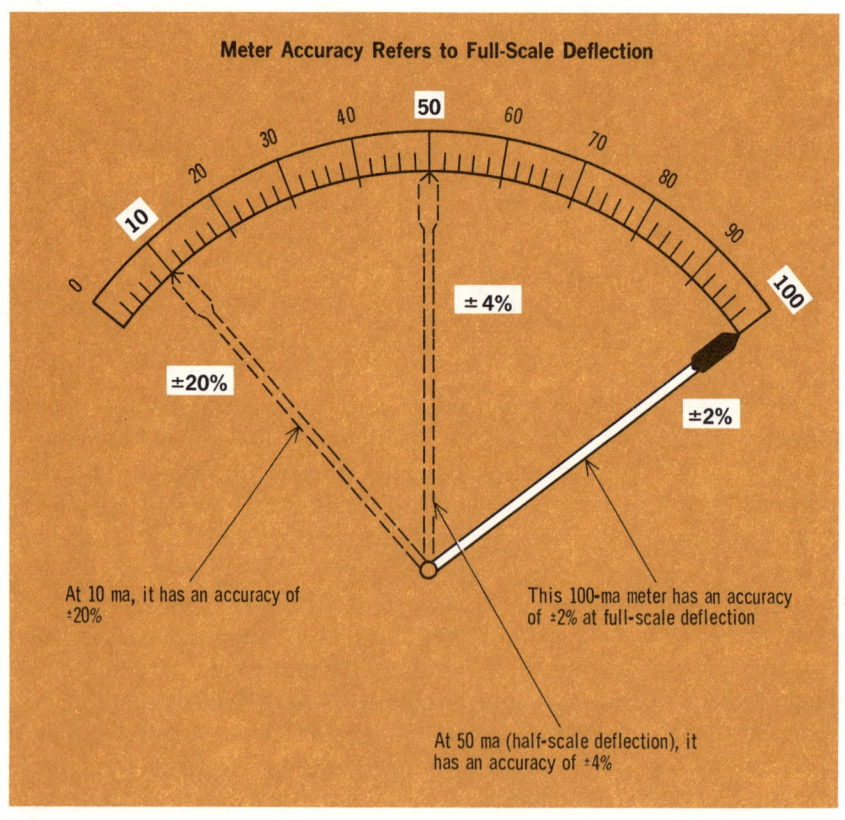

Meter Accuracy Refers to Full-Scale Deflection

At 10 ma, it has an accuracy of ±20%

This 100-ma meter has an accuracy of ±2% at full-scale deflection

At 50 ma (half-scale deflection), it has an accuracy of ±4%

which meters for dc and ac?

All of the basic meter movements discussed so far can be used to measure dc. Of these, however, the moving-coil meter movement is used most often because it is more sensitive and more accurate. Concentric-vane and radial-vane moving-iron meter movements, although they can measure both ac and dc, are generally used to measure low-frequency ac. Actually, even in a-c applications, the moving-coil meter movement is used much more than the other types of movements, but, for it to measure ac, the ac must first be converted to dc and then applied to the meter movement. This type of meter, called a *rectifier meter,* will be discussed shortly.

Thermocouple meters can be used to measure both ac and dc. In electrical and radio work, however, it is used almost exclusively to measure radio-frequency currents. The frequencies of these currents range from a few megacycles to thousands of megacycles, and can only be measured by a thermocouple meter because it operates on the heat produced by the current and is insensitive to frequency. Other meter movements are inaccurate for high-frequency measurements.

D-C AND A-C METER APPLICATIONS

Basic Meter Movement	Measures DC	Measures AC	Remarks
Moving-Coil	Yes	No	Almost all d-c measurements are made with this type of meter movement.
Rectifier-Type Moving-Coil	No	Yes	Basically a moving-coil meter movement with a rectifier added to change ac to dc. It is the most common meter used to measure ac.
Iron-Vane	Yes	Yes	
Thermocouple	Yes	Yes	Used almost exclusively to measure radio-frequency currents because it is the only meter capable of measuring these high frequencies.

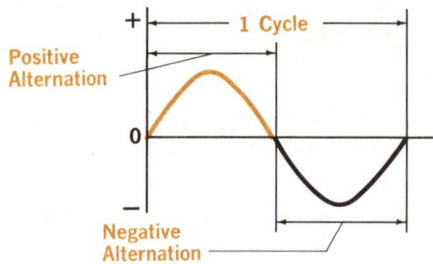

An a-c current completes 1 cycle when it goes from 0 to maximum positive, to 0, to maximum negative, and back to 0. The complete cycle consists of a positive and negative alternation

SINE WAVES

The frequency of this a-c wave is 2 cycles per second because the wave goes from 0 to maximum positive, to 0, to maximum negative, and back to 0 twice in 1 second. This is said to have a frequency of 2 Hz (hertz)

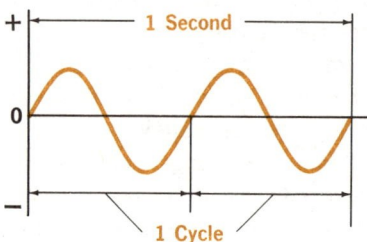

review of ac

Before examining the characteristics of the rectifier a-c meter, it is important that you thoroughly understand the characteristics of a-c waves that were described in Volume 3. Therefore, they will be briefly reviewed here.

Alternating current flows periodically first in one direction and then in the opposite direction. One direction is called a *positive alternation*, and the other direction is called a *negative alternation*. A complete positive and negative alternation is called one cycle. The number of complete cycles that occur each second make up the *frequency*. The frequency is designated in hertz, abbreviated Hz. Therefore, a frequency of one cycle per second is called 1 Hz, and a frequency of 5 cycles per second is called 5 Hz, and so on.

You will recall from Volume 3 that an a-c wave can have many wave shapes; for example, it can be a sine wave, square wave, sawtooth wave, etc. As you will soon see, a-c meters are calibrated on the basis of *sine waves*. When an a-c meter is used to measure nonsinusoidal waveforms, only an approximate indication of values is obtained. Sometimes the indicator could be so far off that the reading is meaningless. Therefore, other measuring instruments such as oscilloscopes should be used instead of a-c meters to measure nonsinusoidal waveforms.

rms and average values of a sine wave

The root-mean-square (rms) value of a sine wave was described in Volume 3, but because of its importance in the study of meters, it will be reviewed briefly.

The basic electrical units, that is, the ampere and the volt, are based on dc. Therefore, a method had to be derived to relate ac to dc. The maximum, or *peak*, value of a sine wave could not be used because a sine wave remains at its peak value for only a very short time during an alternation. Thus, a sine wave with a peak current of 1 ampere is not equal to a d-c current of 1 ampere from an energy standpoint, since the d-c current always remains at 1 ampere.

A relationship based on the heating effects of ac and dc was derived. It was found that a current equal to 0.707 of the peak a-c wave produced the same heat, or lost the same power, as an equal d-c current for a given resistance. For example, a sine wave with a peak value of 3 amperes has a heating effect of 0.707 × 3, or 2.121 amperes of dc.

The value of 0.707 can be derived in the following manner: The heating effect of current is based on the basic power formula you studied in Volume 2; that is, $P = I^2R$, where P is the power dissipated as heat. From the formula, you can see that the heat varies as the square of the current.

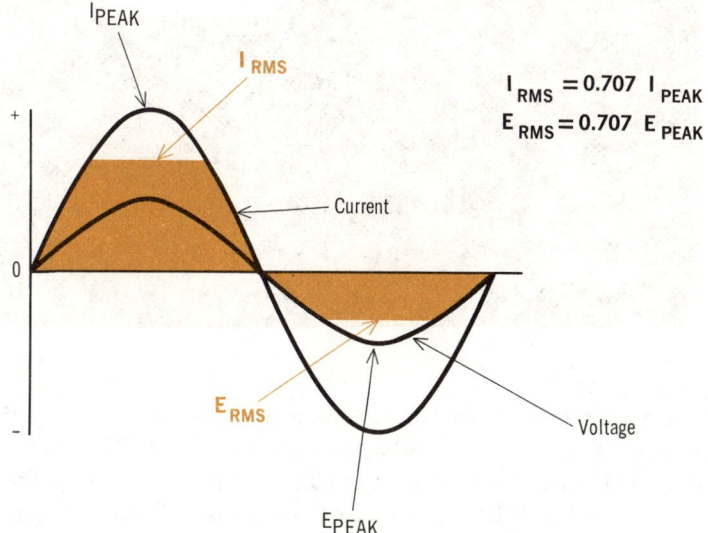

$$I_{RMS} = 0.707 \ I_{PEAK}$$
$$E_{RMS} = 0.707 \ E_{PEAK}$$

rms and average values of a sine wave (cont.)

When a sine wave reaches its peak value, the heat dissipated becomes maximum. Lesser heat values are dissipated for all values of current below the peak value. To find the heat dissipated during an entire sine wave cycle, each *instantaneous* value of *current* is first *squared,* and then *added.* Then the mean (or average) of this sum is found. After this, the *square root* of the *mean* is found, and the answer is called the *root-mean-square* (rms) *value* of the sine wave. Often the rms value of a sine wave is called the *effective value* because 0.707 of the peak value of a sine wave has the same effect as an equal amount of dc.

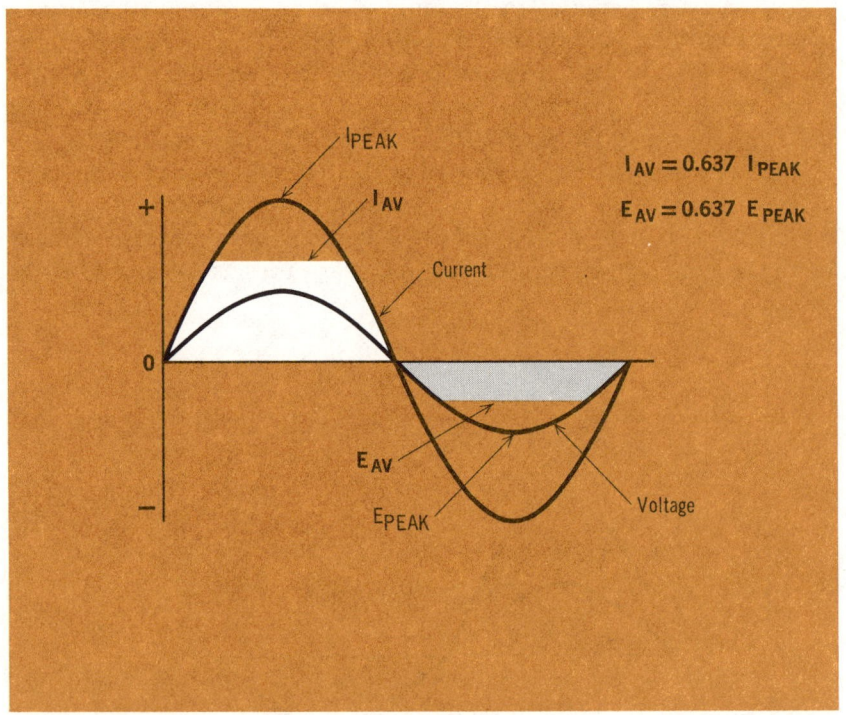

Another sine wave characteristic that is important in the study of meters is the *average value* of the sine wave. The average value is obtained during one alternation, and is equal to 0.637 of the peak value of the sine wave. Both the root-mean-square value and the average value of a sine wave will be discussed again shortly.

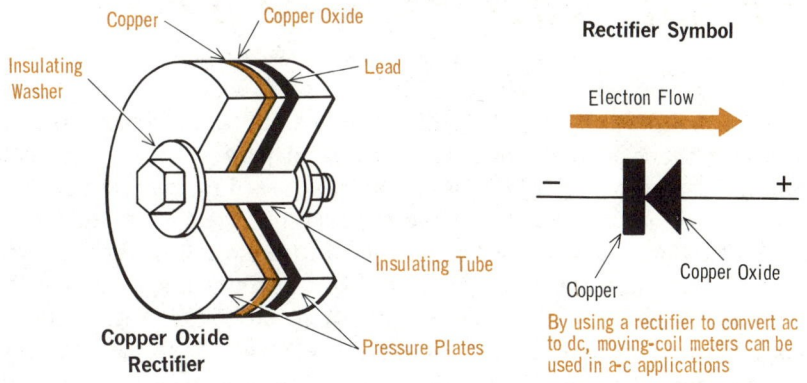

Rectifier Symbol

By using a rectifier to convert ac to dc, moving-coil meters can be used in a-c applications

rectifiers

The moving-coil meter movement is the most sensitive and accurate meter movement that we have discussed. It can only be used to measure dc, however; there is no way that it can be used to measure ac directly. If ac was applied directly to the meter, one half of the cycle would try to make the meter pointer move in one direction and the other half would try to make the pointer move in the opposite direction. Even at very low frequencies, the pointer would not be able to move fast enough to follow the positive and negative alternations of the a-c wave. Therefore, instead of moving across the scale, the pointer would simply vibrate about zero. But, if the ac is first changed to dc before applying it to the meter movement then the moving-coil meter can be used in a-c applications as well as dc.

Ac can be converted to dc by special devices called *rectifiers*. These devices offer a very high opposition to current flow through them in one direction and a very low opposition to current flow in the other direction. Therefore, when a sine wave is applied to a rectifier, it will pass *either* the positive alternation or the negative alternation, depending upon how the rectifier is connected into the meter circuit. In no case, however, will it pass both alternations. Therefore, a rectifier changes a sine wave to a *pulsating d-c wave*.

Copper oxide rectifiers are used most often in meters. They are constructed of a series of copper oxide discs separated by copper discs, and bolted or clamped together. In a rectifier symbol, the arrow points in the direction of high resistance, so current flow through it is in the opposite direction. Rectifier meters also use selenium rectifiers or silicon rectifiers. The principle of operation is the same, however, no matter which kind is used.

the half-wave rectifier meter

You just learned that a rectifier changes ac to dc. There are two basic types of rectifier circuits: the half-wave type and the full-wave type. In the half-wave type, one alternation of current passes through the meter movement and the opposite alternation is bypassed by the rectifier. Even though current through the meter is pulsating, the meter pointer, because of its inertia, will not have sufficient time to follow these fluctuations. Therefore, the meter pointer will rest at the average value of the current flowing through it. The average current for one alternation is 0.637 of peak value, but, for the next alternation, it is zero and that alternation is bypassed by the meter. Therefore, the average current for a complete cycle is the sum of both alternations divided by 2, or $0.637/2 = 0.318$ of peak value. The meter pointer then deflects to the position on the scale that represents 0.318 of the peak value of the current flowing through the meter. But, for the reading to be meaningful the scale is usually calibrated to show the equivalent rms value. Therefore, the points on the scale are calibrated at 0.707 of the equivalent peak values.

Half-Wave Rectifier Meter Operation

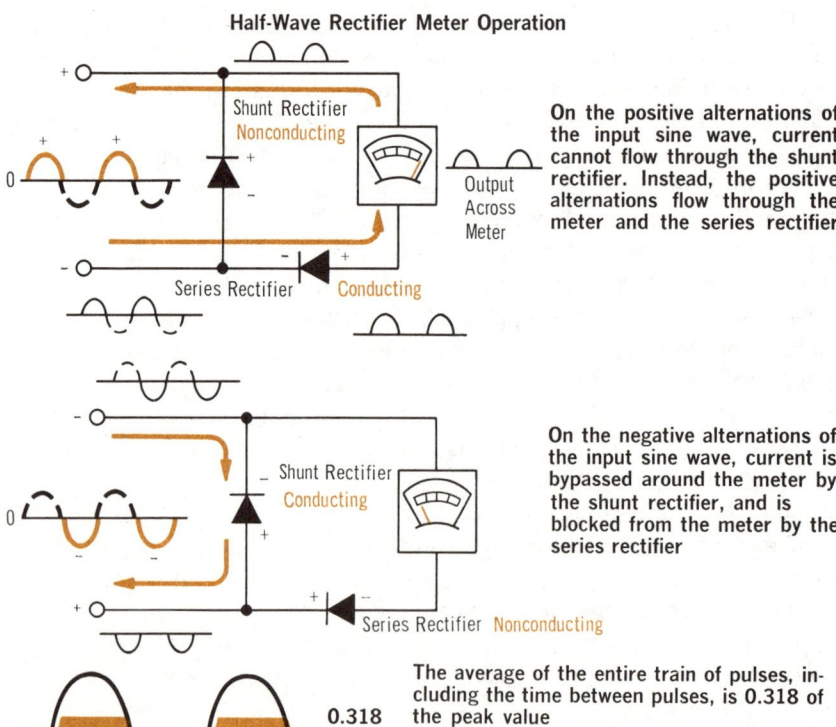

On the positive alternations of the input sine wave, current cannot flow through the shunt rectifier. Instead, the positive alternations flow through the meter and the series rectifier

On the negative alternations of the input sine wave, current is bypassed around the meter by the shunt rectifier, and is blocked from the meter by the series rectifier

The average of the entire train of pulses, including the time between pulses, is 0.318 of the peak value

the full-wave rectifier meter

In the full-wave rectifier meter, current flows through the meter in the *same direction* on both alternations of the a-c sine wave. This is accomplished by using four rectifiers in an arrangement called a *bridge rectifier circuit*. When the input sine wave is positive (positive alternation), current flows from terminal B through rectifier CR_1, *through the meter*, then through rectifier CR_2 to terminal A. On the negative alternation of the input sine wave, current flows from terminal A through rectifier CR_3, *through the meter*, then through rectifier CR_4 to terminal B. You can see, then, that current flows through the meter in the same direction on *both* alternations. The a-c input sine wave was thus changed to pulsating dc. It is pulsating because it is still varying in amplitude and it is dc because it is not changing direction through the meter.

The average current flow through the meter movement is twice as great in the full-wave rectifier arrangement as it is in the half-wave arrangement. In the full-wave arrangement, the average current is 0.637 of peak value because *both halves* of the sine wave flow through the meter. Again the scale is generally calibrated to read in rms values; that is, 0.707 of the peak current being measured.

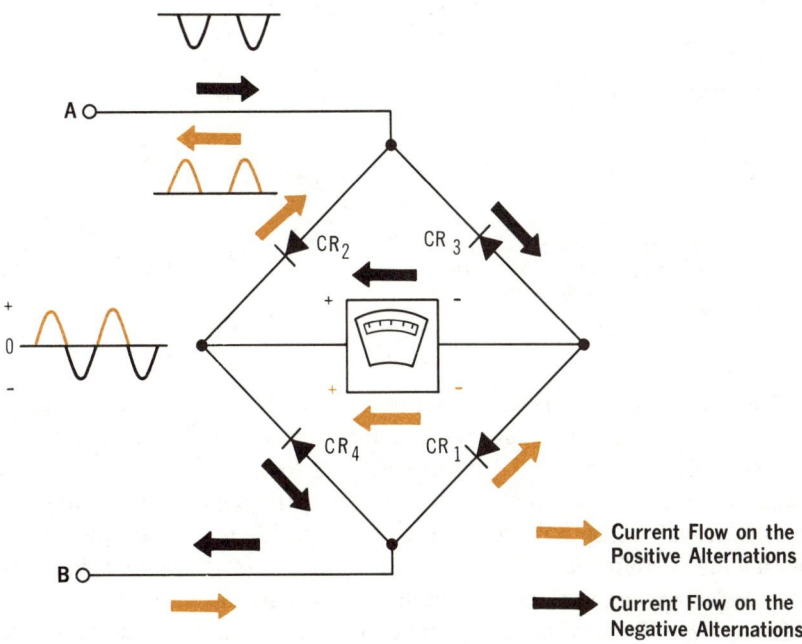

Current Flow on the Positive Alternations

Current Flow on the Negative Alternations

calibrating d-c and a-c meters

Both d-c and a-c meters are calibrated in essentially the same way. To calibrate a d-c meter, a very accurate d-c current source is connected to the meter. The output of the current source must be variable, and some means must be available to monitor the output current of the source. Many sources have a built-in meter for this purpose. The output of the current source is varied in very small steps, and at each step the scale of the meter being calibrated is marked to correspond to the reading on the monitoring device. This procedure is continued until the entire scale of the meter is calibrated.

The same procedure is used to calibrate an a-c meter, except that a 60-Hz sine wave is used mostly. Also, you know that an a-c meter reads the average value of a sine wave, but it is desirable for the meter to indicate rms values. Therefore, the rms equivalents are calculated and marked on the scale.

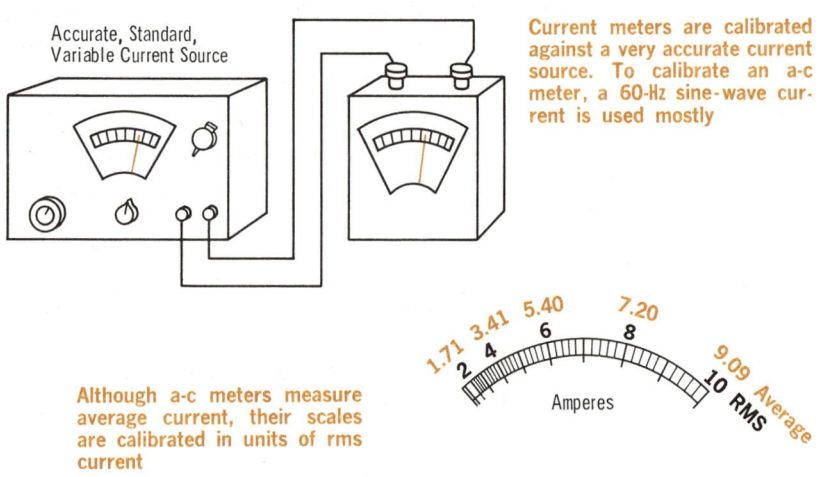

Accurate, Standard, Variable Current Source

Current meters are calibrated against a very accurate current source. To calibrate an a-c meter, a 60-Hz sine-wave current is used mostly

Although a-c meters measure average current, their scales are calibrated in units of rms current

Thermocouple meters are calibrated on the basis of a sine wave, but the calibration is made at the frequency at which the meter will be used. At the extremely high frequencies at which it is used, a phenomenon known as *skin effect* occurs. At these frequencies, the current in a wire travels at the surface of the wire; the higher the frequency, the closer the current moves to the surface of the wire. This effect increases the resistance of the thermocouple heater wire because the diameter of the wire becomes, in effect, smaller. Therefore, since the resistance of the heater wire varies with frequency, thermocouple meters must be calibrated at specific frequencies.

accuracy of d-c and a-c meters

Moving-coil meters designed for general use have an accuracy of about ±2 percent. However, moving-coil meters made for laboratory use or for other special applications often have an accuracy of from ±0.1 to ±0.5 percent (±1/10 to ±1/2 of 1 percent). Keep in mind that meter accuracy, as pointed out earlier, is based on full-scale deflection.

Moving-iron meters, both the concentric-vane and the radial-vane type, have an accuracy of about ±5 percent. The rectifier meter, although basically a moving-coil meter, does not have the accuracy of the basic moving-coil meter. Rectifier inaccuracies add to the inaccuracy of the meter movement itself, and the overall accuracy of the meter is usually about ±5 percent or less.

The use of the thermocouple in a thermocouple meter also adds some error in the meter reading to the error introduced by the basic moving-coil movement itself. The error introduced by the thermocouple usually ranges from about ±1/2 to ±1 percent so that this must be added to the error introduced by the particular moving-coil meter used in the thermocouple meter.

METER ACCURACY	
Meter	Typical Accuracy
Moving-Coil	0.1 to 2%
Moving-Iron	5%
Rectifier-Type Moving-Coil	5%
Thermocouple	1 to 3%

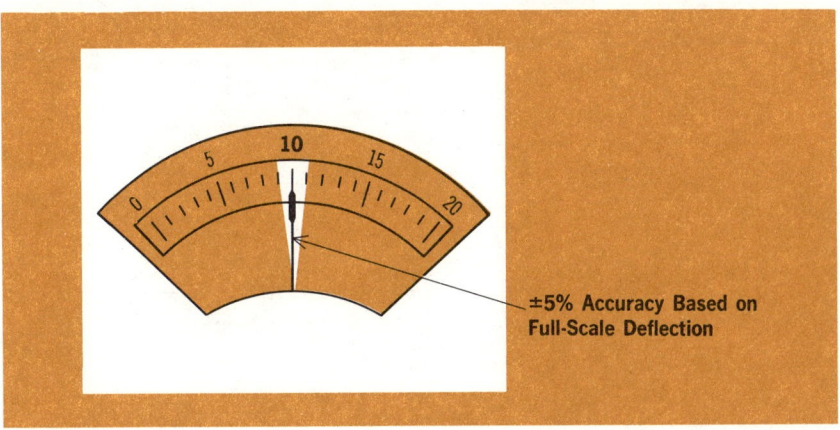

±5% Accuracy Based on Full-Scale Deflection

frequency response

The frequency response of a moving-coil or moving-iron meter is limited mainly by the inductive reactance of the coil. From Volume 3, you will recall that the inductive reactance (X_L) of a coil equals:

$$X_L = 2\pi fL$$

where f is the frequency, and L is the inductance of the coil. Therefore, as frequency increases, the inductive reactance of the coil increases, resulting in a decrease of current through the coil. In addition, moving-iron meters are affected by eddy currents and hysteresis losses in the iron, so that the effect is greater at higher frequencies.

On page 5–50, it was noted that a-c meters are calibrated at 60 Hz. Moving-iron meters can be used up to about 100 Hz without a significant increase in meter error. Above this frequency, however, the readings become very inaccurate. Moving-coil meters have a slightly better frequency response.

The frequency of the rectifier-type meter is limited primarily by the capacitance that is present across the rectifier. For example, a copper oxide rectifier has a capacitance of about 0.009 microfarads, and, as frequency increases, the capacitive reactance of the rectifier decrease's according to the equation:

$$X_C = \frac{1}{2\pi fC}$$

and acts as a low-resistance a-c path across the rectifier. Meter readings become about 1/2 to 1 percent lower for each 1000-Hz increase in frequency. Because of this, rectifier meters generally are not used to measure frequencies above 15,000 Hz. Even at this frequency, the meter reading could be off as much as 15 percent, depending on the basic accuracy of the meter when it was calibrated at 60 Hz.

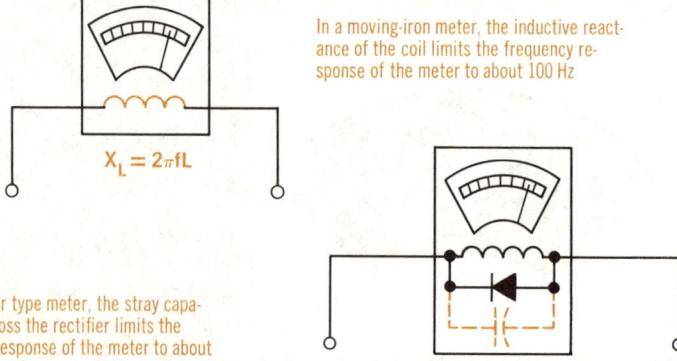

In a moving-iron meter, the inductive reactance of the coil limits the frequency response of the meter to about 100 Hz

$$X_L = 2\pi fL$$

In a rectifier type meter, the stray capacitance across the rectifier limits the frequency response of the meter to about 15,000 Hz maximum

$$X_C = \frac{1}{2\pi fC}$$

summary

☐ Every meter has a certain d-c resistance that depends on the size of the coil wire, and the number of turns in the coil. ☐ The sensitivity of a meter is the current that will cause the pointer to deflect full scale. It is also the maximum current that can be sent through the coil without damaging the meter. ☐ The accuracy of a meter is the percentage error at full-scale deflection. The accuracy becomes progressively poorer the farther the pointer is from full-scale deflection.

☐ Both moving-iron and thermocouple meters can measure ac and dc. ☐ The basic moving-coil meter can only measure dc. It can measure ac, though, by first converting the ac to pulsating dc. Such a meter is called a rectifier meter. ☐ Most moving-coil meters for measuring ac use copper oxide rectifiers. Some use selenium rectifiers. ☐ Half-wave rectifier meters measure only one alternation of the a-c current. The scale is calibrated to read the rms value (0.707 I_{PK}), although the actual deflection is equivalent to 0.318 I_{PK}. ☐ Full-wave rectifier meters measure both alternations of the a-c current. The scale is generally calibrated to read the rms value (0.707 I_{PK}), although the actual deflection is equivalent to 0.637 I_{PK}.

☐ Meters are calibrated by using accurate current sources and monitoring devices. ☐ Generally, moving-coil meters have accuracies of about ±2 percent. ☐ Rectifier meters have accuracies of about ±5 percent. ☐ The accuracy of moving-iron meters is about ±5 percent, and that of thermocouple meters approximately ±3 percent. ☐ The accuracy of all meters decreases at high frequencies.

review questions

1. Define *meter sensitivity*? What is its significance?
2. What is meant by the *peak, average,* and *rms values* of a sine wave?
3. Why can't the basic moving-coil meter measure ac?
4. What is a rectifier meter?
5. What value of the a-c current (peak, average, rms) does a half-wave rectifier meter respond to? What value does it usually indicate?
6. In what type of meter is a bridge rectifier circuit used?
7. Why is it desirable for a-c meters to indicate rms values?
8. What is meant by the *accuracy* of a meter?
9. Why is a rectifier meter less accurate than a basic moving-coil meter?
10. What type of meter is used to measure high-frequency currents?

shunts

Basic meter movements, by themselves, cannot carry large currents. There are very few in use that can measure more than 10 milliamperes. Yet in most of the equipment you will work with, it will often be necessary to measure currents greater than this value. How, then, can these large currents be measured? The simplest way to measure these large currents, and the way that is used in all present-day meters, is to allow only part of the current to flow through the meter movement and divert the rest of it around the movement. This is accomplished by connecting a resistor, called a shunt, in parallel with the meter movement's coil. Because the circuit is designed so that a *specific percentage* of the total circuit current goes through the coil, then the total circuit current is easy to calculate, and can be marked on the meter scale. For example, if the meter only conducts 10 percent of the total circuit current, and the shunt passes the rest, the actual current marked on the scale is 10 times the current the meter passes.

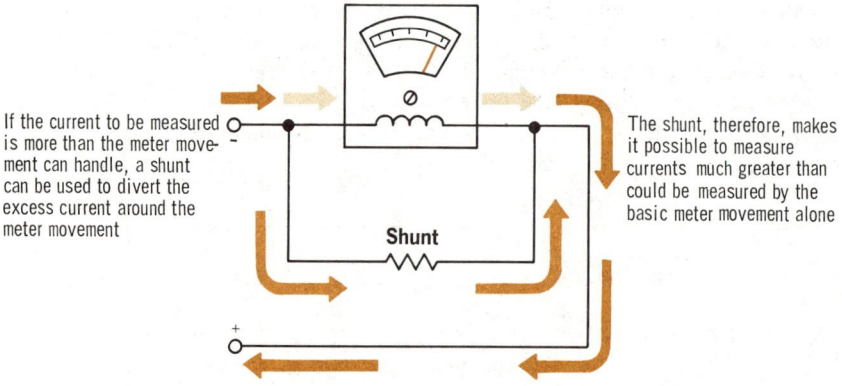

If the current to be measured is more than the meter movement can handle, a shunt can be used to divert the excess current around the meter movement

Shunt

The shunt, therefore, makes it possible to measure currents much greater than could be measured by the basic meter movement alone

Every meter coil has definite d-c resistance. When a shunt is connected in parallel with the coil, the current will divide between the coil and the shunt, just as it does between any two resistors in parallel. By using a shunt of the proper resistance, current through the meter coil will be limited to the value that it can safely handle, and the remainder of the current will flow through the shunt.

Shunts can either be connected inside the current meter case or external to the meter case, depending upon the current the meter will be used to measure. Meters that are designd to measure up to 30 amperes generally have *internal shunts*. Meters designed to measure currents above 30 amperes generally use *external shunts* to prevent damage to the meter movement by the heat that is generated in the shunt.

**Current flow divides between two resistors in parallel
in a ratio inversely proportional to their resistance**

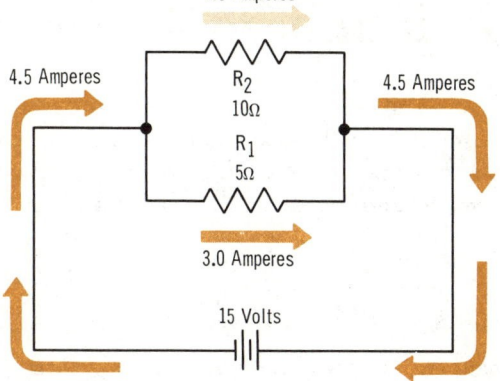

Resistor R_2 is twice as large as resistor R_1. Therefore, the current through R_2 will be one-half the current through R_1

current flow
in parallel circuits

To understand how a shunt can be used to extend the range of a current meter, it is important that you thoroughly understand the behavior of current flow through two resistors connected in parallel. In Volume 2, you learned that current will divide between two resistors in parallel. You also learned that the current through each resistor is *inversely proportional* to its resistance; that is, if one resistor has twice the resistance of another, the current flowing through the larger resistor will be half the current flowing through the smaller one.

Resistor R_2 is three times as large as resistor R_1. Therefore, the current through R_2 will be one-third the current through R_1

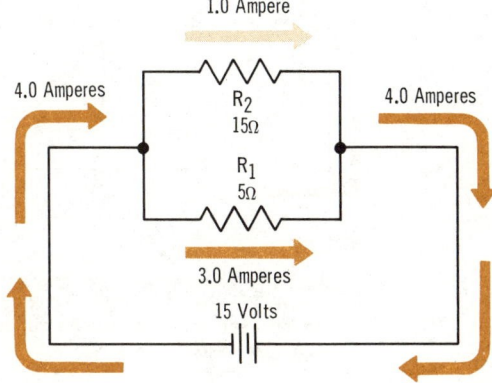

voltage drops
in parallel circuits

Examine the parallel circuit shown. You can see that the voltage across both resistors is the same. You will recall that Ohm's Law states that the voltage across a resistor equals the current through the resistor times the value of the resistor. Therefore, the voltage across R_1 is

$$E_{R1} = I_1 R_1$$

and the voltage across R_2 is

$$E_{R2} = I_2 R_2.$$

However, since the same voltage is across both R_1 and R_2, then

$$E_{R1} = E_{R2}$$

Therefore,

$$I_{R1} R_1 = I_{R2} R_2$$

As you will see next, this simple equation, with very slight modifications, can be used to calculate the value of a shunt for a current meter for any application.

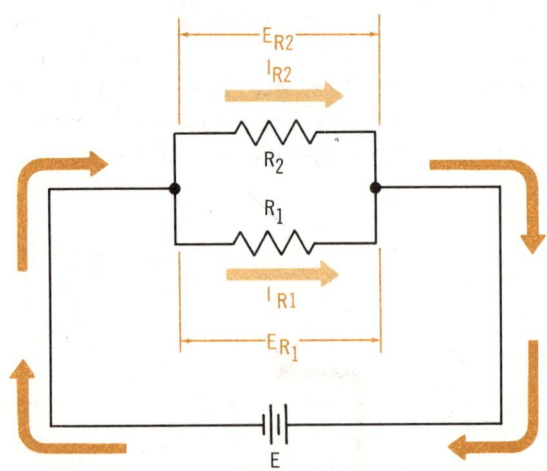

Since the same voltage appears across R_1 and R_2, then $E_{R1} = E_{R2}$. From this, $I_{R1} R_1 = I_{R2} R_2$, which can be used to calculate the shunt needed for a particular current measurement

the shunt equation

A meter and shunt combination is identical to the parallel circuit shown on the previous page. Instead of labeling the top resistor R_2, it can be labeled R_M, which represents the resistance of the meter movement. Resistor R_1 can be labeled R_{SH} to represent the resistance of the shunt. I_{R1} and I_{R2} then become I_{SH} and I_M to indicate current flow through the shunt and through the meter. This means that the equation

$$I_{R1}R_1 = I_{R2}R_2 \qquad \text{can now be written as} \qquad I_{SH}R_{SH} = I_M R_M$$

Therefore, if you know any three of these values, you can calculate the fourth. Since you will be calculating the resistance of the shunt, you will have to know the other three values. Quite often, the meter movement current for full-scale deflection (meter sensitivity), and the meter movement resistance are marked on the face of the meter. If not, look up their values in the meter instruction booklet. The value of current through the shunt (I_{SH}) is simply the difference between the total current you want to measure, and the actual full-scale deflection of the meter movement. For example, if you wish to extend the range of a 1-milliampere meter movement to 10 milliamperes, the I_{SH} will have to be 9 milliamperes, so that the movement itself will conduct no more than 1 milliampere.

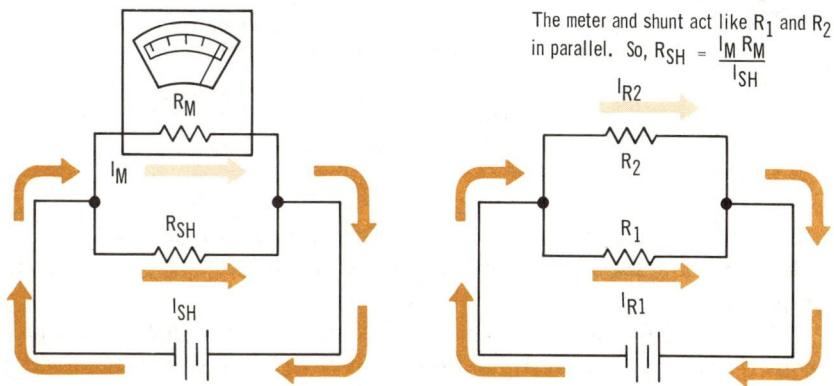

The meter and shunt act like R_1 and R_2 in parallel. So, $R_{SH} = \dfrac{I_M\,R_M}{I_{SH}}$

Since you will always solve for the shunt resistance, set up the equation so that R_{SH} is always the unknown quantity. Therefore, the basic equation $I_{SH}R_{SH} = I_M R_M$ becomes:

$$R_{SH} = \frac{I_M R_M}{I_{SH}}$$

From this equation, shunts can be calculated to extend the range of a current meter to any value.

calculating shunt resistance

Assume that you wish to extend the range of a 1-milliampere meter movement to 10 milliamperes, and the meter movement has a resistance of 27 ohms. Extending the range of the meter to 10 milliamperes means that 10 milliamperes will be flowing in the overall circuit when the pointer is deflected full scale.

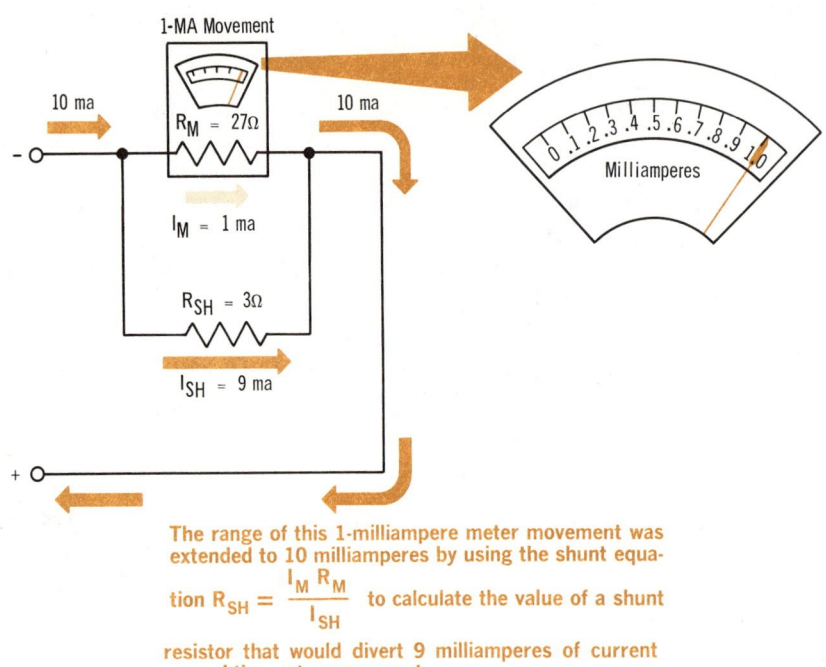

The range of this 1-milliampere meter movement was extended to 10 milliamperes by using the shunt equation $R_{SH} = \dfrac{I_M R_M}{I_{SH}}$ to calculate the value of a shunt resistor that would divert 9 milliamperes of current around the meter movement

Since the meter movement can only carry 1 milliampere at full-scale deflection, the *shunt* must carry the extra current, which is 9 milliamperes.

You now know the meter movement sensitivity (I_M), the meter movement resistance (R_M), and the amount of current that must be diverted through the shunt (I_{SH}). Using the shunt equation, you can now calculate the value of a shunt that will divert 9 milliamperes around the meter movement when 10 milliamperes is measured:

$$R_{SH} = \frac{I_M R_M}{I_{SH}}$$
$$= \frac{0.001 \times 27}{0.009}$$
$$= 3 \text{ ohms}$$

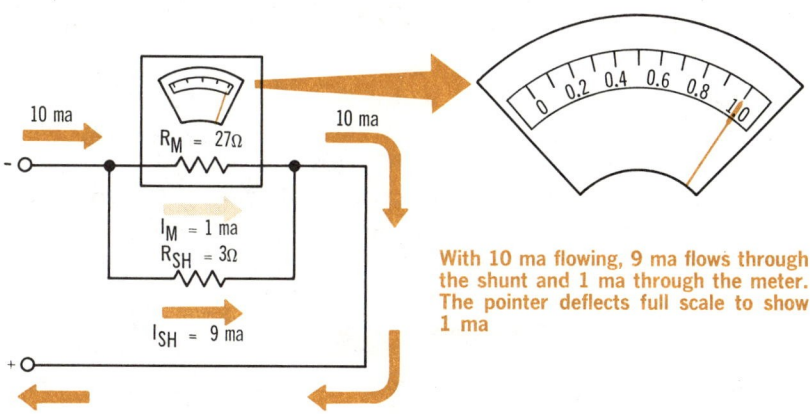

With 10 ma flowing, 9 ma flows through the shunt and 1 ma through the meter. The pointer deflects full scale to show 1 ma

reading a shunted meter

So far, we have only discussed full-scale deflection in a shunted meter. What about the relationship between the current through the meter and the current through the shunt when there is not enough current in the circuit to deflect the pointer full scale?

In the previous example, the shunt resistance is one-ninth that of the meter. Therefore, the shunt current will be nine times greater than that through the meter movement. This is exactly how the current behaves through two resistors in parallel; the current flow through each resistor is inversely proportional to its resistance.

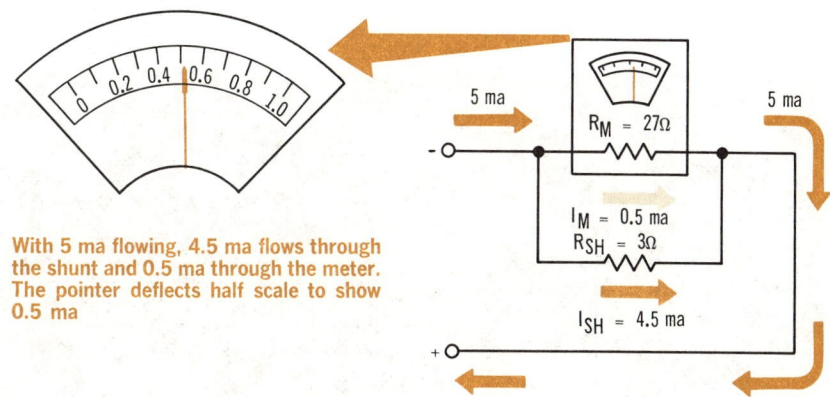

With 5 ma flowing, 4.5 ma flows through the shunt and 0.5 ma through the meter. The pointer deflects half scale to show 0.5 ma

reading a shunted meter (cont.)

If the meter was connected into a circuit carrying only 5 milliamperes, the current would divide in the ratio of nine to one between the shunt and the meter movement. The current through the shunt would be 4.5 milliamperes and the current through the meter would be 0.5 milliampere. Therefore, the pointer would deflect half of full scale.

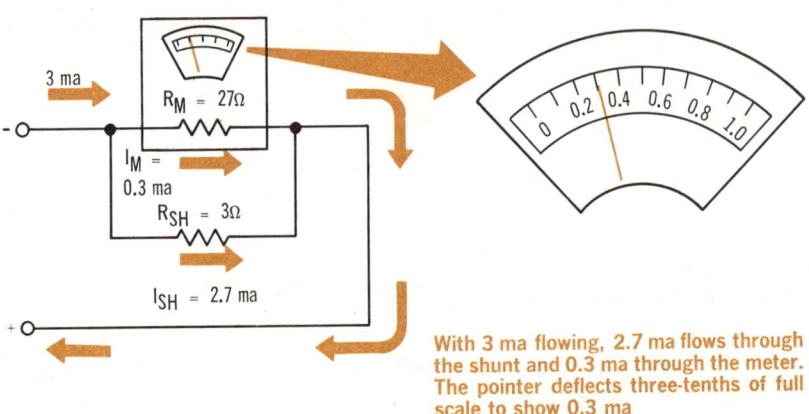

With 3 ma flowing, 2.7 ma flows through the shunt and 0.3 ma through the meter. The pointer deflects three-tenths of full scale to show 0.3 ma

Similarly, if the meter was connected into a circuit carrying 3 milliamperes, the current would again divide in the ratio of nine to one; 2.7 milliamperes would flow through the shunt and 0.3 milliampere through the meter movement. In this case, the meter pointer would deflect three-tenths of full scale. Actually, then, in this case, the total current is always 10 times greater than the meter reading.

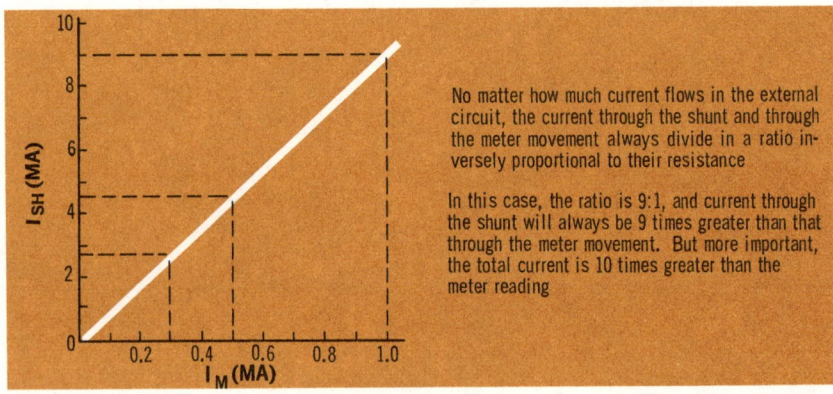

No matter how much current flows in the external circuit, the current through the shunt and through the meter movement always divide in a ratio inversely proportional to their resistance

In this case, the ratio is 9:1, and current through the shunt will always be 9 times greater than that through the meter movement. But more important, the total current is 10 times greater than the meter reading

solved problems

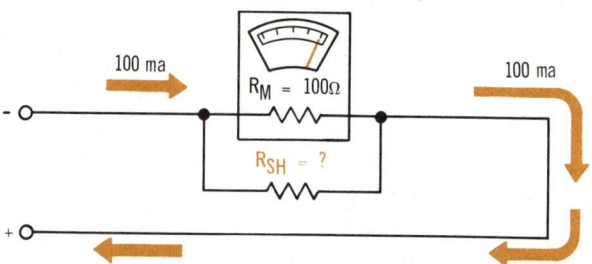

50-Microampere Movement

Problem 1. *If you have a 50-microampere meter movement with a resistance of 100 ohms, what value of shunt is needed to extend the meter range to 100 milliamperes?*

The first step is to find the amount of current that must be diverted through the shunt (I_{SH}). We are given the meter sensitivity and the desired current range; therefore, the shunt current is

$$I_{SH} = \text{desired current range} - \text{meter sensitivity}$$

Be careful, the currents must be expressed in the same units. The meter sensitivity of 50 microamperes is equal to 0.05 milliamperes. Therefore,

$$I_{SH} = 100 \text{ ma} - 0.05 \text{ ma} = 99.95 \text{ ma}$$

Now that you have found the shunt current, and you are given the meter sensitivity and the meter resistance, the next step is to insert these values into the shunt equation to calculate the resistance value of the shunt.

$$R_{SH} = \frac{I_M R_M}{I_{SH}} = \frac{0.05 \times 100}{99.95} = 0.05 \text{ ohms (approx.)}$$

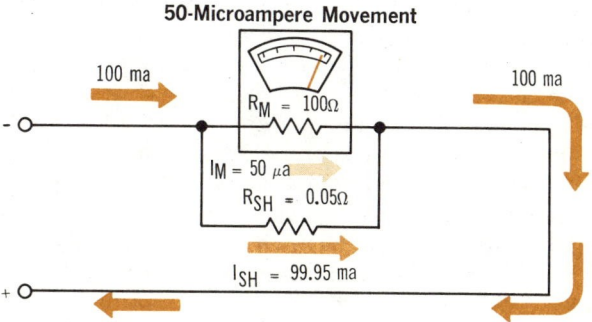

50-Microampere Movement

summary

□ Shunts make it possible for meters to measure large currents. □ A shunt is a resistor connected in parallel with the meter coil. Most of the current being measured is bypassed around the coil by the shunt. □ Shunt circuits are designed so that a specific percentage of the total current being measured goes through the meter coil. □ Shunts that are contained inside of the meter case are called internal shunts. □ Shunts that are connected external to the meter case are called external shunts. □ External shunts are generally used for measuring currents greater than about 30 amperes.

□ As in any parallel resistances, the IR drops across both the shunt and the d-c resistance of the meter coil are equal. So: $I_{SH}R_{SH} = I_MR_M$. □ If the resistance of the coil (R_M), the coil current required for full-scale deflection (I_M), and the maximum current to be measured are known, the required value of the shunt resistance (R_{SH}) can be found by: $R_{SH} = I_MR_M/I_{SH}$. □ The coil resistance and the coil current required for full-scale deflection are often marked on the face of the meter. □ The shunt current (I_{SH}) is the difference between the total current to be measured and the coil current needed for full-scale deflection.

□ The relationship between the current being measured and the amount that flows through the meter is the same throughout the range of the meter. In other words, if the shunt bypasses 90 percent of the current at full scale, it will bypass 90 percent of all other lesser values of measured current.

review questions

1. What is a *shunt?*
2. Why are shunts used?
3. What is the difference between *internal* and *external shunts?*
4. Do shunts have small or large resistances? Why?
5. If a meter will deflect full scale when the current through the coil is 0.5 milliampere, will any current flow through the shunt when a current of 0.1 milliampere is measured?
6. A meter has a sensitivity of 5 milliamperes and a coil resistance of 200 ohms. What shunt resistance is required to extend the range of the meter to 5 amperes?
7. In the meter of Question 6, how much current flows through the shunt when the meter is measuring 1 ampere?
8. If the voltage drop across a meter shunt is 0.8 volt, what is the voltage drop across the meter coil?
9. Does a shunt affect the sensitivity of a meter? Explain.
10. Why are external shunts used with meters designed to measure very large currents?

multirange current meters

In only some applications, it is practical to use a current meter having only one range; for example, only 0–1 milliampere or 0–100 milliamperes, or 0–15 amperes, and so on. Some pieces of electrical and electronic equipment have built-in single-range current meters that measure only one particular current present in the equipment. The output current of a d-c generator and the input current to a radio transmitter are often monitored with their own current meters. In many uses, however, particularly when troubleshooting, it would be impractical to have to use a number of separate current meters to measure all of the currents encountered in a piece of equipment. In these cases, a *multirange current meter* is used.

A multirange current meter is one containing a basic meter movement and several shunts that can be connected across the meter movement. A *range switch* is usually used to select the particular shunt for the desired current range. Sometimes, however, separate terminals for each range are mounted on the meter case.

The range of this 1 milliampere meter movement has been extended to measure 0–10 ma, 0–100 ma, and 0–1 ampere by using multiple shunts

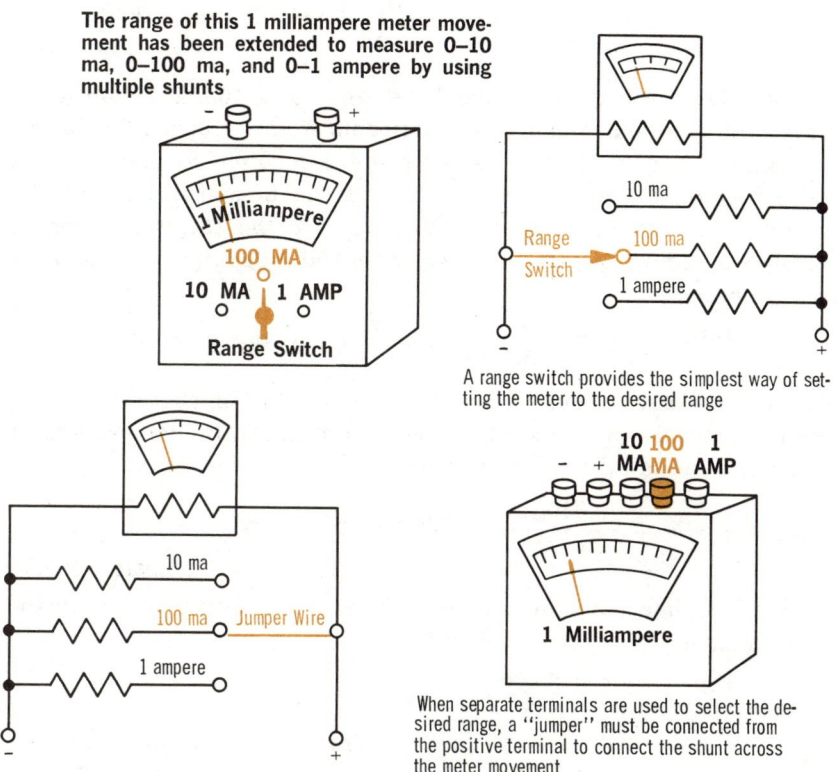

A range switch provides the simplest way of setting the meter to the desired range

When separate terminals are used to select the desired range, a "jumper" must be connected from the positive terminal to connect the shunt across the meter movement

calculating the resistance of multirange shunts

Some multirange current meters have as many as six or seven different current ranges, although most have three or four. The resistance value of the shunt for *each* range is calculated in exactly the same manner as the shunt for a single-range current meter.

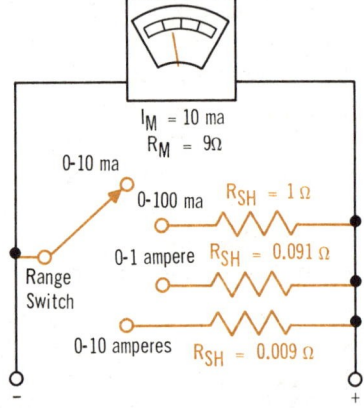

The range of this 0–10-ma, 9-ohm meter movement was extended to measure 0–100 ma, 0–1 ampere, and 0–10 amperes by using shunts for each range, whose values were found by the shunt equation:

$$R_{SH} = \frac{I_M R_M}{I_{SH}}$$

$I_M = 10$ ma
$R_M = 9\,\Omega$

0-10 ma

0-100 ma $R_{SH} = 1\,\Omega$

0-1 ampere $R_{SH} = 0.091\,\Omega$

Range Switch

0-10 amperes $R_{SH} = 0.009\,\Omega$

Assume that you wish to extend the range of a 10-ma, 9-ohm meter movement to permit making measurements from 0–10 ma, 0–100 ma, 0–1 ampere, and 0–10 amperes. No shunt is required for the 0–10-ma range because the meter movement itself can handle up to 10 milliamperes. For the next range, 0–100 ma, use the shunt equation to calculate a shunt resistance that will divert 90 milliamperes around the meter movement:

$$R_{SH} = \frac{I_M R_M}{I_{SH}} = \frac{0.010 \times 9}{0.090} = 1 \text{ ohm} \qquad \text{(for the 0–100-ma range)}$$

Now use the shunt equation to calculate a value of shunt resistance needed for the 0–1-ampere range. For this range, 990 milliamperes must be diverted around the meter movement.

$$R_{SH} = \frac{I_M R_M}{I_{SH}} = \frac{0.010 \times 9}{0.990} = 0.091 \text{ ohm}$$

$$\text{(for the 0–1-ampere range)}$$

Finally, calculate the shunt needed for the 0–10-ampere range in the same manner:

$$R_{SH} = \frac{I_M R_M}{I_{SH}} = \frac{0.010 \times 9}{9.990} = 0.009 \text{ ohm}$$

$$\text{(for the 0–10-ampere range)}$$

solved problems

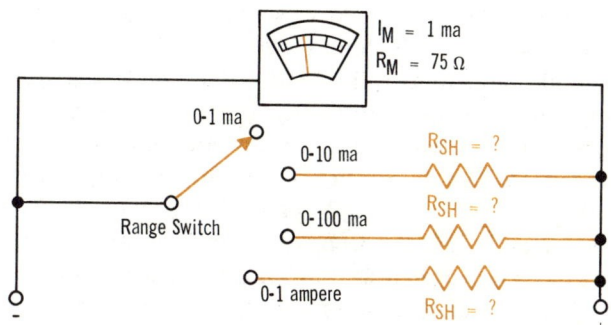

Problem 2. With a 1-ma meter movement with a resistance of 75 ohms, you wish to construct a meter with ranges of 0–1 ma, 0–10 ma, 0–100 ma, and 0–1 ampere. What shunt resistances must be added to the meter circuit?

Calculating the 0–1-ma shunt resistance: The meter movement itself can handle from 0–1 ma, so no shunt is needed for that range.

Calculating the 0–10-ma shunt resistance: The first step is to find the current that must be diverted through the shunt (I_{SH}).

$$I_{SH} = \text{desired current range} - \text{meter sensitivity}$$
$$= 10 \text{ ma} - 1 \text{ ma} = 9 \text{ ma}$$

Insert the three known values of meter current (I_M), meter resistance (R_M), and shunt current (I_{SH}) into the shunt equation, and solve for the shunt resistance (R_{SH}):

$$R_{SH} = \frac{I_M R_M}{I_{SH}} = \frac{0.001 \times 75}{0.009} = 8.333 \text{ ohms}$$

Calculating the 0–100-ma shunt resistance: This is done in the same manner as for the 0–10-ma range:

$$I_{SH} = \text{desired current range} - \text{meter sensitivity}$$
$$= 100 \text{ ma} - 1 \text{ ma} = 99 \text{ ma}$$
$$R_{SH} = \frac{I_M R_M}{I_{SH}} = \frac{0.001 \times 75}{0.099} = 0.758 \text{ ohm}$$

Calculating the 0–1-ampere shunt resistance:

$$I_{SH} = 1 \text{ ampere} - 1 \text{ ma} = 1000 \text{ ma} - 1 \text{ ma} = 999 \text{ ma}$$
$$R_{SH} = \frac{0.001 \times 75}{0.999} = 0.0751 \text{ ohm}$$

ring shunts

Many multirange current meters use shunts in a *ring shunt* arrangement, in which some of the shunt resistors are in series with the meter movement and some in parallel. The values of series resistance and parallel resistance depend upon which range is being used.

As you will learn, ring shunt circuits are somewhat more complicated than the parallel shunt circuits. However, ring shunts have two important advantages over simple parallel shunts:

The resistance of a shunt becomes smaller as you extend the range of a meter movement. When dealing with low shunt resistances of such low values, the resistance of the range switch contacts becomes significant. In the simple parallel shunts just studied, the range switch contacts are in series with the shunts. Therefore a contact resistance of only 0.0001 ohm in series with a 0.001-ohm shunt could cause a significant error in a meter reading. In a ring shunt circuit, however, the contact resistance is external to the shunt circuit, and, therefore, has no effect on the accuracy of the meter reading.

In the simple parallel shunt circuit, the shunt is momentarily *removed* from the circuit when the range switch is moved from one position to another. During this time, the full line current flows through the meter movement, and this surge of current could burn out the meter coil. The ring shunt circuit, though, disconnects the meter from the circuit under test when the range switch is moved from one position to another. A much more complicated switching arrangement would be needed to do the same thing in the simple parallel shunt circuit.

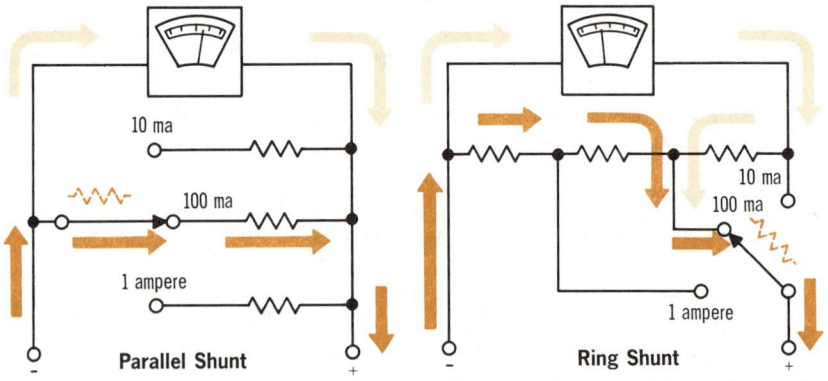

Parallel Shunt	Ring Shunt
The range switch contact resistance adds to the shunt resistance, and may result in a significant error in meter reading. High-surge currents can flow through the meter movement when switching ranges	The range switch contact resistance is external to the shunt and meter circuit, eliminating error found in reading of parallel shunt. High-surge currents are eliminated because meter movement is out of circuit when switching ranges

the ring shunt equation

As in the case of simple parallel shunts, there is a simple equation that can be used to calculate the shunt values for each current range. But, simply memorizing the equation will not result in a thorough understanding of meter operation. A good understanding requires that you be familiar with the derivation of the equation.

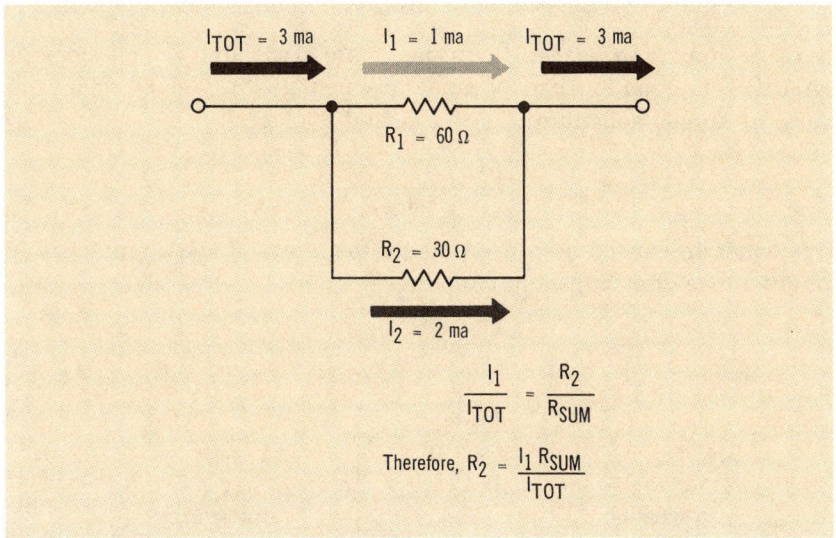

Consider two resistors, R_1 and R_2, in parallel. R_1 is 60 ohms and R_2 is 30 ohms. Both will be used to calculate the values of ring shunts. If a total current of 3 ma is flowing in this parallel circuit, 1/3 of the current will flow through R_1 and 2/3 will flow through R_2. Thus, 1 ma will flow through R_1 and 2 ma will flow through R_2. The current through R_2 (2 ma) is twice the current through R_1 (1 ma). Notice that there is the same ratio between one branch current and the total current (I_{TOT}) as there is between the *other* branch resistance and the *sum total* of the two *branch resistances* ($R_1 + R_2$), which we will call R_{SUM}. In other words, the ratio of I_1 to I_{TOT} is the same as the ratio of R_2 to R_{SUM}. Mathematically,

$$I_1/I_{TOT} = R_2/R_{SUM}$$

This relationship is seen more clearly, if you substitute values into both sides of the equation:

$$I_1/I_{TOT} = 1 \text{ ma}/3 \text{ ma} = 1/3$$

and

$$R_2/R_{SUM} = 30 \text{ ohms}/90 \text{ ohms} = 1/3$$

the ring shunt equation (cont.)

By substituting the resistance and current values into both sides of the equation, you can see that the ratio is 1 to 3 in both cases. Therefore,

$$I_1/I_{TOT} = R_2/R_{SUM}$$

Solving for R_2:

$$R_2 = \frac{I_1 R_{SUM}}{I_{TOT}}$$

Since a shunt and a meter movement are basically two resistors in parallel, this equation can be used to calculate the values of a ring shunt by simply substituting I_M for I_1 and R_{SH} for R_2:

$$R_{SH} = \frac{I_M R_{SUM}}{I_{TOT}}$$

This equation can be used to calculate the value of the shunt resistor for any range of a ring shunt circuit.

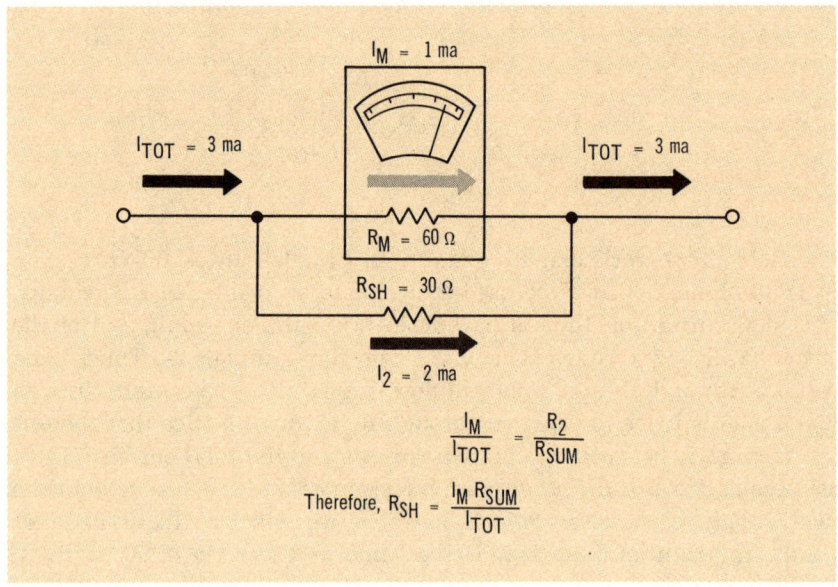

You should note that R_{SUM} is *not* the total *effective resistance* of R_1 and R_2 in parallel. The effective resistance (R_{TOT}) would be less than either R_1 or R_2 because they are in parallel. R_{SUM} is simply R_1 plus R_2, and the only significance it has is that it allows you to set up a mathematical ratio for determining the ring shunt resistance.

calculating the resistance of a ring shunt

On page 5-65 you calculated the values of parallel shunts needed to extend the range of a 1-ma, 75-ohm meter movement to 0–10 ma, 0–100 ma, and 0–1 ampere. Now, you will learn how to extend this meter movement to the same ranges using a ring shunt arrangement.

The first step in solving a ring shunt problem is to find the value of the entire shunt, that is, $R_{SH\,1} + R_{SH\,2} + R_{SH\,3}$. On the 0–10-ma range, all of these resistors are in series, and act as a shunt to the meter movement. Neglecting the other ranges for a moment, you can see that this is actually a simple parallel shunt arrangement. Therefore, the parallel shunt equation can be used to calculate the total shunt resistance $R_{SH\,TOT}$:

$$R_{SH\,TOT} = \frac{I_M R_M}{I_{SH}} = \frac{0.001 \times 75}{0.009} = 8.33 \text{ ohms (approx.)}$$

You now know the total value of $R_{SH\,1}$, $R_{SH\,2}$, and $R_{SH\,3}$, which is also the shunt resistance for the 0–10-ma range. You now have to determine the individual values of $R_{SH\,1}$, $R_{SH\,2}$, and $R_{SH\,3}$ to extend the meter movement to the 0–100-ma and 0–1-ampere ranges. This is where the ring shunt procedure differs from the parallel shunt procedure.

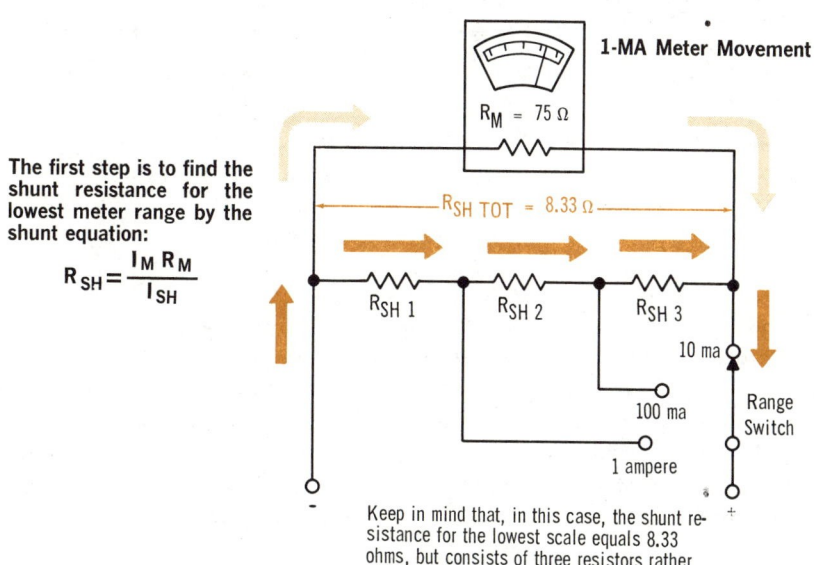

The first step is to find the shunt resistance for the lowest meter range by the shunt equation:

$$R_{SH} = \frac{I_M R_M}{I_{SH}}$$

1-MA Meter Movement

$R_M = 75\,\Omega$

$R_{SH\,TOT} = 8.33\,\Omega$

$R_{SH\,1}$ $R_{SH\,2}$ $R_{SH\,3}$

10 ma
100 ma
1 ampere

Range Switch

Keep in mind that, in this case, the shunt resistance for the lowest scale equals 8.33 ohms, but consists of three resistors rather than a single resistor

calculating the resistance of a ring shunt (cont.)

You have found that the total shunt resistance, $R_{SH\ TOT}$, is 8.33 ohms, and is equal to $R_{SH\ 1} + R_{SH\ 2} + R_{SH\ 3}$. Now you must find the values of each of these shunt resistors.

Assume that you switch to the 1-ampere range. By tracing the current flow, you can see that on this range, $R_{SH\ 2}$ and $R_{SH\ 3}$ are in *series* with the meter movement and $R_{SH\ 1}$ shunts the meter movement and the series resistors. Therefore, now you must calculate the value of $R_{SH\ 1}$, which is the shunt resistor for the 1-ampere range. The ring shunt equation given on page 5-68 provides a simple way to calculate the value of a shunt resistor for any range of a ring shunt meter.

$$R_{SH\ 1} = \frac{I_M R_{SUM}}{I_{TOT}}$$

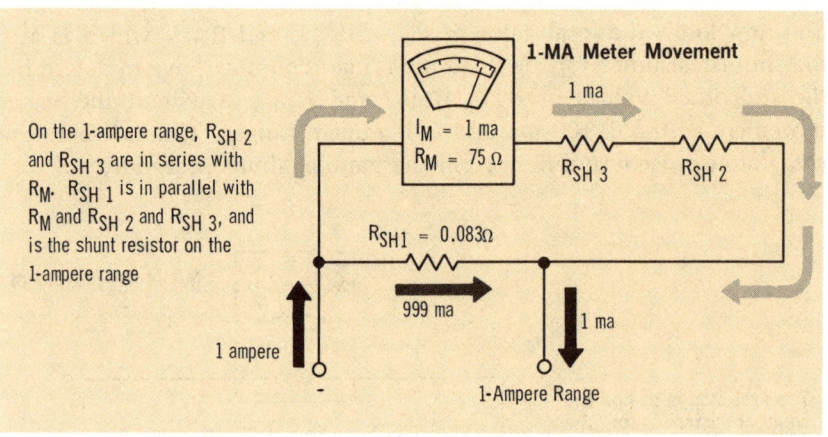

On the 1-ampere range, $R_{SH\ 2}$ and $R_{SH\ 3}$ are in series with R_M. $R_{SH\ 1}$ is in parallel with R_M and $R_{SH\ 2}$ and $R_{SH\ 3}$, and is the shunt resistor on the 1-ampere range

1-MA Meter Movement

$I_M = 1$ ma
$R_M = 75\ \Omega$
$R_{SH\ 3}$
$R_{SH\ 2}$
1 ma

$R_{SH1} = 0.083\Omega$

999 ma
1 ma

1 ampere

1-Ampere Range

You are using a 1-ma meter movement, so you know that I_M equals 1 ma. You are calculating the shunt resistor for the 1-ampere range; therefore, I_{TOT} equals 1 ampere. On page 5-68, it was stated the R_{SUM} is the *sum* of all the resistances in the two branches. You know that the meter resistances (R_M) equals 75 ohms, and on the previous page, you found that the total shunt resistance $R_{SH\ TOT}$ equals 8.33 ohms. Therefore, the total of all resistances in the two branches is 83.33 ohms. You now have all the information needed to calculate the value of $R_{SH\ 1}$:

$$R_{SH\ 1} = \frac{I_M R_{SUM}}{I_{TOT}} = \frac{0.001 \times 83.33}{1} = 0.083 \text{ ohm (approx.)}$$

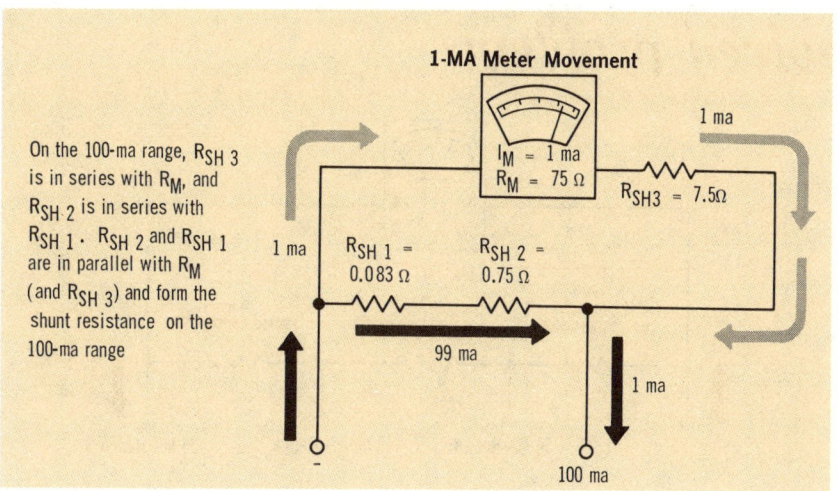

On the 100-ma range, $R_{SH\,3}$ is in series with R_M, and $R_{SH\,2}$ is in series with $R_{SH\,1}$. $R_{SH\,2}$ and $R_{SH\,1}$ are in parallel with R_M (and $R_{SH\,3}$) and form the shunt resistance on the 100-ma range

1-MA Meter Movement

1 ma

$I_M = 1$ ma
$R_M = 75\ \Omega$

$R_{SH3} = 7.5\Omega$

1 ma

$R_{SH\,1} = 0.083\ \Omega$

$R_{SH\,2} = 0.75\ \Omega$

99 ma

1 ma

100 ma

calculating the resistance of a ring shunt (cont.)

So far, you have calculated the total shunt resistance of the ring shunt circuit ($R_{SH\,TOT}$), and the shunt resistance for the 1-ampere range ($R_{SH\,1}$). Now examine the ring shunt circuit with the range switch at the 100-ma position. Tracing the current flow, you can see that $R_{SH\,3}$ is in series with meter resistance R_M. $R_{SH\,1}$ and $R_{SH\,2}$ are in series, and *both* shunt the meter movement (and $R_{SH\,3}$). Therefore, the shunt resistance for the 100-ma range equals $R_{SH\,1}$ plus $R_{SH\,2}$. Using the ring shunt equation, then,

$$R_{SH\,1} + R_{SH\,2} = \frac{I_M R_{SUM}}{I_{TOT}} = \frac{0.001 \times 83.33}{0.100} = 0.83 \text{ ohm (approx.)}$$

This is the resistance value of $R_{SH\,1}$ *plus* $R_{SH\,2}$, but you must find the value of $R_{SH\,2}$ only. Since you know that $R_{SH\,1}$ is 0.083 ohm, and that $R_{SH\,1} + R_{SH\,2} = 0.83$ ohm, then $R_{SH\,2}$ must be the difference of these two values:

$$R_{SH\,2} = 0.83 - R_{SH\,1} = 0.83 - 0.083 = 0.75 \text{ ohm (approx.)}$$

All that remains to complete the design of the ring shunt is to find $R_{SH\,3}$. You have found $R_{SH\,TOT}$, $R_{SH\,1}$, and $R_{SH\,2}$, and you know that

$$R_{SH\,TOT} = R_{SH\,1} + R_{SH\,2} + R_{SH\,3}$$

Therefore,

$$R_{SH\,3} = R_{SH\,TOT} - (R_{SH\,1} + R_{SH\,2})$$

$$= 8.33 - (0.83) = 7.50 \text{ ohms}$$

solved problems

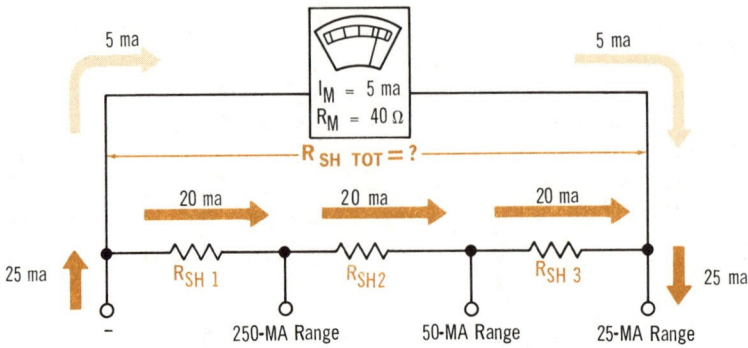

Problem 3. *In the circuit, find* $R_{SH\ TOT}$ *and* $R_{SH\ 1}$.

Calculating $R_{SH\ TOT}$: The parallel shunt equation is used to find the total shunt resistance, which is the shunt resistance for the lowest range (25 ma).

$$R_{SH\ TOT} = R_{SH\ 1} + R_{SH\ 2} + R_{SH\ 3} = \frac{I_M R_M}{I_{SH\ TOT}} = \frac{0.005 \times 40}{0.020} = 10 \text{ ohms}$$

Calculating R_{SUM}: R_{SUM} is the total of all the resistances in the meter and shunt circuits.

$$R_{SUM} = R_{SH\ TOT} + R_M = 10 + 40 = 50 \text{ ohms}$$

Calculating $R_{SH\ 1}$: $R_{SH\ 1}$ is the shunt resistor for the 250-ma range. Using the ring shunt equation,

$$R_{SH\ 1} = \frac{I_M R_{SUM}}{I_{TOT}} = \frac{0.005 \times 50}{0.250} = 1 \text{ ohm}$$

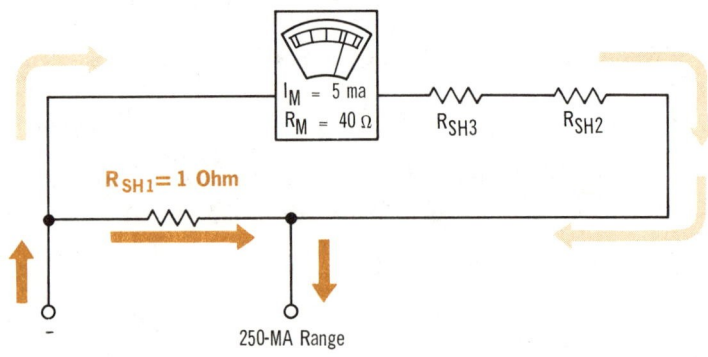

solved problems (cont.)

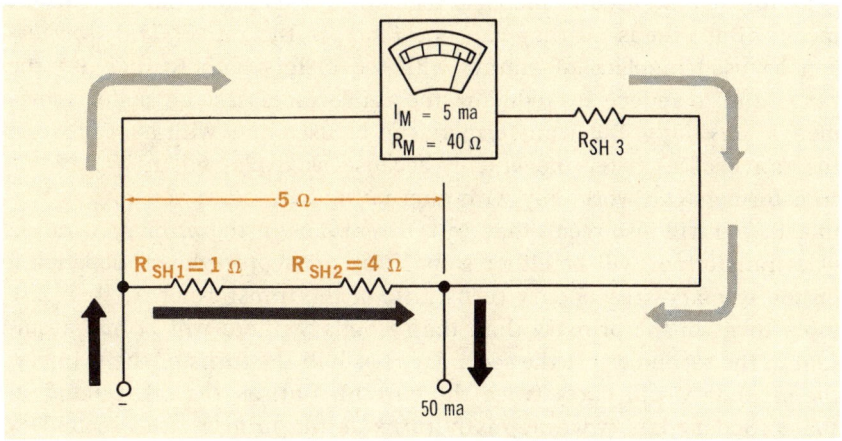

Problem 4. For the circuit on page 5-72, find $R_{SH\,2}$ and $R_{SH\,3}$.

Calculating $R_{SH\,1}$ and $R_{SH\,2}$: $R_{SH\,1}$ and $R_{SH\,2}$ are the shunt resistors for the 50-ma range. Using the ring shunt equation,

$$R_{SH\,1} + R_{SH\,2} = \frac{I_M R_{SUM}}{I_{TOT}} = \frac{0.005 \times 50}{0.050} = 5 \text{ ohms}$$

Calculating $R_{SH\,2}$: Since you solved for $R_{SH\,1}$ in Problem 3,

$$R_{SH\,2} = 5 - R_{SH\,1} = 5 - 1 = 4 \text{ ohms}$$

Calculating $R_{SH\,3}$: Therefore, from the first equation in Problem 3,

$$R_{SH\,3} = R_{SH\,TOT} - (R_{SH\,1} + R_{SH\,2}) = 10 - (5) = 5 \text{ ohms}$$

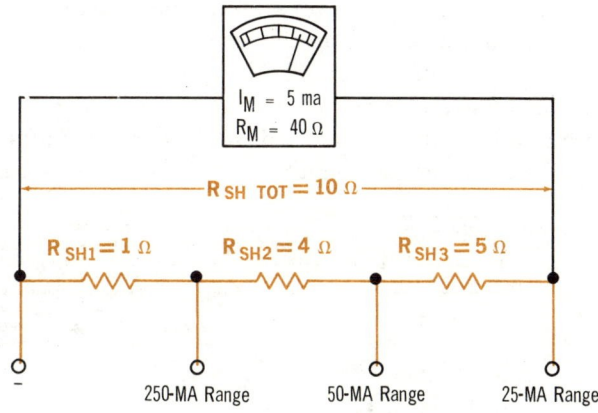

ammeter transformers

Shunts can be used to increase the range of both a-c and d-c current meters. But that is all they can do: *increase* the range. *Transformers* can be used in place of shunts with a-c meters both to increase the range and to reduce it; reducing the range increases the meter movement's *sensitivity*. But transformers can be used *only* with *a-c* to extend the range of a meter movement, because as explained in Volume 3, transformers can work only with alternating current. If you review Volume 3, you will also recall that any current flow in the primary winding of a transformer will be either stepped up or stepped down, depending on the primary-to-secondary turns ratio of the transformer. If there are more turns in the primary than the secondary, there will be more current in the secondary. If the secondary has half the turns of the primary, the secondary will carry twice the current. But, on the other hand, if the secondary has twice as many turns as the primary, the secondary will have half the current. Therefore, transformers can be used to extend the range of the meter movement easily.

If you had a 10-ma meter movement, and wanted it to measure up to 20 ma, you would use a transformer with a 1:2 primary-to-secondary ratio. A current of 20 ma in the primary would then cause 10 ma to flow in the secondary, so that the movement could operate safely in its range. But since you know that the meter is reading only half of the primary currents, you also know that 20 ma is being measured. When the meter is designed in this way, another scale, showing the 20-ma range, is also used. For 30 ma, a 1:3 ratio is used, and so on. In every case, the movement always works with its own rated current, because the current being measured is connected by the transformer.

To make the meter more sensitive than its movement, the transformer would have to step up the current. A 2:1 turns ratio would double the secondary current, so that the range of the movement can be halved. A 3:1 ratio would allow 3⅓ ma in the primary to produce 10 ma in the secondary, to get full scale deflection.

Generally, an autotransformer is used so that by switching to different taps, different ratios can be produced to allow many different ranges. The taps can be in either the primary or the secondary circuit.

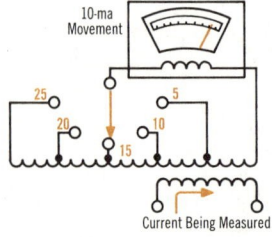

10-ma Movement

25

20

5

15

10

Current Being Measured

Transformers are used to step up or step down the current being measured to increase the range of the movement. When the current is stepped up, the meter is made more sensitive by allowing it to measure smaller currents. When the current is stepped down, currents larger than the movement can normally handle can be measured. In this diagram, the following ratios will cause the 10-ma meter movement to deflect full scale with the currents shown: 1:0.5 = 5 ma
1:1 = 10 ma
1:1.5 = 15 ma
1:2 = 20 ma
1:2.5 = 25 ma

summary

☐ Multirange meters make it possible to measure various ranges of current with a single meter. ☐ A multirange current meter has a basic meter movement and several different shunts that can be connected across the coil. ☐ A range switch is usually used on a multirange meter to select the particular shunt for the desired current range. ☐ The resistance value of each shunt in a multirange meter can be calculated from the same equation used for the shunt value of a single-range meter: $R_{SH} = I_M R_M / I_{SH}$.

☐ Many multirange meters use ring shunts, in which some of the shunt resistors are in series with the meter movement, and some in parallel with it. ☐ In a ring shunt, the range switch contacts are external to the shunt circuit, so the resistance of the contacts do not affect the accuracy of the meter. ☐ A ring shunt disconnects the meter from the circuit being measured when the range switch is moved from one position to another. This protects the meter by preventing the full line current from flowing through the coil.

☐ The resistance of a ring shunt for each range of a multirange meter can be found by: $R_{SH} = I_M R_{SUM} / I_{TOT}$. For the equation, R_{SUM} is found by adding the values of all the ring shunt resistors and the resistance of the meter coil. With a-c measurements, transformers can be used in place of shunts to increase the range or sensitivity of the meter.

review questions

1. What is a *multirange current meter*? What is the advantage of such a meter?
2. What is a *range switch*?
3. Draw a diagram of a meter movement, with three parallel shunt resistors and a range switch.
4. If in the circuit of Question 3, the meter has a coil resistance of 50 ohms and a sensitivity of 2 milliamperes, what shunt values are required to give the meter current ranges of 0–10 milliamperes, 0–100 milliamperes, and 0–1 ampere?
5. If a meter movement has a coil resistance of 100 ohms and a sensitivity of 1 milliampere, what value of shunt is required to give the meter a 0–1 milliampere range?
6. What is a *ring shunt*?
7. What are the advantages of ring shunts over parallel shunts?
8. What is the ring shunt equation?
9. If the ranges of the meter shown on page 5-69 were 0–5, 0–50, and 0–500 milliamperes, what would be the values of the shunt resistors? The meter movement has a sensitivity of 1 milliampere.
10. Which range of a meter requires the smaller shunt resistance: 0–100 milliampere, or 0–1 ampere?

connecting a current meter into a circuit

Current meters must always be *connected in series* with the power source and the load, and *never connected in parallel* with them. A current meter is a very low-resistance device, and the moving coil can be burned out very easily by connecting the meter across a power source, resistor, or other circuit component. Most circuit components have a much higher resistance than the current meter. Should you connect a current meter in parallel with one of these components, you might, in effect, create a short circuit, resulting in very high current flow through the meter. This high current will damage or ruin the meter movement. If you use a multirange current meter, the meter movement, the shunt, or both, could be damaged by the excess current. Therefore, remember: *Always connect a current meter in series with the power source and load.*

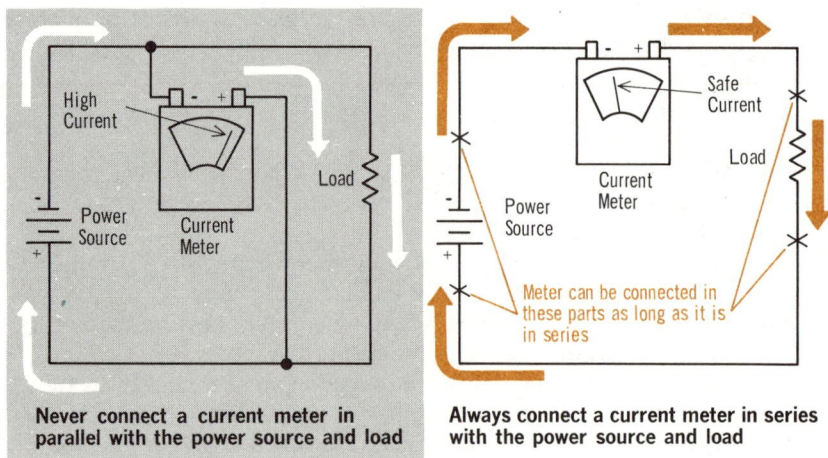

Never connect a current meter in parallel with the power source and load

Always connect a current meter in series with the power source and load

A second important point to remember is that you must observe *polarity* when measuring current in a *d-c current*. In other words, you must connect the negative terminal of the meter to the negative, or low-potential, point in the circuit, and connect the positive terminal of the meter to the high, or positive, potential point in the circuit. Current must flow through the meter from *minus* (−) *to plus* (+). If you connect the meter so that the polarities oppose, the meter coil will move in the opposite direction, and the pointer might strike the left retaining pin. You will not obtain a current reading, and, in some cases, you might bend the pointer.

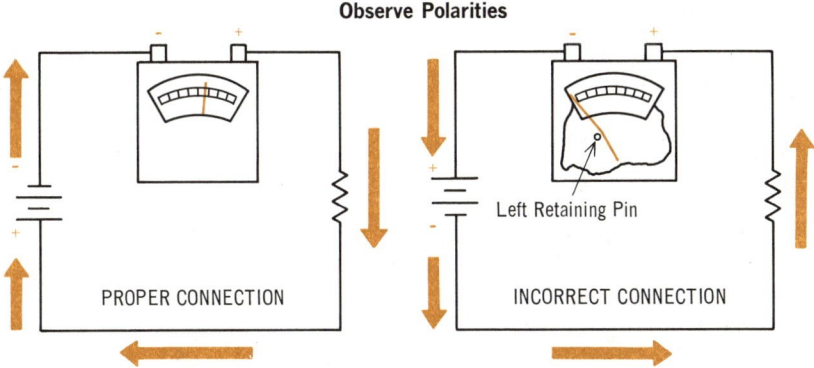

Observe Polarities

PROPER CONNECTION

INCORRECT CONNECTION

Left Retaining Pin

connecting a current meter into a circuit (cont.)

You do not have to observe polarity when using current meters that have their zero position at the center of the scale. When measuring ac, there is no need to observe polarity since the polarity is continually changing through the cycles. Therefore, meters designed for a-c use only do not have plus and minus marked on their terminals.

A third important point to remember is that you should not connect a meter into a circuit unless you know approximately the maximum current flowing in the circuit. Meters are often damaged or ruined because they have to handle currents above their rated meter sensitivity. If you are not sure how much current is flowing, check the circuit diagrams, and, even then, begin by connecting a meter with a range higher than necessary. If a multirange meter is used, set it to its highest scale, and decrease the range until you reach the range that gives a medium deflection. If you start with the lowest range, the pointer could strike the right retaining pin at the high end of the scale.

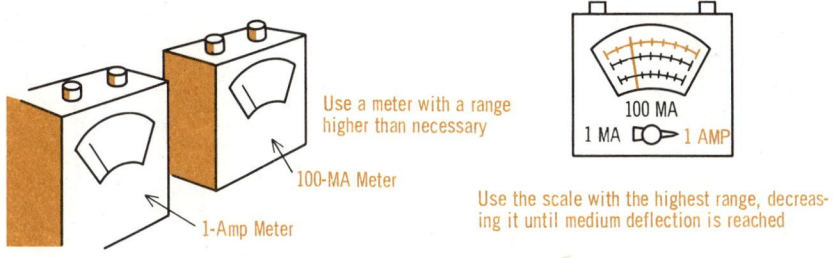

Use a meter with a range higher than necessary

100-MA Meter

1-Amp Meter

100 MA
1 MA 1 AMP

Use the scale with the highest range, decreasing it until medium deflection is reached

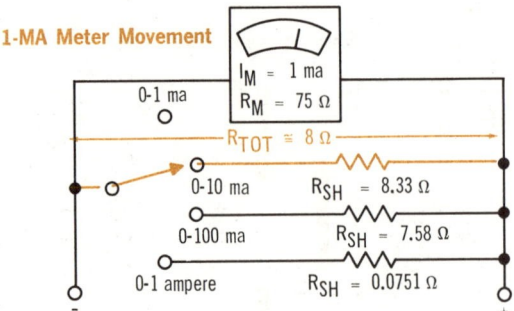

connecting a current meter in parallel

To understand how a current meter can be damaged by connecting it in parallel with a circuit component, let us examine such a case. Assume that the multirange current meter whose circuit is shown is connected by mistake across a 2000-ohm load. According to Ohm's Law, about 5 milliamperes of current should normally be flowing through the load resistor, so you set the meter to the 0–10 ma range. On the 0–10 ma scale, the total resistance of the meter is approximately 8 ohms (8.33-ohm shunt resistance in parallel with a 75-ohm coil resistance equals approximately 8 ohms). Therefore, an 8-ohm resistor has now been connected in parallel with the 2000-ohm load.

With the 8-ohm meter resistance in parallel with the 2000-ohm load resistance, the total resistance of the circuit *drops* to approximately 8 ohms (2000 ohms in parallel with 8 ohms equals approximately 8 ohms).

With only an 8-ohm resistance across a 10-volt power source, 1-1/4 amperes will flow in the circuit, and most of it will flow through the meter because its resistance is so much lower than the load resistance. Since the meter range is only designed to handle 10 milliamperes, the shunt, the meter coil, or both will burn out.

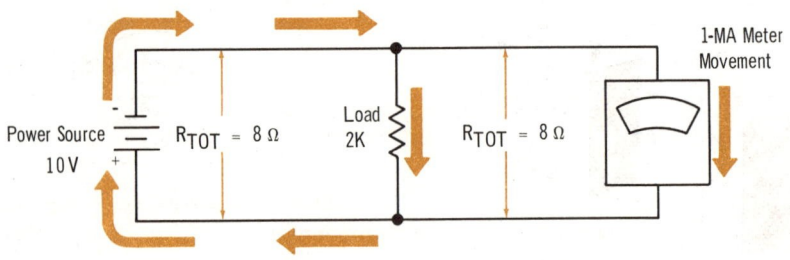

the clamp-on ammeter

You have learned that a current meter must be connected in series with the power source and the load when making a measurement. It is often inconvenient, and, in many cases, almost impossible to do this. The need for a current meter capable of measuring current without breaking the circuit to make a series connection fostered the development of the *clamp-on* ammeter.

The clamp-on ammeter consists basically of an iron core with a coil wound on it and a current meter similar to those you have examined. A triggering device on the core permits opening the core so that one of the conductors of the circuit being measured can be placed inside of the core. This creates a transformer, in which the conductor acts as a one-turn primary winding and the coil on the core acts as the secondary winding. Current through the conductor produces a magnetic field, which in turn, induces current in the secondary winding. Current flows through the meter which is connected across the secondary winding to indicate the current flowing in the circuit being measured. In most clamp-on ammeters, a rectifier-type current meter is used. Since transformer action is required for operation of the clamp-on ammeter, only a-c current can be measured.

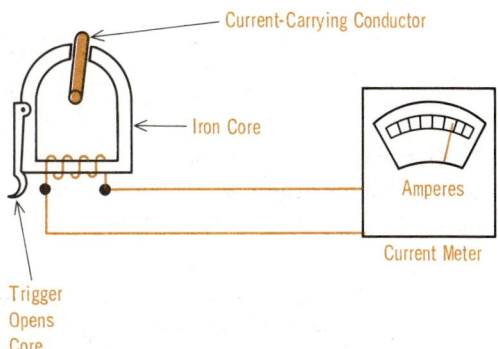

With a clamp-on a-c ammeter, the circuit being tested does not have to be broken to insert the meter. The magnetic field around the conductor induces voltage in the winding, which then provides an induced current to the meter. This meter is particularly useful for high-current measurements

Current-Carrying Conductor

Iron Core

Amperes

Current Meter

Trigger
Opens
Core

The strength of the magnetic field around the conductor is proportional to the number of turns in the conductor and the strength of the current through the conductor. Since, in this case, the conductor has only one turn, the current through one conductor must be high to produce a magnetic field strong enough to cause the meter to operate. Therefore, clamp-on ammeters are normally used to measure current measured in amperes. They are particularly useful in measuring very high currents; for example, hundreds of amperes, because these currents do not flow through the meter movement or meter shunts.

reading the scale

Reading the scale of a single-range current meter like the one shown on page 5-31 is no problem. Since the meter only measures one range of current, only one set of values is required on the scale. However, some multirange current meters also have only one set of values on the scale, even though they measure several ranges of currents. When this is the case, multiply the scale reading by the setting of the range switch. For example, if the scale is calibrated in values of from 0 to 1 milliampere, and the range switch is in the 0–1-milliampere position, read the current directly. However, if the range switch is in the 0–100-milliampere position, multiply the scale reading by 100 to find the amount of current flowing in the circuit.

Some multirange current meters have only one set of values marked on the scale

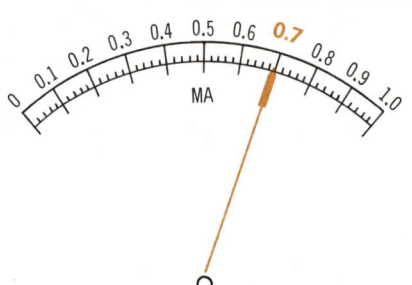

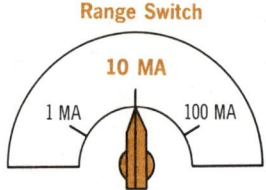

Range Switch

To find the current flowing in the circuit, multiply the scale reading by the range switch setting: current = 0.7 × 10 ma = 7 ma

Other current meters have a separate set of values on the scale for each position of the range switch. In this case, make sure that you read the set of values that corresponds to the setting of the range switch.

Some multirange current meters have a set of values for each range switch position

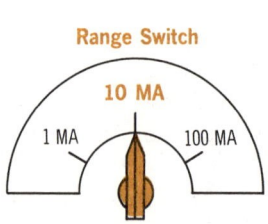

Range Switch

To find the current flowing in the circuit, read the meter scale that corresponds to the position of the range switch: current = 7 ma

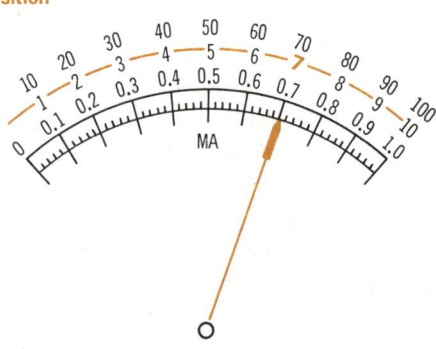

usable part of the scale

Although the current flowing in a circuit may be read anywhere on the scale from zero to full-scale deflection, the closer the pointer is to full-scale deflection, the more accurate will be the reading. Earlier, the accuracy of a meter was specified as the percentage of error at *full-scale deflection*. For example, if the specified accuracy of a 100-milliampere meter is ±2 percent, the meter reading can be off by ±2 milliamperes if it reads full scale. But, the error in a meter is fixed; that is, if it can be off by as much as ±2 milliamperes at 100 ma, it can be off by as much as ±2 milliamperes for any reading below full-scale deflection. Therefore, although the accuracy of a meter is specified at full-scale deflection, keep in mind that the percentage of error becomes progressively greater as you move toward zero.

The greater the deflection, the more accurate the reading

The meter has a full-scale deflection accuracy of ±2%. The error on the 0-100 ma scale is ±2 ma, and the error on the 0-10 scale is ±0.2 ma. Therefore, a more accurate reading will be obtained by making the measurement on the scale that gives the greatest deflection. For a 5-ma reading, the 0-10 ma scale should be used

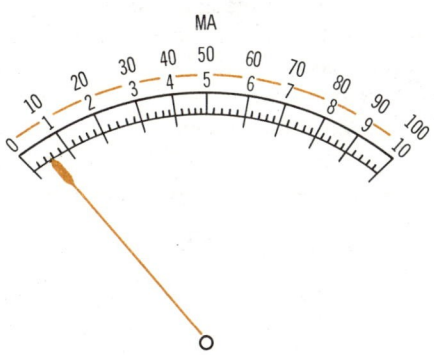

Considering this, if you knew that the current to be measured should be about 5 milliamperes, and you had a meter with a 0–10-, and a 0–100-milliampere range, you should use the 0–10-milliampere range because the reading will fall closer to full-scale deflection. Similarly, if you had two meters, one with a 0–10-milliampere range and the other with a 0–100-milliampere range, you should use the one with the 0–10 milliampere range to get greater deflection, and, therefore, greater accuracy.

Another reason for selecting the scale that gives the greatest deflection is that it is easier to judge the reading on the lower scale if the pointer does not fall on a graduation. For example, as you can see, it would be much easier to judge values between 5 and 6 milliamperes on the 0–10-milliampere scale than it would be on the 0–100'-milliampere scale.

summary

☐ A current meter must always be connected in series with the power source and load. If it is connected in parallel with either of these components, a very high current might flow through the meter and damage the meter movement or shunt. ☐ When a meter is used to measure d-c current, polarity must be observed: it must be connected so that the current flow through the meter is from the negative meter terminal to the positive one. ☐ It is not necessary to observe polarity when using a meter that has the zero position at the center of the scale. ☐ Since the polarity of a-c current continually changes, it is not necessary to observe polarity when measuring ac.

☐ A meter should never be connected into a circuit unless the approximate circuit current is known, and the meter can handle that range. ☐ When using a multirange meter, start with the highest range, and then decrease the range until the most suitable one is reached. ☐ Clamp-on ammeters can be used to make current measurements without breaking the circuit.

☐ Single-range current meters have only one set of values marked on the scale. ☐ Some multirange meters have only a single set of values of their scale. The actual reading is obtained by multiplying the scale reading by the setting of the range switch. ☐ Many multirange meters have a separate set of values on the scale for each current range. ☐ With a multirange meter, the best range for a reading with maximum accuracy and best readability is the one which will give close to full-scale deflection.

---

review questions

1. Why should a current meter never be connected in parallel with a power source or load?
2. What is meant by "observing polarity" when connecting a current meter?
3. For what type of d-c current meter is it unnecessary to observe polarity?
4. Must polarity be observed when measuring ac? Why?
5. If you do not know the approximate value of the current being measured, how should you proceed with a multirange meter?
6. What is a *clamp-on ammeter*?
7. What is the main advantage of a clamp-on ammeter?
8. Can a clamp-on ammeter be used to measure very small currents? Why?
9. Should a 95-milliampere current be measured with the 0–10 milliamperes, 0–100 milliamperes, or 0–1 ampere range of a multirange meter?
10. Explain the reason for your answer to Question 9.

how a current meter movement can measure voltage

A basic current meter movement by itself, whether ac or dc, can be used to measure voltage. You know that every meter coil has a fixed resistance, and, therefore, when current flows through the coil, a voltage drop will be developed across this resistance. According to Ohm's Law, the voltage drop will be proportional to the current flowing through the coil ($E = IR$). For example, assume that you have a 0–1-milliampere meter movement with a coil resistance of 1000 ohms. At full-scale deflection, that is, when 1 milliampere is flowing through the meter coil, the voltage developed across the coil resistance will be:

$$E = I_M R_M = 0.001 \times 1000 = 1 \text{ volt}$$

If only half that current (0.5 milliampere) was flowing through the coil, then the voltage across the coil would be:

$$E = I_M R_M = 0.0005 \times 1000 = 0.5 \text{ volt}$$

You can see, then, that the voltage developed across the coil is proportional to the current flowing through the coil. Also, the current that flows through the coil is proportional to a voltage applied to the coil. Therefore, by calibrating the meter scale in units of voltage instead of in units of current, the voltage in various parts of a circuit can be measured.

Although a current meter movement inherently can measure voltage, its usefulness is limited because the current that the meter movement can handle as well as its coil resistance are very low. For example, the maximum voltage you could measure with the 1-milliampere meter movement examined is 1 volt. In actual practice, you will be required to make voltage measurements higher than 1 volt.

Since the voltage across the meter coil resistance is proportional to the current flowing through the coil, the 1-ma current meter movement can measure voltage directly by calibrating the meter scale in the units of voltage that produce the currents through the coil

$$E = I_R R_M = 0.001 \times 1000 = 1 \text{ volt}$$

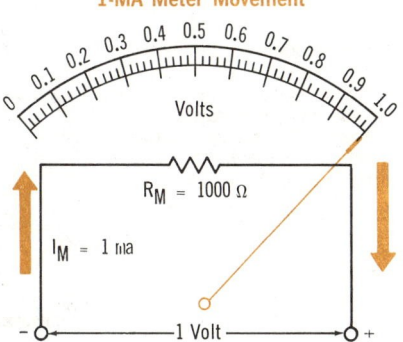

multiplier resistors

Since a basic current meter movement can only measure very small voltages, how can it measure voltages greater than the $I_M R_M$ drop across the coil resistance? The voltage range of a meter movement can be extended by adding a resistor, a *multiplier resistor*, in series. The value of this resistor must be such that, when added to the meter coil resistance, the total resistance limits the current to the full-scale current rating of the meter for any applied voltage. For example, suppose you wished to use the 1-milliampere, 1000-ohm meter movement to measure voltages up to 10 volts. From Ohm's Law, you can see that if you connected the movement across a 10-volt source, 10 milliamperes would flow through the movement and would probably ruin the meter ($I = E/R = 10/1000 = 10$ milliamperes). But, you can limit the meter current to 1 milliampere if you add a multiplier resistor (R_{MULT}) in *series* with the meter resistance (R_M). Since only 1 milliampere maximum can flow through the meter, the total resistance of the multiplier resistor and the meter ($R_{TOT} = R_{MULT} + R_M$) must limit the meter current to 1 milliampere. By Ohm's Law, the total resistance is

$$R_{TOT} = E_{MAX}/I_M = 10 \text{ volts}/0.001 \text{ ampere} = 10,000 \text{ ohms}$$

But this is the *total* resistance needed. The multiplier resistance is

$$R_{MULT} = R_{TOT} - R_M = 10,000 - 1000 = 9000 \text{ ohms}$$

The basic 1-milliampere, 1000-ohm meter movement can now measure 0–10 volts, because 10 volts must be applied to cause full-scale deflection. However, the meter scale must now be recalibrated from 0–10 volts, or, if you use the previous scale, you must multiply all readings by 10.

By connecting a multiplier resistor in series with the meter resistance, the range of a basic meter movement can be extended to measure voltages higher than the $I_M R_M$ voltage drop across the meter coil

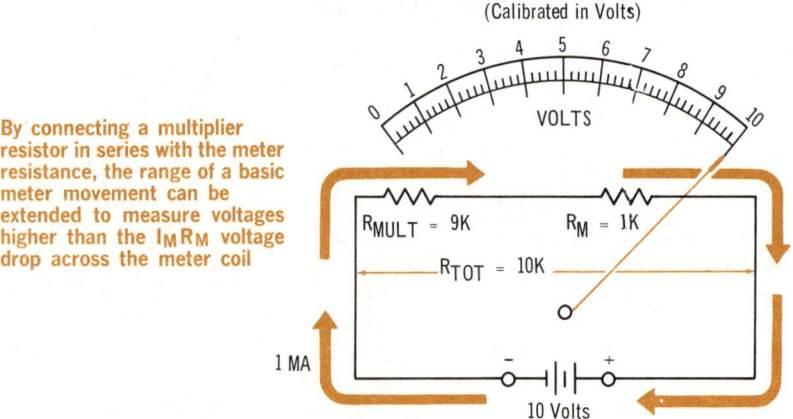

1-MA Meter Movement
(Calibrated in Volts)

solved problems

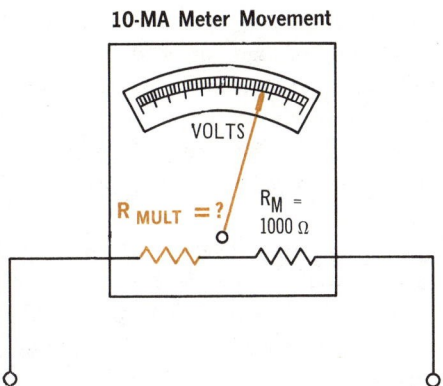

10-MA Meter Movement

VOLTS

R MULT = ?

$R_M = 1000\ \Omega$

Problem 5. *With the basic meter movement shown, what value of multiplier resistor is needed to measure 0–10 volts?*

Calculating Total Resistance (R_{TOT}): The total resistance limits the current across the meter to 10 milliamperes when the voltmeter is connected across 10 volts. From Ohm's Law:

$$R_{TOT} = E_{MAX}/I_M = 10 \text{ volts}/0.01 \text{ ampere} = 1000 \text{ ohms}$$

Calculating Multiplier Resistance (R_{MULT}): Since $R_{TOT} = R_{MULT} + R_M$,

$$R_{MULT} = R_{TOT} - R_M$$

You have solved for R_{TOT} and were given R_M. Therefore,

$$R_{MULT} = 1000 - 1000 = 0 \text{ ohms}$$

The meter, then, can measure 10 volts without a multiplier resistor.

Problem 6. *With the basic meter movement shown, what value of multiplier resistor is needed to measure 0–25 volts?*

Calculating Total Resistance (R_{TOT}): The total resistance limits the current across the meter to 10 milliamperes when the voltmeter is connected across 25 volts. From Ohm's Law:

$$R_{TOT} = E_{MAX}/I_M = 25 \text{ volts}/0.01 \text{ ampere} = 2500 \text{ ohms}$$

Calculating Multiplier Resistance (R_{MULT}): Since $R_{TOT} = R_{MULT} + R_M$:

$$R_{MULT} = R_{TOT} - R_M = 2500 - 1000 = 1500 \text{ ohms}$$

The multiplier resistor that is needed for the voltmeter to measure 25 volts is 1500 ohms.

multirange voltmeters

Previously you discovered that, in many cases, it was not practical to use single-range current meters when working on electrical and electronic equipment. The same is true for voltmeters. In many types of equipment, you will encounter voltages from a few tenths of a volt up to hundreds, and even thousands, of volts. To use single-range meters in these cases will be impractical, and costly. Instead, *multirange voltmeters*—voltmeters that can measure several ranges of voltage—can be used.

A multirange voltmeter contains several multiplier resistors that can be connected in series with the meter movement. Just as with multirange current meters, a range switch is used to connect the proper resistor, or resistors, for the desired range. Also, in some cases, separate terminals for each range are mounted on the meter case.

The range of this 1-ma, 1000-ohm meter movement, which can only measure 0–1 volt, has been extended to measure 0–10, 0–100, and 0–1000 volts by using multiplier resistors for each range

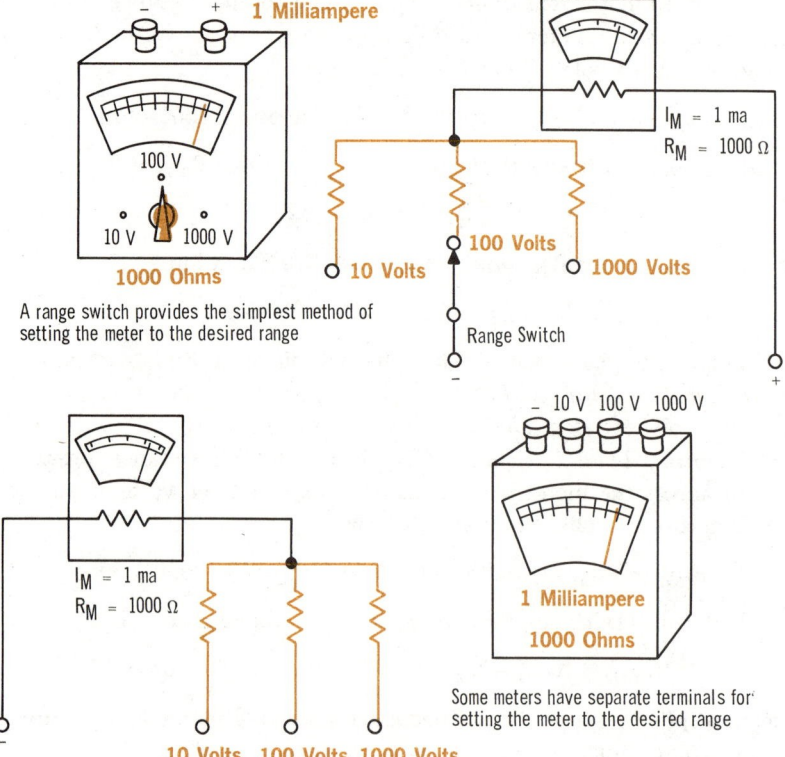

A range switch provides the simplest method of setting the meter to the desired range

Some meters have separate terminals for setting the meter to the desired range

calculating the resistance of multirange multipliers

There are two methods of calculating the values of multiplier resistors for a multirange voltmeter. In the first method, each multiplier is calculated the same as for a single-range voltmeter. Assume that you wish to extend the range of a 1-milliampere movement to measure 0–10, 0–100, and 0–1000 volts, and you also want a 0–1-volt range. Since full-scale deflection equals 1 volt on the 0–1-volt range ($E = I_M R_M = 0.001$ ampere $\times$ 1000 ohms $= 1$ volt), no multiplier is needed. The total resistance (R_{TOT}) needed to limit meter current (I_M) to 1 milliampere on the 0–10-volt range is

$$R_{TOT} = E_{MAX}/I_M = 10 \text{ volts}/0.001 \text{ ampere} = 10{,}000 \text{ ohms}$$

Since the resistance of the meter (R_M) is 1000 ohms, then the multiplier resistance (R_{MULT}) is 9000 ohms.

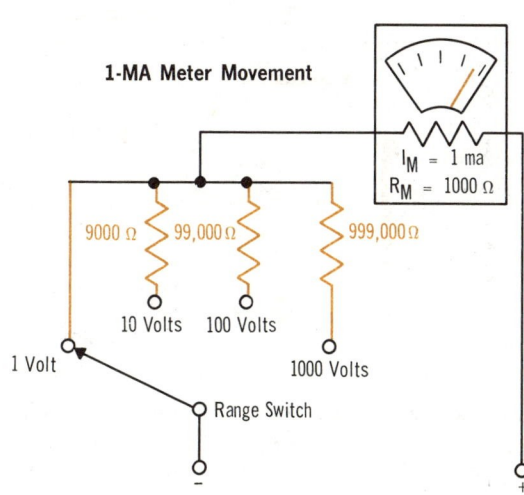

1-MA Meter Movement

$I_M = 1$ ma
$R_M = 1000\ \Omega$

9000 Ω   99,000 Ω   999,000 Ω

10 Volts   100 Volts

1 Volt   1000 Volts

Range Switch

The range of the 0–1 ma, 1000-ohm meter movement was extended to measure 0–10, 0–100, and 0–1000 volts simply by finding the total resistance needed for each range to limit meter current to 1 ma, and then subtracting the meter resistance from that value to obtain the resistance of the multiplier for a particular range

The same procedure is followed to find the multiplier resistor for the 0–100-volt range:

$$R_{TOT} = E_{MAX}/I_M = 100 \text{ volts}/0.001 \text{ ampere} = 100{,}000 \text{ ohms}$$
$$R_{MULT} = R_{TOT} - R_M = 100{,}000 - 1000 = 99{,}000 \text{ ohms}$$

Similarly, for the 0–1000-volt scale:

$$R_{TOT} = E_{MAX}/I_M = 1000 \text{ volts}/0.001 \text{ ampere} = 1{,}000{,}000 \text{ ohms}$$
$$R_{MULT} = R_{TOT} - R_M = 1{,}000{,}000 - 1000 = 999{,}000 \text{ ohms}$$

calculating the resistance
of multirange multipliers (cont.)

A second method of calculating the values of voltmeter multiplier resistors is the *series-multiplier arrangement* in which the multiplier resistors are connected in series. As shown, R_1 is the multiplier resistor for the 0–10-volt range. For the 0–100-volt range, R_1 is in *series* with R_2. Therefore, the value of the multiplier resistance for the 0–100-volt range is equal to R_1 *plus* R_2. Similarly, the multiplier resistance for the 0–1000-volt range is equal to R_1 *plus* R_2 *plus* R_3. By now, you probably realize that the series-multiplier arrangement is similar to the ring shunt arrangement for current meters that you examined earlier.

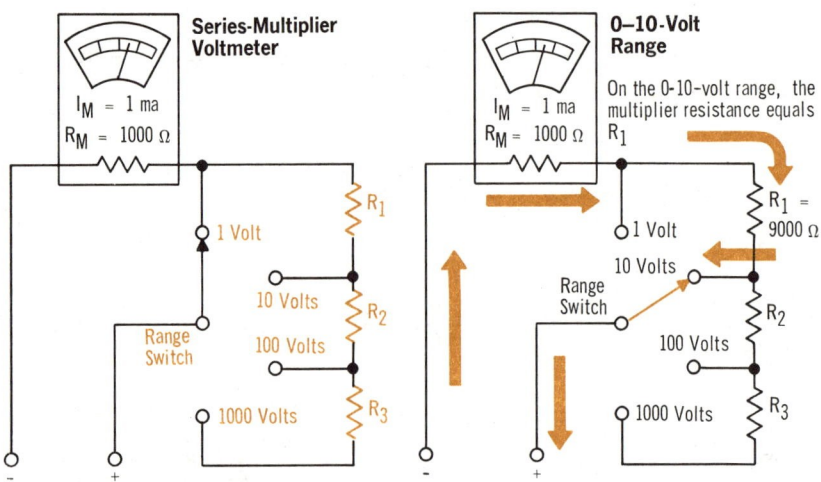

Now, let's calculate the values for a series multipler voltmeter. We will use the same 1-milliampere, 1000-ohm meter movement that we used previously. Since this movement indicates 1 volt for a full-scale deflection, no multiplier resistor is needed for the 0–1-volt range. Therefore, your first step is to calculate the multiplier resistance needed for the 0–10-volt range. Again, using Ohm's Law, find the total resistance (R_{TOT}) needed to limit meter current (I_M) to 1 milliampere at this range:

$$R_{TOT} = E_{MAX}/I_M = 10 \text{ volts}/0.001 \text{ ampere} = 10,000 \text{ ohms}$$

Therefore, multiplier resistor R_1 for the 0–10-volt range equals 10,000 ohms minus the 1000-ohm meter resistance, or 9000 ohms. Thus far, the procedure is the same as in the other method, and the value of the multiplier resistor is the same for the 0–10-volt range.

calculating the resistance of multirange multipliers (cont.)

Having found the series multipliers for the 0–1-and 0–10-volt ranges, let's calculate the total resistance needed for the 0–100-volt range:

$$R_{TOT} = E_{MAX}/I_M = 100 \text{ volts}/0.001 \text{ ampere} = 100,000 \text{ ohms}$$

Subtracting the meter resistance from the total resistance, you find that the multiplier resistance for the 0–100-volt range is 99,000 ohms. Thus far, this method is the same as the previous, but now the multiplier resistance is made up of R_1 *plus* R_2 in series. Therefore, since you need 99,000 ohms for the multiplier resistance and R_1 equals 9000 ohms R_2 must equal 90,000 ohms.

Similarly for the 0–1000-volt range:

$$R_{TOT} = E_{MAX}/I_M = 1000 \text{ volts}/0.001 \text{ ampere} = 1,000,000 \text{ ohms}$$

Thus, $R_{MULT} = R_{TOT} - R_M = 1,000,000 - 1000 = 999,000$ ohms

But $R_{MULT} = R_1 + R_2 + R_3$. Thus,

$$R_{MULT} = 999,000 \text{ ohms} = 9000 + 90,000 + R_3$$

and $\qquad R_3 = 999,000 - 99,000 = 900,000 \text{ ohms}$

No matter which method you use, the value of the multiplier resistance for each range remains the same. However, in the first method, the multiplier is a single resistor, while in the second method, on all but the first extended range, it is made up of a number of resistors in series.

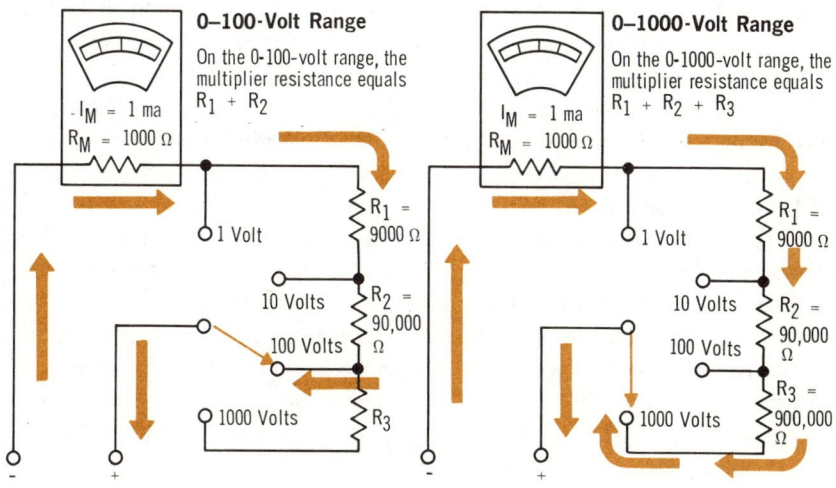

0–100-Volt Range

On the 0-100-volt range, the multiplier resistance equals $R_1 + R_2$

$I_M = 1$ ma
$R_M = 1000 \, \Omega$

1 Volt
10 Volts
100 Volts
1000 Volts

$R_1 = 9000 \, \Omega$
$R_2 = 90,000 \, \Omega$
R_3

0–1000-Volt Range

On the 0-1000-volt range, the multiplier resistance equals $R_1 + R_2 + R_3$

$I_M = 1$ ma
$R_M = 1000 \, \Omega$

1 Volt
10 Volts
100 Volts
1000 Volts

$R_1 = 9000 \, \Omega$
$R_2 = 90,000 \, \Omega$
$R_3 = 900,000 \, \Omega$

ohms-per-volt rating

An important characteristic of any voltmeter is its *impedance,* or ohms-per-volt (ohms/volt) rating. The ohms/volt rating is an indication of the voltmeter sensitivity; and the higher the rating, the more sensitive is the voltmeter. The ohms/volt rating is defined as the resistance ($R_M + R_{MULT}$) required for full-scale deflection. For example, the 1-milliampere, 1000-ohm meter movement we have been using indicates 1 volt at full-scale deflection. Therefore, its ohms/volt rating is 1000/1, or 1000 ohms/volt.

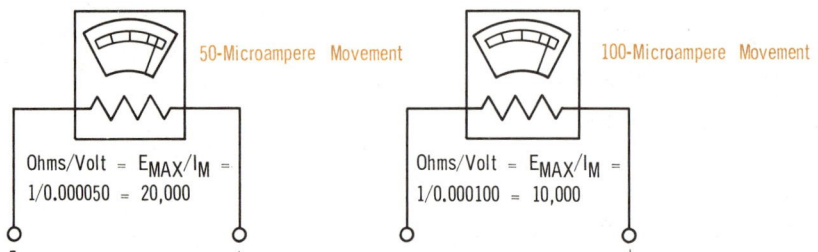

50-Microampere Movement   100-Microampere Movement

Ohms/Volt = E_{MAX}/I_M = 1/0.000050 = 20,000

Ohms/Volt = E_{MAX}/I_M = 1/0.000100 = 10,000

The voltmeter on the left is more sensitive than the one on the right because it offers more resistance to obtain a deflection of 1 volt; or, to put it another way, it requires less current from the circuit it is measuring for full-scale deflection

The ohms/volt rating is an inherent characteristic of the meter movement, and remains the same for all ranges of a multirange voltmeter. For example, when we extended the 1-milliampere, 1000-ohm movement to measure 0–10 volts, a 9000-ohm resistor was connected in series with the 1000-ohm meter resistance. Although total resistance (R_{TOT}) now becomes 10,000 ohms, the voltage for full-scale deflection is now 10 volts instead of 1 volt. Therefore, the ohms/volt rating remains the same (ohms/volt = 10,000/10 = 1000 ohms/volt).

Now, let's examine a 50-microampere meter movement. The meter movement resistance required for a 1-volt full-scale deflection is $R_{TOT} = E_{MAX}/I_M = 1/0.000050 = 20,000$ ohms. Thus, the meter movement is said to have a sensitivity of 20,000 ohms/volt.

While the ohms/volt rating is the same for all ranges of a multirange voltmeter, the total resistance increases as the range increases. The total resistance for any range is the ohms/volt rating of the basic meter movement multiplied by the full-scale deflection at that range. For example, the total resistance on the 0–100-volt range of a 20,000 ohms/volt meter equals 20,000 × 100, or 2,000,000 ohms (2 megohms). A voltmeter rated at 1000 ohms/volt would only have a total resistance of 100,000 ohms on the 0–100-volt range.

low ohms-per-volt rating

When a voltmeter with a low ohms/volt rating is used to measure a voltage across a high resistance, circuit conditions can be so upset that a completely inaccurate voltage reading results. For example, consider two 100K-ohm resistors connected in series across a 60-volt power source. Since the resistors are of equal value, 30 volts will be developed across each one.

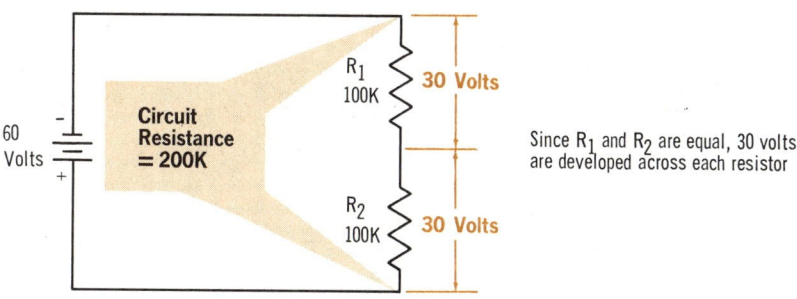

Since R_1 and R_2 are equal, 30 volts are developed across each resistor

Now, connect our 1000 ohms/volt meter across R_2 and set the meter to the 0–100-volt range. As you learned, the total resistance of the meter equals the ohms/volt rating multiplied by the full-scale reading of the range. Therefore, for the 0–100-volt range, the total resistance equals 1000×100, or 100K ohms. When the meter is connected across R_2, they are two 100K-ohm resistors in parallel, and equal 50K ohms. As a result, the total circuit resistance is now only 150K ohms. Resistor R_2 and the meter are now one-third of the total circuit resistance, and, therefore, the meter will only read 20 volts across R_2, which is an error of 33 percent.

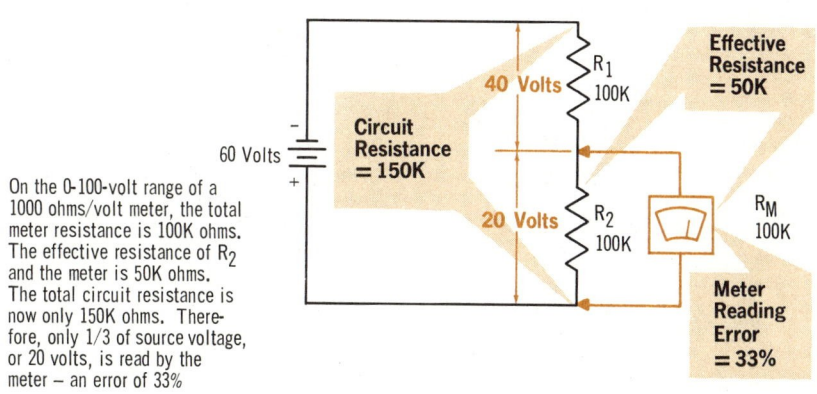

On the 0-100-volt range of a 1000 ohms/volt meter, the total meter resistance is 100K ohms. The effective resistance of R_2 and the meter is 50K ohms. The total circuit resistance is now only 150K ohms. Therefore, only 1/3 of source voltage, or 20 volts, is read by the meter — an error of 33%

low ohms-per-volt rating (cont.)

Now let's connect a 20,000 ohms/volt meter across R_2 and set it to the 0–100-volt range. A 20,000 ohms/volt meter on the 0–100-volt range will have a total resistance of 2 megohms (20,000 × 100 = 2,000,000). The parallel resistance of the 2-megohm meter resistance and the 100K-ohm circuit resistor (R_2) is 95.3K ohms. Therefore, the series resistance of the circuit is affected only slightly, and the meter will read nearly 30 volts across R_2. Actually, it would read about 28.5 volts across R_2, which is an error of about 5 percent.

On the 0-100-volt range of a 20,000 ohms/volt meter, the total meter resistance is 2 megohms. The effective resistance of R_2 and the meter is 95.3K ohms. The total circuit resistance is now 195.3K ohms, nearly the same as if the meter were not connected. The meter reading will be 28.5 volts – an error of 5%

60 Volts

Circuit Resistance = 195.3K

31.5 Volts

28.5 Volts

R_1 100K

R_2 100K

R_M 2 Meg

Effective Resistance = 95.3K

Meter Reading Error = 5%

If you used a meter with a higher ohms/volt rating, for example 100,000 ohms/volt, the voltage error would have been less than 1 percent. Therefore, the ohms/volt rating of the meter you use determines the circuit on which it should be used. The ohms/volt rating should be high for high-impedence circuits. Actually, the total resistance of the meter in the range being used should be about 10 times as much as the resistance of the circuit being checked to get an accurate reading. Otherwise, you should *interpret* the reading based on the meter's effects.

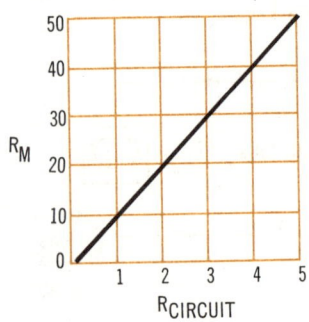

The total resistance of the meter should be at least 10 times as much as that of the circuit

connecting a voltmeter into a circuit

Voltmeters must be in parallel with the circuit component being measured. Unlike the current meter, the voltmeter is less prone to being damaged if it is connected improperly. On the higher ranges, the current flowing through the meter is greatly reduced because of its very high total resistance. However, an erroneous reading will result if a voltmeter is connected in series with a circuit component rather than in parallel with it.

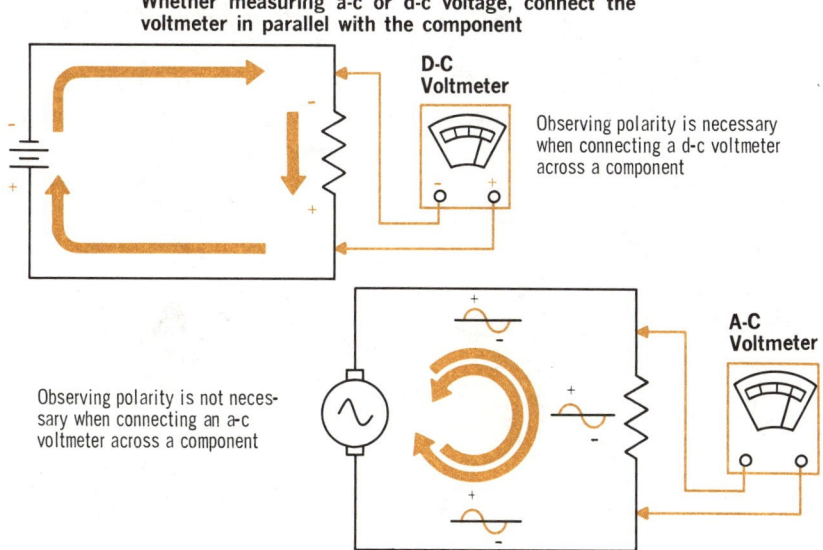

Whether measuring a-c or d-c voltage, connect the voltmeter in parallel with the component

D-C Voltmeter

Observing polarity is necessary when connecting a d-c voltmeter across a component

A-C Voltmeter

Observing polarity is not necessary when connecting an a-c voltmeter across a component

When connecting a d-c voltmeter, always observe the proper polarity. You must connect the negative terminal of the meter to the negative, or low potential, end of the component and the positive terminal to the positive, or high potential, end of the component. As in the case of the current meter, if you connect the voltmeter to the component with opposing polarities, the meter coil will move to the left and the pointer may bend when striking the left retaining pin. These effects, again, do not apply to zero-center meters.

In an a-c circuit, the voltage continually reverses polarity. Therefore, there is no need to observe polarity when connecting a voltmeter across a component in an a-c circuit.

reading a voltmeter scale

Reading a voltmeter scale is as easy as reading a current meter scale. Some multirange voltmeters have only one range of values marked on the scale, and the scale reading must be multiplied by the range switch setting to obtain the correct voltage. Other voltmeters have separate ranges on the scale for each setting of the range switch. When using these meters, make sure that you read the set of values that corresponds to the range switch setting.

When you studied current meters, you learned that the accuracy of the meter was based on full-scale deflection. If the meter was accurate to within ±2 percent, then any reading on the 0–100-volt scale would be off by as much as ±2 volts. Therefore, you were told to use a scale that would give a reading as close as possible to full-scale deflection because the *percentage error* decreases as the reading approaches full scale. However, another factor, the total resistance of the voltmeter on each range, must be considered when making voltage measurements.

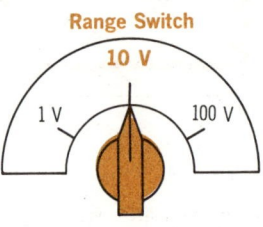

Range Switch

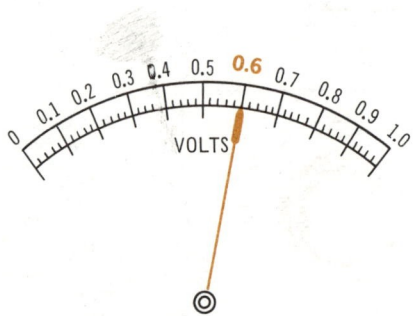

To find the voltage across a component, multiply the scale reading by the setting of the range switch: voltage = 0.6 × 10 = 6.0 volts

Range Switch

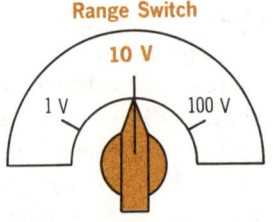

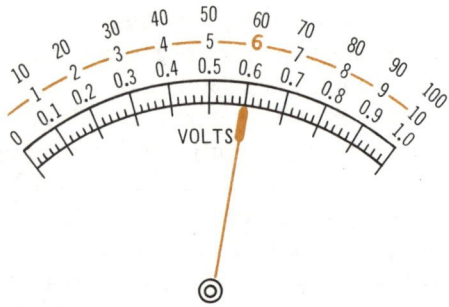

To find the voltage across a component, simply read the meter scale that corresponds to the position of the range switch: voltage = 6 volts

reading a voltmeter scale (cont.)

Since a voltmeter is connected in parallel with a component, the voltmeter resistance should be as high as possible to avoid affecting circuit operation. Therefore, when using lower ohms/volt meters (from 20,000 ohms/volt and lower) in high-resistance circuits, use the *highest* meter range that can be read accurately. For example, for the circuit shown, the voltage developed across a 100K-ohm resistor should be 9 volts, and you will get a more accurate reading using the 0–50-volt scale of a 20,000 ohms/volt meter than you will if you use the 0–10-volt scale.

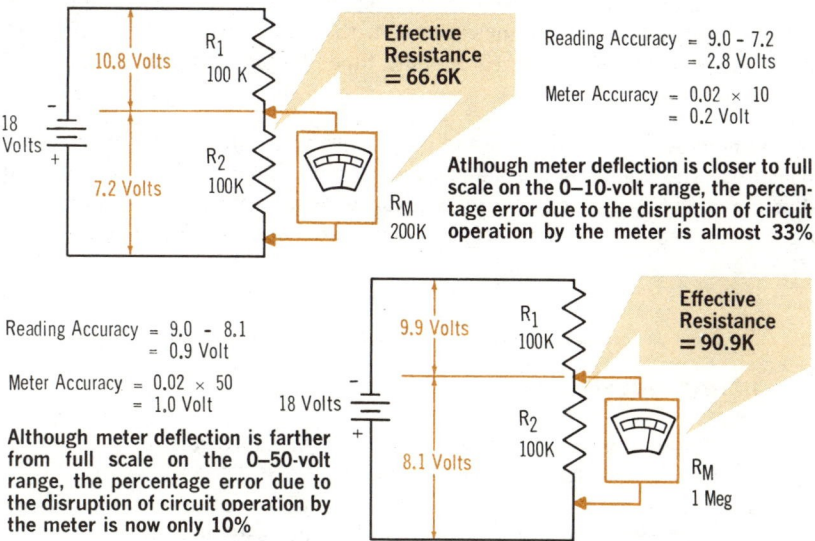

18 Volts

10.8 Volts

R_1 100 K

R_2 100K

7.2 Volts

R_M 200K

Effective Resistance = 66.6K

Reading Accuracy = 9.0 – 7.2
= 2.8 Volts

Meter Accuracy = 0.02 × 10
= 0.2 Volt

Atlhough meter deflection is closer to full scale on the 0–10-volt range, the percentage error due to the disruption of circuit operation by the meter is almost 33%

Reading Accuracy = 9.0 – 8.1
= 0.9 Volt

Meter Accuracy = 0.02 × 50
= 1.0 Volt

Although meter deflection is farther from full scale on the 0–50-volt range, the percentage error due to the disruption of circuit operation by the meter is now only 10%

18 Volts

9.9 Volts

R_1 100K

R_2 100K

8.1 Volts

R_M 1 Meg

Effective Resistance = 90.9K

For a 20,000 ohms/volt meter with an accuracy of ±2 percent on the 0–10-volt range, the effective resistance of R_2 and the meter resistance is 66.6K ohms. Therefore, the voltage developed across R_2 and read by the meter is $18 \times (66.6K/166.6K) = 7.2$ volts. This reading is almost 2 volts lower than the 9 volts that would be developed without the meter connected although the meter accuracy is ±0.2 volt on the 0–10-volt scale.

For the 0–50-volt scale, the effective resistance of R_2 and the meter resistance is 90.9K ohms. Therefore, the voltage developed across R_2 and read by the meter is $18 \times (90.9K/190.9K) = 8.1$ volts. This reading is now 1 volt lower than the 9 volts developed without the meter connected, and within the ±1-volt accuracy on the 0–50-volt scale. Therefore, the 0–50-volt scale disrupts circuit operation much less than the 0–10-volt scale, and a more accurate reading will result.

summary

☐ Because of the Ohm's Law relationship, a current meter scale can be calibrated in volts and used to measure voltage. ☐ Multiplier resistors are added in series with the coil to extend the voltage range of a meter movement. ☐ For a given voltage range, the required value of multiplier resistance can be found by: $R_{MULT} = R_{TOT} - R_M$, where $R_{TOT} = E_{MAX}/I_M$. ☐ Multirange voltmeters contain several multiplier resistors and a range switch.

☐ The sensitivity of a voltmeter is expressed by its ohms/volt rating. The higher this rating, the more sensitive is the meter. ☐ The ohms/volt rating of a multirange voltmeter is the same for all ranges. However, the total resistance of the meter increases as the range increases. ☐ Meters with high ohms/volt ratings should be used for making measurements in high-impedance circuits. ☐ For an accurate measurement, the total resistance of the meter, in the range used, should be 10 or more times larger than the resistance of the circuit being checked.

☐ Voltmeters must be connected in parallel with the circuit component being measured. The voltmeter resistance should be as high as possible to avoid affecting circuit operation. ☐ Proper polarity must be observed when connecting d-c voltmeters. It is not necessary with zero-center d-c voltmeters or a-c voltmeters.

review questions

1. Why can a current meter movement measure voltage?
2. Why are multiplier resistors used?
3. A 1-milliampere meter movement has a coil resistance of 1000 ohms. What value of multiplier resistor is needed to measure 100 volts?
4. The coil and multiplier resistor of a 10-milliampere meter movement have a total resistance of 10K. What is the largest voltage that can be measured with the meter?
5. What is meant by the *ohms/volt rating* of a voltmeter?
6. If a 1-milliampere meter movement requires a combined multiplier and coil resistance of 5000 ohms for a 1-volt full-scale deflection, what is the ohms/volt rating of the meter?
7. What is the significance of a high ohms/volt rating?
8. What is the relationship between a meter's ohms/volt rating and the current required for full-scale deflection?
9. How should a voltmeter be connected in a circuit?
10. Why must the total resistance of a voltmeter be considered when making a voltage measurement?

the ohmmeter

An ohmmeter is a device that measures the resistance of a circuit or component. It can also be used to locate open circuits or shorted circuits.

Basically, an ohmmeter consists of a d-c current meter movement, a low-voltage, d-c power source, and current-limiting resistors, all of which are connected in series. The moving-coil meter movement is the *only* current meter movement used in ohmmeters. A battery is used as the low-voltage, d-c power source.

The series ohmmeter has the power source, meter movement, and current limiting resistor in series

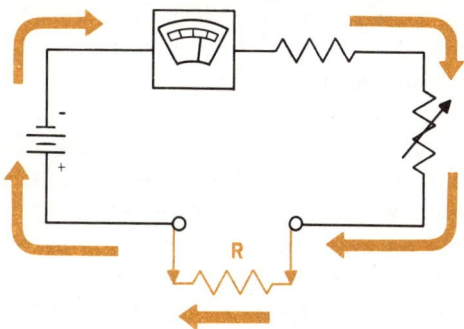

In a series ohmmeter, the resistance to be measured (R) is connected in series with the meter movement

There are two types of ohmmeters: the series type and the shunt type. In the series ohmmeter, the resistance to be measured is connected *in series with the meter movement*. In the shunt ohmmeter, the resistance to be measured is connected *in parallel with the meter movement*. As you will soon learn, each arrangement has advantages over the other for specific applications.

The shunt ohmmeter also has the power source, meter movement, and current limiting resistor in series

In a shunt ohmmeter, the resistance to be measured (R) is connected in parallel with the meter movement

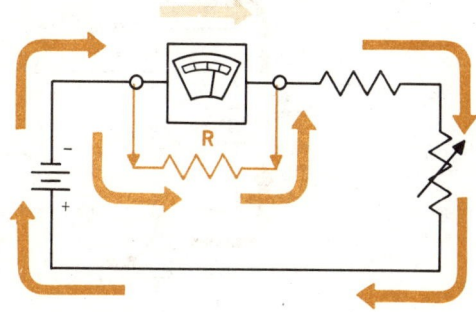

the series ohmmeter

In the series ohmmeter shown, a 1-milliampere, 50-ohm meter movement is connected in series with a 4.5-volt battery, a 4K-ohm fixed resistor (R_1), and a 1K-ohm variable resistor (R_2). If you short the two test leads, current will flow through the meter movement, and the pointer will deflect toward the right. For full-scale deflection, 1 milliampere must flow through the meter movement. To obtain 1 milliampere, the resistance in the circuit must equal 4.5K ohms ($R = E/I = 4.5/0.001 = 4.5K$). Since the resistance of the meter is 50 ohms and the resistance of R_1 is 4K ohms, the value of R_2 must be set to 450 ohms to obtain 1 milliampere of meter current, and, therefore, full-scale deflection. If the resistance of R_2 is set higher than 450 ohms, less than 1 milliampere will flow, and the pointer will not deflect full scale. If R_2 is set less than 450 ohms, more than 1 milliampere will flow, and the pointer will move off scale and hit the right retaining pin.

Notice that shorting the test leads together means that there is *zero resistance* present across the input terminals of the ohmmeter, and when *zero resistance* is present, the pointer deflects *full scale*. Therefore, full-scale deflection of a series ohmmeter indicates zero resistance. You might be wondering why a fixed resistor of 450 ohms is not used in place of the variable resistor that is set at 450 ohms. The reason is simple. You know that batteries age, and as they age their output voltage drops. As the voltage drops, the current through the circuit decreases and the meter will no longer deflect full scale. But, by decreasing the value of R_2, the total circuit resistance decreases, and the current can be brought back to the 1-milliampere level required for full-scale deflection. The variable resistor is usually called the *zero-ohms set control*.

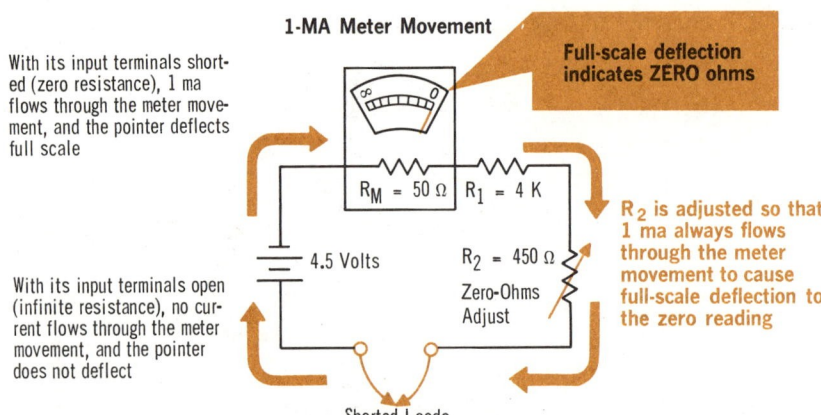

the series ohmmeter (cont.)

If the pointer of a series ohmmeter deflects full scale with zero resistance, what happens when a resistor is connected across the input terminals? Assume that a 9K-ohm resistor is connected across the input terminals of the ohmmeter we discussed on the previous page. Since this resistor is twice the combined resistance of R_M, R_1, and R_2, the total circuit resistance is 13.5K ohms. The meter current, therefore, drops to 1/3 milliampere. Since the meter requires 1 milliampere for full-scale deflection, it will only deflect one-third scale with 1/3 milliampere flowing. Therefore, one-third-scale deflection indicates 9K ohms across the input terminals.

If 4.5K ohms is connected across the meter terminals, the total circuit resistance drops to 9K ohms (twice that of R_M, R_1, and R_2 combined), the meter current increases to 1/2 milliampere, and the pointer deflects half scale. Therefore, half-scale deflection indicates 4.5K ohms across the input terminals.

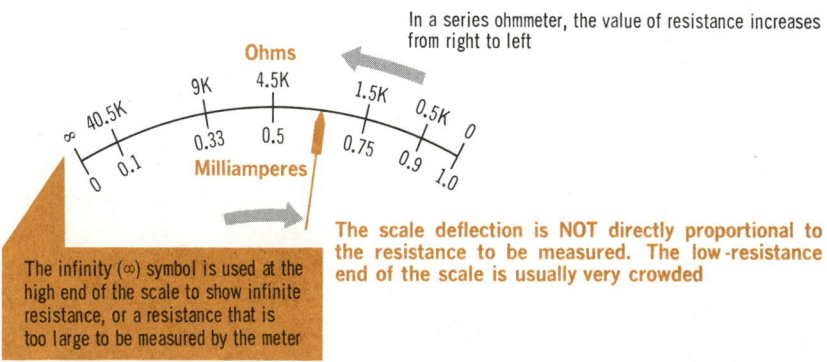

In a series ohmmeter, the value of resistance increases from right to left

The infinity (∞) symbol is used at the high end of the scale to show infinite resistance, or a resistance that is too large to be measured by the meter

The scale deflection is NOT directly proportional to the resistance to be measured. The low-resistance end of the scale is usually very crowded

If 1.5K ohms is connected across the meter terminals, total circuit resistance drops to 6K ohms, meter current increases to 0.75 milliampere, and the pointer deflects three-fourths full scale. If you continue with similar known resistors, you can calibrate the entire scale. You will notice that the ohmmeter does not have a linear scale; that is, the deflection does not increase in direct proportion to the resistance being measured. For example, the 1.5K-ohm resistor caused three-fourths scale deflection; the 4.5K-ohm resistor, three times the value of the 1.5K-ohm resistor, caused half-scale deflection; and the 9K-ohm resistor, six times the value of the 1.5K-ohm resistor, caused one-third scale deflection. Because of nonlinearity, low-resistance readings on the right side of the scale are extremely crowded on a series ohmmeter.

the shunt ohmmeter

Low values of resistance can be measured on a shunt ohmmeter much more accurately than on a series ohmmeter. In the shunt-type ohmmeter, the unknown resistance to be measured is connected in parallel with the meter movement. Therefore, some of the current produced by the battery flows through the meter movement, and some through the unknown resistance.

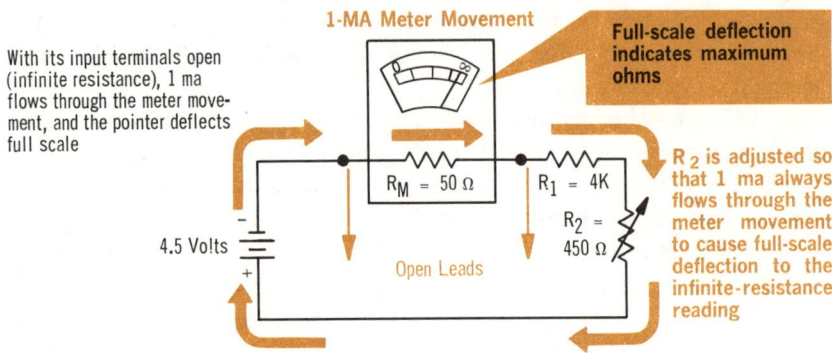

With its input terminals open (infinite resistance), 1 ma flows through the meter movement, and the pointer deflects full scale

1-MA Meter Movement

Full-scale deflection indicates maximum ohms

R_M = 50 Ω R_1 = 4K R_2 = 450 Ω

4.5 Volts

Open Leads

R_2 is adjusted so that 1 ma always flows through the meter movement to cause full-scale deflection to the infinite-resistance reading

Assume that we have a shunt ohmmeter with the same basic meter movement as in the series ohmmeter just discussed; that is, a 1-milliampere, 50-ohm movement. Notice that there is a complete circuit from the negative end of the battery, *through the meter movement,* through R_1 and R_2, to the positive side of the battery. Therefore, 1 milliampere will flow through the meter movement, and the meter pointer will always deflect *full scale* as long as no resistor is connected across the input terminals. Now, assume that you short the input terminals. This causes all of the circuit current to bypass the meter movement, and zero current results in zero deflection. This is exactly the opposite from the series ohmmeter, which deflects full scale when you short the meter input terminals.

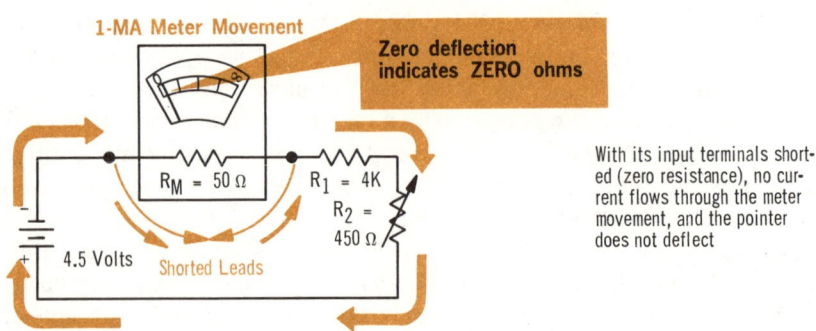

1-MA Meter Movement

Zero deflection indicates ZERO ohms

R_M = 50 Ω R_1 = 4K R_2 = 450 Ω

4.5 Volts

Shorted Leads

With its input terminals shorted (zero resistance), no current flows through the meter movement, and the pointer does not deflect

the shunt ohmmeter (cont.)

The shunt ohmmeter behaves like a current meter with a shunt resistor. The current divides in a ratio inversely proportional to the meter resistance and to the unknown resistance connected across the input terminals. For example, for the 1-milliampere, 50-ohm meter movement we have discussed, suppose we connect a 100-ohm resistor across the input terminals. Twice the current (2/3 milliampere) will flow through the 50-ohm meter resistance than will flow through the 100-ohm resistor (1/3 milliampere). Since the meter requires 1 milliampere for full-scale deflection, it will only deflect two-thirds scale with 2/3 milliampere flowing.

If 50 ohms is connected across the meter terminals, the meter current drops to 1/2 milliampere, and the pointer deflects half scale. If 25 ohms is connected across the meter terminals, the meter current drops to 1/3 milliampere, and the pointer deflects one-third scale. As in the series ohmmeter, you can continue with similar known resistors to calibrate the entire scale.

Shunting the meter resistance with different unknown resistors affects the series resistance of the ohmmeter circuit, but the effect is so slight that it can be ignored. The parallel resistance of the 50-ohm meter resistance and the unknown resistor must always be less than 50 ohms. Since the series resistance is 4500 ohms, variations in resistance up to 50 ohms are insignificant.

The shunt ohmmeter, like the series ohmmeter, does not have a linear scale. For example, the 100-ohm resistor caused two-thirds scale deflection; the 50-ohm resistor, half the value of the 100-ohm resistor, caused half-scale deflection; and the 25-ohm resistor, one-fourth the value of the 100-ohm resistor, caused one-third scale deflection. Because of the nonlinearity, high-resistance readings on the right side of the scale are extremely crowded. However, for the same meter movement, the low-resistance readings of the shunt ohmmeter are less crowded than those of the series ohmmeter, resulting in more accurate low-resistance readings.

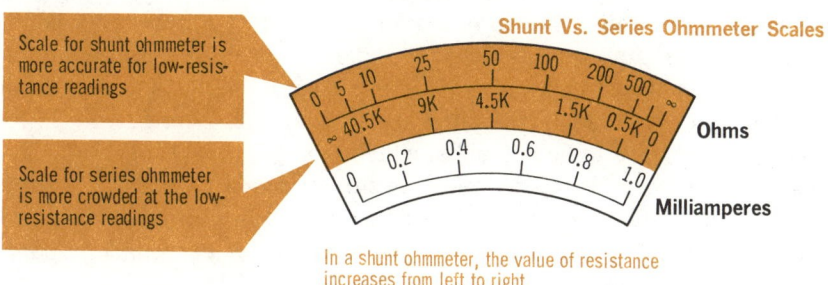

Scale for shunt ohmmeter is more accurate for low-resistance readings

Shunt Vs. Series Ohmmeter Scales

Scale for series ohmmeter is more crowded at the low-resistance readings

Ohms

Milliamperes

In a shunt ohmmeter, the value of resistance increases from left to right

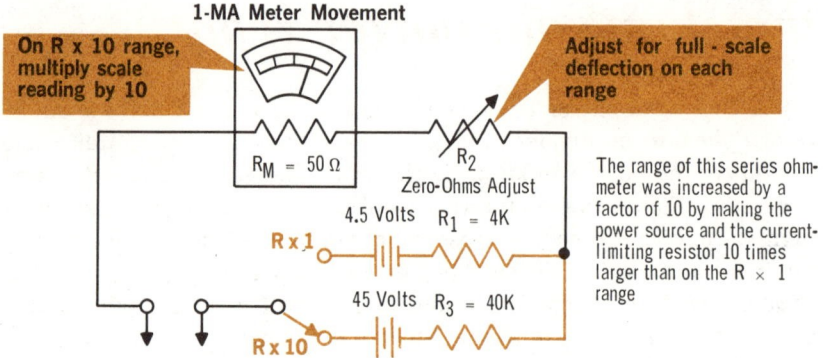

multirange series ohmmeters

You learned that it was often impractical to use single-range current meters and voltmeters when working on electrical equipment. The same holds true for ohmmeters; multirange ohmmeters are more practical.

The range of the series ohmmeter cannot be extended to read higher resistances simply by adding resistors in series or in parallel with the meter movement as is the case for current meters and voltmeters. The only way its range can be extended is to increase the voltage of the power source. Since the current flowing through the ohmmeter circuit and the unknown resistor follows Ohm's Law $(I = E/R)$, as you increase the unknown resistance, a point will be reached at which, essentially, no current will flow and the meter will not deflect. Therefore, for noticeable deflections, a higher voltage is needed to obtain measurable current in the circuit.

For example, if the supply voltage is increased by a factor of ten to 45 volts, the unknown resistance that can be measured is increased by a factor of ten. However, current-limiting resistor R_3 must be ten times greater than R_1. In this arrangement, then, you have an ohmmeter that measures resistance in two ranges: $R \times 1$, the range of the basic series ohmmeter, and $R \times 10$, the range for readings ten times greater than indicated. The range of an ohmmeter is usually extended by multiples of ten, and the range switch is marked $R \times 1$, $R \times 10$, $R \times 100$, etc.

It would be impractical to extend the range of the 1-milliampere series ohmmeter by a factor of more than ten, because a very large voltage supply would be required. For the $R \times 100$ range the power source would have to be 450 volts, which is impractical. For multirange series ohmmeters, therefore, more sensitive meter movements, such as 20,000 ohms/volt, are used. This meter would require only 50 microamperes for full-scale deflection; therefore, a much smaller source voltage can be used.

multirange series ohmmeters (cont.)

You have just learned how to extend the range of a basic series ohm-meter to read higher resistances. The same basic ohmmeter can be modified to measure lower resistances by connecting shunt resistors across the meter movement and its current-limiting resistors, R_1 and R_2. Do not confuse this with the shunt ohmmeter; the unknown resistor is still *in series* with the meter movement.

The basic ohmmeter we examined had a usable range of 0–50K ohms. We will now modify it to include ranges of 0–500 and 0–5000 ohms. Although separate scales could be provided for each range, ohmmeters are usually *calibrated* for the *lowest range*, and the *multiplying factors* ($R \times 10$, $R \times 100$, $R \times 1000$) are used for the *higher ranges*. Therefore, our *lowest* range will be 0–500 ohms, and the pointer deflection must be reduced by a factor of 100 from the basic 0–50K-ohm range. This is accomplished by connecting a shunt resistor across the resistance of the meter circuit ($R_M + R_1 + R_2$). To reduce the deflection by a factor of 100, the shunt resistance must be 1/100 that of the meter circuit so that 1/100 of the current will flow through the meter movement. Since the meter circuit resistance equals 4500 ohms, a 45-ohm shunt resistor must be used on the 0–500-ohm scale. This lowers the meter circuit resistance to about 45 ohms.

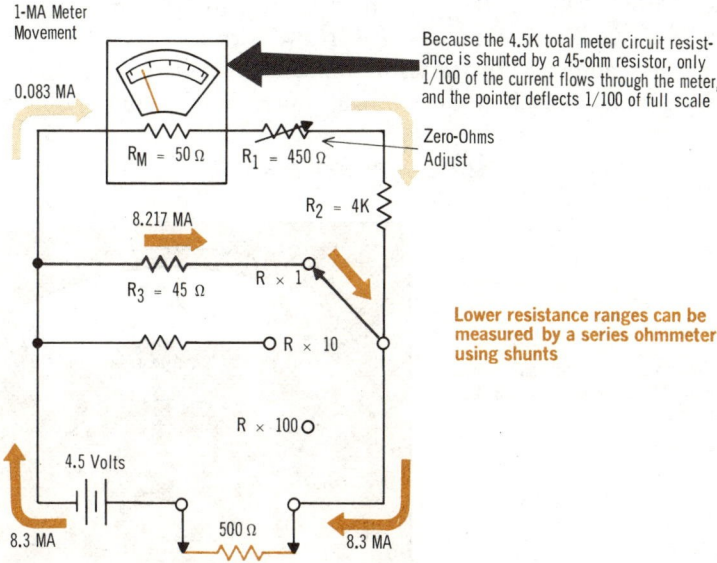

1-MA Meter Movement

0.083 MA

$R_M = 50\ \Omega$ $R_1 = 450\ \Omega$ Zero-Ohms Adjust

$R_2 = 4K$

8.217 MA

$R_3 = 45\ \Omega$ $R \times 1$

$R \times 10$

$R \times 100$

4.5 Volts

8.3 MA $500\ \Omega$ 8.3 MA

Because the 4.5K total meter circuit resistance is shunted by a 45-ohm resistor, only 1/100 of the current flows through the meter, and the pointer deflects 1/100 of full scale

Lower resistance ranges can be measured by a series ohmmeter using shunts

SCALE WITHOUT METER SHUNT

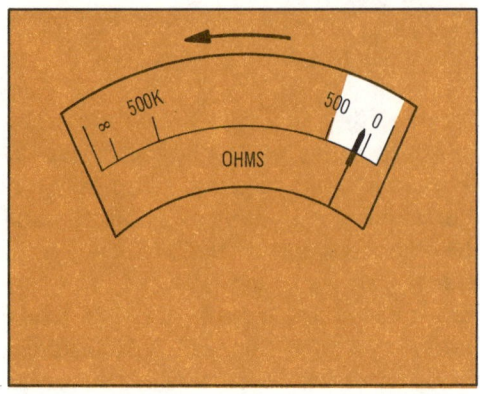

The low-resistance readings on a series ohmmeter are crowded without use of a meter shunt

multirange
series ohmmeters (cont.)

Now, when measuring a 500-ohm resistor, the total circuit resistance becomes 545 ohms, and the total circuit current becomes 8.3 milliamperes ($I = E/R = 4.5/545 = 8.3$ milliamperes). Since the meter current will be 1/100 of the total current, 0.083 milliampere will flow through the meter movement, resulting in 0.083 of full-scale deflection (1 milliampere equals full-scale deflection). On the basic series ohmmeter without shunts, the scale read from 0 to 50K, and the readings from 0 to 500 ohms were *crowded*. Now, with the use of a shunt, the low-range readings of the series ohmmeter are *spread out* so that values between 0 and 500 ohms can be read much more accurately.

SCALE WITH METER SHUNT

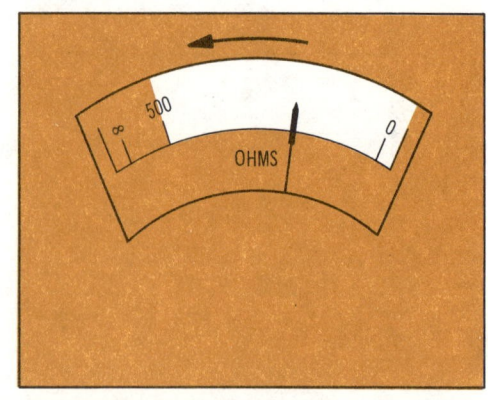

The low-resistance reading on a series ohmmeter are spread out with use of a meter shunt

using the ohmmeter

Not only can an ohmmeter measure the resistance of various circuit parts, but it can check for open or shorted circuit parts and for circuit continuity. In every case, however, to avoid damaging an ohmmeter, make sure that a voltage source is not connected across the ohmmeter prods while you are making a measurement. Resistance readings are only made on deenergized circuits. If the circuit were energized, its voltage could cause a damaging current to flow through the meter.

Simply throwing a switch to the off position does not always prevent an ohmmeter from being connected across a voltage source. Sometimes the switch itself may be defective and voltage will be present throughout the equipment. Or, if you must check the switch itself, a voltage could be applied across the ohmmeter prods even with the switch open.

To Protect an Ohmmeter:

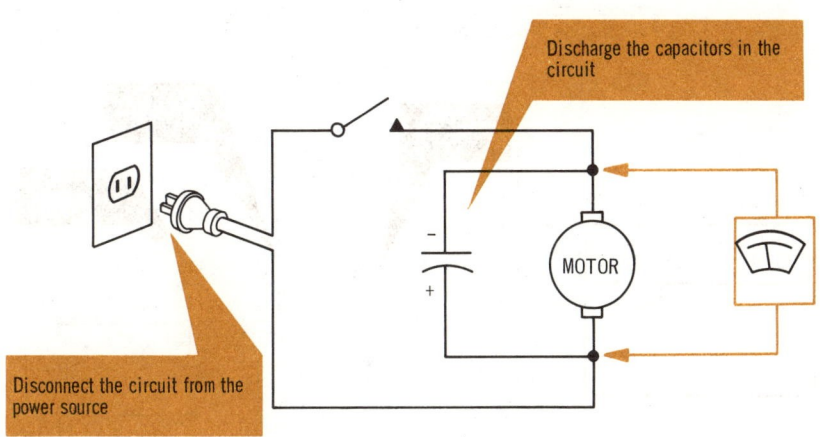

Discharge the capacitors in the circuit

MOTOR

Disconnect the circuit from the power source

The safest way to protect an ohmmeter, then, is to disconnect the equipment from the power source whenever possible. Even with the equipment disconnected from the power source, the ohmmeter is still not completely protected from circuit voltages. This only prevents the possibility of connecting the ohmmeter across a voltage produced by a power supply, battery, or other familiar voltage source. But, it could still be connected across a charged capacitor, which could provide enough current to severely damage the ohmmeter. Therefore, to protect the ohmmeter, you should not only disconnect the equipment from the power source, but you should also discharge any capacitors, particularly electrolytic types, in the circuit being measured.

using the ohmmeter (cont.)

When making resistance measurements in circuits, each part in the circuit can be tested individually by removing the part from the circuit and connecting the ohmmeter prods across it. Actually, the part does not have to be completely removed from the circuit. In most cases, the part can be effectively isolated from the circuit by opening only one of its connections in the circuit. This method takes time, however. Therefore, the manufacturers of many equipments provide charts that list the resistances that you should measure from various *test points* to a *reference point* in the equipment. There are usually many circuit parts between the test point and the reference, or common point, and when an abnormal reading is obtained, you must begin checking smaller groups of parts and individual parts in the circuit to isolate the defective part.

Manufacturers often provide charts to simplify resistance measurements

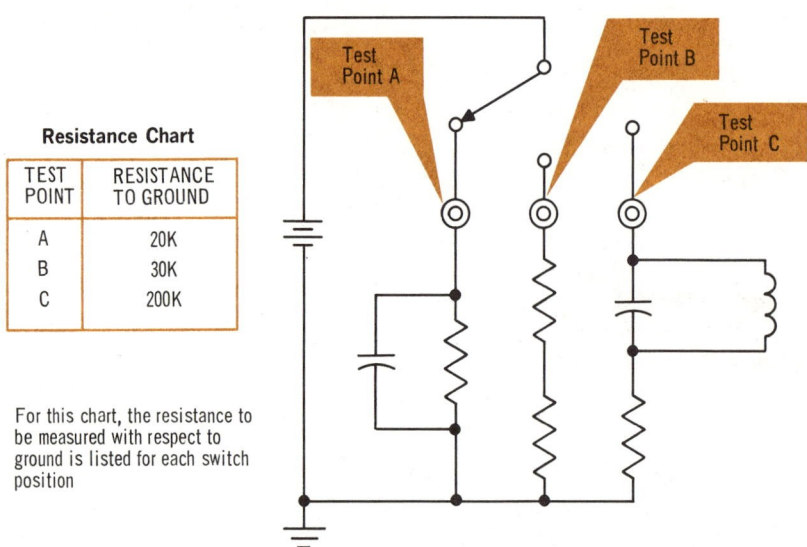

Resistance Chart

TEST POINT	RESISTANCE TO GROUND
A	20K
B	30K
C	200K

For this chart, the resistance to be measured with respect to ground is listed for each switch position

If you do not have resistance charts available, be very careful to make sure that other circuit parts are not connected in parallel with the part. You can check this by examining the schematic and wiring diagram for the particular equipment. If other parts are in parallel with the part you are measuring, you can isolate the part by opening one (or more, if necessary) of its connections in the circuit.

the wheatstone bridge

When extremely accurate resistance measurements are required, a Wheatstone bridge is used. A Wheatstone bridge consists of four resistors connected in a diamond-shaped array. One of the resistors is the unknown one to be measured. A current source is connected to two opposite junctions, and a sensitive galvanometer is connected between the other two junctions. The galvanometer has a center zero reading.

Basic Wheatstone Bridge

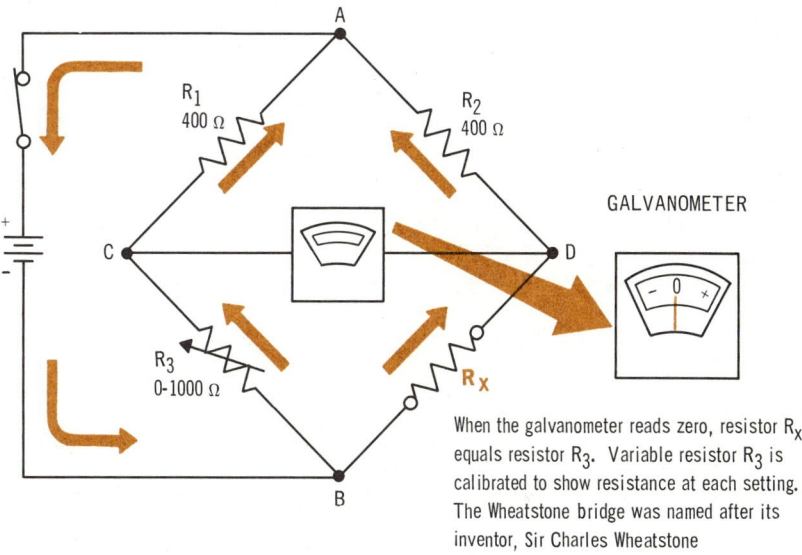

When the galvanometer reads zero, resistor R_X equals resistor R_3. Variable resistor R_3 is calibrated to show resistance at each setting. The Wheatstone bridge was named after its inventor, Sir Charles Wheatstone

To understand how a Wheatstone bridge measures resistance, assume that resistors R_1 and R_2 each equal 400 ohms, and resistor R_3 is variable from 0 to 1000 ohms. Now connect unknown resistor R_X into the bridge circuit and close the switch. You can see that R_1 and R_3 form one divider network, and R_2 and R_X form another divider network. Therefore, since R_1 equals R_2, if R_3 is made equal to R_X, the current and voltage drops in both dividers will be identical. And the potentials at points C and D will be the same, so that no current will flow through the meter. Therefore, when R_3 is adjusted for a zero reading, you know its value equals R_X. Variable resistor R_3 is calibrated to show its exact resistance when adjusted. Therefore, its setting is also the value of unknown resistance R_x. Usually, the Wheatstone bridge contains many components so that different values of R_1, R_2, and R_3 can be switched in to test a wide range of resistances accurately.

capacitor and inductor bridges

The Wheatstone bridge can also be used to measure unknown values of capacitors and inductors in the same way as it does for resistors. However, since capacitors and inductors are reactive devices, an a-c source must be used.

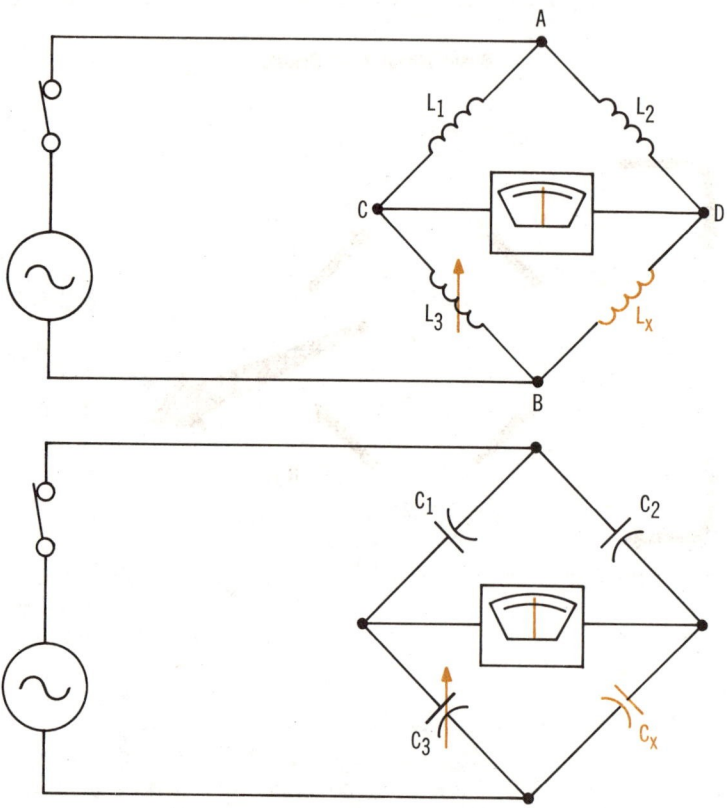

A Wheatstone bridge used with inductors or capacitors works the same as it does with resistors, except that an a-c source must be used

The same diamond-shaped array is used for the bridge, and a sensitive galvanometer is connected between the same opposite junctions. The a-c currents produced by the capacitive or inductive reactances in each leg will be the same when L_1 or C_1 equals L_x or C_x, as the case may be. This will cause points C and D to be at the same potential, and zero current will flow through the galvanometer. The calibrated setting of L_3 or C_3 will then show the value of L_x or C_x.

the megger

Another common resistance-measuring device is the *megger,* which is also called a *megohmmeter* or *megohmer.* The megger measures very high resistances, such as those found in cable insulation, between motor or transformer windings, and so on. These resistances usually range from several hundred to several thousand megohms, and are too high to be measured by an ohmmeter or Wheatstone bridge.

A megger consists basically of a hand crank, a gear box generator, and a meter. When the crank is turned, gears turn the generator at a high speed to generate a voltage of 100, 500, 1000, 2500, or 5000 volts, depending on which megger is used.

The meter is similar to the moving-coil meter that you examined, except that it has two windings. One winding (A) is in series with resistor R_2 across the output of the generator. This winding is wound so that it causes the pointer to move toward the high-resistance end of the scale when the generator is in operation. The other winding (B) is in series with resistor R_1 and the unknown resistance (R_X) to be measured. This winding is wound so that it causes the pointer to move toward the low- or zero-resistance end of the scale when the generator is in operation.

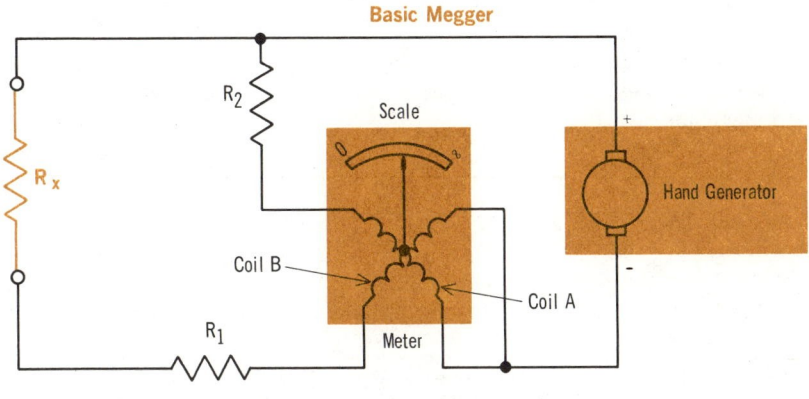

Meggers can measure resistance ranging
from hundreds to thousands of megohms

When an extremely high resistance appears across the input terminals of the megger, such as in the case of an open or near open circuit, current through coil A causes the pointer to read infinity. On the other hand, when a relatively lower resistance appears across the input terminals, current through coil B causes the pointer to deflect toward zero. The pointer stops at a point on the scale determined by the current through coil B, which is governed by R_X.

summary

☐ An ohmmeter is used to measure resistance. It consists of a d-c current meter movement, a low-voltage d-c power source, and current-limiting resistors. ☐ In a series ohmmeter, the resistance to be measured is connected in series with the movement. ☐ With zero resistance present between the input terminals of a series ohmmeter, the pointer deflects full scale. The greater the resistance between the terminals, the less the pointer deflects. ☐ The scale of a series ohmmeter is nonlinear, with the resistance values becoming progressively more crowded at the left side of the scale.

☐ In a shunt ohmmeter, the resistance to be measured is connected in parallel with the movement. The meter behaves like a current meter with a shunt resistor. ☐ In the shunt ohmmeter, the larger the resistance being measured, the more the pointer deflects. ☐ Shunt ohmmeters have nonlinear scales, although not as nonlinear at the low-resistance end of the scale as series ohmmeters. ☐ A multirange ohmmeter uses a different value of source voltage for each range. ☐ Above a certain range, more sensitive meter movements must be used. ☐ For measuring very low resistances, shunt resistors are used across the meter movement and the current-limiting resistors.

☐ Before using an ohmmeter, the circuit under test should be deenergized, and all large capacitors discharged. ☐ A Wheatstone bridge is used for making extremely accurate resistance measurements. It consists of a current source, a galvanometer, and four resistances, one of them the unknown, connected in a diamond-shaped array. ☐ The Wheatstone bridge can also be used to measure capacitors or inductors in the same way. ☐ A megger is used for measuring very high resistances, usually from several hundred to several thousand megohms.

review questions

1. What are the basic parts of an ohmmeter?
2. What is the basic difference between series and shunt ohmmeters?
3. Draw the circuits of simple series and shunt ohmmeters.
4. What type of ohmmeter deflects full scale when measuring zero resistance?
5. What is the purpose of the zero-ohms set control?
6. How is the range extended in multirange series ohmmeters?
7. Why should a circuit be deenergized before resistance measurements are made on it?
8. What are Wheatstone bridges used for?
9. Draw a schematic diagram of a simple Wheatstone bridge.
10. What is a *megger*?

calculating power

If you wish to find the power dissipated in an electrical load, measure any two of the three basic electrical quantities examined: *current, voltage,* and *resistance*. For example, you will recall that power can be calculated by multiplying voltage by current: $P = EI$. Therefore, if you use a voltmeter to measure the voltage across a load, and a current meter to measure the current flowing through the load, insert these values into the power equation. Similarly, you can measure the current through the load and its resistance, and then calculate power by $P = I^2R$. Or you can measure the voltage across the load and its resistance, and then calculate power by $P = E^2/R$.

Voltage and Current Known   **Current and Resistance Known**

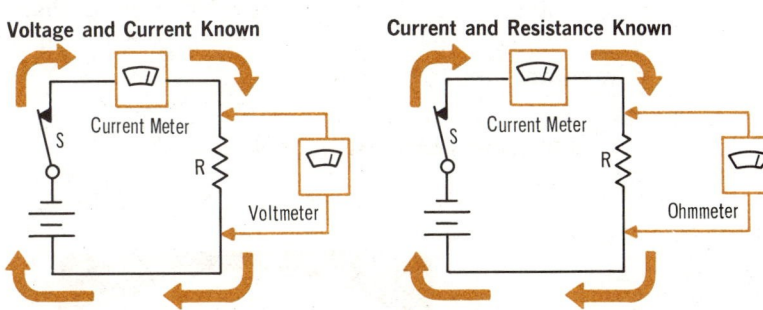

Power dissipated by R can be found by measuring current and voltage and inserting into: $P = EI$

Power dissipated by R can be found by first measuring current with switch closed, and then resistance of R with switch open, and inserting into: $P = I^2R$

Voltage and Resistance Known

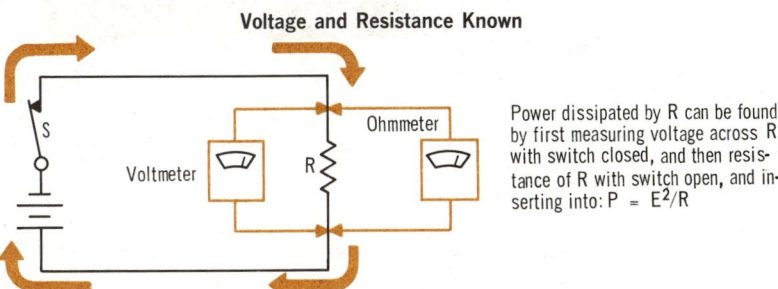

Power dissipated by R can be found by first measuring voltage across R with switch closed, and then resistance of R with switch open, and inserting into: $P = E^2/R$

In actual practice, however, it is often unnecessary to measure two quantities. Usually one, and sometimes two, are known. For example, usually the voltage across a load is known; therefore, all that is necessary is to measure the resistance or the current. Other times, both load voltage and resistance are known; in this case, no measurements are necessary and power can be calculated by $P = E^2/R$.

wattmeters

Rather than perform one or two measurements and then calculate power, a power-measuring meter, called a *wattmeter*, can be connected into a circuit to measure power. The power dissipated can be read directly from the scale of this meter. Not only does a wattmeter simplify power measurements, but it has two other advantages over the previous method of measuring power.

In Volume 3, you learned that often voltage and current in an a-c circuit are not in phase; current sometimes leads or lags voltage (power factor). When this occurs, you learned that simply multiplying voltage by current results in *apparent power* and not *true power*. Therefore, in an a-c circuit, measuring voltage and current and then multiplying one by the other can often result in an incorrect value of power dissipation. However, the wattmeter takes the power factor of a circuit into account, and always indicates true power.

A Wattmeter Always Measures True Power

Some wattmeters are compensated for their own power dissipation so that very accurate power measurements can be obtained

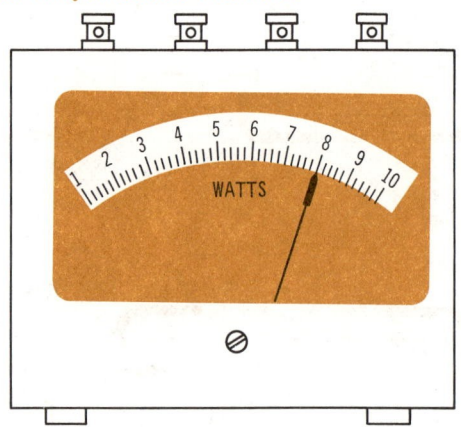

If a wattmeter is not compensated, its power dissipation can easily be determined, resulting in very accurate power measurements

Voltmeters and current meters, themselves, consume power. The amount of power consumed depends upon the levels of the voltage and current in the circuit, and cannot be predicted accurately. Therefore, very accurate power measurements cannot be made by measuring voltage and current, and then calculating power. However, some wattmeters are compensated for their own power losses so that only the power dissipated in the circuit is measured. If the wattmeter is not compensated, the power that is dissipated is sometimes marked on the meter or else can easily be determined so that a very accurate reading can be obtained. Typically, the accuracy of a wattmeter is within 1 percent.

the basic wattmeter

A basic wattmeter consists of two stationary coils connected in series and one moving coil. The moving coil, wound with many turns of fine wire, has a high resistance. The stationary coils, wound with a few turns of a larger wire, have low resistance. For power measurements, the moving coil is connected across the load to the source voltage, which determines the current through the coil and, therefore, the strength of the magnetic fields around it. The stationary coils are connected in series with the load, and the load current builds up magnetic fields about the stationary coils.

The interaction of the two magnetic fields will cause the moving coil and its pointer to rotate in proportion to the voltage across the load and the current through the load. Therefore, the meter indicates E times I, which is power dissipation.

When using a wattmeter, do not exceed its voltage and current ratings. Be careful in interpreting these ratings. For example, a wattmeter with a full-scale reading of 500 watts may be rated at 150 volts and 5 amperes (150 volts × 5 amperes = 750 watts). If the wattmeter is connected into a circuit with 150 volts and 5 amperes, and the circuit has a power factor close to 1.0, then the meter pointer would move off scale to the right, and might bend about the retaining pin. Most wattmeters are rated in this manner because a-c circuits usually have a power factor less than 1.0, and, therefore, the power measured will be less than E × I.

Basic Wattmeter

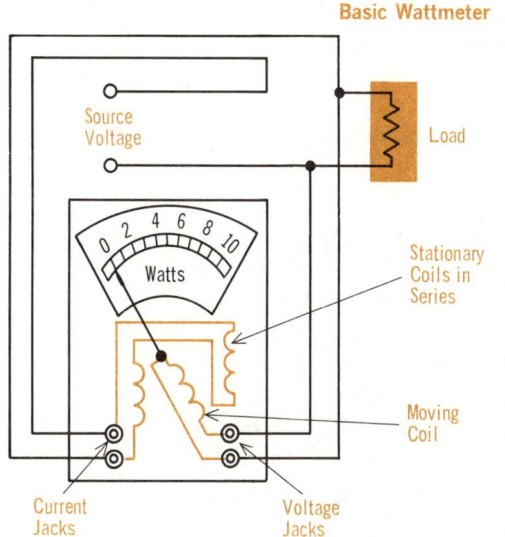

By applying the voltage across the load to the moving coil, and the current through the load to the stationary coil, the voltmeter pointer deflection is proportional to voltage and current

Therefore, since power is proportional to voltage and current, the power dissipated in the load can be read directly from the wattmeter scale

checking wattmeter power loss

The stationary (current) coils and the moving voltage coil of a watt-meter have resistance, resulting in some circuit power loss by the wattmeter. Unless this power loss is considered, incorrect power readings will result.

Some wattmeters are compensated, that is, the power loss was corrected for so that it can be ignored when the meter is used. Many watt-meters that are not compensated have their power loss indicated on the meter itself or in the manufacturer's data. When using a wattmeter of this type, the power loss indicated on the meter, or in the manufacturer's data, must be *subtracted* from the wattmeter scale reading to obtain the true power dissipated by the load.

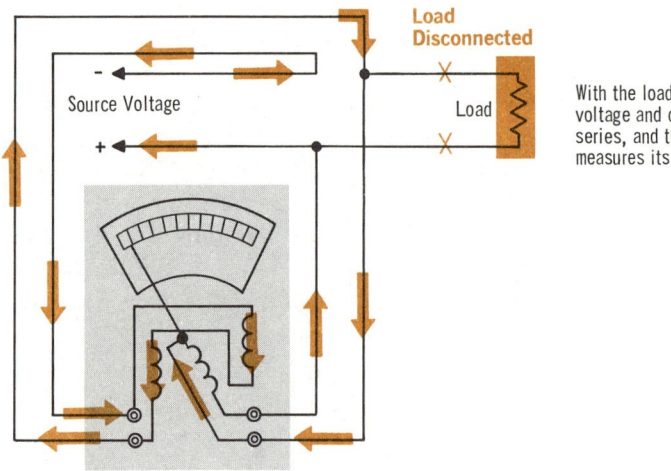

With the load disconnected, the voltage and current coils are in series, and the wattmeter measures its own power loss

Other wattmeters that are not compensated do not have their power loss indicated on the meter or in the manufacturer's data. In this case, the power loss must be determined by simply disconnecting the load from the circuit, but leaving the wattmeter connected. With the load disconnected, the voltage coil is in series with the current coil and the source current flows through both coils. In a sense, the d-c resistance of the coils becomes the load, and since it is this d-c resistance that causes the power loss, the wattmeter will read its own power loss.

This method is used to check the accuracy of a compensated watt-meter or the power loss indicated on the meter or in the manufacturer's data. For accurately compensated meters, the meter reads zero. For accurate ratings and manufacturer's data, the meter reads the power loss of the rating or the data.

the basic multimeter

Probably the three most often-measured electrical quantities are current, voltage, and resistance. You have learned that current can be read on a current meter, voltage on a voltmeter, and resistance on an ohmmeter. But, in many cases, it is impractical, and sometimes almost impossible, for an engineer or technician to carry all of the meters necessary to measure these three quantities. For example, it would be difficult for a field engineer or field technician to carry these separate meters along with all of his other tools and spare parts. There are many other instances where having to use separate current meters, voltmeters, and ohmmeters would be impractical. To overcome this problem, the multimeter was developed.

This completely self-contained multimeter measures voltage, current, and resistance

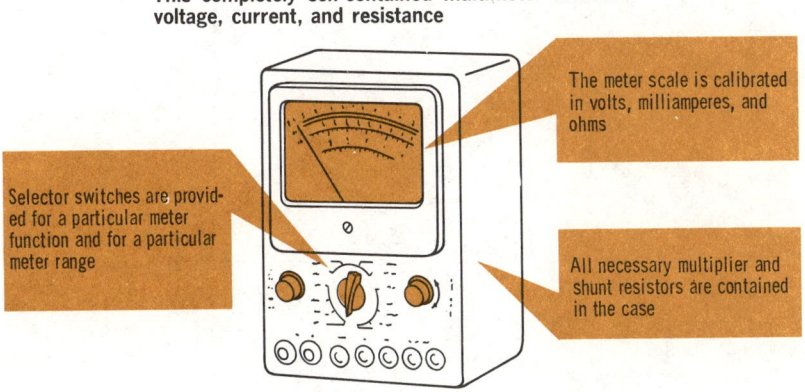

The meter scale is calibrated in volts, milliamperes, and ohms

Selector switches are provided for a particular meter function and for a particular meter range

All necessary multiplier and shunt resistors are contained in the case

The multimeter is probably the piece of test equipment that is used most often in the electrical and electronic industries

Basically, a multimeter consists of a voltmeter, an ohmmeter, and a current meter contained in one case. The meter circuits are almost identical to the ones you studied earlier in this book. However, a multimeter uses a single meter movement, with a scale calibrated in volts, ohms, and milliamperes. The necessary multiplier resistors and shunt resistors are all contained within the case. Front panel selector switches are provided to select a particular meter function and a particular range for that function.

On some multimeters, two switches are used to select a function and a range, and on other multimeters, only one switch is used. Some multimeters do not have switches for this purpose; instead, they have separate jacks for each function and range.

scale and ranges

The voltage, current, and resistance circuits of a multimeter are essentially the same as the ones you studied earlier in this book. A multimeter is basically a multirange voltmeter, multirange current meter, and a multirange ohmmeter combined in one case. Switching circuits or separate jacks are provided to select the proper function and the proper range.

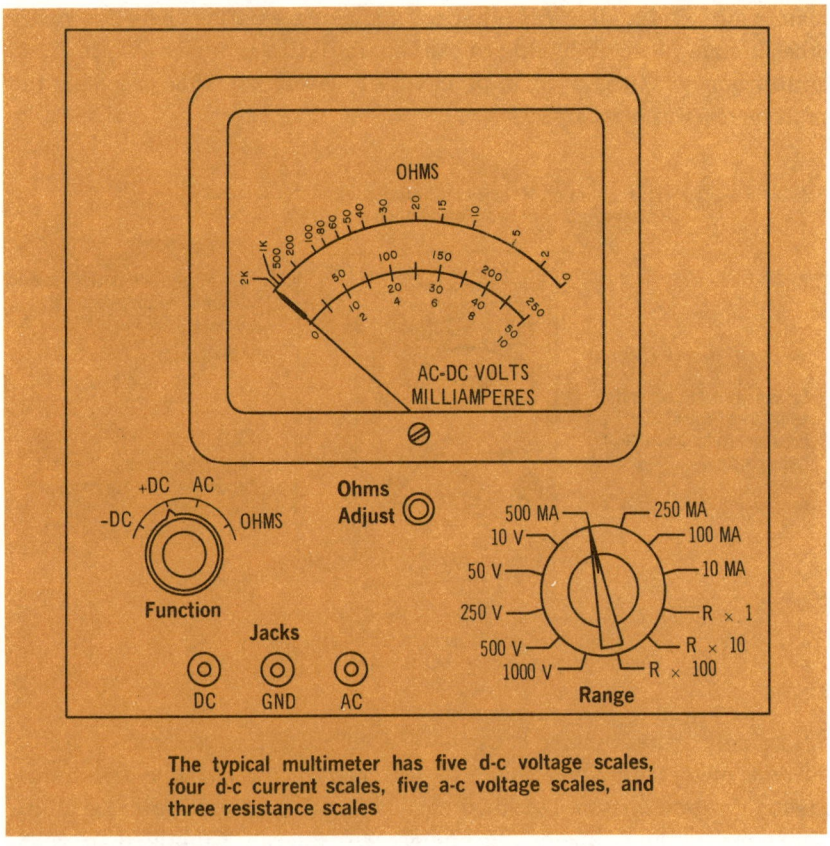

The typical multimeter has five d-c voltage scales, four d-c current scales, five a-c voltage scales, and three resistance scales

Most multimeters have three scales, one calibrated in resistance, one in voltage, and one in current. A typical multimeter may have two selector switches: one to set the circuits to measure either d-c current or voltage, a-c current or voltage, or resistance; and the other to select the range of the quantity to be measured. Sometimes only one switch is used to select the function and range, and other times jacks are provided for this purpose.

meter relays

A *meter relay* is a special type of meter that not only allows the current to be read but also provides built-in contacts so that it can function like a relay when the pointer indicates certain current values. Generally this kind of a relay is used where it is desirable to keep a certain current flow safely within a particular operating range, Two *control pointers* generally exist on the scale face along with the meter pointer. Knobs are available to set the control pointers at any values. One is usually set at the low point of the range, and the other is set at the high point of the range. In normal circumstances, the meter pointer will indicate a current between the two control pointers. If the current changes—goes up or down—the meter pointer will swing toward a control pointer. When it reaches the control pointer, contacts are closed to produce some action, such as setting off an alarm.

There are two types of meter relays: the *contact* type and the *contactless* type. With the contact type, there is generally some direct physical action between the meter and control pointers to close the contacts. This is not usually desirable since the control pointers must then interfere with the free swing of the meter pointer.

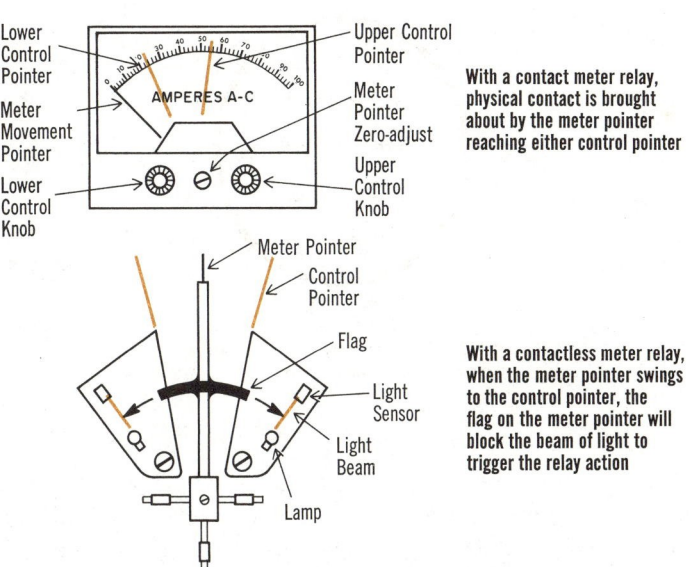

Lower Control Pointer — Upper Control Pointer — Meter Pointer — Meter Movement Pointer — Zero-adjust — Lower Control Knob — Upper Control Knob

With a contact meter relay, physical contact is brought about by the meter pointer reaching either control pointer

Meter Pointer — Control Pointer — Flag — Light Sensor — Light Beam — Lamp

With a contactless meter relay, when the meter pointer swings to the control pointer, the flag on the meter pointer will block the beam of light to trigger the relay action

With the contactless type of meter relay, the close proximity of the meter pointer to the control pointer generally blocks off a beam of light to trigger an electronic relay circuit. There is no physical contact between the meter and control pointers.

the basic electronic multimeter

You learned that the higher the ohms/volt rating of a voltmeter, the less the voltmeter will upset circuit conditions. And the less circuit conditions are upset, the more accurate will be the reading that you obtain. Most of the better voltmeters and multimeters available today are rated at about 20,000 ohms/volt. In some of the very high-resistance circuits found in some present-day equipment, however, even a meter rated at 20,000 ohms/volt will greatly upset circuit conditions, and result in an incorrect reading. To overcome this problem, a device with a very high ohms/volt rating called an *electronic multimeter*, or more commonly, a *vacuum-tube voltmeter* (VTVM), was developed.

A typical VTVM has an ohms/volt rating of 11 megohms. Because of this high rating, the VTVM draws an extremely small current from the circuit under test, and, therefore, has a small effect on circuit conditions. Consequently, the VTVM provides much more accurate voltage readings in high-resistance circuits than do regular voltmeters and multimeters.

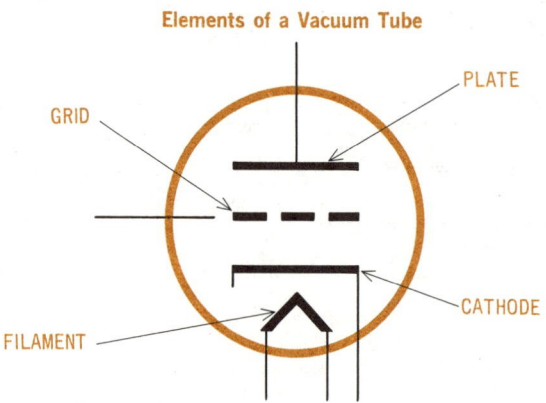

Elements of a Vacuum Tube

The vacuum tube consists of three basic elements: the plate, grid, and cathode. In addition, it contains a filament to heat the cathode

As its name implies, the VTVM makes use of vacuum tubes for operation, and to understand how it works requires a knowledge of these tubes. The most common VTVM is the d-c type. Even though it is a d-c VTVM, it also measures a-c voltage and resistance. The input circuit of a basic VTVM contains a *triode amplifier vacuum tube*, which has three major elements: the *plate*, the *grid*, and the *cathode*. In addition, it contains a *filament* to heat the cathode. However, electronic multimeters can use transistors in place of tubes.

the basic vacuum-tube voltmeter

When the cathode of the triode amplifier vacuum tube is heated, it gives off electrons that are drawn to the plate because the plate is connected to a positive voltage source. The voltage on the grid, which is determined by the grid resistor, controls the number of electrons, or current, that flows through the tube to the plate. Notice that there is a d-c current meter, calibrated in volts, connected to the plate. With only the grid resistor connected into the grid circuit, a certain voltage will be present on the grid, and, therefore, a certain plate current will flow. No matter what this value is, the meter can be adjusted to read zero at that value.

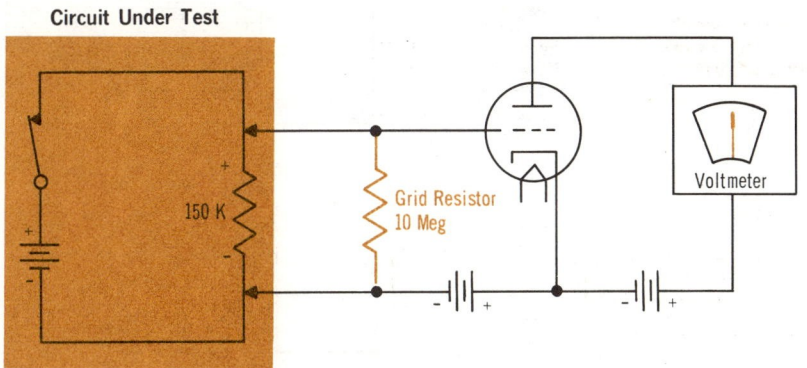

Because this VTVM has a 10-megohm grid resistor across its input, connecting the VTVM across the 150K-ohm resistor will not seriously upset the conditions of the circuit under test (10 megohms in parallel with 150K is effectively 150K)

When the meter probes are connected across a load whose voltage drop is to be measured, that voltage is applied to the grid circuit. This causes the grid voltage to decrease, and, therefore, the plate current will increase. This increase in plate current causes the meter pointer to deflect by an amount proportional to the voltage measured, and the voltage can be read directly on the meter scale.

The fact that the grid resistor of the tube can be made very large, say 10 to 15 megohms, means that the grid circuit will have a minor effect on the circuit under test. This is the major advantage of the VTVM over ordinary voltmeters and multimeters.

The basic VTVM shown is simplified. Actual VTVM's use more than one tube, and a bridge meter circuit for more sensitivity and accuracy.

measuring a-c voltage and resistance

Like the rectifier voltmeter discussed earlier, the d-c VTVM can be used to measure a-c voltage simply by using a rectifier in the meter to convert the a-c voltage to a d-c voltage before applying it to the meter movement. Furthermore, the VTVM can be used to measure a-c voltage at much higher frequencies than the rectifier voltmeter. Some VTVM's can measure a-c voltages up to 250 MHz. To do this, however, a special meter probe called a *radio-frequency* (r-f) *probe* must be used with the VTVM. This probe contains a special crystal rectifier specifically designed to convert very high a-c frequencies to d-c.

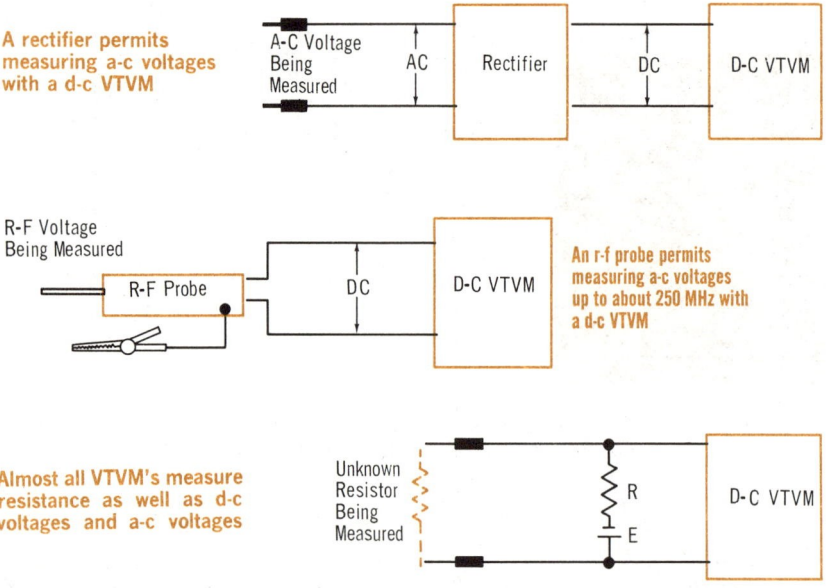

A rectifier permits measuring a-c voltages with a d-c VTVM

An r-f probe permits measuring a-c voltages up to about 250 MHz with a d-c VTVM

Almost all VTVM's measure resistance as well as d-c voltages and a-c voltages

Almost all VTVM's are designed to measure resistance as well as voltage. Just like the basic ohmmeter circuits studied previously, a battery is required in the ohmmeter circuit of a VTVM. In the ohmmeter circuit of the VTVM, a resistor and a battery in series are connected across the test probes. When the test probes are connected across an unknown resistance, current flows through the ohmmeter circuit. This produces a voltage across the unknown resistance, and this voltage is measured by the basic VTVM circuit just as it is measured as described on page 5–119. The value of the unknown resistance is read off the part of the VTVM scale that is calibrated in ohms.

special meters

Most of the information dealt with in this text has been for general-purpose meters, those that can be used in a wide variety of current, voltage, resistance, reactance, and wattage measurements. There are many special-purpose meters that operate similarly to these, but are specially made or have electronic circuitry to allow them to make special electrical or electronic measurements. Some of these are:

Frequency meters to determine the frequency of an a-c signal.

Volume unit or *decibel meters* to indicate sound loudness.

Field-strength meters to determine the strength of radio waves at certain locations.

Null meters, designed specifically to be most accurate at the zero or minimum reading position, which is at center scale.

Primary or *secondary standard meters*, which are very carefully made and well built, and are usually hand calibrated to give reading accuracies which can be used as standards to calibrate ordinary meters.

Recording meters, whose pointer is an ink stylus that writes on a moving sheet of graph paper to record voltage or current fluctuations over any given time period.

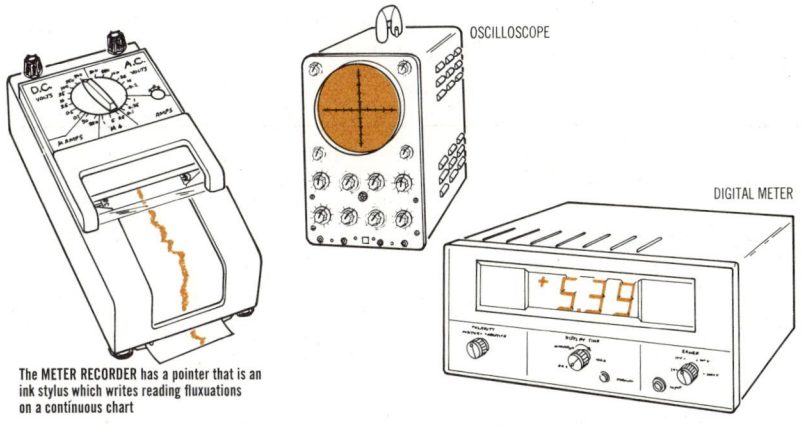

OSCILLOSCOPE

DIGITAL METER

The METER RECORDER has a pointer that is an ink stylus which writes reading fluxuations on a continuous chart

Solid-state meters are electronic multimeters, but use transistors and other semiconductors instead of electron tubes.

Digital meters, which use electronic circuits to produce number displays instead of a pointer and scale to give the reading.

Oscilloscopes, which use a TV-like screen to show what the signal looks like so it can be measured more accurately.

summary

☐ Wattmeters measure power directly. ☐ A wattmeter takes the power factor of a circuit into consideration, and always indicates true power. ☐ Some wattmeters are compensated for their own power loss, and therefore indicate only the power dissipated in the circuit under test. ☐ A basic wattmeter consists of two stationary coils connected in series, and one moving coil. When the meter is connected into a circuit, the moving coil is placed across the source voltage and the stationary coils are placed in series with the load. ☐ If a wattmeter is not compensated, its own power loss must be subtracted from any indicated reading to obtain the actual power measured.

☐ A multimeter consists of a voltmeter, ohmmeter, and current meter contained in one case. ☐ A multimeter has only a single meter movement. The necessary shunts, multiplier resistors, or current-limiting resistors are connected into the circuit by front panel switches.

☐ Electronic multimeters have very high ohms/volt ratings. Because of this, they provide more accurate voltage readings in high-resistance circuits than do regular voltmeters. ☐ Vacuum-tube voltmeters use vacuum tubes. They also use a bridge meter circuit for more sensitivity and accuracy. ☐ VTVM's can be used to measure a-c voltage by having a rectifier in the meter to convert the a-c voltage to d-c. ☐ When a VTVM is used to measure high-frequency voltages, a special high-frequency probe must be used. ☐ Most VTVM's are designed to measure resistance as well as voltage. ☐ There are electronic multimeters that use semiconductors instead of tubes.

---

review questions

1. How else can power be measured, besides by use of a wattmeter?
2. What are the advantages of using a wattmeter to measure power?
3. What is meant by a *compensated wattmeter*?
4. Draw the schematic diagram of a basic wattmeter.
5. If the wattmeter is not compensated and its power loss is not indicated on the scale, how can its power loss be determined?
6. What is a *multimeter*?
7. What is a *VTVM*?
8. What is the advantage of an electronic multimeter over a regular voltmeter?
9. What is an *r-f probe*?
10. What else can a VTVM usually measure besides voltage?

electricity
six

power sources

In previous volumes of this series, you studied what electricity is, how it is put to practical use in electrical circuits, and how it is measured. You will now learn about devices that produce electricity. These devices are called *power sources.*

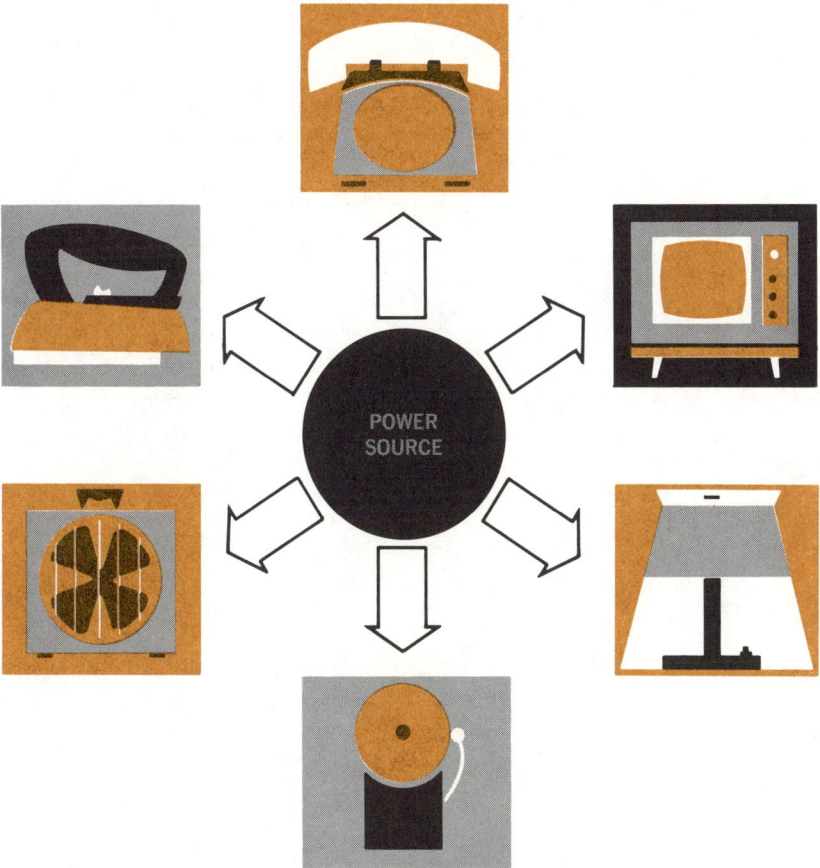

The power source supplies the electricity
needed to operate electrical equipment

When a load is connected to a power source, electric current flows from the source to the load. The power source must supply the amount of voltage and current that the load needs. Any load, such as a lamp or motor, can only work as well as the power source will permit.

BATTERY

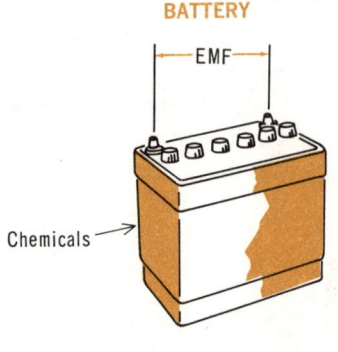

The battery converts chemical energy
to build up a difference of potential

THERMOCOUPLE

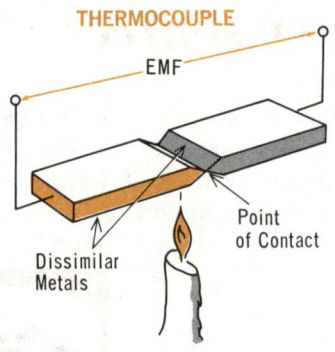

The thermocouple converts heat energy
to develop an electromotive force

PHOTOVOLTAIC CELL

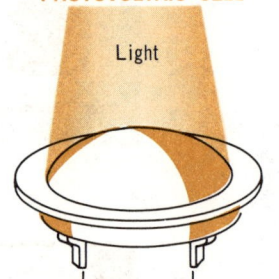

The voltaic cell converts light energy
to develop a voltage

GENERATOR

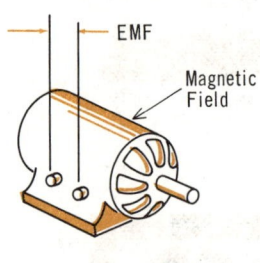

The generator converts magnetic energy
to induce an emf

types

Power sources produce electricity by converting some other form of energy into electrical energy. Power sources supply electrical energy by building up opposite electrical charges on two *terminals*. The *difference of potential*, or *electromotive force* (emf), between these terminals moves the electric current through the load that is connected across the source. The most common power sources you will encounter are the *battery, thermocouple, photovoltaic cell,* and *generator*. They all supply an electric voltage and current, but each converts different forms of energy. The *battery* converts *chemical energy*, the *thermocouple* converts *heat energy*, the *photovoltaic cell* converts *light energy*, and the *generator* converts *magnetic energy*.

the battery

The battery is one of the most important power sources in use today because its energy is *self-contained.* This is one advantage that none of the other power sources have. All the power sources must first be supplied with outside energy, such as heat, light, or mechanical energy, before they can produce electricity. However, the electrical energy of the battery is produced by the chemical energy contained within the battery.

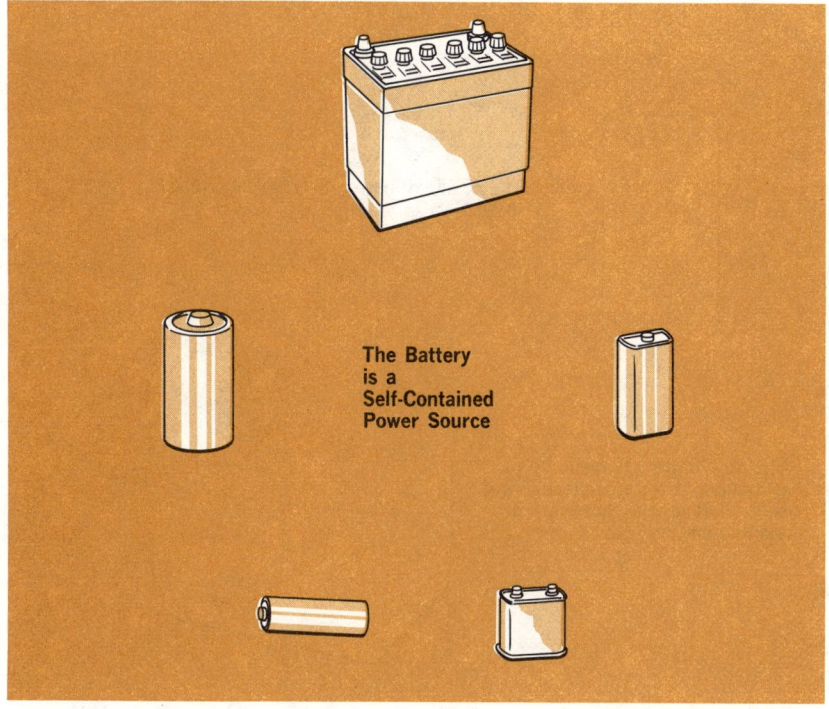

The Battery
is a
Self-Contained
Power Source

The battery is used mostly wherever a *portable* power source is required, such as in flashights and lanterns, in the automobile electrical system, in photography to ignite flash bulbs and photoflash lamps, and in portable radios, measuring instruments, and hearing aids to supply the electronic circuits; they are also used in trains, planes, and ships, toys, special clocks and watches, etc. In fact, the battery is the most versatile power source in use today. The battery is also used as a calibrated source of voltage in a voltage *standard.* Typical of this is the standard cell used by the National Bureau of Standards to establish other units of electricity, such as the ohm and the ampere.

types of batteries

Basically, batteries are classified as *primary* or *secondary* according to the manner in which their chemical energy is converted into electrical energy. The *primary battery* converts chemical energy to electrical energy *directly*, using the chemical materials within the cell to start the action. The *secondary battery* must first be *charged* with electrical energy before it can convert chemical energy to electrical energy. The secondary battery is frequently called a *storage battery,* since it stores the energy that is supplied to it.

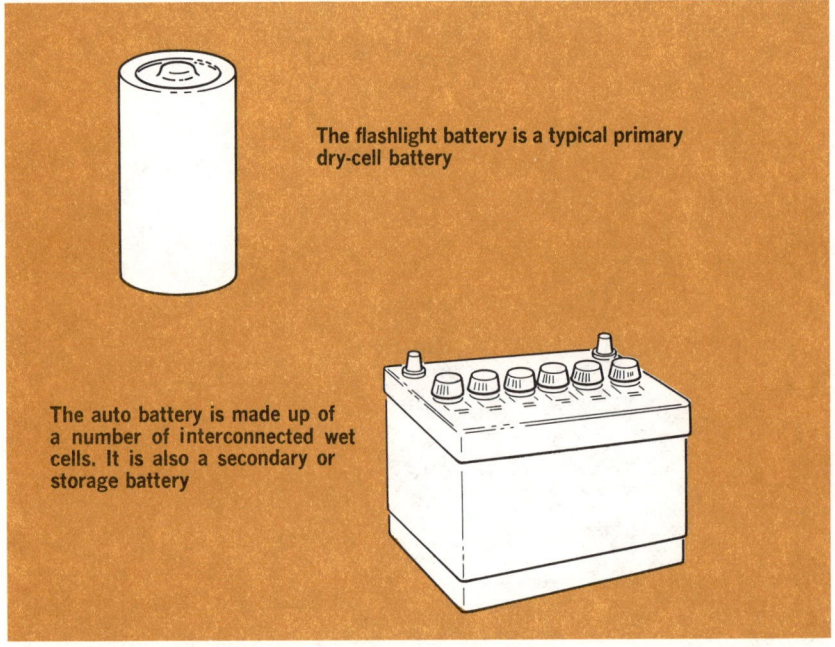

The flashlight battery is a typical primary dry-cell battery

The auto battery is made up of a number of interconnected wet cells. It is also a secondary or storage battery

Batteries are also classified as wet cells or dry cells. The *wet* cell battery uses *liquid* chemicals, while the so-called *dry* cell contains a chemical paste. The *cell* is the basic unit of a battery. A *battery* often consists of a number of cells connected to supply a voltage or current greater than that of a single cell. However, the terms cell and battery are now frequently used interchangeably.

The primary battery is used mostly where a limited current is required. The primary batteries you will probably see most commonly are the dry cells. The secondary battery is generally used where a heavy current is required; secondary batteries are usually wet cells.

early history

Although the discovery of electricity dates back about 2500 years to the ancient Greeks, very little progress in the science of electricity was made until the basic cell was discovered in the late eighteenth century. Up to that time, there was no *convenient* source of electrical energy.

The action of the basic cell was first noticed by Luigi Galvani in 1791, while he was preparing an experiment in anatomy. For the experiment, Galvani had removed dissected frog legs from a salt solution and suspended them by means of a copper wire. He noticed that each time he touched one of the legs with an iron scalpel, the muscles of the leg twitched. Galvani realized that electricity was being produced, but he thought it came from the leg muscles.

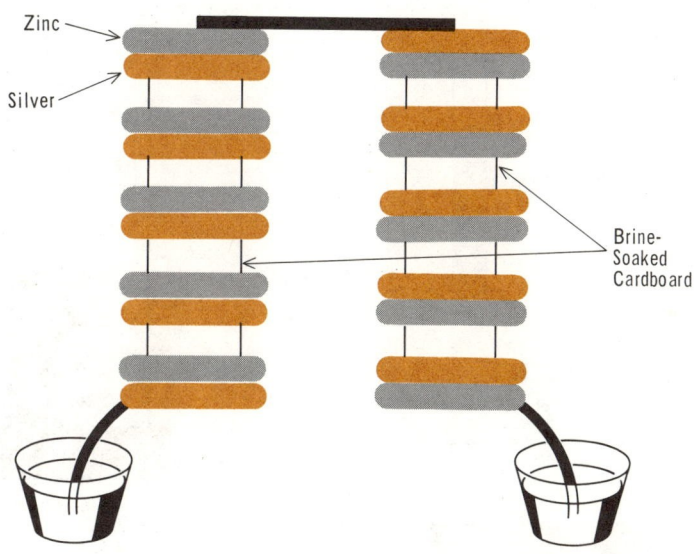

The voltaic pile uses an alternating stack of silver and zinc discs separated by discs of cardboard soaked with a salt solution. This was the first battery

In 1800, Alessandro Volta repeated the experiment, and found that the muscles of the frog did not produce the electricity. Instead, he discovered that the electricity was the result of a chemical action between the copper wire, iron scalpel, and salt solution. Using this knowledge, he built the first practical electric battery, which is known as the *voltaic pile*.

the basic primary wet cell

After Alessandro Volta made his first battery, he continued to experiment with metals and chemicals. He found that by putting two *different metals* in certain chemical solutions, electricity could still be produced. This is the basic primary wet cell. In his honor, it is usually called the *voltaic cell;* however, it is also sometimes referred to as the *galvanic cell,* in honor of Galvani.

The metals in a cell are called the *electrodes,* and the chemical solution is called the *electrolyte.* The electrolyte reacts oppositely with the two different electrodes. It causes one electrode to lose electrons and develop a *positive charge;* and it causes the other electrode to build up a surplus of electrons and develop a *negative charge.* The difference in potential between the two electrode charges is the cell *voltage.*

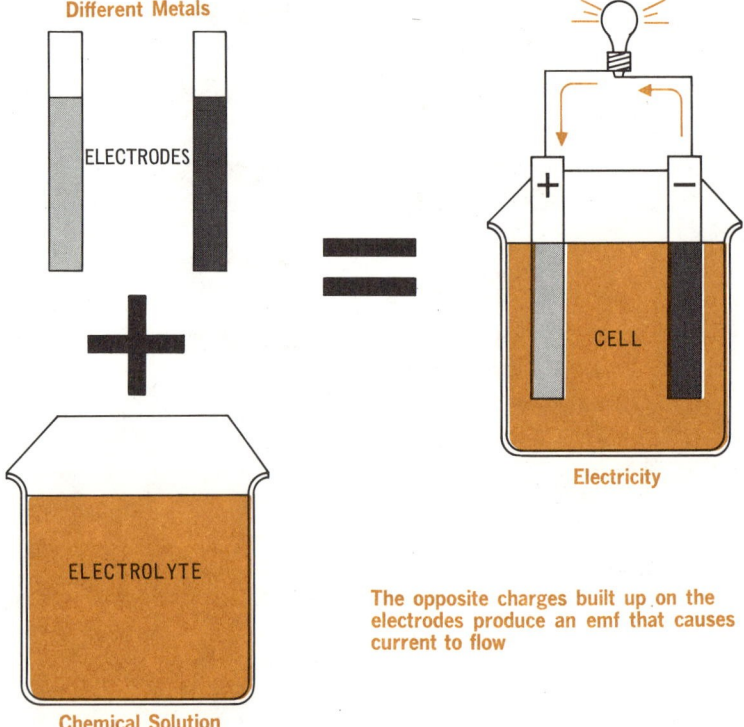

Different Metals

ELECTRODES

CELL

Electricity

ELECTROLYTE

The opposite charges built up on the electrodes produce an emf that causes current to flow

Chemical Solution

In the basic battery you will study on the following pages, copper and zinc will be used for the electrodes, and sulfuric acid mixed with water will be the electrolyte. Actually, as you will learn later, a number of different metals and chemicals can be used.

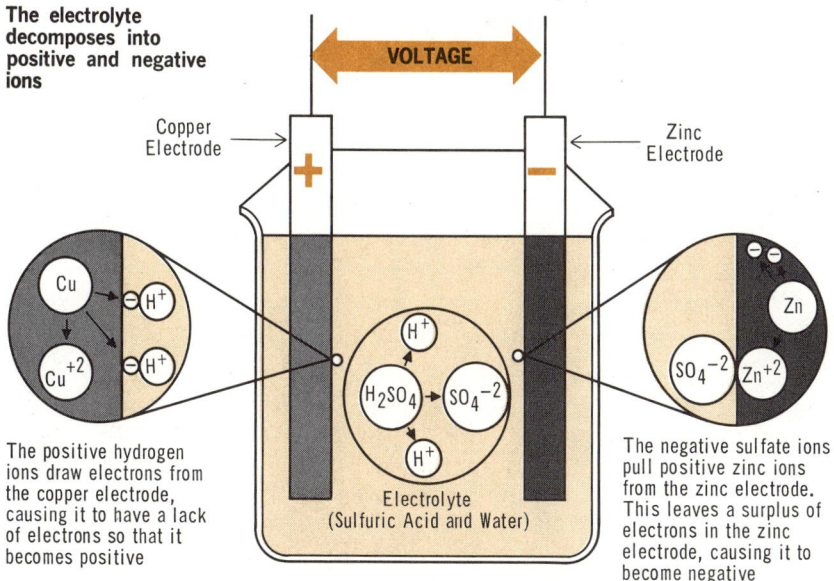

The electrolyte decomposes into positive and negative ions

VOLTAGE

Copper Electrode

Zinc Electrode

The positive hydrogen ions draw electrons from the copper electrode, causing it to have a lack of electrons so that it becomes positive

Electrolyte (Sulfuric Acid and Water)

The negative sulfate ions pull positive zinc ions from the zinc electrode. This leaves a surplus of electrons in the zinc electrode, causing it to become negative

A voltage is developed between the two charged electrodes

developing a voltage

Sulfuric acid mixed with water breaks down into hydrogen ions and sulfate ions. There are two *positive hydrogen ions* produced for each *negative sulfate ion*, but each hydrogen ion has one positive charge (H^+), while each sulfate ion has two negative charges (SO_4^{-2}). Therefore, the entire solution is *neutral*. When the zinc electrode is placed in the solution, the sulfate ions attack the zinc, causing its atoms to give off electrons. The negative sulfate ions attract positive zinc ions (Zn^{+2}) from the electrode, but the electrons released by the zinc atoms remain. So, the *zinc electrode* builds up a *surplus of electrons* and a *negative charge*.

When it develops a sufficient negative charge, the zinc electrode repels the sulfate ions to prevent further activity. The zinc ions then combine with sulfate ions to form neutral zinc sulfate molecules. The electrolyte then has more positive charges than negative charges. As a result, when the copper is placed in the electrolyte, the positive hydrogen ions attract free electrons from the copper. These electrons combine with the hydrogen atoms to neutralize them. This continues until enough hydrogen ions are neutralized to make the electrolyte neutral again. Because of this, the copper electrode develops a *lack of electrons* and a *positive charge*. The *difference in potential* between the positive electrode and the negative electrode produces a *voltage* across the electrodes.

supplying current

The value of voltage developed by the basic cell depends on the materials used for the electrodes and the electrolyte. For the typical zinc-copper cells, this is about 1.08 volts. Once this is reached, the chemical action stops until a load is connected across the electrodes. Then electrons flow from the negative electrode through the load to the positive electrode.

When electrons leave the zinc electrode, its negative charge is reduced, allowing the negative sulfate ions in the electrolyte to again attack the electrode. More zinc atoms release electrons to replenish the supply in the electrode. The new positive zinc ions combine with negative sulfate ions in the electrolyte to again form zinc sulfate. The electrolyte again has an excess of positive hydrogen ions, which are drawn to the copper electrode; they combine with free electrons until the electrolyte is again neutralized.

The action continues in this way: electrons leave the negative zinc electrode and flow through the load to the positive copper electrode; the electrons leaving the zinc are replaced by those that are left behind by the zinc sulfate; and the electrons that enter the copper replace those that leave to neutralize the hydrogen ions. In this manner, the charge on each electrode is kept almost constant, and the terminal voltage remains steady as the cell delivers current.

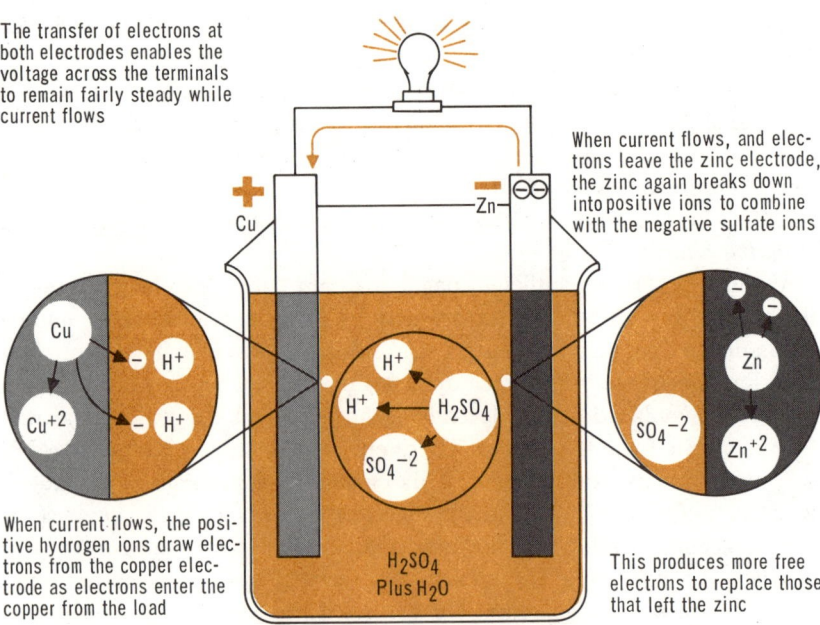

The transfer of electrons at both electrodes enables the voltage across the terminals to remain fairly steady while current flows

When current flows, and electrons leave the zinc electrode, the zinc again breaks down into positive ions to combine with the negative sulfate ions

When current flows, the positive hydrogen ions draw electrons from the copper electrode as electrons enter the copper from the load

This produces more free electrons to replace those that left the zinc

local action

The chemical action in the basic wet primary cell should stop when the cell is not delivering current. This is not always the case, however, because there are some impurities in the electrode material, such as iron (Fe) and carbon (C). These impurities *react* with the zinc and the electrolyte to form many *small cells* that produce local electrical currents about the zinc electrode. This *local action* occurs in the cell whether or not the cell is delivering current. As a result, the electrolyte and electrodes are used unnecessarily, and the battery will not last as long as it should.

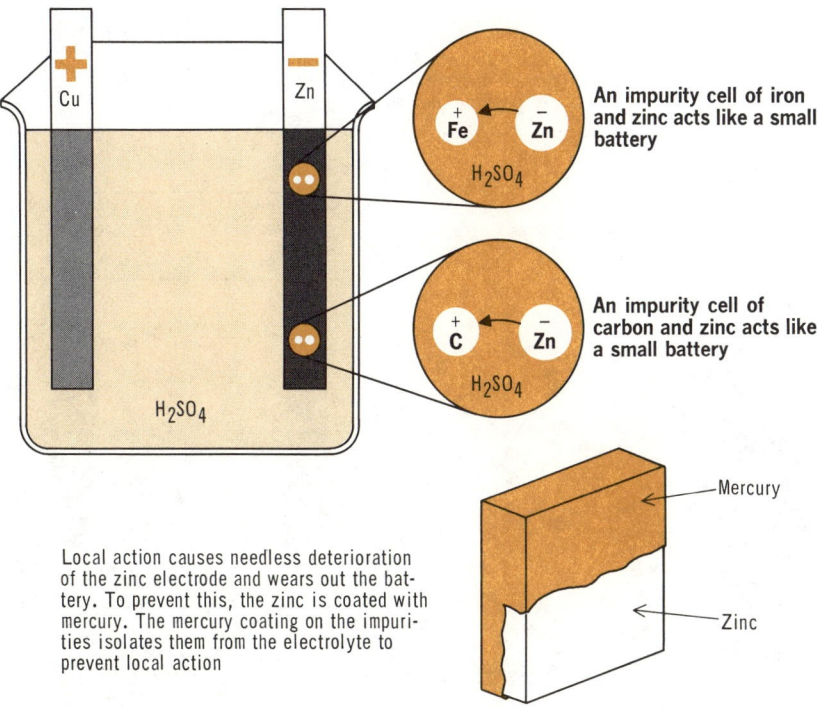

An impurity cell of iron and zinc acts like a small battery

An impurity cell of carbon and zinc acts like a small battery

Mercury

Zinc

Local action causes needless deterioration of the zinc electrode and wears out the battery. To prevent this, the zinc is coated with mercury. The mercury coating on the impurities isolates them from the electrolyte to prevent local action

To reduce the amount of local action that exists, *mercury* is coated on the surface of the zinc electrode. This process is called *amalgamation*. The zinc dissolves in the mercury but is still free to combine with the electrolyte. The impurities, however, do not dissolve in the mercury; instead, they are coated with it, and so are isolated from the electrolyte. Therefore, they cannot combine with the electrolyte to produce local action.

polarization

When the positive hydrogen ions draw electrons from the copper electrodes, neutral hydrogen gas bubbles are formed. Some of these bubbles cling to the copper electrode and produce a layer of nonconducting gas around it. This gas interferes with the action of the cell in two ways. First, the gas bubbles *reduce the effective area* of the copper electrode, so that fewer hydrogen ions can reach it. Second, the positive hydrogen ions tend to collect around these bubbles. This causes a local positive charge to be built up in the electrolyte, which repels the positive hydrogen ions away from the copper electrode. This effect is known as *polarization.*

Polarization is due to the neutral gas bubbles that collect around the positive electrode

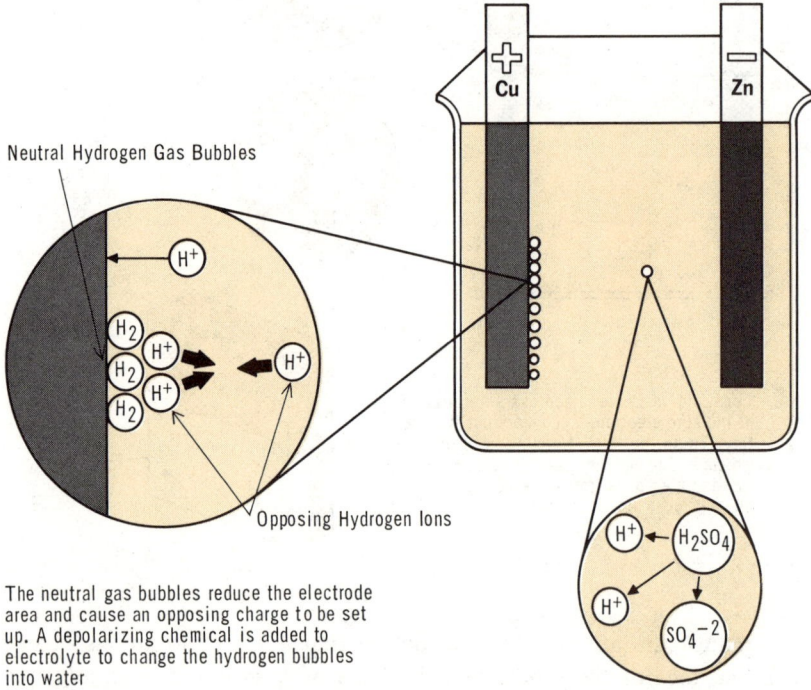

Neutral Hydrogen Gas Bubbles

Opposing Hydrogen Ions

The neutral gas bubbles reduce the electrode area and cause an opposing charge to be set up. A depolarizing chemical is added to electrolyte to change the hydrogen bubbles into water

To remove the neutral hydrogen gas bubbles, a depolarizer is added to the electrolyte. Usually a chemical such as *manganese dioxide* is used. This reacts with the hydrogen gas bubbles to form water. The water then merely mixes with the electrolyte, and polarization is prevented.

the electrolyte

For the basic zinc-copper cell you have studied until now, the electrolyte used was sulfuric acid mixed with water. Actually, many different chemicals can be used. The important functions that the chemicals must perform are

1. Break down into positive and negative ions when they are mixed with water.
2. React chemically with at least one of the electrodes.

Since different chemicals react more or less with different metals, the electrolyte used will determine the actual charges produced on the electrodes, and will determine the amount of voltage and current the battery can supply. Some other electrolytes are ammonium chloride (NH_4Cl), which breaks into positive ammonium ions (NH_4^+) and negative chloride ions (Cl^-); and copper sulfate, which breaks down into positive copper ions (Cu^{+2}) and negative sulfate ions (SO_4^{-2}).

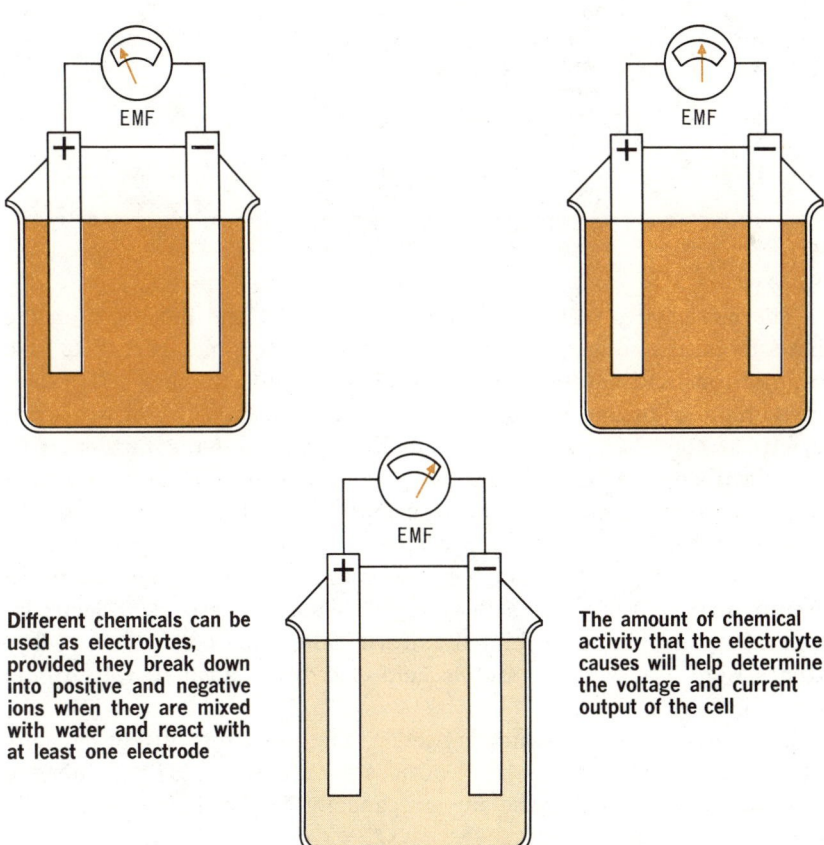

Different chemicals can be used as electrolytes, provided they break down into positive and negative ions when they are mixed with water and react with at least one electrode

The amount of chemical activity that the electrolyte causes will help determine the voltage and current output of the cell

the electrodes

The metals used for the electrodes in the basic cell are chosen so that when they react with the electrolyte, one will give up electrons to develop a *positive* charge, while the other will take on electrons to develop a *negative* charge. The tendency of a metal to give up or take on electrons depends on how *active* the metal is chemically. A special list of metals has been made up to show how active one metal is compared to others. This list is known as the *electromotive* or *galvanic series* of metals, and a partial list is shown.

PARTIAL LISTING OF ELECTROMOTIVE SERIES			
Order of Activity	Metal	Order of Activity	Metal
1	Sodium	10	Nickel
2	Magnesium	11	Tin
3	Aluminum	12	Lead
4	Manganese	13	Antimony
5	Zinc	14	Copper
6	Chromium	15	Silver
7	Iron	16	Mercury
8	Cadmium	17	Platinum
9	Cobalt	18	Gold

The most active metals are highest on the list and the lower on the list a metal is, the less active it is. The more active metals tend to take on electrons and develop a negative charge, while the less active ones give up electrons and produce a positive charge. For a cell to work, one electrode must be more active than the other. And the farther apart the electrodes are on the list, the greater will be the voltage that is developed. For example, for the zinc-copper cell, zinc is fifth on the list and copper is fourteenth. Zinc would be the negative electrode, since it is higher on the list; and copper would be the positive electrode. Approximately 1.08 volts is developed between these two electrodes when they are placed in diluted sulfuric acid. If carbon and zinc electrodes were used with a chromic acid electrolyte, about 2 volts would be developed.

The kinds of electrode metals used do not affect the current capacity of the battery. The size of the electrodes does, however. The greater the electrode size, the higher the current capacity will be. This is more fully explained later in the book.

limitations

The primary wet cell has two basic disadvantages. The first is that its operation relies on the necessity for the electrolyte to attack the negative electrode to produce the chemical action. As a result, as the battery is used, the negative electrode slowly disintegrates. And, when much of it is gone, it cannot supply the amount of current needed. In addition, the electrolyte, too, undergoes a chemical change as it reacts with both electrodes; after a while, the chemical nature of the electrolyte changes to such a degree that it loses much of its electrolytic properties and cannot build up enough of a charge on each electrode. However, the primary wet cell can be restored by replacing the negative electrode and the electrolyte. But this brings us to the second disadvantage.

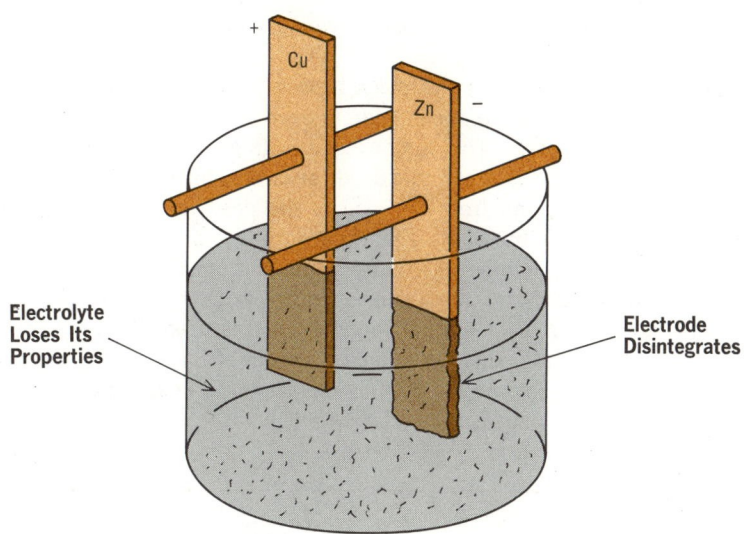

The primary wet cell works in such a way that the negative electrode slowly disintegrates and the electrolyte loses its properties. To make it easy to repair, it is inconvenient to use it commercially

If the wet cell is made so that it is easy to repair, it cannot be convenient, since it uses a liquid electrolyte. If it were made rugged and spillproof so that it would be portable, it would be difficult to repair, and expensive to replace. As a result, most primary wet cells in use today are simply constructed for laboratory use.

For convenient, inexpensive, commercial use, the primary dry battery was developed.

summary

☐ Power sources are devices that produce electricity by converting some other forms of energy into electrical energy. ☐ A battery is a power source that converts chemical energy into electrical energy. ☐ The energy of a battery is self-contained, so batteries are ideally suited for applications requiring a portable source of power. ☐ A primary battery converts chemical energy directly into electrical energy. ☐ A secondary battery first has to be charged with electrical energy before it can produce an electrical output. Secondary batteries are often called storage batteries.

☐ The cell is the basic unit of a battery. ☐ Wet cells use liquid chemicals; while dry cells contain a chemical paste. ☐ Primary batteries are usually of the dry-cell type, and secondary batteries of the wet-cell type. ☐ The basic primary wet cell is usually called the voltaic cell; sometimes it is referred to as the galvanic cell. ☐ A cell consists of two electrodes made of different metals, and a chemical solution, called the electrolyte.

☐ The chemical action that takes place between the electrodes and the electrolyte in a wet cell causes one electrode to develop a positive charge and the other a negative charge. This produces a voltage across the electrodes. ☐ When a cell delivers current to a load, electrons travel from the negative electrode to the load, and from the load to the positive electrode. However, the chemical action of the cell maintains the charge on each electrode constant. ☐ The output voltage of a battery and the amount of current it can supply is determined by the metals used for the electrodes and the chemical used for the electrolyte. ☐ As a primary wet cell is used, the negative electrode slowly disintegrates and the chemical nature of the electrolyte changes. ☐ A primary wet cell can be restored by replacing the negative electrode and the electrolyte. This is usually impractical, though.

review questions

1. What is the difference between a *primary* battery and a *secondary* battery?
2. What is a *wet cell*? A *dry cell*?
3. Name the basic parts of a battery.
4. Briefly describe how a wet cell develops a voltage.
5. What is meant by *local action*?
6. What is a *depolarizer*?
7. What determines the voltage developed by a basic cell?
8. The electrolyte in a cell must perform what two functions?
9. What is the *electromotive series of metals*?
10. What are the disadvantages of the primary wet cell?

the basic primary dry cell

The so-called dry cell actually uses an *electrolytic paste*. The paste is very thick and does not have a tendency to spill or leak. As a result, the dry battery can be made cheaply and light in weight, and are the batteries that are used extensively by the public and industry.

The dry cell works in a manner that is similar to the wet cell. The electrolytic paste reacts with the electrodes to produce a negative charge on one electrode and a positive charge on the other. The difference in potential between the two electrodes is the output voltage.

DRY CELLS CAN BE MADE CHEAPLY IN ALL SIZES AND SHAPES, SO THAT THEY ARE CONVENIENT FOR MANY APPLICATIONS

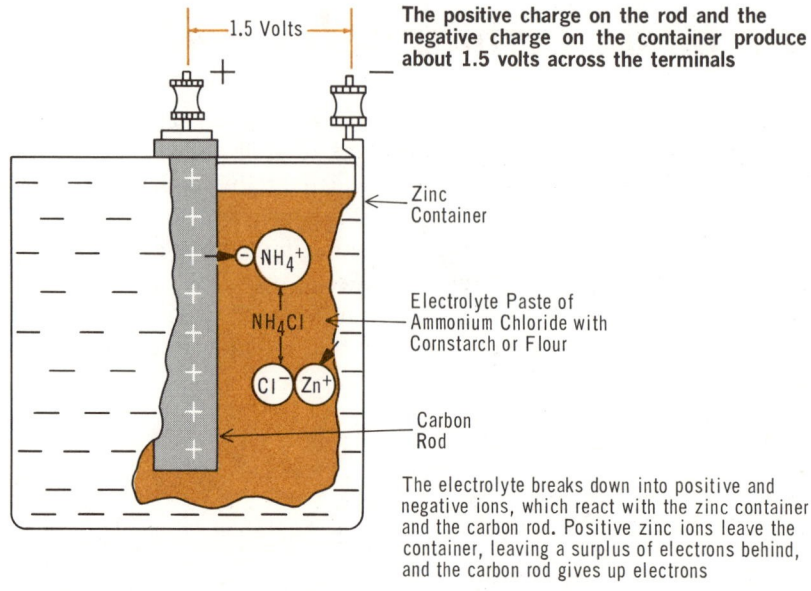

The positive charge on the rod and the negative charge on the container produce about 1.5 volts across the terminals

|←——1.5 Volts——→|

Zinc Container

Electrolyte Paste of Ammonium Chloride with Cornstarch or Flour

Carbon Rod

The electrolyte breaks down into positive and negative ions, which react with the zinc container and the carbon rod. Positive zinc ions leave the container, leaving a surplus of electrons behind, and the carbon rod gives up electrons

the basic zinc-carbon dry cell

The most widely used dry cell is the Leclanche cell. It uses a zinc container as the negative electrode, a carbon rod as the positive electrode, and ammonium chloride as the electrolyte. The ammonium chloride is mixed with starch or flour to make an electrolytic paste. Other materials are also used in the Leclanche cell, but let us begin by studying the basic cell.

When the cell is first made, the electrolyte breaks down into positive ammonia ions (NH_4^+) and negative chloride ions (Cl^-). The positive and negative charges are equal, so that the electrolyte is neutral. The negative chloride ions attack the zinc container causing it to decompose. The zinc atoms give up electrons, and release positive zinc ions (Zn^+) to the electrolyte. The electrons stay behind in the zinc so that the zinc container builds up a surplus of electrons and accumulates enough negative charge to stop the activity. The positive zinc ions combine with and neutralize the chloride ions to form zinc chloride.

Since some of the negative ions in the electrolyte have been neutralized, the electrolyte takes on a positive charge and the ammonia ions attract free electrons from the carbon rod. The carbon rod, then, has a lack of electrons, and builds up a positive charge. This continues until enough ammonia ions take on electrons to make the electrolyte again neutral. The difference in potential between the zinc container and carbon rod in the Leclanche dry cell is about 1.5 volts.

polarization and local action

Dry cells suffer from the effects of polarization and local action in the same way as wet cells, and the conditions are corrected for in a similar manner. The zinc container in a dry cell is usually amalgamated to reduce local action.

For polarization, the Leclanche dry cell uses a mixture of carbon powder, manganese dioxide, and zinc chloride. The carbon powder acts as a pasty binder when mixed with the chemicals.

The dry cell also develops polarization and local action

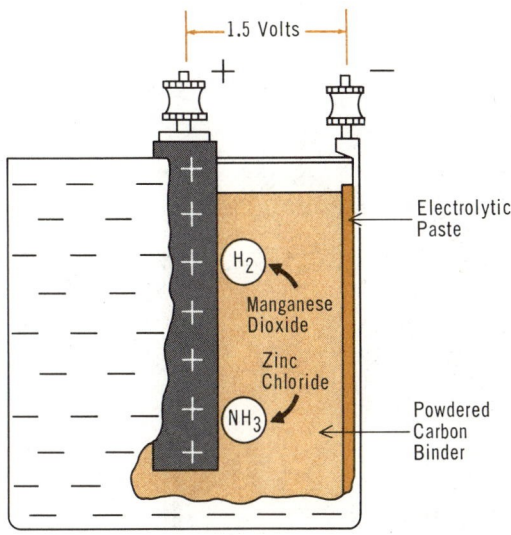

The zinc container is usually amalgamated to keep local action down. And the polarization, due to hydrogen gas (H_2) and ammonia gas (NH_3), is reduced when manganese dioxide and zinc chloride are used. These chemicals are mixed with a powdered carbon binder

When the ammonia ions (NH_4^+) react with the carbon rod, they change into ammonia and hydrogen gas bubbles. The manganese dioxide combines with the hydrogen gas to form water, and the zinc chloride combines with the ammonia gas to form ammonium chloride. This removes the gas bubbles from around the carbon rod, and the resultant chemicals mix with the powdered carbon.

construction of the leclanche dry cell

As you have seen, the dry cell has a considerably different construction than the wet cell. A typical flashlight battery is shown. The cap on the carbon rod is the positive terminal, and the bottom of the zinc container is the negative terminal. Seals are used to make the battery relatively leakproof, and a paper-like cover is generally glued around the zinc container to act as an insulator and to contain information about the battery.

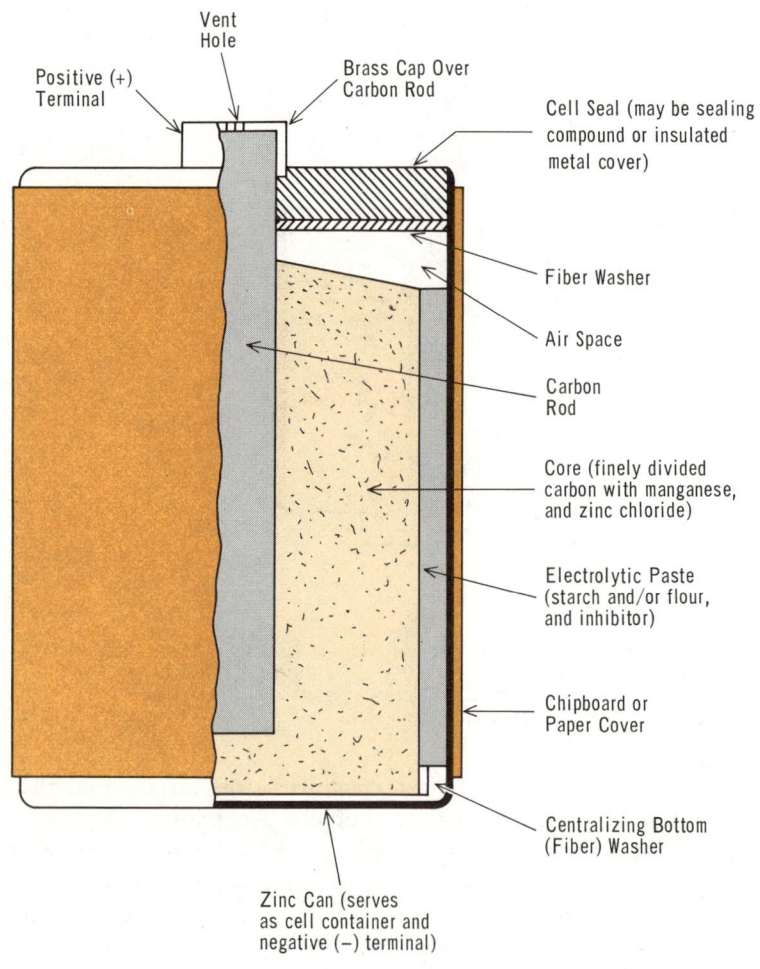

leakproof dry cells

Although the ordinary dry cell uses seals to make it leakproof, it will remain sealed only as long as the zinc container remains intact. However, as you remember, the battery works because the electrolyte attacks the zinc container, and as the battery delivers current, the container is slowly eaten away as it forms zinc chloride with the electrolyte. When the zinc container wears out and develops holes, the battery is no longer leakproof. Moisture can seep in and cause the battery to swell up. And the zinc chloride that forms when the electrolyte reacts with the zinc container can leak out. The white substance that you see on worn out batteries is zinc chloride.

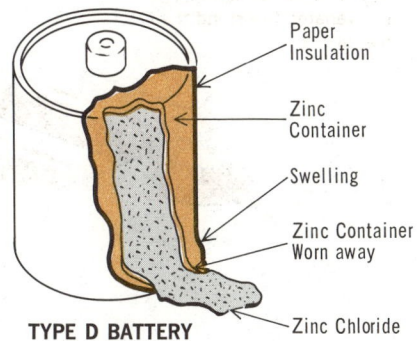

When the zinc container wears out, the ordinary battery is no longer leakproof

Paper Insulation

Zinc Container

Swelling

Zinc Container Worn away

Zinc Chloride

TYPE D BATTERY

To prevent this swelling and leaking, the better batteries are encased in a *steel jacket*. The rest of the battery is made slightly smaller, so that the overall battery size does not change. A paper tube is placed between the steel jacket and zinc container for insulation purposes.

TYPE D BATTERY

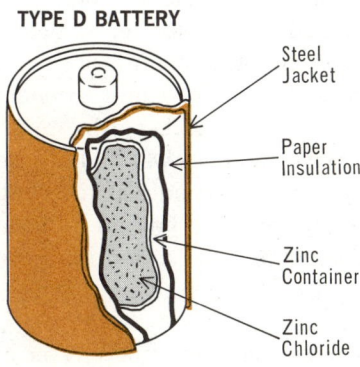

Steel Jacket

Paper Insulation

Zinc Container

Zinc Chloride

When the zinc container deteriorates to the point where holes are developed, the steel jacket prevents moisture from seeping in and chemicals from seeping out. The steel-jacketed batteries are leakproof

rejuvenation

Like the wet cell, the basic primary dry cell deteriorates while it is being used. Zinc atoms from the negative electrode are carried into the electrolyte while the cell is being used, and when much of the electrode is eaten away, the cell becomes useless. Unlike the electrodes in a wet cell, the electrodes in a dry cell cannot be replaced. If there is extensive damage to the electrode, the cell must be replaced. But if the damage is not extensive, the operating life of the dry cell can be extended by a process known as rejuvenation.

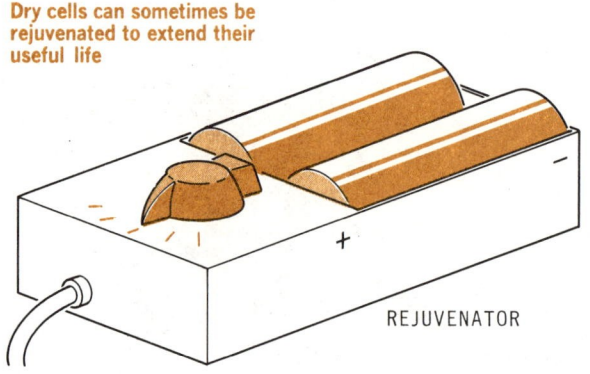

Dry cells can sometimes be rejuvenated to extend their useful life

REJUVENATOR

The electrochemical process that removed particles of zinc and had them combine with the chloride atoms in the electrolyte to form zinc chloride can be reversed. This will cause the zinc chloride molecules to split, and the zinc atoms will be carried back to the negative electrode. Essentially, with a properly applied voltage and current, electroplating occurs to repair the negative electrode. But this can be done only as long as the electrode is not too disintegrated, so that too much electroplating is required. Extensive electroplating cannot be controlled, and the zinc buildup could accumulate in localized areas to cause extended sharp growths that could cause shorts. Essentially, then, the rejuvenation process must be used *before* extensive wear shows, and should be repeated regularly while the battery is in fair condition to get the most extended life. After a while, the rejuvenation process will cause the zinc build up that results in a short. Often, these dry cell rejuvenators are called chargers, but they do not really charge the battery. This is explained for the secondary, or storage, cell.

Rejuvenation of a dry cell is brought about by using a reverse voltage to apply a reverse current through the cell.

dry cell standards

It is possible to design and package the ordinary dry cell, or combinations of cells, in an infinite number of ways, to give any combination of characteristics. To prevent the confusion that might result from any lack of control, the United States of America Standards Institute issues the specifications for standard battery types and sizes, so that the dry cell batteries of different manufacturers can be interchangeable.

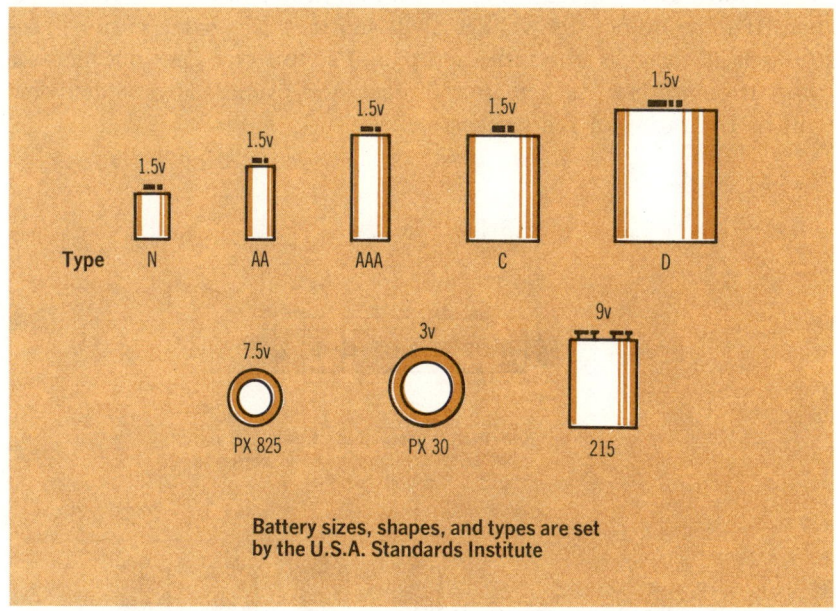

Battery sizes, shapes, and types are set
by the U.S.A. Standards Institute

There are innumerable types, and only the more popular commercial dry cells will be covered here. The Type N, AA, and AAA dry cells are typical of the batteries used in penlights, cameras, portable radios, recorders, etc. The Type C and D batteries are popular for flashlights and electrical toys, but are also used in some of the larger radios and recorders.

The Type PX 30 and PX 825 are button cells designed primarily for use in cameras. The Type 215 battery is typical of the higher voltage batteries used in portable radios, etc.

The size and shape of the battery does not always denote its voltage. For example, the N, AA, AAA, C, and D batteries are all 1.5-volt cells. Their size does, however, give them different current and life ratings, as you will soon learn.

multiple-cell batteries

You learned earlier that although the terms *cell* and *battery* are now used interchangeably, at one time, the term battery was used only to mean a combination of cells. In most applications, the voltage and current that can be supplied by a single cell is not enough, so many batteries have to use *combinations of cells*. When a *higher voltage* is needed, cells are connected in *series*, so that their emf's add. The cells can be connected within the battery, or separate batteries can be combined to obtain the higher voltage. The *polarities* all have to be in the *same direction*. If they are not, the voltages will subtract. The same current goes through all the cells in a series setup, and so this type does not increase the current rating. As a matter of fact, the overall current delivered will be the rating of the weakest cell.

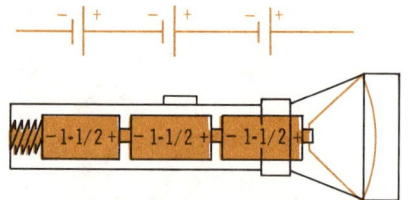

In many flashlights, three 1.5-volt batteries are connected in series to light a 4.5-volt lamp

Four 1.5-volt batteries connected in parallel will produce an emf of 1.5 volts with four times the current rating of a single cell

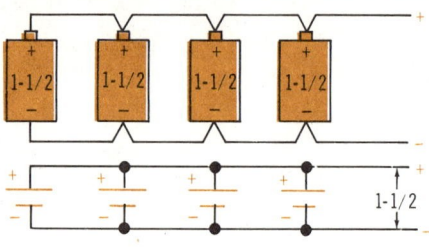

To *increase* the *current rating*, cells must be connected in *parallel*. Then each cell will supply its own current, and the sum of all the cells will be the total current rating. Again, all of the cells should be connected with the same polarity. If not, the cells will supply current to each other and short circuit the combination. All of the cells should have the same voltage rating, otherwise the higher voltage cell will supply current to the lower voltage cell.

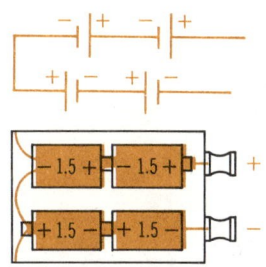

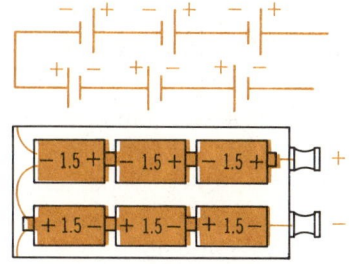

A 6-volt transistor battery contains four 1.5 volt cells in series

A 9-volt transistor radio battery contains six 1.5-volt cells in series

typical multiple-cell batteries

The previous page showed you how ordinary C or D size cells can be used together to increase voltage or current. The batteries that are available with higher voltages actually produce the higher voltage by combining a series of cells within the battery case. A 6-volt transistor battery uses four 1.5-volt cells, and a 9-volt transistor battery uses six 1.5-volt cells. A 45-volt power pack uses thirty 1.5-volt cells in series. A 90-volt pack can use two 45-volt arrangements (60 cells) or, if greater current capacity is desired, four 45-volt arrangements connected in series-parallel (120 1.5-volt cells, connected 60 to each bank). Batteries are available with the following voltages: 1.5, 3, 4.5, 6, 7.5, 9, 12, 15, 22.5, 30, 45, 67.5, 90, and so on up to 500 volts. Those from 30 to 500 volts are mostly for industrial and military use, and are hard to get commercially.

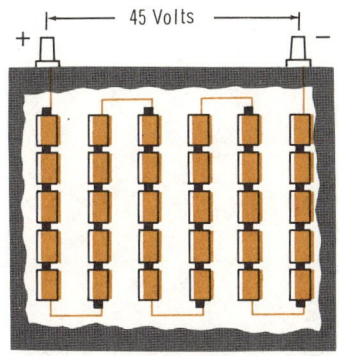

A 45-volt battery uses thirty 1.5-volt cells connected in series

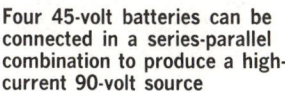

Four 45-volt batteries can be connected in a series-parallel combination to produce a high-current 90-volt source

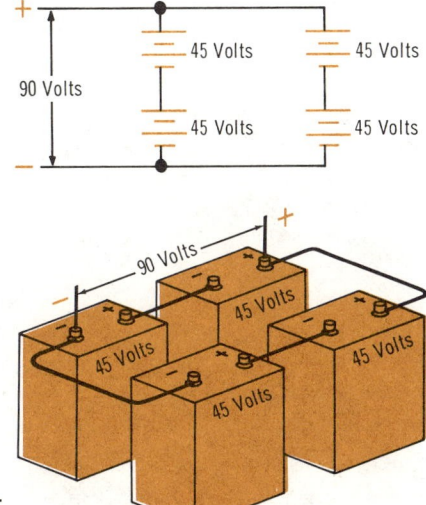

internal resistance

Batteries have two voltage ratings: a no-load rating and a normal load rating. Since the normal load rating is the voltage that actually occurs in normal use, it is the one that is used.

The no-load rating is higher because the battery voltage actually drops when it starts supplying a current. This is because a battery has internal resistance, which results from the opposition that the electrodes and the electrolyte give to the flow of current. When the current flows through this internal resistance, a portion of the battery voltage is dropped and less is available at the output terminals. As the battery wears out, the electrodes and electrolytes in the cell deteriorate and become poorer conductors. The internal resistance increases, and the output voltage decreases further.

With a worn-out battery, the voltage drops to a point where the battery will not function properly. For example, if you had a 9-volt battery in which each of the six cells had a 0.5 ohm internal resistance, the total resistance would be 3 ohms. If the circuit had 10 ma flowing in it, the battery would drop 30 millivolts, having little effect on the 9-volt output. But if each cell deteriorated to 5 ohms, the total would become 30 ohms, and 10 ma would cause 0.9-volt to drop internally. The output would then weaken and go down to 8.1 volts.

The amount of internal resistance a battery has also determines how much peak or flash current it can supply in a short time. A low resistance is needed for high short peaks to keep the output voltage from dropping too far. This is why penlight or flashlight batteries should not be used for photoflash or transistors, even though they have the same type designation. Some typical internal resistances of different Leclanche cells are shown below:

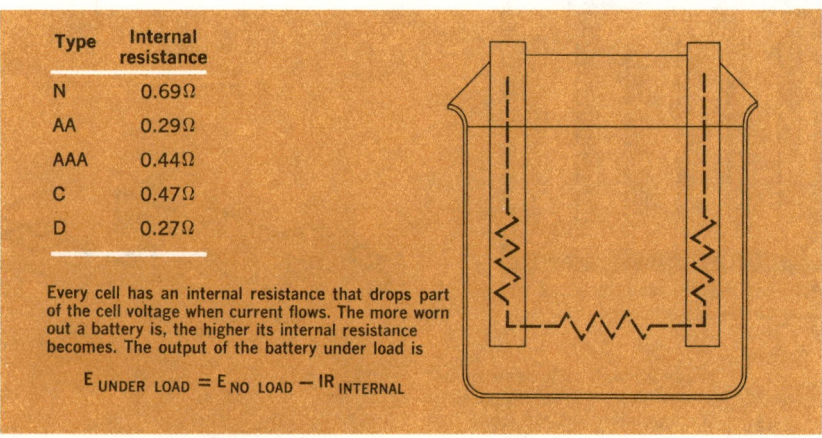

Type	Internal resistance
N	$0.69\,\Omega$
AA	$0.29\,\Omega$
AAA	$0.44\,\Omega$
C	$0.47\,\Omega$
D	$0.27\,\Omega$

Every cell has an internal resistance that drops part of the cell voltage when current flows. The more worn out a battery is, the higher its internal resistance becomes. The output of the battery under load is

$$E_{\text{UNDER LOAD}} = E_{\text{NO LOAD}} - IR_{\text{INTERNAL}}$$

current rating/useful life

The amount of current that a battery can supply depends on the size of the electrodes and the internal resistance of the battery. Small dry cells are limited in how large their electrodes can be, so they generally strive for low internal resistances, particularly if they are designed for photoflash or transistor use.

Also, since the current flowing through the cell is what causes the destructive electrochemical action that ultimately wears out the cell, the amount of current that flows for any given length of time is what determines the useful life of that cell. A battery that supplies a small current over a small duty cycle will last longest. A duty cycle is the ratio of on time to off time. The larger the current or the longer the duty cycle, the shorter will be the life.

As part of their charter for establishing battery standards, the U.S.A. Standards Institute sets drain/life requirements for battery types. Some of these are:

Cell Size	Current Drain (ma)	Life (hours)
N	1.5	275
	7.5	52
	15	24
AA	3	350
	15	40
	30	15
AAA	2	290
	10	45
	20	17
B	5	420
	25	65
	50	25
C	5	430
	25	100
	50	40
D	10	500
	50	105
	100	45

You can see that for each battery type, any significant increase in current drain drastically reduces useful life. In addition to this, these ratings are based on an operating temperature of 70°F and *no more than two hours of operation per day.* If the duty cycle is larger than 2 in 24 hours, useful life drops even more. Ordinary dry batteries of the types listed above are designed not to be used for more than 15 or 20 minutes at a time, to allow them to rejuvenate themselves between uses.

summary

☐ The dry cell uses a thick paste as the electrolyte. ☐ Dry cells operate in a manner similar to wet cells. The electrodes interact with the electrolyte in such a way that a difference of potential is developed between the electrodes. ☐ The Leclanche cell is the most widely used type of dry cell. It has an output voltage of about 1.5 volts. ☐ The zinc container of the Leclanche cell serves as the negative electrode, and a carbon rod is used for the positive electrode. ☐ The electrolyte of the Leclanche cell is made from ammonium chloride mixed with starch or flour to make a paste. ☐ The zinc container of the Leclanche cell is usually amalgamated to reduce local action.

☐ Better types of dry cells are made leakproof by encasement in steel jackets. The jacket keeps the cell leakproof even after the zinc container wears out and develops holes. ☐ Dry cells are made to standard sizes and shapes. ☐ Dry cells can be connected to produce various voltage and current ratings. ☐ A cell's ratings and life depend on its internal resistance.

☐ Although dry cell batteries cannot be repaired as can wet cells, their operating life can be extended by rejuvenation. ☐ After rejuvenation a dry cell is good again for a short period.

review questions

1. Why is a *dry cell* so called?
2. What are the materials used for the electrodes and electrolyte of the Leclanche cell?
3. Draw a sketch of a basic Leclanche cell.
4. What is the output voltage of a Leclanche cell? Of a mercury cell?
5. If you see a white substance on a dry cell, what does it probably indicate?
6. How is a dry cell made leakproof?
7. Who establishes the battery standards?
8. Name four battery type styles.
9. What is *rejuvenation*?
10. Can a dry cell be rejuvenated over and over? Explain.

the secondary (storage) battery

Primary cells have serious limitations because they have short useful lives. The rejuvenation of the dry cell is only a temporary measure. The wet cell can be repaired, but because of this, it is delicate and is usually restricted to laboratory use.

The secondary cell was developed for a *long useful life,* so that it could be ruggedly built for portable applications. The basic difference between the primary and secondary cell is this: The primary cell converts the chemical energy built into it to electrical energy, and in doing so, slowly destroys itself. The secondary cell has no significant electrochemical energy at the start. The energy must first be *charged* into the secondary cell. Then, the cell *stores* the energy until it is used. This is why the secondary cell is also called a *storage cell.*

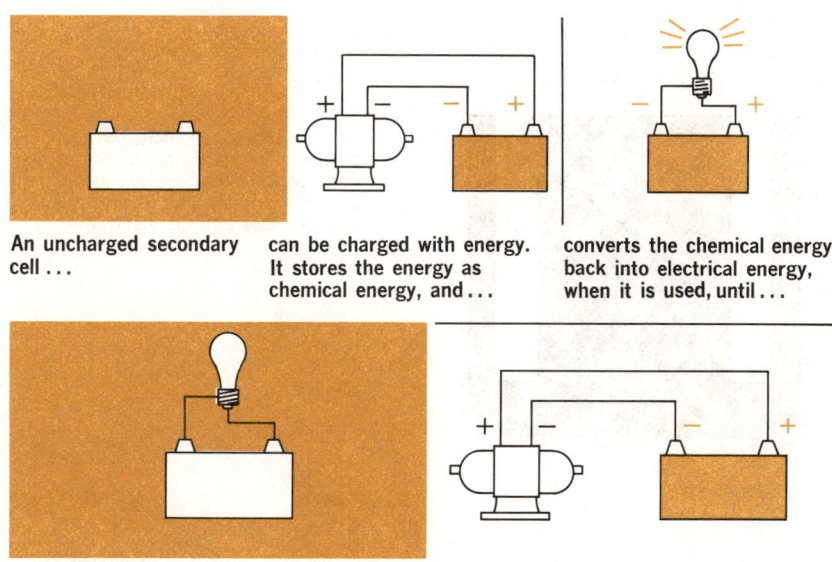

An uncharged secondary cell ...

can be charged with energy. It stores the energy as chemical energy, and ...

converts the chemical energy back into electrical energy, when it is used, until ...

the cell becomes discharged ...

Then it must be charged again

When electrical energy is taken from the storage cell, the cell is said to be *discharging.* When the cell is completely discharged, it no longer can supply electrical energy. But, unlike the primary cell, it can be *recharged.* Basically, the secondary cell *converts* the electrical energy to chemical energy when it is charged. Then, it *reconverts* the chemical energy into electrical energy when it discharges. The most popular storage batteries in use are (1) the lead-acid cell, and (2) various types of alkaline cells.

the basic lead-acid cell

The lead-acid battery can be obtained new, either charged or uncharged, and either with the electrolyte in (wet) or not (dry). For our discussion, let us start with a completely uncharged cell that has the electrolyte in it. The cell, then, consists of two electrodes, both made of lead sulfate ($PbSO_4$), and an electrolyte that is for the most part *distilled* (pure) *water*. There is some sulfuric acid mixed with the water, but in the uncharged state the amount is insignificant. The uncharged lead-acid cell, then, does not meet any of the requirements of a battery: its electrodes are *not* made of *dissimilar metals*, and its electrolyte will *not attack* one of the electrodes.

Lead
Sulfate

Water

The uncharged lead-acid cell cannot generate electrical energy because its electrodes are not dissimilar metals, and its electrolyte will not react with the electrodes

For the lead-acid cell to work, its electrodes must be made dissimilar, and its electrolyte must be made active

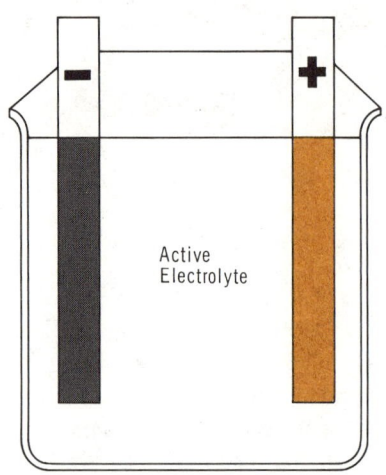

–

+

Active
Electrolyte

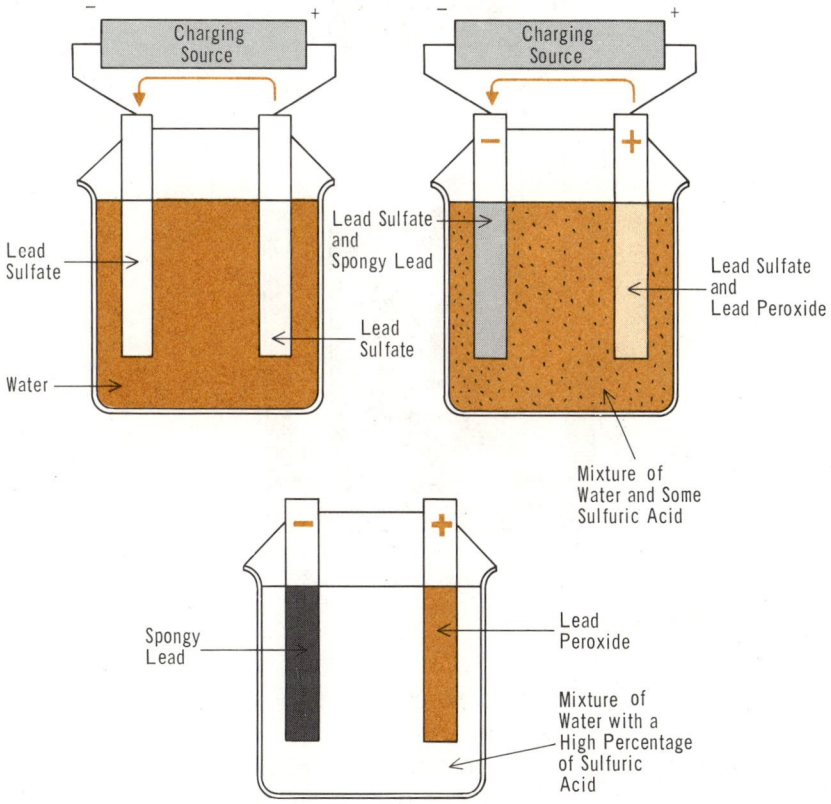

Charging Source (− +)

- Lead Sulfate
- Water
- Lead Sulfate and Spongy Lead
- Lead Sulfate

Charging Source (− +)

- Lead Sulfate
- Lead Sulfate and Lead Peroxide
- Mixture of Water and Some Sulfuric Acid

- Spongy Lead
- Lead Peroxide
- Mixture of Water with a High Percentage of Sulfuric Acid

When the charging current changes the lead sulfate electrodes into spongy lead and lead peroxide electrodes, and produces a high percentage of sulfuric acid in the water, the storage battery is fully charged

charging

For the lead-acid cell to be able to deliver electrical energy, it must have two *different* electrodes, and an *active* electrolyte. This condition exists in the cell when it is charged by an electric current. Because of electrolysis, the water electrolyte breaks down and starts chemical reactions. One lead sulfate ($PbSO_4$) electrode changes into ordinary soft or *spongy lead* (Pb), and the other lead sulfate electrode changes into *lead peroxide* (PbO_2). At the same time, a good part of the water (H_2O) electrolyte becomes *sulfuric acid* (H_2SO_4). When this is done, the battery becomes charged: it has two different metals for electrodes and a chemically active electrolyte. The step-by-step action is shown on the following pages.

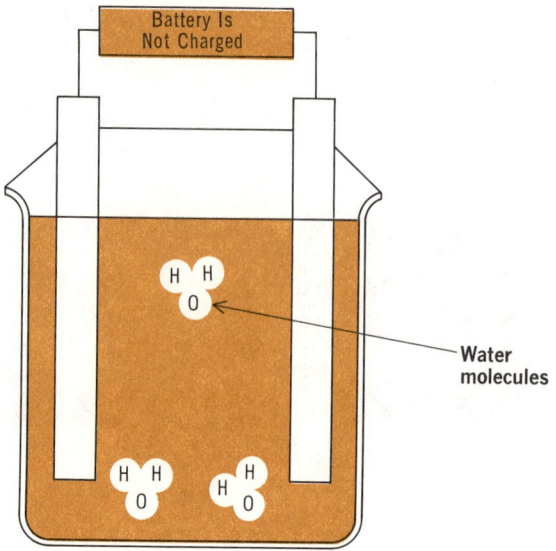

Water
molecules

electrolysis

When charging first starts, the current flowing through the battery causes *electrolysis* of the water. The water molecules (H_2O) begin to break down into their constituent ions. For each *negative* oxygen *ion* (O^{-2}) that is produced, there are two *positive* hydrogen *ions* (H^+), so that the electrolyte is neutral.

Water molecules break
down into positive
hydrogen ions and negative
oxygen ions

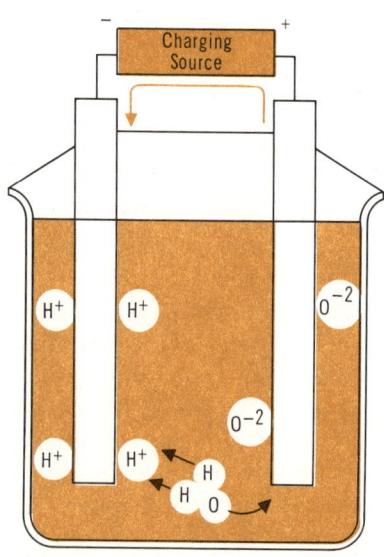

the negative electrode

After electrolysis has broken down the water (H_2O) into positive hydrogen ions (H^+) and negative oxygen ions (O^{-2}), the attraction of the positive hydrogen ions breaks the lead sulfate molecules ($PbSO_4$) into positive lead ions (Pb^{+2}) and negative sulfate ions (SO_4^{-2}). The sulfate ions are attracted out of the electrode by the positive hydrogen ions. This leaves the positive lead ions behind in the electrode.

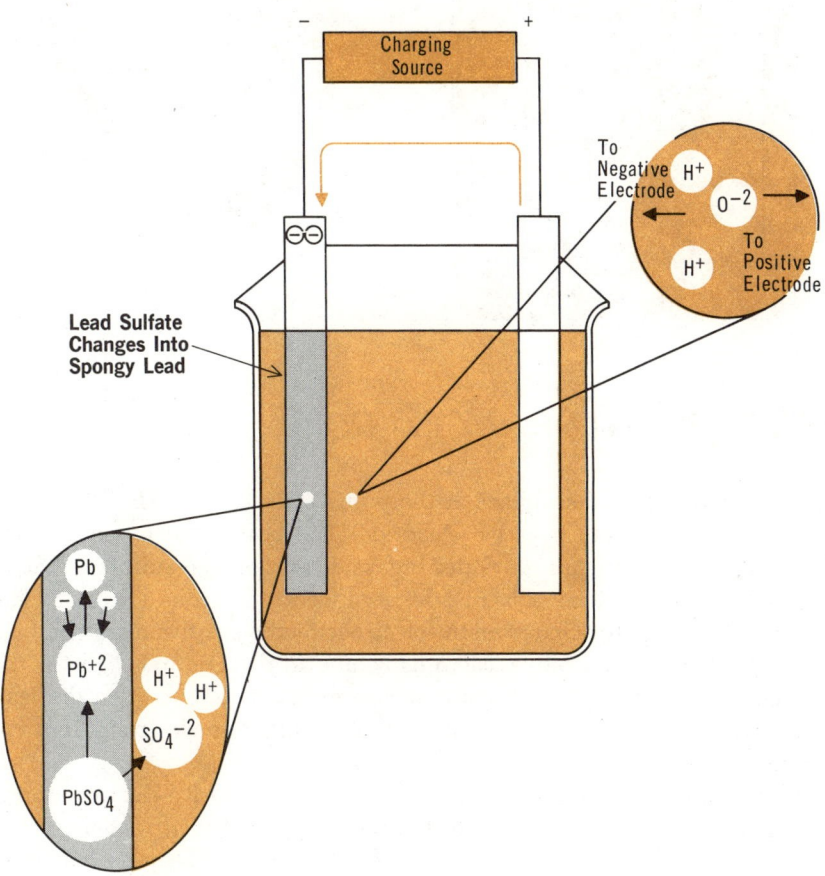

Electrons from the charging current are attracted to the positive lead ions and neutralize them, producing spongy lead (Pb). The action, then, is that the negative electrode begins as lead sulfate, gives off sulfate ions, and then becomes a mixture of lead sulfate and lead while the cell is charging; and continues to give off sulfate ions to become pure spongy lead when the battery is fully charged.

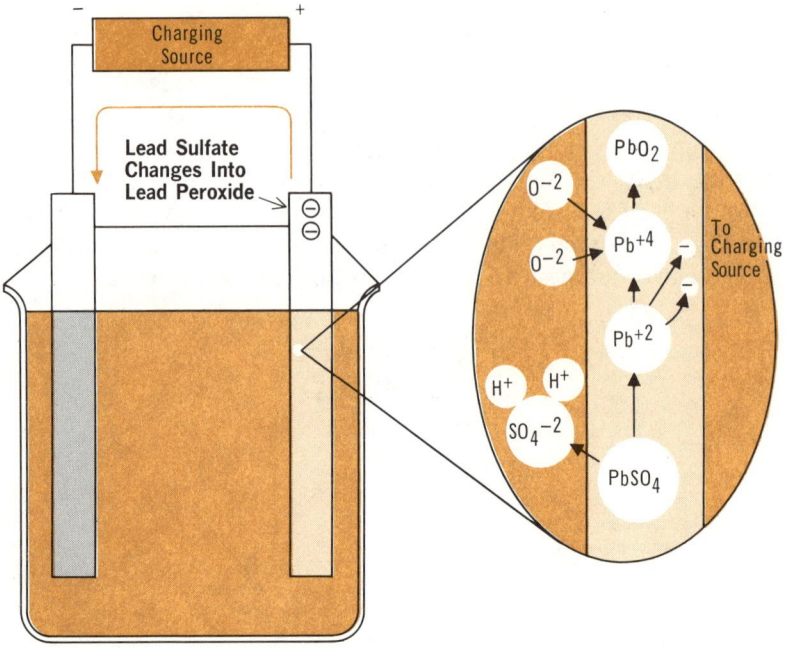

the positive electrode

The action that takes place at the positive electrode is similar to that which occurs at the negative electrode. The positive attraction of the hydrogen ions (H^+) in the water breaks down the lead sulfate ($PbSO_4$) into positive lead ions (Pb^{+2}) and negative sulfate ions (SO_4^{-2}). The sulfate ions are attracted from the electrode by the positive hydrogen ions (H^+). This leaves positive lead ions behind in the electrode. This is the same thing that happened in the negative electrode, but there the lead ions were able to neutralize themselves by taking two electrons from the charging current.

Since the charging current is moving from the positive electrode, however, this cannot be done here. Instead, the positive attraction of the charging source pulls two electrons from the Pb^{+2} to keep the charging current flowing. This produces Pb^{+4}. The Pb^{+4} then attracts two negative oxygen ions (O^{-2}) from the electrolyte and combines with them to become neutral lead peroxide (PbO_2). The action, then, is that the positive electrode begins as lead sulfate, gives off sulfate ions and takes on oxygen ions to become a mixture of lead sulfate and lead peroxide while the cell is charging. The positive electrode continues to give off sulfate ions and to take on oxygen ions until it becomes pure lead peroxide when the battery is fully charged.

the electrolyte

Remember that when charging first started, electrolysis broke down each water molecule (H_2O) into two hydrogen ions (H^+) and one oxygen ion (O^{-2}). The positive hydrogen ions attracted negative sulfate ions (SO_4^{-2}) from each electrode. These combinations produce H_2SO_4, which is *sulfuric acid*. The positive electrode draws negative oxygen ions from the electrolyte. Therefore, as the charging of the cell continues, the electrolyte changes from water to a mixture of water and sulfuric acid. The longer the charging continues, the more the water is changed to sulfuric acid. When the battery is fully charged, the electrolyte contains a high percentage of sulfuric acid as compared to water.

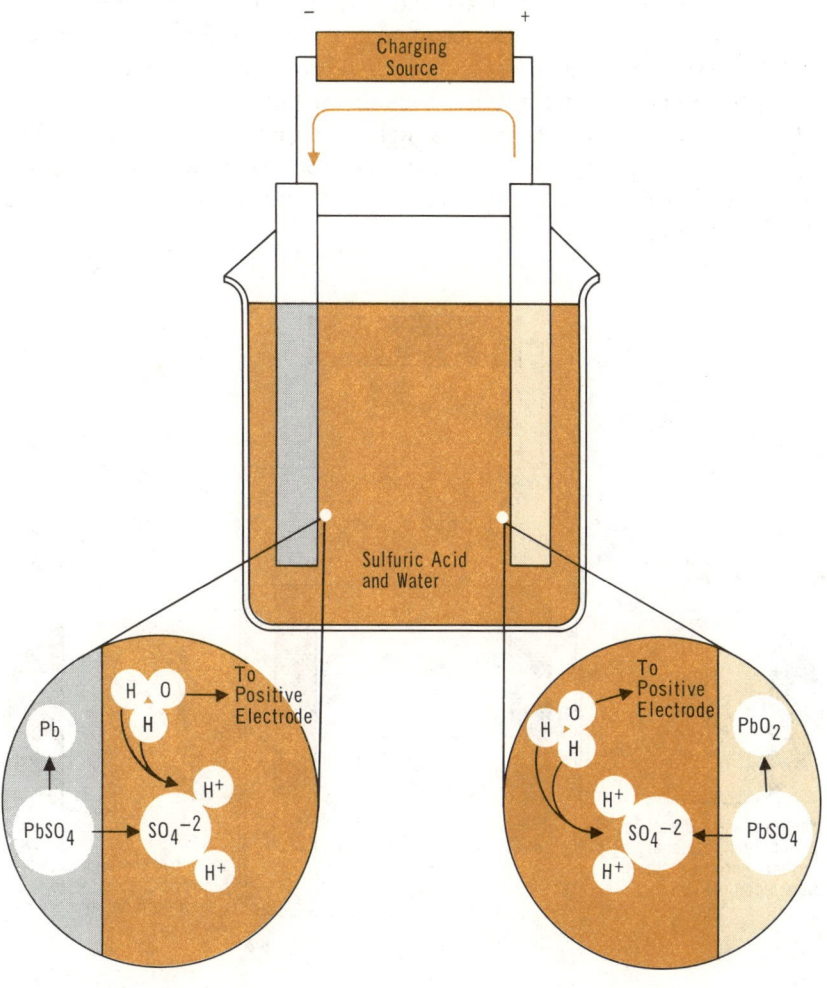

overcharging

When the lead-acid cell becomes fully charged, the negative electrode becomes pure spongy lead (Pb), and the positive electrode becomes pure lead peroxide (PbO_2). These electrodes no longer contain any sulfate ions (SO_4^{-2}) to combine with the positive hydrogen ions (H^+) in the solution, and the positive electrode no longer absorbs the negative oxygen ions (O^{-2}). The hydrogen ions are then attracted to the negative electrode, and the oxygen ions are attracted to the positive electrode.

Electrons supplied by the charging current leave the negative electrode and combine with the hydrogen ions to produce neutral hydrogen gas bubbles. The negative oxygen ions give up electrons to the positive electrode and become neutral oxygen gas bubbles. These gas bubbles accumulate and rise as *gases* to leave the battery through *vent holes*. As a result, further charging does not change the water into more sulfuric acid. Instead, the water is lost as hydrogen and oxygen gases.

The water level will continue to go down until the ratio of sulfuric acid to water is too great, and the high acid content damages the electrodes. The escaping gases are dangerous, too, because they are explosive. When these gases are formed during overcharging, the electrolyte appears to "boil" as the gases rise.

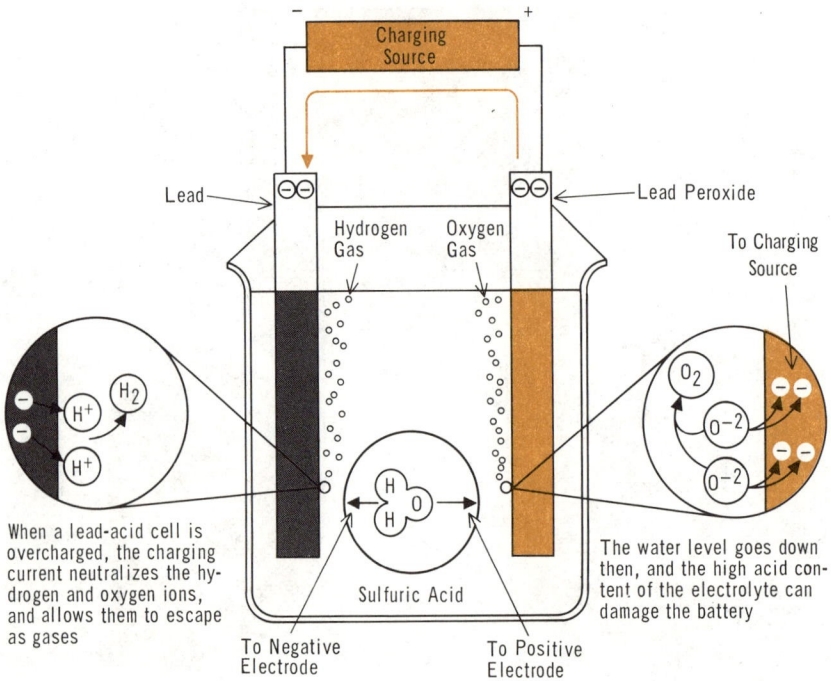

Charging Source

Lead — — Lead Peroxide

Hydrogen Gas　Oxygen Gas

To Charging Source

H^+　H^+　H_2

O_2　O^{-2}　O^{-2}

H O H

Sulfuric Acid

When a lead-acid cell is overcharged, the charging current neutralizes the hydrogen and oxygen ions, and allows them to escape as gases

The water level goes down then, and the high acid content of the electrolyte can damage the battery

To Negative Electrode　　To Positive Electrode

summary

☐ A secondary cell has no significant electrochemical energy until it is charged. ☐ When a secondary cell is being charged, it is converting electrical energy into chemical energy. ☐ When the cell discharges, it converts the chemical energy back into electrical energy. ☐ A secondary cell is also called a storage cell, since after it is charged it stores the energy until it is used. ☐ Unlike the primary cell, a secondary cell can be recharged and used over and over.

☐ An uncharged lead-acid secondary cell has two lead sulfate electrodes and an electrolyte of mostly distilled water. ☐ The uncharged lead-acid cell does not meet any of the requirements of a battery: its electrodes are not made of dissimilar metals, and its electrolyte will not attack one of the electrodes. ☐ When an electric charging current is applied, one electrode changes into spongy lead, and the other into lead peroxide. In addition, most of the water electrolyte becomes sulfuric acid.

☐ Lead-acid cells should never be overcharged. ☐ Overcharging causes explosive hydrogen and oxygen gases to be produced. It also causes the electrolyte to become too highly acid, resulting in damage to the electrodes. ☐ If a lead-acid cell is being overcharged, the electrolyte appears to boil because of the rising hydrogen and oxygen gases.

review questions

1. Why are *secondary cells* often called *storage cells?*
2. What is the principal difference between a *primary cell* and a *secondary cell?*
3. From an energy standpoint, what happens when a secondary cell is charged? When it is discharged?
4. Does an uncharged lead-acid cell meet the requirements of a battery? Explain.
5. What is the composition of the electrodes and electrolyte of an uncharged lead-acid cell?
6. What is the composition of the electrodes and electrolyte of a fully charged lead-acid cell?
7. What is the purpose of vent holes in secondary batteries?
8. Why is overcharging dangerous?
9. How can overcharging be detected visually?
10. Can overcharging damage a secondary battery? Explain.

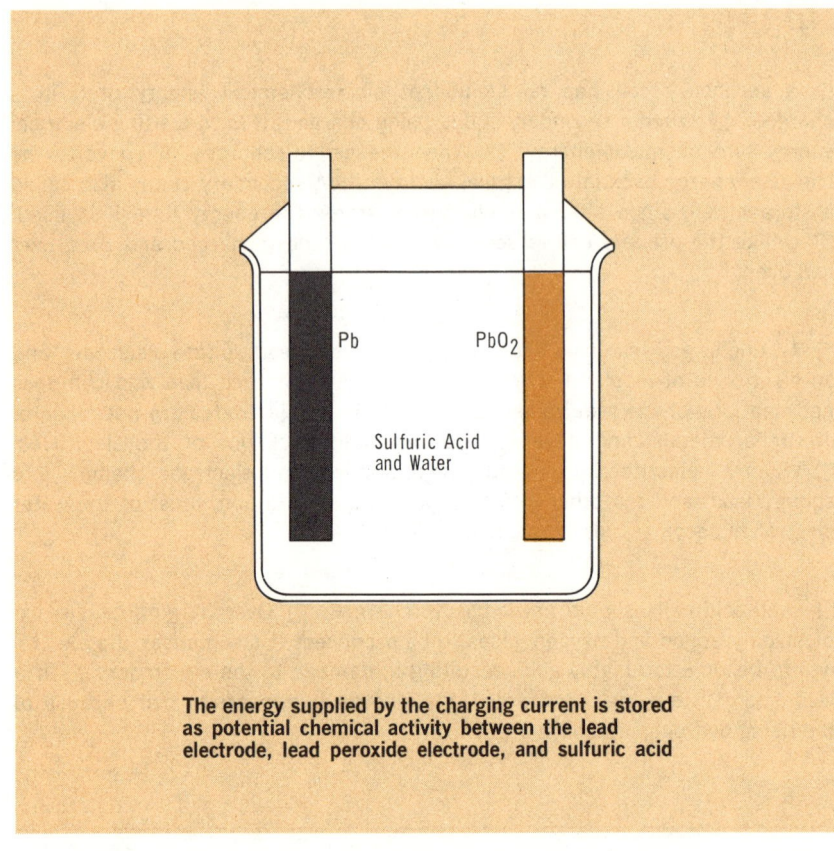

The energy supplied by the charging current is stored as potential chemical activity between the lead electrode, lead peroxide electrode, and sulfuric acid

the charged lead-acid cell

After the lead-acid cell becomes fully charged, one electrode is made of spongy lead (Pb), the other electrode is made of lead peroxide (PbO_2), and the electrolyte is diluted sulfuric acid (H_2SO_4). The cell now has the characteristics of a battery: it has two different electrodes, and an electrolyte that will attack at least one electrode. You may have noticed during the previous explanations, that while the electrodes were being *formed*, the electrodes merely conducted the charging current. They did not build up any charges of their own. As a matter of fact, the electrode atoms were kept neutral by the charging current flowing in the negative terminal and out of the positive terminal. This is the reason why the secondary cell does *not* store *electrical* energy. The energy supplied by the charging current is contained in the *potential chemical activity* of the lead electrode, the lead peroxide electrode, and the sulfuric acid.

developing a negative potential

As soon as charging current stops passing through the lead-acid cell, the sulfuric acid in the cell starts a *reverse chemical action* that is very similar to what happens in the primary wet cell. The electrolyte breaks down into positive hydrogen ions (H^+) and negative sulfate ions (SO_4^{-2}). The negative sulfate ions react with the negative electrode, causing the spongy lead (Pb) to give up two electrons and become a positive ion (Pb^{+2}). The negative sulfate ion combines with the positive lead ion to produce neutral lead sulfate ($PbSO_4$).

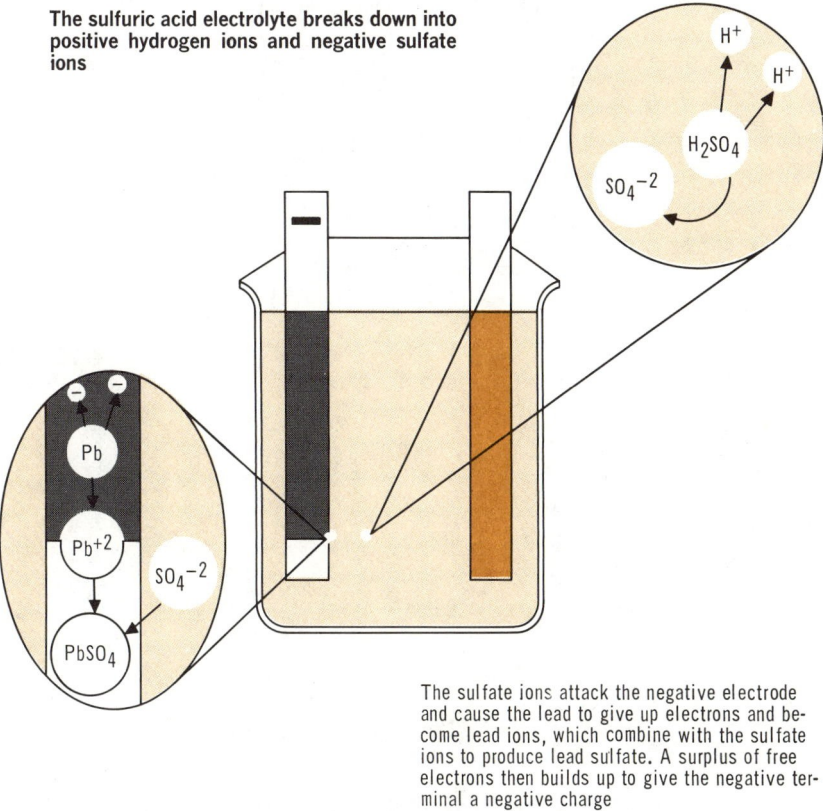

The sulfuric acid electrolyte breaks down into positive hydrogen ions and negative sulfate ions

The sulfate ions attack the negative electrode and cause the lead to give up electrons and become lead ions, which combine with the sulfate ions to produce lead sulfate. A surplus of free electrons then builds up to give the negative terminal a negative charge

In essence, the spongy lead negative electrode starts changing back to lead sulfate, but the two electrons given off by the lead ion remain free. The continued creation of $PbSO_4$ soon allows a surplus of electrons to build up in the negative electrode to give it a negative charge. When enough of a negative charge is built up, the sulfate ions are repelled from the electrode to inhibit further chemical action.

The positive hydrogen ions that are released by the sulfuric acid attract negative oxygen ions from the positive electrode. This leaves Pb^{+4} ions behind, which cause the electrode to build up a positive charge

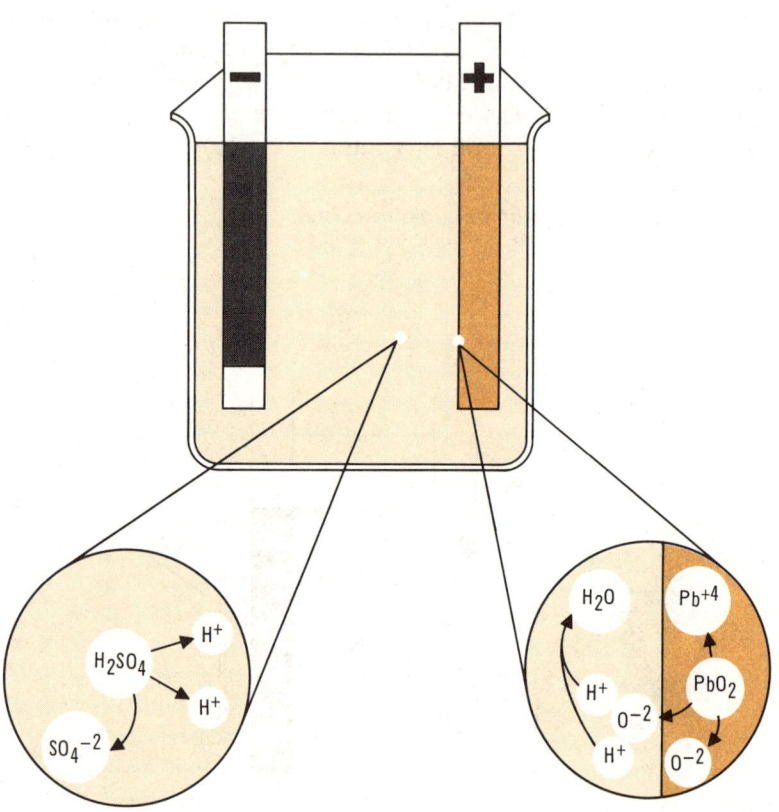

developing a positive potential

The negative sulfate ions (SO_4^{-2}) in the electrolyte also attack the positive electrode. Each lead peroxide molecule (PbO_2) of the positive electrode breaks down into two negative oxygen ions (O^{-2}) and one lead ion (Pb^{+4}). The oxygen ions are attracted into the electrolyte by the positive hydrogen ions (H^+). This leaves the Pb^{+4} ions behind; and when enough oxygen ions leave the electrode, it builds up a positive charge. When this charge becomes sufficient, the positive hydrogen ions are repelled from the positive electrode to inhibit further chemical action.

reducing the electrolyte

At the start of the electrochemical action, the sulfuric acid electrolyte (H_2SO_4) broke down into positive hydrogen ions (H^+) and negative sulfate ions (SO_4^{-2}). The sulfate ions combined with the lead ions of the negative electrode to produce enough lead sulfate so that a negative charge was built up. The positive hydrogen ions attracted enough negative oxygen ions from the positive electrode so that a positive charge was built up. The hydrogen ions and oxygen ions combined to produce water (H_2O), so that in the process of building a difference of potential between the electrodes, part of the sulfuric acid was changed back to water. The action stops when the lead-acid cell develops an emf of about 2.1 volts.

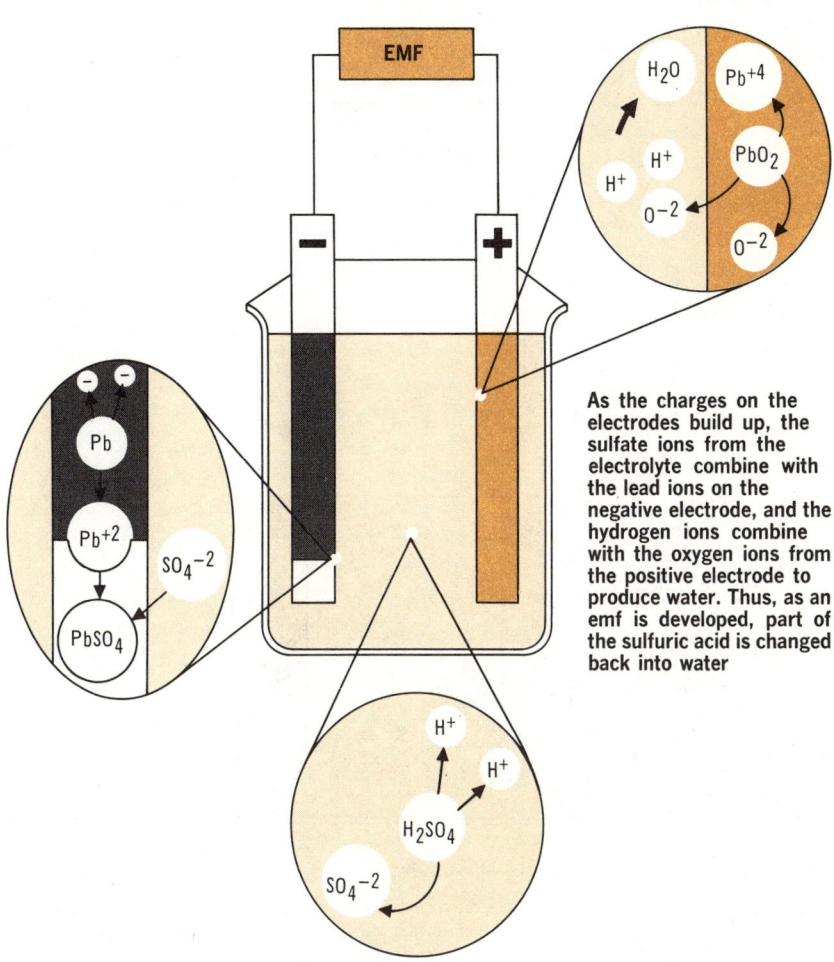

As the charges on the electrodes build up, the sulfate ions from the electrolyte combine with the lead ions on the negative electrode, and the hydrogen ions combine with the oxygen ions from the positive electrode to produce water. Thus, as an emf is developed, part of the sulfuric acid is changed back into water

discharging the lead-acid cell

When the lead-acid cell is connected to a circuit to supply current, the chemical action that produced the emf is continued to replace the electrons that leave the negative terminal, and remove the electrons that enter the positive terminal.

As electrons leave the negative terminal to go to the load, the charge on the negative terminal tends to diminish, allowing the sulfate ions (SO_4^{-2}) in the electrolyte to again react with the negative electrode. The lead atoms of the electrode give up electrons to replace those that have left, and become Pb^{+2} ions. These ions combine with the sulfate ions to produce still more lead sulfate ($PbSO_4$).

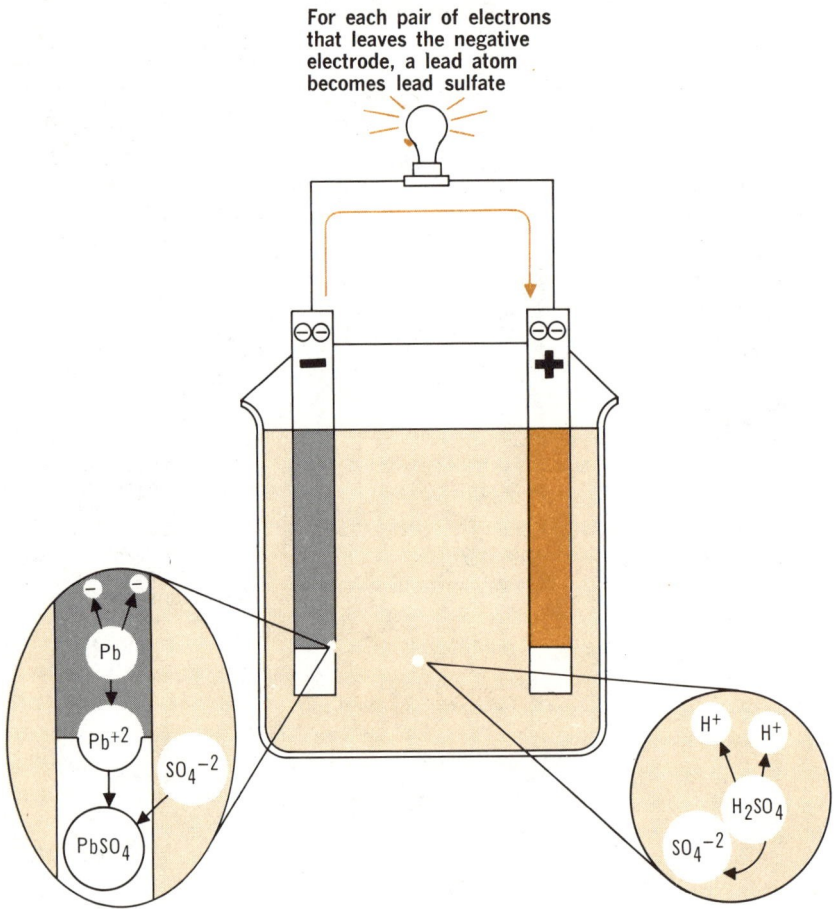

For each pair of electrons that leaves the negative electrode, a lead atom becomes lead sulfate

discharging the lead-acid cell (cont.)

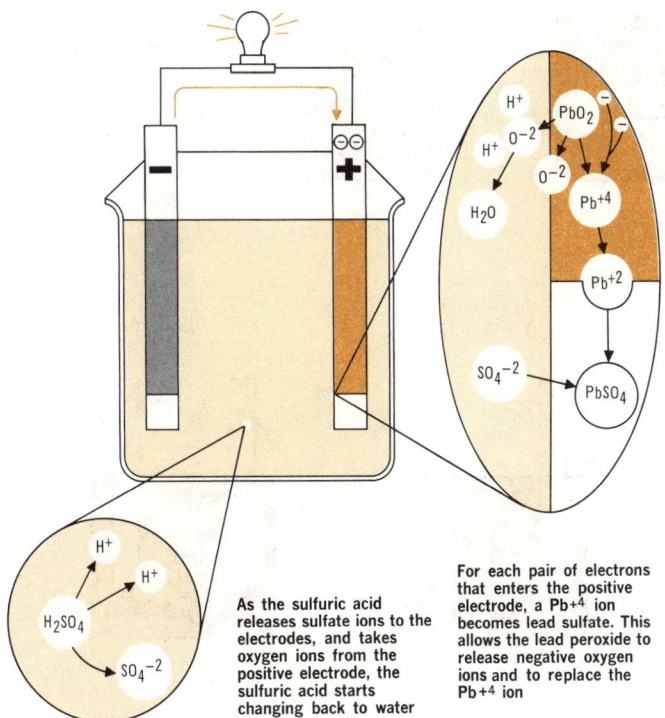

As the sulfuric acid releases sulfate ions to the electrodes, and takes oxygen ions from the positive electrode, the sulfuric acid starts changing back to water

For each pair of electrons that enters the positive electrode, a Pb^{+4} ion becomes lead sulfate. This allows the lead peroxide to release negative oxygen ions and to replace the Pb^{+4} ion

Whenever two electrons leave the negative electrode to go to the load, two electrons are attracted from the load into the positive electrode. These electrons combine with a Pb^{+4} ion, changing it to a Pb^{+2} ion. The Pb^{+2} ion then combines with a negative sulfate ion (SO_4^{-2}) from the electrolyte to deposit lead sulfate on the positive electrode. But the loss of a Pb^{+4} ion to the electrode causes another lead peroxide atom (PbO_2) to break up and give off two oxygen ions (O^{-2}) to the hydrogen ions in the electrolyte. This leaves another Pb^{+4} ion behind to take the place of the one that changed to Pb^{+2} and combined to produce lead sulfate.

In the process of the cell supplying current, the electrolyte gave up more sulfate ions, and absorbed more oxygen ions, so that still more sulfuric acid was changed to water.

While it is supplying current, the emf of the lead-acid cell drops to about 2 volts.

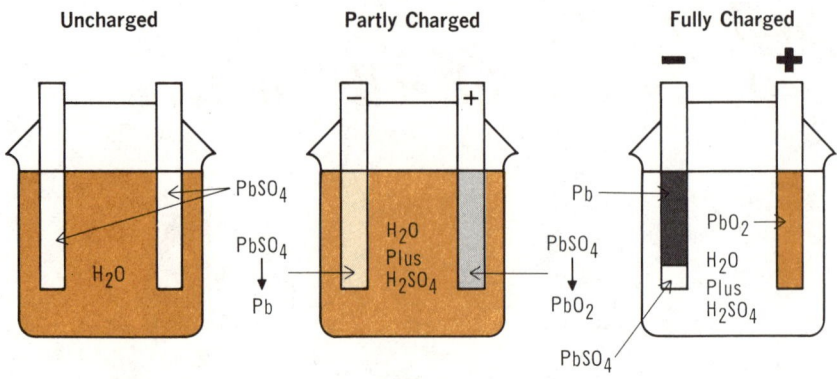

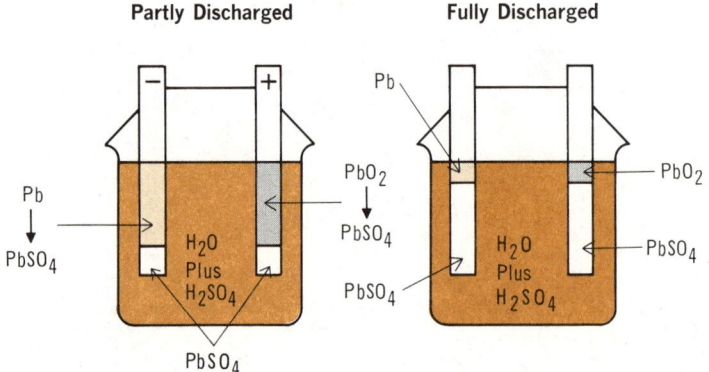

the discharged lead-acid cell

You can see that as the lead-acid cell continues to supply current, the spongy lead (Pb) negative electrode slowly changes into lead sulfate ($PbSO_4$). In a similar manner, the lead peroxide (PbO_2) positive electrode also slowly changes into lead sulfate. And the sulfuric acid in the electrolyte slowly changes into water. As the cell continues to discharge, both electrodes become mostly lead sulfate, and the electrolyte contains very little sulfuric acid. The cell, then, cannot generate sufficient emf, or supply sufficient current to be used; it is then *discharged*. But it can again be charged until the electrodes become spongy lead (Pb) and lead peroxide (PbO_2), and the electrolyte contains more sulfuric acid. It will work again until discharged, but can be used over and over again by periodic *recharging*. The lead-acid cell is considered discharged when its emf drops to 1.75 volts.

specific gravity

You learned that the chemical nature of the electrolyte actually depends on the state of charge that the cell is in. When the cell is fully charged, the electrolyte has a high content of sulfuric acid; and when the cell is discharged, there is very little sulfuric acid in the electrolyte. Therefore, the electrolyte could be tested to determine the state of charge of the cell. To do this chemically would be difficult; but it can be done simply by measuring the *specific gravity* of the electrolyte. Specific gravity is the ratio of the density of a substance to that of water.

Not all materials or liquids have the same *density*. Sulfuric acid is more dense than water, and so also is a mixture of sulfuric acid and water. And, the more acid there is mixed with the water, the more dense the electrolyte would be. The density could be checked by seeing how a *hydrometer* floated in the electrolyte. It would float higher in a dense liquid than it would in a thin liquid.

A syringe draws some of the electrolyte into the hydrometer, and a floating bulb gives indexed readings that show its depth of float. A specific gravity of 1.280 shows a fully charged battery, and a reading of 1.110 shows a discharged battery. The specific gravity of pure water is 1.0. Remember, though, that the density of sulfuric acid changes with temperature, and so its specific gravity is different at different temperatures. Most manufacturers rate their cell readings at 80°F. A good hydrometer will include a thermometer to check the electrolyte temperature, and give compensating adjustments for each 10 degrees of deviation.

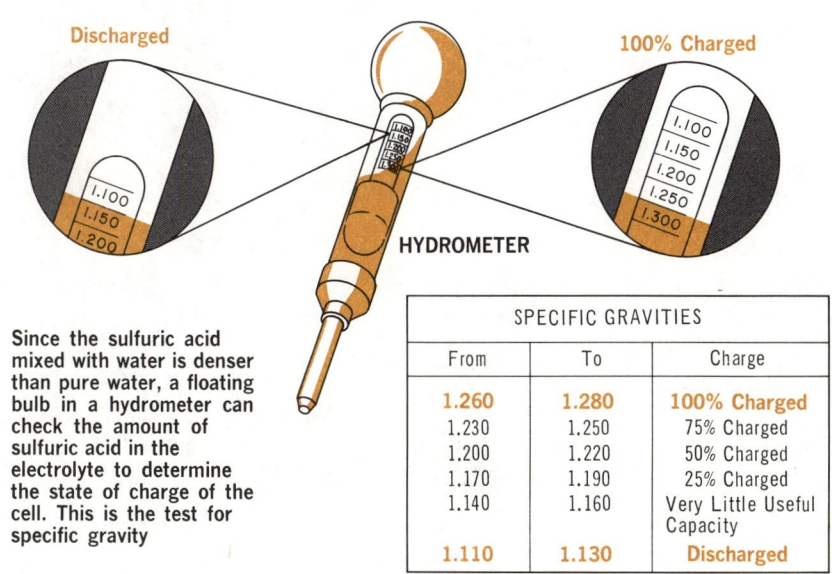

Since the sulfuric acid mixed with water is denser than pure water, a floating bulb in a hydrometer can check the amount of sulfuric acid in the electrolyte to determine the state of charge of the cell. This is the test for specific gravity

SPECIFIC GRAVITIES		
From	To	Charge
1.260	1.280	100% Charged
1.230	1.250	75% Charged
1.200	1.220	50% Charged
1.170	1.190	25% Charged
1.140	1.160	Very Little Useful Capacity
1.110	1.130	Discharged

charging methods

Storage batteries can be charged in various ways. The two basic methods are by *constant-current* and *constant-voltage chargers*. Either method can also be used to give a *high-rate charge*, a *low-rate charge*, or a *trickle charge*. Regardless of the method used, the charging current must be dc. The constant-current method with a low rate is the safest way to charge a battery, but it takes the longest—about 16 to 24 hours with a charging current of about 10 amperes. The constant-voltage, high-rate charger is the fastest method, but it has a tendency to damage batteries that may not be in the best condition. The high current accelerates the chemical action which could deteriorate the electrodes and boil the water, making the electrolyte too strong. This type of charging begins with a current at 50 to 100 amperes, which decreases as the battery becomes charged. This is why this method is also called *taper charging*. A moderate charge can be given with this method in one hour, but a few hours are needed for a full charge. The trickle charger is used to provide a small current to a battery while it is being used, thereby keeping the battery fully charged while it is operating. This, though, has a tendency to overcharge the battery and damage it if charging is not accurately controlled.

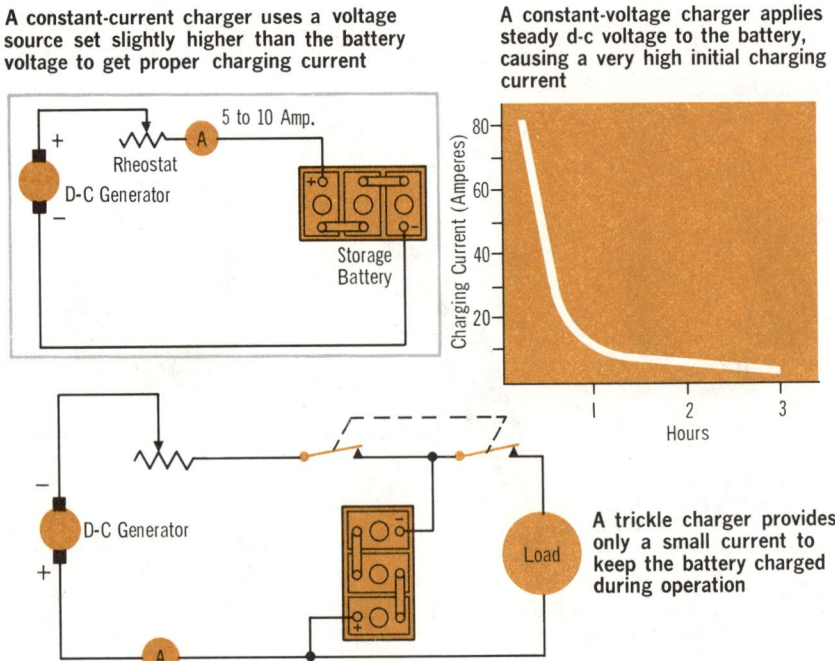

A constant-current charger uses a voltage source set slightly higher than the battery voltage to get proper charging current

A constant-voltage charger applies steady d-c voltage to the battery, causing a very high initial charging current

5 to 10 Amp.

Rheostat

D-C Generator

Storage Battery

Charging Current (Amperes)

D-C Generator

Load

A trickle charger provides only a small current to keep the battery charged during operation

construction

As you learned when you studied the primary wet cell, the amount of current a cell can deliver depends on the surface area of the electrodes. In order to give the electrodes in the lead-acid cell a large effective electrode area so it can deliver large currents, each electrode element is made as a series of *plates*. Then the plates of the negative electrode and those of the positive electrode are *interleaved* so that the negative and positive plates are close together for efficient battery action. Thin sheets of nonconducting porous materials separate the plates to keep them from shorting.

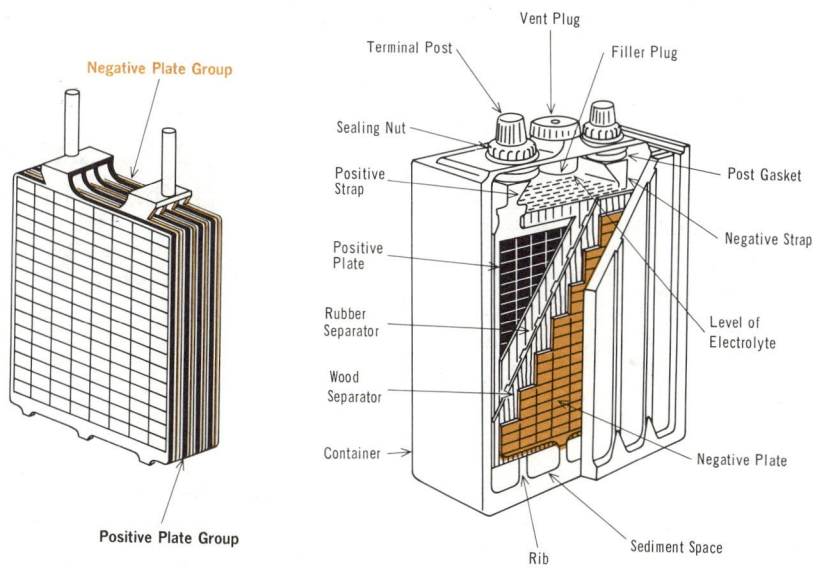

The set of plates for each electrode is connected by a *lead strap* that is attached to the associated lead terminal. The interleaved sets of plates are then usually encased in an acid-resistant molded container. Since the electrode materials are made of forms of lead, they are too soft to stay rigid. Therefore, the plates consist of grids, which have holes to hold the electrode materials. A lead-antimony alloy is generally used for the grid framework. A filler plug is provided at the top of the cell to allow the electrolyte level to be checked, and to allow water to be added when necessary. The filler plug generally has a vent hole to allow gases to escape.

disadvantages

You know that the operation of the lead-acid cell depends on the ability of the lead sulfate on the electrodes to be changed to lead and lead peroxide during the charging cycle. With a battery that is well cared for and *kept* fully charged, this is not a problem. But, when a battery is allowed to remain partially discharged for a long period, the lead sulfate tends to become hard and brittle; this is known as *sulfation*. The sulfated area of an electrode will not react properly with the electrolyte, and so the current capacity of the cell goes down. The sulfation process can continue until so much of the electrode area is lost that the battery becomes too weak to be useful. Also, the brittle sulfate can fall off the electrode. Generally, it drops to a *sediment* area at the bottom, but it could become lodged between the electrode plates and short the cell.

To reduce sulfation, the lead-acid cell must be kept charged, and the electrolyte level must be kept at least 3/8 inch above the plates

Another problem with this cell results from the electrolysis action during charging. Remember that when a cell is overcharged, the excess hydrogen and oxygen ions become gases and escape, reducing the water level of the electrolyte. Even during normal charging this happens, but to a lesser extent. Ultimately, the water content reduces to a point where the electrolyte level is too low, and the relative content of sulfuric acid is too high. This causes sulfation of the exposed areas of the electrodes to accelerate, and the overly strong electrolyte will attack the supporting member of the cell, which should not react with the solution. The electrolyte level must be periodically checked, and distilled water must be added to keep the level about 3/8 inch above the electrode plates.

The electrolyte in this type of cell also has a tendency to freeze in cold weather, particularly when the cell is not fully charged and the specific gravity of the electrolyte is low. The reason for this is that the electrolyte is mostly water, which freezes sooner than sulfuric acid. And when water freezes, it expands and buckles the electrode plates.

current ratings

Although the physical size of a battery has no effect on the emf the battery produces, it does affect the amount of current a battery can supply. The larger the area that the electrodes have, the more current a battery can supply. This is why storage batteries are made with interleaved plates. Both sides of each plate for a given electrode are available to supply current. The interleaving of plates also permits the electrodes to be close together. This reduces the internal resistance of the battery, so that high currents do not decrease the output voltage as much as they might otherwise.

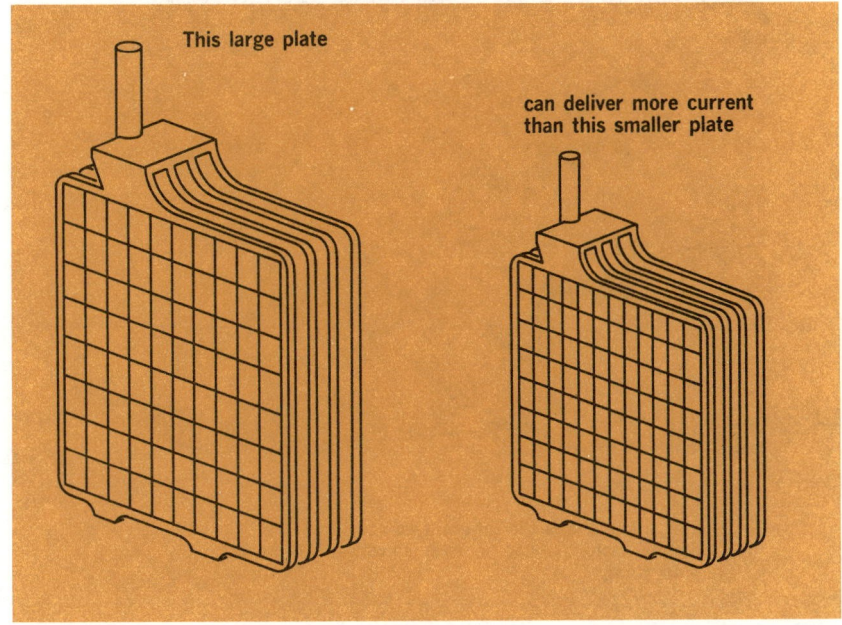

This large plate

can deliver more current than this smaller plate

Batteries are rated according to the amount of *current* they can supply in a given amount of *time*. The rating is given in *ampere-hours*. For storage batteries, the standard time used is generally 20 hours; so, if a storage battery is rated at 100 ampere-hours, it means that it will supply 5 amperes for 20 hours before its emf will drop to the discharged level, which is 1.75 volts per cell. But it can also supply more current for less time, or less current for more time; the ampere-hour rating is the same. For example, it can supply 50 amperes for 2 hours, or 4 amperes for 25 hours. The less current it supplies, the longer it will last.

current ratings (cont.)

In addition, the basic battery rating is given for operation at a temperature of 80°F. At low temperatures, the chemical activity of the battery slows down, and the battery cannot supply as much current. Therefore, batteries that are used outdoors, such as auto batteries, also are rated in ampere-hours for lower temperatures, such as 0°F.

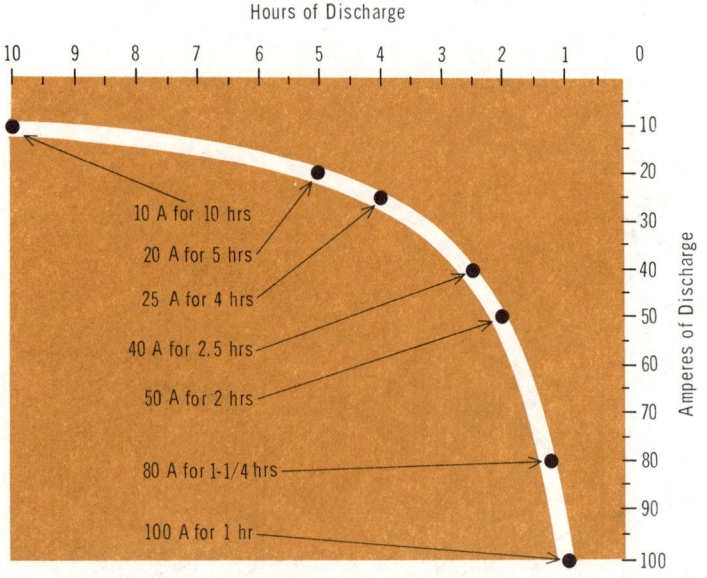

Hours of Discharge

10 A for 10 hrs
20 A for 5 hrs
25 A for 4 hrs
40 A for 2.5 hrs
50 A for 2 hrs
80 A for 1-1/4 hrs
100 A for 1 hr

Amperes of Discharge

This curve shows the different current-time combinations that can be obtained with a 100-ampere-hour rating

Some storage batteries, used for mobile lighting, are rated for four hours of operation, but this rating is still given as *ampere-hours*. Another older method of current rating, which is sometimes also given, is the maximum current that a battery can supply for *20 minutes*. This is just given in *amperes*, and should not be considered with the ampere-hour rating. It is known as the *20-minute rating*.

The ampere-hour rating is specified for *continuous* use over a certain period. Actually, if the battery is used only intermittently, it will last longer than its rating indicates, because it can rejuvenate itself between discharge cycles.

summary

☐ When the charging current is removed from the lead-acid storage cell, a chemical action occurs that is the reverse of that which takes place during charging. This chemical action causes a positive potential to be produced on one electrode and a negative potential on the other. ☐ When fully charged, the lead-acid cell has a no-load output voltage of about 2.1 volts. The output voltage drops to about 2 volts when the cell is supplying current. ☐ The lead-acid cell is considered discharged when its output voltage falls to 1.75 volts.

☐ The state of charge of a storage cell can be determined by measuring the specific gravity of the electrolyte with a hydrometer. ☐ A fully charged cell has a specific gravity of 1.280; a discharged cell 1.110.

☐ Storage batteries can be charged by either constant-current or constant-voltage chargers. Both types can be used to give a high-rate charge, a low-rate charge, or a trickle charge. ☐ The constant-current method with a low rate of charge is the safest charging method, but takes the longest. ☐ The constant-voltage high-rate method is the fastest, but can damage batteries that are not in good condition. This method is also called taper charging, because the charging current decreases as the battery becomes charged. ☐ Sulfation occurs when a lead-acid battery remains partially discharged for a long period. Effectively, sulfation causes part of the electrode to be lost.

review questions

1. What is the output voltage of a lead-acid cell at the instant the charging current is removed? Explain.
2. Explain briefly how a potential is developed on the electrodes of a lead-acid cell.
3. At what output voltage is a lead-acid cell considered discharged?
4. How is the specific gravity of the electrolyte related to the charge of a lead-acid cell?
5. What is the specific gravity of a fully charged lead-acid cell? Of a discharged cell?
6. Can ac be used to charge a storage battery?
7. What is *taper charging*?
8. What is *sulfation*?
9. Why must distilled water periodically be added to a lead-acid storage battery?
10. Why is it good practice to keep storage batteries fully charged at all times?

the alkaline secondary cell

The *alkaline* cell is much more expensive than the lead-acid cell, but it is still finding more and more use because it requires much less attention than the lead-acid cell and has a much longer useful life. Also, its electrodes are made of lighter metals, so that the cell itself is not nearly as heavy as a lead-acid cell. As you will soon learn, the electrolyte in an alkaline cell does not change chemically during charge and discharge, and so it does not tend to freeze as much as the electrolyte in a lead-acid cell. Because of this property, the specific gravity of the alkaline cell does not change, and so cannot be used as a method of checking the state of the battery charge.

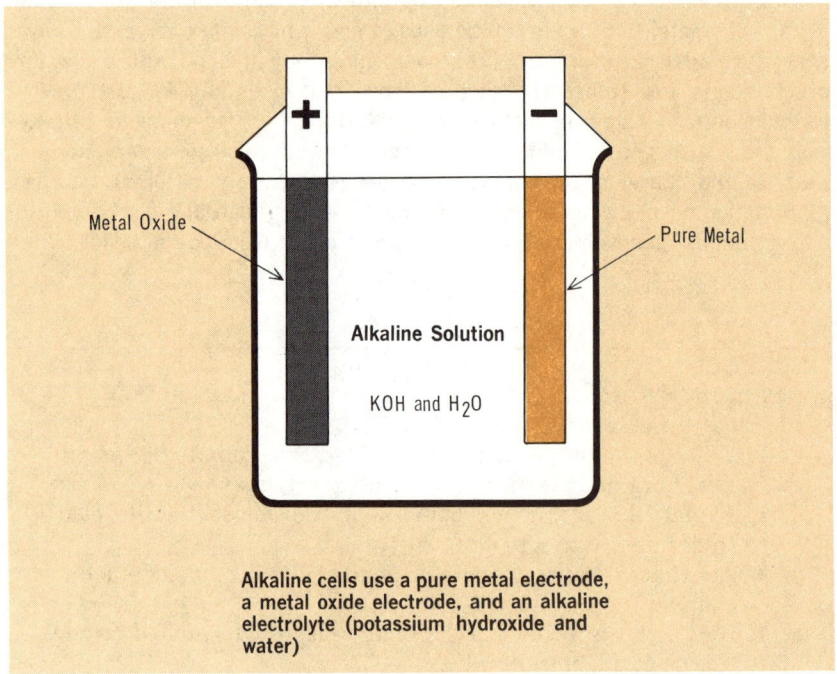

Metal Oxide

Pure Metal

Alkaline Solution

KOH and H_2O

Alkaline cells use a pure metal electrode, a metal oxide electrode, and an alkaline electrolyte (potassium hydroxide and water)

Basically, the alkaline cell is so called because its electrolyte is an alkaline solution rather than an acid. This is just another type of chemical that reacts differently with metals than does an acid. The most common types of alkaline cells are the nickel-iron oxide cell, which was invented by Thomas A. Edison, the nickel-cadmium cell, and the silver-zinc cell. Each of these cells basically use a *pure* metal electrode, a metal oxide electrode, and an *alkaline* electrolyte of potassium hydroxide (KOH) mixed with distilled water (H_2O).

charging

Basically, the alkaline cell works similarly to the acid cell, in that the electrode materials change during the charge and discharge cycles. The electrolyte, though, does not change. All of the alkaline cells work in a similar manner.

To begin, you should know that a metal oxide is the metal combined with oxygen. The more oxygen atoms with which it combines, the more metal oxide molecules are present, and vice versa. When the alkaline cell is uncharged, both electrodes are metal oxides. Since the electrodes are not dissimilar, they do not produce an emf. But when a charging source is connected to the cell, the emf of the source produces electrolysis in the electrolyte. The electrolyte breaks down into an equal number of positive and negative ions. Only the negative ions, which are oxygen ions, react with the electrodes, so we can disregard the rest of the solution. The negative oxygen ions are repelled from the negative electrode, and are attracted to the positive electrode because of the polarities of the charging source. When an oxygen ion reaches the positive electrode, it repels electrons out of a metal oxide molecule, making it positive. The negative oxygen ion is then attracted to the metal oxide ion, and they combine to make the electrode more of an oxide. The electrons released by the oxide molecules are attracted out of the electrode by the charging source, to produce part of the charging current.

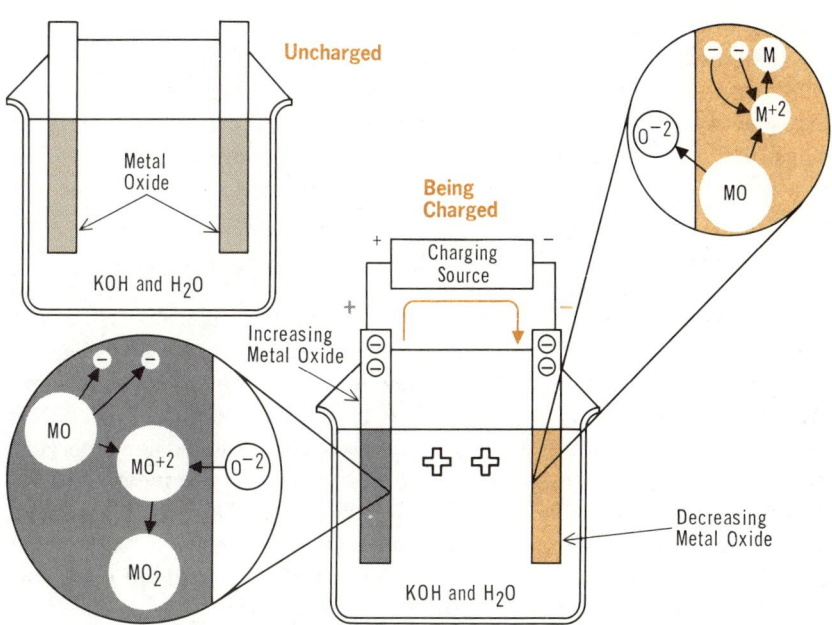

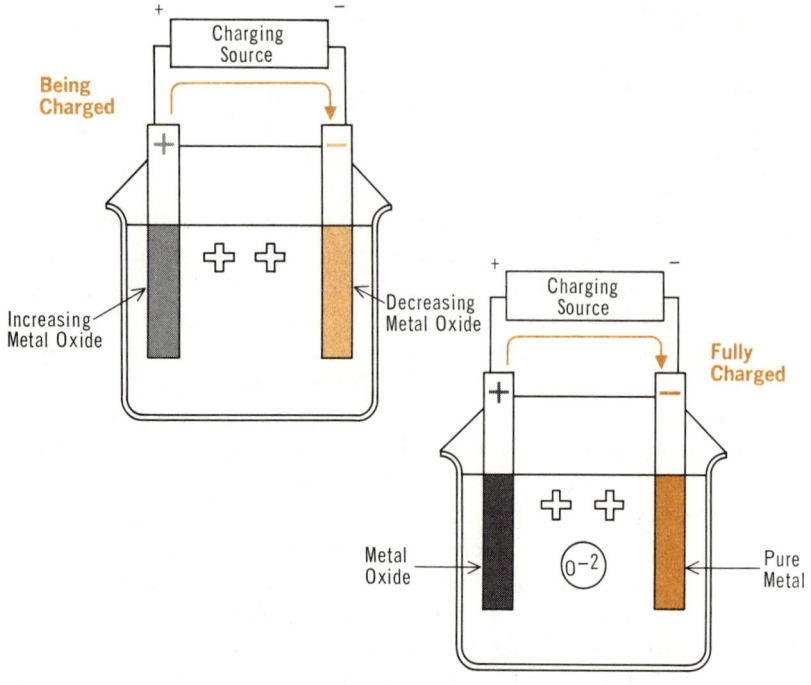

charging (cont.)

When an oxygen ion combines with the positive electrode, the electrolyte becomes electrically unbalanced, and takes on a positive charge. The positive attraction of the electrolyte causes the metal oxide molecules of the negative electrode to break up into positive metal ions and negative oxygen ions. The solution then attracts the oxygen ions into it. And the positive metal ions in the electrode attract electrons out of the charging source to make up the rest of the charging current. Since the negative electrode gave up an oxygen ion, it becomes less of an oxide. This process continues with the positive electrode taking oxygen ions out of the electrolyte to become more of an oxide, and the negative electrode giving up oxygen ions to become less of an oxide. The positive electrode releases electrons and the negative electrode attracts electrons to make up the charging current. This goes on until the negative electrode no longer has oxygen atoms, and so becomes a pure metal, and the positive electrode is fully oxidized: the cell is then completely charged.

Note that for each oxygen ion that the electrolyte gave to the positive electrode, a replacement was received from the negative electrode. Therefore, the nature of the electrolyte *never changes.*

how an emf is developed

You probably noticed that during the charging cycle, positive ions were produced in both electrodes. But in the positive electrode, the positive ions were neutralized when they combined with the negative oxygen ions. And in the negative electrode, the positive ions were neutralized by the electrons from the charging source. Therefore, while it is being charged, the alkaline cell does not build an emf of its own. It stores chemical energy, the same as the lead-acid cell, but as soon as the charging source is removed, the chemical action *reverses*. The negative oxygen ions in the electrolyte are no longer repelled by the pure metal electrode. Instead, they react with the electrode, and cause the metal atoms to give up electrons. The positive metal ions then attract and combine with the negative oxygen ions to produce a small amount of neutral metal oxide. The electrons that are freed accumulate in the pure metal electrode to give it a negative charge. When this negative charge is sufficient, the oxygen ions in the electrolyte are repelled to prevent further oxidation of the electrode.

Since negative oxygen ions have been taken out of the electrolyte, the electrolyte takes on a positive charge. The positive attraction of the electrolyte causes some of the molecules of the metal oxide electrode to break down into positive metal oxide ions and negative oxygen ions, and the oxygen ions are attracted into the electrolyte. This continues until the electrolyte again becomes neutral. The positive ions left behind in the metal oxide electrode build up a positive charge. The difference in potential between the two electrodes is the emf of the cell.

In the process of developing the emf, the pure metal electrode ($-$) became slightly oxidized, and the metal oxide electrode ($+$) became slightly less oxidized. However, since the oxygen ions that were given up to the negative electrode were replaced by those released by the positive electrode, the electrolyte did not change.

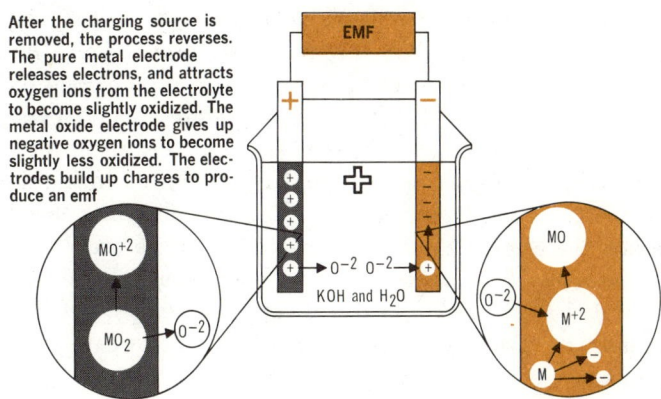

After the charging source is removed, the process reverses. The pure metal electrode releases electrons, and attracts oxygen ions from the electrolyte to become slightly oxidized. The metal oxide electrode gives up negative oxygen ions to become slightly less oxidized. The electrodes build up charges to produce an emf

the discharge cycle

Essentially, when the alkaline cell discharges, it just continues the process that built up the emf. Without current flow, the action stopped because the negative charge on the pure metal electrode prevented further oxidation. But when a load is connected across the cell, and current flows, electrons leave the negative electrode and enter the positive electrode. The number of charges on the electrodes decreases. But

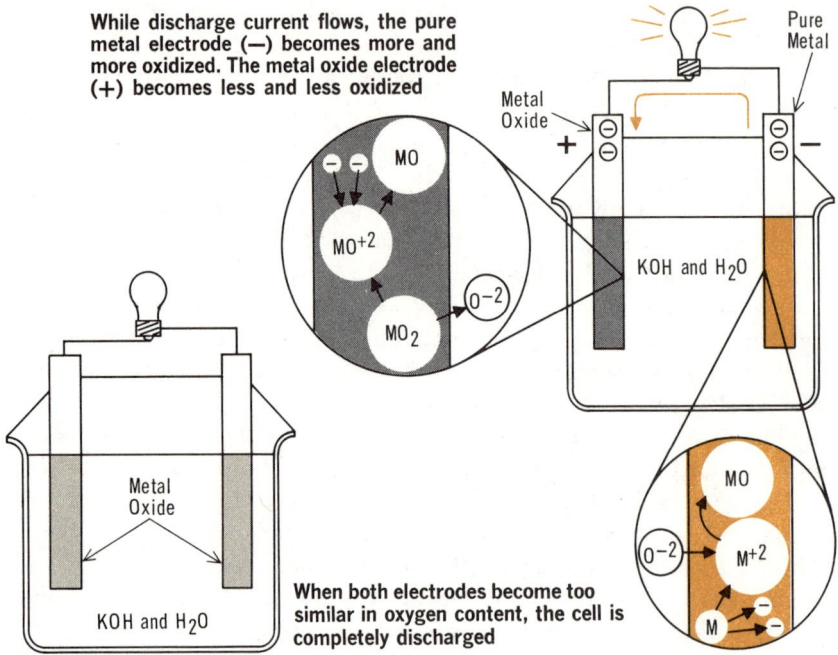

While discharge current flows, the pure metal electrode (−) becomes more and more oxidized. The metal oxide electrode (+) becomes less and less oxidized

When both electrodes become too similar in oxygen content, the cell is completely discharged

this allows negative oxygen ions from the electrolyte to again combine with the metal atoms in the negative electrode to release more free electrons to replace those that became part of the current flow. And, as a result, the metal oxide electrode releases oxygen ions to replace the ones given up by the electrolyte; it, therefore, creates new positive metal ions and negative oxygen ions to replace those neutralized by the current flow. Because of this, as discharge current continues to flow, the pure metal electrode starts changing into more of a metal oxide, and the positive electrode becomes less of a metal oxide.

When the cell is discharged, both electrodes become similar in oxygen content, and the cell is too weak to work. It must be recharged. But throughout it all, the electrolyte did not change; it merely exchanged oxygen ions between the electrodes.

typical alkaline cells

The oldest alkaline cell is the nickel-iron wet cell, called the *Edison cell*. It uses a nickel dioxide (NiO_2) positive electrode, and a pure iron (Fe) negative electrode. During the discharge cycle, the negative electrode changes into iron oxide (FeO_3) and the positive electrode into nickel oxide (Ni_3O_4). Fully charged, this cell produces 1.37 volts, which drops to 1 volt when it is considered discharged. Normal voltage is 1.2 volts. It is used mainly with railroad and telegraph signals, and portable lamps.

Alkaline cells can also be dry primary or secondary cells. A typical alkaline dry cell is the *nickel-cadmium cell*. It uses a positive electrode of nickel dioxide (NiO_2), and a negative electrode of pure cadmium (Cd). The cell also produces 1.2 volts, which remains relatively steady during discharge.

Button Type

Contact Spring / Cover / Gasket / Screen Wrapped Negative Electrode / Separator / Can / Screen Wrapped Positive Electrode / Expanded Metal Spacer

Sintered Plate Type

Negative Pole / Positive Pole / Positive Plates / Negative Plates

The nickel-cadmium alkaline cell has become one of the most popular storage batteries, especially for use in small portable equipment

More recently, the *silver-zinc cell* has a positive electrode of silver oxide (Ag_2O), and a negative electrode of pure zinc (Zn). It delivers 1.86 volts, which drops to 1.6 volts when discharged. It is also available as a rechargeable secondary cell.

alkaline dry cells

The zinc-mercury cell, commonly called the *mercury battery*, is another popular alkaline cell. It uses a negative zinc electrode, and a positive electrode that is a mixture of mercuric oxide and graphite. The electrolyte is a paste of potassium hydroxide and zinc hydroxide. The no-load voltage of this cell is 1.34 volts, which drops to between 1.31 and 1.24 volts when normal current is supplied. It is more expensive than the Leclanche cell but has a higher *constant current* rating. Special versions of the mercury cell can maintain such a steady voltage that they can be used as laboratory *voltage references*. Another variation of the mercury cell is a rechargeable secondary type, and yet another is the zinc-mercury dioxysulfate cell, which uses a zinc sulfate electrolyte.

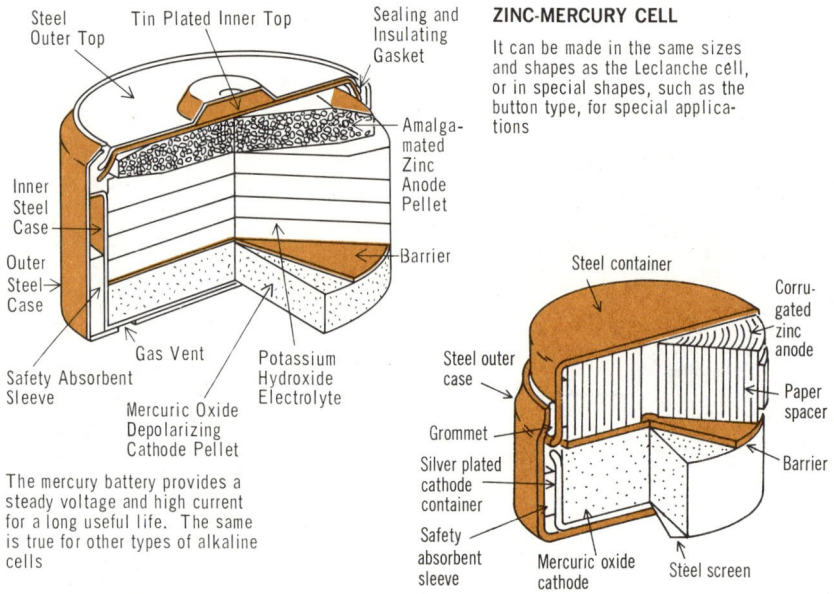

ZINC-MERCURY CELL

It can be made in the same sizes and shapes as the Leclanche cell, or in special shapes, such as the button type, for special applications

The mercury battery provides a steady voltage and high current for a long useful life. The same is true for other types of alkaline cells

The silver oxide-cadmium cell is a secondary cell. It uses a potassium hydroxide electrolyte. It has a no-load voltage of 1.4 volts, which drops to around 1 volt in use.

The zinc-manganese dioxide cell, with a potassium hydroxide electrolyte, is the principal alkaline primary dry cell that has been replacing the Leclanche cell in ordinary commercial applications. Like the Leclanche cell, it produces about 1.5 volts and is available in the typical popular Type N, AA, AAA, C, and D sizes.

summary

☐ Alkaline secondary cells use alkaline solutions as electrolytes instead of acids. ☐ Alkaline cells require less attention and have longer useful lives than lead-acid cells; however, they are considerably more expensive. ☐ The specific gravity of the electrolyte cannot be used to determine the state of charge of an alkaline cell. This is because the electrolyte does not change chemically during charging and discharging.

☐ An uncharged alkaline cell has two metal oxide electrodes and an alkaline electrolyte. ☐ A charging current causes chemical action within the cell which turns one electrode into a pure metal and the other into more of an oxide. ☐ When the charging source is removed, the chemical action within the cell reverses and causes a positive charge to be produced on one electrode and a negative charge on the other.

☐ One type of alkaline cell uses a nickel dioxide positive electrode and a pure iron negative electrode. This nickel-iron cell is also called the Edison cell. ☐ When fully charged, the Edison cell has an output of about 1.37 volts. ☐ The nickel-cadmium cell has a nickel dioxide positive electrode and a pure cadmium negative electrode. When charged, it produces an output voltage of about 1.2 volts. ☐ The silver-zinc cell has a silver oxide positive electrode and a pure zinc negative electrode. Its fully-charged output voltage is about 1.86 volts. ☐ The electrolyte of most alkaline cells is potassium hydroxide mixed with distilled water.

review questions

1. In an alkaline secondary cell, is there a relationship between the specific gravity of the electrolyte and the state of charge of the cell?
2. What are the advantages of an alkaline cell over a lead-acid cell?
3. Is an alkaline cell charged or discharged when both electrodes are metal oxides?
4. What is the function of the electrolyte in an alkaline cell?
5. What is an *Edison cell*?
6. In a charged alkaline cell, is the pure metal electrode positive or negative?
7. What effect does the charging current have on the electrolyte of an alkaline cell?
8. Why doesn't the electrolyte of an alkaline cell tend to freeze as much as the electrolyte of a lead-acid cell?
9. What is a *metal oxide*?
10. What electrolyte is used in most alkaline cells?

voltage ratings

Earlier, you learned that the voltage output of a battery depends only on the materials used for the electrodes and the electrolyte. The amount of material used or the physical size of the battery has nothing to do with the battery's voltage rating. Each specific combination of chemicals produces a specific emf. The only way that the voltage rating can be increased is by using more cells in series.

Batteries are rated with a *no-load voltage.* This is the electromotive force the battery delivers when it is *not* supplying current. When it *is* supplying current, the battery voltage *drops* slightly, up to 0.1 volt or so per cell. The *voltage under load* will drop even more with batteries that are more worn out. This is explained on page 6-24. For this reason, if you test a battery when it is not delivering current, the voltage reading you will get will not be meaningful. You should shunt the battery so that it will supply its rated current to see how much its voltage will drop. If the voltage drops considerably, the battery is worn out. Special *resistance- shunt voltage testers* are available to check storage batteries in this way.

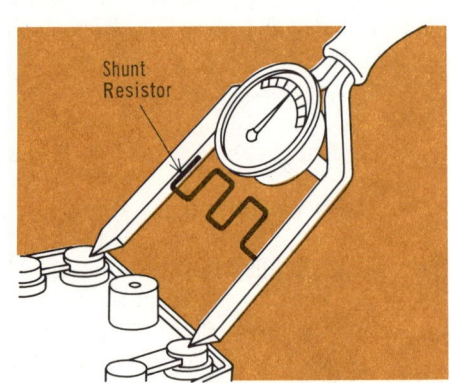

The shunt resistor on this voltmeter causes a high current to flow, so that the voltage of the battery under load can be tested

TYPICAL CELL VOLTAGE RATINGS

Type of Cell	Voltage Ratings		
	No Load	With a Load	Discharge Level
Copper-zinc-sulfuric acid, primary wet cell	1.08	1.008	About 0.8
Carbon-zinc-chromic acid, primary wet cell	2.0	1.9	About 1.7
Carbon-zinc-ammonium chloride, primary dry cell (Leclanche cell)	1.5	1.4	About 1.2
Mercury-zinc-potassium hydroxide, primary dry cell (mercury cell)	1.34	1.31 to 1.24	About 1.0
Lead-acid storage cell	2.1	2.0	About 1.75
Nickel-iron, alkaline storage cell (Edison cell)	1.37	1.3	About 1.0
Nickel-cadmium, alkaline storage cell	1.3	1.2	About 1.0
Silver-zinc, alkaline storage cell	1.95	1.86	About 1.6

other voltaic cells

For the most part, the battery relies on the chemical energy of the electrolyte to cause the electrodes to build up charges. You learned in Volume 1, though, that other forms of energy can be used to produce electric charges. The type of energy most commonly used is magnetism. The devices that produce electrical energy with magnetism fall into a category referred to as d-c and a-c generators, which are explained in the latter part of this volume.

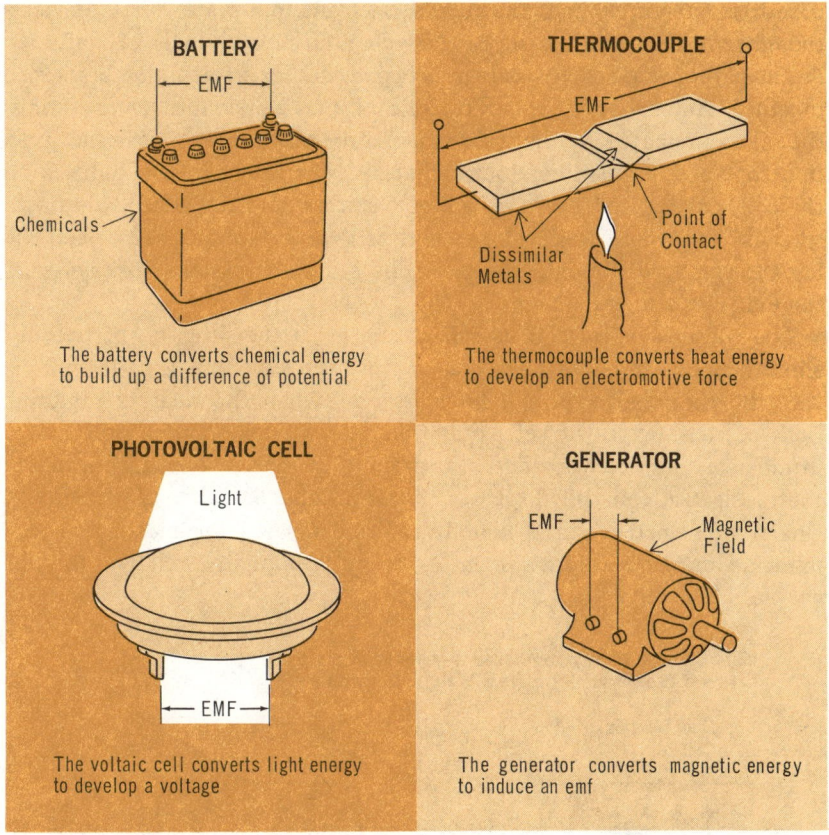

BATTERY

EMF

Chemicals

The battery converts chemical energy to build up a difference of potential

THERMOCOUPLE

EMF

Dissimilar Metals

Point of Contact

The thermocouple converts heat energy to develop an electromotive force

PHOTOVOLTAIC CELL

Light

EMF

The voltaic cell converts light energy to develop a voltage

GENERATOR

EMF

Magnetic Field

The generator converts magnetic energy to induce an emf

Light and heat can also be used by power sources to generate an emf. Sources that use light are known as *photovoltaic cells,* and those that use heat are called *thermoelectric cells.* An emf can also be generated by the triboelectric and piezoelectric effects, but these effects are not usually used in power sources.

photovoltaic and thermoelectric cells

The photovoltaic cell is one of three major categories of photocells. There are also *photoconductive* and *photoemissive* cells. These two types, though, are not power sources. They must be energized by a ·power source before they can work. The photoconductive cell is basically a variable resistive device that permits more current to flow in a circuit when it is struck by light; and a photoemissive cell is usually an electron tube that has an element that emits electrons to a plate. These devices, though, are not voltaic cells because they do not produce an emf.

The *photovoltaic* cell usually uses two dissimilar semiconductors joined together. And when light strikes one of the semiconductor materials, the energy released by the light causes free electrons to cross the junction to the other semiconductor material. Thus, while the light is on the cell, one side of it has a lack of electrons and the other side has a surplus; so an emf is developed between the two materials. When the light is removed, though, the electrons return across the junction and the emf drops to zero.

The *thermoelectric* cell uses heat energy at the junction of two dissimilar metals to cause electrons to cross the junction. Like the photocell, the thermoelectric cell loses its emf when the heat is removed. Both of these types of cells only generate a fraction of a volt, but a number of them can be added in series to increase the emf. A series of thermoelectric cells piled on top of one another is called a *thermopile*. One thermoelectric cell is usually referred to as a *thermocouple*. Lead telluride and germanium silicon perform exceptionally well as a thermocouple.

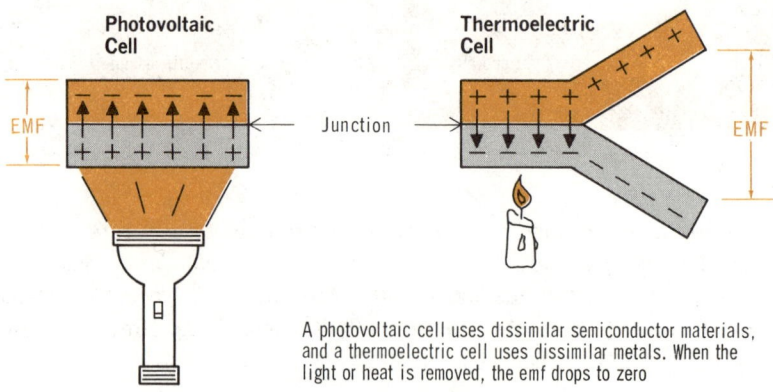

Light or heat energy can cause a transfer of electrons between two dissimilar materials to build up an emf

A photovoltaic cell uses dissimilar semiconductor materials, and a thermoelectric cell uses dissimilar metals. When the light or heat is removed, the emf drops to zero

solar cells

You can see that the basic photovoltaic and thermoelectric cells are limited in their usefulness as power sources because they lose their emf when the heat or light energy disappears. By themselves, these types of cells are usually used as *signal* sources, not as power sources. But, when they are combined with chemical storage cells, they become good power sources. This is generally the way *solar cells,* which are used in portable radios and artificial satellites, make use of the sun's energy.

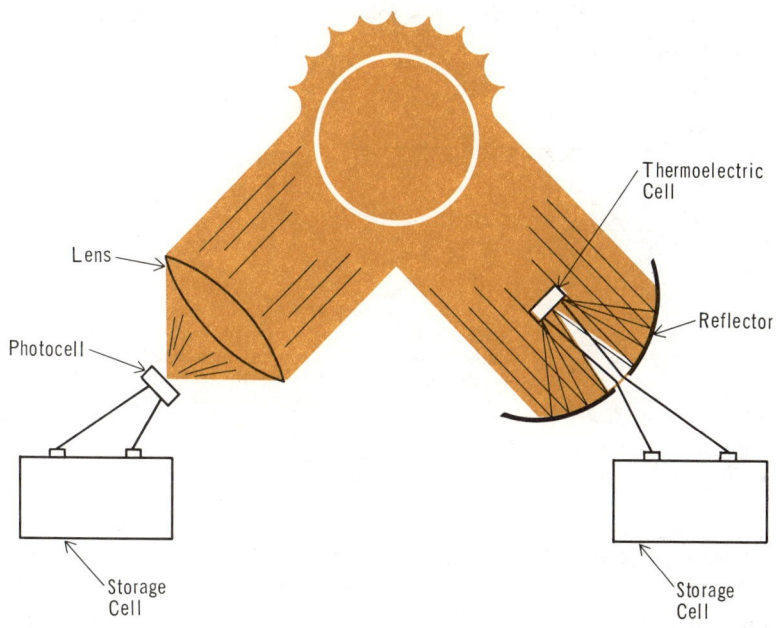

A solar cell uses photovoltaic or thermoelectric cells as primary cells to keep a storage cell charged

The photovoltaic and thermoelectric cells act as **primary sources that** keep a secondary cell charged. Then, as the emf of the primary cells fades when the sun goes down, the secondary cell continues to supply power. The secondary cell is repeatedly recharged whenever the sun's rays strike the primary cells. The photovoltaic cell uses the light energy of the sun directly, but the thermoelectric cell uses the heat from the sun's rays. Lenses of various types are used to concentrate the rays to make the cells more effective, and a number of cells are connected in series to produce the proper emf.

thermionic generator

The *thermionic generator* is one of the more sophisticated developments resulting from our space program research to produce more reliable electrical sources.

A typical example of a thermionic generator is the *cesium cell.* Cesium is a metal that liquifies easily at a little more than room temperature, and also boils over and becomes gaseous at much lower temperatures than other metals. In addition to this, cesium atoms have six shells, with a single valence electron that is so loosely held in orbit that it is easily set free. For this reason, cesium compounds are often used in photocells.

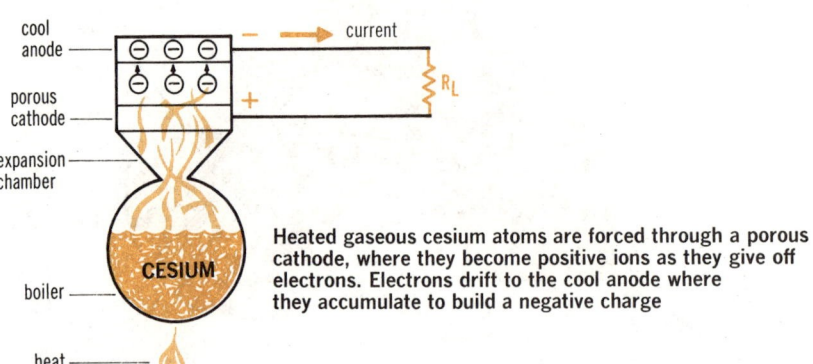

Heated gaseous cesium atoms are forced through a porous cathode, where they become positive ions as they give off electrons. Electrons drift to the cool anode where they accumulate to build a negative charge

These same characteristics make cesium the ideal material to use in a thermionic generator. The cesium is contained in a boiler, which is heated until the cesium liquifies and boils. Hot cesium vapors rise into an expansion chamber and a high vapor pressure builds there. The cesium gas heats a porous cathode, and high energy cesium atoms are forced through the cathode into a vacuum chamber. The high energy atoms give off their valence electrons, and the positive cesium ions form a positive space charge on the cathode. The electrons drift to the cool anode where they accumulate and give off their excess energy. The anode of the cesium cell then takes on a negative charge, and the cathode, a positive charge.

When a load is placed across the cell, electrons flow from the anode to the cathode. Electrons returning to the cell's cathode rejoin cesium ions to neutralize them, but new high energy cesium atoms release electrons to continue the current flow.

The disadvantage of the thermionic generator is, of course, the heating equipment required.

magnetohydrodynamic generator

The *magnetohydrodynamic generator* is another example of America's space-program efforts to produce new, more reliable electrical sources.

This type of generator also uses high energy *plasma* (ionized gas) as the basic energy source. But in this case the plasma is made to interact with a magnetic field to produce the desired electrical charges. A highly conductive plasma is used to provide the source of free electrons.

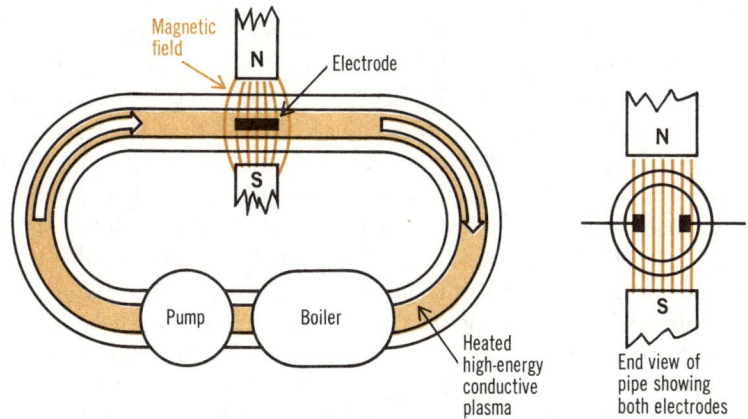

In the magnetohydrodynamic generator, free electrons in the fast-moving plasma are diverted by the magnetic field to the negative electrode for delivery to an electrical circuit. The electrons return to the other electrode to re-enter the plasma stream

The plasma is heated to ionization and pumped in a closed loop so that the ionized gases and free electrons circulate rapidly around the plumbing and through a magnetic field set up by a pair of permanent magnet poles. On either side of the plasma path within the magnetic field are electrodes. As the plasma passes through the field, free electrons are driven at right angles to the field and collect on the negative electrode. When a load is connected between the electrodes, these electrons flow in the circuit around to the positive electrode where they re-enter the flowing plasma stream. The hotter the plasma, the more free electrons there will be available, and the faster the plasma travels through the magnetic field, the greater the accumulation of electrons on the negative electrode.

For proper operation, the magnetic field, plasma flow, and the electrodes must all be at right angles to one another.

You can see that the magnetohydrodynamic generator also requires a good deal of auxiliary equipment.

the fuel cell

The fuel cell is one of the more promising developments of America's space program. Essentially, the fuel cell utilizes the basic principles of the battery cells you studied earlier, but arranges to have a continuous chemical supply to prevent the major disadvantage of the ordinary battery cell. The ordinary battery cell continues to discharge until either the chemical activity of the electrolyte and electrodes becomes impotent, or the electrodes deteriorate. The battery cell, then, has to be recharged or replaced.

In the fuel cell, the electrodes do not take part in the chemical activity that builds up the charge. The electrodes provide only a conducting path for the current. The chemicals that react electrochemically are continuously fed into the fuel cell, and after they produce the current, their impotent products are removed from the cell, and the cell is refueled with active chemicals to continue the action. This takes place continuously, with potent fuel fed in to replace the impotent product that is drained. Therefore, as long as there is a fuel supply, the cell will not deteriorate or become discharged.

The fuel cell uses two chemicals. One is oxygen, and the other is the fuel agent that reacts with the oxygen to produce the impotent product, which is usually water. Three types of fuel are used: (1) hydrogen, (2) hydrocarbon, and (3) a biochemical, which are enzyme and bacteria waste products. At present, hydrogen works best, but it is expensive. The hydrocarbons—gasoline, kerosene, butane, or any other petrochemical—are cheaper, but their carbon content interferes with their activity. The biochemicals are the least reliable, but would be the cheapest. They require the most heat to function properly. You can see that, like the other sophisticated cells, the fuel cell requires plumbing, pumps, and heating equipment to function properly. And the waste product must be disposed of. Also, the flow of fuel must be regulated to match the power drawn by the flow of current. However, this type of cell shows great promise.

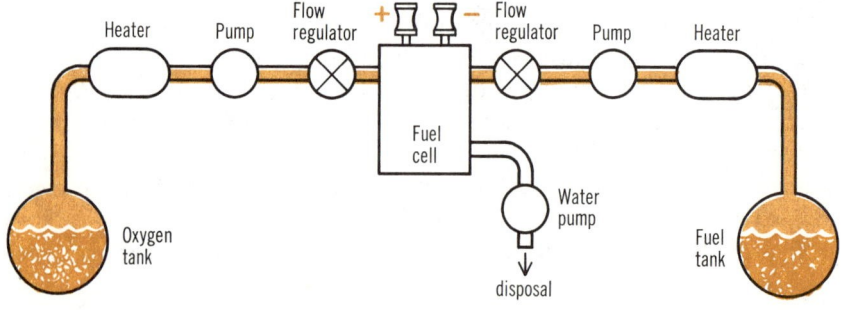

fuel cell operation

The fuel cell is constructed with the electrodes near each side of the cell. The chamber between the electrodes and the cell walls is where the oxygen and hydrogen (fuel) are fed in, each in its own chamber. The large chamber between the electrodes contains an alkaline electrolyte solution of potassium hydroxide (KOH), the same as the alkaline cells you studied earlier. Except for the thin electrodes and the plumbing, the fuel cell looks like the alkaline secondary storage cell.

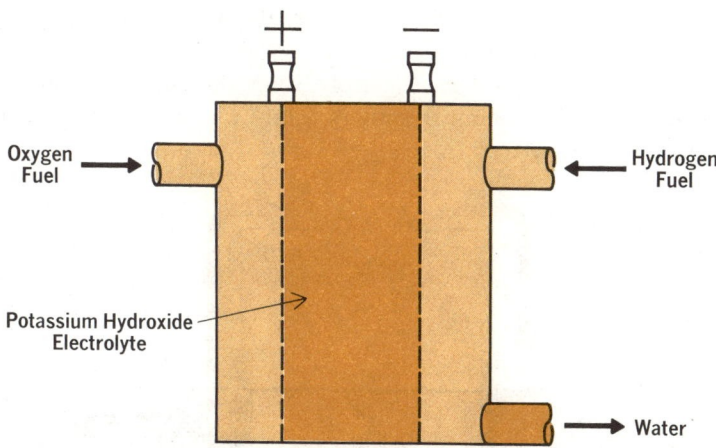

In the fuel cell, however, the chemical gases perform the same function as the electrodes in the earlier cell by reacting with the electrolyte electrochemically. The electrodes here are merely for electrical contact and to contain the chemicals in their respective chambers. The electrodes are made of porous membranes coated with a precious metal, such as platinum. They are porous enough to allow the electrolyte to seep through into each gaseous chamber just enough to react with the gases. The gas pressure keeps the electrolyte near the electrodes.

At the negative electrode, the heated hydrogen atoms ionize, giving off electrons, which accumulate on the electrode to give it a negative charge. These electrons flow to the load. This chemical action prompts a reaction in the water (H_2O) molecules of the electrolyte. The hydrogen atoms give off electrons to the oxygen atoms, creating positive hydrogen ions and negative oxygen ions. The negative oxygen ions are attracted to the positive hydrogen ions of the fuel and combine to produce water, which is drained off. New hydrogen atoms enter the chamber to replace the depleted hydrogen.

fuel cell operation (cont.)

This action leaves positive hydrogen ions in the electrolyte, so that the chemical activity of the hydrogen fuel tends to cause the electrolyte to build up a positive charge. At the oxygen chamber, the heated oxygen atoms ionize by taking electrons from the electrode. The negative oxygen ions are attracted to the positive electrolyte solution, and there react with the positive hydrogen ions to produce H_2O (water), which replaces that which was lost at the negative electrode. The other electrode, in the meantime, which had given up the electrons that ionized the oxygen atoms, has its electrons replaced by the current flow through the load. As the oxygen atoms are used to replenish the electrolyte, more oxygen is pumped into the chamber.

As you can see, neither the electrolyte nor the electrodes change or deteriorate from the chemical activity. Only the oxygen and fuel are used, but they are replaced.

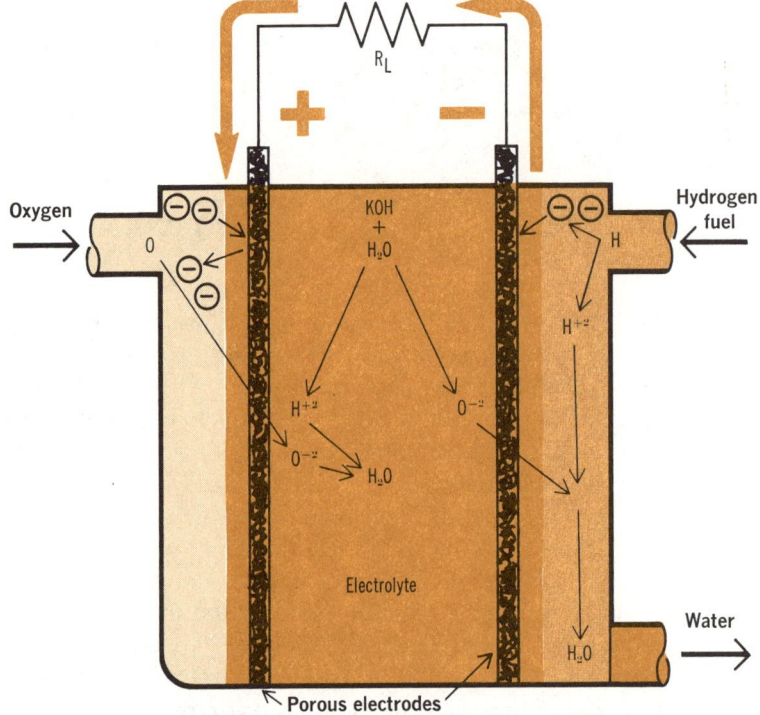

Ionized hydrogen fuel atoms provide the electrons for the current to flow through the load. The depleted hydrogen reacts with hydrogen ions from the electrolyte to produce water which is drained. The electrons from the load ionize the oxygen atoms which enter the electrolyte to replace those that were used. The electrodes and electrolyte do not change or deteriorate. Only the oxygen and fuel are used up

summary

☐ A *thermionic generator* uses heat and chemical activity to generate a charge. Cesium is ideal for this use because it becomes gaseous at lower heat than most other metals and ionizes easily. ☐ A conductive plasma passing through a magnetic field will provide electrons to suitably placed electrodes. This is the principle of a magnetohydrodynamic generator.
☐ In a fuel cell, the electrolyte and electrodes do not deteriorate. Chemical energy is supplied by oxygen and a fuel to produce the electrochemical action.

☐ A photovoltaic cell produces an emf when it is struck by light. When this light is removed, the emf drops to zero. ☐ A thermoelectric cell produces an emf when it is heated. ☐ A series arrangement of thermoelectric cells is called a thermopile. ☐ The solar cell consists of a photovoltaic or thermo-electric cell together with a chemical storage cell.

review questions

1. What is meant by the no-load voltage of a battery?
2. What is another name for a cesium thermionic generator?
3. What is a flowing ionized gas called?
4. How are the plasma, magnetic field, and electrodes in a magnetohydrodynamic generator placed relative to each other?
5. How does a fuel cell differ from a regular battery cell?
6. Name three kinds of fuel used in a fuel cell.
7. What are the disadvantages of a fuel cell?
8. Why are photovoltaic and thermoelectric cells generally not used as power sources?
9. What is a *solar cell*?
10. Two cells, each with an internal resistance of 0.2 ohm, are in parallel. What is the overall internal resistance?

generators

You have seen so far how three different forms of energy are converted into electrical energy. *Cells* convert *chemical energy* into electricity, *photovoltaic cells* convert *light energy* into electricity and *thermocouples* convert *heat energy* into electricity. You will now learn about devices that convert *mechanical energy* into electrical energy. These devices are called *generators*.

Basically, a generator produces electricity by the *rotation* of a group of conductors in a *magnetic field.* The input to a generator, then, is the mechanical energy needed to rotate the conductors. This energy can be supplied by gasoline or diesel engines, steam turbines, electric motors, flowing water, or even atomic reactors. In fact, anything that can be used to make a shaft rotate can be the input to an electrical generator. The output of a generator is the emf induced in the conductors as they move through the magnetic field. Since a generator requires a magnetic field for its operation, a generator might also be defined as a device that converts mechanical energy into electrical energy by means of a magnetic field, or by *magnetic induction.* The principles of magnetic induction were covered in Volumes 1 and 3. Actually, the magnetohydrodynamic generator works on the same principle, but uses a conductive plasma instead of wires.

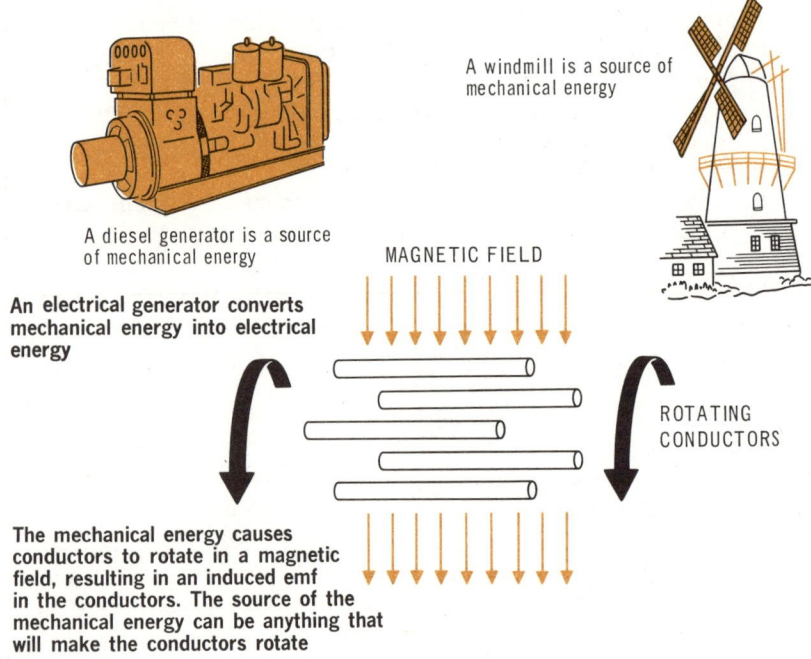

A windmill is a source of mechanical energy

A diesel generator is a source of mechanical energy

MAGNETIC FIELD

An electrical generator converts mechanical energy into electrical energy

ROTATING CONDUCTORS

The mechanical energy causes conductors to rotate in a magnetic field, resulting in an induced emf in the conductors. The source of the mechanical energy can be anything that will make the conductors rotate

generators (cont.)

Although generators are classified in many ways, there are only two basic types: *d-c generators,* which have a d-c voltage output, and *a-c generators,* which have an a-c voltage output. You will find that the principles of operation of the two types are similar in many ways.

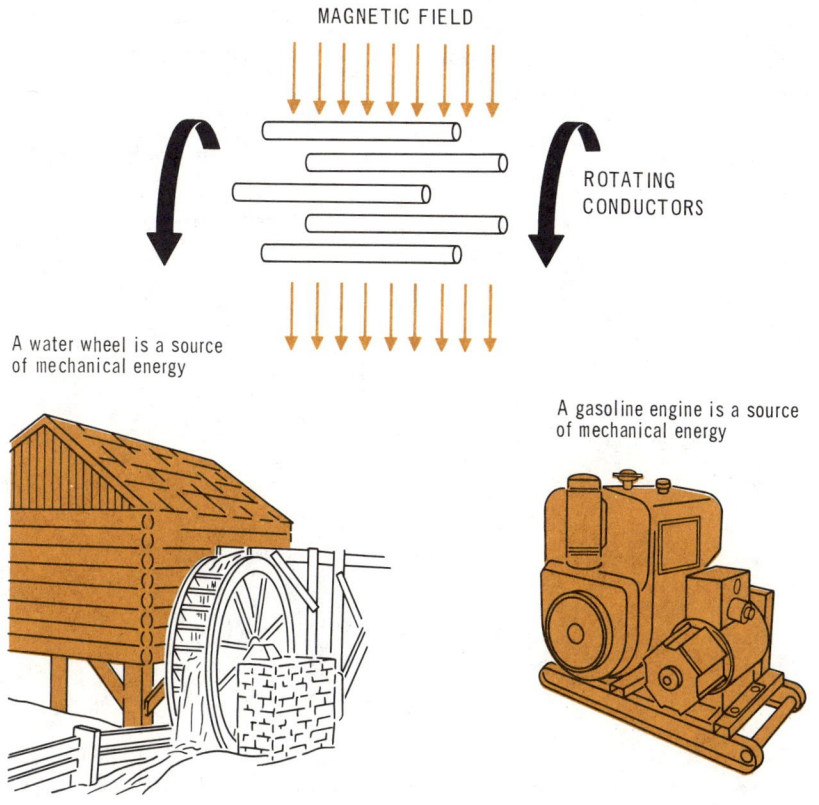

MAGNETIC FIELD

ROTATING CONDUCTORS

A water wheel is a source of mechanical energy

A gasoline engine is a source of mechanical energy

From the standpoint of the total amount of power produced, generators are the number one electrical power source used in the world today. No other practical power source can produce the large amounts of electrical power that generators can. This does not mean, though, that generators are the best power source for all applications. They must be located at or near their source of mechanical energy, and unlike batteries, therefore, cannot be used in cases where portable power sources are required. In addition, they are often uneconomical for producing small amounts of power.

the basic d-c generator

A basic d-c generator has four principal parts: (1) a magnetic field; (2) a single conductor, or loop; (3) a commutator; and (4) brushes. The *magnetic field* can be supplied by either a *permanent magnet* or by an *electromagnet*. For now, we will assume that a permanent magnet is used. As shown, the magnetic field can be pictured as consisting of magnetic flux lines that form a *closed* magnetic circuit. The flux lines leave the north pole of the magnet, cross the air gap between the poles of the magnet, enter the south pole, and then travel through the magnet back to the north pole.

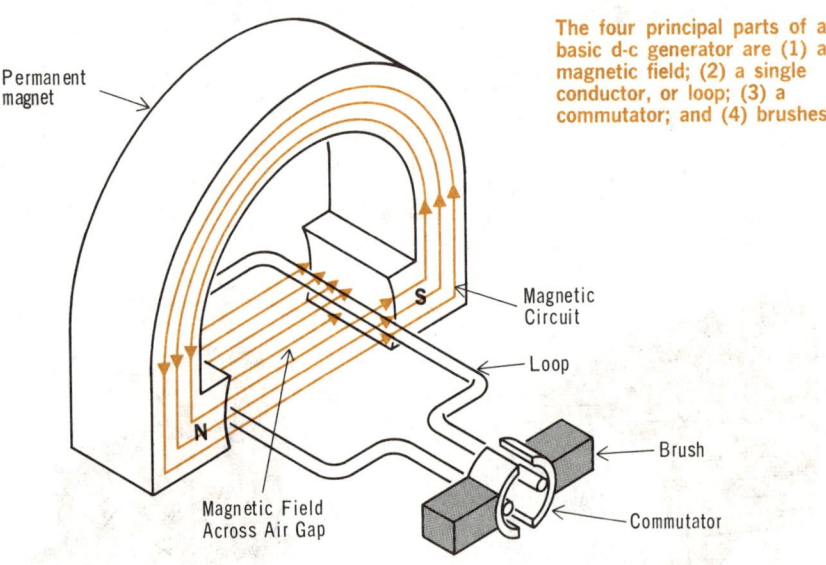

Permanent magnet

The four principal parts of a basic d-c generator are (1) a magnetic field; (2) a single conductor, or loop; (3) a commutator; and (4) brushes

Magnetic Circuit

Loop

Brush

Commutator

Magnetic Field Across Air Gap

The single conductor is shaped in the form of a *loop*, and is positioned between the magnetic poles. The loop is, therefore, in the magnetic field. As long as the loop does not rotate, the magnetic field has no effect on it. But if the loop rotates, it cuts through the lines of magnetic flux, and, as you learned in Volumes 1 and 3, this causes an emf to be *induced* in the loop.

You will learn that for each complete rotation of the loop, the amplitude and direction of the induced emf follows one cycle of a *sine wave*. As the loop rotates, therefore, a sinusoidal, or a-c, voltage is present at the ends of the loop. Since, by definition, d-c generators have d-c outputs, the a-c voltage must be converted to dc. This is done by a *commutator*. The d-c output from the commutator is transferred to an external circuit by *brushes*.

producing voltage

At this point, let us review some basic generator theory introduced in Volumes 1 and 3. This theory applies not only to d-c generators, which you are studying now, but to a-c generators as well.

Whenever there is *relative motion* between a magnetic field and a conductor, and the direction of motion is such that the conductor *cuts* the flux lines of the magnetic field, an emf is induced in the conductor. As far as generators are concerned, the *magnitude* of the induced emf depends mainly on the *strength* of the magnetic field, and the *rate* at which the flux lines are cut. The stronger the field, or the more flux lines cut in a given time, the larger is the induced emf. The direction or polarity of the emf is determined by the *left-hand rule* for generators. According to this rule, extend the thumb, index finger, and middle finger so that they all point at right angles to each other. Then, if you point the index finger in the direction of the magnetic field, and the thumb in the direction of motion of the conductor, the middle finger will be pointing in the direction in which current flows.

Applying the left-hand rule to the basic one-loop generator, you can see from the illustration that two emf's are induced in the loop as it rotates. These are induced on *opposite* sides of the loop, and have *equal amplitudes*. Their directions are such that they are in series with respect to the open ends of the loop. In effect, then, the amplitude of the voltage across the ends of the loop is *twice* the amplitude of the voltage induced in either side of the loop.

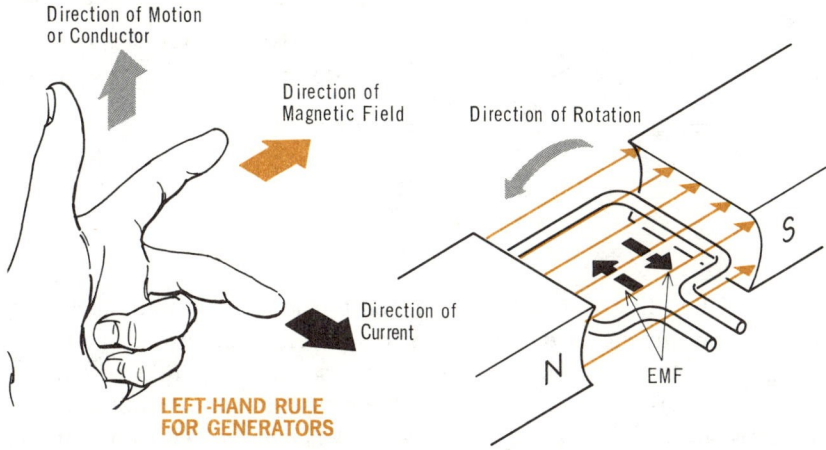

If the left-hand rule is applied to the basic single-loop generator, it shows that an emf is induced in each side of the loop, and that these emf's are in series

You can see that as the loop rotates, the side now moving upward will be moving downward, and vice versa. So the polarity of the induced emf in each side will also reverse

polarities

You have learned that electron current flows from negative to positive in a circuit. However, the generator in itself is not a circuit. It is a power source. So, if you think back to what you learned in Volume 2, you should recall that electron current flow *inside* of power sources goes from positive to negative.

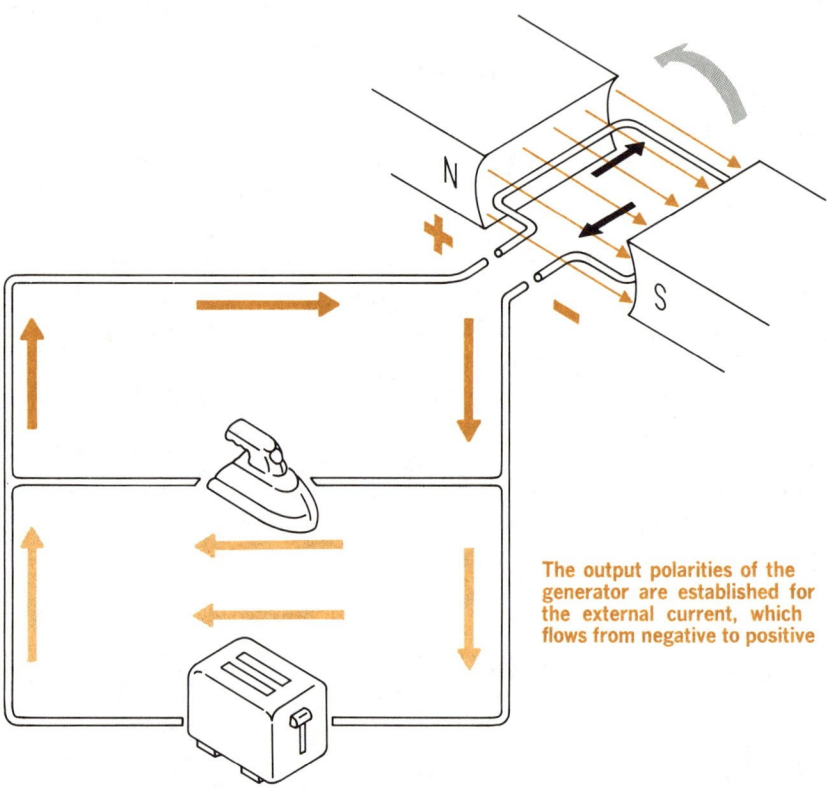

The output polarities of the generator are established for the external current, which flows from negative to positive

Essentially, the polarities are assigned to the generator to show how the electron current that is induced in the generator produces electric charges at the output connections. You can see that the induced current causes electrons to flow in a direction that produces an accumulation of electrons at one output terminal, and a deficiency of electrons at the other output terminal. Thus, the generator *polarities* are labelled according to the *charges* produced. Then, when a load is connected to the generator, current flows through the load from negative to positive.

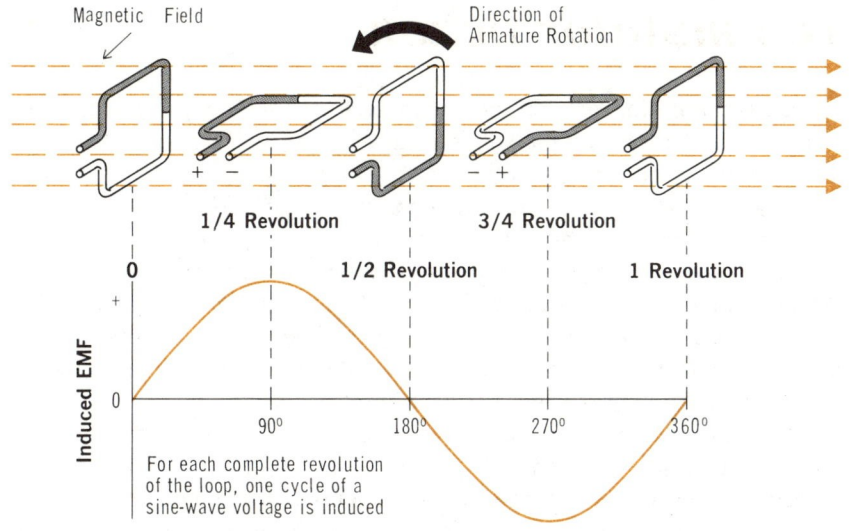

For each complete revolution
of the loop, one cycle of a
sine-wave voltage is induced

producing a sine wave

As was mentioned, the basic d-c generator produces a sine-wave output that is converted to dc by the commutator. Neglecting the commutator for the moment, you can see from the illustration that *one cycle*
of the sine-wave output is generated for each *full rotation* of the loop.
When the plane of the loop is *perpendicular* to the magnetic field, the
sides of the loop are passing *between* the flux lines. Hence, no flux lines
are being cut, so the induced voltage is zero. This happens twice during
each full rotation.

When the plane of the loop is *parallel* to the magnetic field, the sides
of the loop are cutting straight across the flux lines; so the induced voltage is *maximum*, since the rate at which the flux lines are being cut is
maximum. This also occurs twice during each full rotation. However,
in one position of the loop, the maximum emf is in one direction, while
180 degrees later, it is in the opposite direction, following the left-hand
rule.

At all other positions of the loop, the sides of the loop are cutting the
flux lines at an angle. It takes slightly longer for the wire to go from
flux line to flux line, so less lines are cut in a given amount of time.
Therefore, the induced voltage is somewhere between its maximum
value and zero, becoming less as the angle of the loop increases from
parallel to perpendicular. Then as the sides of the loop pass the zero-volt
perpendicular position, they go in the opposite cutting direction, and an
opposite emf is induced, which becomes increasingly greater until the
loop is parallel to the flux lines. Then the process is repeated.

commutator action

The commutator, as you know, converts the a-c voltage generated in the rotating loop into a d-c voltage. However, it also serves as a means of connecting the *brushes* to *the rotating loop*. The way in which it converts ac into dc is directly related to its role of serving as a contact between the brushes and loop.

You will recall that the purpose of the brushes is to connect the generator voltage to an external circuit. To do this, each brush has to make contact with one of the ends of the loop. A *direct* connection is impractical, since the loop rotates. So instead, the brushes are connected to the ends of the loop through the commutator.

The commutator is made of two *semicylindrical* pieces of smooth conducting material separated by insulating material. Each half of the commutator is permanently connected to one end of the loop, and therefore the commutator rotates as the loop rotates. Each brush rests against one side of the commutator, and *slides* along the commutator as both it and the loop rotate. In this way, each brush makes contact with the end of the loop that is connected to the half of the commutator that the brush rests against.

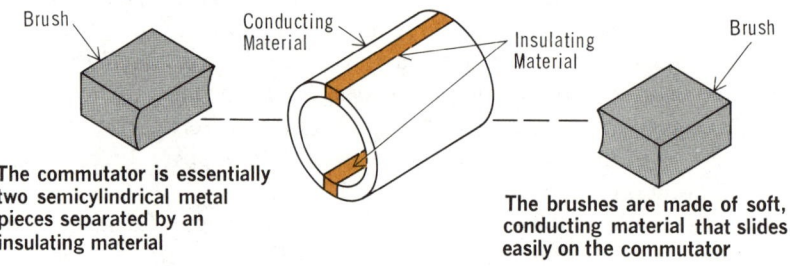

Brush   Conducting Material   Insulating Material   Brush

The commutator is essentially two semicylindrical metal pieces separated by an insulating material

The brushes are made of soft, conducting material that slides easily on the commutator

Since the *commutator rotates* while the *brushes* are *stationary*, each brush first slides along one half of the commutator and then along the other. This means that each brush is first in contact with one end of the loop and then the other. The brushes are positioned on *opposite* sides of the commutator so that they pass from one commutator half to the other at the instant the loop reaches the point in its rotation where the induced voltage reverses polarity. So every time the ends of the loop reverse polarity, the brushes switch from one half, or segment, of the commutator to the other. In this way, one brush is always positive with respect to the other. The voltage between the brushes, therefore, fluctuates in amplitude between zero and some maximum value, but it always has the *same polarity*. The fluctuating d-c voltage, then, is the output of the generator.

commutator action (cont.)

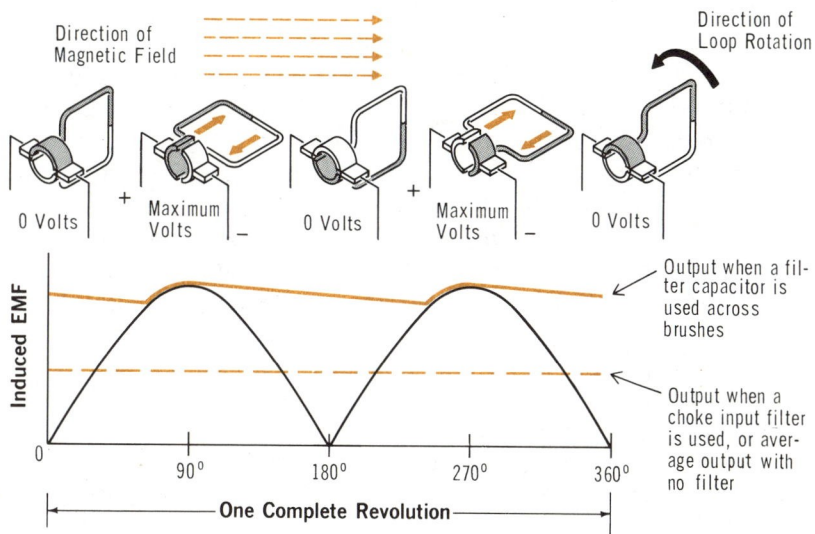

The action of the commutator and brushes in producing a fluctuating d-c output is shown on the following page. An important point to notice is that as each brush passes from one commutator segment to the other, there is an instant during which they contact both segments. The induced voltage at this instant is zero. If it were not, damagingly high currents would flow in the loop, since the brushes are effectively shorting the ends of the loop directly together. The position of the brushes so that they contact both commutator segments when the induced voltage is zero is called the *neutral plane*.

If you study the diagram, you will see that the left-hand brush is always connected to the side of the loop that is moving downward. This makes the left-hand brush always positive, as you can verify by the left-hand rule. Similarly, the right-hand brush is always connected to the side of the loop that is moving upward. This makes the right-hand brush always negative. So instead of the output voltage reversing polarity after one-half revolution, the voltage output for the second half revolution is identical to that of the first half. The commutator and brushes thus convert the induced ac into pulsating dc. If a filter capacitor were connected across the brushes, a more steady d-c voltage close to the peak amplitude would be produced. If a choke input filter were used, the output voltage would be the average level of the fluctuating dc. Even when no filter is used, the average is considered to be the output.

increasing the number of loops

As shown on the previous page, when no filter is used, the output of the basic single-loop generator is a fluctuating d-c voltage that reaches its peak amplitude and falls to zero twice during each full rotation of the loop. This variation in the output voltage is called *ripple,* and makes the output unsuitable for many uses. The variation, or ripple, in the output voltage can be reduced by using *two* rotating loops, positioned at *right angles* to each other. Each end of both loops is connected to a *separate* commutator segment, so the commutator has a total of *four segments.* There are still only two brushes, and they are positioned so that as the loops and the commutator segments rotate, the brushes make contact with the commutator segments first for one loop and then for the other.

For each loop, the brushes and commutator segments perform the same function they do in the single-loop generator. That is, one brush is always in contact with the end of the loop that is negative, and the other brush is always in contact with the end of the loop that is positive. Thus, the a-c voltage induced in the loop is converted to a fluctuating d-c voltage.

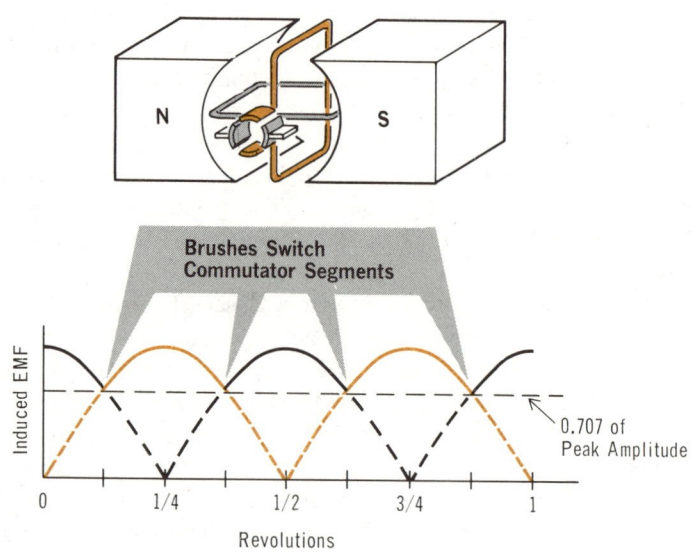

The voltages induced in each loop are equal, but 90° out of phase. When the loops reach the point in their rotation where their voltages are 0.707 of peak, the brushes switch from the commutator segment for the loop with the decreasing voltage to the loop with the increasing voltage

increasing
the number of loops (cont.)

There is one important difference in the two-loop generator, however. It is that one loop is always 90 degrees of rotation behind the other. So when the voltage in one loop is *decreasing,* the voltage in the other is *increasing,* and vice versa. And the position of the brushes is such that as the loops and commutator rotate, the brushes are always in contact with the commutator segments of the loop that has the *greatest* induced voltage. As the voltage in one loop drops below the voltage in the other loop, the brushes pass from the commutator segments of the loop with the decreasing voltage to the segments of the loop with the increasing voltage. This switching occurs *four* times during each full rotation of the two loops, and because of it, the generator output voltage, which appears between the two brushes, never falls below a value 0.707 times the peak amplitude of the voltage induced in either loop. This d-c output needs less filtering than the single-loop generator.

You should notice here that although the use of two *separate* loops decreases the fluctuation in the output voltage, it has no effect on the peak output voltage. The average output, though, is higher.

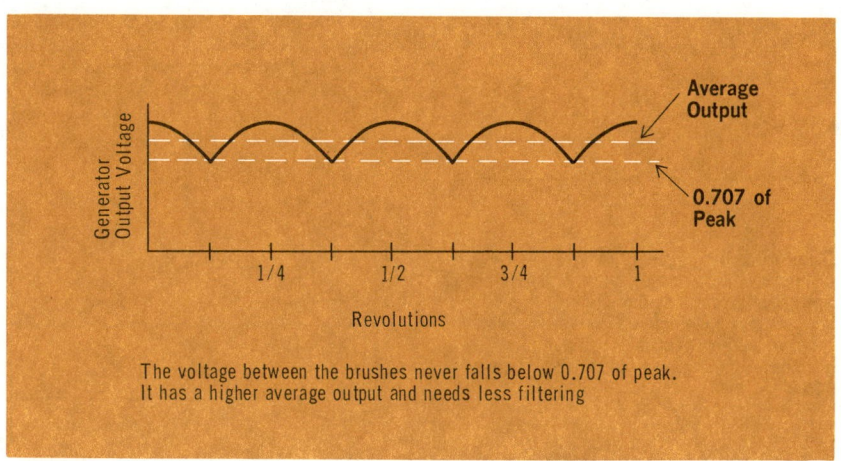

The voltage between the brushes never falls below 0.707 of peak. It has a higher average output and needs less filtering

You have seen how by using two separate loops instead of one, the ripple in the generator output voltage can be reduced. By using more and more separate loops, the ripple can be further reduced, and the output voltage of the generator made very nearly *steady* dc. Little or no filtering would then be needed, and the average output would be almost the peak voltage.

increasing
the number of loops (cont.)

For every separate loop that is added, *two* more commutator segments must also be added; one for each end of the loop. Thus, there is always a *two-to-one ratio* between the number of commutator segments and the number of separate loops. Four segments are needed for two loops, six segments for three loops, and so on. Actual d-c generators contain many separate loops, and twice as many commutator segments. Therefore, if you were to count the number of segments on any generator, you know that there are half as many separate loops.

The output waveform of a generator having four separate loops is shown. Again, you should note that although increasing the number of separate loops decreases the variation between the maximum and minimum output voltage, it does not increase the peak output voltage, only the average.

In practical generators, the loops and the commutator together are usually called the *armature,* and sometimes the *rotor.* The armature, then, in this type of generator is the part that *rotates.*

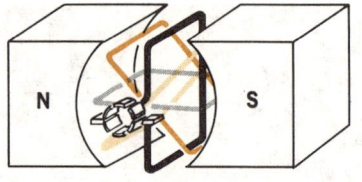

Every additional loop requires two additional commutator segments

The commutator segments and brushes convert the induced a-c voltages into fluctuating d-c voltages, the same as the single-loop generator. However, the brushes are only in contact with the segments for the short period when the voltage in that loop is close to its peak value

Individual Loop Voltages

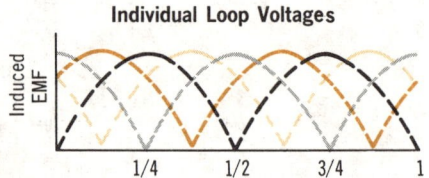

Induced EMF

1/4   1/2   3/4   1

Generator Output Voltage

Generator Output Voltage

1/4   1/2   3/4   1
Revolutions

The average output voltage between the brushes is, therefore, close to the peak value of the individual loops

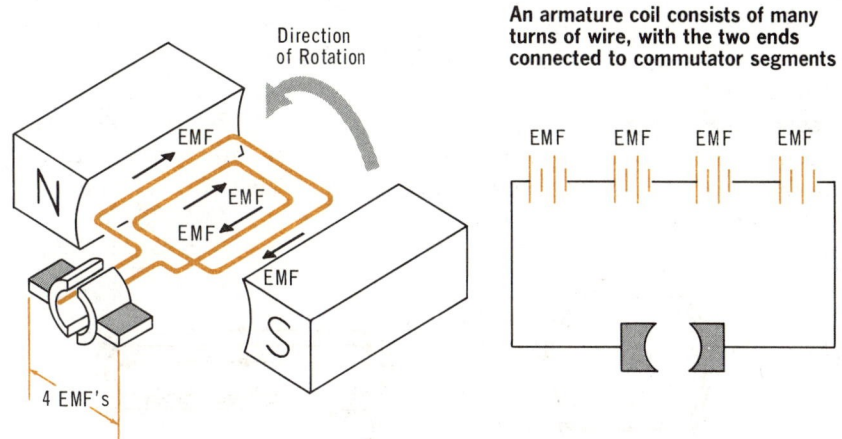

An armature coil consists of many turns of wire, with the two ends connected to commutator segments

The individual voltages induced in each turn are all in series, so their sum appears between the brushes

raising output voltage

In the basic d-c generator described, the output voltage amplitude is the same as that induced in each separate rotating loop, and is *very small.* You will recall that the amplitude of the voltage in each loop is determined by the *rate* at which the loop cuts the magnetic flux lines; and this, in turn, depends on the strength of the magnetic field and the speed at which the loop rotates. You might think, therefore, that the voltage could be increased by increasing the strength of the magnetic field or the speed of rotation, or both. Both of these measures, though, are impractical beyond certain limits.

Instead, the output voltage of a d-c generator can be increased to a usable level by having each rotating loop consist of *many turns* of wire rather than a single turn. These multiturn loops are called *armature coils,* or just coils. Each coil has two ends, and requires two commutator segments, the same as a simple one-turn loop. The total voltage induced in a coil, though is the *sum* of the individual voltages induced in each turn. The armature coils used in actual generators often have many turns, making possible high generator output voltages. You can see, then, that for a given magnetic field and speed of rotation, the *number of turns* in each coil determines the *amplitude* of the generator output voltage, while the number of coils determines the amount of ripple in the output.

producing the magnetic field

As was mentioned, the magnetic field of a d-c generator can be supplied by either a permanent magnet or an electromagnet. Permanent magnets are all right for simple generators, but they have certain limitations for use in most practical generators. Their principal limitations are their size and weight requirements, and their lack of regulation over the generator output.

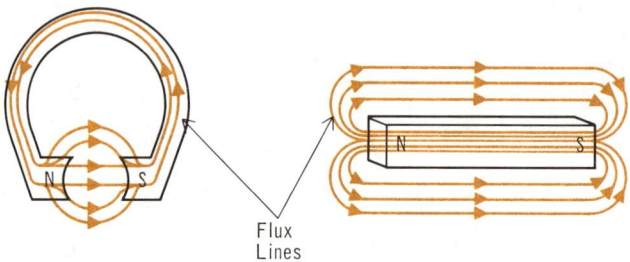

Flux
Lines

To provide magnetic fields of the required strength, permanent magnets must be relatively large and heavy

Since the strength of their magnetic field is fixed, they cannot provide any regulation of the generator output voltage

Electromagnets can produce much stronger magnetic fields than can permanent magnets of the same size and weight. Furthermore, electromagnets have properties that make it possible to regulate the generator output voltage to compensate for changes in load and speed. You will learn more of this later. Because of these factors, the use of permanent magnets in generators is limited. Most of the generators you will encounter will have their magnetic fields supplied by electromagnets. Those generators that do use permanent magnets are usually small and have low outputs. They are often called *magnetos*.

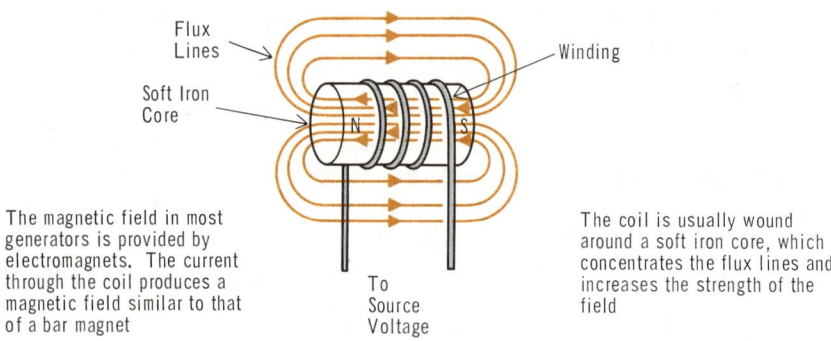

Flux
Lines

Winding

Soft Iron
Core

The magnetic field in most generators is provided by electromagnets. The current through the coil produces a magnetic field similar to that of a bar magnet

To
Source
Voltage

The coil is usually wound around a soft iron core, which concentrates the flux lines and increases the strength of the field

summary

☐ A generator converts mechanical energy into electrical energy by the rotation of a group of conductors in a magnetic field. ☐ Basically, there are two types of generators: dc and ac. ☐ A simple d-c generator consists of a magnetic field, a conductor in the form of a loop, a commutator, and brushes. ☐ The magnetic field of a generator can be produced by either a permanent magnet or an electromagnetic. ☐ Most generators use electromagnets, since they can produce stronger fields than permanent magnets, and also provide a means for regulating the generator output voltage.

☐ As the loop of a simple generator rotates, it cuts the flux lines of the magnetic field. As a result, an emf is induced in the loop. ☐ The emf induced in a single loop is in the form of a sine wave. ☐ The sine-wave voltage induced in the loop is converted to fluctuating dc by the commutator and brushes. ☐ For proper commutation to take place, the brushes must be positioned in the neutral plane. ☐ The amplitude variation in the d-c output voltage of a single-loop generator is called ripple. ☐ Ripple can be reduced by using more than one single rotating loop. ☐ Every loop requires two commutator segments. Thus, there is always a two-to-one ratio between the number of commutator segments and the number of separate loops.

☐ In actual generators, each rotating loop often consists of many turns of wire. These multiturn loops are called armature coils. ☐ The total voltage induced in each armature coil is the sum of the individual voltages induced in each turn. ☐ The number of turns in each coil determines the amplitude of the generator output. ☐ The number of coils determines the amount of ripple in the output.

review questions

1. What is the *left-hand rule for generators*?
2. What is the purpose of the commutator and brushes?
3. What is the *neutral plane*?
4. Which has the smaller ripple: a generator with two armature coils, or one with four armature coils?
5. If a generator has eight commutator segments, how many separate armature coils does it have?
6. What factors determine the output voltage of a generator?
7. What is a *magneto*?
8. Will increasing the number of turns in the armature coils affect the amount of ripple in the output of a d-c generator?
9. Does the number of armature coils affect the number of brushes required for a d-c generator?
10. What would happen if the brushes were not positioned in the neutral plane?

the field winding

The electromagnets used to produce the magnetic field of a generator are called *field coils*. In a simple generator, there are two field coils positioned so that their magnetic fields combine to form one magnetic circuit. As shown, the field coils are wound around cores, called *pole pieces*, that are part of the generator housing. The two pole pieces are separated by a space or gap into which the armature is placed. The closed magnetic circuit is from the "north" pole piece, across the gap to the "south" pole piece, and then through the housing back to the north pole piece. The generator housing, like the pole pieces, is made of material having good magnetic properties, so it helps increase the strength of the magnetic field. There is no electrical connection between the field coil and the pole pieces or generator housing. They form only a *magnetic circuit*.

The field winding supplies the magnetic field required by the generator. It is made up of the individual field coils, and is energized by a d-c voltage source

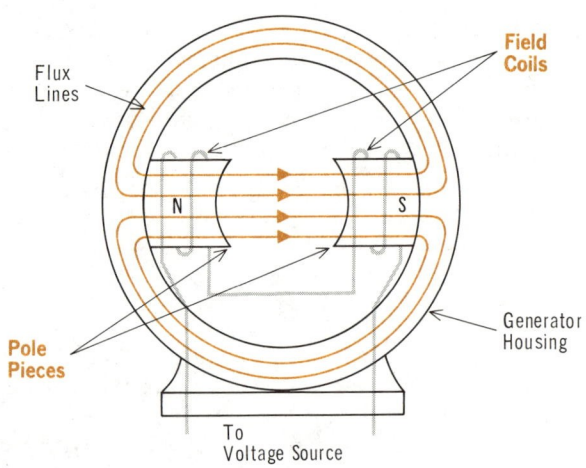

The strength of the magnetic field set up by the field winding depends not only on the physical construction of the field coils, but also on the current level applied to the winding. The greater the current, the stronger will be the field

The two field coils are wound in *series*, and so are energized by the same voltage source. This voltage source is dc, and as a result, the magnetic field produced by the field coils is always in the same direction. Both field coils together are called the *field winding*.

multiple field windings

The simple generator has two field coils, and therefore two poles: one north and one south. This is a two-pole generator. Many actual generators have *four poles, six poles,* and so on. No matter how many field coils there are, the total number of poles is always an *even number,* since for every north pole there must be a south pole.

One reason for having more than two field coils is that by increasing the number of field coils, or poles, the size and weight of the generator can be reduced while its output remains the same. In a two-pole generator, half of the flux lines must pass through the width or thickness of the armature core, which is a magnetic material. The armature core must be thick enough to prevent magnetic *saturation.*

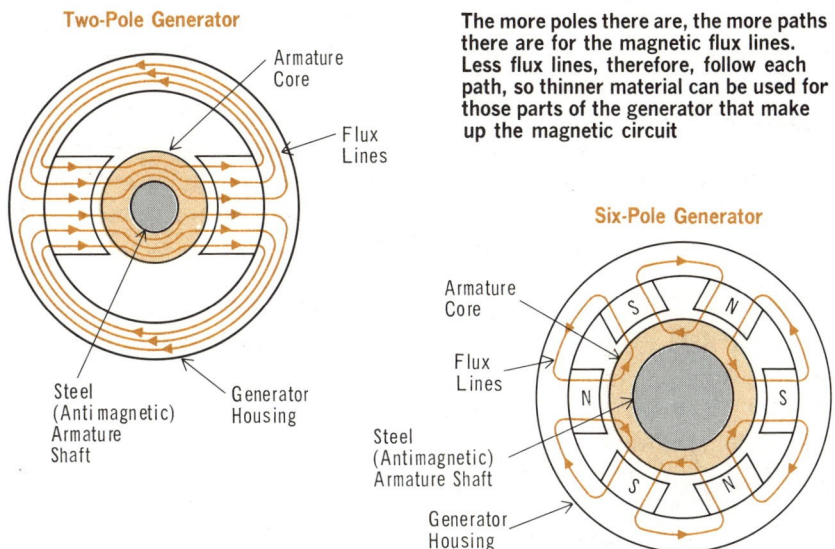

Two-Pole Generator

Armature Core

Flux Lines

Steel (Anti magnetic) Armature Shaft

Generator Housing

The more poles there are, the more paths there are for the magnetic flux lines. Less flux lines, therefore, follow each path, so thinner material can be used for those parts of the generator that make up the magnetic circuit

Six-Pole Generator

Armature Core

Flux Lines

Steel (Antimagnetic) Armature Shaft

Generator Housing

In a six-pole generator, only one-sixth of the flux lines have to pass through the thickness of the armature core in any one spot. But, since there is a series of flux lines, the total number is the same. The core can, therefore, be made substantially thinner and still pass the same total flux lines with little opposition. This also holds true for the generator housing, which serves as part of the path for the flux lines. The more poles there are, the thinner the housing can be. In effect, the more poles there are, the more paths the magnetic flux lines will follow.

Another reason for increasing the number of poles is that with certain types of armature windings, the output voltage of the generator can be increased. This is covered later.

exciting the field winding

Since the field winding is an electromagnet, current must flow through it to produce a magnetic field. For proper operation of a d-c generator, the magnetic field of the field winding must always be in the same direction, so the current through the winding must be dc. This current is called the *excitation current*, and can be supplied to the field winding in one of two ways; it can come from a separate, external d-c voltage source, in which case the generator is called a *separately excited generator;* or it can come from the generator's own output, in which case the generator is called a *self-excited generator*. The excitation source for separately excited generators can be a battery or another d-c generator. When a generator is used, it is called an *exciter*.

In a self-excited generator, the field winding is connected directly to the generator output. It may be connected *across* the output, in *series* with the output, or a *combination* of the two. The way in which it is connected determines many of the generator's characteristics.

Since separate excitation requires a separate battery or generator, it is generally more expensive than self-excitation. As a result, separate excitation is normally used only when self-excitation would be unsatisfactory. This occurs in cases where a generator must respond quickly and precisely to an external control source, or when the generator output voltage must be varied over a wide range during normal operation.

Since the excitation current through the field winding produces the magnetic field for the generator, the larger the excitation current the stronger is the magnetic field. This means that for any given speed of generator rotation, a large excitation current results in a high generator output voltage, while a small excitation current causes a low output voltage. This is true whether the excitation current comes from an outside source, as in a separately excited generator, or from the generator's own output, as in a self-excited generator.

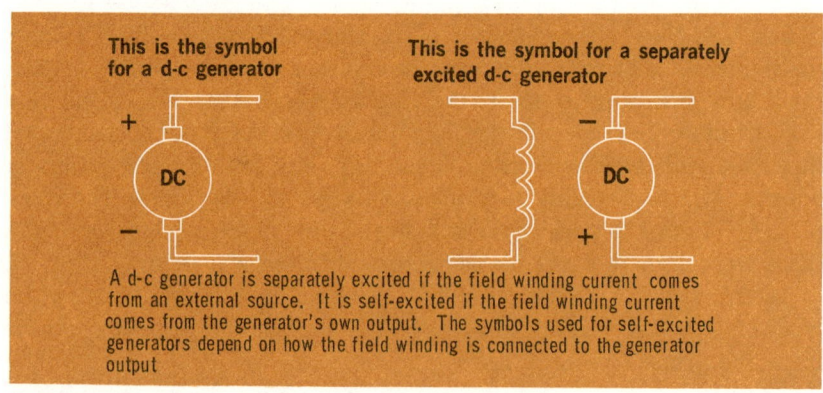

This is the symbol for a d-c generator

This is the symbol for a separately excited d-c generator

A d-c generator is separately excited if the field winding current comes from an external source. It is self-excited if the field winding current comes from the generator's own output. The symbols used for self-excited generators depend on how the field winding is connected to the generator output

series generators

When the field winding of a *self-excited generator* is connected in series with the generator output, the generator is called a *series generator*. The exciting current through the field winding of such a generator is the same as the current the generator delivers to the load. If the load has a high resistance and so draws only a small current from the generator, the excitation current is also small. This means that the magnetic field of the field winding is weak, making the generator output voltage low. Similarly, if the load draws a large current, the excitation current is also large, the magnetic field of the field winding is strong, and the generator output voltage is high. You can see then that in a series generator, changes in load current greatly affect the generator output voltage. A series generator is thus said to have *poor voltage regulation,* and, as a result, series generators are not recommended for fluctuating loads.

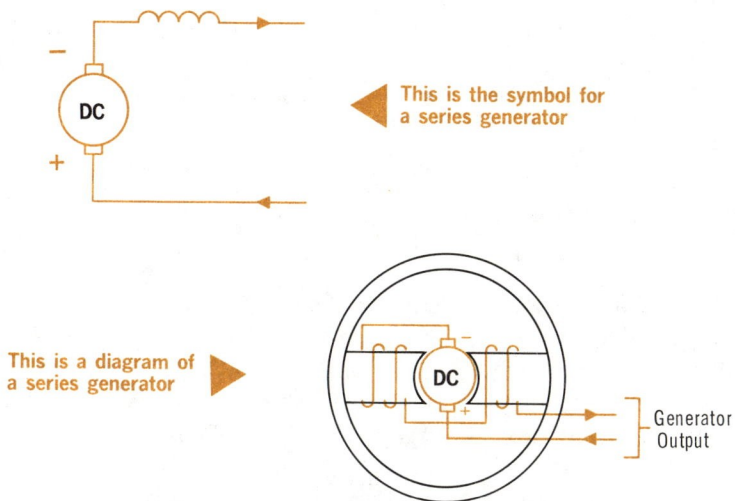

This is the symbol for a series generator

This is a diagram of a series generator

Generator Output

In a series generator, the field winding is connected in series with the generator output. Therefore, the same current that flows through the load flows through the field winding

The graph on the next page shows how the output voltage of a series generator varies with increasing load current. You will notice that as the load current increases, the output voltage also increases, up to a certain point. After that point, further increases in current result in a decrease in the output voltage.

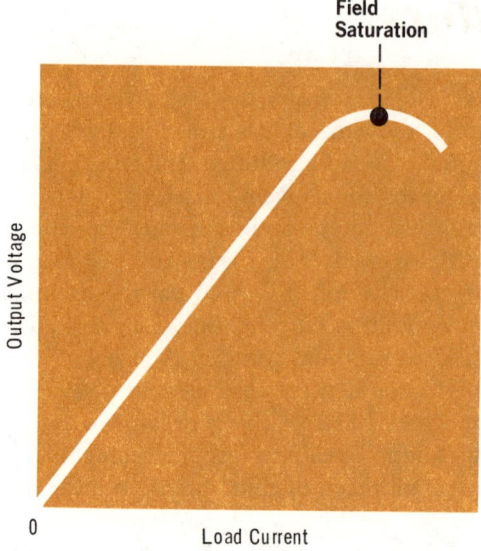

Field Saturation

Up to a certain point, the output voltage of a series generator increases with the load current

After this point, the voltage decreases with increases in load current

Output Voltage

0   Load Current

Since the output voltage varies with load current, a series generator has poor voltage regulation

series generators (cont.)

The point where the voltage no longer increases corresponds to the point of *magnetic saturation* of the field winding. This occurs, as you will recall from Volume 3, when the core material, which is the pole pieces in this case, is completely magnetized. The magnetic flux cannot increase any further, regardless of how much more the current through the winding is increased. The reason why the output voltage drops after this point instead of staying constant at its maximum value is because of the increased voltage drop of the field winding and armature coils. The voltage drop increases because of the increasing current, but the generated voltage stays the same. And since the output voltage equals the generated voltage minus the internal voltage drop, the output voltage must therefore decrease. Another reason for the decrease in the output voltage is that *armature reaction* increases. This will be described later.

A further disadvantage of the series generator, besides its poor voltage regulation, is that the field winding must be wound with wire that can safely carry the entire load current without overheating. This requires wire with a relatively large cross-sectional area.

shunt generators

When the field winding of a self-excited generator is connected in *parallel* with the generator output, the generator is called a *shunt generator*. The value of the excitation current in a shunt generator depends on the output voltage and the resistance of the field winding. Usually, the excitation current is kept somewhere between 0.5 and 5 percent of the total current drawn from the generator.

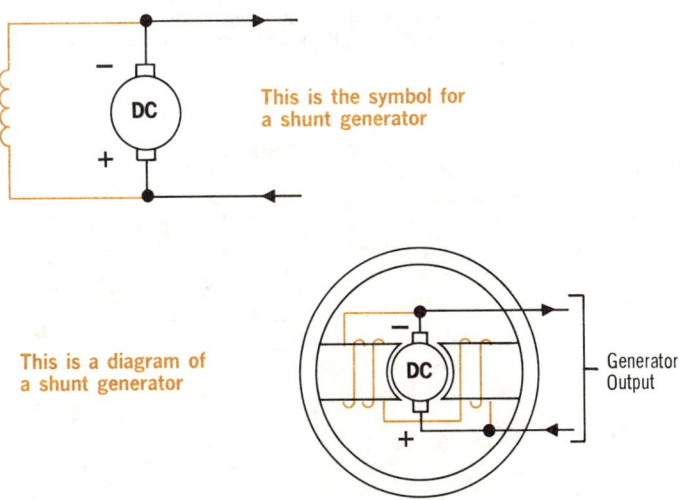

This is the symbol for a shunt generator

This is a diagram of a shunt generator

Generator Output

In a shunt generator, the field winding is connected in parallel with, or across, the generator output. The exciting current, therefore, depends on the value of the output voltage and the resistance of the field winding

The output voltage of a shunt generator running at constant speed under varying load conditions is much more stable than the output voltage of a series generator. However, some change in output voltage still takes place. This change is caused by the fact that when the load current increases, the voltage (IR) drop across the armature coil increases, and this causes the output voltage to decrease. As a result, the current through the field winding decreases, decreasing the magnetic field and thereby causing the output voltage to decrease even more. If the current drawn by the load is much greater than a shunt generator is designed to deliver, the drop in output voltage is extreme. However, for load current changes within the design range, the drop in output voltage with increasing load current is not too severe.

shunt generators (cont.)

The fact that their output voltage drops as load current increases, provides shunt generators with a *self-protective* feature. If the load should be suddenly "shorted," the output voltage would drop to zero. No excitation current, therefore, would flow through the field winding, so the generator would, in effect, be disabled.

The output voltage of a shunt generator drops gradually with increasing load current within its normal operating range

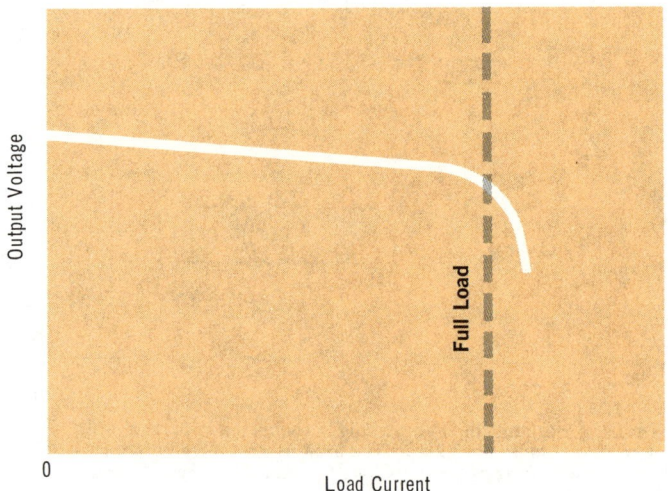

The output voltage drops drastically if the load current increases above the rated full-load value

The change in the voltage output of a shunt generator with changes in load current is also caused to some extent by a change in the *armature reaction,* the same as was previously mentioned for series generators. Armature reaction is covered later.

Compared to series generators, the excitation current in shunt generators is very small. Thin wire can, therefore, be used for the field winding. In actual shunt generators, the field coils consist of many turns of small-diameter wire.

compound generators

Both series and shunt generators have the disadvantage in that changes in their load current from zero to normal full-load cause their output voltage to change also. In a *series generator,* an increase in load current causes an *increase* in output voltage; whereas in a *shunt generator,* an increase in load current causes a *decrease* in output voltage. Many applications in which generators are used require that the generator output voltage be more stable than that supplied by either a series or a shunt generator. One way to supply such a stable voltage is by using a shunt generator with some form of *voltage regulation.* Voltage regulation of generators is covered later. Another means of supplying a stable voltage is by using a *compound generator.*

A compound generator has both series and shunt
field windings

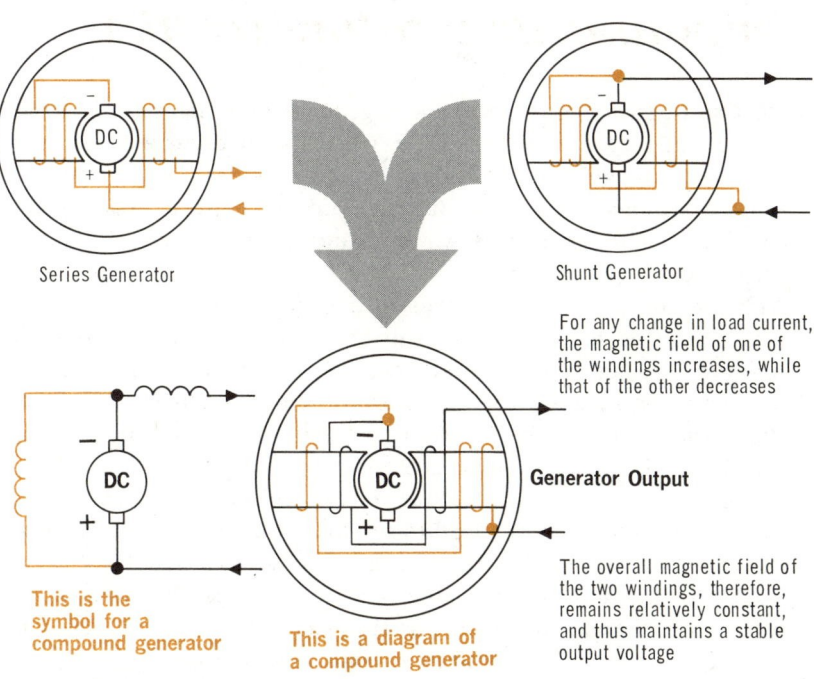

Series Generator

Shunt Generator

For any change in load current, the magnetic field of one of the windings increases, while that of the other decreases

Generator Output

This is the
symbol for a
compound generator

This is a diagram of
a compound generator

The overall magnetic field of the two windings, therefore, remains relatively constant, and thus maintains a stable output voltage

A compound generator has a field winding connected in *parallel* with the generator output, the same as a shunt generator; and it also has another field winding connected in *series* with the generator output, the same as a series generator. Compound generators are sometimes referred to as *series-shunt generators.*

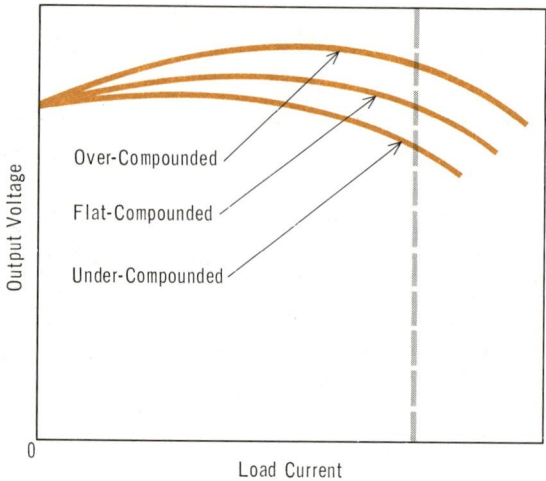

Output Voltage

Over-Compounded

Flat-Compounded

Under-Compounded

0

Load Current

Whether a compound generator is flat-compounded, over-compounded, or under-compounded, the output voltage drops off sharply when the load current exceeds rated full load

compound generators (cont.)

The two windings of the compound generator are made so that their magnetic fields *aid* each other. Thus, when the load current increases, the current through the shunt field winding decreases, reducing the strength of its magnetic field. But the same increase in load current flows through the series field winding, increasing the strength of its magnetic field.

With the proper number of turns in the series winding, the increase in strength in its magnetic field will *compensate* for the decrease in strength of the magnetic field of the shunt winding. The overall strength of the combined magnetic field, therefore, remains almost unchanged, so the output voltage stays constant. Actually, the two field windings cannot be made so that their magnetic fields exactly compensate for one another. Some change in output voltage will take place as the generator current varies from no-load to full-load value. However, as shown in the graph, in practical compound generators, the output voltage at no load and full load is the same, and the change that takes place between no load and full load is less than about 5 percent. A generator with these characteristics is said to be *flat-compounded*.

For some applications, the series winding is made so that it over-compensates for the shunt winding. The generator output voltage then gradually increases with increasing load current over the normal operating range. Such a generator is *over-compounded*. Similarly, the series winding can be made so that it under-compensates for the shunt winding. The output voltage of this type of generator decreases gradually with increasing load current. This type of generator is *under-compounded*.

starting self-excited generators

At the instant any generator is started, both its output voltage and current are zero. You may wonder, therefore, how a self-excited generator can begin to develop an output voltage. With no voltage across the field winding or current through it, how does the winding produce the magnetic field that it must have if the generator is to operate? The answer lies in the *residual magnetism* of the pole pieces. Whenever magnetic materials have been magnetized, they *hold* a small *portion* of this magnetism even after the magnetizing influence is removed.

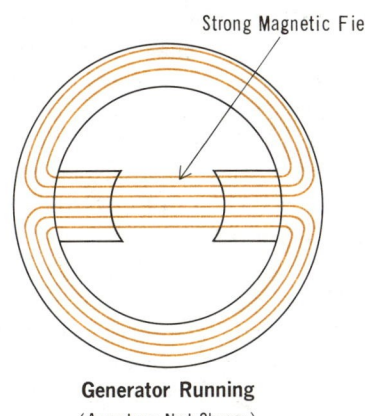

Strong Magnetic Field

Generator Running
(Armature Not Shown)

The residual magnetism of the pole pieces makes it possible to start a self-excited generator even though no exciting current flows through the field windings at the instant the generator is started

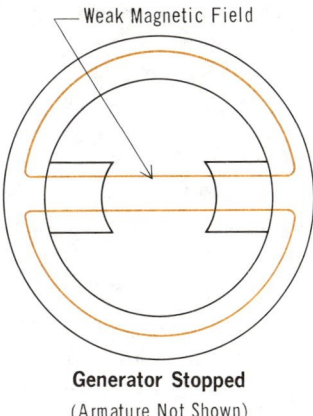

Weak Magnetic Field

Generator Stopped
(Armature Not Shown)

The residual magnetism is caused by the fact that even when the magnetizing current is removed, the pole pieces remain slightly magnetized

Soft iron, of which the pole pieces are made, loses most of its magnetism whenever the exciting current through the field winding drops to zero, such as when the generator is turned off. It retains a small portion of its magnetism, though, and so maintains a *weak* magnetic field in the space between the pole pieces. So, when the generator is started again, the armature coils cut the flux lines of this weak magnetic field. This induces a small voltage in the armature coils, which gives the generator a weak output. The output then causes some exciting current to flow, thereby increasing the strength of the magnetic field, which increases the voltage induced in the armature coils. This action continues until the generator reaches its rated output. In some cases, it can take as long as 30 seconds from the time a self-excited generator is turned on until it reaches its rated output.

summary

☐ Electromagnets used to produce the magnetic field of a generator are called field coils. ☐ Field coils are wound around pole pieces. ☐ A simple generator has two field coils, and is called a two-pole generator. ☐ All of the field coils together are called the field winding. ☐ The current that excites the field winding of a d-c generator must be dc. ☐ In a separately excited generator, the excitation current is supplied by an external d-c voltage source. ☐ In a self-excited generator, the excitation current is supplied by the generator's own output.

☐ There are three types of self-excited generators: series, shunt, and compound. ☐ In a series generator, the field winding is connected in series with the generator output. ☐ A series generator has poor voltage regulation. ☐ In a shunt generator, the field winding is connected in parallel with the generator output. ☐ Shunt generators have better regulation and much smaller excitation current than series generators. ☐ If the load on a shunt generator is shorted, the output voltage will drop to zero. This provides the shunt generator with a self-protective feature. ☐ A compound generator has two field windings: one connected in series with the output, and the other in parallel with it.

☐ A compound generator has good voltage regulation, with the change in output voltage between no load and full load being less than about 5 percent. ☐ A self-excited generator is self-starting because of the residual magnetism of the pole pieces.

review questions

1. What is *excitation current*?
2. Why must the excitation current in a d-c generator be dc?
3. What is a *self-excited generator*? A *separately-excited generator*?
4. Why does a series generator have poor voltage regulation?
5. Why are shunt generators said to be self protective?
6. Why must heavy wire be used for the field windings of series generators?
7. What is a *compound generator*?
8. What is an *over-compounded generator*? An *under-compounded generator*?
9. When a self-excited generator is started, why does it take some time for it to reach its rated output?
10. In a compound generator, do the magnetic fields of the two windings aid or oppose each other?

the armature winding

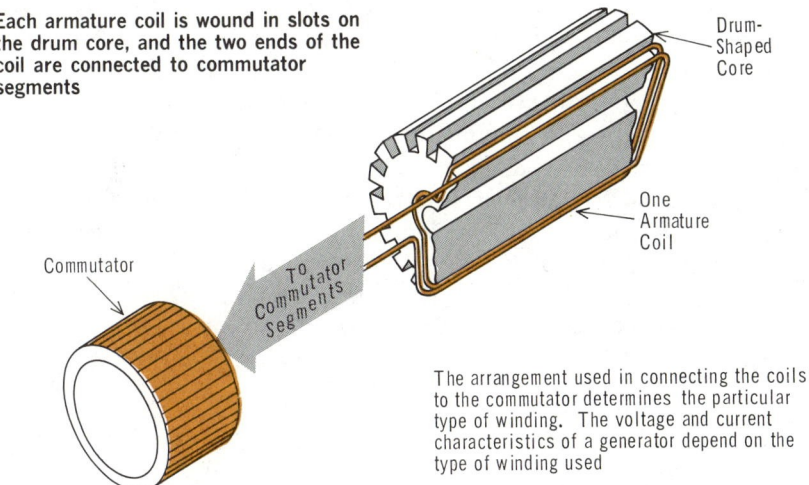

Each armature coil is wound in slots on the drum core, and the two ends of the coil are connected to commutator segments

Drum-Shaped Core

One Armature Coil

Commutator

To Commutator Segments

The arrangement used in connecting the coils to the commutator determines the particular type of winding. The voltage and current characteristics of a generator depend on the type of winding used

Just as the combined field coils make up the field winding, the combined armature coils are called the *armature winding.* The ends of each armature coil are connected to different segments on the commutator, where its emf is picked up by the brushes. The coils of all modern generator armatures are wound on an iron core that is shaped like a drum. The core provides a means for rotating the coils, and, at the same time, is a good low-reluctance path for the flux lines of the magnetic field set up by the field winding. The core has slots along its length, and the coils are wound in these slots. The two sides of each coil are positioned in different slots. A typical drum core with one three-turn armature coil in place is shown.

The ends of the armature coils can be connected to the commutator segments in many different ways. The arrangement used determines to a large extent the voltage and current characteristics of the generator. For practical purposes, all of the different arrangements can be divided into two main types: *lap windings* and *wave windings.* When lap or wave windings are combined, it is called a *frog-leg winding,* because of the shape that is produced.

In any armature winding, either lap or wave, an important point to remember is that each coil is wound on the core in such a way that the two sides of the coil are separated by the same distance that separates a north field pole from a south field pole. Therefore, whenever one side of a coil is at the middle of a north pole, the other side is at the middle of a south pole.

lap windings

Lap windings get their name from their winding diagrams, which represent the connections between the armature coils and the commutator segments. On these diagrams, each coil of a lap winding *overlaps* the previous coil. The two ends of any one coil in a lap winding are connected to *adjacent* commutator segments, and each commutator segment connects the ends of two adjacent coils. This has the effect of placing all those coils under similar pairs of poles in parallel. If the field winding has only one north and one south pole, this means that there are two parallel paths through the armature winding. If there are two north and two south poles, there are four parallel paths. This is a basic characteristic of lap windings: there are the same number of parallel paths through the armature winding as there are field poles.

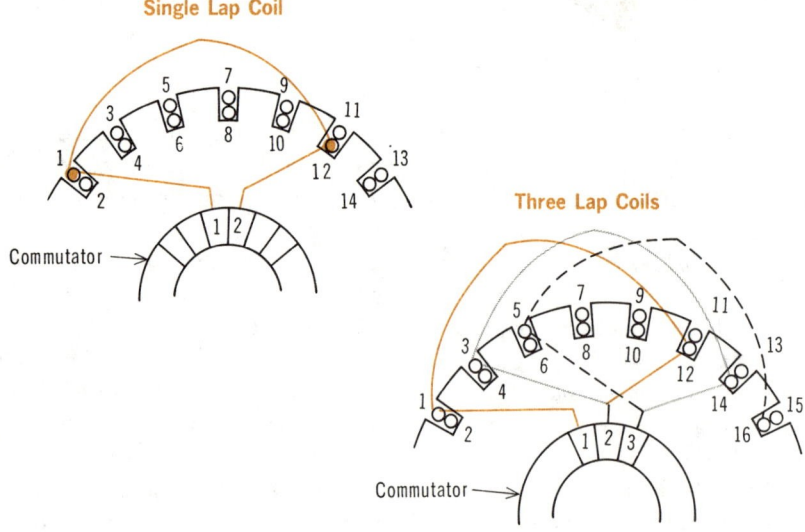

The voltages induced in the coils in each of the parallel paths are equal and have the same polarity, so no circulating current flows among the parallel paths. There is one set of brushes for each two parallel paths, and they are electrically connected (negative to negative and positive to positive) at the generator output. The output voltage is, therefore, equal to the voltage induced in any one of the parallel paths, but the current capacity is large, since the current divides among many paths inside of the generator. Lap windings, therefore, sacrifice voltage output for current capacity.

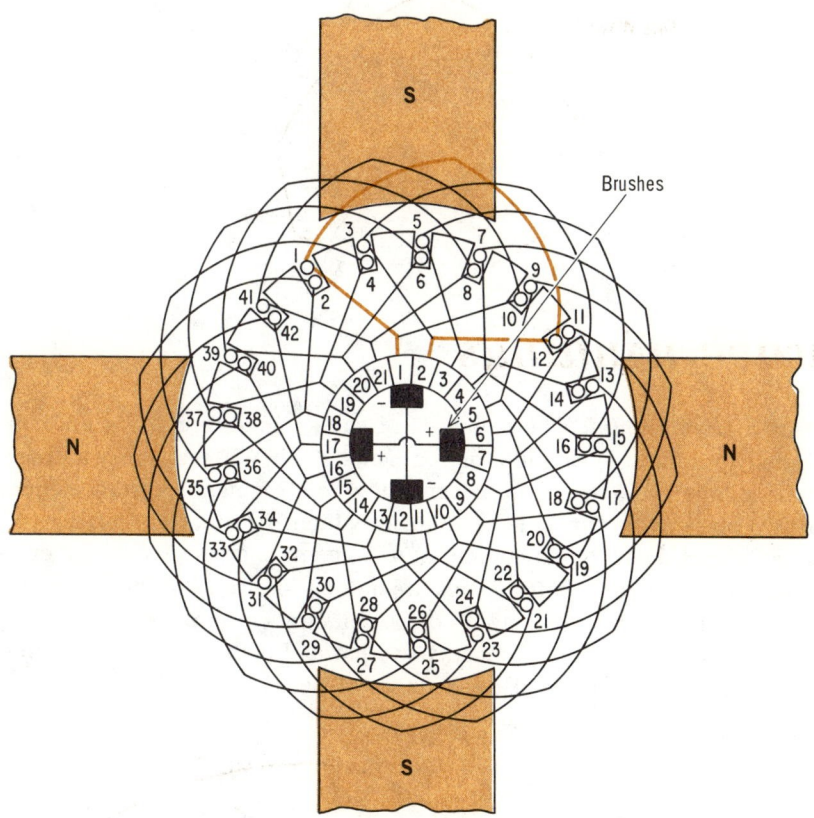

lap windings (cont.)

In summary, lap windings divide the armature winding into as many parallel paths as there are field poles. The voltages induced in the coils in any one of these parallel paths add to produce the total voltage of that path. The total voltages induced in each path are equal and of the same polarity, and so therefore are the currents in each path. There is one set of brushes for each two paths, or, in other words, one brush for each pole, and the combined outputs of each set of brushes make up the generator output. Since the brushes are in parallel, the output voltage is the same as that between any one set of brushes. The output current, though, is the sum of the currents flowing through each set of brushes.

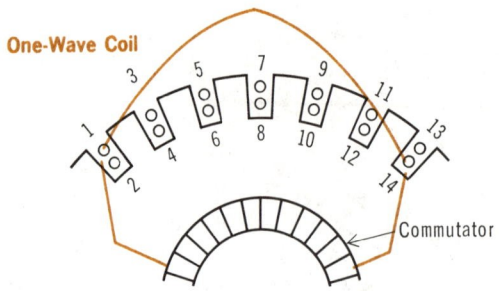

wave windings

In a *wave winding*, the two ends of each armature coil are not connected to adjacent commutator segments as they are in lap windings. Instead, one end of each coil is connected to a segment a distance equal to *two poles* away from the segment to which the other end of the coil is connected; again, each commutator segment is connected to the ends of two different coils, but the coils are on opposite sides of the armature. This has the effect of placing all those coils under similar pairs of poles in *series*. There are, therefore, only two parallel paths through the armature winding, regardless of the number of poles.

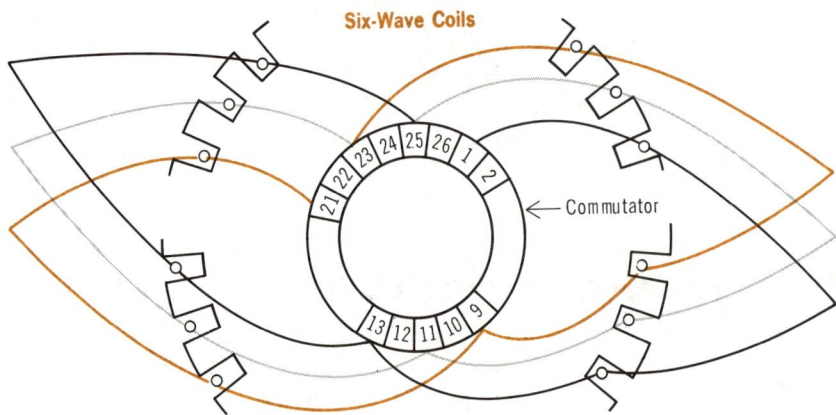

The individual coil voltages add in each path, and since there are only two paths, there are more coils per path than there are in a comparable lap winding. The total voltage induced in each path, therefore, is relatively high. The current capacity of a wave winding, though, is lower than that of a lap winding because there are only two paths for current through the winding. You can see, then, that whereas a lap winding sacrifices output voltage for current-carrying capacity, a wave winding sacrifices current-carrying capacity to achieve high voltage.

wave windings (cont.)

For the same number and size of armature coils, a wave winding will produce a voltage equal to that produced by a lap winding times the number of pairs of poles. But the current capacity is decreased in the same proportion that the voltage is increased. For a simple two-pole generator, it makes no difference whether a lap or wave winding is used. There are two parallel paths through the winding in either case, so the output voltage and current capacity will be the same.

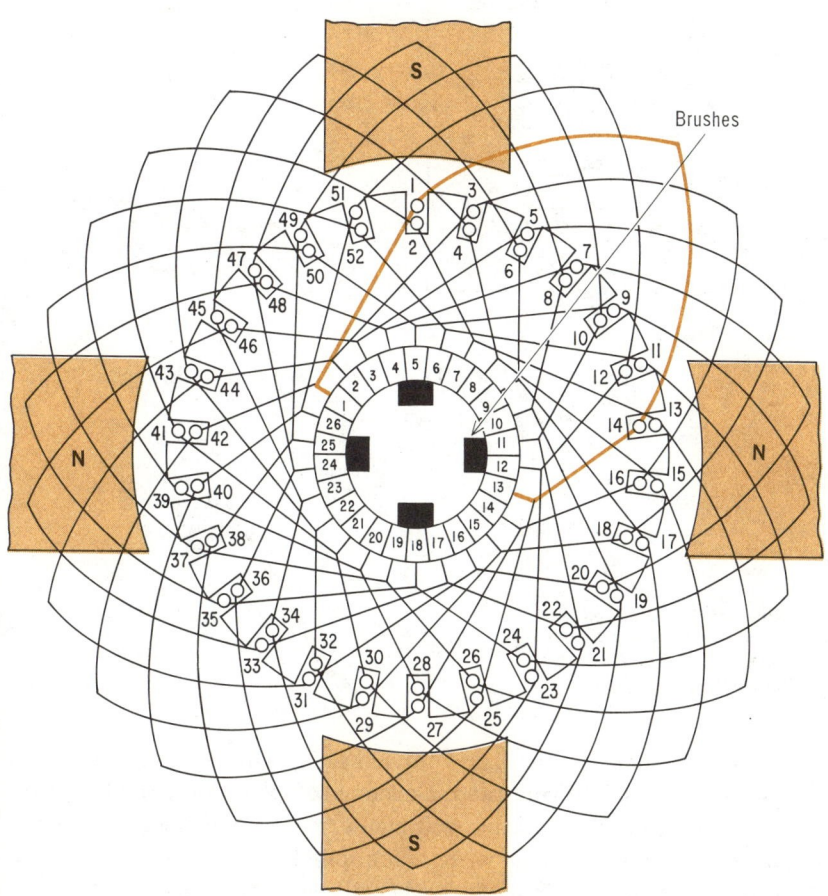

Four-Pole Wave-Wound Armature

the neutral plane

You will recall from the discussion on the basic commutator that whenever a brush makes contact with two commutator segments having a coil connected between them, the brush *short-circuits* the coil. If a voltage was being induced in the coil at this time, a large current would flow through the coil and probably burn it out. To prevent this, a coil must only be shorted by a brush, or *commutated*, when its induced voltage is zero. The points in its rotation where a coil has zero induced voltage lie along what is called the *neutral plane*. As shown for a two-pole generator, the neutral plane is perpendicular to the flux lines and midway between the pole pieces.

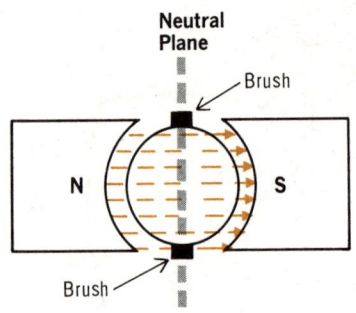

The neutral plane is theoretically where the armature coils cut no flux lines, and so have no voltage induced in them

During operation, the neutral plane tends to shift to a new position. Because of this, the theoretical neutral plane is called the mechanical or geometrical neutral plane

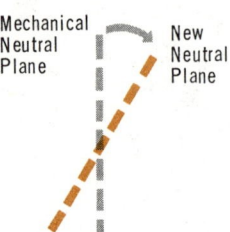

Armature reaction and self-induction of the armature coils tend to shift the position of the neutral plane

The neutral plane is the same for all coils, with each coil passing through it twice as the armature makes one complete rotation. Theoretically, then, perfect commutation will take place if the generator brushes are located in the neutral plane. In actual practice, though, the position of the neutral plane tends to shift when the generator is running. The brushes, therefore, either have to be moved to the new position of the neutral plane, or something must be done to prevent the plane from shifting. The two causes of this shifting of the neutral plane are *armature reaction* and *self-induction* of the armature coils.

armature reaction

When a generator supplies current to a load, the same current that flows through the load also flows through the armature winding. This causes a *magnetic field* to be built up around the conductors of the *armature winding,* since current through any conductor will produce a magnetic field. The magnetic fields around the individual conductors of the armature winding combine to produce an overall magnetic field. There are thus *two* magnetic fields in the space between the generator pole pieces. One is the field caused by the current through the armature, and the other is the main magnetic field produced by the field winding.

It is characteristic of magnetic fields that their flux lines cannot cross; instead, they *combine* to produce a new, total magnetic field. The total field of the two individual fields between the generator pole pieces has a direction as shown. This is the actual field cut by the armature coils as the armature rotates. Zero voltage is still induced in each coil when it cuts no lines of flux, but the two points at which this occurs are no longer in the same place as they were when only the magnetic field of the field winding was considered. The two points have both been *shifted* in the direction of armature rotation. This means, therefore, that the neutral plane has been shifted in the direction of armature rotation.

The neutral plane shift depends on the strength of the magnetic field set up around the armature winding by the load current. The larger the load current, the more the neutral plane shifts. The direction of shift though, is always the same as the direction of rotation. You can see, then, that if the load is constant, the brushes can be positioned at the new location of the neutral plane and left there. But if the load on the generator changes frequently, the brushes must constantly be changed to the new position of the neutral plane if good commutation is to take place. This is obviously a troublesome procedure.

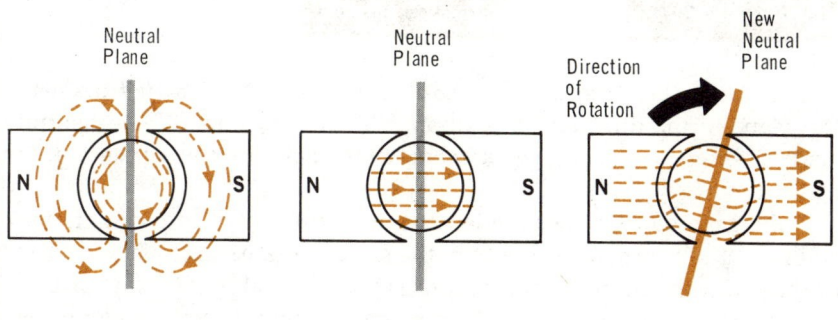

Magnetic field produced by current through armature winding

Magnetic field produced by field winding

Combined magnetic field

self-induction of armature coils

You have seen how the neutral plane is shifted in the direction of rotation by the effect of armature reaction. If you know the exact amount of this shift, and were to set the generator brushes accordingly, you would actually still not get perfect commutation. This is because of *self-induction* of the armature coils.

When the current through a coil tends to drop to zero at the point where the coil is cutting no flux lines, the magnetic field around the coil collapses. You should recall from Volume 3 that this causes a self-induced voltage, which has a polarity that keeps the current flowing, instead of allowing it to drop instantly to zero. This self-induced voltage, then, appears between the commutator segments when the coils are in what is supposedly the neutral plane. Although the self-induced voltage is small, it can cause a large current because of the low resistance of the commutator segments, the brushes, and the coil. This means that the voltage in a coil is not zero until sometime after the coil has passed the point in its rotation where it is cutting no flux lines.

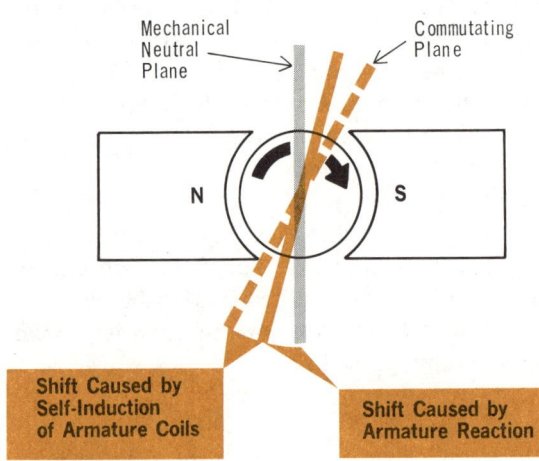

Mechanical Neutral Plane

Commutating Plane

N   S

Shift Caused by Self-Induction of Armature Coils

Shift Caused by Armature Reaction

In addition to the shift of the neutral plane caused by armature reaction, there is a further shift caused by the self-induction of the armature coils. Both shifts are in the direction of rotation

Effectively, then, the self-induced voltage has *further shifted* the neutral plane in the direction of rotation. This new position of the neutral plane is often called the *electrical neutral plane,* or the *commutating plane*.

If the generator brushes are located in this plane, good commutation will result. Like armature reaction, the shift in the neutral plane caused by self-induction of the armature coils is proportional to the load current. Therefore, as was mentioned before, good commutation can only be obtained if the load current is constant, or if the position of the brushes is changed every time the load current changes.

Two-Pole Generator With Interpoles

Direction of Rotation

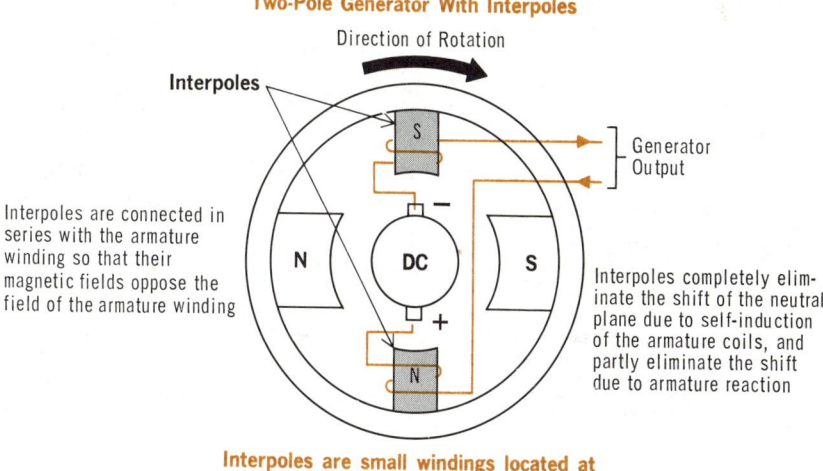

Interpoles

Generator
Output

Interpoles are connected in
series with the armature
winding so that their
magnetic fields oppose the
field of the armature winding

Interpoles completely eliminate the shift of the neutral
plane due to self-induction
of the armature coils, and
partly eliminate the shift
due to armature reaction

**Interpoles are small windings located at
the mechanical neutral plane**

interpoles

When a generator is to supply a variable load current, some provision
must be made to keep the neutral plane from shifting as the load current changes. Without such a provision, constant changing of the position of the brushes would be required. One way of keeping the actual
neutral plane close to the mechanical neutral plane despite changes in
load is by the use of *interpoles*. As shown, interpoles are small windings located at the mechanical neutral plane. The windings are wound
around pole pieces that are part of the generator housing. The interpole
windings are connected in series with the armature winding, so the load
current causes a magnetic field to be set up around each interpole.

The directions of these magnetic fields are such that they *cancel* the
magnetic field around the armature coils in the vicinity of the interpoles. With no magnetic field to collapse, therefore, there is no self-
induction of the armature coils when they reach the neutral plane. So,
no shift of the neutral plane occurs due to self-induction of the armature
coils. To do their job, interpoles must have the correct polarity, which is
opposite to that of the emf of self-induction. If you use the left-hand
rule, you will see that the interpole polarity must be the same as the
next field pole in the direction of rotation.

Interpoles are self-regulating. When the load current increases, the
neutral plane tends to shift in the direction of rotation. But, at the same
time, the field of the interpoles increases in strength, and so opposes
the shift.

compensating windings

Interpoles cannot eliminate armature reaction because they are located only at the neutral plane while armature reaction takes place all around the armature. But, if interpoles were placed all around the armature, armature reaction could be eliminated. In effect, this is what is done by *compensating windings*, which are small windings set in the main pole pieces. The compensating windings are in *series* with each other and with the armature winding.

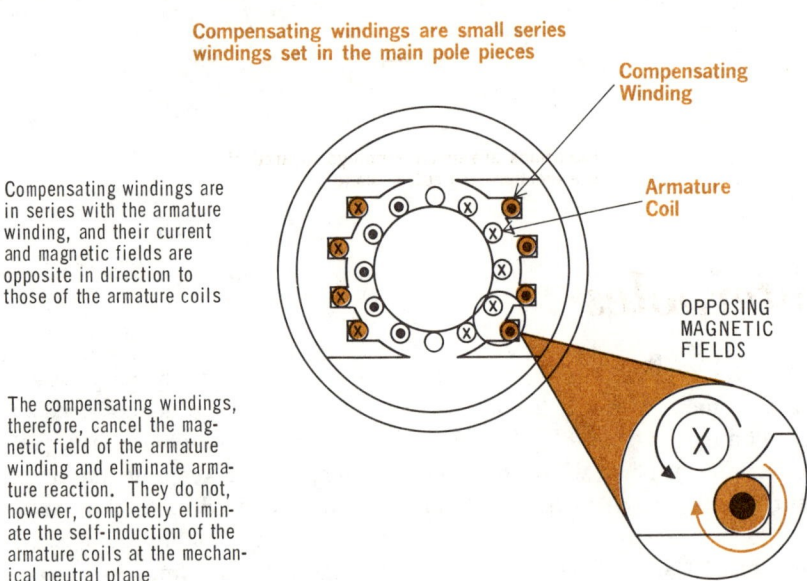

Compensating windings are small series windings set in the main pole pieces

Compensating Winding

Armature Coil

OPPOSING MAGNETIC FIELDS

Compensating windings are in series with the armature winding, and their current and magnetic fields are opposite in direction to those of the armature coils

The compensating windings, therefore, cancel the magnetic field of the armature winding and eliminate armature reaction. They do not, however, completely eliminate the self-induction of the armature coils at the mechanical neutral plane

As shown, the current in the compensating windings flows in the *opposite* direction of the current in the armature coils that face them. Since the currents are opposite, the direction of the magnetic fields of the compensating windings and the armature coils are also opposite. This means that the field of the compensating windings cancels the field of the armature coils. By canceling the magnetic field of the armature coils, the compensating windings eliminate armature reaction, but it does not do away completely with the self-induction of the armature coils.

You can see that both compensating windings and interpoles eliminate most of the shift in the neutral plane that would otherwise be caused by load changes. Some variable load generators use compensating windings, while others use interpoles. On very large generators, or those with wide load variations, compensating windings and interpoles are often used together.

regulating generator voltage

The voltage produced by any d-c generator depends on three factors: (1) the number of conductor loops in the armature that are in series, (2) the speed at which the armature rotates, and (3) the strength of the magnetic field. For a generator to provide a constant voltage output under varying load, at least one of these three factors must be varied to compensate for the voltage changes that the load changes would otherwise cause. During normal operation, neither the number of conductors in the armature, nor their arrangement can be changed. It is also impractical to change the speed at which the armature rotates. The strength of the magnetic field, though, can be changed relatively easily by varying the current through the field winding, and as a result, this is the method used mostly to regulate the output voltage of a generator.

Voltage Regulators Are Often Used With Shunt Generators

Voltage Regulator

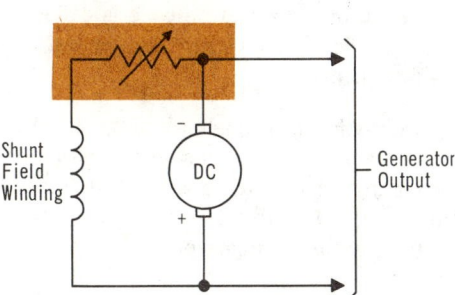

When used with shunt generators, voltage regulators function as variable resistors in series with the field winding. They automatically adjust the field current to compensate for changes in the generator output voltage

You have learned how compound generators provide a relatively constant voltage because their field current automatically adjusts to compensate for changes in load current. By using auxiliary devices or circuits called *voltage regulators,* generators can be made to have output voltages even more stable than those of compound generators. Voltage regulators vary widely in type and design; but they all perform two basic functions: they *sense* the generator output voltage, and they *vary* the current through the field winding in response to the output voltage. Thus, if the regulator senses that the voltage has dropped, it increases the field current to bring the voltage back to normal. Similarly, if the output voltage rises, the regulator decreases the field current. Normally, voltage regulators control the field current either by varying a *resistance* in series with the field, or one in parallel with the field.

In many cases, the regulating resistor is adjusted manually to obtain the right meter reading. For large generators that supply very high currents, special resistance devices are used. One such device is the *carbon pile regulator.*

summary

☐ In a lap winding, the armature winding is divided into as many parallel paths as there are field poles. One set of brushes is used for each two paths. ☐ Lap windings sacrifice voltage output for current capacity. ☐ In a wave winding, all armature coils under similar pairs of poles are in series. There are, therefore, only two parallel paths through the armature. ☐ Wave windings sacrifice current-carrying capacity for high voltage.

☐ The neutral plane is the point in the rotation of an armature where its induced voltage is zero. ☐ When a generator is running, the neutral plane tends to shift as a result of armature reaction and self-induction of the armature coils. ☐ Armature reaction is caused by the magnetic field built up around the armature winding by the load current. ☐ Armature reaction shifts the neutral plane in the direction of armature rotation. ☐ Self-induction of the armature coils prevents the coil voltage from instantly dropping to zero when the coil is cutting no flux lines. ☐ The neutral plane is shifted further in the direction of rotation by the self-induction of the armature coils. ☐ The new position of the neutral plane produced by armature reaction and self-induction of the armature coils is called the electrical neutral plane. ☐ Good commutation results when the generator brushes are located at the electrical neutral plane.

☐ Interpoles are small windings located at the mechanical neutral plane. They are used to compensate for self-induction of the armature coils. ☐ Compensating windings are small windings set in the main pole pieces. They compensate for armature reaction. ☐ Voltage regulators are used to maintain a stable output voltage from a generator.

review questions

1. What are the basic characteristics of a lap winding?
2. In general, does a lap or wave winding have a greater current capacity?
3. What is the *mechanical neutral plane*?
4. What is *armature reaction*?
5. Can armature reaction be eliminated by holding the load current constant?
6. What is the *electrical neutral plane*?
7. How does self-induction of the armature coils affect the electrical neutral plane?
8. What are interpoles used for?
9. What are *compensating windings*?
10. If the output voltage of a generator tends to increase, what should be done to the current through the field winding to keep the voltage constant?

construction

The fundamental principles of operation of d-c generators have been described. You know the basic parts of a generator, the function of each of these parts, and the relationship each part has to the overall operation of the generator. Essentially, what you have learned is the *electrical operation* of a d-c generator. The *physical construction* has been covered only briefly. Very often, you will find that it is important for you not only to know how a generator works electrically, but to be able to recognize its various parts and be familiar with their physical construction.

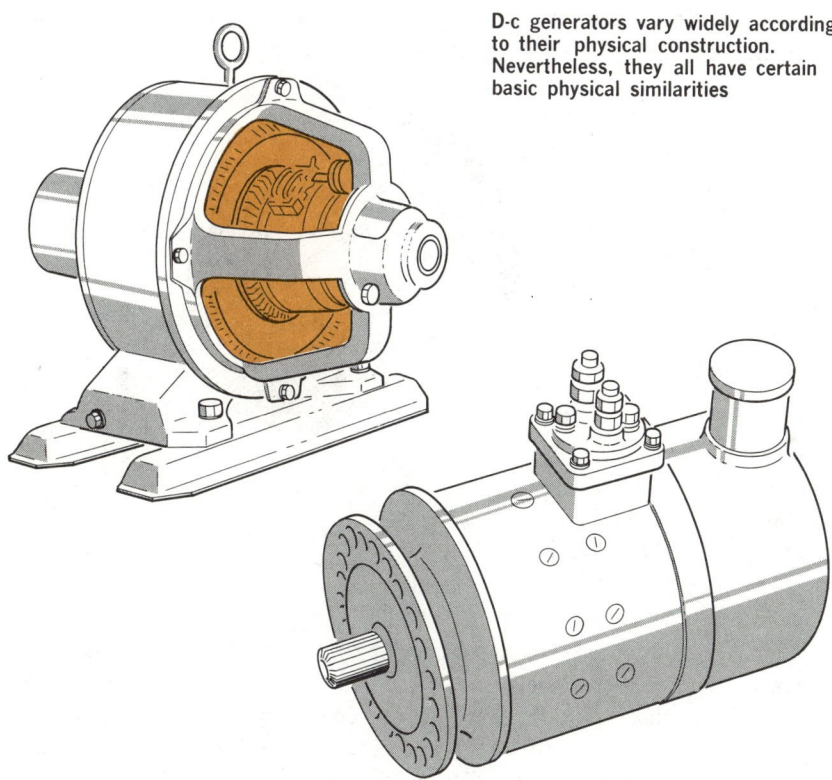

D-c generators vary widely according to their physical construction. Nevertheless, they all have certain basic physical similarities

It is beyond the scope of this volume to describe the construction details of all the many types of d-c generators now in use. However, there are basic physical similarities between practically all generators. Because of this, if you know how a typical generator is made, you will have a good idea of the physical construction of most others. The materials and methods used in the construction of a typical d-c generator, therefore, are described on the following pages.

the armature

The armature, or armature assembly as it is sometimes called, consists of all those generator parts that *rotate*. These parts are the armature *shaft*, the armature *core*, the armature *winding*, and the *commutator*. As shown, the core and the commutator are mounted on the shaft. The winding is wound around the core, in slots, and the ends of the individual coils that make up the winding are connected to the commutator segments. Although not shown, there is one other part that must be mounted on the armature shaft. This is the device or means for connecting the armature to its *driving source*. Usually, this is accomplished by a gear or a belt and pulley arrangement. Or it could be directly attached to the drive shaft of the driving source.

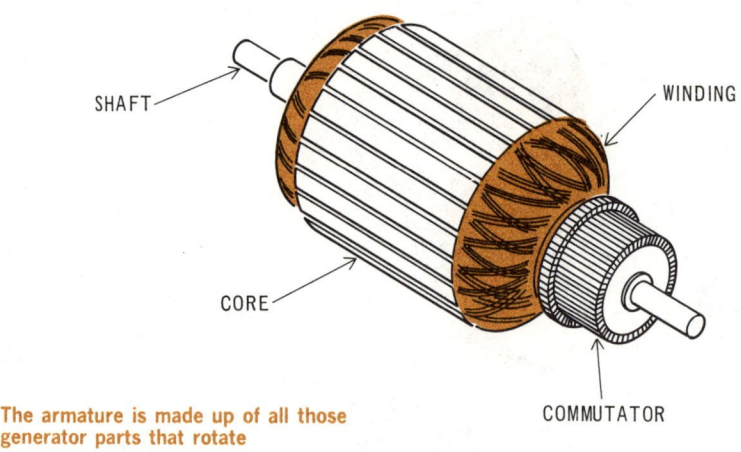

SHAFT

WINDING

CORE

COMMUTATOR

The armature is made up of all those generator parts that rotate

The armature core is drum-shaped, or cylindrical, and is made of soft steel. Instead of being one piece, the core is made of many thin pieces, called *laminations*. The laminations are coated with an insulating varnish, and then are pressed together to form the complete core. Each lamination has notches around its edge, and when the laminations are combined to produce the core, the notches are aligned so that the core has slots along its perimeter. The reason the core is laminated is to reduce losses due to *eddy currents*, which you remember (Volume 3) are circulating currents that are induced in a conducting material when it cuts through magnetic flux lines. The laminations have the effect of greatly reducing the area in which the eddy currents can flow; and this means that the resistance of the material to eddy current flow is greatly increased.

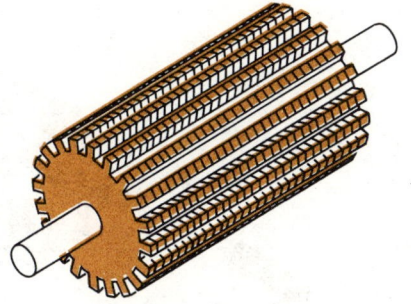

One Core Lamination

**Laminated Core Mounted
on Armature Shaft**

The armature core is laminated to reduce eddy current losses. It is made
of silicon steel to reduce hysteresis losses

the armature (cont.)

To reduce the *hysteresis loss* of the core, which you will recall (Volume 3) occurs because the reversals of the magnetization of the core material actually *lag* the current reversals, practically all generators have cores made of soft silicon steel. This material has an inherently low hysteresis loss.

The generator shaft is a rod made of hard steel, finished with a highly polished bearing surface. The method of mounting the core and the commutator on the shaft varies widely from generator to generator.

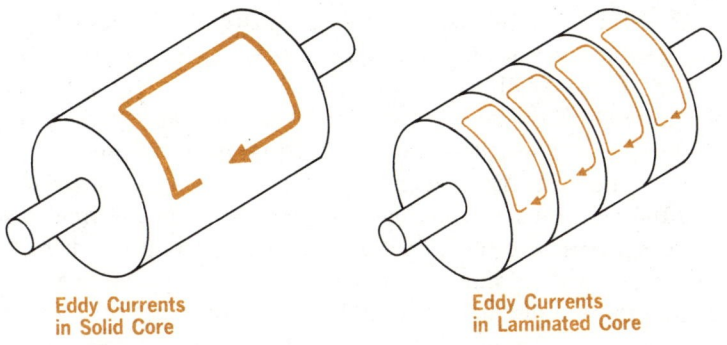

**Eddy Currents
in Solid Core**

**Eddy Currents
in Laminated Core**

By laminating the core, the eddy currents are divided into many small
currents, greatly decreasing the eddy current losses

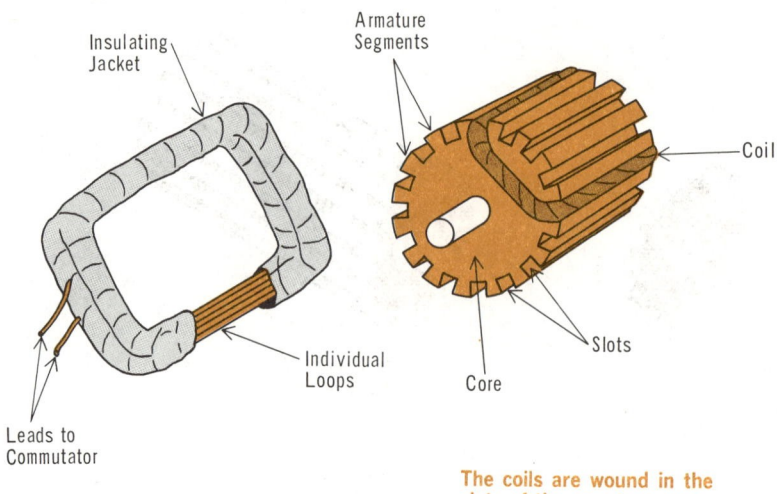

Each coil consists of one or more loops wrapped together in a common insulating jacket

The coils are wound in the slots of the armature core

the armature winding

Each of the coils that make up the armature winding are wound around the armature core, with the sides of the coils set into the slots in the core. In many armatures, the coils are first formed into their final shape, and then placed on the core. This is called *form winding*. All of the loops in a coil are *wrapped together* in a common insulating jacket, and each coil has only *two leads* for connecting to the commutator.

The high portions of the core are called *armature segments,* and the number of armature segments between the two sides of a coil is directly related to the number of poles in the generator. The reason for this is that, as you will recall, for maximum induced voltage, the two sides of a coil must be separated by the same distance that separates adjacent generator poles. Thus, if there are 24 armature segments and the two sides of each coil are 12 segments apart, it is a two-pole generator. Similarly, if there are 24 armature segments and the coil sides are 6 segments apart, it is a four-pole generator. By dividing the total number of armature segments by the number of segments between coil sides, you can determine the number of segments for which the armature was wound.

Wedges made of insulating material are placed in the core slots to hold the coils in place. On some generators, steel bands are also placed around the armature to prevent the coils from being thrown out by centrifugal force.

the commutator

The commutator consists of individual segments made of hand-drawn copper and shaped as shown. The individual segments are assembled in cylindrical form, and are held together by a *clamping flange*. The segments are placed in the wedge-shaped space between the two halves of the clamping flange, and the flange bolts are then tightened, holding the segments rigidly in place. The segments are *insulated* from each other with thin sheets of *mica*. They are also insulated from the clamping flange with rings of mica.

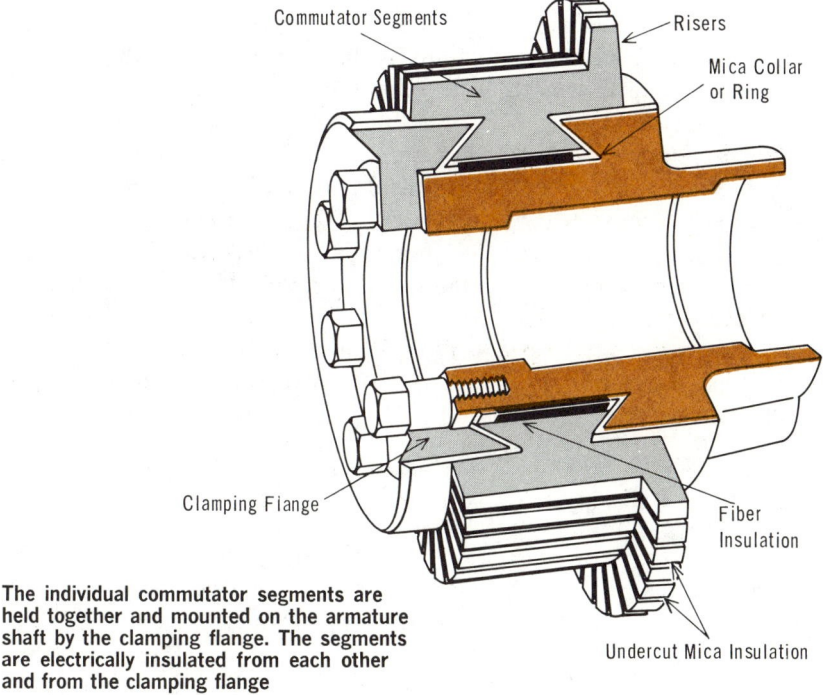

The individual commutator segments are held together and mounted on the armature shaft by the clamping flange. The segments are electrically insulated from each other and from the clamping flange

The leads from the armature coils are connected to the raised portions of the commutator segments, called *risers*. Some commutator segments have no risers. In these cases, the leads from the armature coils are connected to slits in the end of the segments.

The surface of the commutator is cut and ground to a very smooth cylindrical finish. This insures a minimum of friction between the commutator surface and the brushes. In addition, the mica insulation is *undercut* slightly below the surface of the commutator segments so that it does not interfere with the brushes.

the brushes

The brushes transfer the generator output from the commutator to an external circuit. They are usually small blocks of a *carbon* and *graphite* compound. *No lubrication* should be used between the brushes and the commutator, since the graphite in the brushes provides self-lubrication. The brushes are set in holders and held against the commutators by springs. In many generators, the spring pressure can be adjusted. If the pressure is set too high, the brushes will wear out quickly. And if too low a pressure is used, poor electrical contact between the brushes and the commutator results.

In many generators, the brushes are electrically connected to the brush holders by braided copper wires called *pigtails;* but in simple generators, the close fit of the brush inside the holder provides the electrical connection. Connections to an external circuit are then made from the brush holders. The brush holders, although mounted on the generator frame, are electrically insulated from it. In many generators, the brush holders are not easily accessible. The brush holders are then usually connected to *studs* on the outside of the generator housing.

Although the brushes are designed to be long lasting, they are made to wear out faster than the commutator, because it is cheaper and easier to replace the brushes than the armature. The brushes are usually made long so that they can wear down to a fraction of their length before they have to be replaced. The spring keeps moving the brush toward the commutator as it wears.

Brushes Are Usually Made of Carbon and Graphite

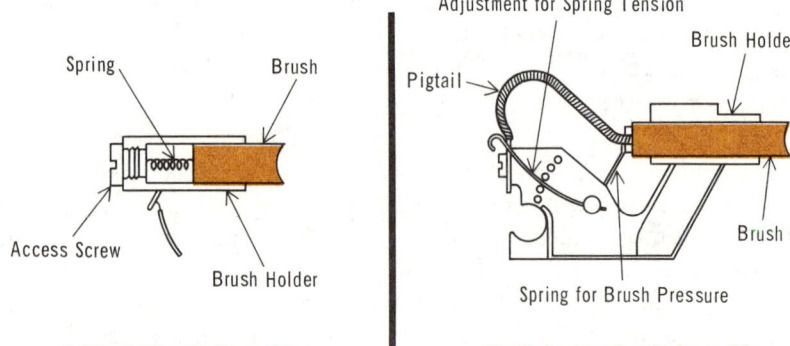

A Simple Brush Assembly   **An Elaborate Brush Assembly**

Brushes are held in holders, and are pressed against the commutator by springs. Good generators have a tension adjustment to allow proper brush contact with minimum wear

the field winding

The field winding of a generator consists of the individual field coils wound around their cores, or pole pieces. The number of field coils depends on how many poles the generator has. In a two-pole generator, there are two field coils; in a four-pole generator, four coils; and so forth. The field coils are mounted on the inside circumference of the generator housing with large countersunk screws that pass through the housing and into the pole pieces.

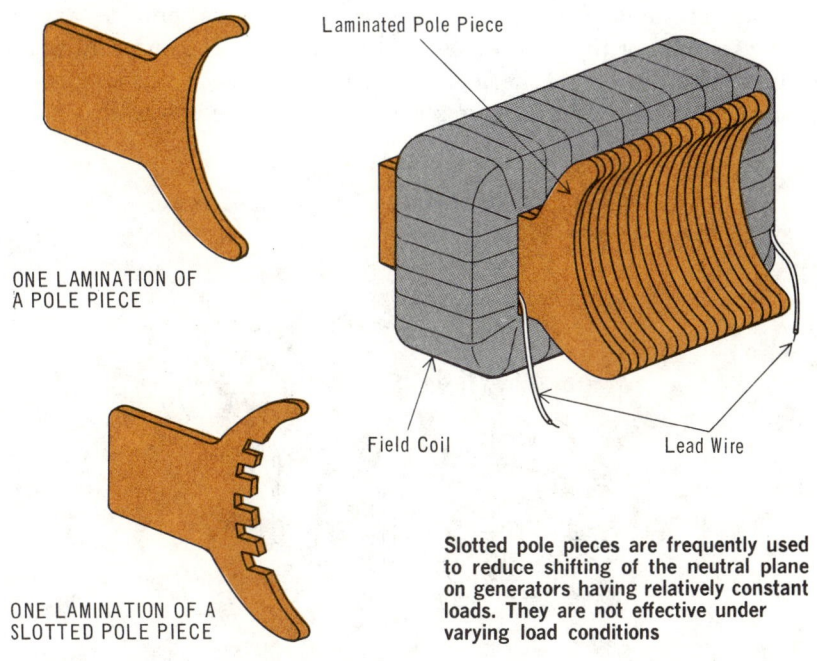

Laminated Pole Piece

ONE LAMINATION OF
A POLE PIECE

Field Coil   Lead Wire

ONE LAMINATION OF A
SLOTTED POLE PIECE

Slotted pole pieces are frequently used to reduce shifting of the neutral plane on generators having relatively constant loads. They are not effective under varying load conditions

The pole pieces are usually made of sheets of steel *laminated* together. This decreases losses due to eddy currents. As shown, the pole pieces are shaped to fit the curvature of the armature. The purpose of this is to keep the *air gap* between the pole pieces and the armature as *small* as possible, since air offers a relatively high reluctance to magnetic flux lines.

The field coils are wound around the pole pieces. *Shunt windings* consist of *many turns* of relatively *small-diameter* insulated copper wire. *Series windings*, on the other hand, consist of a *few turns* of *large-diameter* copper wire, also insulated. The wire used in series windings must be large enough to carry the entire load current without overheating.

the housing and mounting

The generator housing provides *mechanical support* for the parts that make up the generator. On many generators, it also provides protection against outside disturbances such as dust, dirt, and moisture. Most generator housings consist of three parts: a *field frame,* and two *end frames.* The field frame supports the field coils, as well as the interpoles, if they are used. It also serves as part of the magnetic circuit of the field winding. Because of its magnetic function, the field frame is made of iron or steel having good magnetic properties. The thickness of the frame depends on the degree of mechanical support it must provide, as well as the strength of the magnetic field it must carry. The end frames are mounted at either end of the field frame, and are either screwed or bolted to it. The armature bearings are set in recesses in the end frames.

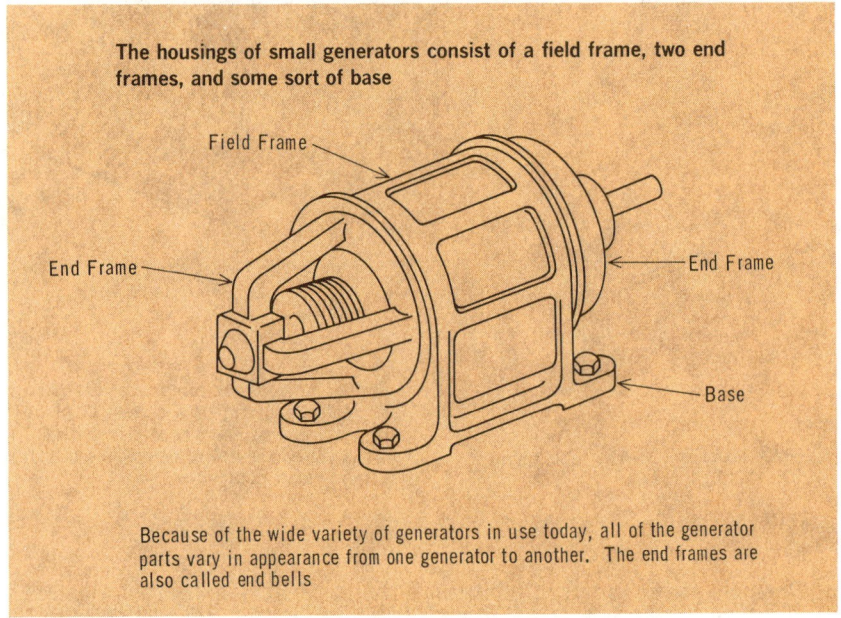

The housings of small generators consist of a field frame, two end frames, and some sort of base

Field Frame

End Frame

End Frame

Base

Because of the wide variety of generators in use today, all of the generator parts vary in appearance from one generator to another. The end frames are also called end bells

There are many methods used for mounting generators, with the method used in any particular case depending on factors such as the size and use of the generator, and the type of source that is driving it. Probably the most common type of mounting is the simple base. This consists of legs or other supports either attached to or part of the field frame, and on which the generator sits. Another common mounting method consists of a flange or plate at the drive end of the generator. The plate has holes in it, and can be bolted to another plate or a panel.

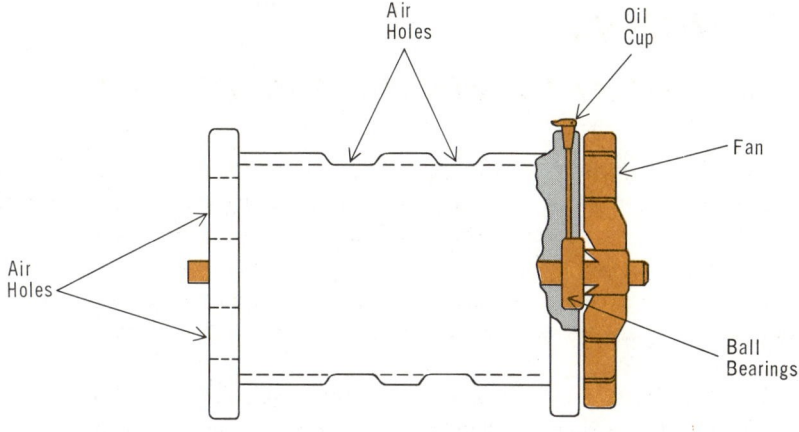

Proper lubrication and cooling are required if a generator is to give long and efficient service

other construction features

Two other important construction features of d-c generators are the *bearings* and the *cooling methods* used. Practically all smaller type generators use ball bearings to provide smooth, high-speed armature rotation with a minimum of friction. The bearings are force-fitted on the ends of the armature shaft, and are set in recesses in the end frames. Thus, when the end frames are screwed or bolted to the field frame, the armature is supported by the end frames. The bearings on some generators are *permanently lubricated,* and sealed during manufacture. No further lubrication of these bearings during use is required, or possible. On generators where the bearings are not sealed, some means is provided for periodic lubrication. Usually, this is in the form of a grease or oil cup in the end frames.

When a generator operates at full capacity, it develops considerable heat. The most common method of dissipating this heat is by means of *air holes* and a *fan.* The air holes are openings in the end frames and in the field frame near the field windings. On larger generators, there are also air holes through the armature. The fan is mounted on one end of the armature shaft. When the armature rotates, the fan forces air through the air holes. The air thus picks up heat from inside the generator housing, and carries it through the air holes to the outside of the housing. Heavy-load generators are often also mounted on a large metal area, referred to as a *heat sink,* which helps conduct the heat away.

regulating
a variable-speed generator

You will recall that regulation is the process of maintaining a *constant* generator output. It is normally accomplished by a device or circuit that senses the generator output, and controls the current through the field winding to compensate for any changes in the output. The two main causes of change, or instability, in a generator's output are changes in the *resistance* of the *load* and changes in the *speed of rotation* of the generator.

The output voltage of shunt and compound generators that are driven at constant speeds does not vary greatly if the load variations are within the design range of the generator. As a result, regulation is only provided for these generators when an extremely stable output is desired, or when the load variations are very great.

Variable-speed generators, on the other hand, usually require regulation. You can understand why by considering the generator used in an automobile. The generator is turned by the automobile engine, so its speed of rotation is very different when the engine is idling than it is when the automobile is traveling at high speed. And yet, despite this wide variation in its speed of rotation, the generator must supply a constant 6 or 12 volts to the electrical system of the automobile. If this voltage should vary, the lights, the horn, in fact the entire electrical system, would be affected. The only way such a generator can supply a stable 6 or 12 volts is by the use of a regulating device. And not only must the regulating device keep the output voltage constant, it must also limit the *output current* of the generator to its maximum rated value to prevent the generator from burning out. Thus, both voltage and current regulation are normally required for a variable-speed generator.

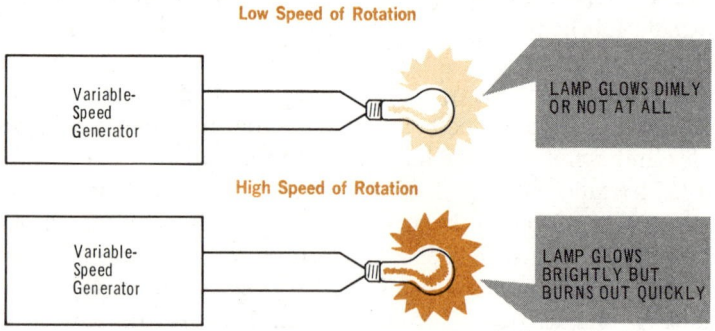

If variable-speed generators did not have their outputs regulated, the loads they supply could not work properly

voltage regulation

A typical circuit used for regulating the generator output voltage is shown. One side of the shunt field winding is connected *directly* to the negative output of the generator. The other side of the field winding is connected to the positive side of the generator output either through resistor R and coil L_2, or through contacts C, which are controlled by the magnetic field of L_1. Thus, the field winding is either directly across the generator output (C closed), or it has R connected in series with it (C open).

The contacts are held closed by a spring so that when the generator is first turned on, the contacts are closed, and the field winding is directly across the generator output. As the generator builds up speed, its output voltage increases, and so also does the current through L_1, which is connected directly across the generator output. When the output voltage reaches a certain point, the magnetic field of L_1 is strong enough to overcome the spring holding C closed, and so C opens.

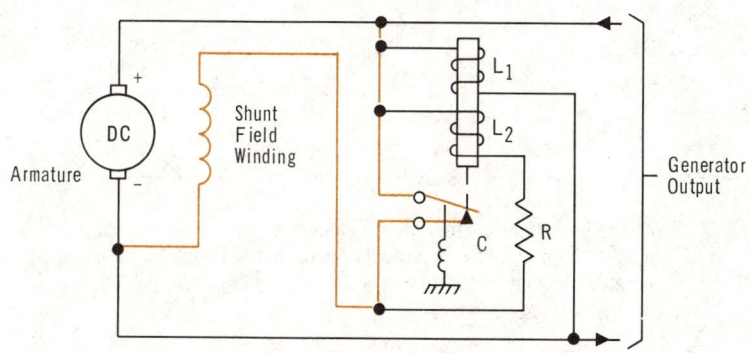

The voltage regulator operates by contacts intermittently inserting a resistance in series with the field winding

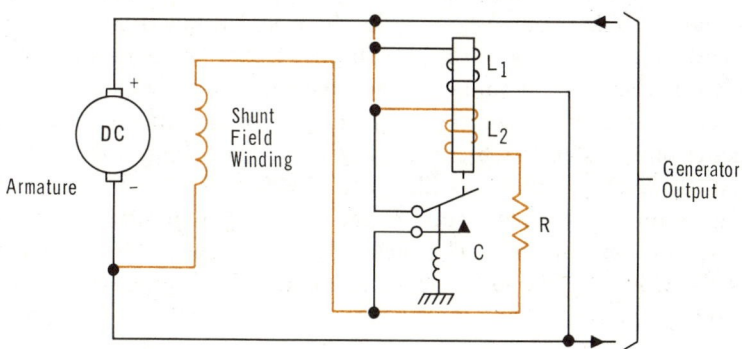

voltage regulation (cont.)

The field current now flows through R and L_2. With the added resistance in the field circuit, the field strength decreases, and the rise in the generator output voltage is checked. The current through L_2 causes a magnetic field around L_2, and this field opposes that around L_1, since the two coils are oppositely wound. This partially neutralizes the magnetic attraction of L_1 on C, and the spring again closes C. As a result, the field winding is again directly across the generator output, so the field current increases, increasing the output voltage, which in turn causes C to open due to the increased magnetic pull of L_1.

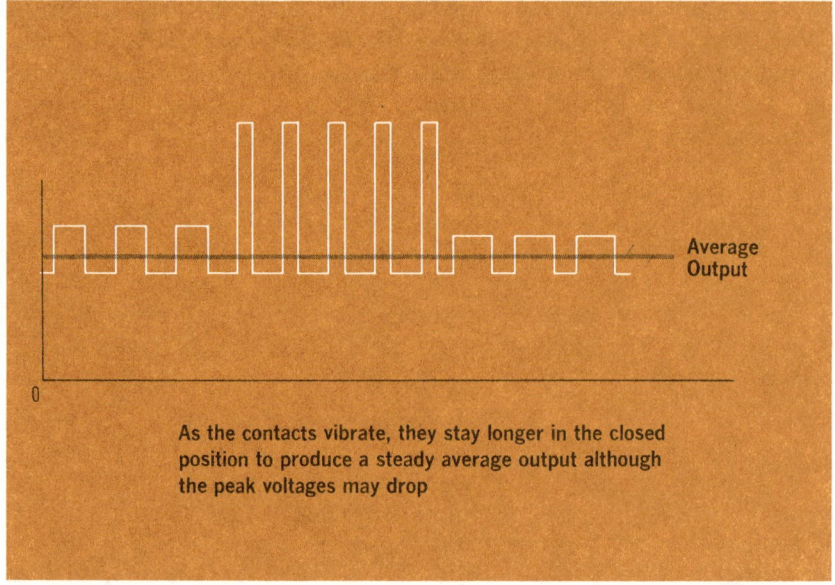

Average Output

0

As the contacts vibrate, they stay longer in the closed position to produce a steady average output although the peak voltages may drop

This cycle occurs very rapidly, many times a second, causing the contacts to vibrate open and closed. The output voltage of the generator thus varies slightly but very rapidly above and below a value determined by the tension of the spring holding C closed. The actual d-c output voltage is the average value between the high and low points. This average depends on whether the contacts stay longer in one position than in the other as they vibrate. If they stay closed longer than they stay open, the average voltage is higher, and vice versa. When the peak voltage goes up, the contacts stay longer in the open position to keep the average output constant. The spring is adjustable, and its setting controls the vibrating action of the contacts and the average generator output voltage.

current regulation

The current regulator, like the voltage regulator, operates by intermittently inserting a resistance in series with the field winding

In the current regulator, the insertion of the resistance is controlled by the load current, whereas in the voltage regulator it is controlled by the output voltage

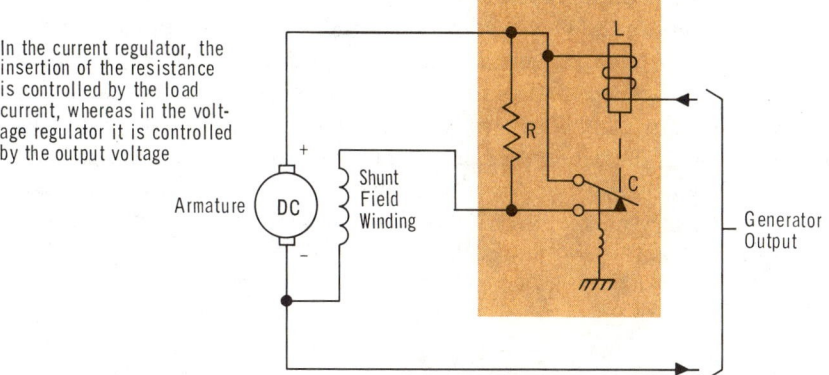

The purpose of regulating the output current of a generator is to prevent the current from exceeding the maximum value that the generator can safely deliver. A commonly used method of current regulation is shown. Essentially, it is very similar to the method for voltage regulation. Depending on whether contacts C are open or closed, the field winding is either directly across the generator output, or it has resistance R in series with it. The opening and closing of C is controlled by coil L, which is in series with the generator output, and so carries the full load current.

Contacts C are normally held closed by a spring. So, when the generator starts, the field winding is directly across the generator output. The output voltage, therefore, builds up, and the load current, which flows through L, also increases. When the current increases to the point where the magnetic attraction of L overcomes the tension on the spring that holds C closed, the contacts open. This inserts resistance R in series with the field winding, and causes the field current, and therefore the output voltage, to decrease.

As a result, the load current also decreases. The decrease in current then lessens the magnetic pull of L, and the spring closes C again. This puts the field winding directly across the generator output, and allows the current to again increase until the magnetic pull of L is sufficient to overcome the spring and open C. As in the voltage regulator, this cycle is continually repeated, and the current varies slightly above and below the average value determined by the spring tension holding C closed.

reverse-current cutout relay

Often a variable-speed d-c generator is used in combination with a *storage battery* to supply power to a load. When the generator's speed, and therefore its output voltage, is low, the battery supplies power to the load. And when the generator comes up to speed and reaches its rated output, it supplies the power to the load, and at the same time *recharges* the battery. In this arrangement, though, some method must be used to disconnect the generator from the battery whenever the generator voltage is less than the battery voltage. Otherwise, the battery would discharge through the generator armature winding and possibly burn it out. A frequently used method for automatically disconnecting the generator from the battery makes use of a *reverse-current cutout relay*. A typical circuit is shown.

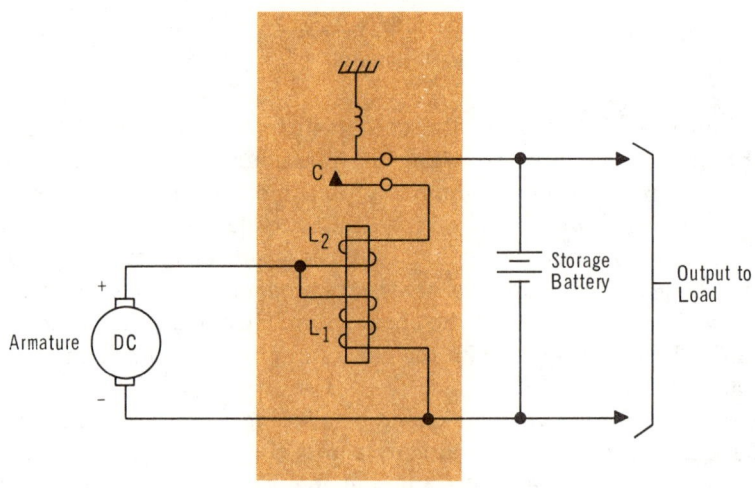

The reverse-current cutout relay connects the generator output to the battery and the load when the generator voltage is greater than the battery voltage. It then disconnects the generator from the battery and the load when the generator voltage drops below that of the battery

The reverse-current cutout relay consists of coils L_1 and L_2, both wound on the same core, and contacts C, which are normally held open by a spring. L_1 is called the voltage winding, and is connected across the generator output. L_2 is called the current winding, and is in series with the generator output.

reverse-current cutout relay (cont.)

The spring holding C open is adjusted so that when the generator voltage is less than the battery voltage, the contacts are open. The generator is therefore disconnected from both the battery and the load, and the battery supplies the output power. At all times, though, the generator output voltage is across L_1, causing current through the coil, and creating a magnetic field. When the generator output voltage rises above the battery voltage, the magnetic pull of L_1 overcomes the spring tension and closes C. This connects the generator to the load, and at the same time allows charging current to flow from the generator to the battery. The generator current flows through L_2, creating a magnetic field which aids that of L_1, and thus keeps the contacts closed tightly.

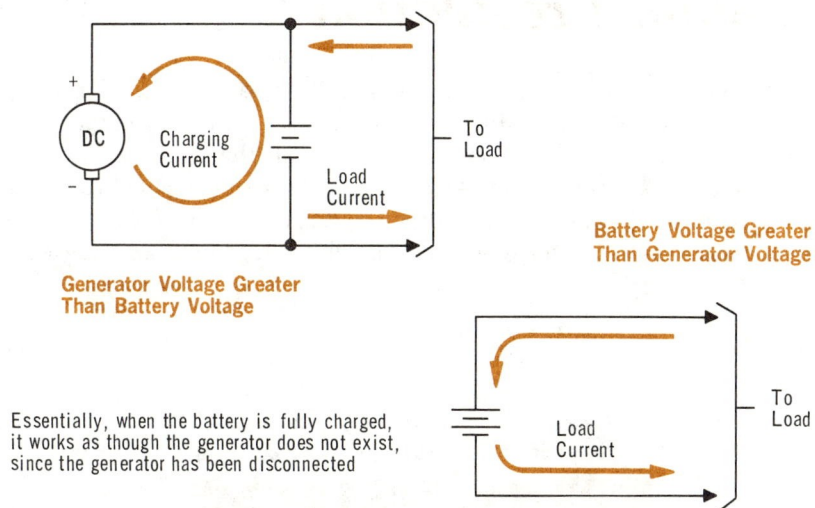

Generator Voltage Greater
Than Battery Voltage

Battery Voltage Greater
Than Generator Voltage

Essentially, when the battery is fully charged,
it works as though the generator does not exist,
since the generator has been disconnected

When the generator voltage is high enough to keep L_1 energized, but its voltage drops below that of the battery, the battery begins to discharge through L_2 and the generator. With the reversal of current through L_2, the magnetic field around it now opposes the magnetic field of L_1. This causes a decrease in the magnetic attraction on C, and as a result, the contacts open, disconnecting the generator from the battery and the load. You should note that the contacts of the reverse-current cutout relay do not vibrate continuously as do the contacts of the voltage and current regulators. They open or close only when the generator voltage either rises above or drops below the battery voltage.

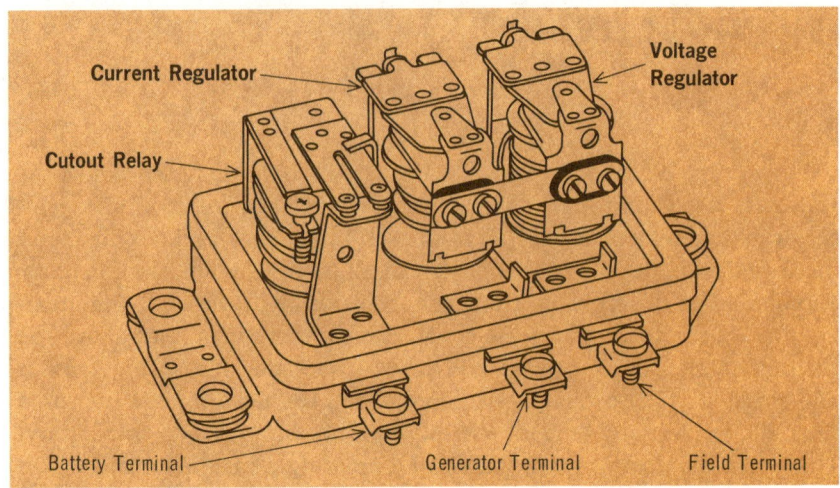

Current Regulator

Voltage Regulator

Cutout Relay

Battery Terminal   Generator Terminal   Field Terminal

combined regulation

Frequently, a voltage and current regulator, as well as a reverse-current cutout relay, are all used together to control the output of a variable-speed generator. When this is done, all three are usually built and installed as a single unit. Actually, this is the unit called the "voltage regulator" in an automobile. Although it is called a voltage regulator, it really consists of a voltage regulator, a current regulator, and a reverse-current cutout relay.

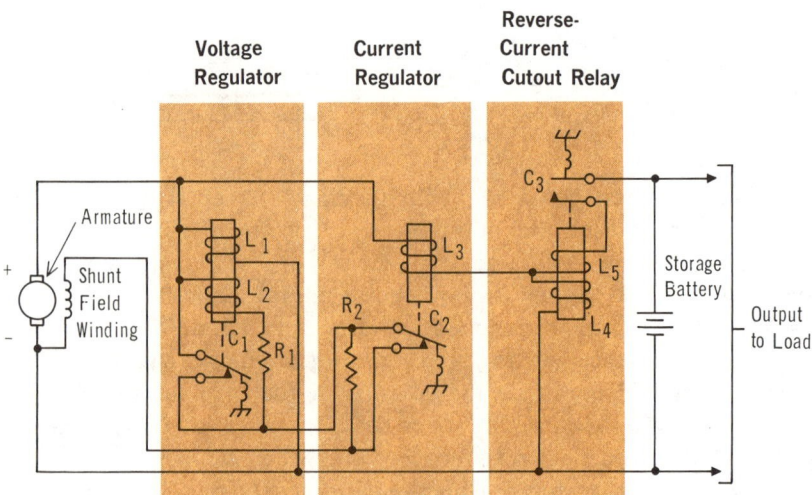

Voltage Regulator   Current Regulator   Reverse-Current Cutout Relay

summary

☐ The armature of a d-c generator consists of all parts that rotate. These are: the armature shaft, the armature core, the armature winding, and the commutator. ☐ The armature shaft is a hard steel rod. ☐ The armature core is made of soft steel laminations insulated from each other. The core is laminated to reduce eddy current losses. ☐ The coils of the armature winding are wound around the core. ☐ The commutator is made of individual copper segments insulated from each other with thin sheets of mica. ☐ Leads from the armature coils are connected to the commutator segments.

☐ Generator brushes are usually small blocks of a carbon and graphite compound. The brushes are mounted in brush holders, which are electrically insulated from the generator frame. ☐ The individual coils of the field winding are wound around the generator pole pieces. ☐ Most generator housings consist of an iron or steel field frame and two end frames. ☐ The armature bearings are set in recesses in the end frames.

☐ Variable-speed generators normally require both voltage and current regulation. The voltage regulation maintains the output voltage essentially constant, while the current regulation prevents the output current from exceeding the maximum value that the generator can safely deliver. ☐ Reverse-current cutout-relay circuits are employed when a d-c generator is used in combination with a battery to supply power to a load. The circuit disconnects the generator from the battery whenever the generator voltage is less than the battery voltage.

review questions

1. Why is the armature core of a generator laminated?
2. Why is soft silicon steel generally used for the armature core of a generator?
3. What is *form winding*?
4. Must the commutator and brushes of a generator be lubricated? Why?
5. Why are pole pieces shaped to fit the curvature of the armature?
6. Physically, what is the difference between shunt and series field windings?
7. Could a plastic field frame be used for a generator? Why?
8. Why do variable-speed generators usually require both voltage and current regulation?
9. What is the purpose of a reverse-current cutout relay?
10. What are the two main causes of instability in a generator's output voltage?

a-c generators (alternators)

Although d-c generators are used extensively for certain applications, they have *inherent limitations* that make them unsatisfactory for many other uses. Some of these limitations are caused by the electrical and physical construction features of the d-c generators themselves, while others are due to the basic nature and properties of d-c electricity. Most of the limitations caused by the generators themselves arise from difficulties in constructing commutators that can carry large outputs reliably and efficiently. A-c generators, as you will learn, do not have commutators, and so in this respect are superior to d-c generators.

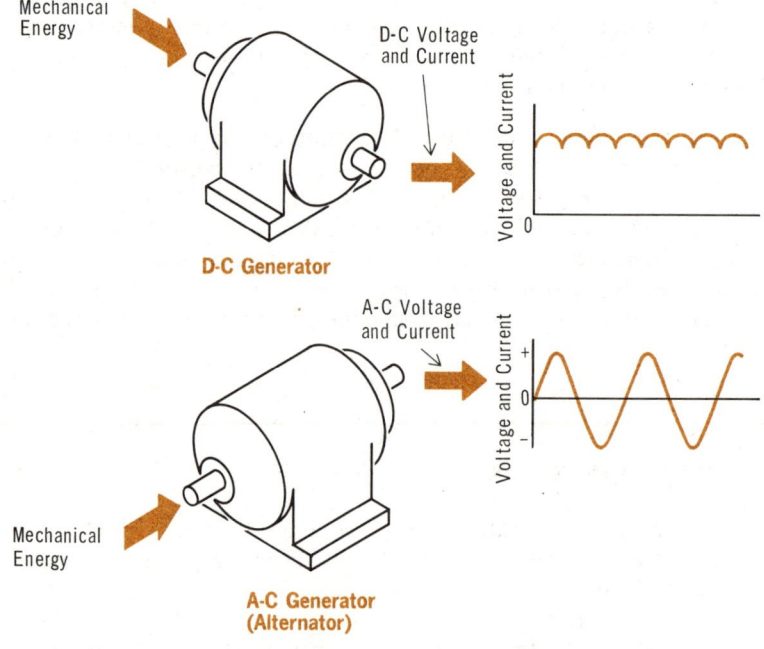

Both a-c and d-c generators convert mechanical energy into electrical energy. However, d-c generators convert mechanical energy into d-c voltages and currents, while a-c generators convert the mechanical energy into a-c voltages and currents

The theory of operation and construction characteristics of a-c generators are described on the following pages. You will find that there are many *basic similarities* between a-c and d-c generators. However, there are also many *significant differences,* as well as some electrical concepts that are entirely new to you. A-c generators are also called *alternators,* since they produce alternating current.

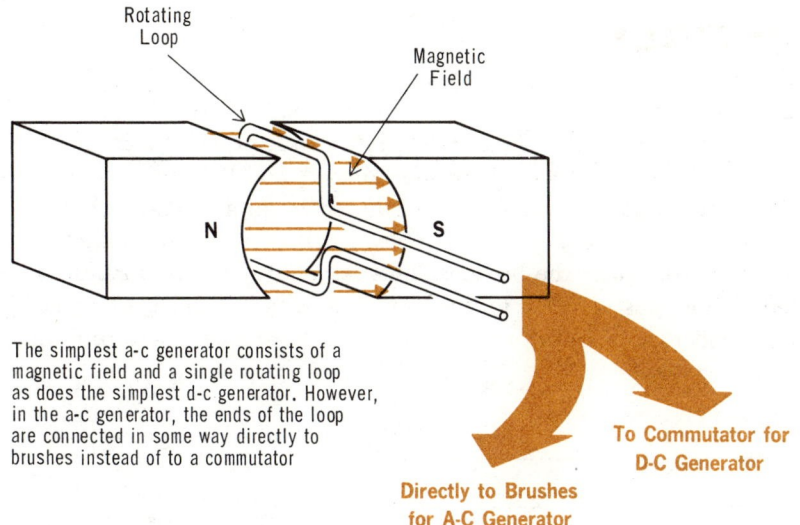

Rotating
Loop

Magnetic
Field

N   S

The simplest a-c generator consists of a
magnetic field and a single rotating loop
as does the simplest d-c generator. However,
in the a-c generator, the ends of the loop
are connected in some way directly to
brushes instead of to a commutator

To Commutator for
D-C Generator

Directly to Brushes
for A-C Generator

the basic a-c generator

You will recall that the simplest d-c generator consists of a single loop of wire rotating in a magnetic field, plus a commutator and brushes. As the loop rotates, an a-c voltage is generated between its two ends. The a-c voltage is then converted to dc by the action of the commutator and brushes. The commutator accomplishes the ac to dc conversion by effectively *switching* the brushes from one end of the rotating loop to the other each time the voltage induced in the loop reverses polarity. This switching is done in such a way that *one brush* is always in contact with the end of the loop that is *positive*, while the *other brush* always contacts the end of the loop that is *negative*. The voltage between the brushes, which is the generator output voltage, is therefore dc.

If the commutator was eliminated, and each brush was permanently in contact with one end of the rotating loop, the voltage between the brushes would be exactly the same as the voltage existing between the ends of the loop. And as you know, this is an a-c voltage. Thus, by eliminating the commutator and in some way permanently connecting the brushes to opposite ends of the loop, the basic d-c generator can be converted to a simple a-c generator.

Of course, a direct connection of the ends of the loop cannot be made to the brushes because the ends must be free to rotate with the loop. If they were not free to rotate, they would become twisted as the loop turned and would break. So, in some way, the brushes must be permanently connected to their loop ends without interfering with their turning. This is covered next.

slip rings

The commutator in a d-c generator performs two functions: (1) it converts the induced a-c voltage to dc, and (2) it provides a means for connecting the induced voltage to the brushes, and thus to an external circuit. No conversion of ac to dc is required in an a-c generator, so in place of the commutator, all that is needed is a means of connecting the induced voltage to the brushes. This is done by *metallic rings* connected to the ends of the rotating loop. One ring is connected to each end of the loop, and both rings rotate as the loop rotates. These rings are called *slip rings*.

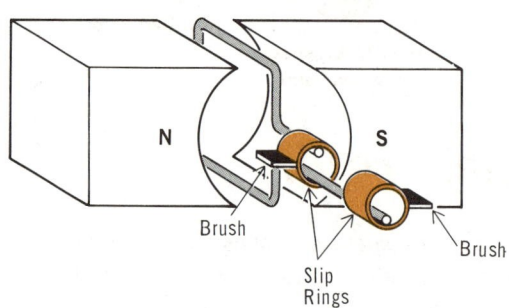

The slip rings are permanently connected to the ends of the loop. The brushes are stationary, and maintain contact with the slip rings as they rotate. The voltage induced in the loop is thus transferred to the brushes, and from there to an external circuit

Brush

Brush

Slip
Rings

Loop and Slip Rings Rotate While Brushes Remain Stationary

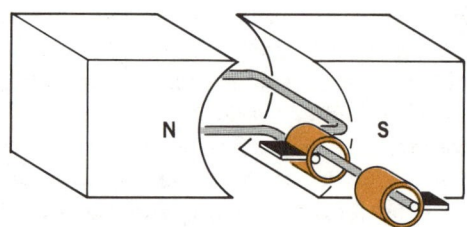

Each slip ring is *permanently* connected to its respective end of the rotating loop, so the voltage induced in the loop appears between the rings. The brushes rest against the slip rings and make electrical contact with them. As the loop rotates, the slip rings slide along the brushes, always maintaining electrical contact with them. Thus, each brush is always in contact with its slip ring, which in turn is permanently connected to one end of the loop. The result is that the a-c voltage induced in the loop appears between the brushes, and can then be applied to an external circuit.

generating a sine-wave output

You can see from the preceding pages that the output of the simple, one-loop a-c generator is identical to the voltage induced in the rotating loop. This voltage is equal to the *sum* of the voltages induced in the *two sides* of the loop as they cut the magnetic flux lines. When *no* flux lines are being cut, the voltage is *zero;* and when the *maximum* number of flux lines are cut, the voltage is *maximum*. As shown, in a two-pole a-c generator, the voltage passes through zero and maximum *twice* during every full rotation of the loop. This variation follows a *sine wave*. So for one full rotation, which corresponds to 360 degrees of rotation, the generated voltage corresponds to 360 electrical degrees.

By comparing the waveform shown with that on page 6-73, you can see that the output voltage of the simple a-c generator is the same as that induced in the rotating loop of a simple d-c generator.

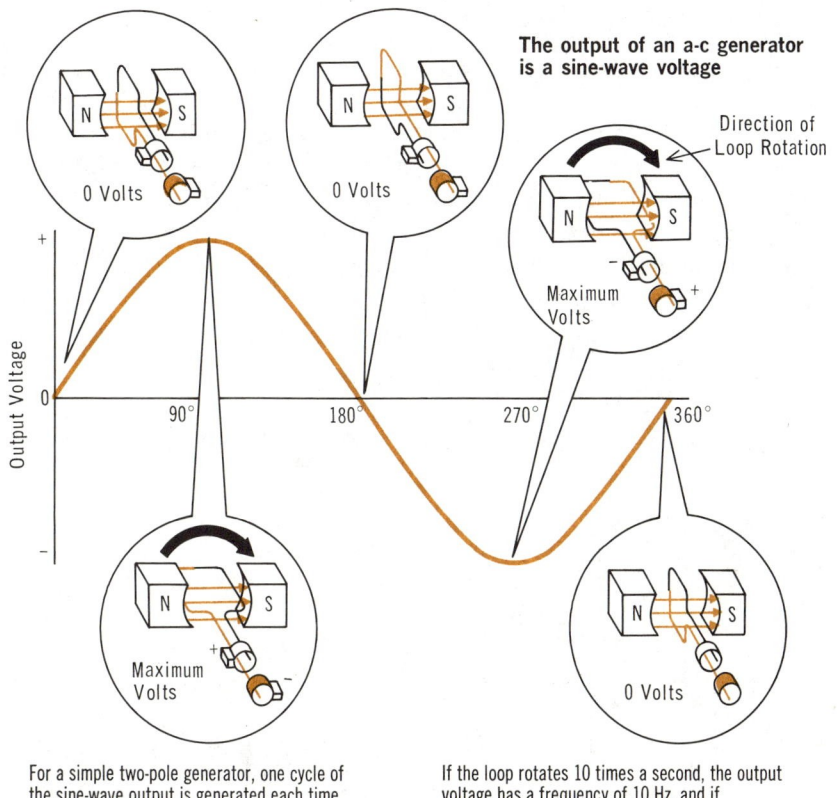

The output of an a-c generator is a sine-wave voltage

Direction of Loop Rotation

0 Volts

0 Volts

Maximum Volts

Maximum Volts

0 Volts

Output Voltage

90° 180° 270° 360°

For a simple two-pole generator, one cycle of the sine-wave output is generated each time the loop makes one full rotation. The frequency of such a generator, therefore, is the same as the speed of rotation of the loop

If the loop rotates 10 times a second, the output voltage has a frequency of 10 Hz, and if the loop rotates 100 times a second, the frequency of the output voltage is 100 Hz

increasing the number of poles

From the waveform on the previous page, you can see that the output voltage of the simple a-c generator is maximum when the sides of the loop pass the centers of the poles. The reason for this is that at these points the sides of the loop are cutting the maximum number of flux lines. If four poles were used instead of two, the output voltage would still reach its maximum value when the sides of the loop passed the centers of the poles. However, since the number of poles was doubled, the voltage would be maximum *four times* during each full rotation of the loop, instead of two times, as it is for a two-pole generator.

If the poles are spaced equally apart, this means that one full sine-wave cycle of the output voltage is generated each time the loop makes one-half, or 180 degrees, of a rotation. The *frequency* of the a-c output voltage is, therefore, *twice* the speed of rotation of the loop. For example, if the loop rotates 30 times a second, the frequency of the voltage is 60 Hz.

It should be obvious that for a given speed of rotation, the more poles there are, the higher will be the frequency of the generator voltage. A general relationship between the speed of rotation of a single loop, the number of poles, and the frequency can be stated as follows: *Frequency is equal to the number of revolutions per second times the number of pairs of poles*. Thus, if six poles are used, the loop rotates 10 times a second, the frequency of the output voltage is 10×3, or 30 Hz. Note that it is the number of *pairs* of poles and not the total number of poles that is used in determining the frequency.

In a 4-pole generator, the sides of the loop cut a maximum number of flux lines four times during each full rotation of the loop

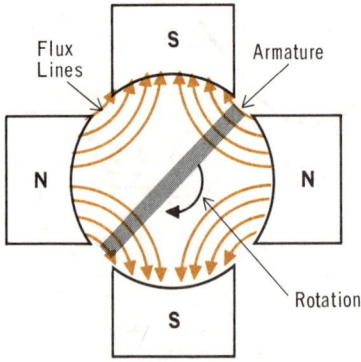

Flux Lines — S — Armature — N — N — S — Rotation

The four points of maximum voltage occur when the sides of the loop pass the centers of the four poles

In a 4-pole generator, two cycles of the output voltage are generated for each full rotation of the loop

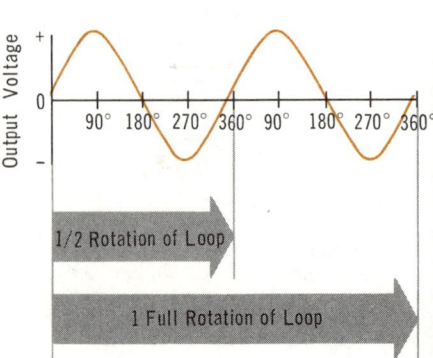

Output Voltage

90° 180° 270° 360° 90° 180° 270° 360°

1/2 Rotation of Loop

1 Full Rotation of Loop

producing the magnetic field

The magnetic field required for the operation of an a-c generator is produced by a *field winding,* the same as it is for d-c generators. You will recall that the field winding is an *electromagnet,* and, therefore, requires current to produce its magnetic field. In a d-c generator, the current for the field winding can be provided by connecting the winding to an external voltage source, in which case the generator is a *separately excited generator.* Or, the exciting current for the field winding can be obtained by connecting the winding to the generator output. This, you will remember, makes it a *self-excited generator.*

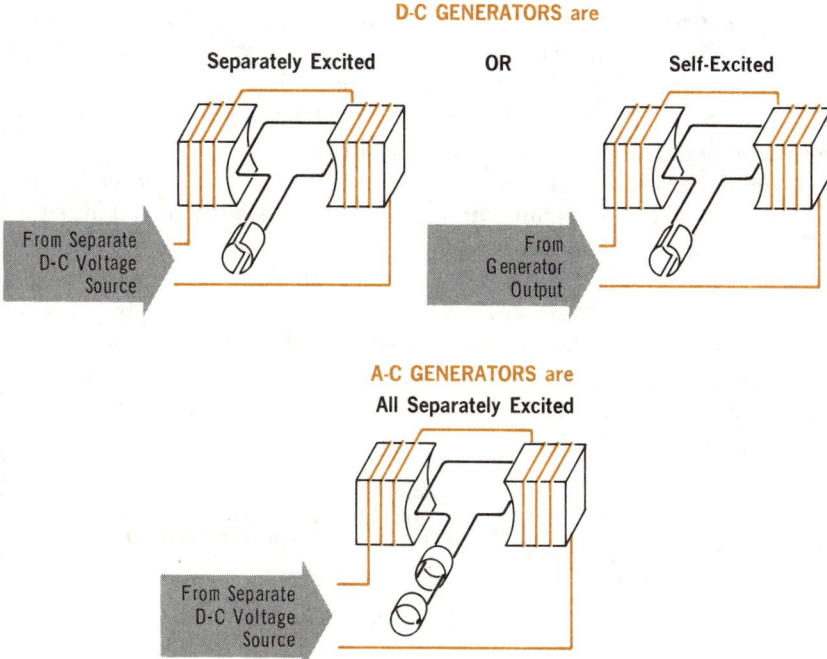

In both cases, though, whether the d-c generator is separately excited or self-excited, the voltage applied to the field winding is dc. This is necessary since a d-c exciting current is required for proper generator operation. As a result, self-excitation *cannot* be used for a-c generators since they have a-c outputs. Separate d-c voltage sources must be used to provide current for the field windings. On many a-c generators, the d-c voltage source for the field winding is a small d-c generator contained in the same housing with the a-c generator.

stationary-armature a-c generators

When an a-c generator delivers a relatively small amount of power, slip rings operate satisfactorily. When large powers are involved, though, it becomes difficult to insulate slip rings sufficiently, and they are therefore a frequent source of trouble. Because of this, most a-c generators have a *stationary armature* and a *rotating field.* In such generators, the armature coils are permanently mounted around the inner circumference of the generator housing, while the field coils and their pole pieces are mounted on a shaft and rotate within the stationary armature. This arrangement of a stationary armature and a rotating field may seem strange at first, but recalling the basic fundamentals of mutual induction, you will see that a voltage is induced in the armature coils whether they cut the flux lines of a stationary magnetic field, or whether the flux lines of a moving magnetic field cut them. *Relative motion* between the magnetic field and the armature coils is all that is required.

With a stationary armature, the generator output can be *directly connected* to an external circuit without the need for slip rings and brushes. This eliminates the insulation difficulties that would otherwise exist if large currents and voltages were delivered to the load through slip rings. Of course, since the field winding rotates, slip rings must be used to connect the winding to its external d-c exciting source. However, the voltages and current involved are small compared to those of the armature, and there is no difficulty in providing sufficient insulation.

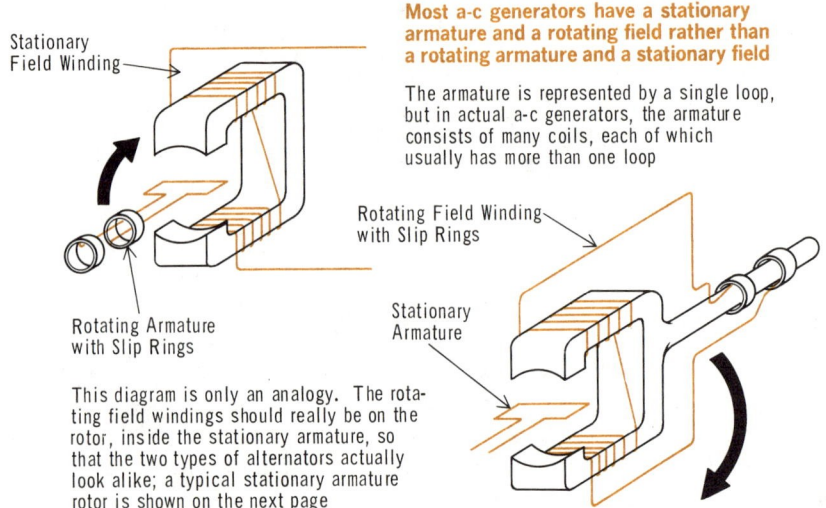

Stationary
Field Winding

Most a-c generators have a stationary armature and a rotating field rather than a rotating armature and a stationary field

The armature is represented by a single loop, but in actual a-c generators, the armature consists of many coils, each of which usually has more than one loop

Rotating Field Winding
with Slip Rings

Rotating Armature
with Slip Rings

Stationary
Armature

This diagram is only an analogy. The rotating field windings should really be on the rotor, inside the stationary armature, so that the two types of alternators actually look alike; a typical stationary armature rotor is shown on the next page

stationary-armature a-c generators (cont.)

Another advantage of using a stationary armature is that it makes possible much higher speeds of rotation, and therefore *higher voltages,* than can be obtained with rotating armatures. The reason for this is again insulating difficulties. At very high speeds of rotation, the large centrifugal force that results makes it difficult to properly insulate the armature winding. No such problem exists when the field winding rotates at high speeds.

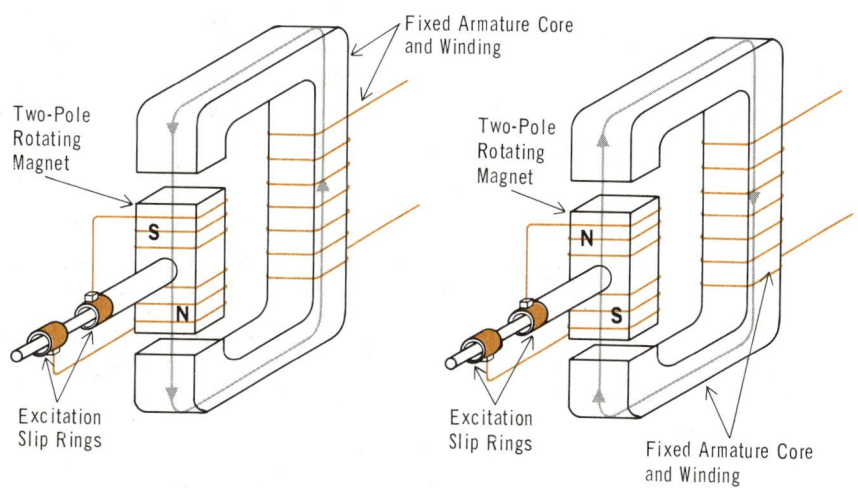

As the field poles in a stationary-armature generator are rotated, the flux lines that go from the N to S poles pass through the fixed-armature core first in one direction and then the other to induce ac in the stationary-armature winding

In summary, then, whereas practically all d-c generators use a rotating armature and a stationary field, most a-c generators have a stationary armature and a rotating field. With a stationary armature, voltages can be produced that are much larger than those possible with rotating-armature generators. The portion of a generator that rotates is often referred to as the *rotor,* while the stationary portion is called the *stator.*

You should note that if a stationary-armature a-c generator uses a *fixed* magnet for the field in the rotor, instead of an electromagnet, no slip rings at all will be required. However, such a generator has a low output, and so is limited in its applications.

single-phase a-c generators

In the discussion of a-c generators, the armature has been represented by a single loop. The voltage induced in such a loop would be very small; so, as in d-c generators, the armature actually consists of a number of coils, each usually having more than one loop. The coils are wound so that the voltages induced in the loops of any one coil *add* to produce the total coil voltage. The coils can be connected in various ways, with the particular method used depending on the desired generator characteristics.

If the armature coils are all connected in series aiding, the generator has a *single output*. The output is sinusoidal, and is equal in amplitude at any instant to the sum of the voltages induced in the individual coils. A generator having its armature wound in this way is a *single-phase generator*. All of the coils connected in series make up the armature winding. Very few practical a-c generators are single phase. Greater efficiency can be gained by connecting the armature coils in other ways.

A single loop rotating in a magnetic field is a single-phase generator, since it has only one output voltage

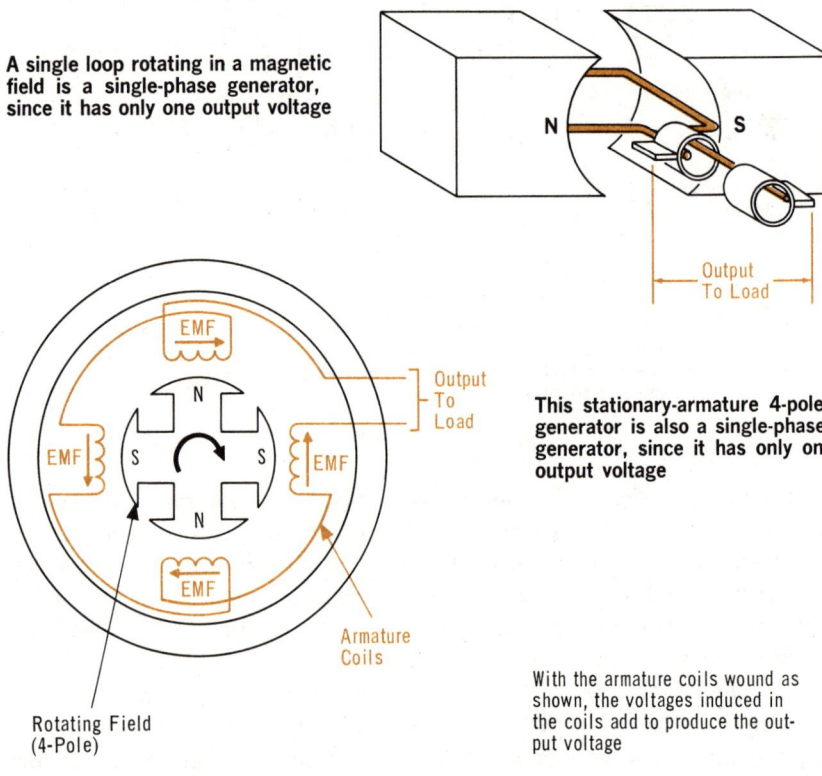

Output To Load

This stationary-armature 4-pole generator is also a single-phase generator, since it has only one output voltage

Armature Coils

Rotating Field (4-Pole)

With the armature coils wound as shown, the voltages induced in the coils add to produce the output voltage

two-phase a-c generators

Rotating-Loop Two-Phase Generator

Two perpendicular rotating loops, each with its own set of slip rings, make up a two-phase generator

The two output voltages are equal in amplitude and 90° out of phase. This, of course, assumes that the loops are the same size

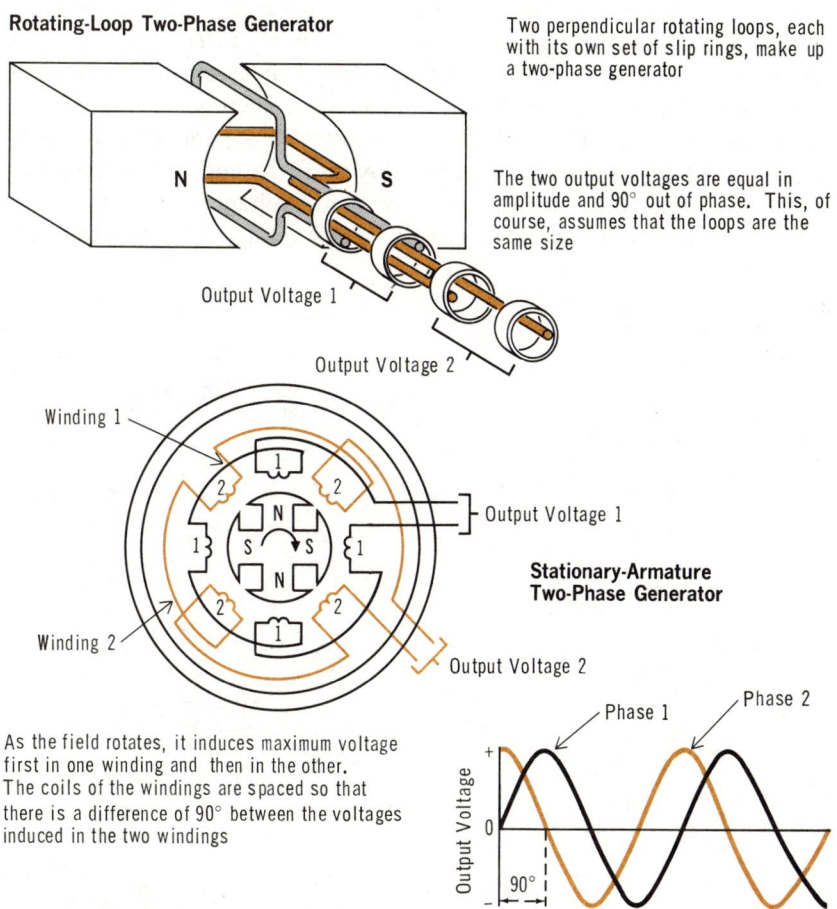

Output Voltage 1

Output Voltage 2

Winding 1

Output Voltage 1

Stationary-Armature Two-Phase Generator

Winding 2

Output Voltage 2

As the field rotates, it induces maximum voltage first in one winding and then in the other. The coils of the windings are spaced so that there is a difference of 90° between the voltages induced in the two windings

Phase 1   Phase 2

Output Voltage

90°

In a two-phase generator, the armature coils are wired so that the generator has two separate output voltages that differ in phase by 90 degrees. A simple, rotating-loop two-phase generator consists of two loops perpendicular to each other, with each loop connected to its own set of slip rings. When the voltage induced in one loop is maximum, the voltage in the other is zero; and vice versa. The voltages taken from the slip rings, therefore, differ in phase by 90 degrees.

The armature coils of a practical two-phase generator having a stationary armature are divided into two single-phase windings, with the individual coils of the two windings spaced so that the voltages induced in the two windings are 90 degrees out of phase.

three-phase a-c generators

Basically, the principles of a three-phase generator are the same as those of a two-phase generator, except that there are *three equally spaced* windings, and three output voltages all *120 degrees out of phase* with each other. A simple rotating-loop three-phase generator with its output waveforms is shown. Physically adjacent loops are separated by an angle equivalent to 60 degrees of rotation. However, the ends of the loop are connected to the slip rings in such a way that voltage 1 leads voltage 2 by 120 degrees, and voltage 2, in turn, leads voltage 3 by 120 degrees.

**Rotating-Loop
Three-Phase Generator**

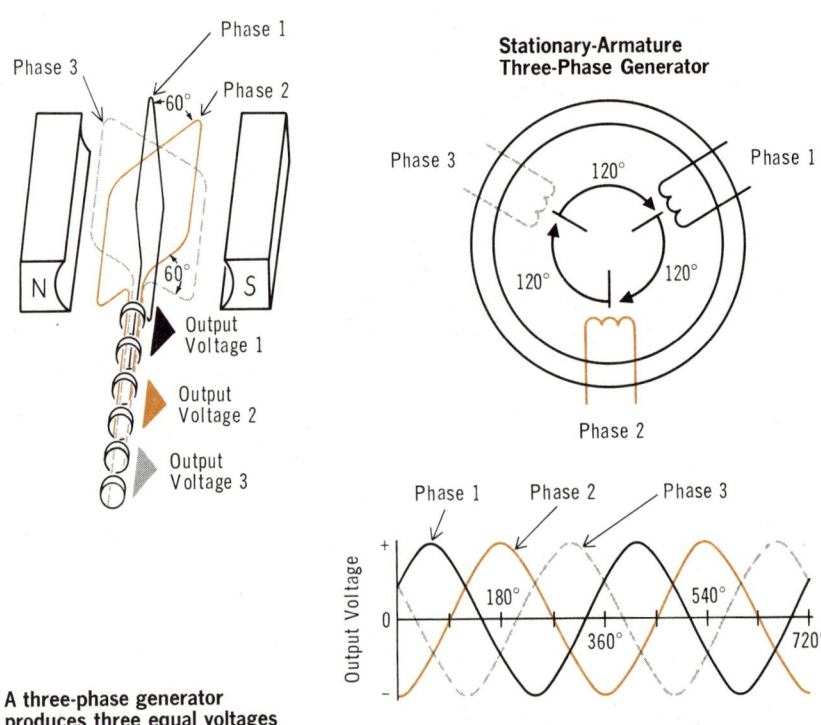

A three-phase generator
produces three equal voltages
120 degrees out of phase

A simplified diagram of a stationary-armature three-phase generator is also shown. On this diagram, the individual coils of each winding are combined, and represented as a single coil. Also, the rotating field is not shown. The significance of the illustration is that it shows that the three-phase generator has three separate armature windings that are 120 degrees out of phase.

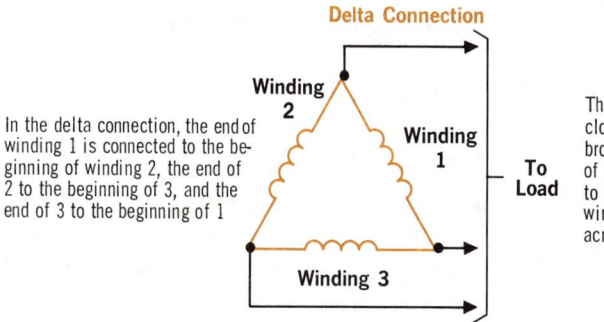

Delta Connection

In the delta connection, the end of winding 1 is connected to the beginning of winding 2, the end of 2 to the beginning of 3, and the end of 3 to the beginning of 1

The three windings thus form a closed circuit. Leads are brought from the three junctions of the windings for connecting to the load. Any 2 of the 3 wires takes the voltage phase across one coil

delta and wye connections

As shown on the previous page, there are six leads from the armature windings of a three-phase generator, and the output voltage is connected to the external load by means of these six leads. In actual practice, this is not the case. Instead, the windings are connected together, and only *three* leads are brought out for connection to the load.

There are two ways that the three armature windings can be connected. Which of the two that is used determines the electrical characteristics of the generator output. In one connection, the three windings are all connected in *series* and form a closed circuit. The load is connected to the three points where two windings are joined. This is called a *delta connection,* since its schematic representation resembles the Greek letter delta (Δ). In the other connection, one of the leads of each winding are connected together, and the other three leads are brought out for connecting the load. This is called a *wye connection,* since schematically it represents the letter Y.

In each case, you can see that the windings are spaced 120 degrees apart, so that each winding will develop a voltage that is 120 degrees out of phase with the other winding voltages.

Wye Connection

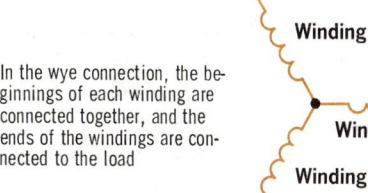

In the wye connection, the beginnings of each winding are connected together, and the ends of the windings are connected to the load

Any two of the 3 wires takes the vector sum of the voltage phases across 2 coils in series

electrical characteristics of delta and wye connections

Since the windings of a delta connection are all connected in series and form a closed circuit, it may seem that a high current will continuously flow through the windings, even when no load is connected. Actually, because of the phase difference between the three generated voltages, negligible or no current flows in the windings under no-load conditions.

The three leads brought out from the delta connection are used for connecting the generator output to the load. The voltage between any two of the leads, called the *line voltage*, is the same as the voltage generated in one winding, which is called the *phase voltage*. Thus, as shown, the three phase voltages are equal and the three line voltages are equal, and they all have the same value. The current in any line, though is $\sqrt{3}$ or approximately 1.73 times the current in any one phase of the winding. You can see, therefore, that a delta connection provides an *increase in current* but *no increase in voltage*.

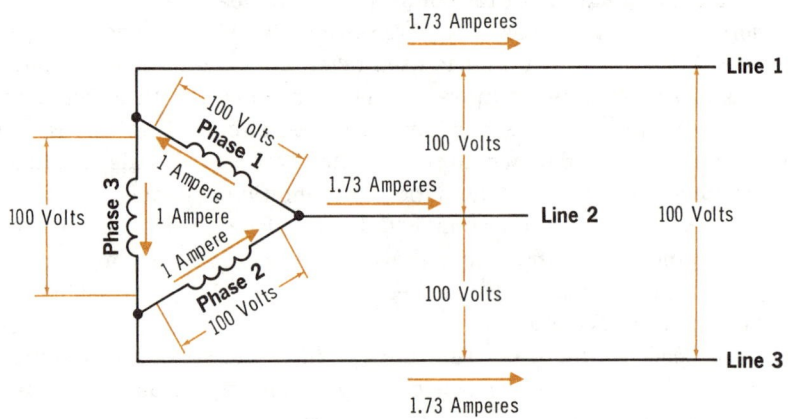

In a delta connection, line voltage equals phase voltage, whereas line current equals $\sqrt{3}$ or 1.73 times phase current

The total *true power* delivered by a delta-connected three-phase generator is equal to $\sqrt{3}$, or 1.73, times the true power in any one line. Remember, though, from Volumes 3 and 4, that the true power depends on the power factor (cos θ) of the circuit. The total true power is therefore equal to 1.73 times the line voltage, times the line current, times the power factor. Or,

$$P_{TRUE} = 1.73 \, E_{LINE} I_{LINE} \cos \theta$$

electrical characteristics of delta and wye connections (cont.)

The voltage and current characteristics of a wye connection are *opposite* to those of a delta connection. The voltage between any two lines of a wye connection is 1.73 times any one phase voltage, while the line currents are equal to the current in any phase winding. This contrasts with the delta connection in which, you will recall, line voltage equals phase voltage and line current equals 1.73 times phase current. Thus, whereas the delta connection provides an increase in current but no increase in voltage, the wye connection gives an increase in voltage but none in current.

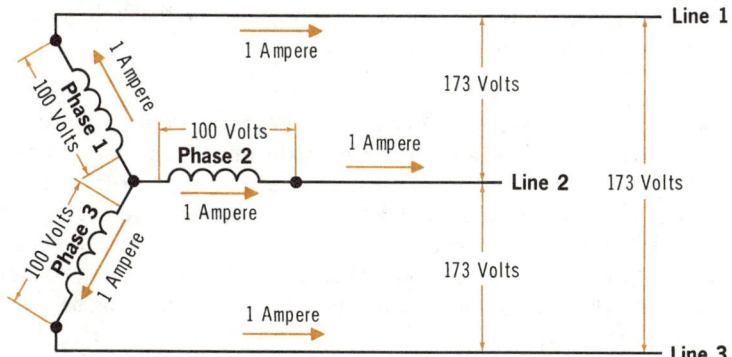

In a wye connection, line voltage equals $\sqrt{3}$ or 1.73 times phase voltage, whereas line current equals phase current

The total *true power* delivered by a wye-connected generator is the same as that of a delta-connected generator. The total true power is therefore equal to:

$$P_{TRUE} = 1.73\ E_{LINE}I_{LINE} \cos \theta$$

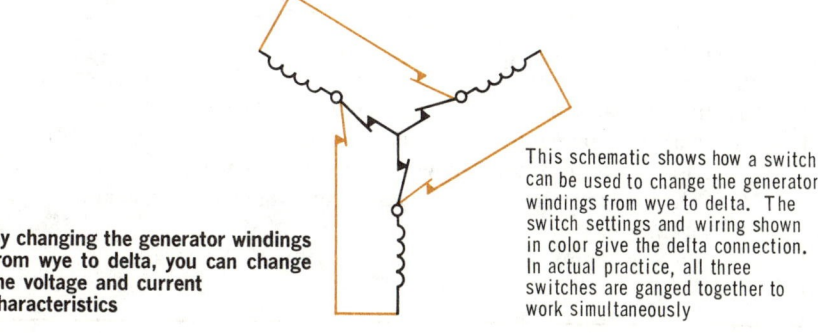

By changing the generator windings from wye to delta, you can change the voltage and current characteristics

This schematic shows how a switch can be used to change the generator windings from wye to delta. The switch settings and wiring shown in color give the delta connection. In actual practice, all three switches are ganged together to work simultaneously

summary

☐ A-c generators are also called alternators. ☐ Since conversion from ac to dc is not required in an a-c generator, a commutator is unnecessary. ☐ The a-c generator uses metallic rings, called slip rings, for connecting the rotating coils to the brushes. ☐ The output frequency of an a-c generator is equal to the frequency of rotation times the number of pairs of poles. ☐ A-c generators cannot be self-excited. Separate d-c voltage sources must be used to provide excitation current.

☐ Most a-c generators have a stationary armature and a rotating field. The generator output is then connected directly to the external circuit. ☐ In a stationary-armature generator, slip rings and brushes are used to connect the rotating field winding to its external d-c exciting source. ☐ Stationary-armature generators can rotate at very high speeds, and therefore, can generate very large voltages. ☐ The portion of an a-c generator that rotates is called the rotor. ☐ The stationary portion is called the stator.

☐ A two-phase a-c generator produces two voltages that differ in phase by 90 degrees. ☐ A three-phase a-c generator produces three voltages that differ in phase by 120 degrees. ☐ In three-phase generators, only three leads are brought out for connection to the load. Either a delta or wye connection can be used. ☐ A delta connection provides an increase in current but no increase in voltage. ☐ A wye connection provides an increase in voltage but no increase in current. ☐ The voltage between any two leads of a three-phase generator is called the line voltage.

review questions

1. Can an a-c generator be self-excited? Why?
2. In an a-c generator, what is the advantage of having a stationary armature and a rotating field?
3. Are slip rings required on a stationary-armature generator?
4. In an alternator, what is the *rotor*? The *stator*?
5. What is a *three-phase a-c generator*?
6. Draw a diagram of a delta connection. A wye connection.
7. In delta and wye connections, what is the line voltage?
8. If a three-phase generator is to supply maximum voltage to a load, would a delta or wye connection be used?
9. What advantages do a-c generators have over d-c generators?
10. Why are commutators unnecessary on a-c generators?

regulating the generator

When the load on an a-c generator changes, the output voltage also tends to change, the same as it does in a d-c generator. The main reason for this is the change in the voltage drop across the armature winding caused by the change in load current. However, whereas in a d-c generator the voltage drop across the armature winding is simply an *IR drop*, in an a-c generator there is an IR drop and an *IX_L drop*, caused by the a-c current flowing through the inductance of the winding. The IR drop depends only on the amount of the load change, but the IX_L drop depends also on the circuit *power factor*. Thus, the output voltage of a-c generators varies with both changes in load current and changes in power factor. As a result, an a-c generator that has satisfactory regulation at one power factor may have very poor regulation at another power factor.

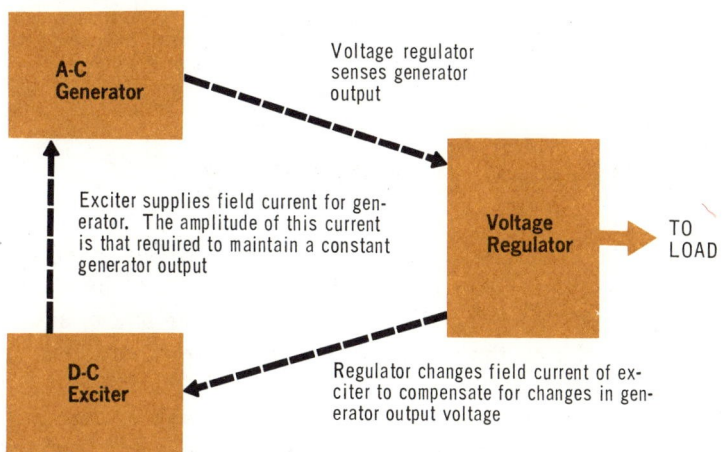

Because of their inherently poor regulation, a-c generators are generally provided with some auxiliary means of regulation. The auxiliary regulators used, whether they are manually operated or work automatically, accomplish their function in basically the same way. They sense the generator output voltage, and when it changes, they cause a corresponding change in the field current of the exciter that supplies field current to the generator. Thus, if the generator output voltage drops, the regulator causes an increase in the exciter field current. The exciter output voltage, therefore, increases, causing the current in the generator field winding to also increase. As a result, the magnetic field of the generator increases in strength and raises the generator voltage to its original amplitude. A similar, but opposite, sequence of events takes place when the regulator senses a decrease in the generator output voltage.

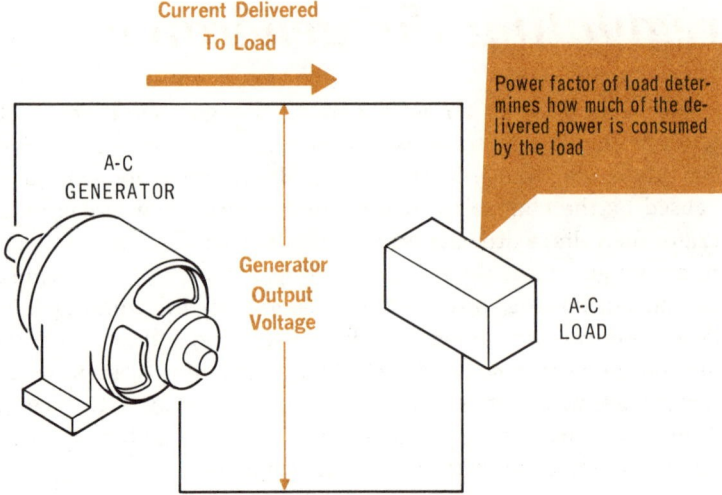

Current Delivered To Load

A-C GENERATOR

Generator Output Voltage

Power factor of load determines how much of the delivered power is consumed by the load

A-C LOAD

A-c generators are rated according to the maximum apparent power they can deliver, regardless of how much of this power is consumed by the load, as determined by the power factor of the load

rating a-c generators

Every d-c generator has a power rating, normally expressed in kilowatts, which indicates the maximum power that can constantly be supplied by the generator. A-c generators, on the other hand, generally cannot be rated in the same way, since the power *consumed* in an a-c circuit depends on the circuit *power factor*. This means that an a-c generator could be supplying a moderate amount of true power for a load, and yet if the power factor of the load was low, the total, or apparent, power actually delivered by the generator could be very large. Such a situation could result in the generator burning out.

For this reason, a-c generators must be rated not on the basis of the maximum allowable power consumption of the load, but on the basis of the *maximum apparent power* they can deliver. This is done by expressing the capacity in *volt-amperes,* or *kilovolt-amperes.* Thus, at any particular output voltage, you know the maximum current that the generator can deliver, regardless of the power factor of the load. For example, if an a-c generator with a rating of 100 kilovolt-amperes has an output of 50 kilovolts, the maximum current that it can safely deliver is 100 kilovolt-amperes divided by 50 kilovolts, or 2 amperes.

Occasionally, a-c generators are designed for use with loads having a constant power factor. The rating of these generators then may be given in watts or kilowatts *at the particular power factor.*

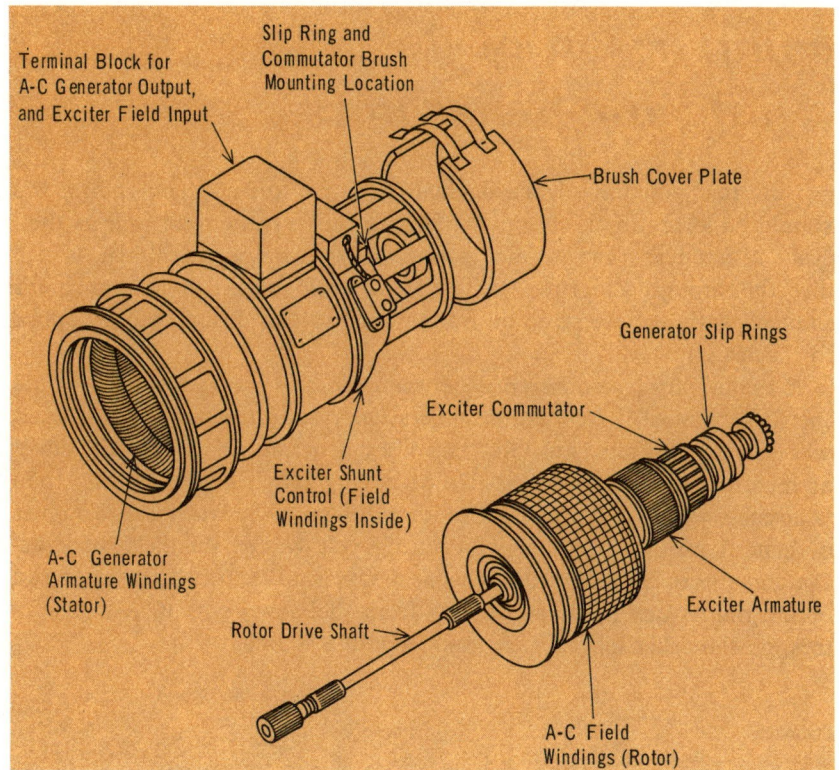

construction of a-c generators

From the standpoint of physical appearance, a-c generators vary considerably. They range from very large turbine-driven types that weigh thousands of pounds to small special-purpose types that weigh only a few pounds, and even less. As has been pointed out, though, practically all a-c generators have stationary armatures and rotating fields. The armature windings are positioned around the inner circumference of the generator housing, and are usually embedded in a laminated iron core. The core and the windings make up the stator.

The field winding and field poles, which make up the rotor, are mounted on a shaft, and turn within the stator. Also mounted on the rotor shaft are the slip rings for the field windings. When the generator contains its own d-c exciter, the exciter armature and commutator are also mounted on the motor shaft. The brush holders for the generator slip rings and the exciter commutator are mounted on the generator housing, as are terminals for making electrical connections to the generator. A typical a-c generator with a self-contained exciter is shown.

comparison of d-c and a-c generators

Now that you have learned about both d-c and a-c generators, you should be aware of the basic *similarities* between them as well as their basic *differences*. In an a-c generator, the induced voltage is fed directly through slip rings to the load, while in a d-c generator, the induced ac is converted to dc by the commutator before being applied to the load.

A major physical difference between d-c and a-c generators is that the field of most d-c generators is stationary and the armature rotates, whereas the opposite is usually true of a-c generators. This has the effect of making a-c generators capable of having much larger outputs than is possible with d-c generators. Another difference between the two types of generators is the source of exciting voltage for the field winding. D-c generators can either use a separate external exciting source, or can obtain the required voltage directly from their own output. A-c generators, on the other hand, must use a separate source.

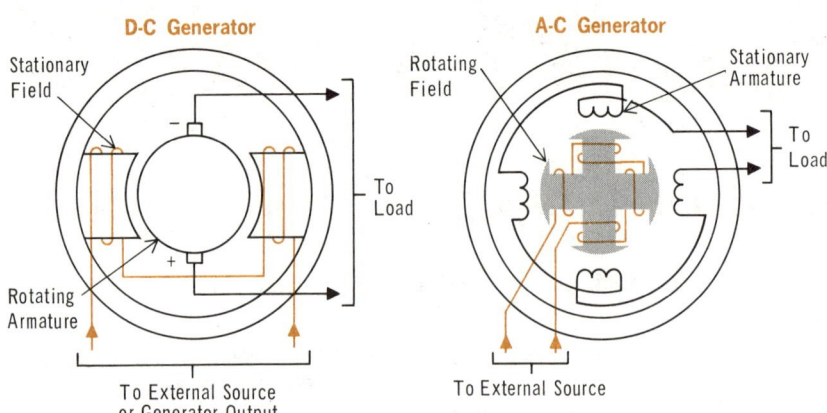

The d-c generator has a rotating armature and a stationary field. Voltage for its field can be obtained either from an external source or from the generator's own output

The a-c generator has a stationary armature and a rotating field. Voltage for its field must come from an external source, called an exciter

In the area of voltage regulation, d-c generators are inherently more stable than a-c generators. One of the reasons for this is that although the output voltages of both types are sensitive to load changes, the output voltage of an a-c generator is also sensitive to changes in the power factor of the load. In addition, a good degree of self-regulation is possible in a d-c generator by using a compound armature winding. This is not possible in a-c generators, since they must be separately excited.

the automobile alternator

The comparison of the advantages of d-c generators and a-c alternators, which you just studied, was based on the accepted categories of basic generators. However, it is possible to combine the advantages of d-c and a-c generators by using additional circuit designs. The automobile alternator does this in a unique way to produce a high-current d-c charging source with an a-c type of generator. It is called an alternator even though it produces a d-c voltage because it is actually a fixed-armature a-c generator that uses *rectifiers* to convert the ac to dc.

Rectifiers are devices that, for the most part, conduct in only one direction. So, the rectifier will pass only one polarity of the a-c voltage to produce a pulsating dc. The typical car alternator produces three-phase ac, so that after the voltage is converted to dc, there is less ripple. A capacitor is then connected at the output to filter out the ripple to get a relatively smooth d-c voltage.

Because the rectifiers oppose current flow in the opposite direction, a reverse-current cutout relay is not needed in the voltage regulator. And, since the alternator is a high-current generator, a current regulator also is not needed. Therefore, the regulator for the alternator is simpler than that for the d-c generator; it uses only one relay circuit to regulate the alternator output voltage by controlling the field current. Notice that although this is an alternator, it *is* self-excited. The more recent alternator circuits use electronic regulators instead of relays. These are actually *semiconductor* devices similar to transistors which can be turned on and off just like a relay. They are sometimes called *solid-state relays* because they control the on-off current flow in the same way. This can be done because the rectified output is *dc*.

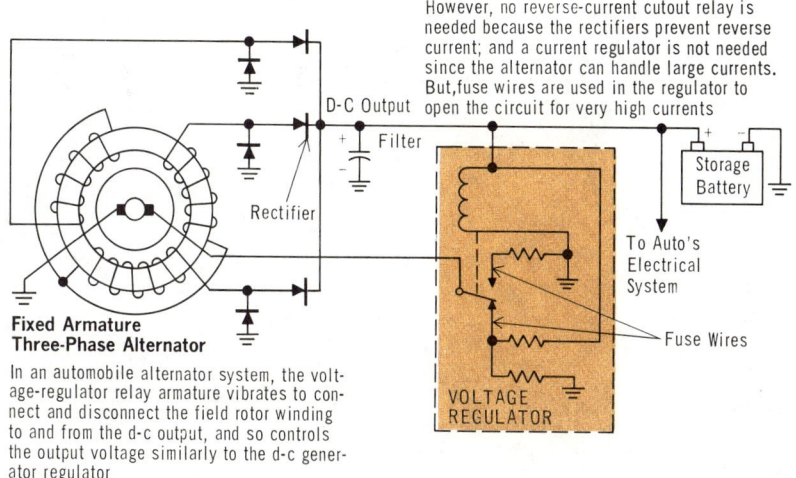

However, no reverse-current cutout relay is needed because the rectifiers prevent reverse current; and a current regulator is not needed since the alternator can handle large currents. But, fuse wires are used in the regulator to open the circuit for very high currents

D-C Output

Filter

+ —

Storage Battery

Rectifier

To Auto's Electrical System

Fixed Armature Three-Phase Alternator

Fuse Wires

VOLTAGE REGULATOR

In an automobile alternator system, the voltage-regulator relay armature vibrates to connect and disconnect the field rotor winding to and from the d-c output, and so controls the output voltage similarly to the d-c generator regulator

auto alternator operation

The three-phase auto alternator uses wye-connected fixed-armature windings, which as you learned, produce one phase voltage between any two output leads. The output of the alternator is a positive voltage with relation to ground. But *no* lead of the wye windings is connected directly to ground because the windings produce ac; all three leads alternately become negative and positive through the a-c cycles. Therefore, each lead must be connected to ground when it is negative, and to the output when it is positive. The rectifiers do this.

The rectifiers act like switches that close for one polarity, and open for the other polarity. Notice that each lead has two rectifiers connected oppositely. One rectifier will connect the lead to the output line when the lead is positive, but disconnect the lead when it is negative. The other rectifier connects the lead to ground when it is negative, and disconnects it when it is positive. The diagram shows how the same two windings are connected for different phase angles of the output voltage. The output, then, is always positive.

If you recall what you learned about the d-c generator, you can see that the commutator was needed to do the same thing whenever the leads switched polarity, since the armature always produces ac. In the alternator, then, the rectifiers act like electronic commutators, and so it is debatable about whether the alternator *is* an alternator, or just another type of d-c generator.

Three-Phase Commutation

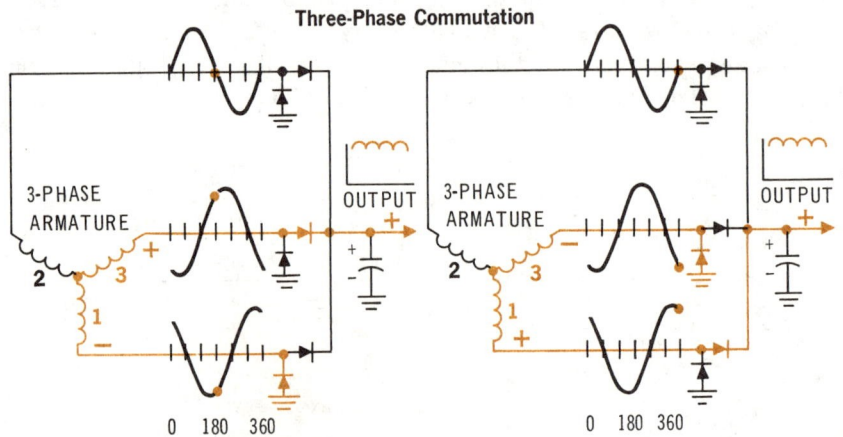

Only two possibilities are shown here. At 120 degrees, winding 2 would be positive and so would be connected to the output instead of winding 3, which would be set at zero volts. Winding 1 would still be negative.

At 60 degrees, winding 3 would be negative and go to ground, while winding 2 would be positive and go to the output. At the intermediate phase angles, it is possible to have two windings connected to the output and one to ground, and vice versa

internal resistance of the generator

In every generator, the load current flows through the armature winding. Like any coil or winding, the armature has *resistance* and *inductance*. The combination of this resistance and the inductive reactance caused by the inductance make up what is known as the *internal resistance* of the generator. When load current flows, it produces a voltage drop across the internal resistance. This voltage drop subtracts from the generator output voltage, and, therefore, represents generated voltage that is *lost* and not available for the load.

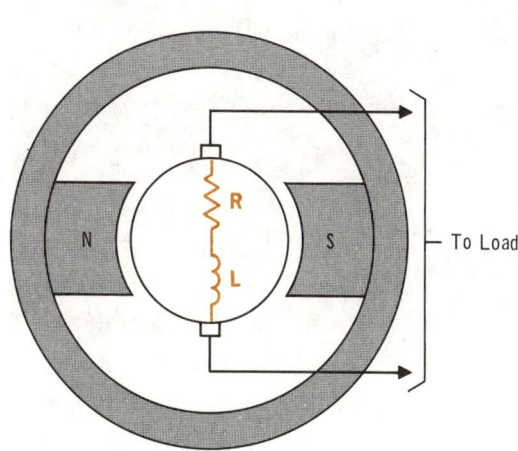

The armature winding of every generator has resistance and inductance. The resistance and the reactance caused by the inductance make up the internal resistance of the generator

To Load

Whenever load current flows, it is opposed by the internal resistance. This opposition produces a voltage drop that subtracts from the generator output voltage

You can see that the larger the internal resistance is, the greater is the portion of the generated voltage that is dropped within the generator and thus lost. In a d-c generator with a given internal resistance, the internal voltage drop is directly proportional to the load current, and is equal to:

$$E = I_{LOAD} R_{INTERNAL}$$

Thus, the greater the load current, the more voltage is dropped across the internal resistance. In an a-c generator, the internal voltage drop also depends on the *frequency* of the generator output voltage, since the inductive reactance of the armature winding changes whenever the frequency does. Inasmuch as the generator speed is one of the factors that determines the frequency, the internal resistance of an a-c generator will change with generator speed.

the motor-generator

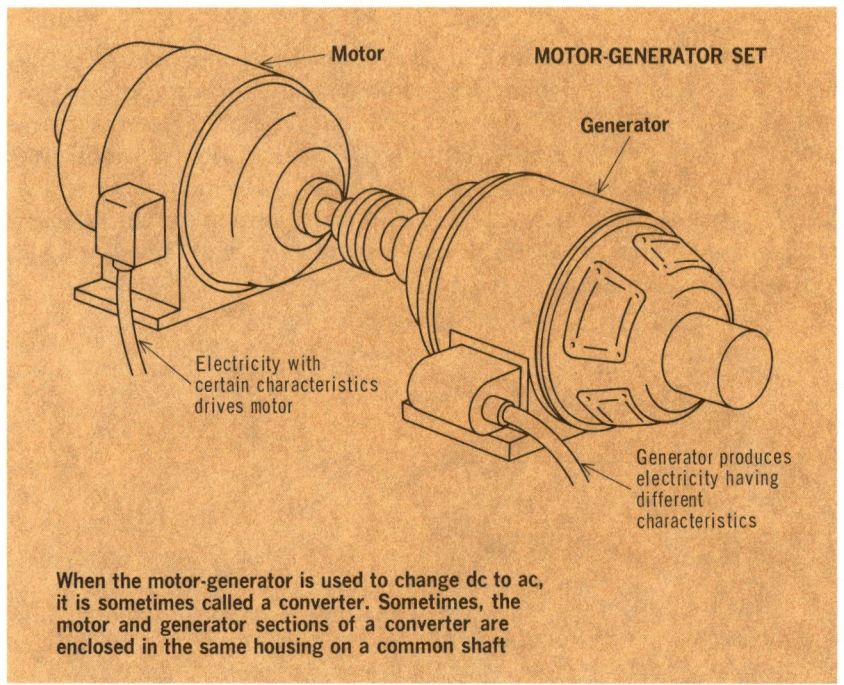

Motor

MOTOR-GENERATOR SET

Generator

Electricity with
certain characteristics
drives motor

Generator produces
electricity having
different
characteristics

**When the motor-generator is used to change dc to ac,
it is sometimes called a converter. Sometimes, the
motor and generator sections of a converter are
enclosed in the same housing on a common shaft**

A motor-generator consists of an electric motor and a generator mechanically connected so that the motor turns the generator. The *motor* thus supplies the *mechanical energy* that the *generator* converts to *electrical energy*. Normally, both the motor and the generator of a motor-generator are mounted on the same baseplate, and are moved about and installed as a single unit.

Motor-generators are usually used to *change* electricity from one voltage or frequency to another voltage or frequency, or to convert ac to dc or dc to ac. Electricity having the characteristic to be changed powers the motor, and the generator is designed to produce electricity having the new desired characteristic. For example, the motor could be driven by a 60-Hz power source, while the generator produces an output having a frequency of 400 Hz. Or, a d-c motor could drive an a-c generator to accomplish dc to ac conversion.

When the device changes ac of one type to ac of another type, or ac to dc, it is referred to as a *motor-generator set*. But, when it is used to convert dc to ac, it is sometimes called a *converter*. Quite often, the converter has the motor and the generator in the same housing.

the dynamotor

In certain respects, the dynamotor is actually a motor-generator. It consists of an electric motor driving a generator. However, a motor-generator set usually uses separate units; in a dynamotor, they are always contained in a *common housing*, similar to the converter, and their armature windings are both wound on the *same shaft*.

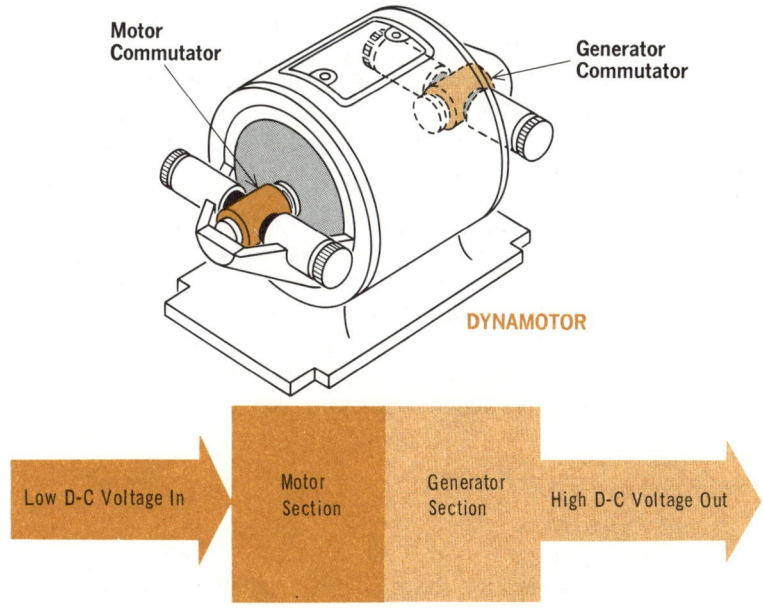

A dynamotor converts a low d-c voltage to a high d-c voltage. Both the motor and the generator portions are mounted on a common shaft and in a common housing

Dynamotors are used for converting *low d-c voltages,* usually supplied by batteries, to *high d-c voltages.* The low voltage drives the motor, turning the generator, which then produces a higher voltage. Dynamotors are frequently used with communications equipment to supply d-c voltages higher than those available from batteries. They are quite common in aircraft, where many kinds of electronic equipment need a few hundred volts of dc to operate, and the main line of the aircraft only supplies 28 volts dc. Some amateur radio equipment, which must operate from the 6- or 12-volt batteries in cars also rely on dynamotors to change the voltage to the required d-c level.

summary

☐ The output voltage of an a-c generator varies with changes in load current and with circuit power factor. ☐ Regulators for a-c generators control the exciter current to the field winding to compensate for changes in output voltage. ☐ A-c generators are rated on the basis of the maximum apparent power they can deliver.

☐ A-c generators vary considerably in appearance. However, practically all have stationary armatures and rotating fields. ☐ The internal resistance of an a-c generator is made up of the d-c resistance and inductive reactance of the armature winding. ☐ The output voltage of a generator is reduced by the voltage drop across the internal resistance. ☐ In a d-c generator, the internal voltage drop is equal to: $E = I_{LOAD} R_{INTERNAL}$. ☐ In an a-c generator, the internal voltage drop also depends on the frequency of the generator output voltage.

☐ A motor-generator consists of an electric motor and a generator mechanically connected so that the motor turns the generator. ☐ If a motor-generator converts ac to another type of ac, or ac to dc, it is called a motor-generator set. ☐ If it converts dc to ac, it is sometimes called a converter. ☐ Dynamotors are used to convert low d-c voltages to high d-c voltages. ☐ A dynamotor consists of a motor and a generator, both contained in the same housing, and with their armature windings wound on the same shaft.

review questions

1. Why do a-c generators have poorer regulation than d-c generators?
2. Why are a-c generators normally not rated in watts or kilowatts?
3. An a-c generator has a rating of 20 kilovolt-amperes and an output of 2 kilovolts. What is the maximum current it can safely deliver?
4. What things affect the internal resistance of an a-c generator?
5. What are motor-generators used for?
6. What is a *converter*?
7. What is a dynamotor used for?
8. If the power factor of the load decreases while the generator output voltage and current remain constant, what happens to the true power delivered by an a-c generator?
9. If the speed of an a-c generator is increased, what happens to the internal resistance?
10. Does generator speed affect the internal resistance of a d-c generator?

electricity
seven

With Volta's development of the first
chemical cell in the 19th century,
man was able to harness the potential
energy in magnetism, eventually in
the form of the electric motor

the electric motor

The electric motor is a machine that can transform electricity into rotary motion to perform useful work. As such, the electric motor represents one of the great advances in harnessing natural forces to do work for mankind.

Man first replaced his own musclepower with that of animals and then with power derived from the available natural forces of wind and water. And when Volta developed the first battery, a new form of power was controlled—electric power. This was the first step in the development of the electric motor.

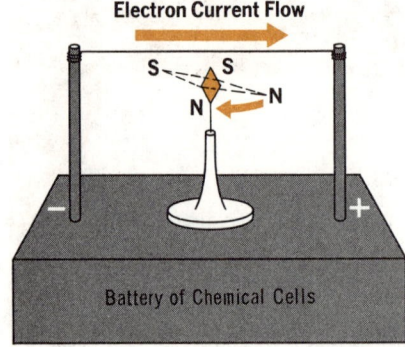

Electron Current Flow

When the current direction is changed, the direction of needle movement is changed

After Volta showed how electricity could be controlled, Oersted discovered that an electric current could cause a magnetic needle to move, and thus proved that the flow of electricity created a magnetic field

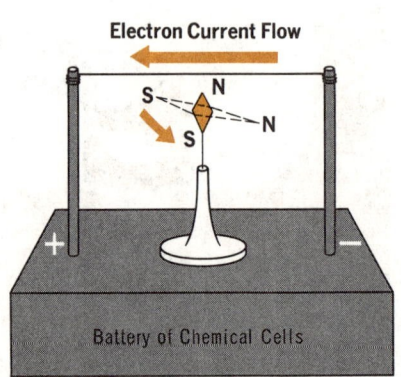

Electron Current Flow

motion from electricity

Once the chemical battery became generally available, scientists were able to conduct experiments to see if they could control the energy in electric current. One of the most important discoveries made at this time concerned the *magnetic effects* of electricity. Hans Christian Oersted, in 1819, noticed that when he placed a freely pivoted magnetic needle near a wire that conducted current, the needle moved until it was at right angles to the wire. And, when he reversed the current, the needle turned again and aligned itself pointing in the opposite direction, still at right angles to the wire. The conclusions drawn from Oersted's experiment and their results were as follows:

1. A wire carrying an electric current produced a magnetic field; thus electricity could be converted into magnetism. Using this information, scientists of Oersted's day developed the *electromagnet*.
2. The magnetic field created by electric current could interact with the field of a magnet to produce *motion*. Therefore electrical energy could be changed into mechanical energy. The electric motor was developed as a direct result of this observation.

the faraday motor

Faraday used Oersted's discovery, that electricity could be used to produce motion, to build the world's first electric motor, in 1821. Ten years later, using the same logic in reverse, Faraday discovered the electric generator principle.

Faraday was interested in getting the motion produced by Oersted's experiment to be continuous, rather than just a rotary shift in position. In his experiments, Faraday thought in terms of magnetic *lines of force*. He visualized how flux lines existed around a current-carrying wire and a bar magnet. He was then able to produce a device in which the different lines of force could interact to produce continuous rotation. The basic Faraday motor uses a free-swinging wire that circles around the end of a bar magnet. The bottom end of the wire is in a pool of mercury, which allows the wire to rotate while keeping a complete electric circuit.

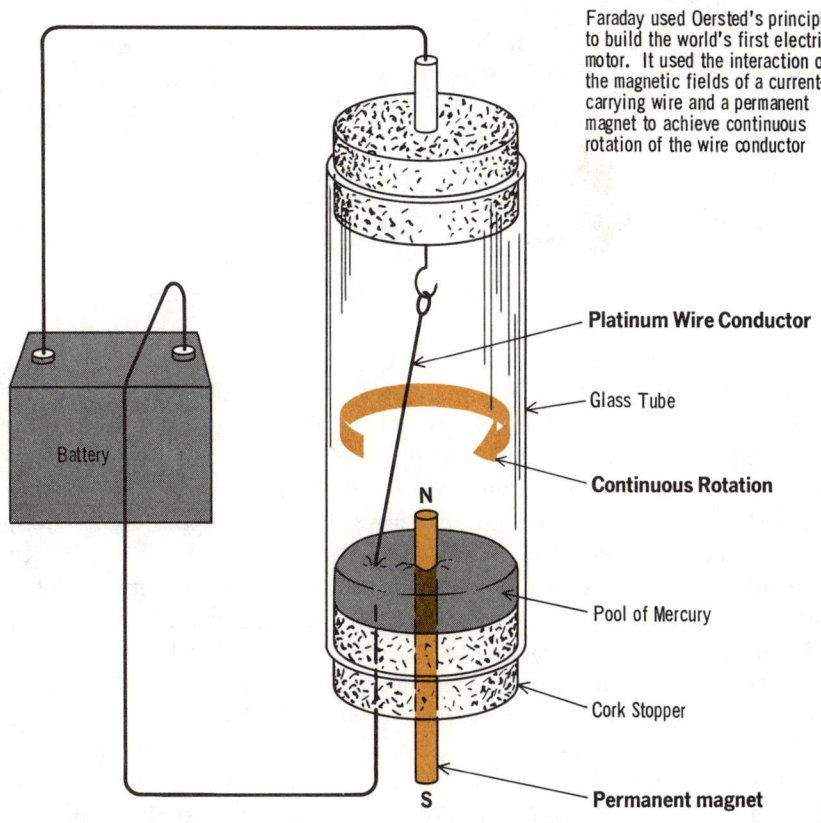

Faraday used Oersted's principle to build the world's first electric motor. It used the interaction of the magnetic fields of a current-carrying wire and a permanent magnet to achieve continuous rotation of the wire conductor

Battery

Platinum Wire Conductor

Glass Tube

Continuous Rotation

N

Pool of Mercury

Cork Stopper

S

Permanent magnet

basic motor action

Although Faraday's motor was ingenious, it could not be used to do any practical work. This is because its drive shaft was enclosed and it could only produce an internal *orbital* motion. It could not transfer its mechanical energy to the outside for driving an external load. However, it did show how the magnetic fields of a conductor and a magnet could be made to interact to produce continuous motion. Faraday's motor orbited its wire rotor on the outside of the field of the magnet. In a practical motor the rotor must pass *through* the magnet's lines of force.

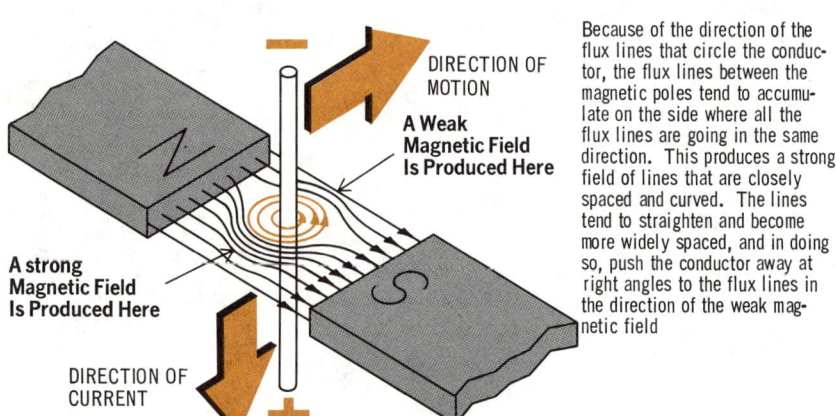

DIRECTION OF MOTION

A Weak Magnetic Field Is Produced Here

A strong Magnetic Field Is Produced Here

DIRECTION OF CURRENT

Because of the direction of the flux lines that circle the conductor, the flux lines between the magnetic poles tend to accumulate on the side where all the flux lines are going in the same direction. This produces a strong field of lines that are closely spaced and curved. The lines tend to straighten and become more widely spaced, and in doing so, push the conductor away at right angles to the flux lines in the direction of the weak magnetic field

When a current is passed through the wire, circular lines of force are produced around the wire. Those flux lines go in a direction described by the *left-hand rule*. The lines of force of the magnet go from the N pole to the S pole. You can see that on one side of the wire, the magnetic lines of force are going in the same direction as the circular field around the wire. On the other side, though, they are going in the opposite direction; as a result, the wire's flux lines oppose the magnet's flux lines. Since flux lines take the path of least resistance, more lines concentrate on the other side of the wire. Because of the concentration of lines on one side of the conductor, the lines are bent and are very closely spaced. The lines tend to straighten and be wider spaced. Because of this, the denser, curved field pushes the wire in the opposite direction.

The direction in which the wire is moved is determined by the *right-hand rule*. If the current in the wire went in the opposite direction, the direction of its flux lines would reverse, and the wire would be pushed the other way.

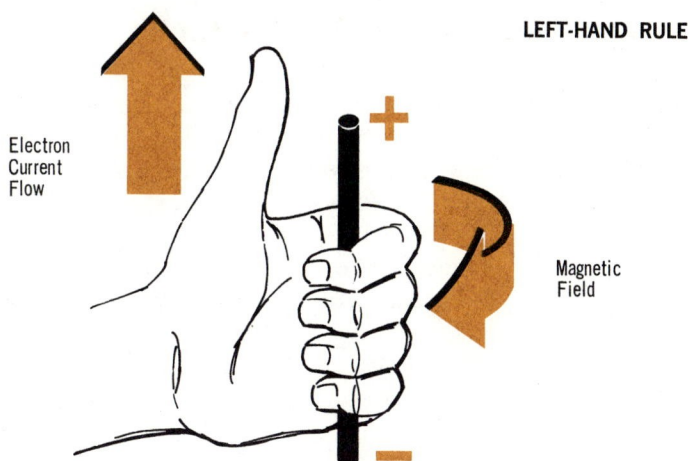

LEFT-HAND RULE

Electron
Current
Flow

Magnetic
Field

rules for motor action

The *left-hand rule* shows the direction of the flux lines around a wire that is carrying current. When the thumb points in the direction of the electron current, the fingers will point in the direction of the magnetic lines of force.

The *right-hand rule* for motors shows the direction that a current-carrying wire will be moved in a magnetic field. When the forefinger is pointed in the direction of the magnetic field lines, and the center finger is pointed in the direction of the current in the wire, the thumb will point in the direction that the wire will be moved.

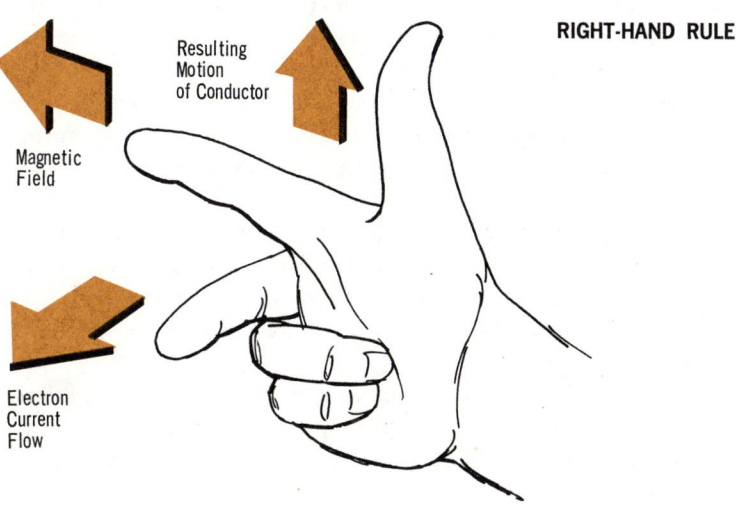

RIGHT-HAND RULE

Resulting
Motion
of Conductor

Magnetic
Field

Electron
Current
Flow

torque and rotary motion

In the basic motor action you just studied, the wire only moves in a straight line, and stops moving once out of the field, even though the current is still on. A practical motor must develop *continuous rotary motion*. To produce this, we must develop a basic twisting force called *torque*.

By bending the straight wire of the basic motor into a sort of loop, we can see how torque is produced. If the loop is connected to a battery, current flows in one direction on one side of the loop, and in the opposite direction on the other. Therefore, the concentric lines of force developed about the loop are in opposite directions on the two sides.

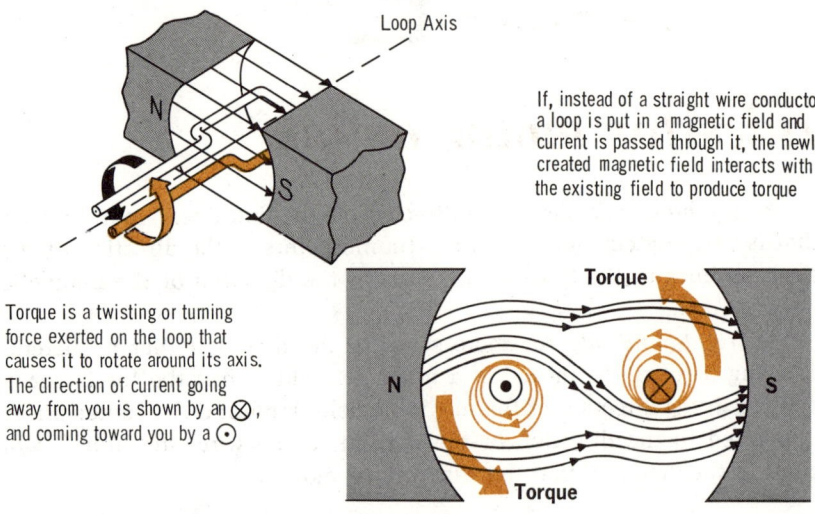

Loop Axis

If, instead of a straight wire conductor, a loop is put in a magnetic field and current is passed through it, the newly created magnetic field interacts with the existing field to produce torque

Torque is a twisting or turning force exerted on the loop that causes it to rotate around its axis. The direction of current going away from you is shown by an ⊗, and coming toward you by a ⊙

Torque

Torque

If we mount the loop in a fixed magnetic field and supply the current, the flux lines of the field and both sides of the loop will interact, causing the loop to act like a lever with a force pushing on its two sides in opposite directions. The combined forces result in a turning force, or *torque*, because the loop is arranged to pivot on its axis. In a motor, the loop that moves in the *field* is called an *armature* or *rotor*. The overall turning force on the armature depends on several factors, including *field* strength, armature *current* strength, and the physical *construction* of the armature, especially the distance from the loop sides to the axis lines. Because of the lever action, the forces on the sides of the armature loop will increase as the loop sides are further from the axis; thus, larger armatures will produce greater torques.

In the practical motor, the torque determines the energy available for doing useful work. The greater the torque, the greater the energy. If a motor does not develop enough torque to pull its load, it *stalls*.

producing continuous rotation

You learned that the armature turns when torque is produced, and torque is produced as long as the fields of the magnet and armature interact. When the loop reaches a position perpendicular to the field, the interaction of the magnetic fields stops. This position is known as the *neutral plane*. In the neutral plane, no torque is produced and the rotation of the armature should stop; however, *inertia* tends to keep a moving object in motion even after the prime moving force is removed, and thus the armature tends to rotate past the neutral plane. But when the armature continues on, the sides of the loop start to swing back into the flux lines, and compress them again. As a result, the flux lines apply a force to push the sides of the loop back, and a torque is developed in the opposite direction. Instead of a continuous rotation, an oscillating motion is produced until the armature stops in the neutral plane.

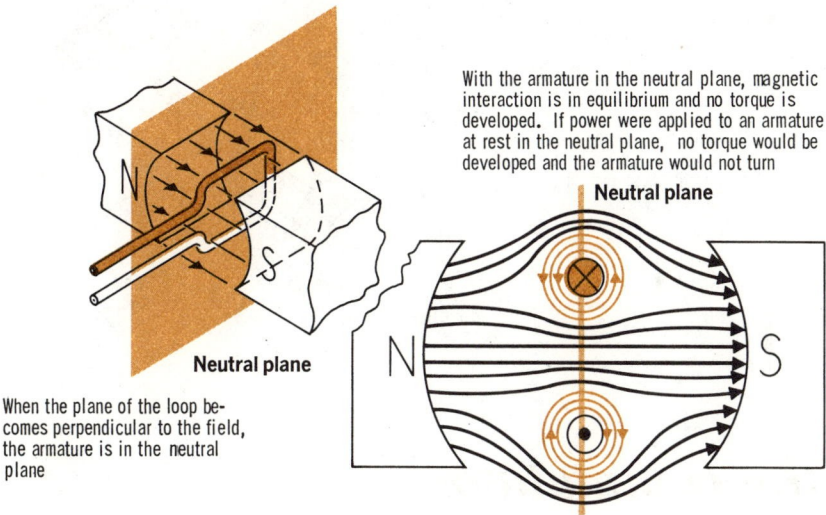

With the armature in the neutral plane, magnetic interaction is in equilibrium and no torque is developed. If power were applied to an armature at rest in the neutral plane, no torque would be developed and the armature would not turn

Neutral plane

Neutral plane

When the plane of the loop becomes perpendicular to the field, the armature is in the neutral plane

To get continuous rotation, we must keep the armature turning in the same direction as it passes through the neutral plane. We could do this by reversing either the direction of the pole field or the direction of the current flow through the armature at the instant the armature goes through the neutral plane. Current reversals of this type are normally the job of circuit-switching devices. Since the switch would have to be synchronized with the armature, it is more logical to build it into the armature than into the field. The practical switching device which can change the direction of current flow through an armature to maintain continuous rotation is called a *commutator*.

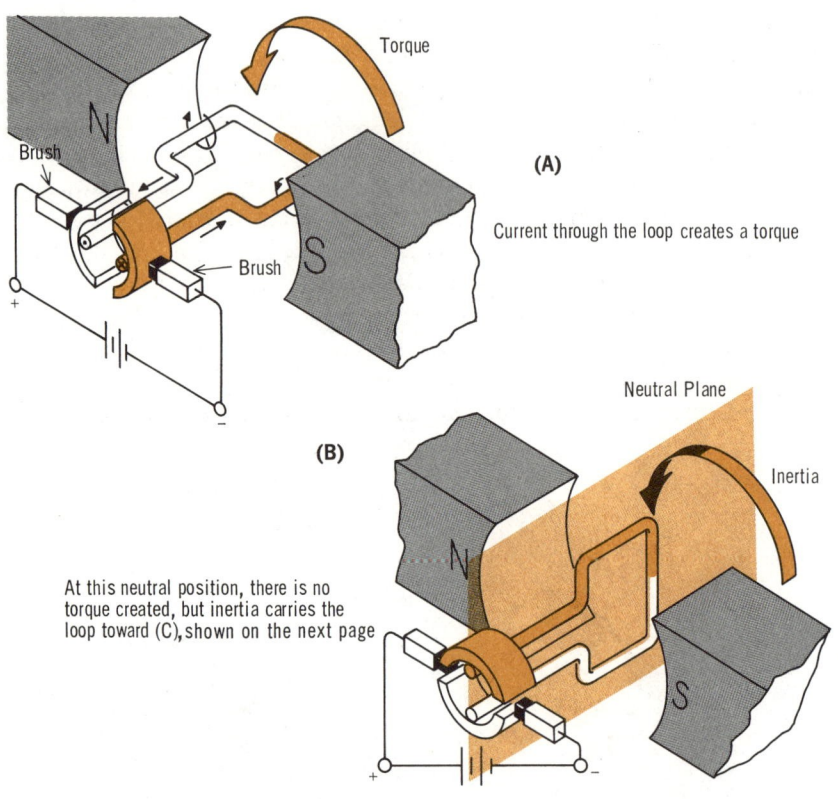

(A)

Current through the loop creates a torque

(B)

At this neutral position, there is no torque created, but inertia carries the loop toward (C), shown on the next page

the commutator

For the single-loop armature, the commutator is simple. It is a con-ducting ring that is split into two segments, with each segment con-nected to an end of the armature loop. Power for the armature from an external power source, such as a battery, is brought to the commutator segments by means of brushes. The arrangement is almost identical to that for the basic d-c generator you learned about in Volume 6.

The logic behind the operation of the commutator is easy to see in the figures. You can see in figure A that current flows into the side of the armature closest to the south pole of the field and out of the side closest to the north pole. The interaction of the two fields produces a torque in the direction indicated, and the armature rotates in that direction.

the commutator (cont.)

The armature is shown in the neutral plane in figure B on page 7-8; theoretically, no torque is produced, but the armature continues to rotate past the neutral plane due to inertia. Notice that at the neutral position, the commutator disconnects from the brushes. Once the armature goes past neutral, as shown in figure C, the sides of the loop reverse positions. But the switching action of the commutator keeps the direction of current flow through the armature the same as it was in figure A. Current still flows into the armature side that is now closest to the south pole.

Since the magnet's field direction remains the same throughout, the interaction of fields after commutation keeps the torque going in the original direction; thus, the same direction of rotation is maintained.

As you can see in figure D, inertia again carries the armature past neutral to the position shown in figure A on page 7-8, while commutation keeps the current flowing in the direction that continues to maintain rotation. In this way, the commutator keeps switching the current through the loop, so that the field it produces always interacts with the pole field to develop a continuous torque in the same direction.

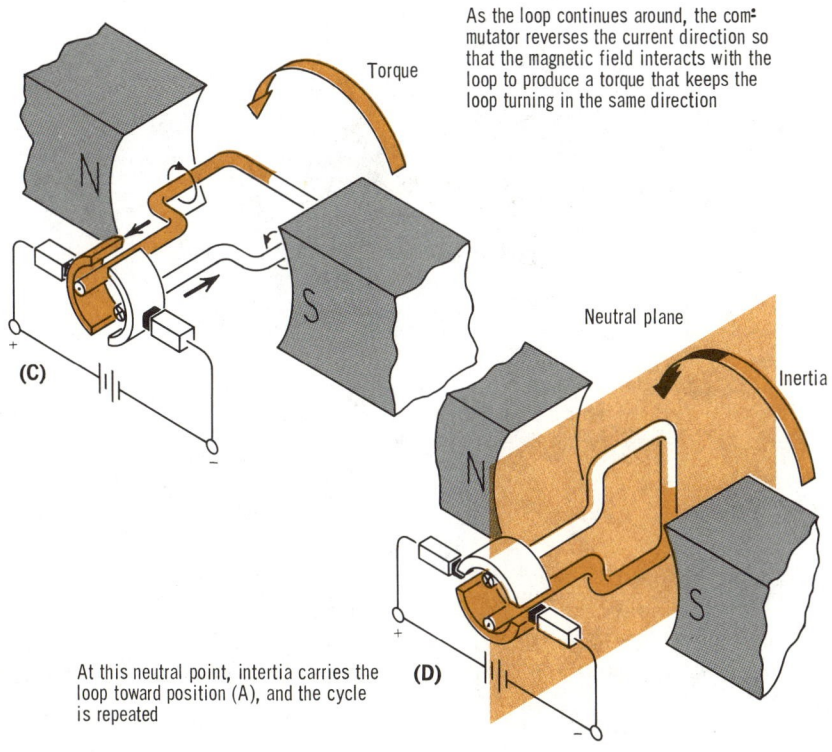

As the loop continues around, the commutator reverses the current direction so that the magnetic field interacts with the loop to produce a torque that keeps the loop turning in the same direction

At this neutral point, intertia carries the loop toward position (A), and the cycle is repeated

the elementary d-c motor

At this point, you have been introduced to the four principal parts that make up the elementary d-c motor. These parts are the same as those you met in your study of the basic d-c generator in Volume 6: a magnetic field, a movable conductor, a commutator, and brushes. In practice, the magnetic field can be supplied by a permanent magnet or by an electromagnet. For most discussions covering various motor operating principles, we will assume that a permanent magnet is used. At other times, when it is important for you to understand that the field of the motor is developed electrically, we will show that an electromagnet is used. In either case, the magnetic field itself consists of magnetic flux lines that form a closed magnetic circuit. The flux lines leave the north pole of the magnet, extend across the air gap between the poles of the magnet, enter the south pole, and then travel through the magnet itself back to the north pole. The movable conductor, usually a loop, called the *armature,* is placed in the air gap between the magnet poles. The armature, therefore, is in the magnetic field.

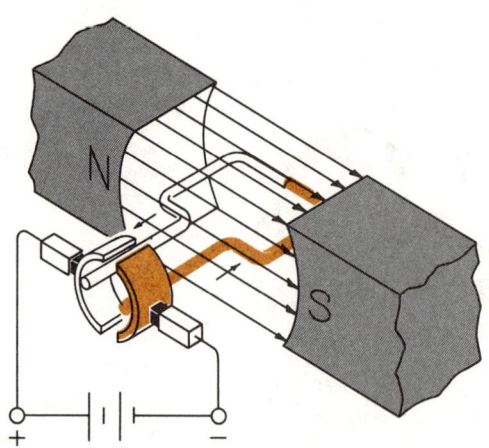

A single loop in a magnetic field with power supplied via brushes through a split-ring commutator is the most elementary form of d-c motor

When d-c power is supplied to the armature through the brushes and commutator, magnetic flux is also built up around the armature. It is this armature flux that interacts with the magnetic field in which the armature is suspended to develop the torque that makes the motor operate.

summary

☐ An electric motor transforms electricity into rotary motion to perform useful work. ☐ The first electric motor was built by Faraday in 1821. Although it was an ingenious device, it could not perform practical work. ☐ Basic motor action arises from the interaction of two magnetic fields: one produced around a current-carrying conductor, and the other, a fixed magnetic field. ☐ The left-hand rule for motors shows the direction of the magnetic field around a current-carrying conductor. ☐ The right-hand rule for motors shows the direction that a current-carrying conductor will move in a magnetic field.

☐ A practical motor must produce continuous rotary motion. It does this by developing a turning force, or torque, on a current-carrying conductor formed into a loop. ☐ The greater the torque, the more useful work can be done by the motor. ☐ The neutral plane of a motor is the position of the rotating loop when it is perpendicular to the flux lines of the fixed magnetic field. ☐ No torque is produced when the rotating loop is in the neutral plane. ☐ To keep a d-c motor turning continuously, instead of oscillating about the neutral plane, a commutator is used. The commutator reverses the direction of the current in the rotating loop each time the loop passes through the neutral plane.

☐ The four principal parts of an elementary d-c motor are a magnetic field, a movable conductor, a commutator and brushes. ☐ The magnetic field can be supplied by either a permanent magnet or an electromagnet. ☐ The movable conductor is a loop, and is called the armature. ☐ D-c current is supplied to the armature through the brushes and commutator.

---

review questions

1. What was learned from Oersted's experiments on the magnetic effects of electricity?
2. Why was Faraday's first electric motor impractical?
3. What is the *right-hand rule for motors*?
4. What is the *left-hand rule for motors*?
5. What is *torque*? Why is it important in motors?
6. What is the *neutral plane*?
7. What is a *commutator*?
8. Why are commutators necessary?
9. Draw a sketch of an elementary d-c motor.
10. Briefly describe the operation of the motor drawn in Question 9.

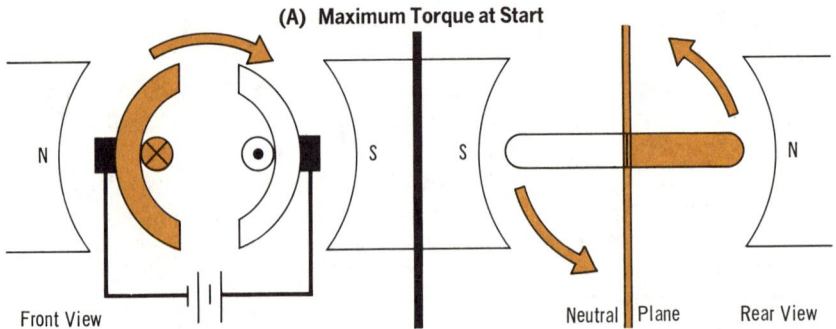

(A) Maximum Torque at Start

Front View   Neutral ‖ Plane   Rear View

shortcomings of the elementary d-c motor

Although an elementary d-c motor of the type you learned about can be built and operated, it has two serious shortcomings which prevent it from being useful. First, such a motor cannot always start by itself, and once started, it operates very irregularly.

The elementary d-c motor with a single-loop armature cannot start by itself when the armature begins in the neutral plane. In the neutral plane, no current flows through the armature since the brushes are disconnected from the commutator. However, even if the current were made to flow in the armature, remember that in the neutral plane no flux interaction takes place. As a result, no torque could be produced and inertia would maintain the motor at rest.

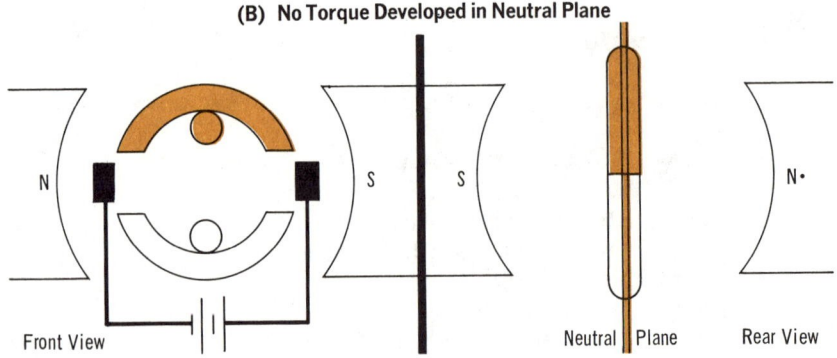

(B) No Torque Developed in Neutral Plane

Front View   Neutral ‖ Plane   Rear View

The elementary d–c motor develops torque erratically. Thus, it is ineffective against realistic loads. Since no torque is developed when the loop is in the neutral plane, the motor will not start by itself if the loop is there when power is first applied

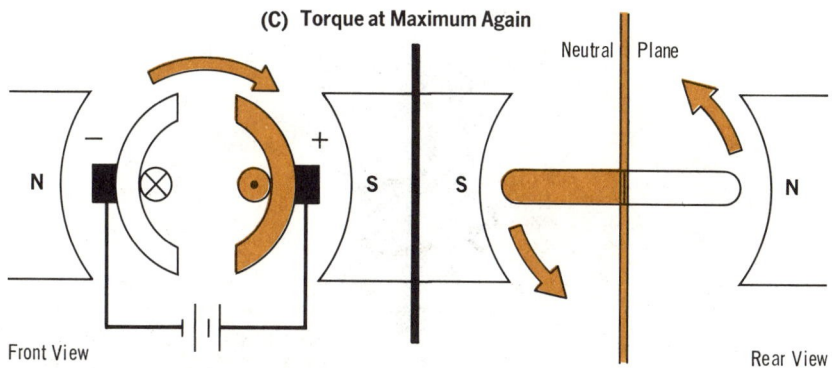

(C) Torque at Maximum Again

shortcomings of the elementary d-c motor (cont.)

To get the motor started, it is necessary to move its armature out of the neutral plane. In any other position, the brushes are reconnected to the commutator, current flows in the armature, and torque will be developed. Once started this way, the motor continues to run until power is removed.

This brings us to the second shortcoming. When the elementary d-c motor runs, its operation is erratic because it produces torque irregularly. Maximum torque is produced only when the plane of the single loop armature is parallel with the plane of the field. This is the position at right angles to the neutral plane. Once the armature passes this plane of maximum torque, less and less torque is developed, until, of course, when it arrives at the neutral plane again, no torque is developed. Inertia carries the armature past the neutral plane, and so the motor continues to turn. Its irregularity in producing torque, however, prevents the single-loop elementary d-c motor from being used for practical jobs.

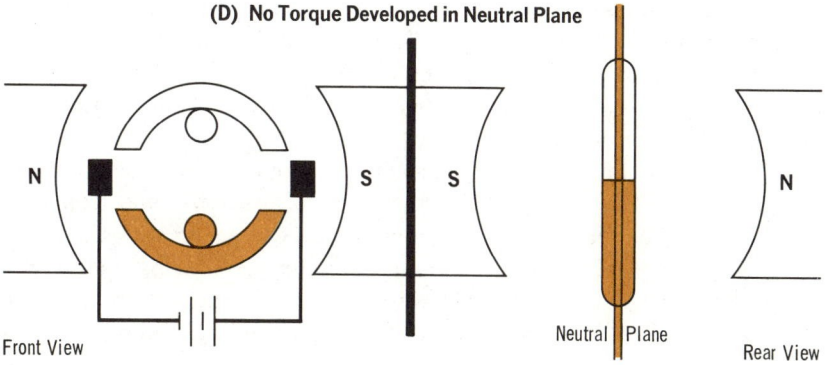

(D) No Torque Developed in Neutral Plane

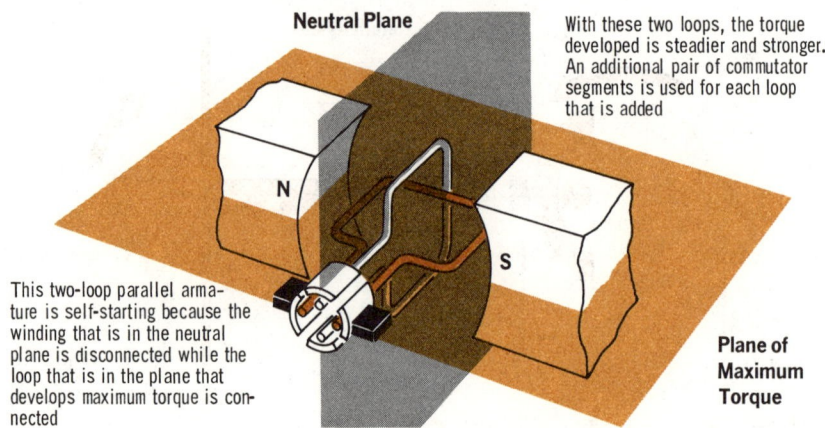

Neutral Plane

With these two loops, the torque developed is steadier and stronger. An additional pair of commutator segments is used for each loop that is added

This two-loop parallel arma-ture is self-starting because the winding that is in the neutral plane is disconnected while the loop that is in the plane that develops maximum torque is con-nected

Plane of Maximum Torque

the two-loop armature d-c motor

We can make the basic d-c motor start by itself by giving it an arma-ture that contains two or more loops. In a two-loop armature, the loops are placed at right angles to each other; when one loop lies in the neutral plane, the other is in the plane of maximum torque. The commutator in this case is split into two pairs, or four segments, with one segment associated with each end of each armature loop. This gives us two *parallel loop circuits*. Only one loop at a time is ever connected if we supply power through one pair of fixed brushes to one set of ring segments.

In this multiloop armature, the commutator serves two functions: it maintains current flow through the armature in the same direction at all times; and it switches power to the armature loop approaching the maximum torque position. Thus, in the two-loop armature, as loop 2 approaches the maximum torque position, loop 1 is approaching the minimum torque position and the commutator acts to switch loop 1 out and switch loop 2 in. Current through loop 2 is maintained in the direction that supports continuous rotation. A short time later as loop 2 approaches minimum torque while loop 1 again approaches maximum, the commutator switches power from loop 2 back to loop 1. This time it also switches the direction of current through loop 1 to maintain the original direction of rotation.

With this type of parallel loop armature, the motor will be self-starting but will still operate somewhat erratically because only one loop at a time provides the torque that drives the motor. Of course, with two loops the motor is driven with twice as much much torque during one revolu-tion as in a single-loop motor. However, the added dead weight of the second loop offsets the full advantage of the additional torque.

improving armature operation

As you have learned, adding loops to the armature of an elementary d-c motor insures only that it will be self-starting, not necessarily that the motor will operate smoothly enough to work effectively under a load. In an *elementary* motor with only *one* pair of brushes, no matter how many independent loops there are, only one loop at a time is supplied with current and is developing torque to drive the motor. In a three-loop motor, for example, the one loop that develops torque must carry the dead weight of the two remaining loops. To gain a real improvement in operation, current must be supplied to all the loops in the armature at the same time; except, of course, to any loops that happen to be in the neutral plane.

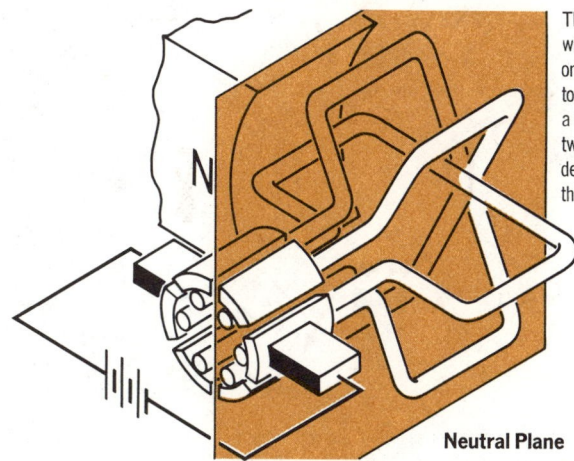

The three-loop parallel wired armature that uses one set of brushes develops torque in only one loop at a time. As a result, the two other loops represent dead weight that makes this type of motor inefficient

Neutral Plane

One seemingly simple solution would be to provide a three-loop motor with a brush for each commutator segment. With six brushes it would be possible to supply current to all armature loops at the same time. As a result all loops would provide torque at the same time, improving motor operation. The use of six brushes, however, is not a practical solution. This motor would be costly, cumbersome, and maintenance would be quite complicated. Therefore, a way to overcome the problem while still using one pair of brushes is the more desirable approach. By connecting the armature loops in a series circuit arrangement it becomes possible to use a single pair of brushes to supply current to all the loops at the same time. As a result, the windings all create torque at the same time to aid motor operation. As you will learn, most practical d-c motors make use of one set of brushes and many armature loops connected in various series-parallel arrangements.

the four-loop armature d-c motor

A practical four-loop armature is shown in the form of a schematic diagram. The schematic will help you understand the layout of the electric circuit. Notice first that the loops are connected between adjacent commutator segments. Also notice that the connections are arranged so that the combination of loops and commutator segments make up *one* large *series* circuit *loop*. In the 2-loop parallel armature you learned about, each end of the loop was assigned to opposite segments on the commutator. As a result, where we needed *one pair* of segments per loop, we now only need *one segment per loop*.

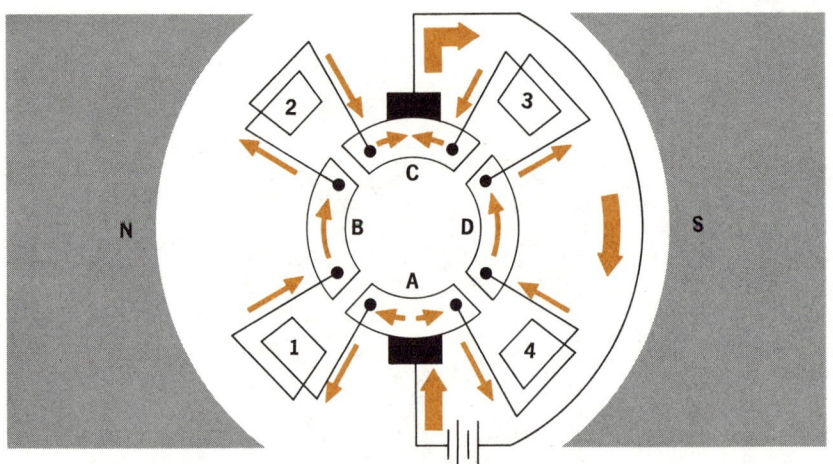

With the brushes on either pair of commutator segments, current flows through all the loops at the same time. In this wiring arrangement, no loop is ever switched out of the circuit in favor of another loop. All loops, therefore, work together to produce a maximum torque

In the case of the four-loop armature this means we only need 4 commutator segments for the series winding. The parallel winding would require eight segments. If we now add brushes to our commutator at opposite segments A and C, we will have divided the armature into two series circuits in parallel: one series circuit consisting of segment A, loop 1, segment B, loop 2, and segment C in parallel with the series circuit consisting of segment C, loop 3, segment D, loop 4, and segment A. Thus, when current flows in the circuit as shown, all four windings will carry the current and will be able to contribute torque to help the motor operate. With this type of armature, the brushes are wider than the gaps between the commutator segments, so that the circuit is never opened as the brushes pass from segment to segment.

operation

If you examine in detail the illustration of the practical d-c motor, you will see that the armature is shown in a position where none of its windings is in the neutral plane; therefore, all the windings can contribute torque. Of course, for all the windings to produce torque, they all have to be supplied with current at the same time. If you trace the path to be described here, you will see how this is done. Beginning at the negative side of the power supply, current flows through the negative brush through the two parallel sets of series windings, over to the positive brush and back to the positive side of the power supply. Notice that a complete circuit for current flow exists through all the windings just as it did in the schematic diagram on the previous page. Actually the circuit here is identical to that on the previous schematic.

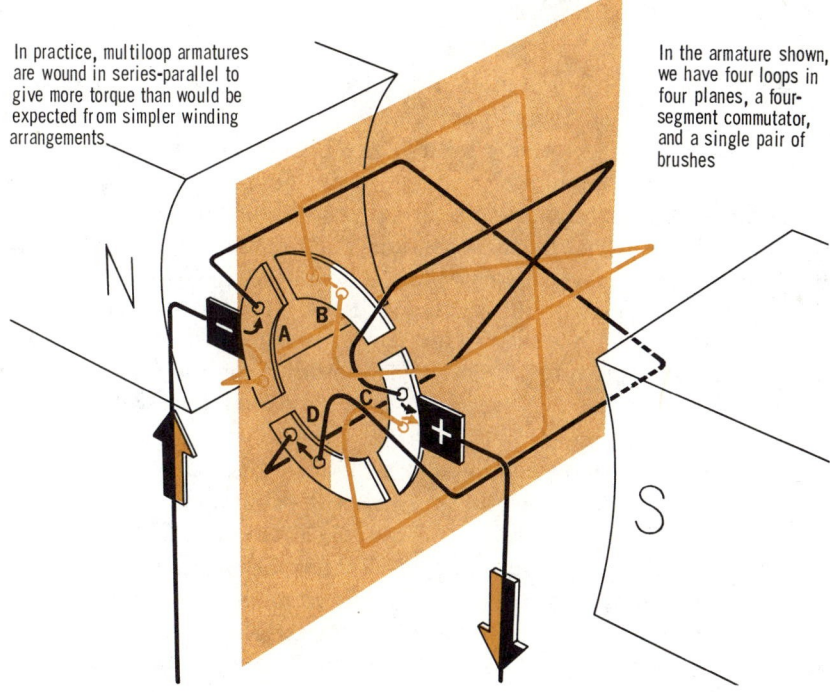

In practice, multiloop armatures are wound in series-parallel to give more torque than would be expected from simpler winding arrangements.

In the armature shown, we have four loops in four planes, a four-segment commutator, and a single pair of brushes

When torque is developed, the armature turns and is soon brought into a position where one of its loops enters the neutral plane. At this point commutation must take place. As you will soon learn, the commutation technique in the practical d-c motor differs slightly from that you learned about for the elementary d-c motor.

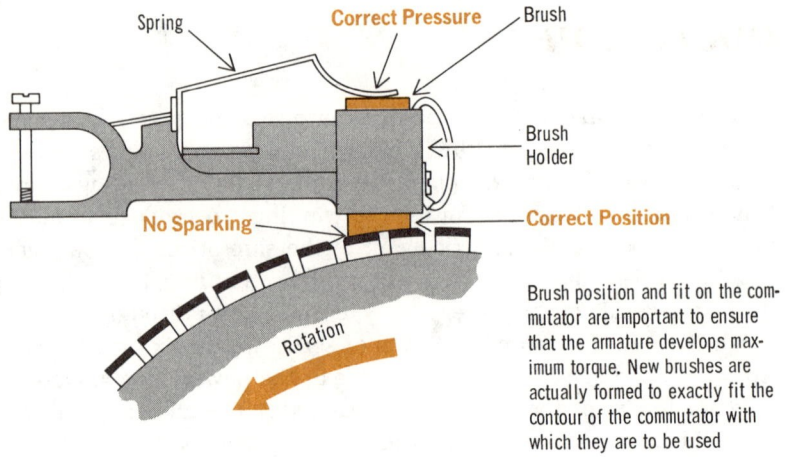

Spring   Correct Pressure   Brush

Brush Holder

No Sparking   Correct Position

Rotation

Brush position and fit on the commutator are important to ensure that the armature develops maximum torque. New brushes are actually formed to exactly fit the contour of the commutator with which they are to be used

commutation

In your study of commutation in the elementary single-loop d-c motor, you learned that the commutator and brush assembly operated as an opening and closing reversing switch. The brushes first became *disconnected* from the commutator segment as the loop entered the neutral plane before being *reconnected* to opposite segments as the loop swung past the neutral plane. In that operation, a current-carrying circuit was first broken and then was made complete again. These switching actions cause an arc to form at the switching point. Thus, in the elementary motor, there is usually much arcing and sparking between brushes and commutator, causing burned spots on the commutator and rapid wearing down of the brushes. The useful life of such a motor between repairs would be shortened.

In practical d-c motors commutator switching eliminates current from a loop passing through the neutral plane with minimum arcing. In part this is accomplished by using a brush to short circuit that loop for the instant it lies within the neutral plane. Because the loop is in the neutral plane during commutation and there is minimum or no interaction with the field flux, the potential difference across the loop is also at a minimum. This means that when the brush short circuits the loop in the neutral plane and then reestablishes the circuit with current flow in the new direction, the make and break operations involve points of near equal potential, so that commutation takes place with a minimum of arcing and sparking. By keeping the shorting period to an absolute minimum, we insure that torque is lost only for the instant the loop is in neutral and that the motor performs at its highest possible operating speed.

commutation in the four-loop armature

Commutation in a practical d-c motor with a four-loop armature is carried out by having the brushes short circuit adjacent commutator segments to stop the flow of current in an associated armature loop passing through the neutral plane. When commutation is being carried out properly, the motor operates at highest possible speed with the least amount of sparking. In the system used in practical motors, as shown, when a pair of loops, say 1 and 3, arrives at the neutral plane, they are shorted out so they stop carrying current. At the same time, the other two loops, 2 and 4, are still connected in this circuit and continue to carry current.

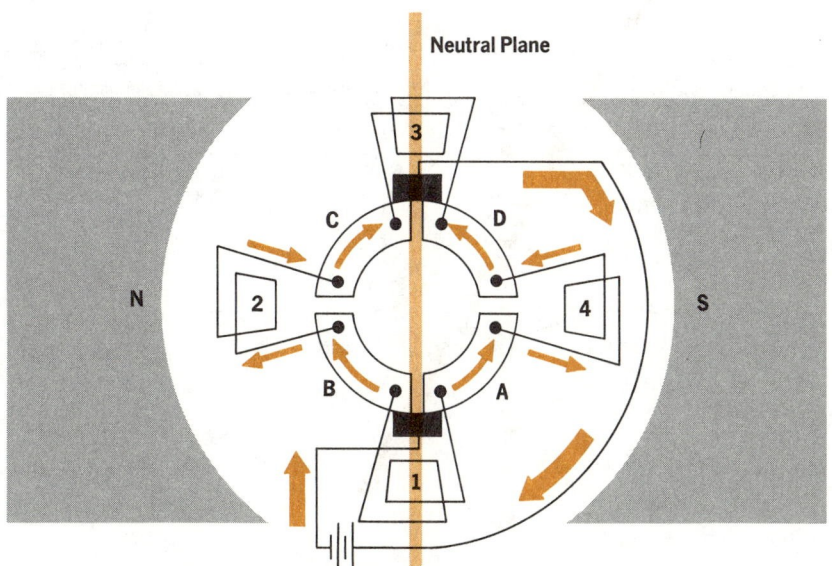

Brushes short circuit the commutator segments that feed the loops in the neutral plane. These loops, therefore, carry no current at the point where they cannot produce any torque. The other loops carry their normal currents and provide the torque to keep the armature turning

Notice that because of the symmetrical arrangement and action of the commutator and brushes, two of the armature windings are always short circuited at one time. As a result, in practice, armatures are intentionally wound as shown so that pairs of windings affected by brush action arrive at the neutral plane at the same time.

the neutral plane

You will recall from the discussion on commutation in the practical d-c motor, that arcing during motor operation is kept to a minimum because there is a minimum potential difference across an armature loop when it is in the neutral plane. As shown for a two-pole motor, the neutral plane should be in the axis at right angles to the field flux lines and midway between the pole pieces. This axis is sometimes called the *geometric neutral plane* of the motor. The neutral plane in the motor is the same for all the loops of the armature. Each loop passes through the neutral plane twice as the armature makes one complete revolution.

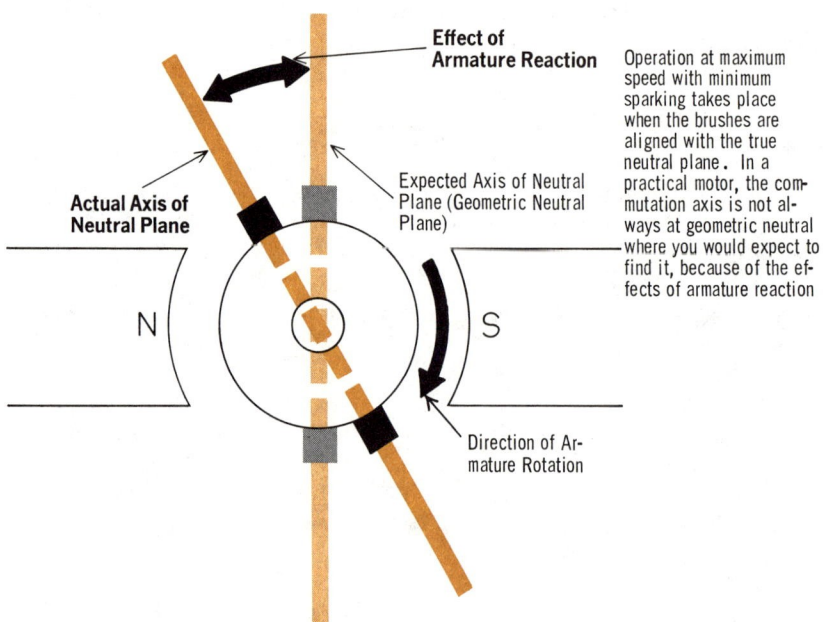

Effect of Armature Reaction

Actual Axis of Neutral Plane

Expected Axis of Neutral Plane (Geometric Neutral Plane)

N   S

Direction of Armature Rotation

Operation at maximum speed with minimum sparking takes place when the brushes are aligned with the true neutral plane. In a practical motor, the commutation axis is not always at geometric neutral where you would expect to find it, because of the effects of armature reaction

Theoretically, then, to achieve perfect commutation, the plane in which the brushes of the motor are positioned, called the *brush axis*, should ideally coincide with the axis of the neutral plane of the motor. In practice, however, the position of the actual neutral plane tends to shift from the geometric neutral axis when the motor is operated.

The shift is affected by the speed and direction of operation of the motor. Therefore, either the brush axis has to be moved to the new position of the neutral plane or something has to prevent the plane from shifting. The principal cause behind this shifting of the neutral plane when the motor operates is an effect known as *armature reaction*.

armature reaction

When a motor armature is supplied with current, a magnetic flux is built up around the conductors of the armature winding. There are thus two magnetic fields in the space between the field pole pieces: the main magnetic field and the field produced by the armature. These two fields combine to produce a new *resultant magnetic field*. In this case, the resultant field is distorted as to shift it opposite to the direction of rotation of the armature. This distortion of the original main field is called *armature reaction*. Since the neutral plane of the motor is at right angles to the field of flux, we find that it too is shifted opposite to the direction of rotation of the armature.

The amount of armature reaction determines how far the neutral plane gets shifted. Armature reaction varies depending on the amount of armature current flowing. The larger the current, the more the neutral plane shifts from geometric neutral. Similarly, the direction of the shift depends on the direction of current flow through the armature.

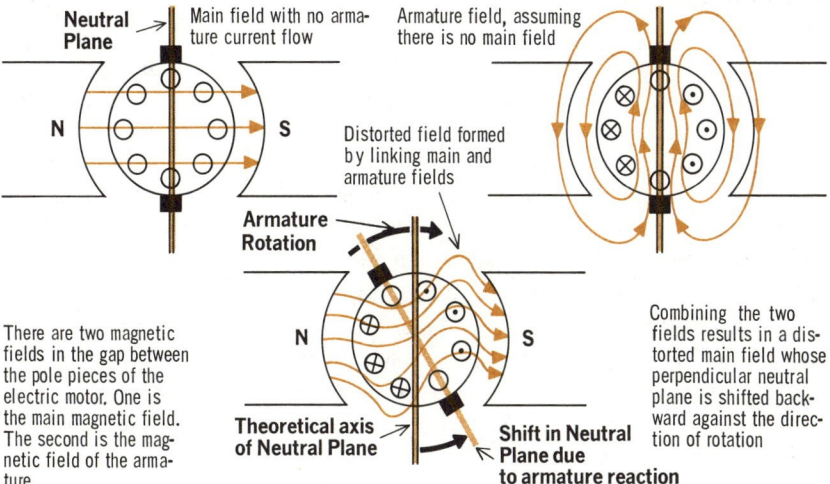

Neutral Plane

Main field with no armature current flow

Armature field, assuming there is no main field

Distorted field formed by linking main and armature fields

N S

Armature Rotation

N S

There are two magnetic fields in the gap between the pole pieces of the electric motor. One is the main magnetic field. The second is the magnetic field of the armature

Theoretical axis of Neutral Plane

Shift in Neutral Plane due to armature reaction

Combining the two fields results in a distorted main field whose perpendicular neutral plane is shifted backward against the direction of rotation

If the motor is to be operated at a constant speed in a single direction, the brushes can be positioned at the new location of the neutral plane and left there to get effective commutation. But if the motor is to be operated at various speeds, directions, and against varying loads, current through the armature will vary considerably. As a result, armature reaction will vary and so will the position of the neutral plane. This means that for effective commutation, the brush would have to be changed in position each time the neutral plane changes. Obviously this is a most troublesome procedure.

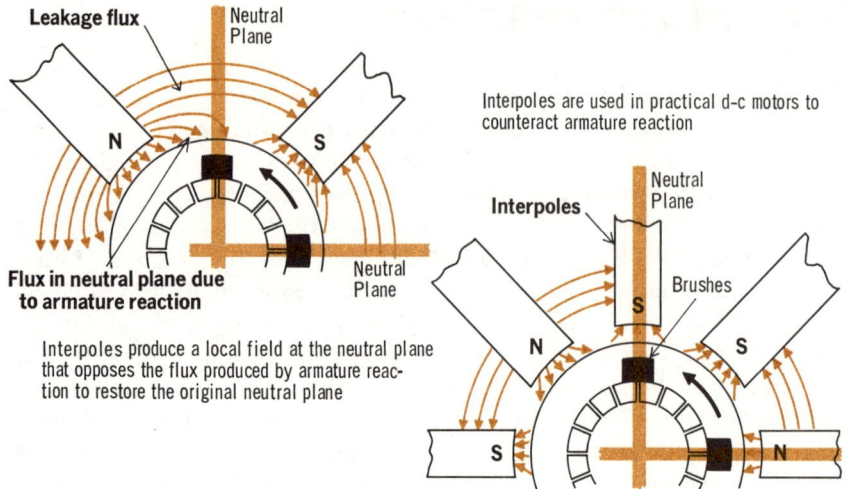

Interpoles are used in practical d-c motors to counteract armature reaction

Interpoles produce a local field at the neutral plane that opposes the flux produced by armature reaction to restore the original neutral plane

interpoles

When a motor is operated at various speeds, in different directions, or against variable loads, armature current and armature reaction also vary. To operate such a motor with efficient commutation would require a change in brush positions to account for each resultant shift in the neutral plane. Since such constant realignment of brushes is unrealistic, practical motors need some provisions to keep the neutral plane from shifting as a result of armature reaction. One solution to this problem is the use of special windings called *interpoles.*

Interpoles are special electromagnet pole pieces placed in line with the neutral plane between main pole pieces. Interpole windings are connected in series with the armature winding, so the armature current causes magnetic fields to be set up around them. The directions of these fields are such that they cancel the magnetic fields around the armature coils in the vicinity of the interpoles, and thus counteract the tendency for armature reaction to shift the neutral plane. As a result, the neutral plane is kept reasonably close to geometric neutral for all modes of motor operation.

The fact that the interpole windings are in series with the armature makes their action self-regulating: the interpoles will apply the proper amount of cancellation field for each different set of conditions. For example, at high armature currents where armature reaction is high and the tendency to shift the neutral plane is great, the interpole field cancelling the shift will also be strong. The reverse is true at low armature current levels.

summary

☐ The elementary d-c motor will not start by itself when the armature loop is in the neutral plane, because no flux interaction occurs there. ☐ Because of the loss of torque that occurs whenever the loop passes through the neutral plane, the elementary d-c motor runs erratically after it is started. ☐ When an armature is made with two loops at right angles to each other, at least one loop always develops torque to make such a motor self-starting; however, this type of motor still runs erratically because of the effect of the neutral plane. ☐ When an armature with three or more loops is used, and the loops are connected in series so that every loop, except the one in the neutral plane, conducts current, a steadier torque is produced.

☐ Practical motors use multiple-loop armatures with series-parallel arrangements to bring about smooth operation. ☐ To eliminate arcing at the brushes and to produce the highest motor speed, practical motors have commutation take place with the loop passing through the neutral plane; the brushes short the loop to remove it from the circuit.

☐ The brush axis cannot always be positioned along the geometric neutral plane because of the effect of armature reaction, which tends to shift the neutral plane. ☐ With relatively constant-speed motors, the brushes can be positioned along the shifted neutral plane; but with varying-speed motors, some means, such as interpoles, must be used to keep the neutral plane from shifting.

review questions

1. In the elementary single-loop motor, why doesn't the armature flux interact with the field flux in the neutral plane?
2. Why is a two-loop armature motor self-starting?
3. Does a basic two-loop armature motor run smoothly? Why?
4. If you add loops, how would you arrange the loop circuits with one set of brushes to get smooth operation? Why?
5. What would you have to do to get smooth operation with a multiple-loop motor if a series-circuit loop arrangement were not used?
6. What is the disadvantage of using the brushes to energize the loops away from the neutral plane? Why does this occur?
7. Why is it safe for a loop to be shorted in the neutral plane?
8. What is *armature reaction,* and why does it shift the neutral plane?
9. In what direction is the neutral plane shifted by armature reaction?
10. How do interpoles counteract armature reaction?

Cutaway View of Practical D-C Motor

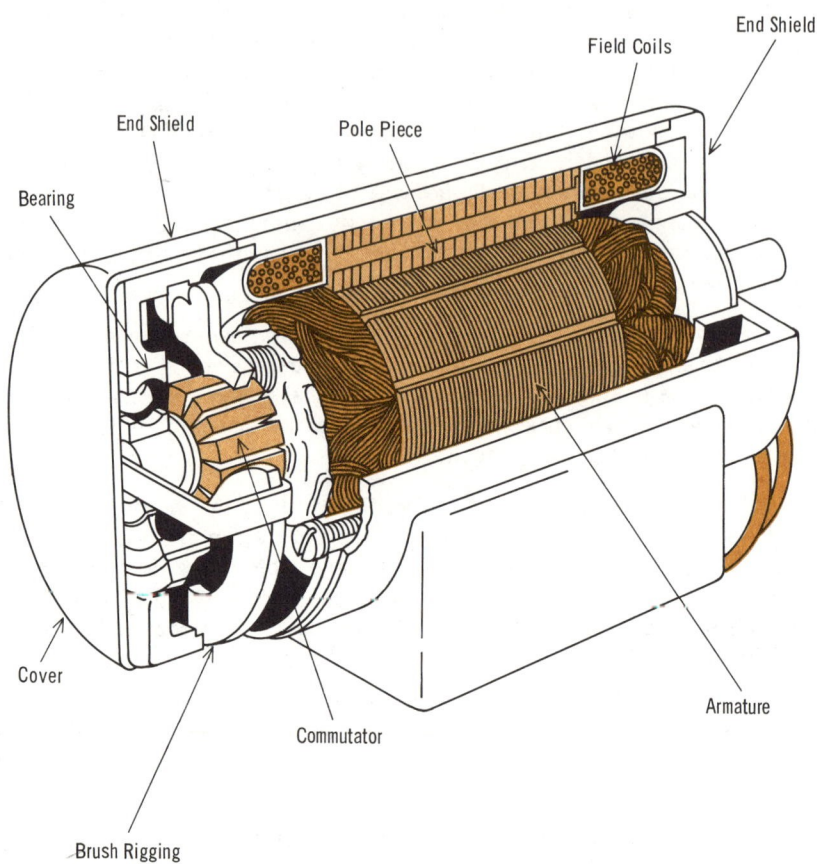

construction

In your study so far, you have been introduced to the principles of operation of d-c motors. In learning about the electrical operation of the d-c motor, you have also been introduced to most of the major physical features of the d-c motor. These features included the armature and commutator, the brush assembly, and the field magnet. At this point, you will be taught the details about the physical construction of d-c motors to enable you to recognize the various parts of actual motors. You will also learn how a motor is made in terms of the material parts and methods used in its construction. Finally, you will be given an appreciation for some of the operations involved in motor maintenance.

armature core and shaft

The term *armature*, or *rotor*, is applied to the part of the motor that rotates. When you look at an operating motor, you usually see its shaft turning. The shaft is an external extension of the armature through the motor housing and frame, the side opposite the commutator end of the motor. The commutator is described later.

A typical armature core looks like a solid metal slotted cylinder. Actually the core is made up of sandwiched, or *laminated*, thin pieces of soft steel. The laminations, each with notches around its edge, are coated with an insulating varnish and pressed together to form the core. In the forming process, the notches are lined up so that the finished core has a series of longitudinal slots all around its perimeter. Laminations are used to form the core in order to reduce losses due to eddy currents.

Eddy currents are the circulating currents induced in a conducting material when it cuts magnetic flux lines (see Volume 3). The laminations reduce the area in which eddy currents can flow and thus the relative resistance of the material goes up; hence, power losses due to eddy current flow go down. The use of soft steel in the core material reduces *hysteresis losses*, which occur, you will recall from Volume 3, when reversals of the magnetization of the core material lag the actual current reversals.

The slots in the formed core are used for positioning the copper wire loops or windings of the armature. The armature core is mounted on the shaft of the motor, which is usually a hard steel rod with a highly polished bearing surface. The method of mounting the core on the shaft varies widely from motor to motor.

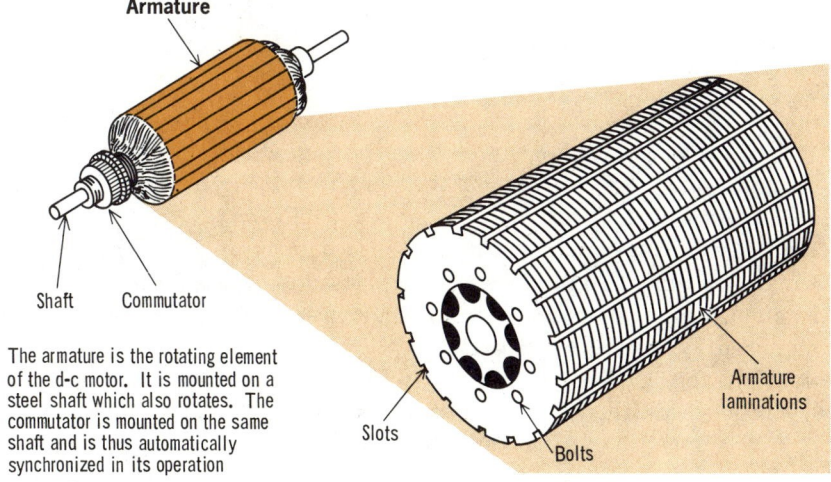

Armature

Shaft   Commutator

The armature is the rotating element of the d-c motor. It is mounted on a steel shaft which also rotates. The commutator is mounted on the same shaft and is thus automatically synchronized in its operation

Slots

Bolts

Armature laminations

gramme-ring armature windings

Armature winding is a science unto itself. At this point you will be introduced to some of the important aspects of the subject which will be of benefit later on in appreciating the basic differences and similarities between d-c and a-c motors.

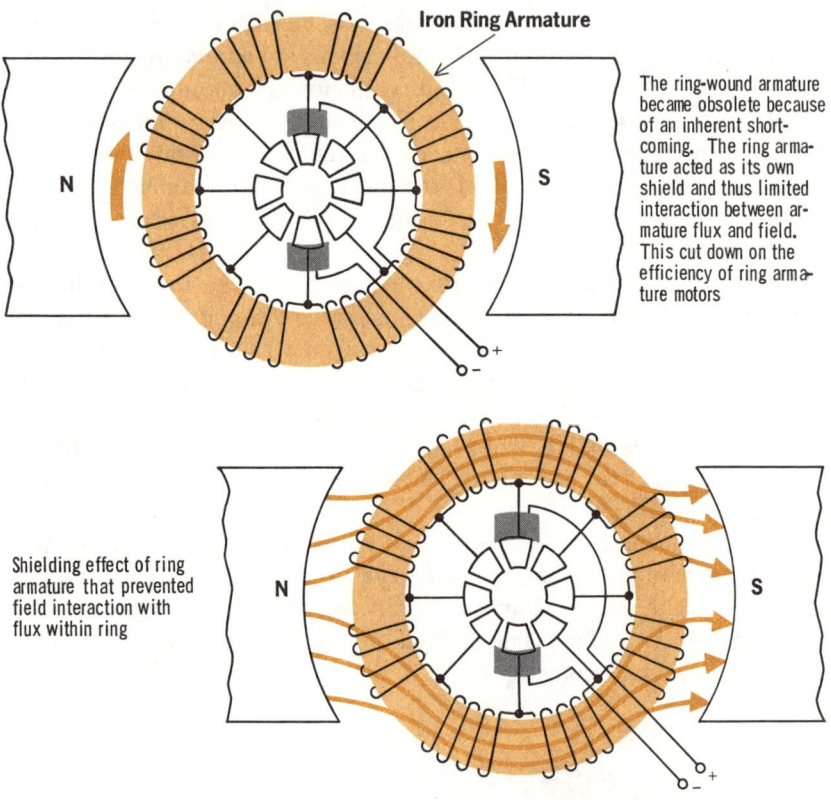

Iron Ring Armature

The ring-wound armature became obsolete because of an inherent short-coming. The ring armature acted as its own shield and thus limited interaction between armature flux and field. This cut down on the efficiency of ring armature motors

Shielding effect of ring armature that prevented field interaction with flux within ring

D-c armature windings are classified as either *ring windings* or *drum windings*, depending on the shape of the armature core. The earliest practical motors used the so-called *Gramme-ring winding*. In the Gramme-ring winding, the armature core is an iron ring around which is wound a continuous winding closed upon itself and connected at regular intervals to commutator segments. This type of armature is inefficient because the conductors on the inner half of the ring are magnetically shielded by the iron and thus cannot interact with the field. Because of this and other disadvantages, the Gramme-ring armature is now little more than a laboratory curiosity.

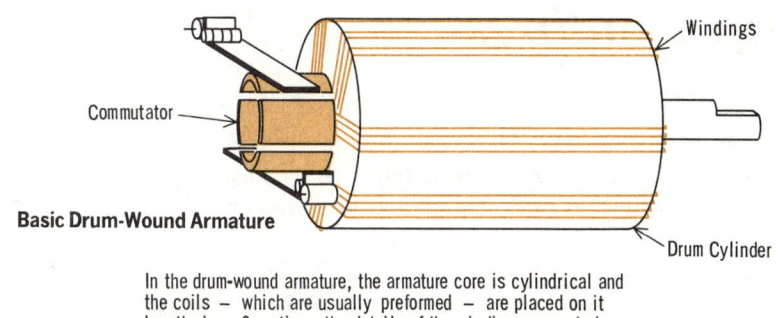

Basic Drum-Wound Armature

In the drum-wound armature, the armature core is cylindrical and the coils — which are usually preformed — are placed on it lengthwise. Sometimes the details of the windings are not obvious from merely looking at the completed armature

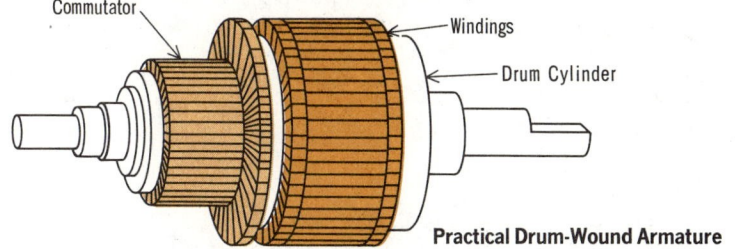

Practical Drum-Wound Armature

the drum-wound armature

In place of the Gramme ring, most modern motors use the drum-shaped armature core you learned about previously.

The loops or coils that make up the drum-wound armature are wound around the armature core by setting the sides of the coils into the slots in the core. The slots are usually insulated with "fish" paper to protect the windings. In many armatures, the coils are first preformed into their final shapes and then placed in the slots of the core. This is called *form winding* and is carried out either by shaping the coils over a wooden form or by bending them in a press before placing them on the drum. Each winding is always the same as every other on the armature and the final wound armature should always be perfectly symmetrical.

After the armature coils have been placed on the core, wedges of insulating material are placed in the core slot to hold the coils in place. After this, adhesive and steel bands are used to secure the coils in order to insure they will not be thrown out by the centrifugal force developed during armature rotation.

There are two basic types of drum winding arrangements in use: *lap windings* and *wave windings*. The lap winding is used for low-voltage, high-current motors. The wave winding is used in motors with high-voltage, low-current requirements.

lap windings

The lap winding gets its name from the way the winding laps back and forth on the drum armature. Here, one coil element goes forward and lies under a north field pole and then laps back upon itself to return under a south field pole. You can better understand the lap wiring by using the wiring diagram on this page to trace the wiring as follows:

Begin with the heavy-lined coil which begins its run from commutator segment 1, through slot 1, and comes back through slot 4 to commutator segment 2. Notice that the coil begins and ends at adjacent commutator segments, but that in doing so, it runs through slots under opposite poles. Check the other coils and you will find this is true of all of them.

Now look at the schematic diagram and notice that there are four parallel paths through the armature.

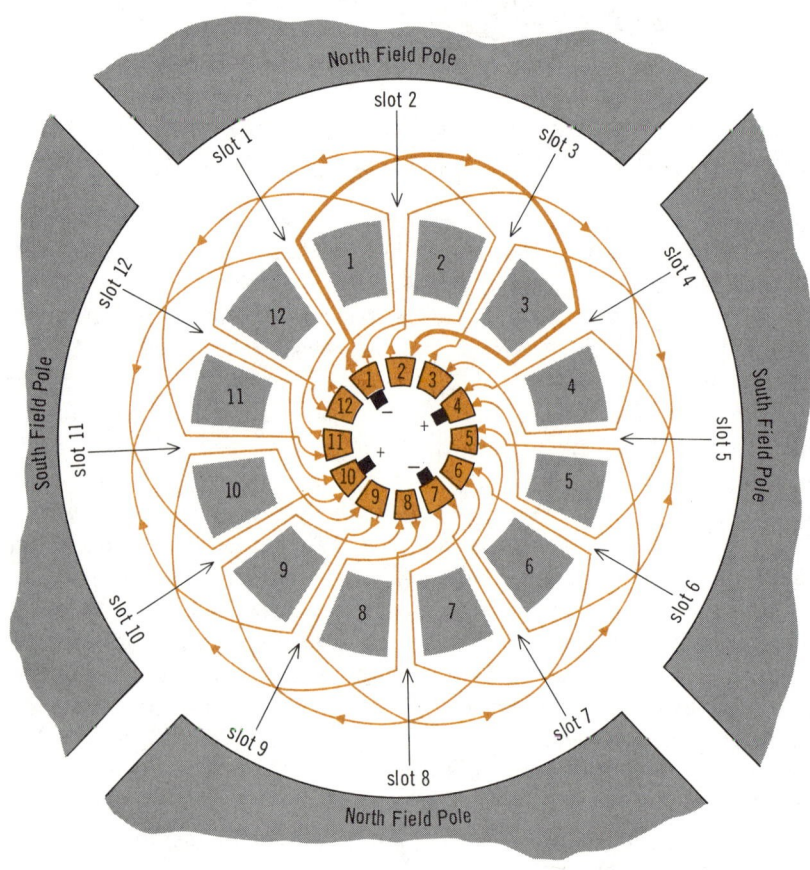

lap windings (cont.)

The four parallel paths through the armature are:
1. From the brush at 1 on the negative side, through the windings in the path 2-3, and on to the brush at 4 on the positive side.
2. From the brush at 1 on the negative side, through the windings in the path 12-11, and on to the brush at 10 on the positive side.
3. From the brush at 7 on the negative side, through the windings in the path 6-5, and on to the brush at 4 on the positive side.
4. From the brush at 7 on the negative side, through the windings in the path 8-9, and on to the brush at 10 on the positive side.

This shows that all twelve armature windings are carrying current at the same time and thus all are contributing to motor action and operation at the same time.

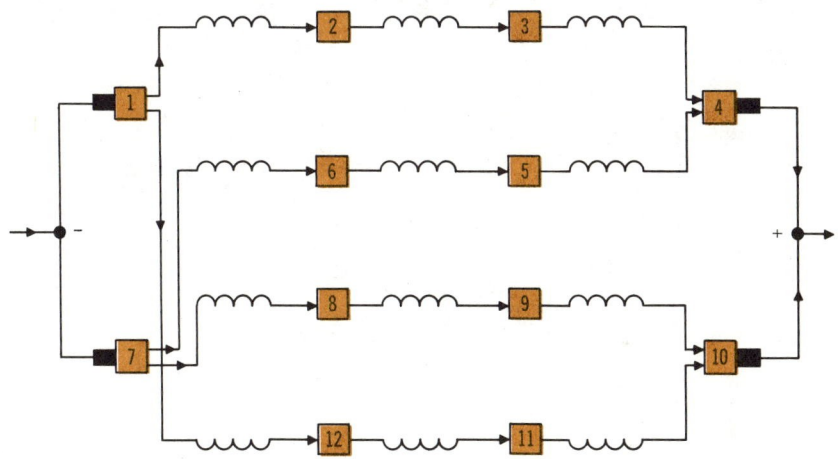

A four-pole motor with lap windings on its armature has four brushes. It also has four parallel paths through the windings, each of which carries current and contributes torque for motor operation

Although there are many, many variations of the lap winding in use, two characteristics are an aid to recognizing them in motors you may meet:
1. Motors with lap-wound armatures generally have as many brushes as there are poles.
2. Motors with lap-wound armatures generally have as many parallel paths through the armature as there are poles.

You can verify these statements, using the illustrations on these two pages.

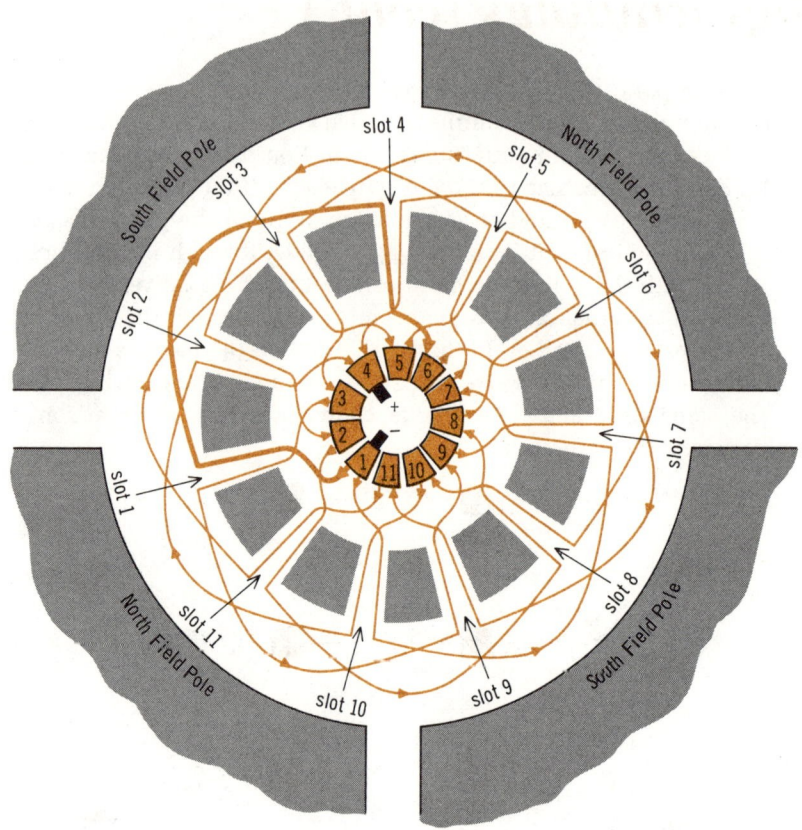

Wiring Diagram

wave windings

The *wave winding* gets its name from the appearance of the windings on the drum. As in the case of the lap winding, many variations of the wave windings are possible. However, two characteristics are common to all:

1. Motors with wave-wound armatures require only the minimum of two brushes. However, in some variations, they may have as many brushes as there are poles, just like the motor with a lap-wound armature.

2. In wave-wound armatures, there are only two parallel paths through the armature in one complete wave winding, regardless of the number of brushes or poles used.

wave windings (cont.)

The schematic and wiring diagrams of the wave-wound armature bear out these points. By tracing through the schematic diagrams you will see that there are two parallel paths through the winding as follows:

1. From the brush at 1 on the negative side, through the windings in the path 6-11-5-10, and on to the brush at 4 on the positive side.
2. From the brush at 1 on the negative side, through the windings in the path 7-2-8-3-9, and on to the brush at 4 on the positive side.

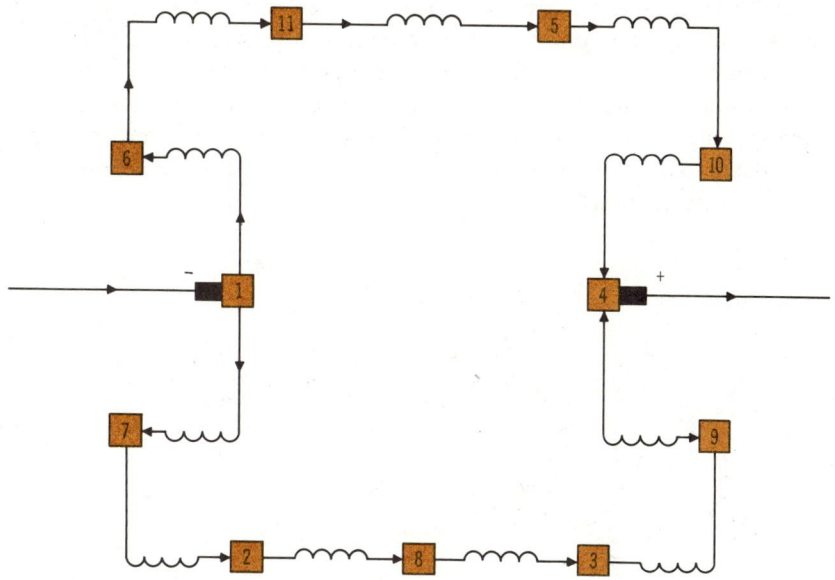

Motors with wave-wound armatures that have multiple sets of poles, need only one set of brushes. This fact is an aid to recognizing whether a wave-type armature winding is being used in a particular d-c motor

On tracing the arrangement of the wave windings in the wiring diagrams, you can find out about the most significant difference between lap and wave windings. Using the heavy-lined coil as an example, notice that the ends of the coil terminate on commutator segments 1 and 4, which are rather *widely separated*. Remember that in the lap-wound armature the ends of a winding terminated on *adjacent segments*.

Another item of interest is the fact that for a four-pole motor, only 11 wave windings can be used, whereas 12 lap windings could be accommodated by comparison.

the commutator

The commutator consists of individual conducting *segments* made of copper insulated from each other with thin sheets of *mica*. The individual segments with their mica separators are assembled in a cylindrical form and are held together by a clamping *flange*. The segments are insulated from the clamping flange with rings of mica.

The leads from the armature coils are connected to the raised portions of the commutator segments known as *risers*. Some commutator segments are manufactured without risers, and instead have slits in their ends to which the armature coil leads are connected. After assembly, the surface of the commutator is cut perfectly cylindrical in a lathe and ground to a very smooth finish. This insures that there will be a minimum of friction between the commutation surface and the brushes. Finally, and of great importance, the mica insulation between segments is undercut slightly below the surface of the commutator segments so that it does not interfere with the operation of the brushes.

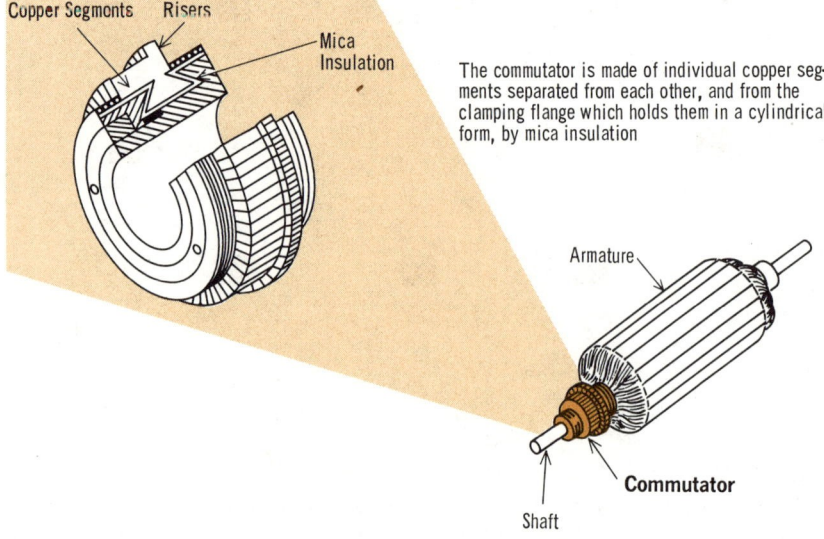

The commutator is made of individual copper segments separated from each other, and from the clamping flange which holds them in a cylindrical form, by mica insulation

After a motor has been in service for a time, the copper surface of the commutator usually wears somewhat. Whenever wear brings the copper down to the level of the mica, it is necessary to again undercut the mica if the motor is expected to perform satisfactorily. At the same time it is usually also necessary to turn the commutator on a lathe again to insure that its shape is that of a true cylinder.

the brush assembly

The *brush assembly* consists of the brushes, their holders, and brush springs. The brushes themselves are made of a soft carbon material containing a large amount of graphite. This material serves two purposes: it is soft enough to cause only minimum wear to the commutator, while hard enough not to wear down too quickly in operation. No lubrication should ever be used between the brushes and commutator. Whatever lubrication may be required is supplied by the graphite in the brushes.

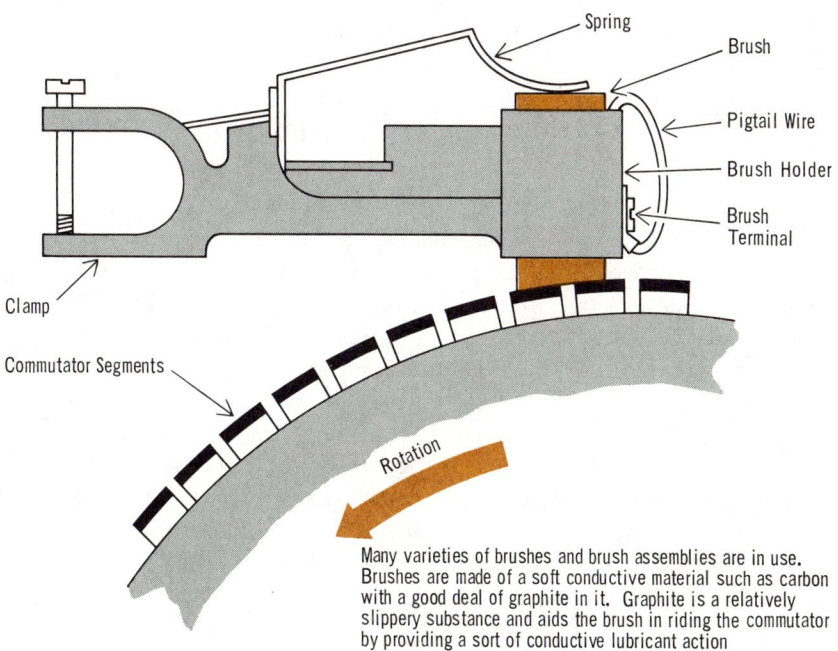

Many varieties of brushes and brush assemblies are in use. Brushes are made of a soft conductive material such as carbon with a good deal of graphite in it. Graphite is a relatively slippery substance and aids the brush in riding the commutator by providing a sort of conductive lubricant action

The brushes are usually set into separate assemblies known as *brush holders*. The brush holders are held in a fixed position supported by, but insulated from, the motor housing. The brushes are set loosely in their holders and held against the commutator by springs. The loose fit and spring action allow the brushes to move somewhat freely in their holders so that they can adjust to follow small irregularities in the commutator surface.

In many motors, the spring pressure can be adjusted to the correct specification of the motor manufacturer. If the pressure is too great, the brushes will wear out quickly; if too low, poor electrical contact will result. Poor contact will also cause sparking and erratic motor operation.

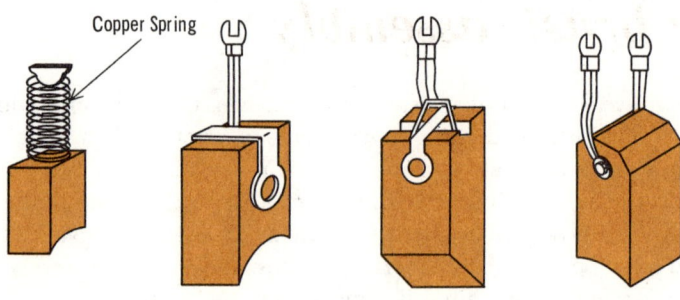

BRUSH TYPES

the brush assembly (cont.)

In most motors, the electrical connection between the brushes and the external power source is accomplished as follows: the brushes themselves are electrically connected to their respective brush holders by means of braided copper wires called pigtails. The brush holders, in turn, are connected to *studs* on the outside of the motor housing. The holders and studs are both insulated from the housing itself. The studs provide the tie points to which power leads may be connected to the motor.

Although brushes are designed to be long lasting, they are always made to wear out faster than the commutator, because it is cheaper and easier to replace brushes than to service an armature. The brushes are usually made long so that they can stay in service for a relatively long time before they finally wear to the point where replacement is necessary. The spring keeps the brush pressed firmly against the commutator all during its useful life.

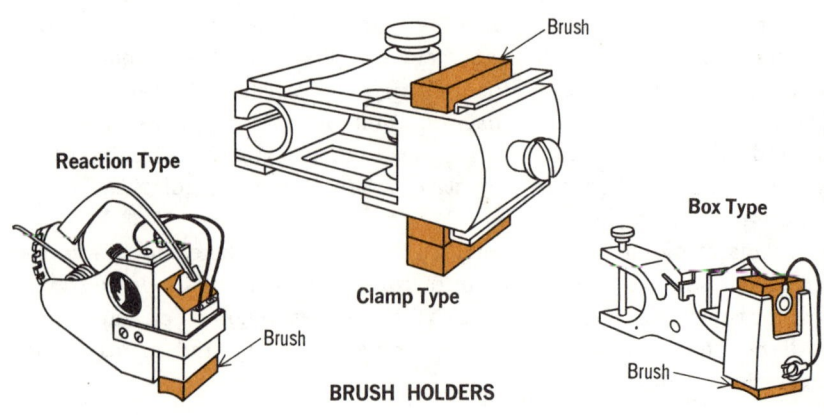

BRUSH HOLDERS

the field winding

The *field* is a common name for the polar magnetic field in which the armature turns. The field can be supplied by a permanent magnet, or, as in most practical motors, by an electromagnet. The electromagnet current has the same power source as the armature current.

The field assembly consists of *pole pieces* and *field coils*. The field pole pieces are usually bolted to the inside circumference of the housing and made of sheets of soft steel *laminated* together to decrease losses due to eddy currents. The pole pieces are usually shaped to fit the curvature of the armature to keep the air gap between the pole pieces and the armature as small as possible, since air offers a relatively high reluctance to magnetic flux lines.

Most d-c motors have individual pole pieces, called *salient poles*, which project inward toward the armature area. The field winding of a salient pole motor consists of all the individual field coils wound around their cores, the pole pieces. The number of individual·field coils determines the number of poles in the motor. A two-pole motor has two field coils, a four-pole motor has four field coils, and so on.

In some motors, the field winding is not formed on discrete salient poles, but is distributed around the field frame; the windings are arranged and power is applied to create fixed magnetic poles. Distributed field windings of this type are found in the so-called ac-dc motors discussed later in this volume.

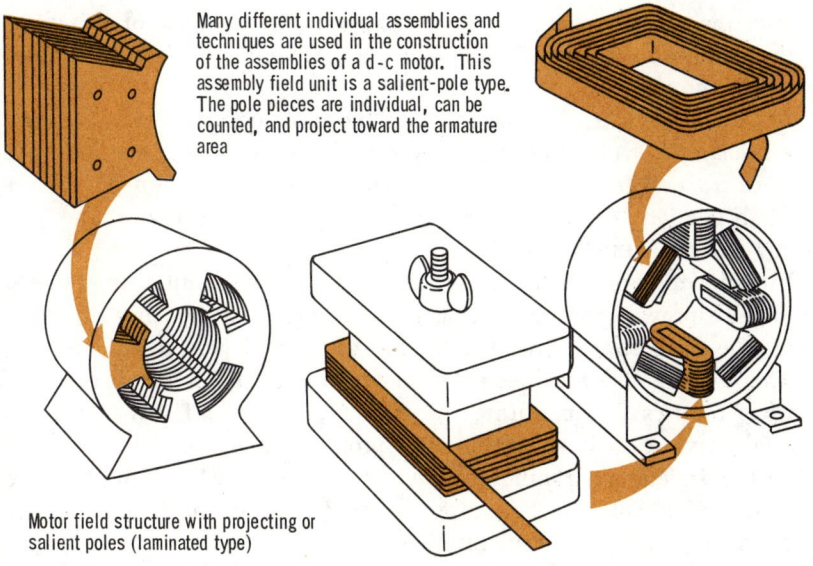

Many different individual assemblies and techniques are used in the construction of the assemblies of a d-c motor. This assembly field unit is a salient-pole type. The pole pieces are individual, can be counted, and project toward the armature area

Motor field structure with projecting or salient poles (laminated type)

the housing and mounting

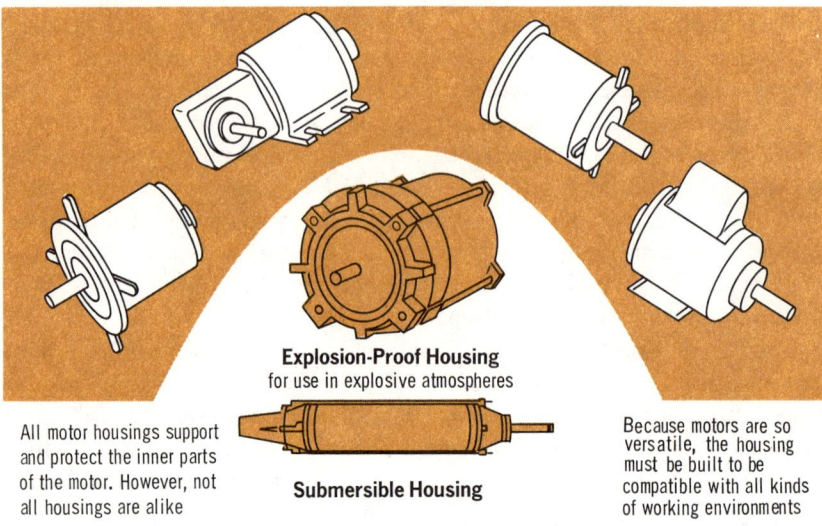

Explosion-Proof Housing
for use in explosive atmospheres

Submersible Housing

All motor housings support and protect the inner parts of the motor. However, not all housings are alike

Because motors are so versatile, the housing must be built to be compatible with all kinds of working environments

    The motor housing provides the mechanical support for the various parts that make up the motor. It also provides protection for the moving parts from outside influences such as dust, dirt and water. Most motor housings consist of three parts: a *field frame* and two *end frames*. The field frame supports the field coils and interpoles, if any. It also forms part of the magnetic circuit of the field winding. Because of this magnetic function, the field frame is made of iron and steel with good magnetic properties.

    The armature must be suspended in the field so as to be free to turn. The housing provides this basic support via bearings set in recesses in each of the end frames to allow for rotation with minimum power loss through friction. The end frames are mounted at each end of the field frame and are bolted to it.

    In some motors, oil holes or grease fittings are built into the end frames to provide for lubricating the bearings. The housing also provides the fixed support needed for the brush and field assemblies.

    The housing also provides external support for the motor. It contains various fixtures for mounting the motor, and is itself constructed in special ways to best withstand and protect the motor working parts from the various environments in which the motor will be used. There are usually air holes in the end frame, and often in the field frame, to allow air circulation to cool the motor.

bearings and cooling

Two other important construction features of electric motors are the *bearings* and the built-in devices to provide for *cooling*. Bearings support the armature and provide for smooth high-speed rotation with a minimum of friction. Where *ball bearings* are used, the bearings are force-fitted on the ends of the armature shaft and are set in recesses in the end frames. Where *ring bushing bearings* are used, these are merely set into the recesses in the end frames and the polished armature shaft is set into the bushing. When the end frames are bolted to the field frame in assembling the housing, the armature is automatically supported by the bearings in the end frame.

The bearings of some motors are made with built-in permanent lubrication to last for the life of the motor. In other motors, some means is provided for periodically lubricating the bearings. Usually, grease fittings or oil cups are built into the end frames for this purpose. Regularly lubricating a motor is perhaps the most important part of motor maintenance.

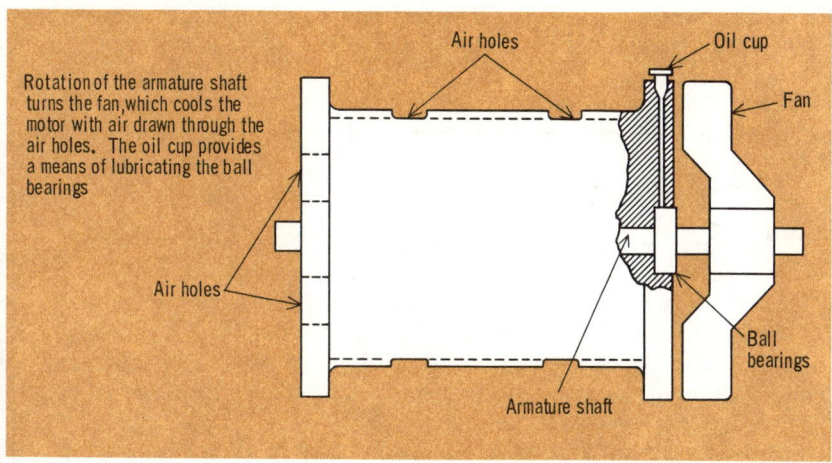

When a motor operates it normally develops a considerable amount of heat. This heat must be rapidly dissipated if the motor is to have a long service life. The most common method of dissipating this heat is by means of air holes and a built-in *fan* unit. The air holes are openings in the end frames and in the field frame to provide for ventilating the internal current-carrying parts. The built-in fan is a wheel with fan vanes, usually mounted on one end of the armature shaft. When the armature rotates, the fan draws air through the air holes and in this way carries heat from within the housing with it.

summary

☐ The rotor or armature core is made of laminated soft steel to minimize eddy current and hysteresis losses. ☐ The armature core is made with slots to hold the loop windings. ☐ Either ring or drum windings can be used, but the latter is used in most motors. The windings can be wound directly on the armature core, or form windings, which are preshaped, can be used. ☐ Fish paper is used in the slots for insulation, and insulation wedges, adhesives, and steel bands hold the coils in place.

☐ The armature coils are connected to the commutator in lap or wave, or both, winding arrangements. ☐ Lap windings are used for low-voltage, high-current motors; and wave windings, for high-voltage, low-current motors. ☐ The commutator is made of copper segments separated by undercut sheets of mica. ☐ Each end of every loop winding is connected to its own segment riser, so there are two segments for every loop. ☐ Current is supplied to the smoothly polished commutator by graphite brushes that slide from segment to segment.

☐ The field winding can be mounted on laminated salient poles or can be distributed around the field frame. ☐ The motor housing consists of the field frame and two end frames. ☐ The end frames contain the bearings that support the armature shaft, and the field frame carries the field winding.

review questions

1. How are the armature core and field poles made to reduce eddy current and hysteresis losses? How does this help?
2. How are form windings made?
3. When are lap or wave windings used?
4. Which type of winding has its ends connected to adjacent commutator segments? How does this differ from the other type of winding?
5. How do lap and wave windings differ in their relationship between the number of brushes and poles? The parallel current paths?
6. What is the relationship between the number of commutator segments and armature windings?
7. Why is the mica insulation between the segments undercut?
8. Why are the brushes made of soft graphite? Why is the brush pressure important?
9. What are *salient poles*?
10. Why is the armature shaft made of highly polished hard steel?

comparison of motors and generators

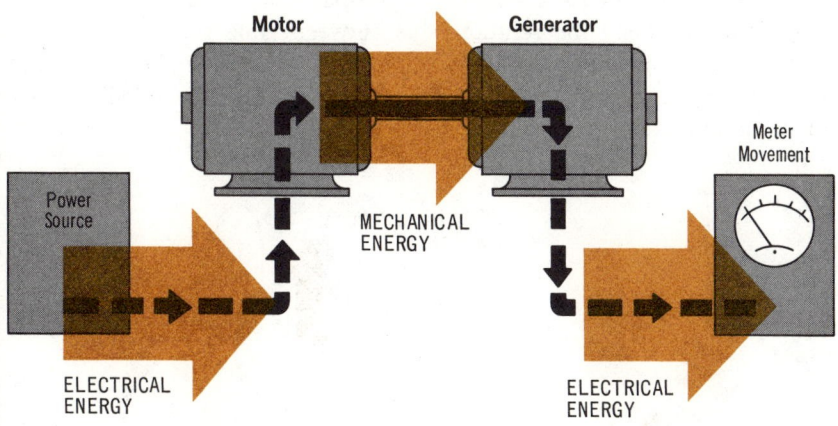

Motors and generators are opposites. If you feed electricity into a motor, the motor produces mechanical energy. If you drive a generator with that mechanical energy, it produces electricity

From everything you have learned to this point about d-c motors, you have probably concluded that the d-c motor we have been discussing is closely related to the d-c generator you learned about in Volume 6. The motor is a machine for converting electricity into motion while the generator is just the opposite—a machine for converting motion into electricity.

To perform their opposite jobs the d-c motor and generator use the same essential parts—a fixed field winding, a rotating armature with commutator, and a brush assembly. In fact, if you supply electric current to a generator, it will operate as a motor and drive a mechanical load. By the same token, if you turn the shaft of a d-c motor with a suitable source of mechanical energy, such as a steam turbine or gasoline engine, it will operate as a generator.

Although they are similar in general construction, some of the physical features of motors and generators differ in some practical details because they are used differently. Methods of mounting are one notable difference. The generator is usually located next to its source of mechanical energy and is most often seated in a convenient horizontal position. Motors, on the other hand, are used in so many different applications that the number of conceivable positions for mounting them is unlimited.

left-hand rule for generators and right-hand rule for motors

The left-hand rule for generators is compared with the right-hand rule for motors to show the relationships between generators and motors. For both rules, the forefinger gives the direction of the field, the thumb gives the direction of motion of the conductor, and the middle finger gives the direction of current flow.

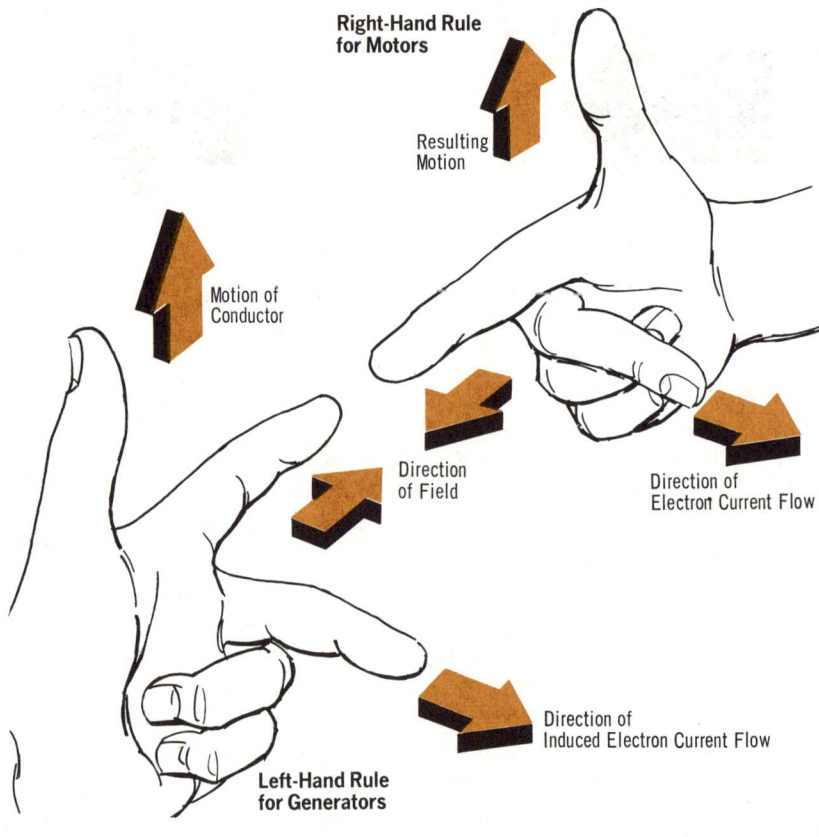

The interesting conclusion to be drawn from studying the rules is that if any two of the quantities are the same in a generator and motor, the third will be the opposite. In the illustration, motion and current flow are in the same direction in both; therefore, field directions in the two are opposite.

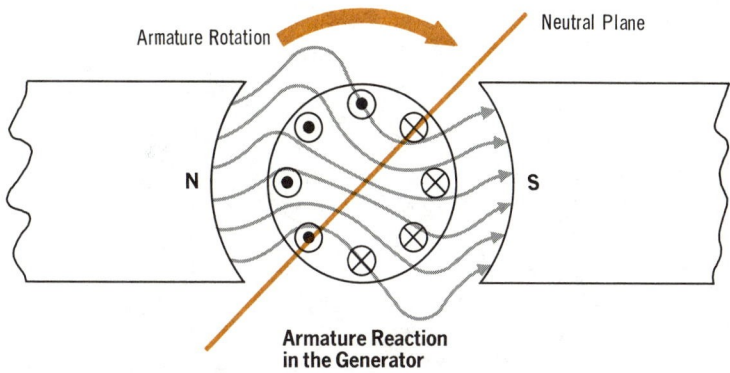

**Armature Reaction
in the Generator**

Armature reaction in a generator is opposite to
that of a motor. The neutral plane is shifted forward in the direction of rotation

armature reaction

Although our comparison of motors and generators indicated that
their operation is convertible, one does not often hear of using a motor
as a generator and vice versa. This is because armature reaction in
motors shifts the neutral plane *backwards*, opposing the direction of
rotation, while in *generators* the neutral plane is shifted *forward* in the
direction of rotation. As a result, trying to use one as the other will
result in inefficient operation accompanied by much sparking at the
commutator. In special cases, where one must be converted to the other,
this can be done by either shifting the brush axis and/or rewiring any
interpoles to compensate for the new direction of shift of the neutral
plane.

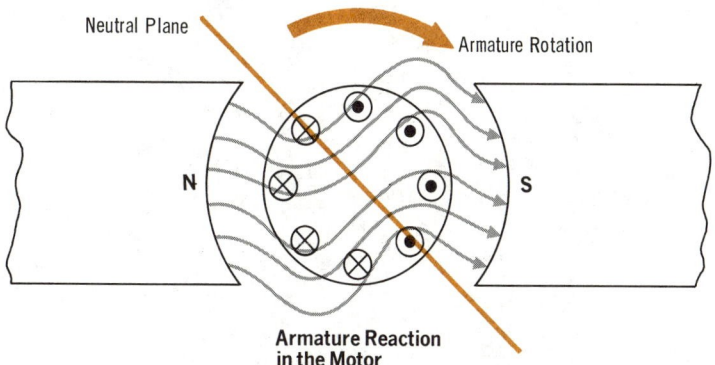

**Armature Reaction
in the Motor**

Armature reaction in a motor shifts the neutral
plane backwards, opposing the direction of
rotation

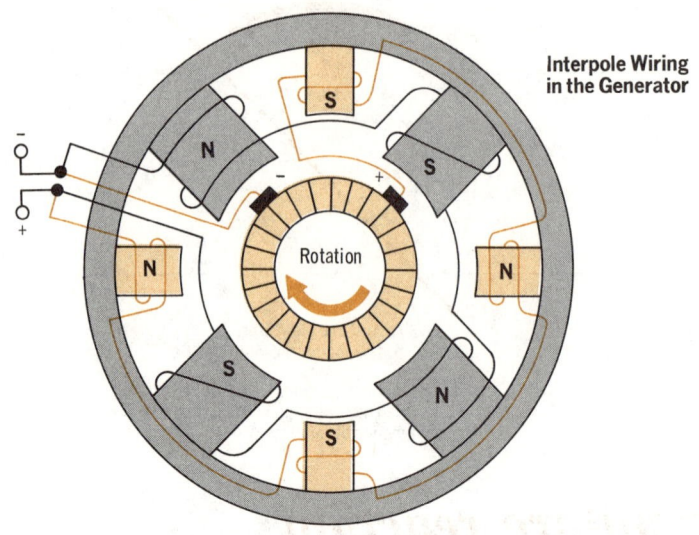

**Interpole Wiring
in the Generator**

interpole wiring

The polarities of interpoles in the generator are the reverse of the polarities of interpoles in motors. Notice that, in each case, the interpoles are wired in series with the armature so as to be able to respond to changes in armature reaction at various speeds.

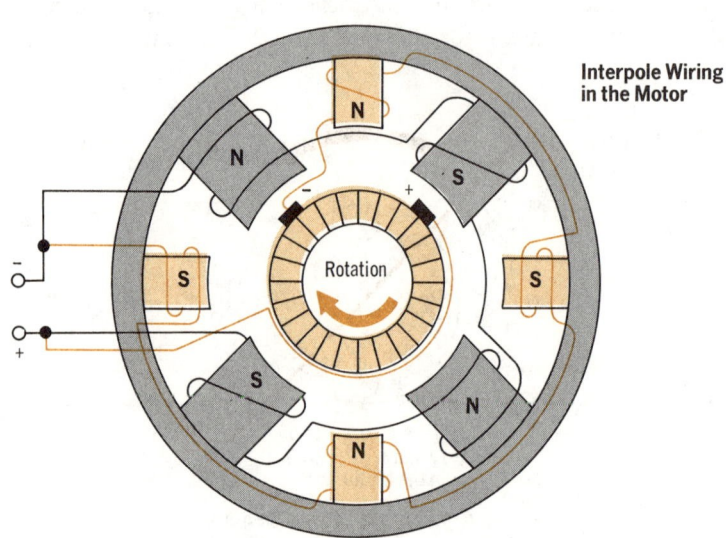

**Interpole Wiring
in the Motor**

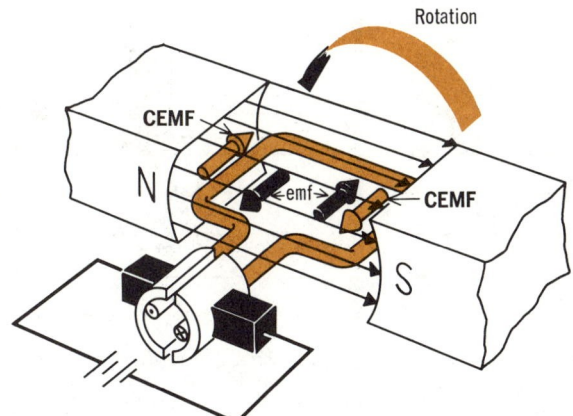

Rotation

When a motor is used as a generator, the left-hand rule for generators states that the current flow in the armature is the reverse of current flow in the motor for the same direction of rotation. This means that the emf induced in the armature of an operating motor by generator action opposes the emf applied to the motor armature. For this reason, it is called a counterelectromotive force, or cemf

counterelectromotive force

Generator action is always taking place in an *operating motor*. As the armature of the motor turns, its conducting loops cut the magnetic flux lines of the field. These are the conditions for electromagnetic induction. Therefore, an emf is always being induced into the turning motor armature during normal motor operation.

To understand the effects of this induced emf we will return to the single-loop elementary d-c motor. Current to start the armature turning flows in the direction determined by the applied emf. Immediately after rotation starts, the moving conductor cuts flux lines. Applying the left-hand rule for generators, the emf induced produces a current in the opposite direction. Thus, the emf induced in a motor armature due to its rotation in the field *opposes* the applied line voltage. This is consistent with a statement derived from Lenz's Law, which you learned about in Volume 3: The current produced by an induced emf will be in a direction which will oppose the motion of the conductor producing the induced emf.

The emf induced as a result of normal motor operation is called a *counterelectromotive force* or cemf. The actual emf acting in an armature circuit always equals the applied voltage less the cemf.

$$\text{emf}_{\text{actual}} = \text{emf}_{\text{applied}} - \text{cemf}$$

To develop armature current flow for operating a motor, the voltage applied to the armature circuit must be greater than the cemf that will be developed after the motor starts turning. The amount of cemf produced in a motor is always proportional to the field flux strength and the speed of rotation of the armature (the amount of flux and the rate at which flux lines are cut).

(A)

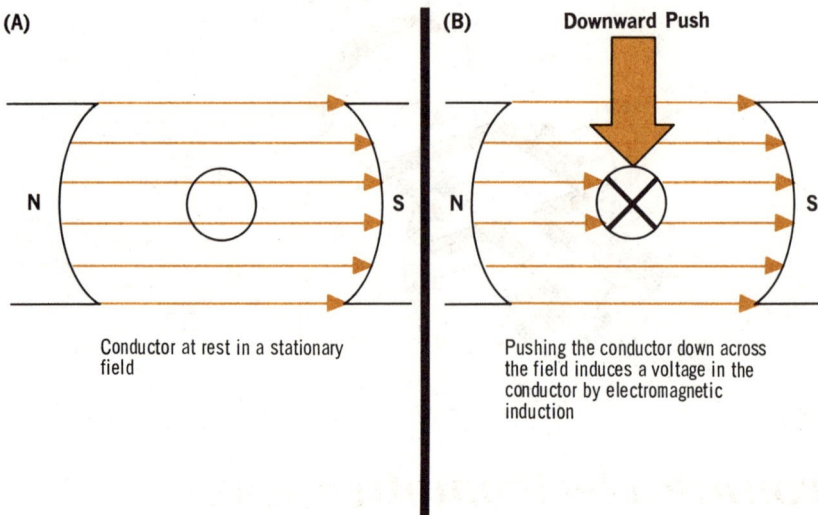

Conductor at rest in a stationary field

(B) **Downward Push**

Pushing the conductor down across the field induces a voltage in the conductor by electromagnetic induction

review of lenz's law

Whenever you move a conductor across a magnetic field, the current induced in the conductor sets up a field about the conductor that tends to move the conductor against the initial direction of motion. Stated another way, Lenz's Law tells us that motor action is produced as a result of the normal operation of a generator and vice versa.

(C)

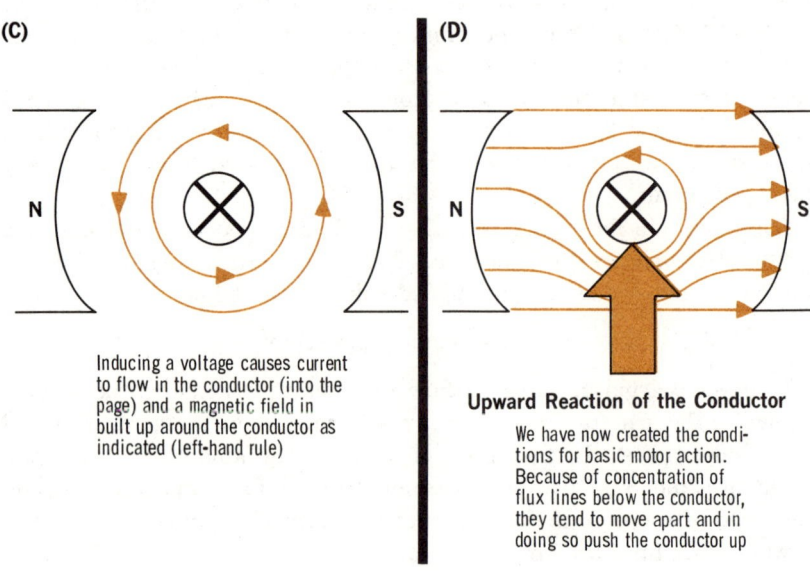

Inducing a voltage causes current to flow in the conductor (into the page) and a magnetic field in built up around the conductor as indicated (left-hand rule)

(D)

Upward Reaction of the Conductor

We have now created the conditions for basic motor action. Because of concentration of flux lines below the conductor, they tend to move apart and in doing so push the conductor up

cemf and counter torque

The generator action in the motor which produces *counter electromotive force,* or cemf, is comparable to the motor action in the generator which produces *counter torque.*

The initial driving motion imparted to the armature in *the generator* pushes a conductor through the magnetic field of the pole pieces and thereby induces a voltage in the conductor. This voltage, however, sets up a magnetic field around the conductor, and the interaction between this field and the field of the pole pieces results in a force opposing the initial driving motion. This force is known as *counter torque.*

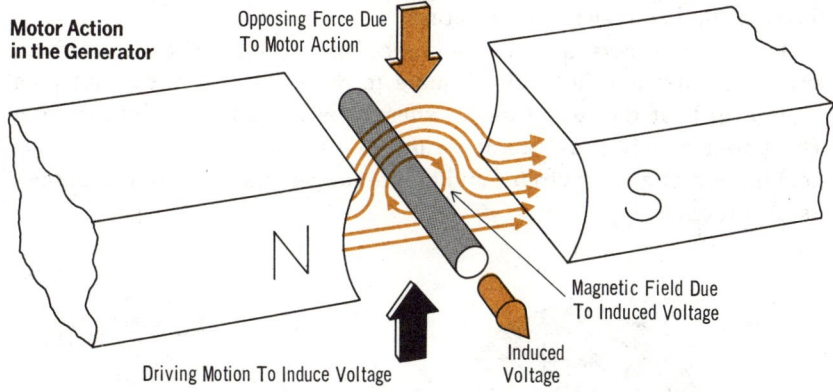

Motor Action in the Generator

Opposing Force Due To Motor Action

Magnetic Field Due To Induced Voltage

Driving Motion To Induce Voltage

Induced Voltage

In the *motor,* the initial emf imparted to the conductor sets up a magnetic field around it which interacts with the field of the pole pieces to cause motion of the conductor. This motion through the field induces a voltage, the cemf, in the conductor opposing the original emf.

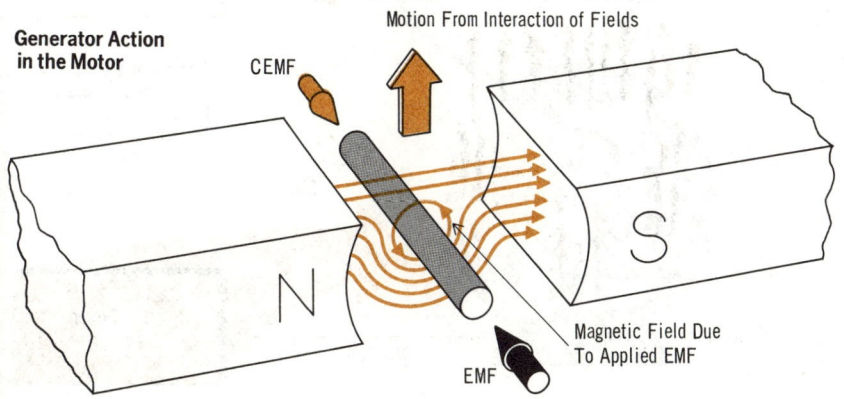

Generator Action in the Motor

CEMF

Motion From Interaction of Fields

Magnetic Field Due To Applied EMF

EMF

role of cemf

Since the armature windings of a motor are made of copper, the resistance of the armature is very small. In fact, it is frequently less than one ohm. Applying almost any voltage to such a 1-ohm armature at rest, causes a huge short-circuit current to flow through it. If the armature were jammed so that it was impossible for it to turn, it would heat up and probably even burn out from all that current. However, remember that the armature is free to rotate. With the application of current, the armature starts turning, cuts lines of force, and an induced cemf is built up. Since the cemf opposes the line voltage, it acts to cut down the short-circuit current which the line voltage could potentially cause through the low-resistance armature.

Hence, cemf acts as an automatic current limiter which reduces armature current to a level adequate to drive the motor but not great enough to heat the armature to where it is in danger of burning out. The power supply thus operates into the armature as a true electrical load instead of a short circuit and the motor gets useful power to support its operation.

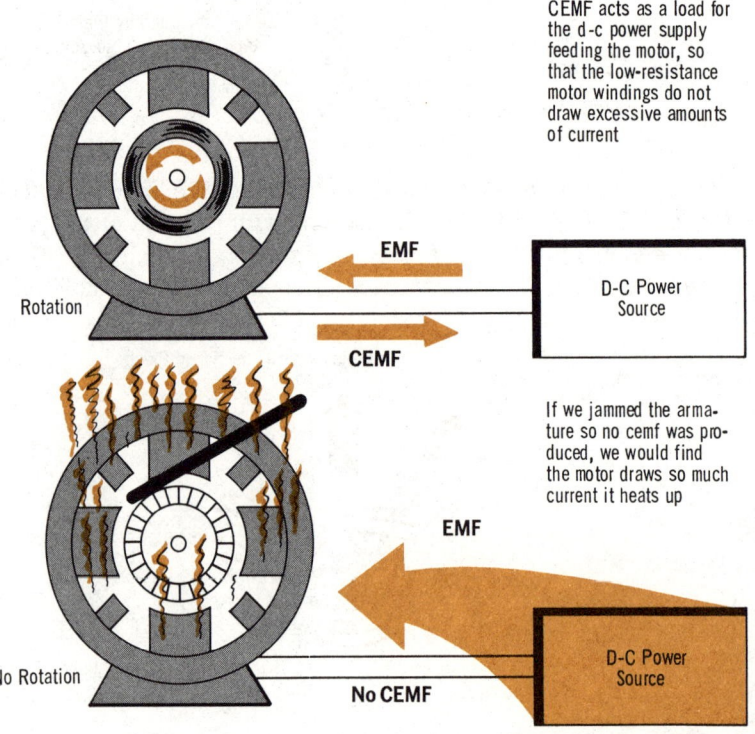

CEMF acts as a load for the d-c power supply feeding the motor, so that the low-resistance motor windings do not draw excessive amounts of current

EMF

D-C Power Source

Rotation

CEMF

If we jammed the armature so no cemf was produced, we would find the motor draws so much current it heats up

EMF

No Rotation

No CEMF

D-C Power Source

cemf in motor speed regulation

You learned that the *speed* of operation of the electric motor is determined by the *torque* produced by the motor and the *load* the motor has to drive. Torque is the force developed when the magnetic flux lines of the field and armature interact, and is dependent on the strength of the field and the strength of the armature flux. Assuming no external load on a motor and a constant field, by increasing the armature current, the armature flux strength increases, increased torque is developed, and the speed of the motor increases.

You also learned that when the speed of a motor is increased, its field flux lines are being cut at an increased rate by the armature. As a result, a greater cemf is produced in the armature. This means that whenever armature current is increased to speed a motor up, the corresponding increase in cemf acts to cut down on the original increase in armature current. As a result, the final motor speed developed in this case is usually not much greater than the initial speed. This type of action on the part of cemf gives electric motors a built-in system for regulating their own speed.

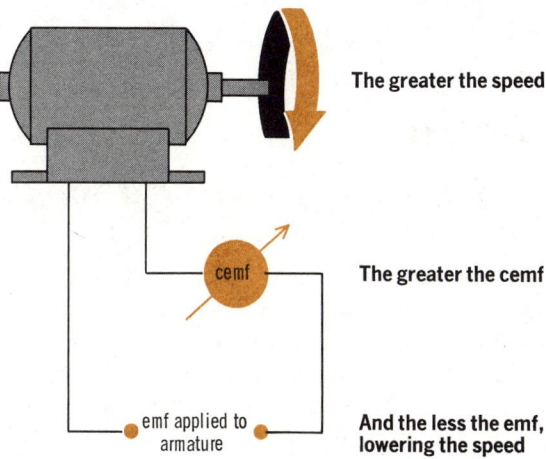

When the motor turns too fast, more cemf is generated to lower the emf that reaches the motor. This lowers the motor speed to normal. The reverse is also true, so that cemf tends to regulate motor speed

The greater the speed

cemf

The greater the cemf

emf applied to armature

And the less the emf, lowering the speed

The d-c motor, therefore, generally tends to run at a natural speed where the sum of the voltage drop across the armature as a result of its built-in resistance plus the cemf developed just equals the applied line voltage. Once established at this speed, the d-c motor will tend to regulate its own operation at this speed. It does this by adjusting the value of cemf developed in order to change armature current to where the final torque developed is just sufficient to maintain the natural speed of operation.

When a load is applied to the motor, the motor tends to slow down. But this reduces cemf, which increases the emf supplied to the motor to bring it back to the right speed

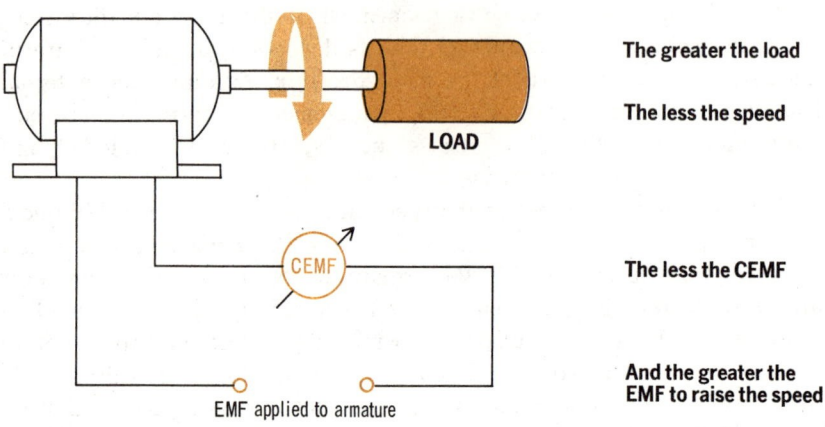

LOAD

EMF applied to armature

The greater the load

The less the speed

The less the CEMF

And the greater the EMF to raise the speed

The opposite is true when a load is removed from the motor. The motor tends to speed up; but cemf goes up to reduce the emf supplied to the motor to slow it back to the right speed

the load
in motor speed regulation

If we begin with a motor operating at its natural speed without a load, and suddenly apply a load to it, the immediate effect of the load is always to reduce the speed of the motor. Slowing down the motor reduces the rate at which field flux lines are being cut by the armature and as a result the cemf drops. This drop in cemf enables the armature to draw additional current from the power supply. With the additional armature current there is an increase in torque and a corresponding increase in motor speed. Generally, the motor speed builds up to meet the demands of the load. Of course, as the speed builds up, so does the cemf, and once again there is the tendency for the motor to regulate its own speed. Eventually the motor stabilizes itself at an operating speed in line with the requirements of the load placed on it.

These relationships between the load placed on a motor and the corresponding speed of operation of the motor make up an important factor in motor design and application called the *load-speed characteristic.*

field strength in motor speed regulation

You learned that, in general, increasing armature current tends to increase armature flux, which produces greater torque and speeds the motor up. However, increasing the strength of the field has the exact opposite effect. When field strength is increased, more flux lines are put in the path of the turning armature, generator action is greater, and the cemf induced is increased. Assuming there is a constant load on the motor, the armature current will decrease and the motor will slow down. At this point some speed regulation takes place in that cemf is also decreased, allowing some additional armature current to flow. This new value of armature current is always slightly lower than it was initially. Therefore, the speed of the motor is slightly less than it was before the field strength was increased.

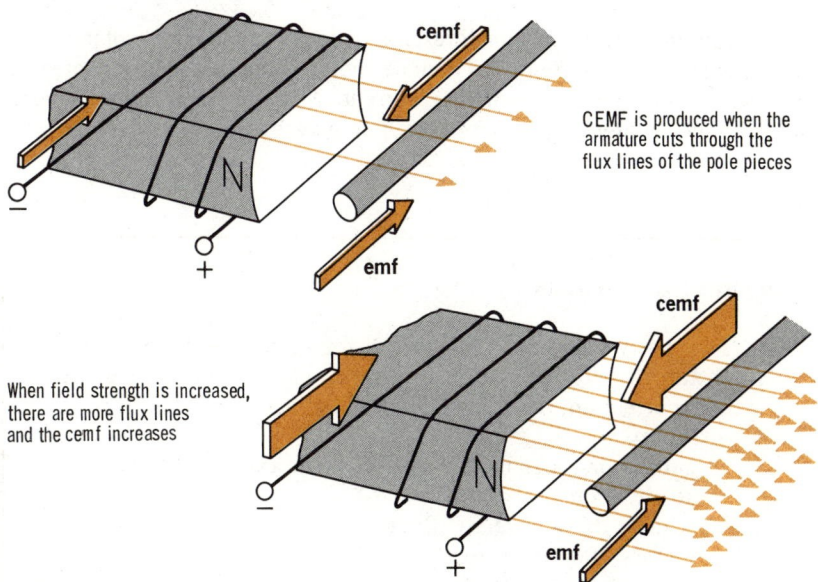

cemf

CEMF is produced when the armature cuts through the flux lines of the pole pieces

emf

When field strength is increased, there are more flux lines and the cemf increases

cemf

emf

By the same reasoning, reducing the field strength of a motor, decreases the value of induced cemf, causing a greater current flow in the armature, and thus increasing motor speed. Again, assuming a constant load, as the motor speed increases, the induced cemf builds up and the armature current is brought to a value just high enough to handle the load. However, since the load is now being driven at a higher speed, it requires more power; and the value of armature current is higher than it was initially.

summary

☐ Motors and generators are similar mechanically and electrically. However, since they are devices that operate on opposite principles, certain of their characteristics are opposite, which prevent their being used interchangeably. ☐ Although the right-hand rule is used for motors, the left-hand rule is used for generators. ☐ Armature reaction in the motor shifts the neutral plane backwards, whereas in the generator, it shifts the plane forward. Therefore, brush positioning and interpole wiring in the motor and generator are opposite.

☐ Because a motor has windings passing through a magnetic field, it always acts somewhat like a generator, and so produces a cemf. This cemf opposes the applied voltage to reduce armature current. ☐ Since cemf is proportional to the armature speed, it tends to regulate the motor speed. Whenever the armature speeds up, a greater cemf is produced which lowers armature current to reduce the armature speed. ☐ When a load is suddenly applied to the d-c motor, its armature will slow down, but then cemf will go down, causing armature current to go up and increase the motor speed to normal.

☐ Although an increase in armature current tends to increase motor speed, and increase in field current will reduce motor speed because it increases cemf, which then reduces armature current. ☐ Cemf in a motor is comparable to the counter torque that is set up in a generator.

review questions

1. For the two different rules used for motors and generators, if any two corresponding fingers on each hand point in the same direction, the third ones will point in the _____ direction.
2. In generators, armature reaction shifts the neutral plane in what direction?
3. How does the answer to Question 2 affect interpole wiring in motors and generators?
4. Why does a motor produce cemf?
5. Cemf allows the armature windings to have _____ resistance.
6. Why does an increase in applied voltage to the armature have little effect on motor speed?
7. How does cemf counteract load changes?
8. Whose law describes cemf? State the law.
9. What can you do to change the speed of a d-c motor?
10. What does the load-speed characteristic of a motor describe?

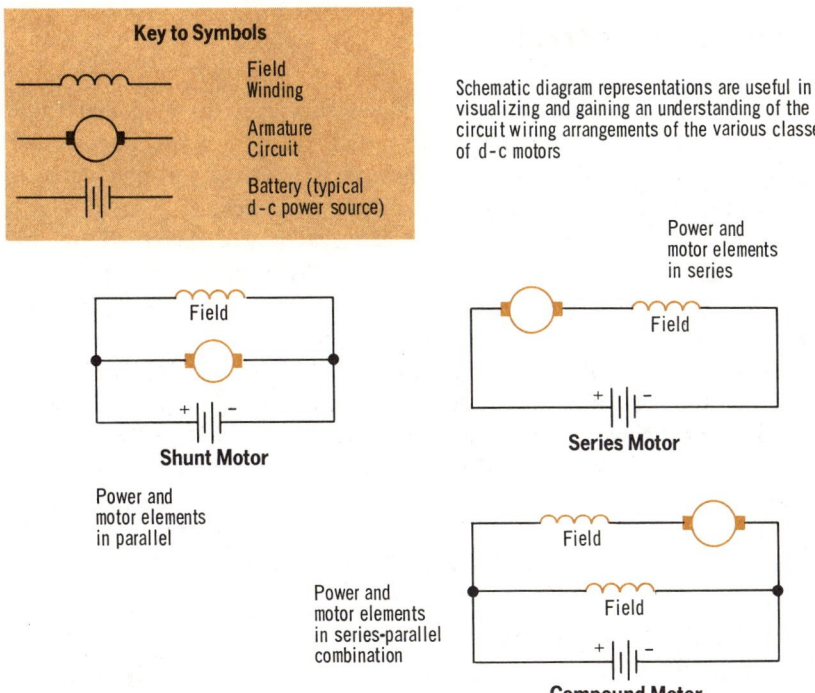

Key to Symbols

Field Winding

Armature Circuit

Battery (typical d-c power source)

Schematic diagram representations are useful in visualizing and gaining an understanding of the circuit wiring arrangements of the various classes of d-c motors

Power and motor elements in series

Field

Series Motor

Field

Shunt Motor

Power and motor elements in parallel

Power and motor elements in series-parallel combination

Field

Field

Compound Motor

classification of d-c motors

D-c motors have come to be classified electrically by the way in which their field windings are connected to the source of electrical power that drives them. The descriptive names, *shunt, series,* and *compound* identify the three principal d-c motor types. In the shunt-type motor, the field winding and armature winding are connected in parallel across the input circuit. In the series-type motor the field winding and armature winding are connected in series to the input circuit.

As the name implies, the compound-type motor indicates that series and shunt connections of the field winding and the armature are combined in a single motor.

D-c motors have come to be classified mechanically by the type of housing in which they come, and by their so-called load-speed characteristic.

Motor housings come in the open, semi-enclosed, drip-proof, waterproof, submersible, and explosion-proof types.

Load-speed characteristics of motors identify them as constant-speed, multispeed, adjustable-speed, varying-speed and adjustable varying-speed types.

motor ratings

Motors are rated in terms of the load they can drive, called their *power output;* by the electrical power they take from the line, called their *power input;* and by how well they perform in converting electrical energy into mechanical energy, called their *efficiency.*

Power output is a measure of mechanical power delivered by the motor at full load and is given in *horsepower* ratings. Horsepower is a way of telling how much work a motor can do in a given period of time as compared to a horse. One horsepower is the same as doing 33,000 foot-pounds of work per minute. The amount of foot-pounds of work a motor does is the product of its torque times its speed of operation. The horsepower of any motor can be computed if its torque and speed are known. Torque can be measured directly by a device called a *prony brake,* and speed can be found by using a tachometer or revolutions-per-minute-counter and stopwatch. The rated horsepower of a motor is usually stamped on the nameplate of the motor.

The prony brake test is used to measure the torque of a motor

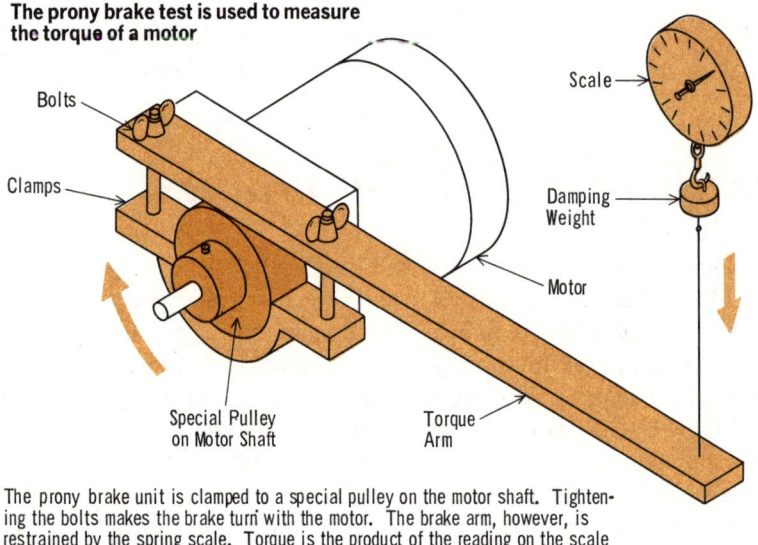

The prony brake unit is clamped to a special pulley on the motor shaft. Tightening the bolts makes the brake turn with the motor. The brake arm, however, is restrained by the spring scale. Torque is the product of the reading on the scale (force in lbs.) times the length of the torque arm in feet

The electrical power taken by a d-c motor from the line supply is a simple product of the current and voltage and is expressed in watts.

$$\text{Power} = \text{Voltage} \times \text{Current}$$

$$P_{watts} = E_{volts} \times I_{amps}$$

motor ratings (cont.)

Unfortunately, not all the power put into a motor comes out as useful mechanical power for driving the load. Some of the power is used up as heat in the field winding; another part is used up as heat in the armature; and still another part is used up in overcoming the mechanical loading effects of friction, air resistance, etc. In each case, a quantity of power is used up that is not transferred to the load as mechanical energy. This wasted power is called *loss*. The power *output* of a motor is always *equal* to its power *input minus* all the power *losses*.

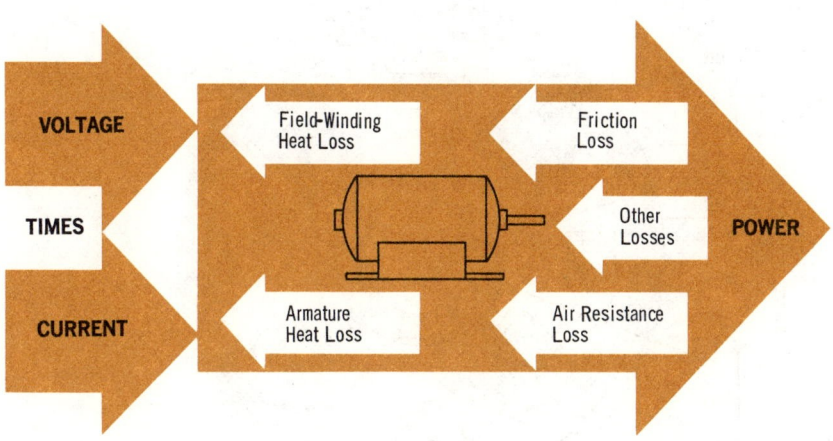

Motor efficiency is a measurement of how well power input is converted to power output. If losses are low, efficiency is said to be high. Efficiency is computed by dividing the power input into the power output and multiplying by 100 to yield a percent figure. Input in *watts* can be divided into output in *horsepower* because there is a definite relationship between the two as follows:

$$1 \text{ horsepower} = 746 \text{ watts}$$

Rating of information on the motor nameplate usually provides enough information to enable you to have or calculate all you need to know about the motor.

Often, motors are rated with the input given in horsepower. With such motors, then, a 1/4-hp motor will have an output something less than 1/4 hp because of the losses.

shunt motors

The shunt motor gets its name from the fact that its field winding is connected across the power supply line in parallel with the armature winding. This means that there is an independent path for current flow through each winding. In a shunt motor, the field current can be held constant and the armature circuit alone can be used to control the motor. One of the main features of such a motor is the fact that it can hold a constant speed while serving a changing load, and the load can be entirely removed without danger to the motor.

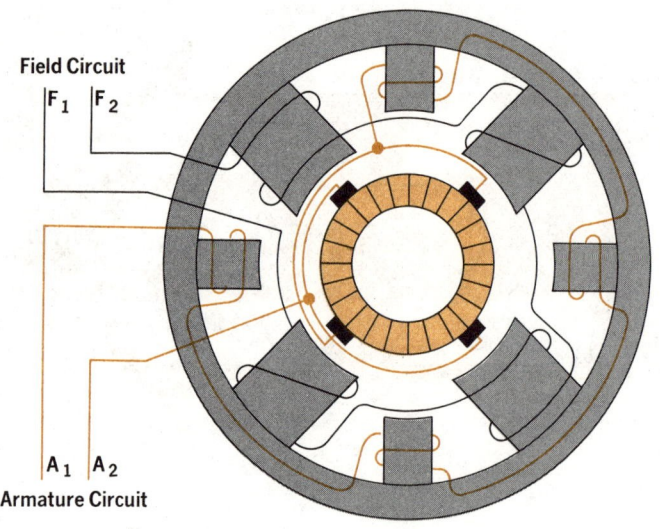

Field Circuit
F_1 F_2

A_1 A_2
Armature Circuit

Tracing the wiring will indicate that the field circuit and the armature circuit are independent. Therefore, this is a shunt motor. You will notice that the interpoles are in series with the armature circuit to enable them to respond to armature current changes

Suppose the load on a shunt motor increased. The immediate effect of this is to reduce armature speed. Slowing down the armature reduces the cemf, producing an increase in the amount of armature current flowing This has the effect of increasing torque to bring the armature speed right back up. The reverse happens when the load on a shunt motor is decreased.

speed control

The shunt motor can be operated at various speeds by using a *rheostat* control either in series with the field winding, the armature winding, or in both places. The use of a rheostat in series with the field winding is the most common method of varying the speed of a shunt motor. It is preferable to the use of an armature rheostat because the field current is less than the armature current, hence the power loss across the rheostat is much less in the field circuit. As a result more current is made available for actual motor operation.

By inserting more resistance in series with the field, less field current flows, field strength drops, and the motor speeds up. This happens because when fewer flux lines are cut by the rotating armature, there is a tendency for cemf to drop.

This permits more current to flow in the armature, resulting in increased torque out of proportion to the amount actually required. As a result the motor quickly speeds up, increasing the cemf to where the current is reduced to a value that produces the correct amount of torque.

By inserting less resistance in series with the field, the field strength increases and the motor slows down.

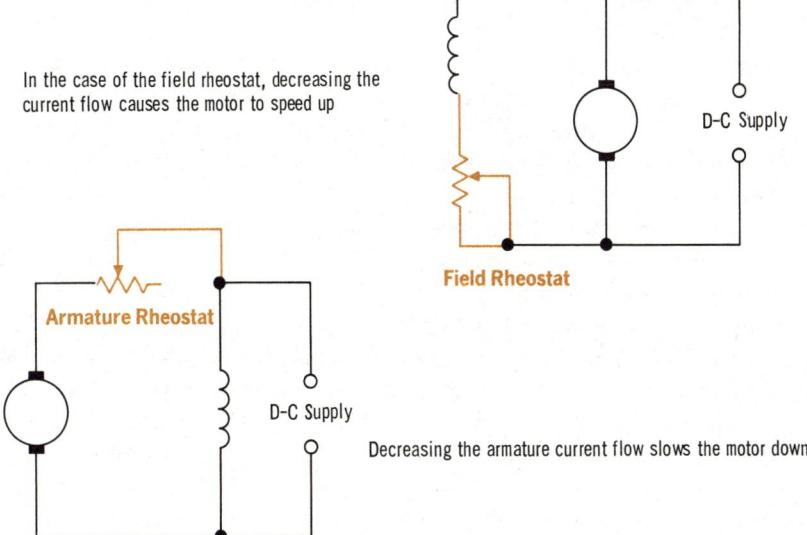

With any part of the rheostat in the circuit, a portion of d–c supply voltage is dropped across it and less current flows through the motor element in series with it

In the case of the field rheostat, decreasing the current flow causes the motor to speed up

D-C Supply

Field Rheostat

Armature Rheostat

D-C Supply

Decreasing the armature current flow slows the motor down

**Residual Magnetism
Maintains a Weak Field**

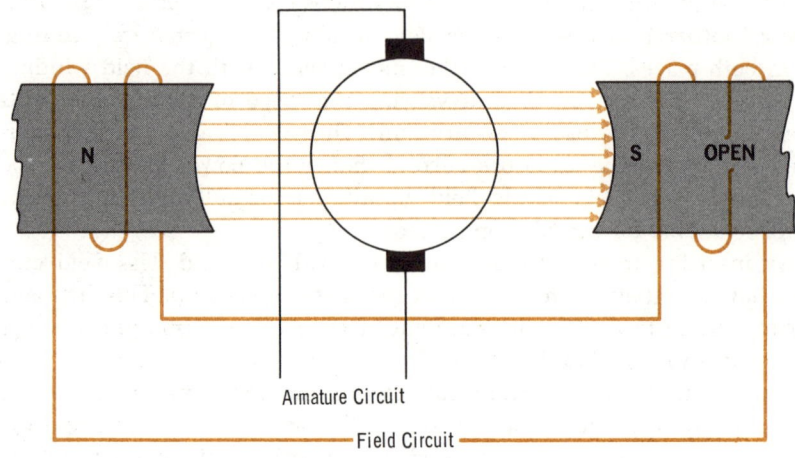

Armature Circuit

Field Circuit

With a field winding open, the shunt motor tends to run away
rather than stop because residual magnetism of the pole pieces
maintains a weak field. The armature circuit, of course, can
still draw current because the field winding and the armature are
in parallel across the power supply

runaway characteristic

Remember that one of the outstanding characteristics of the shunt motor is that it speeds up as current through the field winding is decreased. We use this feature to control the speed of the shunt motor with a series field rheostat. The increase in speed with the decrease in field strength is due to a lowering of cemf, and a corresponding increase in armature current. As it happens, this increase in armature current produces a torque increase that is way out of proportion to the reduction of field strength.

Thus, if the field strength is suddenly made extremely weak, as when a field winding opens and only *residual magnetism* sustains the field, the shunt motor will tend to operate so fast we say it *runs away*. Runaway operation may completely destroy the motor, which, in most cases, has not been built to withstand the physical stress of such speed In a like manner with the field circuit of the shunt motor completely opened, the motor might also possibly burn up because of the high armature current drawn.

series motors

The series motor gets its name from the fact that its field winding is connected in series with the armature. This means that a common current flows through both windings. Whatever happens to armature current as a result of driving a load is automatically felt in the field winding.

Let's examine what effect this has on motor operation. Assume the load on a series motor is increased. We know from the shunt motor that this slows the motor down, thus dropping the cemf. The armature then draws additional current to increase the torque, as a first step in building up speed and cemf. However, in the series motor, this additional armature current is also present in the series field winding. Thus the field flux increases and restores the cemf. This action then prevents the motor speed from building up as was the case in the shunt motor.

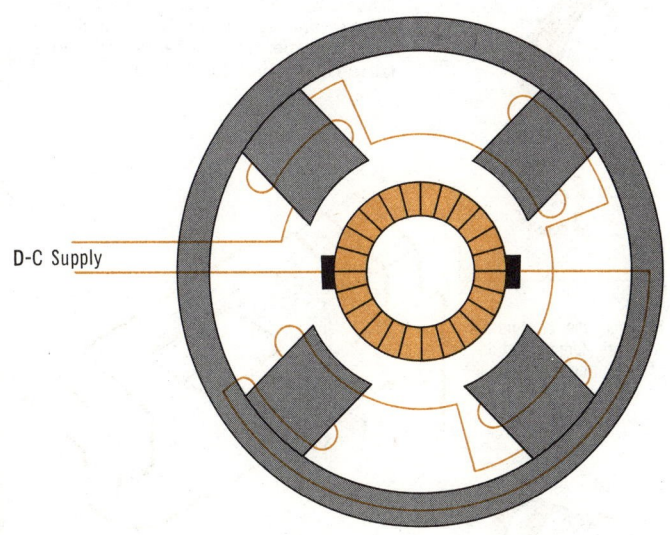

D-C Supply

Tracing the wiring diagram will indicate that the field windings and armature circuit are in series; therefore, this is a series motor

The series motor then, is not a constant-speed motor. In the shunt motor when torque increased, speed increased, and vice versa. In the series motor the reverse is true—that is, torque and speed are inversely proportional. This means that when torque is high, speed is low; and when torque is low, speed is high.

no-load characteristic

The load-speed property of the series motor just discussed gives the series motor a tendency to "run away" if either started or allowed to run without a load. This is due to the fact that without a load, a low torque is required to turn the armature. Speed of the series motor thus characteristically increases in an attempt to build up cemf to reduce armature current in order to keep the torque low.

Unfortunately, as the motor runs faster to reduce the armature current, the field flux also drops and so does the cemf. So the motor runs still faster to build up still more cemf. The speed continues to increase until the sheer physical force of rotation wrecks the motor through a combination of heating through friction and centrifugal force.

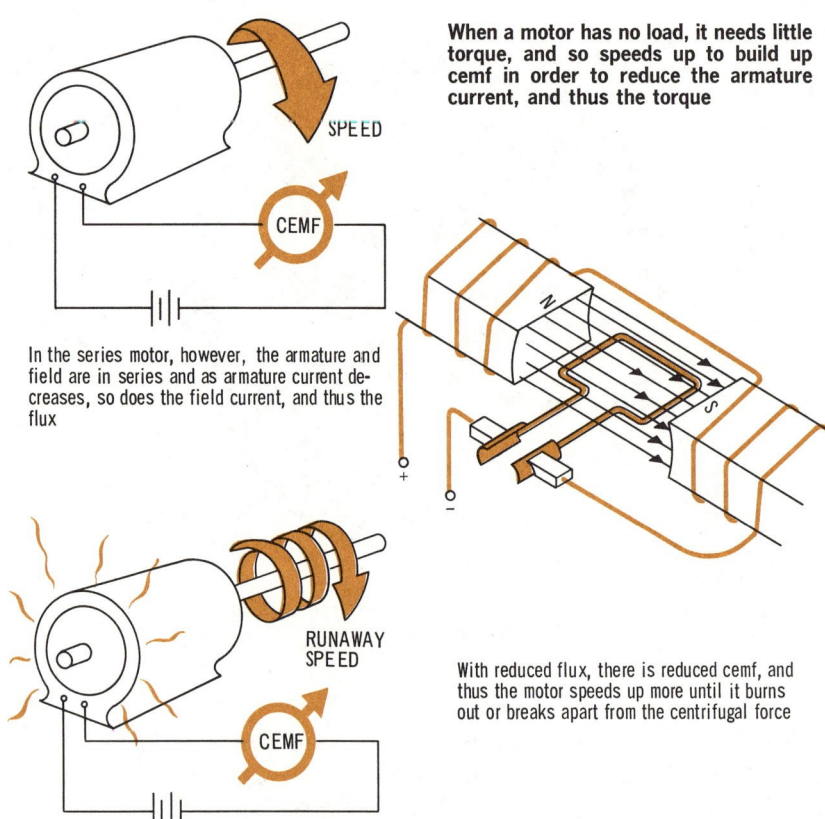

When a motor has no load, it needs little torque, and so speeds up to build up cemf in order to reduce the armature current, and thus the torque

In the series motor, however, the armature and field are in series and as armature current decreases, so does the field current, and thus the flux

With reduced flux, there is reduced cemf, and thus the motor speeds up more until it burns out or breaks apart from the centrifugal force

torque characteristic

The series motor can provide very high starting torque and can thus meet the high torque demands of suddenly imposed heavy overloads. This is because the torque of a series motor varies approximately as the square of the current through it. If the current in the armature were suddenly tripled as a result of an overload, the current through the field and hence flux strength would be automatically tripled as well. Since torque is a product of armature current and flux strength, the torque that results will be 9 times the original value.

Series motors are useful where sudden changing loads are put on the motor. In industry, series motors are used to do heavy work on cranes, hoists, subway trains, electric locomotives, and similar applications

Because of this characteristic, series motors are used wherever a high starting torque is needed against heavy loads which are to remain coupled to it during an entire operation. Series motors are especially effective where a sudden very heavy overload may be encountered while operating. The series motor is a poor choice in application where a relatively constant speed is required from no-load to full-load conditions.

summary

☐ D-c motors can be classified in many ways: by their field winding arrangements; by their housings; or by their load-speed characteristics. ☐ Motors can be rated by power input, power output, and power efficiency. ☐ Power can be described electrically in watts or mechanically in horsepower.

☐ In shunt motors, armature current is independent of the parallel field winding; so cemf regulates motor speed for varying loads. ☐ A rheostat can be used in series with the field winding to control motor speed. Increased resistance will reduce field current and increase speed. Zero field current will cause a runaway motor.

☐ In series motors, the same current goes through the field and armature windings. Cemf, then, cannot regulate motor speed for varying loads. If a load reduces armature speed, cemf goes down, and armature current goes up to increase field current. The increased field current increases cemf before the motor speed can increase to build up the cemf. ☐ The motor speed stays down for heavy loads, and goes up for light loads. ☐ Without a load, the series motor will run away. ☐ The series motor provides a high starting torque, but has a varying load-speed characteristic. ☐ Compound motors combine the advantage of series and shunt motors.

review questions

1. How do the field windings in shunt and series motors differ?
2. Which motor would you use if you wanted constant speed with varying loads: series or shunt?
3. Which motor would you use for high starting torque: series or shunt?
4. How many watts produce 1 horsepower?
5. Which is stronger: a motor that has a 186.5-watt input or one that has a 1/4-hp output?
6. If you had a 1-hp output motor rated at 75% efficiency, how much current would it draw from a 110-volt line?
7. How would you make a shunt motor into an adjustable-speed motor?
8. What happens to a series motor when its field winding opens? A shunt motor?
9. What happens to a shunt motor when the load is removed? A series motor?
10. What is a *compound motor*?

reversing practical d-c motors

To reverse the direction of rotation of a d-c motor, you have to reverse *either* the field *or* the armature current, but *not* both at the same time. In the illustration, this principle is reviewed. In (A), we find conditions that support counterclockwise rotation. In (B), we have reversed the field current only, and the motor now turns clockwise.

In (C), we have reversed the armature current instead, and the motor runs in a clockwise direction. Finally, in (D), we have reversed both the field and armature current, and so the motor continues to run counterclockwise.

(A)   **(B)**

Starting with the initial condition shown in (A), you must reverse either the field polarity, as shown in (B), or the armature current flow, as shown in (C), to go from counterclockwise to clockwise

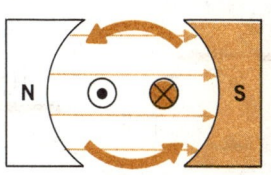 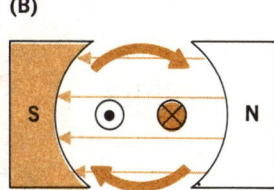

(C)   **(D)**

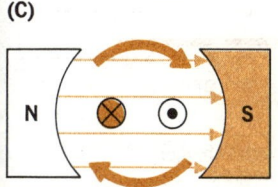 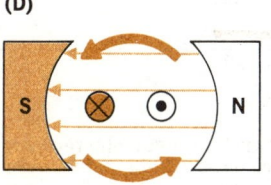

If the initial field polarity and armature current of (A) are both reversed at the same time, as in (D), the motor continues to rotate in the initial counterclockwise direction

Understanding this principle is important for understanding why reversal cannot take place in the practical d-c motor by merely reversing the polarity of the power supply. In most practical motors the field is produced by an electromagnet, and both field winding and armature are fed from the same two d-c power terminals. Hence, reversing the polarity of the power supply changes current direction in *both* the armature and field *simultaneously*. Thus, the motor continues to turn in its original direction. As you will learn, to reverse the rotation of a practical motor, a reversing switch has to be used that changes the direction of current flow through either the armature or field circuits, but not both.

reversing a series d-c motor

The most important element for reversing direction in a practical series d-c motor is a switch. The type of switch we will use is called a double-pole, double-throw switch because it makes it possible to switch two wires at a time into one of the two circuits.

The schematic shows what is happening in the circuit. First, to prove we have a series motor, trace the electron flow in (A) from the negative side of the power supply to the positive side. Notice that you pass through both field winding and armature, in turn, proving they are in series. You also pass through the switch; its two poles are on contacts 1 and 2. With current flowing in the direction shown in (A), the armature rotates clockwise, as indicated.

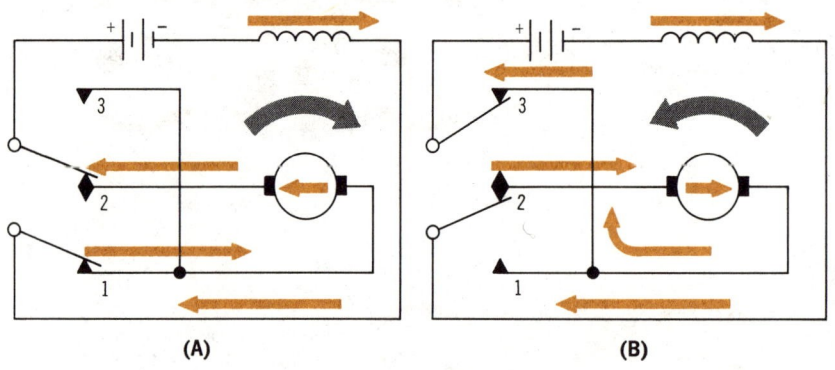

(A)   (B)

Throwing the switch in the series motor reversing circuit changes the current direction through the armature while it keeps the same current direction through the field. As a result, the direction of rotation of the motor is changed whenever the switch is thrown

Now look at the circuit as shown in (B). First, quickly check that the field and armature are still in series. Now trace in detail the path of electron flow. Notice that current goes through the field winding in the same direction as it did in (A). However, this time the pole of the switch that was on contact 1 before, is now on contact 2; and the pole which was on contact 2, is now on contact 3. As a result, as you trace the electron path further, you find that the current passing through the armature is in a new direction in going to the positive side of the power supply. With the field current flowing in the same direction as in (A), and armature current reversed, the motor rotates in the opposite (counterclockwise) direction, as indicated.

reversing a shunt d-c motor

Apply the experience you've just gained in circuit tracing to the circuits shown. First, quickly trace the current path of both circuits. Begin at the negative side of the battery and follow the arrows through the field winding, then through the armature. Notice that there is an independent parallel path for current through each, making the motor a shunt type.

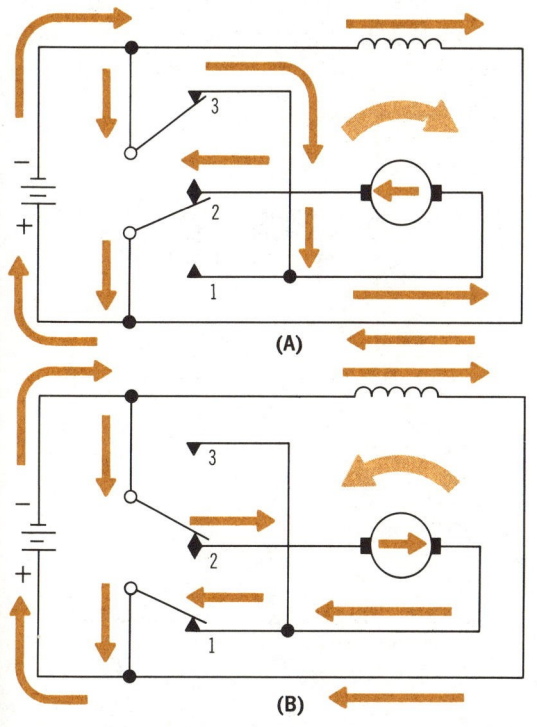

(A)

(B)

Current through the field winding remains the same for motor rotation in either direction. The reversing switch changes the direction of current flow only through the armature in order to reverse the direction of motor rotation

Now retrace the circuit (A) in detail to establish direction first through the field and then through the armature. Notice that the switch is in the armature circuit only. With the switch making contact at 2 and 3, as shown in (A), the motor rotates clockwise.

Now examine the circuit in (B). The switch is now in its alternate position where it makes contact at 1 and 2. The field winding circuit is undisturbed, and current still passes through the field in the same direction as in (A). However, in tracing the armature path, you find that, because of the new switch position, current passes through the armature in a new direction. As a result the motor rotates in the opposite (counterclockwise) direction.

compound motors

Shunt and series motors each have some special features not found in the other. For example, the constant-speed characteristic of the shunt motor is not found in the series motor, and the excellent high-torque characteristic of the series motor is not found in the shunt motor. Clearly, a combination of the plus features of each in a single motor would be desirable. These characteristics can be combined by giving a single motor two field wirings, one in series with the armature, the second shunted across it. We call this type of d-c motor a *compound motor.*

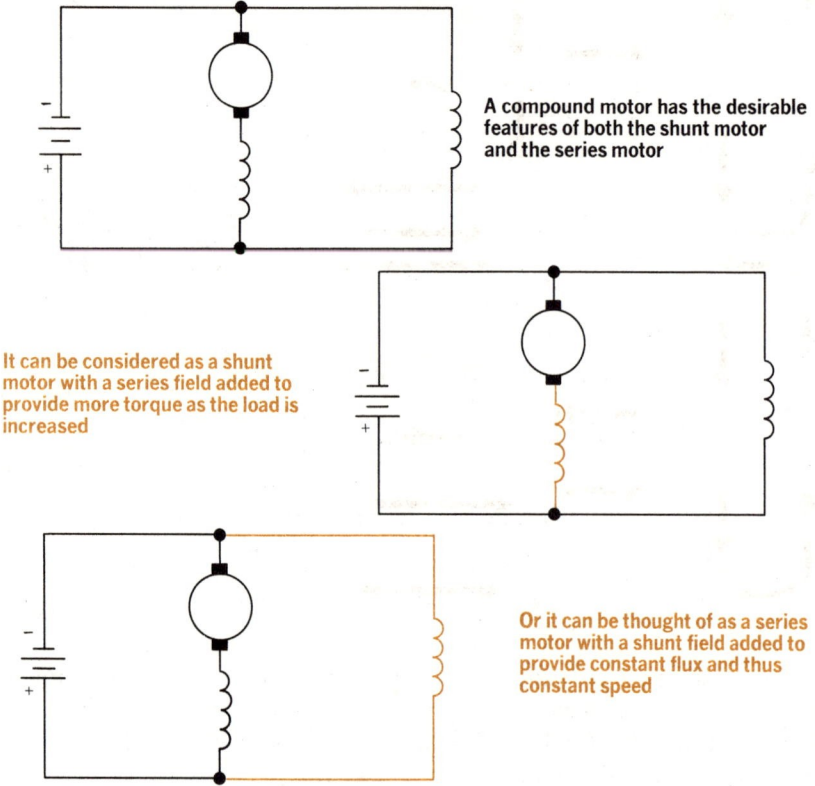

A compound motor has the desirable features of both the shunt motor and the series motor

It can be considered as a shunt motor with a series field added to provide more torque as the load is increased

Or it can be thought of as a series motor with a shunt field added to provide constant flux and thus constant speed

To understand how the compound motor acts, consider a shunt motor with an extra series field. As the load on such a motor increases and the motor slows, the resultant increase in armature current also boosts the strength in the series field winding. Since we get more total interacting flux, torque increases. Hence, we have added some series motor benefits to the shunt motor through compounding.

compound motors (cont.)

Now, consider a series motor to which we have added a shunt field. The ordinary series motor tended to run away at no load, in part because of a constantly decreasing field flux. Now that we have added a constant-flux shunt field, the motor speed tends to limit itself to a reasonable value, as in the case of the regular shunt motor.

By arranging the compounded field windings so that the strength of one will be greater than the other, we can make the compound motor more like either the series or the shunt motor. This control over the relative strength of the two field windings results in the classification of compound motors as *cumulative-compound* or *differential-compound* types. Most compound motors are of the cumulative-compound type.

If we connect the shunt field across both the armature and its series field, we have made a *long-shunt cumulative-compound motor*. If we connect the shunt field across only the armature, we have made a *short-shunt cumulative-compound motor*.

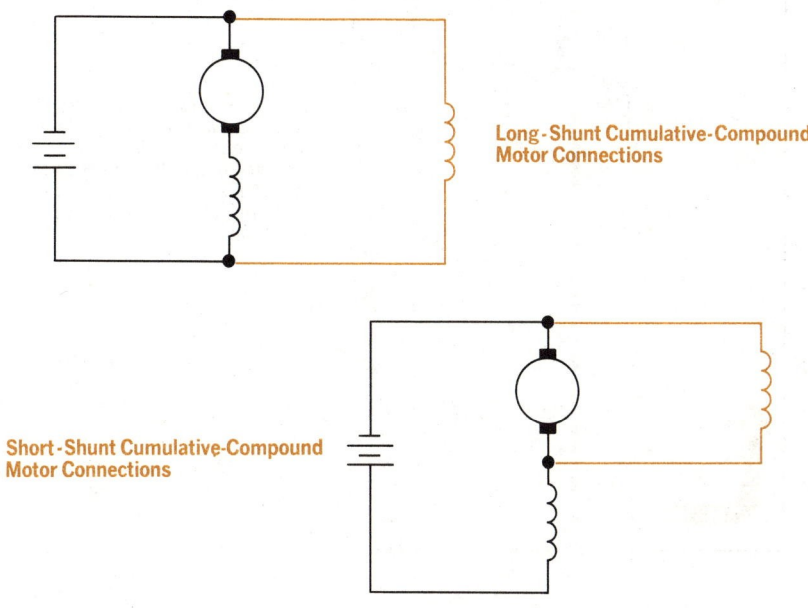

Long-Shunt Cumulative-Compound Motor Connections

Short-Shunt Cumulative-Compound Motor Connections

Most practical compound motors in use today are of the long- and short-shunt cumulative-compound types. Differential-compound motors are very rarely used

cumulative-compound motors

If we make a compound motor with its series field winding and shunt field winding wound in the same direction, *both* windings will tend to aid each other in generating magnetic flux. In this case, we have made a *cumulative-compound motor*. The word *cumulative* tells us that the fluxes combine to form an overall stronger field.

Many of these cumulative-compound types are wound to give preference to the superior torque characteristic of the series motor. Motors of this type are frequently series motors with a few shunt turns to prevent them from running away under no-load conditions. Sometimes such motors are referred to as *series-shunt motors*.

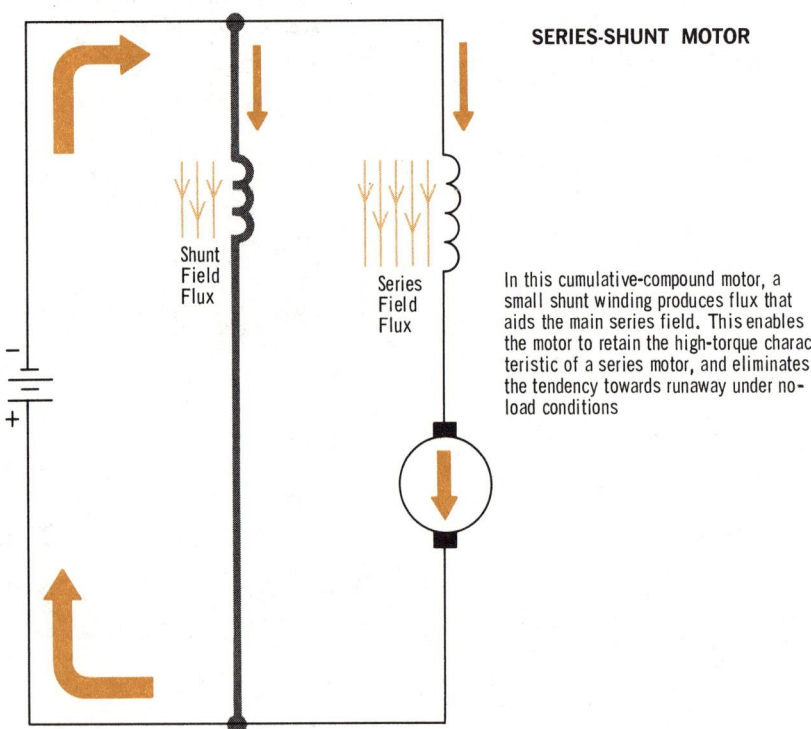

SERIES-SHUNT MOTOR

Shunt
Field
Flux

Series
Field
Flux

In this cumulative-compound motor, a small shunt winding produces flux that aids the main series field. This enables the motor to retain the high-torque characteristic of a series motor, and eliminates the tendency towards runaway under no-load conditions

On the other hand, there are cumulative-compound motors in which the series characteristic of high starting torque is used just to start the motor. Once it is up to normal running speed, a switch disconnects the series winding and the motor operates favoring the speed-regulating characteristics of the shunt motor.

differential-compound motors

The *differential-compound motor* is essentially a shunt motor with a series field winding. In this motor, the field windings are wound in *opposite* directions so that the series flux *subtracts* from the shunt flux. The word *differential* is used to tell us that the resultant flux is the *difference* between the two.

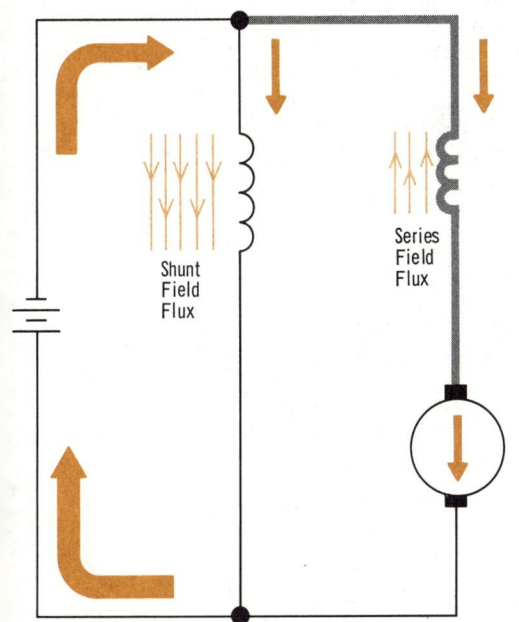

Shunt Field Flux

Series Field Flux

The differential-compound motor has a small series field winding that produces flux in opposition to the main shunt field winding. This allows the motor to operate at practically a constant speed under varying load conditions

Because the series field is overridden by the shunt field, the differential-compound motor does not have the good starting torque characteristic of the typical series motor. The series field here acts instead to make the motor more sensitive to load changes so that it gives better constant-speed regulation than the normal shunt motor.

There is better constant-speed regulation because when an increase in load slows the motor down, in addition to the normal process in which the decrease in cemf causes more armature current to flow (in an attempt to bring motor speed up), we also have an increase in the current in the series field windings. This increase in series flux opposes and hence reduces the dominant shunt field flux. This resultant decrease in field strength then also causes the motor to speed up to build up the cemf. The result is that the motor reacts much more quickly to maintain its speed. Therefore, we say that the differential-compound motor is more of a sensitive constant-speed regulator.

comparison of d-c motors

Motor Type	Features	Uses
SHUNT	1. Fair starting torque. 2. Fairly constant speed. 3. Easy to control speed. 4. Self-regulating. 5. Runs away if field is opened.	Very popular for industrial applications where reasonably constant speed is important
SERIES	1. High starting torque. 2. Speed varies with load. 3. Runs away under no-load conditions.	Popular for applications where the load is always connected but varies over wide limits
CUMULATIVE-COMPOUND	1. Excellent starting torque. 2. Good constant speed characteristic; better than in shunt motor. 3. Will not run away under no-load conditions.	Popular in applications where high starting torque is desirable but where runaway characteristics of series motors are undesirable
DIFFERENTIAL-COMPOUND	1. Very poor starting torque. 2. Absolutely constant speed under varying loads over small range.	Very few applications probably because of even better characteristics of this type available from a-c motors

summary

☐ The direction of rotation in a d-c motor can be reversed if the current in either the field or the armature is reversed, but not both. ☐ The rotation can be reversed in both the series and shunt motors by using a switch to reverse the wire connections of the field or the armature to the applied power.

☐ Compound motors use series and shunt fields to improve the characteristics of motors. ☐ When the flux strengths of the series and shunt windings add, the motor is called a cumulative-compound type, and is designed to give high starting torque with moderate speed regulation so the motor will not run away with no load. ☐ Long-shunt or short-shunt cumulative-compounding can be used to get various characteristics.

☐ When the series field is wound to cancel part of the flux of the shunt field, the motor is called a differential-compound type. It works like a shunt motor with low starting torque, but has a higher degree of speed regulation. ☐ Some compound motors have their fields switched in and out so that they start as series motors for high starting torque and then become shunt motors for speed regulation.

review questions

1. Does the differential compound motor have good starting torque?
2. What is the difference between a long-shunt and short-shunt cumulative-compound motor?
3. What characteristic is the cumulative-compound motor usually designed to improve?
4. What effect would you have on the direction of rotation of a d-c motor if you reversed the polarity of the power connections to the motor while it was running?
5. What is the difference between cumulative-compound and differential-compound motors?
6. Does the cumulative-compound motor have better speed regulation than the differential-compound motor?
7. Does the cumulative-compound motor have better speed regulation than the shunt motor?
8. The differential-compound motor is essentially a _____ motor?
9. How can you reverse the direction of rotation of a series motor? A shunt motor?
10. _____ compound motors are also called series-shunt motors.

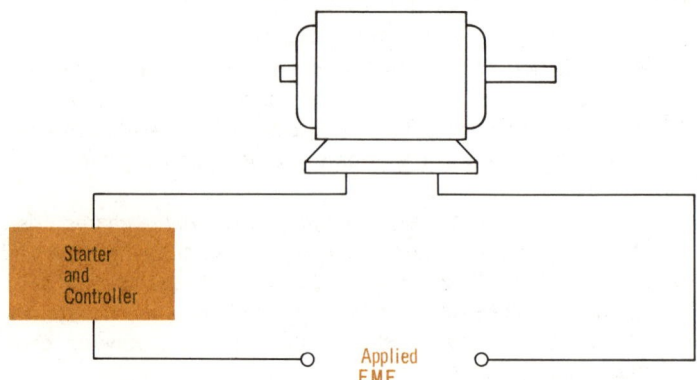

Starters are needed to limit the armature current that flows when the motor is first turned on. The starter is used to bring the motor up to its normal speed, and is then switched out of the circuit. The controller then adjusts the speed of the motor as needed

starters and controllers

Our discussion of cemf showed that when power is first applied to a d-c motor, an almost unlimited amount of current flows in the armature. As soon as the motor starts turning, cemf builds up and the armature current is limited to a reasonable value. High current flows at the instant of starting, and hence great stress is placed on the armature windings and the brushes and commutator. The high current flow can cause brush and commutator burns and could possibly cause the armature winding itself to burn open.

To avoid this, special devices, called *d-c motor starters*, aid in safely accelerating the motor to its normal operating speed. Although there are many varieties of starters in use, they are all basically some form of variable resistance or rheostat placed in series with the armature.

It is important to understand that a motor starter is only what its name implies and serves no other function once the motor is brought to operating speed. However, another device, called a *d-c motor controller*, combines the functions of starting and variable speed control in a single physical box. It contains a starting rheostat used to bring the motor to normal speed and an additional rheostat connected to allow for varying motor speed in actual use. We have already discussed how the speed of a d-c motor may be varied. The control rheostat is merely the physical means for doing the job.

Both starters and controllers contain a mechanism for automatically disconnecting the motor from the power line should the line voltage fail.

classification

Starters and controllers have been devised to match the needs of the many varieties of d-c motors in use. For instance, small d-c motors can use a relatively simple across-the-line switch for starting while large d-c motors require more elaborate setups.

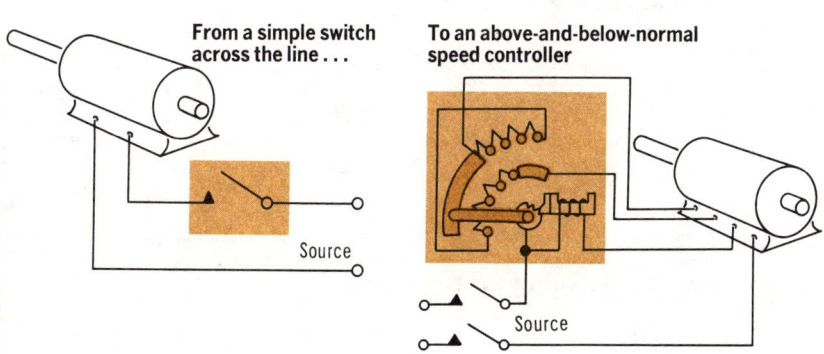

Starters and controllers range in complexity . . .

From a simple switch across the line . . .

To an above-and-below-normal speed controller

Source

Source

You will find starters and controllers classified (1) by how they are operated: as either manual or automatic types; (2) by how they are built: as faceplate and drum types; and (3) by how they are enclosed: as open type or protected (drip proof, watertight, etc.). In addition, starters and controllers are classified according to the number of terminals they possess that must be connected to the motor: as two-point, three-point, and four-point starters.

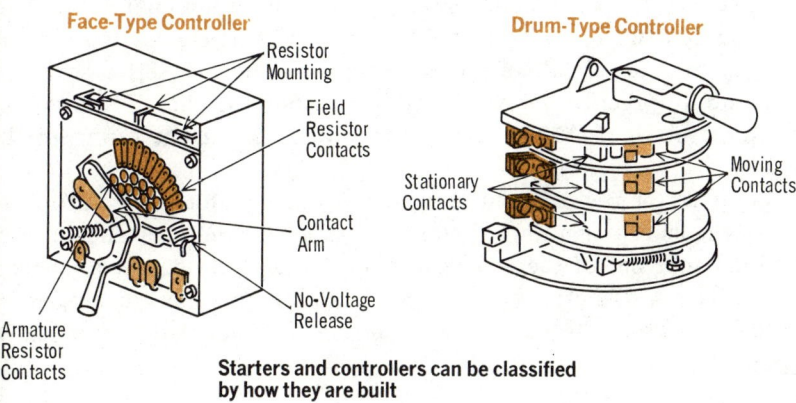

Face-Type Controller

Resistor Mounting

Field Resistor Contacts

Contact Arm

No-Voltage Release

Armature Resistor Contacts

Drum-Type Controller

Stationary Contacts

Moving Contacts

Starters and controllers can be classified by how they are built

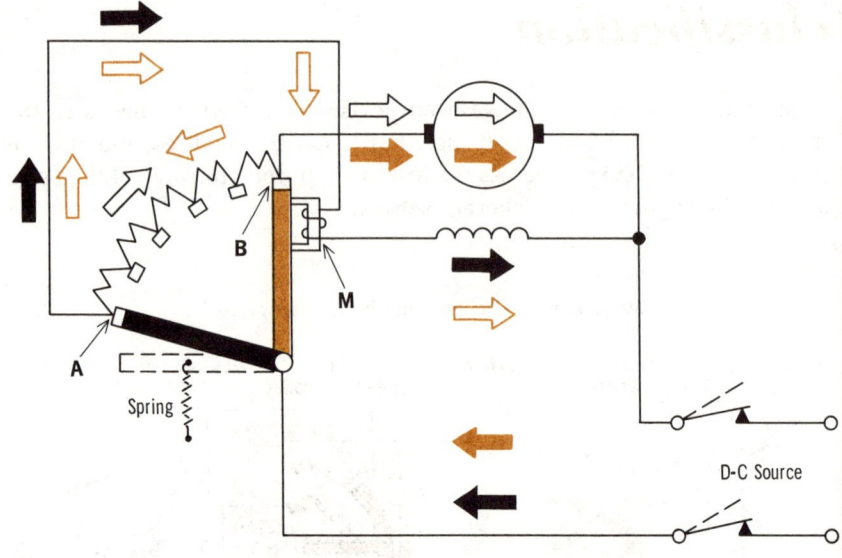

The three-point starter gets its name from the three connections that must be made between it and the motor it will be used to start. Note the different distributions of the current at the different positions of the arm

three-point starters for shunt and compound motors

The three-point starter for shunt motors shown, is of the *face type* and is manually operated. The resistor element of the rheostat is tapped by six contact buttons. The movable arm of the rheostat has a spring reset, and is arranged so that it can be moved over the contact buttons to bypass sections of the tapped resistor.

After closing the line switch, the operator manually positions the movable arm from the stop position to the first button contact, A. This connects the full line voltage to the shunt field, energizes the holding magnet, and places all of the starting resistance in series with the armature. In practice, the value of this resistance is chosen to limit the starting current to 150 percent of the full-load current rating of the armature.

As the motor begins to build up speed, the operator gradually moves the spring-loaded arm toward contact B. The starting resistance is thus gradually cut out of the armature circuit and is placed in series with the field circuit where, because of its very low resistance compared to the field resistance, it has practically no effect on field strength or motor speed.

three-point starters for shunt and compound motors (cont.)

When the arm of the three-point starter we have been discussing is on B, the armature is connected directly across the line and the motor is considered to be at normal running speed. Holding magnet M then holds the arm at the B position, opposing any tendency for the spring to bring the arm back to the stop position. Since the holding magnet is in series with the shunt field, it senses any variations that develop in the field winding.

In the shunt motor, as the field strength is decreased, the armature tends to speed up. Since it is possible to reach a runaway point if the field strength drops greatly, the holding magnet is designed to deenergize at a particular reduced value of field current. At this point, the spring-loaded arm automatically flies back to the stop position. This same arrangement causes the arm to immediately fly back to stop if the line voltage is interrupted for any reason. This then, requires an operator to go through the starting cycle again in order to return the motor to service after power is restored.

The same three-point starter used for the shunt motor can be used with a cumulative-compound motor. The illustration shows that the only difference between the two setups is the addition of the series field winding in the case of the compound motor.

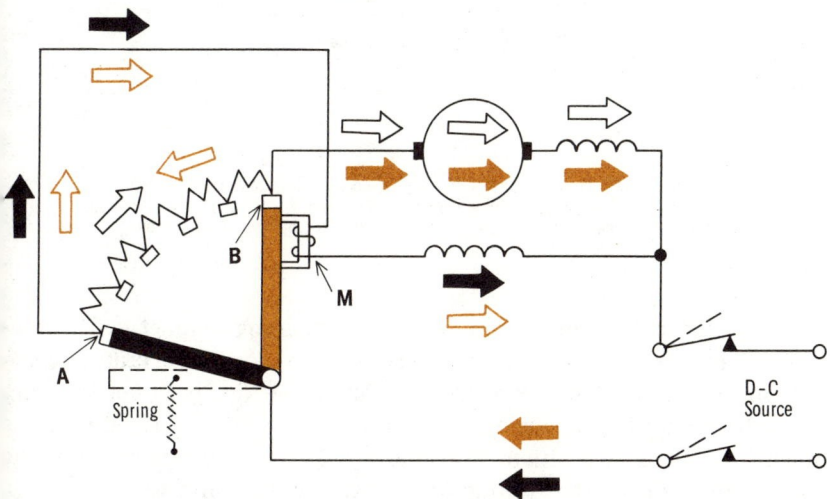

The three-point starter for a compound motor is the same as for a shunt motor. Note the spring arranged to return the starter to its original position each time the motor is stopped for any reason. This arrangement forces the operator to repeat the starting operation each time the motor is started

the three-point starter
for series motors

The three-point starter for series motors serves the same purpose as the starters you have learned about that are used with shunt and compound motors.

The three-point starter for series motors illustrated, features *low-voltage protection*. This means that if the source voltage drops to a very low value or is lost, the motor will be removed from the circuit.

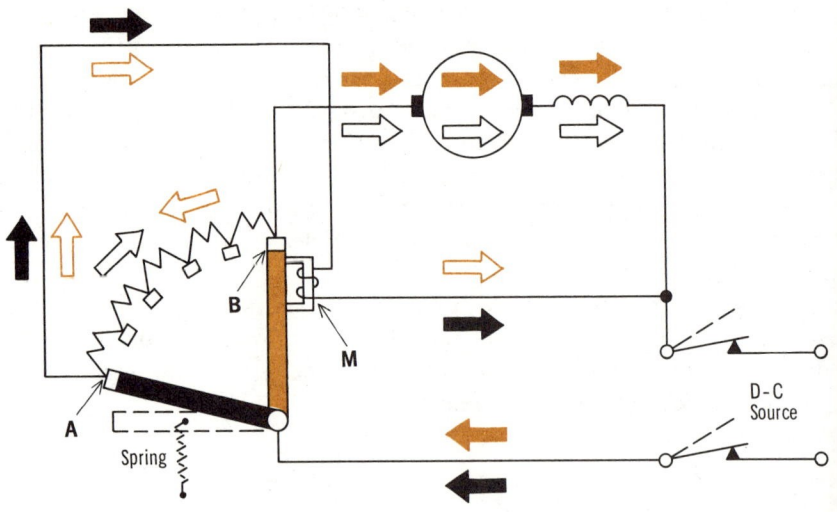

Although the three-point series motor starter requires three connections to the motor and closely resembles the three-point starter for shunt and compound motors, its internal construction and wiring differ considerably from the other starters

Note that in this three-point starter, the holding magnet coil is connected across the source voltage. To start the motor, the operator gradually moves the movable arm from the off to the run position, stopping at each contact button for one or two seconds, until the rated speed is reached at the run position. The holding electromagnet then keeps the arm in the run position, acting against the reset spring tension.

If the source voltage should drop, the holding magnet deenergizes and releases the movable arm, which is then rapidly sprung back to the off position, thus protecting the motor from possible damage.

the two-point starter for series motors

The two-point starter for series motors is one which features *no-load protection*. This means that if the load is suddenly removed during motor operation, the starter will remove power from the motor to keep it from running away.

Note that in the two-point starter, the holding magnet coil is connected in series with the power supply, motor armature, and field winding. To start the motor, the operator gradually moves the movable arm from the off to the run position, stopping on each rheostat contact button for one or two seconds. Finally, the arm is kept in the run position against the reset spring tension by the action of the holding magnet.

If the load on the motor drops, the corresponding drop in armature current is sensed by the series holding coil, which drops out. As a result, the movable arm is released and returned to the off position by the action of the reset spring. This feature prevents the series motor from damage as a result of high-speed operation under light and no-load conditions.

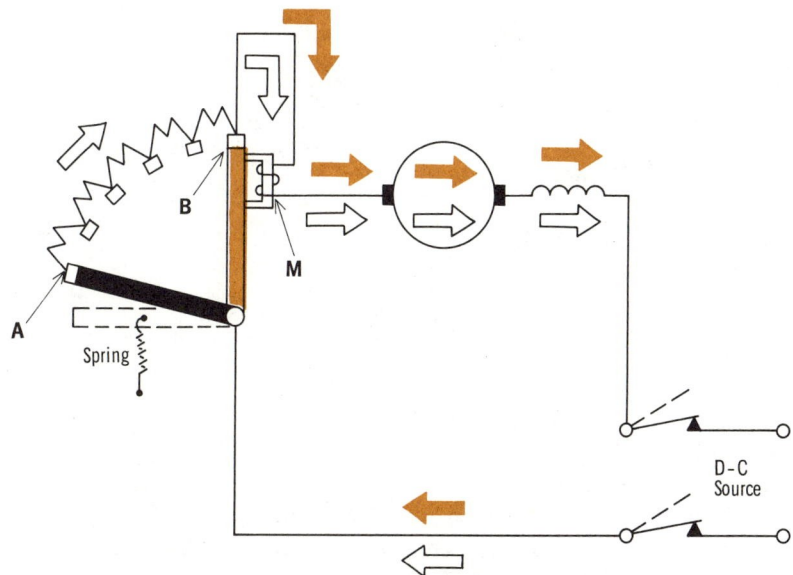

The two-point starter overcomes the no-load runaway characteristic of the series motor. Thus, in addition to its starting function, the two-point starter acts as a monitor for sensing no-load conditions. Under such conditions, the starter cuts out power to the motor and returns to the off position

four-point starters for shunt and compound motors

The four-point starters for shunt and compound motors serve the same basic functions as their three-point counterparts, and in addition, provide for the use of a field rheostat with the motors to obtain above-normal speeds.

A four-point starter used with a shunt motor is shown. The holding coil is not connected in series with the shunt field as it was in the three-point starter. Instead, the holding coil and a series resistor are connected directly across the source voltage. In this manner, the holding coil current is made independent of the field current, which will be varied to change motor speed. Nevertheless, the holding coil is still arranged to release under low- or no-source voltage conditions.

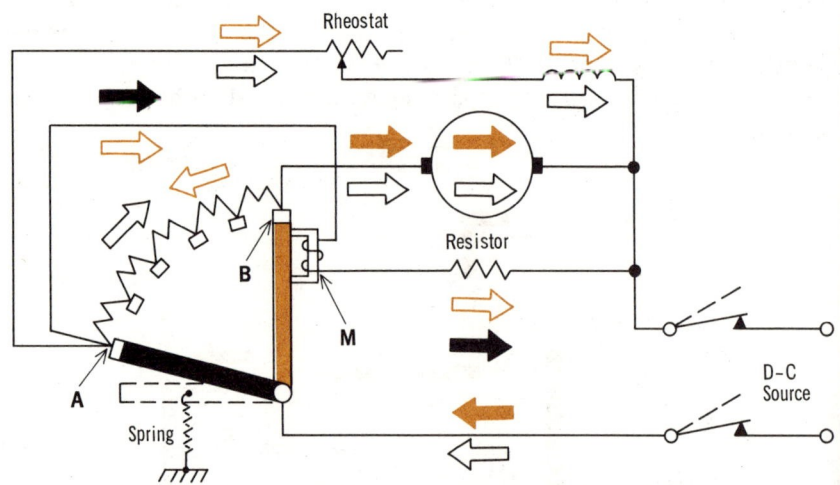

The four-point starter is essentially the same as the three-point starter previously described. A separate series rheostat is shown in the circuit. Since the rheostat here is separate, the starting device is a true starter and not a controller

The four-point starter is used to start the motor in a manner exactly as described for the three-point starter. Once the arm is in the run position, the separate field rheostat in series with the shunt field is used to adjust the motor speed to a desired value. When the motor is to be stopped, the operator will usually reset the field rheostat so that all resistance is cut out and the motor speed is reduced to its normal value. This ensures that the next time the motor is started, a strong field, and thus maximum starting torque, will be available.

summary

☐ The d-c resistance of armature coils is very low. So, when a motor is first energized, the armature draws a very high current until the armature speed becomes great enough to produce enough cemf to limit the current. ☐ To protect the motor against such high damaging currents, starters and controllers are used to limit the current to a safe value until the motor reaches normal speed.

☐ The three-point starter used with a shunt or cumulative-compound motor is first set to place no resistance in series with the field, and maximum resistance in series with the armature. ☐ As the armature builds up speed, the starter arm is moved across sections of a tapped resistor to slowly reduce the resistance in series with the armature and increase the resistance in series with the field; until finally, all of the resistance is switched from the armature circuit to the field circuit. ☐ Field current through a holding magnet keeps the starter arm in the run position. If field current drops too low, the magnet will release the arm to shut down the motor to prevent runaway.

☐ The three-point starter for series motors works in a similar manner to that of the shunt or cumulative-compound motor, except that the holding magnet is energized directly by the line voltage. ☐ In a two-point starter for series motors, the holding magnet is in series with the armature and field, and so can sense when motor current drops when the load is removed. ☐ The magnet will release the arm to shut down the motor to prevent no-load runaway. ☐ Four-point starters can be used with shunt and compound motors to give the added features of speed control.

review questions

1. Why is a motor starter needed?
2. What do all starters use to control motor current?
3. What do the points in a 2-, 3-, or 4-point starter signify?
4. Generally, to what percentage of rated full-load current does the three-point starter limit starting current?
5. How does the three-point starter prevent shunt-motor runaway?
6. Why is the holding magnet of a three-point starter connected differently with series and shunt motors?
7. What kind of protection does the holding magnet in a three-point starter give to series motors?
8. What starter gives runaway protection to a series motor?
9. How does the four-point starter differ from the three-point starter for shunt motors?
10. Does the operator usually move the starter arm with a steady, continuous motion? Why?

motor controllers

As you have learned previously, motor controllers combine the functions of motor starting and speed control in a single device. Up to this point, we have considered starters only. In our discussions of the four-point starter, you were introduced to the use of field rheostats to achieve motor speed control. In that case the field rheostat was independent of the starter box. Here you will learn about controllers which combine both the start and control functions.

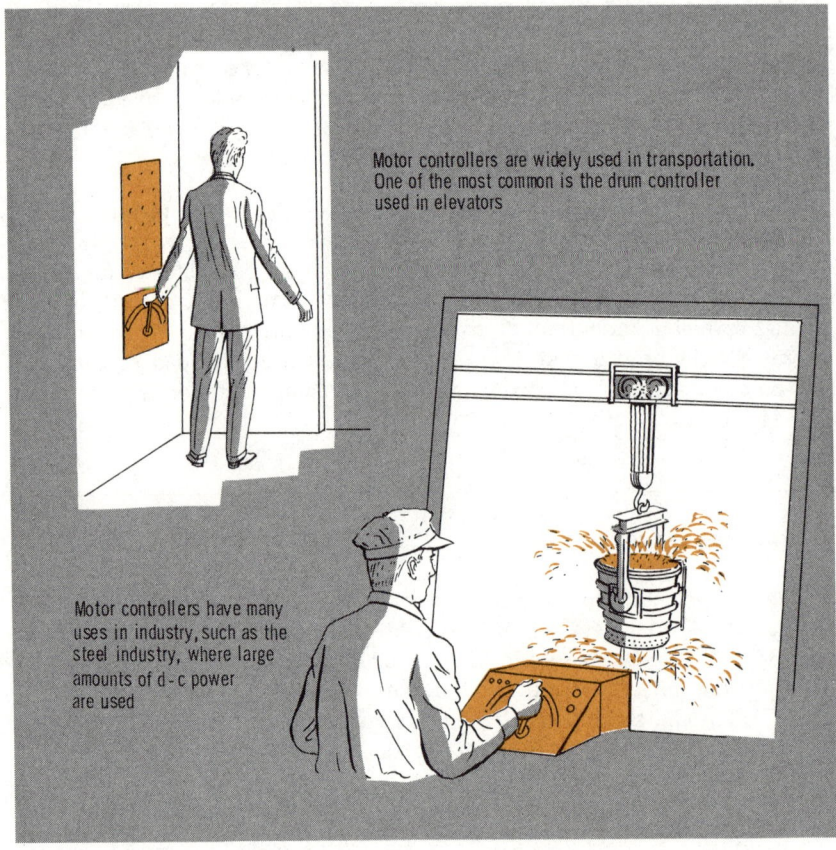

Motor controllers are widely used in transportation. One of the most common is the drum controller used in elevators

Motor controllers have many uses in industry, such as the steel industry, where large amounts of d-c power are used

In general, controllers are used exclusively with shunt and cumulative-compound motors. For the sake of simplicity, we will deal with the shunt motor in our discussion. There are two functional types of controllers in use today—the *above-normal speed controller;* and the *above-and-below-normal speed controller.*

start operation of the above-normal speed controller

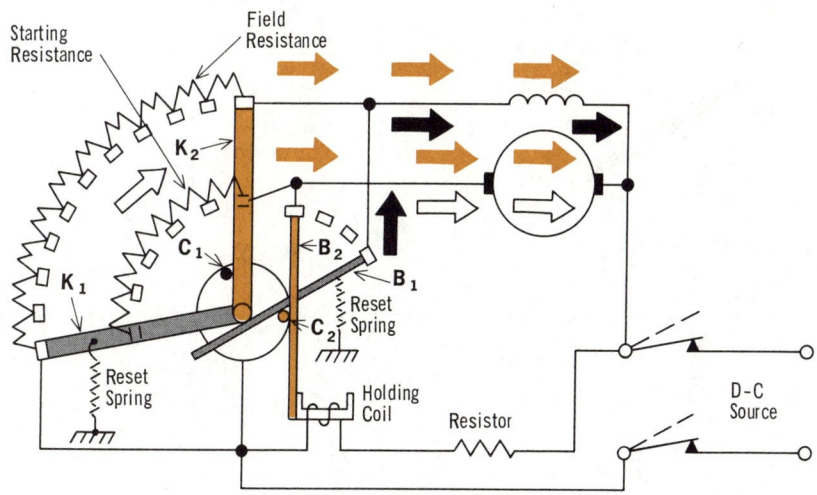

Although the controller resembles the starters previously studied, it includes, in addition to a starting mechanism, a built-in mechanism for controlling the motor speed

The above-normal speed controller combines the functions of the starter and field rheostat in a single open-face box. The starting resistance is used in the armature circuit of the shunt or compound motor only during the starting period. This limits the armature current while the motor builds up to normal operating speed. After the motor is at operating speed, the field resistance circuit becomes effective.

In the controller illustrated, there are two rows of contacts. The top row of small contacts is connected to the series field resistance. The bottom row of large contacts is connected at the starting resistance in series with the armature. Control arm K is connected to both sets of contacts.

In the start position as shown, auxiliary arm B bypasses the field resistance and the full supply voltage is applied to the shunt field. As arm K is moved clockwise, starting resistance is cut out of the circuit and the motor accelerates. As arm K approaches the run position (K_2), pin C pushes auxiliary arm B counterclockwise. Arm B contacts the holding coil at B_2 and is kept against it by magnetic action. By this time the motor will have accelerated to normal speed.

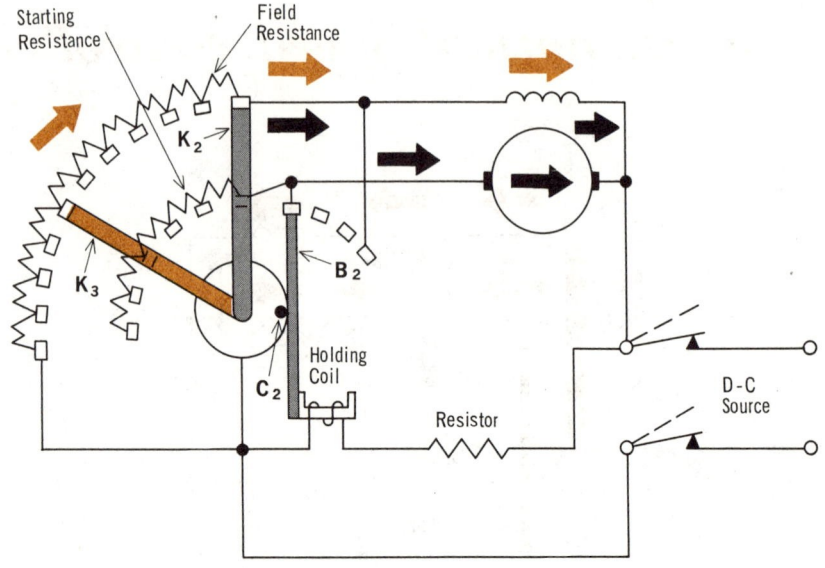

Once the controller is taken through the start cycle and the starter arm is in the run position, the speed-controlling section of the controller can be used

run operation of the above-normal speed controller

The above-normal speed controller in the run position is shown. Note that auxiliary arm B has been removed from the field circuit so that it no longer serves as a short across the *field* resistance. Arm B is now bypassing the *starting* resistance, thus providing a circuit path from the source voltage to the armature.

With arm B in the position indicated, arm K can now be moved without any effect on the armature current, but with a resultant effect on the resistance of the field circuit. If arm K is moved counterclockwise, resistance is inserted in series with the field winding and the speed of the motor is increased to some value above normal running speed. Arm K may be adjusted to any position to obtain a desired above-normal running speed to meet operating requirements.

To stop the motor, the line switch is opened. This causes the holding magnet to release arm B which is sprung back to original start position by a reset spring. Pin C is thus released, which permits a reset spring to return arm K to its original off position. With both arms reset in this manner, the controller is ready once again to start the motor.

above-and-below-normal speed controller

For a very wide range of speed control, including the ability to run our motor at above and below normal speed, we use the above-and-below-normal speed controller. As shown, it is set for below-normal speed operation. The movable arm is connected to two rows of contacts, the lower row connected to the starting resistance and the upper row to the series field resistance.

When the movable arm is moved clockwise, with the line switch closed, the motor starts and gradually accelerates, first because resistance is removed from the armature circuit, and then because resistance is introduced in the field circuit. Normal speed is attained when the arm contacts radial conductors A and D simultaneously, at the upper end of the starting resistance and point B, and full line voltage is applied to both armature and field circuits.

Before that position is reached, with resistance still in series with the armature and with full line voltage applied to the field, since the arm is in contact with radial conductor D, motor speed will be below normal.

The gear and latch system, operated with the holding coil, will lock the movable arm at any contact point, providing a variety of below-normal speeds. In the other starters, a reset spring would return the arm to off if it was set between off and run.

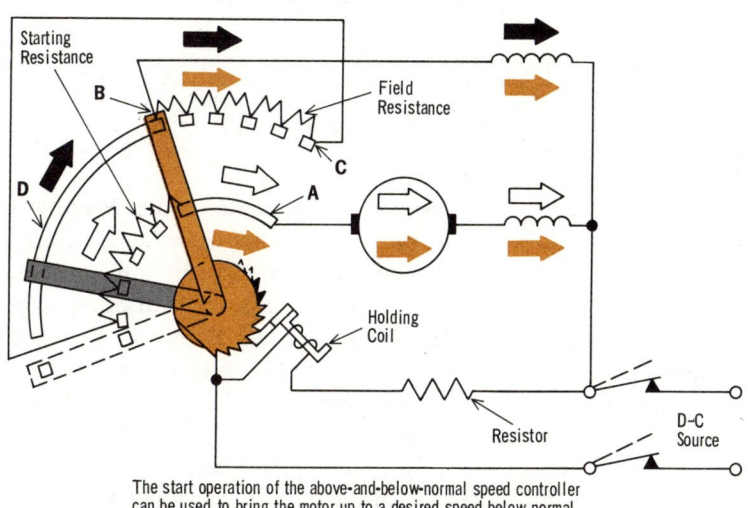

The start operation of the above-and-below-normal speed controller can be used to bring the motor up to a desired speed below normal and to leave it there. From that point, the motor may be further slowed down, or brought up to normal speed or beyond, as the operator desires

above-and-below-normal speed controller (cont.)

The above-and-below-normal speed controller can also be set to give the desired above-normal speed operation, as shown.

Remember that at normal speed full line voltage was applied to both the armature and field circuits. This was achieved when the movable arm contacted radial conductors A and D simultaneously in position B. For below-normal speed operation, we inserted resistance into the armature circuit by moving the arm into a position on the inner, or starting, resistance.

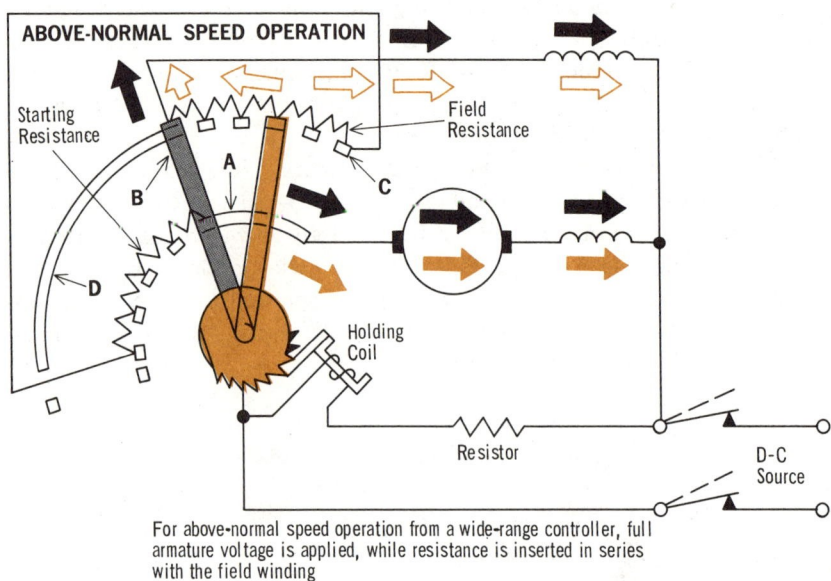

For above-normal speed operation from a wide-range controller, full armature voltage is applied, while resistance is inserted in series with the field winding

For above-normal speed operation we do the opposite. We move the arm to contact some point on radial conductor A, while contacting a tap on the field resistance. Under these conditions full line voltage is applied to the armature while resistance has been inserted in series with the field winding. The weaker field causes the motor to speed up.

As in the previous case, the arm can be latched in place on any one of the taps to provide a variety of above-normal speeds. With the arm at point C, maximum speed is attained.

When the line switch is opened to stop the motor, the holding coil releases the latch and the reset spring returns the arm to off position. The controller is thus in the position for starting the motor again the next time power is applied.

drum controllers

The controllers you have learned about to this point are not generally used with motors in applications that require very frequent starting, stopping, and changing of speed. Yet many of the motors you are familiar with, such as those used on elevators, cranes, machine tools, etc., require the constant attention of an operator who starts, speeds up, slows down, stops, and reverses them as needed. To accomplish this job a manually operated controller, which is rugged enough to withstand all the wear and tear that goes with constant and continual use, is employed. This type of device is called a *drum controller*.

The drum controller makes it easy to operate a motor-driven device that needs to be constantly changed in direction and speed

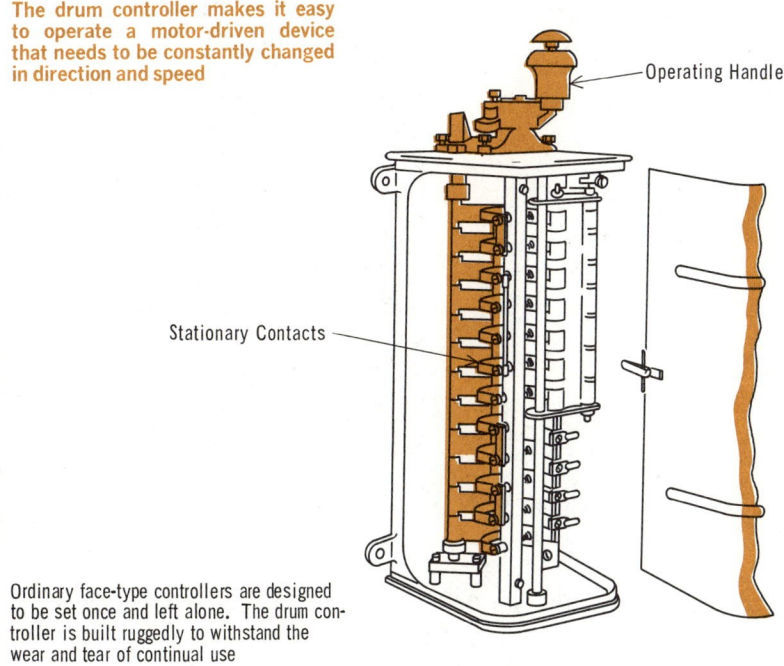

Operating Handle

Stationary Contacts

Ordinary face-type controllers are designed to be set once and left alone. The drum controller is built ruggedly to withstand the wear and tear of continual use

The drum controller is usually housed in a box shaped like a cylinder, which has an operating handle mounted on one end plate. Inside the box are heavy-duty switch contacts, which are mounted on a movable drum cylinder, and a set of matching stationary contact fingers. The drum cylinder itself is mounted on, but insulated from, a central drive shaft, which ends up on the outside, connected to the operating handle.

The operating handle can be moved in either a clockwise or counter-clockwise direction to give a range of speed control for either direction of rotation.

mechanical operation
of the drum controller

The detent action of the drum controller relieves the operator from constant control. For constant speed and direction, the operator positions the handle as required and the detent action keeps the setting in operation until the operator has to change it

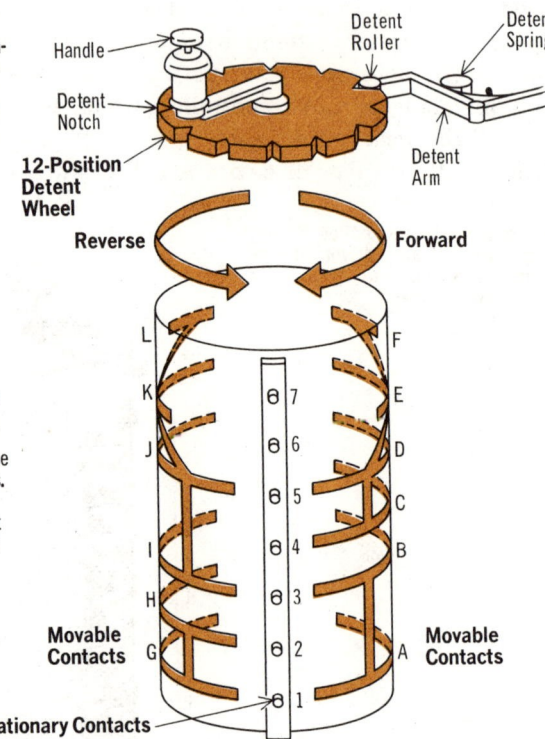

The operation of the contacts on a drum controller as shown on this simplified diagram will be discussed for forward and reverse operation on the following pages. The stationary contacts are actually fingers that brush against the movable contacts to connect the circuits

In the drum controller we find a drum cylinder insulated from the shaft it is mounted on, and from the shaft operating handle. Contacts are mounted on the movable drum, and stationary contacts, or *fingers,* are arranged to come in contact with the movable drum contacts as the operator turns the handle. Although the stationary fingers are insulated from each other, they are connected to starting resistances, reversing switches, and other starting and controlling circuit elements.

The drum controller is arranged so that once the operating handle is set, it is kept in position until the operator moves the handle again. This is accomplished by a roller and notched wheel assembly called a *detent.* The detent is keyed to the central shaft. When the handle is turned, a spring forces the roller into one of the notches on the wheel to hold the drum in the selected position.

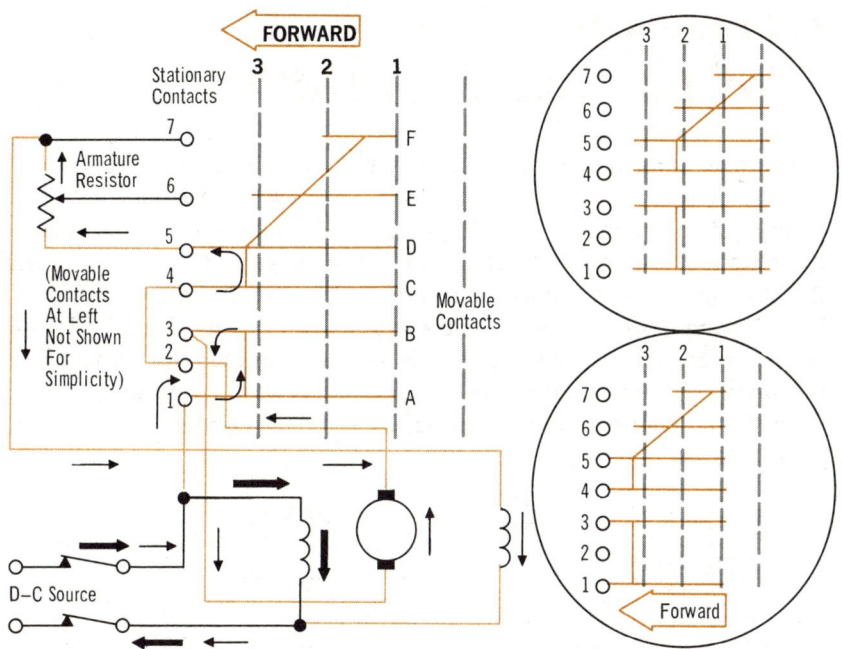

electrical operation of the drum controller

A drum controller used with a typical compound motor is connected as indicated in the diagram. When the operator positions the handle for forward operation, the set of contacts on the movable drum shown on the right contact the stationary contacts in the center. This controller offers three forward and three reverse speeds.

In the first forward position, all of the external armature resistance is in series with the armature. Movable contacts A, B, C, and D touch stationary contacts 1, 3, 4, and 5. Current flows from the source to 1, from 1 to A, from A to B, from B to 3, and from 3 to the armature.

After the current has passed through the armature winding, it continues to stationary contact 2, from 2 to 4, from 4 to C, from C to D, from D to 5, and from 5 to the resistor and then to the series field and back to the source.

In the second forward position, part of the resistance is cut out by the connection from D to E to 6. In the third forward position, all of the resistance is bypassed by the connection from E to F to 7, and the armature is connected directly across the line.

reverse operation of the drum controller

If the operator returns the control handle to neutral and then turns it in the opposite direction, we enter the first reverse position, in which all resistance is again inserted in series with the armature. The current flows from the source to 1, from 1 to G, from G to H, from H to 2, from 2 to the armature winding, from the armature to 3, from 3 to I, from I to J, from J to 5, from 5 to the resistor, the series field, and back to the source. Notice that current is flowing through the armature in the *opposite* direction than in the case of forward operation. Current through the shunt and series fields, however, is in the *same* direction as before.

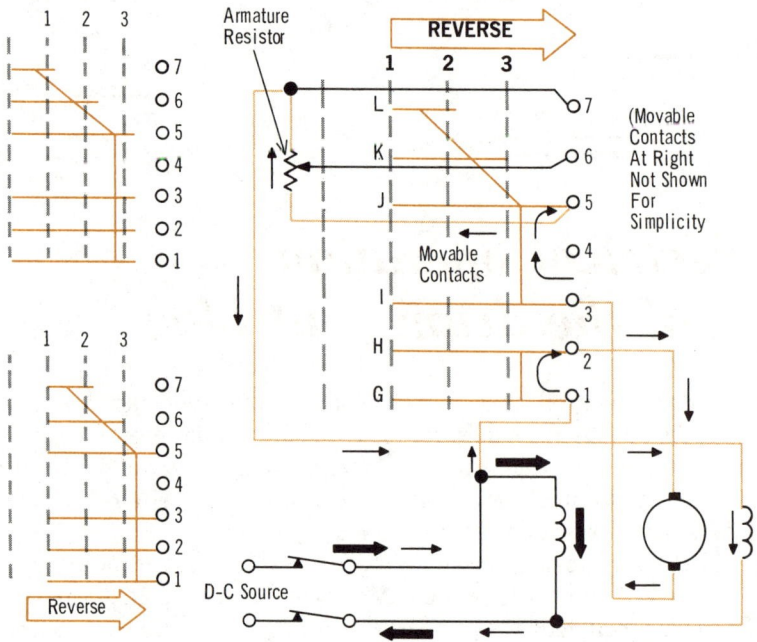

As you have learned, if we reverse the flow of current through either the armature or field of a d-c motor, but not both, we will produce a reversal of direction of rotation.

If the operator positions his handle in the second reverse position, part of the resistance is cut out of the armature circuit by the connection from J to K to 6. And, as before, placing the system in the third reverse position places the armature directly across the line by the connection K to L to 7.

automatic controllers

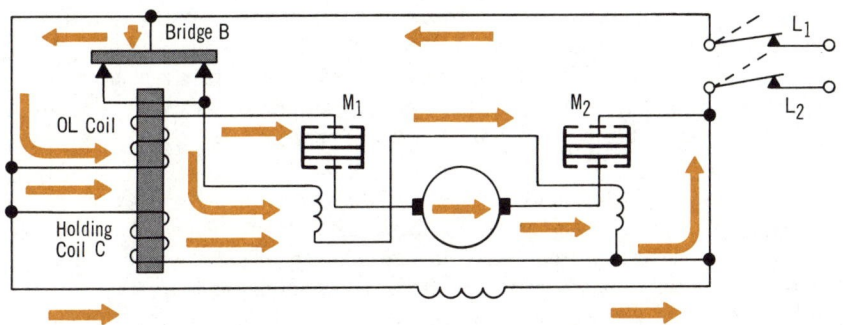

The magnetic contactor is the heart of all automatic controllers.
The ultimate purpose of the contactor is to connect the motor to the
power line. Contactor operation is usually initiated by throwing a
remote switch, or pressing a pushbutton

Push-button devices used to start and control motors are the remote
control elements of motor control systems commonly called *automatic
controllers,* which are convenient, and reduce the possibility of damage
arising from reliance on human judgment.

The key element in automatic controllers is a switching device, the
magnetic contactor, which ultimately connects the motor to the power
line. This switch is magnetic or relay-operated.

An elementary magnetic contactor system is shown connected to a
shunt motor. When the circuit is closed at L_1 and L_2, contactors M_1 and
M_2 are energized and closed. The path of current through them includes
circuit B, a so-called *bridge element,* consisting of two pins connected
to the coils of M_1 and M_2. A metallic plate lying across the tip of these
two pins is connected electrically to the L_1 side of the power line, and is
held on the pins by gravity only. Holding coil C prevents the plunger
from disturbing this setup during normal operation.

Closing contacts M_1 and M_2 energizes the motor. Current flows from
L_1 through overload coil OL, contact M_1, the armature, contact M_2 and
on to power L_2. The field coil is energized in a parallel path.

Once energized, the controller keeps the motor operating under normal conditions. In the event of an overload, the armature draws excessive current and overload coil OL, in series with the armature, energizes
and overcomes the effect of holding coil C; it moves the plunger through
the coil to contact and raise bridge B. Thus, the current path through
the coils of M_1 and M_2 is broken, the contacts of M_1 and M_2 open, and
the motor is shut down.

summary

☐ Motor controllers provide both starting and speed control for shunt and cumulative-compound motors. ☐ The above-normal-speed controller uses two resistances: one for the armature circuit, and one for the field circuit. ☐ On starting, the arm is first moved up to slowly reduce armature resistance until normal speed is reached. Then, the arm can be moved back to any position to add resistance in the field circuit to increase speed.

☐ The above-and-below-normal speed controller also has two sets of resistances, but the control arm connects to both resistances simultaneously. ☐ As the arm is moved up from stop, it reduces the armature resistance and increases the field resistance to give from below-normal to normal speed control. ☐ At normal speed, the arm switches all resistance out of the armature and field circuits. ☐ As the arm is moved further, resistance is added only to the field circuit to give above-normal speed control.

☐ Drum controllers are used with motors that require frequent starting, stopping, reversing, and speed changing. ☐ A drum is used with surface contacts that are arranged to make or break with fixed sliding contacts, as the drum is turned, to bring about circuit switching that can apply power, add resistances, and reverse wiring to give starting and speed control in the forward and reverse directions.

review questions

1. How do motor controllers differ from starters?
2. With the above-normal speed controller, in what direction must the arm be moved to add resistance in the field circuit?
3. Referring to Question 2, will that have any effect before normal speed is reached? Why?
4. What kinds of motors are controllers generally used with?
5. How many sets of resistances does the above-and-below-normal speed motor controller have?
6. With the above-and-below-normal speed controller, when is the armature resistance changed? The field resistance?
7. With the above-normal speed controller, how are the resistances tapped for normal speed?
8. When are drum controllers more useful?
9. What extra motor control does a drum controller give?
10. When the controller arm is set in the desired position, how does the way in which the drum controller holds the arm in that position differ from the other controllers?

a-c motors

In an a-c motor, as in the d-c motor, electrical energy is changed to mechanical energy. As its name implies, alternating current is used to power a-c motors instead of direct current. Since most commercial power is ac, a-c motors are easier to use than d-c motors, which require special conversion equipment.

The basis of the a-c motor is the rotation of the stator's magnetic field

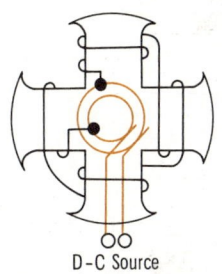
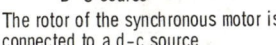

D-C Source
The rotor of the synchronous motor is connected to a d–c source

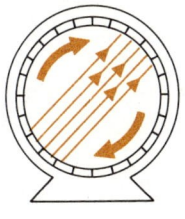

The rotor of the induction motor has no external connections

Because parts of the United States are still supplied with dc, appliance manufacturers interested in selling a uniform product from coast-to-coast pioneered the development of a special motor that could operate equally well on ac and dc. This ac-dc, or *universal,* motor is found in vacuum cleaners, electric drills, etc., and is closer in concept to the d-c motor than to the a-c motor. The universal motor, therefore, like the d-c motor, has some inherent disadvantages which can be avoided in purely a-c motors; chiefly, the need for commutation.

Commutation requires that parts of the motor rub against each other as the motor operates, so that the motor literally wears itself out. The pure a-c motor does not depend on commutation for its operation.

There are several different classes of a-c motors. In Volume 6 you studied the a-c generator or alternator; in this volume you learned that generators can be used as motors. When an alternator is connected for use as a motor, it becomes still another class of a-c motor called a *synchronous motor.* Perhaps the purest form of a-c motor is the *induction motor,* which uses no physical connection between its rotating member, or *rotor,* and stationary member, or *stator.* A third class is the a-c commutator motor, including the ac-dc universal motor.

basic operation

Since a-c power is supplied to the stator winding, the field generated between the poles *alternates* with the applied alternating power. In alternating, the field builds up from zero to a peak in one direction, falls, arrives at zero again, and then repeats the cycle in the opposite direction. The *rotor* in the basic a-c motor is shown as a permanent magnet.

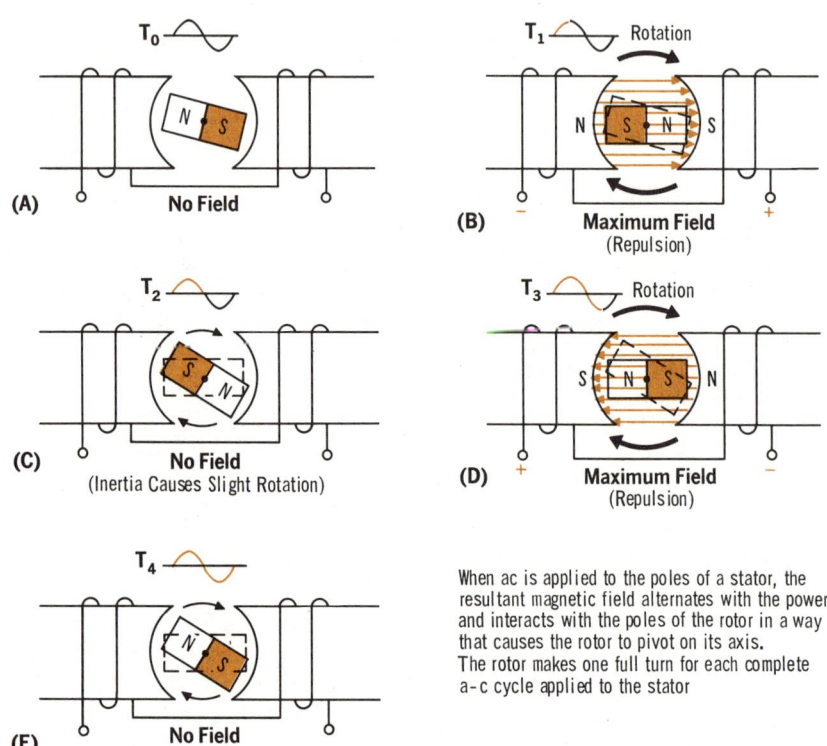

(A) T$_0$ No Field

(B) T$_1$ Rotation Maximum Field (Repulsion)

(C) T$_2$ No Field (Inertia Causes Slight Rotation)

(D) T$_3$ Rotation Maximum Field (Repulsion)

(E) T$_4$ No Field

When ac is applied to the poles of a stator, the resultant magnetic field alternates with the power and interacts with the poles of the rotor in a way that causes the rotor to pivot on its axis.
The rotor makes one full turn for each complete a-c cycle applied to the stator

When an alternating current is first applied to the electromagnetic stator winding at time T_0, no field is developed between the stator poles since there is zero current.

However, in the time between T_0 and T_1, a field builds up as the applied current does. The stator starts developing magnetic poles. Since like poles repel, the rotor is first repelled by the magnetic field. Then, since opposite poles attract, the rotor will continue to rotate until its north and south poles come under the opposite poles of the stator.

basic operation (cont.)

If the polarity of the stator current did not change, the rotor would be held in the position shown in B on page 7-90. However, since ac is being used, the field current starts to drop after time T_1, and the rotor continues to turn because of *inertia*. At time T_2, when the applied current returns to zero, the stator magnetic field is returned to zero; as shown in C, the rotor is carried on due to its inertia. However, between T_2 and T_3, the power alternation builds up in the opposite direction. The polarity of the stator magnetic poles is reversed, and the rotor is again repelled.

The rotor is rotated clockwise until it reaches position D, where it again would be held stationary by the attractive force of the stator if the a-c current would not fall and allow inertia to carry it past to position A. Back at position A, the a-c power supplied to the field again alternates to reverse the field and the cycle is repeated to keep the rotor turning.

You will notice that in A and C the rotor is slightly offset from the maximum flux positions of B and D. This slight rotation produced by the inertia of the rotor is important because it allows the motor action to continue.

If the rotor was exactly parallel to the field position of B and D, its rotation would not be possible, because the magnetic repulsion would be equal for both directions of rotation; thus the rotor might not move in either direction. This means that the basic a-c motor shown will not reliably start by itself. Also, in the illustration, the rotor turned clockwise only because the rotor was offset in that direction at the beginning. If the rotor was offset in the other direction, it would have turned counterclockwise.

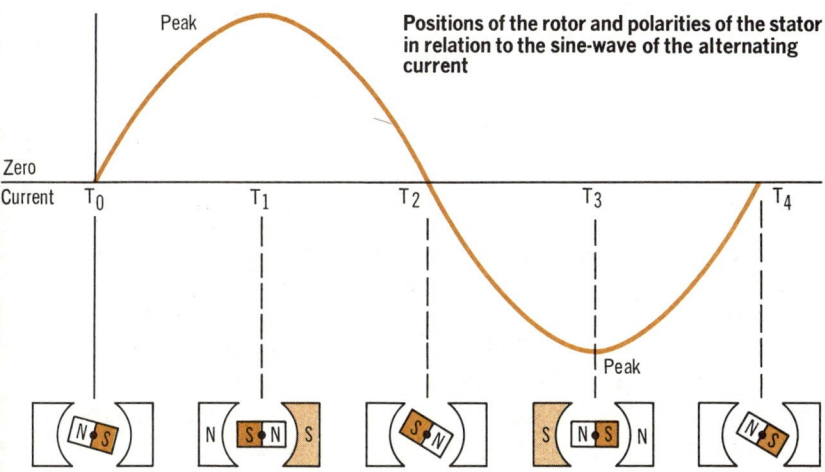

Positions of the rotor and polarities of the stator in relation to the sine-wave of the alternating current

rotating the stator field

The basic disadvantages of the simple a-c motor you just studied are that it might not start by itself and if it did, it might not run in the right direction. Both of these depend on the position of the rotor when you first apply power. As a matter of fact, this rotor might also not start no matter what position it is in because the field only alternates back and forth and does so at such a fast rate—60 times a second—that the rotor might not have enough time to follow it. To start the motor then, you would have to turn it by hand in the direction you want until it was going fast enough to follow the alternating field.

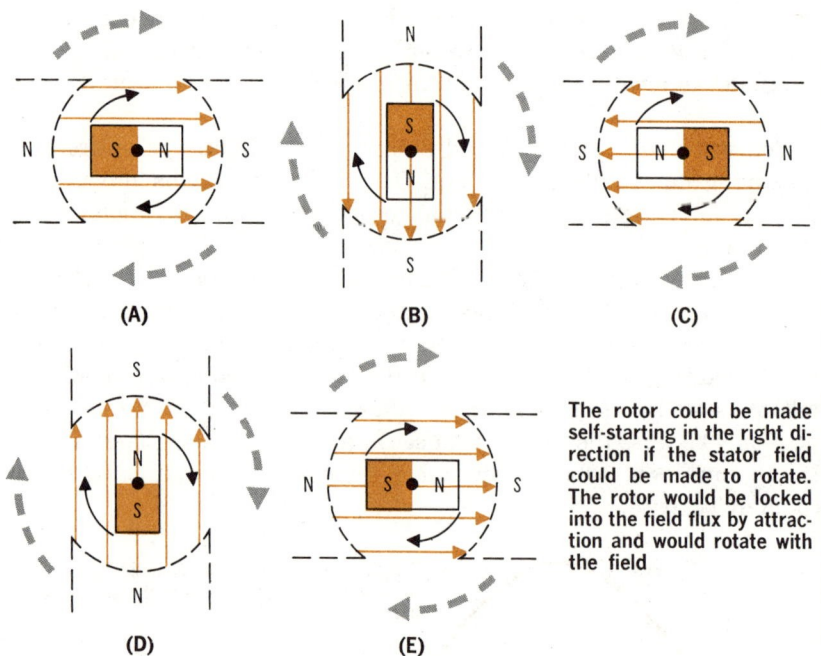

(A)   (B)   (C)

(D)   (E)

The rotor could be made self-starting in the right direction if the stator field could be made to rotate. The rotor would be locked into the field flux by attraction and would rotate with the field

The best way to overcome these disadvantages is to make the stator's magnetic field *rotate* instead of just alternate. Then, as the field turned, its revolving poles would attract the opposite poles of the rotor; the rotor would be locked in position by the magnetic attraction and it would rotate with the field.

In practical a-c motors, many different methods are employed to electrically rotate the stator field. In single-phase a-c motors, the effect is created through phase splits and shifts of the a-c power applied to the stator winding; while in polyphase a-c motors, the natural phase differences between the different a-c voltages are used.

rotating the magnetic field

Alternating current has some special properties that make it possible to use it to create a rotating magnetic field in the stator of an a-c motor. Previously, you learned that alternating current has cycles of build up and fall, and that these cycles follow a pattern called a *sine wave*. One full sine-wave cycle of 360 degrees recurs at the a-c frequency. Two different alternating currents of the same frequency can be described as being *in phase,* or *out of phase.*

Out-of-phase alternating currents are further described in terms of their relative phase angle difference at a given instant of time. The two can thus be said to be *in phase* when they rise and fall simultaneously, and 180 degrees *out of phase* when one begins to rise past zero as the other simultaneously begins to fall as it passes zero. When currents are 90 degrees out of phase, one current reaches a peak while the other is at zero.

By splitting an a-c input into two 90-degree, out-of-phase alternating currents in a specially arranged stator winding of an a-c motor, it becomes possible to alternately produce a series of electromagnetic poles that create the effect of a rotating magnetic field.

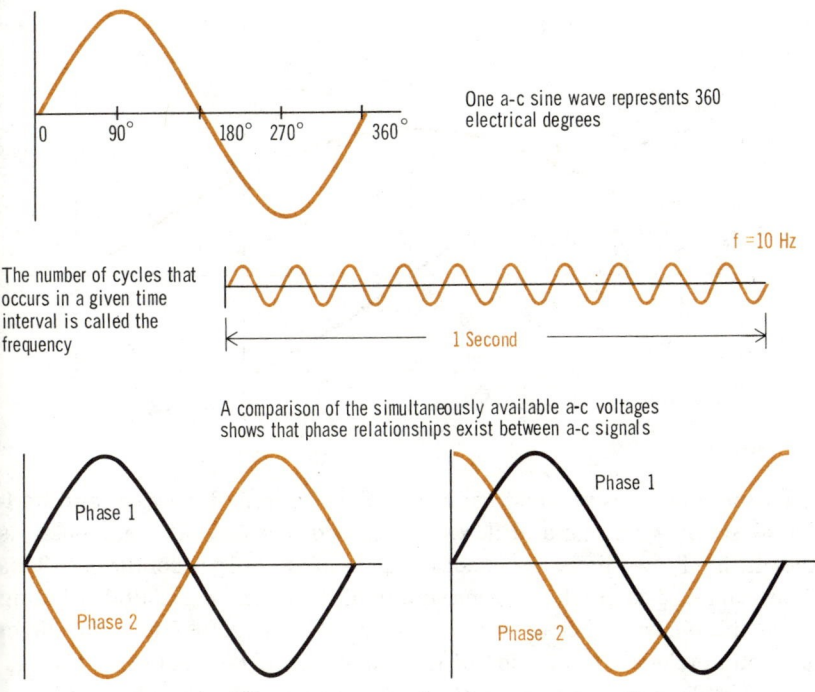

One a-c sine wave represents 360 electrical degrees

f = 10 Hz

The number of cycles that occurs in a given time interval is called the frequency

1 Second

A comparison of the simultaneously available a-c voltages shows that phase relationships exist between a-c signals

Phase 1

Phase 2

1 and 2 are 180° out of phase

Phase 1

Phase 2

1 and 2 are 90° out of phase

principle of the
rotating magnetic field

Knowing how alternating currents can exist with different phase angle relationships to one another, you are ready to learn how a rotating magnetic field may be created in the stator of an a-c motor.

At T_0 in the illustration, alternating current of phase 1 is being supplied to the vertical stator windings, while alternating current 90 degrees out of phase (phase 2) is being supplied to the horizontal windings. At this instant of start, phase 1 produces a maximum vertical magnetic field, while phase 2 produces no horizontal field.

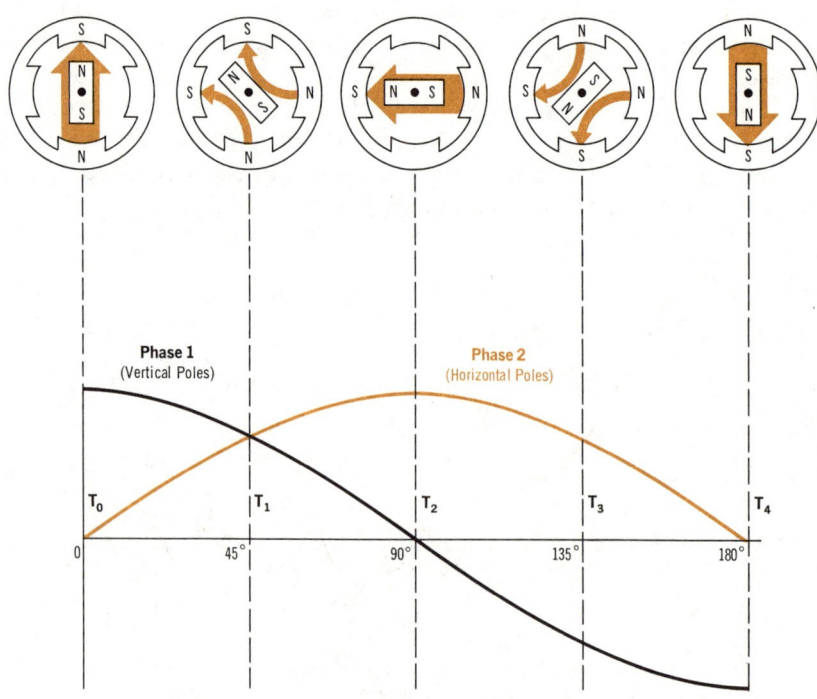

By the time T_1 is reached, current is flowing in both vertical and horizontal stator windings and flux is developed between adjacent poles as shown. By T_2, we have the exact opposite of the initial conditions. This time, phase 2 is producing maximum flux in the horizontal winding while the phase 1 flux at the vertical winding is zero. The total effect is a counterclockwise rotation of the magnetic field we begin with at T_0 so that by T_2 the pole lies in a plane at right angles to the plane at T_0.

principle of the rotating magnetic field (cont.)

Between T_0 and T_2, the strength of the initial vertical pole is gradually diminishing, while the strength of the horizontal pole is building up. Between T_2 and T_4, the process continues. By T_3, the strength of the horizontal pole has diminished, while the strength of the vertical pole has again begun to build up. This time, the vertical field is in a new direction, to account for the fact that the alternating current is building up in a new direction. By T_4, the horizontal field has diminished to zero while the vertical field has built up to maximum in the new direction.

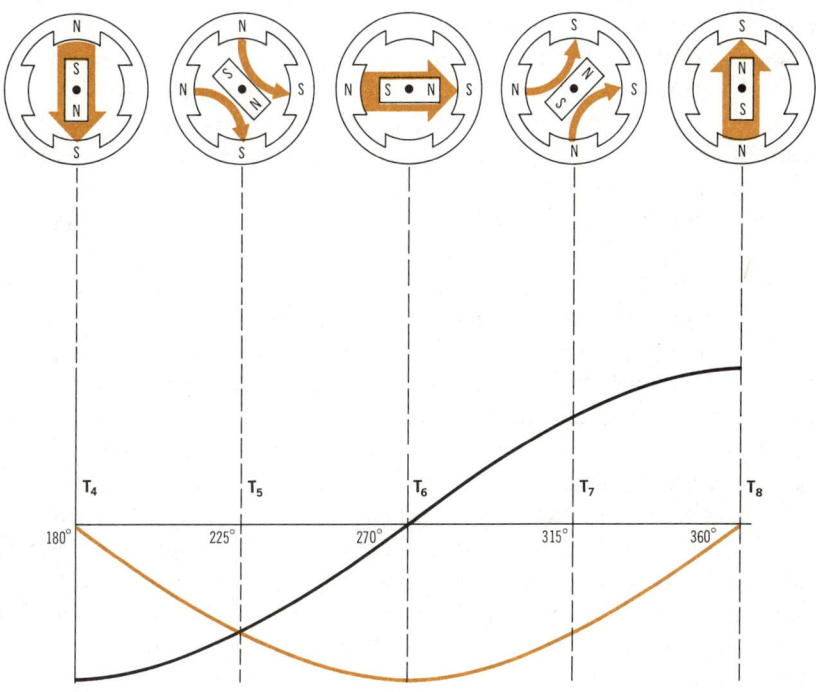

Reviewing the action between T_0 and T_4 as a continuous process, you can see that the magnetic pole has gradually been rotated 180 degrees. Between T_4 and T_8 the process continues in the same direction until at T_8 the magnetic pole is back to the original phase of T_0. For one alternating-current cycle, the magnetic field has been rotated 360 degrees. Incidentally, the natural rate at which the magnetic field rotates about the stator is called the *synchronous speed*.

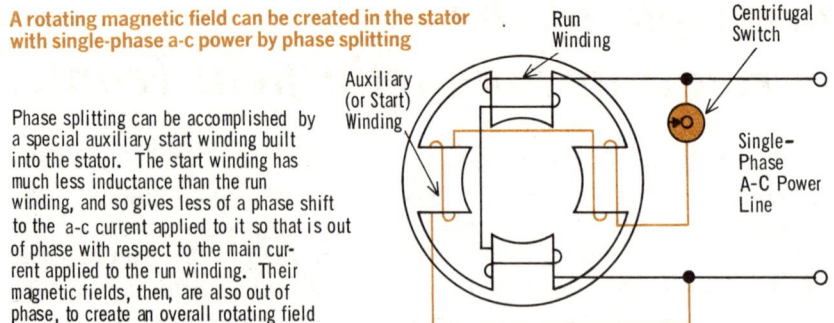

A rotating magnetic field can be created in the stator with single-phase a-c power by phase splitting

Phase splitting can be accomplished by a special auxiliary start winding built into the stator. The start winding has much less inductance than the run winding, and so gives less of a phase shift to the a-c current applied to it so that is out of phase with respect to the main current applied to the run winding. Their magnetic fields, then, are also out of phase, to create an overall rotating field

the split-phase rotating field

Since commercial power is ordinarily supplied to homes by *single-phase* ac, some means must be provided, in either the home electrical circuit or in the motor, for getting two phases from the available single-phase power, if it is to be used to start and run an a-c motor. The process of deriving two phases from one is known as *phase splitting*. Usually the means for splitting single-phase ac into two phases is built into the stator circuit of the a-c motor. Once the phase has been properly split, we can use the two phases derived to create the rotating magnetic field.

One means is a special auxiliary winding built into the stator called the *start winding* to differentiate it from the actual *run winding* of the stator. In most split-phase a-c motors, the start winding is used only for starting the motor. It has a high resistance and low inductive reactance, while the run winding has low resistance and high reactance; the two windings have different electrical characteristics. When power is first applied, both windings are energized. Because of their different inductive reactances, the run winding current lags the start winding current, creating a phase difference between the two. Ideally, the phase difference should be 90 degrees; but in practical motors, it is much less, Nevertheless, the windings develop fields that are out of phase. This creates a rotating magnetic field in the stator which applies torque to the rotor, thereby starting the motor.

Once the motor gets up to about 80 percent of its normal operating speed, the rotor is able to follow the alternations of the magnetic field created by the run winding. To minimize energy losses, the start winding is then switched out of the circuit by a mechanical device called a *centrifugal switch*, because it is operated by the centrifugal force created by the rotor revolutions. The direction of a split-phase rotating field can be reversed by reversing the connections to the start winding. This changes the direction of the initial phase shift, which means that the overall magnetic field is rotated in the opposite direction.

reactor-start motor

As explained for the split-phase rotating field, which uses a special start winding, the ideal phase angle that should be created between the run and start windings should be 90 degrees. However, because of the resistance that exists in the run winding, the actual phase difference between the windings is much less than 90 degrees. The starting torque produced by a rotating field that is produced by a small phase angle is low, causing the motor to start inefficiently and slowly, particularly under a heavy load. Therefore, it is important to try to increase the phase angle between the windings.

Normally, with a split-phase motor the run winding is designed to have as little resistance as possible, and the start winding much more. However, there is a limit as to how high the resistance of the start winding can be, since enough current must be allowed to flow through the start winding for it to produce a usable magnetic field. Also, the run winding has to be designed to give the best performance while the motor is running. So the design of the windings for the best starting characteristics may not be the same as that for the best running characteristics.

To alleviate this problem, special auxiliary devices can be used to modify the winding characteristics during the start operation, allowing the run winding to be designed for the best run operation. One such device is the *reactor* connected in *series* with the run winding to increase the total inductance of the run circuit. This increases the lag between the magnetic fields.

The centrifugal switch in this circuit has two jobs to perform. First, before the motor comes up to running speed, the switch allows the reactor to be in series with the run winding, and also connects in the start winding. Second, when the motor starts to get close to running speed, the switch disconnects the start winding and shorts out the reactor, so that only the run winding is in the circuit.

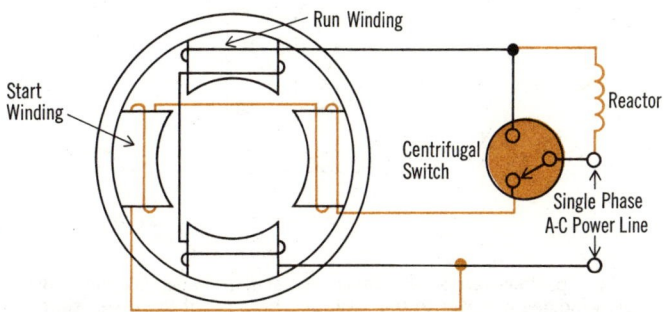

The reactor increases the phase angle between the start and run windings. When the motor comes up to speed, the centrifugal switch disconnects the start winding, and at the same time shorts out the reactor so that only the run winding is energized

the capacitor-start rotating field

Although the use of a reactor in series with the run winding improves the phase lag of the winding, the phase difference still falls far short of 90 degrees. Therefore, even though the reactor helps, the starting torque it produces still leaves much to be desired.

A phase shift closer to the ideal 90 degrees is possible through the capacitor-start system for creating a rotating stator field. This system is a modification of the split-phase system; a high-value starting capacitor is placed in series with the start winding of the stator to provide a phase shift of approximately 90 degrees for the start current. As a result, the starting torque that results is greatly improved over the standard split-phase system.

The start winding of the capacitor-start stator usually has a lower resistance and a greater number of turns than the standard split-phase type so that it is more efficient. In some capacitor motors, the starting winding with capacitor is left in the circuit even after starting, to provide improved motor operation. In most of the more common motors, however, the capacitor and start winding are taken out of the circuit by a centrifugal switch as in the case of the standard split-phase motor. A simple method of reversing the direction of rotation of a capacitor motor is the same as in the case of the split-phase motor, that is, to reverse the connections to the start winding leads.

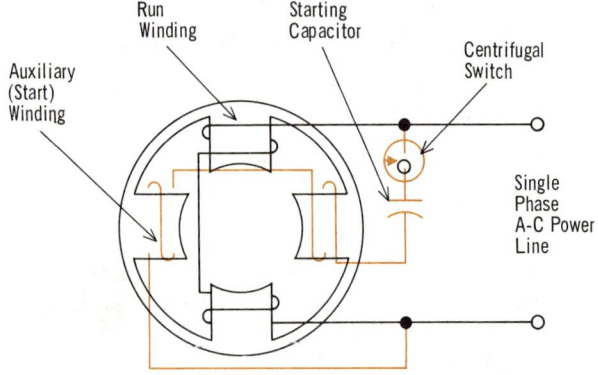

In the capacitor-start stator, the phase shift between the run and start windings is closer to the ideal 90° because of the phase-shift action of the capacitor

the capacitor-run motor

It was mentioned earlier that some capacitor-start motors do not remove the auxiliary winding after the motor comes up to speed. This is because the capacitor is able to provide a phase angle so close to 90 degrees that the use of the efficient, low-resistance auxiliary winding allows much more torque to be developed while the motor is running. This is not so with the ordinary split-phase and reactor-start motors because they cannot produce enough of a phase difference for the motor to run efficiently after it comes up to speed. Also, since their auxiliary windings have high resistance, much power is lost in the form of heat (I^2R loss).

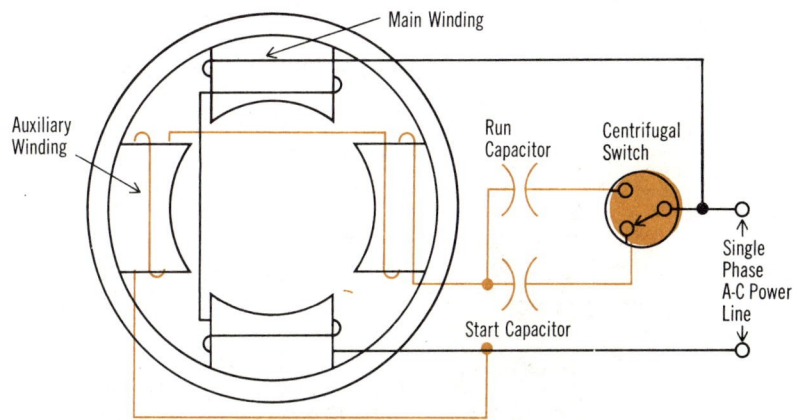

With the two-capacitor motor, the centrifugal switch cuts out the starting capacitor and switches in the run capacitor when the motor comes up to speed. The permanent-split capacitor motor does not use a centrifugal switch, and the same capacitor is used for both start and run

The capacitor motors that also *run* with the capacitor are able to have their auxiliary windings designed to act as efficiently as the main run winding. There are two types of *capacitor-run motors*. One type uses two capacitors and the other uses only one. With two capacitors, a centrifugal switch connects a high-value capacitor during start to provide higher starting current to the auxiliary winding for greater starting torque. Then, when the motor comes up to speed, the switch disconnects the starting capacitor, and connects in the *run capacitor*.

The capacitor motor that uses only one capacitor continuously during run and start is called a *permanent-split capacitor motor*. The capacitance value of this capacitor is a compromise between what is best for starting and running.

starting capacitors

Capacitors used to start a-c motors are not the same as ordinary capacitors. Since they must have very high capacitance values, they are *electrolytics*. Ordinary electrolytic capacitors are designed for d-c operation, and if they are used for a-c, they will blow up. Capacitors for starting a-c motors are *a-c electrolytics*. These were discussed in Volume 3.

Essentially, an a-c electrolytic capacitor is made of two d-c capacitors connected back-to-back in the same shell. Because of this, it is larger and bulkier than an ordinary d-c electrolytic of the same value. A-c electrolytics do not have polarity markings, and their ratings for motor use are more critical than for ordinary electrolytic applications.

In many ordinary applications, one electrolytic can usually be replaced with another having a close value, without any apparent effects. But in a motor, if the value is changed, starting torque can be adversely affected, and resonance with the auxiliary winding can result.

Motor-starting capacitors are a-c electrolytics, and must not be replaced with ordinary d-c electrolytics. In addition, their capacitance values are more critical and have special duty-cycle ratings

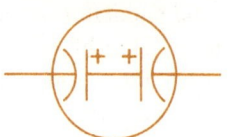

A-c electrolytics are actually two capacitors connected back-to-back to conduct in both directions

Unlike ordinary electrolytics, motor starting capacitors are also specially rated for *duty cycle*. The *equivalent* of 20 three-second periods in an hour is the typical rating.

Capacitors used to *run* motors are usually oil-filled capacitors. These, too, were covered in Volume 3.

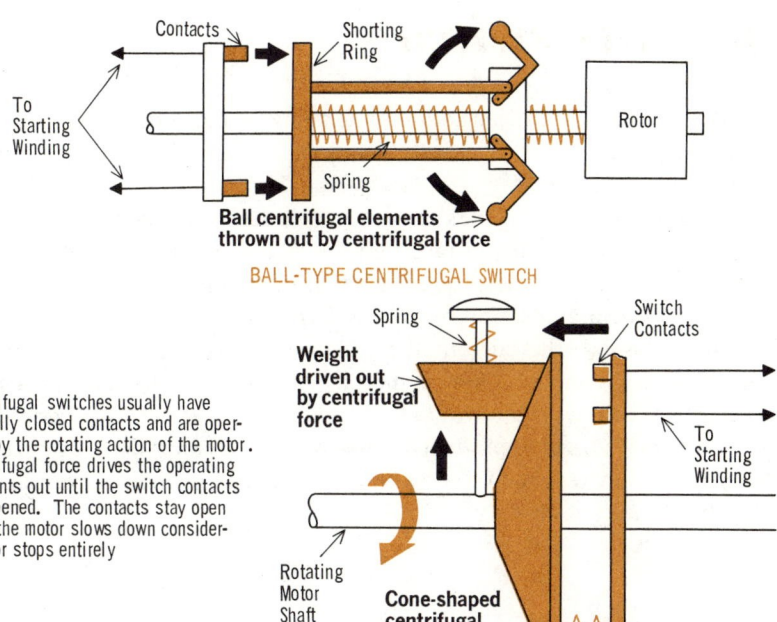

BALL-TYPE CENTRIFUGAL SWITCH

Centrifugal switches usually have normally closed contacts and are operated by the rotating action of the motor. Centrifugal force drives the operating elements out until the switch contacts are opened. The contacts stay open until the motor slows down considerably or stops entirely

CONE-TYPE CENTRIFUGAL SWITCH

the centrifugal switch

Although many varieties of centrifugal switches are in use to control the connection of start windings and starting capacitors, they all consist of two basic parts: a switch arm and a switch contact. Part of the centrifugal switch is mounted on the motor's rotor or rotor shaft. The specific arrangement and type of arm and contacts used depend on the particular application. The switch arm is usually loaded or weighed down in some way and arranged so that the switch contacts are held normally closed by spring tension. This means that before starting, the start winding and/or the starting capacitor are always connected.

As the motor gains speed after starting, centrifugal force throws out the loaded arm which in turn overcomes the spring tension and opens the switch contacts to disconnect the start winding and capacitor. During operation, the switch remains in the open position; when the motor is stopped, the springs return the centrifugal switch to its original starting position, reconnecting the start winding and capacitor.

Two types of centrifugal switches are shown. Of the two, the cone type is commonly found in home appliance motors. The ball type is used mostly in large industrial motors.

the relay switch

Some motors use an electromagnetic relay in place of the centrifugal switch to cut the starting winding out of the circuit when the motor reaches rated speed.

Since the run winding has a very low resistance, the operation of a *cutout relay* depends on the fact that the main or run winding draws a great deal of current when the motor is first energized. However, as the rotor starts turning, a counterelectromotive force is generated in the run winding to oppose the current flow. This was discussed earlier for d-c motors. As the rotor turns faster, more cemf is produced, and the current in the run winding goes down. Therefore, as more speed is produced in the motor, more cemf is produced, and the run winding has less current. When the motor reaches rated speed, the current in the run winding is at its minimum.

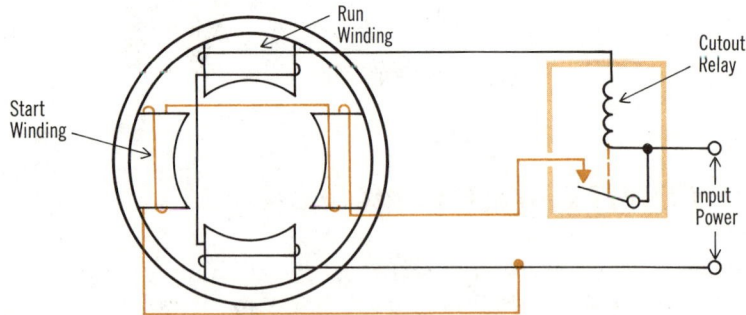

The high current flowing through the run winding energizes the relay to connect in the start winding. But at rated speed, the run winding current drops to deenergize the relay, and the start winding is disconnected

The cutout relay uses the high starting current and low running current to switch the starting winding in or out of the circuit.

Ordinarily, the cutout relay's contacts are open. When the power is first applied to the motor, a high current flows through the relay winding in series with the run winding of the motor. This energizes the relay, causing its contacts to close. The closed contacts connect the start winding to the power source. Then, as the motor picks up speed, the current in the run winding that flows through the relay winding diminishes, but is still large enough to keep the relay energized until the motor gets close to rated speed. When this happens, the run current is too weak to keep the relay energized. Its contacts open to disconnect the start winding.

the shaded-pole rotating field

Another way of creating the effect of a rotating magnetic field is to divide each pole of the stator of a simple, single-phase a-c motor into two sections, one encircled with a heavy *shorting conductor*, such as a shorted turn of copper ring. This *shading-coil* arrangement displaces the axis of the shaded poles from the axis of the main poles. When power is applied to the stator, the flux in the main part of the pole induces a voltage in the shading coil, which acts as a transformer secondary winding.

Since the current in the secondary winding of a transformer is out of phase with the current in the primary winding (Vol. 3), the current in the shading coil is out of phase with the current in the main field winding; thus, the flux of the shading pole is out of phase with the flux of the main pole. At one time, then, the field flux will go between the main pole sections, and then between the shaded sections. At other times, the flux between the poles interact to produce a distorted field that follows a revolving pattern. Since the field does not follow a smooth 360-degree rotation, *wrap-around* poles are usually used to make the shifting flux pattern more effective.

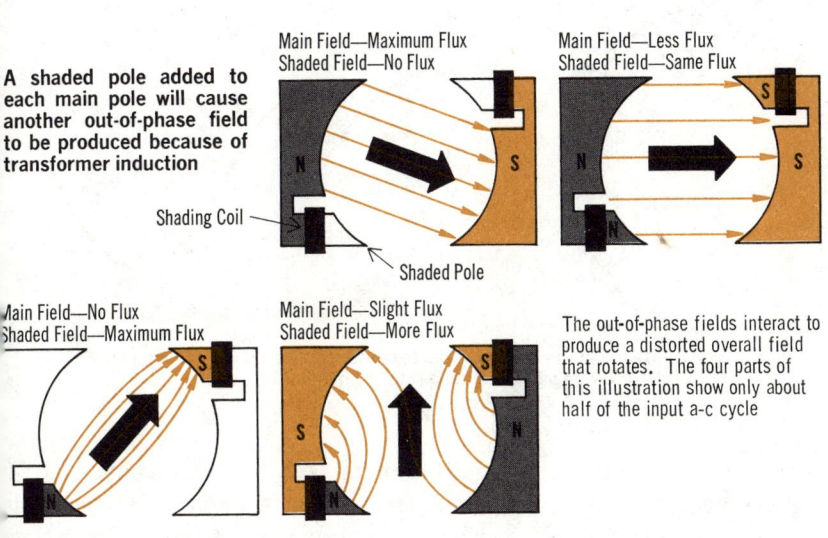

A shaded pole added to each main pole will cause another out-of-phase field to be produced because of transformer induction

Shading Coil

Shaded Pole

Main Field—Maximum Flux
Shaded Field—No Flux

Main Field—Less Flux
Shaded Field—Same Flux

Main Field—No Flux
Shaded Field—Maximum Flux

Main Field—Slight Flux
Shaded Field—More Flux

The out-of-phase fields interact to produce a distorted overall field that rotates. The four parts of this illustration show only about half of the input a-c cycle

Shaded-pole motors can be constructed to rotate either clockwise or counterclockwise. This is controlled by the placement of the shading coil. In addition, these motors can be built to be reversible by fitting each pole piece with a pair of shading coils. One or the other of the pair is connected by an external switch to select the direction of rotation.

the two-phase rotating field

In studying the split-phase and capacitor-start methods of creating a rotating stator field, you learned that the object of the system was to produce within the stator circuit two alternating currents 90 degrees out of phase, beginning with single-phase a-c power available from home outlets. If instead of single-phase a-c power, two-phase a-c power 90 degrees out of phase was available, it would be unnecessary to build the stator with provisions for phase splitting. Instead, the stator could be set up as shown, with a separate a-c phase applied to each of two pairs of stator windings. This is the equivalent of having the phase split take place at the generating station before the power is applied to the motor.

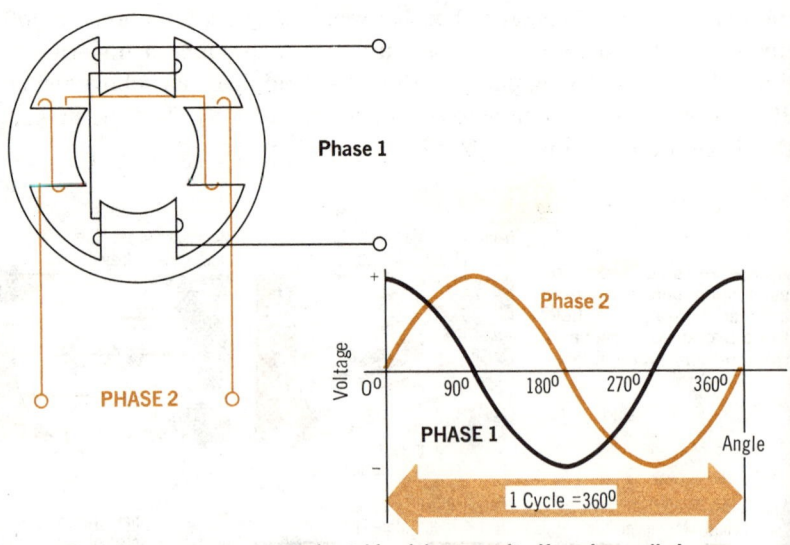

Two-phase a-c power created outside of the stator itself can be applied
to the stator directly to generate a rotating magnetic field

In addition, since the phase difference is a perfectly controlled 90 degrees with an individual power generator source for each line, it is possible not only to use the two-phase power to start the motor, but to run it as well. Operation is efficient, as there is no loss across a capacitor or as a result of a less than ideal 90-degree phase displacement. The two-phase a-c motor thus has good starting torque and running torque characteristics. As an interesting historical note, Nikola Tesla, who invented the a-c induction motor, conceived and built his first motor to run off two-phase a-c power.

the three-phase rotating field

You have so far been introduced to rotating stator fields produced by single-phase and two-phase a-c power. Single phase a-c power is in a class by itself, but two-phase power falls into the larger *polyphase* (meaning more than one phase) power category.

Electric companies in the United States normally generate and transmit three-phase a-c power. Single-phase a-c power for the individual home is obtained from one phase of the three-phase a-c power lines. Three-phase motors are common in industry because electric companies supply three phase a-c power lines to industrial users on order.

The creation of a rotating stator field using three-phase power is similar to the principle of the split-phase or two-phase system. In the three-phase system, a rotating magnetic field is generated in three phases instead of two. The three phases of alternating current in the illustration are each displaced by 120 degrees. Each supplies one of three separate pair of poles. At the first position, phase 1 has the greatest magnitude, placing the field between pole 1 and pole 4 in control. By the second position, phase 2 has the greatest magnitude and the field between poles 2 and 5 is in control. At the third position, phase 3 has the greatest magnitude and the field of poles 3 and 6 is in control.

The direction of rotation of the field in a three-phase motor can be changed by reversing any two of the leads to the stator. This displaces the phases in a manner that causes the stator's magnetic field to rotate in an opposite direction.

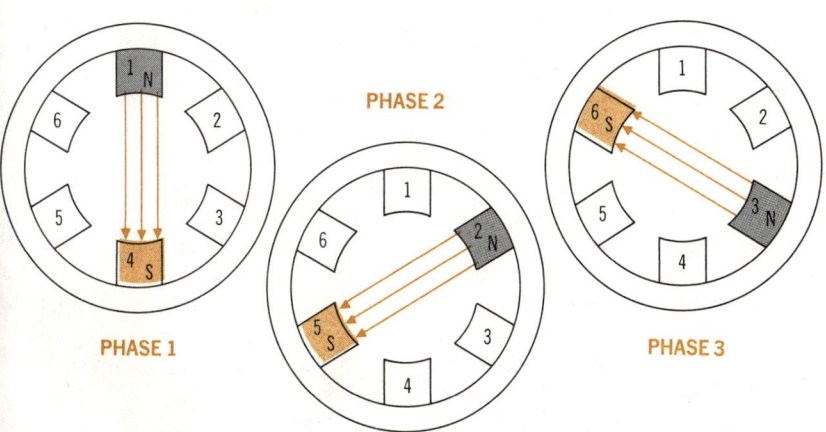

PHASE 1   PHASE 2   PHASE 3

summary

☐ Since most commercial power is ac, a-c motors are easier to use than d-c motors. ☐ The basic a-c motor, which uses a permanent magnet that follows the alternations of a field produced by ac, has difficulty starting, may not run in the right direction, and runs erratically.

☐ A practical a-c motor has a rotating magnetic field that carries the permanent magnet armature around with it. ☐ The speed at which the field rotates is called the synchronous speed. ☐ A simple method of producing a rotating field is with phase splitting. This can be done with single-phase ac with two windings, one of which produces more inductive current than the other, or the run winding can be supplied with a reactor or the start winding with a capacitor to give a phase shift. ☐ Shaded poles can also be used to produce interacting fields that result in rotation.

☐ Two-phase input power can also be used to energize two windings to rotate the field, or three-phase power can be applied to three windings to get rotation. ☐ Polyphase a-c motors are common in industrial applications. ☐ Often, split-phase motors use two windings only to start the motor. A start winding is used with a run winding until the motor gets up to speed in the right direction. Then a centrifugal switch or relay cuts out the start winding. Sometimes an auxiliary winding is used for both starting and running.

review questions

1. In the basic a-c motor, what position would the rotor have to be in to prevent the motor from starting?
2. Why can the basic motor run in either direction?
3. What kind of a field is necessary to make the motor practical?
4. In the capacitor motor, one winding conducts capacitive current, and the other winding conducts _____ current.
5. With a split-phase motor that does not use a capacitor, what kind of current does each winding conduct?
6. Why is the capacitor motor better than the simple two-winding split-phase motor?
7. What are the positions of the field windings in the split-phase and two-phase motors?
8. How are the windings positioned for a three-phase motor?
9. How does the shaded-pole motor work?
10. How does the centrifugal switch work?

synchronous speed in the two-pole stator

When you first studied the principle of the rotating magnetic field, you learned that the natural rate of rotation of the magnetic field of the stator is called the *synchronous speed*.

Since the rotor will follow this rotation of the stator's magnetic field, obviously the speed that the motor can attain will be tied to this natural rate, or synchronous speed, at which the field rotates. The synchronous speed, in turn, is dependent on the frequency of the applied alternating current, and the number of poles in the stator.

One-Phase, Two-Pole Stator

Two-Phase, Two-Pole Stator

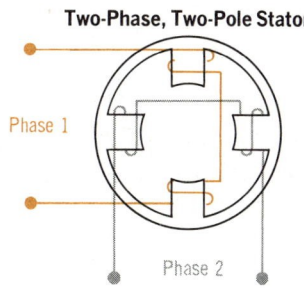

Three-Phase, Two-Pole Stator

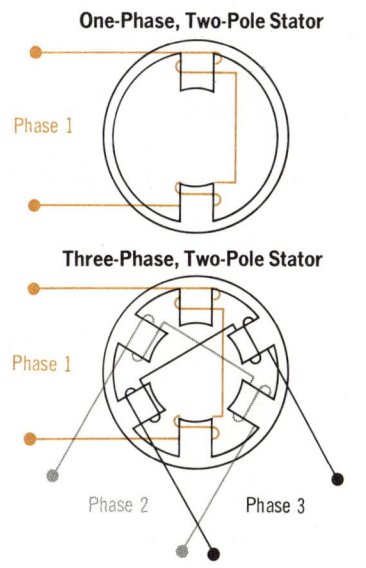

The one-phase, two-phase, and three-phase stators covered to this point are all classed as two-pole stators because there is one main N–S pole pair for each phase of power applied. As a result, the synchronous speed of the rotating magnetic field of all these stators is the same

All of the motor stators you have been introduced to up to this point are classed as two-pole stators because, by definition, in each of them, the opposites side of the stator coil that make up one north and one south pole, or a two-pole set, are *180 physical degrees* apart. In these motors there are only *two main poles*. The other poles are really only auxiliary poles used to produce a phase shift for a rotating field. And the field makes one full revolution for each time the main poles' polarities make a full alternation. Actually, you can determine the number of main poles in a motor by counting the number of actual poles and dividing that number by the phases used. Therefore, a four-pole, two-phase motor is really a two-pole motor, and so is a six-pole, three-phase motor.

synchronous speed in the two-pole stator (cont.)

In going through one a-c cycle, the field in the main plane of a two-pole motor is, in effect, reversed once, and then restored to its original state. This means that in the time of one a-c cycle, the stator's magnetic field would have made one complete revolution. If we consider that a 60-Hz a-c input is applied, we can say that the synchronous speed is 60 revolutions per second. Since motor speeds are more often rated in revolutions per minute, we multiply 60 revolutions per second by 60 seconds per minute to get 3600 revolutions per minute.

We can express this as a general equation:

$$\text{Synchronous speed (rpm)} = \frac{\text{a-c frequency (Hz)} \times 120}{\text{number of poles/number of phases}}$$

If we assume that the frequency of the applied ac was the same for all of the a-c motor stators you learned about up to now, and we know that by definition they were all two-pole stators, you can see that all their fields operate at the same synchronous speed of 3600 rpm.

The speed an a-c motor can attain depends on the speed of the rotor

Since the rotor follows the rotation of the magnetic field of the stator, its speed depends on the rotating magnetic field

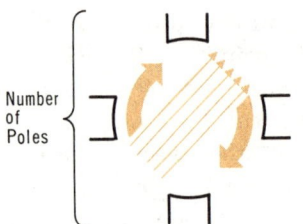

Number of Poles

The natural rate of rotation of the field or synchronous speed, depends on the frequency of the applied ac, the number of phases, and the number of poles

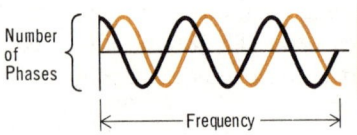

Number of Phases

← Frequency →

Synchronous speed =
$$\frac{\text{a-c frequency (Hz)} \times 120}{\text{number of poles/number of phases}}$$

synchronous speed in the four-pole stator

A-c motors can also be made with more than two main poles. When more than two main poles are used, the fields can be made stronger, and can rotate more smoothly for better operation. The four-pole stator is wound so that the opposite coil sides, which form a complementary north and south pole set, are *90 physical degrees* apart. This means that the effects of the four-pole stator are also felt in the planes between the adjacent poles.

Four-pole stators have their complementary pole pieces 90° apart

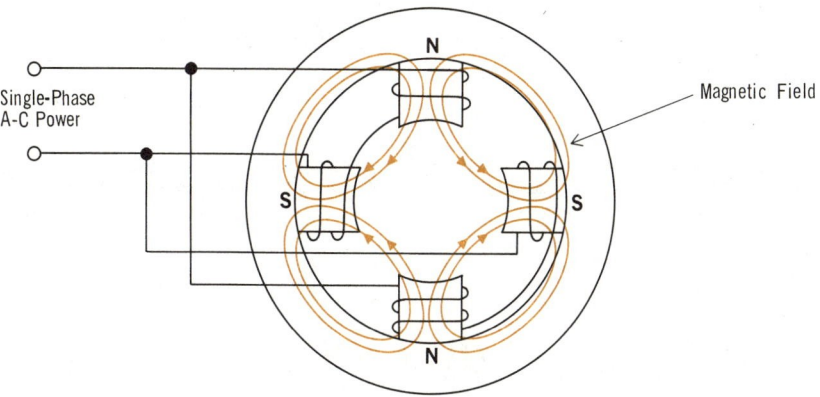

Because the complementary pole pieces are 90° apart, the rotating magnetic field takes twice as long to revolve around the stator as it does in the two-pole stator. The speed of the rotating field of the four-pole stator is thus half that of the two-pole stator

In the two-pole stator, with each a-c alternation, the magnetic field rotates a full 360 degrees. In the four-pole stator, however, at the end of one full 360-degree a-c sine wave, the field rotates through only 180 degrees. This means that in the motor with a four-pole stator, the synchronous speed is half that of the two-pole motor, or 1800 rpm. Substituting in the formula you saw previously proves this:

$$\text{Synchronous speed (rpm)} = \frac{\text{a-c frequency (Hz)} \times 120}{\text{number of poles/number of phases}}$$

$$1800 = \frac{60 \times 120}{4}$$

In a like manner, it is possible to calculate the synchronous speed of the rotating magnetic field of motors with any number of main poles.

The alternator and the synchronous motor can be thought of as equal opposites. The two are difficult to tell apart physically

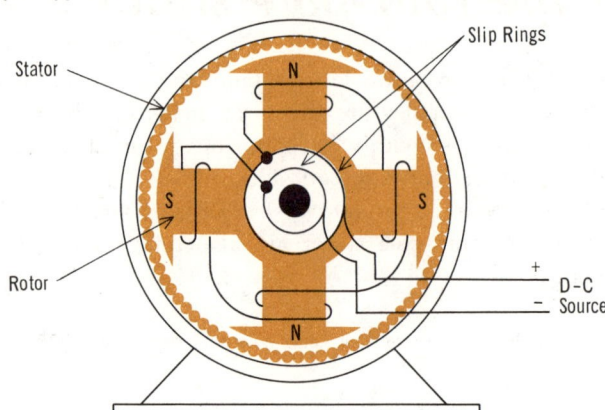

The synchronous motor can be used as an alternator and vice versa. This parallels the case of being able to use the d–c motor as a generator and the generator as a motor

synchronous motors

As you have learned, it is possible to operate a d-c generator as a d-c motor and vice versa. When an a-c generator or alternator is driven by electrical energy, it also develops mechanical power. Under these conditions, the alternator operates as a *synchronous motor*.

The synchronous motor gets its name from the term *synchronous speed,* which was used to describe the natural speed of the rotating magnetic field of the stator. In the synchronous motor, a rotating magnetic field is created and is used to react with a specially created rotor field. In the motor action that results, the rotor is actually *locked in step* with the rotating stator field and is stiffly dragged along at the synchronous speed of the rotating magnetic field. As you have learned in your study of the rotating magnetic field, its speed of rotation is controlled strictly by the frequency of the applied a-c power and the number of main poles. Since power frequency is regulated by the power companies, synchronous motors tend to maintain their speed with a high degree of accuracy. For this reason synchronous motors find important application in electric clocks and other timing devices.

the basic synchronous motor

A basic synchronous motor of the type you might find in a typical electric clock is shown. The stator, you will notice, includes shaded poles to insure that a rotating magnetic field will be created. The rotor shown is a soft iron slug into which the rotating stator field magnetically induces poles of opposite polarity. As a result, a magnetic attraction force is set up between the corresponding north and south poles of the rotating field and the induced south and north poles of the rotor. As the stator field rotates, the rotor is dragged along at this synchronous speed. When a light load is placed on the rotor, such as the hands of a clock, the rotor may fall out of step with the rotating field for an instant, but will then again follow the rotation of the rotor field, remaining at the same speed as long as the load remains unchanged.

Soft-iron rotor becomes magnetized by the field's flux lines

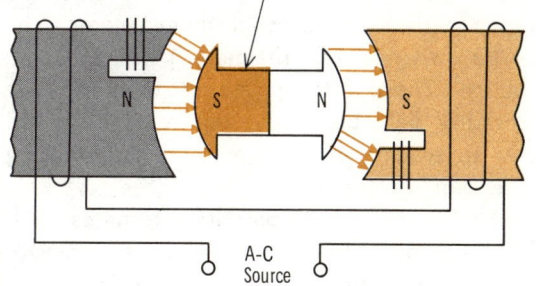

In the basic synchronous motor, the rotor is locked in position by the attractive force of the stator field, and is turned at a synchronous speed with the stator's rotating magnetic field

It happens that in the simple motor of the type shown, the actual magnitude of the poles induced in the rotor is small and there is only a weak interaction with the rotating stator field. As a result, the motor does not develop enough torque to start by itself. Motors of this type in use in electric clocks usually have to be started by hand. In doing this, the user actually gives the rotor a slight spin to overcome starting inertia. Once started in this manner, the motor then develops sufficient torque to keep itself operating.

There are synchronous motors used in electric clocks and other timing devices which are self starting. These motors use so-called *damper windings* on their rotor, which produce an effective increase in the magnetic induction between stator and rotor and result in an improvement in stator-rotor interaction to where sufficient torque is developed to start the motor.

the three-phase synchronous motor

Although the synchronous motor and the alternator are equal opposites, the basic synchronous motor in the electric clock just explained bears little, if any, resemblance to the alternator in Volume 6. The electric clock's motor falls into a special class of *fractional horsepower synchronous motors* generally operated from single-phase a-c power. The synchronous motor matching the alternator is generally a large motor which has a wound rotor and slip rings just as the alternator has. In addition, these motors are usually operated on three-phase power.

In the three-phase synchronous motor, polyphase ac is applied to the stator to produce the required rotating magnetic field in the manner you learned about earlier. If dc is then supplied as excitation voltage to the rotor winding, as in the alternator, a most important difference becomes evident between the single-phase basic synchronous motor as found in electric clocks and the three-phase synchronous motor. The rotor of the latter develops its own powerful magnetic poles from the separate d-c excitation voltage. These rotor poles are then strongly attracted to the stator's rotating field poles and the rotor is locked rigidly in position to rotate in exact synchronism with the rotating field, at the synchronous speed of the motor.

The rotor polarity is fixed while the stator polarity alternates as it rotates. Thus, on starting, if the rotor first approaches an attractive force (opposite pole), it tends to move in one direction; but an instant later, it approaches a repulsive force (like pole) which tends to move it in another direction. The net effect is to maintain the three-phase synchronous motor at a standstill. Thus, like the basic single-phase synchronous motor, the three-phase synchronous motor develops no torque when power is first applied and hence is not self-starting.

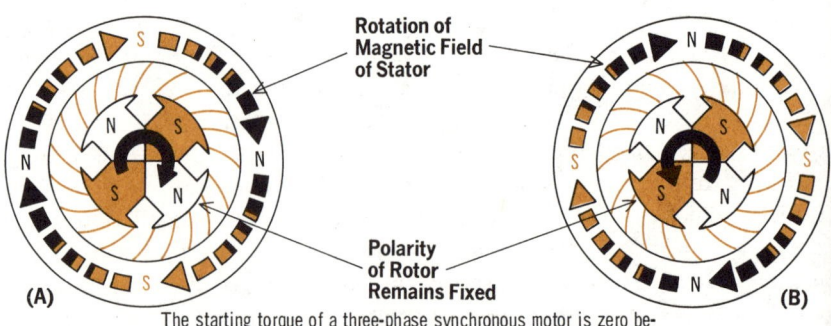

(A)   Rotation of Magnetic Field of Stator   Polarity of Rotor Remains Fixed   (B)

The starting torque of a three-phase synchronous motor is zero because for every tendency for the rotating magnetic field of the stator to pull the rotor around with it, there is an attracting force which tends to pull the rotor in the opposite direction

starting the three-phase synchronous motor

The three-phase synchronous motor is not naturally self-starting. Therefore, these motors have to incorporate some type of starting device or system into their design. In the single-phase synchronous clock motor, which had the same problem, the motor could be started by either giving the rotor a spin by hand or through the use of damper windings on the rotor. The case of the three-phase synchronous motor is similar. Of course, since the three-phase synchronous motor is a huge, powerful device, starting by hand is out of the question. Instead, a practical system is in use by which another motor, a-c or d-c, is used to bring the rotor of the large synchronous motor up to about 90 percent of its synchronous speed. The starting motor is then disconnected, the rotor locks in step with the rotating magnetic field of the stator, and the motor continues to develop operating torque and runs at synchronous speed.

Synchronous motors are often started by a special exciter motor provided just for that purpose

Exciter Motor

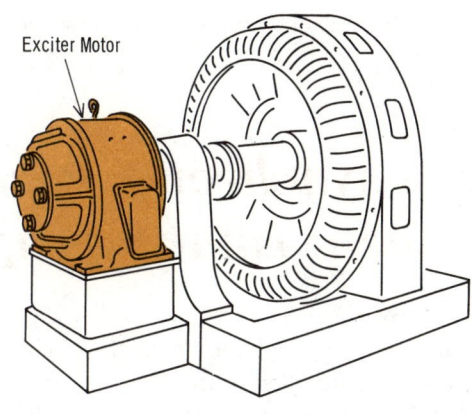

Damper Winding

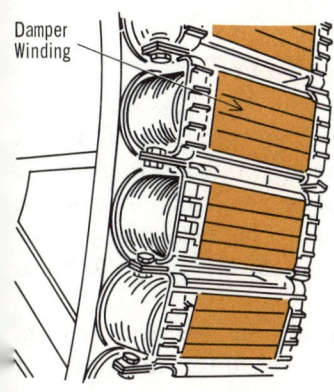

Synchronous motors can also be made self-starting by providing them with a damper winding over their regular wound rotor

In a second system for starting the three-phase synchronous motor, a damper winding is placed over the windings of the rotor. The damper winding allows inductive self starting to take place. You will learn more about this concept later when you study induction motors. When the motor is started by this method, d-c excitation is first removed. As the synchronous motor approaches operating speed, the d-c winding is excited and the rotor locks into step with the rotating magnetic field.

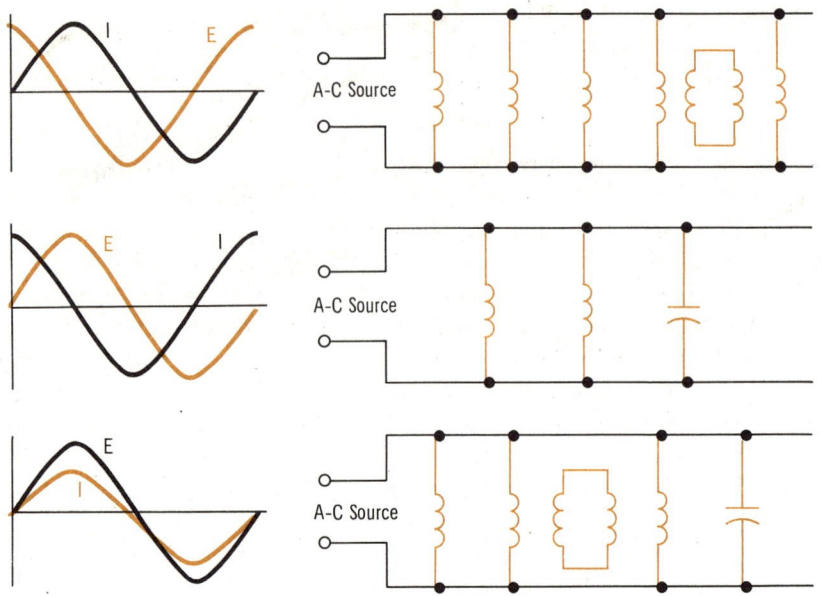

The inductive system formed when many induction motors and transformers are used on the same line has a poor power factor. By adding capacitance to the line in the form of an overexcited synchronous motor, the power factor is corrected and mechanical power is made available as well

the synchronous motor
as a rotating capacitor

A very special characteristic makes synchronous motors useful in industrial applications employing many induction motors. If more d-c current is supplied to the field of a synchronous motor than is required, the motor is said to be *overexcited* and will take a phase-leading current from the power line like a capacitor connected to the line in an otherwise resistive circuit.

Where many transformers and induction motors are used, for example, in large industrial plants, the currents drawn tend to lag the impressed voltage, as is common with inductive circuits. The greater the lag, the poorer the power factor. Thus, power derived from the power lines is not doing the work required. We can improve the power factor by adding capacitance to the load circuits. It thus becomes practical to replace the induction motors in some of the plant machinery with synchronous motors, which help improve the power factor while furnishing useful mechanical power. A synchronous motor used in this way is often said to be operating as a *synchronous capacitor*.

construction

There is practically no difference in construction between the synchronous motor and the alternator. The synchronous motor, therefore, will have a stator and a rotor outfitted with a slip ring and brush arrangement just as the alternator.

The construction of the rotor usually depends on how heavy a load the motor must drive and how fast it has to run. Most synchronous motors are built for low and moderate speeds. These rotors have *salient* or *projecting poles* which are wound on cores bolted to a heavy, cast steel spider ring with good magnetic properties. Motors designed for higher speed service have their cores dovetailed and rigidly locked to the spider ring, which is generally made of laminated steel. In all motors, the rims of the spider ring are usually made extra-heavy to provide the best design for overcoming inertia to insure the easiest possible starting operation.

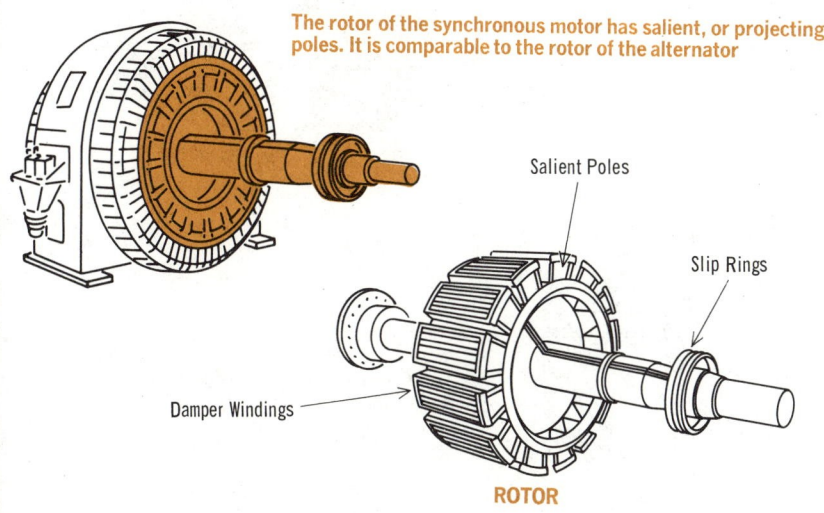

The rotor of the synchronous motor has salient, or projecting poles. It is comparable to the rotor of the alternator

Salient Poles

Slip Rings

Damper Windings

ROTOR

In synchronous motors that use damper windings for starting, the pole faces are slotted to accept the copper bars of the damper structure. The bars set in the slots are then all short-circuited together at each end by means of a conducting ring.

The rotor windings are generally made of copper wire wound on each core so that bare wire is exposed to air to facilitate cooling the structure. In very large motors, the coils are made of copper strapping wound on edge, and asbestos or other heat resistant material must be used to insulate each turn.

construction (cont.)

The stator of the synchronous motor generally consists of the cast iron or welded steel frame which supports a slotted ring of laminated soft steel. The laminations are insulated from each other, and the slots in them are lined with a horn-fiber material. The stator winding coils are set into the open lined slots of the steel ring. All the coils, which are individually insulated, are alike and interchangeable. With this type of construction it is easy to remove and install new coils when repairs are necessary. The stator incorporates spacing blocks which provide an open ended radial air duct in the structure through which cooling air can circulate.

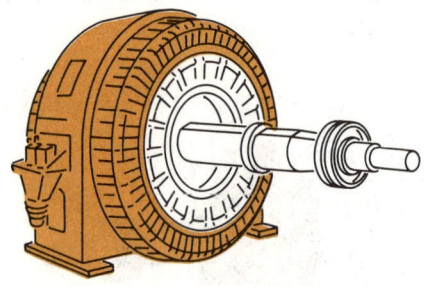

STATOR

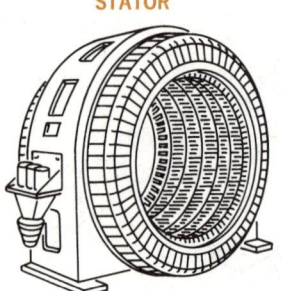

The construction of the stator of the synchronous motor is similar to the stator construction of the alternator. The stator has distributed windings, so that its poles are not discernible

The magnetic poles in the stator are not discernible because no pole pieces are used, as they are on the rotor. Remember, the magnetic polarities of the stator rotate to produce a rotating field. The poles exist at any instant in those coils that conduct the proper current and since the current phases differ in the coils, the poles shift from coil to coil to follow the rise and fall phase shift from coil to coil. Thus, the stator is said to have *distributed* poles.

summary

☐ The natural rate that the stator's magnetic field rotates at is known as the motor's synchronous speed. ☐ The synchronous speed depends on the frequency of the applied power and the number of main poles in the motor:

$$\text{Synchronous speed (rpm)} = \frac{\text{a-c frequency (Hz)} \times 120}{\text{number of poles / number of phases}}$$

☐ A two-pole motor that works on 60-Hz power has a synchronous speed of 3600 rpm. ☐ A four-pole motor driven by 60-Hz power has a synchronous speed of 1800 rpm.

☐ An a-c motor that runs with its rotor locked in step with the rotating field, so that it operates at exactly the synchronous speed, is called a synchronous motor. ☐ The electric clock has a synchronous motor, which uses a soft-iron rotor that is magnetized by the field's flux so that the rotor locks into the rotating flux. ☐ The larger industrial three-phase synchronous motors use a d-c excitation current to magnetize the rotor, so that greater magnetic attraction and torque are developed.

☐ Synchronous motors are not naturally self-starting. ☐ The smaller single-phase motors can be started by hand, or they can be made with damper windings on their rotor to be self-starting. ☐ The larger three-phase industrial motor can be made with damper windings, too, but often an auxiliary d-c motor is used to drive the synchronous motor up to speed. ☐ The larger synchronous motors can operate as synchronous capacitors to improve a plant's power factor.

review questions

1. Define a *synchronous motor*.
2. The basic three-phase motor has six salient poles. What is its synchronous speed on 60-Hz power?
3. The basic split-phase motor has four projecting poles. How fast will it turn on 60-Hz power?
4. If a synchronous motor had four shaded poles, how fast would its rotor turn?
5. Why is the rotor in the simple single-phase synchronous motor made of soft iron?
6. Why is current used to magnetize the rotors in the larger synchronous motors?
7. How would a synchronous motor be designed to be self-starting?
8. How can you make the synchronous motor improve a plant's power factor?
9. A three-phase synchronous motor resembles an _____.
10. What are *distributed poles*?

induction motors

When you learned about the basic single-phase synchronous clock motor, you were told that the rotor becomes magnetized because of magnetic induction by the field poles. You learned about this in Volume 1. A soft temporary magnet placed in a flux field has its molecules oriented so that the metal becomes magnetized.

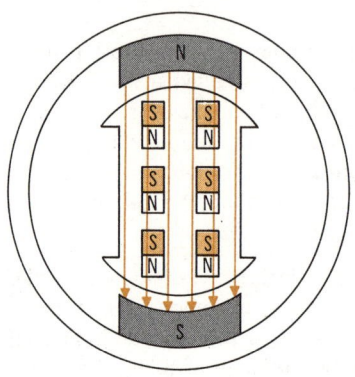

Synchronous motors use magnetic induction to magnetize their rotors

The flux lines of the field orient the magnetic molecules in the soft iron rotor

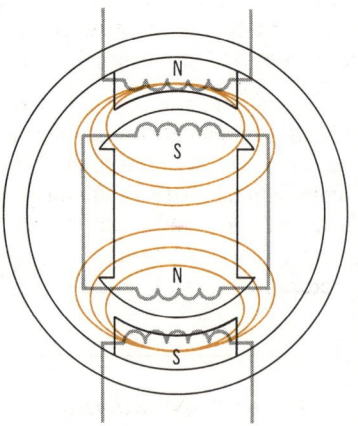

Induction motors use electromagnetic induction to magnetize their rotors

The stator acts like the primary winding of a transformer, and the rotor acts like the secondary winding

The induction motor is similar to the synchronous motor in that it depends on the magnetic field for its operation. However, here the similarity ends. The induction motor does not depend on magnetic induction. Instead, it relies on *electromagnetic induction*. The stator and rotor act like the primary and secondary windings of a transformer. The rotating magnetic field of the stator induces high currents in the rotor; and these currents, in turn, produce their own magnetic fields, which interact with the main field to make the rotor turn.

Because the simple synchronous motor used magnetic induction, the rotor's field was weak. But because the induction motor uses electromagnetic induction, the rotor's field is strong. Therefore, the induction motor can be self-starting and produce enough torque without slip rings, commutators, or brushes. The only power supplied to the rotor is through electromagnetic induction from the stator.

construction

In the induction motor, as in the synchronous motor, the stationary member is called a *stator* and the rotating member is called a *rotor*. The stator of the induction motor is the only member to which power is applied. All of the various stators you learned about in your study of how a rotating magnetic field is created are also used in induction motors. In fact, induction motors are usually named according to the manner in which the rotating field of the stator is created. Therefore, you hear of *shaded-pole, split-phase, capacitor-start,* and *polyphase induction motors.*

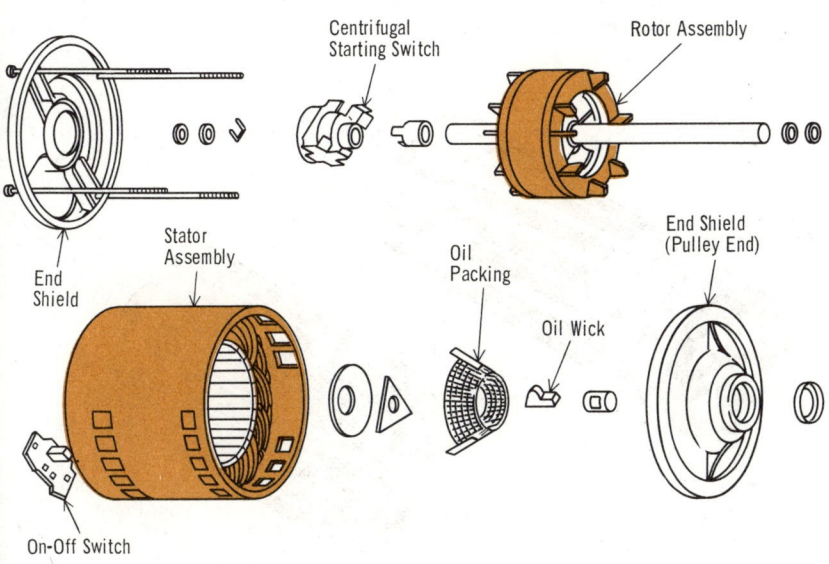

In the a–c induction motor, there are usually no electrical connections made to the rotor by brushes as in synchronous and d–c motors. Therefore, this motor has long trouble-free service and is most widely used of all a–c motors

Although you learned about two-pole, four-pole, etc. stators, it is not always easy to count a motor's poles. Instead of the easily distinguishable *salient* (projecting) poles used to illustrate the stators, many induction motor stators have *distributed windings* similar to the type described for synchronous motors. This means that if you looked at the stator windings of some induction motors, you could not count the poles. Instead you would have to rely on the manufacturer's information on the nameplate of the motor to get this information.

construction (cont.)

The simplest and most widely-used rotor for induction motors is the so-called *squirrel-cage rotor*, from which the *squirrel-cage induction motor* gets its name. The squirrel-cage rotor consists of a laminated iron core which is slotted lengthwise all around its periphery. Solid bars of aluminum, copper, or other conductors are tightly pressed or embedded into the slots. At both ends of the rotor, short-circuiting rings are welded or brazed to the bars to make a solid structure. The short-circuited bars, because of their very low resistance with respect to the core, do not have to be specially insulated from the core. In some rotors the bars and end rings are cast as a single integral structure for placement on the core. The short circuiting elements actually form *shorted turns* that have high currents included in them by the field flux.

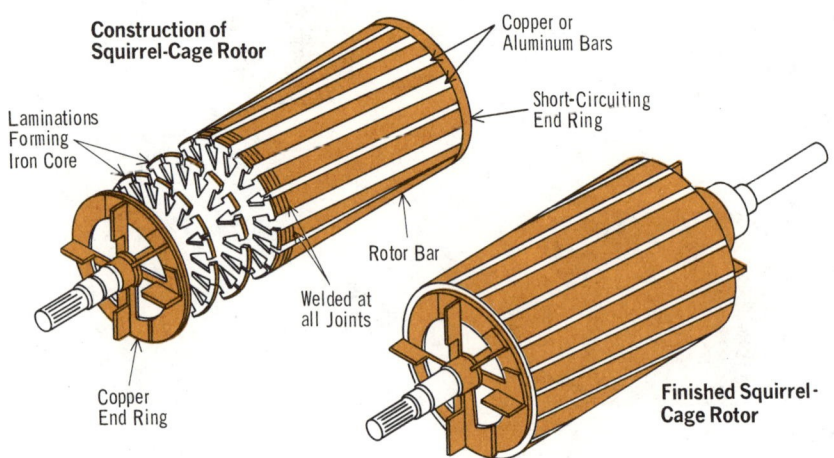

The squirrel-cage rotor used in a–c induction motors is extremely simple in construction when compared, for example, to the armature of the d–c motor, with its complicated winding arrangements

Compared to the intricately wound and arranged rotor of the synchronous motor or the armature of the d-c motor, the squirrel-cage rotor is relatively simple. It is easy to manufacture and is essentially troublefree in actual service.

In an assembled squirrel-cage induction motor, the periphery of the rotor is separated from the stator by a very small air gap. The width of this air gap in fact is as small as mechanical clearance needs will permit. This insures that the strongest possible electromagnetic induction action will take place.

motor action

You have learned how a rotating magnetic field can be created in a stator and know a little about the squirrel-cage rotor. You are therefore ready to learn about basic motor action in the induction motor.

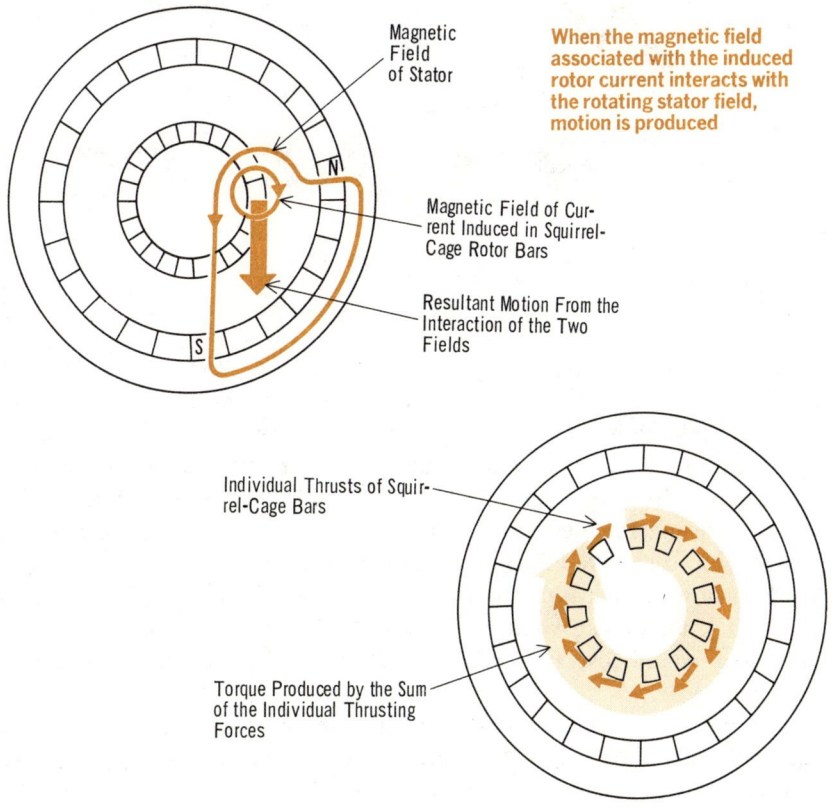

Magnetic
Field
of Stator

When the magnetic field associated with the induced rotor current interacts with the rotating stator field, motion is produced

N

Magnetic Field of Current Induced in Squirrel-Cage Rotor Bars

Resultant Motion From the Interaction of the Two Fields

S

Individual Thrusts of Squirrel-Cage Bars

Torque Produced by the Sum of the Individual Thrusting Forces

When power is applied to the stator of a practical induction motor, a rotating magnetic field is created by any one of the means you learned about. As the field begins to revolve, its flux lines cut the shorted turns embedded around the surface of the squirrel-cage rotor and generate voltages in them by electromagnetic induction. Because these turns are short circuits with very low resistance, the induced voltages cause high currents to circulate in the rotor bars. The circulating rotor currents then produce their own strong magnetic fields. These local rotor flux fields produce their own magnetic poles, which are attracted to the rotating field. Thus, the rotor revolves with the main field.

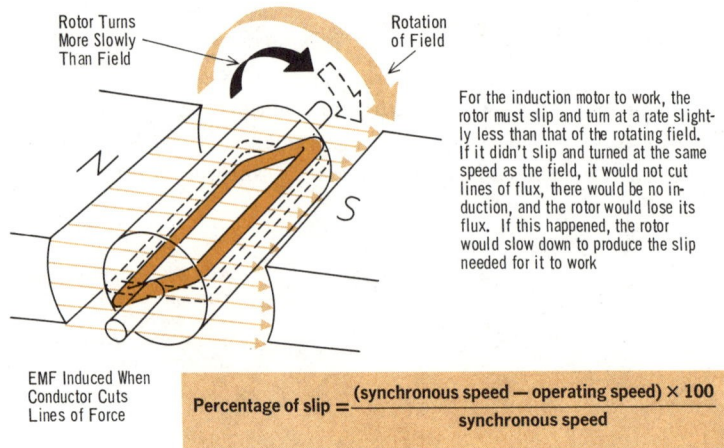

Rotor Turns More Slowly Than Field

Rotation of Field

For the induction motor to work, the rotor must slip and turn at a rate slightly less than that of the rotating field. If it didn't slip and turned at the same speed as the field, it would not cut lines of flux, there would be no induction, and the rotor would lose its flux. If this happened, the rotor would slow down to produce the slip needed for it to work

EMF Induced When Conductor Cuts Lines of Force

$$\text{Percentage of slip} = \frac{(\text{synchronous speed} - \text{operating speed}) \times 100}{\text{synchronous speed}}$$

slip

You just learned how the electrically created rotary motion of the magnetic field of the stator created mechanical rotation of the rotor, through induction, in the same direction as the field of the stator. If there were no load on the motor and no friction or resistance to motion, you might think the motor speed would build to the synchronous speed of the rotating magnetic field. However, if the rotor turned exactly at this speed, there would be no relative difference in motion between field and rotor, the copper bars in the rotor would not cut the field flux lines, and no current would be induced in the rotor. Without induction, the rotor would lose its own magnetic field and no torque would be produced.

For the induction motor to work, the rotor must run at a different speed than the rotating stator field. Actually, the rotor speed is slightly slower than the rotating stator field. This difference is known as *slip*. In practical motors, rotors run from 2 to 10 percent behind the synchronous speed at no load; as the load increases, the percentage of slip increases.

In a four-pole motor with a synchronous speed of 1800 rpm, if the rotor actually turns at 1764 rpm as measured with a *tachometer,* there is a difference of 36 rpm between synchronous and rotor speeds. Expressed as percentages this would be $36/1800 \times 100$, reflecting 2 percent slip. In a two-pole motor with a 36 rpm difference, slip would be less, because, you will recall, at 60 Hz the synchronous speed of two-pole motor is 3600 rpm. Substituting in the formula:

$$\frac{3600}{36} \times 100 = 1\% \text{ slip}$$

starting torque characteristics

The starting torque of the basic squirrel-cage induction motor is low, because at rest the rotor has a relatively large inductive reactance (X_L) with respect to its resistance (R). Under these conditions, we would expect the rotor current to lag rotor voltage by 90 degrees. We thus say that the power factor in the circuit is low. This means that the motor is inefficient as a load and cannot derive really useful energy for its operation from the power source.

Despite the inefficiency, a torquing force is developed and the motor does begin to turn. As it starts turning, the difference in speed between rotor and rotating field, or slip, goes from a maximum of 100 percent to some intermediate value, say 50 percent. As the slip decreases in this manner, the frequency of the voltage induced in the rotor decreases, because the rotating field cuts conductors at a decreased rate; this, in turn, causes the overall inductive reactance in the circuit to decrease. As inductive reactance decreases, the power factor begins to increase and there is an improvement in the manner in which the motor uses the power applied to it. This improvement is reflected as an increase in torque and a subsequent increase in speed.

When the slip drops to some value between 2 and 10 percent, the motor speed stabilizes. This stabilization occurs because every tendency for the motor speed to increase to where slip will drop below 2 percent is naturally offset by the fact that, as the rotor approaches within 2 percent of the synchronous speed, the effects of reduced induction overcome the previous tendency to increase torque as the motor is speeded up from start. Thus, the a-c motor exhibits an automatic speed control characteristic similar to that of the d-c shunt motor.

CLASSES OF SQUIRREL-CAGE MOTORS
(According to Starting Characteristics)

Class	Starting Characteristics
A	Moderate torque, moderate starting current.
B	Moderate torque, low starting current.
C	High torque, low starting current.
D	High slip.
E	Low starting torque, moderate starting current.
F	Low starting torque, low starting current.

operating torque characteristics

You have learned that after some difficulty in getting started because of a poor starting torque characteristic, the rotor of an induction motor will eventually reach a stable speed at from 2 to 10 percent slip. This is the no-load operating speed of the motor. At this speed, with no-load conditions, very little torque is required to keep the motor going.

If a load is placed on the operating motor, the rotor slows down and the slip increases. As a result, the rotating field cuts the copper rings at an increased rate, inducing higher rotor currents to provide increased torque for the motor to handle the load.

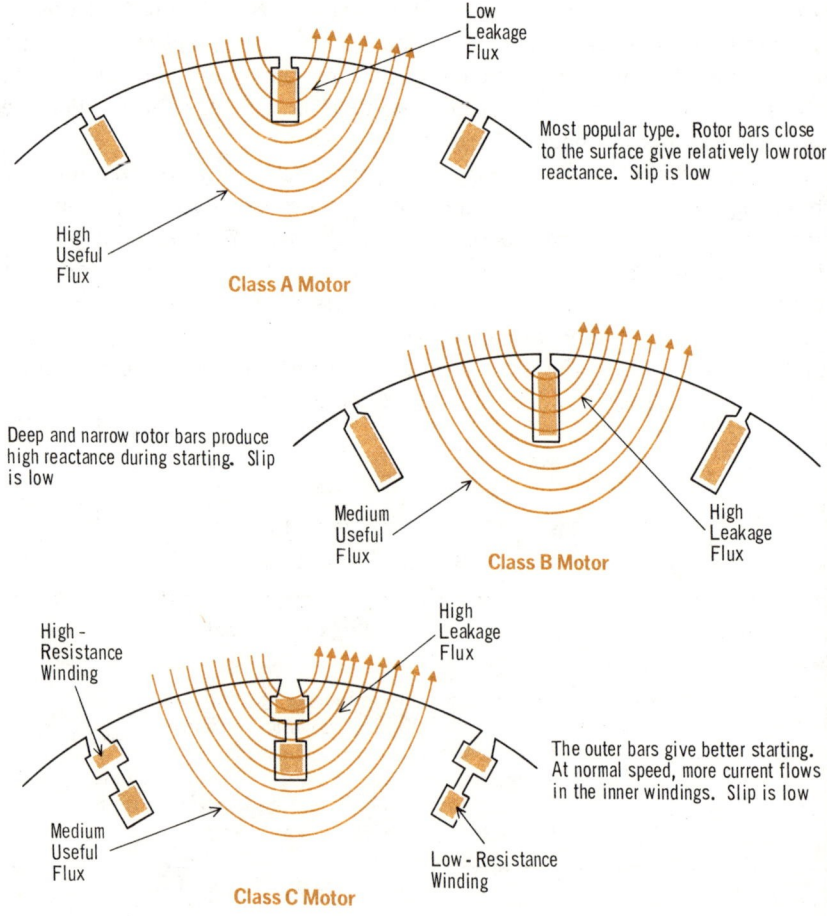

Low Leakage Flux

Most popular type. Rotor bars close to the surface give relatively low rotor reactance. Slip is low

High Useful Flux

Class A Motor

Deep and narrow rotor bars produce high reactance during starting. Slip is low

Medium Useful Flux

Class B Motor

High Leakage Flux

High - Resistance Winding

High Leakage Flux

The outer bars give better starting. At normal speed, more current flows in the inner windings. Slip is low

Medium Useful Flux

Low - Resistance Winding

Class C Motor

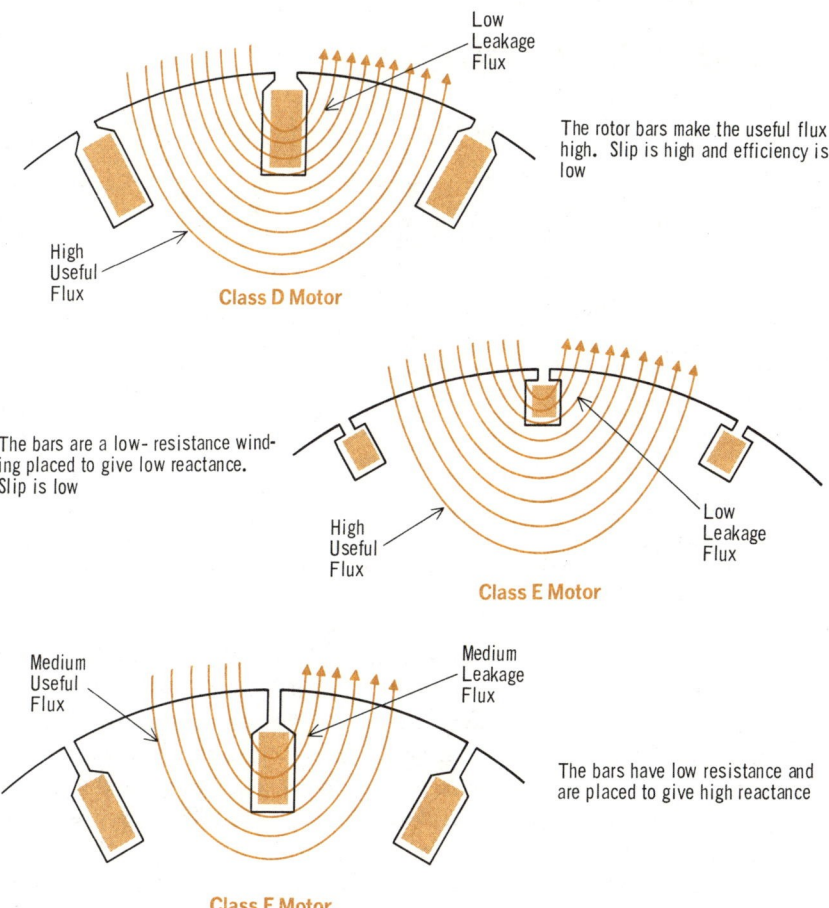

The rotor bars make the useful flux high. Slip is high and efficiency is low

Low Leakage Flux

High Useful Flux

Class D Motor

The bars are a low- resistance winding placed to give low reactance. Slip is low

High Useful Flux

Low Leakage Flux

Class E Motor

Medium Useful Flux

Medium Leakage Flux

The bars have low resistance and are placed to give high reactance

Class F Motor

operating
torque characteristics (cont.)

If the load on the motor is increased to the point at which slip is increased and hence motor speed is decreased, a reversal of the starting conditions for the motor occurs. That is, by decreasing the rotor speed, we decrease torque by lowering the power factor so that the tendency for inductive action to increase torque is overcome. With the torquing force thus exhausted, the motor is brought to a stop or is stalled by the load. Under these conditions, we say that the *breakdown torque* of the motor has been reached.

the double-squirrel-cage induction motor

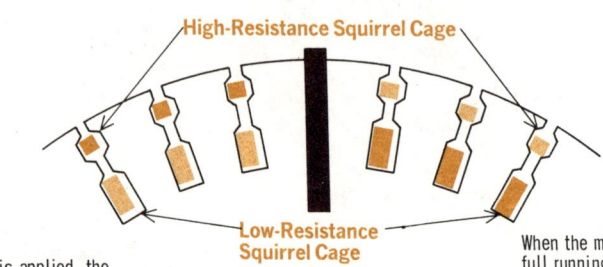

High-Resistance Squirrel Cage

Low-Resistance Squirrel Cage

When current is applied, the high-resistance cage carries most of the current induced in the rotor, giving high starting torque with low starting current

When the motor has reached full running speed, the low-resistance inner cage current is much higher than that of the high-resistance outer cage, and high-efficiency operation results

The double-squirrel-cage rotor provides higher starting torque than the conventional squirrel-cage rotor. High-resistance bars set in the rotor make this possible. It is difficult to distinguish between squirrel-cage rotors, since construction differences are internal

The poor starting torque of the standard squirrel-cage induction motor comes from its relatively high inductive reactance at rest compared to its resistance. Since inductance and resistance properties are designed into the rotor and cannot be changed for different applications, the standard squirrel-cage motor is limited to loads requiring variable torque at approximately constant speed with high full-load efficiency, such as blowers, centrifugal pumps, and motor-generator sets.

For a high starting torque we need a special high-resistance rotor, which has two sets of rotor bars. A set of low-resistance bars is embedded deep in the core slots. A second set, of higher resistance bars, is embedded in the core slots close to the surface of the rotor.

Most of the current flows through the high resistance bars on starting, because the inductive reactance of the low-resistance bars is higher on starting than the resistance of the high-resistance bars. Since the power factor is better in the high-resistance circuit, starting torque is improved with respect to the standard rotor. When the motor reaches operating speed, most of the current is carried by the lower bars because inductive reactances are equalized and ohmic value alone is the controlling factor.

Under such conditions with no load, the double-squirrel-cage motor is indistinguishable from the standard squirrel-cage motor. However, under varying load conditions, the current automatically divides between sets of bars in proportions required to produce the required amount of torque for the given load condition.

the wound-rotor induction motor

Up to this point in our study of a-c induction motors we have considered only motors with squirrel-cage rotors. It is also feasible to construct an a-c induction motor which has windings around the rotor core in place of the conducting bars of the squirrel-cage motor. In this case, currents are induced in the windings just as they would be in shorted turns. However, the advantage of using windings is that the wires can be brought out through slip rings so that the resistance, and therefore the current through the windings, can be controlled. The so-called *form-wound rotor* is similar in basic appearance to the wound armature of the d-c motor.

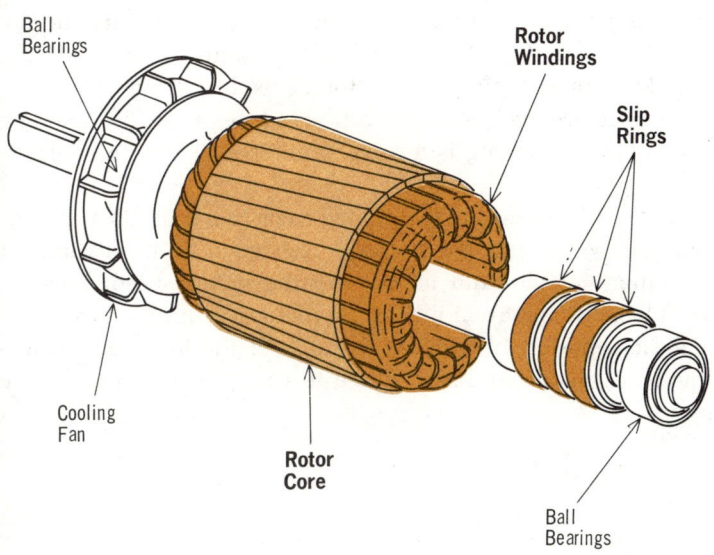

The wound rotor is generally used in larger induction motors that are operated from three-phase a-c power

By and large, the squirrel-cage induction motor finds almost universal application. Where the standard squirrel-cage motor cannot do because of starting torque requirements, the double-squirrel-cage motor is usually adequate. In fact, for applications using single-phase a-c power, the squirrel-cage family of a-c motors is used almost exclusively. The wound-rotor induction motor is used only in special applications and is always operated on three-phase a-c power.

the wound-rotor induction motor (cont.)

The three ends of the three-phase rotor windings are brought out to three slip rings mounted on the rotor shaft. The brushes bearing on the slip rings play an important role in realizing maximum advantage from the wound-rotor motor. By connecting the brushes through rheostats, it becomes possible to develop a higher starting torque than is possible with either the single- or double-squirrel-cage motors. On starting, the full resistance of the rheostats is maintained in the rotor circuit, thus providing the very maximum starting torque, since the resistance values are chosen to be equal to or greater than the reactance of the rotor at standstill.

As the motor approaches normal operating speed, the rheostat resistance is gradually reduced until it is out of the circuit entirely at full speed. Ironically, although its starting torque is higher, it is not as efficient as the squirrel-cage motors at full speed, because the inherent resistance of the rotor winding is always more than that of a squirrel-cage rotor.

An obvious special feature of the wound-rotor motor is its variable-speed capability. By varying the rheostat resistance, it is possible to vary the percentage of slip and hence vary the motor speed. In such cases, below full-speed operation means the motor is running at reduced efficiency and horsepower. In addition, because of the high rotor resistance, the motor is made more susceptible to variation in speed as the load changes.

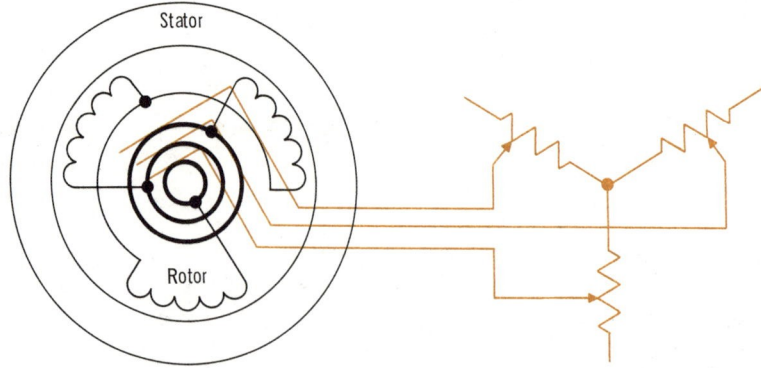

The wound-rotor motor differs from other induction motors in that it uses resistance in series with its rotor windings through a slip ring and brush arrangement

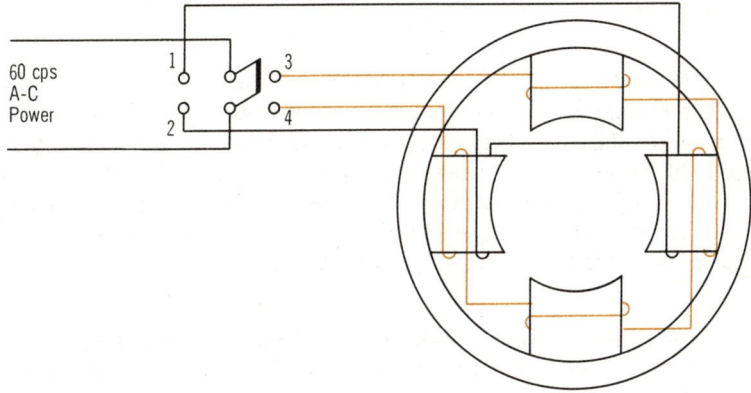

In this multiple-winding motor, if wires 1 and 2 are connected to the power source, it is a two-pole motor that runs at about 3600 rpm. But if wires 3 and 4 are connected, it becomes a four-pole motor that runs at about 1800 rpm

multispeed squirrel-cage induction motors

Since the variable-speed wound-rotor motor is limited to three-phase applications for the reasons just discussed, you may be wondering how more than one speed can be provided by a squirrel-cage induction motor. How, for example, do we get several speeds in such household appliances as the multiple-speed fan or mixer, which use squirrel-cage motors?

You have learned that the speed of an induction motor depends on the power supply frequency and the number of pairs of poles used in the motor. Obviously, to alter motor speed it is merely necessary to change one of these two factors. By far the most common method used involves changing the number of poles, generally at some type of external controller. There are two types of multispeed squirrel-cage induction motors in common use: the multiple-winding motor and the consequent-pole motor. Both feature poles that may be changed as required by reshifting key external connections, and in this way provide for operating the motor at a limited number of different speeds.

In the multiple distributed-pole-winding motor, two or more separate windings are placed in the stator core slots, one over the other. For example, an eight-pole winding can be laid in the core slots and a four-pole winding placed on top of it. The windings are, of course, insulated from each other and are arranged so that only one winding at a time can be energized.

summary

☐ The stator and rotor of an induction motor act like windings of a transformer. The rotating stator field induces current in the rotor to produce the rotor's magnetic field. ☐ The stator in the induction motor can produce a rotating field by any method used in the basic a-c motor. ☐ The induction motor rotor, though, is designed for efficient electromagnetic induction. ☐ The squirrel-cage rotor, which is widely used, has solid conducting bars connected to form shorted turns to conduct large induced currents.

☐ Since the induction motor relies on electromagnetic induction, its rotor cannot run at the synchronous speed. It must slip to cut the field's lines of force. ☐ The induction motor usually slips from 2 to 10 percent behind synchronous speed.

☐ The standard induction motor has a poor starting torque. But starting torque can be improved by using a double-squirrel-cage rotor. ☐ Some special three-phase induction motors use wound rotors with a brush and slip ring arrangement, so that resistance can be added to control rotor current and increase starting torque, or even increase slip to control the speed. ☐ Multispeed motors can be made with selectable multiple-pole stators, or with consequent-pole stators.

review questions

1. The rotors of synchronous motors become magnetized through _____ induction.
2. The rotors of induction motors become magnetized through _____ induction.
3. Compare the speeds of an induction motor and a synchronous motor that are similar in design.
4. What is the equation for slip?
5. How different are the stator windings of an induction motor from that of a synchronous motor?
6. Why does the squirrel-cage rotor use shorted turns?
7. How fast will a single-phase, four-pole induction motor turn if it has 5 percent slip?
8. What kind of induction motors can have good starting torque?
9. What is *breakdown torque*?
10. What kind of stators can a motor have to give multiple speeds?

the consequent-pole motor

In the consequent-pole motor we get two speeds, for example, by arranging the motor to have a certain number of poles for high-speed operation and then by a switching action doubling this number of poles to give low-speed operation. The switching action is illustrated by use of a **two**-phase motor to simplify the explanation. If you trace the wiring in A, you can see how the system is phased so that both magnetic north and south poles are produced at the winding projections as indicated. With **two**-phase power applied to the four-pole system, we get a rotating magnetic field at **3600 rpm**.

Consequent poles get their name because they are formed as a direct consequence of having formed monopole north poles

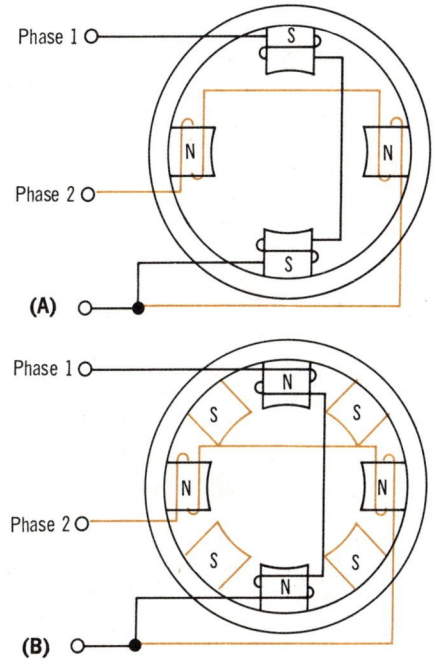

The combination of consequent poles and actual monopoles gives double the number of actual poles that can be counted. By doubling the number of poles, the speed of the motor is reduced by one-half

In B, we have changed the connections so that the system is phased to produce four magnetic north poles at the winding projections. Since for every north pole there must be a south pole, south poles, known as *consequent poles* because they are formed as a consequence of having formed north poles, are produced between the projecting north poles as indicated. This means that in B, we have twice as many pole groups as in A. We, therefore, get an eight-pole rotating magnetic field at 1800 rpm.

the practical single-phase consequent-pole motor

You just learned how an opposite pole is automatically formed as a consequence of having created an apparent monopole. Although we used a **two**-phase motor as an example, the consequent-pole system is more often found in single-phase induction motors.

Part A of the illustration shows what we will call the *short jumpering arrangement* of the motor. In this, all windings are in series and we produce alternate north and south poles. To produce consequent poles, we replace the series connection shown in A with a parallel connection accomplished by the *long jumpering arrangement* shown in B. By connecting the motor in this manner, we produce four salient monopoles and, as a result, create four opposite consequent poles. In the practical consequent-pole motor, all necessary internal connection rearrangements are accomplished at an external control panel.

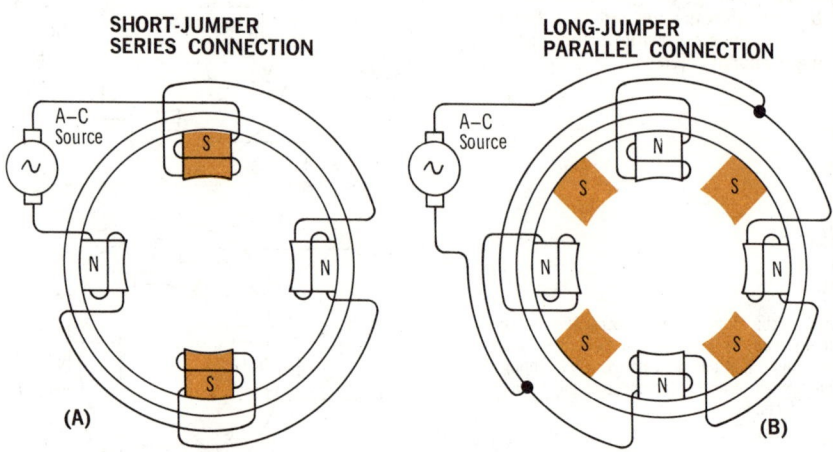

In the single-phase, consequent-pole motor, series and parallel methods of interconnecting pole-phase groups are used. The complicated wiring rearrangements are controlled from an external panel to which all motor leads are connected

Consequent-pole motor characteristics depend on their intended application. In the *constant-horsepower motor,* torque varies *inversely* with speed. It is used for driving machine tools, lathes, etc. In the constant-torque motor, horsepower varies directly with speed. It is used to drive pumps, air compressors and in constant pressure blowers. In the *variable-torque, variable-horsepower motor,* both torque and horsepower *increase* with increases in speed. This is the type of motor found in household fans and air conditioners.

the reluctance motor

The reluctance motor is a fractional-horsepower synchronous motor that uses a special type of magnetic squirrel-cage rotor. In the reluctance motor, special grooves are cut into an otherwise conventional squirrel-cage rotor. These grooves are then used to wind salient poles **of soft iron** that correspond to the number of poles in the stator. This arrangement is sort of the reverse of the wound rotor with a squirrel-cage like damper winding for starting on top.

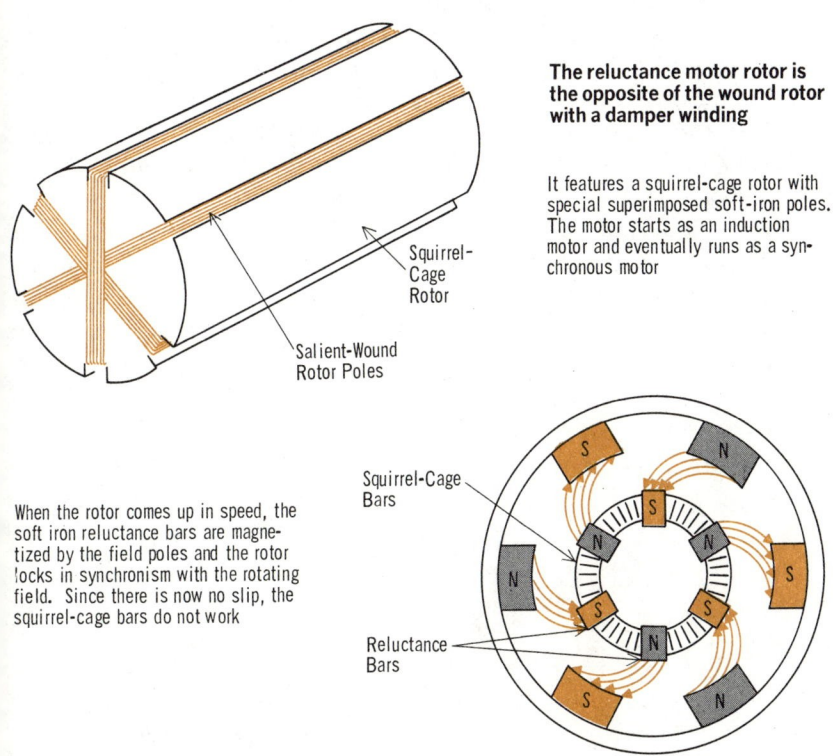

Squirrel-Cage Rotor

Salient-Wound Rotor Poles

The reluctance motor rotor is the opposite of the wound rotor with a damper winding

It features a squirrel-cage rotor with special superimposed soft-iron poles. The motor starts as an induction motor and eventually runs as a synchronous motor

Squirrel-Cage Bars

Reluctance Bars

When the rotor comes up in speed, the soft iron reluctance bars are magnetized by the field poles and the rotor locks in synchronism with the rotating field. Since there is now no slip, the squirrel-cage bars do not work

The reluctance motor starts as a squirrel-cage induction motor. As the reluctance motor comes to synchronous speed, those soft iron poles on the rotor gradually slip less and less. At some point in the operation, the relative difference in speed between the rotating magnetic field and the rotor is sufficiently slow to cause the poles of the rotor to be magnetized by the stator poles. Then, the rotor locks into step with the rotating field. At this point, the rotor tends to run like a synchronous motor at synchronous speed.

the hysteresis motor

The *hysteresis motor* is a special form of synchronous motor in which the magnetic properties of the material used in the rotor are used to create a large torque and to achieve synchronous action without the need for external d-c excitation.

From your study of magnetism, you will recall that when an iron bar was placed in a magnetic field, the molecules of iron arranged themselves in an orderly way. In doing so, the iron bar became a magnet. The north pole of the bar pointed to the south pole of the field, and the south pole pointed to the north pole of the field.

If such a bar is rotated clockwise in the field, each molecule in the bar remains in its original position, aligned with the external magnetic field. In other words, the molecules themselves were rotated counterclockwise with respect to the physical position of the bar. This change in the position of the molecules with respect to the position of the bar consumes power.

We say that this use of power is due to *hysteresis effects,* and it is from this that we get the name hysteresis motor. In the hysteresis motor, the large amount of power consumed in the rotor due to hysteresis losses, provides us with a large amount of effective torque for motor operation.

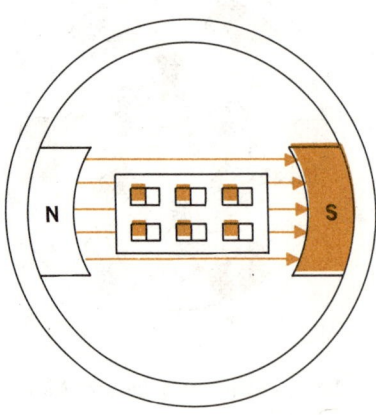

Hysteresis losses are created when the molecules of an iron bar remain aligned in an external magnetic field after the position of the bar in the field has been changed

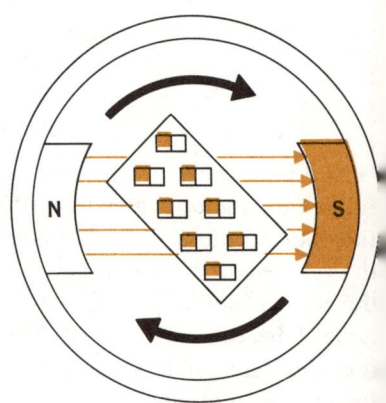

These losses are consumed as power by the rotor to provide torque

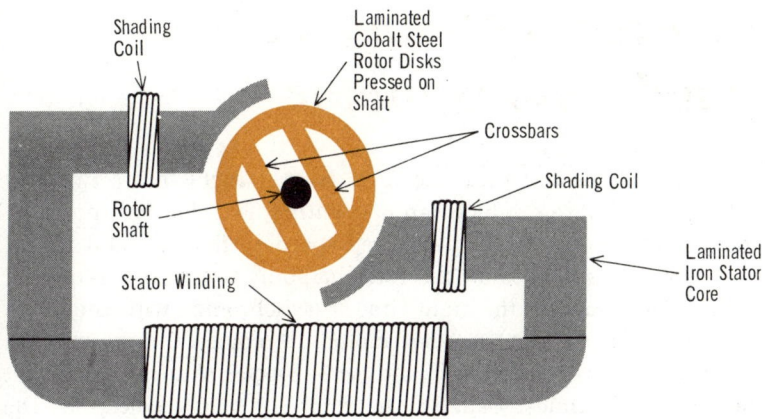

Hysteresis losses represent power being consumed. In the hysteresis motor, we take advantage of this principle. Hysteresis losses in the rotor of a hysteresis motor cause a very large torque to be developed

The hysteresis motor starts as a shaded-pole induction motor and runs as a synchronous motor

the hysteresis motor (cont.)

A typical hysteresis motor uses cobalt steel in its rotor. This material has good magnetic retentivity and is highly permeable to a magnetic field. The rotor consists of several flat disks of cobalt steel mounted on a shaft. The rotor assembly is mounted between two shaded stator poles made of laminated iron. The effect of a rotating field created by the shaded poles induces a current in the rotor circuit which starts the motor as an induction motor. However, as the motor develops speed, a comparatively large hysteresis loss develops in the disk rotor, creating a larger torque than is available from a conventional motor. A path of minimum reluctance exists along the axis of the two rotor crossbars.

The induced current flow in the rotor thus causes permanent magnetic poles to be developed along the crossbar axis. Because the cobalt steel is retentive, the poles are maintained through the effects of residual magnetism. The poles formed in this manner then lock into synchroism with the rotating field, and the motor operates as a synchronous motor. The rotor system continues to consume power in the form of hysteresis loss, and the large torque developed remains. This type of motor can maintain a constant speed very effectively, even with erratic load changes.

Hysteresis motors are well-known for their application in high-fidelity phonograph and tape recording mechanisms where constant speed is a must. They are also used in clock motors and other timing devices.

the permanent-magnet synchronous motor

When the a-c motor was first discussed, it was taught that a two-pole field with a single phase produces an *alternating* field that will not turn the rotor unless it is first run up to rated speed so that it could follow the alternations. This is because the rotor had only two poles which had to be in the right place at the right time to synchronize with the alternating field. The permanent-magnet motor overcomes this with extra rotor poles, all the same polarity, and with special shapes to interact with the field at all times. Usually, the rotor has many poles, but the one shown here is simplified.

The diagram shows a two-pole, single-phase motor with three rotor poles, all permanently magnetized *South*. The rotor shape is such that when the upper pole is North, a rotor S pole is attracted into alignment with it. The lower field pole is South, and repels the S rotor on either side of the field pole. When the field polarity alternates, the lower field pole becomes North, and attracts a rotor pole in alignment with it, thus rotating the rotor a distance equal to one-half its pole pitch. At this time, the upper field pole is South, and holds two rotor S poles on either side of it. At the next alternation, the upper field pole becomes North again to rotate the rotor another half a pole pitch to align a rotor pole with it. This continues, with each alternation rotating the rotor one-half a pole pitch.

Since the rotor moves one *full pole pitch* for two alternations (one *cycle*), each complete revolution of the rotor requires a full cycle for each rotor pole. Thus, if we use 60 Hz power, which is equivalent to 3600 rpm, the rotor will turn 3600 ÷ 3 poles, or 1200 rpm.

Each alternation of the power moves the rotor a distance equal to half its pole pitch

Two alternations in a 2-pole permanent-magnet synchronous motor

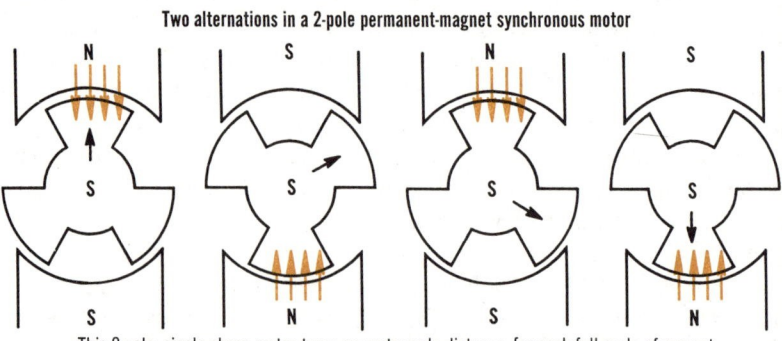

This 2-pole, single-phase motor turns one rotor pole distance for each full cycle of current. Since it requires three cycles for one revolution, its synchronous rpm is 1200

the permanent-magnet synchronous motor (cont.)

The number of rotor poles is what determines the rotor speed. A single-phase, four-pole field motor with a six-pole rotor works in the same way as the two-pole motor: the rotor turns half a pole pitch for each alternation, and a full pole pitch for a complete cycle. Since there are six rotor poles, the motor speed for 60 Hz power is 3600 ÷ 6, or 600 rpm.

Unlike other a-c motors, this motor can turn at relatively low speeds. The more poles there are in the rotor, the lower the rpm will be. Having more rotor poles also helps the motor run more smoothly. Since there is no rotating field, the rotor actually "steps" for each alternation. The more poles there are, the more the stepping or pulsing action becomes smooth.

Any permanent-magnet synchronous motor that uses single-phase power must be first turned in the desired direction to make sure it will run in that direction. However, if two-phase, or split-phase, power is used, the motor will always turn in the proper direction. The four-pole, two-phase motor is an example of this. Notice that for the eight field poles, there are ten rotor poles. For one phase of excitation, the rotor poles are aligned with their respective field poles. In this condition, the second-phase field poles are only half aligned with the rotor poles they face. When the first phase drops off, and the second phase comes up, its field poles move the rotor *one-quarter* pole pitch distance to align them. This puts the pole pieces half in alignment with the second winding of the first phase, so that when the second phase drops off and the first phase alternates, the rotor will again move *one-quarter* pole pitch distance. This continues with the rotor moving one pole distance for two alternations (one cycle) of *any one phase*. Thus, ten poles on the rotor require ten cycles, so that with 60 Hz power, the rotor speed will be 3600 ÷ 10, or 360 rpm.

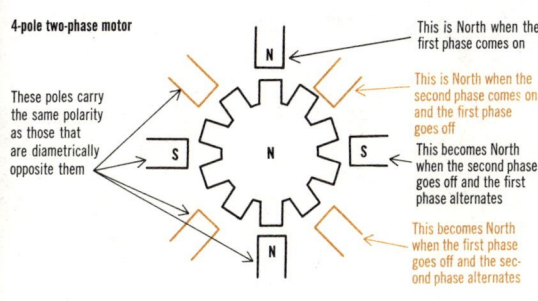

4-pole two-phase motor

These poles carry the same polarity as those that are diametrically opposite them

This is North when the first phase comes on

This is North when the second phase comes on and the first phase goes off

This becomes North when the second phase goes off and the first phase alternates

This becomes North when the first phase goes off and the second phase alternates

The first phase of the field poles is shown in black, and the second phase is shown in color. When the first phase drops off, and the second phase comes on, the rotor will move a quarter pole pitch in alignment with the North pole of the second phase. When the second phase drops off, and the first phase alternates, the rotor will move another quarter pitch. So, like the single-phase motor, the rotor will move half a pole pitch for one alternation of any one phase

subfractional permanent-magnet synchronous motors

The permanent-magnet synchronous motor has become very popular for use as a small timing motor because it is capable of allowing very low rotor speeds simply by using more rotor poles. A typical example of one such motor uses a rotor with 48 poles. Such a rotor will give a synchronous motor speed of 3600 ÷ 48 = 75 rpm.

A simplified example of how such a motor will work is shown in the diagram. The 48-toothed rotor is set between two specially shaped field poles. The upper pole has pole teeth that have the same pitch as the rotor teeth, and all its teeth will align with the rotor teeth when it has the proper magnetic polarity, say North if the rotor is South. The lower pole piece also has pole teeth that match the rotor pitch, but this pole

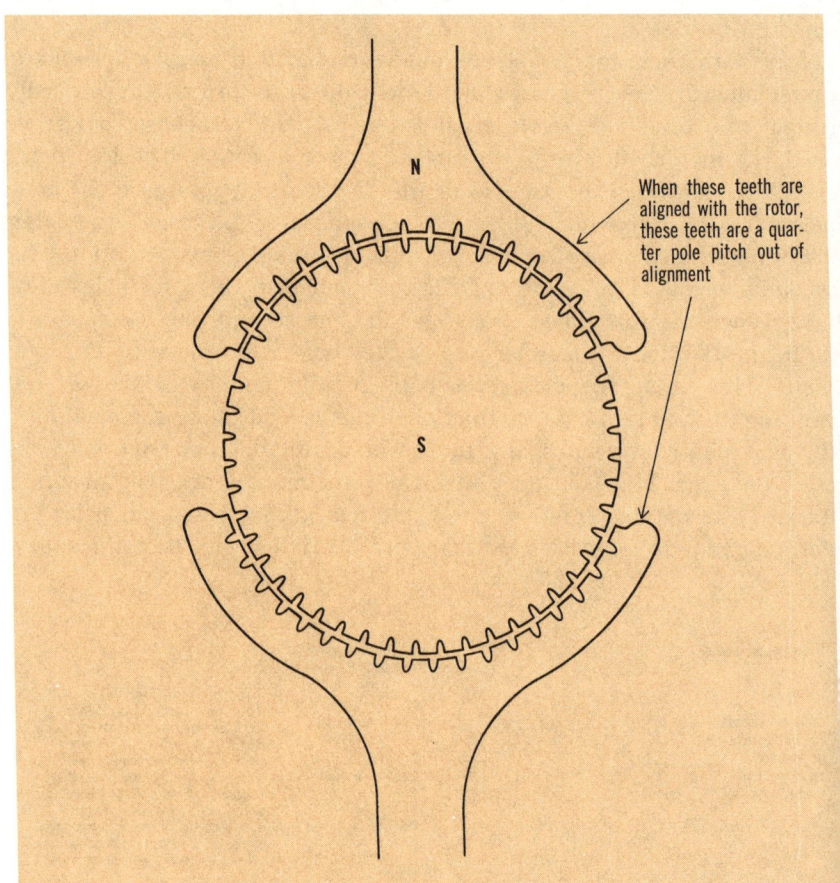

When these teeth are aligned with the rotor, these teeth are a quarter pole pitch out of alignment

subfractional permanent-magnet synchronous motors (cont.)

piece has teeth *displaced* one-quarter pole pitch, so that when the rotor teeth are aligned with the upper field pole, they are only partially aligned with the lower field pole. Then, when the field alternates, the rotor will turn one-half pitch to align its poles with the lower field poles. This displaces the teeth at the upper pole. But at the next alternation, the rotor again turns one-half pitch to align with the upper poles, and so on. Again, since each cycle (two alternations) moves the rotor one pole pitch, 48 cycles are needed for one full revolution of the rotor, or, with 60 Hz power, the rotor will turn 3600 ÷ 48 = 75 rpm.

The motor just described is an inefficient version of how these motors are actually made. Since each field pole acts on only part of the rotor at a time, less than full torque can be developed. In the actual motor, the stator or field poles are made of two complete circular discs with all their pole teeth aligned. The first disc is one polarity, and the other disc is the opposite polarity. Also, the rotor is made in two layers, one disc aligned with the first field, and the second disc aligned with the second field. The two sets of pole teeth on the rotor are displaced one-quarter pitch from each other, so that only one set at a time can be attracted into alignment with its corresponding stator. Thus, full torque is developed since each entire stator section acts on each entire rotor section at once.

Actually, these motors are usually two-phase motors, so that there are four stator sections and four rotor sections, two sets for each phase.

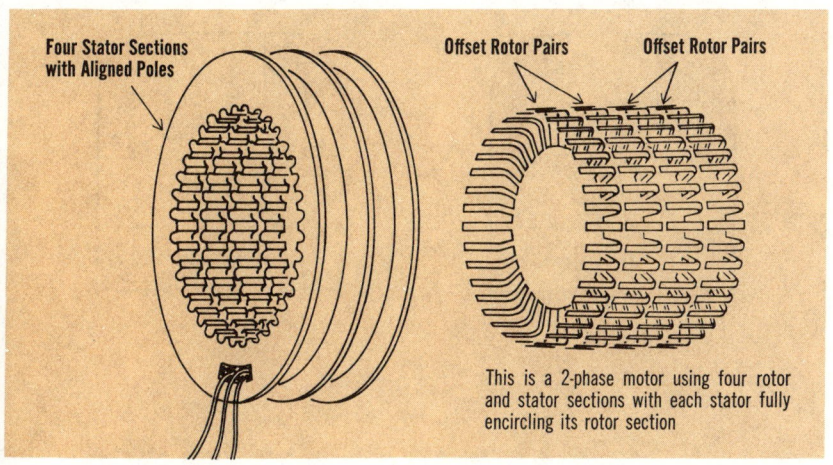

Four Stator Sections with Aligned Poles

Offset Rotor Pairs   Offset Rotor Pairs

This is a 2-phase motor using four rotor and stator sections with each stator fully encircling its rotor section

stepper motors

A *stepper motor* is one that can be stepped through its revolutions under controlled conditions. It may be stepped in only one direction, or it may be stepped back and forth, not always at the same distance. It can be used for continuous revolving, but it is usually applied in places for counting or indicating position. Some stepper motors are controlled by polyphase power, and some by pulses. You have probably deduced that the permanent-magnet synchronous motor is a form of stepper motor. If, instead of using regular a-c power to drive the motor, pulses were used, which could be turned on and off, the rotor would step around following the pulses. However, the synchronous motor is designed to operate best as a rotating motor. Stepper motors can use a similar principle, but work best in a sequencing function.

Stepper motors are designed to be excited by pulsed a-c power or pulsed d-c power. Those that use d-c pulses use the same principle as a single-phase permanent-magnet synchronous motor, with a positive pulse to establish one magnetic polarity, and a negative pulse to alternate the poles to move the rotor one-quarter pitch. Thus, the number of poles on the rotor determines how far the rotor will move for each pulse.

A-c stepper motors use two-, three-, and four-phase field windings to control the rotor. Each winding is energized in sequence to step the rotor around. A wide variety of stepper motors are made. Those that are not polyphase may use split-phase windings to control the rotor direction,

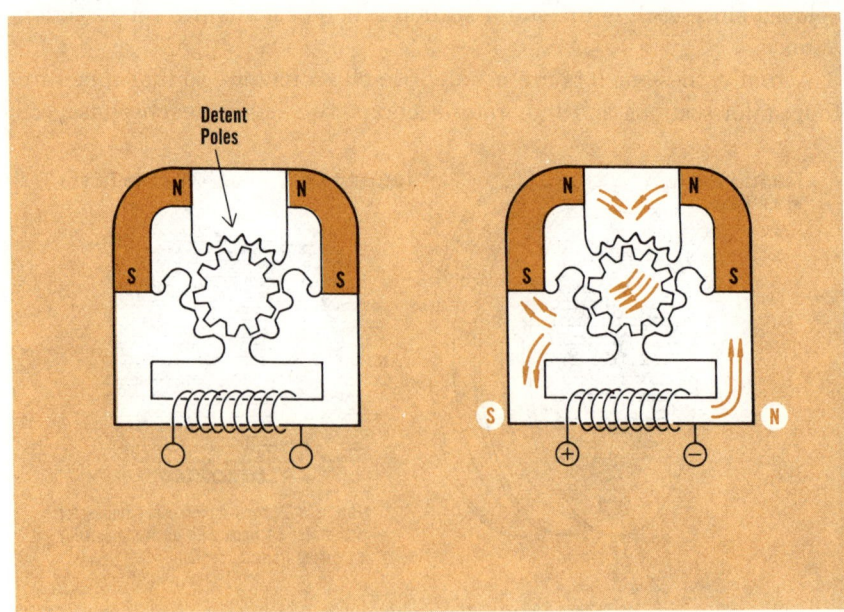

stepper motors (*cont.*)

while others may use shaded poles. The polyphase types use reluctance rotors. They are available with just a few degrees of rotation per step to 90 degrees per step. Some motors are designed to be driven by pulse logic circuits which automatically provide the precise number of pulses needed to get any precise rotor positioning needed.

The Sigma Cyclonome® stepper motor is an example of one that functions similarly to the permanent-magnet synchronous motor. Two permanent magnets establish the field. When the field winding is not energized, the detent pole section, which is magnetized by the North ends of the field magnets, holds the rotor by aligning with three of its pole teeth. When a pulse of current is sent through the coil, as shown in the second figure, the flux travels through the rotor to the left field pole teeth, and the rotor turns to align its teeth with them. When the pulse is stopped, the detent poles hold the rotor there. Then, when an opposite current pulse is sent through the coil, the flux lines go to the other side, and the rotor turns in the same direction to align its teeth. Again, when the pulse is stopped, the detent pole holds the rotor. Notice that alternately negative and positive pulses are needed here to step the rotor around in the same direction. Not all stepper motors act this way.

®Registered Trademark of Sigma Instruments Incorporated.

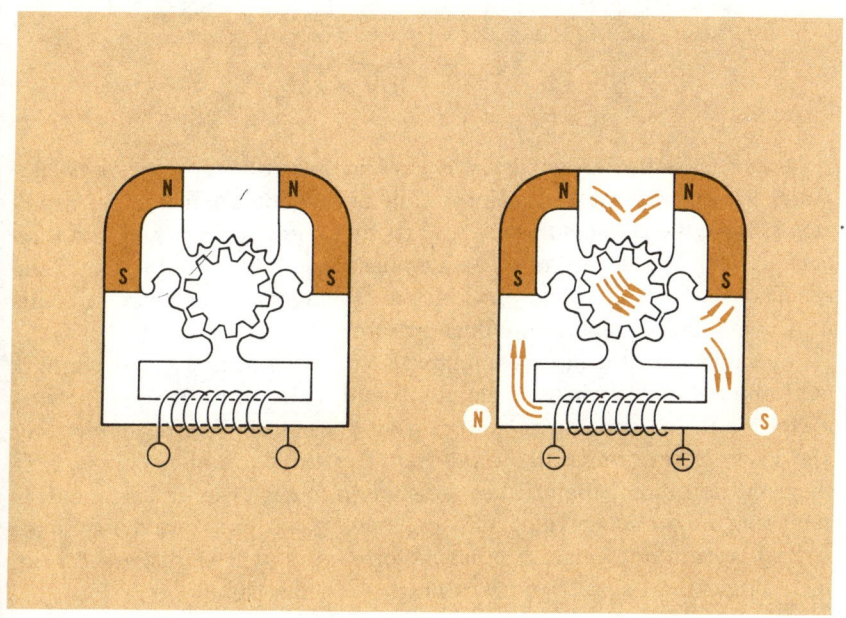

a-c commutator motors

In your study of a-c motors so far you have learned about induction motors and synchronous motors. There is a third class of a-c motors, known as *a-c commutator motors*. This motor is built very much like the d-c motor, in that it features brushes and a commutator. A-c commutator motors include *repulsion motors,* and *a-c series motors*.

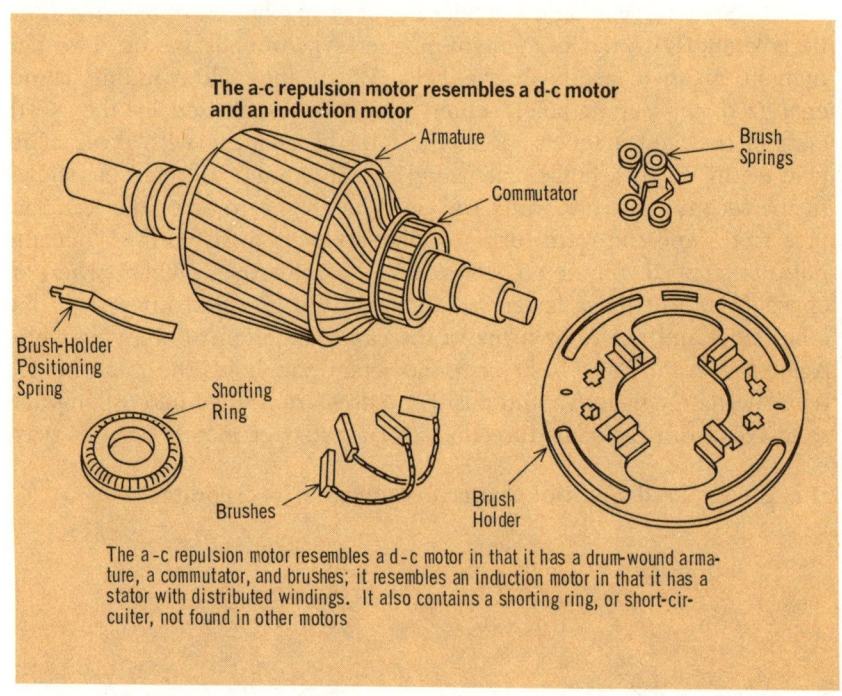

The a-c repulsion motor resembles a d-c motor and an induction motor

Armature

Commutator

Brush Springs

Brush-Holder Positioning Spring

Shorting Ring

Brushes

Brush Holder

The a-c repulsion motor resembles a d-c motor in that it has a drum-wound armature, a commutator, and brushes; it resembles an induction motor in that it has a stator with distributed windings. It also contains a shorting ring, or short-circuiter, not found in other motors

Motors classified as *repulsion motors* include the *standard repulsion motor,* which starts and runs by so-called magnetic repulsion; the *repulsion-start induction motor,* which starts by magnetic repulsion and then converts to operation as a single-phase induction motor; and the *repulsion-induction motor,* which starts by magnetic repulsion and runs as a combination repulsion-induction motor.

A repulsion motor has a laminated stator with a distributed winding very much like that of the induction motor. The rotor of the repulsion motor, on the other hand, is a drum-wound armature with a commutator and brushes very much like those found in the d-c motor. However, in the repulsion motor the brushes are shorted to each other. The standard repulsion motor is operated off single-phase a-c power and has some special feature not found in other motors, which will be discussed next, including a short-circuit and brush-lifting mechanism.

operation of the repulsion motor

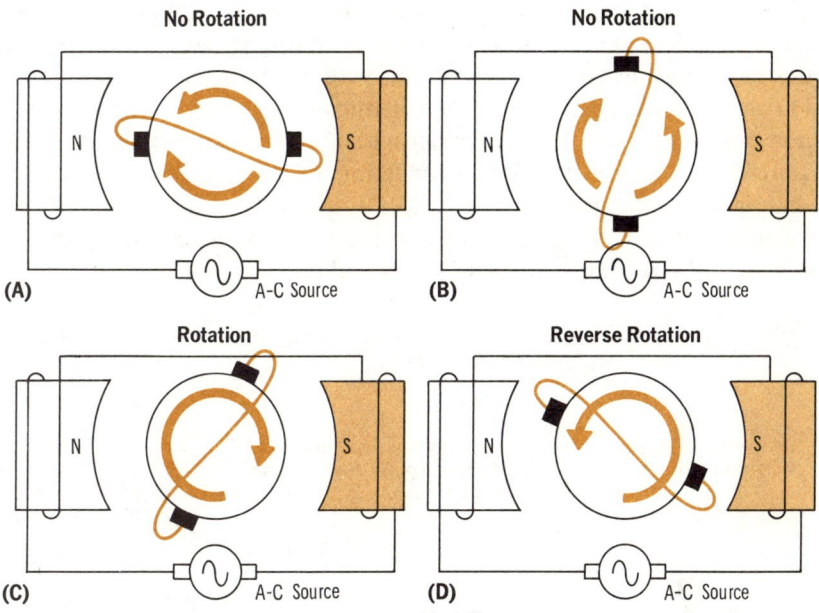

Though it resembles both induction and d-c motors, the repulsion motor operates uniquely by the principle of magnetic repulsion. To simplify the explanation of this principle, salient (or projecting) poles are shown in the illustration. In A, with the brush axis aligned with the poles, the stator induces equal and opposite currents in the two halves of the rotor windings. The net effect cancels the torque and the motor does not run.

With the brush axis perpendicular to the poles, as in B, the voltages induced in the rotor neutralize each other, producing no voltage at the brushes, no armature current flow, and hence, no torque. However, with the brushes anywhere between the position of A and B, as in C, there is a resultant voltage and current flow in the armature, creating a field which produces like poles between rotor and stator. This creates a magnetic repulsion force (torque) that turns the rotor in the direction of the brush shift.

The operating characteristics of the repulsion motor are very similar to those of the d-c series motor. It has high starting torque and can operate at relatively high speeds under light loads. To produce reversal, the brush position is shifted to the opposite side of the neutral plane, as in D. The repulsion motor always rotates in the direction of the brush shift from neutral.

repulsion-start induction motors

The repulsion-start induction motor starts just like a standard repulsion motor, but when it is near normal operating speed it converts and runs as a single-phase induction motor with a wound rotor. The conversion takes place when a special centrifugally operated device called a *short-circuiter* is driven into contact with the inner face of the commutator. The short-circuiter is a conducting ring that short circuits the commutator segments, thus cutting off all current flow through the brushes. At this point, a conventional rotating magnetic field system is created and the rotor is turned by this action.

The short-circuiter and brush-lifting mechanisms of the repulsion-start induction motor are operated by centrifugal force

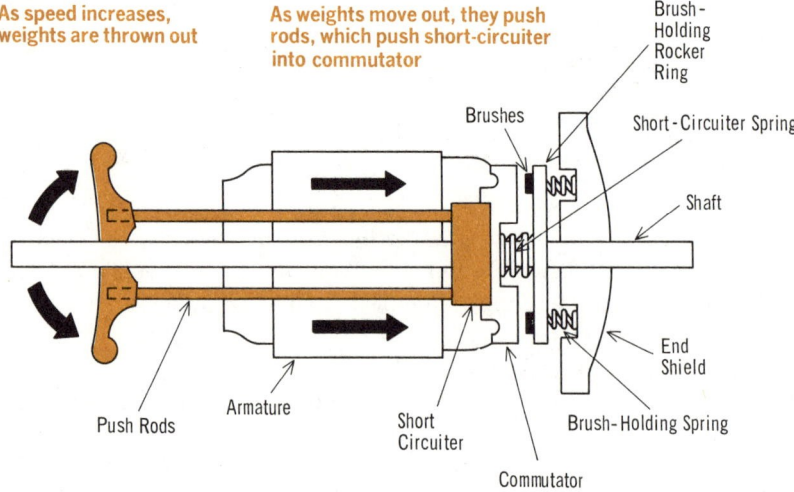

The arrangement is similar to the one used to operate the centrifugal switch of the capacitor-start induction motor

There are two classes of repulsion-start induction motors: the *brush-riding* and the *brush-lifting* types. In the brush-riding type, the brushes maintain contact with the commutator even after the short circuiter has activated and the motor is operating as an induction motor. In the brush-lifting type, a mechanism in the motor lifts the brushes off the commutator when the short-circuiter is activated. This action prevents wear of the brushes and commutator during the time the motor is operating as a normal induction motor.

the repulsion-induction motor

The repulsion-induction motor is unique because its rotor combines a squirrel-cage construction, and a drum winding with a commutator and short-circuited brushes. When single-phase a-c power is applied, the drum winding produces the starting torque by magnetic repulsion. As the armature rotates, torque is also produced by the squirrel-cage winding. This torque adds to that produced by repulsion.

As a result, at no load, the speed of operation is slightly above the synchronous speed, due to the action of the repulsion winding, while at full load, its speed is slightly below synchronism, the same as in an induction motor.

It is important to realize that both windings are always in operation when the armature rotates. The repulsion-induction motor does not have a short-circuit or brush-lifting mechanism.

The repulsion-induction motor can be reversed by changing the rigging of the brushes with respect to the neutral plane, as in the case of the standard repulsion motor.

The repulsion-induction motor resembles the reluctance motor

The repulsion-induction motor features a squirrel-cage rotor acting as the drum for the wound armature. In this respect, it resembles the reluctance motor
The repulsion–induction motor also features a commutator and shorted brushes, which make it unlike the reluctance motor

a-c series motors

Think back to your study of d-c motors. If we took a d-c motor and connected it to a-c power what would happen? Remember, that with armature and field connections constant, the direction of rotation of the d-c motor is fixed regardless of the polarity of the applied voltage. With the motor connected to ac, the polarity would change every half cycle, but the flux created by armature and field would tend to create a torque that acts in the same direction for each half cycle. This means we might expect the d-c motor to operate satisfactorily on ac.

However, if we consider a shunt motor, we find that the relatively high inductance of the field winding causes the field current to lag the armature current by a large phase angle. Hence the shunt motor develops very little torque. In some cases, the amount of torque developed might not even be enough to overcome inertia and frictional forces and the motor will not even start. In the case of the series motor, however, the field and armature current are the same, and an adequate torque can theoretically be developed, and, in fact, the series d-c motor can be operated from ac.

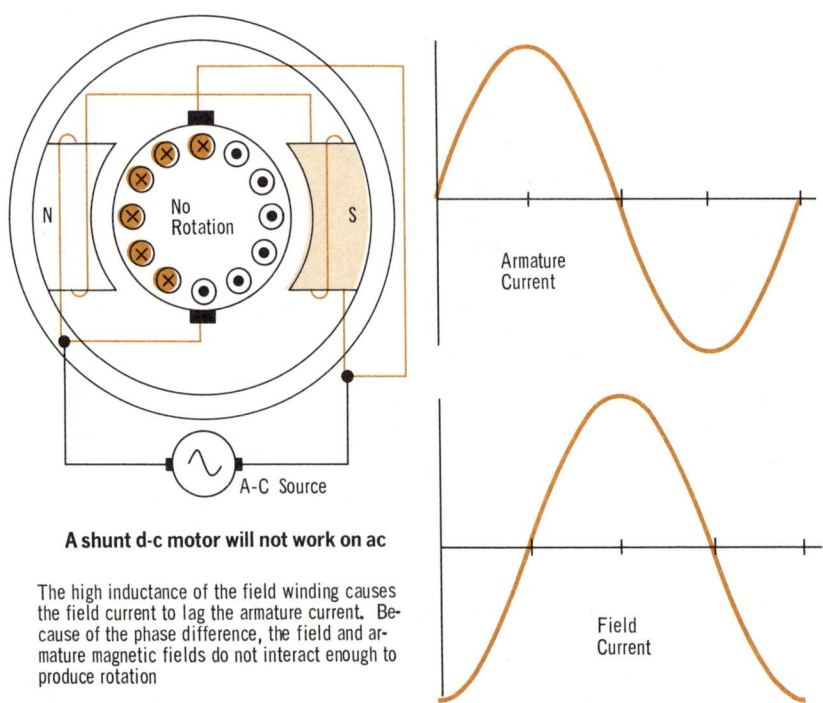

A shunt d-c motor will not work on ac

The high inductance of the field winding causes the field current to lag the armature current. Because of the phase difference, the field and armature magnetic fields do not interact enough to produce rotation

When an alternating current is applied to the input terminals of an a–c series motor, because of the series winding arrangements, both the field flux and the armature current reverse simultaneously every half cycle. Because both change at the same time, the direction of rotation remains the same for each cycle.

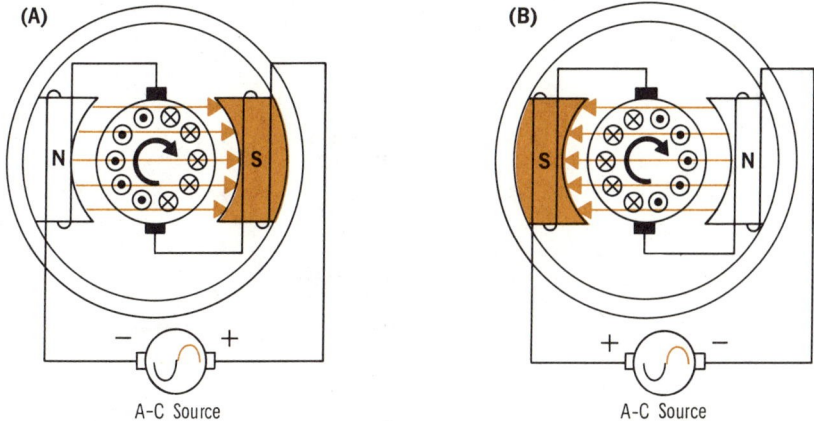

(A)   **(B)**

A-C Source   A-C Source

Although its basic diagram resembles the d–c series motor, practical a–c series motors have special design features to facilitate their use with a–c power

a-c series motors (cont.)

In the practical a-c series motor, the armature and field are wound and connected in the manner of the d-c series motor. When we apply ac to the terminals, the field flux and armature current both reverse simultaneously every half cycle. Since the field and armature windings are connected in series, the direction of rotation remains the same for each a-c half cycle.

There are certain limitations to operating series d-c motors on ac. Most notable, alternating field flux causes large eddy current and hysteresis losses; self-induction of field and armature windings produces a low power factor, hence inefficiency; and armature reaction causes excess sparking at the commutator. If some way could be found to overcome these limitations, a d-c motor could be built to work successfully on ac.

In the a-c series motor, the limitations are overcome by adjusting the design of the d-c motor to include the following:

1. The field poles and core are laminated to minimize eddy currents and hysteresis losses.
2. Power factor is improved by reducing field winding reactance through the use of as few turns as possible on the field. A high field flux is obtained by using a low-reluctance magnetic circuit.
3. Armature reactance is reduced by using a compensating winding in the pole faces.

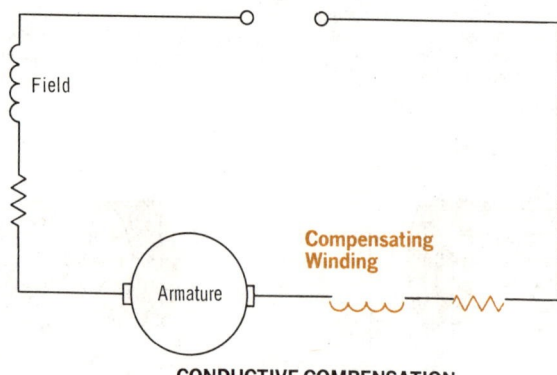

CONDUCTIVE COMPENSATION

Compensating windings must be conductively connected in series with the field and armature windings for ac–dc operation

compensation for
a-c series motors

A main feature of a-c series motors is the use of compensating windings to reduce armature reaction. The most common means of compensation involve the use of distributed compensating windings embedded into the laminated pole faces of the motor. If the a-c series motor is to be used for both a-c and d-c applications, the compensating winding is always connected in series with the armature, and the motor is said to be *conductively* compensated. If the compensating winding is short-circuited on itself, the motor is said to be *inductively* compensated.

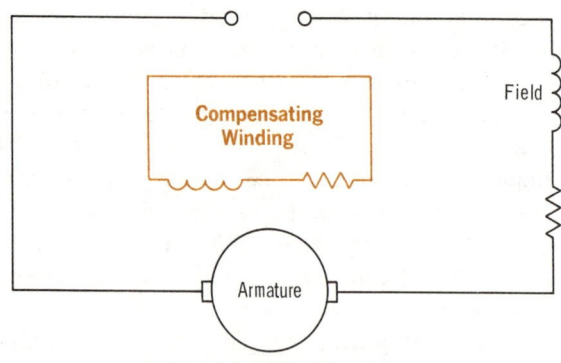

INDUCTIVE COMPENSATION

The compensating windings may also be inductively connected by short-circuiting it on itself to form a transformer coupling loop

universal motors

Universal motors are fractional horsepower a-c series motors specially designed for use on either d-c or a-c power. These motors provide the same speed and torque characteristic when operated on either ac or dc. In general, small universal motors do not require compensating windings because the number of turns used on their armature is low and so, therefore, is their armature reactance. As a result, motors below 3/8 horsepower are generally built without compensation. The cost of uncompensated universal motors is relatively low and so their use is widespread in light household appliances, such as vacuum cleaners, hand drills, malted milk mixers, etc.

The large-type universal motors use some form of compensation. This normally is the compensating winding of the in-series motor, or a distributed field winding specially wound to take care of armature reaction problems.

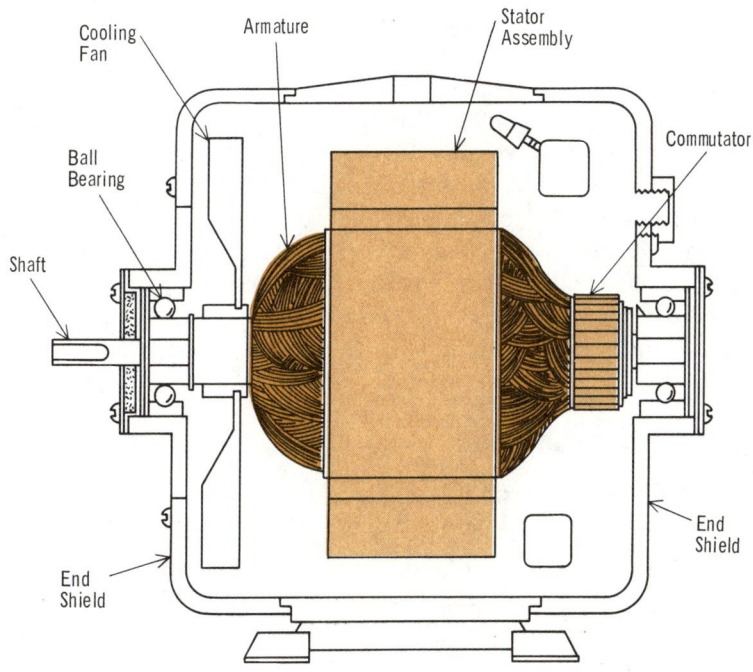

The universal motor operates equally well on ac and dc. Small fractional-horsepower universal motors perform well on ac without the need for compensation because their armature resistance is low

reversing motors

Some motors run in only one direction because of the way they are made, but many motors are designed for two-way operation or can be so modified. D-c motors can be reversed as shown on pages 7–62 and 7–63. Reversing the wiring (current flow) to *either* the field wniding *or* armature, but *not both*, will reverse a d-c motor. To reverse the rotation of any a-c motor that uses a start winding, such as the split-phase, capacitor, or reactor motors, only the leads (current flow) to the start winding need be shifted. With two-phase motors, only one winding need be reversed. But with the three-phase a-c motor any two out of the three windings have to be reversed in order to reverse the motor direction.

With the a-c commutator repulsion motors, the motor will run in the direction of the brush shift (from neutral). But not all motors will permit the brushes to be shifted this way.

Most small single-phase a-c motors use shaded poles to determine the direction of rotation. As shown on page 7–103, the direction of flux rotation is toward the shaded pole, so that it is the position of the shaded pole that controls the rotor direction. Unless the motor is made to react differently, the shaded pole motor can turn in only one direction. Reversible shaded pole motors can be made in two ways. The first method uses two different and independent field windings, each having its shaded poles on opposite sides. You use one field winding for one direction, and the other field winding for the opposite direction. (Two of the same motors connected back-to-back will do the same thing.) The second, and less expensive, method to make a reversible shaded pole motor is to use *shading coils* on either side of the pole pieces. The coils, however, are not shorted to themselves. Instead, their leads come out, and can be connected to a switch that will short either one pair or the other to control the flux direction of rotation.

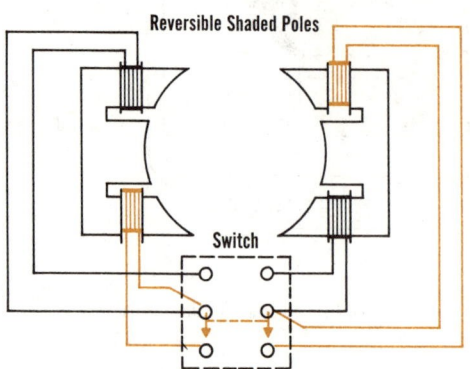

Reversible Shaded Poles

Switch

With the switch connected as shown, the flux field will rotate counterclockwise. With the switch connected to the other terminals, the other two shading coils are shorted to rotate the flux field clockwise

dual voltage motors

Many of the larger motors, ¼-horseower and above, are designed to function on two different line voltages. Although most voltages at the power outlet are 115 volts a-c, the main power line coming into the service box is a center-tapped 230-volt line, from which two different 115-volt lines are taken off to feed most appliance and equipment needs. However, if you recall from your earlier studies (Volume 3), the same amount of power can be obtained from different combinations of voltage and current, since P (watts) = E × I. You also learned that high currents cause high I²R losses, which are always in the form of heat. For many larger motor applications, which tend to draw a great deal of current to obtain their horsepower rating, it would be more economical to run them on 230 volts with half the current than at 115 volts with twice the current. But since 230 volts is not always conveniently available, many motors are made to work on both 115/230 volts.

These motors are made with two field windings, each rated at 115 volts, and both windings are usually wound together so that they both give very similar torque characteristics. For 230-volt operation, the windings are connected in series and carry the same current. But for 115-volt operation, the windings are connected in parallel so that each of their currents is added, doubling the total field current drawn from the power line.

Usually, the start or auxiliary winding is a single 115-volt winding, since it does not draw as much current as the field windings. When the field windings are connected in parallel, the auxiliary winding is merely connected across them to the same 115-volt line. But when 230 volts is used, and the field windings are in series, the auxiliary winding is connected to the center of the two windings across one of the fields to get 115 volts. All winding wires are generally brought out to a terminal board, where they can be connected for either 115- or 230- volt operation.

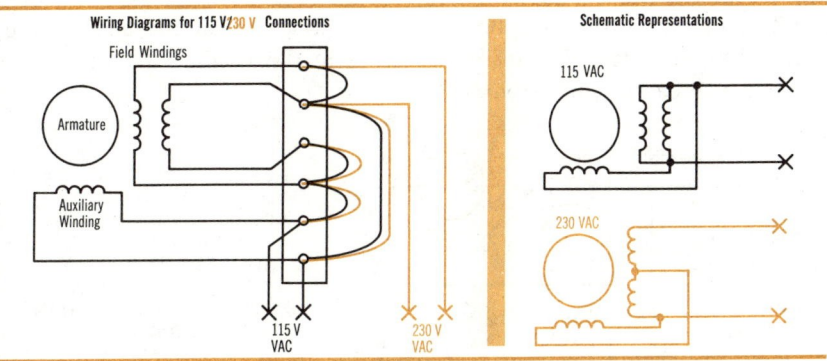

motor protection

Many motors operate under changing or heavy loads, or in warm environments. Also, because of the heavy currents drawn by some motors, they tend to get hot. Too much heat in a motor can cause its windings to expand, or its insulation to deteriorate. Expanding metal at the shafts or bearings can interfere with free rotation and cause excessive bearing wear.

In addition to this, turns in the windings can short, causing excessive damaging currents to flow. But even more important, intermittent or sudden heavy loads on the motor can slow it down excessively. If you remember what you learned earlier about cemf, you will recall that the motor speed is what produces the cemf, which limits the current flow through the windings. Without cemf, or without enough cemf for too long a time, too much current will flow and damage the windings or excessive heat will cause other damage.

To prevent this, larger motors generally have built-in protective devices to shut off the motor temporarily if it gets too hot or draws too much current. These devices are called current or thermal or combination *cutouts*.

Generally, the cutouts are connected in series with the power line so that the entire current to the motor passes through it. If it gets too hot, either from ambient temperature or the current passing through it, it will open the line and shut off the motor. When it cools down, it will close automatically to turn the motor on again. There are many types of cutouts. Some cutouts use a metal strip that bends with heat to open the contacts; some use a bimetallic strip that bends with heat; and some use heat coils or bellows.

Many motors use only the thermal cutout in the start winding, since it is usually during startup time, when the load is too great, that trouble can develop. Motors that cut out during startup, often repeatedly, are not always defective, as long as they soon stay on and operate at normal speed. Air conditioner motors are good examples of this. When the start winding cutout is used, the heating coil of the thermal cutout is in the start winding circuit, but its contacts are in the main line so that the entire motor can be shut down.

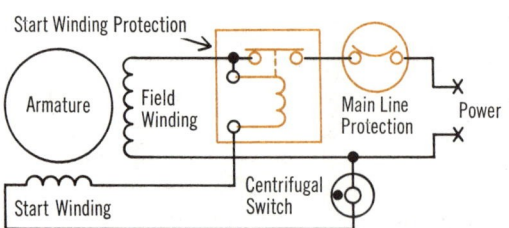

The thermal cutout can be either in the main line to both windings or only in the line to the start winding. When it is in the start winding circuit, it protects the motor only during startup, not while it is running. In either case, the main line is opened when trouble develops

summary

□ Consequent-pole motors provide multiple speeds by rearranging the wiring to each set of poles so that they act as monopoles; consequently, virtual opposite poles are created between adjacent like poles. □ This, in effect, doubles the number of poles to halve the synchronous speed.

□ The reluctance motor rotor has squirrel-cage shorted turns so that it starts as an induction motor, and soft-iron salient poles so that it runs as a synchronous motor. □ The hysteresis synchronous motor uses a cobalt-steel rotor that, through hysteresis effects, develops a greater torque. □ The permanent-magnet synchronous motor uses a rotor to follow the field alternations. The more pole teeth the rotor has, the more slowly it will turn. Stepper motors are similar, and are designed to step around when pulsed.

□ The a-c commutator motor is built like the d-c type. □ The a-c repulsion motor uses commutator shorting brushes to choose shorted turns that develop magnetic polarities which are repelled by the rotating field. □ The repulsion-start induction motor begins as a repulsion motor, but runs as a wound-rotor motor. □ The repulsion-induction motor starts with magnetic repulsion, but runs on both principles. □ A-c series motors are designed to minimize eddy current and hysteresis losses, field winding reactance, and armature reaction. □ Some motors are designed for dual 115/230-volt operation. Thermal cutouts are used to protect motors when they have trouble starting, are running too low, or are running too hot.

review questions

1. How is a four-pole motor wired to produce four consequent poles?
2. How fast would the motor in Question 1 run on 60 Hz if it were a synchronous motor?
3. What would you add to a squirrel-cage induction motor to make it a reluctance motor? What material would you use?
4. How does the material in Question 3 differ in magnetic properties from that used in the hysteresis motor?
5. What is the basic difference in operation between the wound-rotor induction motor and the a-c repulsion motor?
6. What is done to the repulsion-start motor to make it run as an induction motor?
7. Describe the rotor of the repulsion-induction motor.
8. Why can some d-c series motors also run on ac?
9. What design considerations must be given to a series motor if it is to be used on ac?
10. Do fractional-horsepower series motors, used in household appliances, usually need compensation? Why?

cumulative index

cumulative index